# COMPREHENSIVE POLYMER SCIENCE

## IN 7 VOLUMES

# COMPREHENSIVE POLYMER SCIENCE

## *The Synthesis, Characterization, Reactions & Applications of Polymers*


CHAIRMAN OF THE EDITORIAL BOARD

SIR GEOFFREY ALLEN, FRS

*Unilever Research and Engineering, London, UK*

DEPUTY CHAIRMAN OF THE EDITORIAL BOARD

JOHN C. BEVINGTON

*University of Lancaster, UK*


## Volume 4

Chain Polymerization II


VOLUME EDITORS

GEOFFREY C. EASTMOND

*University of Liverpool, UK*

ANTHONY LEDWITH

*Pilkington plc Group Research, UK*

SAVERIO RUSSO

*Università di Sassari, Italy*

PIERRE SIGWALT

*Université Pierre et Marie Curie, Paris, France*



**PERGAMON PRESS**

**OXFORD · NEW YORK · BEIJING · FRANKFURT**

**SÃO PAULO · SYDNEY · TOKYO · TORONTO**

| | |
|---|---|
| U.K. | Pergamon Press plc, Headington Hill Hall, Oxford OX3 0BW, England |
| U.S.A. | Pergamon Press, Inc., Maxwell House, Fairview Park, Elmsford, New York 10523, U.S.A. |
| PEOPLE'S REPUBLIC OF CHINA | Pergamon Press, Room 4037, Qianmen Hotel, Beijing, People's Republic of China |
| FEDERAL REPUBLIC OF GERMANY | Pergamon Press GmbH, Hammerweg 6, D-6242 Kronberg, Federal Republic of Germany |
| BRAZIL | Pergamon Editora Ltda, Rua Eça de Queiros, 346, CEP 04011, Paraiso, São Paulo, Brazil |
| AUSTRALIA | Pergamon Press (Australia) Pty Ltd., P.O. Box 544, Potts Point, NSW 2011, Australia |
| JAPAN | Pergamon Press, 5th Floor, Matsuoka Central Building, 1-7-1 Nishishinjuku, Shinjuku-ku, Tokyo 160, Japan |
| CANADA | Pergamon Press Canada Ltd., Suite No. 271, 253 College Street, Toronto, Ontario, Canada M5T 1R5 |



First edition 1989

**Library of Congress Cataloging-in-Publication Data**

Comprehensive polymer science.

Includes index.
Contents: v. 1. Polymer characterization/volume editors, Colin Booth & Colin Price—v. 2. Polymer properties/volume editors, Colin Booth & Colin Price—[*etc.*]—v. 7. Specialty polymers & polymer processing/volume editor, Sundar L. Aggarwal.
1. Polymers and polymerization. I. Allen, G. (Geoffrey), 1928– II. Bevington, J. C.
QD381.C66 1988 547.7 88-25548

**British Library Cataloguing in Publication Data**

Comprehensive polymer science.

Vol. 4: Chain polymerization

1. Polymers
I. Allen, Geoffrey II. Bevington, John C.
III. Eastmond, Geoffrey C.
547.7

ISBN 0-08-036208-7 (vol. 4)
ISBN 0-08-032515-7 (set)

*Printed in Great Britain by A. Wheaton & Co. Ltd., Exeter*

# Contents

# Preface

It is only 60 years since Staudinger's model of the molecular nature of a polymer was becoming universally accepted and the physical states of rubbers, plastics and fibres understood. Unfortunately, for some time many academic chemists continued not to appreciate the full significance of polymerization reactions and physicists tended to regard polymeric materials as inevitably being of indeterminate composition and unamenable to study by conventional physical methods.

Nevertheless, in the 1930s the foundations were laid for the understanding of the main polymerization mechanisms. An industry based on synthetic rubbers, plastics and fibres was soon established. In World War II it played a major strategic role and afterwards grew to be one of the main elements of the heavy chemicals industry. It became recognized that synthetics may be superior to natural materials in their properties and that they may be used for completely new purposes.

Alongside the production of well-defined materials there grew the ability to characterize the structure of polymer molecules and to understand the relationships between methods of preparation and subsequent treatment, structure and properties, both chemical and physical. As a result, a vast literature of polymer science and technology has been generated and four Nobel prizes awarded specifically for contributions to polymer science. Add to this the fact that many biological molecules, including polypeptides, enzymes, antibodies, carbohydrates and so on, are polymers of varying degrees of complexity, then the universality of polymers in the physical and biological sciences and technologies forms a dominant modern theme.

*Comprehensive Polymer Science* is a series of volumes designed to set down the structure of this vast subject in such a way that researchers and teachers of polymer science and workers in associated fields can find an authoritative and comprehensive account of the topic of immediate interest. That topic is set out in a framework of related subjects. The text is focused on synthetic polymers with little reference to biological macromolecules *per se* but the science underpins both physical and biological systems.

To ensure that the wide coverage is maintained at an authoritative level, more than 250 authors from 20 countries have been enlisted. Their contributions have been organized into a series of major themes:

| | |
|---|---|
| Volume 1 | Polymer Characterization |
| Volume 2 | Polymer Properties |
| Volumes 3–5 | Polymerization Mechanisms |
| Volume 6 | Polymer Reactions |
| Volume 7 | Specialty Polymers & Polymer Processing |

Because of the wide coverage the editors were presented with a particularly difficult decision with regard to symbols and nomenclature. The latter does not follow strictly the recommendations of IUPAC nor are symbols consistent throughout the whole work. However, usage in a particular chapter is consistent with the practice in the current literature. Thus a reader will be able to frame new publications in the context of the information presented in this series of volumes.

We should like to acknowledge the way in which the staff at the publisher, particularly Dr Colin Drayton (who initially proposed the project), Dr Helen McPherson and their editorial team, have supported the editors and authors in their endeavour to produce a text that is both complete and up-to-date and that will appeal to industrial and academic researchers alike. *Comprehensive Polymer Science* is a milestone in the literature of the subject in terms of coverage, clarity and a sustained high level of presentation.

GEOFFREY ALLEN
*London*

JOHN C. BEVINGTON
*Lancaster*

# Contributors to Volume 4

Professor H. R. Allcock
Department of Chemistry, College of Science, 152 Davey Laboratory, The Pennsylvania State University, University Park, PA 16802, USA

Dr A. J. Amass
Speciality Polymer Group, Department of Molecular Sciences, University of Aston, Aston Triangle, Birmingham B4 7ET, UK

Professor B. M. Baysal
Department of Chemistry, Faculty of 'Fen-Edebiyat', Technical University of Istanbul, Maslak-Istanbul, Turkey

Dr I. G. Berry
Department of Chemistry, University of Manchester Institute of Science and Technology, PO Box 88, Sackville Street, Manchester M60 1QD, UK

Dr A. Bolognesi
Istituto di Chimica delle Macromolecole del CNR, via E Bassini 15, I-20133 Milano, Italy

Dr V. Busico
Dipartimento di Chimica, Università di Napoli, via Mezzocannone 4, I-80134 Napoli, Italy

Professor G. B. Butler
Department of Chemistry, University of Florida, Gainesville, FL 32611, USA

Dr F. Candau
Institut Charles Sadron (CRM-EAHP), 6 rue Boussingault, F-67083 Strasbourg Cedex, France

Dr M. Catellani
Istituto di Chimica delle Macromolecole del CNR, via E Bassini 15, I-20133 Milano, Italy

Professor P. Corradini
Dipartimento di Chimica, Università di Napoli, via Mezzocannone 4, I-80134 Napoli, Italy

Dr G. Costa
Centro di Studi Chimico-Fisici di Macromolecole Sintetiche e Naturali, Corso Europa 30, I-16132 Genova, Italy

Professor J. M. G. Cowie
Department of Chemistry, University of Stirling, Stirling FK9 4LA, UK

Professor J. V. Dawkins
Department of Chemistry, University of Loughborough, Loughborough, Leicestershire, LE11 3TU, UK

Dr S. Destri
Istituto di Chimica delle Macromolecole del CNR, via E Bassini 15, I-20133 Milano, Italy

Professor G. Di Silvestro
Dipartimento di Chimica Organica e Industriale, Università di Milano, via Venezian 21, I-20133 Milano, Italy

Dr A. S. Dunn
Department of Chemistry, University of Manchester Institute of Science and Technology, PO Box 88, Sackville Street, Manchester M60 1QD, UK

Professor W. J. Feast
Department of Chemistry, University of Durham, South Road, Durham DH1 3LE, UK

Professor J. L. Garnett
Department of Chemistry, University of New South Wales, Kensington, New South Wales, Australia 2033

Dr A. Giarrusso
Dipartimento di Chimica Industriale e Ingegneria Chimica Giulio Natta, Politecnico di Milano, Piazza Leonardo da Vinci 32, I-20133 Milano, Italy

Dr R. G. Gilbert
Department of Theoretical and Physical Chemistry, University of Sydney, Sydney, New South Wales 2006, Australia

Dr G. Guerra
Dipartimento di Chimica, Università di Napoli, via Mezzocannone 4, I-80134 Napoli, Italy

Dr G. D. Guerra
Facoltà di Ingegneria, CNR Pisa, 2 via Diotisalvi, I-56100 Pisa, Italy

Professor A. Guyot
CNRS-LMO, Laboratoire de Materiaux Organiques, BP 24, F-69390 Vernaison, France

Dr Y. Iriyama
Materials Research Center, Institute for Thin Films Processing Science, University of Missouri-Rolla, Rolla, MO 65401-0249, USA

Dr T. C. Kendrick
Research and Development Department, Dow Corning Ltd., Barry, S. Glamorgan CF6 7YL, UK

Professor D. H. Napper
Department of Theoretical and Physical Chemistry, University of Sydney, Sydney, New South Wales 2006, Australia

Professor S. P. Pappas
Polymers and Coating Department, North Dakota State University, Fargo, ND 58105, USA

Dr B. M. Parbhoo
Dow Corning Research Group, School of Molecular Sciences, University of Sussex, Brighton BN1 9QJ, UK

Professor L. Porri
Dipartimento di Chimica Industriale e Ingegneria Chimica Giulio Natta, Politecnico di Milano, Piazza Leonardo da Vinci 32, I-20133 Milano, Italy

Professor J. Silverman
University of Maryland, College Park, MD 20742, USA

Dr D. Y. Sogah
Central Research and Development Department, EI du Pont de Nemours & Co. Inc., Experimental Station, Building 328, Wilmington, DE 19898, USA

Dr P. Sozzani
Dipartimento di Chimica Organica e Industriale, Università di Milano, via Venezian 21, I-20133 Milano, Italy

Professor V. T. Stannett
Department of Chemical Engineering, North Carolina State University, PO Box 7905, Raleigh, NC 27695-7905, USA

Dr P. J. T. Tait
Department of Chemistry, University of Manchester Institute of Science and Technology, PO Box 88, Sackville Street, Manchester M60 1QD, UK

Mr D. J. Walbridge
Paints Division, ICI plc, Wexham Road, Slough SL2 5DS, UK

Dr N. D. Watkins
Department of Chemistry, University of Manchester Institute of Science and Technology, PO Box 88, Sackville Street, Manchester M60 1QD

Dr O. W. Webster
Central Research and Development Department, EI du Pont de Nemours & Co. Inc., Experimental Station, Building 328, Wilmington, DE 19898, USA

Dr J. W. White
Dow Corning Research Group, School of Molecular Sciences, University of Sussex, Brighton, BN1 9QJ, UK

Professor H. Yasuda
Materials Research Center, Institute for Thin Films Processing Science, University of Missouri-Rolla, Rolla, MO 65401-0249, USA

# Contents of All Volumes

**Volume 2 Polymer Properties**



**Volume 3 Chain Polymerization I**

**Volume 4 Chain Polymerization II**

**Volume 5 Step Polymerization**

**Volume 6 Polymer Reactions**



**Volume 7 Specialty Polymers & Polymer Processing**

# 1

# Monoalkene Polymerization: Ziegler–Natta and Transition Metal Catalysts

PETER J. T. TAIT
*University of Manchester Institute of Science and Technology, UK*



## 1.1 INTRODUCTION

One of the most important achievements in the field of synthetic polymer chemistry during the past 50 years has been the discovery in 1953 by Ziegler and his co-workers[1,2] that aluminum alkyl compounds used along with many transition metal halides in an inert hydrocarbon medium could polymerize ethylene under conditions of normal temperatures and pressures to yield a solid polymer of high molecular weight and of linear structure. The developments which led to the discovery of this catalyst have been described by Ziegler.[2] For some time Ziegler's group had studied the 'aufbau' or growth reaction in which ethylene undergoes oligomerization when heated with an aluminum alkyl compound. Although the insertion step proceeds at temperatures of 100–120 °C the displacement reaction requires temperatures of greater than 120 °C. By accident Holzkamp, one of Ziegler's students, discovered that the displacement reaction could be catalyzed at lower temperatures by traces of colloidal nickel inadvertently left in a reactor. Subsequently, during wider investigations using other metals, Breil discovered that zirconium acetylacetonate, used with aluminum triethyl, was an exceedingly active catalyst with the autoclave being filled with polyethylene. Other transition metal compounds from Groups IV to VI when used with aluminum alkyl compounds gave similar results, with titanium compounds showing a high activity. Characterization studies on the polyethylenes produced by the new catalyst systems showed them to have linear unbranched structures and high molecular weights.

Ziegler and his co-workers were mainly concerned with the polymerization of ethylene and its copolymerization with α-alkenes, and it was left to Natta to extend Ziegler's discovery to α-alkenes. Natta disclosed his results to Montecatini, and it was agreed that research could be carried out on the polymerization of α-alkenes. Using a combination of titanium tetrachloride and aluminum triethyl Natta and co-workers[3,4] established that propylene could be polymerized to give a product which could be separated into largely crystalline and amorphous fractions. Subsequently Natta and his co-workers[5] extended the type of transition metal halide used to solid crystalline transition metal salts in a lower valence state, *e.g.* $TiCl_3$, $VCl_3$, *etc.* It was also established that stereoregular polymers of butene-1, styrene, dienes, *etc.*, could be prepared.

Ziegler's discovery, together with the ensuing work into the polymerization of propylene and other α-alkenes by Natta, has given rise to the development of new classes both of catalysts and polymers. In the case of ethylene, the essential advance which was achieved was the preparation of a more linear type of polyethylene, whilst in the case of propylene and other α-alkenes, it was the control of the propagation step so as to give linear unbranched stereoregular molecules. The commercial advantage was that this control could be achieved in the polymerization of such readily available and cheap monomers as ethylene and propylene. In recognition of their contributions the 1963 Nobel Prize for Chemistry was awarded to Ziegler and to Natta.

Independent and concurrent research carried out in the USA in several industrial laboratories during the early 1950s led to the discovery of a related class of polymerization catalysts, which are formed from oxides of transition metals from Groups V–VII of the periodic table. To increase catalytic activity the transition metal oxides were supported on high area substrates such as silica, alumina, silica–alumina, *etc.* These catalysts constitute some of the earliest low pressure catalysts for alkene polymerization. The silica-supported chromium trioxide catalyst developed by the Phillips Petroleum Company[6,7] is the most important transition metal oxide catalyst of this class and is used for the polymerization of ethylene, producing linear, high density polymers. The discovery of these catalysts, their high activity, and the particular methods of their preparation have had profound effects on subsequent developments in Ziegler–Natta catalysis.

Due to their commercial and scientific importance Ziegler–Natta and related catalysts have been extensively reviewed in the scientific literature and in publications of symposia proceedings.[8–31]

## 1.2 DEFINITION OF THE ZIEGLER–NATTA CATALYST

In the broadest sense Ziegler–Natta catalysts are formed from reactions involving certain transition metal compounds (more commonly halides) of Group IV–VIII such as titanium, vanadium, chromium, zirconium, *etc.*, with alkyl, aryl or hydrides of Groups I–IV (equation 1). These reactions being carried out in an inert solvent, or diluent, and under inert conditions.

$$\underset{\text{(Group IV–VIII)}}{\text{Transition metal compound}} + \underset{\text{(Group I–IV)}}{\text{Metal alkyl, aryl, or hydride}} \rightarrow \text{Active Ziegler–Natta catalyst} \qquad (1)$$

This definition is too broad since not all possible combinations of the two components will form active and stereoregulating catalysts for the polymerization of ethylene and α-alkenes.

The type of solvent of diluent should not be omitted in reporting a catalyst system. Ziegler–Natta polymerizations are usually carried out in inert solvents, *e.g.* heptane or toluene. The use of polar solvents can drastically alter the reaction mechanism. Both monomer and diluent or solvent must be clearly specified in defining a particular Ziegler–Natta catalyst system.

In an attempt to recognize a common mechanistic feature believed to operate in these polymerization systems the use of the term 'coordination polymerization' is sometimes adopted. This description has the obvious advantage of highlighting the relationship between Ziegler–Natta polymerization processes and alkene metathesis and oligomerization. A possible disadvantage is that such a procedure leaves the definition of Ziegler–Natta catalysts too general and vague.

## 1.3 POLYMER STRUCTURES

Stereoregularity arises because of order in the spatial structures of polymer chains. Such ordered spatial structures are determined by two phenomena. During the polymerization step the monomer units are joined together in a regular sequence determined by the catalyst used and by the polymerization conditions. Regularity of this type cannot be altered or changed during subsequent physical treatment of the polymer. Subsequently these regularly constructed polymer chains spontaneously arrange themselves into helical structures.

If the backbone of a polymer chain is drawn in a flat zigzag form in the plane of the paper, the patterns shown in Figure 1 can be easily envisaged in the case of monosubstituted vinyl units. It should be noted that in (a) all the substituent R groups lie uniformly on the same side of the zigzag plane. Natta called this structure isotactic. In (b) the substituent R groups occupy positions alternatively above and below the backbone plane. Such a structure is termed syndiotactic. In (c) there is no regular arrangement of the substituent R groups and this structure is called atactic.

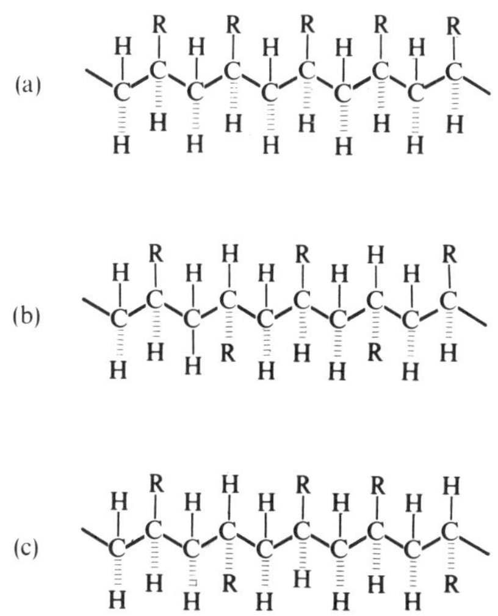

**Figure 1** Planar representations of chains of poly(α-alkenes): (a) isotactic; (b) syndiotactic; and (c) atactic (—, in plane of paper; — above plane of paper; ||||||, below plane of plane)

The tertiary carbon atoms in polyalkene chains are not strictly asymmetric since two substituents are constructed from infinite monomer sequences, and are thus similar, and so the symbols of asymmetric carbon atoms, *R* and *S*, can not be usefully applied. Normally the symbols D and L are used to designate the two possible configurations of the tertiary carbon atoms in polyalkene chains: DDDDDDD or LLLLLLL, an isotactic sequence; DLDLDLDL, a syndiotactic sequence; DLDDLDDDLLDLLLDDDDLL, an atactic sequence.

In reality the polymer chain never adopts the flat extended forms shown in Figure 1 but is coiled in a helix since such a structure results in the least crowding between the hydrogen atoms. The form of the helix depends on the substituent R. Helical structures for isotactic and syndiotactic polypropylene are shown in Figure 2. Normally only one helical form is stable under given conditions but sometimes different helical forms can coexist depending on the melt crystallization conditions, *e.g.* isotactic polybutene-1.

The polymerization of propylene by heterogeneous Ziegler–Natta catalysts produces a mixture of isotactic and atactic polymer, and one of the many significant contributions made by Natta and his

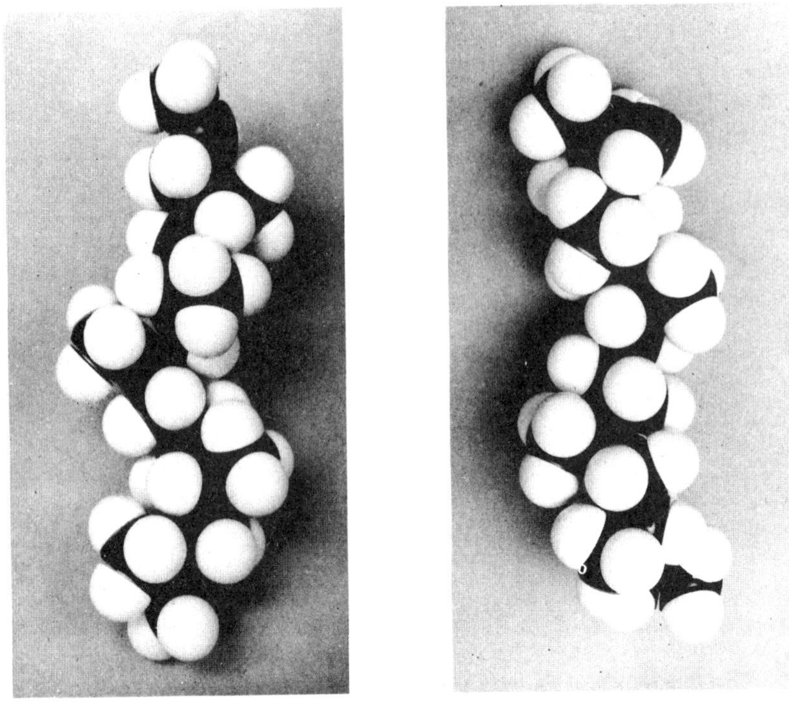


**Figure 2** Chain models of helices of (a) isotactic and (b) syndiotactic polypropylene (reproduced from ref. 21)

co-workers[32] was a rapid method they devised for catalyst characterization which was based on simple stereoisometric fractionation procedures involving the use of boiling solvents, *e.g.* ether, heptane, xylene, *etc.* However, different solvents tend to extract varying fractions of lower molecular weight material, as well as removing atactic polymer. The most common procedure adopted in industrial laboratories is extraction using boiling *n*-heptane, and the percentage of polymer insoluble in this solvent under these conditions is normally referred to as the Isotactic Index (II). It should, however, be realized that the term atactic is used loosely since the soluble fraction may contain chains made up of blocks of isotactic or syndiotactic placements in addition to purely atactic material.

## 1.4 POLYMERIZATION CATALYSTS

As has been remarked the history of the development of high activity catalysts for ethylene and α-alkene polymerization falls conveniently into three periods.[33] During the early period, from 1955 and onwards, developments were concerned with solid $TiCl_3$-based catalysts and the so-called first generation of industrial catalysts was produced. During a subsequent period higher activity catalysts were formulated. These catalysts were based on the use of appropriate additives and used with considerable success for the polymerization of propylene. Finally, during a third period, catalysts for the polymerization of propylene of spectacular activity, based on the use of titanium tetrachloride, magnesium dichloride and electron donors, have been discovered. As will be appreciated these three periods overlap each other considerably.

Although it is common practice at the present time to refer to the particular generation of a given catalyst it is considered of more scientific value to divide Ziegler–Natta and related catalysts into four general types:[20] (i) heterogeneous Ziegler–Natta catalysts; (ii) homogeneous Ziegler–Natta catalysts; (iii) modified Ziegler–Natta catalysts; and (iv) transition–metal alkyl catalysts.

### 1.4.1 Heterogeneous Catalysts

#### *1.4.1.1 Non-supported heterogeneous catalysts*

*(i) First generation titanium trichloride catalysts*

Ziegler's discoveries[2] were concerned with the use of an *in situ* prepared catalyst which was produced as a precipitate by the reaction of a soluble transition metal halide with a soluble metal alkyl, the reaction being carried out in an inert diluent. Natta, however, found it to be advantageous to use *preformed* solid transition metal halides, usually in lower valence states (Scheme 1). Preformed solid catalysts of this type were found to produce higher yields of isotactic propylene. The establishment by Natta *et al.*[34] that titanium trichloride could exist in four crystalline modifications had a significant effect on the course of catalyst development since the interconversion procedures have been used frequently in subsequent catalyst preparations.

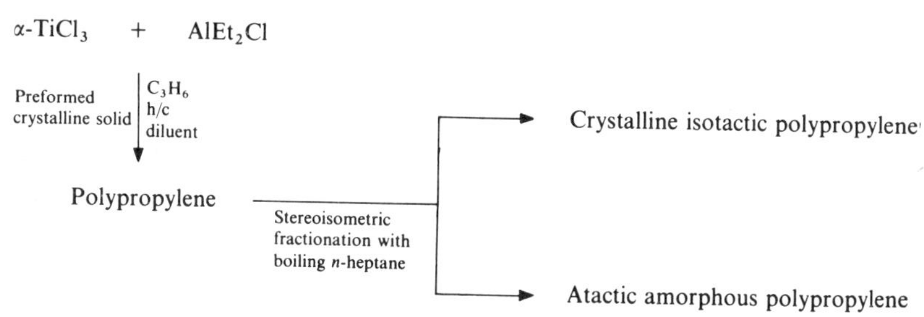


**Scheme 1**

The $\alpha$-form can be prepared by reduction of $TiCl_4$ with $H_2$ at temperatures above 400 °C, usually at 800 °C.[34] Reduction with Al in hydrocarbon medium at 250 °C produces $\alpha$-$TiCl_3 \cdot 0.33\ AlCl_3$ in which the $AlCl_3$ is cocrystallized with the $\alpha$-$TiCl_3$. $\beta$-$TiCl_3$ is obtained by reducing $TiCl_4$ with Al or aluminum alkyl compounds at low temperatures (usually in the range 0–100 °C) or by reduction with $H_2$.[35] $\gamma$-$TiCl_3$ is obtained by heating the $\beta$-form above 150 °C or reducing $TiCl_4$ at temperatures in the range 150–200 °C, reduction of $TiCl_4$ with Al yields a product having the $TiCl_3 \cdot 0.33\ AlCl_3$ composition.[36] $\delta$-$TiCl_3$ is obtained by prolonged grinding of $\alpha$- or $\gamma$-$TiCl_3$. The $\alpha$- and $\gamma$-forms have lower activities than the $\delta$-form whilst the $\beta$-form produces polymer which is essentially amorphous.[37]

Most first generation non-supported catalysts are of the solid solution type, $\delta$-$TiCl_3 \cdot 0.33\ AlCl_3$ and owe their origins to the discoveries of Wilchinsky, Looney and Tornqvist[38–41] who discovered that ball milling the product obtained from the reduction of $TiCl_4$ by aluminum metal or aluminum alkyl produced a more active catalyst than pure $\alpha$-$TiCl_3$. Wilchinsky, Looney and Tornqvist[41] concluded from their studies that during the milling process both the $\alpha$- and $\gamma$-forms were converted into the $\delta$-form which displayed double-layer stacking. The catalyst preparation procedure can be illustrated as in Scheme 2, in which the reactions are carried out under carefully controlled conditions so as to regulate the size and shape of the catalyst particles.

Most first generation commercial catalysts are of this solid solution type and have been dry milled and heat treated and have specific surface areas in the range of 40 $m^2\,g^{-1}$.

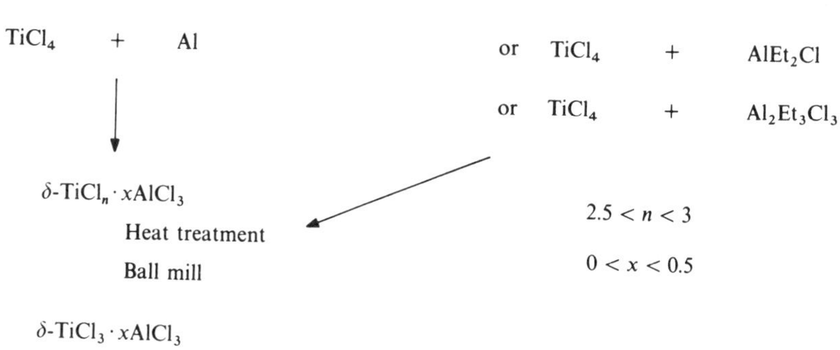


**Scheme 2**

The morphology of $\delta$-$TiCl_3$ has important effects on the catalyst activity and kinetic behaviour. The primary particles of $\delta$-$TiCl_3$ are 0.03–0.7 $\mu$m in diameter[42,43] and agglomerate to form secondary particles of 20–40 $\mu$m.[43] These secondary particles are loosely constructed with voids between the primary particles. During the initial stages of polymerization these voids are rapidly filled with polymer leading to disintegration into primary particles.[41,43,44] The rate at which polymer is formed in the initial period determines whether the separate primary particles either disperse in the slurry or remain loosely bound into aggregates having the shape of the secondary particles but greatly increased in size.

The activity and stereospecificity of these catalysts also depends on the type of aluminum alkyl compound used as cocatalyst. Use of triethylaluminum gives a much higher activity but a much lower stereospecificity, and for this reason diethylaluminum chloride is used for the polymerization of propylene and other $\alpha$-alkenes, whilst a trialkylaluminum, *e.g.* $AlEt_3$, $Al(Bu^i)_3$, *etc.*, is used for ethylene polymerization.

### (ii) Donor modified titanium trichloride catalysts

It was observed early in the development of Ziegler–Natta catalysts that the addition of certain Lewis base compounds could increase the stereospecificities of propylene polymerization systems. A very wide range of donors such as ethers, esters, ketones, amines, amides, phosphines, organophosphorus compounds, polymeric derivatives, *etc.*, can be used and are often ball milled with the $\delta$-$TiCl_3 \cdot 0.33\ AlCl_3$ catalyst. Most commercial companies involved in alkene polymerization have their own preferred recipes.

The mode of addition of the donor and the conditions of concentration and temperature used in the preparation stages as well as the intrinsic nature of the donor are all important in determining the success or otherwise of a particular complexing agent.

Whilst the use of donors often produces catalysts with higher stereospecificities for $\alpha$-alkene polymerization they often adversely affect catalyst activity. However, this is not always the case and catalysts not only with higher stereospecificities but also with higher activities can sometimes be prepared by the judicious addition of an appropriate donor.

Whilst the deliberate addition of electron donor type compounds to catalyst systems from which all trace impurities such as $H_2O$, $O_2$, CO, allene, propadiene, *etc.*, have been carefully excluded may seem paradoxical there are good fundamental principles which govern the selective reactions of selected donors. The mechanism by means of which electron donors may affect activity and stereospecificity are complex but may be formulated as follows.

(*a*) Complex formation of donor (D) with active centres (S), which can lead to the formation of hindered centres, or centres of increased stability and chirality, or indeed blocked centres, depending on the concentration and nature of the donor D. A particularly important reaction, and one which is believed to operate in many donor-modified catalyst systems, is that of the donor with more exposed and less stereospecific centres, leading to an increased stereospecificity and a reduced activity.

(*b*) Complex formation of donor with aluminum alkyl (A)

$$A + D \rightleftharpoons A \cdot D \tag{2}$$

This type of reaction will lead to a reduced alkylation ability and a reduction in the reducing capacity of the aluminum alkyl as well as a reduction in chain transfer to aluminum alkyl.

(*c*) Strong complex formation with aluminum alkyl dihalides, produced during catalyst formation or surface alkylation reactions, and leading to increased catalyst activity.

(*d*) Catalyst modification reactions such as the removal of $AlCl_3$ from $\delta$-$TiCl_3 \cdot 0.33\ AlCl_3$ type catalysts.

Table 1 lists some typical examples of the use of electron donors as additives in catalyst preparation.

An important landmark in the development of catalysts with both increased activity and stereospecificity was the discovery in 1973 by Solvay & Cie of ether treated $TiCl_3$-based catalysts.[45,46] Their patents describe in a most elegant and accurate manner the necessary details of the catalyst preparation, and such a method is given in the following paragraph.

$TiCl_4$ in an inert diluent such as *n*-heptane is reacted under mild conditions with $AlEt_2Cl$ at a temperature of $0 \pm 2$ °C over a period of about 3 h to give the solid product, $\beta$-$TiCl_3 \cdot x'(AlCl_3) \cdot z(AlEtCl_2)$, where $x' - 0.15$ and $z - 0.20$. This product is then treated with diisoamyl ether at 35 °C for 1 h when much of the $AlCl_3$ and $AlEtCl_2$ is removed from the solid matrix. The resulting solid is again isolated and heated with $TiCl_4$ at 65 °C for 2 h to yield a violet solid of composition

**Table 1** Some Selected Lewis Bases Used as Catalyst Additives

| *Company* | *Catalyst System* | *Patent Number* |
|---|---|---|
| Montecatini | $TiCl_3/AlEt_3$ + pyridine | US 3 139 418 (1964) |
| Eastman Kodak | $TiCl_3/AlEt_3$ + HMPT | US 3 186 977 (1965) |
| Phillips Petroleum Co. | $TiCl_3 \cdot 0.33AlCl_3/AlEt_2Cl$ + aromatic esters | US 3 317 498 (1967) |
| Mitsui Petrochem. Ind. | $TiCl_4$ reduced with aluminum alkyl; $TiCl_3$ produced ground with organonitrogen or organophosphorus compound | Jpn. 79 016 479 (1972) |
| Kazan Kirov | $TiCl_3/AlEt_2Cl$ + organophosphorous compound | USSR 434 746 (1972) |
| Toho Titanium KK | $TiCl_3 \cdot 0.33AlCl_3$ milled with $(EtO)_4Si$ and methyl methacrylate | Jpn. 50 103 495 (1974) |
| Plastipolimer Res. | $TiCl_3/AlEt_2Cl$ + diisopropyl ether | USSR 495 326 (1975) |
| BASF AG | $TiCl_3 \cdot 0.33AlCl_3$ milled with $Bu_3P/AlEt_2Cl$ | Ger. 2 519 582 (1975) |
| Shell Int. Res. Mij. BV | —$TiCl_3$ prepared from $TiCl_4 + AlEt_3$ heated to 120 °C with BuCl to give violet $TiCl_3$ | Belg. 845 791 (1975) |
| Stauffer Chemical Co. | $TiCl_3 \cdot 0.33AlCl_3$ milled with $Ph_3PO$ or benzophenone | Ger. 2 657 917 (1975) |
| Mitsubishi Chem. Ind. KK. | $TiCl_3$ treated with thioether as used with *n*-propyl or *n*-hexyl aluminum alkyl | Ger. 2 905 330 (1978) |
| BASF AG | $TiCl_3 \cdot nAlCl_3$ ($n = 0.01–1$) ball milled with an ester, $RO\text{-}COR_2$ or $R_1OCOCR_2CHR_2$ and reacted with an alcohol, *e.g.* $TiCl_3 \cdot 0.33AlCl_3$ and ethyl phenyl acetate, ball milled and reacted with $Bu^nOH$ | Ger. 3 021 297 (1980) |
| Dart Industries Inc. | Al reduced $TiCl_3$, ball milled with an electron donor such as $R_1OR_2$, and heat treated so as to remove $R_1Cl$ from the catalyst, *e.g.* $TiCl_3 \cdot 0.33AlCl_3$ ball milled with anisole | US 4 382 018 (1979) |
| Northern Petrochem. Co. | $TiCl_3 \cdot 0.33AlCl_3$ ball milled with benzoic acid ester, *e.g.* butyl benzoate | Br. 2 109 800 (1981) |

$\alpha\text{-}TiCl_3 \cdot Al(R_nX_{3-n}) \cdot C_y$, where C is the complexing agent (diisoamyl ether) and $0 > n > 2$ and $0.11 > y > 0.009$. The best results are reported when $n = 1$. The catalyst produced is of high porosity and has a high surface area, usually greater than $150\,m^2\,g^{-1}$, and may be used in conjunction with $AlEt_2Cl$ to polymerize propylene and other α-alkenes.

Ether-treated catalysts of this type have a five-fold increase in activity in propylene polymerization over conventional $\delta\text{-}TiCl_3 \cdot 0.33\ AlCl_3$ catalysts and excellent stereospecificities, producing only between 2 and 5% amorphous polypropylene. In slurry type polymerizations these catalysts produce polymer particles with very good particle size distribution and shape characteristics due to the rapid disintegration of the catalyst matrix on treatment with $AlEt_2Cl$ and monomer.

It is emphasised that the development of this class of catalysts represented an important stage in catalyst development. The prescribed preparation involves two highly innovative and significant reactions which are performed sequentially on the same catalyst system, *viz.* treatment with an ether (or similar compound) so as to preferentially extract most of the $AlCl_3$ and complexed $AlCl_2$ and leaving behind a porous and weakly bonded matrix, and the transformation by heat treatment at 65 °C of the $\alpha\text{-}TiCl_3$ into a more active form.

The behaviour of catalysts of this type have been analyzed by Nielson.[47] A selection of catalysts is listed in Table 2.

### *1.4.1.2 Supported Ziegler–Natta catalysts*

A parallel course of catalyst investigation has led to exciting and new catalysts for the polymerization of both ethylene and propylene. The use of supports in heterogeneous catalysis was well understood by 1950 and was highlighted by the discovery in 1958 by Hogan and Banks[6] of highly active chromium trioxide catalysts supported on silica which could polymerize ethylene under mild conditions of temperature and pressure to produce predominantly linear, high density polyethylene.

**Table 2** Selected Metal Alkyl Reduced Catalysts — Donor Treated

| *Company* | *Catalyst System* | *Patent Number* | *Chemical Abstract* |
|---|---|---|---|
| Mitsubishi Chem. Ind. KK | Solvay-type preparation: $I_2$ added at the beginning to increase stereospecificity | Jpn. 54 093 691 (1974) | **91**, 212 024 |
| Solvay & Cie | Solvay preparation: final $TiCl_4$ treatment replaced by $CCl_4$ | Fr. 2 320 309 (1975) | **87**, 202 358 |
| Shell Int. Res. Mij. BV | $TiCl_4$ reduced with $AlEt_3$ at $-100\,^\circ C$ to give —$TiCl_3$; washed with toluene | Neth. 7 602 920 (1975) | **86**, 122 054 |
| Exxon Res. & Eng. | Solvay-type preparation: use of diisopentyl ether | Br. 1 557 329 (1975) | **92**, 147 519 |
| Naphthachemie SA | $TiCl_4$ reduced with $AlEt_2Cl$ in the presence of diisopentyl ether | Ger. 2 807 361 (1977) | **89**, 164 197 |
| Sumitomo Chem. KK | $TiCl_4$ in toluene reduced with $AlEt_2Cl$ in the presence of $Bu_2O$; treated with $I_2$ at $100\,^\circ C$ | Belg. 104 670 (1977) | **90**, 104 670 |
| Mitsubishi Chem. Ind. KK | Reduce $TiCl_4$ with aluminum alkyl, treated with poly(dimethylsiloxane) in toluene | Jpn. 54 079 195 (1977) | **91**, 124 686 |
| Tokuyama Soda Co. | Solvay-type preparation: $TiCl_4$ reduced with $AlEt_2Cl$; treated with diisopentyl ether then $TiCl_4$ polymerization stage | Jpn. 54 127 995 (1977) | **92**, 127 995 |
| Chisso Corp. | Solvay-type preparation: $TiCl_4$ reduced with $AlEt_2Cl$; product treated with diisopentyl ether and $TiCl_4$; treated with methyl hydrogen polysiloxane | Jpn. 55 009 657 (1978) | **93**, 027 093 |
| Montedison Spa | $TiCl_4$ reduced with aluminum alkyl in the presence of diatomaceous support; treated with dibutyl ether | Eur. 5824 (1978) | **92**, 199 063 |
| Gulf Oil Corp. | $TiCl_4$ reduced with $Al_2Et_3Cl_3 + AlE_2Cl$; treated with di-*n*-butyl ether then $CO_2$ | US 4 330 652 (1980) | — |
| Standard Oil (Amoco) | $TiCl_4$ reduced with $AlEt_2Cl$; product treated with propylene; reacted with di-*n*-butyl ether and diisoamyl ether, then $TiCl_4$ | Belg. 890 678 | — |
| Chisso Corp. | $TiCl_4$ added to $AlEt_2Cl$ diisoamyl ether; propylene added, product treated with diisoamyl ether + $TiCl_3$, used with $AlEt_2CO$ | Belg. 839 913 (1982) | — |

At about the same time Natta and co-workers[48] demonstrated that only a small percentage of the titanium atoms (< 1%) in first generation preformed catalysts was active. It was evident that higher catalyst activities were likely to be achieved through the use of transition metal compounds supported on appropriate matrices. Over the years a wide range of supported catalysts have been successfully developed and used for the commercial production of ethylene, and more recently of propylene also.

The term *supported catalyst* is used in a very wide sense and includes not only systems in which the transition metal compound is linked to the substrate by means of a chemical bond but also systems in which the transition metal atom may occupy a position in a lattice structure, or where complexation, absorption, or even occlusion may take place.

### *(i) Catalysts based on reaction products of hydroxyl-containing compounds with transition metal compounds*

The use of $SiO_2$, $Al_2O_3$, *etc.* to chemically anchor transition metal compounds has been widespread since the early 1960s. Heat treatment of $SiO_2$ can control the number and type of surface hydroxyl groups and indirectly the amount and distribution of transition metal atoms which are anchored to the surface (equation 3).[54–60]

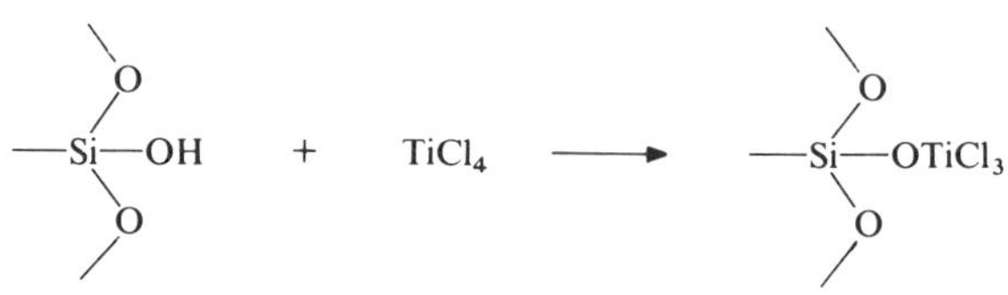


(3)

Chemically anchored transition metal catalysts have been used industrially for the preparation of high density linear polyethylene and ethylene–α-alkene copolymers. Karol[61] has divided such catalysts into three classes: (i) metal oxide, particularly $CrO_3/SiO_2$; (ii) Ziegler, particularly $Mg(OH)Cl/TiCl_4 + AlR_3$; and (iii) organotransition metal, particularly $(C_5H_5)_2Cr/SiO_2$.

Many of these catalyst systems exhibit high activity in the polymerization of ethylene, but are not, however, useful for the polymerization of propylene and α-alkenes. Typical examples are listed in Table 3.

### *(ii) Catalysts based on reaction products of magnesium alkoxides with transition metal compounds*

Superactive catalysts for ethylene polymerization have been prepared by reacting titanium tetrachloride with magnesium alkoxides. The original structure of the alkoxide is usually destroyed during reaction and new species of increased surface area produced.

**Table 3** Catalysts Based on Reaction Products of Surface Hydroxyl-containing Compounds with Transition Metal Compounds

| *Catalyst* | *Company* | *Patent Number* |
|---|---|---|
| Transition metal compounds reacted with $SiO_2$, $Al_2O_3$ $SiO_2$–$Al_2O_3$, $ZrO_2$, $TiO_2$, ThO, MgO, *etc* | Cabot | Br. 969 761 (1960–1963)<br>969 767<br>916 132<br>1 038 882 |
| Transition metal compound reacted with a hydroxy chloride of a bivalent metal preferably Mg(OH)Cl | Solvay & Cie | Belg. 650 679 (1963)<br>Br. 1 024 336 (1963) |
| Transition metal halide reacted with $Mg(OH)_2$ | Mitsui Petrochem. Ind.<br>Solvay & Cie<br>Montecatini<br>Hoechst AG | Jpn. 70 40 295 (1967)<br>Belg. 726 839 (1968)<br>Belg. 728 002 (1968)<br>Belg. 735 291 (1968) |
| Transition metal benzyl and allyl compounds reacted with $SiO_2$, $Al_2O_3$ $Al_2O_3$–ZnO, $Al_2O_3$–MgO | Imperial Chemical Industries | Br. 1 265 747 (1969)<br>Br. 1 314 828 (1969) |
| Bis(triphenylsilyl) chromate reacted with $SiO_2$–$Al_2O_3$ or $SiO_2$ | Union Carbide | US 3 324 095 (1967) |

Böhm[62] has described the use of a highly active catalyst containing 8.5% by weight Ti, derived from $Mg(OEt)_2$ and $TiCl_4$ and using $AlEt_3$ as cocatalyst. The solid catalyst produced from $Mg(OEt)_2$ and $TiCl_4$ was amorphous and had a surface area of $60\,m^2\,g^{-1}$ and a high porosity, 61% of the catalyst particles consisting of cavities. These hollow-type structures would account for the complete disintegration of the catalyst particles observed during polymerization. Final catalyst residues were reported to be smaller than 0.0005 $\mu$m in diameter.

High ratios of Al:Ti are required for maximum activities. Böhm reports the use of Al:Ti of 80–100:1. These catalysts show no settling periods and reach their high activity right at the start of the polymerization.

Some typical examples are listed in Table 4.

**Table 4** Selected Catalysts Based on Reaction Products of Magnesium Alkoxides with Transition Metal Compounds

| *Catalyst* | *Company* | *Patent Number* |
|---|---|---|
| 1. Transition metal halide reacted with $Mg(OR)_2$, *etc.* | Hoechst AG | Belg. 737 778 (1968)<br>US 3 644 318 (1968) |
| 2. Transition metal halide reacted with $Mg(OR)_2$, *etc.* | Solvay & Cie | Belg. 743 325 (1969) |
| 3. Transition metal compound reacted with $Mg(OR)_2$ and acid halide | Hoechst AG | Belg. 780 530 (1971) |
| 4. $Ti(OPr^i)_4 + Mg(OEt)_2 + SiCl_4$ | Hoechst AG | Br. 1 357 474 (1974) |
| 5. $TiCl_4 + Mg(OEt)_2 +$ THF, *etc.* | Hoechst AG | Ger. 3 231 031 (1984) |
| 6. $(Mg(OEt)_2 + AlEt_3) + TiCl_4 +$ 1,2-dichloroethane + EB | Standard Oil (Amoco) | US 4 567 155 (1986) |
| 7. $Mg(OEt)_2 + TiCl_4 +$ EB | Shell Int. Res. Mij. BV | Eur. 0 159 736 (1985)<br>US 4 535 068 (1985) |
| 8. $(Mg(OEt)_2 + SiHCl_3) + TiCl_4$ | Toa Nenryo Kogyo KK | Eur. 0 227 455 (1987) |
| 9. $Ti(OBu)_4 + Mg(OSiMeBuPh)_2$ | Conoco | US 4 440 869 (1984) |
| 10. $Mg[OCH(CF_3)_2]_2 + TiCl_4 + HSiCl_3$ | Phillips Petroleum Co. | US 4 559 317 (1985) |

### *(iii) Catalysts based on reaction products of magnesium alkyls and titanium compounds*

The reactions between magnesium alkyls and titanium compounds have been used as a means of preparing highly active catalyst systems. Stamicarbon[6] have prepared highly active catalysts using $AlEt_2Cl$, $Bu_2Mg$ and $TiCl_4$. Shell International Research[65,66] have used the reduction of $TiCl_4$ with organomagnesium compounds to prepare highly active catalysts for ethylene polymerization. These catalyst systems have been extensively investigated by Haward and co-workers,[67–69] as well as by Radenkov *et al.*[70–73] Characterization studies have revealed that the catalysts contain appreciable amounts of magnesium chloride and that they have nodular structures. The size of the primary particles is believed to be less than 0.05 $\mu$m.

### *(iv) Catalysts based on reaction products of magnesium chloride with transition metal compounds*

Some of the most promising catalysts, especially for propylene polymerization, have resulted from the use of magnesium chloride as supports. An early patent in 1960 described the use of ground $MgCl_2$ with electron donors in propylene polymerization.[74] However, it was the discoveries by Montecatini–Edison Co.[75] and Mitsui Petrochemicals Ind.[74] that catalysts, prepared from $MgCl_2$, $TiCl_4$ and electron donors, and activated by a mixture of trialkylaluminum and an electron donor, could polymerize propylene with a high yield [$>$ 50 kg PP $(g\,Ti)^{-1}$] and with good stereospecificity (isotactic index $>$ 90%), that have set the scene for much of the present explosion in catalyst development. The initial patents describe two basic routes for catalyst preparation.

Typically a 30-fold molar excess of very carefully dried anhydrous $MgCl_2$ is ball milled with a $TiCl_4$–electron donor complex, such as $TiCl_4\cdot$ethyl benzoate, washed with *n*-heptane, and dried. The catalyst is then activated using a cocatalyst mixture consisting of trialkylaluminum (TAA) and ethyl benzoate (EB) such that TAA:EB:Ti = 300:10:1.

Alternatively, dried anhydrous $MgCl_2$ is ball milled for 20 h at 0–5 °C with EB ($MgCl_2$:EB = 1:0.15) and then treated with neat $TiCl_4$ at 80–130 °C for 2 h, washed with *n*-heptane, and finally dried, yielding a pale yellow solid catalyst containing typically 1–5% Ti and 5–20% EB. This catalyst is then activated by treatment with a mixture of trialkylaluminum and an electron donor (*e.g.* ethyl benzoate, *p*-ethyl anisate, *p*-methyl toluate, *etc.*).

Since 1968 many hundreds of related catalysts have been claimed in the patent literature for ethylene, propylene and other α-alkene polymerization. Some selected examples are listed in Table 5 with additional examples to be found in the literature.

**Table 5** Some Selected Magnesium Chloride Catalysts

| *Company* | *Catalyst System* | *Patent Number* |
|---|---|---|
| Phillips Petroleum Co. | $MgCl_2$ milled with $TiCl_4$ ethyl benzoate complex and durene; used with $AlEt_3$ + ethyl anisate | Belg. 843 224 (1975) |
| Montedison Spa | $MgCl_2$ milled with $TiCl_4$ ethyl benzoate complex; used with $AlEt_3$ + methyl *p*-toluate | Belg. 845 593 (1975) |
| Mitsui Petrochem. Ind. KK | $Mg(OPh)_2$ milled with ethyl benzoate; reacted with $TiCl_4$; used with $AlEt_3$ and *p*-toluate | Belg. 856 189 (1976) |
| Mitsui Toatsu Chem. Inc. | $MgCl_2$ milled with $AlCl_3$ ethyl benzoate; treated with $TiCl_4$; used with $AlEt_3$ | Ger. 2 734 652 (1976) |
| Mitsui Petrochem. Ind. KK | $MgCl_2$ milled with ethyl benzoate and silicone oil; *p*-cresol added then $AlEt_3$; then $TiCl_4$; used with $AlEt_3$ + methyl *p*-toluate | Ger. 2 809 318 (1977) |
| Exxon Res. and Eng. Co. | $MgCl_2$ milled with $TiCl_4$ ethyl benzoate complex; product treated with $TiCl_4$; used with $(Me_3SiCH_2)_2AlEt$ | Eur. 4739 (1978) |
| BASF AG | $Mg(OEt)_2$ milled with PhCOCl; treated twice with $TiCl_4$; used with $Et_3Al$ + ethyl anisate | Ger. 2 831 830 (1978) |
| Montedison Spa | MgO treated with $SOCl_2$; product treated with ethyl benzoate and EtOH and then $TiCl_4$ | Belg. 875 494 (1978) |

'Activated' $MgCl_2$, a key ingredient of these catalyst systems, has been known for some time, being mentioned by Bryce-Smith[77] in 1959, and described by Kamienski[78] in 1967. A number of preparation methods are available and include; (i) treatment of $MgCl_2$ with activating agents such as electron donors; (ii) ball milling $MgCl_2$; and (iii) reaction of compounds such as Grignard reagents with chlorinating agents. All three procedures have been found useful in catalyst preparation.

*(a) Effect of ball milling.* Magnesium dichloride seems almost uniquely suitable as a support in propylene polymerization, other possible substitutes including calcium hydride[79] and manganese dichloride. The reason for this unique behaviour is believed to be associated with similarities between the ionic radii of $Ti^{4+}$ and $Mg^{2+}$ which are 0.68 and 0.65 Å respectively. Magnesium dichloride may exist as a cubic close-packed structure of double chlorine layers with interstitial $Mg^{2+}$ ions in six-fold coordination.[80] This layer structure is of the $CdCl_2$ type and leads to a characteristic X-ray diffraction spectrum with a strong reflection 104 at $d = 2.56$ Å.[81] Additionally magnesium dichloride can exist in a hexagonal close-packed structure. This structure is less stable and is similar to that of α-$TiCl_3$ with a strong reflection 104 at $d = 2.78$ Å.

The introduction of structural faults is believed to explain the activity of ball-milled magnesium dichloride.

Giannini has reported that during ball milling the X-ray reflection at $d = 2.5$ Å gradually disappears and is replaced by a broad halo at $d = 2.65$ Å, a position intermediate between reflections 104 for cubic and hexagonal structures, and Giannini has attributed this behaviour as arising from stacking faults induced by the rotation of Cl—Mg—Cl layers.[81] From their extensive studies on ball-milling Galli *et al.*[82] have concluded that the best model describing the structural disorder of activated $MgCl_2$ (in the presence of $TiCl_4$) is similar to that proposed for δ-$TiCl_3$.[83] Additional factors such as surface area and porosity are also believed to be important in determining activity. However, studies by Gerbasi *et al.*[84] would indicate that activity is related to the introduction of defects alone.

Representations of commercial and ball-milled $MgCl_2$ crystallites are shown in Figure 3.

Goodall[85] has proposed the existence of three different types of surface magnesium ions in milled $MgCl_2$: (i) type 1, present on lateral faces, with single vacancy Mg ions, having an effective charge of

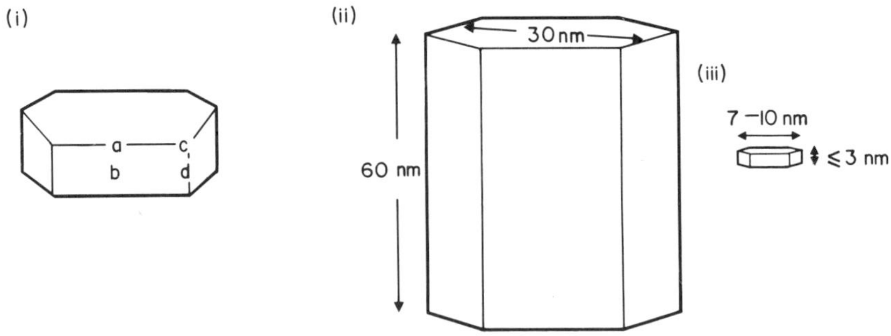


**Figure 3** Influence of milling on magnesium chloride: (i) $MgCl_2$ crystallite, in which a = basal edge, b = lateral facet, c = corner, d = lateral edge; (ii) commercially available $MgCl_2$; and (iii) ball milled $MgCl_2$

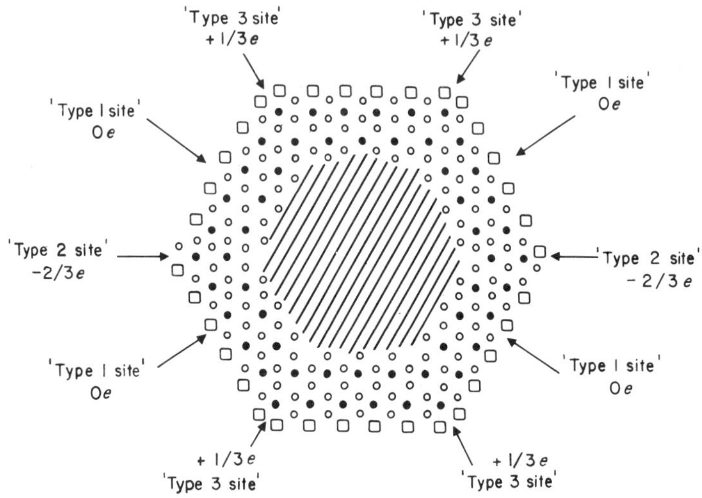


**Figure 4** $MgCl_2$ crystallite (two-dimensional view), in which ● = Mg, ○ = Cl and □ = vacancy

0*e*; (ii) type 2, present on corners, with single vacancy Mg ions having an effective charge of $-2/3e$; and (iii) type 3, present on corners, with double vacancy Mg ions having an effective charge of $+1/3e$. The various types of surface magnesium ions are shown in Figure 4.

As the size of the crystallite decreases, the fraction of double vacancy Mg ions increases more rapidly than that of single vacancy ions.

During the ball-milling procedure of $MgCl_2$ in the presence of ethyl benzoate, the agglomerates are broken down to yield primary crystallites, when the freshly cleaved surfaces are coated with ethyl benzoate preventing reaggregation (*cf.* protected colloids). At the same time absorption of ethyl benzoate between layers in the $MgCl_2$ structure, or even solid solution formation can take place.

*(b) Interaction of $MgCl_2$ with electron donors.* The use of aromatic esters as internal donors seems the only choice for high stereospecificities although some sterically hindered alcohols offer distinct possibilities. Keszler *et al.*[86] have concluded from thermal studies that rapid adsorption of ethyl benzoate occurs followed by slower complex formation. Chien *et al.*[87] have investigated the effects of ball milling $MgCl_2$ with ethyl benzoate by means of BET, mercury porosity and X-ray techniques, and have concluded that the milled material (60 h) has higher surface areas, (5.1–7.3 $m^2\,g^{-1}$), smaller pore radii, higher pore surface area and smaller crystallite dimensions than unmilled material.

The interaction between $MgCl_2$ and ethyl benzoate has been extensively studied using IR techniques. Goodall[85] has reported shifts in the C═O stretching frequency from 1721 $cm^{-1}$ in the free ester to 1678–1688 $cm^{-1}$ in the $MgCl_2$ complex. Kashiwa[88] and Chien[89] report similar observations. Goodall[85] concluded that virtually all the surface sites were complexed with ethyl benzoate. Absorption at around 1680 $cm^{-1}$ was assigned to molecules bound *via* their carbonyl groups to the basal edges and corners of crystallites, whilst that around 1700 $cm^{-1}$ was assigned to

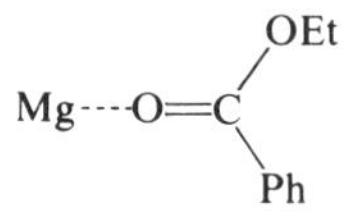

**(1)**

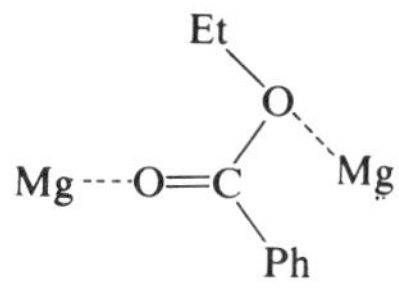

**(2)**

molecules adsorbed at lateral faces. Chien *et al.*[89] have proposed complexes (**1**) and (**2**) to explain the observed shifts.

*(c) Effect of $TiCl_4$ treatment.* The second step in the catalyst preparation involves treatment of the solid products from ball milling with hot neat $TiCl_4$ at 80–130 °C. During treatment with $TiCl_4$ some ethyl benzoate is extracted by the $TiCl_4$ and a typical catalyst will contain only 5–20% ethyl benzoate after treatment. Ethyl benzoate is lost not only from weakly bonded sites on the lateral faces (1700 $cm^{-1}$ absorption) but also from the more strongly coordinating corner sites (1650 $cm^{-1}$ absorption). It is believed that the $TiCl_4$ coordinates to these sites by replacing the ethyl benzoate molecules.

Electron micrographs of polymer growth when using $MgCl_2$ supported catalysts show[90] that growth is more evident on the edges and corners rather than on the basal faces, which is consistent with the belief that these sites may accommodate Ti atoms with higher activity.

The oxidation state of the titanium complexed to $MgCl_2$ has been investigated by several authors.[91] Baulin *et al.*[92] have concluded from electron paramagnetic and elemental analytical studies that all the titanium exists as $Ti^{IV}$. Chien's CW catalyst is more complex since the procatalyst preparation involved $AlEt_3$ treatment.[89] Figures of 8% $Ti^{II}$, 38% $Ti^{III}$ and 54% $Ti^{IV}$ have been reported for this catalyst. On the basis of their results Chien and Wu[93] have proposed structures (**3**), (**4**) and (**5**) for the surface $Ti^{III}$ species. Activation of the procatalyst by a cocatalyst system involves reduction of $Ti^{IV}$ to lower valence states.

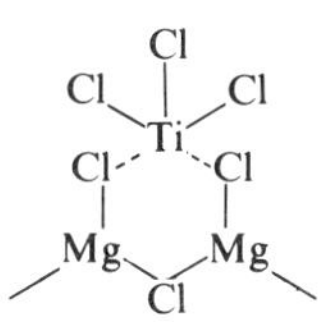

**(3)**

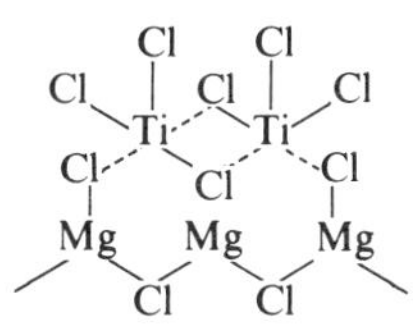

**(4)**

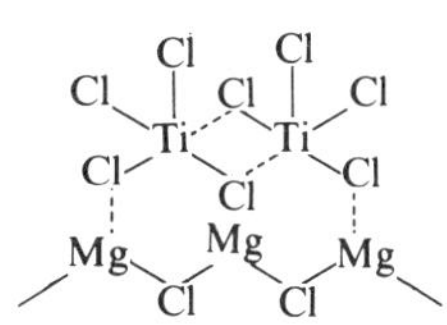

**(5)**

*(d) The cocatalyst system.* These $MgCl_2$/EB/$TiCl_4$ catalyst systems may be activated with trialkylaluminum compounds and used for the polymerization of ethylene. However, for propylene polymerization the addition of an electron donor (external donor) to the trialkylaluminum solution is necessary in order to achieve high stereospecificity. Benzoates, toluates, anisates and various amines have been successfully used with $MgCl_2$/EB/$TiCl_4$ catalyst systems, whilst various di- and tri-alkoxysilanes have been used with $MgCl_2$/phthalate/$TiCl_4$ systems.

The role of the external donor is complex and includes the following reactions.[33] (i) Complex formation with the aluminum alkyl compound leading to a reduction in its capacity to reduce the titanium entity. (ii) Deactivation of the specific centres by selective complex formation or even conversion of specific centres into isospecific centres. (iii) Alkylation and reduction of the donor compounds; these reactions have been extensively investigated and it has been suggested that the products may themselves act as catalyst modifiers.

The exact role of external electron donors is still unresolved but their presence is a crucial feature of these polymerizations, affecting both catalyst activity and stereospecificity. Under experimental conditions where extensive chemical reaction between the donor and aluminum alkyl takes place[85,94] the use of large concentrations of alkyl are necessary. Hence the use of modifiers already built into the procatalyst system itself may obviate the need to use high alkyl concentrations.

*(e) Super high activity catalysts.* A new class of super high activity $MgCl_2$-supported catalysts (SHAC) have been developed by Shell[85,97] which are believed to have only one type of active centre. Activities of up to 2000 kg PP (g Ti)$^{-1}$ are claimed (*cf.* 500–1000 kg PP (g Ti)$^{-1}$ for more conventional based catalyst systems) together with high stereospecificities (isotacticity around 95%). In addition lower concentrations of activator are required and the usual inverse relationship between isotactic index and polymer yield is virtually absent.

*(f) The use of phthalate esters as internal donors.* A major disadvantage of $MgCl_2$/EB/$TiCl_4$ catalyst systems is the rapid decrease in activity with time which is normally observed. More stable kinetic rate–time profiles are often shown by catalysts incorporating phthalate esters.

Preparation of phthalate-based catalysts may involve conventional ball-milling procedures or the use of magnesium chloride–alcoholate supports. The latter procedure produces completely spherical catalyst particles and so offers a distinct commercial advantage. A wide range of phthalate esters is available but the diisobutyl and diisooctyl compounds have been found to be the most useful. Longer chain esters may form colloids, whilst distearyl phthalate causes fragmentation of the $MgCl_2$–alcoholate support.

The patent literature is particularly rich with examples of the use of phthalates.[98–100] Details from these refs. are shown in Schemes 3, 4 and 5 respectively.

$MgCl_2$ + 2-ethylhexanol in decane
↓ 120 °C, 2 h
↓ $TiCl_4$ in decane added dropwise, −10 °C, 1 h
↓ 120 °C, 2 h
↓ Diisobutyl phthalate
↓ Stir 120 °C, 2 h
↓ Filter
↓ $TiCl_4$ (neat), 120 °C, 2 h
↓ Filter and wash with hexane
(Catalyst, 2.2–2.5% w/w Ti)

**Scheme 3**[98]

$MgCl_2$ + 2-ethyl hexanol in decane
↓ 130 °C, 2 h
↓ Phthalic anhydride
↓ $TiCl_4$ added dropwise, −20 °C, 20 min
↓ Stir, −20 °C, 1 h
↓ 120 °C
↓ Dioctyl phthalate
↓ Stir, 2 h
↓ Filter
↓ $TiCl_4$ (neat)
↓ 120 °C, 2 h
↓ Filter
(Catalyst, 2.0% w/w Ti)

**Scheme 4**[99]

Mg $(OEt)_2$ (spray dried)
↓ $TiCl_4$, 20 °C
↓ 85 °C
↓ Diisooctyl phthalate
↓ 110 °C, 1 h
↓ Filter
↓ $TiCl_4$, 110 °C, 1 h
↓ Filter, wash
(Catalyst, 2% w/w Ti)

**Scheme 5**[100]

Alkoxysilanes such as dimethoxydiphenylsilane and phenyltriethoxysilane have been used with particular success as external donors for $MgCl_2$/phthalate/$TiCl_4$ catalyst systems.

Although it is apparent that external donors need to be matched with particular internal donors in order to achieve high stereoregularity, the principles which determine donor selection are still poorly understood.

*(g) Catalyst replication effects.* It has been established that the polymer granules produced in propylene polymerization replicate the shape of the catalyst particles.[101–105] This effect is shown in Figure 5.

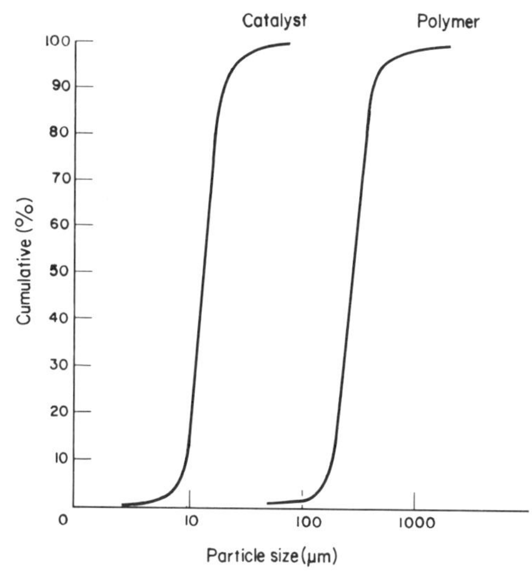


**Figure 5** Particle size distributions for a $MgCl_2$ supported catalyst

There are distinct commercial advantages, particularly for gas phase polymerization, in using spherical catalyst particles of narrow particle size distribution, and as a result there has been considerable effort in recent years to prepare catalysts in this form. The support material typically consists of spheres of 10–40 $\mu$m in diameter, having a porous brittle structure.

An interesting preparative procedure for obtaining spherical catalyst particles has been described by Union Carbide.[106] $TiCl_4$ is added dropwise to a mixture of anhydrous $MgCl_2$ in tetrahydrofuran (THF) and the products heated to dissolve all solids. Pure *n*-hexane is then added slowly when a yellow solid is precipitated which is filtered, washed with *n*-hexane, and then dried to yield the catalyst precursor. This precursor is then activated using a multi-step procedure. First the precursor is mixed with $SiO_2$ (in spherical form) in an aliphatic diluent, and then treated with sufficient aluminum alkyl compound (*e.g.* $AlEt_3$) to achieve partial activation (Al:Ti = 4.8:1). The resulting slurry is then dried under an inert atmosphere. This catalyst may then be activated with $AlEt_3$ (Al:Ti = 10–400:1) and used for the polymerization of ethylene.

## 1.5 HOMOGENEOUS ZIEGLER–NATTA CATALYSTS

Homogeneous Ziegler–Natta catalysts are formed when a transition metal halide and a metal alkyl interact to produce a soluble catalyst. Usually the transition metal halide is itself soluble in the solvent used. However, the polymer is very often precipitated as it is formed.

Commercial use of soluble catalysts has been limited so far to the field of diene polymerization and ethylene–propylene copolymerization. Nevertheless many of these catalyst and polymerization systems have received considerable academic attention in the hope that they might provide models for heterogeneous systems since it was believed that problems associated with surface properties and particle sizes could be avoided. The behaviour of many soluble polymerization systems is very complicated. Complex formation reactions often occur with much of the transition metal being locked up in inactive complexes. In addition, systems may only be colloidally dispersed, whilst in others precipitation of the catalyst as well as the polymer may occur.

### 1.5.1 Homogeneous Vanadium Catalysts

#### *1.5.1.1 Catalysts for syndiotactic polypropylene*

The chief interest in these catalysts is that they can be used under appropriate conditions to prepare syndiotactic polypropylene.[107] Thus $VCl_4$–$AlEt_3$, $VCl_4$–$AlEt_2Cl$, $VCl_4$–$AlEt_2Cl$/anisole and $V(acac)_3$–$AlEt_2X$ (acac = acetylacetonate, X = halogen) polymerize propylene to give syndiotactic polypropylene. It is necessary that the catalyst and cocatalyst are reacted at temperatures below $-40\,^\circ C$, preferably at $-78\,^\circ C$, and also that the polymerization is carried out at a low temperature. For catalyst activity either the vanadium compound or the organometallic compound must contain at least one halogen atom. Furthermore syndiotactic polymer can only be prepared if the Al:V ratios are within certain ranges, *e.g.* Al:V = 3–10:1 for $VCl_4$.

#### *1.5.1.2 Catalysts for 'living' syndiotactic polypropylene*

$V(acac)_3$ activated by $AlEt_2Cl$ has been used[108–112] polymerize propylene at $-78\,^\circ C$ to give a 'living'-type polymer with a narrow molecular weight distribution ($\bar{M}_w/\bar{M}_n = 1.19 \pm 0.05$). $^{13}C$ NMR spectroscopy demonstrated that the polypropylene produced was predominately syndiotactic.

#### *1.5.1.3 Catalysts for ethylene–propylene copolymers*

Homogeneous Ziegler–Natta catalysts based on vanadium compounds have found widespread industrial applications for the manufacture of ethylene–propylene copolymers and ethylene–propylene–diene terpolymers (see Chapter 4 of Volume 4).

### 1.5.2 Soluble Metallocene Catalysts

As early as 1955 it was discovered[113,114] that biscyclopentadienyltitanium dichloride, $Cp_2TiCl_2$, activated by aluminum alkyls (or chloroalkyls) could polymerize ethylene but not propylene. Since that time many catalyst systems have been developed and studied. A feature of these catalysts has been their relatively low activity, *viz.* 50–200 g PE (mmol Ti)$^{-1}$ h$^{-1}$ atm$^{-1}$. However, the addition of small amounts of water was found to lead to increased activities.

One of the most fascinating developments in the field of transition metal polymerization during recent years has been the discovery by Sinn, Kaminsky and co-workers[115–117] that homogeneous catalysts such as $Cp_2ZrMe_2$, or $Cp_2ZrCl_2$ in toluene solution, when activated with methylaluminoxane, can polymerize ethylene with activities in excess of $25 \times 10^6$ g PE (g Zr)$^{-1}$ h$^{-1}$. A large excess of aluminoxane is required for high activity (Al:Zr = 1000:1). Molecular weight distributions, $\bar{M}_w/\bar{M}_n$, of 1.6 to 2.4 are reported. These catalysts, in contrast to those in which water is added directly to the aluminum alkyl compound, are able to polymerize propylene and to copolymerize ethylene with propylene, hexene-1, *etc.* Only atactic polypropylene is formed from the use of non-chiral compounds such as $Cp_2TiCl_2$. Aluminoxanes are usually prepared from the hydrolysis of aluminum trimethyl with $CuSO_4 \cdot 5H_2O$ or $Al_2SO_4 \cdot 16H_2O$ in toluene.

Natural substances such as starch, cellulose or lignin, due to their free OH groups, can form aluminoxane structures on addition of $AlMe_3$, and it has been reported that starch grains or cellulose fibres when treated in this manner show considerable polymerization activity when suspended in toluene in the presence of the metallocene catalyst. In this way it is possible to cover cellulose, lignin or inorganic materials with polyethylene or other polyalkenes.

Recently the racemic mixture of ethylene bis(indenyl)zirconium dichloride or ethylene bis(tetrahydroindenyl)zirconium dichloride activated by methylaluminoxane has been used[118–120] to polymerize propylene to yield highly isotactic polypropylene. The structures of the 1-*R*-enantiomers of (a) *racemo*-$Et(Ind)_2HfCl_2$ and (b) *racemo*-$Et(IndH_4)_2HfCl_2$ are shown in Figure 6.

This important discovery clearly indicates that stereospecific polymerizations are dependent on the participation of chiral centres, and that provided stable chiral transition metal compounds can be used isotactic polypropylene can be prepared even in homogeneous systems. Typical experimental details and results are listed in Tables 6 and 7.

Similar catalysts based on the use of hafnium as the transition metal have been successfully prepared and used in polymerization and copolymerization studies. Some of these are reported to have improved performances[121] with respect to activities and molecular weights.

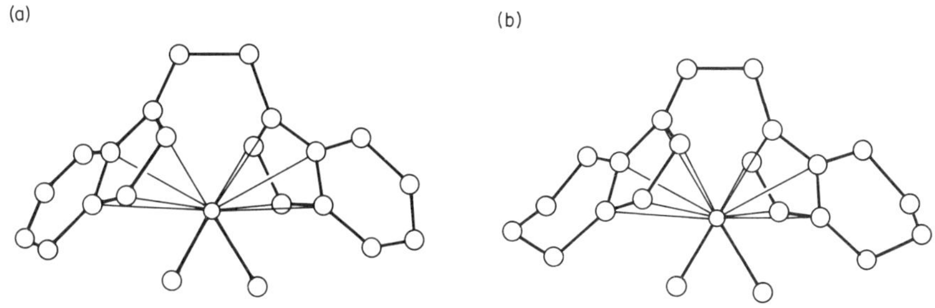


**Figure 6** Molecular structure of a chiral hafnium complex (a) *racemo*-$Et(Ind)_2HfCl_2$ and (b) *racemo*-$Et(IndH_4)_2HfCl_2$

**Table 6** Propylene Polymerizations with Ethylene Bis(indenyl)-zirconium Dichloride and Methylaluminoxane

| | |
|---|---|
| Activity (35 °C, 4 bar) | = 43 000 kg PP (mol Zr)$^{-1}$ h$^{-1}$ |
| [Zr] | = $5 \times 10^{-6}$ mol dm$^{-3}$ |
| [AlMe—O] | = $3 \times 10^{-2}$ mol dm$^{-3}$ |
| Molecular weight of the polypropylene | = 45 000 |
| Turnover time | = $3.5 \times 10^{-3}$ s |

**Table 7** Some Characteristics of Isotactic Polypropylene Produced with Ethylene Bis(tetrahydroindenyl)zirconium Dichloride and Methylaluminoxane

| *Temperature* (°C) | $10^{-3}\bar{M}_w$ | $\bar{M}_w/\bar{M}_n$ | *Toluene soluble (atactic) polymer* |
|---|---|---|---|
| −10 | 305 | 2.6 | 0.25% |
| 0 | 144 | 2.4 | 0.2% |
| 15 | 62 | 2.0 | 0.7% |
| 20 | 45 | 1.8 | 1.0% |

## 1.6 MODIFIED ZIEGLER–NATTA CATALYSTS

It has become somewhat more difficult to distinguish and identify this particular class of catalysts due to the widespread use of electron donors in attempts to increase catalyst activities and stereospecificities. However, in this case the objective is quite different in that it is to prepare Ziegler–Natta catalysts which may be used for the polymerization of polar-type monomers, such as vinyl chloride, methyl methacrylate, *etc.* Success has been very limited. Most modified Ziegler–Natta catalysts are characterized by low activities and lack of stereospecificity, producing polymers with properties which are very similar to those of polymers produced by more conventional procedures. In many cases there continues to be doubt whether or not these polymerizations operate *via* coordination polymerization mechanisms, and for this reason these catalysts are only briefly outlined in this article. Additional information can be found elsewhere.[20]

### 1.6.1 Catalysts Modified by Additives

Catalysts modified by the addition of ethers, esters, amines, ketones and nitroaromatics have been claimed for the polymerization of vinyl chloride.[122] In particular tetrahydrofuran (THF), diethyl ether, triethylamine and vinyl acetate have been used as additives. Many of these catalyst systems are soluble and are based on vanadium compounds. In particular the soluble catalyst system $VOCl_2 \cdot 2THF–AlBu^i_3$ has been used for the polymerization of vinyl chloride and vinyl fluoride.[123]

The role of the additive is believed to be that of complexation with the aluminum alkyl, so preventing attack of the halogen in the monomer.

### 1.6.2 Catalysts Derived from Metal Alkoxides

A number of modified Ziegler–Natta catalysts have been prepared from the use of either a transition metal alkoxide and an aluminum alkyl or chloroalkyl, or a transition metal halide and an aluminum alkoxide or chloroalkoxide. Whilst some of these catalyst systems have been shown to be active for the polymerization of vinyl chloride, vinylidene chloride, vinyl fluoride, *etc.*, there is considerable uncertainty about the mechanism of polymerization. Many polymerization systems are believed to proceed by radical mechanisms.[20]

## 1.7 TRANSITION METAL CATALYSTS

### 1.7.1 Heterogeneous Transition Metal Catalysts

#### *1.7.1.1 The Phillips catalyst*

Catalysts of the type first discovered by Hogan and Banks[6,7] still continue to make an important contribution to the production of HDPE. This catalyst has been described by Hogan[124] as 'an amorphous silica gel, finely divided and containing a small amount of $Cr^{VI}$ oxidic species'.

In practice $CrO_3$ is dissolved in water and the resulting orange solution is used to impregnate the silica support. After mixing in sufficient aqueous $CrO_3$ solution to slightly more than fill all the pores and to provide a final concentration of about 1%, the catalyst is heated until it is again free flowing. The catalyst is then activated using a fluidized bed system by passing dried air upwards through the bed and progressively increasing the temperature to 500–1000 °C. Then the maximum temperature is held for a few hours before the catalyst is allowed to cool and stored under a positive pressure of dry air or nitrogen. The activation process results mainly in water removal and the fixation of some of the Cr as a surface silyl chromate or dichromate species.

Various types of silica can be used in the preparation of commercial catalysts.[124] Surface area, pore size and pore size distribution have been shown to be important catalyst parameters, bringing about differences in catalyst activity, molecular weight and molecular weight distribution of the polymers produced.[31] Thus as pore size is increased the molecular weight of polyethylene decreases. Most commercial catalysts contain about 1% total Cr and have particle sizes of 30–150 $\mu$m.

Typically, ethylene polymerization is carried out in hydrocarbon diluents, *e.g.* hexane or butane, at pressures up to 40 atm and at temperatures from 80 °C to above 100 °C. The polymerization temperature is the primary control of molecular weight.

Copolymerization of ethylene with various $\alpha$-alkenes can be readily performed using the Phillips catalyst.[124] Butene-1 and hexene-1 have been extensively used for the production of copolymers.

Numerous modifications of the basic Phillips catalyst have been carried out, and include the addition of fluoride and titania[125–127] and the use of special calcining procedures.[127,128] At the present time the use of aluminum phosphate as a support has received particular attention because of its close similarity to silica, being isoelectronic and isostructural. Details of such catalyst systems are given in recent papers by McDaniel and co-workers.[129–131]

#### *1.7.1.2 Chromate and chromocene catalysts*

When bis(triphenylsilyl) chromate is deposited from hydrocarbon solution onto a silica–alumina of large surface area, very active catalysts for ethylene polymerization without the need for a cocatalyst or secondary treatment are produced.[132] Catalysts derived from chromocene deposited on silica supports also show very high activity for the polymerization of ethylene.[133] The formation of the catalyst has been formulated by Carrick as in Scheme 6.

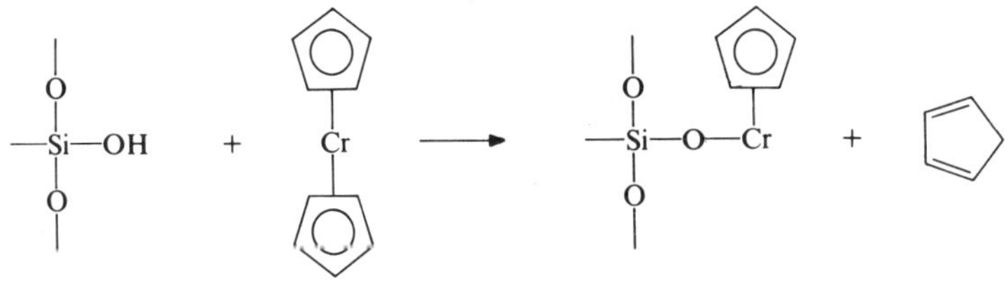


**Scheme 6**

Patents to Union Carbide claim very high activities for $SiO_2/Cp_2Cr$ catalysts (250 kg PE (mol $Cp_2Cr)^{-1}$ $atm^{-1}$ $h^{-1}$).[134]

### 1.7.1.3 *Transition metal alkyl and benzyl compounds*

Transition metal alkyl and benzyl compounds can be reacted with hydroxyl groups on pre-dried silica and alumina to give highly active polymerization catalysts of long lifetime.[135,136]

Typically $Zr(allyl)_4$ reacts with dried silica (220 °C) as in equation (4).

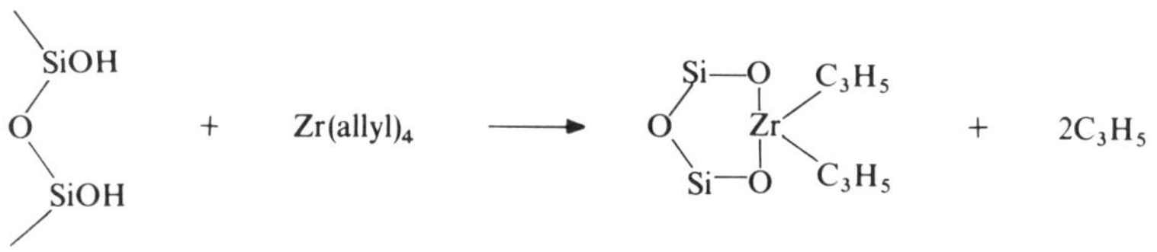

(4)

A variety of supports such as $Al_2O_3$, $Al_2$–ZrO, $Al_2O_3$–MgO may also be used. Ethylene, propylene, 4-methylpentene-1, butadiene may all be polymerized.

These supported catalyst systems show much higher activities than the original transition metal alkyl compounds.

## 1.7.2 Homogeneous Transition Metal Catalysts

### 1.7.2.1 *Titanium and zirconium alkyl catalysts*

Early investigations[137,138] showed that methyltitanium halides and titanium tetramethyl showed only weak activity. Giannini *et al.*[139] have demonstrated that the benzyl derivatives of Ti and Zr will polymerize ethylene. $TiBz_4$ (Bz = benzyl) and $TiBz_3Cl$ have been claimed[139,140] to produce isotactic polypropylene although polymerization could have occurred on the surface of colloidal decomposition products.

Extensive studies on catalysts such as $ZrBz_4$, $Zr(4\text{-}Me\text{-}Bz)_4$, *etc.*, have been reported by Ballard.[141,142] Polymerization of ethylene, propylene, acrylonitrile, styrene, *p*-bromostyrene and methyl methacrylate have all been reported.

## 1.8 CATALYST TARGETS

Industrial and economic factors have greatly influenced the development of more efficient and selective catalysts for ethylene and α-alkene polymerization, and for copolymerization. These factors can be summarized as follows: (i) the elimination of catalyst deactivation procedures and of steps for the removal of catalyst residues from the polymer; (ii) the need for increased yields of stereoregular and high molecular weight polymer and the elimination of the production of wax deposits within the reactor; (iii) better control of molecular weight and molecular weight distribution of polymer products; (iv) the production of polymers with increased ease of processability; (v) the development of industrial processes of greater flexibility so allowing the use of different types of catalysts and comonomers in the same reactor; and (vi) the production of polymer of controlled particle size and bulk density.

Many of the requirements necessary for the economic production of polyethylene and polypropylene have been partially achieved. However, catalysts of greater activity and of greater selectivity in the production of polymers and copolymers for specialist applications can be anticipated.

## 1.9 POLYMERIZATION PROCESSES

Solution, slurry and gas phase processes have been developed for the production of polyethylene and polypropylene. Discussion in this chapter is necessarily brief and full details should be consulted elsewhere.[143]

### 1.9.1 Solution Polymerization Processes

The 'term' solution should not be regarded strictly since the polymer produced is often present in a molten state.

An early process developed by Du Pont for the production of HDPE uses mixed transition metal catalysts and operates at temperatures above 200 °C and pressures above 70 bar.

Other processes of this type have been developed by Mitusi, Dow and Stamicarbon NV amongst others. In the Stamicarbon process, a solution of ethylene in hexane is cooled to −40 °C and then fed into a polymerization reactor where polymerization is carried out at temperatures higher than 130 °C. Stamicarbon have filed patent for the use of titanium tetrachloride supported on magnesium chloride supports for use in such processes.

A solution process for propylene polymerization has been used by Eastman Kodak for many years using modified first generation catalysts. The flow diagram shown in Figure 7 fully illustrates the problems associated with the industrial use of these catalysts as far as catalyst removal and polymer isolation procedures are concerned.

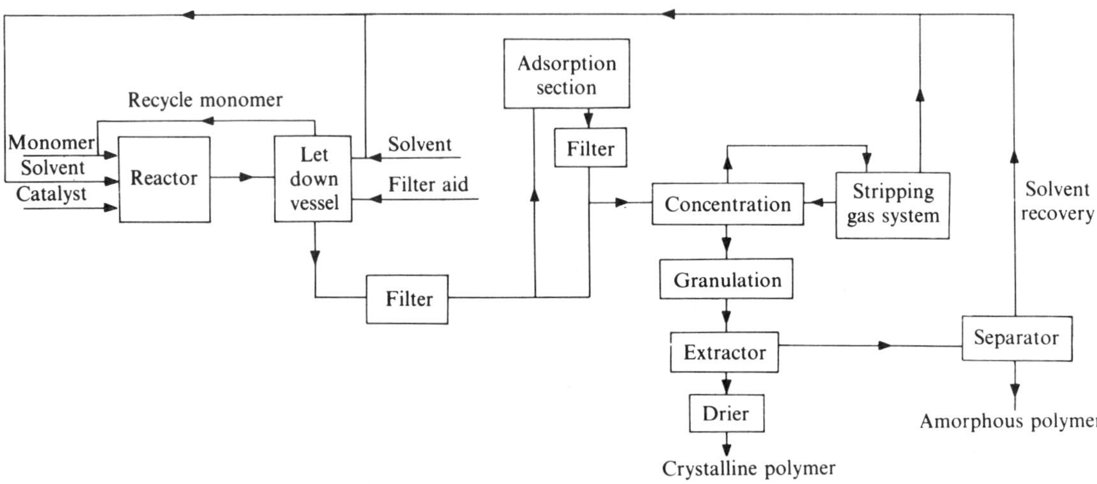


**Figure 7** Representation of Eastman Kodak process

### 1.9.2 Slurry Polymerization Processes

The use of slurry processes is widespread in propylene polymerization. Normally a poor solvent for the polymer is used and dispersion of the catalyst maintained.

For propylene polymerization an inert hydrocarbon, *viz.* heptane, pentamethylheptane (EC180), *etc.*, is used and polymerizations effected at temperatures between 55–70 °C and pressures of about 10 atm. A diagrammatic scheme is shown in Figure 8.

Removal of catalyst residues can be achieved by the addition of alcohol, followed by recovery and extraction of catalyst residues. The polymer is freed from diluent by centrifuging and then dried. Atactic polypropylene remains in the diluent.

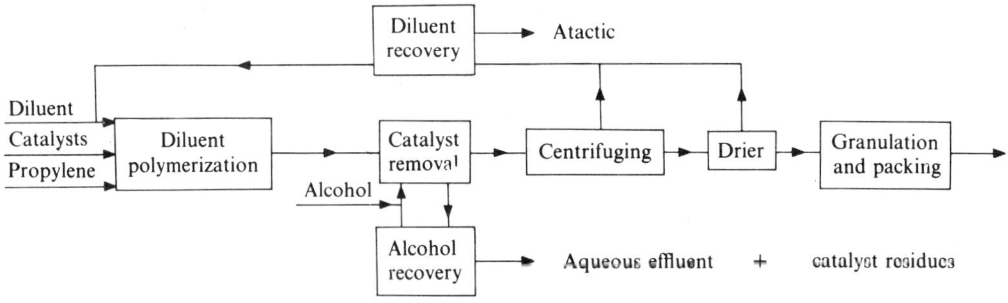


**Figure 8** Representation of slurry polypropylene process

Dart Industries have developed a process in which the monomer forms the reaction diluent. Pressures of approximately 22 atm (1 atm = $1.01 \times 10^5$ Pa) are used to maintain the propylene in a liquid form. Hydrocarbon diluent and isopropyl alcohol may be used to remove catalyst residues when the atactic polymer is extracted in the diluent.

Solvay & Cie ether-treated catalysts and $MgCl_2$/donor/$TiCl_4$ systems may be used with these processes.

Low cost slurry processes for polyethylene production have been developed by many producers. These include Hoechst AG, Mitsubishi Chemical Ind., Montecatini–Edison, Solvay & Cie and Veba Chemie. Most producers use high activity catalysts.

These new processes are claimed to be very flexible and to produce polymers of varying melt indexes, densities and molecular weight distribution. Flexibility and the capacity to produce specialize types of copolymers now seems established as being a vital component for the future success of any new commercial process.

### 1.9.3 Gas Phase Polymerization Processes

The successful development of such processes represents one of the greatest technological successes in this field.

#### *1.9.3.1 Union Carbide Unipol gas phase process*

A gas phase, fluidized bed process (Unipol) was developed during the 1960s by the Union Carbide Corporation for the production of HDPE and first commercialized in 1968. The process was based initially on the use of chromium-type silica supported catalysts but may also be used with other types of high activity supported catalysts including those of a $SiO_2/MgCl_2/TiCl_4$/donor type. The process operates at a nominal reaction pressure of 18 atm and at temperatures between 85 and 100 °C. A full range of HDPE polymers can be prepared.

The versatility of the original process led to a new gas phase fluidized bed process for the commercial production of LLDPE in 1975 by copolymerizing ethylene with α-alkenes such as butene-1. The essential key to this new process is the development of new high activity catalysts that operate at low temperatures, and at pressures of only 7–20 atm. Molecular weight is controlled by polymerization temperature and the use of hydrogen.

The Unipol processes have been licensed with more than 25 licenses and over 40 reactors are operated in 15 countries.

#### *1.9.3.2 BP gas phase process*

In the BP Chemicals gas phase process polymerization is carried out under mild conditions of temperature and pressure, typically at temperatures of between 60 and 100 °C and pressures between 15 and 30 bar, in a fluid bed reactor. The growing polymer particles are maintained in a fluidized condition by means of a gas stream containing ethylene, the α-alkene (in copolymerizations) and hydrogen (for molecular weight control). The gas stream from the top of the reactor is cooled in a heat exchange and then recycled through a compressor. Polyethylene granules leave the reactor as a free-flowing powder. Homogeneity in the fluid bed and temperature control are ensured by the use of high activity titanium and magnesium catalysts.

The process has a high degree of flexibility and can produce a broad range of products, using the same catalyst.

A diagrammatic representation of the process is shown in Figure 9.

#### *1.9.3.3 BASF Novolen gas phase process*

A gas phase process (Novolen) for the production of polypropylene was first developed by BASF and operated by Rheinische Olefinwerke at Wesseling in West Germany in 1969. Other groups such as ICI and USI (Norchem) operate similar units and El Paso has used its own version for some years. Additionally Himont uses gas phase technology in a process commercialized along with Mitsui Petrochemical.

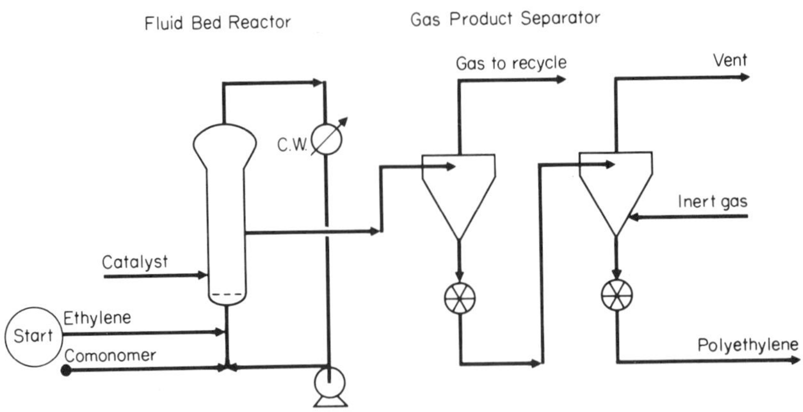


**Figure 9** Representation of BP gas phase process

In the Novolen process the reactor consists of a stirred polymer bed into which liquid propylene and catalyst are introduced at the bottom and activator at the top. Gas phase polymerization is achieved by maintaining a quantity of high molecular weight polypropylene in the reactor.

A variety of catalysts have been and are being used, and include donor modified $\delta$-$TiCl_3 \cdot 0.33$ $AlCl_3$ catalysts activated by $AlEt_2Cl$ and also a variety of high activity catalysts such as $MgCl_2$/donor/$TiCl_4$ activated by an $AlEt_3$/donor system. The use of $MgCl_2$/phthalate/$TiCl_4$ catalysts of controlled particle shape and size offers exciting possibilities for use in these systems.

### *1.9.3.4 Union Carbide Unipol polypropylene gas phase process*

Union Carbide extended their Unipol technology to the production of polypropylene in 1985 using Shell's high activity SHAC catalyst. The reactor consists of a vertical fluidized bed containing granular polypropylene and catalyst. Reaction gas is continuously circulated through the reactor by means of a gas blower and provides agitation which absorbs the heat of polymerization.

A single reactor system is used to make homopolymers and random copolymers. Two reactors are operated in series for the production of impact copolymers as are shown in Figure 10.

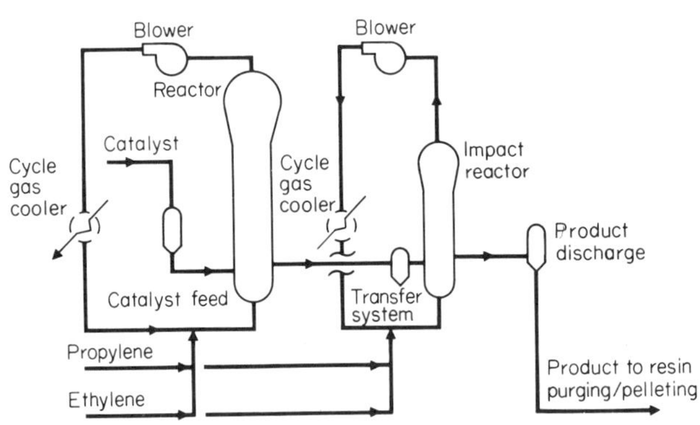


**Figure 10** Representation of Union Carbide Unipol process

### *1.9.3.5 Amoco gas phase process*

A horizontal stirred reactor has been developed by Amoco and used for polypropylene production.

### *1.9.3.6 Advantages of gas phase technology*

The development of gas phase technology must represent a major advance for the commercial production of polyalkenes. The commercial advantages of these processes for new plants is significant and can be summarized as follows: (i) elimination of costs associated with storage and handling of large amounts of solvent leading also to improved safety; (ii) simplification of polymerization systems; (iii) reduction of environmental hazards; and (iv) simple and flexible production of block copolymers of propylene and ethylene — a very important consideration at the present time.

## 1.10 REFERENCES


1. K. Ziegler, *Belg. Pat.* 533 362 (1953).
2. K. Ziegler, E. Holzkamp, H. Breil and H. Martin, *Angew. Chem.*, 1955, **67**, 541.
3. G. Natta, P. Pino, P. Corradini, F. Danusso, E. Mantica, G. Mazzanti and G. Moraglio, *J. Am. Chem. Soc.*, 1955, **77**, 1708.
4. G. Natta, *J. Polym. Sci.*, 1955, **16**, 143.
5. G. Natta, P. Pino and G. Mazzanti, *Ital. Pat.* 526 101 (1954).
6. J. P. Hogan and R. L. Banks, *US Pat.*, 2 825 721 (1954).
7. A. Clark, J. P. Hogan, R. L. Banks and W. C. Lanning, *Ind. Eng. Chem.*, 1956, **48**, 1152.
8. G. Natta and I. Pasquon, *Adv. Catal.*, 1959, **11**.
9. J. Boor, *Macromol. Rev.*, 1967, **2**, 115.
10. D. F. Hoeg, in 'Stereochemistry of Macromolecules', ed. A. D. Ketley, Dekker, New York, 1967, vol. 1.
11. D. O. Jordan, in 'Stereochemistry of Macromolecules', ed. A. D. Ketley, Dekker, New York, 1967, vol. 1.
12. M. N. Berger, G. Boocock, and R. N. Haward, *Adv. Catal.*, 1969, **19**, 24.
13. J. Boor, *Ind. Eng. Chem. Prod. Res. Dev.*, 1970, **9(4)**, 437.
14. T. Keii, in 'Kinetics of Ziegler–Natta Polymerization', Chapman and Hall, London, 1972.
15. I. Pasquon and L. Porri, *MTP Int. Rev. Sci.; Macromol, Sci.*, 1972, **8**, 159.
16. J. C. W. Chien (ed.), 'Coordination Polymerization, A Memorial to K. Ziegler', Academic Press, New York, 1975.
17. P. Pino and U. U. Suter, *Polymer*, 1976, **17**, 11.
18. W. Cooper, in 'Comprehensive Chemical Kinetics', ed. C. H. Bamford and C. F. H. Tipper, Elsevier, Amsterdam, 1976, p. 133.
19. A. D. Caunt, *Catalysis (London)*, 1977, **1**.
20. P. J. T. Tait, in 'Developments in Polymerization', ed. R. N. Haward, Applied Science, London, 1979, vol. 2, p. 81.
21. J. Boor, Jr., 'Ziegler–Natta Catalysis and Polymerization', Academic Press, New York, 1979.
22. R. W. Lenz and F. Ciardelli (eds.), 'Preparation and Properties of Stereoregular Polymers', Reidel, Dordrecht, 1980.
23. P. J. T. Tait, *Spec. Period. Rep.: Macromol. Chem.*, 1980, **1**, 3.
24. R. P. Quirk (ed.), *MMI Press Symp. Ser.*, 1981, **4**.
25. P. J. T. Tait, *Spec. Period. Rep.: Macromol. Chem.*, 1984, **3**, 3.
26. F. J. Karol, *Catal. Rev. Sci. Eng.*, 1984, **26 (3/4)**, 557.
27. T. Keii and K. Soga (eds.), 'Catalytic Polymerization of Olefins', in 'Studies in Surface Science and Catalysis', Kodansha and Elsevier, Tokyo and Amsterdam, 1986, vol. 25.
28. Y. V. Kissin, 'Isospecific Polymerization of Olefins', Springer-Verlag, New York, 1985.
29. A. Munoz-Escalona, *MMI Press Symp. Ser.*, 1983, **4**, 323.
30. Yu. I. Yermakov, B. N. Kuznetsov and V. A. Zakharov, 'Catalysis by Supported Complexes', in 'Studies in Surface Science and Catalysis', Elsevier, Amsterdam, 1981, vol. 8.
31. M. P. McDaniel, *Adv. Catal.*, 1985, **33**, 47.
32. G. Natta, P. Pino and G. Mazzanti, *Gazz. Chim. Ital.*, 1957, **87**, 528.
33. P. J. T. Tait, in 'History of Polyolefins', ed. R. B. Seymour and T. Cheng, Reidel, Dordrecht, 1986, p. 213.
34. G. Natta, P. Corradini and G. Allegra, *J. Polym. Sci.*, 1961, **51**, 399.
35. G. Natta, P. Corradini, I. W. Bassi and L. Porri, *Atti Accad. Naz. Lincei, Cl. Sci. Fis., Mat. Nat., Rend.*, 1951, **24**, 121.
36. G. Natta, P. Corradini and G. Allegra, *Atti Accad. Naz. Lincei, Cl. Sci. Fis., Mat. Nat., Rend.*, 1959, **26**, 155.
37. G. Natta, I. Pasquon, A. Zambelli and G. Gatti., *J. Polym. Sci.*, 1961, **51**, 387.
38. E. Tornqvist, C. W. Seelback and A. W. Langer, *US Pat.* 3 128 252 (1964).
39. E. G. M. Tornqvist, J. T. Richardson, Z. W. Wilchinsky and R. W. Looney, *J. Catal.*, 1967, **8**, 189.
40. E. Tornqvist and A. W. Langer, *US Pat.* 3 032 510 (1962).
41. Z. W. Wilchinsky, R. W. Looney and E. G. M. Tornqvist, *J. Catal.*, 1973, **28**, 352.
42. E. J. Arlman, *J. Catal.*, 1964, **3**, 89.
43. G. Guidetti, R. Zanetti, D. Ajo, A. Marigo and M. Vidali, *Eur. Polym. J.*, 1980, **15**, 1007.
44. V. W. Buls and T. L. Higgins, *J. Polym Sci., Part A-1*, 1970, **8**, 1025.
45. J. P. Hermans and P. Henrioulle, *Ger. Pat.* 2 213 086 (1972) (*Chem. Abstr.*, 19??, **??**, ?? ???).
46. J. P. Hermans and P. Henrioulle, *US Pat.* 3 769 233 (1973) (*Chem. Abstr.*, 19??, **??**, ?? ???).
47. R. P. Nielson, *MMI Press Symp. Ser.*, 1983, **4**, 47.
48. G. Natta, *J. Polym. Sci.*, 1959, **34**, 21.
49. *Br. Pat.* 841 822 (1955).
50. *US Pat.* 3 153 634 (1956).
51. *Belg. Pat.* 650 679 (1963).
52. *Belg. Pat.* 726 839 (1968).
53. *Belg. Pat.* 735 290 (1968).
54. K. Weissermel, H. Cherdon, J. Berthold, B. Diedrich, K. D. Keil, K. Rust, H. Strametz and T. Toth, *J. Polym. Sci., Polym. Symp.* 1975, **51**, 187.
55. J. C. W. Chien and J. T. T. Hsieh, *J. Polym. Sci., Polym. Chem. Ed.*, 1976, **14**, 1915.
56. V. A. Zakharov and Y. I. Yermakov, *Catal. Rev. Sci. Eng.*, 1979, **19**, 63.
57. M. J. Sharp and J. A. Hockey, *J. Catal.*, 1970, **18**, 52.
58. D. D. Eley, D. A. Keir and R. Rudham, *J. Chem. Soc., Faraday Trans.*, 1976, **1**, 1685.

59. A. Simon and G. Grobler, *J. Polym. Sci., Polym. Chem. Ed.*, 1980, **18**, 1311.
60. K. Soga, R. Ohnishi and Y. Doi, *Polym. Bull. (Berlin)*, 1983, **9**, 299.
61. F. J. Karol, 'International Symposium on Transition Metal Catalyzed Polymerizations', Akron, June 16–20, 1986.
62. L. L. Böhm, *Polymer*, 1978, **19**, 553.
63. *Ger. Pat.* 2 003 075 (1970).
64. *Belg. Pat.* 751 315 (1969).
65. *Br. Pat.* 1 299 862 (1970).
66. *Belg. Pat.* 776 301 (1970).
67. R. N. Haward, A. N. Roper and K. Fletcher, *Polymer*, 1973, **14**, 365.
68. D. G. Boucher, I. W. Parsons and R. N. Haward, *Makromol. Chem.*, 1974, **175**, 3461.
69. K. Gardner, I. W. Parsons and R. N. Haward, *J. Polym. Sci.*, 1978, **16**, 1683.
70. P. Radenkov, T. Petrova, L. Petkov and D. Jelyazkova, *Eur. Polym. J.*, 1976, **11**, 313.
71. P. Radenkov, L. Petrov, S. Karaenev and R. Kyrtcheva, *Eur. Polym. J.*, 1976, **12**, 427.
72. L. Petrov, R. Kyrtcheva, P. Radenkov and D. Dobreva, *Polymer*, 1978, **19**, 567.
73. L. Petkov, P. Radenkov and R. Kyrtcheva, *Polymer*, 1978, **19**, 570.
74. Mitsui Petrochemicals Ind., *Ger. Pat.* 904 510 (1960).
75. Montecatini–Edison Co., *Br. Pat.* 1 286 807 (1968).
76. *Ital. Pat.* 912 345 (1968).
77. D. Bryce-Smith, *Ger. Pat.* 955 807 (1959).
78. C. W. Kamienski, 'Synthesis and Properties of Diorganomagnesium Compounds', Ph.D. Thesis, University of Tennessee, 1967.
79. E. Kinkelin, G. Fink and B. Bogdanov, *Makromol. Chem., Rapid Commun.*, 1986, **7**, 85.
80. G. Bruni and A. Ferrari, *Atti Accad. Naz. Lincei, Cl. Sci. Fis., Mat. Nat., Rend.*, 1925, **2**, 457.
81. U. Giannini, *Makromol. Chem., Suppl.*, 1981, **5**, 216.
82. P. Galli, P. Barbe, G. Guidetti, R. Zannetti, A. Martorana, A. Marigo, M. Bergozza and A. Fichera, *Eur. Polym. J.*, 1983, **19**, 19.
83. G. Allegra and I. W. Bassi, *Gazz. Chim. Ital.*, 1980, **110**, 437.
84. R. Gerbasi, A. Marigo, A. Martorana, R. Zannetti, G. P. Guidetti and G. Baruzzi, *Eur. Polym. J.*, 1984, **20**, 967.
85. B. L. Goodall, *MMI Press Symp. Ser.*, 1983, **4**, 355.
86. B. Keszler, G. Bodor and A. Simon, *Polymer*, 1980, **21**, 1037.
87. J. C. W. Chien, J.-C. Wu and C.-I. Kuo, *J. Polym. Sci., Polym. Chem. Ed.*, 1983, **21**, 725.
88. N. Kashiwa, *MMI Press Symp. Ser.*, 1983, **4**, 379.
89. J. C. W. Chien, J.-C. Wu and C.-I. Kuo, *J. Polym. Sci., Polym Chem. Ed.*, 1982, **20**, 2019.
90. P. C. Barbe, L. Noristi, G. Baruzzi and E. Marchetti, *Makromol. Chem., Rapid Commun.*, 1983, **4**, 249.
91. N. Kashiwa and J. Yoshitake, *Makromol. Chem.*, 1984, **185**, 1984.
92. A. A. Baulin, Ye. I. Novikova, G. Ya. Mal'kova, V. L. Maksinov, L. I. Vyshinskaya and S. S. Ivanchev, *Polym. Sci. USSR (Engl. Transl.)*, 1980, **22**, 205.
93. J. C. W. Chien and J. C. Wu, *J. Polym. Sci., Polym. Chem. Ed.*, 1982, **20**, 2461.
94. J. C. W. Chien and J. C. Wu, *J. Polym. Sci., Polym. Chem. Ed.*, 1982, **20**, 2445.
95. J. C. W. Chien, *Catal. Rev. Sci. Eng.*, 1984, **26 (3/4)**, 613.
96. S. Weber, J. C. W. Chien and Y. H., in 'Transition Metals and Organometallics as Catalysts for Olefin Polymerization', ed., W. Kaminsky and H. Sinn, Springer-Verlag, Berlin, 1988, p. 45.
97. Shell, *Eur. Pat.* 29 623 (1979).
98. Mitsui, *Eur. Pat.* 0 086 288 (1983).
99. *Aust. Pat.* 181 018/81 (1983).
100. BASF, *Ger. Pat.* 3 342 039 (1985), 3 411 197 (1985).
101. P. Galli, IUPAC International Symposium on Macromolecules, Florence, Sept. 7–12, 1981.
102. P. Galli, in 'Structural Order in Polymers', ed. F. Ciardelli and P. Giusti, Pergamon Press, Oxford, 1981, p. 63.
103. G. Di Drusco and L. Luciani, *J. Appl. Polym. Sci., Appl. Polym. Symp.*, 1981, **36**, 95.
104. P. Galli, L. Luciani and E. Cecchin, *Angew. Makromol. Chem.*, 1981, **94**, 63.
105. P. Galli, P. C. Barbe and L. Noristi, *Angew. Makromol. Chem.*, 1984, **120**, 73.
106. *Eur. Pat.* 0 004 651 (1979).
107. G. Natta, I. Pasquon and A. Zambelli, *J. Am. Chem. Soc.*, 1962, **84**, 1488.
108. Y. Doi, S. Ueki and T. Keii, *Makromol. Chem.*, 1979, **180**, 1359.
109. Y. Doi, S. Ueki and T. Keii, *Macromolecules*, 1979, **12**, 814.
110. Y. Doi, S. Suzuki, F. Nozawa and K. Soga, *Stud. Surf. Sci. Catal.*, 1986, **25**, 257.
111. Y. Doi, S. Ueki, S. Tamura, S. Nagahara and T. Keii, *MMI Press Symp. Ser.*, 1983, **4**.
112. G. G. Evens and E. M. J. Pijpers, *MMI Press Symp. Ser.*, 1983, **4**, 245.
113. G. Natta, P. Pino, G. Mazzanti and R. Lanzo, *Chim. Ind. (Milan)*, 1957, **39**, 1032.
114. G. Natta, P. Pino, G. Mazzanti and U. Giannini, *J. Inorg. Nucl. Chem.*, 1958, **8**, 612.
115. A. Andreson, H. G. Cordes, J. Herwig, W. Kaminsky, A. Merk, R. Mottweiler, J. H. Sinn and H.-J. Vollmer, *Angew. Chem., Int. Ed. Engl.*, 1976, **15**, 630.
116. W. Kaminsky, *MMI Press Symp. Ser.*, 1983, **4**, 225.
117. W. Kaminsky, *Stud. Surf. Sci. Catal.*, 1986, **25**, 293.
118. W. Kaminsky, K. Külper, H. H. Brintzinger and F. R. W. P. Wild, *Angew. Chem.*, 1985, **97**, 507.
119. W. Kaminsky, A. Bark, R. Spiehl, N. Möller-Lindenhof and S. Niedoba, in 'Transition Metals and Organometallics as Catalysts for Olefin Polymerization', ed. W. Kaminsky and H. Sinn, Springer-Verlag, Berlin, 1988, p. 281.
120. H. H. Brintzinger, in 'Transition Metals and Organometallics as Catalysts for Olefin Polymerization', ed. W. Kaminsky and H. Sinn, Springer-Verlag, Berlin, 1988, p. 249.
121. J. A. Ewen, L. Haspeslagh, M. J. Elder and J. L. Atwood, in 'Transition Metals and Organometallics as Catalysts for Olefin Polymerization', ed. W. Kaminsky and H. Sinn, Springer-Verlag, Berlin, 1988, p. 249.
122. *US Pat.* 834 937 (1957).
123. A. G. Chesworth, R. N. Haszeldine and P. J. T. Tait, *J. Polym. Sci.*, 1974, **12**, 1703.

124. J. P. Hogan and R. R. Witt, *Prepr., Div. Pet. Chem., Am. Chem. Soc.*, 1979, **24**, Part 2, 377.
125. T. J. Pullukat, R. E. Hoff and M. Shida, *J. Polym. Sci., Polym. Chem. Ed.*, 1980, **18**, 2857.
126. T. J. Pullukat, R. E. Hoff and M. Shida, *J. Appl. Polym. Sci.*, 1981, **26**, 2927.
127. M. P. McDaniel, M. B. Welch and M. J. Dreiling, *J. Catal.*, 1983, **82**, 118.
128. M. P. McDaniel and M. B. Welch, *J. Catal.*, 1983, **82**, 98, 110.
129. M. P. McDaniel, T. T. P. Cheung and M. M. Johnson, 'International Symposium on Transition Metal Catalyzed Polymerizations', Akron, June 16–20, 1986.
130. M. P. McDaniel and M. M. Johnson, *J. Catal.*, 1986, **101**, 446.
131. T. T. P. Cheung, K. W. Willcox, M. P. McDaniel and M. M. Johnson, *J. Catal.*, 1986, **102**, 10.
132. W. L. Carrick, R. J. Turbett, F. J. Karol, G. L. Karapinka, A. S. Fox and R. N. Johnson, *J. Polym. Sci., Part A*, 1972, **10**, 2609.
133. F. J. Karol, G. L. Karapinka, C. Wu, A. W. Dow, R. N. Johnson and W. L. Carrick, *J. Polym. Sci., Part A*, 1972, **10**, 2621.
134. F. J. Karol, B. E. Wagner, I. J. Levine, G. L. Goeke and N. Noshay, 190th ACS National Meeting, Chicago, Sept. 8–13, 1985.
135. *Br. Pat.* 1 314 828 (1969).
136. D. G. H. Ballard, E. Jones, R. J. Wyatt, R. T. Murray and P. A. Robinson, *Polymer*, 1974, **15**, 169.
137. C. Beerman and H. Bestian, *Angew. Chem.*, 1959, **71**, 618.
138. H. J. Berthoid and G. Groh, *Z. Anorg. Chem.*, 1963, **319**, 230.
139. U. Giannini, U. Zucchini and E. Albizzati, *J. Polym. Sci., Part B*, 1970, **8**, 405.
140. D. G. H. Ballard, *Adv. Catal.*, 1973, **23**, 263.
141. D. G. H. Ballard and P. W. Van Lienden, *Makromol. Chem.*, 1972, **154**, 177.
142. D. G. H. Ballard, J. V. Dawkins, J. M. Key and P. W. Van Lienden, *Makromol. Chem.*, 1973, **165**, 173.
143. J. N. Short, *MMI Press Symp. Ser.*, 1983, **4**, 651.
144. N. F. Brockmeier, *MMI Press Symp. Ser.*, 1983, **4**, 671.

# 2

# Monoalkene Polymerization: Mechanisms

PETER J. T. TAIT and NEIL D. WATKINS
*University of Manchester Institute of Science and Technology, UK*

Because of the very late delivery of the manuscript, this chapter now appears at the end of this Volume, commencing on page 533. This is in order to minimize any delay in publishing 'Comprehensive Polymer Science' as a whole.

# 3

# Monoalkene Polymerization: Stereospecificity

PAOLO CORRADINI, VINCENZO BUSICO and GAETANO GUERRA
*Università di Napoli, Italy*



## 3.1 INTRODUCTION

We will discuss here a few concepts which are useful in the following discussion of stereospecific Ziegler–Natta (ZN) polymerizations of 1-alkenes.

In a polymerization reaction, a prochiral 1-alkene molecule will add to a growing polymer chain in one of four possible modes (Scheme 1).

Addition modes (a) and (b) on the one hand, and (c) and (d) on the other, lead to monomeric units with equal constitution (*1,2-addition* and *2,1-addition* respectively, signifying that in cases (a) and (b) the 1-alkene molecule has linked to the growing chain end at C-2, but in cases (c) and (d) at C-1), but opposite configurations at the newly formed tertiary C atom.

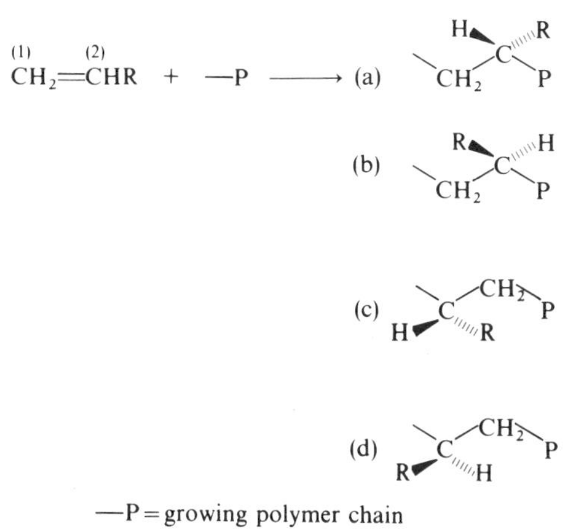


**Scheme 1** Possible addition modes of a 1-alkene molecule to a growing poly-1-alkene chain

Macromolecules resulting from successions of (prevailingly) equal additions (either 1,2 or 2,1, independent of the choice of configuration) are said to be *regioregular* (from a *regiospecific* polymerization process); if there is regularity in the succession of configurations of the tertiary C atoms, the macromolecules are said to be *stereoregular* (from a *stereospecific* polymerization process).

In the description of a poly-1-alkene chain stereostructure, two successive monomeric units having the same configuration are said to constitute a *meso (m)* diad, and two having opposite configurations a *racemo (r)* diad. A chain ideally resulting from the regular succession of *m* diads is defined as *isotactic*, and one from the regular succession of *r* diads as *syndiotactic*, but in the absence of stereoregularity, the chain is said to be *atactic* (Scheme 2).

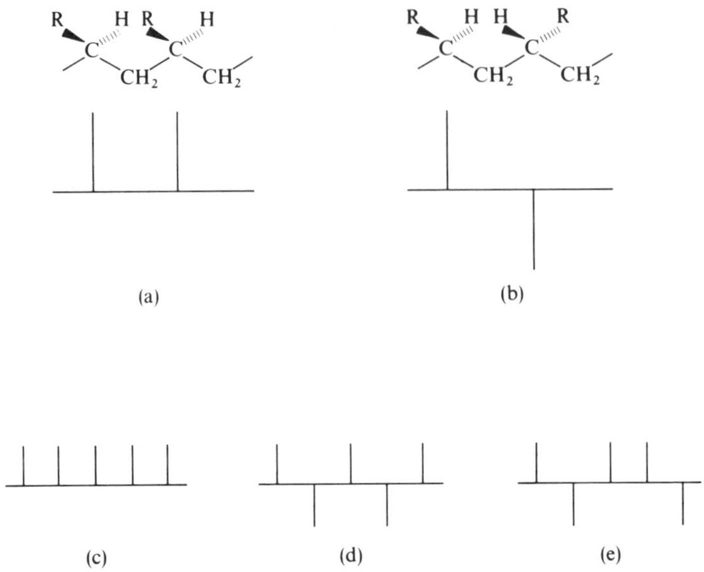


**Scheme 2** Planar zigzag projection of the backbone carbon atoms of poly-1-alkene chain segments: *meso* (a) and *racemo* (b) diads, both in the saw-horse (upper) and in an adapted Fischer (lower) notation; isotactic (c), syndiotactic (d) and atactic (e) sequences in an adapted Fischer notation

In general, ZN catalysts are highly regiospecific in the polymerization of 1-alkenes, but differ widely in stereospecificity. Until recently, it was customary to classify stereospecific ZN catalysts into *heterogeneous isospecific* and *homogeneous syndiospecific*.[1,2] In 1984, the first report[3] appeared in the literature of homogeneous catalyst systems which are able to polymerize 1-alkenes to isotactic polymers. This was soon followed by others which began to throw light on what is probably the most significant scientific achievement in the field of ZN catalysis since its discovery. In this review, which is perhaps the first to be written after such reports, the new class of *homogeneous isospecific* ZN catalysts was added to the two traditional ones.

This chapter is structured in two parts. In Section 3.2, experimental data on the stereospecificity of polymerization of 1-alkenes are presented; these are not intended to provide an exhaustive collection of data on stereospecific ZN catalysts (excellent reviews on this are refs. 1 and 4), but rather a selection of results useful—in the authors' opinion—in the comprehension of the principles of catalyst specificity. In Section 3.3, possible reasons for the origins of regio- and stereo-specificity, and models of catalytic sites are discussed.

## 3.2 SELECTED EXPERIMENTAL RESULTS

### 3.2.1 General Remarks

It is widely accepted that, in the presence of ZN catalysts, poly-1-alkene chain growth consists of successive insertions of 1-alkene molecules into a metal–carbon bond.[1,2,4]

A *cis* opening of the 1-alkene double bond has been proved conclusively.[3,5–7]

An insertion leading to the formation of a M–$CH_2$–CHR–P sequence (where M = metal; P = growing polymer chain) is said to be *primary* (1,2 addition of Scheme 1), while one leading to the formation of a M–CHR–$CH_2$–P sequence is said to be *secondary* (2,1 addition of Scheme 1). Regiospecificity implies the non-equivalence of the two insertion modes, for electronic and/or steric reasons.

Stereospecificity, which for a given insertion mode entails the ability of the catalyst to discriminate between the two faces of the prochiral monomer, requires at least one chiral species to participate in the polymerization mechanism. The steric control may be dictated by the configuration of the asymmetric tertiary C atom of the last inserted monomer unit, as well as by the chirality of the organometallic complex of which the active site is part (in case the optically inactive catalyst consists of a racemic mixture of enantiomorphic sites). Both possibilities (*e.g. chain end stereocontrol* and *enantiomorphic site stereocontrol*) have been verified, depending on the catalyst system (*vide infra*).

No ZN catalyst is totally stereospecific. Terms such as 'isotactic' and 'syndiotactic', when referred to real poly-1-alkene chains, should be intended as 'mainly isotactic' or 'mainly syndiotactic', to account for the occurrence of failures in the catalyst stereocontrol.

The distribution of steric defects along the polymer chains may be indicative of which kind of stereocontrol is operating. Bernoullian statistics have been shown[8] to be consistent with a chain end stereocontrol, while non-Bernoullian distributions originate from an enantiomorphic site stereocontrol (Table 1).[9,10]

**Table 1** Stereochemical Sequence Distributions for Bernoullian (Chain End) and Enantiomorphic Site Stereocontrol

| *Sequence* | *Probability (Bernoullian chain end)* | *Probability (Enantiomorphic site)* |
|---|---|---|
| *m* | $P$ | $1-2b(1-b)$ |
| *r* | $1-P$ | $2b(1-b)$ |
| *mm* | $P^2$ | $1-3b(1-b)$ |
| *mr* | $2P(1-P)$ | $2b(1-b)$ |
| *rr* | $(1-P)^2$ | $b(1-b)$ |
| *mmmm* | $P^4$ | $5b^4-10b^3+10b^2-5b+1$ |
| *mmmr* | $2P^3(1-P)$ | $-6b^4+12b^3-8b^2+2b$ |
| *rmmr* | $2P^2(1-P)^2$ | $b^4-2b^3+b^2$ |
| *mmrr* | $2P^3(1-P)$ | $-6b^4+12b^3-8b^2+2b$ |
| *mrmm + rmrr* | $P^2(1-P)^2+2P(1-P)^3$ | $(2b^4-4b^3+2b^2)+(2b^4-4b^3+2b^2)$ |
| *rmrm* | $2P^2(1-P)^2$ | $2b^4-4b^3+2b^2$ |
| *rrrr* | $2P(1-P)^3$ | $b^4-2b^3+b^2$ |
| *rrrm* | $P^2(1-P)^2$ | $2b^4-4b^3+2b^2$ |
| *mrrm* | $(1-P)^4$ | $-3b^4+6b^3-4b^2+b$ |

The difference can be readily realized for isotactic propagation. Intuitively, in the case of chain end stereocontrol, an occasional change in the configuration of the last inserted monomer unit tends to be perpetuated (as in Scheme 3a), while in the case of enantiomorphic site stereocontrol the same occasional change, having no effect on site chirality, tends to remain isolated (as in Scheme 3b).[2–4]

(a) 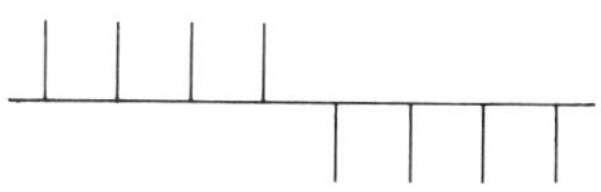 (b) 

**Scheme 3** Typical steric defects in a (mainly) isotactic poly-1-alkene chain (adapted Fischer notation)

It should be realized that more than one type of active site may be present (as is the case particularly with heterogeneous catalysts), and that, even for a single type of active site, both kinds of stereocontrol may operate.[2,4]

The extent of chain regularity is determined primarily with spectroscopic methods. For a long time IR spectroscopy has been the only available tool. The main drawback of the IR technique, *i.e.* the lack of generally accepted procedures for the quantitative evaluation of the spectra, has been overcome by the development of NMR spectroscopy, particularly $^{13}$C NMR. For most poly-1-alkenes of practical importance, the chain stereostructure can be quantitatively determined at least at the triad level. In the following, IR data are reported only in the absence of more reliable $^{13}$C NMR results.

Polymer chains differing in regularity have different solubilities in organic solvents (less stereoregular chains being more soluble), and solvent fractionation methods have been proposed to provide indirect estimates of chain regularity. Although it is well recognized that solubility depends on average molecular weight as well (shorter macromolecules being more soluble), these methods are so widely and routinely used that they cannot be ignored. In particular, for polypropene obtained from heterogeneous isospecific catalysts, the weight fraction insoluble in *n*-heptane at its boiling point is largely identified with the relative amount of isotactic polymer in a sample, and for this reason denoted *isotactic index (I.I.)*. When dealing with the complicated mixture of macromolecules of variable regularity typically resulting from such catalysts (*vide infra*), this practice is justified by the need of an (admittedly arbitrary) borderline between what is to be considered isotactic polymer and what is not.

The specificity of a ZN catalyst system is somewhat dependent on polymerization conditions (temperature, solvent, concentrations of components and of monomer). A detailed description of such conditions for each cited example would have rendered this presentation too burdensome; for details, the interested reader is referred to the original papers. An exception is made for polymerization temperature in the presence of homogeneous catalysts, which is always indicated for the dramatic dependence of stereospecificity on this parameter.

Due to the overwhelming importance of polypropene (polypropylene), most literature data on stereospecific ZN polymerizations refer to propene (propylene); in the following, significant results for other 1-alkenes will be given when available.

### 3.2.2 Heterogeneous Isospecific Ziegler–Natta Catalysts

In 1954, Natta and co-workers found that the catalyst disclosed by Ziegler in 1953[11] for the low pressure polymerization of ethylene, *e.g.* the product of the reaction between $TiCl_4$ and an Al trialkyl in hydrocarbon solvent, is able to polymerize propene to a partly crystalline polymer.[12]

An ingenious combination of solvent fractionation and X-ray diffraction methods revealed that the raw polymer was a mixture of atactic and isotactic polypropene (the finding was so revolutionary that even the terms 'isotactic' and 'atactic' had to be coined for the occasion).[12–16]

What was defined as 'isotactic polypropene', *i.e.* the polymer fraction insoluble in boiling *n*-heptane, amounted to 20–40% by weight of the total polymer, depending on the experimental conditions.[17] From later $^{13}C$ NMR spectra, a *mm* triad content well over 90% of the total triad distribution was evaluated for this fraction.[18]

It was soon recognized that, in polymerization conditions, $TiCl_4$ and Al trialkyls react to form a brown modification of $TiCl_3$ (prevailingly $\beta$-$TiCl_3$), which is insoluble in the reaction medium.[19] This prompted Natta and co-workers to explore the catalytic activity of preformed modifications of $TiCl_3$, prepared either by reduction of $TiCl_4$ (*e.g.* with $H_2$ or Al) or from different crystalline forms of $TiCl_3$ by mechanical or thermal treatments. A dramatic improvement in stereospecificity was obtained in propene polymerization using suspensions of a layered (violet) modification of $TiCl_3$ ($\alpha$, $\gamma$ or $\delta$) in hydrocarbon medium, in a mixture with Al trialkyls or Al dialkyl halides: I.I.'s in the range 80–95% were achieved.[20]

These findings opened the way to a whole generation of $TiCl_3$-based catalysts for the polymerization of 1-alkenes to linear isotactic polymers of wide industrial application.

In spite of large differences in catalytic activity, resulting from treatments to increase surface areas and from modifications with organic and/or inorganic compounds,[21–24] a similar stereospecificity was shown by such catalysts in propene polymerization (see Table 2(a) where data for the catalyst system $TiCl_4/AlEt_3$ are added for comparison).

The non-transition metal co-catalyst has limited influence on polymer stereoregularity, though its effect has been sometimes overemphasized for an incorrect interpretation of solvent fractionation data. As an example, it is seen from Table 3 that the I.I. of polypropene produced in the presence of the catalyst system $\alpha$-$TiCl_3/MR_n$ decreases with increasing chain transfer ability of $MR_n$ from $BeEt_2$ to $ZnEt_2$, the effect being primarily due to a decrease in average molecular weight rather than in polymer regularity. This has been taken as an indication that the active site is a transition metal–carbon bond.[1]

Data on the polymerization of 1-alkenes other than propene in the presence of $TiCl_3$-based catalysts are given in Table 4.

Stereospecific catalyst systems based on titanium(III) halides other than $TiCl_3$ as well as on different transition metal halides are known, though they are generally of lower activity and/or stereospecificity (Table 2b). A layered crystal structure similar to that of violet $TiCl_3$ is presented by most of these metal halides.[25]

**Table 2** Isotactic Index (I.I) and Degree of Isotacticity of the *n*-Heptane Insoluble Fraction (D.I.) of Polypropene Samples Obtained in the Presence of Different Heterogeneous Isospecific ZN Catalysts

| | *Catalyst system* | *I.I.* (%) | *D.I.* (%) | *Ref.* |
|---|---|---|---|---|
| (a) | $TiCl_4/AlEt_3$ | 26 | 91[a] | c |
| | $\alpha$-$TiCl_3/AlEt_3$ | 87 | 95[b] | d |
| | $\delta$-$TiCl_3/AlEt_3$ | 75 | 92[b] | d |
| | $\alpha$-$TiCl_3/AlEt_2Cl$ | 95 | 94[b] | d |
| | $\delta$-$TiCl_3/AlEt_2Cl$ | 89 | 90[a] | e |
| (b) | $\alpha$-$TiBr_3/AlEt_3$ | 76 | 87[b] | d |
| | $VCl_3/AlEt_2Cl$ | 61 | 80[a] | e |
| | $CrCl_3/AlEt_2Cl$ | 98 | 94[a] | e |
| (c) | $MgCl_2/TiCl_4-AlEt_3$ | 31 | 91[a] | f |
| | $MgCl_2/EB/TiCl_4-AlEt_3$ | 50 | 93[a] | f |
| | $MgCl_2/TiCl_4-AlEt_3/EB$ | 88 | 93[a] | f |
| | $MgCl_2/EB/TiCl_4-AlEt_3/EB$ | 95 | 93[a] | f |

Abbreviation: EB = ethyl benzoate.
[a] % mm triad as evaluated by $^{13}C$ NMR. [b] Evaluated by IR. [c] Y. Doi, *Makromol. Chem.*, 1979, **180**, 2447. [d] Y. V. Kissin, V. I. Tsvetskova and N. M. Chirkov, *Eur. Polym. J.*, 1972, **8**, 529. [e] C. Wolfsgruber, G. Zannoni, E. Rigamonti and A. Zambelli, *Makromol. Chem.*, 1975, **176**, 2765. [f] Authors' data.

**Table 3** Isotactic Index (I.I.), IR Degree of Isotacticity (D.I.) and Intrinsic Viscosity $[\eta]$ of Polypropene Samples from $\alpha$-$TiCl_3/MR_n$ Catalyst Systems

| $MR_n$ | *I.I.* (%) | *D.I.* (%) | $[\eta]$ (dl $g^{-1}$)[a] | *Ref.* |
|---|---|---|---|---|
| $BeEt_2$ | 91–98 | 93–98 | 2–11.8 | b, c |
| $AlEt_3$ | 77–80 | 85–92 | 3–4 | b, c |
| $ZnEt_2$ | 46–66 | 90–95 | 0.1–0.4 | b, c |

[a] For the *n*-heptane insoluble fraction. [b] A. P. Firsov, B. G. Kashporov, Y. V. Kissin and N. M. Chirkov, *J. Polym. Sci.*, 1962, **62**, s-104. [c] Y. V. Kissin, V. I. Tsvetskova and N. M. Chirkov, *Eur. Polym. J.*, 1972, **8**, 529.

**Table 4** Degree of Isotacticity (D.I.) of Different Poly-1-alkene Samples from $TiCl_3$-based Catalyst Systems

| *Poly-1-alkene* | *Catalyst system* | *D.I.* (%) | *Ref.* |
|---|---|---|---|
| Poly-1-butene | $\gamma$-$TiCl_3/AlEt_2Cl$ | 95[a] | c |
| Poly-4-methyl-1-pentene | $\delta$-$TiCl_3/AlBu^i_3$ | 77[b] | d |
| Polystyrene | $\alpha$-$TiCl_3/AlEt_3$ | 96[b] | d |

[a] % by weight insoluble in diethyl ether. [b] Evaluated by IR. [c] J. Boor, *J. Polym. Sci., Part C*, 1963, **1**, 237. [d] Y. V. Kissin, 'Isospecific Polymerization of Olefins', Springer-Verlag, New York, 1985, chap. IV.

A more effective utilization of the catalytically active species has been achieved by a new generation of catalysts, rapidly taking over the market, in which a Ti compound (mostly $TiCl_4$) is supported on a crystalline matrix. It may be of interest to remark that, up to now, stereospecific catalysts have been obtained using supports with a layered crystal structure of $TiCl_3$-type, primarily $MgCl_2$.

In the absence of third components (*vide infra*), average catalyst activities of two orders of magnitude higher than the maximum activity of conventional $TiCl_3$-based catalysts can be reached with $MgCl_2/TiCl_4$ catalysts, but their stereospecificity is poor: I.I. for polypropene is in the range 20–50%.[26,27] Nevertheless, the polymer fraction insoluble in *n*-heptane is highly stereoregular, as is the homologous fraction from $TiCl_3$-based catalysts (*cf.* Tables 2a and c). This suggests the existence, in both cases, of (basically) two classes of active sites — highly stereospecific and non-stereospecific — in variable ratios.

It seemed reasonable to suppose that for these two classes of sites — being part of different organometallic complexes — there is a different chemical reactivity. This assumption justified the search for third components able to 'poison' selectively the non-stereospecific sites. Indeed, several families of electron donor compounds were found able to improve the stereospecificity of $MgCl_2$/$TiCl_4$ catalysts, *e.g.* esters of aromatic carboxylic acids, hindered amines, alkylalkoxysilanes. I.I.'s of 95% and over can be obtained in propene polymerization, with an appropriate choice of electron donors as components of the solid catalyst ('internal' donor) and of the co-catalyst mixtures with Al trialkyls ('external' donor) (Table 2c).[27–30]

Although the first industrial patents date back to the early 1970's,[31] systematic studies on the mechanism of catalyst modification with electron donors came much later.[27,32] It has been shown that both internal and external donors are necessary to achieve high stereospecificities without depressing catalyst productivities too much. In particular, according to a recent investigation,[27] the internal donor, which is contacted with $MgCl_2$ *prior* to $TiCl_4$, prevents the formation of non-stereospecific sites relatively insensitive to selective poisons (external donors) acting either by complexation or by some other chemical reaction.[27,33] Sound evidence has been presented in favour of the view that the electron donors *do not* participate in the organometallic complex of which the active site is part.[27,34]

Tables 5 and 6 present more detailed spectroscopic data on the regio- and stereo-regularity of polypropene samples obtained in the presence of a number of heterogeneous isospecific catalyst systems. A similar picture emerges for $TiCl_3$-based and $MgCl_2$-supported catalysts.

**Table 5** Regioregularity (Evaluated by $^{13}C$ NMR) of Polypropene Samples from Different Heterogeneous Isospecific ZN Catalysts

| *Catalyst system* | *Polymer fraction* | *Amount of regioirregular placements* (mol %) | *Ref.* |
|---|---|---|---|
| $TiCl_4/AlEt_3$ | i7 | n.d. | a, b |
| | s7 | 4–5 | |
| $\delta$-$TiCl_3/AlEt_3$ | i7 | n.d. | a, b |
| | s7 | 1 | |
| $VCl_3/AlEt_2Cl$ | i7 | n.d. | c |
| | s7 | 1–2 | |
| $MgCl_2/TiCl_4$–$AlEt_3$ | i7 | n.d. | d |
| | s7 | $\simeq$0.1 | |

Abbreviations: i7 = insoluble in *n*-heptane; s7 = soluble in *n*-heptane; n.d. = not detected
[a] Y. Doi, *Makromol. Chem.*, 1979, **180**, 2447. [b] Y. Doi, *Makromol. Chem. Rapid Commun.*, 1982, **3**, 635. [c] C. Wolfsgruber, G. Zannoni, E. Rigamonti and A. Zambelli, *Makromol. Chem.*, 1975, **176**, 2765. [d] Authors' data.

Table 5 reports, for different polymer fractions, the relative amounts of regioirregular placements of monomer units. These are virtually absent in the isotactic (*n*-heptane insoluble) fraction; 0.1–5% regioirregular placements, on the other hand, are found in the non-stereoregular (*n*-heptane soluble) fractions. The spectroscopic analysis of the chain end groups pointed out that, in most cases, the preferred insertion mode is primary.[34–37]

Table 6 gives the pentad stereosequence distribution (as determined *via* $^{13}C$ NMR) for various polymer fractions. The data for the isotactic (*n*-heptane insoluble) polymer are consistent with the hypothesis of an enantiomorphic site stereocontrol (*cf.* Table 1).[9,10,38]

Such a conclusion is independently confirmed by the $^{13}C$ NMR analysis of isotactic polypropene chains with low amounts of copolymerized ethylene. If the steric control was due to the chirality of the last inserted monomer unit, the insertion of a propene unit following that of an achiral ethylene unit would be non-stereospecific (both cases (a) and (b) of Scheme 4 would be possible), whereas virtually complete stereospecificity has been observed experimentally (case (a) of Scheme 4).[2,39]

Table 6 deserves a further comment. It is seen from the pentad stereosequence distribution that what is cumulatively denoted as non-stereoregular (atactic) polypropene (*i.e.* the fraction soluble in *n*-heptane) is a complicated mixture of macromolecules of variable stereostructure. In all cases, the distribution is not Bernoullian, but short isotactic stereoblocks alternate with atactic sequences and (with a smaller amount of) short syndiotactic stereoblocks.[2]

**Table 6** Pentad Stereosequence Distribution (Evaluated by $^{13}C$ NMR) for Polypropene Samples from Different Heterogeneous Isospecific ZN Catalysts.

| *Catalyst system* | *Polymer fraction* | % | $^{13}$*C NMR pentad stereosequence distribution (%)* | | | | | | | | | *Ref.* |
|---|---|---|---|---|---|---|---|---|---|---|---|---|
| | | | *mmmm* | *mmmr* | *rmmr* | *mmrr* | *mmrm + rrmr* | *mrmr* | *rrrr* | *mrrr* | *mrrm* | |
| β-$TiCl_3$/$AlEt_2Cl$ | i7 | 47.5 | ≃90 | ≃4 | — | ≃4 | — | — | — | — | ≃2 | a |
| | s7–i6 | 9 | 52 | 8 | 10 | 9 | 8 | ≃2 | 6 | ≃2 | ≃3 | |
| | s6–i5 | 12.5 | 24 | 9 | ≃2 | 15 | 16 | 6 | 14 | 10 | ≃4 | |
| | s5 | 31 | 16 | 11 | ≃3 | 16 | 16 | 6 | 18 | 10 | ≃4 | |
| δ-$TiCl_3$/$AlEt_2Cl$ | i7 | 91 | ≃93 | ≃3 | — | ≃3 | — | — | — | — | ≃1 | b, c |
| | s7 | 9 | 25 | 13 | 2 | 17 | 13 | 2 | 14 | 9 | 5 | |
| $VCl_3$/$AlEt_2Cl$ | i7 | 60.5 | 74 | 6 | — | 7 | 6 | ≃1 | ≃2 | ≃1 | ≃3 | a |
| | s7–i6 | 5.5 | 41 | 9 | ≃2 | 12 | 12 | ≃2 | 13 | 5 | ≃4 | |
| | s6 | 34 | 15 | 10 | ≃4 | 13 | 23 | 5 | 14 | 12 | ≃4 | |
| $CrCl_3$/$AlEt_2Cl$ | i7 | 98.5 | ≃90 | ≃4 | — | ≃4 | — | — | — | — | ≃2 | a |
| | s7 | 1.5 | 32 | 9 | ≃2 | 11 | 10 | ≃2 | 18 | 9 | 7 | |
| $MgCl_2$/EB/$TiCl_4$–$AlEt_3$ | i7 | 71.7 | 90 | 4 | — | 4 | — | — | — | — | 2 | d |
| | s7–i6 | 8.6 | 68 | 10 | — | 10 | 3 | — | 3 | 2 | 4 | |
| | s6 | 19.7 | 28 | 12 | 3 | 16 | 11 | 3 | 15 | 8 | 4 | |

Abbreviations: i = insoluble; s = soluble; 7 = *n*-heptane; 6 = *n*-hexane; 5 = *n*-pentane

[a] C. Wolfsgruber, G. Zannoni, E. Rigamonti and A. Zambelli, *Makromol. Chem.*, 1975, **176**, 2765. [b] A. Pavan, A. Provasoli, G. Moraglio and A. Zambelli, *Makromol. Chem.*, 1977, **178**, 1099. [c] Y. Doi, *Makromol. Chem., Rapid Commun.*, 1982, **3**, 635. [d] Y. Doi, E. Suzuki and T. Keii, *Makromol. Chem., Rapid Commun.*, 1981, **2**, 293.

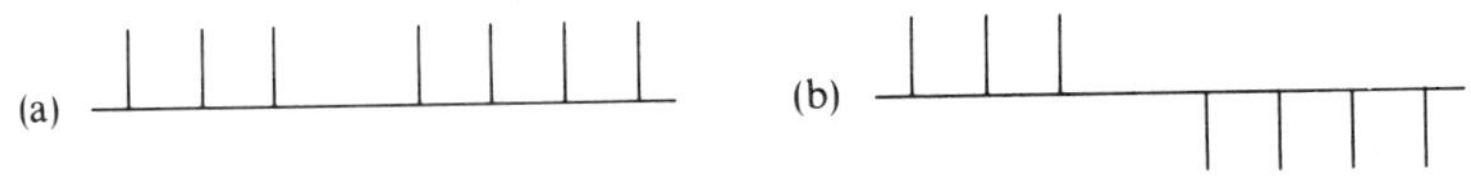

**Scheme 4** Possible isotactic poly-1-alkene chain propagation after the copolymerization of an achiral ethylene unit (adapted Fischer notation)

Scheme 5 illustrates recent results of a high resolution $^{13}C$ NMR investigation[40] of the specificity of a $\delta$-$TiCl_3$ catalyst in the first two insertions of the monomer (either propene or 1-butene) into an initial M–alkyl bond of a growing chain, the alkyl group deriving from a selectively $^{13}C$-enriched Al or Zn alkyl.

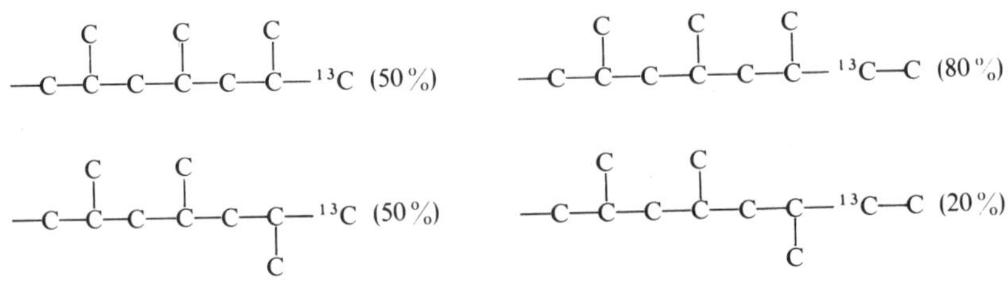

**Scheme 5** Structure and relative occurrence of polypropene chain ends for polymer samples obtained from the catalyst systems $\delta$-$TiCl_3/Al(^{13}CH_3)_3/Zn(CH_3)_2$ (left) and $\delta$-$TiCl_3/Al(^{13}CH_2CH_3)_3/Zn(C_2H_5)_2$

The spectra confirm a primary insertion mode, as already discussed. For both propene and 1-butene, the first insertion is *non*-stereospecific when the active site is a M–$^{13}CH_3$ bond, and only partly (*ca.* 80%) stereospecific when the active site is a M–$^{13}CH_2CH_3$ bond, while the second insertion is in all cases highly (over 90%) isospecific. These results prove once again that stereospecificity is dictated by the intrinsic chirality of the organometallic complex at the active site, and not by the chirality of the growing polymer chain (no asymmetric C atoms being present in ethyl or isobutyl groups), but they also reveal that its extent is dependent on the size of the alkyl group bound to the metal in the active site; in particular, stereospecificity appears when the alkyl group has at least two C atoms. The implications of this latter finding on the mechanism of stereocontrol are discussed in Section 3.3.

We end with a brief account of the polymerization of chiral 1-alkenes. In the presence of heterogeneous isospecific catalysts, it has been shown[41] that a racemic mixture of a chiral monomer is converted into a racemic mixture of macromolecules each containing prevailingly one of the two enantiomers (*stereoselective* polymerization).

The degree of stereoselectivity strongly depends on the position of the asymmetric C atom with respect to the double bond,[42] being relatively high (*ca.* 70–90% for 3-methyl-1-pentene in the presence of the catalyst system $\delta$-$TiCl_3/AlMe_3$, according to recent $^{13}C$ NMR investigations[43]) when this is in the $\alpha$ position, moderate when in the $\beta$ position, and virtually absent when in the $\gamma$ position or further apart.

In principle, stereoselectivity—as with stereospecificity—may arise from the chirality of the last inserted monomer unit as well as from that of the active site. Spectroscopic investigations have pointed out, in this case too, an enantiomorphic site stereocontrol.[43,44] It has also been shown, on the other hand, that stereospecificity and stereoselectivity are independent phenomena, arising from different interactions between the 1-alkene monomer and the organometallic complex of which the active site is part.[43]

### 3.2.3 Homogeneous Syndiospecific Catalysts

Propene and—very recently—styrene are the only two 1-alkenes which have been polymerized with ZN catalysts to syndiotactic polymers with high steric purity.

Low amounts of syndiotactic polymer are obtained when polymerizing propene in the presence of some heterogeneous isospecific catalysts;[45] even more frequent is the formation of more or less stereoirregular macromolecules containing short syndiotactic stereoblocks.[2,16] This proves the existence in such catalysts of small fractions of active sites having—at least temporarily—a syndiospecific stereocontrol.

Highly syndiotactic polypropene as a primary product, on the other hand, can be obtained—to our present knowledge—only in the presence of a few V-based catalyst systems[47,48] (Table 7) operating in the homogeneous phase at low temperature (below *ca.* $-50\,^{\circ}C$, as the polymerization becomes non-stereospecific at higher temperatures).

**Table 7** Regularity (Evaluated by $^{13}C$ NMR) of Polypropene Samples from Different Homogeneous Syndiospecific ZN Catalysts

| *Catalyst system* | $^{13}C$ *NMR triad distribution* % *mm* | % *mr* | % *rr* | *Fraction of regioirregular placements* (mol%) | *Ref.* |
|---|---|---|---|---|---|
| $VCl_4/AlEt_2Cl$, $-78\,^{\circ}C$ | — | 24 | 76 | 2.1 | b |
| $VCl_4/AlEt_2Br$, $-78\,^{\circ}C$ | 22 | 25 | 53 | 5.1 | b |
| $VCl_4/AlEt_2I$, $-78\,^{\circ}C$ | 46 | 31 | 23 | 15.0 | b |
| $V(acac)_3/AlEt_2Cl$, $-78\,^{\circ}C$ | 3 | 32 | 65 | 3.6 | c |
| $VCl_4/AlEt_2Cl$, $41\,^{\circ}C$[a] | 59 | 26 | 15 | 4.6 | b |

Abbreviations: acac = acetylacetonate.
[a] data added for comparison. [b] Y. Doi, M. Takada and T. Keii, *Makromol. Chem.*, 1979, **180**, 57; [c] Y. Doi, *Macromolecules*, 1979, **12**, 1012.

From Table 7, it appears that both regio- and stereo-regularity are lower than those typical of isotactic polypropene; improvements in syndiospecificity have been obtained by catalyst modification with specific electron donor compounds (typically anisole).[47,48]

A lot of spectroscopic evidence exists[50,51] that propene insertion is prevailingly secondary. The insertion mode appears to be mainly dictated by steric factors; indeed it has been shown that propene insertion into an initial M–ethyl bond,[51] as well as after an ethylene insertion in ethylene/propene copolymerization,[52] is substantially non-regiospecific.[53] Regiospecificity would thus arise in consequence of propene secondary insertion.

The observed distribution of steric defects (Table 8) obeys Bernoullian statistics.[54,55] This suggests that syndiospecificity results from a chain end stereocontrol (unlike (ul) 1,3 asymmetric induction of the last inserted propene unit). In agreement with this hypothesis, it has been shown that, in ethylene/propene copolymerization, propene insertion after an ethylene insertion is substantially non-stereospecific.[38]

**Table 8** Pentad Stereosequence Distribution (Evaluated by $^{13}C$ NMR) for a Typical Polypropene Sample from the Homogeneous Syndiospecific ZN Catalyst System $VCl_4/AlMe_2Cl$/Anisole at $-78\,^{\circ}C$[a]

| *Pentad* | % |
|---|---|
| *mmmm* | 1.1 |
| *mmmr* | 3.3 |
| *rmmr* | 3.1 |
| *mmrr* | 3.7 |
| *mmrm* + *rmrr* | 23.9 |
| *rmrm* | 10.9 |
| *rrrr* | 25.4 |
| *mrrr* | 21.7 |
| *mrrm* | 6.9 |

[a] A. Zambelli, P. Locatelli, A. Provasoli and D. R. Ferro, *Macromolecules*, 1980, **13**, 267.

Table 7 shows that syndiospecificity is affected by the choice of the Al alkyl co-catalyst. This has been taken as an indication that the Al alkyl participates in the catalytic complex.[55]

A preliminary report has recently appeared in the literature on the first polymerization of styrene to a syndiotactic polymer.[56] An exceedingly high steric purity (*rr* triad content over 98%) was achieved. No detail was given about the catalyst system, generically described as a ZN catalyst comprising of a Ti compound and an organo–Al compound.

In a subsequent independent paper,[57] the synthesis of syndiotactic polystyrene has been claimed at 50 °C in the presence of the catalyst system $Cp_2TiPh_2$/methylalumoxane, which is also known to polymerize propene to an isotactic polymer at low temperature (below *ca.* −40 °C, *vide infra*).

### 3.2.4 Homogeneous Isospecific Catalysts

Homogeneous catalyst systems able to polymerize 1-alkenes to isotactic polymers were first described by Giannini *et al.* in 1970.[58] Isotactic polypropene and poly-4-methyl-1-pentene were obtained with catalysts originating from benzyl derivatives of Ti and Zr (typically $MBz_4$, $MBz_3X$ with M = Ti or Zr; X = halogen), even in the absence of a non-transition metal organometallic co-catalyst. The exceedingly low catalytic activities discouraged further research on these catalyst systems.

Much more impact was produced by a recent (1984) paper by Ewen,[3] describing homogeneous catalysts able to polymerize propene to isotactic polymer with high productivities, and consisting of titanocene compounds in combination with alkylalumoxanes.

$Cp_2TiR_2/AlR'_3$ catalyst systems (Cp = cyclopentadienyl; R = alkyl or aryl; R′ = alkyl) for ethylene polymerization have long been known. In the early 1970s, substantial improvements in productivities were obtained by controlled addition of $H_2O$, which reacts with the Al alkyl to yield mixtures of linear and cyclic oligomeric alkyl alumoxanes. At the usual polymerization temperatures (20–70 °C), such catalysts were also known to polymerize propene with lower productivities to purely atactic polymer.[18,59]

Quite unexpectedly, Ewen discovered[3] that the catalyst system $Cp_2TiPh_2/AlMe_3/H_2O$ becomes increasingly isospecific in propene polymerization with decreasing temperature (Table 9 and Figure 1).

**Table 9** Diad and Triad Distributions in Polypropene Samples Obtained in the Presence of the Catalyst System $Cp_2TiPh_2/AlMe_3/H_2O$ at Various Temperatures[a]

| *Temperature* | % *m* | % *r* | % *mm* | % *mr* | % *rr* |
|---|---|---|---|---|---|
| 50 | 49 | 51 | 23 | 53 | 24 |
| 25 | 50 | 50 | 24 | 52 | 24 |
| 0 | 67 | 33 | 45 | 44 | 11 |
| −15 | 78 | 22 | 62 | 33 | 5 |
| −30 | 83 | 17 | 69 | 28 | 3 |
| −45 | 85 | 15 | 73 | 25 | 2 |
| −60 | 85 | 15 | 73 | 24 | 3 |
| −75 | 85 | 15 | 73 | 25 | 2 |
| −85 | 84 | 16 | 72 | 25 | 3 |

[a] J. A. Ewen, *J. Am. Chem. Soc.*, 1984, **106**, 6355.

The observed Bernoullian distribution of steric defects in the polypropene chains (Table 9) is consistent with a chain end stereocontrol.[3] The spectroscopic analysis of the chain end groups suggested that chain propagation takes place *via* primary insertion; no regioirregular placements of propene units were detected by $^{13}$C NMR. These conclusions were subsequently confirmed by a $^{13}$C NMR analysis of the stereostructure of the chain end groups resulting from the first propene and 1-butene insertions into an initial M–alkyl bond, the alkyl group being derived from a selectively $^{13}$C-enriched Al alkyl.[60]

Remarkably, this is the first known case of isotactic chain propagation dictated by the chirality of the last inserted monomer unit (like (lk) 1,3 asymmetric induction).

An unusual decrease of catalyst stereospecificity has been pointed out[57] with increasing size of the alkyl substituent at the double bond in the polymerization of different 1-alkenes: the *m* diad content in the resulting polymer, which is over 80% for polypropene, drops to *ca.* 60% for poly-1-butene and approaches 50% for poly-4-methyl-1-pentene. This has been taken as a further indication for the existence of a chain end stereocontrol.[57]

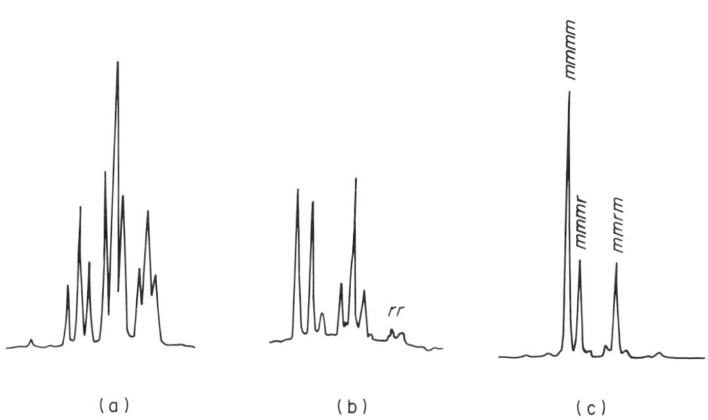


**Figure 1** Comparison of $^{13}$C NMR spectra of the methyl pentad region for polypropylene obtained with $Cp_2TiPh_2$/methylalumoxane at three different polymerization temperatures: (a) 25 °C; (b) 0 °C; (c) −45 °C (reproduced by permission of American Chemical Society from *J. Am. Chem. Soc.*, 1984, **106**, 6355)

Ewen also showed[3] that stereorigid *chiral* titanocenes (*e.g. rac*-(ethylene)bis(indenyl)$TiX_2$ with X = halogen, alkyl or aryl, Figure 2), in mixture with alkylalumoxanes, polymerize propene to an isotactic polymer with a non-Bernoullian distribution of steric defects consistent with an enantiomorphic site stereocontrol, as with that observed for the heterogeneous isospecific catalysts. In this case too, chain propagation was shown to proceed *via* primary insertion with very high regiospecificity. The $^{13}$C NMR investigation of the stereostructure of chain end groups resulting from the first propene and 1-butene insertions into initial M–($^{13}$C-enriched)-alkyl bonds[61] confirmed that the steric control is due to the chirality of the active site, and pointed out a dependence of stereospecificity on the size of the alkyl group bound to the metal very similar to that observed for the heterogeneous isospecific catalysts[40] (*cf.* Scheme 5).

Noticeably, stereorigid *achiral* titanocenes (*e.g.* meso-(ethylene)bis(indenyl)$TiX_2$), in combination with alkylalumoxanes, polymerize propene to purely atactic polymer.[3]

Ewen's pioneering work has been subsequently extended by Kaminski to homogeneous catalyst systems based on the homologous or strictly related zirconocenes, which turned out to be characterized by similar specificity but much higher catalytic activity in propene and 1-butene polymerization.[62] Productivities, expressed in moles of polymerized monomer per mole of Zr an hour, analogous to or even higher than those typical of $MgCl_2$-supported catalysts have been achieved, though a number of still unsolved problems preclude — at the moment — the possibility of short-term industrial applications.

### 3.2.5 Summary

Table 10 summarizes the main conclusions derived from over three decades of research on the stereochemistry of polymerization of 1-alkenes in the presence of stereospecific ZN catalysts.

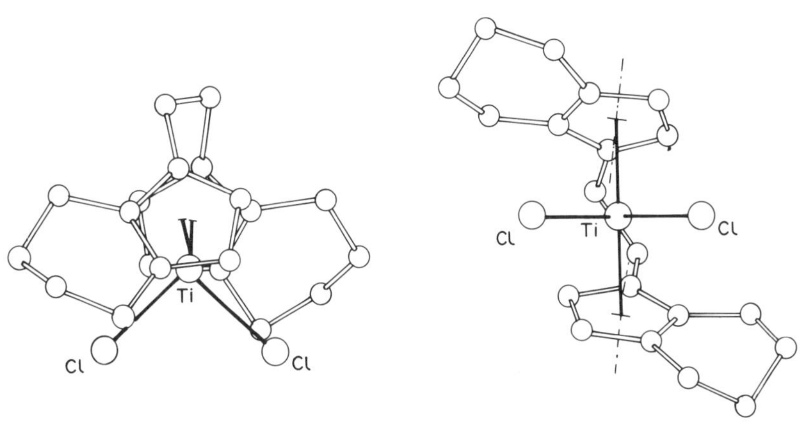


**Figure 2** Molecular structure of racemic ethylene-bis(4,5,6,7-tetrahydro-1-indenyl)titanium dichloride (reproduced by permission of Elsevier from *J. Organomet. Chem.*, 1982, **232**, 233)

**Table 10** Stereochemistry of Propene Polymerization in the Presence of Stereospecific ZN Catalysts

| *Catalyst type* | *Stereochemistry of addition to the double bond* | *Monomer insertion* | *Type of stereocontrol* |
|---|---|---|---|
| Heterogeneous isospecific | *cis* | primary | enantiomorphic site |
| Homogeneous syndiospecific | *cis* | secondary | chain end |
| Homogeneous isospecific, achiral | *cis* | primary | chain end |
| Homogeneous isospecific, chiral | *cis* | primary | enantiomorphic site |

This frame of experimental *facts* partly obviates the lack of direct observations of the catalytic species, providing a basis from which reasonable models of such species can be devised, as is discussed in Section 3.3.

## 3.3 MECHANISMS FOR STEREOSPECIFICITY IN ZIEGLER–NATTA POLYMERIZATION

### 3.3.1 General Remarks

In this section, we review a number of mechanisms proposed in the literature, in order to account for the stereospecific behaviour of the Ziegler–Natta catalysts, in the heterogeneous as well as in the homogeneous polymerization of monoalkenes.

It is largely accepted, on the ground of specific experimental evidence (*cf.* also Section 3.2) and of basic considerations of organometallic chemistry, that polymer chain growth implies successive insertions into a transition metal–carbon bond of previously coordinated 1-alkene molecules, through *cis* openings of the alkene double bond. Only mechanisms will be considered which conform to this view and take explicitly into consideration the origin of the stereospecificity, as arising from possible diastereoisomeric situations.

The main elements of chirality possibly present in the intermediates and transition states, which can be hypothesized in this framework, are as follows.

Firstly, upon coordination, a prochiral 1-alkene molecule, such as propene, gives rise to non-superposable *si* and *re* coordinations[63,64] as indicated in Figure 3. According to the considered mechanisms, an isotactic polymer is generated by a long series of insertions of all *si* or all *re* coordinated monomers; a syndiotactic polymer by alternate insertions of *si* and *re* coordinated monomers.

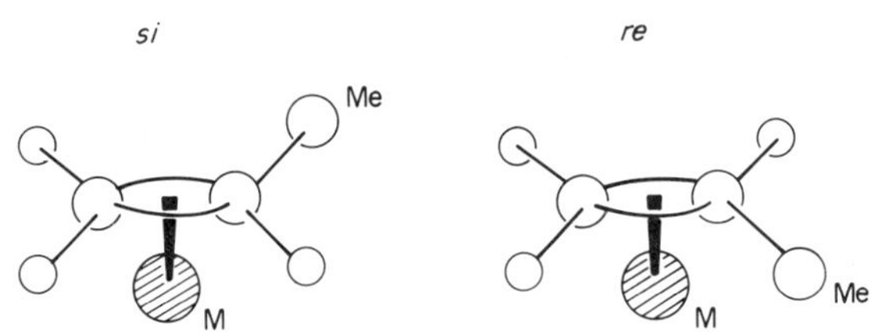


**Figure 3** The two possible chiral coordinations to a metal atom (M) of a propene molecule (*re* and *si*). The metal atom in the figure is below the propene molecule

A second element of chirality is the configuration of the tertiary carbon atoms in the growing chain, and in particular of that in the last inserted monomer unit.

Last but not least, the catalytic site may be chiral itself. An example for the heterogeneous catalysts, first proposed by Arlman[65] and common to other models[66–69] is shown in Figure 4: a metal atom is bridge-bonded through halogen atoms to two more metal atoms and two coordination positions are available to the monomer and to the growing chain. The chirality of these sites can be labelled with the symbols $\Lambda$ and $\Delta$,[67] defined for octahedral coordination compounds with at least two bidentate chelating agents.[70]

In this framework the stereospecificity of polymerization is to be connected with the energy differences between the diastereoisomeric situations which originate from the combination of two or

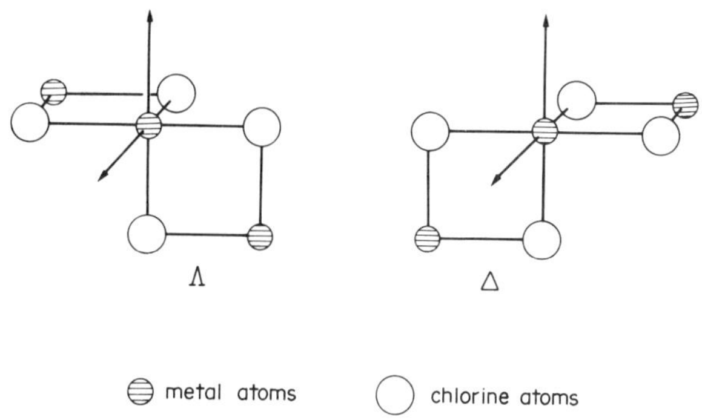


**Figure 4** Chiral model sites for the heterogeneous catalysts constituted by a metal atom bridge-bonded to two other metal atoms. Their chirality can be labelled with the Λ and Δ notation, defined for coordination compounds. The two positions at the model sites available for the incoming monomer and the growing chain are indicated by arrows

more of the above elements of chirality. All of the literature studies on the subject consider the energy differences between diastereoisomeric intermediates, instead of transition states, since it is easier to assume reasonable internal coordinates. Of course, the relevance of these studies is based on the assumption that such intermediates be energetically and geometrically close (much more than reagents and products) to the transition states; in any case, the comparison between structures of suitable diastereoisomeric intermediates may give an indication of the relative height of the energy barriers to the products.

### 3.3.2 Heterogeneous Catalysis

As discussed in Section 3.2, it is well established that the steric control of the heterogeneous Ziegler–Natta catalysts is due to the chirality of the catalytic site and not to the configuration of the last inserted monomer unit.

The hypothesis that the steric control is due to the structure of catalytic sites on the border of crystal layers of $TiCl_3$ was first put forward by Natta[71] and thoroughly developed by Arlman and Cossee.[65,72,73]

The structure of violet titanium trichlorides, as well as that of other transition metal trichlorides effective as components of Ziegler–Natta catalysts (*e.g.* $VCl_3$, $CrCl_3$), is built up of layers of the kind shown in Figure 5, piled one on top of the other according to a close packing of the chlorine atoms. The various known crystalline modifications[74–76] correspond to different ways in which the piling may be obtained. Within each layer, the metal atoms occupy in an ordered arrangement two thirds of the octahedral positions. As a result, neighbouring metal atoms (bridged by two Cl atoms) have opposite chirality,[66] which may be designated with the symbols Λ and Δ.

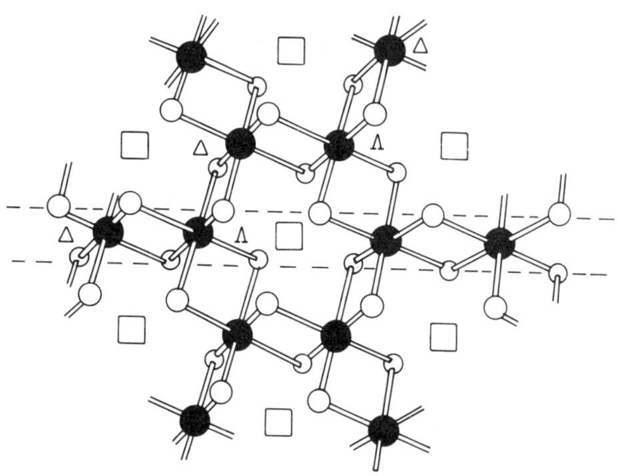


**Figure 5** Schematic drawing of a structural layer of violet $TiCl_3$. The titanium atoms (full circles) are each bonded to six chlorine atoms (empty circles) and occupy only two thirds of the octahedral positions ; the vacant octahedral positions are indicated by squares. The chirality of some of the metal atoms are also labelled. The dashed lines correspond to the lateral cut of Figure 6

On the basis of considerations of crystallochemistry and taking into account electron microscopy observations of the surface of crystals upon which some polymer was formed,[77] Arlman and Cossee[72,73] concluded that the active sites are located on crystal surfaces different from the basal 001 ones and parallel to the direction $\vec{a}-\vec{b}$ of the unit cell defined as in ref. 76.

Figure 6 illustrates that, if we cut a $TiCl_3$ layer parallel to the direction defined above, which corresponds to the line connecting two bridged Ti atoms, electroneutrality conditions impose that each Ti atom at the surface of the cut be bonded to five Cl atoms only. Of these, four are bridged to further metal atoms and are then more strongly bonded; the fifth, non-bridged Cl atom may be replaced by an alkyl group. The octahedral position which is still left free may $\pi$-coordinate an alkene. In general for this kind of model sites, the two positions accessible to the growing chain and to the monomer are nonequivalent: as shown in Figure 6, one (indicated as inward) is further in than the other (indicated as outward) with respect to the lateral cut.

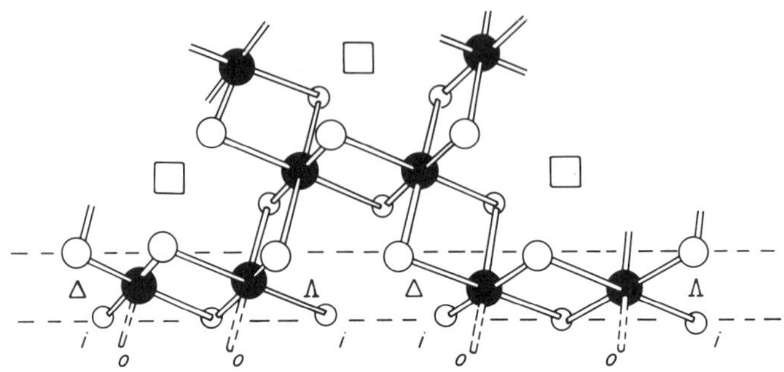


**Figure 6** Schematic drawing of a lateral cut of a $TiCl_3$ layer. The chirality of the Ti atoms on the termination of the layer is indicated. Electroneutrality conditions impose that the Ti atoms at the surface of the cut are bonded to four Cl atoms bridge-bonded to further metal atoms and one Cl atom not bridge-bonded, and that a coordination position is left free (dashed bonds). The two nonequivalent coordination positions, inward and outward, are indicated by *i* and *o* respectively (see text)

Of the set of crystal faces parallel to $\vec{a}-\vec{b}$ of indices (*hkl*), only the 110 face of $\alpha$-$TiCl_3$ was considered explicitly and an inward coordination of the alkene was supposed by Arlman and Cossee.

In these authors' opinion, the environment of the chlorine atoms would force the alkyl group of the alkene (coordinated with the double bond parallel to the Ti–chain bond, in such a way as to give rise to a primary insertion) to protrude out of the crystal. For a given site, such a condition can be realized for only one of the two possible coordinations of the alkene. However, as we shall see, this does not seem to be relevant in the explanation of stereospecificity.

The possibility that active sites be located on $TiCl_3$ layers protruding above the two adjacent ones was discussed by Allegra.[66] For the particular termination of the layer considered by Allegra, both octahedral sites of coordination at a Ti atom are equivalent for alkene complexation because the surface atoms with relevant non-bonded interactions at the catalytic site are locally related by a two-fold axis (Figure 7).

An argument in favour of such a situation was presented as follows. The Cossee reaction mechanism implies, as originally published, that at the end of each polymerization step the growing chain occupies the coordination site previously occupied by the alkene. Since, on the surface proposed by Cossee, inward and outward coordinations of the chain would prefer opposite

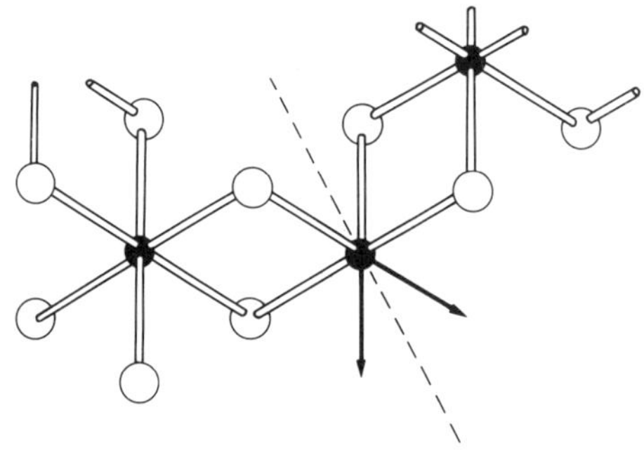

**Figure 7** The particular $TiCl_3$ layer termination, proposed by Allegra,[66] for which the two coordination positions available for the polymerization reaction (indicated by arrows) are equivalent. In fact they are locally related by a two-fold axis (dashed line)

coordinations of the alkene,[73] an isotactic polymerization would imply a back jump of the Ti–chain bond to the original position before the subsequent reaction step. The equivalence just discussed between both octahedral positions (as in the Allegra protruding layer) allows the back jumping of the Ti–chain bond to be considered an unnecessary step. However, quantum mechanical computations,[78,79] as well as chemical intuition, indicate that the best coordination of the growing chain at the catalytic site in the absence of the alkene (as at the end of each polymerization step) should be intermediate between inward and outward positions, so that the equivalence between the inward and outward positions does not appear as a determining argument in favour of a particular catalytic site.

Alternative octahedral sites on the basal or lateral surfaces of violet $TiCl_3$ crystals, presenting two vacancies accessible to the incoming monomer, were also proposed.[4,80] The main reason for these suggestions was, in the proponents' opinion, that in the Cossee model the access of the monomer to the only vacancy at the metal atom would be largely hindered by the growing chain. These model sites, besides the unlikelihood of two chlorine ligands being removed from the same metal atom, require the helicity of the chain as an element of chirality in order to account for the stereospecificity, which is definitely in contrast with many experimental observations.[1]

The model proposed by Arlman and Cossee has been more recently examined in some detail by Corradini, Guerra *et al.* by calculating the non-bonded energies for all sets of possible internal coordinates of the atoms at the catalytic site.[67,81] The main internal coordinates which were varied in the calculations are shown in Figure 8. They are the dihedral angle $\theta_0$ to be associated with the rotation of the alkene around the axis connecting the metal to the center of the double bond and the internal rotation angles $\theta_1$, $\theta_2$, $\theta_3$, . . . , around the first bonds of the chain.

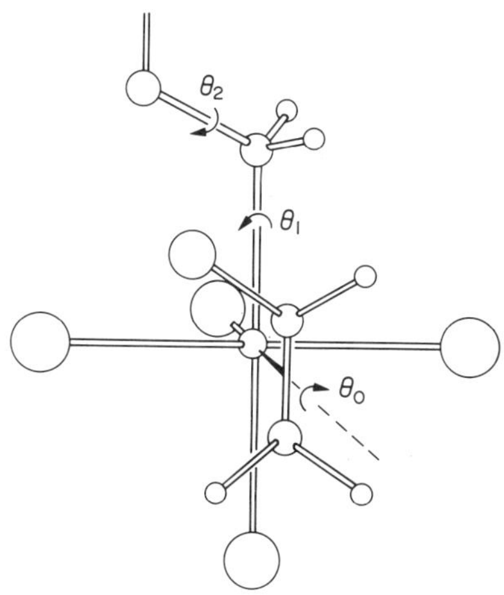


**Figure 8** The main internal coordinates which were varied in the non-bonded energy calculations for the model catalytic sites by Corradini *et al.*[67,81]

According to the calculations, in agreement with the conclusions of Arlman and Cossee, if the catalytic sites are thought to reside on the bulk of lateral (110) α-$TiCl_3$ surfaces, the alkene would prefer to coordinate in the inward position and the steric hindrance readily explains the observed regiospecificity; however, no evidence for the preference of a specific alkene coordination (*re* or *si*) at a site of given chirality was found. On the other hand, due to the interaction with the surface, the orientational freedom of the first carbon–carbon bond of the growing chain appeared to be strongly restricted; this 'chiral' orientation was suggested to result in different activation energies for the enantiomeric coordinated alkenes (*re* or *si*) in the successive insertion step. The stereospecificity could thus find an explanation.

On the other hand, calculations of the non-bonded interactions for the bulk of possible lateral surfaces of γ-$TiCl_3$ showed that the interaction energies are always much higher than those for the corresponding situations in α-$TiCl_3$.[81] Hence, if the catalytic sites are thought to reside on plain lateral surfaces, it seems rather difficult to explain the experimental observation that all layered modifications of $TiCl_3$ have a quite analogous stereospecificity irrespective of the particular stacking of the structural layers (*cf.* Section 3.2).

Instead, the computations suggested that, in agreement with previous hypotheses,[66,82,83] catalytic sites with similar behaviour for all layered modificiations of $TiCl_3$ can be located on some kind of terminal layers or on layers protruding above both neighbouring ones. These sites in relief involve much lower steric hindrances than sites on plain lateral surfaces and their behaviour is unified by the same general model: lower steric repulsions are obtained for a coordination of the growing polymer chain in the more hindered (inward) position, which, in turn, causes a chiral orientation of the first C–C bond of the chain (for a Λ site, values for $\theta_1$ near $+90°$ would be allowed while values near $-90°$ would be forbidden) (Figure 9). This orientation was identified as an essential factor in determining the isospecificity: indeed, as shown in Figure 9, the alkyl group of the 1-alkene and the second carbon atom of the growing chain, for the *si* monomer coordination, are on opposite sides with respect to the plane defined by the Ti–C bonds, while for the *re* coordination they are on the same side. This would determine a lower activation energy and henceforth favour the insertion of the *si* with respect to the *re* coordinated alkene on a Λ site. The opposite, of course, holds for a Δ site.

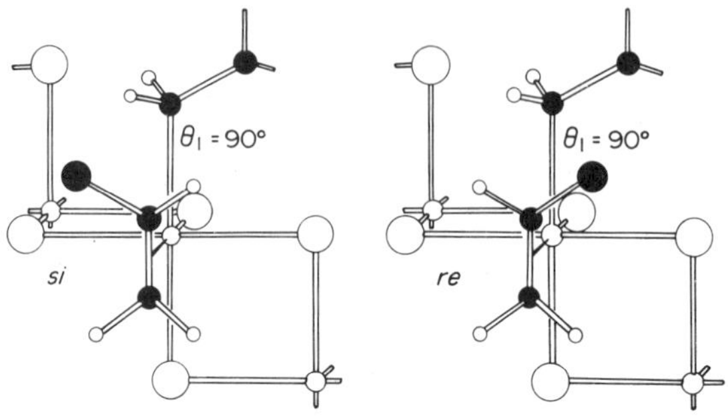


**Figure 9** Perspective drawing of the two possible chiral coordinations of propene (*si* and *re*), suitable for a primary insertion, on a Λ site of the kind of Figure 6. In both cases the conformation $\theta_1 \approx 90°$ of the growing chain corresponds to an energy minimum

The analysis by $^{13}$C NMR techniques of the chain end groups resulting from the initiation steps on $^{13}$C-enriched Ti–alkyl bonds, performed some years later[40] and already described in Section 3.2.2, pointed out that the chiral site is able to discriminate between *si* and *re* insertions of the monomer only in cases where alkyl groups with at least two C atoms are initially bound to the Ti atom, thus confirming the essential role of the chain in conferring the isospecific behaviour. Indeed specific calculations, relative to the steric control in the first steps of propene polymerization,[84] showed that for the model site proposed by Corradini *et al.* the insertion of the first monomer unit is non-stereospecific into an initial Ti–Me bond, partially stereospecific into an initial Ti–Et bond and totally stereospecific into an initial Ti–Bu$^i$ bond, in agreement with what observed experimentally.

At least for single vacancy sites in relief with respect to lateral surfaces, calculations in the presence of a long polypropene chain[85] showed that this, though able to discriminate between the two faces of the prochiral alkene, does not hinder severely the access of the monomer to the vacancy, thus overcoming the previously discussed criticism to the single vacancy sites.[4]

As recently discussed by Corradini *et al.*, catalytic sites in relief with respect to lateral surfaces of the layered crystals are also suitable to account for the similarity of the general behaviour of $TiCl_3$-based and $MgCl_2$-supported catalysts.[68] In fact, magnesium dichloride has crystal structures very similar to those of violet titanium trichlorides, built up of layers piled one on top of the other according to a close packing of the chlorine atoms, with the magnesium atoms occupying all the octahedral positions within each structural layer (Figure 10); the crystal lattice has also similar dimensions. This dictates the possibility of an epitactic coordination of $TiCl_4$ and, after reduction, $TiCl_3$ units on lateral, coordinatively unsaturated, faces of $MgCl_2$ crystals, giving rise to reliefs crystallographically coherent with the matrix.

Possible lateral cuts of $MgCl_2$ which correspond to 100 and 110 faces are also shown in Figure 10. Both have been observed by optical and electron microscopy on $MgCl_2$ crystals.[86] Electroneutrality conditions impose an average coordination number 5 for Mg atoms on 100 cuts, 4 for Mg atoms on 110 cuts. This causes differences in the surface coordination of single mononuclear $TiCl_4$ or $TiCl_3$ units. In any case, calculations suggested that the reduced $TiCl_3$ species formed after activation of these complexes by Al alkyl would behave as non-stereospecific sites.[68]

ESR studies on $MgCl_2$-supported catalysts,[87] on the other hand, showed that a significant fraction of $Ti^{3+}$ is ESR silent. This was explained by the formation of complexes with bridged Ti, *e.g.* through chlorine atoms as in crystalline $TiCl_3$.

Bridged dinuclear $Ti_2Cl_8$ species can easily coordinate to the lateral cuts of $MgCl_2$;[68] activation by the Al alkyl would change them into $Ti_2Cl_6$ reliefs on the $MgCl_2$ support. The epitactic placement

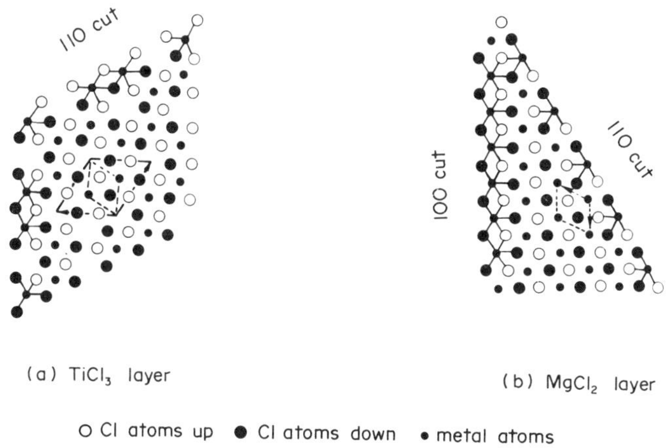


**Figure 10** Schematic drawing of structural layers of (a) violet $TiCl_3$ and (b) $MgCl_2$. Possible lateral cuts of the layers are indicated. The bonds of the metal atoms belonging to the lateral cuts are also drawn to point out the different coordination numbers

of a $Ti_2Cl_6$ unit on the 100 face of $MgCl_2$ deserves special attention, since it results in reliefs (Figure 11b) very similar to those proposed[81] as active sites on lateral 110 cuts of $TiCl_3$ crystals (Figure 11a).

It is seen from Figure 11 that the environment of the Ti atoms is chiral. Extensive calculations of non-bonded interactions showed that these sites can have stereoregulating ability, the main factor determining their stereospecific behaviour being the fixed, chiral orientation into which the growing polymer chain appears to be forced.[68]

As reported in Section 3.2, in the absence of Lewis bases, the stereospecificity of $MgCl_2$-supported catalysts is low. This has been explained assuming that, upon coordination of $TiCl_4$ to $MgCl_2$, mononuclear and dinuclear species are formed (as suggested by ESR[87]) both at 100 and 110 faces, only dinuclear species on the 100 faces leading to the formation of stereospecific sites.[68] Possible roles of the Lewis bases in increasing catalyst specificity have been discussed in Section 3.2.

The main general drawback of catalytic sites in relief with respect to the lateral surfaces is that no steric hindrance is present in the coordination step which can account for the observed regiospecificity.[88] The origin of regiospecificity, however, could be traced back to electronic factors[89] and, at least in part, to non-bonded interactions in the insertion step.[88]

The Cossee mechanism of polymerization, when completed with the chiral orientation of the growing polymer chain, is also able to account for the stereoselective behaviour of the isospecific heterogeneous catalysts.[90] In particular, the model proved able to explain the experimental results relative to the first insertion of a chiral alkene into an initial Ti–methyl bond,[43] *i.e.* the absence of enantioselectivity (discrimination between *si* and *re* insertions) and the presence of diastereoselectivity (preference for *S* (*R*) enantiomer upon *si* (*re*) insertion). Upon si (*re*) coordination of the two

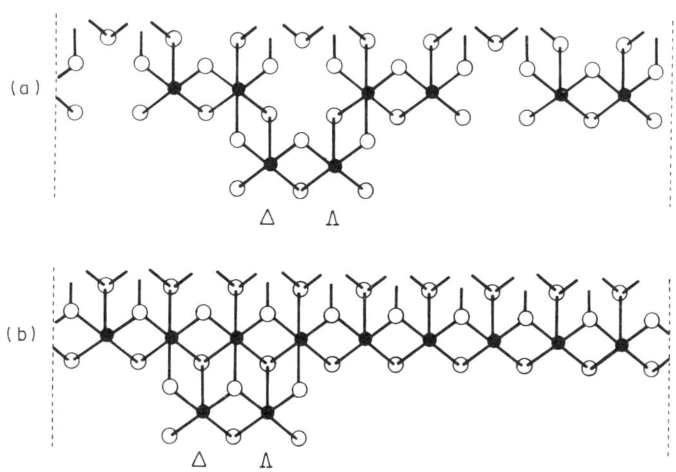


**Figure 11** Detailed view of (a) a $Ti_2Cl_6$ group in relief with respect to the 110 cut of $TiCl_3$ and (b) a $Ti_2Cl_6$ group epitactically placed on a 100 cut of $MgCl_2$. Chiral environments of metal atoms are explicitly labelled

enantiomers of 3-methyl-1-pentene to the proposed catalytic model, it was calculated that energy minima only occur when the conformation relative to the single C–C bond adjacent to the double bond, referred to the hydrogen atom bonded to the tertiary carbon atom, is nearly anticlinal minus, $A^-$ (anticlinal plus, $A^+$). Thus one can postulate the reactivity only of the $A^-$ conformations upon *si* coordination, and of the $A^+$ conformations upon *re* coordination (Figure 12). In other words, upon *si* coordination, only the *Z* synperiplanar methyl conformation would be accessible to the *S* enantiomer and only the (less populated) *Z* synperiplanar ethyl conformation to the *R* enantiomer; this would favour the *si* attack of the *S* enantiomer with respect to the same attack of the *R* enantiomer, independently of the chirality of the catalytic site. This result is in agreement with a previous hypothesis of Zambelli and co-workers, based only on the experimental reactivity ratios of the different faces of C-3 branched 1-alkenes.[43]

It is also worthy to note that, although the mechanism for stereospecificity of the chiral orientation of the chain was developed in the hypothesis of a $\sigma$ bond between the metal and the first carbon atom of the chain, it would remain valid in the hypothesis of carbenic intermediates.[91]

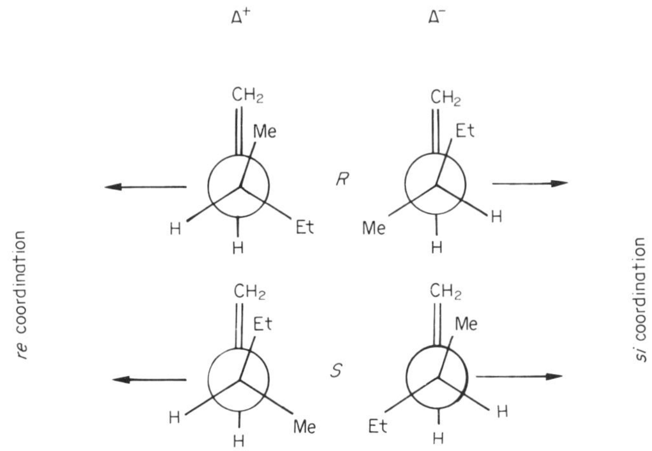


**Figure 12** Newman projections of the energy minimum conformations for (*R*)-3-methyl-1-pentene and (*S*)-3-methyl-1-pentene upon coordination on a site of the kind of Figure 6. The energy minimum conformations for the *si* coordination are shown on the right while those for the *re* coordination on the left

### 3.3.3 Homogeneous Syndiospecific Catalysis

The problem of the origin of stereoregulation in the syndiospecific polymerization of propene in the presence of homogeneous V-based catalyst systems has been relatively little investigated up to now.

As previously discussed (*cf.* Section 3.2.3), there are experimental evidences that the polymerization is not completely regiospecific and that the only syndiospecific step is propene insertion into a metal–secondary carbon bond with formation of a new secondary metal–carbon bond. A chain end stereocontrol is also well established.

All the proposed models for syndiotactic propagation suppose that the active center is a vanadium–carbon bond and that the monomer first coordinates to the metal; moreover, all of them attribute the stereospecificity to steric factors. However, different driving forces for the syndiospecificity have been proposed.

According to an early model,[73] there are two adjacent accessible positions at the catalytic site, each favouring the coordination of the prochiral monomer with one of its two faces; if the growing polymer chain alternates between the two positions at each insertion step, syndiotactic propagation is ensured. An analogous model had been proposed for the ring-opening polymerization of norbornene derivatives by $ReCl_5$, where the propagation species is a metallacarbene complex.[92] Due to the successive finding of a chain end stereocontrol, this model has to be rejected.

Most authors[2,93,50] favoured the hypothesis that syndiotacticity arises from steric repulsions between the methyl group of the complexed propene and the last inserted monomer unit of the growing chain. One of these models[93] is no more suitable, since it assumed a primary monomer insertion. It was suggested that such steric repulsions could occur in an activated four-center complex, resulting from propene secondary insertion into a metal–secondary carbon bond (Figure 13),[2,95] though this model has not been subjected to quantitative examination.

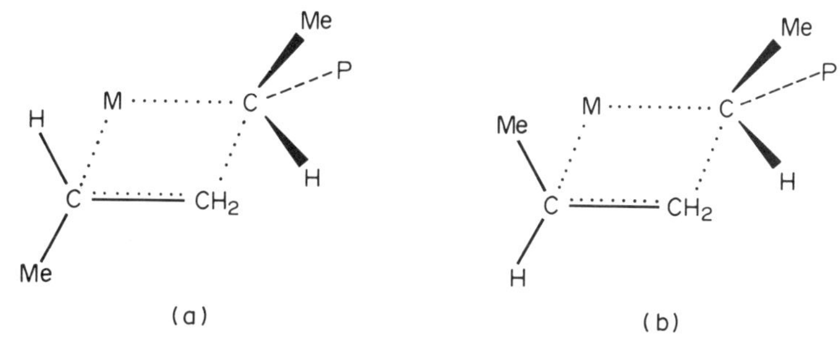


**Figure 13** Supposed four-center activated complexes for secondary insertion of a propene molecule (reproduced by permission of Springer–Verlag from *Adv. Polym. Sci.*, 1974, **15**, 31)

A detailed catalytic model, with a pentacoordinated metal (V) atom was proposed by Zambelli and Allegra,[94] able to account for several experimental observations concerning the homo- and copolymerization of propene and other 1-alkenes in the presence of the syndiospecific catalyst systems under consideration. However, the authors pointed out that the steric repulsions between the methyl group of a coordinated propene molecule and the last inserted propene unit of the growing chain do not generate energy differences between diastereoisomeric situations in the hypothesized catalytic complex, so that steric interferences with the growing chain, during monomer approach to the catalyst, were suggested as the source of the syndiospecific control.

More recently a further model of the origin of the syndiospecificity has been advanced,[69] consisting of a hexacoordinated metal (V) atom surrounded by four chlorine atoms assumed to be bridge bonded to other metal (*i.e.* Al) atoms (Figure 14). The catalytic site is chiral and analogous to that proposed for the isospecific heterogeneous catalysts by the same research group,[88] but, unlike the heterogeneous model site, interconversion between enantiomeric complexes is assumed to be possible, after each insertion step, when the metal atom is pentacoordinated. The analysis of the nonbonded interactions at the catalytic site showed that a *si* insertion of the last monomer unit (which generates what was called a *si* chain) favours the formation of the Λ complex, which in turn favours the *re* coordination and insertion of the successive monomer unit, this assuring the syndiospecific propagation of the chain. In summary, the chirality of the growing chain (*e.g.* the configuration of the last added unit) influences the chirality of the coordinated monomer (hence the configuration of the successive inserted unit) by influencing directly the chirality of the metal atom.

None of the last three models can be discarded on the basis of the available experimental data.

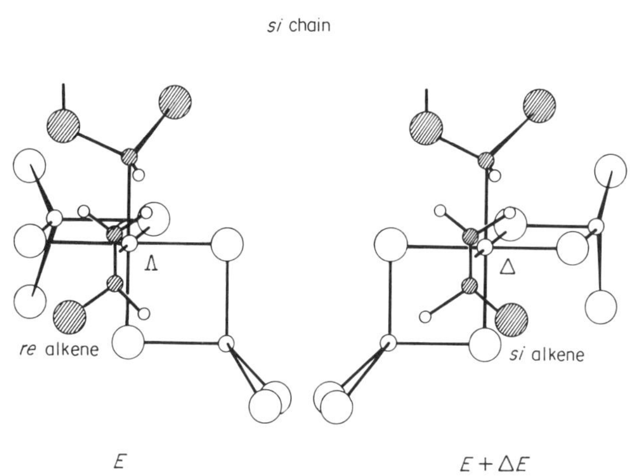


**Figure 14** Possible diastereoisomeric catalytic complexes for the homogeneous syndiospecific catalysts, with a *si* chain, presenting suitable orientation of the chain and of the alkene. The corresponding minimum energies are indicated below the models

### 3.3.4 Homogeneous Isospecific Catalysis

As already discussed (*cf.* Section 3.2.4), a class of homogeneous isospecific Ziegler–Natta catalysts has recently been disclosed, based on titanocenes and zirconocenes in combination with alkylalumoxanes. The stereocontrol has been attributed in some cases to the chirality of the catalytic complex and in some other cases to the chirality of the last inserted monomer unit in the growing chain, depending on the ligands of the metal.

A possible model for the chiral site-controlled catalysis has already been proposed.[95] It refers to stereorigid chiral complexes, with an ethylenebis(1-indenyl) ligand (Figure 15). The coordination to the metal of three more ligands in addition to the bis(indenyl) ligand was regarded as unlikely on the basis of calculations of non-bonded interactions. In consequence, it was assumed that the catalytic complex presents only two ligands (coordinated monomer and growing chain) besides the indenyl groups in the stage preceding monomer insertion. This is possible for instance if the active catalyst has an ion pair nature, in analogy to the cationic character of the catalytic species assumed for similar systems:[96–99]

$$[\text{bis(indenyl ligand)Ti(propene)}(P_n)]^+\ [R(AlMeO)_x]^- \rightarrow [\text{bis(indenyl ligand)Ti}P_{n+1}]^+\ [R(AlMeO)_x]^-$$

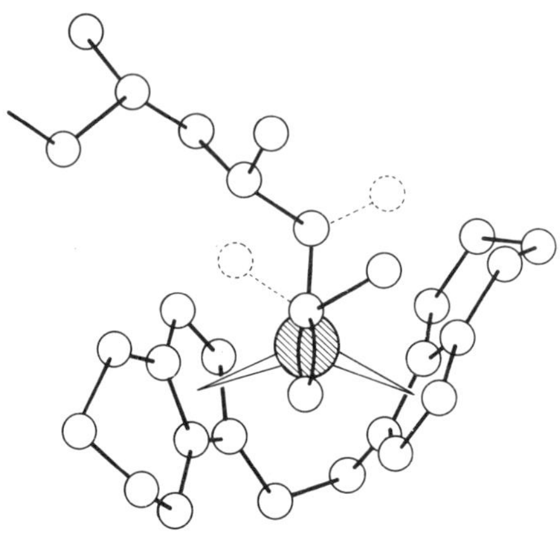

**Figure 15** A possible model site for the homogeneous isospecific catalysis: a sterically rigid chiral complex, in which, besides the monomer and the growing chain, an ethylenebis(4,5,6,7-tetrahydro-1-indenyl) ligand is present. The broken lines indicate the conformation of the growing chain which is forbidden, and the corresponding favoured coordination of the monomer

The calculations indicated that the chiral environment of the metal atom forces the growing chain to a chiral orientation (for instance, in Figure 15 the orientation of the chain indicated with full lines is allowed, that indicated with broken lines is forbidden). This in turn favours the coordination of the prochiral monomer with one of its two faces (for the sketched complex, the *re* coordination would be favoured). The similarity between this model and that for the heterogeneous isospecific catalysts proposed by the same research group[88] is pointed out by a comparison between Figure 9 and Figure 15. In both cases the chiral environment of the metal atom forces the growing chain to assume a chiral orientation, which in turn discriminates between *si* and *re* coordinated alkenes.

The predicted behaviour of this model catalytic site seems to be in agreement with the available experimental data. In particular, the model is nonstereospecific for the monomer insertion into a metal–methyl bond (*cf.* Section 3.2.4). Moreover, the model is in agreement with the elegant results of the analysis and optical activity measurements on the saturated propene oligomers obtained, under suitable conditions, with this kind of catalysts, for which the *re* insertion of the monomer has been verified to be favoured in the case of the (*R*,*R*) chirality of coordination of the bis(indenyl) ligand.[100]

At the moment, no model has yet been proposed for the isospecific homogeneous Ziegler–Natta catalysts with a chain end stereocontrol.

## 3.4 REFERENCES


1. J. Boor, Jr., 'Ziegler–Natta Catalysts and Polymerizations', Academic Press, New York, 1979.
2. A. Zambelli and C. Tosi, *Fortschr. Hochpolym.-Fortsch.*, 1974, **15**, 31.
3. J. A. Ewen, *J. Am. Chem. Soc.*, 1984, **106**, 6355.
4. Y. V. Kissin, 'Isospecific Polymerization of Olefins', Springer-Verlag, New York, 1985.
5. G. Natta, M. Farina and M. Peraldo, *Chim. Ind. (Milan)*, 1960, **42**, 255.
6. T. Miyazawa and T. Ideguchi, *J. Polym. Sci. Part B*, 1963, **1**, 389.
7. A. Zambelli, M. G. Giongo and G. Natta, *Makromol. Chem.*, 1968, **112**, 183.
8. F. A. Bovey and G. V. D. Tiers, *J. Polym. Sci.*, 1960, **44**, 173.

9. R. A. Sheldon, T. Fueno, T. Tsunetsugu and J. Furukawa, *J. Polym. Sci., Part B*, 1965, **3**, 23.
10. Y. Doi and T. Asakuru, *Makromol. Chem.*, 1975, **176**, 507.
11. K. Ziegler, *Belg. Pat.* 533 362 (1953).
12. G. Natta, P. Pino and G. Mazzanti, (issued to Montecatini), *US Pat.* 3 715 344 (1973) (*Chem. Abstr.*, 1973, **78**, 148 866).
13. G. Natta, P. Pino, P. Corradini, F. Danusso, E. Mantica, G. Mazzanti and G. Moraglio, *J. Am. Chem. Soc.*, 1955, **77**, 1708.
14. G. Natta, P. Pino and G. Mazzanti, *Chim. Ind. (Milan)*, 1955, **37**, 927.
15. G. Natta and P. Corradini, *Atti Accad. Naz. Lincei, Cl. Sci. Fis., Mat. Nat., Rend.*, 1955, **4** (Ser. 8), 73.
16. G. Natta, P. Pino and G. Mazzanti, *Gazz. Chim. Ital.*, 1957, **87**, 528.
17. G. Natta, P. Pino, G. Mazzanti and P. Longi, *Gazz. Chim. Ital.*, 1957, **87**, 549.
18. P. Pino and R. Muelhaupt, *Angew. Chem., Int. Ed. Engl.*, 1980, **19**, 857.
19. G. Natta, P. Corradini, I. W. Bassi and L. Porri, *Atti. Accad. N az. Lincei, Cl. Sci. Fis., Mat. Nat., Rend.*, 1958, **24** (Ser. 8), 121.
20. G. Natta, P. Pino and G. Mazzanti, (issued to Montecatini), *Ital Pat.* 526 101 (1954).
21. E. Tornqvist, W. Seelbach and A. W. Langer, Jr., (issued to Esso Res. Eng. Co.), *US Pat.* 3 032 510 (1962) (*Chem. Abstr.*, 1964, **60**, 14 630).
22. Mitsubishi Petrochem. Corp., *Br. Pat.* 1 092 390 (1966) (*Chem. Abstr.*, 1968, **68**, 13 557).
23. R. L. McConnel, M. A. McCall, G. O. Cash, Jr., F. B. Joyner and H. W. Coover, Jr., *J. Polym. Sci., Part A*, 1965, **3**, 2135.
24. J. P. Hermans and P. Henrioulle, *Ger. Pat.* 2 256 789 (1973); J. P. Hermans and P. Henrioulle, (issued to Solvay), *Ger. Pat* 2 213 086 (1972) (*Chem. Abstr.*, 1973, **78**, 16 791).
25. R. Colton and J. H. Canterford, 'Halides of the First Row Transition Metals', Wiley, London, 1969.
26. P. Longi, U. Giannini and A. Cassata (issued to Montedison), *Belg. Pat.* 774 600 (1971).
27. V. Busico, P. Corradini, L. De Martino, A. Proto, V. Savino and E. Albizzati, *Makromol. Chem.*, 1985, **186**, 1279; V. Busico, P. Corradini, L. De Martino, A. Proto and E. Albizzati, *Makromol. Chem.*, 1986, **187**, 1115.
28. U. Giannini, E. Albizzati and S. Parodi, (issued to Montedison) *US Pat.* 4 277 589 (1976) (*Chem. Abstr.*, 1978, **89**, 25 087); G. Cecchin and E. Albizzati, (issued to Montedison), *US Pat.* 4 294 721 (1977) (*Chem. Abstr.*, 1979, **90**, 138 426).
29. L. Luciani, N. Kashiwa, P. Barbé and A. Toyota, (issued to Montedison and Mitsui Petrochemical), *US Pat.* 4 226 741 (1975) (*Chem. Abstr.*, 1977, **87**, 68 893).
30. A. W. Langer, T. J. Burkhardt and J. J. Steger, Proceedings of the MMI International Symposium on 'Transition Metal Catalyzed Polymerization: Unsolved Problems', Midland, MI, 1981, Part A, p. 97.
31. U. Giannini, A. Cassata, P. Longi and R. Mazzocchi, (issued to Montedison), *Belg. Pat.* 785 332 (1972) (*Chem. Abstr.*, 1973, **98**, 98 293).
32. J. C. W. Chien, J. C. Wu and C. J. Kuo, *J. Polym. Sci., Polym. Chem. Ed.*, 1982, **20**, 2019; J. C. W. Chien and J. C. Wu, *J. Polym. Sci., Polym. Chem. Ed.*, 1982, **20**, 2445, 2461; J. C. W. Chien, J. C. Wu and C. J. Kuo, *J. Polym. Sci., Polym. Chem. Ed.*, 1983, **21**, 725.
33. A. Zambelli, L. Oliva and P. Ammendola, *Gazz. Chim. Ital.*, 1986, **116**, 259.
34. A. Zambelli, P. Locatelli, M. C. Sacchi and I. Tritto, *Macromolecules*, 1982, **15**, 831.
35. P. Longi, G. Mazzanti, A. Roggero and M. P. Lachi, *Makromol. Chem.*, 1963, **61**, 63.
36. Y. Takegami, T. Suzuki and T. Okazaki, *Bull. Chem. Soc. Jpn.*, 1969, **42**, 1060.
37. A. Zambelli, P. Locatelli and E. Rigamonti, *Macromolecules*, 1979, **12**, 156.
38. A. Zambelli, G. Gatti, M. C. Sacchi, W. O. Crain, Jr. and J. D. Roberts, *Macromolecules*, 1971, **4**, 475; A. Zambelli, P. Locatelli, G. Zannoni and F. A. Bovey, *Macromolecules*, 1978, **11**, 923; A. Zambelli, G. Bajo and E. Rigamonti, *Makromol. Chem.*, 1978, **179**, 1249.
39. W. O. Crain, Jr., A. Zambelli and J. D. Roberts, *Macromolecules*, 1971, **4**, 330.
40. A. Zambelli, M. C. Sacchi, P. Locatelli and G. Zannoni, *Macromolecules*, 1982, **15**, 211.
41. P. Pino, *Adv. Polym. Sci.*, 1965, **4**, 363; P. Pino, F. Ciardelli and G. Montagnoli, *J. Polym. Sci., Part C*, 1968, **6**, 3265; P. Pino, F. Ciardelli, G. P. Lorenzi and G. Natta, *J. Am. Chem. Soc.*, 1962, **84**, 1487; P. Pino, G. Montagnoli and E. Benedetti, *Makromol. Chem.*, 1966, **93**, 158.
42. P. Pino, F. Ciardelli, G. Montagnoli and O. Pieroni, *J. Polym. Sci., Part B*, 1967, **5**, 307; F. Ciardelli, E. Benedetti and O. Pieroni, *Makromol. Chem.*, 1967, **103**, 1.
43. A. Zambelli, P. Ammendola, M. C. Sacchi, P. Locatelli and G. Zannoni, *Macromolecules*, 1983, **16**, 341; A. Zambelli, P. Ammendola, P. Longo and A. Grassi, *Gazz. Chim. Ital.*, 1987, **117**, 579.
44. F. Ciardelli, P. Locatelli, M. Marchetti and A. Zambelli, *Makromol. Chem.*, 1974, **175**, 923; P. Pino, F. Ciardelli and G. P. Lorenzi, *J. Polym. Sci., Part C*, 1964, **4**, 21.
45. G. Natta, I. Pasquon, P. Corradini, M. Peraldo, M. Pegoraro and A. Zambelli, *Atti. Accad. Naz. Lincei. Cl. Sci. Fis., Mat. Nat., Rend.*, 1960, **28** (Ser. 8), 452; G. Natta, P. Corradini, I. Pasquon, U. Pegoraro and M. Peraldo, (issued to Montecatini), *US Pat.* 3 258 455 (1966); E. J. Addink and J. Beintema, *Polymer*, 1961, **2**, 185.
46. Y. Doi, *Makromol. Chem., Rapid Commun.*, 1982, **3**, 635.
47. G. Natta, I. Pasquon and A. Zambelli, *J. Am. Chem. Soc.*, 1962, **84**, 1488.
48. A. Zambelli, G. Natta and I. Pasquon, *J. Polym. Sci., Part C*, 1963, **4**, 411.
49. A. Zambelli, I. Pasquon, R. Signorini and G. Natta, *Makromol. Chem.*, 1968, **112**, 160.
50. Y. Takegami and T. Suzuki, *Bull. Chem. Soc. Jpn.*, 1969, **42**, 848; 1970, **43**, 1484.
51. Y. Doi, F. Nozawa, M. Murata, S. Suzuki and K. Soga, *Makromol. Chem.*, 1985, **186**, 1825.
52. A. Zambelli, C. Tosi and M. C. Sacchi, *Macromolecules*, 1972, **5**, 649.
53. *Cf.* ref. 2; primary propene addition to a growing chain ending with an ethylene unit is slightly favoured.
54. A. Zambelli, P. Locatelli, A. Provasoli and D. R. Ferro, *Macromolecules*, 1980, **13**, 267.
55. Y. Doi, *Macromolecules*, 1979, **12**, 248.
56. N. Ishihara, T. Seimija, M. Kuramoto and M. Uoi, *Macromolecules*, 1986, **19**, 2465.
57. P. Ammendola, C. Pellechia, P. Longo and A. Zambelli, *Gazz. Chim. Ital.*, 1987, **117**, 65.
58. U. Giannini, U. Zucchini and E. Albizzati, *J. Polym. Sci. Polym. Lett. Ed.*, 1970, **8**, 405.
59. H. Sinn and W. Kaminski, *Adv. Organomet. Chem.*, 1980, **18**, 99.
60. A. Zambelli, P. Ammendola, A. Grassi, P. Longo and A. Proto, *Macromolecules*, 1986, **19**, 2703.
61. P. Longo, A. Grassi, L. Pellecchia and A. Zambelli, *Macromolecules*, 1987, **20**, 1015.

62. W. Kaminski, S. Kuelper and S. Niedoba, *Makromol. Chem. Macromol. Symp.*, 1986, **3**, 377; W. Kaminski, Preprints of International Symposium on 'Transition Metal Catalyzed Polymerizations', Akron, 1986.
63. P. Corradini, G. Paiaro and A. Panunzi, *J. Polym. Sci., Part C*, 1967, **16**, 2905.
64. K. R. Hanson, *J. Am. Chem. Soc.*, 1966, **88**, 2731.
65. E. J. Arlman, *J. Catal.*, 1964, **3**, 89.
66. G. Allegra, *Makromol. Chem.*, 1971, **145**, 235.
67. P. Corradini, V. Barone, R. Fusco and G. Guerra, *Eur. Polym. J.*, 1979, **15**, 133.
68. P. Corradini, V. Barone, R. Fusco and G. Guerra, *Gazz. Chim. Ital.*, 1983, **113**, 601.
69. P. Corradini, G. Guerra and R. Pucciariello, *Macromolecules*, 1985, **18**, 2030.
70. Nomenclature of Inorganic Chemistry, *Pure Appl. Chem.*, 1971, **28**, 77.
71. G. Natta, *J. Inorg. Nucl. Chem.*, 1958, **8**, 589.
72. E. J. Arlman and P. Cossee, *J. Catal.*, 1964, **3**, 99.
73. P. Cossee, in 'The Stereochemistry of Macromolecules', ed. A. D. Ketley, Marcel Dekker, New York, 1967, vol. 1, chap. 3.
74. G. Natta, P. Corradini, I. W. Bassi and L. Porri, *Atti. Accad. Naz. Lincei, Cl. Sci. Fis., Mat. Nat., Rend.*, 1958, **24** (Ser. 8), 121.
75. G. Natta, P. Corradini and G. Allegra, *J. Polym. Sci.*, 1961, **51**, 399.
76. G. Allegra, *Nuovo Cimento*, 1962, **23**, 502.
77. B. Hargitay, L. Rodriguez and M. Miotto, *J. Polym. Sci.*, 1959, **35**, 599.
78. D. R. Armstrong, P. G. Perkins and J. J. P. Stewart, *J. Chem. Soc., Dalton Trans.*, 1972, 1972.
79. G. Giunchi, E. Clementi, M. E. Ruiz-Vizcaya and O. Novaro, *Chem. Phys. Lett.*, 1977, **49**, 8.
80. Yu. V. Kissin and N. M. Chirkov, *Eur. Polym. J.*, 1970, **6**, 525.
81. P. Corradini, G. Guerra, R. Fusco and V. Barone, *Eur. Polym. J.*, 1980, **16**, 835.
82. L. A. Rodriguez and A. Gabant, *J. Polym. Sci., Part C*, 1964, **4**, 125.
83. J. Boor, Jr., *J. Polym. Sci., Part C*, 1963, **1**, 257.
84. P. Corradini, V. Barone and G. Guerra, *Macromolecules*, 1982, **15**, 1242.
85. P. Corradini, V. Barone and G. Guerra, *Eur. Polym. J.*, 1984, **20**, 1177.
86. U. Giannini, G. Giunchi, E. Albizzati and P. C. Barbé, 'Frontiers in Polymerization Catalysis and Polymer Synthesis', NATO Advanced Research Work, Feb. 1–6, 1987, in press.
87. J. C. W. Chien and J. C. Wu, *J. Polym. Sci., Polym. Chem. Ed.*, 1982, **20**, 2461.
88. P. Corradini, V. Barone, R. Fusco and G. Guerra, *J. Catal.*, 1982, **77**, 32.
89. J. J. Eisch, 'Proceedings, International Symposium on Transition Metal Catalyzed Polymerization', Akron, 1986, in press.
90. P. Corradini, G. Guerra and V. Villani, *Macromolecules*, 1985, **18**, 1401.
91. K. J. Ivin, J. J. Rooney, C. D. Steward, M. L. H. Green and R. Mahtab, *J. Chem. Soc., Chem. Commun.*, 1978, 604.
92. J. G. Hamilton, K. J. Ivin and J. J. Rooney, *J. Mol. Catal.*, 1985, **28**, 255.
93. J. Boor, Jr. and E. A. Youngman, *J. Polym. Sci., Part A-1*, 1966, **4**, 1861.
94. A. Zambelli and G. Allegra, *Macromolecules*, 1980, **13**, 42.
95. P. Corradini, G. Guerra, M. Vacatello and V. Villani, *Gazz. Chim. Ital.*, in press.
96. F. S. Dyachkowsky, A. J. Shilova and A. E. Shilov, *J. Polym. Sci., Part C*, 1967, **16**, 2333.
97. J. W. Lauher and R. Hoffman, *J. Am. Chem. Soc.*, 1976, **98**, 1729.
98. E. Giannetti, G. M. Nicoletti and R. Mazzocchi, *J. Polym. Sci., Polym. Chem. Ed.*, 1985, **23**, 2117.
99. J. J. Eisch, A. H. Piotrovsky, S. K. Brownstein, E. J. Gabe and F. L. Lee, *J. Am. Chem. Soc.*, 1985, **107**, 7219.
100. P. Pino, P. Cioni and J. Wei, *J. Am. Chem. Soc.*, 1987, **109**, 6189.

# 4

# Monoalkene Polymerization: Copolymerization

PETER J. T. TAIT and IAN G. BERRY
*University of Manchester Institute of Science and Technology, UK*

---

Because of the very late delivery of the manuscript, this chapter now appears at the end of this Volume, commencing on page 575. This is in order to minimize any delay in publishing 'Comprehensive Polymer Science' as a whole.

# 5

# Conjugated Diene Polymerization

LIDO PORRI and ANTONINO GIARRUSSO
*Politecnico di Milano, Italy*

## 5.1 INTRODUCTION

The polymerization of conjugated dialkenes with transition metal catalysts began in 1954, soon after the first results on the polymerization of propylene had been obtained. The first stereoregular dialkene polymer obtained was a polyisoprene with a structure similar to that of natural rubber.[1] An isoprene polymer with a structure identical to that of natural gutta percha was obtained soon after,[2] along with stereoregular polymers of butadiene and pentadiene.[2–4]

The polymerization with transition metal catalysts shows two distinctive features with respect to other methods used for polymerizing 1,3-dialkenes: (1) it is characterized by a high degree of chemoselectivity, *i.e.* it can yield polymers made up almost exclusively of one type of monomeric unit, 1,4, 1,2 or 3,4; and (2) it shows a high degree of stereoselectivity, *i.e.* it can yield polymers with a high configurational regularity when the monomeric units contain sites of steric isomerism (internal double bonds, tertiary or asymmetric carbon atoms).

No other method of polymerizaton so far known shows the high degree of chemo- and stereo-selectivity of transition metal catalysts. Radical initiators give, from the most common dialkenes, polymers which are amorphous by X-ray at room temperature, because they are made up of monomeric units of different types, in variable proportions.[5] Cationic catalysts give irregular, crosslinked polymers with a lower degree of unsaturation than that expected.[6] A higher chemo- and stereo-selectivity is shown by catalysts based on alkali metals. The Alfin catalyst system (a combination of sodium allyl, sodium isopropoxide and sodium chloride) gives a polybutadiene consisting of about 70% *trans*-1,4 units, which is crystalline by X-ray at room temperature, due to the presence in each chain of long sequences made up of *trans*-1,4 units only.[7–9]

A good degree of chemoselectivity is shown by lithium alkyls. They give a polybutadiene consisting of more than 90% 1,2 units in ethers,[10] and of about 90–93% 1,4 units in hydrocarbon solvents.[11] However, both these polymers are amorphous, the first because it lacks configurational order at the tertiary carbon atoms, the second because it has a mixed *cis/trans* structure.

Lithium alkyls show a good degree of chemo- and stereo-selectivity only in the polymerization of isoprene, giving a polymer consisting of 92–93% *cis*-1,4 units.[12]

In conclusion, the methods of polymerization other than those based on transition metal catalysts give only two stereoregular polymers from the most common 1,3-dialkenes: the Alfin *trans*-polybutadiene and the lithium *cis*-polyisoprene. Transition metal catalysts have dramatically changed the situation in the field of dialkene polymerization. They are effective for the polymerization of various types of monomers: $CH_2{=}CHCH{=}CH_2$; $CH_2{=}CRCH{=}CH_2$; $CH_2{=}CHCH{=}CHR$; $CH_2{=}CRCR{=}CH_2$; $CH_2{=}CRCH{=}CHR$; $CH_2{=}CHCR{=}CHR$; $CH_2{=}CHCH{=}CR_2$; and $CHR{=}CHCH{=}CHR$ (R = alkyl or aryl group).

Stereoregular polymers have been obtained from each of the above classes, which shows the uniqueness of this type of polymerization for conjugated dialkenes. Extensive reviews in this field have appeared.[13–19]

## 5.2 STEREOISOMERISM IN POLYMERS OF 1,3-DIALKENES

Butadiene and symmetric monomers of type $CH_2{=}CRCR{=}CH_2$ can give *cis*-1,4, *trans*-1,4 or 1,2 monomeric units. Four stereoregular isomers are therefore possible for polymers from these monomers: *cis*-1,4, *trans*-1,4, 1,2 isotactic and 1,2 syndiotactic. Isoprene or more generally monomers $CH_2{=}CRCH{=}CH_2$ can give in addition 3,4 units, hence, in principle, 3,4 isotactic and 3,4 syndiotactic polymers are possible from these monomers.

Isomerism becomes more complex in polymers from terminally substituted butadienes. A monomer $CH_2{=}CHCH{=}CHR$ can give 1,4, 1,2 or 3,4 monomeric units, each of which, independently of its structure, contains two sites of isomerism.

A 1,4 monomeric unit (**1**) contains a double bond and an asymmetric carbon atom, which can assume two opposite configurations. For this reason a *cis*-1,4 or a *trans*-1,4 polymer of $CH_2{=}CHCH{=}CHR$ can exist in an isotactic, a syndiotactic or an atactic form. The isotactic polymers can assume two enantiomeric forms; isotactic polymers consisting predominantly of one enantiomeric form have actually been obtained from 1,3-pentadiene using an optically active catalyst (see Section 5.4.3.2).

A 1,2 unit (**2**) contains a double bond in the side group, hence two types of isotactic and two types of syndiotactic polymers made up of units (**2**) are possible, each type differing for the configuration of the double bonds, *cis* or *trans*. A 3,4 unit (**3**) contains two tertiary carbon atoms. Therefore three types of stereoregular polymers with 3,4 structure are possible: *threo* diisotactic, *erythro* diisotactic and disyndiotactic.

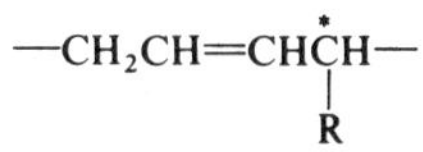

**(1)**

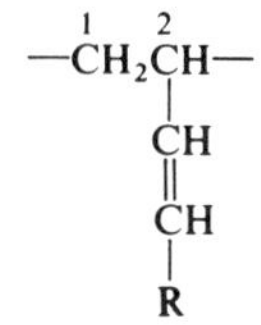

**(2)**

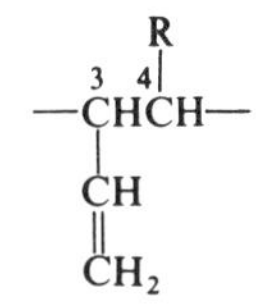

**(3)**

Terminally disubstituted butadienes, CHR=CHCH=CHR, give monomeric units each containing three types of isomerism. A 1,4 unit contains a double bond and two asymmetric carbon atoms. Units of this type (either *cis* or *trans*) can exist in two isomeric forms, (**4a**) and (**4b**), in which the two asymmetric carbon atoms have respectively the same and opposite chirality. Each isomer can exist in the form of two enantiomers, as indicated. As a consequence, eight stereoregular isomeric forms are foreseeable for 1,4 polymers of CHR=CHCH=CHR: *cis*-1,4 (or *trans*-1,4) *threo* diiso; *cis*-1,4 (or *trans*-1,4) *threo* disyndio; *cis*-1,4 (or *trans*-1,4) *erythro* diiso; *cis*-1,4 (or *trans*-1,4) *erythro* disyndio.

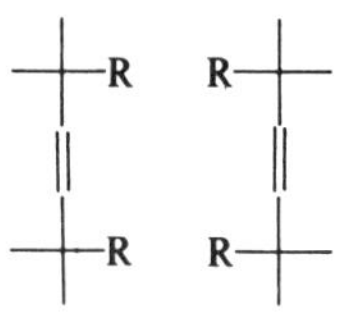

**(4a)**

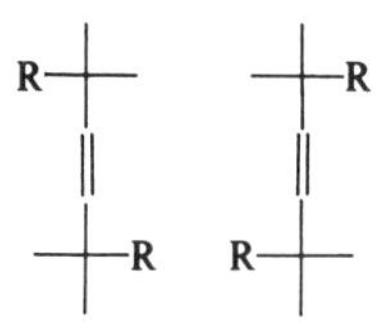

**(4b)**

Polymers having some of these structures have actually been obtained (see Section 5.4.4). Polymers of CHR=CHCH=CHR consisting of 1,2 units show the same type of isomerism as the polymers made up of units (**3**). However, due to the presence of a double bond in the side group, two types of *threo* diisotactic, *erythro* diisotactic or disyndiotactic polymers are foreseeable, each type differing for the configuration of the double bond. A more detailed discussion of stereoisomerism phenomena in dialkene polymers can be found in a recent publication.[20]

## 5.3 POLYMERIZATION CATALYSTS

### 5.3.1 General Considerations

The first transition metal catalysts used for the polymerization of conjugated dialkenes were those that had been found to be effective for the polymerization of ethylene and propylene, *i.e.* $TiCl_4$ and $TiCl_3$ in combination with aluminum alkyls (1954).[1,2] Following intensive work which was carried out in the following years and has continued up to the present, a vast number of catalyst systems have been discovered, which give stereoregular polymers of conjugated dialkenes. These catalysts are usually a combination of aluminum alkyls and compounds of transition metals of groups IV–VIII and of the lanthanide series. At the beginning of the 1960s it was found that some simple derivatives of transition metals are effective for the polymerization of dialkenes, without requiring an aluminum alkyl. In 1961 it was reported that some rhodium salts ($RhCl_3 \cdot 3H_2O$, $Rh(NO_3)_3 \cdot 2H_2O$) polymerize butadiene to *trans*-1,4 polymers in protic solvents such as water and ethanol.[21,22] In 1964 it was reported that allylnickel halides can polymerize butadiene to *cis*-1,4 or *trans*-1,4 polymers.[23] Allyl derivatives of other transition metals were later reported to be effective catalysts.

Transition metal catalysts known at present for 1,3-dialkene polymerization can be grouped into the following classes: (a) catalysts derived from aluminum alkyls or aluminum hydrides and transition metal compounds (Ziegler–Natta catalysts); (b) catalysts derived from precursors not containing metal–carbon bonds; and (c) catalysts based on allyl derivatives of transition metals.

### 5.3.2 Ziegler–Natta Catalysts

The catalysts of this class are at present the most important both with regard to activity and stereospecificity, and are the only ones now used for industrial manufacture of dialkene polymers. Some of them are soluble, whereas others are insoluble in the polymerization medium. Various derivatives of transition metals can be used, but the most common are halides, halides complexed with Lewis bases, alcoholates, acetylacetonates and salts of organic acids. Aluminum alkyls ($AlR_3$, $AlR_2Cl$, $AlRCl_2$; R = alkyl or aryl group) are the preferred cocatalysts, but aluminum hydrides are also effective. Other metal alkyls such as LiR, $MgR_2$, $CaR_2$ and $ZnR_2$ can be used in some cases, but are of minor interest.

A detailed examination of the most important Ziegler–Natta catalysts is given in the following sections.

### 5.3.2.1 Titanium catalysts

#### (i) Catalysts based on $TiCl_4$

The catalyst system $AlR_3$–$TiCl_4$ was the first used[1] for the polymerization of 1,3-dialkenes and still is one of the most important for the preparation of high-*cis* polyisoprene. The reaction between $AlR_3$ and $TiCl_4$ has been extensively investigated.[24–38] A wide variety of aluminum alkyls can be used, the most common of which are $AlEt_3$ and $Al(Bu^i)_3$. The preferred method for catalyst preparation is gradual addition of $AlR_3$ to a solution of $TiCl_4$ in an aliphatic solvent (pentane, heptane), at room temperature. The reactions that take place are shown in Scheme 1.

$$AlR_3 + TiCl_4 \longrightarrow AlR_2Cl + TiCl_3R \quad (1)$$

$$TiCl_3R \longrightarrow \beta\text{-}TiCl_3 + R\cdot \quad (2)$$

$$TiCl_3R + AlR_3 \longrightarrow TiCl_2 + AlR_2Cl + 2R\cdot \quad (3)$$

$$\beta\text{-}TiCl_3 + AlR_3 \longrightarrow TiCl_2R + AlR_2Cl \quad (4)$$

$$TiCl_2R \longrightarrow TiCl_2 + R\cdot \quad (5)$$

**Scheme 1**

The species $TiCl_3R$ is unstable at room temperature when $R \geqslant C_2$ and decomposes with formation of a brown precipitate. This is essentially a crystalline modification of $TiCl_3$, called $\beta$-$TiCl_3$,[39] with Al present in small amounts. The solid product appears identical, by X-ray analysis, with that obtained by action of dark discharges on a gaseous mixture of $H_2$ and $TiCl_4$.[40,41]

At Al/Ti molar ratios > 1, the precipitate turns gradually black and the Cl/Ti ratio in the solid becomes lower than three. Changes in composition of the precipitate on varying the $AlR_3/TiCl_4$ ratio are shown in Table 1.

**Table 1** Analysis of the Solid from the Reaction of $AlEt_3$ with $TiCl_4$ at 15 °C

| *$AlEt_3/TiCl_4$ mole ratio* | *Cl/Ti mole ratio* | *Al/Ti mole ratio* | *Average valence of Ti in the solid* | *Ref.* |
|---|---|---|---|---|
| 0.5 | 3.4 | 0.24 | | a |
| 1.2 | 3.08 | 0.07 | 3 | a,b |
| 2.0 | 2.36 | 0.33 | 2.31 | a,b |
| 2.5 | 2.15 | 0.30 | | |
| 3.0 | 2.0 | 0.32 | 2.04 | a,b |

[a] G. Natta, L. Porri, A. Mazzei and D. Morero, *Chim. Ind. (Milan)*, 1959, **41**, 398 (*Chem. Abstr.*, 1960, **54**, 9353).
[b] E. Schoenberg, H. A. Marsh, S. Y. Walters and W. M. Saltman, *Rubber Chem. Technol.*, 1977, **52**, 526.

Some authors separated the solid precipitates and the liquid portions of the catalysts prepared at various Al/Ti ratios by filtration.[27,30,42,43] Neither the liquid nor the solid from the Al/Ti = 1 catalyst are capable of initiating polymerization. Only the combination of them is catalytically active. The thoroughly washed inactive solid can be reactivated with a small amount of $AlR_3$ (*ca.* Al/Ti = 0.1) or a larger amount of $AlR_2Cl$ (Al/Ti > 1).[43,44]

The precipitate from the Al/Ti = 2 reaction product, separated by filtration and resuspended in fresh heptane, shows moderate catalytic activity, which probably can be attributed to some alkylated species (*e.g.* $TiCl_2R$) present in the solid, along with $TiCl_2$.[30,43]

$AlR_2Cl$ and $AlRCl_2$ give $\beta$-$TiCl_3$ on reaction with $TiCl_4$ at room temperature, even at Al/Ti > 1.[45]

The kinetics of the $TiCl_4$ and $Al(Bu^i)_3$ reaction have been examined.[46] At room temperature, reactions (1) and (2) of Scheme 1 are equal in rate, whereas reaction (4) is much slower, suggesting that $TiCl_2$ is predominantly formed by side reaction (3) rather than by reactions (4) and (5). Reaction of $AlR_2Cl$ with $TiCl_4$ is also very slow compared with that of $AlR_3$.

Reaction of $AlR_3$ with $TiCl_4$ at temperatures > 80 °C yields a violet modification of $TiCl_3$, called $\gamma$-$TiCl_3$.[38,47,48] This modification shows a different stereospecificity than does $\beta$-$TiCl_3$ for the polymerization of dialkenes.[45]

Compounds containing Al—H bonds (alanes) have been used as reducing agents instead of aluminum alkyls.[49–51] Examples of these Al hydrides are $AlHCl_2 \cdot OEt_2$, $AlHCl_2 \cdot NMe_3$, $AlH_2Cl \cdot NMe_3$, $AlH_2NMe_2$, $AlH_3 \cdot NMe_3$ and poly(iminoalane)s of formula $\text{—}[\text{Al(H)N(R)}]_n\text{—}$.

These alanes do not react or react very slowly with the monomer at room temperature,[52] but they react very rapidly with $TiCl_4$, with formation of a precipitate. At low Al/Ti molar ratio (from *ca.* 0.9 to 1.5, depending on the alane) the precipitate is crystalline by X-ray and consists essentially of $\beta$-$TiCl_3$; at higher ratios it is amorphous, with a Cl/Ti ratio lower than three. Therefore the reaction between alanes and $TiCl_4$ is similar to that with $AlR_3$.

For the polymerization of dialkenes, only the catalysts containing $\beta$-$TiCl_3$ are of interest, *i.e.* $AlR_3$–$TiCl_4$ at Al/Ti = 1 and the alane systems at Al/Ti = 0.9–1.5. In these catalysts, the active species is believed to be a complex formed on the surface of $TiCl_3$ crystals, containing Ti—R or Ti—H bonds according to the reaction

$$AlR_2Cl \quad + \quad \beta\text{-}TiCl_3 \longrightarrow AlRCl_2 \cdot TiCl_2R \qquad (6)$$

The concentration of active centers in the $AlR_3$–$TiCl_4$ system is estimated at only a small percent of the Ti present. Using inhibitors, it was found that less than 1% of Ti is active.[53,54]

### *(ii) Catalysts based on $TiCl_3$*

There are four modifications of $TiCl_3$: three violet modifications, called $\alpha$, $\gamma$ and $\delta$, and a brown modification, $\beta$-$TiCl_3$. The reaction between an aluminum alkyl and $TiCl_3$ follows a different course depending on the modification used. The violet modifications apparently do not react with $AlR_3$ or $AlR_2Cl$.[45] Presumably an alkylation reaction occurs at the surface of the crystals, as shown in equation (6), with formation of catalytic complexes containing Al and Ti, in which a Ti—C bond is present.

$\beta$-$TiCl_3$ can be easily prepared by reaction of $TiCl_4$ with the complex $2AlCl_3 \cdot TiCl_2 \cdot C_6H_6$ and has been used, in combination with $AlR_3$ or $AlR_2Cl$, for the polymerization of isoprene or butadiene.[45,55] It is apparent from the previous paragraph that $AlR_3$–$\beta$-$TiCl_3$ (Al/Ti = 1) and $AlR_2Cl$–$\beta$-$TiCl_3$ are equivalent to $AlR_3$–$TiCl_4$ (Al/Ti = 2) and $AlR_3$/$TiCl_4$ (Al/Ti = 1), respectively.

### *(iii) Titanium catalysts containing iodine*

Aluminum trialkyls give, in combination with $TiI_4$, catalysts that behave differently from $AlR_3$–$TiCl_4$, in that they give polymers of butadiene with a higher *cis* content, while they are poorly active in the polymerization of isoprene. $AlR_3$–$TiI_4$ was the first system used for the synthesis of a high-*cis* polybutadiene.[56,57]

The reaction between $AlR_3$ and $TiI_4$ gives rise to a precipitate containing reduced Ti, analogously to the $AlR_3$–$TiCl_4$ reaction. However, a higher Al/Ti ratio is necessary in order to have the same halogen/Ti ratio in the precipitate. Since $TiI_4$ is only slightly soluble in hydrocarbon solvents, alternative systems have been proposed, *e.g.* $AlR_3$–$TiCl_2I_2$,[59–63] $AlR_3$–$AlEt_2I$–$TiCl_4$[64] and $AlR_3$–$I_2$–$TiCl_4$.[58,65] In the last system $I_2$ reacts with $AlR_3$ to give $AlR_2I$, which on reaction with $TiCl_4$ gives a precipitate similar to that obtained with $AlR_3$–$TiI_4$.[58,59] Tetraalkylaluminoxanes[66] and LiBu[67] have been used instead of $AlR_3$.

Several aluminum hydrides have been used as cocatalysts instead of $AlR_3$, *e.g.* $AlH_3 \cdot NMe_3$, $AlH_2Cl \cdot NMe_3$, $AlH_2NMe_2$, $AlHCl_2 \cdot OEt_2$. Some of the systems reported are $AlHCl_2 \cdot OEt_2$–$TiCl_4$–$AlI_3$,[68] $AlHI_2 \cdot OEt_2$–$TiCl_4$,[69] $AlH_2NMe_2$–$TiCl_2I_2$,[69] $LiAlH_4$–$AlEt_2I$–$TiCl_4$,[64] $LiAlH_4$–$I_2$–$TiCl_4$.[69] The equivalence of all the above systems with regard to the stereospecificity is attributable to the fact that iodine, independently of the form in which it is initially present, ends up on titanium. The composition of the precipitate in the various systems is therefore very similar.[58,64,69]

The percentage of Ti active in the polymerization has been determined for the $TiCl_2I_2$ system and it was found to be less than 5%.[54]

Although the catalyst-forming reaction in the above systems gives rise to a precipitate, evidence was reported that the polymerization occurs in a homogeneous phase.[57,62,63]

*(iv) Catalysts based on titanium tetraalkoxides*

Soluble catalysts result from the reaction at room temperature between $AlR_3$ and a titanium tetraalkoxide $Ti(OC_nH_{n+1})_4$ in hydrocarbon solvents. These systems give polymers with a predominant 1,2 structure from butadiene and a 3,4 structure from isoprene. Use of Ti compounds with $n$ varying from one to nine in combination with $AlR_3$ (mostly $AlEt_3$ or $Al(Bu^i)_3$) indicated that the change of orthotitanate component of the catalyst has practically no effect on polymer microstructure.[70,71]

The catalyst-forming reaction consists of an alkylation of $Ti(OR)_4$ with formation of Ti species in a lower oxidation state, as shown by the evolution of gas.[14] Polarographic studies have been effected on the systems based on $Ti(OMe)_4$ or $Ti(OC_9H_{19})_4$ and $Al(Bu^i)_3$ (Al/Ti from 2 to 12) and have shown that these systems contain $Ti^{IV}$, $Ti^{III}$ and $Ti^{II}$ species, in a ratio dependent on the Al/Ti ratio and on the size of the alkoxy radical attached to Ti. Most active was the catalyst containing the largest concentration of $Ti^{III}$ compounds.[72] IR investigation also suggests that the reactive polymerization species contain $Ti^{III}$.[73] In the catalyst system $Al(Bu^i)_3$–$Ti(OR)_4$ (R = Bu, nonyl) less than 1% of Ti is active in the polymerization, according to kinetic data.[74]

When $Ti(OR)_4$ is used in combination with alkylaluminum chlorides ($AlEt_2Cl$, $Al_2Et_3Cl_3$, $AlEtCl_2$, Al/Ti > *ca.* 4) heterogeneous catalysts are obtained. An exchange reaction between Cl and OR groups occurs with formation of a precipitate consisting of $TiCl_3$. $Ti(OR)_4$ in which R is a linear group gives the $\beta$ form of $TiCl_3$ in reactions at room temperature; the systems based on these Ti compounds have therefore the same stereospecificity as $AlR_3$–$TiCl_4$ at Al/Ti = 1.[75–80] $Ti(OR)_4$ in which R is a branched group behaves differently,[80] as indicated in Section 5.5.2.1.

### 5.3.2.2 *Vanadium catalysts*

Vanadium halides, *e.g.* $VCl_3$, $VOCl_3$ and $VCl_4$, in conjunction with aluminum alkyls ($AlR_3$, $AlR_2Cl$), give heterogeneous systems that are highly stereospecific for the *trans* polymerization of conjugated dialkenes.[81,82] $VCl_3$ is insoluble in hydrocarbon solvents and apparently does not react with $AlR_3$ or $AlR_2Cl$. Actually a reaction occurs on the surface of the crystals, analogous to reaction (6). $VOCl_3$ and $VCl_4$ are soluble in hydrocarbon solvents, but they give a precipitate on reaction with an aluminum alkyl. The precipitate contains vanadium in a lower oxidation state, along with Al.[82] The catalysts from $VCl_3$ are more stereospecific than those from $VOCl_3$ or $VCl_4$, but give poor polymer yields. These can be improved by supporting $VCl_3$ on an inert support (kaolin, clay, *etc.*).[83]

Several soluble vanadium catalysts are known. Those obtained from $AlR_3$ and $V(acac)_3$ or $VO(OR)_3$ have interest for the polymerization of butadiene to 1,2 syndiotactic polymer. It does not seem that these systems have been used for other dialkenes. Catalysts prepared from $V(acac)_3$ or other soluble V compounds and $AlR_2Cl$ or $AlRCl_2$ are also soluble, but have a different stereospecificity, since they yield *trans* polymers from butadiene.[84,85] In these systems when prepared at high Al/V ratio (at least five) the chlorine ends up eventually on V and the active complex contains the species $Cl_2VR$ complexed with the aluminum alkyl.[86] This is the reason why $AlR_2Cl$–$VCl_3 \cdot 3THF$ is practically equivalent to $AlR_2Cl$–$V(acac)_3$. The V—C bond is rather unstable at room temperature and homolytic breakage of this bond results in the formation of inactive $V^{II}$ species; therefore, these systems are preferably used at lower temperatures. However, in the presence of Cl donors (*e.g.* $Cl_3CCO_2H$, $C_5Cl_6$) capable of reoxidizing $V^{II}$ to $V^{III}$, the polymerization can be carried out even at 80 °C.[87] Soluble vanadium catalysts that give high-*trans* polymers from isoprene have been obtained from $Ti(OR)_4$, $VCl_3$ and $AlR_3$.[83] Little is known about the nature of these systems.

### 5.3.2.3 *Chromium and molybdenum catalysts*

Catalysts based on metals of Group VI have not been extensively studied, but have some interest for the 1,2 polymerization of butadiene. Several soluble chromium complexes [$Cr(acac)_3$, $Cr(CO)_5py$ (py = pyridine) and $Cr(CNPh)_6$] give apparently homogeneous systems on reaction with $AlR_3$.[88,89] ESR analysis of $AlEt_3$–$Cr(acac)_3$ showed the formation of low-valent Cr species.[90] Several soluble catalysts based on Mo were reported, *e.g.* $AlEt_3$–$Mo(acac)_3$ (or $MoO_2(acac)_2$),[48] $AlEt_2(OEt)$–$MoCl_5$, in the presence of ethyl acetate[91] or sulfides[92] as activators; $Al(Bu^i)_2$-

OPh–$MoCl_5$,[93] $Al(Bu^i)_2OPh$–$MoCl_{4-n}$.[94,95] ESR analysis of the $MoCl_{4-n}(OR)_n$ systems showed that $Mo^{III}$ is present in the catalytic complex.[95]

### *5.3.2.4 Cobalt and nickel catalysts*

The reaction at room temperature between an alkylaluminum chloride ($AlR_2Cl$, $AlRCl_2$, $AlR_2Cl + AlRCl_2$ in various proportions) and a cobalt compound affords hydrocarbon-soluble catalysts that are effective for the *cis* polymerization of butadiene. Cobalt compounds of various types can be used, such as cobalt chloride complexed with bases (pyridine, alcohol, *etc.*), cobalt salts of organic acids, $Co(acac)_2$, cobalt tetrachloroaluminate $[Co(AlCl_4)_2]$, complexes of $Co^I$ [*e.g.* $\eta^3$-$C_8H_{13}Co(C_4H_6)$] and of $Co^0$ (*e.g.* $Co_2(CO)_8$).[96–112]

In addition to the above aluminum alkyls the product of the reaction between $H_2O$ and $AlEt_2Cl$ (molar ratio 1:2) was also used as a catalyst. This product, aluminoxane, is a mixture of compounds of different molecular weights, containing Al—O—Al bonds.

The reaction between aluminum alkyl chlorides and several $Co^{II}$ complexes ($Co(acac)_2$, $CoCl_2 \cdot 2py$, cobalt octanoate, *etc.*) was studied by various authors.[111,115–117] The reaction occurs through the following steps: (1) at the Al/Co ratio normally used ($>50$) the anionic ligands are removed from the cobalt and replaced by the Cl atoms of the aluminum compound,[103,115,116] explaining why catalysts prepared from different cobalt compounds have a similar behavior; (2) formation of an Al/$Co^{II}$ complex, the existence of which was proved by spectroscopic methods;[111,115] (3) alkylation of the $Co^{II}$ species, to give an unstable Co—C bond, with eventual formation of $Co^I$ complex. The presence of a $Co^I$ species is proved by the fact that the complex disproportionates with bases to give $Co^0$ and $Co^{II}$ in equivalent amounts.[103,111,117]

According to some authors,[111] the complex of $Co^{II}$ formed in step (2) above is the active species; others have reported that the complex of $Co^I$ can also give an active catalyst.[14,117] A crystalline complex of $Co^I$, active in the polymerization of butadiene to *cis* polymer, was isolated from the reaction between $AlPh_2Cl$, $AlPhCl_2$ and $CoCl_2$ in benzene.[116] It seems therefore that active species can be derived either from $Co^I$ or $Co^{II}$ complexes; both complexes are precursors of the catalyst and presumably both can give, on reaction with the monomer, the cobalt–allyl bond, on which the polymer chain grows.[14,118]

Determinations of the number of active centers have been carried out on the system $AlEt_2Cl$–$Co(acac)_2$ in the polymerization of butadiene. The technique of termination with tritiated methanol was used.[119] This method is not very accurate,[118] but the data reported indicate that at least 60–70% of Co is catalytically active.[120]

Cobalt catalysts supported on polymers[121,122] or on inorganic carriers[123–125] have been reported. These catalysts give polymers with a *cis* content as high as that reported for the homogeneous systems, but their activity is higher.

Catalysts derived from aluminum trialkyls and cobalt compounds show different chemo- and stereo-selectivity than those from aluminum alkyl chlorides, since they give polymers of butadiene with a 1,2[113,126–129] or a mixed 1,2/*cis*-1,4[98,130–132] structure. Only the reaction between $AlEt_3$ and $Co(acac)_3$ was studied and it was found that in the presence of butadiene the complex (5-methylhepta-2,6-dien-1-yl)(butadiene)cobalt, ($\eta^3$-$C_8H_{13}$)–$CoC_4H_6$, is formed.[128,129] This $Co^I$ complex is believed to be the active species in the $AlR_3$–cobalt compound system.

Soluble nickel catalysts are obtained from the reaction of nickel compounds such as $Ni(acac)_2$, $Ni(OCOR)_2$, $NiCl_2 \cdot 2py$ and aluminum alkyl chlorides. These catalysts are less stereospecific and give polymers of lower molecular weight than the corresponding Co catalysts.[133–135] Polymers of higher molecular weight and higher *cis* content are obtained with fluorine-containing systems such as $AlR_3$–$Ni(OCOR)_2$–$BF_3 \cdot OEt_2$ and $AlR_3$–$Ni(OCOR)_2$–HF[136] (or $HF \cdot OEt_2$).[137] Several variants of these systems were reported, in which fluorine-containing compounds of various types (*e.g.* acyl fluorides + $H_2O$,[138] $PhCF_3 + CF_3CO_2H$, $CH_2ClCHClCHF_2$)[139] are used instead of $BF_3 \cdot OEt_2$ or HF.

Systems containing lithium alkyls as alkylating agent were also reported: nickel diisopropylsalicylate–$BF_3 \cdot OEt_2$, lithium butyl[140,141] or lithium tetrabutylboron.[142]

The nature of the active center in Ni catalysts has not been much investigated. ESR examination of the systems $Al(Bu^i)_3$–Ni(naphthenate)–$BF_3 \cdot OEt_2$[143] and $Al(Bu^i)_2Cl$–Ni(naphthenate)[144] indicated that $Ni^{II}$ was reduced to $Ni^I$ and $Ni^0$. It is not clear, however, whether the species containing $Ni^I$ or that containing $Ni^0$ is the effective catalyst. This should contain a Ni–allyl bond and formation of a bond of this type was actually observed by IR investigation of the system $AlEt_3$–$Ni(acac)_2$–$P(OPh)_3$–$BF_3 \cdot OEt_2$ in the presence of butadiene.[145] Most probably the active

Al–Ni catalyst is an ionic species containing $Ni^{II}$, of the type formed in the reaction between $[(allyl)NiBr]_2$ and $AlBr_3$.[146,147]

The efficiency of $AlR_3$–$Ni(OCOR)_2$–$BF_3 \cdot OEt_2$ was reported to be rather low: 6.8% maximum of Ni was found to be active in the polymerization.[148] Data for other fluorine-containing nickel systems, which are reported[136] to be more active than the $BF_3$ system, are not available.

### 5.3.2.5 *Lanthanide catalysts*

Catalytic compositions based on Ce for the *cis* polymerization of 1,3-dialkenes are reported in a patent of 1963.[149] Some papers on the polymerization of 1,3-dialkenes with catalysts based on Ce, Nd, Pr and Gd have appeared in the following years.[150–152] The use of catalysts based on U is reported in a patent of 1971[153] and in the following papers.[154–156]

The complexes $(allyl)_3UX$ (X = Cl, Br, I) were found to be good initiators of the polymerization of butadiene to polymers with *ca.* 99% *cis* structure. The activity was greatly enhanced by addition of a Lewis acid such as $AlEtCl_2$, $AlCl_3$, $TiCl_4$, $BCl_3$ (Lewis acid/U molar ratio about 0.3).[154] The $AlEtCl_2$–$U(OMe)_4$ system has the same activity as the sysems from $(allyl)_3UX$ and $AlEtCl_2$, but is much simpler to prepare.[156,157] All the U catalysts can be used in aliphatic or aromatic solvents and the polymerization apparently takes place in a homogeneous phase.

Although uranium catalysts have shown a very high activity and stereospecificity for the *cis* polymerization of butadiene and isoprene, no practical application has ensued. The polymers prepared with U catalysts, after purification according to the standard procedures, still contain about 40–50 p.p.m. of uranium and this represents a problem for the commercialization of the product.

Cerium catalysts also have some disadvantages, due to the fact that the cerium residues in the polymer catalyze the oxidation of the product.

In the second half of the 1970s research work focused on neodymium systems. These do not show the disadvantages of uranium and cerium catalysts and are more active than the catalysts based on other elements of the series (*e.g.* Pr, Gd, Yb, *etc.*).[158,159] At present, two types of Nd systems are used: (a) the ternary systems $AlEt_2Cl$–$Nd(OCOR)_3$–$Al(Bu^i)_3$ (2.7–3:1:20–30) where the catalyst is usually prepared by reacting first the Nd salt of an organic acid ($Nd_2$ ethylhexanoate, Nd naphthenate) or also Nd alkoxides with $AlEt_2Cl$ in a hydrocarbon solvent and then by adding $Al(Bu^i)_3$ to the precipitate so formed;[157,158,160] and (b) the binary systems derived from $AlR_3$ and an Nd complex of the type $NdCl_3 \cdot nL$, where the ligand L can be THF,[160] alcohols,[158,162,163] phenols,[164] pyridine, ethylenediamine,[165] DMSO, $Ph_2SO$,[166] organic *N*-oxides,[167] *etc.*

The type of aluminum alkyl has some influence on the catalytic activity, which decreases in the order: $Al(Bu^i)_3 = Al(Bu^i)_2H > AlEt_3 \gg AlMe_3$.[158,159,168,169]

The catalysts from $NdCl_3$ are practically insoluble in hydrocarbon solvents. More soluble catalysts can be obtained using complexes such as $Nd(OCOR)_nCl_{3-n}$ or $Nd(OR)_nCl_{3-n}$, where R is a long hydrocarbon group; these complexes can be formed *in situ* by reacting $Nd(OCOR)_3$ or $Nd(OR)_3$ with an amount of $AlEt_2Cl$ (or other Cl-donating compound) insufficient to form $NdCl_3$. A soluble catalyst was obtained from $AlEt_3$ and $(CF_3CO_2)_2NdCl \cdot EtOH$. A crystalline complex of composition $(CF_3CO_2)EtNdCl \cdot Et_2AlH$ was isolated from this homogeneous system. This complex functions as the active species in the polymerization of butadiene and isoprene, just as does the $(CF_3CO_2)_2NdCl$–$AlEt_3$ system, to give the same polymer regularity and structure.[170–174]

The presence of Cl in the catalyst system seems to be essential for the formation of a *cis* polymer. The systems obtained by reacting $Al(Bu^i)_3$ with $Nd(OR)_3$ or $Nd(OCOR)_3$ give, in fact, *trans*-1,4-polybutadiene;[175] the same type of polymer was obtained with $MgR_2$–$Nd(OCOR)_3$.[176,177] Addition of $AlR_nCl_{3-n}$ to these systems results in the formation of a catalyst stereospecific for the *cis* polymerization of butadiene or isoprene. A similar effect of Cl was observed with $Nd(CH_2Ph)_3$: this complex gives *trans*-1,4-polybutadiene,[178] but $Nd(CH_2Ph)Cl_2$ (formed *in situ* from $Nd(CH_2Ph)_3 + 2HCl$) gives *cis*-1,4-polybutadiene.[179]

Little is known about the nature of the Nd systems. Spectrophotometric studies indicated that $Nd^{III}$ is present in the system Nd naphthenate–$Al(Bu^i)_3$–$Al_2Et_3Cl_3$.[158,180] Presumably the catalyst contains an alkylated $Nd^{III}$ species complexed with the Al compound.[181,182]

Catalysts were also prepared from compounds of $Nd^{II}$ ($NdCl_2$ complexed with THF,[183,184] $PhNdCl$[185]) and even from an Nd–benzene complex, $(C_6H_6)_3Nd_2$,[186] in combination with aluminum alkyls. Nothing is known about the oxidation state of Nd in these catalysts; it is possible that in these systems also $Nd^{III}$ is present, as a result of disproportionation reactions and/or of oxidative addition of the monomer to Nd in the catalytic complex.

Determinations of the number of active centers have been carried out on the systems $AlEt_2Cl$–$Nd(OCOR)_3$–$Al(Bu^i)_3$ and $NdCl_3 \cdot EtOH$–$AlR_3$; in catalysts prepared at room temperature, 5–6% maximum of Nd was found to be active in the polymerization.[159,160] A higher efficiency was found in catalysts prepared at −18 °C and aged at this temperature for some weeks. It appears that catalyst efficiency depends on two factors: (a) the level of alkylation of the Nd compound, which, at a given temperature, depends on the nature of $AlR_3$ and on the $AlR_3$/Nd ratio;[158,159,168,169,188] and (b) on the stability of the Nd—C bond, which depends on the temperature.[182]

Polymer-bound Nd catalysts have been reported. Their activity and stereospecificity is comparable to that of typical Nd catalyst systems and are maintained without significant loss when the catalyst is recycled.[189–191]

### 5.3.3 Catalysts Derived from Precursors Not Containing Metal–Carbon Bonds

Several systems not containing preformed metal–carbon bonds are effective for the polymerization of conjugated dialkenes. Rhodium salts ($RhCl_3 \cdot 3H_2O$, $Rh(NO_3)_3 \cdot 2H_2O$) are active in aqueous emulsions or in protic solvents and give butadiene polymers with a *trans*-1,4 structure.[21,22,192–196] Halides of some transition metals, subjected to thermal treatment under vacuum at 100–300 °C, can induce the stereospecific polymerization of butadiene.[197–199] Active catalysts were also obtained from haloacetates of Ni after thermal activation.[200,201] (Cycloocta-1,5-diene)NiX (X = Br, I) gives, although in low yields, polymers of butadiene with a *cis* (X = Br) or a *trans* (X = I) structure.[202] Activity in the polymerization of butadiene to *cis* polymers is shown even by metallic nickel alone or by metallic nickel in combination with benzyl chloride, benzoyl chloride or Lewis acids[203,204] and by U powder.[205]

Several bimetallic systems not containing preformed metal–carbon bonds have been reported. The following give *cis*-polybutadiene: $(NiC_8H_{12})_2$–$MoCl_5$ (or $TiCl_4$, $SbCl_5$),[206] $Ni(CO)_4$–$WCl_6$, $Ni(CO)_4$–$VCl_4$,[207] $Ni(PCl_3)_4$–$TiCl_4$,[208] $CoCl_2$–$AlCl_3$, $CoCl_2$–$AlCl_3$–Al[209–211] and $CoCl_2$–$AlCl_3$–thiophene.[212] The $Co_2(CO)_8$–$MoCl_5$ system gives instead 1,2-polybutadiene.[207]

Most of the above systems are only of theoretical interest, but a few of them have activity and stereospecificity comparable with that of Ziegler–Natta systems.

### 5.3.4 Catalysts Based on Allyl Derivatives of Transition Metals

These systems include simple allyl derivatives of transition metals and more complex catalysts derived from the reaction of the allyl complexes with electron donors, electron acceptors or Lewis acids. The systems are of interest because they are good models of the active site structure and therefore provide information which is useful to elucidate the catalysis of the polymerization. Some of them, however, have activity and stereospecificity comparable to that of Ziegler–Natta catalysts.

The first examples of stereospecific polymerization induced by simple allyl derivatives of transition metals were reported in 1964. A predominantly 1,2 polybutadiene was obtained with $(allyl)_3Cr$ and a predominantly *cis* polymer with $(allyl)_2CoI$.[213,214] A crystalline polybutadiene consisting almost exclusively of *trans*-1,4 units and a predominantly *cis* polymer were obtained with (allyl)NiBr in different solvents.[23] Allyl derivatives of other transition metals (Ti, Zr, Mo, W, Ru, Rh, U, Nd) were later found to be effective catalysts.[198,215–217]

Table 2 reports the microstructure of the polybutadienes obtained with some allyl complexes. Activity and stereospecificity depend both on the metal and the type of ligand. $(\eta^3\text{-Allyl})_3Cr$ gives 1,2-polybutadiene, but replacement of an allyl group with another ligand (*e.g.* Cl, OR, $OCOCL_3$) results in the formation of catalytic species that give predominantly *cis*-1,4-polybutadiene.[218–223]

$(\eta^3\text{-Allyl})_4Mo$ is a rather poor catalyst, but complexes of the type $(\eta^3\text{-allyl})_3MoX$ (X = Cl, I, Br, $OCOCF_3$, $OCOCCl_3$), which can be obtained by reacting $(allyl)_4Mo$ *in situ* with $I_2$, HI, HCl, $Ph_3CCl$ or $CH_2{=}CHCH_2I$, are highly active for the polymerization of butadiene to predominantly 1,2 polymers.[224–226] Analogously, $(allyl)_4W$ is practically inactive, whereas $(allyl)_3WX$ (X = Cl, OR, $OCOCl_3$) are good catalysts for the 1,2 polymerization of butadiene.[227] Change in activity and stereospecificity by substitution of an allyl group with anionic ligands such as Cl, I, $OCOCF_3$ has also been observed with allyl derivatives of Ti, Zr, Nb, Co and Rh.[218,219,228]

Extensive work has been carried out with nickel complexes of the type (allyl)NiX. These complexes give 1,4 polymers from butadiene, but both polymer microstructure and catalyst activity strongly depend on the nature of X (Table 3). High-*trans* polymers are obtained with (allyl)-NiI,[146,229–235] whereas high-*cis* polymers are obtained when X is Cl[146,229,230,236–238] or a ligand

**Table 2** Polymerization of Butadiene with Allyl Derivatives of Transition Metals

| *Catalyst* | *Polymer microstructure* % *cis* | % *trans* | % *1,2* | *Ref.* |
|---|---|---|---|---|
| $(\eta^3\text{-}C_4H_7)_{3-4}Ti$ | — | 17 | 83 | a |
| $(\eta^3\text{-}C_4H_7)_3Cr$ | | 19 | 81 | a |
| $(\eta^3\text{-}C_4H_7)_3Cr + \frac{1}{2}O_2$ | | 92.5 | 7.5 | b |
| $(\eta^3\text{-}C_4H_7)CrCl_2$ | 90 | 5.5 | 4.5 | a |
| $(\eta^3\text{-}C_4H_7)_2CrOCOCl_3$ | 93 | 4 | 3 | a |
| $(\eta^3\text{-}C_3H_5)_2CrOBu^n$ | 12 | 14 | 74 | c |
| $(\eta^3\text{-}C_4H_7)_3Nb$ | — | — | 100 | a |
| $(\eta^3\text{-}C_4H_7)NbCl$ | 91 | 5.5 | 3.5 | a |
| $(\eta^3\text{-}C_3H_5)_4Mo + Cl_3CCO_2H$ | | 2 | 98 | d |
| $(\eta^3\text{-}C_3H_5)MoCl$ | — | — | 90 | e |
| $(\eta^3\text{-}C_3H_5)_4Mo + I_2$ | | 2 | 98 | d |
| $(\eta^3\text{-}C_5H_5)_4Mo + CH_2{=}CHCH_2I$ | | 0.5 | 99.5 | d |
| $(\eta^3\text{-}C_5H_5)W + CH_2{=}CHCH_2I$ | | 5 | 95 | a |
| $(\eta^3\text{-}C_4H_7)_3Co + 2HCl$ | 91.5 | 1.5 | 7 | a |
| $(\eta^3\text{-}C_8H_{13})(C_4H_6)Co$ | 1 | 1 | 98 | f |
| $(\eta^3\text{-}C_{12}H_{18})RuCl_2$ | 54 | 28 | 18 | g |
| $(\eta^3\text{-}C_4H_7)_3Rh$ | | 94 | 6 | a |
| $Li(\eta^3\text{-}C_5H_5)_4Nd \cdot$ dioxane | 2.6 | 92.8 | 4.6 | h |
| $(\eta^3\text{-}C_3H_5)_3UCl$ | 99 | 0.7 | 0.3 | i |

[a] B. A. Dolgoplosk, *Vysokomol. Soedin., Ser. A*, 1971, **13**, 325; *Polym Sci. USSR (Engl. Transl.)*, 1971, **13**, 367.
[b] I. A. Oresklin, E. I. Tinyakova and B. A. Dolgoplosk, *Vysokomol. Soedin., Ser. A*, 1969, **11**, 1840; *Polym. Sci. USSR (Engl. Transl.)*, 1969, **11**, 2096.
[c] B. A. Dolgoplosk, E. I. Tinyakova, N. N. Stefanovskaya, I. A. Oreshkin and V. L. Shmonina, *Eur. Polym. J.*, 1974, **10**, 605.
[d] B. A. Dolgoplosk and E. I. Tinyakova, *Vysokomol. Soedin., Ser. A*, 1977, **19**, 2441; *Polym. Sci. USSR (Engl. Transl.)*, 1977, **19**, 2805.
[e] P. W. Jolly, C. Krueger, C. C. Romao and M. J. Romao, *Organometallics*, 1984, **3**, 936.
[f] G. Ricci and L. Porri, in press.
[g] K. Hirachi and H. Hirai, *Macromolecules*, 1970, **3**, 382.
[h] A. Mazzei, *Makromol. Chem., Suppl.*, 1981, **4**, 61.
[i] G. Lugli, A. Mazzei and S. Poggio, *Makromol. Chem.*, 1974, **175**, 2021.

**Table 3** Polymerization of Butadiene by (Allyl)NiX: Influence of the Anion X and of Electron Donors and Acceptors on Polymer Microstructure

| *Allyl compound* | *Electron donor or acceptor* | *Polymer microstructure* % *cis* | % *trans* | % *1,2* | *Ref.* |
|---|---|---|---|---|---|
| $(\eta^3\text{-}C_4H_7)NiCl$ | | 92 | 6 | 2 | a |
| $(\eta^3\text{-}C_4H_7)NiBr$ | | 72 | 25 | 3 | a |
| $(\eta^3\text{-}C_4H_7)NiI$ | | 0 | 97 | 3 | b |
| $(\eta^3\text{-}C_3H_5)NiOCOCCl_3$ | | 94 | 4 | 2 | b |
| $(\eta^3\text{-}C_3H_5)NiOCOCHCl_2$ | | 93 | 5 | 2 | b |
| $(\eta^3\text{-}C_3H_5)NiOCOCH_2Cl$ | | 92 | 6 | 2 | b |
| $(\eta^3\text{-}C_4H_7)NiOCOCF_3$ | | 97 | 2 | 1 | a |
| $(\eta^3\text{-}C_3H_5)NiOC_6H_2(NO_2)_3$ | | 94 | 4 | 2 | a |
| $(\eta^3\text{-}C_3H_5)NiOC_6H_2Br_3$ | | 0 | 97 | 3 | b |
| $(\eta^3\text{-}C_{12}H_{19})NiOCOCF_3$ | | 98 | 2 | — | a |
| $(\eta^3\text{-}C_4H_7)NiOCOCF_3$ | $CF_3CO_2H$ | 50 | 49 | 1 | c |
| $(\eta^3\text{-}C_4H_7)NiI$ | *p*-Fluoranil | 97 | 2 | 1 | b |
| $(\eta^3\text{-}C_4H_7)NiCl$ | $CCl_3CHO$ | 95 | 4 | 1 | b |
| $(\eta^3\text{-}C_4H_7)NiCl$ | EtOH | 30 | 70 | 0 | a |
| $(\eta^3\text{-}C_4H_7)NiCl$ | THF | 54 | 41 | 5 | a |
| $(\eta^3\text{-}C_4H_7)NiCl$ | $Et_2O$ | 75 | 23 | 2 | a |
| $(\eta^3\text{-}C_3H_5)NiOCOCF_3$ | $PhNO_2$ | 50 | 50 | 0 | d |
| $(\eta^3\text{-}C_{12}H_{19})NiOCOCF_3$ | EtOH | 0 | 96 | 4 | c |
| $(\eta^3\text{-}C_{12}H_{19})NiOCOCF_3$ | $P(OPh)_3$ | 0 | 96 | 4 | c |
| $[(\eta^3\text{-}C_3H_5)NiP(OPh)_3]PF_6$ | | 5 | 94 | 1 | e |
| $[(\eta^3\text{-}C_3H_5)Ni(SbPh_3)]PF_6$ | | 88 | 9 | 3 | e |

[a] F. Dawans and Ph. Teyssié, *Ind. Eng. Chem. Prod. Res. Dev.*, 1971, **10**, 261.
[b] B. A. Dolgoplosk, *Vysokomol. Soedin., Ser. A*, 1971, **13**, 325; *Polym. Sci. USSR (Engl. Transl.)*, 1971, **13**, 367.
[c] J. P. Durand, F. Dawans and Ph. Teyssié, *J. Polym. Sci., Part A-1*, 1970, **8**, 979.
[d] J. C. Maréchal, F. Dawans and Ph. Teyssié, *J. Polym. Sci., Part A-1*, 1970, **8**, 1993.
[e] R. Taube, J. P. Gehrke and U. Schmidt, *J. Organomet. Chem.*, 1985, **292**, 287.

having a high electron-withdrawing power, *e.g.* $CF_3CO_2$ or $CCl_3CO_2$.[218,232,239,240] Even the nature of the substituents on the anionic ligand can affect the microstruture: (allyl)Ni(picrate) gives 97% *cis*-polybutadiene,[232] whereas allylnickel 2,4,6-tribromophenolate gives a 96% *trans* polymer;[240] allylnickel *p*-toluenesulfonate gives instead a polymer composed of nearly 50% *trans* and 50% *cis* units.

The rate of polymerization is also strongly influenced by the nature of X; *e.g.* allylnickel trifluoroacetate polymerizes butadiene much faster than allylnickel halides. A systematic investigation of the polymerization with allylnickel haloacetates has shown that the rate increases with the electron-withdrawing power of the anionic ligand: $CF_3CO_2 > CCl_3CO_2 > CHCl_2CO_2 > CH_2ClCO_2 > MeCO_2$.[215,218,233,234,242]

The catalytic activity of allylnickel halides and other allyl derivatives of transition metals is strongly enhanced by addition of various types of electron acceptors such as $CF_3CO_2H$,[243–246] halogen-containing quinones (fluoranil, chloranil, *etc.*),[247–250] $Cl_3CCHO$, $Cl_3CCOCl$, $CCl_3COCCl_3$,[250–252] haloacetates of Co, Ni, Zn or Mn,[220,253,254] $I_2$, $SnI_4$,[230] $SnCl_4$,[204,223,240,255] Ni halides,[204,223,236,240] $TiCl_4$,[204,240,256–260] $TiCl_3Me$,[261] $AlCl_3$ or $AlBr_3$,[146,204,223,240,255,256] $AlEtCl_2$[262,264] and halides of V, W or Mo.[223,255,256]

In addition to enhancing the activity, the electron acceptors affect the stereospecificity, favoring the formation of *cis* units. (Crotyl)NiCl and (crotyl)NiBr give butadiene polymers consisting respectively of 93% and 88% *cis* units.[230,265] On addition of $Cl_3CCO_2H$, $Cl_3CCHO$, $Cl_3CCOCl$ or chloranil these systems give polymers composed of 96–98% *cis* units.[243,265] The effect of electron acceptors on the stereospecificity is particularly evident in the case of (allyl)NiI. This complex alone gives a polybutadiene consisting of > 95% *trans* units; on addition of $I_2$ or $SnI_4$ a polymer with *ca.* 85% *cis* units is obtained,[230] while in the presence of chloral, $Cl_3CCOCl$ or fluoranil the polymer *cis* content is 92–95%.[231,247,248] Typical Lewis acids have the same effect: a combination of (allyl)NiI and $TiCl_4$ or $AlCl_3$ gives predominantly *cis* polymers from butadiene.[146,204,240,256]

The reaction between allylnickel halides and Lewis acids leads to the formation of ionic complexes; *e.g.* allylnickel bromide and $AlBr_3$ in benzene give a precipitate of formula $[(C_3H_5)Ni(C_6H_6)]AlBr_4$.[146] The reaction with quinones, $Cl_3CCHO$ or $CCl_3COCCl_3$ results in the formation of charge-transfer complexes (CTCs), in which the positive charge on the transition metal is increased. The formation of a CTC was confirmed by IR spectroscopy from the shift of the carbonyl group oscillation frequency.[248,266]

While the addition of electron acceptors favors the formation of *cis* polymers, the addition of electron donors (ethers, $H_2O$, alcohols, phosphites, *etc.*) causes a decrease in the *cis* content and an increase in the *trans* content; *e.g.* allylnickel chloride, which gives predominantly *cis*-polybutadiene in hydrocarbon solvents, gives *trans* polymers in the presence of $H_2O$, EtOH or THF;[146,215,236,240] allylnickel trifluoroacetate, which gives *cis*-polybutadiene in cyclohexane, gives *trans*-polybutadiene in the presence of $P(OPh)_3$.[234]

Other ligands such as $CF_3CO_2H$ or benzene cause the formation of polymers composed of *cis* and *trans* units in a practically 1:1 ratio (equibinary polymers).[234,245,267–269] Allylnickel trifluoroacetate can therefore give three types of polybutadiene, depending on the solvent or ligand used (Scheme 2).

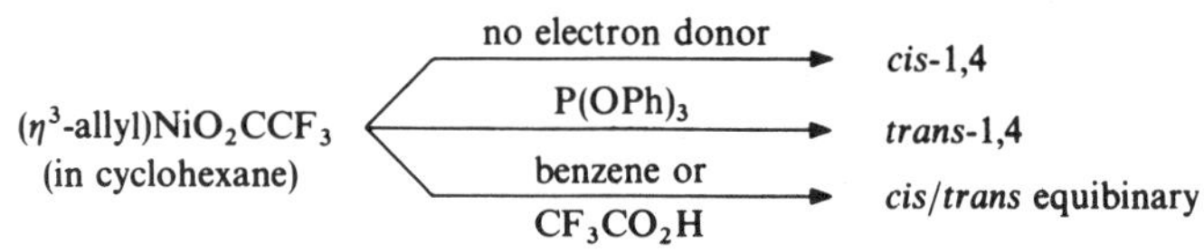


**Scheme 2**

Ionic complexes of the type $[(allyl)NiL_2]PF_6$ ($L = P(OR)_3$, $PPh_3$, CN—Bu$^t$, $AsPh_3$, $SbPh_3$) are also effective catalysts for the 1,4 polymerization of butadiene. The activity and stereospecificity of these systems is strongly influenced by the nature of the ligand L; some of them show a noticeable activity and give predominantly *cis*-polybutadiene.[270–277]

Although most of the work has concerned the polymerization of butadiene, isoprene and other dialkenes were polymerized with allyl derivatives of transition metals.[228,235,278–282]

Catalysts based on allyl derivatives of transition metals supported on alumina, silica or silica–alumina have been reported.[217,221,222,283–290] The activity is generally enhanced with respect to the unsupported catalysts, while the stereospecificity was found to vary with the nature of the support, *e.g.* $(allyl)_3Cr$, which gives 1,2-polybutadiene only, becomes a stereospecific catalyst for the formation of *trans* units of butadiene and isoprene when supported on silica or silica–alumina and for the formation of *cis* units on alumina.[222]

## 5.4 SYNTHESIS OF THE POLYMERS

Most of the work on the polymerization of 1,3-dialkenes has concerned butadiene and isoprene, due to the practical significance of the polymers from these monomers. Some work has been done on the polymerization of 1,3-pentadiene, due to the phenomena of stereoisomerism shown by its polymers. Other monomers have been less examined, usually by application of a single catalyst, without systematic investigation of the experimental conditions.

Some catalysts give products consisting of macromolecules of the same microstructure, others give a mixture of stereoregular and amorphous polymers. The latter are of little interest and usually have not been well characterized. In the following sections the syntheses of the most interesting polymers are examined, with particular regard to those of butadiene, isoprene and pentadiene.

### 5.4.1 Butadiene Polymers

By the use of transition metal catalysts it is possible to obtain from butadiene all four foreseeable stereoregular polymers: *trans*-1,4, *cis*-1,4, 1,2 isotactic and 1,2 syndiotactic. All these polymers have been obtained with a high degree of steric purity and hence are highly crystalline or crystallizable. Polymers consisting predominantly of only two types of monomeric units in a practically 50:50 ratio (*cis*-1,4/*trans*-1,4; *cis*-1,4/1,2) were also obtained. These have been called 'equibinary' polybutadienes. In addition, several types of irregular polymers have been obtained, consisting of different monomeric units in various proportions. Some of the polybutadienes obtained with transition metal catalysts have practical significance: the most important is the *cis*-polybutadiene, which exhibits excellent elastomeric properties.

#### *5.4.1.1* cis-*1,4-Polybutadiene*

This polymer can be obtained with catalysts from titanium, cobalt, nickel and lanthanides (Table 4).

*(i) Titanium catalysts*

The $AlR_3$–$TiCl_4$ system was the first one used for the polymerization of butadiene. At an Al/Ti molar ratio of about unity it gives polymers with a *cis* content of 65–70%; at $Al/Ti > 2$ it gives polymerization products consisting of a mixture of predominantly *cis* polymers and of crystalline

**Table 4** Polymerization of Butadiene to *Cis*-1,4 Polymer

| *Type of catalyst and polymerization conditions* | *Concentration of transition metal* (mmhm)[a] | *Polymer structure* % cis | % trans | % 1,2 | *Ref.* |
|---|---|---|---|---|---|
| $Al(Bu^i)_3$–$TiI_4$ (4–5:1) in benzene at 30 °C | 1–2 | 92–3 | 2–3 | 4–6 | b |
| $AlEt_3$–$AlEt_2I$–$TiCl_4$ (4:4:1) at −5 °C | 0.35–3 | 92.7 | 3.3 | 4 | c |
| $AlHCl_2 \cdot OEt_2$–$TiCl_4$–$AlI_3$ (5.5:1:1.3) in toluene at 5 °C | 0.34–2.2 | 93 | 2.5 | 4.5 | d |
| $AlEt_2Cl$–$CoCl_2 \cdot py$–$H_2O$ (1000:1:100) in benzene at 5 °C | 0.02 | 98 | 1 | 1 | e |
| $AlEt_3$–Ni(octanoate)–HF (12–15:1:30 to 70) in benzene at 50 °C | 0.05 | 97 | 2 | 1 | f |
| $AlEt_3$–Ni(octanoate)–$BF_3 \cdot OEt_2$ (17:1:15) in benzene at 50 °C | 0.92 | 96–97 | 2–3 | 1 | g |
| $Al(Bu^i)_3$–$NdCl_3 \cdot nL$ in cyclohexane at 50 °C (L = electron donor) | 0.06–6 | 97 | 2.7 | 0.3 | h |

[a] mmhm = mmol per 100 g of monomer.
[b] W. Cooper, in 'The Stereo Rubbers', ed. W. M. Saltman, Wiley, New York, 1977, p. 21.
[c] W. Marconi, M. Araldi, A. Béranger and M. De Maldé, *Chim. Ind. (Milan)*, 1963, **45**, 522.
[d] W. Marconi, A. Mazzei, A. Araldi and M. De Maldé, *J. Polym. Sci., Part A*, 1965, **3**, 735.
[e] M. Gippin, *Ind. Eng. Chem. Prod. Res. Dev.*, 1962, **1**, 32.
[f] M. C. Throckmorton and F. S. Farson, *Rubber Chem. Technol.*, 1972, **45**, 268.
[g] W. Saltman and L. J. Kuzma, *Rubber Chem. Technol*, 1973, **46**, 1055.
[h] H. L. Hsieh and H. C. Yeh, *Rubber Chem. Technol.*, 1985, **58**, 117.

*trans* polymers.[14,82] Polybutadiene with a higher *cis* content (80–85%) was obtained with $AlR_3$–$TiBr_4$.[291] However, this system, as with that from $TiCl_4$, has a tendency to produce gelled polymer and also low molecular weight oligomers.

Only systems containing iodine give high-*cis* polybutadiene of high molecular weight, which is essentially gel free. $AlR_3$–$TiI_4$ gives polymers with a 92–94% *cis* content.[56] Practically identical results are obtained with catalysts prepared from $TiCl_2I_2$ or $TiCl_3I$.[60,292] Various types of $AlR_3$ were examined, although $Al(Bu^i)_3$ is the most commonly used. The main features of these systems are as given below.[54,57,61–63,66]

(1) The catalysts become active at an Al/Ti ratio of about two. Yields increase rapidly with the Al/Ti ratio, reach a maximum, the position of which depends on the particular $AlR_3$ used, then decrease.[54,58,61] Figure 1 shows some data for the $Al(Bu^i)_3$–$TiI_4$ system.

(2) The *cis* content is maximum at Al/Ti ratios corresponding to maximum activity, then it decreases slowly with increasing Al/Ti ratio, while the content of 1,2 units increases (about 4% at $Al/Ti = 4$; 10% at $Al/Ti \approx 10$).[61]

(3) The *cis* content of the polymers depends on the concentration of the catalyst and increases with decrease in concentration. At low concentration (about 1 mmol of Ti per 100 g $C_4H_6$) the *cis* content may reach 94–95%, which is probably the highest value attainable with Ti catalysts.[57,61]

(4) At fixed monomer concentration, the polymer molecular weight is highly influenced by the catalyst concentration. Polymers with molecular weight ranging from $30–40 \times 10^3$ to $10^6$ can be obtained by varying the monomer and catalyst concentration.[57,61] The latter can be varied within a wide range, from about 0.3 to more than 3 mmol of Ti per 100 g of monomer (mmhm), but in order to obtain a processable material ($[\eta] = 2–3$) the level of Ti should be 1–2 mmhm.[16] The polymer has a narrow MWD at the early stages of polymerization, but becomes highly polydispersed as the reaction proceeds, in runs carried out over the range 5 °C to 50 °C.[62,63,293]

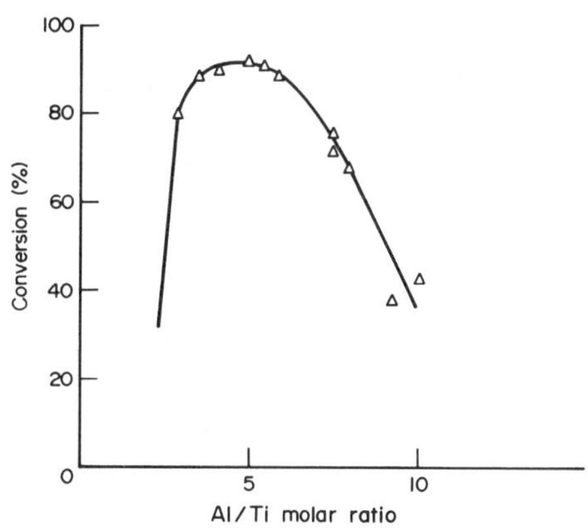


**Figure 1** Polymerization of butadiene with $Al(Bu^i)_3$–$TiI_4$. Monomer conversion as a function of the Al/Ti molar ratio[58]

As indicated earlier (Section 5.3.2.1), several ternary systems have been proposed. These exhibit characteristics similar to $AlR_3$–$TiI_4$. In the ternary systems, high activity and stereospecificity are obtained only for compositions varying within a limited range; *e.g.* the optimum composition for $Al(Bu^i)_3$–$I_2$–$TiCl_4$ is 8/1.5/1, while for $AlEt_3$–$AlEt_2I$–$TiCl_4$ it is 4/4/1.[16,64]

Some work has been carried out with ternary systems containing metal hydrides as reducing agents. The systems obtained from $TiCl_4$ with $AlI_3$ and one of the compounds $AlHCl_2 \cdot OEt_2$, $AlH_2NMe_2$, $AlH_2Cl \cdot NMe_3$ or $AlH_3 \cdot NMe_3$ are particularly active and give a polymer about 95% *cis*.[68,294]

Titanium catalysts were the first used for the synthesis of high-*cis* polybutadiene and have been the basis for commercial processes in Europe and the USA. Now other systems with higher activity and stereospecificity are available.

### *(ii) Cobalt catalysts*

Although several Co compounds can be used for the preparation of the catalyst, salts of organic acids with a long chain, *e.g.* Co octanoate, are particularly suitable for their solubility in aromatic or aliphatic solvents.

$AlEt_2Cl$, $Al_2Et_3Cl_3$, $AlEtCl_2$ or their mixtures can be used as cocatalysts. The alkyl to chloride ratio in the alkylaluminum compound plays an important role in the polymerization activity, stereospecificity and polymer molecular weight.[80,102]

The presence of $H_2O$ is essential to the catalytic activity of the $AlEt_2Cl$ catalyst.[80,99–101,103,106,295–299] The activity, which is negligible in the absence of $H_2O$, is at a maximum for a ratio of $H_2O$ to $AlEt_2Cl$ of about 0.03 to 0.2. The *cis* content is also affected by the presence of $H_2O$ and reaches a value of 95–96% at $H_2O/AlEt_2Cl$ ratios higher than 0.1. The effect of $H_2O$ is attributable, at least in part, to the formation of aluminoxanes, $[Et(Cl)Al]_2O$.[114] Besides $H_2O$, other substances can act as activators: peroxides, hydroperoxides, alcohols, $O_2$, $AlCl_3$,[99,100] $SnCl_4$[104] and HCl.[110] Catalysts prepared from $AlEtCl_2$ or $AlEt_{1.5}Cl_{1.5}$ or, more generally, from an aluminum compound in which the Cl/Et ratio is $> 1$, do not require $H_2O$ or any other activator.[80,99,100,102,104]

Aromatic hydrocarbons are the best polymerization solvents. The rate of polymerization decreases in the order benzene > toluene > xylenes, inversely with the basicity of the solvent.[117,296] Hexamethylbenzene, a strong electron donor, is capable of inactivating the polymerization if present in a suitable amount. Aliphatic solvents are also suitable, but normally they are used in a mixture with some benzene or toluene (10–20%).[80,99,100,102] *cis*-2-Butene is an excellent solvent for catalysts based on aluminoxanes; the polymerization rate in this solvent is higher than in benzene or in heptane.

The *cis* content of the polymers can be as high as 99.5%. Several parameters influence the *cis* content, *e.g.* the polymerization temperature, the Cl/Al ratio in the aluminum alkyl and to a lesser extent the nature of the solvent. All the catalysts are practically equivalent at low temperature (0 to +5 °C) with regard to polymer *cis* content (98% minimum). The *cis* content decreases with increasing polymerization temperature,[96,99,100,109] but the effect of temperature becomes more pronounced with an increase of the alkyl/Cl ratio in the aluminum compound (Figure 2).

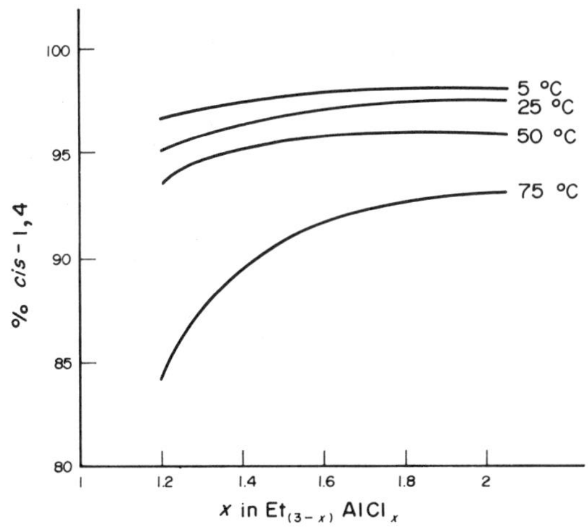


**Figure 2** Polymerization of butadiene with aluminum alkyl chloride–cobalt octanoate. Effect of chloride content on percentage of *cis*-1,4 structure at various temperatures[80]

The polymer molecular weight is affected by several variables. such as (a) the type of solvent, (b) the monomer and cobalt concentration, (c) the concentration of the aluminum alkyl, (d) the polymerization temperature and (e) the monomer conversion.

(a) In aromatic solvents, the polymer molecular weight decreases with increasing basicity of the solvent.[14] In mixed aromatic/aliphatic solvents, the polymer molecular weight decreases with increasing proportion of the aliphatic component.[80,99,100,102,296,300]

(b) The polymer molecular weight increases with the monomer concentration[14,80,99–101,106,301] (Figure 3) while it decreases with increasing cobalt concentration.[98,101,106,107,109]

(c) At constant Al level, polymer molecular weight (MW) increases with the Cl/alkyl ratio in the aluminum alkyl chloride.[80,102,104] The effect of the concentration of Al seems to depend on the type of aluminum compound used. With $AlEt_2Cl$ as cocatalyst the concentration of Al has no significant effect on polymer MW at 20% monomer conversion,[301] while at higher conversions MW decreases slightly with increasing Al concentration.[109,134,302] By contrast, polymer MW has been reported to

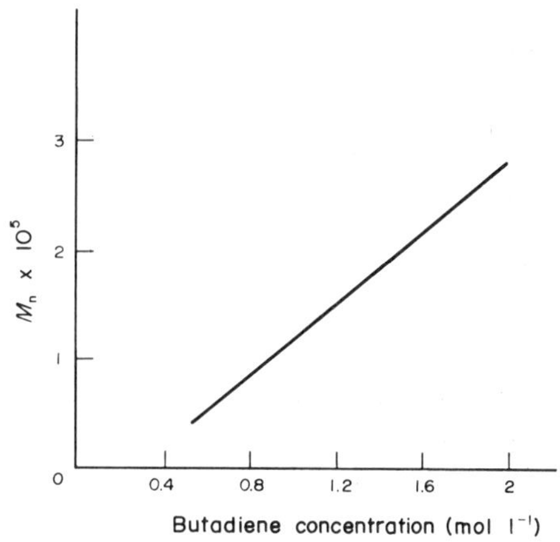


**Figure 3** Polymerization of butadiene to the *cis* polymer with $Al_2Et_3Cl_3$–cobalt octanoate. Effect of monomer concentration on polymer MW[80]

increase with Al concentration when Al compounds in which the Cl/Et ratio is higher than 0.5 are used as cocatalyst.[80]

(d) At fixed Al/Co ratio, polymer MW is strongly dependent upon the polymerization temperature (Figure 4). The effect is different for catalysts prepared from different aluminum alkyls (Figure 5). In benzene, with $AlEtCl_2$/Co octanoate and a monomer concentration of 0.7 mol l$^{-1}$, $\bar{M}_n$ is 750 000 at 5 °C, 400 000 at 50 °C, and about 200 000 at 70 °C.[80]

(e) Figure 5 shows the change in molecular weight with conversion. The molecular weight increases during the first 30% conversion, reaches a plateau, then drops slightly.[80] The decline may be accounted for by the relationship shown in Figure 3.

Several substances can act as transfer agents in the polymerization of butadiene by Co catalysts. Allenes are particularly efficient, but ethylene, propylene and also $H_2$ can be used to reduce the polymer molecular weight. The data concerning some transfer agents are listed in Table 5.[303] The molecular weight distribution (MWD) of polybutadiene prepared with Co catalysts varies with the degree of monomer conversion and is strongly dependent on the level of water in the catalyst systems. Values of $\bar{M}_w/\bar{M}_n$ varying from two to about nine, depending on the experimental conditions, have been reported.[103,109,302]

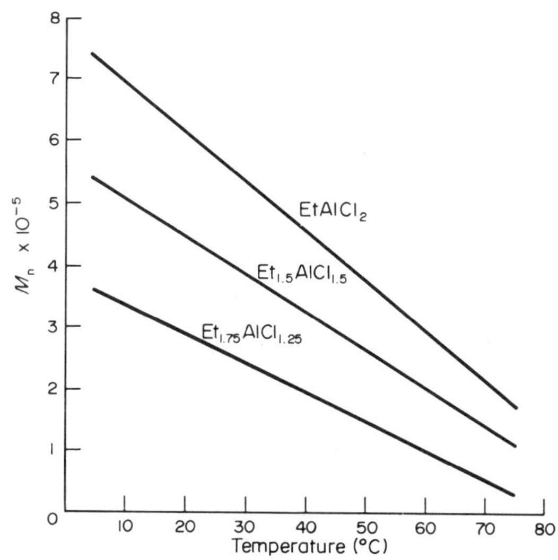


**Figure 4** Polymerization of butadiene with aluminum alkyl chloride–cobalt octanoate. Effect of temperature on polymer MW with various Al compounds[80]

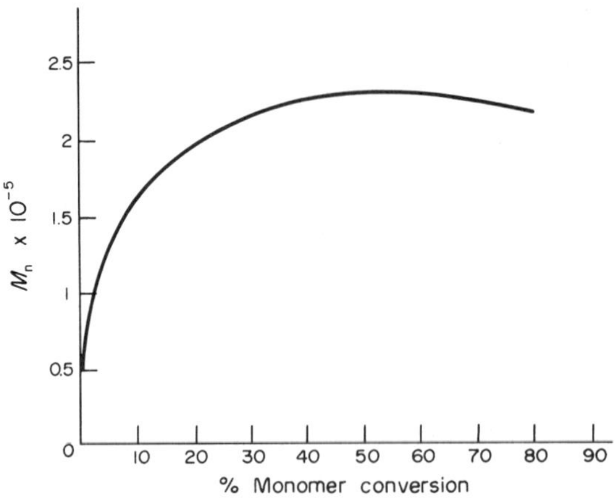


**Figure 5** Polymerization of butadiene with $AlEt_{1.1}Cl_{1.9}$–Co(octanoate). Dependence of polymer MW on conversion[80]

**Table 5** Effect of Transfer Agents on Intrinsic Viscosity of *cis*-Polybutadiene Obtained with $AlEt_2Cl$–$CoCl_2 \cdot 2py$ in Benzene[a,b]

| *Transfer agent* | $[\eta]^c$ ($dl\,g^{-1}$) |
|---|---|
| None | 5.57 |
| Ethylene (2.1 mol/100 mol $C_4H_6$) | 2.72 |
| Ethylene (5.08 mol/100 mol $C_4H_6$) | 1.78 |
| Propylene (72.2 mol/100 mol $C_4H_6$) | 3.43 |
| 1,2-Butadiene (0.09 mol/100 mol $C_4H_6$) | 3.12 |
| 1,2-Butadiene (0.223 mol/100 mol $C_4H_6$) | 2.10 |
| Hydrogen (0.5 MPa) | 2.95 |
| Hydrogen (2 MPa) | 1.37 |

[a] C. Longiave, R. Castelli and M. Ferraris, *Chim. Ind. (Milan)*, 1962, **44**, 725 (*Chem. Abstr.*, 1962, **57**, 16 825).
[b] $CoCl_2 \cdot 2py$ 0.09 mmol; Al/Co (molar ratio) = 400; benzene 1000 ml; $C_4H_6$ 100 g; 15 °C.
[c] In toluene at 26 °C.

### *(iii) Nickel catalysts*

Several catalysts based on nickel give polybutadiene with a high *cis* content. Some of them have activity and stereospecificity comparable with those from Co and have become of industrial interest.

The catalysts prepared from aluminum alkyl chlorides and soluble nickel compounds ($Ni(acac)_2$, Ni napthenate) give polymers with a lower *cis* content and lower molecular weight than the corresponding Co systems; *e.g.* $AlEt_2Cl$–$Ni(acac)_2$ gives at room temperature an oily product with a *cis* content of about 85%.[304] The *cis* content may be increased to values >90% by lowering the polymerization temperature and also by adding water to the catalyst system. Addition of water also increases the molecular weight, which, however, remains lower than with Co catalysts.[134] Variants of the above Al–Ni systems containing fluoride are of high activity and give polymers of 97–98% *cis* structure. Intrinsic viscosities are within the range 3 to 5 $dl\,g^{-1}$, depending on the experimental conditions. The following three-component systems have been particularly studied: $AlEt_3$–Ni(carboxylate)–$BF_3 \cdot OEt_2$ (7–8:1:7.3–7.7)[148,298,305] and $AlR_3$–Ni(carboxylate)–HF (12–15:1:30–55).[136] The presence of fluoride is essential to produce high-*cis* high molecular weight polymer.

With the above two systems the molecular weight increases rapidly at the beginning of the reaction, until the conversion has reached a value of 40–50%, then remains constant and finally drops slightly.

Transfer to monomer is the main process governing the molecular weight. The catalyst level has some effect on the molecular weight, higher values being obtained at lower catalyst concentration. The polymerization temperature is usually 30–50 °C.

Polybutadiene with a *cis* content of 97–98% can be obtained with (allyl)$NiOCOCF_3$ in heptane.[233,234,242,268,306] The activity of this catalyst is, however, lower than that of the fluoride-containing Al–Ni systems.[135,296]

*(iv) Lanthanide catalysts*

High-*cis* polybutadiene can be obtained with several catalysts from halides of lanthanides in combination with $AlR_3$. Only a few of these systems have been examined in detail. Good yields of polymer were obtained with the three-component system $AlEt_2Cl$–Ce salt–$AlR_3$.[152] *Cis* content was 97–98% for polymers of high molecular weight. Cerium catalysts have not been further investigated because of problems connected with the presence of cerium residues in the polymers.

Catalysts based on uranium have been investigated much more intensively.[154–156] The allyl complexes $(allyl)_3UX$ (X = Cl, Br, I) are effective initiators for the polymerization of butadiene. Polymers having a *cis* content up to 99% are obtained, independently of the nature of X. When a Lewis acid is added ($AlEtCl_2$, $BCl_3$, $TiCl_4$, $AlCl_3$; Lewis acid/U molar ratio 0.2 to 0.3) the catalytic activity is much improved. *Cis* content is always high, about 99%, independently of the nature of the halogen in the uranium compound and in the Lewis acid. The systems prepared from uranium alkoxides, in particular $U(OMe)_4$, and $AlBr_3 + AlEt_3$ ($AlBr_3/U = 0.5$; $AlEt_3/U = 15$) give results equivalent to $AlEtCl_2$–$U(allyl)_3X$, both with regard to catalyst activity and polymer properties.

Uranium polybutadiene has high stereoregularity, excellent processability and gives elastomers with very good properties.[155,156] Unfortunately it cannot be commercialized for the reasons indicated in Section 5.3.2.5.

The catalysts based on Nd appear now as the most interesting among those prepared from lanthanides for the manufacture of *cis*-polybutadiene. They have a good activity and stereospecificity and do not have the disadvantages of Ce and U catalysts. The two systems that are more extensively used are $AlEt_2Cl$–Nd(carboxylate)–$Al(Bu^i)_3$ (2.5 to 3:1:30) and $Al(Bu^i)_3$–$NdCl_3 \cdot nL$ (*ca.* 30:1), where L is an electron donor (see Section 5.3.2.5). The two systems are practically equivalent in the polymerization of butadiene.

Nd catalysts give a polymer which is essentially gel free. The *cis* content can be as high as 99%. A relationship exists between polymer molecular weight and *cis* content: this decreases with the molecular weight. 1,2 units are always in the range 0.1 to 0.8%.[163]

Several types of solvents can be used, such as aliphatic or aromatic hydrocarbons, chlorobenzene, tetrachloroethylene and alkenes. The rate of polymerization is higher in aliphatic solvents than in arenes. In the latter, the rate decreases with the basicity of the solvent: benzene > toluene > xylenes. Alkenes are in general equivalent to heptane or cyclohexane, but some of them, *e.g.* (*Z*)-2-butene, give even higher rates. Since Nd catalysts have no cationic activity, alkenes such as isobutene or isopentylenes do not affect the polymerization.[163,307,308]

The polymer molecular weight increases continuously with the polymerization time.[163] This is one of the most significant differences between Nd catalysts and those from Ni or Co. With the latter catalysts, the polymer molecular weight remains practically constant after a 40–50% conversion, transfer with monomer being the main termination process.[80,305] In the polymerization with Nd catalysts, transfer with monomer is much less pronounced.

The most important transfer process with these catalysts is that between the growing chain and $AlR_3$.[158,163,308] As a consequence of this transfer reaction, polymer molecular weight decreases as the $AlR_3$/Nd ratio is increased. $Al(Bu^i)_2H$ is more effective than $Al(Bu^i)_3$ in reducing the molecular weight. The effect of the $AlR_3$/Nd ratio on $[\eta]$ at different polymerization temperatures is shown in Figure 6.

The polymerization temperature and catalyst concentration also affect the molecular weight. At fixed $AlR_3$/Nd ratio and Nd concentration, an increase in the polymerization temperature results in a decrease in the molecular weight. An increase in the catalyst level produces an effect in the same direction. Operating on the above indicated variables, polymers having $[\eta]$ in the range 0.5 to 10 can be obtained. $\bar{M}_w/\bar{M}_n$ is always rather high, in the range five to eight. The catalyst level is 0.1 to 0.6 mmol per 100 g of monomer.[163]

### 5.4.1.2 trans-*1,4-Polybutadiene*

Crystalline polybutadiene with a high-*trans* structure has been obtained using catalysts containing Ti, V, Cr, Rh, Ir, Co, Ni and Nd. Only a few of these catalysts have been studied in detail, probably because of the low commercial interest of the polymer.

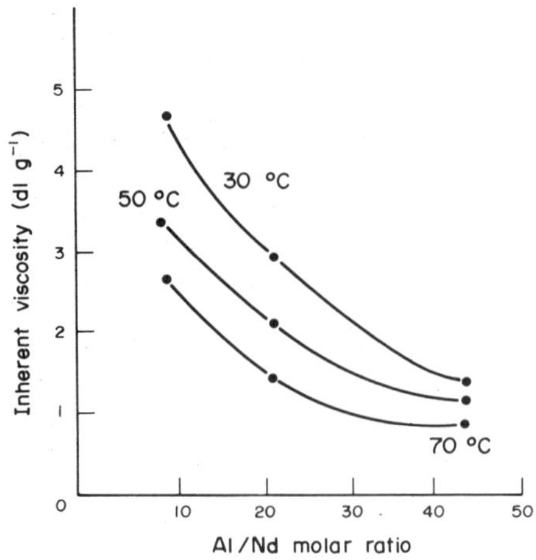


**Figure 6** Polymerization of butadiene to the *cis* polymer with Nd catalyst. Effect of Al/Nd molar ratio on polymer inherent viscosity at various temperatures[163]

*(i) Vanadium catalysts*

These catalysts are by far the most important for preparing high molecular weight *trans*-polybutadiene. At room temperature the heterogeneous system $AlEt_3$–$VCl_3$ gives products with a *trans* content of 99% minimum. These contain a very small proportion (1–2%) of low molecular weight products, soluble in diethyl ether. About 20–30% of the crude product is extractable with boiling benzene; this fraction is crystalline by X-rays and $[\eta]$ is in the range 2.5–3 dl $g^{-1}$ (in toluene at 30 °C). The residue after benzene extraction is also highly crystalline by X-rays and its insolubility is probably attributable to its high molecular weight.[81]

Catalysts derived from $VCl_4$ and $VOCl_3$, in combination with $AlR_3$, give products containing a higher proportion of diethyl ether soluble fraction, as compared with $VCl_3$ catalysts at the same polymerization temperature.[82]

Polymer yield with $VCl_3$ catalysts depends markedly on the degree of subdivision of $VCl_3$ and is generally lower than with $VCl_4$ or $VOCl_3$ catalysts. Higher yields can be obtained when $VCl_3$ is dispersed on an inert carrier such as kaolin or $SiO_2$.[83] Another method for increasing the activity of $VCl_3$ is to form this compound in a coprecipitate with $TiCl_3$ by reacting blends of $TiCl_4$ and $VCl_4$ at 170 °C with $AlEt_3$ (Al/Ti approximately 3/1).[102] Presumably, in the coprecipitate the $VCl_3$ is highly dispersed and a higher proportion is active in the polymerization.

High-*trans* polybutadiene can also be obtained with soluble vanadium catalysts. The $AlEt_2Cl$–$VCl_3 \cdot 3THF$[84] and $AlEt_2Cl$–$V(acac)_3$[86] systems give in benzene or toluene a polymer containing more than 99% *trans*-1,4 units. The polymer has lower molecular weight than that obtained with the heterogeneous systems from $VCl_3$, $VCl_4$ or $VOCl_3$. The soluble vanadium catalysts are used at temperatures $<20$ °C, because at higher temperatures the catalyst activity drops rapidly. However, in the presence of Cl donors (*e.g.* $Cl_3CCO_2H$), which reoxidize $V^{II}$ to $V^{III}$, they are active even at 80 °C.[87]

Some data concerning the polymerization of butadiene with vanadium catalysts are shown in Table 6.

*(ii) Titanium catalysts*

Crystalline *trans*-1,4-polybutadiene has been obtained with several catalysts based on titanium. These are less stereospecific than vanadium catalysts and usually give products containing a large proportion of amorphous non-*trans* polymers.

$AlEt_3$–$\alpha$-$TiCl_3$ was the first catalyst used for the synthesis of high-*trans* polybutadiene.[2] It gives products containing 20–40% of amorphous polymers with a mixed *cis–trans* structure. These can be removed by extraction with diethyl ether; the residue has a *trans* content of *ca.* 97–98% and is highly crystalline by X-rays. Catalysts from $\gamma$-$TiCl_3$ give similar results.[45,102]

The $\beta$ form of $TiCl_3$ gives a predominantly *cis* (about 60–65%) polybutadiene in combination with a small amount of $AlR_3$ (Al/Ti about 0.1) or larger amounts of $AlEt_2Cl$ (Al/Ti = 0.1 to 3–4). At an Al/Ti ratio of two, the $AlR_3$–$\beta$-$TiCl_3$ system gives a product consisting of a mixture of crystalline

**Table 6** Polymerization of Butadiene to *Trans*-1,4 Polymers by Vanadium-based Catalysts

| *Catalyst (molar ratio)* | *Solvent* | *Polymerization temperature* (°C) | *Polymer yield* ($g\,mmol^{-1}\,h^{-1}$) | *Polymer structure* (% *trans*-1,4) | *Ref.* |
|---|---|---|---|---|---|
| $AlEt_3$–$VCl_3$[a] (2:1) | Heptane | 15 | 0.14 | 99–100 | c |
| $AlEt_3$–$VCl_3$[a] (2:1) | Heptane | 60 | 1.25 | 89 | c |
| $AlEt_3$–$VCl_3$[b] (3:1) | Benzene | 60 | 0.9 | >95 | d |
| $AlEt_2(OEt)$–$VCl_3$[b] (3:1) | Benzene | 60 | 8.4 | >90 | d |
| $AlEt_3$–$VCl_4$ (1.8:1) | Heptane | 15 | 2.3 | 95 | e |
| $AlEt_2Cl$–$VCl_3 \cdot 3THF$ (50:1) | Toluene | −20 | 43 | 99–100 | f |
| $AlEt_2Cl$–$V(acac)_3$–$Cl_3CCO_2H$ (244:1:60) | Toluene | 80 | 1000 | — | g |

[a] $VCl_3$ Fluka.
[b] $VCl_3$ prepared at 80 °C from $AlEt_3$ and $VCl_4$.
[c] G. Natta, L. Porri, P. Corradini and D. Morero, *Chim. Ind. (Milan)*, 1958, **40**, 362 (*Chem. Abstr.*, 1959, **53**, 195).
[d] G. F. van Amerongen, in 'Elastomer Stereospecific Polymerization', ed. B. L. Johnson and M. Goodman, American Chemical Society, Washington, DC, 1966, p. 136.
[e] G. Natta, L. Porri and A. Mazzei, *Chim. Ind. (Milan)*, 1959, **41**, 116 (*Chem. Abstr.*, 1959, **53**, 15 619).
[f] G. Natta, L. Porri and A. Carbonaro, *Atti Accad. Naz. Lincei, Cl. Sci. Fis., Mat. Nat., Rend.*, 1961, **31** (5), 189 (*Chem. Abstr.*, 1962, **57**, 4848b).
[g] A. Carbonaro (SNAM Progetti SpA), *Ger. Pat.* 2 534 495 (1976) (*Chem. Abstr.*, 1976, **84**, 151 265).

*trans*-1,4 polymers and of amorphous polymers with a predominantly (about 70%) *cis* structure. Similar results are obtained with $AlR_3$–$TiCl_4$ at Al/Ti ratios of about two.[45]

Polymers of high *trans* content are obtained from $LiAlH_4$ and $TiI_4$ (Al/Ti = 0.9 to 3) in aromatic or aliphatic solvents, at about 50 °C.[309] The crude polymerization product has a *trans* content of 80 to 85%. It has not been reported whether it consists of macromolecules of low stereoregularity or if it is a mixture of macromolecules with different *trans* contents.

Polymerization products consisting of 93 to 94% *trans*-1,4 units, crystalline by X-rays, have been obtained with $AlEtCl_2$–$Ti(OBu)_4$, Al/Ti ≈ 10, at 30 °C in heptane.[75]

### *(iii) Cobalt catalysts*

Polybutadiene with 80 to 95% *trans* content can be obtained with the system $AlEt_2Cl$–cobalt compound–$NMe_3$ (or $NEt_3$). In the absence of the amine, the system gives high-*cis* polymers. The effect of amine is to decrease the yield of polymer, the molecular weight and the *cis* content, until at a ratio 8–9:1:10 of $NR_3$:Co:Al a *trans* polymer is produced. Methanol (about 2 mol% on $AlEt_2Cl$) acts as an activator. With a cobalt concentration of $1.9 \times 10^{-2}$ $mol\,l^{-1}$ and a butadiene concentration of about 100 $g\,l^{-1}$, a polymer conversion of 80% was obtained in 30 h at 20 °C, after an induction period of 6 h. Small variations in catalyst composition and concentration result in some change in structure, from 70 to 95% *trans*, the remainder being almost exclusively 1,2 units.[120]

### *(iv) Rhodium catalysts*

Simple salts of rhodium, such as $Rh(NO_3)_3 \cdot 2H_2O$ or $RhCl_3 \cdot 3H_2O$, in aqueous or alcoholic solution, are capable of polymerizing butadiene to *trans* polymer. No alkylating cocatalyst (*e.g.* aluminum alkyl, lithium alkyl) is necessary.[21,22] The nitrate is effective in water or alcohol, but $RhCl_3$ requires an emulsifier (usually dodecyl sulfate or sodium dodecylbenzenesulfonate). The polymerization rate depends on the nature of the rhodium salt, on the type of emulsifier and on pH. The polymers obtained are highly stereoregular (99% *trans* minimum) and have a melting point of 145 °C. Molecular weights are rather low; $[\eta]$ is 0.37–0.5 $dl\,g^{-1}$ in aqueous systems, whereas it is 0.14 to 0.16 in methanol or ethanol.

Addition of cyclohexadiene (CHD) to the rhodium salt (CHD/Rh molar ratio 16–20:1) increases the rate of polymerization. CHD probably reduces the valence of the metal, thus favoring the formation of the active species.[192,194,310]

Catalysts from rhodium salts and aluminum alkyls have been examined. They have moderate activity and give polymers with low molecular weight ($[\eta] = 0.1$–$0.2$).[311]

#### *5.4.1.3 1,2-Polybutadiene*

Three types of butadiene polymers with a 1,2 structure have been prepared with transition metal catalysts: (a) crystalline syndiotactic polymers; (b) crystalline isotactic polymers; and (c) amorphous polymers, some of which most probably are not completely atactic.

Several catalyst systems have been found for the syndiotactic polymer, but only chromium catalysts have given the isotactic stereoisomer. The crystalline 1,2 polymers have not become of commercial interest and their synthesis has not been intensively studied.

##### *(i) Syndiotactic polymer*

This polymer has been obtained with catalysts based on Ti, V, Cr and Co (Table 7). Normally alcoholates or acetylacetonates are used, but other compounds (complexes with carbon monoxide, with nitriles, isonitriles, pyridine or other Lewis bases) give, in combination with an aluminum trialkyl, seemingly homogeneous catalysts effective for the preparation of syndiotactic 1,2-polybutadiene.

**Table 7** Catalysts for the Preparation of 1,2 Syndiotactic Polybutadiene

| *Catalyst system type* | *Ageing time* | *Syndiotactic polymer* % *in crude polymer* | *m.p.* (°C) | dl $g^{-1}$ |
|---|---|---|---|---|
| $AlR_3$–$Ti(OR)_4$ (*ca.* 7:1) in heptane at 0–15 °C | 2–3 | — | — | — |
| $AlH_2NMe_2$–$Ti(NEt_2)_4$ (8:1) in benzene at 15 °C | 15–20 | 50–55 | — | 1.5 |
| $AlEt_3$–$V(acac)_3$ in benzene at 15 °C | | | | |
| Al/V = 6 | 180 | 40 | 156 | 1.8–2 |
| Al/V = 10 | 180 | 60 | 156 | 1.8–2 |
| $AlR_3$–$Mo(acac)_3$ (4–5:1) in benzene at 15 °C | 2–3 | 80 | — | — |
| $AlR_3$–$Co(acac)_3$–$CS_2$ (50:1) in benzene at 40 °C | 2–3 | 99–100 | 206–216 | 2.7 |
| $(\eta^3\text{-}C_8H_{13})Co(C_4H_6)$–$CS_2$ in benzene | | 99–100 | 200–205 | 1.7 |
| $AlEt_3$–$Co(acac)_3$–fumaric acid in benzene at 16 °C (Al/Co = 1) | — | — | 150 | — |
| $Al(Bu^i)_3$–$CoBr_2 \cdot 2PPh_3$–$H_2O$ | — | — | 75–90 | — |
| $(Et_2Al)_2SO_4$–$Co(SCN)_2 \cdot 2PPh_3$ in toluene at 80 °C (Al/Co = 2–4) | — | — | 105–115 | — |

The $AlEt_3$–$Ti(OR)_4$ system (Al/Ti $\approx$ 7) was the first catalyst used for the preparation of syndiotactic polybutadiene.[3] It gives a polymer of moderate crystallinity which is partially crosslinked, especially when obtained at high monomer conversion.[70] Gel free polymers with a good degree of crystallinity have been obtained with catalysts based on $Ti(NEt_2)_4$, in combination with $AlR_3$ or aluminum hydrides. A typical system of this class is $AlH_2NMe_2$–$Ti(NEt_2)_4$, with an Al/Ti ratio of 6–20 in benzene.[312]

The $AlEt_3$–$V(acac)_3$ system gives polymerization products consisting of 80–86% 1,2 structure. These can be separated by successive extractions with solvents into amorphous or slightly crystalline polymers (soluble in boiling diethyl ether) and highly crystalline polymers, with 1,2 content of 95 to 97%, melting point 155 °C (residue after diethyl ether extraction). Both the Al/V ratio and catalyst ageing influence the stereospecificity.[313]

Catalysts showing a higher degree of stereospecificity can be obtained from $AlR_3$ and various Mo compounds, such as $MoO_2(OR)_2$, $MoO_2(acac)_2$, $Mo(acac)_3$ and $MoCl_5$. The crystalline fraction (residue after ether extraction) is about 80%.[48]

Several catalysts based on cobalt have been proposed. $AlR_3$ in combination with various cobalt compounds was reported to give high yields of syndiotactic polybutadiene.[126] In a variant of this system, the reaction product of $AlEt_3$ with $H_2O$ (Al:$H_2O$ = 1:0.5) or with a primary amine such as aniline (Al:amine = 1:0.5) was used as cocatalyst. The compounds $Et_2AlOAlEt_2$ and $Et_2AlN(Ph)AlEt_2$ are presumably formed in these reactions.[113] According to some authors, $H_2O$ is essential to the activity of the $AlR_3$–Co compound systems.[314]

Syndiotactic polybutadiene with a 1,2 content > 99% and a crystallinity of about 79% has been obtained with $AlR_3$–$Co(acac)_2$–$CS_2$ (66:1:2.9). The melting point of the polymer is 210 to 216 °C. The presence of $CS_2$ is essential for the stereospecificity. The $AlR_3$–$Co(acac)_2$–$CS_2$ system is by far the most active and stereospecific of those so far reported for the preparation of syndiotactic polybutadiene.

The simple allyl complex $(\eta^3\text{-}C_8H_{13})Co(C_4H_6)$ also gives, in conjuction with $CS_2$, a catalyst effective for the preparation of a highly stereoregular polymer, with a melting point of 210–216 °C.[127–129,315]

Several cobalt catalysts that give a polymer of lower stereoregularity have been reported. The $AlEt_3$–diethyl fumarate–$Co(acac)_3$ system (1/0.33/1) gives a polymer with m.p. about 150 °C.[316–318] A polymer with m.p. 105–110 °C, 1,2-content about 95% and $[\eta] = 0.54\ dl\,g^{-1}$ was obtained with $(Et_2Al)_2SO_4$–$Co(SCN)_2 \cdot 2PPh_3$ in toluene at 80 °C,[319] while a less crystalline polymer (m.p. 75–90 °C) was obtained with $CoBr_2 \cdot 2PPh_3$–$Al(Bu^i)_3$–$H_2O$.[311] This polymer was developed as a thermoplastic resin in Japan. It appears from the above that the catalysts from cobalt can afford syndiotactic polybutadiene with different degrees of stereoregularity depending on the particular catalyst composition used.

*(ii) Isotactic polymer*

This isomer has been obtained with catalysts prepared from aluminum trialkyls and soluble Cr compounds, such as $Cr(acac)_3$, $Cr(CNPh)_6$, $Cr(CO)_6$ and $Cr(CO)_3py_3$. These apparently homogeneous systems give either syndiotactic or isotactic polymers or a mixture of them depending on the preparation conditions of the catalysts. At low Al/Cr ratios the unaged catalysts give the syndiotactic polymer or mixtures of predominantly syndiotactic and isotactic polymer. Upon increasing the Al/Cr ratio and/or ageing the catalyst, the proportion of isotactic polymer increases until it becomes the only crystalline isomer.[88]

The crystalline polymers are always obtained in a mixture with amorphous polymers, which have 80 to 85% 1,2 content. The proportion of isotactic polymer in the crude products depends on the catalyst used, but is generally low: only 2–2.5% with catalysts from $Cr(acac)_3$, but 40–45% with catalysts from $Cr(CO)_3py_3$ or $Cr(CNPh)_6$ (Table 8).

**Table 8** Polymerization of Butadiene by Catalysts from $AlEt_3$ and Cr Compounds[a,b]

| *Cr compound* | *Al/Cl molar ratio* | *Time of ageing of catalyst* (min) | *Polymer* *g* | *Diethyl ether insoluble fraction* % | *X-ray examination* |
|---|---|---|---|---|---|
| $Cr(CNPh)_6$ | 1.49:2 | 2 | 0.3 | — | Syndio |
| | 1.49:5 | 2 | 14.5 | 14 | Syndio(> 50%) + iso |
| | 1.49:5 | 240 | 9.7 | 32.5 | Iso |
| | 0.745:7.5 | 2 | 6.4 | 15.3 | Syndio(> 50%) + iso |
| | 0.745:7.5 | 120 | 5.15 | 25 | Iso |
| | 0.745:15 | 2 | 7.2 | 13.5 | Iso(> 50%) + syndio |
| | 0.745:15 | 120 | 6.6 | 24 | Iso |
| | 0.745:15 | 240 | 6.8 | 34 | Iso |
| $Cr(CO)_5py$ | 1.84:3 | 2 | 5.4 | 31 | Syndio + iso |
| | 1.84:5 | 2 | 5.75 | 15 | Iso |
| $Cr(acac)_3$ | 1.42:3 | 2 | 6 | 15 | Syndio |
| | 1.42:10 | 2 | 9.85 | 2 | Iso |

[a] Benzene, 80 g; monomer, 30 g; 16 h at 15 °C.
[b] G. Natta, L. Porri, G. Zanini and A. Palvarini, *Chim. Ind. (Milan)*, 1959, **41**, 1163 (*Chem. Abstr.*, 1961, **55**, 20 906).

The behavior of Cr catalysts indicates that the reaction betwen $AlR_3$ and the Cr compounds proceeds stepwise, leading to the formation of complexes of different stereospecificity. First a complex is formed which is stereospecific for the synthesis of the syndiotactic polymer. On further reaction with $AlR_3$, this complex is transformed into a new one, which gives the isotactic polymer.

Although pure isotactic polybutadiene has been obtained only with Cr catalysts, the presence of this polymer in very small amounts has been detected by X-ray analysis also in the polymerization products obtained with $V(acac)_3$ or $Mo(acac)_3$ systems, when $AlR_2H$ or $AlR_3$ containing $AlR_2H$ are used as cocatalyst.[88] It is likely that V or Mo compounds require strong reducing agents for the formation of catalysts for isotactic polybutadiene.

### 5.4.1.4 Other butadiene polymers

Polymers of butadiene consisting almost exclusively of *cis* and *trans* units in practically 1:1 ratio have been synthesized using as catalyst allylnickel trifluoroacetate in the presence of various ligands such as $CF_3CO_2H$, benzene and *o*-dichlorobenzene.[234,241,245,267–269] Allylnickel *p*-toluenesulfonate also gives equibinary *cis/trans* polybutadiene.[24] The polymer molecular weight can vary over a wide range, from a few thousands to about 70 000, depending on the experimental conditions.[269]

NMR studies on the distribution of *cis* and *trans* units along the chain were carried out by various authors.[321–325] These studies have shown that the polymers do not possess a regular alternating *cis/trans* structure, which is consistent with the finding that the polymers are amorphous by X-ray. Some effect of the solvent and other experimental conditions on the distribution of *cis* and *trans* units was observed.[321]

Another type of equibinary polybutadiene, consisting of approximately the same amount of *cis* and 1,2 structures and of almost no *trans* units, has been prepared by various catalyst systems: $AlEt_3$–$Co(acac)_3$–$H_2O$,[132] $AlR_3$–$MoCl_3(OR)_2$ at an Al/Mo molar ratio of about 10,[326] $AlEt_3$–$MoCl_5$ at an Al/Mo molar ratio between two and ten,[326] $AlEt_2Cl$–$MoO_2(acac)_2$ at Al/Mo $>9$,[326] and $Al(Bu^i)_3$–$Fe(acac)_3$–1,10-phenanthroline.[327] $^{13}C$ NMR analysis of the polymers showed that *cis* and 1,2 units are randomly distributed along the chains.[326–328]

### 5.4.2 Isoprene Polymers

Two highly stereoregular polymers have been obtained from isoprene with transition metal catalysts: a polymer consisting of 99–100% *trans*-1,4 units, of structure equal to that of balata or gutta percha, and a polymer consisting of *ca.* 97% *cis* units, very similar to natural rubber. Various polymers of mixed *cis*-3,4 structure, amorphous by X-ray, were also obtained.

#### *5.4.2.1* cis-*1,4-Polyisoprene*

Two families of catalysts are now available for the synthesis of *cis*-1,4-polyisoprene, one based on titanium and the other based on neodymium. The $AlR_3$–$TiCl_4$ system was the first used, in 1954, for the polymerization of isoprene to a high-*cis* polymer. Several variants of this system have been proposed since. Neodymium catalysts have been developed more recently and have been less studied, but they have activity and stereospecificity comparable with those based on Ti.

##### *(i)* $AlR_3$–$TiCl_4$ *system*

This is the system most intensively studied. The influence on the activity of the mode of catalyst preparation and the effect of polymerization variables have been thoroughly investigated.

*(a) Mode of catalyst preparation.* The following variables have been examined: (a) influence of the monomer on the catalyst-forming reaction (preformed catalysts *vs.* '*in situ*' catalysts); (b) influence of Al/Ti ratio; (c) influence of reaction temperature; (d) influence of the R group in the $AlR_3$ compound; (e) effect of catalyst ageing; and (f) effect of electron donors.[329]

Preformed catalysts from $AlR_3$ and $TiCl_4$ are more active and give more reproducible results than the optimum catalysts prepared '*in situ*'. Catalysts are preferably prepared by adding $AlR_3$ slowly to the stirred $TiCl_4$ solution in aliphatic hydrocarbon, but simultaneous introduction of the two reagents in the solvent at a constant Al/Ti ratio of 1.0 gives equivalent results.[47]

The $AlR_3/TiCl_4$ molar ratio is the most important variable in $TiCl_4$ catalysts. Figure 7 shows the effect of the Al/Ti ratio on conversion. The maximum activity and stereospecificity for the *cis*-1,4 polymerization of isoprene is obtained at an Al/Ti ratio close to unity. The polymer obtained under these conditions is 96–97% *cis*-1,4. At lower ratios, the yield decreases, until at ratios of 0.3–0.5 the product is a crosslinked powdery substance, probably attributable to cationic type polymerization. At Al/Ti ratios higher than unity the yield of solid polymer (still 96–97% *cis*) decreases rapidly and increasing amounts of oil (extractable) are formed. The oils, which have an odor characteristic of terpenes, are linear or cyclic oligomers of isoprene, some of which have been identified.[27,330–332]

It appears that $AlR_3$–$TiCl_4$ is highly active and stereospecific for the *cis* polymerization of isoprene over only a narrow Al/Ti ratio range. This range corresponds to the formation of $\beta$-$TiCl_3$.

The temperature of preparation of the catalyst has some influence on the activity. The catalysts prepared at low temperature ($-20\,^{\circ}C$ or below) are the most active. Increasing preparation temperature results in a gradual decrease in activity.[47] Figure 8 shows conversion/time plots for catalysts prepared at three different temperatures ($-20$, $+20$, $+40\,^{\circ}C$) and aged for four months at room temperature. Catalysts prepared at $+125\,^{\circ}C$ or above are practically inactive for the formation of *cis*-polyisoprene.

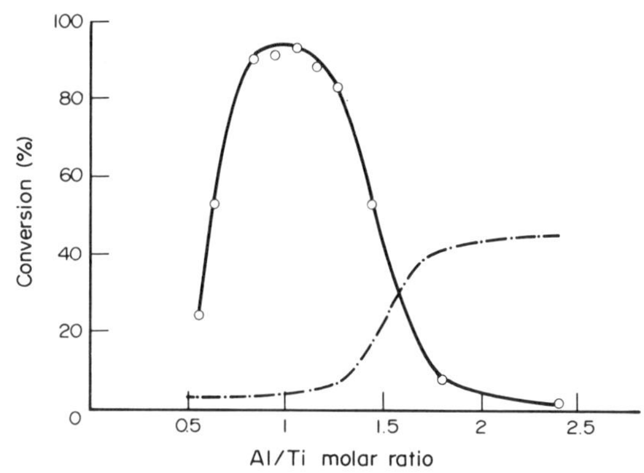


**Figure 7** Polymerization of isoprene to the *cis* polymer with $AlR_3$–$TiCl_4$. Monomer conversion to polymer as a function of Al/Ti molar ratio[47] (——— solid polymer; –·–·–· oligomers)

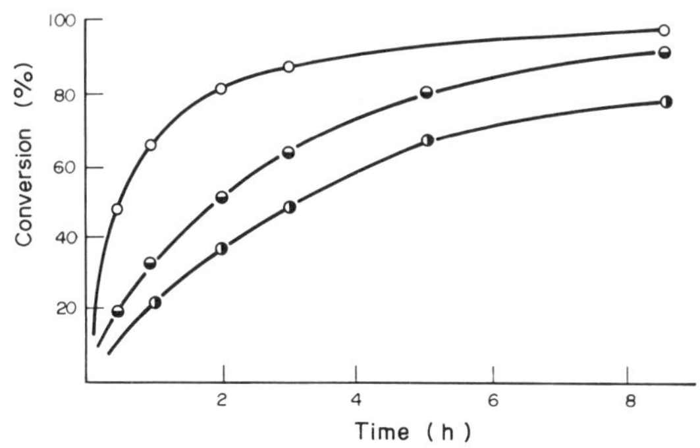


**Figure 8** Polymerization of isoprene to the *cis* polymer with $Al(Bu^i)_3$–$TiCl_4$. Effect of catalyst preparation temperature on monomer conversion. Catalyst preparation temperature (°C): ○, – 20; ○, 20; ●, 40. Al/Ti = 1.0 (ref. 47)

With increasing temperature, the Al/Ti ratio at which optimum polyisoprene conversion is obtained shifts gradually to lower values, *e.g.* for catalysts prepared at 90 °C, the optimum Al/Ti ratio is 0.6–0.7.[47] This effect can be interpreted by considering that the optimum Al/Ti ratio corresponds to the formation of $\beta$-$TiCl_3$. This ratio is lower than unity when the reaction is carried out at 80–90 °C, because at these temperatures $AlR_2Cl$ participates in the reduction of $TiCl_4$. The temperature of catalyst preparation also influences some polymer properties (Table 9). Catalysts

**Table 9** Properties of *cis*-1,4-Polyisoprene Obtained with Catalysts Prepared by Adding $AlR_3$ to $TiCl_4$ at Different Temperatures[a]

| *Catalyst preparation temperature* (°C) | *Al/Ti* | *Inherent viscosity* ($dl\,g^{-1}$) | *Gel* (%) | *cis-1,4* (%) |
|---|---|---|---|---|
| – 20 | 0.98 | 1.8 | 31 | 96.9 |
| – 20 | 1.08 | 2.5 | 9 | |
| 0 | 0.98 | 2.8 | | 97.4 |
| 40 | 0.80 | 2.3 | 9 | 95.5 |
| 40 | 0.90 | 2.1 | 6 | 96.5 |
| 60 | 0.80 | 3.1 | 10 | 96.4 |
| 60 | 1.00 | 3.0 | 7 | 96.9 |
| 75 | 0.81 | 3.1 | 9 | 94.6 |
| 90 | 0.59 | 5.0 | 23 | 96.4 |
| 90 | 0.68 | 5.7 | 12 | 95.0 |

[a] E. Schoenberg, D. L. Chalfant and R. H. Mayor, *Rubber Chem. Technol.*, 1964, **37**, 103.

prepared at higher temperature give polymers of increased intrinsic viscosity, while the microstructure remains unchanged. In addition, these catalysts give products containing only a very small amount of oils.

The nature of the aluminum alkyl has some influence on catalyst activity. Catalysts prepared at 20 °C from various $AlR_3$, where R was a normal or branched alkyl group from $C_1$ to $C_8$, were examined. The optimum Al/Ti ratio was 0.9–1 for all the catalysts, except for $AlMe_3$, which requires a slightly higher Al/Ti ratio. The activity increased significantly with the length of the alkyl group, while microstructure and other properties of the polymer prepared with the various aluminum alkyls were similar.[44, 333, 334]

The effect of catalyst ageing was examined by several authors, but conflicting results were reported.[44, 330, 335] It seems, however, that aged catalysts prepared from $AlR_3$ with $C_4$ or higher alkyls have both improved activity and stereospecificity and give polymerization products having a reduced amount of oily oligomers.[44, 334, 336] Inherent viscosities of the resultant polymers were also moderately increased.[44]

Some electron donors, such as ethers and amines, have beneficial effects on the polymerization of isoprene with $AlR_3$–$TiCl_4$. Ethers or amines, at low levels (< 0.7–0.9 mol per mol $AlR_3$) reduce the amount of gel in the polymer. Probably their effect is attributable to the suppression of some cationic activity of the catalyst system.[329]

Particularly interesting are the catalysts prepared from $AlR_3$ complexed with aromatic ethers.[337, 338] These systems have higher activity than the corresponding non-etherate systems and give polymers with a reduced content of gel and of isoprene oligomers. The polyisoprene obtained with systems from $Ph_2O \cdot AlR_3$ was reported to give vulcanizates with improved properties with respect to non-etherate systems.[338] $AlR_3$ complexed with aromatic ethers were reported to have a greater alkylating or reducing power than the non-complexed aluminum alkyls, but the mode by which aromatic ethers affect polymerization remains obscure.[337, 339–341]

Addition of $CS_2$ to $AlR_3$–$TiCl_4$ ($CS_2$/Ti ≈ 0.1) suppresses the formation of isoprene oligomers and enhances the formation of *cis*-polyisoprene. Carbon disulfide is the most effective suppressant of oligomer formation of those so far proposed.[332] Addition of aldehydes is also reported to increase catalyst activity and to reduce oligomer formation.[342]

*(b) Effect of polymerization variables.* Polymerization temperature and catalyst level are the two most important polymerization variables. Molecular weight is slightly influenced by catalyst concentration.[80, 338] This behavior is characteristic of polymerizations by heterogeneous catalysts. The polymerization temperature influences the polymer microstructure only slightly in the range − 30 to + 70 °C, but strongly affects molecular weight. An approximate doubling of DSV is obtained by altering the polymerization temperature from − 50 to 0 °C.[329–331]

Monomer conversion is directly related to catalyst level, the other conditions being constant. At fixed Ti concentration, the molecular weight increases rapidly at the beginning of the polymerization, reaches a maximum at 30–50% conversion, then remains practically constant as the polymerization proceeds.[33, 335, 338] At conversion above 80%, a drop in $M_v$ is observed, whereas $M_n$ does not change.[338]

The molecular weight distribution of the *cis*-polyisoprene obtained with $AlR_3$–$TiCl_4$ was found to be characterized by symmetry and a comparatively small dispersion.[336, 343, 344]

Data on the influence of the solvent have been reported. Aliphatic solvents are preferred, although aromatic solvents give products with a slightly lower amount of gel.[331]

### *(ii) Catalysts based on $Ti(OR)_4$*

Catalyst systems effective for the polymerization of isoprene to the *cis* polymer can be prepared from $Ti(OBu^n)_4$ and various aluminum alkyl chlorides ($AlEt_2Cl$, $AlEt_2Cl + AlCl_3$ (or $AlBr_3$), $Al_2Et_3Cl_3$, $AlEtCl_2$), at an Al/Ti ratio of 4–5.5.[75–78, 80, 345] At room temperature β-$TiCl_3$ is formed, which is responsible for the stereospecificity of the catalyst. Addition of amines, such as $N(Bu^n)_3$ (N/Ti about 0.6), reduces the cationic activity. Aliphatic or aromatic solvents can be used.

The stereospecificity of the system $Al(Bu^i)Cl_2$–$Ti(OR)_4$ changes when R is an iso- or tertiary alkyl group. With $Al(Bu^i)Cl_2$–$Ti(OPr^i)_4$, *cis*-polyisoprene is obtained in pentane, whereas *trans*-polyisoprene is obtained in benzene. Catalysts from $Ti(OBu^t)_4$ give *trans* polymers either in benzene or heptane. The difference in stereospecificity was attributed to the fact that, depending on the structure of $Ti(OR)_4$ and on the nature of the solvent, different modifications of $TiCl_3$ are formed (β-$TiCl_3$ or violet modification).[80, 345]

*(iii) Alane catalysts*

An important group of catalysts for the *cis* polymerization of isoprene is based on alanes and $TiCl_4$.[49–51] Several alanes can be used, but the most active and stereospecific systems are those prepared from poly(*N*-alkyliminoalane) (PIA).[50] Particularly interesting results have been obtained with PIA having an alkyl group branched in the positions $\alpha$ or $\beta$ to the nitrogen, *e.g.* that derived from isopropylamine, $(HAlN(Pr^i))_6$. While alanes such as $AlH_2Cl \cdot NMe_3$ or $AlH_2NMe_2$ give catalysts that require an aromatic solvent, the catalysts from PIA give good results also in aliphatic solvents.

The PIA catalysts show the following features, with respect to those from $AlR_3$: (1) they give polymers with a higher molecular weight, probably attributable to the absence of transfer reactions between the growing chain and the cocatalyst; and (2) they give rise to negligible branching or crosslinking reactions, thus allowing the formation of linear polymers, practically free of gel. Branching and crosslinking reactions occur with conventional $AlR_3$–$TiCl_4$ systems, because of the cationic action of soluble reaction products, such as $AlRCl_2$. In the case of PIA systems, the chlorinated species formed in the reaction with $TiCl_4$ have a lower acidity and lower solubility, thus preventing cationic type reactions. As a consequence, PIA systems do not need an aromatic ether or other chemicals to reduce the tendency of gel formation.[50,51]

*(iv) Lanthanide catalysts*

Several catalysts based on lanthanides can polymerize isoprene to high-*cis* polymer. Catalysts from Ce, Th and U are reported in some patents and papers. However, the catalysts based on Nd are the most interesting and so far the only ones of practical value for the reasons indicated in Section 5.3.2.5. The polymer *cis* content obtained is usually $> 95\%$.[158,161,163] Some dependence of the *cis* content on the polymerization temperature has been observed, polymers with a higher *cis* content being obtained at low temperature.

$^{13}C$ NMR analysis shows that neodymium *cis*-polyisoprene has a regular head-to-tail structure of the monomeric units. The polymers are free of gel and of oily oligomers, which is probably related to the absence of cationic activity in Nd catalysts.

Molecular weight can be varied within a wide range and can be as high as $10^7$. The $AlR_3/Nd$ ratio is the most important parameter for regulating molecular weight. Molecular weight distribution is usually broad. The amount of catalyst normally used is about 0.1–0.6 mmol of Nd per 100 g of monomer, with Al/Nd ratio 10–40.[163]

Physical and mechanical properties of vulcanizates from neodymium polyisoprene are reported to be practically equal to those of titanium polyisoprene.[163]

### 5.4.2.2 *trans-1,4-Polyisoprene*

The synthesis of a polyisoprene with practically 98–100% *trans*-1,4 structure was reported in 1955, using $AlEt_3$–$\alpha$-$TiCl_3$ as catalyst.[2] The crystalline polymers were obtained along with amorphous products, which were removed by extraction with diethyl ether.

The synthetic polymer of high-*trans* content is very similar in its properties to the natural polymers balata and gutta percha. The IR, X-ray and NMR spectra of the synthetic and natural polymers are identical. The molecular weight of the synthetic polymer is normally higher than those of the natural ones.

Several systems can give a high-*trans* polyisoprene, but those based on vanadium are the most active and stereospecific.

*(i) Vanadium catalysts*

All vanadium halides give, in combination with $AlR_3$, high-*trans* polyisoprene, but $VCl_3$ gives the most stereoregular polymer. Polymerization conditions are not critical. The $AlR_3/VCl_3$ ratio can be varied within a wide range, 3–6 being the preferred range. Aliphatic hydrocarbons are used as solvent and the polymerization temperature is 20–50 °C. Above 50 °C the polymerization rate drops rapidly.[16,81] Since $VCl_2$ is inactive for the polymerization, the active species in the $AlR_3$–$VCl_3$ system is presumably an allylvanadium dichloride, $Cl_2V$(allyl), complexed with the aluminum alkyl. Vanadium–carbon bonds are unstable at high temperature and this accounts for the rapid drop of activity at temperatures higher than 50 °C.

The unmodified $AlR_3$–$VCl_3$ system gives rather poor polymer yields.[45,81] These can be markedly increased by addition of an ether or by depositing $VCl_3$ on an inert support such as kaolin or titanium dioxide. Addition of diisopropyl ether [$O(Pr^i)_2/V = 5:1$ at $Al/V = 2:1$] produces a fourfold increase in polymer yield, while it has no effect on polymer microstructure. The ether causes an increase in the number of active centers, which is probably attributable to the fact that it facilitates the cleavage of the $VCl_3$ crystals at lattice defects.[16,346,347]

Deposition of $VCl_3$ on an inert support is even more effective than diisopropyl ether in increasing polymer yields. Usually the deposition is made by suspending the support in a heptane solution of $VCl_4$ and by irradiating with UV or by heating the suspension to decompose $VCl_4$ to $VCl_3$.[16,83]

The heterogeneous systems from $VCl_4$ or $VOCl_3$ and $AlR_3$ give products that are usually over 90% *trans*, but the polymers are of lower molecular weight and of lower stereoregularity than those obtained with the $VCl_3$ catalyst.[16,83,348,349]

*(ii) Mixed vanadium–titanium catalysts*

These catalysts are the most active for the *trans* polymerization of isoprene. Two types of systems have been proposed, one heterogeneous, in which a blend of $TiCl_3$ and $VCl_3$ is used,[102] and the other homogeneous, resulting from the reaction between $Ti(OR)_4$, $AlR_3$ and $VCl_3$, preferably dispersed on a support.[83]

The heterogeneous system is prepared by reacting $AlEt_3$ at 100–170 °C with a solution of $VCl_4$ and $TiCl_4$ in liquid paraffin ($Ti/V = 4:1$; $Al/V = 1:3$) and by activating the mixed chlorides so produced with further addition of $AlEt_3$ to give a final $Al/(Ti+V)$ ratio of about five.[102]

In a variant of this system, $VCl_3$ deposited on $TiCl_3$ is used. $AlEt_3$ is first reacted with $TiCl_4$ at 170 °C in liquid paraffin to give $\gamma$-$TiCl_3$, then $VCl_4$ is added, which is decomposed to $VCl_3$ by heating or by reaction with $AlEt_3$. The $VCl_3$–$TiCl_3$ mixed chlorides are separated from the liquid paraffin and activated with $AlEt_3$ ($Al/(Ti+V) = 5$–$10$).[16,350] The $VCl_3$–$TiCl_3$ systems are more active than those from $VCl_3$ alone by a factor of about 50. Polymer molecular weights are rather high ($[\eta] = 4$–$5\ dl\ g^{-1}$). One role of $TiCl_3$ in these systems is to disperse $VCl_3$, thus increasing the proportion of $VCl_3$ active in the polymerization.[102] Presumably, $TiCl_3$ gives little contribution to polymer formation, in consideration of the fact that catalysts from $\gamma$-$TiCl_3$ alone are much less active than those from $VCl_3$ in the *trans* polymerization of isoprene.[45]

The Ti–V homogeneous systems have been reported to be more active than the heterogeneous ones. $Ti(OR)_4$ of different types have been used ($R = Bu^n$, 2-ethylhexyl, $Bu^t$), usually in combination with $Al(Bu^i)_3$ as the aluminum alkyl. Supported $VCl_3$ was reported to give efficient catalysts,[83] but unsupported $VCl_3$ or even $VCl_4$ are reported in some patents.[351,352] Typical compositions are: $Al:V:Ti = 20:2:1$ (molar ratio). Hydrogen, $ZnEt_2$ or $CdEt_2$ have been proposed as molecular weight regulators.

The nature of these catalysts is not clear, but probably their high activity is attributable to the fact that a high fraction of vanadium is active in the polymerization.

*(iii) Titanium catalysts*

Catalysts prepared from violet $TiCl_3$ and $AlR_3$ give polymerization products (*ca.* 85% *trans*-1,4) that contain crystalline *trans* polymers in mixtures with products of lower stereoregularity.[45] With these catalysts, the rate of polymerization and polymer yields are rather low. Deposition of $TiCl_3$ on a support and addition of an ether increase markedly the rate, which, however, remains much lower than with vanadium catalysts.[353]

In some early patents it was reported that high-*trans* polyisoprene can be obtained with $AlR_3$–$TiCl_4$, at $Al/Ti = 0.5$. The product obtained under these conditions is a resinous material which has IR absorptions corresponding to *trans*-1,4 units, but it is dissimilar from balata or gutta percha. These polymers are highly branched and contain gel as a result of the cationic activity of the catalyst system.[16]

*trans*-Polyisoprene can be obtained with $AlRCl_2$–$Ti(OBu^t)_4$ in aromatic or aliphatic solvents and with $AlRCl_2$–$Ti(OPr^i)_4$ in aromatic solvents. There is little information on the properties of the polymer obtained with these systems.[80,345]

#### 5.4.2.3 Other isoprene polymers

Polymers of isoprene consisting of *cis*-1,4 and -3,4 units in various proportions have been obtained with several catalysts. The aluminum alkyl chloride–cobalt compound system gives polymers containing from 70 to 80% *cis*-1,4 units, depending on the type of the aluminum alkyl, the solvent and the amount of water.

$AlR_3$–$Ti(OR)_4$ (Al/Ti = 7 to 8) gives polymers consisting predominantly of 3,4 units, from 60 to about 70%, depending on the polymerization temperature. The polymer obtained in toluene at 0 °C consists of 64% 3,4 and 36% *cis* units, as determined by $^{13}C$ and $^{1}H$ NMR.[354] All these polymers are amorphous by X-ray at room temperature.

Two types of equibinary polyisoprene were reported. Polymers consisting of equal amounts of *cis*-1,4 and -3,4 units were obtained with PhMgBr–$CoX_2$–MeOH (X = F, Cl, Br, I, acac; the maximum yields were obtained with X = I or F)[267,355,356] and $Et_2AlOEt$–$CoX_2$ (X = I, Br, Cl, F).[357] Polymers consisting of equal amounts of 1,2 and 3,4 units were obtained with PhMgBr–$CoF_2$–HMPA.[356] All the above polymers are amorphous by X-ray.

### 5.4.3 1,3-Pentadiene Polymers

The polymerization of this monomer is of particular interest due to the complex phenomena of isomerism shown by its polymers. 1,3-Pentadiene exists in two isomeric forms, (*E*) and (*Z*). Both isomers are polymerized by some catalysts, whereas with others only the (*E*) isomer is polymerized. Of the various stereoregular polymers that can be predicted for 1,3-pentadiene, only four have been prepared so far: *trans*-1,4 isotactic, *cis*-1,4 isotactic, *cis*-1,4 syndiotactic and 1,2 syndiotactic. In addition to these stereoregular polymers, various types of amorphous polymers have been obtained, consisting of different monomeric units in variable proportions.

#### *5.4.3.1* Trans-*1,4 isotactic polypentadiene*

This polymer was first obtained with $AlEt_3$–α-$TiCl_3$ in heptane at about 60 °C, using a mixture of (*E*) and (*Z*) isomers as the monomers.[2] The crude product consisted of a mixture of amorphous polymers, soluble in $Et_2O$, and of crystalline polymers, insoluble in $Et_2O$. Much more active and stereospecific is the heterogeneous system $AlEt_3$–$VCl_3$, which gives polymerization products containing only a small proportion of amorphous fraction, either from the (*E*) or the (*Z*) isomer.[81,358] When mixtures of the two isomers are used, the product contains a higher percentage of amorphous fraction. In addition, the crystalline fraction has a lower melting point than that from the pure isomers (89 *vs.* 95 °C), an indication of a lower stereoregularity. Polymer yields with the $VCl_3$ catalyst are rather low, but since the polymer has no practical interest, little work has been done to improve its preparation (Table 10).

**Table 10** Polymerization of 1,3-Pentadiene to *Trans*-1,4 Isotactic Polymer with $AlEt_3$–$VCl_3$[a,b]

| *Monomer* (*E*) *isomer* (%) | *Polymerization*[c] *Time* (h) | *Conversion* (%) | *Crystalline fraction*[d] (%) | *m.p.* (°C) |
|---|---|---|---|---|
| 100 | 14 | 45.5 | 87 | 95 |
| 100 | 70 | 75.5 | 85 | 95 |
| 75 | 14.3 | 25 | 66 | 92 |
| 50 | 16 | 32.2 | 64 | 89–90 |
| 28 | 14.3 | 25.1 | 69 | 89 |
| 0 | 16 | 35.1 | 94 | 95 |
| 0 | 76 | 59.2 | 92 | 95 |

[a] Percentage and melting point of the crystalline fraction given as a function of the isomer monomer composition.
[b] G. Natta, L. Porri, P. Corradini, G. Zanini and F. Ciampelli, *J. Polym. Sci.*, 1961, **51**, 463.
[c] $VCl_3$ 0.3 g; Al/V = 2.5; monomer 4 ml; heptane 15 ml; 18 °C.
[d] Residue after extraction with boiling diethyl ether; about 95% soluble in boiling benzene; 1.3–1.5 dl $g^{-1}$ (in toluene at 30 °C).

In addition to $AlR_3$–$VCl_3$, which is the most stereospecific system so far known for the synthesis of *trans*-1,4 isotactic polypentadiene, other heterogeneous catalysts were reported which give this polymer, *e.g.* $AlR_3$–$VCl_4$, $AlR_3$–$TiCl_4$[117] and $AlR_3$–$TiCl_4$–$VCl_4$.[359] At present, no homogeneous catalyst based on transition metals is known which will give *trans*-1,4 isotactic polymers from 1,3-pentadiene. The homogeneous systems that give crystalline high-*trans* polymers from butadiene normally do not give crystalline high-*trans* polymers from pentadiene. Thus, the systems from

$AlEt_2Cl$ and $VCl_3 \cdot 3THF$ or $V(acac)_3$, which give 99–100% *trans*-polybutadiene,[84] give from (*E*)-pentadiene amorphous polymers consisting of 55–60% *trans*-1,4 and 40–45% 1,2 units.[360] The homogeneous system $AlEt_3$–$Ti(OR)_4$–$VCl_3$, which was reported to give crystalline *trans*-1,4 polymers from isoprene[83] and 2-methyl-1,3-pentadiene,[361] does not seem to have been used for the polymerization of 1,3-pentadiene.

#### 5.4.3.2 *Cis-1,4 isotactic polypentadiene*

This polymer was first obtained with the homogeneous $AlR_3$–$Ti(OR)_4$ system, at an Al/Ti ratio of about seven (Table 11). The crude products had a *cis* content of about 70%, the remainder being predominantly 1,2 and 3,4 units, By removing the fraction soluble in methyl ethyl ketone (MEK), a residue was obtained, crystalline by X-ray at room temperature, with a *cis* content 90–92%, $[\eta] = 5$–$7\ dl\ g^{-1}$.[362,363] Both the (*E*) and (*Z*) isomers were polymerized with $AlR_3$–$Ti(OR)_4$, but only the (*E*) isomer gave crystalline polymers. It has been shown that the (*Z*) isomer is isomerized to the (*E*) one by the catalyst, since when using pure (*Z*) isomer, the monomer recovered after about 10% conversion was found to be a mixture of the two isomers, with (*E*) being predominant. Therefore the polymerization product from the pure (*Z*) isomer may be a copolymer of the two monomer isomers.

By using optically active systems, optically active *cis*-1,4 isotactic polypentadiene was obtained.[364,365]

Much more active and stereospecific for the synthesis of *cis* isotactic polypentadiene are the catalysts based on lanthanides. The $AlEt_2Cl$–$Nd(OCOR)_3$–$Al(Bu^i)_3$ system gives highly crystalline polymers from the (*E*) isomer (the (*Z*) isomer gives only very poor yields).[366–368] The *cis* content of the product obtained depends on the polymerization temperature: it is about 85% at 20 °C, 90% at 0 °C, and 95% at − 18 °C. The non-*cis* units are predominantly 1,2 with no trace of 3,4 units (which are instead present in the $Ti(OR)_4$ polymer). The amount of *trans*-1,4 units is negligible.[369,370] Polymer yields with Nd catalysts are much higher than with $AlR_3$–$Ti(OR)_4$, by a factor of 400–500. This difference is at least in part attributable to the fact that the number of active centers is higher in Nd than in Ti catalysts.[369]

Isotactic polypentadiene with a *cis* content of 88% was obtained with a cerium catalyst at 50 °C.[359] It is likely that at a lower temperature this system also will give a polymer with a higher *cis* content.

#### 5.4.3.3 *Cis-1,4 syndiotactic polypentadiene*

This polymer was obtained with two types of catalysts, based on cobalt and nickel respectively. Only the (*E*) isomer is polymerized (Table 11). The system $AlEt_2Cl$–$Co(acac)_2$–$H_2O$ (Al/Co = 300; $H_2O/Al = 0.1$–$0.5/1$) gives benzene polymerization products consisting of 85–90% *cis*-1,4 units, crystalline by X-ray.[116,371] The presence of $H_2O$, which has the function of producing '*in situ*' ethylaluminoxanes, $O(AlEtCl)_2$, is essential for stereospecificity.[372,373]

The effect of aluminoxanes is shown by the results of polymerizations in heptane using $Co(acac)_3$ in combination with $AlEt_2Cl$ and preformed aluminoxanes respectively. The catalyst from $AlEt_2Cl$ gives the 1,2 polymer (about 90%), and that from aluminoxanes a 90% *cis* polymer. Addition of thiophene to the catalyst system (thiophene/Al = 0.5–5) is required in order to suppress cationic activity, which is responsible for the formation of low molecular weight polymers with a predominantly 1,2 structure.

One of the best solvents for the polymerization of pentadiene to *cis* syndiotactic polymer with aluminoxanes–$Co(acac)_3$ is *cis*-2-butene. The polymerization in this solvent is faster than in heptane or benzene and gives polymers with a higher molecular weight.[114]

Crystalline syndiotactic polypentadiene was obtained with $AlEt_3$–$Ni(OCOR)_2$–$BF_3 \cdot$ phenolate. The *cis* content of the polymer obtained at 50 °C in heptane was about 88%. Probably a polymer of higher stereoregularity can be obtained at a lower temperature with this system.[359]

#### 5.4.3.4 *1,2 Syndiotactic polypentadiene*

Polymerization of 1,3-pentadiene with $AlEt_2Cl$–$Co(acac)_3$ in heptane at 20 °C yields a polymer consisting of about 90% 1,2 units (Table 11). The polymer is amorphous by X-ray in the unstretched

**Table 11** Polymerization of 1,3-Pentadiene to *Cis*-1,4 and 1,2 Stereoregular Polymers

| *Catalyst system* | *Polymerization temperature* (°C) | *Type of tacticity* (%) | *Cis-1,4* (%) | *Polymer structure Trans-1,4* (%) | *1,2* (%) | *3,4* (%) | *Ref.* |
|---|---|---|---|---|---|---|---|
| $AlR_3$–Ti(OBu$^n$)$_4$ (7:1) in heptane | 0<br>−18 | *cis*-1,4 iso | 73.4<br>77.2 | —<br>— | 16.4<br>13.9 | 10.2<br>8.9 | b |
| $AlEt_2Cl$–Nd(OCOR)$_3$–Al(Bu$^i$)$_3$ | 20<br>−18 | *cis*-1,4 iso<br>*cis*-1,4 iso | 84.2<br>93 | —<br>— | 15.8<br>7 | —<br>— | c |
| Al(Bu$^i$)$_2$H–Ce(OCOR)$_3$–HBr (24:1:2) in hexane | 50 | *cis*-1,4 iso | 76 | 4 | 20 | — | d |
| $AlEt_2Cl$–Co(acac)$_2$–$H_2O$ [or $(Et_3Sn)_2O$] (600:1:300) in benzene, heptane or *cis*-butene[a] | 0 to 18 | *cis*-1,4 syndio | 92–94 | <1 | 5–7 | — | e |
| $AlEt_3$–Ni octanoate–$BF_3 \cdot$phenolate (10:1:1.5) in heptane | 50 | *cis*-1,4 syndio | 88 | 7 | 5 | — | d |
| $AlEt_2Cl$–Co(acac)$_3$ (600:1) in heptane | 18 | 1,2 syndio | 6 | 3–5 | 89–91 | — | e |
| $AlEt_3$–Fe octanoate–BuSCN (3:1:1) in heptane | 50 | 1,2 syndio | 11 | 4 | 81 | 4 | d |

[a] $AlEt_2Cl$ was first reacted with $H_2O$ or $(Et_3Sn)_2O$ and the chloroethylaluminoxanes so obtained were used as cocatalyst.
[b] G. Natta, L. Porri, A. Carbonaro and G. Stoppa, *Makromol. Chem.*, 1964, **77**, 114.
[c] L. Porri and G. Ricci, in press.
[d] D. H. Beebe, C. E. Gordon, R. N. Thudium, M. C. Throckmorton and T. L. Hanlon, *J. Polym. Sci., Polym. Chem. Ed.*, 1978, **16**, 2285.
[e] L. Porri, A. Di Corato and G. Natta, *Eur. Polym. J.*, 1969, **5**, 1.

state, but a moderate crystallinity appears on stretching. From the fiber spectrum an identity period of 5.1 Å was measured, which is consistent with a 1,2 syndiotactic structure. The stereoregularity of the polymer, however, is rather low.[374]

A polymer of pentadiene with a prevailingly 1,2 structure (81%) was obtained with the system $AlEt_3$–Fe octanoate–BuSCN (3:1:1) in heptane at 50 °C. Nothing was reported about the stereoregularity of this polymer.[359]

### 5.4.4 Polymers of Other 1,3-Dialkenes

In addition to butadiene, isoprene and pentadiene, several other 1,3-dienes have been polymerized with transition metal catalysts (Tables 12 and 13). Amorphous and crystalline polymers of various structures were obtained and some of them showed a high degree of stereoregularity. The practical interest of these polymers is low and probably it is for this reason that they have not been the subject of detailed study. However, the results obtained with these monomers are of interest for the information they give on some aspects of the mechanism of 1,3-diene polymerization.

#### *5.4.4.1 Monomers $CH_2$=CHCH=CHR*

Crystalline polymers with a *trans*-1,4 isotactic structure have been obtained from $CH_2$=CHCH=CHR monomers, with R = Et, $Pr^i$, $Bu^n$ and $Bu^i$, using the $AlR_3$–$VCl_3$ system. The polymerization products are always a mixture of crystalline and amorphous polymers. The latter can be removed by extraction with diethyl ether. Either the (*E*) or (*Z*) isomers of each monomer may be polymerized.[375]

With linear monomers (R = Et, $Pr^n$, $Bu^n$), the crystalline fractions obtained from either isomer are identical by X-ray and NMR examination and consist only of *trans*-1,4 isotactic polymers. With the non-linear monomers (R = $Pr^i$, $Bu^i$), the (*E*) isomers give a crystalline fraction (30 to 45% of the total) consisting only of *trans*-1,4 isotactic polymers. The (*Z*) isomers give instead a crystalline fraction (60 to 70% of the total), from the fiber spectra of which two identity periods have been measured, 4.85 and 6.5 Å. The first is that of *trans*-1,4 isotactic polymers, the other is that expected for a polymer with a 1,2 isotactic structure. The two polymers have not been isolated, but X-ray and IR evidence indicates the presence of both *trans*-1,4 and 1,2 isotactic sequences.[375]

**Table 12** Polymerization of Some Substituted Butadienes

| *Monomer* | *Catalyst system* | *Crystalline polymer* | |
|---|---|---|---|
| | | *Type* | *m.p.* (°C) |
| 1,3-Hexadiene[a] | $AlR_3$–$VCl_3$ | *trans*-1,4 iso | 82 |
| 1,3-Heptadiene[a] | $AlR_3$–$VCl_3$ | *trans*-1,4 iso | 85 |
| 1,3-Octadiene[a] | $AlR_3$–$VCl_3$ | *trans*-1,4 iso | 87 |
| (*E*)-5-Methyl-1,3-hexadiene | $AlR_3$–$VCl_3$ | *trans*-1,4 iso | 88 |
| (*Z*)-5-Methyl-1,3-hexadiene | $AlR_3$–$VCl_3$ | *trans*-1,4 iso + 1,2 iso | |
| (*E*)-6-Methyl-1,3-heptadiene | $AlR_3$–$VCl_3$ | *trans*-1,4 iso | 119 |
| (*Z*)-6-Methyl-1,3-heptadiene | $AlR_3$–$VCl_3$ | *trans*-1,4 iso + 1,2 iso | |
| (*E*)-5-Methyl-1,3-heptadiene (racemic) | $Al(Bu^i)_2$–$TiCl_3$HRA[b] | *trans*-1,4 iso | 165–170 |
| (*E*)-5-Methyl-1,3-heptadiene [(*S*)-enantiomer] | $Al(Bu^i)_3$–$TiCl_3$HRA[b] | *trans*-1,4 iso | 175–180 |
| 2-Ethyl-1,3-butadiene | $AlR_3$–$TiCl_4$ | *cis*-1,4 | — |
| 2-Propyl-1,3-butadiene | alanes–$TiCl_4$ | *cis*-1,4 | — |
| 2-Propyl-1,3-butadiene | $AlR_3$–$VCl_3$ | *trans*-1,4 | *ca.* 42 |
| 2-*t*-Butyl-1,3-butadiene | alanes–$TiCl_4$ | *cis*-1,4 | 103–106 |
| 2-Phenyl-1,3-butadiene | $AlHCl_2 \cdot OEt_2$–$AlI_3$–$TiCl_4$ | *cis*-1,4 | — |
| | $AlR_3$–$TiCl_4$, Al/Ti *ca.* 0.5 | *trans*-1,4 | *ca.* 270 |
| 2,3-Dimethylbutadiene | $AlR_3$–$TiCl_4$, Al/Ti ≥ 1; | *cis*-1,4 | 195 |
| | $AlEt_2Cl$–$Nd(OCOR)_3$–$Al(Bu^i)_3$ | *cis*-1,4 | 195 |
| 2-Methyl-1,3-pentadiene | $Al(Bu^i)_3$–$Ti(OBu^n)_4$–$VCl_3$ | *trans*-1,4 | 182–192 |
| 2-Methyl-1,3-pentadiene | $AlEt_2Cl$–$Nd(OCOR)_3$–$Al(Bu^i)_3$ | *cis*-1,4 | 168 |
| 4-Methyl-1,3-pentadiene | $AlEt_3$–$TiCl_3$ | 1,2 isotactic | 166 |

[a] Both (*E*) and (*Z*) isomers give *trans*-1,4 isotactic polymers.
[b] HRA: hydrogen reduced and activated.

**Table 13** Structure of the Polymers Obtained from some 1,4-Disubstituted Butadienes

| *Monomer* | *Catalyst system* $AlR_3$–$VCl_3$ | $AlR_2Cl$–$Co(acac)_2$–$H_2O$ | $AlEt_2Cl$–$Nd(OCOR)_3$–$Al(Bu^i)_3$ | *Ref.* |
|---|---|---|---|---|
|  | *trans-threo* | *cis-threo* | | a |
| 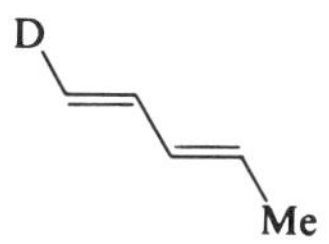 | *trans-threo*-diiso | *cis-threo*-disyndio | *cis-erythro*-diiso | b, c |
| 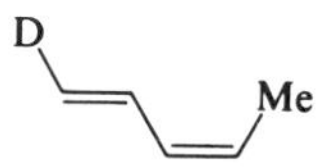 | *trans-erythro*-diiso | | | b |
| 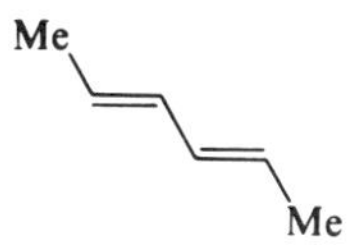 | | *trans-threo*-diiso | *trans-threo*-diiso | d, e |

[a] L. Porri and M. Aglietto, *Makromol. Chem.*, 1976, **177**, 1465.
[b] S. Destri, G. Gatti and L. Porri, *Makromol. Chem., Rapid Commun.*, 1981, **2**, 605.
[c] S. Destri, A. Bolognesi, L. Porri and F. Wang, *Makromol. Chem., Rapid Commun.*, 1982, **3**, 187.
[d] S. Destri, M. C. Gallazzi, A. Giarrusso and L. Porri, *Makromol. Chem., Rapid Commun.*, 1980, **1**, 293.
[e] F. Wang, A. Bolognesi, A. Immirzi and L. Porri, *Makromol. Chem.*, 1981, **182**, 3617.

### 5.4.4.2 5-Methyl-1,3-heptadiene (stereoselective polymerization)

Each isomer, (*E*) and (*Z*), of this monomer can exist in two enantiomeric forms, (*R*) and (*S*), due to the presence of an asymmetric carbon atom: CH═CHCH═CH$\overset{*}{C}$H(Me)Et. Only the polymerization of the (*E*) isomer was examined, using the $Al(Bu^i)_3$–$TiCl_3HRA$ system.[376] Crystalline polymers with a *trans*-1,4 isotactic structure were obtained, in a mixture with amorphous polymers. The optically active monomer [(*S*) enantiomer] gives a crystalline fraction (55–60% of the total), from which two types of macromolecules were isolated by solvent fractionation. The two types of macromolecules differ in the configuration of the asymmetric carbon atom of the main chain, which is predominantly (*R*) in one type (soluble in diethyl ether) and (*S*) in the other as shown from the results of oxidation (Figure 9). The latter type is largely predominant; (*S*):(*R*) = 9:1. The racemic monomer gives a polymerization product containing a lower amount of crystalline fraction (30–35% of the total), but this shows the same X-ray and IR spectra as the crystalline fraction from the (*S*) enantiomer and on oxidation gives the same diastereoisomer of 3-carboxy-2-methylhexanoic acid. This means that the polymerization of the racemic monomer is stereoselective, *i.e.* each macromolecule is made up predominantly of units derived from one enantiomer. The melting point of the crystalline polymer from the racemic monomers is slightly lower than that from the (*S*) monomer (165–170 °C *vs.* 170–175 °C, for the most stereoregular fractions). This is indicative of a lower stereoregularity, attributable to the fact that the polymerization is not 100% stereoselective.

### 5.4.4.3 Monomers $CH_2$═CRCH═$CH_2$

Polymers with a predominantly *cis*-1,4 structure were obtained from the monomers $CH_2$═CRCH═$CH_2$ (R = Et, $Pr^n$, $Bu^t$, Ph) using $TiCl_4$ and $AlR_3$ or aluminum hydrides. While the polymers from 2-ethyl-[377] and 2-propyl-butadiene[378] are rubbery products at room temperature, those from 2-*t*-butylbutadiene are highly crystalline, with m.p. = 103–106 °C, identity period 15.3 Å.[379] The polymers from 2-phenylbutadiene are of low molecular weight, with a *cis* content in the range 80–95%. Some of them show a very low degree of crystallinity by X-ray.[380,381]

A polymer with a prevailingly *trans*-1,4 structure was obtained from 2-propylbutadiene, with $AlR_3$–$VCl_3$.[378] The polymer is crystalline, m.p. *ca.* 42 °C, identity period 9.2 Å. Attempts to

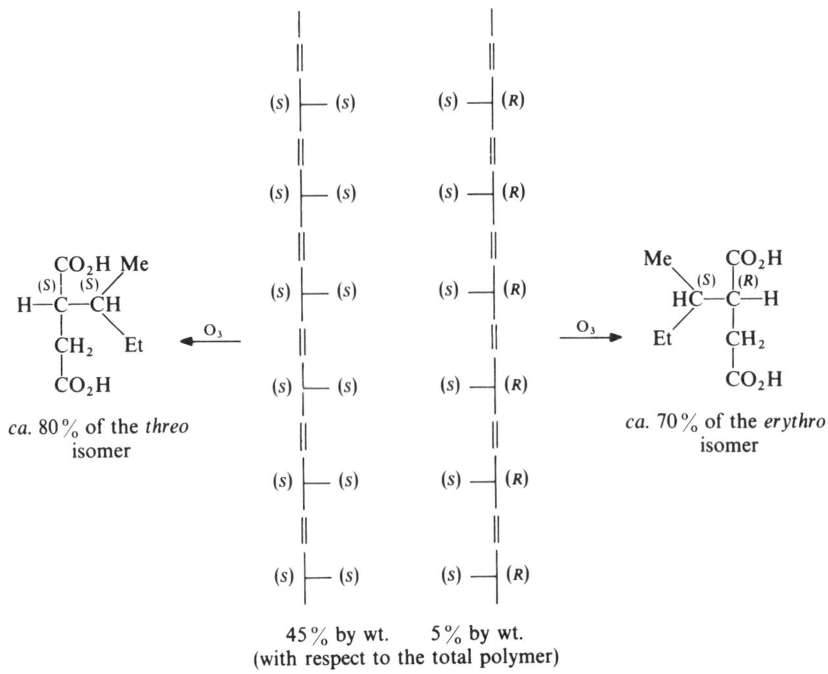

**Figure 9** Polymerization of (+) (*S*)-5-methyl-1,3-heptadiene to *trans*-1,4 isotactic polymer with $Al(Bu^i)_3$–$TiCl_3HRA$. Absolute configuration and percentage with respect to the total polymer of the isotactic diastereoisomeric macromolecules

prepare a *trans*-1,4 polymer from 2-*t*-butylbutadiene failed, presumably because for this monomer the transoid conformation is energetically unfavorable (see Section 5.8.1)

### *5.4.4.4 2,3-Dimethylbutadiene*

This monomer was polymerized with catalysts based on Ti, Co and Nd, and two crystalline polymers, having a *trans*-1,4 and a *cis*-1,4 structure, were obtained. The *trans* polymer (*trans* content >95%; m.p. about 270 °C) was prepared with $AlEt_3$–$TiCl_4$, at an Al/Ti molar ratio of about 0.5. The same system, at Al/Ti = 1 to 2, gives a high-*cis* (>97%) polymer (m.p. 162 to 190 °C).[382–384] A polymer consisting of 99–100% *cis* units was obtained with $AlEtCl_2$–Nd naphthenate–$Al(Bu^i)_3$ and other Nd systems.[385,386] Probably, other lanthanide catalysts will give high-*cis* polymer from 2,3-dimethylbutadiene. A polymer with a high-*trans* structure was obtained with $[(crotyl)NiCl]_2$ plus chloranil.[281]

The cobalt catalysts that are highly stereospecific for the *cis*-polymerization of butadiene (*e.g.* $AlEt_2Cl$–$Co(acac)_2$–$H_2O$) give a polymer of 2,3-dimethylbutadiene consisting of about 80% *cis*-1,4 and 20% 1,2 units.[387,388]

### *5.4.4.5 2-Methyl-1,3-pentadiene*

Two crystalline polymers were obtained from this monomer [(*E*) isomer], one with a *trans*-1,4 and the other with a *cis*-1,4 structure. The *trans* polymer (m.p. = 182–192 °C) was prepared with the seemingly homogeneous system $Ti(OBu^n)_4$–$VCl_3$–$Al(Bu^i)_3$ (1:2:20),[361] the *cis* polymer with the two different systems: $Al(Bu^i)_3$–$TiCl_4$ (Al/Ti = 1; aged for 30 min) and $AlEt_2Cl$–Nd octanoate–$Al(Bu^i)_3$ (2.7–3:1:30).[361,369] The titanium system gives polymerization products consisting of about 90% *cis*-1,4 units. Extraction with MEK at room temperature leaves as residue an essentially *cis* polymer with m.p. = 164.5 °C. The neodymium system gives a crude product with a *cis* content of 99–100%. This product has been fractionated by successive extractions with different solvents and NMR analysis has shown the various fractions to have a different degree of stereoregularity. The most stereoregular fractions have m.p. = 168 °C.[369]

The 1,4 polymers of 2-methyl-1,3-pentadiene have one asymmetric carbon atom per monomeric unit and can therefore have an isotactic or a syndiotactic structure. No information has been reported on the structure of the crystalline polymers so far obtained. The catalysts used for the preparation of the *cis* polymers give *cis*-1,4 isotactic polymers from 1,3-pentadiene. It seems likely that the polymers of 2-methyl-1,3-pentadiene also have a *cis*-1,4 isotactic structure.

#### *5.4.4.6 4-Methyl-1,3-pentadiene*

This monomer was polymerized with several catalyst systems based on Ti or V. The polymers obtained are made up of 1,2 units only, independent of the catalyst used. The particular behavior of this monomer is attributable to the presence of two substituents in position 4, which makes the formation of 1,4 units difficult.

The heterogeneous systems $AlEt_3$–α-$TiCl_3$ and $AlEt_3$–$TiCl_4$ (Al/Ti = 1.2) give polymerization products containing a high proportion (70–80%) of crystalline 1,2 isotactic polymers. $AlEt_3$–$VCl_3$ gives a lower percentage of the crystalline fraction. The soluble systems from $Ti(OR)_4$ or $VO(OR)_3$ and $AlEt_3$ give only amorphous 1,2 polymers. The 1,2 isotactic poly(4-methyl-1,3-pentadiene) is a highly stereoregular polymer, with m.p. about 166 °C.[389]

#### *5.4.4.7 1,4-Disubstituted butadienes*

Only a few monomers of this class were polymerized, *i.e.* 2,4-hexadiene and the following deuterated monomers: (*Z*,*Z*)-1,4-$d_2$-1,3-butadiene; (*E*,*E*)-1-$d_1$-1,3-pentadiene; and (*E*,*Z*)-1-$d_1$-1,3-pentadiene.

2,4-Hexadiene was polymerized with several catalyst systems, *e.g.* $Al(Bu^i)_3$–$TiCl_4$, $AlEt_2Cl$–$Ti(acac)_3$, $AlEt_2Cl$–$Co(acac)_3$[390,391] and $AlEt_2Cl$–Nd octanoate–$Al(Bu^i)_3$.[392] Crystalline polymers with a *trans*-1,4 *threo* diisotactic structure were obtained from the (*E*,*E*) isomer, whereas amorphous polymers were obtained from the (*E*,*Z*) isomer. The (*Z*,*Z*) isomer does not polymerize with all the systems used. It is interesting that a poly(2,4-hexadiene) with a *trans* structure is obtained with catalysts that give predominantly *cis*-1,4 polymers from butadiene, isoprene and other simple dialkenes. The reason for this change of stereospecificity is discussed in Section 5.8.1

(*Z*,*Z*)-1,4-$d_2$-butadiene was polymerized with several catalyst systems. *Trans*-1,4 stereoregular polymers were obtained with $AlEt_3$–$VCl_3$ or $AlEt_2Cl$–$VCl_3 \cdot 3THF$, and *cis*-1,4 stereoregular polymers with $AlEt_2Cl$–$Co(acac)_2$–$H_2O$.[393] Several stereoregular polymers with a 1,4 structure were also obtained from the two deuterated pentadienes.[394,395] The results are summarized in Table 13. The mechanistic implications of the results obtained with these deuterated monomers are discussed in Section 5.8.

### 5.5 THE STAGES OF POLYMERIZATION

The polymerization of 1,3-dienes with transition metal catalysts is an insertion polymerization, as is that of 1-alkenes with the same catalysts; that is, the polymer chain grows by monomer insertion into the metal–carbon bond between the growing chain and the transition metal of the catalyst complex. The fundamental similarity between the polymerization of the two classes of monomers is shown by the fact that several monoalkene/dialkene copolymers can be prepared (*e.g.* ethylene/-butadiene and propene/butadiene alternating copolymers). There are, however, several differences in the polymerization of dialkenes from that of 1-alkenes. These are due to the fact that in diene polymerization the growing chain is bonded to the transition metal by an $\eta^3$-allyl bond. This distinctive feature of dialkene polymerization was recognized soon after the first results were obtained using simple allyl derivatives of transition metals as polymerization catalysts.[23,118,214,215] Work on NMR investigation of the allyl complexes and of their reaction with dialkenes has allowed full elucidation of the individual steps of chain growth.[396–406]

*The first stage.* As shown in Section 5.3.1, three classes of catalysts are effective for 1,3-diene polymerization: (a) systems derived from aluminum alkyls and transition metal compounds; (b) mono- or two-component systems not containing preformed metal–carbon bonds; and (c) systems based on allyl derivatives of transition metals.

The first stage of the polymerization of butadiene with simple allyl derivatives of nickel has been elucidated by $^1H$ NMR. Reaction of $[(\eta^3\text{-}C_4H_7)NiI]_2$ with perdeuterobutadiene causes a gradual disappearance of the resonances of the allylic protons, while the signals of the free group —$CH_2CH{=}CHMe$ appear simultaneously (equation 7). On the other hand, when

$[(\eta^3\text{-}C_4D_7)NiI]_2$ is reacted with butadiene, resonances due to the allylic protons appear[396] (equation 8).

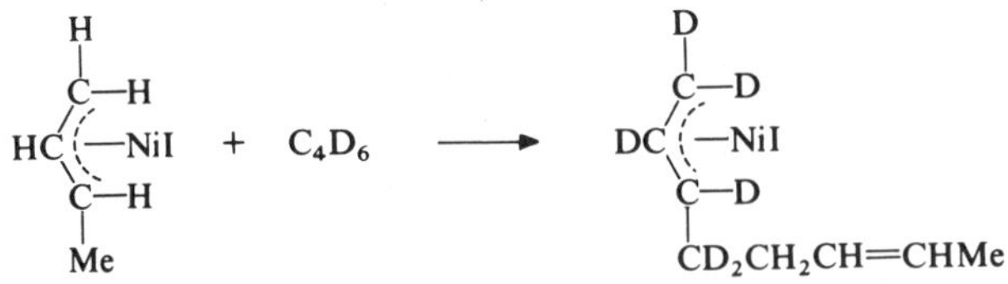

(7)

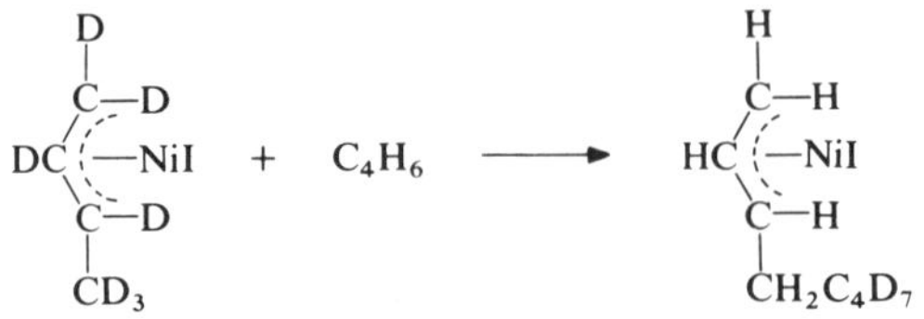

(8)

These results indicate that insertion of the first monomer molecule occurs on the metal–allyl bond, with re-formation of a new $\eta^3$-butenyl group. Similar conclusions were reached using other allyl derivatives of nickel.[268,269,398,405] More detailed information on the insertion of butadiene on a metal–allyl bond was obtained by NMR examination of the reaction between butadiene and $[(\eta^3\text{-}C_4H_7)PdCl]_2$. The first step is the formation of an $\eta^1$-allylic species as a result of the coordination of one or more monomer molecules. The $\eta^1$-allyl species is in rapid equilibrium with the $\eta^3$-species (equation 9). Such $\eta^3 \rightleftharpoons \eta^1$ equilibria of allyl ligands under the influence of donor ligands are well known.[407,408]

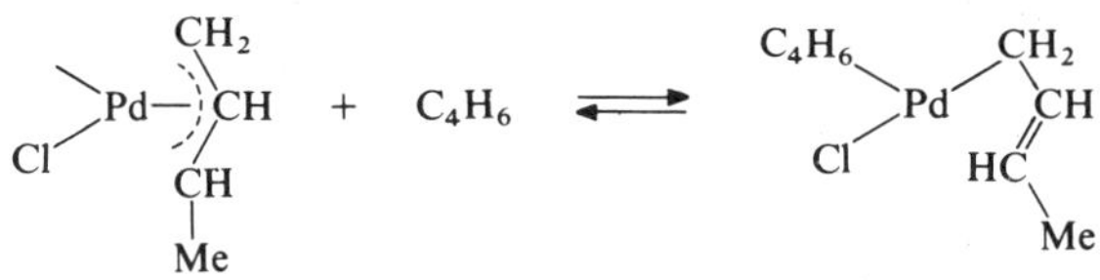

(9)

The second step is the actual insertion reaction, which presumably consists of a transfer of the Mt—C bond of the $\eta^1$-species to the coordinated monomer.[409,410]

In the case of the system obtained from aluminum alkyls and transition metal compounds, monomer insertion takes place on the Mt—C (Mt = metal) bond that is produced in the catalyst-forming reaction. When aluminum alkyls containing $^{14}C$ were used, $^{14}C$ was found in the polymer.[70] The insertion of the first monomer molecule results in the formation of an $\eta^3$-butenyl group. In the case of catalyst systems that do not contain preformed metal–carbon bonds, it is assumed that a metal–allyl bond is formed on reaction of the complex with the monomer. There are several catalyst systems of this type, *e.g.* $RhCl_3$ in $H_2O$,[21,22] $AlCl_3$–$CoCl_2$,[209–211] $Ni(CO)_4$–$VCl_4$,[207] *etc.* The formation of the Mt–allyl bond may occur in different ways, *i.e.* by addition of the monomer to a Mt—H bond (this is probably the case for $RhCl_3$ in $H_2O$ or EtOH), through an oxidative addition of the monomer to a complex containing Mt in a low oxidation state, or even by monomer insertion into a Mt—Cl bond. In the case of Rh catalysts, the formation of Rh–allyl bonds in the initiation reaction has been demonstrated;[411] it was also shown that the polymer chain grows on a Rh—C bond.[412]

*The growth reaction.* The growth of the polymer molecule occurs by insertion of the monomer into the $\eta^3$-butenyl bond between the transition metal and the last polymerized unit. Using as catalyst simple allyl derivatives of Ni, it was possible to investigate the initial stages of the polymerization and to demonstrate that a new $\eta^3$-allyl bond is re-formed after each monomer insertion.[268,396,398,400,402,413] NMR examination of the polymerization of the deuterated monomer $CD_2{=}CHCH{=}CD_2$ using as catalyst the complex $[(\eta^3\text{-}C_4D_6H)NiI]_2$, (**5**), allows observation of the successive formation of species (**6**), (**7**) and (**8**) from the appearance of the resonance of the protons $H^a$, $H^b$, $H^c$, $H^d$ and $H^e$ (Scheme 3).[400]

NMR examination of the polymerization of monomers of the type $CH_2{=}CRCH{=}CH_2$ (R = Me, Et, Pr$^i$, Bu$^t$) with $[(\eta^3\text{-}C_4H_7)NiI]_2$ has shown in addition that an *anti* butenyl group is first formed, which then isomerizes to the *syn* form.[401,402]

In the case of polymerizations with catalysts from aluminum alkyls and transition metal compounds NMR examination of the polymer growth reaction was not carried out. However, there are several indirect pieces of evidence suggesting that with these systems also a new $\eta^3$-allyl bond is

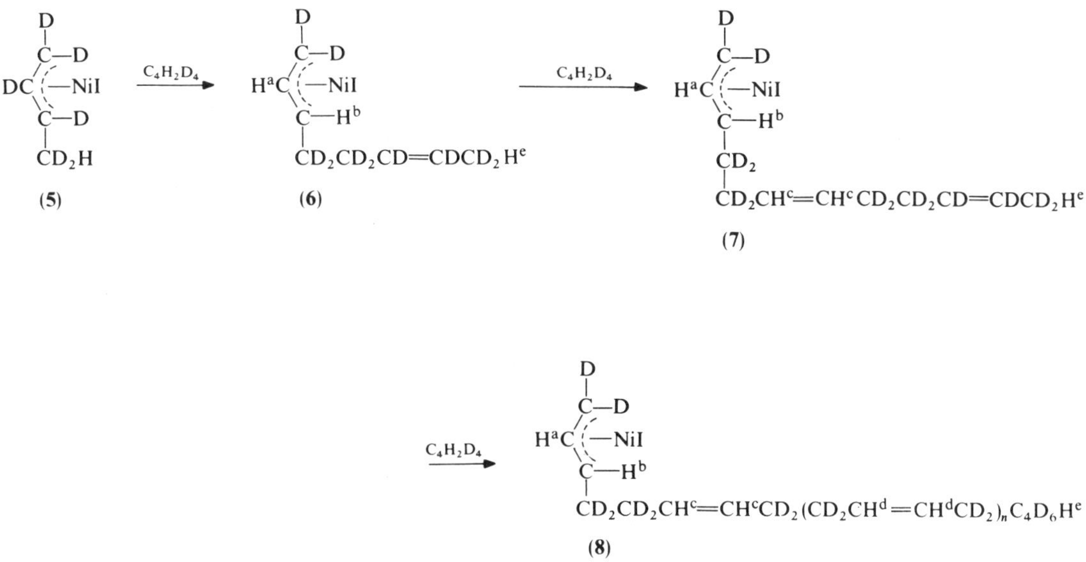

Scheme 3

re-formed after each individual insertion step. Several systems (*e.g.* $AlEt_2Cl$–$Co(acac)_2$–$H_2O$; $AlR_3$–$Ni(OCOR)_2$–$BF_3 \cdot OEt_2$) do not polymerize monoalkenes, or give only oligomers from them, whereas they give high polymers from 1,3-dialkenes. The different behavior is attributable to the higher stability of the Mt–allyl bond as compared with Mt—C $\sigma$ bonds. The copolymerization of monoalkenes with dialkenes is rather difficult compared with the facility of copolymerizing an alkene with another alkene or a dialkene with a dialkene. In addition, steric control in 1,3-diene polymerization is interpreted far better by assuming that the last polymerized unit is $\eta^3$- rather than $\eta^1$-bonded to Mt.

*Chain termination.* Several processes can bring about the termination of a growing polymer, *e.g.* transfer with the aluminum alkyl or with the monomer, transfer with a suitable molecule introduced in the reaction medium, monomolecular or bimolecular terminations. These processes were studied in detail mostly for polymerization of butadiene or isoprene to *cis* polymers, due to the commercial interest of these products.

*Transfer with aluminum alkyls.* Formation of a dead chain by reaction of the growing polymer with the aluminum alkyl has been observed in several cases. In the polymerization of 1,3-dialkenes to *cis* polymers with neodymium catalysts, the polymer molecular weight decreases with increasing Al/Nd molar ratio,[158,163,308] as already shown in Figure 6. The transfer reaction is represented in reaction (10). Reaction (10) appears as the most practicable method for regulating polymer molecular weight with Nd catalysts. Transfer with the aluminum alkyl has been observed in several other cases, *e.g.* the polymerization of butadiene with $Al(Bu^i{}_3$–$TiI_4$–$O(Pr^i)_2$,[57] $MoCl_4OC_8H_7$–$Al(Bu^i)_2OPh$[414] or with $AlEt_3$–$Cr(acac)_3$[301] and in the polymerization of isoprene with $AlR_3$–$VCl_3$[346] or $AlR_3$–$Ti(OR)_4$.[70] Probably transfer reactions of this type occur with several catalyst systems, but only a few cases have been investigated.

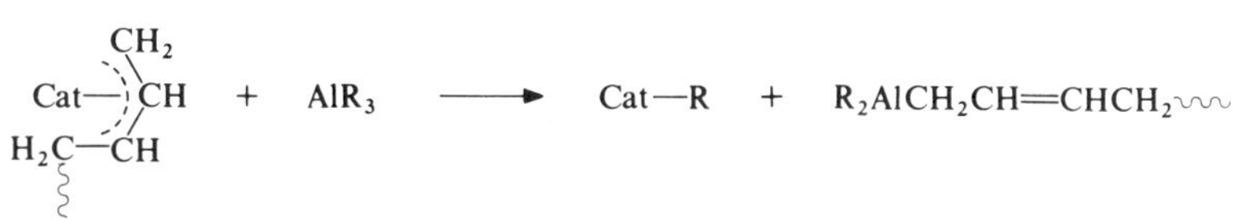 (10)

*Transfer to monomer.* This type of transfer was observed in the polymerization of butadiene or isoprene with several systems, *e.g.* $AlEt_3$–$Ti(OR)_4$,[415] $AlR_3$–$TiCl_4$,[335] $MoCl_4OC_8H_7$–$Al(Bu^i)_2$-$OPh$,[414] Ni stearate–$AlEt_2Cl$[134] and monometallic Rh catalysts.[411] Most probably a transfer of this type occurs in the *cis* polymerization of butadiene with $AlEt_2Cl$–$Co(acac)_2$–$H_2O$. In this polymerization polymer molecular weight increases initially with conversion, then remains practically constant. Since transfer to aluminum alkyl is practically absent in this case,[301] transfer to monomer seems to be the main process regulating molecular weight. Probably this type of transfer occurs to a greater or lesser extent in many other polymerizations.

*Reaction with transfer agents.* Several substances such as ethylene, 1-alkenes, $H_2$ and allenes act as transfer agents in the polymerization of butadiene with $AlEt_2Cl$–$CoCl_2 \cdot 2py$ as already shown in Table 5. Reaction with $C_2H_4$, which is the most active among the monoalkenes, occurs as indicated in Scheme 4.[303,416]

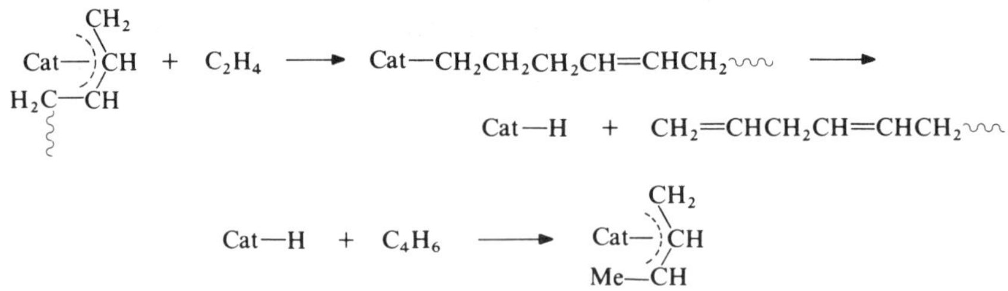

**Scheme 4**

Reaction with $H_2$ gives rise to the species Cat—H (Cat = catalyst), which re-forms a new $\eta^3$-allyl group on reaction with the monomer. Allenes are more effective than alkenes or $H_2$, but it has not been elucidated by which mechanism the transfer occurs.

*Mono- and bi-molecular terminations.* Two types of molecular reactions can lead to the formation of a dead polymer chain: (1) hydrogen abstraction from the growing chain, with formation of a Mt—H bond (equation 11). The Mt—H species re-forms a Mt–allyl bond on reaction with the monomer; and (2) homolytic breakage of the Mt–allyl bond, with formation of a chain containing a free radical, while the metal assumes a lower oxidation state (equation 12). The radical chain will form a dead polymer molecule by coupling or a disproportionation reaction or by hydrogen abstraction from the solvent.

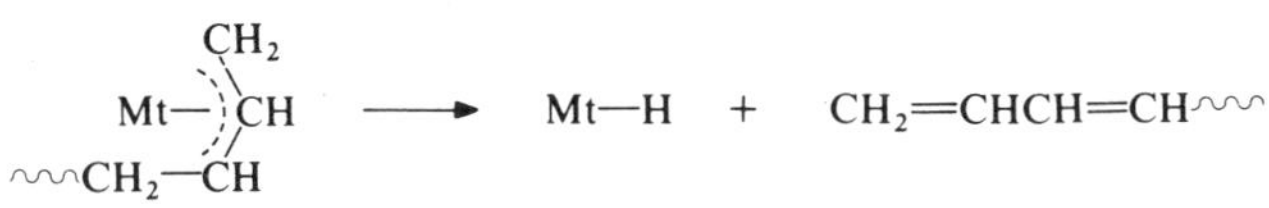 (11)

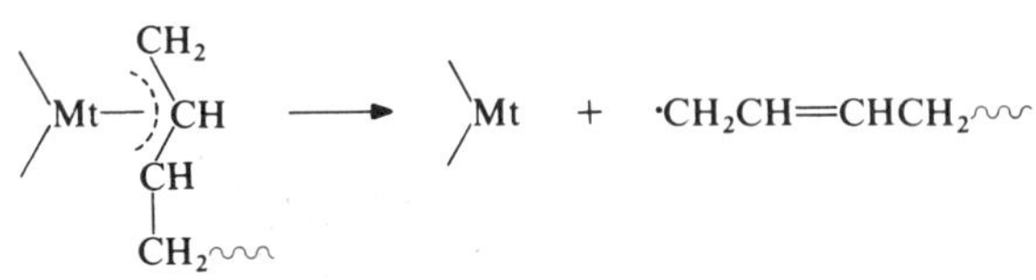 (12)

The dimerization of butadiene to 5-methyl-1,3,6-heptatriene using as catalyst ($\eta^3$-$C_8H_{13}$)($C_4H_6$)Co in heptane at 65–70 °C occurs according to reaction (11).[417] Most probably the same type of reaction occurs in the polymerization of butadiene to 1,2 syndiotactic polymer using as catalyst ($\eta^3$-$C_8H_{13}$)($C_4H_6$)Co in toluene, heptane or $CH_2Cl_2$ at room temperature or below.[418] Formation of dead chains according to reaction (11) was also observed in the polymerization of butadiene with (allyl)NiI[419–421] or (allyl)NiOCOMe.[269] In the latter case, NMR spectra recorded during the polymerization indicated the presence of small amounts of (crotyl)$NiOCOCF_3$ even at high monomer conversion; its formation presumably occurs through reaction of butadiene with the Ni—H intermediate formed according to equation (11).

Formation of dead chains according to reaction (12) occurs with some vanadium systems ($AlR_3$–$VCl_3$, $AlEt_2Cl$–$VCl_3 \cdot 3THF$ or V(acac)$_3$) for which a drop of activity was observed during polymerization at room temperature or higher.[84,346]

Bimolecular terminations have been observed with some soluble Ziegler–Natta catalysts, *e.g.* in the polymerization of butadiene with $AlR_3$–$TiI_4$–O(Pr$^i$)$_2$.[57]

## 5.6 KINETIC ASPECTS

Kinetic studies on the polymerization of butadiene and isoprene have been performed with various catalyst systems. With most of them the rate of polymerization is first order with respect to both monomer and catalyst concentration, but with some a different dependence is observed. The results reported by the various authors are sometimes difficult to compare, because the experimental conditions are not exactly identical.

The kinetics of the *cis* polymerization of butadiene with Ti catalysts was investigated with the systems $TiI_4$–$Al(Bu^i)_3$–$O(Pr^i)_2$,[57] $TiI_2Cl_2$–$Al(Bu^i)_3$[63,422] and $TiCl_4$–$AlI_3$–$AlHCl_2 \cdot OEt_2$.[294] The kinetics with the $TiI_4$ and $TiCl_2I_2$ systems follows a rather complex scheme due to the occurrence of reversible bimolecular terminations. The rate with the $TiCl_4$ system has been reported to be first order in monomer and Ti concentration when the catalyst is prepared in the presence of the monomer. When the catalyst is prepared in the absence of monomer, the catalyst is, at least in part, heterogeneous and the rate is still first order in monomer concentration but the dependence on Ti is no longer linear.

For the *cis* polymerization of isoprene with the $AlR_3$–$TiCl_4$ system the rate is represented by equation (13).

$$R_p = K_p[M][Al][Ti] \tag{13}$$

The overall activation energy $E_a$ is 60.2 kJ mol$^{-1}$. The first order dependence of the rate on both Ti and Al concentrations derives from the fact that the catalyst system affords a *cis* polymer only at an Al/Ti molar ratio of 1.[27]

Much work has been carried out on the *cis* polymerization of butadiene with Co catalysts.[99,101,109,295,301,302] In benzene or toluene the rate was found to be first order with respect to monomer; a second order dependence was observed only in heptane, with Co carboxylate–$Al_2Et_3Cl_3$.[111] This presumably reflects the different association of the catalyst in aliphatic solvents. The rate dependence on Co was also generally found to be linear, in the range of concentrations usually employed (0.01 to 0.09 mmol l$^{-1}$). Most authors have reported that the rate of polymerization is independent of the concentration of the Al compound; a linear dependence of the rate on Al was however reported for the system Co carboxylate–$AlEt_{1.1}Cl_{1.9}$ in a mixed benzene/aliphatic solvent (benzene 10% by volume).[80] A 0.5 order rate dependence on Al was observed with $Al(Bu^i)_2Cl$–$CoCl_2 \cdot EtOH$–$H_2O$, but only at concentrations of $Al(Bu^i)_2Cl$ lower than 12 mmol l$^{-1}$.[295] However, most of the evidence indicates that for concentrations higher than this value, the Al compound has practically no effect on the rate. The overall activation energy $E_a$ was reported to be 49.6 kJ mol$^{-1}$.[301]

The butadiene polymerization to the *cis*-1,4 polymer was reported to be first order with respect to monomer concentration with $Ni(stearate)_2$–$AlEt_2Cl$,[134] and first order in both butadiene and catalyst concentration with $Ni(carboxylate)_2$–$BF_3 \cdot Et_2O$–$AlR_3$ (R = Et, Bu$^i$), the total activation energy being 46.9 kJ mol$^{-1}$.[305] The butadiene polymerization rate in the presence of nickeldiethyldithiocarbamate–$AlEt_2Cl$ was reported to be first order in butadiene and nickel concentration, while the order with respect to $AlEt_2Cl$ was found to decrease asymptotically from two, at low Al/Ni ratios, to zero as the Al concentration increased.[423]

The kinetics of the *cis* polymerization of butadiene and isoprene with lanthanide catalysts was examined by various authors; the rate was found to be first order with respect to monomer with all the systems examined.[163,424–427] The polymerization of butadiene with Nd octanoate–$Al(Bu^i)_2H$–EtCl and $NdCl_3$ $3Pr^iOH$–$AlEt_3$ was reported to be first order with respect to Nd, whereas the order with respect to the Al alkyl was reported to be 0.33 for the former and 0.55 for the latter system.[424,425] The rate dependence on the Al alkyl is due to the fact that the activity of the system increases with the concentration of Al, which is probably attributable to an increase in the number of active centers. The isoprene polymerization with Nd naphthenate–$Al(Bu^i)_3$–$Et_3Al_2Cl_3$ was reported to be of order 1.7 with respect to catalyst concentration (Al + Nd).[426,427] $E_a$ was 34.9 kJ mol$^{-1}$ for the polymerization of butadiene and 34.3 kJ mol$^{-1}$ for that of isoprene.[163]

Only a few data have been reported on the *trans* polymerization of butadiene and isoprene; a first order dependence of the rate on monomer was observed with the $VOCl_3$–$Al(Bu^i)_3$ system.[428]

The kinetics of the polymerization of butadiene to the 1,2 polymer was investigated with the following systems: $Ti(OBu^n)_4$–$AlEt_3$,[415] $MoCl_2(OC_8H_{17})_2$–$Al(Bu^i)_2OPh$,[429] $MoCl_4(OC_8H_{17})_2$–$Al(Bu^i)_2OPh$[430] and $MoCl_4$–$Al(Bu^i)_2OC_6H_4Me$.[431] The rate was found to be first order in monomer and catalyst concentration. For the Mo catalysts $E_a$ was reported to be 77.4, 71.5 and 72.8 kJ mol$^{-1}$ respectively.

Kinetic data have been reported for the polymerization of butadiene with various allyl derivatives of transition metals. Table 14 reports some data concerning the most common systems. The different order with respect to metal concentration probably reflects the different degrees of association of the complexes in the polymerization conditions. The different order with respect to monomer seems to indicate a different mechanism in the participation of the monomer in the formation of the active species.

**Table 14** Kinetic Data of Butadiene Polymerization with some Allyl Complexes

| *Catalyst* | $E_a$ (kJ mol$^{-1}$) | *Reaction order with respect to the concentration of* *Metal* | *Monomer* | *Ref.* |
|---|---|---|---|---|
| $(\eta^3\text{-}C_3H_5)NiI$ | | 0.5 | 1 | a |
| $(\eta^3\text{-}C_4H_7)NiX$ (X = Cl, Br, I) | 60.7–66.9 | 0.5 | 1 | b |
| $(\eta^3\text{-}C_4H_7)NiOCOCl_3$ | | 1 | 1.6 | b |
| $(\eta^3\text{-}C_3H_5)NiOCOCF_3$ | | 1 | 1 | c |
| $(\eta^3\text{-}C_4H_7)NiCl$ + chloranil | 25.1 | 1 | 1 | b |
| $(\eta^3\text{-}C_3H_5)Ni + CCl_3CHO$ | | 2 | 2 | d |
| $(\eta^3\text{-}C_4H_7)NiCl + CCl_3CHO$ (or $CCl_3CO_2H$) | 8.4 | 1 | 2 | d, e |
| $(\eta^3\text{-}C_4H_7)NiCl + SO_2Cl$ | | 2[f] | 1 | g |
| $(\eta^3\text{-}C_3H_5)_4Mo + HCl$ | | 1 | 1 | h |
| $(\eta^3\text{-}C_3H_5)_4Mo + C_3H_5I$[i] | 50.2 | 1 | 0 | l |
| $(\eta^3\text{-}C_3H_5)PdX$ (X = Cl, Br) | | 0.5 | 1 | m |
| $(\eta^3\text{-}C_4H_7)PdCl$ | | 0.5 | 1 | n |
| $(\eta^3\text{-}C_3H_5)_3Cr$ | 79.5 | 2 | 0 | b |

[a] G. Henrici-Olivé, A. Olivé and E. Schmidt, *J. Organomet. Chem.*, 1972, **39**, 201.
[b] B. A. Dolgoplosk, *Vysokomol. Soedin., Ser A*, 1971, **13**, 325; *Polym. Sci. USSR (Engl. Transl.)*, 1971, **13**, 367.
[c] J. C. Maréchal, F. Dawans and Ph. Teyssié, *J. Polym. Sci., Part A-1*, 1970, **8**, 1993.
[d] E. S. Novikova, O. P. Parenago, V. M. Frolov and B. A. Dolgoplosk, *Kinet. Katal.*, 1976, **17**, 928; *Kinet. Catal. (Engl. Transl.)*, 1976, **17**, 809.
[e] V. M. Gorelik, E. S. Novikova, O. P. Parenago, V. M. Frolov and B. A. Dolgoplosk, *Kinet. Katal.*, 1972, **13**, 1428 (*Chem. Abstr.*, 1973, **78**, 85 513).
[f] At low catalyst concentration; the order drops to zero at high catalyst concentration.
[g] E. S. Novikova, O. P. Parenago and V. M. Frolov, *Kinet. Katal.*, 1978, **19**, 268; *Kinet. Catal. (Engl. Transl.)*, 1978, **19**, 215.
[h] V. S. Stroganov, E. S. Novikova and V. M. Frolov, *Kinet. Katal.*, 1982, **23**, 995; *Kinet. Catal. (Engl. Transl.)*, 1982, **23**, 845.
[i] The rate also shows a first order dependence on allyl iodide.
[l] R. I. Ter-Minasyan, O. P. Parenago, V. F. Chirkova, V. M. Frolov and B. A. Dolgoplosk, *Kinet. Katal.*, 1976, **17**, 935; *Kinet. Catal. (Engl. Transl.)*, 1976, **17**, 815.
[m] L. Ya. Al't, A. I. Lazutkina, A. M. Lazutkin and Yu. I. Ermanov, *Kinet. Katal.*, 1971, **12**, 1410 (*Chem. Abstr.*, 1972, **76**, 128 384).
[n] G. M. Khvostik, I. Ya. Poddubnyi and V. N. Sokolov, *Dokl. Akad. Nauk SSSR*, 1969, **186**, 894 (*Chem. Abstr.*, 1969, **71**, 71 656).

## 5.7 REGIO- AND CHEMO-SELECTIVITY OF THE INSERTION REACTION

Butadiene can give only one type of butenyl group on insertion into a metal–carbon bond. This group can exist in two forms, *syn* and *anti*, (**9a**) and (**9b**). A non-symmetric 1,3-dialkene, such as $CH_2{=}CRCH{=}CH_2$, $CH_2{=}CHCH{=}CHR$ or $CH_2{=}CHCH{=}CR_2$, can give instead two different positional isomers. Isoprene, for example, can give the allylic groups (**10**) and (**11**), each of which can exist in the *syn* form (as shown) or in the *anti* one.

1,3-Pentadiene and 4-methyl-1,3-pentadiene can give the groups (**12**)–(**13**) and (**14**)–(**15**) respectively (only the *syn* form is shown).

Information on the regioselectivity of the insertion reaction has been obtained with NMR examination in the case of the polymerization of various monomers with $[(\eta^3\text{-butenyl})NiI]_2$.[402]

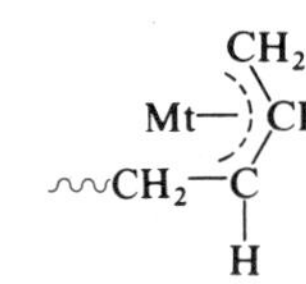

*anti*

(**9a**)

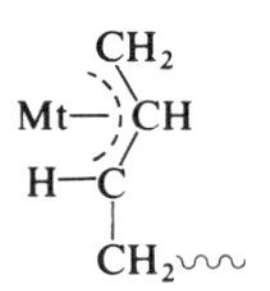

*syn*

(**9b**)

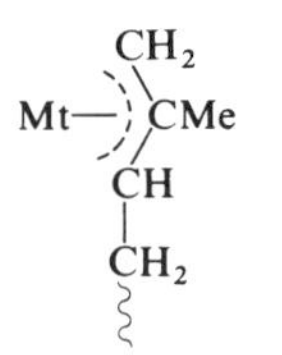

(**10**)

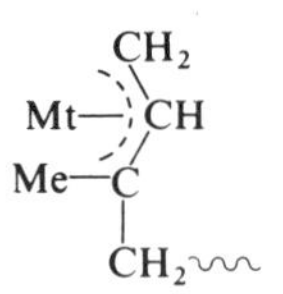

(**11**)

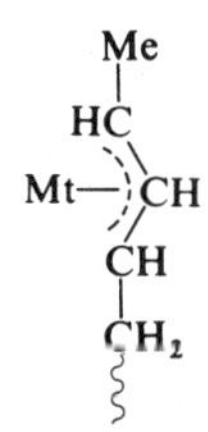

(**12**)

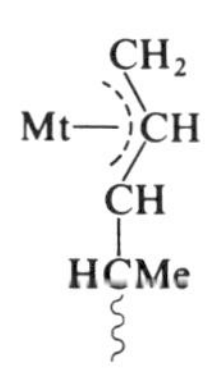

(**13**)

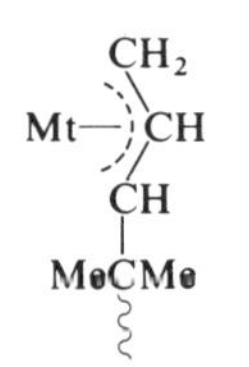

(**14**)

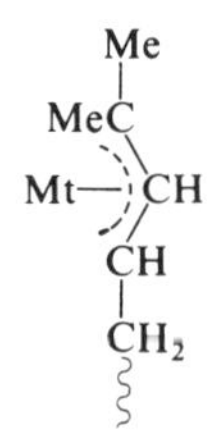

(**15**)

This catalyst is homogeneous, and monomer incorporation is sufficiently slow to be followed by NMR techniques. Isoprene gives predominantly (about 90%) unit (**10**), while pentadiene and 4-methylpentadiene give exclusively (**13**) and (**15**) respectively.

NMR examination has not been carried out on other catalyst systems, but information on the type of allylic unit formed can be derived indirectly from the nature of non-1,4 units present in the polymer, as will be shown in the following.

An $\eta^3$-butenyl group has two reactive points, C(1) and C(3). There are several examples in the literature of insertion reactions of simple molecules (*e.g.* carbon monoxide) into Mt—C(1) or Mt—C(3) of an allylic complex. In the polymerization of 1,3-dienes the chemoselectivity, that is the formation of 1,2 *vs.* 1,4 units, depends on whether insertion of the new monomer occurs at C(3) or at C(1) of the last polymerized unit (Scheme 5).

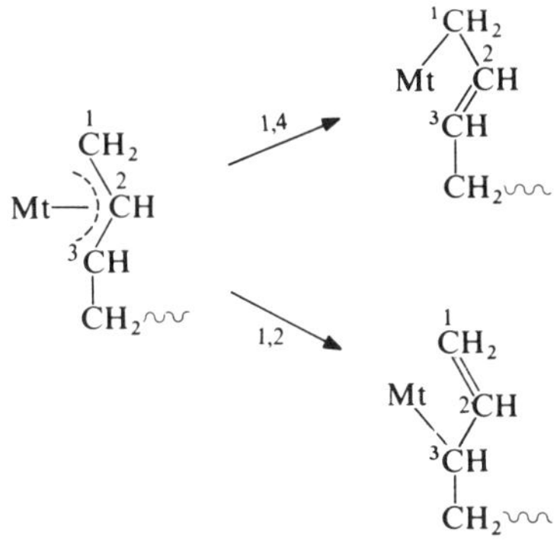

**Scheme 5**

Both butenyl groups of the *syn* form (as shown) or the *anti* form can give 1,2 units. It is clear that in the polymerization of non-symmetric monomers the type of unit that results from reaction at C(3) will depend on the type of the allylic group. Thus, group (**10**) will give a 3,4 isoprene unit, whereas group (**11**) will give a 1,2 isoprene unit. Analogously, (**12**) will give a 1,2 pentadiene unit and (**13**) a 3,4 unit. In the polymers of isoprene and pentadiene obtained with the most common catalysts (*e.g.* $AlR_3$–$TiCl_4$, $AlR_3$–$VCl_3$, $AlEt_2Cl$–$Co(acac)_2$–$H_2O$ and $AlR_3$–$NdCl_3 \cdot nD$, where D is a donor) the non-1,4 units are exclusively 3,4 and 1,2 respectively. These can be generated only by groups (**10**) and (**12**).

The case of 4-methyl-1,3-pentadiene is particularly simple because this monomer gives only 1,2 units in polymerizations with transition metal catalysts.[389] Units of this type can derive only from (**15**). It appears from the above that the nature of non-1,4 units can give an indication of the regioselectivity of monomer insertion reactions.

On the other hand, in stereoregular *cis*-1,4 or *trans*-1,4 polymers of isoprene or pentadiene head-to-head inversions are negligible, if any, according to NMR examination.[163,432] This means that the same type of allylic group is formed throughout all the polymerization, since, if butenyl groups (**10**) and (**11**) or (**12**) and (**13**) respectively would form on the same catalytic center, inversions should be present in the polymers.

With regard to the chemoselectivity, the factors that determine reaction of the new monomer at C(3) rather than at C(1) have not been fully clarified. Two cases have been examined in detail, the formation of 1,2-poly(4-methylpentadiene) and the formation of 1,2-polybutadiene.

4-Methylpentadiene has been polymerized with several catalyst systems, homogeneous or heterogeneous, based on Ti and V. Only 1,2 polymers have been obtained, with exclusion of 1,4 units, even with catalysts (*e.g.* $AlR_3$–$TiCl_4$, $AlR_3$–$VCl_3$) that give 1,4 polymers from butadiene or isoprene.[389] The particular behavior of 4-methylpentadiene has been interpreted on the basis of steric effects.[360] In the butenyl groups formed from butadiene or isoprene, (**9**), (**10**) and (**11**), C(1) is less substituted and hence more reactive than C(3). In the butenyl group (**15**) C(1) is more substituted than C(3) and hence less reactive. This may be the reason why catalyst systems that give 1,4 polymers from butadiene or isoprene give exclusively 1,2 polymers from 4-methylpentadiene.

Steric factors, however, cannot explain the formation of 1,2 units from butadiene, since in (**9a**) and (**9b**) C(3) is more substituted than C(1). For the formation of 1,2-polybutadiene with some soluble catalyst systems based on $AlR_3$ and alcoholates or acetylacetonates of Ti, V, Cr, Mo and Co an interpretation has been reported, which is based on the following observations:[360] (i) in the polybutadienes obtained with the above systems the non-1,2 units are predominantly *cis*-1,4;

(ii) some of the above systems give high-*cis* polymers or polymers with a mixed *cis*–1,2 structure from monomers other than butadiene; and (iii) catalysts based on Mo or Co are known that give polybutadienes with a 50/50 *cis*–1,2 structure.

The interpretation proposed is shown in Figure 10. The last polymerized unit is below the plane of the figure, the transition metal Mt is on the plane and the incoming monomer is above the plane. According to this scheme a 1,2 or a *cis*-1,4 unit can be obtained depending on whether the new Mt—C bond is formed at C(1′) or at C(4′) of the new monomer. Small variations in the position of the latter with respect to the last polymerized unit, due to the nature of the anionic ligand or the nature of the monomer itself, can determine the formation of one or other of the monomeric units. The scheme of Figure 10 can explain the formation with $AlR_3$–$Ti(OR)_4$ of 1,2 polymers from butadiene[70] and *cis*-1,4 polymers from pentadiene.[363] It can also account for the formation, with some systems, of equibinary *cis*-1,4/1,2 polybutadienes.[132,326,327] In these systems, probably due to the nature of the anionic ligands, the formation of 1,2 and *cis*-1,4 units becomes equally probable. It will be shown in Section 5.8.2 that the scheme proposed also accounts for the stereospecificity of the polymerization (the 1,2 polymers have a syndiotactic and the *cis*-1,4 an isotactic structure).

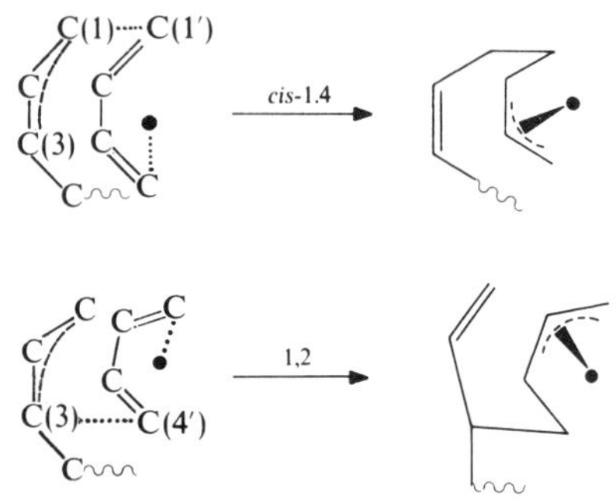


**Figure 10** Mode of formation of a *cis*-1,4 *vs.* a 1,2 monomeric unit from butadiene

The above interpretation is not valid in general. There are cases in which the formation of 1,2 units depends on other factors, not yet elucidated. The polybutadiene obtained with $AlEt_2Cl$–Co-$(acac)_2$–$H_2O$ has a 97–99% *cis*-1,4 structure, but the polymers of isoprene or 2,3-dimethylbutadiene obtained with the same system contain respectively 25–30% of 3,4 and 19–20% of 1,2 units. When butadiene is copolymerized with isoprene or dimethylbutadiene with the Co system the butadiene units in the polymer become partly 1,2, as shown in Section 5.9. Evidently, the presence of the methyl groups in isoprene and dimethylbutadiene affects in some way, not yet clarified, the chemoselectivity of the insertion reaction, both in homo- and co-polymerization.

## 5.8 STEREOSELECTIVITY OF THE INSERTION REACTION

### 5.8.1 *Cis–Trans* Isomerism in 1,4 Polymers

The formation of a *cis*-1,4 or a *trans*-1,4 unit is related to the structure, *anti* or *syn*, of the last polymerized unit (Scheme 6).

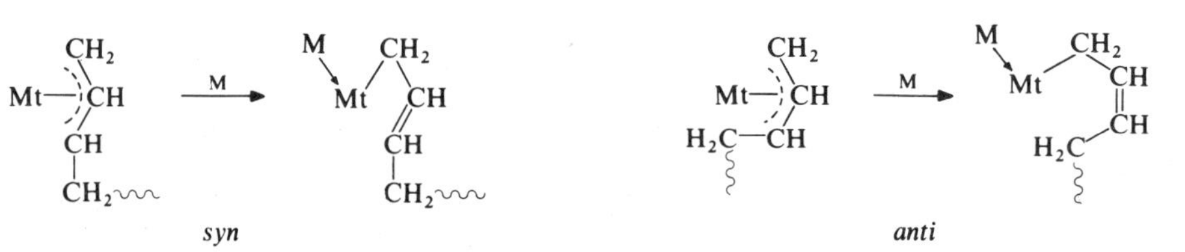


**Scheme 6**

The factors that determine the formation of a *syn*-butenyl group with some catalysts and of an *anti* one with others are not completely clear. There is a general agreement, however, that the mode of coordination of the monomer is the main factor in determining the structure of the butenyl group. A 1,3-diene can coordinate to a transition metal with only one double bond, as an *s-trans*-$\eta^2$ ligand, or

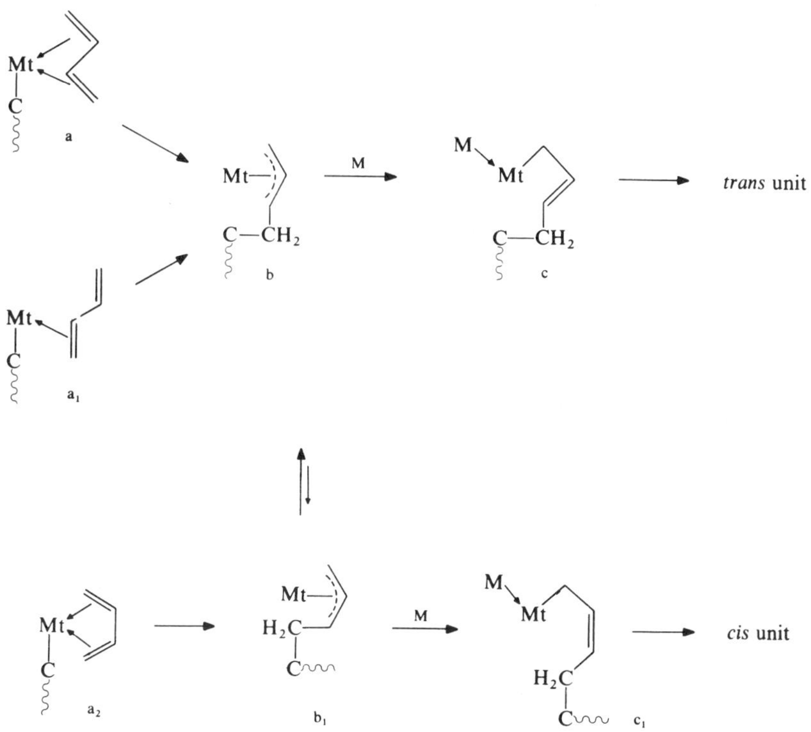


**Figure 11** Scheme for the formation of a *cis*-1,4 or *trans*-1,4 monomeric unit

with the two double bonds, as an *s-cis-*$\eta^4$ or an *s-trans-*$\eta^4$ ligand.[433] Figure 11 shows a scheme for the insertion of butadiene into the metal–carbon bond Mt—C between the transition metal and the last polymerized unit (the latter is in the $\eta^1$-allylic form).

The *s-trans-*$\eta^4$ coordination, although less common than the *s-cis* one, was first observed in some Zr–diene complexes[434–436] and has recently also been found in some Hf[437,438] and Mo[439,440] complexes. Therefore, one cannot exclude its intermediacy in the polymerization with some transition metal catalysts. The current interpretation is that a *syn*-butenyl group can derive either from an *s-trans-*$\eta^2$ or an *s-trans-*$\eta^4$ coordination, whereas an *anti*-butenyl group derives from an *s-cis-*$\eta^4$ coordination (Figure 11).

The *anti* and *syn* forms of a butenyl group are in equilibrium. It is known that when no bulky substituent is present at C(2), the equilibrium at room temperature is completely shifted towards the *syn* form, which is thermodynamically much more stable.[441,442] A *trans*-1,4 unit can therefore be obtained either *via* a (or $a_1$) → b → c or *via* $a_2 \rightarrow b_1 \rightarrow$ b → c of Figure 11. A *cis*-1,4 unit is obtained *via* $a_2 \rightarrow b_1 \rightarrow c_1$.

The formation of a *syn* butenyl group *via anti* → *syn* isomerization was clearly demonstrated in the polymerization of monomers $CH_2{=}CRCH{=}CH_2$ (R = Me, Et, $Pr^i$, $Bu^t$) with $[(\eta^3\text{-}C_4H_9)NiI]_2$,[427] and also in the polymerization of butadiene with some allyl nickel complexes.[277] Another case in which a *trans* polymer is formed *via anti* → *syn* isomerization is the polymerization of 2,4-hexadiene with $AlEt_2Cl$–$Co(acac)_2$ or $AlEt_2Cl$–$Nd(OCOR)_3$–$Al(Bu^i)_3$. These systems afford a *cis*-polymer from butadiene, but give a *trans* polymer from hexadiene.[391,392] The polymerization of this monomer is much slower than that of butadiene, probably due to the steric effect of the methyl groups. Since the insertion step is slow, the *anti* butenyl group has time to isomerize to the more stable *syn* structure before monomer incorporation. This interpretation is in accord with the results of the butadiene/hexadiene copolymerization with Co or Nd catalysts.[443] In copolymers containing a large amount of butadiene (> 90%) the hexadiene units are isolated between the butadiene units and have a *cis* structure. Insertion of butadiene into a hexadiene unit is more rapid than insertion of hexadiene into the same unit and occurs more rapidly than the *anti* → *syn* isomerization. In other words, the formation of a *trans vs.* a *cis* unit depends on the relative rate of two processes: (a) the insertion process of the new monomer on the allylic group; and (b) the *anti* → *syn* isomerization of the allylic group. If the insertion process is more rapid than the isomerization process, a *cis*-1,4 unit will result, otherwise isomerization occurs with consequent formation of a *trans* unit according to Scheme 7.

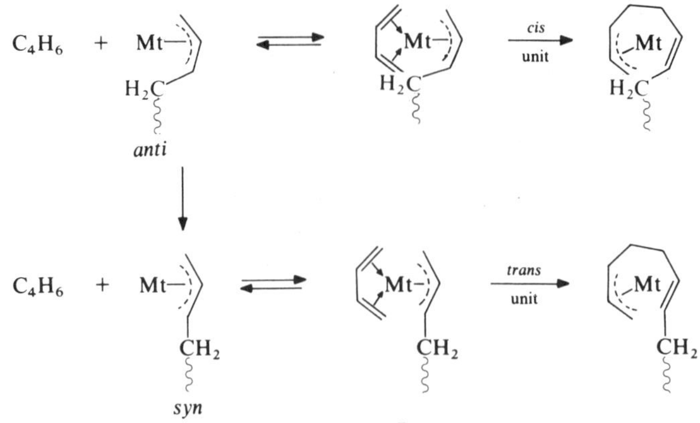

**Scheme 7**

As a result of *anti* → *syn* isomerization, a decrease in the *cis* content and a simultaneous increase in the *trans* content were observed in the polymerization of butadiene with some Ti, Co and Ni systems, at very low monomer concentration.[444,445] Under these conditions, the rate of polymerization is low and a certain amount of *trans* units are formed *via* $a_2 \rightarrow b_1 \rightarrow b \rightarrow c$ of Figure 11.

The evidence available indicates that with various catalysts (in particular with the soluble ones) *trans*-1,4 polymers are formed *via* $a_2 \rightarrow b_1 \rightarrow b \rightarrow c$ of Figure 11. However, *trans*-1,4 polymers have also been obtained, using heterogeneous systems, from the (*Z*) isomers of monomers $CH_2{=}CHCH{=}CHR$, for which the *s-cis*-$\eta^4$ coordination is unfavored.[358,375] These polymers are probably formed *via* a (or $a_1$) → b → c in Figure 11.

The scheme of Figure 11 also accounts for the effect of electron donors on polymer microstructure. As indicated in previous paragraphs, some systems, typically for the *cis* polymerization of butadiene, give a *trans* polymer on addition of electron donors. For example $AlEt_2Cl$–$Co(acac)_2$ and $(allyl)NiOCOCF_3$, both highly stereospecific for the formation of *cis* polybutadiene, give a *trans* polymer in the presence respectively of $NEt_3$[120] and $P(OPh)_3$ or EtOH.[234] A plausible interpretation is that the electron donor occupies a coordination site, thus obliging the new monomer to coordinate as an *s-trans*-$\eta^2$ ligand, as in $a_1$ of Figure 11. When the electron donor D has a basicity comparable to that of the monomer, a competitive monomer/donor coordination takes place as shown in equation (14).

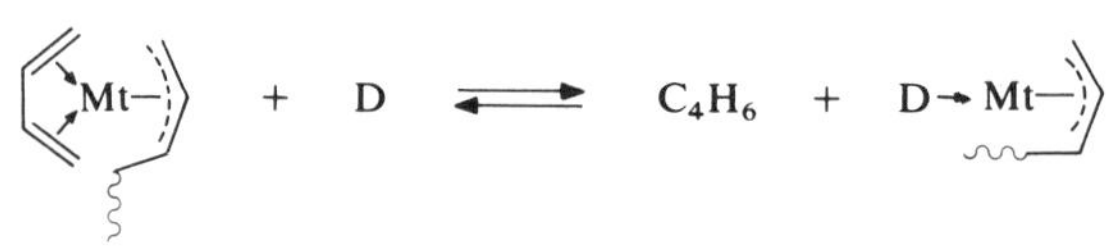

(14)

**(16)** **(17)**

Equilibrium (14) may affect polymer microstructure in different ways. If D occupies only one site of coordination, species (**17**) will coordinate the monomer as in $a_1$ of Figure 11 to give *trans* units. A polymer with a mixed *cis*/*trans* structure will result, the length of the *cis* or *trans* sequences depending on the lifetimes of species (**16**) and (**17**) respectively. If D occupies more than one coordination site (*e.g.* as arenes do), species (**17**) becomes inactive, since it cannot coordinate a new monomer. Depending on the nature of the metal Mt and of the anionic ligands, an *anti* → *syn* isomerization of (**17**) may occur. Also in this case addition of D will result in the formation of a polymer with a mixed *cis*/*trans* structure, according to a mechanism similar to that outlined in Scheme 7.

Equibinary *cis*/*trans* polybutadienes have been prepared by addition of ligands such as benzene or $CF_3CO_2H$ to $(allyl)NiOCOCF_3$ under specific conditions. The formation of these polymers is presumably interpretable on the basis of the above rationalization.

The scheme of Figure 11 also accounts for the effect of electron acceptors on (allyl)Ni halides. The formation of ionic or charge transfer complexes favors the coordination of the monomer, as in $a_2$ of Figure 11.

Support for the validity of the scheme of Figure 11 has come from the polymerization of 1,4-disubstituted butadienes. Figure 12 reports a scheme, analogous to that of Figure 11, in which the stereochemistry of insertion of a monomer (*E*,*E*)-$CHX{=}CHCH{=}CHX$ is indicated.[391,446] If the insertion occurs *via* a *cis* addition (*i.e.* if it consists of a transfer of the Mt—Ç bond of the $\eta^1$-allylic

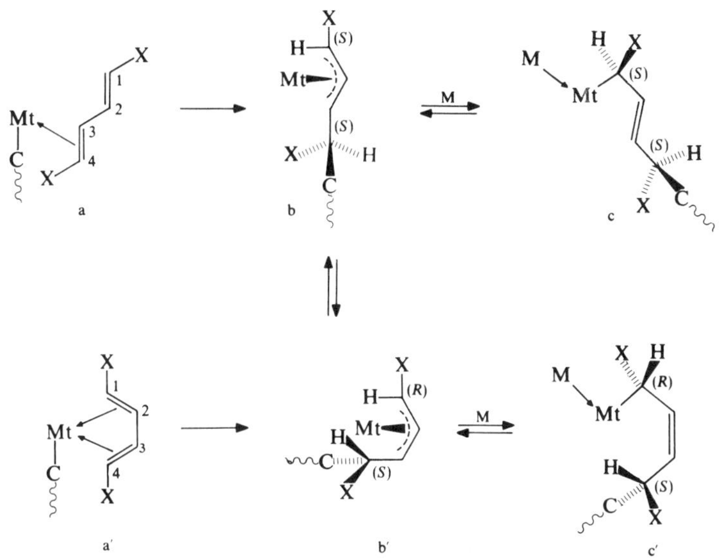

**Figure 12** Stereochemistry of insertion of a monomer $(E,E)$-1,4-$X_2$-1,3-butadiene

unit to the coordinated monomer) in the *syn* complex b, C(1) and C(4) have the same chirality, whereas in the *anti* complex b′, they have an opposite chirality.

A *syn* complex can also be obtained *via anti* → *syn* isomerization. The stereochemistry of this rearrangement occurs as shown in the scheme of Figure 13. The intermediate is an $\eta^1$-allylic species and the rearrangement occurs through rotation around the C(2)—C(3) bond and reattachment of the double bond to Mt, with the other face, to yield a new $\eta^3$-allylic complex.[441,442] The net result is: (1) inversion of configuration at C(3), from *anti* to *syn*; and (2) inversion of chirality at C(1). Hence, in a *syn* complex C(1) and C(4) have the same chirality independently of whether the *syn* complex is

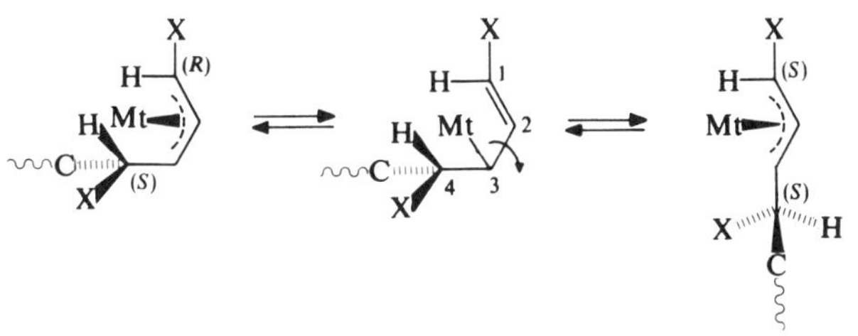

**Figure 13** Stereochemistry of the *anti–syn* isomerization of an $\eta^3$-butenyl group

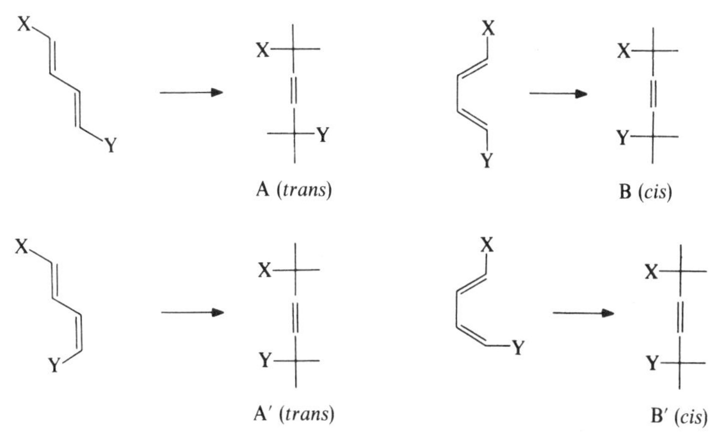

**Figure 14** Fischer projections of *cis* and *trans* monomeric units derived from the $(E,E)$ and $(E,Z)$ isomers of a monomer XCH═CHCH═CHY, according to the scheme of Figure 12

obtained *via* a → b or *via* a′ → b′ → b of Figure 12. When a new monomer is incorporated, the *syn* complex gives a *trans* unit in which C(1) and C(4) have the same chirality, the *anti* one gives a *cis* unit in which C(1) and C(4) have opposite chirality. More generally, the *trans* and *cis* units that can be obtained from two generic (*E*,*E*)- and (*Z*,*E*)-1,4-disubstituted butadienes are those represented as Fischer projections in Figure 14.[446]

Several stereoregular *cis*-1,4 or *trans*-1,4 polymers have been obtained from 1,4-disubstituted butadienes, as shown in Table 13. The structures of the polymers are in agreement with the structures of the *cis* and *trans* units indicated in Figure 14.

An implication of the above structures of *cis* or *trans* units is that only some types of 1,4 stereoregular polymers can be obtained from a 1,4-disubstituted butadiene. For example, if a monomer (*E*,*E*)-CHX=CHCH=CHX yields *trans* units A and *cis* units B, the stereoregular 1,4 polymers that can be obtained from it are those indicated in Figure 15.[391,446]

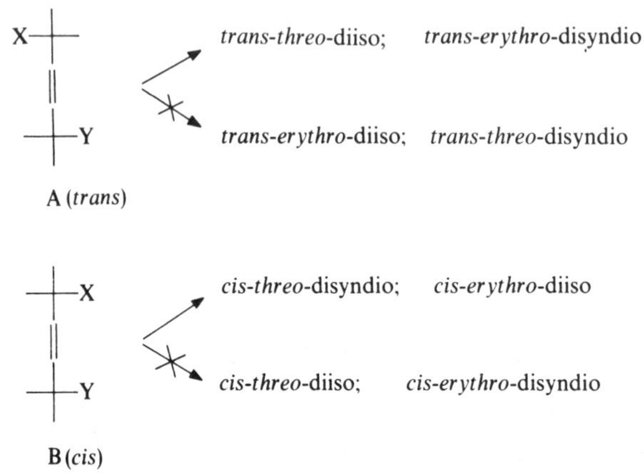


**Figure 15** Types of stereoregular 1,4 polymers that can be obtained from a monomer XCH=CHCH=CHY

The results so far obtained from 1,4-disubstituted butadienes are in agreement with the above scheme. It appears from the above that the use of 1,4-disubstituted butadienes provides stereochemical evidence that *cis* and *trans* units derive from *anti* and *syn* butenyl groups respectively.

### 5.8.2 Isotactic–Syndiotactic Isomerism

It has been shown in Section 5.8.4 that several polymers having an isotactic structure have been obtained from 1,3-dienes, *e.g.* 1,2 iso- or syndio-tactic polybutadiene, *cis*-1,4 iso- or syndio-tactic polypentadiene, *trans*-1,4 isotactic polypentadiene and *trans*-1,4 *threo* diisotactic poly(2,4-hexadiene). The factors that determine stereoregulation in 1,3-diene polymerization are more complex than in monoalkene polymerization. As shown in Figure 11, a dialkene can coordinate in different ways to the catalytic complex. In addition, as will be shown later, the same mode of presentation of the monomer to the active center can give rise to an isotactic or a syndiotactic placement, depending on whether a 1,2 or a 1,4 unit is formed. This explains why some catalysts give isotactic polymers from one monomer and syndiotactic polymers from a different monomer. For example, $AlR_3$–$Ti(OR)_4$ gives *cis*-1,4 isotactic polypentadiene and 1,2 syndiotactic polybutadiene; $AlEt_2Cl$–$Co(acac)_2$–$H_2O$ gives *cis*-1,4 syndiotactic polypentadiene and *trans*-1,4 *threo* diisotactic poly(2,4-hexadiene).

A metal–butenyl group, *syn* or *anti*, is a chiral group, independently of the substituents; (**18a**) and (**18b**) represent the two enantiomers of an *anti* metal–butenyl group derived from butadiene.

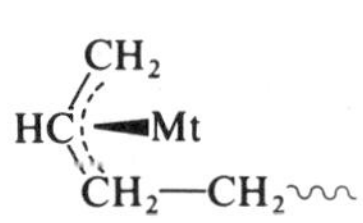


(**18a**)

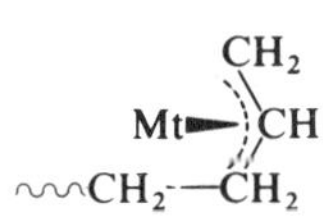


(**18b**)

Insertion of a new monomer will give a butenyl group of the same chirality as the previous one (and hence an isotactic diad) or of opposite chirality (and hence a syndiotactic diad) depending on the orientation of the new monomer with respect to the last polymerized unit.

Only the formation of a few iso- or syndio-tactic polymers have been examined. One of these is the formation of *cis*-1,4 isotactic and *cis*-1,4 syndiotactic polypentadiene.[147, 395,447] The schemes proposed are shown in Figure 16. These schemes show the last polymerized unit, in the *anti* configuration, the transition metal Mt and the incoming monomer. The latter can assume two possible orientations with respect to the butenyl group, as shown in a and b respectively. a′ and b′ represent the same situation as a and b, but observed from the right to the left along an axis that contains the butenyl group. With the orientation as in a, the new monomer will give on insertion a butenyl group of the same chirality as the preceding one (hence, an isotactic diad), and orientation as in b will give a new butenyl group of opposite chirality (hence a syndiotactic diad). According to the schemes of Figure 17, the monomer (*E*, *E*)-CHD=CHCH=CHMe should give a *cis*-diisotactic polymer with an *erythro* structure, and a *cis*-disyndiotactic polymer with a *threo* structure.

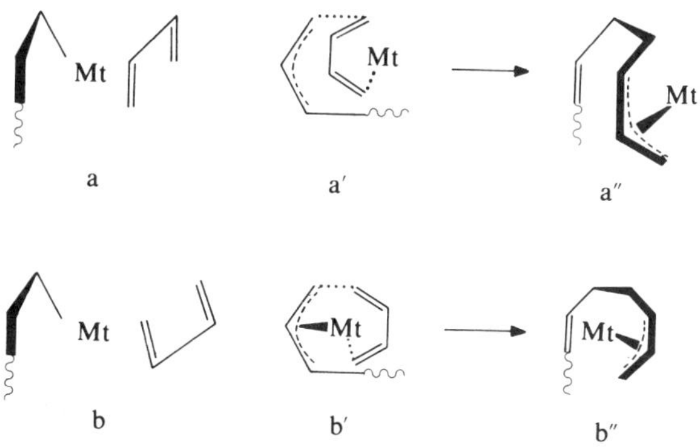


**Figure 16** Formation of a butenyl group of the same (a′ a″) or of opposite (b′ b″) chirality of the preceding one, depending on the orientation of the incoming monomer

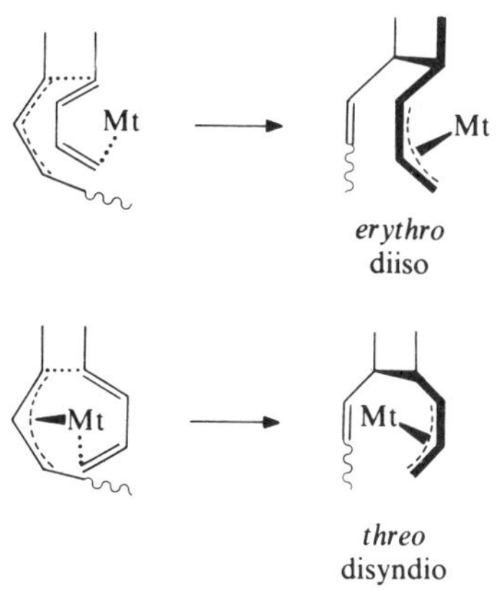


**Figure 17** Scheme for the formation of an *erythro* diisotactic and of a *threo* disyndiotactic *cis*-1,4 polymer from a terminally substituted butadiene

It was actually found that the polymers of (*E*,*E*)-CHD=CHCH=CHMe obtained with $AlEt_2Cl$–$Nd(OCOR)_3$–$Al(Bu^i)_3$ and $AlEt_2Cl$–$Co(acac)_2$–$H_2O$ have the structures expected, which supports the validity of the schemes presented (Table 13).

Another case that has been examined is the formation of 1,2 syndiotactic polybutadiene.[360,446] The interpretation reported for the formation of this polymer is based on the scheme of Figure 10. The last polymerized unit has an *anti* configuration, the new monomer has an *s-cis*-$\eta^4$-coordination. It has been shown in Section 5.7 that a scheme of this type can lead to a 1,2 or a *cis*-1,4 unit, depending on whether the new carbon–carbon bond is formed between C(1) and C(1′) or C(3) and C(4′). The formation of a *cis* unit according to this scheme leads to a butenyl group of the same chirality as the preceding one, as shown in the previous paragraph. The formation of a 1,2 unit

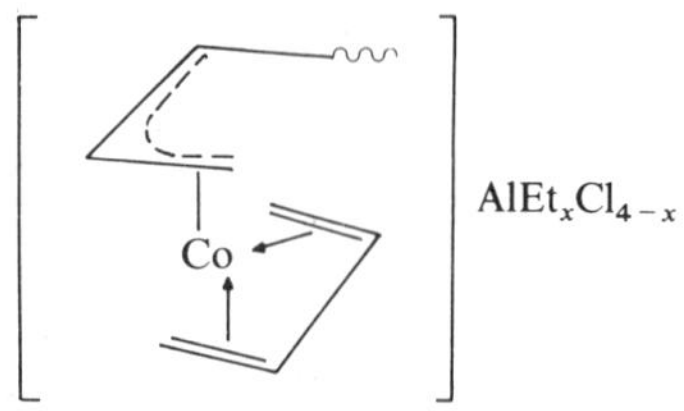


**Figure 18** Possible structure for the catalytic complex derived from a cobalt compound and an aluminum alkyl chloride

instead gives rise to a butenyl group of opposite chirality and hence to a syndiotactic polymer. The scheme accounts for the formation, with $AlR_3$–$Ti(OR)_4$, of a 1,2 syndiotactic polymer from butadiene and of a *cis*-1,4 isotactic polymer from pentadiene.

An interpretation of the factors that determine orientation of the incoming monomer as in a or in b of Figure 16 has been reported.[147] In the system $AlEt_2Cl$–$Co(acac)_2$–$H_2O$ the active catalyst is an ionic species which can be visualized as in Figure 18. In this species no anionic ligand is bonded to Co. The orientation of the new monomer seems to be essentially determined by the interaction between the last polymerized unit and the incoming monomer. Orientation as in b of Figure 16 presumably minimizes such an interaction.

In the systems $AlR_3$–$Ti(OR)_4$ and $AlR_3$–$NdCl_3 \cdot n$D, which give *erythro cis*-1,4 isotactic polymers from $(E,E)$-CHD=CHCH=CHMe, anionic ligands are bonded to Ti and Nd of the catalytic complexes. The $Ti(OR)_4$ system gives an optically active *cis*-1,4 isotactic polymer from pentadiene when R is an optically active group.[364] The asymmetric induction exerted by the catalyst can be accounted for only by assuming that at least one OR group is bonded to Ti. The active species in the Nd system contains $Nd^{III}$ (see Section 5.3.2.5), hence Cl atoms are bonded to Nd in the active species. The difference between the Co system and those based on Ti and Nd consists of the presence in the latter of anionic ligands bonded to the transition metal. It is presumably the presence of those ligands that forces the new monomer to coordinate as in a of Figure 16.

This interpretation is supported by the results obtained with $AlEt_2Cl$–$Co(acac)_2$–$H_2O$ in the presence of some donors. Addition of THF or $CS_2$ to a toluene solution of the Co system results in a new catalytic species, which gives a 1,2 syndiotactic polybutadiene. This means that coordination of the monomer with $AlEt_2Cl$–$Co(acac)_2$–$H_2O$ occurs as in b of Figure 16, whereas addition of THF or $CS_2$ causes the new monomer to coordinate as in a. This emphasizes the importance of ligands in determining stereospecificity in dialkene polymerization.

## 5.9 DIALKENE COPOLYMERIZATION

The copolymerization of dialkenes with transition metal catalysts has been extensively investigated. For most of the copolymerizations the monomer reactivity ratios have been determined.[16]

In most of the cases the microstructure of each comonomer unit is the same as in the homopolymers but in some cases it differs.

Butadiene/isoprene copolymerization has been examined with several catalyst systems. The copolymers obtained with $AlR_3$–$TiCl_4$ are essentially *cis*-1,4 and the microstructure of butadiene and isoprene units does not differ substantially from that in the homopolymers.[448–450] The same phenomenon has been observed in butadiene/isoprene copolymers obtained with Nd catalysts, which are $>95\%$ *cis*-1,4.[163,168,451,452] By contrast, systems based on Co give copolymers in which the microstructure of the comonomer units markedly depends on copolymer composition.[392,453–455] The $AlEt_2Cl$–$Co(acac)_2$–$H_2O$ system gives homopolymers of butadiene which are *ca.* 97% *cis*. In the copolymers the *cis* content of the butadiene fraction decreases with increasing isoprene content. For a 50/50 copolymer only 70% of the butadiene fraction has a *cis* structure. On the other hand, the isoprene units, which are *ca.* 70% *cis* in the homopolymer, are *ca.* 90% *cis* in a 50/50 copolymer.[453] A similar phenomenon, but much more pronounced, has been observed in butadiene/2,3-dimethylbutadiene copolymers prepared with $AlEt_2Cl$–$Co(acac)_2$–$H_2O$. In 50/50 copolymers the butadiene units are 60% 1,2, whereas the dimethylbutadiene units, which are *ca.* 81% *cis* and 19% 1,2 in the homopolymer, are more than 98% *cis* in copolymers containing more than 50% butadiene.[388]

The copolymerization of butadiene with 1,3-pentadiene,[308,456] 2-phenylbutadiene[457] and other dienes[443,458–460] with Ti or Nd catalysts has been reported.

Copolymers with a predominantly *trans*-1,4 structure have also been prepared, mainly using vanadium-based catalysts.[461] The butadiene/pentadiene copolymerization with the soluble system $AlEt_2Cl–V(acac)_3$ has been particularly studied.[86,462,463] The butadiene units in the copolymer are all *trans*-1,4, while the pentadiene units are 60–65% *trans*-1,4 and 35–40% 1,2. The copolymers containing up to *ca.* 30% by weight of pentadiene are crystalline by X-ray, those containing 30–40% of pentadiene are amorphous at rest but crystallize on stretching. The elastomeric properties of these copolymers have been evaluated.[464] The copolymerization of butadiene with isoprene, pentadiene, 2,3-dimethylbutadiene and 1,3-cyclohexadiene with allyl complexes of Ni and Cr has been reported.[221,465,466]

The alkene/dialkene copolymerization has attracted much interest. Random copolymers of ethylene and/or propylene with butadiene, isoprene or pentadiene have been prepared with various catalysts, mainly based on Ti or V.[467–477]

Copolymers of butadiene with isobutene have also been prepared.[478] Homopolymers of isobutene (due to cationic activity) and/or of butadiene are usually present in the reaction product, along with the copolymer. The best yields of the copolymer (*ca.* 80% of the crude polymerization product) were obtained with $AlH(Bu^i)_2–TiCl_4$–benzophenone (3:1:1.5), at −40 °C. Copolymers containing up to 30% of isobutene were isolated by fractionation. In the copolymers, isobutene units are present only in BIB sequences (B = butadiene; I = isobutene).

Of particular interest are the alternating alkene/dialkene copolymers. Ethylene/butadiene alternating copolymers were first obtained with $Al(Bu^i)_3–Al(Bu^i)_2Cl–VCl_4$–PhOMe;[479] other catalysts based on V or Ti were reported in subsequent patents.[480–482] Ethylene/isoprene[483–485] and ethylene/pentadiene[486] alternating copolymers were also prepared.

Much work has been carried out on the preparation of alternating propylene/butadiene copolymers.[487–491] The following are some of the systems used: $VOCl_3–VO(OEt)_3–Al(Bu^i)_3$; $TiCl_4$–PhOMe–$Al(Bu^i)_3$; $VOCl_3–Al(OPr^i)_3–Al(Bu^i)_3$; and $VO(OCH_2CMe_3)_2Cl–Al(Bu^i)_3$. Variants of these systems have been reported in recent literature and in several patents.[474,492–497] $^{13}$C NMR analysis has shown that the copolymers have a perfectly alternating monomer sequence distribution.[498] In the copolymers obtained with V catalysts, all the butadiene units have a *trans*-1,4 structure, while in the Ti copolymers a small amount of *cis* structure is also present. The butadiene/propylene alternating copolymers have good elastomeric properties.[489,490]

## 5.10 REFERENCES


1. S. E. Horne, Jr., J. P. Kichl, J. J. Shipman, V. L. Folt and C. T. Gibbs, *Ind. Eng. Chem.*, 1956, **48**, 784.
2. G. Natta, L. Porri and G. Mazzanti (Montecatini SpA), *Ital. Pat.* 536 631 (1955) (*Chem. Abstr.*, 1959, **53**, 3756).
3. G. Natta and L. Porri (Montecatini SpA), *Ital. Pat.* 538 453 (1956) (*Chem. Abstr.*, 1958, **52**, 5032).
4. G. Natta, L. Porri and A. Palvarini (Montecatini SpA), *Br. Pat.* 854 615 (1960) (*Chem. Abstr.*, 1961, **55**, 10 977).
5. J. L. Binder, *Ind. Eng. Chem.*, 1954, **46**, 1727.
6. J. P. Kennedy, 'Cationic Polymerization of Olefins: a Critical Inventory', Wiley, New York, 1975.
7. A. A. Morton, *Ind. Eng. Chem.*, 1950, **42**, 1488.
8. A. A. Morton, I. Nelidow and E. Schoenberg, *Rubber Chem. Technol.*, 1957, **30**, 326.
9. A. A. Morton, in 'Advances in Catalysis', ed. A. Farkas, Academic Press, New York, 1957, vol. 9, p. 743.
10. L. E. Forman, in 'Polymer Chemistry of Synthetic Elastomers', ed. J. P. Kennedy and E. Tornqvist, Wiley-Interscience, New York, 1969, part II, p. 492.
11. I. Kuntz and A. Gerber, *J. Polym. Sci.*, 1960, **42**, 299.
12. H. Hsieh and A. V. Tobolski, *J. Polym. Sci.*, 1957, **25**, 245.
13. W. Marconi, in 'Stereochemistry of Macromolecules', ed. A. D. Ketley, Dekker, New York, 1967, vol. 1, p. 239.
14. G. Natta and L. Porri, in 'Polymer Chemistry of Synthetic Elastomers', ed. J. P. Kennedy and E. Tornqvist, Wiley-Interscience, New York, 1969, part II, p. 597.
15. Ph. Teyssié, M. Julémont, J. M. Thomassin, E. Walkiers and R. Warin, in 'Coordination Polymerization', ed. J. C. W. Chien, Academic Press, New York, 1975, p. 327.
16. W. Cooper, in 'The Stereo Rubbers', ed. W. M. Saltman, Wiley, New York, 1977, p. 21.
17. Ph. Teyssié and F. Dawans, in 'The Stereo Rubbers', ed. W. M. Saltman, Wiley, New York, 1977, p. 79.
18. J. Boor, Jr., 'Ziegler Natta Catalysts and Polymerizations', Academic Press, New York, 1979.
19. B. A. Dolgoplosk, *Soc. Sci. Rev., Sect. B*, 1980, **2**, 203 (*Chem. Abstr.*, 1980, **93**, 239 967)
20. M. Farina, *Top. Stereochem.*, 1987, **17**, 1.
21. R. E. Rinehart, H. P. Smith, H. Witt and H. Romeyn, Jr., *J. Am. Chem. Soc.*, 1961, **83**, 4864.
22. R. E. Rinehart, H. P. Smith, H. Witt and H. Romeyn, Jr., *J. Am. Chem. Soc.*, 1962, **84**, 4145.
23. L. Porri, G. Natta and M. C. Gallazzi, *Chim. Ind. (Milan)*, 1964, **46**, 428 (*Chem. Abstr.*, 1964, **61**, 1944).
24. H. N. Friedlander and K. Oita, *Ind. Eng. Chem.*, 1957, **49**, 1885.
25. G. Natta, P. Pino, G. Mazzanti and P. Longi, *Gazz. Chim. Ital.*, 1957, **87**, 549 (*Chem. Abstr.*, 1957, **51**, 15 993).
26. H. Martin and J. Stedefeder, *Justus Liebigs Ann. Chem.*, 1958, **618**, 17.
27. W. M. Saltman, W. E. Gibbs and J. Lal, *J. Am. Chem. Soc.*, 1958, **80**, 5615.
28. C. Beerman and H. Bestian, *Angew. Chem.*, 1959, **71**, 618.

29. M. L. Cooper and J. B. Rose, *J. Chem. Soc.*, 1959, 765.
30. G. Natta, L. Porri, A. Mazzei and D. Morero, *Chim. Ind. (Milan)*, 1959, **41**, 398 (*Chem. Abstr.*, 1960, **54**, 9353).
31. Y. P. Tepenitsyna, M. I. Faberov, A. M. Kutin and G. S. Sevskaia, *Vysokomol. Soedin.*, 1959, **1**, 1148 (*Chem. Abstr.*, 1960, **54**, 20 604).
32. A. G. Pozamantir, A. A. Korotkov and I. S. Lishanskii, *Vysokomol. Soedin.*, 1959, **1**, 1207 (*Chem. Abstr.*, 1960, **54**, 22 313).
33. M. H. Jones, U. Martins and M. P. Thorne, *Can. J. Chem.*, 1960, **38**, 2303.
34. R. J. Kern and H. G. Hurst, *J. Polym. Sci.*, 1960, **44**, 272.
35. A. Simon and G. Ghymes, *J. Polym. Sci.*, 1961, **53**, 327.
36. E. J. Arlman and J. R. de Jong, *Recl. Trav. Chim. Pays-Bas*, 1960, **79**, 910 (*Chem. Abstr.*, 1961, **55**, 4222).
37. E. J. Arlman and J. R. de Jong, *Recl. Trav. Chim. Pays-Bas*, 1960, **79**, 1319 (*Chem. Abstr.*, 1961, **55**, 16 404).
38. E. J. Arlman, J. R. de Jong, J. Beintema and L. L. van Reijen, *Recl. Trav. Chim. Pays-Bas*, 1961, **80**, 1129 (*Chem. Abstr.*, 1962, **56**, 15 120).
39. G. Natta, P. Corradini, I. W. Bassi and L. Porri, *Atti Accad. Naz. Lincei, Cl. Sci. Fis., Mat. Nat., Rend.*, 1958, **24**, 121 (*Chem. Abstr.*, 1958, **52**, 15 183).
40. F. Bock and L. Moser, *Monatsh. Chem.*, 1912, **33**, 1407.
41. F. Bock and L. Moser, *Monatsh. Chem.*, 1913, **34**, 1825.
42. K. Ziegler, E. Holzkamp, H. Breil and H. Martin, *Angew. Chem.*, 1955, **67**, 541.
43. W. M. Saltman, *J. Polym. Sci., Part A*, 1963, **1**, 373.
44. E. Schoenberg, *Adv. Chem. Ser.*, 1966, **52**, 46.
45. G. Natta, L. Porri and L. Fiore, *Gazz. Chim. Ital.*, 1959, **89**, 761 (*Chem. Abstr.*, 1960, **54**, 22 335).
46. G. M. Tolstopyatov, B. F. Raitses, R. A. Korina and R. S. Bresler, *Kinet. Katal.*, 1977, **18**, 1387; *Kinet. Catal. (Engl. Transl.)*, 1977, **18**, 1136.
47. E. Schoenberg, D. L. Chalfant and R. H. Mayor, *Rubber Chem. Technol.*, 1964, **37**, 103.
48. G. Natta, *Chim. Ind. (Milan)*, 1960, **42**, 1207.
49. W. Marconi, A. Mazzei, S. Cucinella and M. De Maldè, *Makromol. Chem.*, 1964, **71**, 118.
50. A. Mazzei, S. Cucinella and W. Marconi, *Makromol. Chem.*, 1969, **122**, 168.
51. A. Balducci, M. Bruzzone, S. Cucinella and A. Mazzei, *Rubber Chem. Technol.*, 1975, **48**, 736.
52. W. Marconi, A. Mazzei and L. Gatto, *Eur. Polym. J.*, 1969, **5**, 747.
53. Yu. B. Monakov, N. K. Minchenkova and S. R. Rafikov, *Dokl. Akad. Nauk SSSR*, 1977, **236**, 1151 (*Chem. Abstr.*, 1978, **88**, 7511).
54. S. R. Rafikov, Yu. B. Monakov, G. A. Tolstikov, N. G. Marina, N. Kh. Minchenkova and I. G. Savelieva, *J. Polym. Sci., Polym. Chem. Ed.*, 1983, **21**, 2697.
55. G. H. Smith and D. C. Perry, *J. Polym. Sci., Part A-1*, 1969, **7**, 707.
56. Phillips Petroleum Co., *Br. Pat.* 848 065 (1960) (*Chem. Abstr.*, 1961, **55**, 15 982).
57. J. F. Henderson, *J. Polym. Sci., Part C*, 1963, **4**, 233.
58. W. M. Saltman and T. H. Link, *Ind. Eng. Chem. Prod. Res. Dev.*, 1964, **3**, 199.
59. P. H. Mayer, *J. Polym. Sci., Part A*, 1965, **3**, 209.
60. Phillips Petroleum Co., *Br. Pat.* 863 337 (1961) (*Chem. Abstr.*, 1961, **55**, 20 479).
61. P. H. Mayer and M. H. Lehr, *J. Polym. Sci., Part A*, 1965, **3**, 217.
62. S. L. Bresler, V. A. Grechanovskii, A. Muzhai and I. Ya. Poddubnyi, *Vysokomol. Soedin., Ser. A*, 1969, **11**, 1165; *Polym. Sci. USSR (Engl. Transl.)*, 1969, **11**, 1325.
63. S. L. Bresler, V. A. Grechanovskii, A. Muzhai and I. Ya. Poddubnyi, *Makromol. Chem.*, 1970, **133**, 111.
64. W. Marconi, M. Araldi, A. Beranger and M. De Maldè, *Chim. Ind. (Milan)*, 1963, **45**, 522 (*Chem. Abstr.*, 1963, **59**, 6593).
65. Firestone Tire and Rubber Co., *Ger. Pat.* 1 112 834 (1961) (*Chem. Abstr.*, 1961, **56**, 10 351).
66. V. A. Krol and Ye. K. Khrennikova, *Vysokomol. Soedin., Ser. A*, 1972, **14**, 2365; *Polym. Sci. USSR (Engl. Transl.)*, 1972, **14**, 2757.
67. M. H. Lehr and P. H. Mayer, *J. Polym. Sci., Part A*, 1965, **3**, 231.
68. W. Marconi, A. Mazzei, M. Araldi and M. De Maldè, *J. Polym. Sci., Part A*, 1965, **3**, 735.
69. W. Marconi, A. Mazzei, M. Araldi and M. De Maldè, *Chim. Ind. (Milan)*, 1964, **46**, 245 (*Chem. Abstr.*, 1964, **60**, 14 709).
70. G. Natta, L. Porri and A. Carbonaro, *Makromol. Chem.*, 1964, **77**, 126.
71. Yu. B. Monakov, G. R. Rafikov, A. M. Ivanova, A. A. Pozdeeva, E. E. Zaev, Yu. G. Grechishnikov, G. S. Igoshkina and M. S. Miftakhov, *Dokl. Akad. Nauk SSSR*, 1974, **218**, 353 (*Chem. Abstr.*, 1975, **82**, 43 849).
72. A. M. Ivanova, R. Z. Lukmanova and Yu. G. Grechishnikov, *Khim. Vysokomol. Soedin. Neftekhim.*, 1975, 65 (*Chem. Abstr.*, 1976, **85**, 63 423).
73. Yu. B. Monakov, S. R. Rafikov, A. M. Ivanova, A. A. Panasenko, G. A. Tolstikov, A. A. Pozdeeva, E. E. Zaev, R. Z. Lukmanova and G. S. Igoshkina, *Vysokomol. Soedin., Ser. A*, 1975, **17**, 2631; *Polym. Sci. USSR (Engl. Transl.)*, 1975, **17**, 3025.
74. Yu. B. Monakov, N. G. Marina, I. G. Savel'eva and S. R. Rafikov, *Vysokomol. Soedin., Ser. A*, 1981, **23**, 50 (*Chem. Abstr.*, 1981, **94**, 140 246).
75. S. Cucinella, A. Mazzei, W. Marconi and C. Busetto, *J. Macromol. Sci., Chem.*, 1970, **4**, 1549.
76. S. Nakatomi, *Nippon Kagaku Kaishi*, 1976, (2), 29 (*Chem. Abstr.*, 1976, **85**, 6105).
77. S. Nakatomi, K. Kamma, T. Inoue and Y. Kai, *Nippon Kagaku Kaishi*, 1976, (5), 814 (*Chem. Abstr.*, 1976, **85**, 34 321).
78. S. Nakatomi and T. Inoue, *Nippon Kagaku Kaishi*, 1976, (5), 820 (*Chem. Abstr.*, 1976, **85**, 34 322).
79. S. Nakatomi, *Nippon Kagaku Kaishi*, 1976, (2), 296 (*Chem. Abstr.*, 1976, **85**, 6141).
80. S. E. Horne, Jr., in 'Transition Metal Catalyzed Polymerizations', ed. R. P. Quirk, Harwood Academic, New York, 1983, part B, p. 527.
81. G. Natta, L. Porri, P. Corradini and D. Morero, *Chim. Ind. (Milan)*, 1958, **40**, 362 (*Chem. Abstr.*, 1959, **53**, 195).
82. G. Natta, L. Porri and A. Mazzei, *Chim. Ind. (Milan)*, 1959, **41**, 116 (*Chem. Abstr.*, 1959, **53**, 15 619).
83. J. S. Lasky, H. K. Garner and R. H. Ewart, *Ind. Eng. Chem. Prod. Res. Dev.*, 1962, **1**, 82.
84. G. Natta, L. Porri and A. Carbonaro, *Atti Accad. Naz. Lincei, Cl. Sci. Fis., Mat. Nat., Rend.*, 1961, **31** (5), 189 (*Chem. Abstr.*, 1962, **57**, 4848).
85. H. O. Froehlich and H. Kaholdt, *Z. Chem.*, 1975, **15** (9), 365 (*Chem. Abstr.*, 1975, **83**, 206 623).
86. L. Porri, A. Carbonaro and F. Ciampelli, *Makromol. Chem.*, 1963, **61**, 90.
87. A. Carbonaro (SNAM Progetti SpA), *Ger. Pat.* 2 534 495 (1976) (*Chem. Abstr.*, 1976, **84**, 151 265).

88. G. Natta, L. Porri, G. Zanini and A. Palvarini, *Chim. Ind. (Milan)*, 1959, **41**, 1163 (*Chem. Abstr.*, 1961, **55**, 20 906).
89. C. H. Bawn, A. M. North and J. S. Walker, *Polymer*, 1964, **5**, 419.
90. K. Hiraki and H. Hirai, *J. Polym. Sci., Polym. Chem. Ed.*, 1973, **11**, 901.
91. F. Dawans and Ph. Teyssié, *Bull. Soc. Chim. Fr.*, 1967, (8), 2930 (*Chem. Abstr.*, 1967, **67**, 100 447).
92. L. A. Kokhanenko, R. I. Ter-Minasyan and V. M. Frolov, *Vysokomol. Soedin., Ser. B*, 1982, **24**, 538 (*Chem. Abstr.*, 1982, **97**, 145 343).
93. H. Ikeda and A. Kogure (Japan Synthetic Rubber Co., Ltd.), *Jpn. Pat.* 75 154 389 (1975) (*Chem. Abstr.*, 1976, **84**, 151 913).
94. L. Xu, S. Zhao and X. Tang, *Hecheng Xiangjiao Gongye*, 1986, **9** (4), 258 (*Chem. Abstr.*, 1986, **105**, 98 932).
95. S. Wang, H. Song, Y. Yan and X. Tang, *Gaofenzi Tongxun*, 1984, (4), 292 (*Chem. Abstr.*, 1985, **102**, 132 508).
96. C. Longiave, R. Castelli and G. F. Croce, *Chim. Ind. (Milan)*, 1961, **43**, 625 (*Chem. Abstr.*, 1961, **55**, 24 086).
97. V. N. Zgonnik, B. A. Dolgoplosk, V. A. Kropachev and N. I. Nikolaiev, *Vysokomol. Soedin.*, 1962, **4**, 1000 (*Chem. Abstr.*, 1963, **58**, 14 098).
98. B. A. Dolgoplosk, B. L. Erussaliniskii, E. N. Kropatcheva and E. I. Tinyakova, *J. Polym. Sci.*, 1962, **58**, 1333.
99. M. Gippin, *Ind. Eng. Chem. Prod. Res. Dev.*, 1962, **1**, 32.
100. M. Gippin, *Ind. Eng. Chem. Prod. Res. Dev.*, 1965, **4**, 160.
101. V. N. Zgonnik, B. A. Dolgoplosk, N. I. Nikolaiev and V. A. Kropachev, *Vysokomol. Soedin.*, 1965, **7**, 308; *Polym. Sci. USSR (Engl. Transl.)*, 1965, **7**, 338.
102. G. J. van Amerongen, *Adv. Chem. Ser.*, 1966, **52**, 136.
103. L. V. Gavrilova, V. A. Grechanovski, Ye. N. Kropacheva and B. A. Dolgoplosk, *Vysokomol. Soedin., Ser. A*, 1968, **10**, 2001; *Polym. Sci. USSR (Engl. Transl.)*, 1968, **10**, 2324.
104. G. V. Timofeyeva, N. A. Kokorina and S. S. Medvedev, *Vysokomol. Soedin., Ser. A*, 1969, **11**, 596; *Polym. Sci. USSR (Engl. Transl.)*, 1969, **11**, 677.
105. S. S. Medvedev, L. A. Volkov, V. S. Byrikhin and G. V. Timofeyeva, *Vysokomol. Soedin., Ser. A*, 1971, **13**, 1388; *Polym. Sci. USSR (Engl. Transl.)*, 1971, **13**, 1561.
106. L. A. Volkov, Ye. F. Dudko and S. S. Medvedev, *Vysokomol. Soedin., Ser. A*, 1973, **15**, 2366; *Polym. Sci. USSR (Engl. Transl.)*, 1973, **15**, 2681.
107. V. S. Byrikhin, V. A. Tverskoi, S. S. Potapov, N. N. Luzina and N. P. Fedorov, *Vysokomol. Soedin., Ser. A*, 1974, **16**, 1059; *Polym. Sci. USSR (Engl. Transl.)*, 1974, **16**, 1222.
108. V. S. Byrikhin, N. N. Luzina, V. A. Tverskoi, N. P. Fedorov, M. O. Vasyshak and A. N. Pravednikov, *Vysokomol. Soedin., Ser. A*, 1975, **17**, 665; *Polym. Sci. USSR (Engl. Transl.)*, 1975, **17**, 767.
109. C. C. Hsu and L. Ng, *AIChE J.*, 1976, **22**, 66.
110. L. A. Volkov, I. D. Chernova, I. L. Radushnova, V. T. Robysheva and V. I. Lysova, *Vysokomol. Soedin., Ser. A*, 1976, **18**, 1486; *Polym. Sci. USSR (Engl. Transl.)*, 1976, **18**, 1701.
111. J. H. Howell and W. G. Curtis, in 'Transition Metal Catalyzed Polymerizations', ed. R. P. Quirk, Harwood Academic, New York, 1983, part B, p. 509.
112. A. Takahashi and S. Kambara, *J. Polym. Sci., Part B*, 1965, **3**, 279.
113. C. Longiave and R. Castelli, *J. Polym. Sci., Part C*, 1963, **4**, 387.
114. P. Racanelli and L. Porri, *Eur. Polym. J.*, 1970, **6**, 751.
115. K. Matsuzaki and T. Yasukawa, *Polym. Lett.*, 1965, **3**, 393.
116. L. Porri and A. Carbonaro, *Makromol. Chem.*, 1963, **60**, 236.
117. G. Natta and L. Porri, *Adv. Chem. Ser.*, 1966, **52**, 24.
118. G. Natta, L. Porri, A. Carbonaro and A. Greco, *Makromol. Chem.*, 1964, **71**, 207.
119. W. Cooper, D. E. Eaves and G. Vaughan, *Makromol. Chem.*, 1963, **67**, 229.
120. W. Cooper, D. E. Eaves and G. Vaughan, *Adv. Chem. Ser.*, 1966, **52**, 46.
121. N. D. Golubeva, A. D. Pomogailo, A. I. Kuzaev, A. N. Ponomarev and F. S. D'yachkovskii, *J. Polym. Sci., Polym. Symp.*, 1980, **68**, 33.
122. I. N. Ivleva, S. B. Echmaev, N. D. Golubeva and A. D. Pomogailo, *Kinet. Catal. (Engl. Transl.)*, 1983, **24**, 558.
123. K. Soga and K. Yamamoto, *Polym. Bull. (Berlin)*, 1982, **6**, 263 (*Chem. Abstr.*, 1982, **96**, 123 405).
124. K. Soga, K. Yamamoto and S. Chen, *Polym. Bull. (Berlin)*, 1981, **5**, 1 (*Chem. Abstr.*, 1981, **95**, 116 789).
125. R. P. Chaplin, R. P. Burford, G. J. Tory and S. Kirby, *Polymer*, 1987, **28**, 1418.
126. E. Susa, *J. Polym. Sci., Part C*, 1963, **4**, 399.
127. H. Ashitaka, H. Ishikawa, H. Ueno and A. Nagasaka, *J. Polym. Sci., Polym. Chem. Ed.*, 1983, **21**, 1853.
128. H. Ashitaka, K. Jinda and H. Ueno, *J. Polym. Sci., Polym. Chem. Ed.*, 1983, **21**, 1951.
129. H. Ashitaka, K. Jinda and H. Ueno, *J. Polym. Sci., Polym. Chem. Ed.*, 1983, **21**, 1989.
130. W. L. Yang and C. C. Hsu, *J. Appl. Polym. Sci.*, 1983, **28**, 145.
131. K. K. Li and C. C. Hsu, *Ind. Eng. Chem. Prod. Res. Dev.*, 1981, **20**, 624.
132. J. Furukawa, K. Haga, E. Kobayashi, Y. Iseda, T. Yashimoto and K. Sakamoto, *Polym. J.*, 1971, **2**, 371.
133. B. A. Dolgoplosk, E. N. Kropacheva, E. K. Khrennikova, E. I. Kuznetsova and K. G. Golodova, *Dokl. Akad. Nauk SSSR*, 1960, **135**, 847 (*Chem. Abstr.*, 1961, **55**, 13 292).
134. D. H. Lee and C. C. Hsu, *J. Appl. Polym. Sci.*, 1980, **25**, 2373.
135. C. Dumas and C. C. Hsu, *Chem. Eng. Commun.*, 1983, **24**, 37.
136. M. C. Throckmorton and F. S. Farson, *Rubber Chem. Technol.*, 1972, **45**, 268.
137. M. C. Throckmorton and W. M. Saltman (Goodyear Tire and Rubber Co.), *Ger. Pat.* 2 257 137 (1973) (*Chem. Abstr.*, 1973, **79**, 67 631).
138. M. C. Throckmorton (Goodyear Tire and Rubber Co.), *Ger. Pat.* 2 113 527 (1971) (*Chem. Abstr.*, 1972, **76**, 60 719).
139. Y. Yagi, H. Sato, S. Narisawa, S. Yasui, A. Kobayashi, M. Hino and K. Hata (Sumitomo Chemical Co., Ltd.) *Ger. Pat.* 2 164 414 (1972) (*Chem. Abstr.*, 1973, **78**, 17 346).
140. C. Dixon, E. W. Duck, D. P. Grieve, D. K. Jenkins and M. N. Thornber, *Eur. Polym. J.*, 1970, **6**, 1359.
141. E. W. Duck, D. K. Jenkins, D. P. Grieve and M. N. Thornber, *Eur. Polym. J.*, 1971, **7**, 55.
142. C. Dixon, E. W. Duck and D. K. Jenkins, *Eur. Polym. J.*, 1972, **8**, 13.
143. S. Hua and X. Tang, *Kao Fen Tzu T'in Hsun*, 1982, (5), 331 (*Chem. Abstr.*, 1983, **98**, 107 819).
144. B. S. Turov, T. A. Nuzhdina and G. N. Shilova, *Vysokomol. Soedin., Ser. A*, 1980, **22**, 2577 (*Chem. Abstr.*, 1981, **94**, 84 606).
145. A. G. Azizov, G. A. Mamedaliev and S. M. Aliev, *Azerb. Khim. Zh.*, 1977, (2), 26 (*Chem. Abstr.*, 1977, **87**, 136 454).

146. L. Porri, G. Natta and M. C. Gallazzi, *J. Polym. Sci., Part C*, 1967, **16**, 2525.
147. L. Porri, M. C. Gallazzi, S. Destri and A. Bolognesi, in 'Transition Metal Catalyzed Polymerizations', ed. R. P. Quirk, Harwood Academic, New York, 1983, p. 555.
148. R. Sakata, J. Hosono, A. Onishi and K. Ueda, *Makromol. Chem.*, 1970, **139**, 73.
149. W. C. Von Dohlen, T. P. Wilson and E. G. Caflisch (Union Carbide Corp.), *Belg. Pat.* 644 291 (1964) (*Chem. Abstr.*, 1965, **63**, 5874).
150. C. C. Shen, C. Y. Kung, C. C. Chung and Y. C. Ou, *K'o Hsueh T'ung Pao*, 1964, **4**, 335 (*Chem. Abstr.*, 1964, **61**, 12 091).
151. C. C. Shen, C. Y. Kung and Y. C. Ou, *Kao Fen Tzu T'ung Hsun*, 1965, **7**, 193 (*Chem. Abstr.*, 1966, **64**, 11 320).
152. M. C. Throckmorton, *Kautsch. Gummi Kunstst.*, 1969, **22**, 293 (*Chem. Abstr.*, 1969, **71**, 50 995).
153. G. Lugli, A. Mazzei and G. Modini (Snam Progetti SpA), *Ger. Pat.* 2 257 786 (1973) (*Chem. Abstr.*, 1973, **79**, 54 072).
154. G. Lugli, A. Mazzei and S. Poggio, *Makromol. Chem.*, 1974, **175**, 2021.
155. A. De Chirico, P. C. Lanzani, E. Raggi and M. Bruzzone, *Makromol. Chem.*, 1974, **175**, 2029.
156. M. Bruzzone, A. Mazzei and G. Giuliani, *Rubber Chem. Technol.*, 1974, **47**, 1175.
157. A. Mazzei, *Makromol. Chem. Suppl.*, 1981, **4**, 61.
158. Z. Shen, J. Ouyang, F. Wang, Z. Hu, F. Yu and B. Qian, *J. Polym. Sci., Polym. Chem. Ed.*, 1980, **18**, 3345.
159. Yu. B. Monakov, N. G. Marina, I. G. Savel'eva, L. E. Zhiber, V. G. Kozlov and S. R. Rafikov, *Dokl. Akad. Nauk SSSR*, 1982, **265**, 1431 (*Chem. Abstr.*, 1983, **98**, 54 523).
160. G. Silvester and J. Witte (Bayer A. G.), *Ger. Pat.* 2 848 964 (1980) (*Chem. Abstr.*, 1980, **93**, 96 555).
161. J. H. Yang, M. Tsutsui, Z. Chen and E. Bergbreiter, *Macromolecules*, 1982, **15**, 2303.
162. S. Yang, J. Hu, S. Feng, E. Pan, D. Xie, C. Zhong and J. Ouyang, *Sci. Sinica*, 1980, **2/3**, 734 (*Chem. Abstr.*, 1980, **93**, 155 054).
163. H. L. Hsieh and H. C. Yeh, *Rubber Chem. Technol.*, 1985, **58**, 117.
164. G. Yu, W. Chen, Y. Wang and H. Guan, *Kao Fen Tzu T'in Hsun*, 1985, (6), 452 (*Chem. Abstr.*, 1986, **105**, 79 404).
165. J. Yang, S. Pang, Y. Li and J. Ouyang, *Cuihva Hsueh Pao*, 1984, **5**, 291 (*Chem. Abstr.*, 1985, **102**, 25 110).
166. S. Pang, Y. Li, W. Ding, J. Xue and J. Ouyang, *Ying Yung Chieh Mien Hua Hsueh*, 1984, **1**, 50 (*Chem. Abstr.*, 1985, **102**, 62 595).
167. Asahi Chemical Industry Co., Ltd., *Jpn. Pat.* 59 113 003 (1984) (*Chem. Abstr.*, 1985, **102**, 7256).
168. Z. Shen, X. Song, S. Xiao, J. Yang and X. Kan, *J. Appl. Polym. Sci.*, 1983, **28**, 1585.
169. N. G. Marina, Kh. K. Gadeleva, Yu. B. Monakov and S. R. Rafikov, *Dokl. Akad. Nauk SSSR*, 1984, **274**, 641 (*Chem. Abstr.*, 1984, **101**, 55 556).
170. Y. Jin, X. W. Li, Y. Sun and J. Ouyang, *K'o Hsueh T'ung Pao*, 1982, (1), 32 (*Chem. Abstr.*, 1982, **96**, 124 251).
171. Y. Jin, X. W. Li, Y. Sun and J. Ouyang, *Kao Fen Tzu T'in Hsun*, 1984, (5), 351 (*Chem. Abstr.*, 1985, **102**, 185 561).
172. Y. Jin, X. W. Li, Y. Sun and J. Ouyang, *Kao Fen Tzu T'in Hsun*, 1984, (5), 358 (*Chem. Abstr.*, 1985, **102**, 185 562).
173. Y. Jin, Y. Sun, X. Liu, X. Li and J. Ouyang, *Fen Tzu K'o Hsueh Yu Hua Hsueh Yanjin*, 1984, **4**, 247 (*Chem. Abstr.*, 1984, **101**, 131 149).
174. Y. Jin, X. Li, Y. Sun and J. Ouyang, *K'o Hsueh T'ung Pao*, 1985, **30**, 1047 (*Chem. Abstr.*, 1986, **104**, 69 213).
175. N. N. Chigir, O. K. Sharaev, E. I. Tinyakova and B. A. Dolgoplosk, *Vysokomol. Soedin., Ser. B*, 1983, **25**, 47 (*Chem. Abstr.*, 1983, **98**, 126 695).
176. D. K. Jenkins, *Polymer*, 1985, **26**, 147.
177. D. K. Jenkins, *Polymer*, 1985, **26**, 152.
178. N. N. Chigir, I. Sh. Guzman, O. K. Sharaev, E. I. Tinyakova and B. A. Dolgoplosk, *Dokl. Akad. Nauk SSSR*, 1982, **263**, 375 (*Chem. Abstr.*, 1982, **97**, 56 281).
179. E. L. Vollershtein, N. N. Glebova, S. B. Gol'shtein, E. N. Zavadovskaya, O. K. Sharaev, V. A. Yakovlev, E. I. Tinyakova and B. A. Dolgoplosk, *Dokl. Akad. Nauk SSSR*, 1985, **284**, 140 (*Chem. Abstr.*, 1986, **104**, 130 300).
180. F. Wang, X. J. Zhao, C. D. Jiang and Z. Gong, *Hua Hsueh Hsueh Pao*, 1979, **37**, 111 (*Chem. Abstr.*, 1979, **91**, 211 891).
181. A. Ruan and B. Li, *Hua Hsueh Hsueh Pao*, 1983, **41**, 783 (*Chem. Abstr.*, 1983, **99**, 212 962).
182. G. Ricci, S. Italia, F. Cabassi and L. Porri, *Polym. Commun.*, 1987, **28**, 223.
183. A. A. Sanyagin and V. A. Kormer, *Dokl. Akad. Nauk SSSR*, 1985, **283**, 1209 (*Chem. Abstr.*, 1986, **104**, 51 150).
184. J. Yang, S. Pang and J. Ouyang, *Kao Fen Tzu T'in Hsun*, 1984, (1), 73 (*Chem. Abstr.*, 1984, **100**, 210 508).
185. S. Qi, X. Liu, Y. Chang and W. Chen, *Ying Yung Hua Hsueh*, 1986, **3** (3), 73 (*Chem. Abstr.*, 1986, **105**, 209 457).
186. S. Qi, X. Gao, S. Xiao and W. Chen, *Ying Yung Hua Hsueh*, 1986, **3** (1), 63 (*Chem. Abstr.*, 1986, **104**, 186 909).
187. E. Pan, J. Hu and C. Zhon, *Kao Fen Tzu T'in Hsun*, 1985, (2), 125 (*Chem. Abstr.*, 1985, **103**, 215 850).
188. J. Witte, *Angew. Makromol. Chem.*, 1981, **94**, 119.
189. Y. Li, S. Pang, D. Xue and J. Ouyang, *Kao Fen Tzu T'in Hsun*, 1985, (2), 111 (*Chem. Abstr.*, 1985, **103**, 255 848).
190. Y. Li, D. Han and J. Ouyang, *Cuihua Hsueh Pao*, 1986, **7** (3), 276 (*Chem. Abstr.*, 1986, **105**, 227 405).
191. D. E. Bergbreiter, L. B. Chen and R. Chandran, *Macromolecules*, 1985, **18**, 1055.
192. R. Dauby, F. Dawans and Ph. Teyssié, *J. Polym. Sci., Part C*, 1967, **16**, 1989.
193. E. Ochiai, H. Hirai and S. Makishima, *J. Polym. Sci., Part B*, 1966, **4**, 1003.
194. M. Morton, I. Piirma and B. Das, *Rubber Plast. Age*, 1967, **46**, 404.
195. M. Morton and B. Das, *J. Polym. Sci., Part C*, 1969, **27**, 1.
196. V. N. Sokolov, B. D. Babitskii, V. A. Kormer and I. Ya. Poddbunyi, *J. Polym. Sci., Part C*, 1969, **16**, 4345.
197. W. S. Anderson, *J. Polym. Sci., Part A-1*, 1967, **5**, 429.
198. B. A. Dolgoplosk, *Vysokomol. Soedin., Ser. A*, 1971, **13**, 325; *Polym. Sci. USSR (Engl. Transl.)*, 1971, **13**, 367.
199. S. A. Radzinskii, A. P. Sheinker and A. D. Abkin, *Vysokomol. Soedin., Ser. A*, 1979, **21**, 608 (*Chem. Abstr.*, 1979, **90**, 187 416).
200. F. Dawans, J. P. Durand and Ph. Teyssié, *J. Polym. Sci., Part B*, 1972, **10**, 493.
201. I. Y. Ostrovskaya and K. Makovetskii, *Dokl. Akad. Nauk SSSR*, 1973, **208**, 1110 (*Chem. Abstr.*, 1973, **79**, 19 190).
202. L. Porri, M. C. Gallazzi and G. Vitulli, *J. Polym. Sci., Part B*, 1967, **5**, 629.
203. T. Otsu, M. Yamaguchi, *J. Polym. Sci., Part A-1*, 1969, **7**, 387.
204. E. I. Tinyakova, A. V. Alferov, T. G. Golenko, B. A. Dolgoplosk, I. A. Oreshkin, O. K. Sharaev, G. N. Chernenko and V. A. Yakovlev, *J. Polym. Sci., Part C*, 1967, **16**, 2625.
205. T. C. Wen, C. C. Chang, Y. D. Chuang, J. P. Chin and C. T. Chang, *J. Am. Chem. Soc.*, 1981, **103**, 4576.
206. F. Dawans and Ph. Teyssié, *Polym. Lett*, 1965, **3**, 1045.

207. B. D. Babitskii, V. A. Kormer, I. M. Lapuk and V. I. Skoblikova, *J. Polym. Sci., Part C*, 1968, **16**, 3219.
208. D. K. Jenkins, D. G. Timms and W. E. Duck, *Polymer*, 1966, **7**, 419.
209. H. Scott, R. E. Frost, R. F. Belt and D. E. O'Reilly, *J. Polym. Sci., Part A*, 1964, **2**, 3233.
210. H. Scott, *J. Polym. Sci., Part B*, 1966, **4**, 105.
211. D. E. O'Reilly, C. P. Poole, Jr., R. E. Belt and H. Scott, *J. Polym. Sci., Part A*, 1964, **2**, 3257.
212. J. G. Balas, H. E. De La Mare and D. O. Schussler, *J. Polym. Sci., Part A*, 1965, **3**, 2243.
213. G. Wilke (Studiengesellschaft Kohle), *Fr. Pat.* 1 410 430 (1963) (*Chem. Abstr.*, 1966, **64**, 9928).
214. G. Wilke, B. Bogdanovic, P. Hardt, P. Heimbach, W. Keim, M. Kromer, W. Oberkirch, K. Tanaka, E. Steinrucke, D. Walter and H. Zimmerman, *Angew. Chem., Int. Ed. Engl.*, 1966, **5**, 151.
215. B. A. Dolgoplosk and E. I. Tinyakova, *Izv. Akad. Nauk SSSR, Ser. Khim.*, 1970, **2**, 344 (*Chem. Abstr.*, 1970, **73**, 25 939).
216. F. Dawans and Ph. Teyssié, *Ind. Eng. Chem. Prod. Res. Dev.*, 1971, **10**, 261.
217. B. A. Dolgoplosk and E. I. Tinyakova, *Vysokomol. Soedin., Ser. A*, 1977, **19**, 2441; *Polym. Sci. USSR (Engl. Transl.)*, 1977, **19**, 2805.
218. B. A. Dolgoplosk, I. A. Oreshkin, E. I. Tinyakova and V. A. Yakovlev, *Izv. Akad. Nauk SSSR, Ser. Khim.*, 1967, (9), 2130 (*Chem. Abstr.*, 1967, **67**, 109 053).
219. I. A. Oreshkin, I. Ya. Ostrovskaya, V. A. Yakovlev, E. I. Tinyakova and B. A. Dolgoplosk, *Dokl. Akad. Nauk SSSR*, 1967, **173**, 1349 (*Chem. Abstr.*, 1967, **67**, 74 299).
220. I. A. Oreshkin, E. I. Tinyakova and B. A. Dolgoplosk, *Vysokomol. Soedin., Ser. A*, 1969, **11**, 1840; *Polym. Sci. USSR (Engl. Transl.)*, 1969, **11**, 2096.
221. V. L. Shmonina, N. N. Stefanovskaia, E. I. Tinyakova and B. A. Dolgoplosk, *Vysokomol. Soedin., Ser. A.*, 1973, **15**, 647; *Polym. Sci. USSR (Engl. Transl.)*, 1973, **15**, 731.
222. B. A. Dolgoplosk, E. I. Tinyakova, N. N. Stefanovskaia, I. A. Oreshkin and V. L. Shmonina, *Eur. Polym. J.*, 1974, **10**, 605.
223. I. A. Oreshkin, G. M. Chernenko, E. I. Tinyakova, and B. A. Dolgoplosk, *Dokl. Akad. Nauk SSSR*, 1966, **169**, 1102 (*Chem. Abstr.*, 1966, **65**, 15 630).
224. R. Ter-Minasyan, O. P. Parenago, V. M. Frolov and B. A. Dolgoplosk, *Dokl. Akad. Nauk SSSR*, 1974, **214**, 824 (*Chem. Abstr.*, 1974, **81**, 26 896).
225. R. Ter-Minasyan, O. P. Parenago, V. G. Chirkova, V. M. Frolov and B. A. Dolgoplosk, *Kinet. Katal.*, 1976, **17**, 935; *Kinet. Catal. (Engl. Transl.)*, 1976, **17**, 815.
226. V. S. Stroganov, E. S. Novikova and V. M. Frolov, *Kinet. Catal.*, 1984, **25**, 487; *Kinet. Catal. (Engl. Transl.)*, 1984, **25**, 407.
227. I. A. Oreshkin, L. I. Redkina, K. L. Makovetskii, B. A. Dolgoplosk and E. I. Tinyakova, *Vysokomol. Soedin., Ser. B*, 1972, **14**, 384 (*Chem. Abstr.*, 1972, **77**, 75 570).
228. V. A. Vasil'ev, V. A. Kormer and N. A. Kalinicheva, *Vysokomol. Soedin., Ser. B*, 1972, **14**, 212 (*Chem. Abstr.*, 1972, **77**, 21 094).
229. B. D. Babitskii, B. A. Dolgoplosk, V. A. Kormer. M. I. Lobach, E. I. Tinyakova and V. A. Yakovlev, *Dokl. Akad. Nauk SSSR*, 1965, **161**, 583 (*Chem. Abstr.*, 1965, **63**, 3142).
230. O. K. Sharayev, A. V. Alferov, E. I. Tinyakova and B. A. Dolgoplosk, *Izv. Akad. Nauk SSSR*, 1967, (5), 1170 (*Chem. Abstr.*, 1968, **68**, 69 380).
231. O. K. Sharayev, A. V. Alferov, E. I. Tinyakova and B. A. Dolgoplosk, *Izv. Akad. Nauk SSSR*, 1967, (11), 2584 (*Chem. Abstr.*, 1968, **68**, 30 171).
232. J. P. Durand, F. Dawans and Ph. Teyssié, *Polym. Lett.*, 1967, **5**, 785.
233. J. P. Durand, F. Dawans and Ph. Teyssié, *J. Polym. Sci., Part B*, 1968, **6**, 757.
234. J. P. Durand, F. Dawans and Ph. Teyssié, *J. Polym. Sci., Part A-1*, 1970, **8**, 979.
235. L. M. Vardanyan, Yu. V. Korshak and B. A. Dolgoplosk, *Vysokomol. Soedin., Ser. B*, 1971, **13**, 1922 (*Chem. Abstr.*, 1971, **74**, 126 152).
236. E. V. Mushina, T. K. Vydrina, E. V. Sakharova, V. A. Yakovlev, E. I. Tinyakova and B. A. Dolgoplosk, *Vysokomol. Soedin., Ser. B*, 1967, **9**, 784 (*Chem. Abstr.*, 1968, **68**, 13 827).
237. T. Matsumoto, J. Furukawa and H. Morimura, *J. Polym. Sci., Part B*, 1968, **6**, 869.
238. T. Matsumoto, J. Furukawa and H. Morimura, *J. Polym. Sci., Part B*, 1969, **7**, 541.
239. S. V. Lebedev (All-Union Research Institute of Synthetic Rubber), *Fr. Pat.* 1 550 097 (1968) (*Chem. Abstr.*, 1969, **71**, 40 048).
240. V. A. Yakovlev, B. A. Dolgoplosk, E. I. Tinyakova and O. N. Yakovleva, *Vysokomol. Soedin., Ser. A*, 1969, **11**, 1965; *Polym. Sci. USSR (Engl. Transl.)*, 1969, **11**, 1869.
241. V. A. Yakovlev, B. A. Dolgoplosk, K. L. Makovetskii and E. I. Tinyakova, *Dokl. Akad. Nauk SSSR*, 1969, **187**, 354 (*Chem. Abstr.*, 1969, **71**, 102 859).
242. F. Dawans and Ph. Teyssié, *J. Polym. Sci., Part B*, 1969, **7**, 111.
243. O. K. Sharayev, A. V. Alferov, E. I. Tinyakova, B. A. Dolgoplosk, *Izv. Akad. Nauk SSSR*, 1967, (11), 2583 (*Chem. Abstr.*, 1968, **68**, 30 172).
244. V. M. Gorelik, O. P. Parenago, V. M. Frolov and B. A. Dolgoplosk, *Kinet. Katal.*, 1973, **14**, 1446 (*Chem. Abstr.*, 1974, **81**, 50 755).
245. J. P. Durand and Ph. Teyssié, *J. Polym. Sci., Part B*, 1968, **6**, 299.
246. E. A. Mushina, T. K. Vydrina, E. V. Sakharova, E. I. Tinyakova and B. A. Dolgoplosk, *Dokl. Akad. Nauk SSSR*, 1967, **177**, 361 (*Chem. Abstr.*, 1968, **68**, 40 722).
247. A. G. Azizov, O. K. Sharayev, E. I. Tinyakova and B. A. Dolgoplosk, *Vysokomol. Soedin., Ser. B*, 1969, **11**, 746 (*Chem. Abstr.*, 1970, **72**, 32 935).
248. G. Lugli, W. Marconi, A. Mazzei and N. Palladino, *Inorg. Chim. Acta*, 1969, **3**, 151.
249. A. G. Azizov, O. K. Sharayev, E. I. Tinyakova and B. A. Dolgoplosk, *Dokl. Akad. Nauk SSSR*, 1970, **190**, 582 (*Chem. Abstr.*, 1970, **72**, 133 806).
250. A. G. Azizov, *Dokl. Akad. Nauk SSSR*, 1982, **265**, 362 (*Chem. Abstr.*, 1982, **97**, 198 600).
251. R. I. Ter-Minasyan, O. P. Parenago, V. M. Frolov and B. A. Dolgoplosk, *Dokl. Akad. Nauk SSSR*, 1970, **194**, 1372 (*Chem. Abstr.*, 1971, **74**, 32 485).
252. E. S. Novikova, O. P. Parenago, V. M. Frolov and B. A. Dolgoplosk, *Kinet. Katal.*, 1976, **17**, 928; *Kinet. Catal. (Engl. Transl.)*, 1976, **17**, 809.

253. E. I. Tinyakova, A. D. Litmanovich and R. V. Rabovskaya, *Vysokomol. Soedin., Ser. A*, 1978, **20**, 2552 (*Chem. Abstr.*, 1979, **90**, 55 349).
254. R. V. Rabovskaia, E. I. Tinyakova and G. A. Parfenova, *Vysokomol. Soedin., Ser. A*, 1977, **19**, 2352; *Polym. Sci. USSR (Engl. Transl.)*, 1977, **19**, 2701.
255. B. D. Babitskii, B. A. Dolgoplosk, V. A. Kormer, M. I. Lobach, E. I. Tinyakova and V. A. Yakovlev, *Dokl. Akad. Nauk SSSR*, 1965, **161**, 583 (*Chem. Abstr.*, 1965, **63**, 3142).
256. B. D. Babitskii, B. A. Dolgoplosk, V. A. Kormer, M. I. Lobach, E. I. Tinyakova, N. N. Chesnokova and V. A. Yakovlev, *Vysokomol. Soedin., Ser. A*, 1964, **6**, 2202; *Polym. Sci. USSR (Engl. Transl.)*, 1964, **6**, 2436.
257. E. V. Kristal'nyi, N. V. Orekhov, Ye. V. Zabolotskaya, A. R. Gantmakher and S. S. Medvedev, *Vysokomol. Soedin., Ser. A*, 1970, **12**, 836; *Polym. Sci. USSR (Engl. Transl.)*, 1970, **12**, 947.
258. N. I. Pakuro, E. V. Zabolotskaya and B. A. Dolgoplosk, *Dokl. Akad. Nauk SSSR*, 1973, **209**, 631 (*Chem. Abstr.*, 1973, **79**, 53 861).
259. N. I. Pakuro and E. V. Zabolotskaya, *Vysokomol. Soedin., Ser. B*, 1973, **15**, 185 (*Chem. Abstr.*, 1973, **79**, 42 904).
260. N. I. Pakuro and E. V. Zabolotskaya, *Vysokomol. Soedin., Ser. B*, 1974, **16**, 457 (*Chem. Abstr.*, 1974, **81**, 169 963).
261. E. V. Kristal'nyi, N. I. Galkina, E. V. Zabolotskaya and A. R. Gantmakher, *Vysokomol. Soedin., Ser. B*, 1974, **16**, 64 (*Chem. Abstr.*, 1974, **81**, 64 896).
262. A. I. Kadantseva, V. S. Byrikhin and S. S. Medvedev, *Vysokomol. Soedin., Ser. A*, 1972, **14**, 2668; *Polym. Sci. USSR (Engl. Transl.)*, 1972, **14**, 3104.
263. S. S. Potapov, V. S. Byrikhin, M. F. Margaritova, V. A. Vashkevich, V. A. Isaeva and V. A. Tverskoi, *Vysokomol. Soedin., Ser. A*, 1974, **16**, 2515; *Polym. Sci. USSR (Engl. Transl.)*, 1974, **16**, 2924.
264. V. S. Byrikhin and A. I. Kadantseva, *Vysokomol. Soedin., Ser. A*, 1975, **17**, 878; *Polym. Sci. USSR (Engl. Transl.)*, 1975, **17**, 1011.
265. O. K. Sharayev, A. V. Alferov, E. I. Tinyakova, B. A. Dolgoplosk, V. A. Kormer and B. D. Babitskii, *Dokl. Akad. Nauk SSSR*, 1967, **177**, 140 (*Chem. Abstr.*, 1968, **68**, 30 165).
266. E. N. Zavadovskaya, M. P. Teferina, O. K. Sharayev, A. G. Azizov, T. K. Vydrina, E. I. Tinyakova and B. A. Dolgoplosk, *Dokl. Akad. Nauk SSSR*, 1969, **88**, 822 (*Chem. Abstr.*, 1970, **72**, 26 633).
267. Ph. Teyssié, F. Dawans and J. P. Durand, *J. Polym. Sci., Part C*, 1968, **22**, 221.
268. J. C. Marechal, F. Dawans and Ph. Teyssié, *J. Polym. Sci., Part A-1*, 1970, **8**, 1993.
269. J. M. Thomassin, E. Walkiers, R. Warin and Ph. Teyssié, *J. Polym. Sci., Polym. Chem. Ed.*, 1975, **13**, 1147.
270. J. F. Harrod and A. Navarre, *Macromolecules*, 1984, **17**, 2455.
271. N. Stransky, R. Herzschu, J. P. Gehrke and R. Taube, *J. Organomet. Chem.*, 1984, **270**, 357.
272. R. Taube, J. P. Gehrke and U. Schmidt, *J. Organomet. Chem.*, 1985, **292**, 287.
273. R. Taube and U. Schmidt, *Z. Chem.*, 1977, **17**, 349 (*Chem. Abstr.*, 1978, **88**, 7470).
274. R. Taube and J. P. Gehrke, *Z. Chem.*, 1983, **23** (12), 437 (*Chem. Abstr.*, 1984, **100**, 139 680).
275. R. Taube, U. Schmidt, J. P. Gehrke and U. Anaeker, *J. Prakt. Chem.*, 1984, **326**, 1 (*Chem. Abstr.*, 1984, **100**, 192 348).
276. R. Taube, J. P. Gehrke and R. Radeglia, *J. Organomet. Chem.*, 1985, **291**, 101.
277. R. Taube, J. P. Gehrke and U. Schmidt, *Makromol. Chem., Makromol. Symp.*, 1986, **3**, 389.
278. J. P. Durand and F. Dawans, *J. Polym. Sci., Part B*, 1970, **8**, 743.
279. T. V. Soboleva, V. A. Yakovlev, E. I. Tinyakova and B. A. Dolgoplosk, *Dokl. Akad. Nauk SSSR*, 1973, **212**, 893 (*Chem. Abstr.*, 1974, **80**, 84 422).
280. V. A. Kormer, V. A. Vasil'ev, N. A. Kalinicheva and O. I. Belgorodkaya, *J. Polym. Sci., Polym. Chem. Ed.*, 1973, **11**, 2557.
281. S. K. Ngo, Y. V. Korshak, G. N. Bondarenko, M. P. Teterina and B. A. Dolgoplosk, *Vysokomol. Soedin., Ser. B*, 1973, **15**, 375 (*Chem. Abstr.*, 1973, **79**, 66 863).
282. V. A. Vasil'ev, T. I. Sokoryanskaya and V. A. Kormer, *J. Polym. Sci., Polym. Chem. Ed.*, 1978, **16**, 2159.
283. N. N. Stefanovskaya, V. L. Shmonina, E. I. Tinyakova and B. A. Dolgoplosk, *Dokl. Akad. Nauk SSSR*, 1973, **211**, 862 (*Chem. Abstr.*, 1974, **80**, 38 047).
284. Y. I. Ermakov, B. N. Kutznetzov, Y. P. Grabovskii, A. N. Startsev, A. M. Lazatkin, V. A. Zakharov and A. I. Lazatkina, in 'Proceedings of the International Symposium on Catalysis, 1974', ed. B. Delmon and G. Jannes, Elsevier, Amsterdam, 1975, p. 145.
285. Y. I. Ermakov, V. A. Zakharov, Y. P. Grabovskii, G. D. Bukatov and N. B. Chumaevskii, *Vysokomol. Soedin., Ser. B*, 1975, **17**, 668 (*Chem. Abstr.*, 1975, **83**, 206 646).
286. Y. I. Ermakov, Y. P. Grabovskii, A. M. Lazutkin and V. A. Zakharov, *Kinet. Katal.*, 1975, **16**, 911; *Kinet. Catal. (Engl. Transl.)*, 1975, **16**, 787.
287. Y. I. Ermakov, Y. P. Grabovskii and A. M. Lazutkin, *React. Kinet. Catal. Lett.*, 1974, **1**, 345 (*Chem. Abstr.*, 1975, **83**, 59 412).
288. Y. I. Ermakov, A. M. Lazutkin, E. A. Demin, Y. P. Grabovskii and A. V. Zakharov, *Kinet. Katal.*, 1972, **13**, 1422 (*Chem. Abstr.*, 1973, **78**, 98 902).
289. N. N. Stefanovskaya, V. L. Shmonina, E. I. Tinyakova and B. A. Dolgoplosk, in 'Symposium on Mechanism of Hydrocarbon Reactions, 1973', ed. F. Marta and D. Kallo, Elsevier, Amsterdam, 1975, p. 497.
290. W. Skupinski, M. Zawartke and S. Mulinowski, *React. Kinet. Catal. Lett.*, 1980, **13**, 319 (*Chem. Abstr.*, 1980, **93**, 186 869).
291. Chemische Werke Hüls Akt. Ges., *Br. Pat.* 824 201 (1959) (*Chem. Abstr.*, 1960, **54**, 25 977).
292. Polymer Corp. Ltd., *Belg. Pat.* 610 400 (1962) (*Chem. Abstr.*, 1962, **57**, 12 685).
293. I. Ia. Poddubnyi, V. A. Grechanovskii and E. G. Ehrenburg, *Makromol. Chem.*, 1966, **94**, 268.
294. A. Mazzei, M. Araldi, W. Marconi and M. De Maldè, *J. Polym. Sci., Part A*, 1965, **3**, 753.
295. A. I. Diaconescu and S. S. Medvedev, *J. Polym. Sci., Part A*, 1965, **3**, 31.
296. F. P. van de Kamp, *Makromol. Chem.*, 1966, **93**, 202.
297. I. Yu. Kirchovskaya and S. S. Medvedev, *Dokl. Akad. Nauk SSSR*, 1964, **158**, 1116 (*Chem. Abstr.*, 1965, **62**, 2826).
298. W. M. Saltman and L. J. Kuzma, *Rubber Chem. Technol.*, 1973, **46**, 1055.
299. L. A. Volkov and S. S. Medvedev, *Dokl. Akad. Nauk SSSR*, 1969, **184**, 125 (*Chem. Abstr.*, 1969, **70**, 83 314).
300. A. Guyot, M. Rochina and Y. Trambouze, *Bull. Soc. Chim. Fr.*, 1962, 1893 (*Chem. Abstr.*, 1963, **58**, 4651).
301. C. E. H. Bawn, *Rubber Plast. Age*, 1965, **46**, 510.
302. J. A. J. Honig, R. P. Burford and R. P. Chaplin, *J. Polym. Sci., Polym. Chem. Ed.*, 1983, **21**, 2559.
303. C. Longiave, R. Castelli and M. Ferraris, *Chim. Ind. (Milan)*, 1962, **44**, 725 (*Chem. Abstr.*, 1962, **57**, 16 825).

304. B. Schleimer and H. Weber, *Angew. Makromol. Chem.*, 1971, **16/17**, 253.
305. T. Yoshimoto, K. Komatskii, R. Sakata, K. Yamamoto, Y. Takeuchi, A. Onishi and K. Ueda, *Makromol. Chem.*, 1970, **139**, 61.
306. P. Hadjiandreou, M. Julémont and Ph. Teyssié, *Macromolecules*, 1984, **17**, 2455.
307. G. Ricci, G. Boffa and L. Porri, *Makromol. Chem., Rapid Commun.*, 1986, **7**, 355.
308. L. Yang and X. Zhu, *Hecheng Xiangjiao Gongye*, 1984, **7**, 356 (*Chem. Abstr.*, 1985, **102**, 80 021).
309. Phillips Petroleum Co., *Br. Pat.* 848 067 (1960) (*Chem. Abstr.*, 1961, **55**, 15 982).
310. Ph. Teyssié and R. Dauby, *J. Polym. Sci., Part B*, 1964, **2**, 413.
311. J. Zachoval and B. Vernovic, *Polym. Lett.*, 1966, **4**, 965.
312. A. Mazzei, D. Cucinella, W. Marconi and M. De Maldè, *Chim. Ind. (Milan)*, 1963, **45**, 528 (*Chem. Abstr.*, 1963, **59**, 4043).
313. G. Natta, L. Porri, G. Zanini and L. Fiore, *Chim. Ind. (Milan)*, 1959, **41**, 526 (*Chem. Abstr.*, 1960, **54**, 1258).
314. K. Matsuzaki and T. Yasukawa, *J. Polym. Sci. Part A-1*, 1967, **15**, 511.
315. H. Ashitaka, K. Inaishi and H. Ueno, *J. Polym. Sci., Polym. Chem. Ed.*, 1983, **21**, 1973.
316. K. Mori, T. Taketomi and F. Imaizumi, *Nippon Kagaku Kaishi*, 1975, (11), 1982 (*Chem. Abstr.*, 1976, **84**, 74 676).
317. C. L. Parris and L. S. Rieve (Allied Chemical Corp.), *Ger. Pat.* 2 007 052 (1970) (*Chem. Abstr.*, 1971, **74**, 23 218).
318. J. Zymonas, E. R. Santee, Jr. and H. J. Harwood, *Macromolecules*, 1973, **6**, 129.
319. M. Iwamoto and S. Yuguchi, *J. Polym. Sci., Part B*, 1967, **5**, 1007.
320. Y. Takeuchi, A. Sekimoto and M. Abe, *ACS Symp. Ser.*, 1964, **4**, 15.
321. M. Julémont, E. Walkiers, R. Warin and Ph. Teyssié, *Makromol. Chem.*, 1974, **175**, 1673.
322. J. Furukawa, E. Kobayashi, T. Kawagoe, N. Katsuki and M. Imanari, *J. Polym. Sci., Polym. Lett. Ed.*, 1973, **11**, 239.
323. K. Hatada, Y. Terawaki, K. Okuda, Y. Tanaka and H. Sato, *J. Polym. Sci., Polym. Lett. Ed.*, 1974, **12**, 305.
324. F. Conti, A. Segre, P. Pini and L. Porri, *Polymer*, 1974, **15**, 5.
325. J. M. Thomassin, E. Walkiers, R. Warin and Ph. Teyssié, *J. Polym. Sci., Polym. Lett. Ed.*, 1973, **11**, 299.
326. J. Furukawa, E. Kobayashi and T. Kawagoe, *Polym. J.*, 1973, **5**, 231.
327. Z. Y. Zhang, H. J. Zhang, H. M. Ha and Y. Wu, *J. Mol. Catal.*, 1982, **17**, 65.
328. K. F. Elgert, G. Quack and B. Stutzel, *Polymer*, 1974, **15**, 612.
329. E. Schoenberg, H. A. Marsh, S. Y. Walters and W. M. Saltman, *Rubber Chem. Technol.*, 1977, **52**, 527.
330. H. E. Adams, R. S. Stearns, W. A. Smith and J. L. Binder, *Ind. Eng. Chem.*, 1958, **50**, 1507.
331. C. F. Gibbs, S. E. Horne, Jr., J. H. Macey and H. Tucker, *Rubber World*, 1961, **144**, 69.
332. D. C. Perry, F. S. Farson and E. Schoenberg, *J. Polym. Sci., Polym. Chem. Ed.*, 1975, **13**, 1071.
333. Y. B. Monakov, S. R. Rafikov, A. M. Ivanova, N. K. Minchenkova and I. G. Saveleva, *Dokl. Akad. Nauk SSSR*, 1976, **229**, 667 (*Chem. Abstr.*, 1976, **85**, 124 449).
334. S. E. Bresler, M. I. Mosevitskii, I. Ya. Poddubnyi and G. I. Shi, *Vysokomol. Soedin.*, 1961, **3**, 1591; *Polym. Sci. USSR (Engl. Transl.)*, 1962, **3**, 1033.
335. N. Yamazaki, T. Suminoe and S. Kambora, *Makromol. Chem.*, 1963, **65**, 157.
336. S. E. Bresler, M. I. Mosevitskii, I. Ya. Poddubnyi and G. I. Shi, *Vysokomol. Soedin.*, 1961, **3**, 820; *Polym. Sci. USSR (Engl. Transl.)*, 1962, **3**, 670.
337. M. Gippin, *J. Appl. Polym. Sci.*, 1970, **14**, 1807.
338. W. M. Saltman, F. S. Farson and E. Shoenberg, *Rubber Plast. Age*, 1965, **46**, 502.
339. R. E. Cunningham and R. A. Dove, *J. Polym. Sci., Part A*, 1968, **6**, 1751.
340. V. P. Mardykin, P. N. Gaponik, Z. U. Rogoshkina and V. V. Bolotov, *Vestsi Akad. Navuk BSSR, Ser. Khim. Navuk*, 1970, (4), 112 (*Chem. Abstr.*, 1970, **74**, 13 942).
341. V. P. Mardykin, V. K. Bradaev and P. N. Gaponik, *Izv. Vyssh. Uchebn. Zaved. Khim. Khim. Technol.*, 1967, **10**, 1258 (*Chem. Abstr.*, 1968, **68**, 92 609).
342. W. Y. Chiang, *Ta T'ung Hsueh Pao*, 1976, **6**, 225 (*Chem. Abstr.*, 1977, **86**, 140 553).
343. S. E. Bresler, M. I. Mosevitskii, I. Ya. Poddubnyi and N. N. Chesnokova, *Rubber Chem. Technol.*, 1960, **33**, 689.
344. S. E. Bresler, M. I. Mosevitskii, I. Ya. Poddubnyi and G. I. Shi, *J. Polym. Sci.*, 1961, **52**, 317.
345. S. E. Horne, Jr. and C. J. Carman, *J. Polym. Sci., Part A-1*, 1971, **9**, 3039.
346. W. Cooper, D. E. Eaves, G. D. T. Owen and G. Vaughan, *J. Polym. Sci., Part C*, 1964, **4**, 211.
347. W. Cooper, R. K. Smith and A. Stokes, *J. Polym. Sci., Part B*, 1966, **4**, 309.
348. Yu. B. Monakov, S. R. Rafikov, N. K. Minchenkova, I. R. Mullagaliev and K. S. Minsker, *Dokl. Akad. Nauk SSSR*, 1981, **258**, 892 (*Chem. Abstr.*, 1981, **95**, 169 868).
349. B. J. Pantukh, V. A. Rozentsvet, Yu. B. Monakov and S. R. Rafikov, *Dokl. Akad. Nauk SSSR*, 1982, **265**, 1186 (*Chem. Abstr.*, 1982, **97**, 163 588).
350. Dunlop Rubber Co., Ltd., *Fr. Pat.* 1 348 983 (1964) (*Chem. Abstr.*, 1964, **61**, 7209).
351. T. Makimoto and K. Fuji (Kuraray Co., Ltd.), *Br. Pat.* 1 387 135 (1975) (*Chem. Abstr.*, 1975, **84**, 44 972).
352. T. Makimoto and K. Fuji (Kuraray Co., Ltd.), *Jpn. Pat.* 74 31 554 (1974) (*Chem. Abstr.*, 1975, **82**, 58 551).
353. W. Cooper and G. Vaughan (Rubber Co., Ltd), *Br. Pat.* 1 007 646 (1965) (*Chem. Abstr.*, 1966, **64**, 903).
354. W. Gronski, N. Murayama, H. J. Cantow and T. Miyamoto, *Polymer*, 1976, **17**, 358.
355. G. Delheye and F. Dawans, *Makromol. Chem.*, 1966, **98**, 164.
356. F. Dawans and Ph. Teyssié, *Makromol. Chem.*, 1967, **109**, 68.
357. F. Dawans and Ph. Teyssié, *Eur. Polym. J.*, 1969, **5**, 541.
358. G. Natta, L. Porri, P. Corradini, G. Zanini and F. Ciampelli, *J. Polym. Sci.*, 1961, **51**, 463.
359. D. H. Beebe, C. E. Gordon, R. N. Thudium, M. C. Throckmorton and T. L. Hanlon, *J. Polym. Sci., Polym. Chem. Ed.*, 1978, **16**, 2285.
360. M. C. Gallazzi, A. Giarrusso and L. Porri, *Makromol. Chem., Rapid Commun.*, 1981, **2**, 59.
361. D. Cuzin, Y. Chauvin and G. Lefebvre, *Eur. Polym. J.*, 1969, **5**, 283.
362. G. Natta, L. Porri, G. Stoppa, G. Allegra and F. Ciampelli, *J. Polym. Sci., Part B*, 1963, **1**, 67.
363. G. Natta, L. Porri, A. Carbonaro and G. Stoppa, *Makromol. Chem.*, 1964, **77**, 114.
364. G. Natta, L. Porri and S. Valenti, *Makromol. Chem.*, 1963, **67**, 225.
365. G. Costa, P. Locatelli and A. Zambelli, *Macromolecules*, 1973, **6**, 653.
366. A. A. Panasenko, V. N. Odinokov, Yu. B. Monakov, L. M. Khalilov, A. S. Bezgina, V. K. Ignatynk and S. R. Rafikov, *Vysokomol. Soedin., Ser. B*, 1977, **19**, 656 (*Chem. Abstr.*, 1977, **87**, 185 152).

367. D. M. Xie, C. G. Zhong, N. A. Yuan, Y. F. Sun, S. X. Xiao and J. Ouyang, *Kao Fen Tzu T'in Hsun*, 1979, **4**, 233 (*Chem. Abstr.*, 1980, **92**, 77 008).
368. A. Bolognesi, S. Destri, L. Porri and F. Wang, *Makromol. Chem., Rapid Commun.*, 1982, **3**, 187.
369. F. Cabassi, S. Italia, G. Ricci and L. Porri, in 'Proceedings of a Meeting on Transition Metal Catalyzed Polymerization', Akron, June 1986, in press.
370. D. Xie and C. Zhong, *Kao Fen Tzu T'in Hsun*, 1983, (3), 220 (*Chem. Abstr.*, 1983, **99**, 159 096).
371. G. Natta, L. Porri, A. Carbonaro, F. Ciampelli and G. Allegra, *Makromol. Chem.*, 1962, **51**, 229.
372. L. Porri, A. Di Corato and G. Natta, *J. Polym. Sci., Part B*, 1967, **5**, 321.
373. L. Porri, A. Di Corato and G. Natta, *Eur. Polym. J.*, 1969, **5**, 1.
374. G. Natta, L. Porri and G. Sovarzi, *Eur. Polym. J.*, 1965, **1**, 81.
375. G. Natta, L. Porri and M. C. Gallazzi, *Chim. Ind. (Milan)*, 1964, **46**, 1158 (*Chem. Abstr.*, 1965, **62**, 5346).
376. L. Porri and D. Pini, *Chim. Ind. (Milan)*, 1973, **55**, 196.
377. R. Ohno, Y. Tanuka and M. Kawakami, *Polym. J.*, 1973, **4**, 56.
378. W. Marconi, A. Mazzei, S. Cucinella, M. Cesari and E. Paoluzzi, *J. Polym. Sci., Part A*, 1965, **3**, 123.
379. W. Marconi, A. Mazzei, S. Cucinella and M. Cesari, *J. Polym. Sci., Part A*, 1964, **2**, 4261.
380. J. K. Stille and E. D. Vessel, *J. Polym. Sci.*, 1961, **49**, 419.
381. W. Marconi, A. Mazzei, S. Cucinella, M. Cesari and E. Paoluzzi, *J. Polym. Sci., Part C*, 1967, **16**, 805.
382. T. F. Yen, *J. Polym. Sci.*, 1959, **35**, 533.
383. T. F. Yen, *J. Polym. Sci.*, 1959, **38**, 272.
384. S. Grassman, A. Yamada and O. Vogl, *J Macromol. Sci., Chem.*, 1981, **16**, 897.
385. H. L. Hsieh and H. C. Heh, *Ind. Eng. Chem. Prod. Res. Dev.*, 1986, **25**, 456.
386. F. Wang and A. Bolognesi, *Kao Fen Tsu T'in Hsun*, 1982, (3), 238 (*Chem. Abstr.*, 1982, **97**, 182 930).
387. M. Zimmerman and G. Marwede (Bayer A. G.), *Ger. Pat.* 2 541 004 (1977) (*Chem. Abstr.*, 1977, **86**, 156 808).
388. F. Cabassi, S. Italia, A. Giarrusso and L. Porri, *Makromol. Chem.*, 1986, **187**, 913.
389. L. Porri and M. C. Gallazzi, *Eur. Polym. J.*, 1966, **2**, 189.
390. M. Kamachi, N. Wakabayashi and S. Murahashi, *Macromolecules*, 1974, **7**, 744.
391. S. Destri, M. C. Gallazzi, A. Giarrusso and L. Porri, *Makromol. Chem., Rapid Commun.*, 1980, **1**, 293.
392. F. Wang, A. Bolognesi, A. Immirzi and L. Porri, *Makromol. Chem.*, 1981, **182**, 3617.
393. L. Porri and M. Aglietto, *Makromol. Chem.*, 1976, **177**, 1465.
394. S. Destri, G. Gatti and L. Porri, *Makromol. Chem., Rapid Commun.*, 1981, **2**, 605.
395. S. Destri, A. Bolognesi, L. Porri and F. Wang, *Makromol. Chem., Rapid Commun.*, 1982, **3**, 187.
396. M. I. Lobach, V. A. Kormer, I. Yu. Tsereteli, G. R. Kondratenkov, B. D. Babitskii and V. I. Klepikova, *J. Polym. Sci., Polym. Lett. Ed.*, 1971, **9**, 71.
397. V. A. Kormer and M. I. Lobach, *Polym. Lett.*, 1972, **10**, 177.
398. R. Warin, Ph. Teyssié, P. Bourdandur and F. Dawans, *Polym. Lett.*, 1973, **11**, 177.
399. V. I. Klepikova, G. P. Kondratenkov, V. A. Kormer, M. I. Lobach and L. A. Churlayeva, *J. Polym. Sci., Polym. Lett. Ed.*, 1973, **11**, 193.
400. V. I. Klepikova, G. B. Erusaliniskii, M. I. Lobach, L. A. Churlayeva and V. A. Kormer, *Macromolecules*, 1976, **9**, 217.
401. V. A. Kormer, M. I. Lobach, V. I. Klepikova and B. D. Babitskii, *J. Polym. Sci., Polym. Lett. Ed.*, 1976, **14**, 317.
402. V. A. Kormer and M. I. Lobach, *Macromolecules*, 1977, **10**, 572.
403. N. N. Druz, A. V. Zak, M. I. Lobach, P. P. Shpakov and V. A. Kormer, *Eur. Polym. J.*, 1977, **13**, 875.
404. N. N. Druz, A. V. Zak, M. I. Lobach, V. A. Vasiliev and V. A. Kormer, *Eur. Polym. J.*, 1978, **14**, 21.
405. R. Warin, M. Julémont and Ph. Teyssié, *J. Organomet. Chem.*, 1980, **185**, 413.
406. M. Julémont, R. Warrin and Ph. Teyssié, *J. Mol. Catal.*, 1980, **7**, 523.
407. K. Vrieze, C. Maclean, P. Cossee and C. W. Hilbers, *Recl. Trav. Chim. Pays-Bas*, 1966, **85**, 1077 (*Chem. Abstr.*, 1967, **66**, 24 250).
408. W. Brenner, P. Heimbach, H. Hey, E. W. Muller and G. Wilke, *Justus Liebigs Ann. Chem.*, 1969, **727**, 161.
409. V. N. Sokolov, G. M. Kvostic, I. Ya. Poddubnii and G. P. Kondratenkov, *J. Organomet. Chem.*, 1971, **29**, 313.
410. R. P. Hughes, T. Jack and J. Powell, *J. Am. Chem. Soc.*, 1972, **94**, 7723.
411. J. Zachoval, F. Mikes, J. Krepelka, O. Prouzova and O. Pradova, *Eur. Polym. J.*, 1972, **8**, 397.
412. V. A. Kormer, B. D. Babitskii, M. I. Lobach and N. N. Chesnokova, *J. Polym. Sci., Part C*, 1969, **16**, 4351.
413. L. A. Churlayaeva, V. I. Valuev, T. S. Dmitrieva, M. I. Lobach, V. N. Sokolov, R. A. Shlyachter and V. A. Kormer, *Macromolecules*, 1976, **9**, 214.
414. N. Shaoru, X. Gao and X. Tang, *Kao Fen Tzu T'in Hsun*, 1983, (5), 328 (*Chem. Abstr.*, 1984, **100**, 139 644).
415. D. H. Dawes and C. A. Winkler, *J. Polym. Sci., Part A*, 1964, **2**, 3029.
416. M. Dubini, C. Longiave and R. Castelli, *Chim. Ind. (Milan)*, 1963, **45**, 923.
417. G. Allegra, F. Lo Giudice, G. Natta, U. Giannini, G. Fagherazzi and P. Pino, *Chem. Commun.*, 1967, 1263.
418. G. Ricci *et al.*, *Polym. Commun.*, submitted for publication.
419. G. Henrici-Olivé, S. Olivé and E. Schmidt, *J. Organomet. Chem.*, 1972, **39**, 20.
420. J. T. Harrod and L. R. Wallace, *Macromolecules*, 1969, **2**, 449.
421. J. T. Harrod and L. R. Wallace, *Macromolecules*, 1972, **5**, 686.
422. A. Muzhai, L. S. Bresler, V. A. Grechanovskii and I. Ya. Poddubnyi, *Dokl. Akad. Nauk SSSR*, 1968, **180**, 920 (*Chem. Abstr.*, 1968, **69**, 52 478).
423. V. S. Aliev, A. G. Azizov, F. A. Nasirov, T. A. Ismailov and F. M. Agaev, *Deposited Doc.*, 1984, *VINITI* 3495-84 (*Chem. Abstr.*, 1985, **103**, 178 671).
424. T. Sun, J. Yang, S. Pang and J. Ouyang, *Ying Yung Chieh Mien Hua Hsueh*, 1985, **2**, 47 (*Chem. Abstr.*, 1985, **103**, 142 442).
425. J. Hu, C. Zhou and J. Ouyang, *Chung-Kuo K'o Hsueh Yuan Ying Yung Hua Hsueh Yen Chiu So Chi K'an*, 1982, **19**, 63 (*Chem. Abstr.*, 1984, **101**, 153 246).
426. X. Zhu, Y. Qiao and D. Han, *Kao Fen Tzu T'in Hsun*, 1984, **3**, 207 (*Chem. Abstr.*, 1985, **102**, 25 087).
427. Z. Gong, B. Li, G. Yang and F. Wang, *Kao Fen Tzu T'in Hsun*, 1983, **2**, 116 (*Chem. Abstr.*, 1983, **99**, 176 315).
428. B. I. Pantukh, V. A. Rozentsvet, Yu. B. Monakov and S. R. Rafikov, *Acta Polym.*, 1983, **34**, 732 (*Chem. Abstr.*, 1984, **100**, 86 972).

429. S. Xia, C. Wang, L. Xu and X. Tang, *Gaodeng Xuexiao Huaxue Xuebao*, 1985, **6**, 945 (*Chem. Abstr.*, 1986, **104**, 149 467).
430. S. Ni, X. Gao and X. Tang, *Kao Fen Tzu T'in Hsun*, 1983, (4), 241 (*Chem. Abstr.*, 1984, **100**, 103 922).
431. Y. Yang, Y. Ling, H. Xin and X. Tang, *Hecheng Xiangjiao Gongye*, 1986, **9**, 179 (*Chem. Abstr.*, 1986, **105**, 60 992).
432. P. Aubert, J. Sledz and F. Schué, *J. Polym. Sci., Polym. Chem. Ed.*, 1981, **19**, 955.
433. H. Yasuda and A. Nakamura, *Angew. Chem., Int. Ed. Engl.*, 1987, **26**, 723.
434. G. Erker, J. Wicher, K. Engel, S. Rosenfeldt, W. Dietrich and C. Kruger, *J. Am. Chem. Soc.*, 1980, **102**, 6344.
435. G. Erker, J. Wicher, K. Engel and C. Kruger, *Chem. Ber*, 1982, **115**, 3300.
436. G. Erker, K. Engel and P. Vogel, *Angew. Chem., Suppl.*, 1982, 1746.
437. R. Benn and G. Schroth, *J. Organomet. Chem.*, 1982, **228**, 71.
438. W. H. Hersh, F. J. Hollander and R. G. Bergman, *J. Am. Chem. Soc.*, 1983, **105**, 5834.
439. A. D. Hunter, P. Legzdins, C. R. Nurse, F. W. B. Einstein and A. C. Willis, *J. Am. Chem. Soc.*, 1985, **107**, 1791.
440. A. D. Hunter, P. Legzdins, F. W. B. Einstein, A. C. Willis, B. E. Bursten and M. G. Gatter, *J. Am. Chem. Soc.*, 1986, **108**, 3843.
441. P. W. N. M. van Leeuwen and A. Praat, *J. Organomet. Chem.*, 1970, **21**, 501.
442. J. Lukas, P. W. N. M. van Leeuwen, H. C. Volger and A. P. Kouwenhoven, *J. Organomet. Chem.*, 1973, **47**, 153.
443. A. Bolognesi, S. Destri, Z. Zinam and L. Porri, *Makromol. Chem.*, 1984, **5**, 679.
444. B. A. Dolgoplosk, K. L. Makovetskii, L. I. Red'kina, T. V. Soboleva, E. I. Tinyakova and V. A. Yakovlev, *Dokl. Akad. Nauk SSSR*, 1972, **205**, 387 (*Chem. Abstr.*, 1972, **77**, 140 604).
445. K. L. Makovetskii and L. I. Red'kina, *Dokl. Akad. Nauk SSSR*, 1974, **215**, 1380 (*Chem. Abstr.*, 1975, **83**, 61 116).
446. L. Porri, in 'Structural Order in Polymers', ed. F. Ciardelli and P. Giusti, Pergamon Press, Oxford, 1981, p. 51.
447. L. Porri, M. C. Gallazzi, S. Destri and A. Bolognesi, *Makromol. Chem., Rapid Commun.*, 1983, **4**, 485.
448. J. Furukawa, T. Saegusa, K. Irako, N. Hirooka and T. Narumiya, *Kogyo Kagaku Zasshi*, 1962, **65**, 2074 (*Chem. Abstr.*, 1963, **58**, 11 553).
449. L. S. Bresler, B. A. Dolgoplosk, M. F. Kolechkova and E. N. Kropacheva, *Rubber Chem. Technol.*, 1964, **37**, 121.
450. T. Suminoe, K. Sasaki, N. Yamazachi and S. Kambara, *Kobunshi Kagaku*, 1964, **21**, 9 (*Chem. Abstr.*, 1964, **61**, 801).
451. Yu. B. Monakov, N. G. Marina, N. V. Duvakina and S. R. Rafikov, *Dokl. Akad. Nauk SSSR*, 1977, **236**, 617 (*Chem. Abstr.*, 1977, **87**, 185 045).
452. C. Shan and J. Ouyand, *Kao Fen Tzu T'in Hsun*, 1983, **3**, 238 (*Chem. Abstr.*, 1984, **99**, 176 330).
453. I. Pasquon, L. Porri, A. Zambelli and F. Ciampelli, *Chim. Ind. (Milan)*, 1961, **43**, 509 (*Chem. Abstr.*, 1961, **55**, 22 901).
454. J. Furukawa, T. Saegusa, T. Naruyama and S. Kurahashi, *Kogyo Kagaku Zasshi*, 1962, **65**, 2082 (*Chem. Abstr.*, 1963, **58**, 14 263).
455. H. Weber, B. Schleimer and H. Winter, *Makromol. Chem.*, 1967, **101**, 320.
456. E. A. Delaney, N. J. Pinkowski and W. Nudenberg (Uniroyal Inc.), *Ger. Pat.* 2 513 206 (1975) (*Chem. Abstr.*, 1976, **84**, 32 343).
457. W. Marconi, A. Mazzei, G. Lugli and M. Bruzzone, *J. Polym. Sci., Part C*, 1967, **16**, 805.
458. N. G. Marina, N. V. Duvakina, Yu. B. Monakov, U. M. Dzhemilev and S. R. Rafikov, *Vysokomol. Soedin., Ser. A*, 1985, **27**, 1203 (*Chem. Abstr.*, 1985, **103**, 178 676).
459. S. R. Rafikov, R. K. Kudashev, Yu. B. Monakov, N. M. Vlasova, E. A. Glukhov, R. I. Khusnutdinov and U. M. Dzhemilev, *Acta Polym.*, 1984, **35**, 558 (*Chem. Abstr.*, 1984, **101**, 132 161).
460. Yu. B. Monakov, E. A. Glukhov, I. T. Farkhieva, R. Kh. Kudashev, K. S. Minsker, U. M. Dzhemilev, G. A. Tolstikov and S. R. Rafikov, *Vysokomol. Soedin., Ser. A*, 1980, **22**, 385 (*Chem. Abstr.*, 1980, **92**, 181 370).
461. R. Kh. Kudashev, M. M. Blasova, Yu. B. Monakov, K. S. Minsker and S. S. Rafikov, *Prom-st Synt. Kauch.*, 1982, **8**, 4 (*Chem. Abstr.*, 1982, **97**, 164 291).
462. G. Natta, L. Porri, A. Carbonaro and G. Lugli, *Makromol. Chem.*, 1962, **53**, 52.
463. A. Carbonaro, V. Zamboni, G. Novajra and G. Dall'Asta, *Rubber Chem. Technol.*, 1973, **46**, 1274.
464. M. Bruzzone, A. Carbonaro and L. Gargani, *Rubber Chem. Technol.*, 1978, **51**, 907.
465. B. A. Dolgoplosk, S. I. Beilin, Yu. V. Korshak, G. M. Chernenko, L. M. Vardanyan and M. P. Teterina, *Eur. Polym. J.*, 1973, **9**, 895.
466. T. V. Soboleva, V. A. Yakovlev, E. I. Tinyakova and B. A. Dolgoplosk, *Dokl. Akad. Nauk SSSR*, 1976, **228**, 619 (*Chem. Abstr.*, 1976, **85**, 109 042).
467. M. Bruzzone, A. Carbonaro and C. Corno, *Makromol. Chem.*, 1978, **179**, 2173.
468. J. Furukawa, E. Kobayashi and K. Haga, *J. Polym. Sci., Polym. Chem. Ed.*, 1973, **11**, 629.
469. J. Furukawa, K. Haga and E. Kobayashi (Bridgestone Tire Co., Ltd.), *Ger. Pat.* 2 123 802 (1971) (*Chem. Abstr.*, 1972, **76**, 73 497).
470. T. Takahashi and M. Yamaguchi (Japanese Bureau of Industrial Technology), *Jpn. Pat.* 72 15 696 (1967) (*Chem. Abstr.*, 1972, **77**, 89 763).
471. M. Ichikawa, Y. Takeuchi, K. Makino and M. Endo (Japan Synthetic Rubber Co., Ltd.), *Ger. Pat.* 2 117 616 (1971) (*Chem. Abstr.*, 1972, **76**, 60 720).
472. Y. Ishizuka and K. Irako (Bridgestone Tire Co., Ltd.), *Jpn. Pat.* 71 41 375 (1971) (*Chem. Abstr.*, 1972, **76**, 142 114).
473. K. Shirayama, S. Kida, M. Katata and M. Kashitani (Sumitomo Chemical Co., Ltd.), *Jpn. Pat.* 73 34 983 (1973) (*Chem. Abstr.*, 1973, **79**, 67 626).
474. A. Kawasaki, H. Veda and I. Maryuama (Maruzen Petrochemical Co., Ltd.), *Jpn. Pat.* 72 23 416 (1972) (*Chem. Abstr.*, 1972, **77**, 153 531); *Jpn. Pat.* 72 19 682 (1972) (*Chem. Abstr.*, 1972, **77**, 141 189); *Jpn. Pat.* 72 19 693 (1972) (*Chem. Abstr.*, 1972, **77**, 141 190); *Jpn. Pat.* 72 19 694 (1972) (*Chem. Abstr.*, 1972, **77**, 141 191); *Jpn. Pat.* 72 16 185 (1972) (*Chem. Abstr.*, 1972, **77**, 89 766); *Jpn. Pat.* 72 16 183 (1972) (*Chem. Abstr.*, 1972, **77**, 89 767); *Jpn. Pat.* 72 16 184 (1972) (*Chem. Abstr.*, 1972, **77**, 89 768); *Jpn. Pat.* 72 08 613 (1972) (*Chem. Abstr.*, 1972, **77**, 49 808).
475. K. Yamaguchi, N. Kanoh, I. Ito and N. Enokido (Mitsubishi Chemical Industries Co., Ltd.), *Ger. Pat.* 2 320 848 (1973) (*Chem. Abstr.*, 1974, **80**, 134 054).
476. K. Hayashi, A. Kawasaki and I. Maryuama (Maruzen Petrochemical Co., Ltd.), *Jpn. Pat.* 74 40 631 (1974) (*Chem. Abstr.*, 1975, **83**, 194 304); *Jpn. Pat.* 74 40 632 (1974) (*Chem. Abstr.*, 1975, **82**, 171 737).
477. S. Shiga and T. Yamao (Sumitomo Chemical Co., Ltd.), *Jpn. Pat.* 73 67 380 (1973) (*Chem. Abstr.*, 1974, **81**, 26 956).
478. A. Priola, C. Corno, G. Ferraris and S. Cesca, *Makromol. Chem.*, 1979, **180**, 2859.

479. G. Natta, A. Zambelli, I. Pasquon and F. Ciampelli, *Makromol. Chem.*, 1964, **79**, 161.
480. A. Kawasaki and M. Katsuta (Maruzen Petrochemical Co., Ltd.), *Jpn. Pat.* 73 25 087 (1973) (*Chem. Abstr.*, 1973, **79**, 67 630); *Jpn. Pat.* 73 74 586 (1973) (*Chem. Abstr.*, 1974, **80**, 60 414).
481. A. Kawasaki and I. Maryuama (Maruzen Petrochemical Co., Ltd.), *Jpn. Pat.* 73 02 937 (1973) (*Chem. Abstr.*, 1974, **80**, 4097).
482. K. Hayashi, A. Kawasaki and I. Maryuama (Maruzen Petrochemical Co., Ltd.), *Jpn. Pat.* 74 26 070 (1974) (*Chem. Abstr.*, 1975, **82**, 58 589); *Ger. Pat.* 2 113 164 (1971) (*Chem. Abstr.*, 1972, **76**, 60 117); *Jpn. Pat.* 72 34 791 (1972) (*Chem. Abstr.*, 1973, **78**, 137 082); *Jpn. Pat.* 72 34 792 (1972) (*Chem. Abstr.*, 1973, **78**, 137 080); *Ger. Pat.* 2 128 804 (1971) (*Chem. Abstr.*, 1972, **76**, 114 513); *Jpn. Pat.* 72 42 882 (1972) (*Chem. Abstr.*, 1973, **78**, 137 096); *Jpn. Pat.* 73 11 232 (1973) (*Chem. Abstr.*, 1974, **80**, 134 064).
483. L. A. Myagkova, E. N. Kropacheva and A. S. Khachaturov, *Vysokomol. Soedin., Ser. A*, 1984, **26**, 954. (*Chem. Abstr.*, 1984, **101**, 55 721).
484. A. Kawasaki and I. Maryuama (Maruzen Petrochemical Co., Ltd.), *Jpn. Pat.* 73 25 086 (1973) (*Chem. Abstr.*, 1973, **79**, 67 616).
485. K. Hayashi, A. Kawasaki and M. Katsuta (Maruzen Petrochemical Co., Ltd.), *Jpn. Pat.* 72 34 980 (1972) (*Chem. Abstr.*, 1973, **78**, 112 553).
486. A. Kawasaki and I. Maryuma (Maruzen Petrochemical Co., Ltd.), *Jpn. Pat.* 73 25 085 (1973) (*Chem. Abstr.*, 1973, **79**, 67 617; *Jpn. Pat.*, 73 70 791 (1973) (*Chem. Abstr.*, 1974, **81**, 50 322).
487. A. Kawasaki, I. Maryuama, M. Taniguchi, R. Hirai and J. Furukawa, *J. Polym. Sci., Polym. Lett. Ed.*, 1969, **7**, 613.
488. J. Furukawa, R. Hirai and M. Nakamiwa, *J. Polym. Sci., Polym. Lett. Ed.*, 1969, **7**, 671.
489. J. Furukawa, *Angew. Makromol. Chem.*, 1972, **23**, 189.
490. J. Furukawa, *J. Polym. Sci., Polym. Symp.*, 1974, **48**, 19.
491. W. Wieder and J. Witte, *J. Appl. Polym. Sci.*, 1981, **26**, 2603.
492. N. A. Venediktova, L. V. Smirnova and E. N. Kropacheva, *Vysokomol. Soedin., Ser. A*, 1985, **27**, 1452 (*Chem. Abstr.*, 1986, **104**, 6207).
493. L. A. Yatsenko, A. G. Boldyrev, I. G. Zhuchikhina and E. N. Kropacheva, *Kinet. Katal.*, 1985, **26**, 1416; *Kinet. Catal. (Engl. Transl.)*, 1985, **26**, 1225.
494. L. V. Smirnova, N. A. Venediktova and E. N. Kropacheva, *Vysokomol. Soedin., Ser. A*, 1985, **27**, 1676 (*Chem. Abstr.*, 1986, **104**, 34 363).
495. K. Hayashi, A. Kawasaki and I. Maryuama (Maruzen Petrochemical Co., Ltd.), *Jpn. Pat.* 72 34 792 (1972) (*Chem. Abstr.*, 1973, **78**, 137 080).
496. A. Kawasaki and I. Maryuama (Maruzen Petrochemical Co., Ltd.), *Jpn. Pat.* 72 47 861 (1972) (*Chem. Abstr.*, 1974, **80**, 60 409); *Jpn. Pat.* 74 26 073 (1974) (*Chem. Abstr.*, 1975, **82**, 58 584).
497. N. A. Venediktova, E. N. Kropacheva, L. V. Smirnova, L. I. Vyshinskaya and S. Ya. Timoshenko, *Vysokomol. Soedin., Ser. A*, 1980, **22**, 977 (*Chem. Abstr.*, 1980, **93**, 26 894).
498. C. J. Carman, *Macromolecules*, 1974, **7**, 789.

# 6

# Metathesis Polymerization: Chemistry

ALLAN J. AMASS
*Aston University, Birmingham, UK*



## 6.1 AIMS AND OBJECTIVES

In this chapter concerned with the ring-opening polymerization of cycloalkenes, bicycloalkenes and alkynes, it is hoped that the reader will acquire an understanding of the metathesis reaction, particularly as it applies to polymerization reactions. In part, it will be necessary to refer to the metathesis of linear alkenes but this will only be done in order to assist in an explanation of the chemistry involved in metathesis because much work of a fundamental nature has been carried out with linear alkenes.

This chapter deals with a brief review of the nature and scope of the alkene metathesis reaction as applied to linear and cyclic alkenes and establishes the mechanistic similarities between the two areas. After dealing with the range of monomers and initiators that are available for metathesis ring-opening polymerization, a discussion will be undertaken of the kinetic chain nature of the reaction that has been developed to explain metathesis. Detailed discussion of the stereochemistry of the propagation reaction as applied particularly to the polymerization of bicyclic monomers gives much insight into the chemistry of metathesis reactions in general. The chapter deals then with the kinetic and thermodynamic aspects of the metathesis polymerization, which in some respects is typical of ring-opening polymerization in general, but in others shows significant differences.

## 6.2 NATURE AND SCOPE OF METATHESIS POLYMERIZATION

The alkene metathesis reaction is a catalyzed reaction that can be summarized quite simply, as shown in equation (1), and when applied to simple linear alkenes[1] the substituents $R^1$ and $R^2$ may be either linear or branched-chain alkyl groups. They are not normally structures that have or will generate conjugated double bonds because these usually lead to deactivation of the catalysts. When the alkene is cyclic,[2] polymerization of the monomer takes place by a ring-opening process that converts an unsaturated monomer into an unsaturated polymer. Thus the polymerization of cyclopentene may be described by equation (2).

$$2\,R^1CH{=}CHR^2 \rightleftharpoons R^1CH{=}CHR^1 + R^2CH{=}CHR^2 \qquad (1)$$

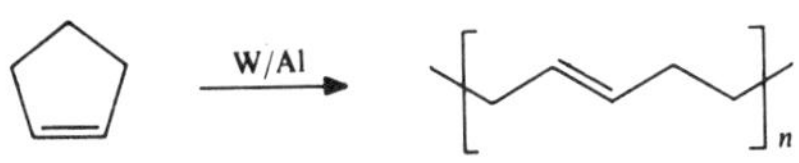


(2)

Natta[3] first suggested that the most likely site for cleavage of the ring was the carbon–carbon single bond $\alpha$ to the double bond. However, it was Calderon and Ofstead[4] who recognized that the ring-opening polymerization reaction is a special case of the more general alkene metathesis reaction (1). Both reactions are catalyzed by the product of the reaction between $WCl_6$ or $WOCl_4$ and $AlEtCl_2$. In the reaction between 2-butene and perdeuterated 2-butene only one new alkene is formed, 1,1,1,2-tetradeutero-2-butene. This is consistent with cleavage of the carbon–carbon double bond and a transalkylidenation reaction (equation 3), whereas transalkylation (equation 4), an exchange of alkyl groups, would have resulted in the production of two new alkenes.

$$CH_3CH{=}CHCH_3 + CD_3CD{=}CDCD_3 \rightleftharpoons 2\,CH_3CH{=}CDCD_3 \qquad (3)$$

$$CH_3CH{=}CHCH_3 + CD_3CD{=}CDCD_3 \rightleftharpoons CH_3CH{=}CHCD_3 + CD_3CD{=}CDCH_3 \qquad (4)$$

Confirmation that the ring-opening polymerization of a cycloalkene also proceeds by cleavage of the carbon–carbon double bond was provided by Dall'Asta and Motroni.[5] Cyclooctene was copolymerized with cyclopentene which was $^{14}C$-labelled at the carbon–carbon double bond. A random copolymer of the two monomers was obtained and the product of copolymerization was ozonized and reduced to a mixture of diols. If cleavage of the ring had taken place at the single bond to the double bond then the radioactivity would be distributed as shown in structure (**1**), whereas cleavage of the double bond itself would give structure (**2**). Ozonolysis of the copolymer and reduction of the products with $NaBH_4$ gave only one radioactive product, 1,5-pentanediol, confirming that polymerization by ring-opening is also a reaction involving cleavage of the carbon–carbon double bond.

$$\text{—}CH{=}CH(CH_2)_6C^*H{=}C^*H(CH_2)_3CH{=}CH(CH_2)_6\text{—}$$

$\leftarrow C_8\,\text{unit} \rightarrow \leftarrow C_5\,\text{unit} \rightarrow$

(**1**)

$$={CH(CH_2)_6CH{=}C^*H(CH_2)_3C^*H{=}CH(CH_2)_6CH}=$$

$\leftarrow C_8\,\text{unit} \rightarrow \leftarrow C_5\,\text{unit} \rightarrow$

(**2**)

## 6.3 MONOMERS FOR METATHESIS POLYMERIZATION

### 6.3.1 General Aspects

In common with the polymerization of 1-alkenes by Ziegler–Natta catalysts, the ring-opening metathesis polymerization of cycloalkenes and bicycloalkenes and the metathesis of acyclic alkenes are catalyzed by the products of the reaction between a transition metal compound and an alkylmetal or Lewis acid.[6] In the ensuing discussion of the mechanism of reaction, the initial step of the propagation reaction will involve the complexation of a carbon–carbon double bond to a transition metal carbene.[7] Consequently, the substrates that may be used as monomers for polymerization are severely restricted in structure, and those monomers that possess functional groups that would be expected either to react with the transition metal carbene or to form strong complexes with the transition metal atom are excluded. Monomers that possess functional groups such as amino, hydroxyl, carbonyl and carboxyl groups and esters, ethers, amides and thiols are not normally suitable for polymerization.

### 6.3.2 Monocyclic Monomers

The principal monomers used in metathesis ring-opening polymerization are the monocyclic cycloalkenes. Prominent among these monomers are cyclopentene,[8] cyclooctene[9] and 1,5-cyclooctadiene.[10] The cycloalkene may also possess an alkyl substitutent on the ring[11] and the only restrictions that should be applied to this class of monomers is that they should not: (i) be cyclohexene, or its derivatives if the $C_6$ ring is strain-free, (ii) possess an alkyl substituent at the carbon–carbon double bond or (iii) be a conjugated diene, such as cyclopentadiene or 1,3-cyclooctadiene.

A recent publication by Patton and McCarthy[12] suggests that the ring size of the cycloalkene is not a factor that determines the polymerization of the ring, but on thermodynamic grounds one would not expect cyclohexene to polymerize except at extremely low temperatures or exceedingly high pressures.

There is no kinetic data available to allow comparison of the polymerization of unsubstituted cycloalkenes. Even if such data were available, the extraction from it of unequivocal values for the rate constants for propagation would be difficult because: (i) the efficiencies of catalyst systems, for the most part unknown, appear to be extremely low when estimated from molecular weight data and (ii) catalyst systems may well be produced from the reaction not only between tungsten halides and alkylaluminums but also by reaction of the transition metal compound and the cycloalkene. The fraction of catalyst actually converted to active sites may depend upon the nature of the cycloalkene.

Systems recently described by Grubbs[13] and Ivin[14] as 'living' metathesis catalysts might provide catalysts to be used in such studies, because under these conditions it will at least be possible to estimate the concentration of propagating ends. This would be particularly useful in those cases in which the only sequence of reactions to be considered is coordination of the monomer to an active centre followed by metathetic cleavage of the double bond. Unfortunately, this is not the only reaction that has to be considered, particularly in the case of monocyclic monomers of low ring-strain energy, in which case consideration has to be taken of: (i) intramolecular back-biting reactions[15] that lead to the establishment of a thermodynamic equilibrium between open-chain polymer, monomer and cyclic oligomers; and (ii) reaction of the propagating end with an in-chain carbon–carbon double bond by an intermolecular reaction which would result in molecular weight randomization reactions.[16]

Both of these factors will be discussed in detail later, but essentially, from a kinetic point of view, the instantaneous concentration of active centres participating in propagation reactions is reduced by the occurrence of this type of reaction.

Monocyclic alkenes that possess an alkyl substituent may be polymerized by metathesis catalysts provided that the alkyl substituent is not at the carbon–carbon double bond. This correlates with the decreasing order of activity towards metathesis in the series of alkenes $RCH{=}CH_2 > R^1CH{=}CHR^2 > R^1R^2C{=}CHR^3 > R^1R^2C{=}CR^3R^4$. Feast[17] has studied the activity of fluorinated monomers towards polymerization. When $WCl_6/Ph_4Sn$ is used as catalyst, 1*H*,2*H*-perfluorocyclopentene will not polymerize but its presence does not inhibit the polymerization of cyclopentene. The effect of the fluorine substituents is to withdraw the electrons from the $\pi$ system and reduce the coordinating ability of the monomer. If the monomer is bicyclic it will not undergo metathesis polymerization if the fluorine substituent is present in the same ring as the double bond. If the substituent is part of a different ring then the substituent effect is negligible or non-existent and the monomer will polymerize. Thus polymers **(3)**–**(8)** have been obtained from fluorinated bicyclic monomers.

### 6.3.3 Bicyclic Monomers

Bicyclic and polycyclic monomers, particularly monomers possessing a double bond susceptible to metathesis, are readily polymerized by a wide range of catalysts.[18] Monomers of this type have

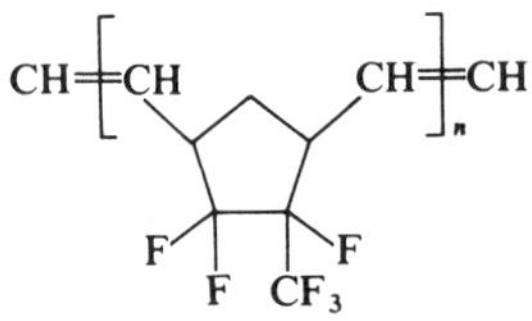

**(3)**

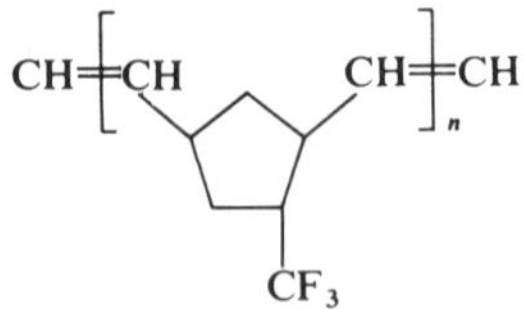

**(4)**

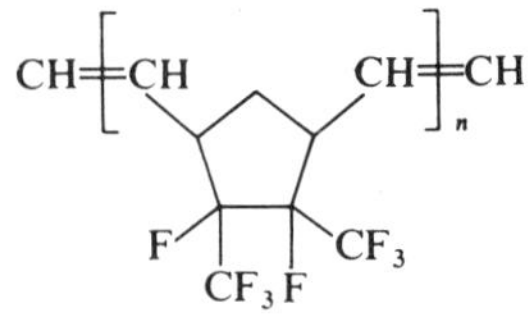

**(5)**

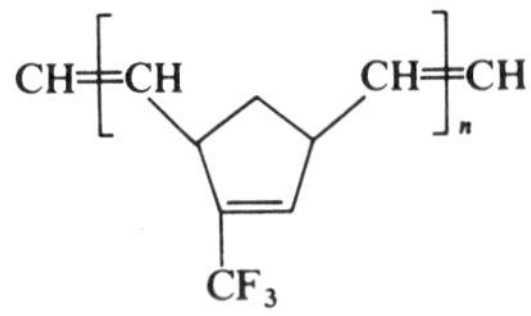

**(6)**

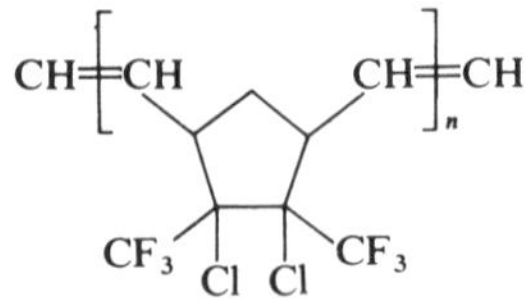

**(7)**

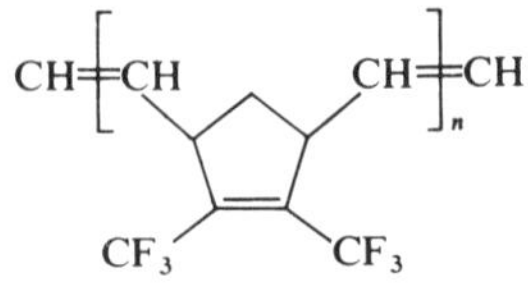

**(8)**

been particularly useful in the synthesis of precursors to polyacetylene by the 'Durham' route[19] and in the study of the stereochemistry of metathesis reactions.

Volume 4, Chapter 7 of this work will discuss in greater detail the chemistry of the application of alkene metathesis to the synthesis of polyacetylene. Suffice it to say here that a highly strained monomer such as 7,8-bis(trifluoromethyl)tricyclo[4.2.2.0$^{2,5}$]deca-3,7,9-triene (**9**) will ring open at the C(3) double bond which is the most highly strained double bond in the molecule. The polymer produced will be (**10**; equation 5). Polyacetylene may then be obtained by the thermal elimination of perfluoroxylene from the backbone (equation 6).

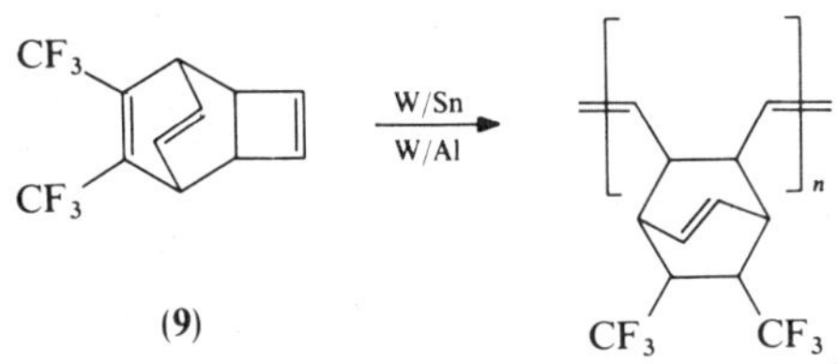

**(9)** **(10)** (5)

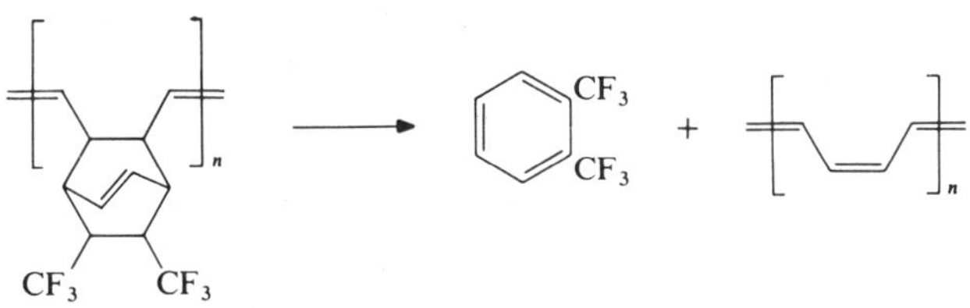

**(10)** (6)

This is a particularly convenient route to the synthesis of polyacetylene because the rigid polyacetylene chain may be obtained from a flexible precursor polymer that can be fabricated in a number of forms. The monomer may be obtained by a Diels–Alder-type reaction between perfluoro-2-butyne and cyclooctatetraene, which, although a cyclic alkene, does not polymerize readily by a ring-opening metathesis polymerization.

Recent work published by Grubbs,[20] however, describes a catalyst system that provides a route to polyacetylene from cyclooctatetraene. This catalyst is (**11**) and the polymerization may be described by equation (7). The substituent groups are principally ether groups. The efficient cometathesis of cyclopentene with an unsaturated ester has also been reported for a $WCl_6$ system when the ester contains an electron-withdrawing group, the trichloroacetate group.[85]

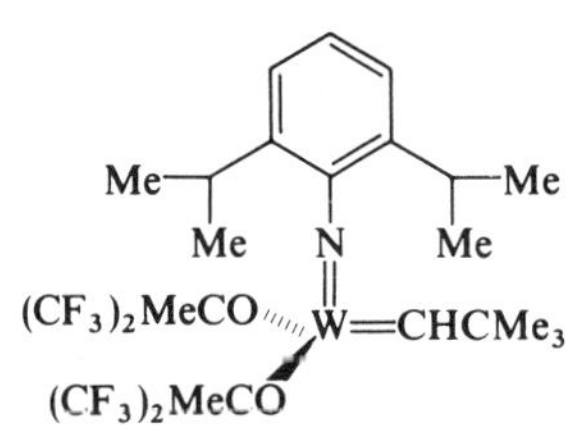

**(11)**

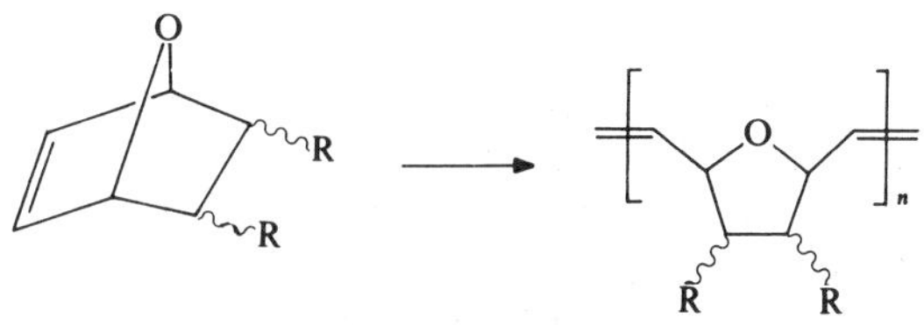

(7)

Bicyclic monomers such as the derivatives of bicyclo[2.2.1]heptene (norbornene) provide a class of polymers that are not susceptible to intramolecular or intermolecular back-biting reactions (equation 8).[21] The carbon–carbon double bond in the polymer does not participate in secondary metathesis reactions. The most likely explanation for this is that steric crowding around the double bond prevents coordination of the double bond to the active centre. The principal advantage of this lack of reactivity of the in-chain double bond is that the stereochemistry generated in the propagation step is frozen into the system, unlike polypentenylene where the in-chain double bond may undergo isomerization by metathesis and the stereochemistry generated in the propagation step may be lost.[22] Furthermore, the orientation of the ring with respect to the rest of the chain may be regarded as a form of tacticity, which, when combined with information about the *cis* or *trans* nature of the double bond, gives startling information about the reproducibility of propagation steps. These features will be discussed in greater detail in Section 6.6.

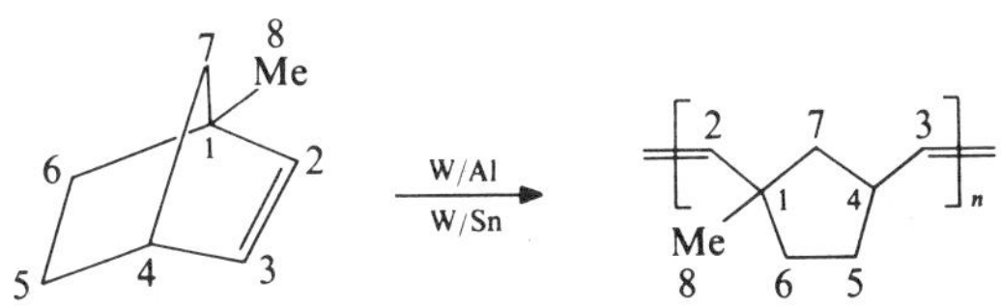


(8)

### 6.3.4 Metathesis and Polymerization of Alkynes

Alkynes should be included in a discussion of metathesis and ring-opening polymerization as they display activity towards heterogeneous and homogeneous metathesis catalysts, in one of three possible reactions (Scheme 1). Polymerization of 1-alkynes[23,24,25] can be effected using conventional alkene metathesis catalysts, but the question arises as to whether the reaction is simple addition polymerization initiated by a metal carbene or whether the polymerization proceeds by a route that is accepted as a metathesis route. The fact that Fischer- and Casey-type carbenes, $Ph(MeO)C{=}W(CO)_5$ and $(3,5\text{-}Me_2C_6H_3)_2C{=}W(CO)_5$, initiate the polymerization of 1-butyne as described by Katz, and that copolymers of phenylacetylene and cyclopentene have also been prepared, suggests that polymerization of the alkyne is a metathesis-type reaction. A similar report by Amass[26] has shown that acetylene may be polymerized by a $WCl_6/AlEtCl_2$ catalyst, and that soluble block copolymers of cyclopentene and acetylene are obtained from the addition of acetylene to an active cyclopentene polymerization.[27]

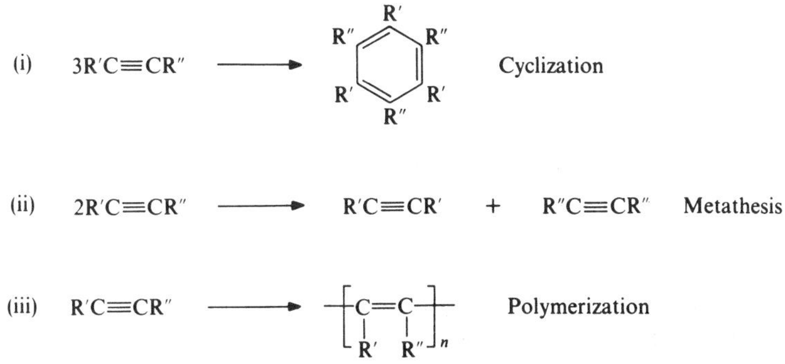


**Scheme 1**

In general, the heterogeneous catalysts, in particular $WO_3$[28] or $MoO_3$[29] adsorbed on silica, are inclined to promote metathesis, as are the carbyne complexes (**12**), equivalent to the Casey and Fischer carbenes, prepared by Schrock.[30] For the most part, the homogeneous catalysts tend to promote the polymerization reaction of alkynes. The polymerization is not restricted to the aliphatic series of alkynes, as phenylacetylene is a particularly reactive monomer in this respect. In the copolymerization of phenylacetylene with disubstituted alkynes, 2-butyne or 4-octyne, phenylacetylene is by far the more reactive monomer.[31]

$$W{\equiv}CCMe_3(OCH_2CMe_3)_3$$

**(12)**

Hasegawa[32] has studied the effects of substituents on the reactivity of substituted phenylacetylenes towards metathesis polymerization. Copolymerization of phenylacetylene ($M_1$) with a series of substituted phenylacetylenes ($M_2$) yields various values of the reactivity ratios ($r_1, r_2$) for the comonomers. Since the value of $r_1$ is given by $r_1 = k_{p11}/k_{p12}$, $1/r$ may be used as a measure of the reactivity of the substituted phenylacetylene. If a Hammett plot is then made of log ($r_1^{-1}$) against $\sigma^+$, the Hammett factor for the substituent group, a straight line with a negative slope ($\rho^+$) is obtained. This dependence is in agreement with the chain carrier in the polymerization, *i.e.* the site at which the alkyne is coordinated, being electron deficient in nature. Similar effects are found in the cationic polymerization of styrene, but their magnitudes are much greater for that polymerization than for the metathesis polymerization of phenylacetylene.

The polymers obtained by the polymerization of phenylacetylene[33] are highly coloured, unlike, for example, poly(1-phenylpropyne). The colour depends upon the method used to prepare the sample, but, in general, the polymers do not appear to be black as in the case of polyacetylene. This would indicate that there is a limited conjugation, restricted by the steric hindrance of the phenyl group. Indeed, poly(3-hexyne) is white, in the case of the disubstituted alkyne, as the freedom to rotate about the carbon–carbon single bond is so restricted by non-bonded steric interactions that the chain is prevented from adopting a completely delocalized structure.

Recent work has shown that acetylene itself may be polymerized using metathesis catalysts, even though the yields of polymer are extremely low.[26] Catalysts derived from $WCl_6$ and alkylaluminums, alkyl halides and butyllithium have been used. The activities of these catalysts for the polymerization of acetylene are likely to be low because the propagating centres are encapsulated by intractable insoluble polymer, preventing monomer from diffusing to the active site. These catalysts also form the basis of the formation of block copolymers of polyacetylene with either polystyrene or polypentenylene materials that are 'soluble' in many organic solvents. Schrock[34] has also shown that living catalysts can be generated for metathesis systems, and that the copolymerization of an alkyne with a cyclic alkene can lead to the formation of block copolymers.

## 6.4 CATALYST SYSTEMS FOR RING-OPENING METATHESIS POLYMERIZATION

### 6.4.1 General Aspects of Catalysis

A wide variety of transition metal compounds may be used as the catalyst component, but the most widely used, in order of importance, are tungsten, molybdenum, rhenium and ruthenium compounds. Other transition metals have been used in special studies, particularly those related to mechanistic studies of linear alkene metathesis, and these include niobium and tantalum compounds. The catalyst systems are conveniently classified as:

(i) Heterogeneous catalysts, which are normally transition metal oxides or carbonyls adsorbed on to a high surface-area support such as alumina or silica. These supports may have varying Brønsted acidities and this may determine the applicability of a support for a particular purpose. These catalysts are activated under an inert atmosphere and are used in conjunction with a promoter and/or a cocatalyst for the metathesis of linear alkenes and unsaturated esters.[35] Heterogeneous catalysts are very rarely used for the metathesis polymerization of cycloalkenes and bicycloalkenes.

(ii) Homogeneous catalysts, which are prepared as complexes of transition metal compounds or derived from the reaction of a transition metal compound, normally a halide, with an alkylmetal or a Lewis acid. There are obvious similarities between these systems and the conventional Ziegler–Natta catalysts for the the polymerization of 1-alkenes. Like Ziegler–Natta catalysts, the metathesis catalysts are susceptible to reaction with traces of moisture and other impurities and hence it is necessary to work under dry anaerobic conditions, although traces of moisture or oxygen-containing compounds have often been used as activators for the homogeneous catalysts.[36] Although these catalysts are referred to as homogeneous or soluble, it is feasible that the soluble precursors react to give finely divided heterogeneous catalysts in some cases. The homogeneous catalysts are by far the most important class of catalysts for the ring-opening polymerization of cyclic and bicyclic alkenes.

### 6.4.2 Tungsten-based Catalysts

The most widely used initiator systems for the polymerization of a variety of monocyclic and bicyclic alkenes are based upon $WCl_6$[37] as a catalyst; the catalyst normally, but not exclusively, requires activation by a cocatalyst. In a typical reaction, the components of the catalyst system may either be added sequentially as solutions to a solution of the monomer or the components can be reacted together prior to addition to the cyclic alkene. There is no doubt that when the monomer is a highly strained ring, such as norbornene,[38] rapid polymerization ensues when pre-mixed catalysts are used, but when the monomer is a monomer with only moderate ring-strain, then the most active catalysts are obtained when the components are added separately to the solution of monomer. It is possible that this results from the fact that when the monomer is highly strained only a trace of active centres is required to effect a high rate of polymerization.

If the polymerization is catalyzed by separate additions of the components to the monomer solution, it is more than likely that the catalyst does not remain as $WCl_6$ in a solution of monomer. On addition of $WCl_6$ to the solution of monomer there is an immediate change from dark blue to brown.[39] Although this might be ascribed to a change in the solvent, Amass has shown that this immediate colour change is followed by a further series of spectral changes indicating that reaction takes place between the alkene and the catalyst. Accompanying this reaction there is a reduction in oxidation state[40] of the tungsten atom and the formation of by-products; hydrogen chloride and 3-chlorocyclopentene are among these. A mechanism for this reduction has recently been proposed by Amass *et al.*[41] which is shown in Scheme 2 (□ = vacant coordination site). Coordination of the monomer to the reduced tungsten halide may prevent $WCl_4$ from forming completely a polymeric structure, incapable of acting as a catalyst.

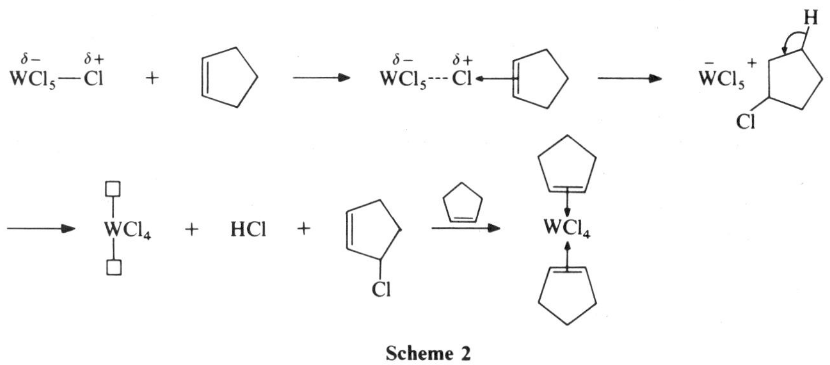


**Scheme 2**

In addition to the catalysts based on tungsten halides ($WCl_6$, $WF_6$, $WOCl_4$) other tungsten catalysts include the Casey[42] and Fischer[43] carbenes, particularly for the polymerization of highly strained monomers such as norbornene. Ivin has also investigated the use of tricarbonyl(mesitylene)tungsten[44] as a catalyst with triethylaluminum in the presence of *exo*-2,3-epoxybicyclo[2.2.1]-hept-5-ene.

The most recent developments, however, in the area of catalysis have come with the synthesis of tungsten compounds that are reportedly living catalysts. Schrock had previously reported that tungsten–carbene complexes, propagating the metathesis of acyclic alkenes, could be detected using $^1H$ NMR.

Kress[45] used catalysts derived from complexes formed between $GaBr_3$ and $Br_2W{=}CHCMe_3(OCH_2Bu^t)_2$. The chain ends associated with the polymerization of a number of norbornene derivatives were detected by investigating the $^1H$ NMR of a polymerizing system; monomers **(13)**–**(20)** were polymerized. The living nature of this catalyst permits the synthesis of block copolymers by much the same procedure as that for the anionic polymerization of styrene and 1,3-butadiene. If one monomer is allowed to polymerize to completion before a second monomer is added, then, provided the active centre initiates the polymerization of the second monomer, a block copolymer is formed; there would not appear to be the restrictions associated with the synthesis of block copolymers by anionic means.

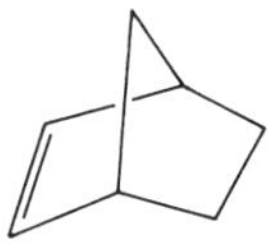

**(13)**

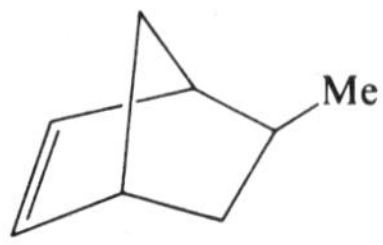

**(14)**

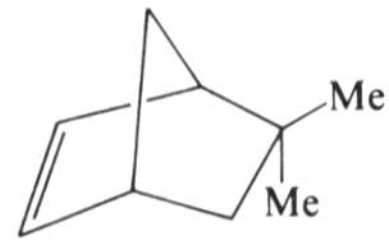

**(15)**

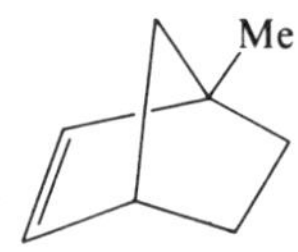

**(16)**

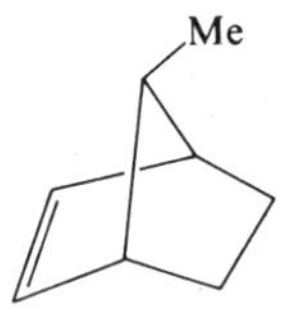

**(17)**

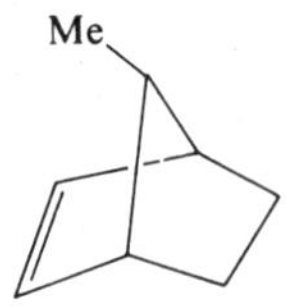

**(18)**

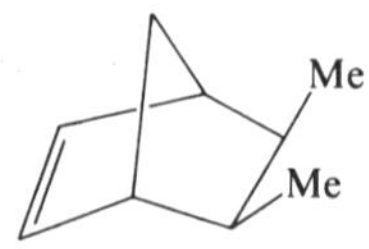

**(19)**

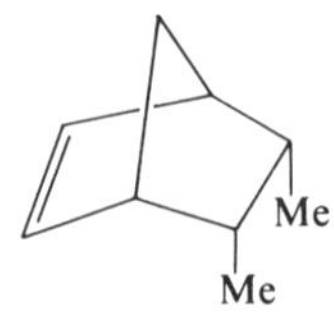

**(20)**

The catalyst (**21**) will polymerize norbornene to produce a living polymer and even a living polymer with cyclopentene ends has been reported.[46] Functionalized monomers such as 7,8-bis(acyloxy)tricyclo[$4.2.2^{1,6}0^{2,5}$]deca-3,7-diene (**22**) may be polymerized by the more strained cyclobutene ring (equation 9). The reactivity, as measured by the rate of polymerization, depends on the alkoxy ligand in the catalyst; reactivities that range from high rates of polymerization at $-50\,^{\circ}C$ when the ligand is 1,1-bis(trifluoromethyl)ethoxy to moderate rates at room temperature when the ligand is *t*-butoxy. These remarkable catalysts appear to remain unchanged for long periods of time at ambient temperature. Block copolymers can again be synthesized with these catalyst systems when a second monomer is added to the living system generated by the first monomer. In addition to monoene cyclic compounds, cyclooctatetraene may be used as a monomer yielding efficient conversion to polyacetylene chains. Copolymerizations involving cyclooctatetraene may be either of a block type, if the comonomers are added sequentially, or statistical, if the second monomer added has a reactivity similar to that of 1,5-cyclooctadiene. Molecular weight studies show that these monometallic catalysts are 100% efficient and narrow molecular weight distribution polymers are obtained. This would appear to be additional evidence in favour of the metal carbene as the active site in metathesis. Furthermore, the greater activity of the bis(trifluoromethyl)ethoxy ligand over the *t*-butoxy group would indicate that the electron-withdrawing effect of the trifluoromethyl group activates the transition metal towards complexing the alkene.

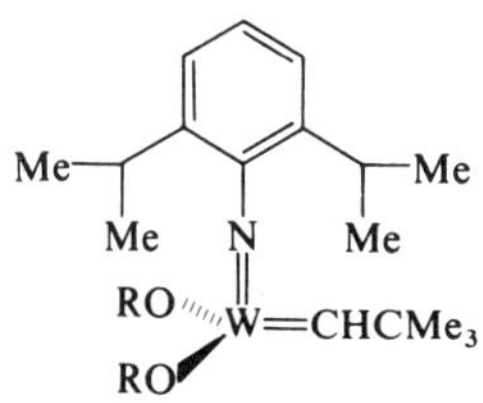

**(21)**

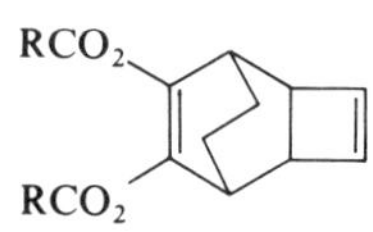

**(22)**

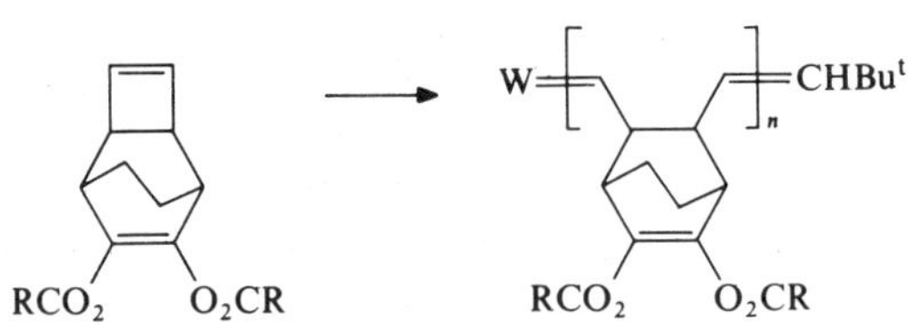

(9)

### 6.4.3 Molybdenum-based Catalysts

In common with tungsten compounds, molybdenum catalysts can be formed either as homogeneous or heterogeneous systems. The heterogeneous catalysts, such as $MoO_3$ adsorbed on a $\gamma$-$Al_2O_3$[47] surface, are particularly useful as catalysts for the metathesis of linear alkenes. Indeed, when used for the metathesis of alkenes, such molybdenum catalysts show a greater specificity for retention of stereochemistry and higher rate of reaction than the corresponding tungsten catalysts.

Initially it was thought that *cis* stereospecificity was a feature of molybdenum catalysts; examples of this specificity are shown by: (i) $(py)_2MoCl_2(NO)_2/EtAlCl_2$[48] gives initially 89% *cis*-2-butene and 73% *cis*-3-hexene on the metathesis of *cis*-2-pentene; (ii) the intramolecular metathesis of *cis,cis*-2,8-decadiene with $MoO_3/CoO/\gamma\text{-}Al_2O_3$[49] similarly gave 78% *cis*-2-butene; and (iii) $MoCl_5/AlEt_3$[3] polymerizes cyclopentene to *cis*-polypentenylene at −35 °C.

However, many catalyst systems that produce initially high *cis* contents in the alkene tend eventually to produce a product mixture that is thermodynamically controlled.[50–52]

Despite the fact that initial reports of the polymerization of cycloalkenes included molybdenum and tungsten compounds, there is no doubt that the catalysts based on molybdenum are not nearly so efficient and 'clean' as those based upon tungsten compounds. Eleuterio first reported that hydrogen-reduced $MoO_3/\gamma\text{-}Al_2O_3$[53] would polymerize cyclopentene when activated by $LiAlH_4$. The first homogeneous catalysis of the ring-opening polymerization of cyclopentene disclosed by Natta in 1964 included the use of $MoCl_5/AlEt_3$[3] for the production of *cis*-polypentenylene. Apart from other isolated instances, such as the report by Laverty[54] that $MoCl_5$ is a single-component catalyst for the polymerization of norbornene, there are few instances where molybdenum compounds have been used for the polymerization of cyclic or bicyclic alkenes.

### 6.4.4 Miscellaneous Catalyst Systems

Besides tungsten catalysts (the most important) and molybdenum systems, catalysts based on ruthenium and rhenium compounds are by far the most interesting. Ivin and Rooney[55,56] have used $ReCl_5$ alone or in conjunction with an alkylmetal to polymerize monomers such as norbornene and derivatives of norbornene, because, together with $RuCl_3$[57] catalysts, it is possible to generate a number of catalysts with tendencies towards differing stereospecificity, both carbon–carbon bond specificity and tacticity of the cyclopentyl ring. Although rhenium catalysts necessitate the use of a cocatalyst, the initiation of polymerization using $RuCl_3$ is somewhat different, because polymerizations initiated by $RuCl_3$ do not require the use of a Lewis acid or alkylmetal as cocatalyst; rather, it is necessary to use $RuCl_3$ either as a complex with water or cyclooctadiene.

### 6.4.5 The Role of the Cocatalyst

An exhaustive survey of the catalyst systems that are used for the metathesis of alkenes or the polymerization of cyclic alkenes would show that the cocatalysts used in bimetallic systems are (i) organometallic compounds and/or (ii) Lewis acids.

Immediately there is the possibility to be considered that the alkylmetal is necessary in a normal catalyst system for the formation of the metal carbene that is the active centre for the metathesis reaction. Grubbs, using $WCl_6/Me_4Sn$,[58] and Muetterties, using $WCl_6/Me_2Zn$,[59] both showed that methane is a product of the reactions. By inference it was suggested that methane was one of the disproportionation products of a methylated tungsten derivative, the other product being the catalytically active methylene complex (equation 10).

$$WCl_6 + SnMe_4 \longrightarrow Cl_5WMe \longrightarrow Cl_4W{=}CH_2 + CH_4 + WCl_6 \qquad (10)$$

Bencze[60] has also shown that $WCl_6$ reacts with $Ph_4Sn$ to form the metathetically active $PhWCl_3$, but in this case the formation of an active metal carbene must involve other reactions such as reaction of the W—C(phenyl) bond with an alkene. Disproportionation of phenyltungsten is obviously not a possible route to the active centre in this case.

Evidence for the direct participation of the alkylmetal in the formation of the active centre has recently been put forward by Amass[61] for both the polymerization of cycloalkenes and the polymerization of acetylene. The catalyst system ($WCl_6$/BuLi) first described by Wang and Menapace[62] for the metathesis of 2-pentene was modified by Amass for the polymerization of cyclopentene. (Polystyryl)lithium with a low molecular weight, $M_n$ 8000, was prepared as a solution in cyclohexane using either *n*-butyl- or *s*-butyl-lithium as initiator. This solution of living (polystyryl)lithium was then used as a cocatalyst for the polymerization of cyclopentene, using $WCl_6$ as the catalyst. The final product of the polymerization reaction could be separated into two fractions, one soluble in hexane and rich in polypentenylene and the other insoluble in hexane and rich in polystyrene. Gel permeation chromatographic analysis of the product showed the conventional bimodal molecular weight distribution, normally associated with polyalkenylenes prepared by metathesis ring-opening polymerization. Comparison of the gel permeation chromatograms produced by the differential refractometer and the UV/visible detector ($\lambda = 254$ nm) showed that both the high and low molecular weight materials contained materials absorbing at 254 nm; polystyrene but not polypentenylene absorbs at this wavelength. It was concluded that the product contained at least some block copolymeric material and inferences were drawn relating to the mechanism of initiation in metathesis polymerization reactions.

The synthesis of block copolymers by this anionic-to-metathesis transformation technique suggested that the organometallic cocatalyst played a vital role in generating the initial metal carbene. A reaction sequence of metallation followed by rearrangement to the metal carbene was suggested as the initiation mechanism (Scheme 3).

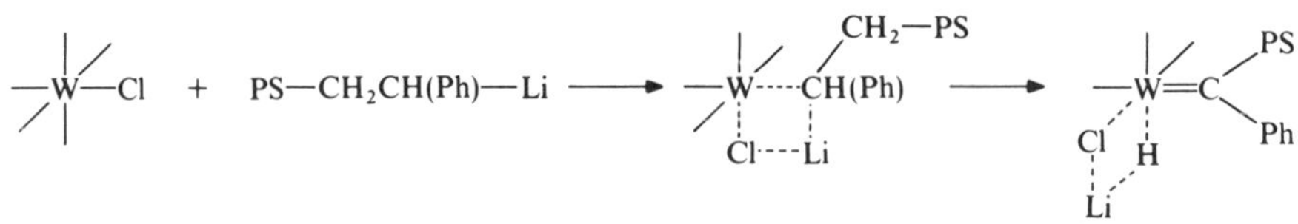

PS = Polystyrene

**Scheme 3**

## 6.5 MECHANISM OF REACTION

### 6.5.1 The Kinetic Chain Reaction

Prior to the proposal by Hérisson and Chauvin[63] that the mechanism of polymerization of cycloalkenes was a non-pairwise mechanism, a number of pairwise schemes had been proposed, but the evidence is so substantially in favour of a non-pairwise mechanism that these other mechanisms will not be discussed here. The reader is referred to a number of previous reviews for a detailed discussion of these, now largely defunct, mechanisms.

The non-pairwise mechanism as proposed by Hérisson and Chauvin may be summarized for the ring-opening polymerization of a monomer such as cyclopentene as shown in Scheme 4. The propagation reaction involves a transition metal carbene as the active species which must possess a vacant coordination site. During propagation the carbon–carbon double bond of the cycloalkene coordinates at this vacant site to form a typical metal $\pi$ complex. This complexation step is followed by the formation of a metallacyclobutane intermediate that is unstable and can cleave to form a new metal carbene and a new carbon–carbon double bond that decoordinates to regenerate the propagating metal carbene.

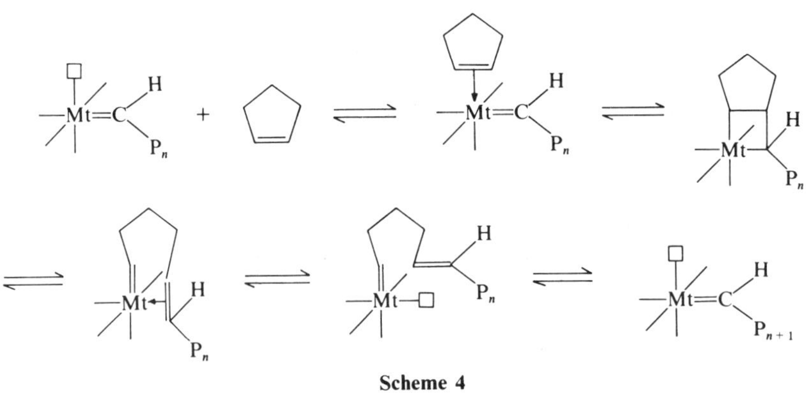

**Scheme 4**

The evidence that led to the proposal of this kinetic chain type mechanism came from studies of the 'cross-metathesis' of cycloalkenes with linear, particularly unsymmetrical, alkenes. For example, when cyclopentene is reacted with 2-pentene, analysis of the products formed in the initial stages of the reaction showed that not only is $MeCH{=}CH(CH_2)_3CH{-}CHEt$ present, which would be expected from a pairwise mechanism, but the two symmetrical products $MeCH{=}CH(CH_2)_3$-$CH{=}CHMe$ and $EtCH{=}CH(CH_2)_3CH{=}CHEt$ are found.

The latter cannot be accounted for by a pairwise mechanism such as equation (11) unless the reaction between two molecules of 2-pentene to form a mixture of 3-hexene and 2-butene takes place preferentially in the earliest stages before reaction of any alkene with cyclopentene can take place.

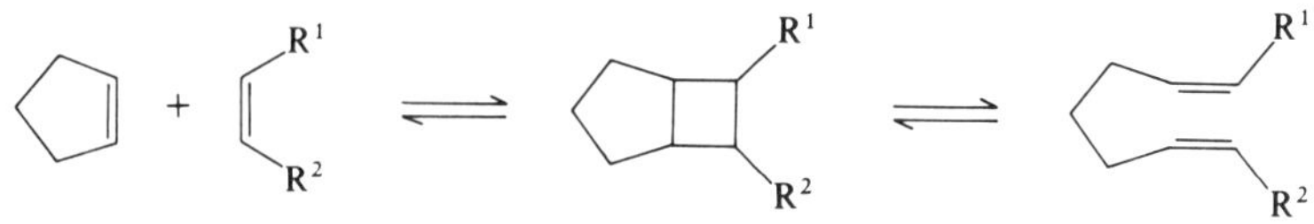


(11)

Subsequently, a number of studies were carried out by various workers to substantiate the non-pairwise scheme but perhaps the most conclusive were those carried out by Katz.[64] Cyclooctene was reacted with a mixture of 4-octene and 2-butene. The possible products of the combined cross-metathesis reactions are shown in Scheme 5. The product ratio $(C_{14}/C_{12}) \times (C_{14}/C_{16})$ was determined as a function of reaction time and extrapolated to zero time, when, for a pairwise mechanism, the product ratio would be expected to be zero. Contrary to this observation the product of the ratio was found to be $4.05 \pm 0.05$ for *cis* products and $4.1 \pm 0.09$ for *trans* alkenes, experimentally close to the predicted value of 4.0 for a non-pairwise kinetic chain reaction such as that promoted by the metal carbene.

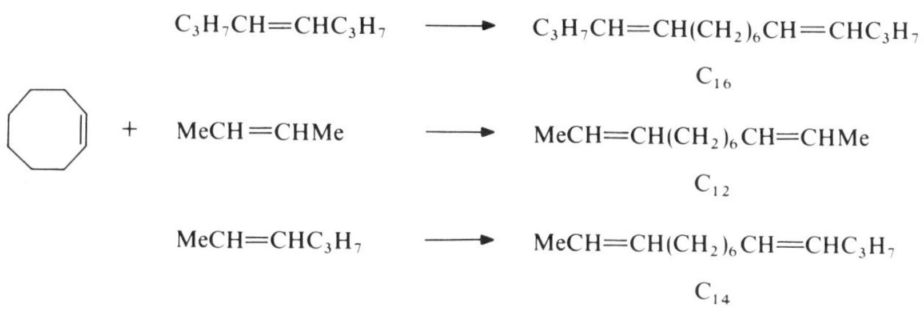


**Scheme 5**

## 6.5.2 Metal Carbene Initiators

If a transition metal carbene is formed as a product of the reaction between the catalyst components in the ring-opening polymerization of cycloalkenes or the metathesis of alkenes, then metal carbenes would be expected to act as active catalysts. Transition metal carbenes are, for the most part, not particularly stable under normal reaction conditions. However, Fischer has shown that transition metal carbenes such as (**23**) may be prepared as stable compounds at low temperatures; Casey has also prepared the diphenylmethylene equivalent of the Fischer compound (**24**). These structures are stabilized by the delocalizing effects of the benzene rings and methoxy groups. Furthermore, they would be expected to form $\pi$ complexes with the reacting alkene.

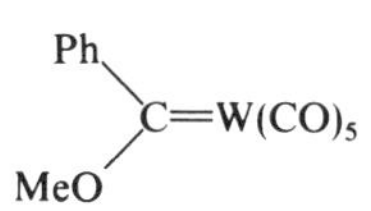


(**23**) Fischer carbene

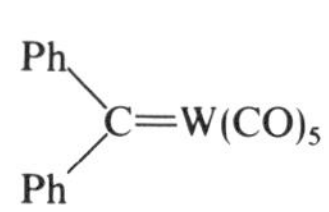


(**24**) Casey carbene

Reaction of $(CO)_5W{=}C(OMe)C_6H_4$-*p*-Me with 3-buten-1-ol resulted in replacement of the methoxy group by the 3-buten-1-oxy group.[65] This compound slowly eliminated carbon monoxide and formed a metal carbene–alkene complex (**25**).

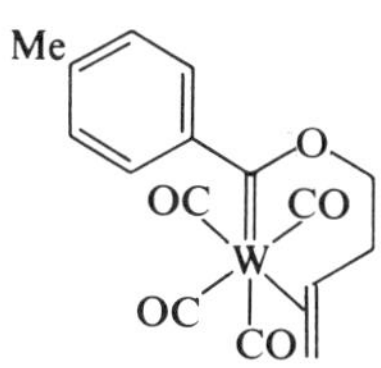


(**25**)

The participation of transition metal–carbene complexes in the metathesis reaction scheme was established by Casey[66] who reacted $(CO)_5W{=}CPh_2$ with isobutene. The principal product of the reaction, in the early stages, was 1,1-diphenylethylene; also formed was 1,1-dimethyl-2,2-diphenylethylene. Both these products are in agreement with the proposed reaction scheme (equation 12).

$$(CO)_5W{=}C(Ph)_2 \quad + \quad Me_2C{=}CH_2 \longrightarrow W(CO)_6 \quad + \quad Ph_2C{=}CH_2 \quad +$$ 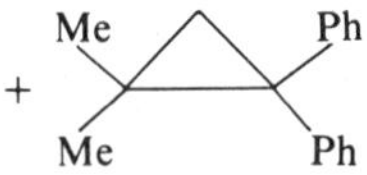 (12)

Cyclopropanes result from the decomposition of the intermediate metallacyclobutane. Schrock[67] has established that bisneopentyl complexes of tantalum and niobium metathesize 1-butene and that any complexes formed during the metathesis do not rearrange; for example, the propylidene complexes do not rearrange according to equation (13). However, the rearrangement of the metallacyclobutane may well be an important termination step along with the bimolecular decomposition of tungsten–methylene complexes. Second-order termination reactions have also been observed for the polymerization of cyclopentene,[68] although in the case of polymerization reactions using conventional bimetallic catalyst systems it would appear that the termination reaction produces a complex of tungsten that promotes further termination. This is particularly prevalent in polymerizations that have a high initial rate of polymerization.

$$Mt{=}C(H)(CH_2R) \longrightarrow Mt \leftarrow \begin{matrix} CHR \\ \| \\ CH_2 \end{matrix} \qquad (13)$$

### 6.5.3 Metallacyclobutanes

An integral part of the overall mechanism of metathesis reactions is the metallacyclobutane intermediate. In the case of active metathesis catalysts it is highly unlikely that such structures are to be found because they are so reactive. However, Grubbs[69] has studied the synthesis of metallacyclobutane derivatives using titanocene compounds as precursors. The reaction for the formation of bimetallic complexes between titanocene dichloride with trimethylaluminum was first described by Tebbe (equation 14).[70] Under carefully controlled conditions this may be reacted with an alkene (2-butene) to produce a 'stable' metallacyclobutane (equation 15).

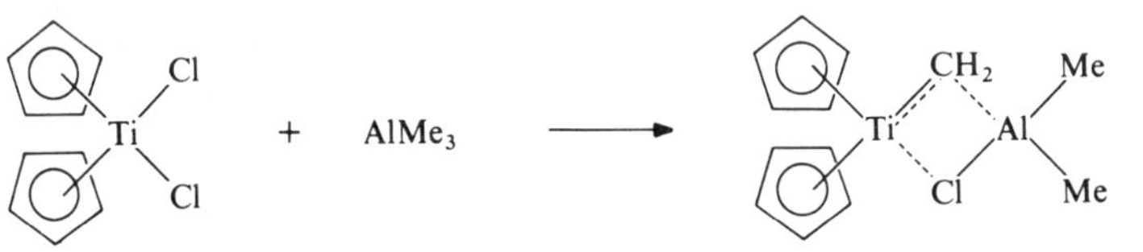


(14)

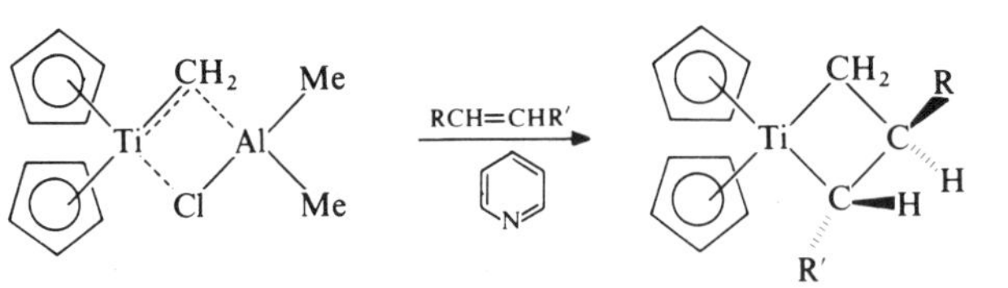


(15)

The *cis* isomer formed by the reaction of the bimetallic complex with 2-butene in the presence of pyridine is less stable than the corresponding *trans*-dimethyltitanacyclobutane but there is substantial evidence, particularly NMR, to corroborate the structure of these titanacyclobutanes. Furthermore, the metathesis of *cis,cis*- and *trans,trans*-2,8-decadiene is catalyzed in low yield with 1,1-dicyclopentadienyl-*cis*-2,3-dimethyltitanacyclobutane. The polymerization of norbornene[71] is initiated by the titanium methylene generated during the reaction of the bimetallic complex with an alkene and pyridine (Scheme 6).

An interesting polymerization that can be effected in this way is the polymerization of 3,4-dipropylidenecyclobutene (equation 16). This highly delocalized polymer may be doped with iodine to generate a conducting polymer.

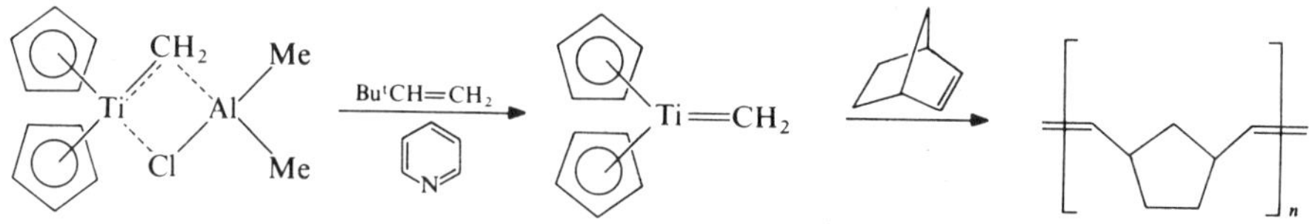


**Scheme 6**

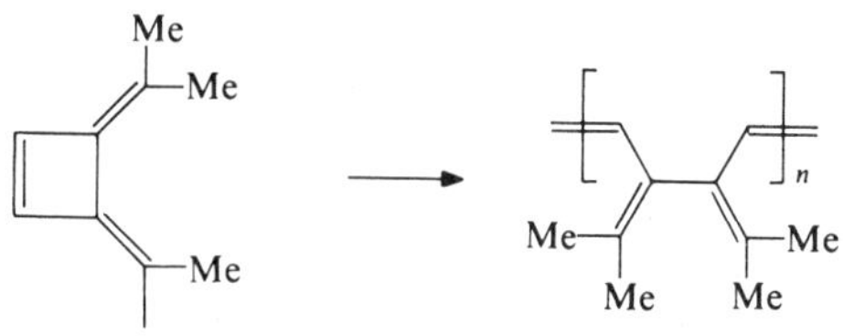

(16)

There appears, therefore, to be substantial evidence to suggest that metallacyclobutanes are essential intermediates in the metathesis reaction. In the proposed reaction sequence involving metal carbenes and metallacyclobutanes, in the case of tungsten compounds, the more stable species is the carbene, whereas, in the case of titanium complexes such as those prepared by Grubbs, the more stable form is that of the metallacyclobutane. However, there is also in this work evidence to suggest that the structure of the intermediate in a conventional catalyst system is not a simple metallacyclobutane because these isolable intermediates are not as active as the catalyst centres in normal polymerization systems.

The differences in activity between these model systems and normal polymerization systems or alkene metathesis systems could well lie in the nature and distribution of ligands around the transition metal atom together with the fact that the most active catalysts may exist in metastable oxidation states.

## 6.6 STEREOCHEMISTRY OF POLYMERIZATION REACTIONS

### 6.6.1 Stereoisomerism in Polymers by Ring-opening Polymerization

Polymerization of cyclic alkenes by ring-opening polymerization gives rise to the occurrence of a number of types of stereoisomerism during propagation. The first of these relates to the structure of the carbon–carbon double bond as it appears in the polymer as a *cis* or *trans* structure and applies to all cyclic monomers. Additionally, when the monomer is bicyclic, the orientation of the resulting in-chain cycloalkyl ring with respect to the chain may generate a polymer that is syndiotactic, isotactic or atactic. The latter type of stereoisomerism is shown for the syndiotactic and isotactic polymers of norbornene in structures (**26**) and (**27**).

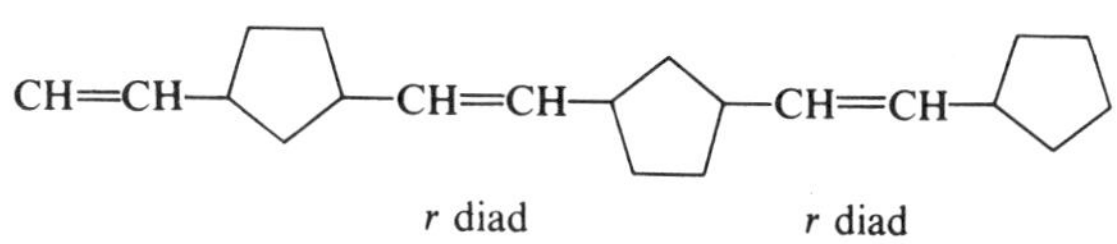


*r* diad *r* diad

(**26**) syndiotactic

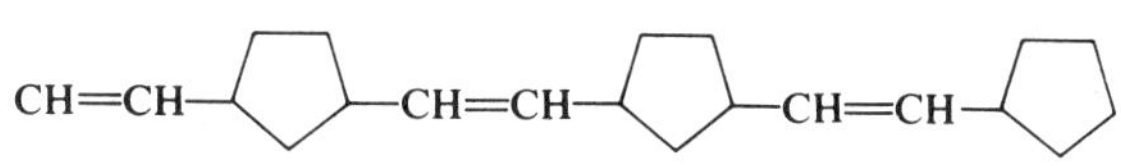


*m* diad *m* diad

(**27**) isotactic

It was recognized by many workers that the structures of the catalyst, cocatalyst and indeed the molar ratios of these components in a polymerization system could affect dramatically the stereochemistry of the products.

### 6.6.2 Stereoisomerism of Polyalkenylenes

The dependence of the stereochemistry of the polymer on the molar ratio of the catalyst components was demonstrated by Pampus *et al.*[11] for the polymerization of cyclopentene using the catalyst $WF_6/Al_2Et_3Cl_3$. At Al:W molar ratios up to 1:1 high *cis* polymers are obtained, whereas the product is of increasingly higher *trans* content when the molar ratio is increased above this value. No attempt was made to interpret this effect except to recognize that the structure of the propagating intermediate had a controlling effect upon the stereochemistry of the propagating reaction. Subsequently, Ivin[22] demonstrated the problems associated with the interpretation of structural data obtained for simple polyalkenylenes because the carbon–carbon double bond in the polymer retains its activity towards metathesis and the structure of this bond may change during the course of a secondary metathesis reaction. It has been reported that during the copolymerization of cyclopentene with norbornene the fraction of *cis* double bonds ($\rho_c$) related to pentenylene-containing diads decreases with conversion over a wide range of monomer compositions. Accordingly, a determination of the structure of polymers of the polyalkenylene type may not necessarily yield reliable information concerning the structures resulting from the propagation step. In contrast, the structure of the double bonds associated with the norbornene units of the copolymer remain constant with conversion. Unlike cyclopentene and similar monomers, norbornene and its derivatives generally form polymers that do not participate in secondary metathesis reactions. This makes them more satisfactory monomers for the study of the stereochemistry of metathesis polymerizations and metathesis reactions in general.

### 6.6.3 Stereoisomerism in Norbornene Polymers

$^{13}C$ NMR spectroscopy has proved a powerful tool in the analysis of the ring-opening polymerization of norbornene and its derivatives. Not only can the technique yield information on the fraction of double bonds that possess a *cis* structure ($\rho_c$) but also the importance of sequences of *cis* or *trans* double bonds in the polymer and the frequency of *cis,trans* (*c,t*) [and *trans,cis* (*t,c*)]diads. In this way analysis will yield a distinction between a random distribution of double bonds and any tendency for the alternation or blockiness in a polymer that contains both *cis* and *trans* double bonds.

$\rho_c$ may be derived from the intensities of a number of peaks in the spectrum of polynorbornene, including peaks arising from the carbon atoms in *cis* and *trans* double bonds, which have chemical shifts, respectively, of 133.88 and 133.10 p.p.m. downfield from TMS. The chemical shifts of the chiral carbon atoms (*) depend not only on the carbon–carbon double bond to which they are bonded but also the structure of the next nearest carbon–carbon double bonds. Thus these atoms may be labelled as shown in structure (**28**).

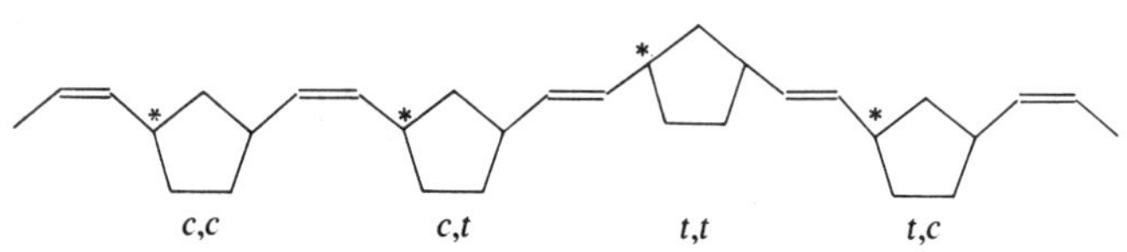


(**28**)

Similarly, the distribution of diads may be determined from the C(3) resonance, C(3) being derived from the bridgehead carbon in the monomer. Now, because the *c,t* and *t,c* arrangements are equivalent, there are only three resonances that exist for this carbon atom depending on whether it is flanked by *cis* or *trans* double bonds.[18] The chemical shifts of these atoms are shown in Table 1.

Homopolymerization of norbornene may be considered to yield a copolymer of *cis* and *trans* comonomer repeat units. It is then possible to define reactivity ratios, in common with other copolymerizations, that give rise to the structures shown in Table 1. The reactivity ratios are $r_t = (2)t,t/(2)t,c$ and $r_c = (2)c,c/(2)c,t$. A tendency to blockiness is then shown when $r_t r_c > 1$ and a random distribution of *cis* and *trans* units when $r_t r_c = 1$.

Ivin *et al.*[73] have studied the polymerization of norbornene using a variety of catalysts and cocatalysts and in so doing have prepared polymers with $\rho_c$ values ranging from 0.2 to 0.8. The data show that if the catalyst generates a polymer with $\rho_c < 0.35$ then the distribution of *cis* and *trans* double bonds is close to random, whereas when $0.5 < \rho_c < 0.85$ there is a pronounced increase in the value of $r_t r_c$ indicating a strong tendency to produce blocky polymers. There is only a slight tendency for this to happen when $0.35 < \rho_c < 0.50$. There is a tendency therefore for the *cis* units to occur in blocks; values of $r_c$ as high as 5 are obtained when the *cis* content is high. In the main, fast-acting catalysts produce low $\rho_c$ polymers whilst slow rates of polymerization are obtained with high *cis* structures.

**Table 1** Chemical Shifts of C(2) and C(3) in Polynorbornene

| *Designation* | *Chemical Shift (TMS)* |
|---|---|
| (2)*t,c* | 43.41 |
| (2)*t,t* | 43.15 |
| (2)*c,c* | 38.67 |
| (2)*c,t* | 38.47 |
| (3)*t,t* | 41.40 |
| (3)*c,t* = (3)*t,c* | 42.11 |
| (3)*c,c* | 42.83 |

### 6.6.4 Effects of Catalyst Structures on the Structure of the Polymer

Detailed studies show that it is now possible to propose general schemes to account for the changes in polymer structure that are brought about by changes in the structure of the catalyst composition. Clearly, the transition metal alone is not responsible for the control of the polymer structure because it is possible to produce a range of polymer structures with the same transition metal. Within any class of transition metal compounds, it is also obvious that the cocatalyst has a significant effect upon the resulting polymer structure, *e.g.* $WCl_6$/(allyl)$_4$Sn yields polymers of high *cis* content ($\rho_c = 0.53$–$0.7$) and blockiness ($r_tr_c = 5$–$6.4$), whereas $WCl_6/Ph_4Sn$ gives polymers with a relatively low $\rho_c$ (0.41) and a tendency towards randomness ($r_cr_t = 1.4$). $WCl_6/AlEtCl_2$ generally produces polymers in the medium range of $\rho_c$ (0.43–0.53) but, in common with the molybdenum-based systems, particularly $MoCl_5$, the *cis* content can be increased by the incorporation of a third component such as ethyl acrylate. This probably changes the ligand environment of the transition metal and/or the oxidation state which together determine the stereocontrol of the propagation reaction.

To explain these results, Ivin[74] invoked the existence of at least two kinetically distinct propagation species, **(29)** and **(30)**. These two species may be in chemical equilibrium but the rates of the participating exchange reactions in the equlibrium change according to the value of $\rho_c$ for the catalyst. Thus at low $\rho_c$ values these propagation species can interchange rapidly and hence the polymer has a lesser tendency to be blocky, whereas at higher values of $\rho_c$ the rates of exchange are considerably reduced; thus, once an active centre has been formed in a *cis* generating mode, then it continues to produce *cis* polymer until disturbed from that mode.

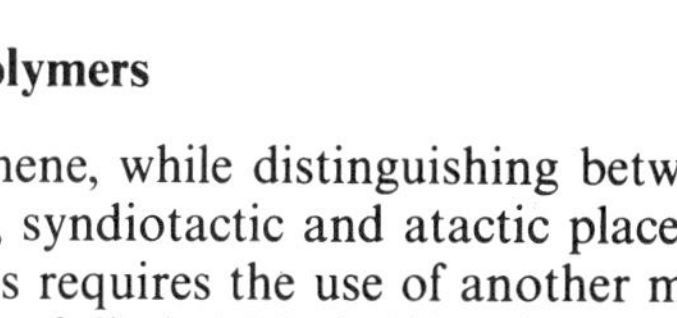

**(29)** **(30)**

### 6.6.5 Substituted Norbornene Polymers

The polymerization of norbornene, while distinguishing between *cis* and *trans* structures, does not distinguish between isotactic, syndiotactic and atactic placements of the cyclopentyl ring. To distinguish between *m* and *r* diads requires the use of another monomer in which there is a ready distinction between the two types of diad. 1-Methylbicyclo[2.2.1]hept-2-ene (**31**; MNBE) is such a monomer.[73]

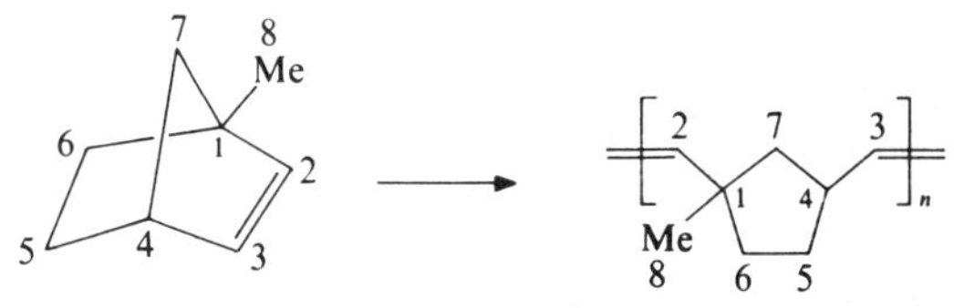

(17)

**(31)**

During the polymerization of MNBE (equation 17), in addition to the isomerism associated with the arrangement of substituents about the double bond and the position of the cycloalkyl ring, there are the possibilities of head-to-head, head-to-tail and tail-to-tail arrangements of the repeat units. These may be shown schematically as illustrated in structure (**32**). The $^{13}C$ NMR spectrum of this polymer is now extremely complex. The alkenic carbon chemical shift depends not only upon the nature of the double bond (*cis* or *trans*) but also on the arrangements of the two repeat units from which the double bond is derived, *i.e.* HH, HT (TH) or TT. Experiments have shown that all arrangements, with the exception of *cis*-HH, can be formed and that it is now necessary to define a term $B$, the bias of polymerization, as shown in equation (18).

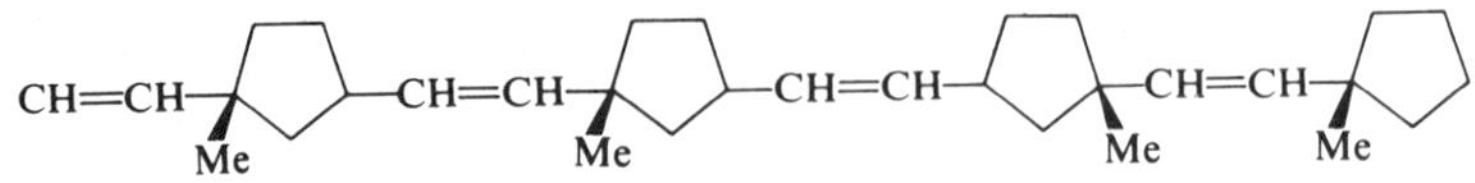


tail-to-head(TH) tail-to-tail(TT) head-to-head(HH)

(**32**)

$$B = \frac{\text{HT} + \text{TH}}{\text{HH} + \text{TT}} \qquad (18)$$

The bias is a measure of the tendency of the polymerization to repeat head-to-tail placements, considered to be normal placements in a polymerization. The number of plausible propagation steps in the polymerization is now eight. There are two propagating species, one in which the 1-methyl group is adjacent to the carbene ($P_H$) and the other in which it is removed from the carbene ($P_T$) *viz.* (**33**) and (**34**). If it is supposed that the metal centre is chiral then each of $P_H$ and $P_T$ may exist in two enantiomeric forms (labelled + and −); it is also postulated that the carbon–metal bond is polarized with the negative end of the dipole located on the metal atom.

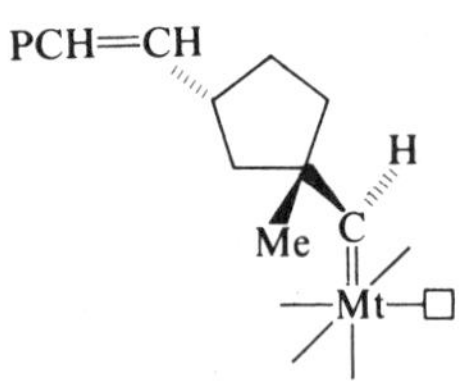


(**33a**) $P_H^+$

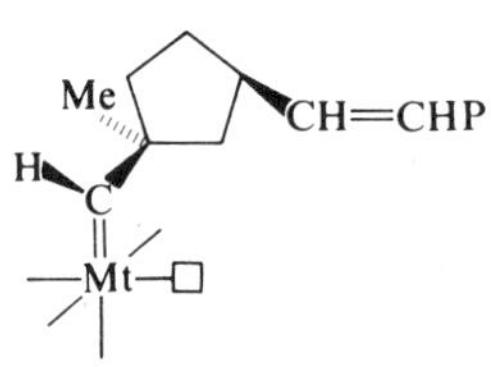


(**33b**) $P_H^-$

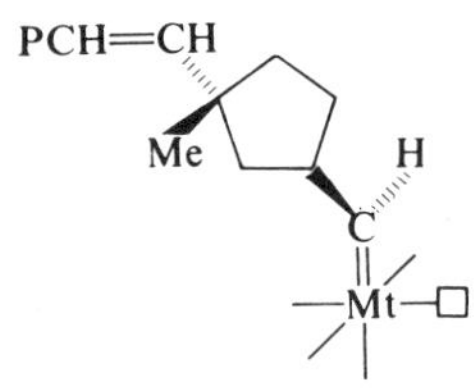


(**34a**) $P_T^+$

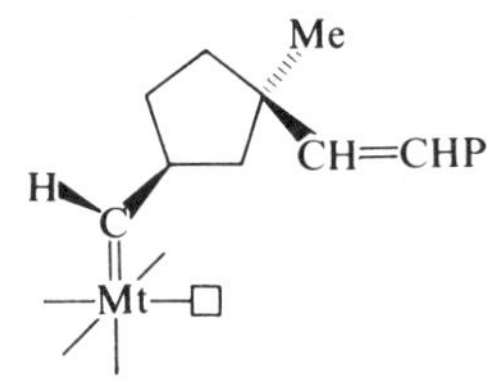


(**34b**) $P_T^-$

Inspection of the structure of the monomer shows that it also exists as a racemic mixture of two enantiomers which may be arbitrarily assigned the (+) and (−) forms (structures **35** and **36**).

The most interesting polymers are those that are generated by high *cis*-forming catalysts. Analysis of the NMR spectra of these polymers indicates that not only are these polymers blocky but that they also show a high syndiotactic structure with a high bias towards head-to-tail arrangements. *Trans* double bonds are normally formed in these polymers in association with head-to-head or tail-to-tail ring-opening. Experimental evidence suggests that the head-to-head reaction is usually preceded by a head-to-tail (*trans*) placement and followed immediately by a tail-to-tail reaction that may be either *cis* or *trans* generating.

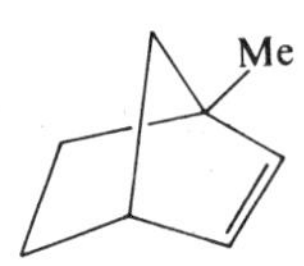


(**35**) (+)MNBE

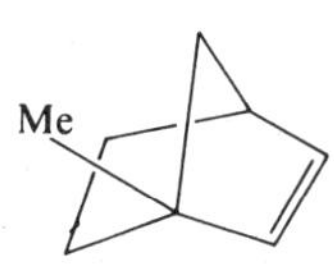


(**36**) (−)MNBE

If one considers the reaction between a pure enantiomer with a racemic mixture of another substrate, it is usually found that the pure enantiomer will react with only one of the enantiomers in the racemate. Such principles can be applied to the generation of all *cis*-HT syndiotactic polymers which may be considered to be obtained by the alternating copolymerization of the (+) and (−) enantiomers at the (+) and (−) enantiomeric catalyst sites respectively. If one presupposes that 'like reacts with like', then the reaction scheme for the formation of *cis* polymer with a preponderance of syndiotactic structures can be summarized as shown in Scheme 7. On the basis of this mechanism the polymer is predicted to be syndiotactic and the propagating species must alternate in structure between the (+) and (−) forms. If this is so, both enantiomers must be present in the reaction mixture if propagation is to proceed, and in fact Rooney[78] has observed that catalysts that polymerize racemic mixtures of MNBE to high *cis* syndiotactic polymer will not polymerize a pure enantiomer. One would expect that some of the catalyst species present will react in one step with their conjugate enantiomer, but having undergone such a reaction propagation comes to a halt because the other enantiomer, essential for further propagation, is absent.

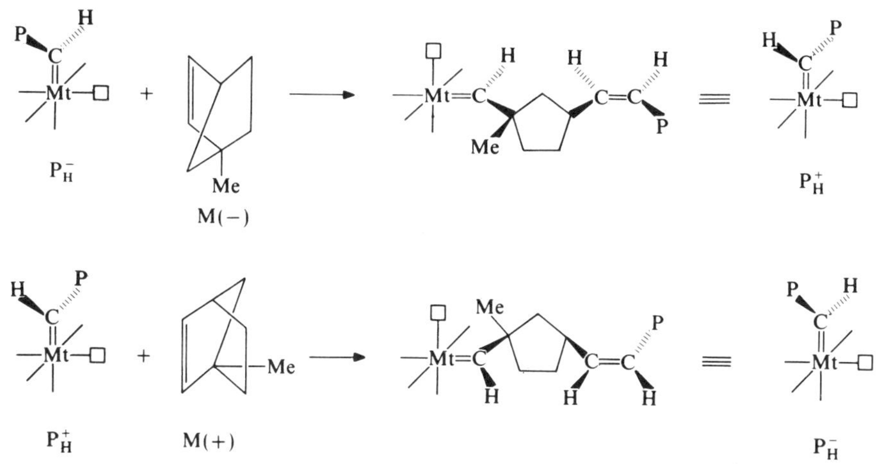

**Scheme 7**

If either form of $P_H$ is to react to form the *trans* double bond by a head-to-tail reaction then it must do so by reacting unfavourably with the non-conjugate enantiomeric form of the monomer. For example, if $P_H^-$ is to form a *trans* double bond it must do so by reacting with M(+) (Scheme 8). Such reactions lead to a retention of configuration of the catalyst site. Other reactions that might conceivably lead to *trans* double bonds are shown in Scheme 9. These are reaction pairs that will lead to head-to-head and tail-to-tail sequences if they arise in the system. The predominance of the reaction sequences that lead to 'like' enantiomers reacting together predicts that *trans* double bonds should be associated with head-to-head and tail-to-tail structures rather than those reactions that would lead to head-to-tail structures with a retention of configuration. This elegant work by Ivin and Rooney has demonstrated the importance of stereoisomerism in the propagation reaction. It is doubtful whether such factors are important in the polymerization of molecules without asymmetric structures.

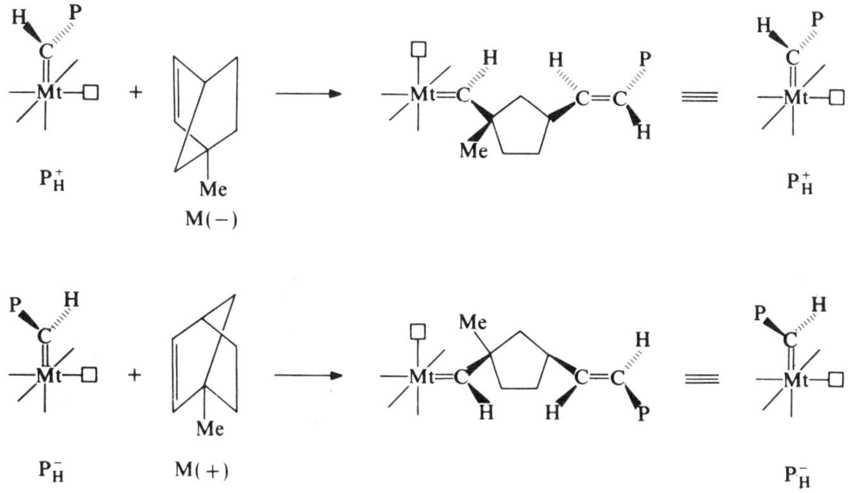

**Scheme 8**

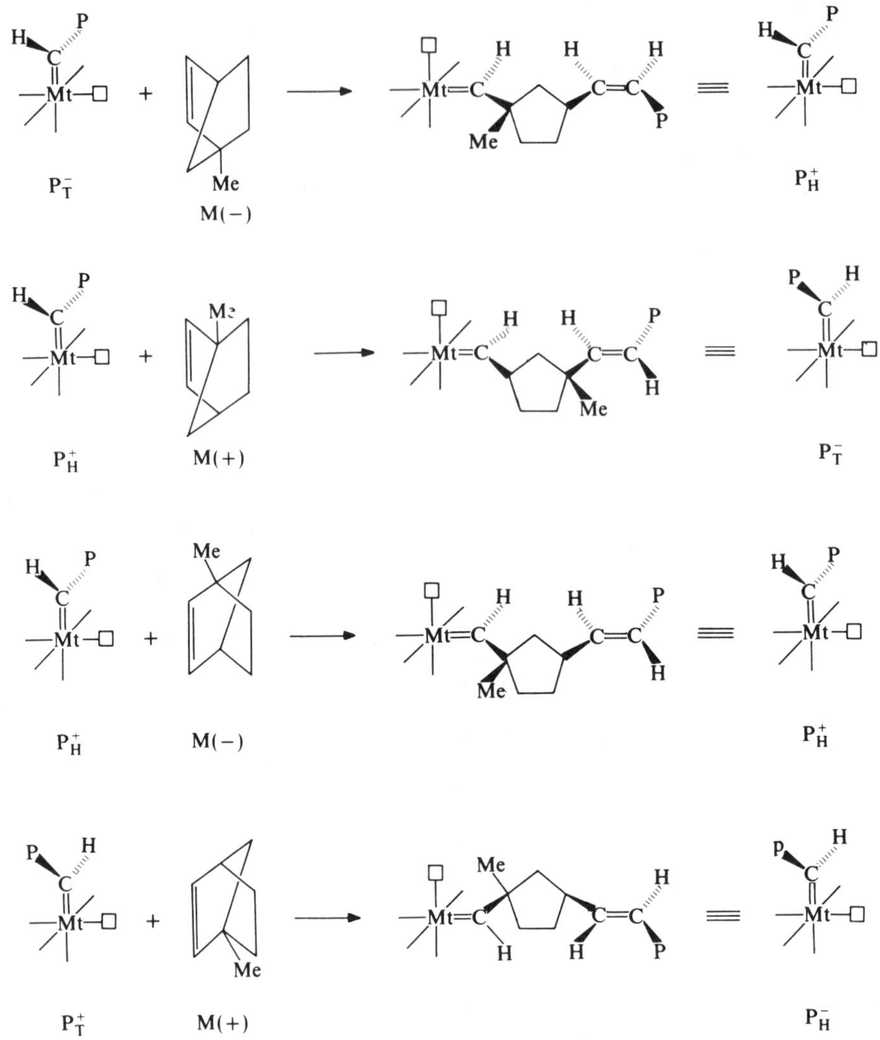

Scheme 9

## 6.7 KINETICS OF POLYMERIZATION

It has to be appreciated that the metathesis ring-opening polymerization of cycloalkenes is unlike the majority of other polymerization systems in that the functional group present in the monomer, the carbon–carbon double bond, is also a feature of the polymer chain, although there may be a change in the stereochemistry around the double bond, *i.e. cis* → *trans*. This double bond maintains an activity towards metathesis, and indeed Hummel[75] has used this fact to great effect in the study of the degradation of unsaturated polymers with linear alkenes (equation 19).

$$P_mCH{=}CHP_n \ + \ RCH{=}CHR \longrightarrow P_mCH{=}CHR \ + \ P_nCH{=}CHR \qquad (19)$$

The kinetics of polymerization of cycloalkenes will thus be complicated by the fact that any active polymerization site in the system will not only be able to coordinate the monomer to lead to a propagation step but can also coordinate a double bond from a polymer chain. If coordination of the in-chain carbon–carbon double bond is an intramolecular reaction then a cyclic molecule is eliminated and a new propagating centre of identical structure is formed (equation 20). Detailed discussions of the distribution of ring sizes in cyclic oligomers are included in Section 6.9. Coordination of a carbon–carbon double bond by an intermolecular mechanism will lead to the metathesis reaction shown in equation (21).

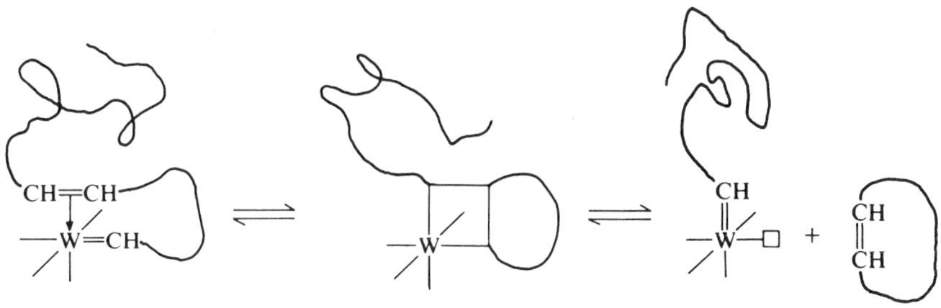

(20)

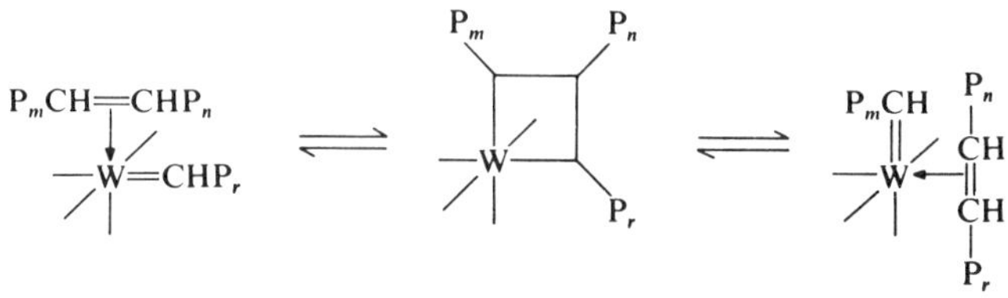

(21)

If metathesis is carried out on a narrow molecular weight distribution polyalkenylene then the molecular weight distribution will broaden to that of a polymer produced by a normal metathesis polymerization, including the production of oligomers, as shown by Höcker.[76] These reactions will therefore be expected to take place in an active polymerizing system, and it must be expected that the kinetics of the polymerization reaction will be affected because a fraction of the active sites that should be available as propagation sites are involved in these non-productive metathesis reactions.

If the propagation reaction is described by equation (22) then the rate of polymerization ($R_p$) will be given by equation (23), where [M] is the concentration of monomer. If $f$ is the fraction of propagation centres involved in non-productive reactions with in-chain carbon–carbon double bonds of polymer molecules then one obtains equation (24) where [W*] is the concentration of all active centres.

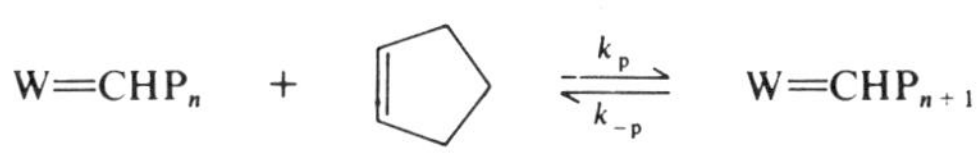

(22)

$$R_p = -d[M]/dt = k_p[M]\,[\text{Active centres}] \quad (23)$$

$$-d[M]/dt = k_p[M]\,[W^*]\,(1 - f) \quad (24)$$

Rooney[77] has considered that propagation occurs in a number of steps, firstly involving reversible coordination of the monomer to the active centre to form a complex that reacts in a first-order manner to regenerate the propagating centre with one more monomer unit in the chain (equation 25).

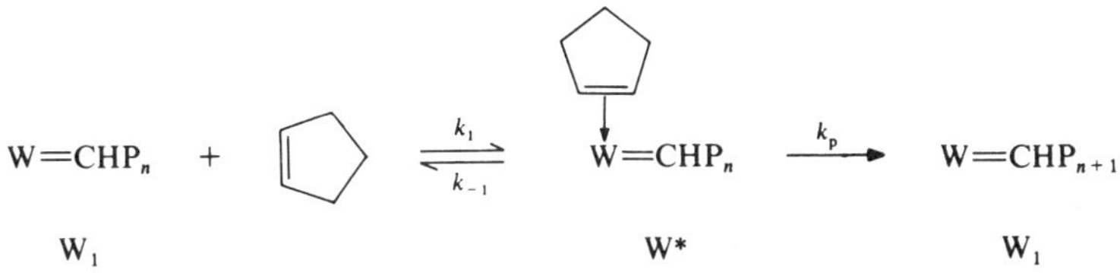

(25)

Michaelis–Menten kinetics can now be applied to this system, in which case

$$d[W^*]/dt = k_1[W_1][M] - k_{-1}[W^*] - k_p[W^*] \quad (26)$$

If $d[W^*]/dt = 0$, then

$$[W^*] = \frac{k_1[M]\,[W_1]}{k_{-1} + k_p} = \frac{[M]\,[W_1]}{K} \quad (27)$$

where $K$ is a Michaelis–Menten constant $K = (k_{-1} + k_p)/k_1$. The overall concentration of active sites $[W]_0$ may be given by

$$[W]_0 = [W_1] + [W^*] \quad (28)$$

$$[W]_0 = [W_1] + [M]\,[W_1]/K \quad (29)$$

where

$$[W_1] = \frac{K[W]_0}{K + [M]}, \quad [W^*] = \frac{[M]\,[W]_0}{K + [M]} \quad (30)$$

Since the rate of propagation is given by

$$R_p = -d[M]/dt = k_p[W^*] \tag{31}$$

substituting for $[W^*]$ gives

$$-d[M]/dt = \frac{k_p[W]_0[M]}{K + [M]} \tag{32}$$

Integration and rearrangement of this equation leads to

$$\Delta[M] = \frac{k_p[W]_0[M]_0\, t}{K + [M]_0} \tag{33a}$$

*i.e.* the conversion of polymerization is decreased by a factor $(K + [M]_0)$.

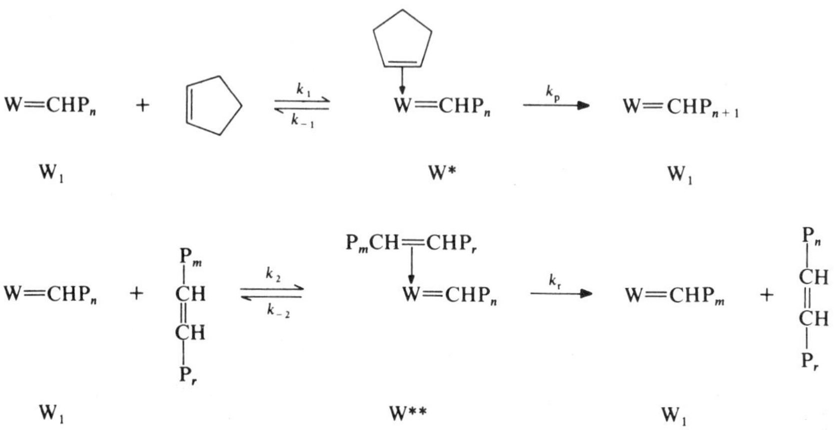


**Scheme 10**

However, this treatment does not take account of the fact that the in-chain carbon–carbon double bond can participate in metathesis also, effectively removing the active centre from the reaction. One should therefore consider the two sequences of reactions shown in Scheme 10. Now $d[W^*]/dt = k_1[W_1][M] - k_{-1}[W^*] - k_p[W^*]$ and $d[W^{**}]/dt = k_2[W_1]\ ([M]_0 - [M]) - k_2[W^{**}] - k_r[W^{**}]$ since the concentration of in-chain double bonds that participate in non-productive metathesis must be identical to the decrease in concentration of monomer. If stationary-state kinetics are applicable to both active species then

$$[W^*] = \frac{k_1[M]\,[W_1]}{k_{-1} + k_p}, \quad [W^{**}] = \frac{k_2[W_1]\,([M]_0 - [M])}{k_{-2} + k_r} \tag{33b}$$

Two Michaelis–Menten constants may now be defined as

$$K_1 = \frac{k_{-1} + k_p}{k_1}, \quad K_2 = \frac{k_{-2} + k_r}{k_2} \tag{34}$$

Then $[W^*] = ([M]\ [W_1])/K_1$ and $[W^{**}] = \{[W_1]\ ([M]_0 - [M])\}/K_2$ and $[W]_0 = [W_1] + [W^*] + [W^{**}]$. Thus

$$[W]_0 = [W_1] + \frac{[M]\,[W_1]}{K_1} + \frac{[W_1]([M]_0 - [M])}{K_2} \tag{35}$$

which can be solved for $[W_1]$, $[W^*]$ and $[W^{**}]$ to give

$$[W_1] = \frac{K_1K_2[W]_0}{K_1K_2 + (K_2 - K_1)[M] + K_1[M]_0} \tag{36}$$

$$[W^*] = \frac{K_2[M]\,[W]_0}{K_1K_2 + (K_2 - K_1)[M] + K_1[M]_0} \tag{37}$$

$$[W^{**}] = \frac{K_1[W]_0([M]_0 - [M])}{K_1K_2 + (K_2 - K_1)[M] + K_1[M]_0} \tag{38}$$

The rate of polymerization is again given by

$$R_p = -d[M]/dt = k_p[W^*] \tag{39}$$

and when suitable substitution is made for [W*]

$$R_p = \frac{k_pK_2[M][W]_0}{K_1K_2 + (K_2 - K_1)[M] + K_1[M]_0} \tag{40}$$

This equation is integrated and rearranged to give

$$\Delta[M] = \frac{k_p[M]_0[W]_0 t}{K_1 + [M]_0} \tag{41}$$

when $-\ln [M]/[M]_0 = \Delta[M]/[M]_0$. However, at higher conversions when this is not true

$$-\ln\left(\frac{[M]}{[M]_0}\right) = \frac{k_pK_2[W]_0t - (K_2 - K_1)\Delta[M]}{K_1(K_2 + [M]_0)} \tag{42}$$

Hence the deviation of the conversion from the expected value increases with conversion of monomer to polymer (Δ[M]).

Amass and Zurimendi[68] studied the kinetics of polymerization of cyclopentene using $WCl_6$/$AlBu^i_3$ catalysts. They observed that the rate of polymerization decreased with conversion of monomer to polymer at a rate faster than could be accounted for by the consumption of monomer alone. They suggested that the decrease in rate of polymerization might be accounted for by the occurrence of termination reactions, but evidence shows that the plot of $-\ln [M]/[M]_0$ is not a straight line at higher conversions of monomer to polymer. Separation of this effect from the reduction in rate polymerization caused by termination reactions is very difficult. If a metathesis ring-opening polymerization is allowed to proceed to zero rate of polymerization, the further addition of an aliquot of catalyst will cause polymerization to recommence but the rate is lower than would have been expected if complexation of the active site with monomer alone takes place.

The kinetics of polymerization of cyclic alkenes with catalysts derived from $WCl_6$ and alkylmetals has other complications that have to be considered. Considerable evidence has been amassed to suggest that reactions take place between not only the catalyst components but between the monomer and particularly the tungsten compound to produce intermediates that react with the cocatalyst to produce the active centres. Amass and Tuck[78] showed that the initial rate of polymerization of cyclopentene depends on the time that elapses between the separate additions of the two catalyst components. Recent kinetic studies of the reaction between $WCl_6$ and cyclopentene have suggested that the active centre is produced in the reduction of $WCl_6$ by cyclopentene. Product analysis indicated the formation of 3-chlorocyclopentene and HCl as principal products of the reaction of $WCl_6$, as discussed in Section 6.4. A careful detailed study of the kinetics of polymerization is still required if the steric and electronic effects of structure are to be assessed in a quantitative manner.

## 6.8 THERMODYNAMICS OF POLYMERIZATION

The thermodynamics of the ring-opening polymerization of cyclic alkenes is a typical example of the thermodynamics of ring-opening polymerization in general. In the reaction of linear alkenes, the metathesis reaction is characterized by the cleavage and formation of an equal number of essentially identical chemical bonds and the enthalphy changes associated with such reactions are zero or at most very small. Calderon[1] pointed out that the position of equilibrium in metathesis reactions is determined by a progress towards a state of maximum entropy in the system, generally represented by a statistical distribution of the alkylidene groups between the reactants and the products. Thus the metathesis of 2-pentene leads to a mixture at equilibrium that comprises 2-pentene, 2-butene and 3-hexene in the molar ratio of 2:1:1 (equation 43). The occurrence of *cis–trans* isomerism in these reactions may upset this position of ideal equilibrium but such disturbances are expected to be very small.

$$\underset{\text{2 moles}}{\mathrm{EtCH{=}CHMe}} \rightleftharpoons \underset{\text{1 mole}}{\mathrm{MeCH{=}CHMe}} + \underset{\text{1 mole}}{\mathrm{EtCH{=}CHEt}} \tag{43}$$

While the metathesis of linear alkenes leads to products with structures essentially similar to those of the reactants, the metathesis of cyclic alkenes is accompanied by cleavage of a ring to form a linear molecule, which is accompanied by the release of the associated ring-strain energy. The enthalpy of polymerization is not expected therefore to be zero except in the case of a strain-free monomer such as cyclohexene, for which the equilibrium lies far to the side of the monomer. The ring-strain energy is the principal factor to be considered in the polymerizability of cyclic alkenes and it is found that the cycloalkenes conform to the expectations for ring–chain equilibria discussed in detail by Dainton and Ivin[79] in their celebrated review of the thermodynamics of polymerization.

The equilibrium that applies in the case of the polymerization of cycloalkenes may be written as equation (44), if it is assumed that a metal carbene is the propagating centre. If the principles of kinetic and thermodynamic equilibria are applied to this system, then $k_p[M]_{eq} = k_{-p}$ or $K_p = k_p/k_{-p} = 1/[M]_{eq}$ and since $\Delta G = \Delta H - T\Delta S = -RT \ln K_p$ then $\Delta H - T\Delta S = RT \ln [M]_{eq}$.

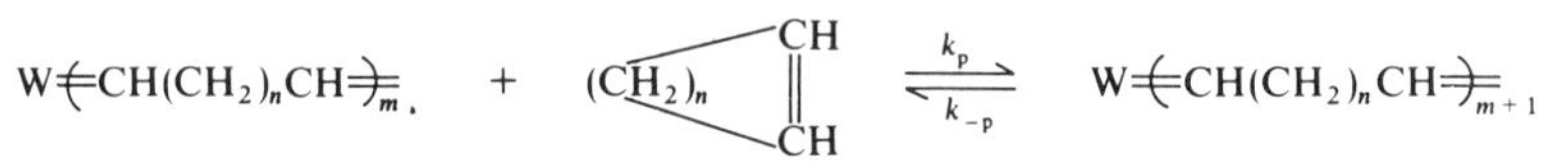
(44)

Dainton and Ivin calculated the theoretical enthalpies and entropies of polymerization of a series of cycloalkanes varying in ring size from three to eight carbon atoms. Only cyclohexane showed the combination of positive enthalpy and negative entropy of polymerization, indicative of a monomer that is always much more stable thermodynamically than the corresponding polymer. From the data supplied by Dainton and Ivin it is possible to estimate ceiling temperatures and the equilibrium concentrations of the pure monomers at any given temperature. These properties and their dependences on ring size are shown in Table 2.

**Table 2** Thermodynamics of the Polymerization of Cycloalkanes[a]

| *Ring size, n* | $\Delta H^\circ$ (kcal mol$^{-1}$) | $\Delta S^\circ$ (cal mol$^{-1}$ K$^{-1}$) | $\Delta G^\circ$ (kcal mol$^{-1}$) |
|---|---|---|---|
| 3 | −27 | −16.5 | −22.1 |
| 4 | −25.1 | −13.2 | −21.2 |
| 5 | −5.2 | −10.2 | −2.2 |
| 6 | 0.7 | −2.5 | 1.4 |
| 7 | −5.1 | −0.7 | −4.9 |
| 8 | −8.3 | 8.9 | −11.0 |

[a] Temperature = 298 K; 1 cal = 4.184 J.

Comparison of these data with those available for the polymerization of cycloalkenes shows that the presence of the carbon–carbon double bond has the most significant effect upon the polymerizability of cyclopentene. The slightly increased ring-strain brought about by the inclusion of a double bond increases the ceiling temperature of the polymerization to 390 K. The enthalpy of polymerization was found to be $-20.5$ kJ mol$^{-1}$ and the entropy of polymerization $-60.25$ J K$^{-1}$ mol$^{-1}$, which are significantly higher than the corresponding figures for cyclopentane polymerization.

The existence of a true equilibrium in metathesis systems has been demonstrated by a number of workers. Hoecker[16] and Calderon[80] have independently demonstrated the existence of kinetically attainable equilibria by reacting polyalkenylenes with active metathesis catalysts. Narrow molecular weight distribution polymers of cyclooctene and 1,5-cyclooctadiene both react with metathesis catalysts to re-establish the molecular weight distribution associated with the polymerization of these monomers. The equilibrium molecular weight distribution of the polymer shows bimodal character, whether obtained from monomer or polymer, and it was Calderon[80] who showed that the lower molecular weight polymers were cyclic structures. In the case of polymerization of 1,5-cyclooctadiene, not only did the equilibrium distribution of products include dimer, trimer and tetramer *etc.*, but also $C_{12}$, $C_{20}$ and $C_{28}$ ring structures. It has to be concluded, therefore, that an equilibrium exists not only between the monomer and the polymer but also between the polymer and the cyclic oligomers.

## 6.9 MOLECULAR WEIGHT DISTRIBUTIONS IN METATHESIS POLYMERIZATION

A considerable amount of information concerning the course of a polymerization reaction may be obtained from a study of the molecular weight and molecular weight distribution changes that take place in the polymerization of a monomer. It is well known that the number- and weight-average molecular weights of the products of polymerization may increase or remain relatively constant during the course of a polymerization, depending upon the kinetics and mechanism of reaction. Some of the earliest reports by Pampus[81] and Amass[82] on the molecular weight changes in the polymerization of cyclopentene catalyzed by $WCl_6/AlBu^i_3$, showed that the reaction is accompanied by an unexpected decrease of number-average molecular weight with increase in conversion. Pampus showed that, as the number-average molecular weight of the polymer appears to decrease, there is a commensurate decrease in the *cis* content of the polymer from 15% to 9% of the double bonds. Pampus concluded that metathesis catalysts are not only responsible for the ring-opening polymerization of cycloalkenes but also for the isomerization of double bonds and reactions that lead to the degradation of polymer chains. Without reference to any specific data, it was concluded that the molecular weight distribution of the polymer does not change significantly during conversion.

Gel permeation chromatography data[82] show that the number-average molecular weight changes that occur during polymerization can be ascribed to the changes that take place in the molecular weight distribution as a function of reaction time. The products of the ring-opening polymerization of cyclopentene have bimodal molecular weight distributions, and samples taken at various times from a polymerizing system show that the weight ratios of the two components of the distribution change with reaction extent. The generation of a bimodal distribution at the outset of the polymerization was accounted for by the presence of more than one type of active centre. These result from the reaction of $WCl_6$ with the cycloalkene in reactions that lead to formation of at least two products, and when the cocatalyst is added to the reaction these are converted to active species with different propagation and termination rate constants. Moreover, these species are converted by reactions in the system to inactive compounds. Eventually the molecular weight distribution approaches an equilibrium distribution that is dependent principally on the concentration of the monomer. The initial molecular weight distribution is therefore kinetically controlled, whereas, in the later stages of the reaction, control of the final molecular weight distribution is thermodynamic.

Although Calderon[10] had first disclosed the existence of macrocylic rings in the products of the polymerization of cyclooctene and 1,5-cyclooctadiene, Hoecker[16] investigated the thermodynamically controlled molecular weight distribution in polyoctenylene. Because the polymerization of a cycloalkene results in the formation of a polymer with an identical functional group to the reactant, it is possible for that functional group to react with the catalyst after polymerization has taken place. This reaction might be either an intermolecular or an intramolecular reaction.

If the secondary metathesis reaction is of an intermolecular type, then except for the fact that there is a randomizing process taking place which leads to a statistical distribution of the molecular weight, the number-average degree of polymerization remains unchanged by reactions of the type shown in equation (45). It is this type of secondary metathesis reaction that was proposed by Pampus[11] to account for the degradation of the polyalkenylene and also the *cis–trans* isomerism reaction.

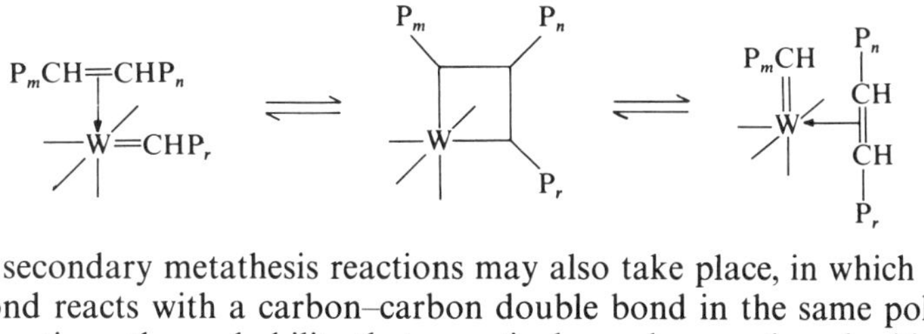

(45)

Intramolecular secondary metathesis reactions may also take place, in which case the tungsten–carbon double bond reacts with a carbon–carbon double bond in the same polymer chain as the carbene. In such reactions, the probability that a particular carbon–carbon double bond participates in the reaction depends on the distance of that double bond from the reaction site. The products of this reaction are: (i) a shortened polymeric metal carbene and (ii) a cyclic molecule.

When the polymerization of cyclooctene is carried out at low concentrations of monomer ($<0.25$ mol $l^{-1}$) then, in addition to the high molecular weight product, a homologous series of oligomers is observed, irrespective of the nature of the catalyst used.[16] Mass spectroscopic evidence, particularly of the hydrogenated oligomers, suggests that these compounds are cyclic and are thus formed by a back-biting process (equation 46).

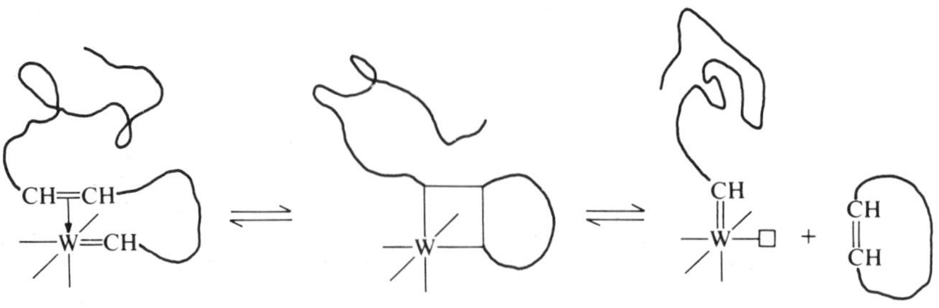


(46)

Above the equilibrium concentration of monomer for polymer formation, the relative concentrations of the oligomers are independent of the initial concentrations of the monomer and the catalyst; only a slight dependence on temperature is observed, and above the equilibrium concentration of monomer the [polymer]/[oligomers] ratio increases with increasing concentration of monomer.

The equilibrium between open-chain polymer and each cyclic oligomer ($M_x$) may be summarized as

$$A{=}M_y{=}B \rightleftharpoons A{=}M_{y-x}{=}B + M_x \tag{47}$$

The equilibrium constant ($K_x$) for the formation of cyclic $x$-mer ($M_x$) may be written as

$$K_x = \frac{[M_x][{=}M_{y-x}{=}]}{[M_y]} \tag{48}$$

From simple statistical theory, if $p$ represents the probability that a double bond has undergone metathesis polymerization, then

$$\frac{[{=}M_{y-x}{=}]}{[M_y]} = \frac{p^{y-x}}{p^y} = \frac{1}{p^x} \tag{49}$$

and

$$K_x = \frac{[M_x]}{p^x} \tag{50}$$

which for high conversions of monomer to polymer may be approximated by

$$K_x = [M_x] \tag{51}$$

Because the equilibrium described above is an example of ring–chain equilibria, the equilibrium concentration of oligomer may be described by the Jacobson–Stockmayer equation, which predicts that the equilibrium concentration of the ring oligomer containing $x$ monomer units is inversely proportional to mean distance ($\langle r^2 \rangle$) between the two units in the chain that are required to form such a ring. This leads to the celebrated equation

$$K_x = \left[\frac{3}{2\pi}\right]^{3/2} \frac{1}{xL} \left[\frac{1}{\langle r^2 \rangle}\right]^{3/2} \tag{52}$$

The value for $\langle r^2 \rangle$ may be replaced by a characteristic ratio for that ring oligomer ($c_x$) which is given by

$$c_x = \langle r^2 \rangle / vxl^2 \tag{53}$$

where $v$ is the number of bonds in the monomer unit and $l$ is the average bond length. Then

$$[M_x] = \left[\frac{3}{2\pi}\right]^{3/2} \frac{1}{L} \left[\frac{1}{vl^2 c_x}\right]^{3/2} x^{-5/2} \tag{54}$$

Plots of ln $[M_x]$ against ln $x$ for a number of cycloalkenes show that there is agreement with the Jacobson–Stockmayer equation[84] with slopes of $-5/2$. The intercepts of such plots also show the predicted dependence on ring size.

## 6.10 REFERENCES


1. N. Calderon, N. Y. Chen and K. W. Scott, *Tetrahedron Lett.*, 1967, 3327.
2. G. Natta, G. Dall'Asta, G. Mazzanti and G. Motroni, *Makromol. Chem.*, 1963, **56**, 163.
3. G. Natta, G. Dall'Asta and G. Mazzanti, *Angew. Chem., Int. Ed. Engl.*, 1964, **3**, 723.
4. N. Calderon, E. A. Ofstead, J. P. Ward, W. A. Judy and K. W. Scott, *J. Am. Chem. Soc.*, 1968, **90**, 4133.
5. G. Dall'Asta and G. Motroni, *Eur. Polym. J.*, 1971, **7**, 707.
6. F. Haas, K. Nuetzel, G. Pampus and D. Theisen, *Rubber Chem. Technol.*, 1970, **43**, 1126.
7. C. P. Casey and T. J. Burkhardt, *J. Am. Chem. Soc.*, 1974, **96**, 7808.
8. K. Nuetzel, K. Dinges and F. Haas, *Ger. Pat.* 1 720 797 (1968) (*Chem. Abstr.*, 1970, **72**, 79 656M).
9. N. Calderon and M. C. Morris, *J. Polym. Sci., Part A-2*, 1967, **5**, 1283.
10. N. Calderon, E. A. Ofstead and W. A. Judy, *J. Polym. Sci., Part A-1*, 1967, **5**, 2209.
11. P. Guenther, F. Haas, G. Marwede, K. Nuetzel, W. Oberkirch, G. Pampus, N. Schoen and J. Witte, *Angew. Makromol. Chem.*, 1970, **14**, 87.
12. P. A. Patton and T. J. McCarthy, *Macromolecules*, 1987, **20**, 778.
13. L. R. Gillom and R. H. Grubbs, *J. Am. Chem. Soc.*, 1986, **108**, 733.
14. J. Kress, J. A. Osborne, R. M. E. Green, K. J. Ivin and J. J. Rooney, *J. Chem. Soc., Chem. Commun.*, 1985, 874.
15. K. W. Scott, N. Calderon, E. A. Ofstead, W. A. Judy and J. P. Ward, *Adv. Chem. Ser.*, 1969, **91**, 399.
16. H. Hoecker, W. Reiman, L. Reif and K. Riebel, *Recl. Trav. Chim. Pays-Bas*, 1977, **96**, M47.
17. W. J. Feast and B. Wilson, *J. Mol. Catal.*, 1980, **8**, 277.
18. K. J. Ivin, D. T. Laverty, J. H. O'Donnell, J. J. Rooney and C. D. Stewart, *Makromol. Chem.*, 1979, **180**, 1989.
19. J. H. Edwards and W. J. Feast, *Polymer*, 1980, **21**, 595.
20. B. M. Novak and R. H. Grubbs, *J. Am. Chem. Soc.*, 1988, **110**, 960.
21. K. J. Ivin, D. T. Laverty and J. J. Rooney, *Makromol. Chem.*, 1977, **178**, 1545.
22. K. J. Ivin, J. H. O'Donnell, J. J. Rooney and C. D. Stewart, *Makromol. Chem.*, 1979, **180**, 1975.
23. T. Masuda, K. Hasegawa and T Higashimura, *Macromolecules*, 1974, **7**, 728.
24. T. J. Katz, S. J. Lee and M. A. Shippey, *J. Mol. Catal.*, 1980, **8**, 219.
25. T. J. Katz and S. J. Lee, *J. Am. Chem. Soc.*, 1980, **102**, 7940.
26. A. J. Amass, M. S. Beevers, T. R. Farren and J. A. Stowell, *Makromol. Chem., Rapid Commun.*, 1987, **8**, 119.
27. T. R. Farren, A. J. Amass, M. S. Beevers and J. A. Stowell, *Makromol. Chem.*, 1987, **188**, 2535.
28. F. Pennella, R. L. Banks and G. C. Bailey, *J. Chem. Soc., Chem. Commun.*, 1968, 1548.
29. A. Mortreux and M. Blanchard, *Bull. Soc. Chim. Fr.*, 1972, 1641.
30. J. Sancho and R. R. Schrock, *J. Mol. Catal.*, 1982, **15**, 75.
31. T. J. Katz and S. J. Lee, *J. Am. Chem. Soc.*, 1980, **102**, 422.
32. K. Hasegawa, T. Masuda and T. Higashimura, *Macromolecules*, 1975, **8**, 255.
33. T. Masuda, N. Sasaki and T. Higashimura, *Macromolecules*, 1975, **8**, 717.
34. R. R. Schrock, R. T. DePue, J. Feldman, C. J. Schaverien, J. C. Dewan and A. H. Liu, *J. Am. Chem. Soc.*, 1988, **110**, 1423.
35. C. Boelhouwer and C. J. Mol, *Prog. Lipid Res.*, 1985, **24**, 243.
36. F. W. Kupper and R. Streck, *Makromol. Chem.*, 1974, **175**, 2055.
37. G. Natta, G. Dall'Asta, I. W. Bassi and G. Carella, *Makromol. Chem.*, 1966, **91**, 87.
38. K. J. Ivin, D. T. Laverty, J. H. O'Donnell, J. J. Rooney and C. D. Stewart, *Makromol. Chem.*, 1979, **180**, 1989.
39. A. J. Amass and J. A. Zurimendi, *J. Mol. Catal.*, 1980, **8**, 243.
40. B. A. Dolgoplosk, I. A. Oreshkin, K. L Makovetsky, E. I. Tinyakova, I. Ya. Ostrovskaya, I. L. Kershenbaum and G. M. Chernenko, *J. Organomet. Chem.*, 1977, **128**, 339.
41. A. J. Amass, M. Lotfipour, C. Thompson, B. J. Tighe and J. A. Zurimendi, *Makromol. Chem.*, 1987, **188**, 2121.
42. T. J. Katz, J. McGinnis and C. Altus, *J. Am. Chem. Soc.*, 1976, **98**, 606.
43. Y. Chauvin, D. Commereuc and G. Zaborowski, *Recl. Trav. Chim. Pays-Bas*, 1977, **96**, M131.
44. K. J. Ivin, B. S. R. Reddy and J. J. Rooney, *J. Chem. Soc., Chem. Commun.*, 1981, 1062.
45. M. Szwarc, *Science (Washington, D.C.)*, 1970, **170**, 23.
46. R. R. Schrock, R. H. Grubbs, J. Feldman and L. F. Cannizzo, *Macromolecules*, 1987, **20**, 1169.
47. R. Thomas and J. A. Moulijn, *J. Mol. Catal.*, 1982, **15**, 157.
48. W. B. Hughes, *J. Chem. Soc., Chem. Commun.*, 1969, 431.
49. R. H. Grubbs and S. Swetnick, *J. Mol. Catal.*, 1980, **8**, 25.
50. M. Leconte, Y. Ben. Taarit, J. L. Bilhou and J. M. Basset, *J. Mol. Catal.*, 1980, **8**, 263.
51. M. Leconte and J. M. Basset, *J. Am. Chem. Soc.*, 1979, **101**, 7296.
52. M. Leconte and J. M. Basset, *Ann. N. Y. Acad. Sci.*, 1980, **333**, 165.
53. H. S. Eleuterio, *US Pat.* 3 074 918 (1957) (*Chem. Abstr.*, 1961, **55**, 16 005)
54. D. T. Laverty, M. A. McKervey, J. J. Rooney and A. Stewart, *J. Chem. Soc., Chem. Commun.*, 1976, 193.
55. K. J. Ivin, D. T. Laverty and J. J. Rooney, *Makromol. Chem.*, 1977, **178**, 1545.
56. K. J. Ivin, D. T. Laverty and J. J. Rooney, *Makromol. Chem.*, 1978, **179**, 253.
57. F. W. Michelotti and W. P. Keaveney, *J. Polym. Sci., Part A*, 1965, **3**, 895.
58. R. H. Grubbs and C. R. Hoppin, *J. Chem. Soc., Chem. Commun.*, 1977, 634.
59. E. L. Muetterties, *Inorg. Chem.*, 1975, **14**, 951.
60. L. Bencze, L. Marko, R. Optiz and K. H. Thiele, *Hung. J. Ind. Chem.*, 1976, **4**, 15.
61. A. J. Amass and D. Gregory, *Br. Polym. J.*, 1987, **19**, 263.
62. J. L. Wang and H. R. Menapace, *J. Org. Chem.*, 1968, **33**, 3794.
63. J. L. Herisson and Y. Chauvin, *Makromol. Chem.*, 1970, **141**, 161.
64. T. J. Katz and J. McGinnis, *J. Am. Chem. Soc.*, 1977, **99**, 1903.
65. C. P. Casey and A. J. Shusterman, *J. Mol. Catal.*, 1980, **8**, 1.
66. C. P. Casey, H. E. Tuinstra and M. C. Saeman, *J. Am. Chem. Soc.*, 1976, **98**, 608.
67. R. R. Schrock, S. Rocklage, J. Wengrovius, G. Rupprecht and J. Fellman, *J. Mol. Catal.*, 1980, **8**, 73.
68. A. J. Amass and J. A. Zurimendi, *Eur. Polym. J.*, 1981, **17**, 1.
69. T. R. Howard, J. B. Lee and R. H. Grubbs, *J. Am. Chem. Soc.*, 1980, **102**, 6876.

70. F. N. Tebbe, G. W. Parshall and G. Reddy, *J. Am. Chem. Soc.*, 1978, **100**, 3611.
71. K. C. Ott, J. B. Lee and R. H. Grubbs, *J. Am. Chem. Soc.*, 1982, **104**, 2942.
72. J. G. Hamilton, K. J. Ivin, G. M. McCann and J. J. Rooney, *Makromol. Chem.*, 1985, **186**, 1477.
73. R. M. E. Green, J. G. Hamilton, K. J. Ivin and J. J. Rooney, *Makromol. Chem.*, 1986, **187**, 619.
74. K. J. Ivin, G. Lapiensis, J. J. Rooney and C. D. Stewart, *J. Mol. Catal.*, 1980, **8**, 203.
75. K. Hummel, P. Heiling, C. Karaoulis, W. Kathan and F. Stelzer, *Makromol. Chem.*, 1980, **181**, 3041.
76. H. Hoecker and R. Musch, *Makromol. Chem.*, 1974, **175**, 1395.
77. H. T. Ho, B. S. R. Reddy and J. J. Rooney, *J. Chem. Soc., Faraday Trans. 1*, 1982, **78**, 3307.
78. A. J. Amass and C. N. Tuck, *Eur. Polym. J.*, 1978, **14**, 817.
79. F. S. Dainton and K. J. Ivin, *Q. Rev., Chem. Soc.*, 1958, **12**, 61.
80. E. A. Ofstead and N. Calderon, *Makromol. Chem.*, 1972, **154**, 21.
81. G. Pampus, J. Witte and M. Hoffmann, *Rev. Gen. Caoutch. Plast.*, 1970, **47**, 1343.
82. A. J. Amass and T. A. McGourtey, in 'Proceedings of Catalyse par Coordination (Villeurbanne)', 1974, p. 181.
83. A. J. Amass and J. A. Zurimendi, *Polymer*, 1982, **23**, 211.
84. H. Jacobson and W. H. Stockmayer, *J. Chem. Phys.*, 1950, **18**, 1600.
85. A. J. Amass, A. L. Dale and B. J. Tighe, *Makromol. Chem.*, 1988, **189**, 515.

# 7

# Metathesis Polymerization: Applications

W. JAMES FEAST
*University of Durham, UK*



## 7.1 INTRODUCTION

The first observation of the transition metal-catalyzed reaction, now known as alkene metathesis, was reported from Du Pont's laboratories.[1] In that work it was established that bicyclo[2.2.1]heptene was polymerized by a catalyst derived from the reaction of titanium tetrachloride with ethylmagnesium bromide, the structure of the product was not recognized immediately but is now known to have been poly(1,3-cyclopentylene vinylene) (Scheme 1a). Subsequently the same company filed a patent describing the ring-opening polymerization of cyclopentene (Scheme 1b).[2] Shortly after these initial observations on the polymerization of cyclic alkenes, novel reactions of acyclic alkenes were also recognized and described variously as alkene exchange, disproportionation or dismutation. Thus, for example, it was shown that propene could be catalytically converted to a mixture of ethene and 2-butene (Scheme 1c);[3] the reaction was reversible and the product distribution usually statistical.

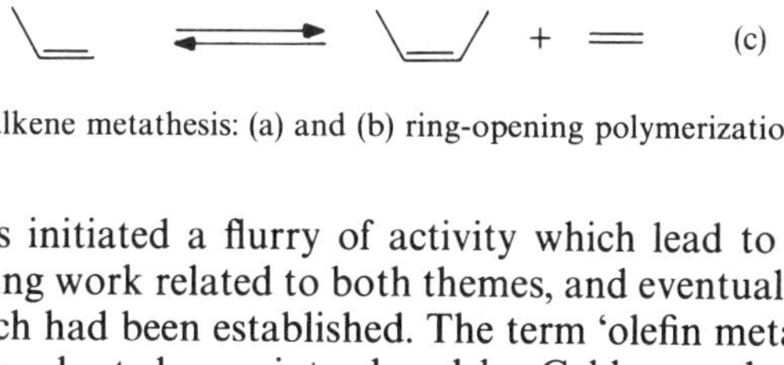


**Scheme 1** Early examples of alkene metathesis: (a) and (b) ring-opening polymerization, (c) alkene disproportionation

These early observations initiated a flurry of activity which lead to the publication of further patents and papers describing work related to both themes, and eventually to commerical processes utilizing the chemistry which had been established. The term 'olefin metathesis' (alkene metathesis) which has been universally adopted, was introduced by Calderon who recognized that both the alkene exchange and ring-opening polymerization processes were manifestations of the same chemical reaction.[4] This remarkable and somewhat unexpected reaction has attracted a considerable research effort; the chemical and mechanistic aspects of the reaction have been reviewed in the previous chapter and here we shall consider some of its applications in polymer science.

Those examples of alkene metathesis which have been subjected to detailed mechanistic investigation are known to proceed *via* a chain reaction in which the chain-carrying species is a metallocarbene, and the propagation reaction involves alternating formation and scission of a metallocyclobutane. The reaction can, in principle at least, be used in two distinct ways for the synthesis of polymers.

(*a*) *Chain growth polymerization.* When monocyclic or polycyclic alkenes, such as those exemplified in Scheme 1(a and b), are used as monomers the polymer grows by a chain growth process and the microstructure of the polymer backbone provides a record of the propagation steps in the chain reaction by which it was produced.

(*b*) *Step growth polymerization.* It is possible to envisage a step growth polymerization process in which, for example, a divinyl monomer undergoes polymerization *via* the stepwise elimination of ethylene initiated by a metathesis catalyst (see Scheme 2). In such processes, while the metathesis reaction has the characteristics of chain reaction, the growth of the polymer backbone occurs in a stepwise fashion.

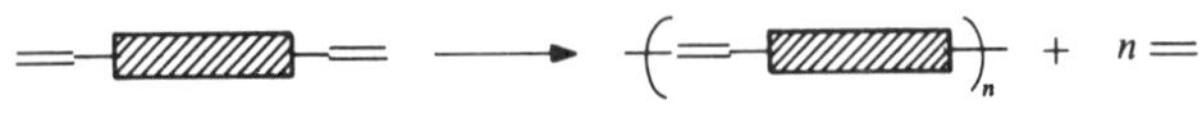

**Scheme 2** Step-growth polymerization *via* an alkene metathesis process

Up until now this step growth approach to polymer synthesis using alkene metathesis does not appear to have attracted as much attention as chain growth ring-opening polymerization. Thus, although it has been shown that by selection of appropriate catalysts and reaction conditions the simplest non-conjugated diene, 1,4-pentadiene, can be converted in 70% yield to octa-1,4,7-triene ($n=1$, in the Table 1),[5] or to the telomer distribution shown in Table 1,[6] there are very few reports of attempts to apply this reaction to polymer synthesis. It is possible, and indeed seems quite likely, that careful manipulation of this kind of reaction might yield genuine high polymers.

**Table 1** Distribution of Telomers in the Metathesis of 1,4-Pentadiene

| $n$ | 1 | 2 | 3 | 4 | 5 | 6 |
|---|---|---|---|---|---|---|
| Weight (%) | 15.7 | 7.9 | 4.1 | 1.8 | 1.6 | 0.6 |

i, $WCl_6$: $(Et)_3Al_2Cl_3$: $MeCCl(OH)CH_2Cl$: $P(Ph)_3$, 1:2:1:1, at 0°C for 6 hours.

At this point it is appropriate to mention some applications of the alkene metathesis reaction which, while they are not concerned with polymerization, are important in the synthesis of monomers, intermediates and polymer ancillaries. Several companies have been active in this area and a few examples of the many possible processes and products available are shown in outline in Scheme 3. The Phillips' 'Triolefin Process' is exemplified in Scheme 3(a) by the conversion of propene to ethylene and 2-butene, which can be dehydrogenated to the more valuable 1,3-butadiene. The first step in this scheme can equally well be run in the reverse direction depending on market demand. Shell have established their 'SHOP' and 'FEAST' processes for the manufacture of alkenes, polyenes and derived chemicals; examples of the products available are shown in Scheme 3(b), 3(c) and 3(d). The initial step involves a cross metathesis between ethylene and cyclooctadiene and, since the ethylene can be replaced by isobutene in this process, a range of branched products is also available. As well as the difunctional compounds shown in Scheme 3(d), it is possible to produce a range of 6-substituted 1-hexenes. Monsanto have reported a route to styrene from toluene and ethylene (Scheme 3e). This list is illustrative and whether a route is used or not will depend on the exigencies of the market place. Other examples have been collated and discussed in some detail.[7-9]

Another application of the alkene metathesis reaction in polymer science involves the degradation of unsaturated polymers *via* cross metathesis with acyclic alkenes;[10] the idea is illustrated in Scheme 4. Separation and identification of the fragments provides information about the structure of the original network. The process is essentially a post synthesis modification and is dealt with elsewhere in these volumes.

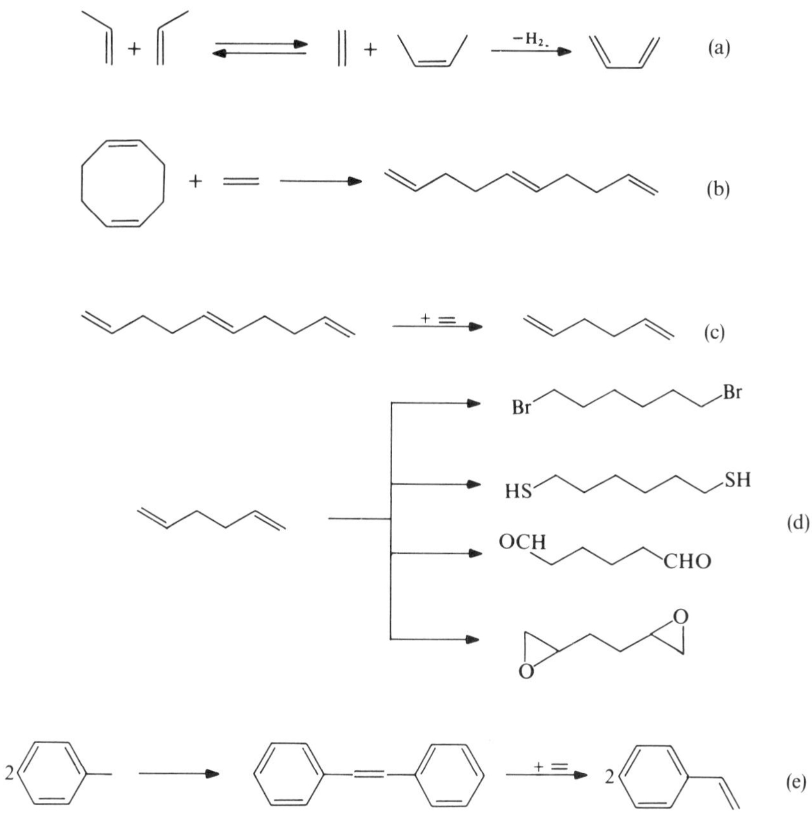


**Scheme 3** Examples of applications of alkene metathesis in monomer synthesis

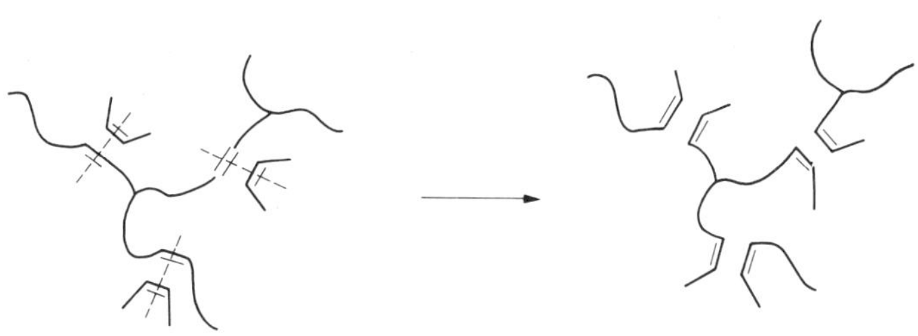

**Scheme 4** Diagrammatic representation of degradation of a part of an unsaturated network *via* alkene metathesis with an acyclic alkene

Undoubtedly, the simple ring-opening polymerization of cycloalkenes has generated the most interest and activity as far as applications of metathesis polymerization are concerned. A very wide range of potential monomers have come under scrutiny in both industrial and academic research laboratories and many interesting materials have been obtained, some of which are manufactured commercially at the time of writing. We shall consider the polymers from monocyclic unsubstituted alkenes in the next section, with subsequent sections on polymers derived from polycyclic and substituted monomers. The features of polymerizations of this kind which make them interesting and potentially useful can be summarized as follows: (i) the monomer's alkenic unsaturation is retained in the polymer; (ii) there are novel opportunities for isomeric incorporation of monomer units and hence novel stereoregularities are, in principle, available; (iii) a variety of functional substituent groups can be tolerated; (iv) copolymerization is possible; and (v) the availability of true living metathesis allows good control of molecular size and block length in block coplymers.

The synthesis of polymers with alkenic unsaturation along the backbone has focused attention in industry on the possibility of manufacturing elastomers by this route, indeed speciality elastomers are the major commercial success of metathesis polymerization so far.[7,8,11]

## 7.2 POLYMERS FROM HYDROCARBON MONOCYCLIC ALKENES AND POLYENES

Thermodynamic considerations lead to the expectation that ring-opening polymerization reactions of three, four- and eight- or larger membered rings will be associated with a large negative free energy change and consequently favoured, providing a mechanism is available; whereas the opening of five- and seven-membered rings is associated with significantly smaller negative free energies and in this situation small changes in structure or reaction conditions may completely inhibit polymerization; six-membered rings are predicted to be unpolymerizable by ring-opening.[12] These expectations are fulfilled in practice in the case of ring-opening metathesis although there is no evidence concerning the polymerizability of cyclopropenes by this method.

Cyclobutene, cyclooctene and larger cycloalkenes are readily polymerized to high polymers by a range of conventional metathesis catalysts. The polymers from cyclobutene and 1-methylcyclobutene are, of course, poly-1,4-butadiene and polyisoprene respectively and the polymerization of 1-methylcyclooct-1,5-diene gives an alternating copolymer of 1,4-butadiene and isoprene units. All these reactions have been established and thoroughly investigated but none of them are operated commercially since equivalent products may be obtained more cheaply by established technologies.

The exact nature of the initiator in metathesis ring-opening polymerizations is unknown for the vast majority of commonly used catalyst formulations. Many catalyst systems have been found to be sensitive to a wide range of common organic structural units which appear to act as inhibitors or poisons for these common catalysts. Consequently a kind of 'folk lore' concerning what constitutes a polymerizable monomer in metathesis developed and, for example, it was widely held that groups containing heteroatoms and conjugated dienes were polymerization inhibitors in this field. A reappraisal of these views is made necessary by the results of recent work, particularly from the research groups of Osborne, Grubbs and Schrock, on structurally, well-defined catalysts of the kind shown by (**1**), (**2**) and (**3**).[13–15]

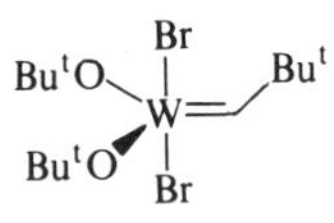


(**1**)

(**2**)

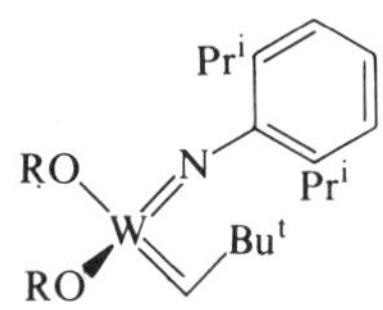


(**3**)

It is now clear that, with the appropriate initiating system, the range of polymerizable monomers is very wide. Thus, for example, Grubbs has reported that the novel conjugated diene, 3,4-diisopropylidenecyclobutene, may be polymerized using a bis(cyclopentadienyl)titanacyclobutane derivative (**2**) as initiator, (see Scheme 5).[16]

**Scheme 5** Ring-opening polymerization of 3,4-diisopropylidenecyclobutene[16]

The linear cross conjugated polymer which is the product of this reaction is colourless and soluble but rather susceptible to oxidative degradation, none of which is particularly surprising. Films can be cast from solution which, in their pristine state, are insulators but which display an electrical conductivity of up to 200 $(\Omega\ cm)^{-1}$ after exposure to iodine. At the time of publication of this result it was well known that polymers containing long sequences of conjugated double bonds could be oxidatively modified from insulators to conductors, but the observation that this cross-conjugated system behaved in a similar fashion was a surprise in many quarters.

More recently this group have reported that the tungsten carbene complex (**3**) will polymerize cyclooctatetraene to give polyacetylene which is similar to that obtained *via* the interfacial polymerization of acetylene developed by Shirakawa.[17,18]

Although cyclobutene derived polymers have not yet merited commercialization, cyclopentene provided what appeared to be a rather more attractive possibility for a successful synthetic elastomer *via* metathesis. Cyclopentene is readily available *via* hydrogenation of cyclopentadiene which is a major component of the C-5 refinery stream. The ring-opening polymerization of cyclopentene has been investigated in great detail. The product, generally referred to as polypentenamer, is a good general purpose rubber which can be vulcanized to give products comparable with those obtained from polyisoprene and polybutadiene. As would be expected, the detailed physical properties of the raw gum depend on the frequency and distribution of *cis* and *trans* double bonds, an indication of these effects is shown in Table 2. *Cis*-polypentenamer has the lowest glass transition temperature known for an hydrocarbon polymer.

**Table 2**

| —CH=CH(Me)$_3$— | $T_g$ (°C) | $T_m$ (°C) |
|---|---|---|
| High *trans* | −90 | 20 |
| Natural rubber | −70 | 30 |
| High *cis* | −114 | −41 |

Several companies, including Goodyear, Bayer and Japan Synthetic Rubber, have taken the development of material produced by this method to the pilot plant manufacture and tyre fabrication and testing stage, however the materials produced have not been marketed since the technology available makes them marginally more expensive than traditional well-established materials.[9]

Although methyl substitution at either the vinylic or the allylic position of cyclobutene does not inhibit its polymerization, the more marginal polymerizability of cyclopentene results in greater sensitivity to substituent effects. Thus, the 1-methyl isomer cannot be polymerized whereas a methyl substituent can be tolerated at the 3-position but an isopropyl group can not. Polymerization of cyclopentene is readily reversed and the reaction constitutes a good example of equilibrium polymerization. This has been the subject of much discussion (see ref. 7), and Schrock recently reported that polymerization of this monomer under living conditions can be regulated to allow the production of narrow molecular weight distribution polypentenamer at −60 °C which can be completely reversed to give a mixture of initiator and monomer at + 60 °C.[19]

As expected cyclohexene can not be readily polymerized by metathesis catalysts.

There have been few studies of the polymerization of the cycloheptene, presumably because of its cost. It seems unlikely to have sufficiently novel properties to find application even in speciality markets.

Cyclooctene, by contrast with the two previously mentioned monomers, is both readily available (*via* butadiene dimerization and partial hydrogenation) and easily polymerized. Since 1980 the product of its polymerization has been successfully commercialized by Huls AG as a speciality elastomer under the name Vestenamer.[9] This material has a relatively low molecular weight (*ca.* 60 000), it has a $T_g$ of − 65 °C but is a hard elastomer at room temperature due to the presence of crystallinity (*ca.* 33%), and it melts at 55 °C to a low viscosity fluid, a fact which aids the processing of blends of which it forms a part. Its main outlet appears to be in blends with more common elastomers, these products find specialist applications.[20] At the time of writing Vestenamer is probably the most successful metathesis polymer in terms of production tonnage with Huls A.G. operating a plant of 12 000 tonnes per annum capacity.[9]

## 7.3 POLYMERS FROM POLYCYCLIC ALKENES AND POLYENES

Polycyclic monomers are generally more easily polymerized than their monocyclic alkene analogues, an observation which is usually ascribed to increased ring strain,[7] although this rationalization has been disputed.[21] In practice it is observed that virtually any and all metathesis catalysts will effect the polymerization of bicyclo[2.2.1]heptene, whereas cyclopentene is somewhat more fastidious in the selection of catalysts to which it responds. Therefore it is not surprising that bicyclo[2.2.1]heptene was the first monomer shown to undergo metathesis ring-opening polymerization,[1,2] the monomer is also relatively cheaply obtained *via* the Diels–Alder reaction of ethylene with cyclopentadiene, yet despite this the polymer was not commercialized until 1976, that is, about 20 years after its discovery. A polymer having *ca.* 90% *trans* double bonds is sold by CdF Chimie under the name Norsorex. This material has a high molecular weight (greater than $10^6$ g mol$^{-1}$), a glass transition temperature of 35 °C, a melting point in excess of 170 °C, and an unusually large capacity to absorb oil and plasticizers (up to five time its own mass). It is highly compatible with fillers and, since it has two tertiary allylic C—H bonds per repeat unit, it is susceptible to oxidative degradation and requires antioxidants. The highly extended and vulcanized compounds obtained from Norsorex show good tear strength and a high dynamic damping which makes them useful in noise and vibration applications.

Many other bicyclo[2.2.1]heptene derivatives have been investigated as monomers in metathesis polymerizations and many patents have been issued; this is undoubtedly due to the relatively low cost and ease of synthesis of such monomers together with their ease of polymerization. In a recent review[9] Streck reported that in a five year period two Japanese companies issued 127 patents in this field. Poly(5-cyanonorbornene) appears to have been the most energetically pursued product of this activity (Showa Denko K.K.) and was reported to show promise as a thermoplastic which has not, as yet, been fulfilled in the market place.[22] Speciality materials containing silyl and tributylstannyl substituents have been reported to impart improved adhesive bonding and biocidal activity respectively.[23]

The polymer obtained by metathesis ring opening from bicyclo[2.2.1]heptene (Scheme 1a) has, in the allylic carbons, two chiral centres per repeat unit. If the centres on either side of the vinylene have the same chirality, a racemic diad is generated, whereas centres with opposite chirality give a *meso* diad. If the monomer is unsymmetrically substituted a further complication is added in that head–head, tail–tail and head–tail incorporation of repeat units is possible. Thus, ring opening of a simple substituted norbornene can, in principle, give rise to a very wide range of microstructures defined by the frequency and distribution of: *cis* and *trans* vinylenes; *meso* and racemic diads; and HH, TT or HT placements.

In the event the outcome of a particular polymerization depends on a combination of factors including the structure of the monomer, the nature of the catalyst, and the reaction conditions (solvent, temperature and concentrations). The elucidation of the structures of these polymers has depended heavily on analysis of high resolution $^{13}$C NMR spectra, with Ivin being the major contributor to this field. In favourable cases total specification of all aspects of microstructure can be achieved and several totally stereoregular polymers have been characterized.[7,24] A particularly striking example of this area of synthesis is provided by the polymerization of racemic 1-methylnorbornene, initiated by rhenium pentachloride, which gives a polymer with an all head–tail *cis*-syndiotactic microstructure; this result requires that the monomer enantiomers are incorporated into the polymer in a strictly alternating sequence (Scheme 6).

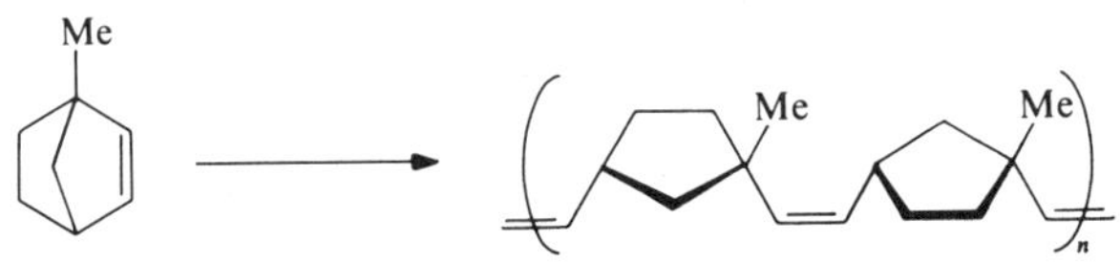


**Scheme 6** Polymerization of 1-methylnorbornene giving a totally stereoregular alternating copolymer of enantiomers[24]

The location and nature of a substituent on the norbornene skeleton has a marked effect on the polymerizability of the monomer. It is not surprising, since approach of catalyst to monomer occurs from the *exo* face, that substituents at C-7 which are *syn* to the double bond prevent polymerization whereas *anti*-C-7 substituents are tolerated. Somewhat more surprising is the observation that *endo* substituents at C-5 and C-6 often inhibit polymerization whereas *exo* substituents do not, this has been reported for both anhydride and chlorine.[25,26] However, it is necessary to avoid making dogmatic classifications of polymerizable and non-polymerizable monomers since the continuing development of new initiator systems seems likely to invalidate them.

Examples of the substituents which are tolerated by the easily prepared conventional metathesis catalysts include fluorine, giving rise to the possibility of making stereoregular fluoropolymers, aryl, esters (including carbonates), and fulvene derivatives.[27–30]

Tricyclo[5.2.1.0(2,6)]deca-3,8-diene, cyclopentadiene dimer (DCP), is a cheap potential monomer which has attracted a lot of attention. In 1982 the linear polymer obtained by ring opening exclusively at the vinylene in the bicyclo unit was reported to be about to be launched as a thermoplastic by B.F. Goodrich under the name Telene,[31] however this did not occur. In 1985 Hercules established a production unit using DCP in a metathesis reaction injection moulding (RIM) process (Scheme 7).

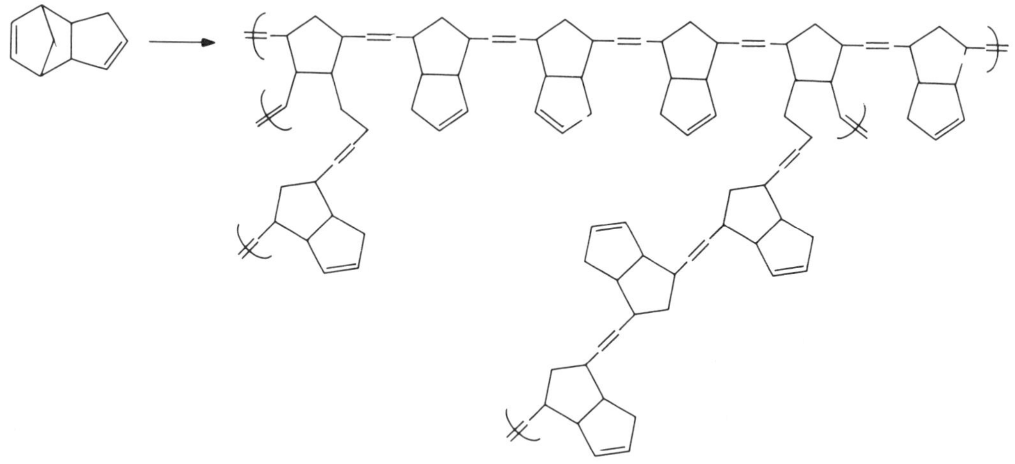

**Scheme 7** Network formation by metathesis RIM with dicyclopentadiene[32]

The catalyst which initiates the opening of the vinylenes in both the bicyclo[2.2.1]- and the 3,4-disubstituted cyclopentene units of DCP is formed when a transition metal compound and an activator are combined as two monomer streams and mixed immediately prior to injection into the mould. The solutions of the individual catalyst components in monomer are stable, and the delay between mixing and initiation of network formation can be controlled which allows the kinetics of the process to be adjusted to the technical requirements of the particular job in hand. The reaction is exothermic, cycle times for the production of large items can be low, and the polydicyclopentadiene product has an unusual and advantageous balance of mechanical properties.[32] Inspection of the structure drawn in Scheme 7 shows a very high concentration of tertiary allylic C–H bonds and, as would be expected, these mouldings develop an oxidatively crosslinked surface fairly quickly after their formation. It turns out that the initially formed oxidation layer serves two useful purposes: firstly, it acts as a barrier to ingress of oxygen into the bulk of the polymer; and secondly it provides a surface to which paint adheres very well. It is not difficult to imagine other compounds containing two or more vinylene units capable of participating in metathesis polymerization and it will be interesting to see if other metathesis RIM processes are developed in the future.

Earlier in this section the retention of monomer unsaturation in the product polymer was emphasised as one of the potentially useful features of metathesis polymerization. Conjugated polymers have been an interesting target for polymer synthesis since the first theoretical studies of their electrical properties were published in the 1930s. Experimentally the physics of conjugated polymers such as polyacetylene proved rather difficult to study in a satisfactory way because the material is air sensitive, insoluble and infusible, and consequently not easily purified or processed. A standard solution to the problem of access to an intractable but desirable polymer is to proceed *via* a processable precursor. In this particular case an unsaturated precursor provides a start *en route* to the desired objective and hence a metathesis polymerization becomes an attractive option. A successful solution to the polyacetylene synthesis problem is outlined in Scheme 8.

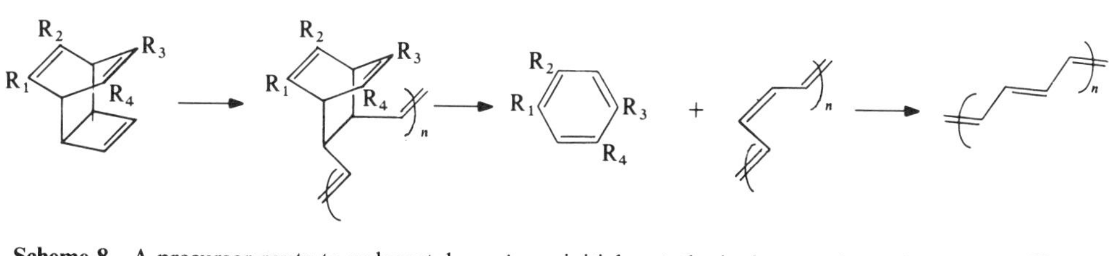

**Scheme 8** A precursor route to polyacetylene *via* an initial metathesis ring-opening polymerization[33]

The tricyclictriene monomer undergoes ring opening exclusively at the cyclobutene double bond to give the soluble precursor polymer which can be purified and characterized (GPC, spectroscopy, *etc.*) prior to its conversion, *via* a symmetry-allowed elimination, to polyacetylene. The process has been the subject of detailed study and provides access to material of high purity and in a variety of morphologies.[33]

## 7.4 THE FUTURE

Attempting to predict the future of a topic is notoriously difficult but during the preparation of this manuscript some exciting developments have been occuring in this field. These result chiefly from the attention of skilled transition metal chemists to the problem of defining the structure and mechanism of action of metathesis catalysts. Three examples of well-defined catalysts were cited earlier. The papers of Grubbs and Schrock[17,19,34–36] demonstrate that true living polymerization is possible providing access to narrow molecular weight distribution and block copolymers; switching from metathesis to other propagation mechanisms has been demonstrated; and it is clear that a wider range of monomers are polymerizable than was previously thought to be the case. One confident prediction can be made, namely that many interesting papers, patents and (possibly) processes concerning metathesis polymerization will appear in the immediate period after the publication of this work. There is a penalty to be paid for the intending participant in this activity in as much as the synthesis and use of these newer catalysts places experimental skill at a premium.

## 7.5 REFERENCES


1. A. W. Anderson and N. G. Merkling, *Chem. Abstr.*, 1955, **50**, 3008.
2. H. S. Eleuterio, *US Pat.*, 1957 3 074 918 (*Chem. Abstr.*, 1961, **55**, 16 005).
3. E. F. Peters and B. L. Evering, *US Pat.*, 1960 2 963 447 (*Chem. Abstr.*, 1961, **55**, 5810).
4. N. Calderon *et al.*, *Chem. Eng. News.*, 1967, **45**, 51.
5. G. Doyle, *J. Catal.*, 1973, **30**, 118.
6. G. Dall'Asta, *Pure Appl. Chem.*, 1973/4, **1**, 133.
7. K. J. Ivin, 'Olefin Metathesis', Academic Press, New York, 1983.
8. V. Dragutan, A. T. Balaban and M. Dimonie 'Olefin Metathesis and Ring-Opening Polymerization of Cyclo-Olefins', Editura Academiei/Wiley, New York, 1985.
9. R. Streck, 'Olefin Metathesis in Industry—A Status Report', paper presented at the 7th International Symposium on Olefin Metathesis, Hull, August 1987 and to be published in *J. Mol. Catal.*
10. K. Hummel, *Pure Appl. Chem.*, 1982, **54 (2)**, 351.
11. G. Dall'Asta, *Rubber Chem. Tech.*, 1974, **47 (3)**, 511.
12. K. J. Ivin, in 'Reactivity, Mechanism and Structure in Polymer Chemistry', eds. A. D. Jenkins and A. Ledwith, Wiley, London 1974, chap. 16.
13. J. Kress, M. Wesolak and J. A. Osborn, *J. Chem. Soc., Chem. Commun.*, 1982, 514.
14. T. R. Howard, J. B. Lee and R. H. Grubbs, *J. Am. Chem. Soc.*, 1980, **102**, 6877; and D. A. Straus and R. H. Grubbs, *J. Mol. Catal.*, 1985, **28**, 9.
15. C. Schaverien, J. Dewan and R. R. Schrock, *J. Am. Chem. Soc.*, 1986, **108**, 2771.
16. T. M. Swager and R. H. Grubbs, *J. Am. Chem. Soc.*, 1986, **108**, 2771.
17. R. H. Grubbs, paper presented at the 7th International Symposium on Olefin Metathesis, Hull, August 1987; and F. L. Klavetter and R. H. Grubbs, *Polym. Prepr., Am. Chem. Soc., Div. Polym. Chem.*, 1987, **28** (2), 425.
18. T. Ito, H. Shirakawa and S. Ikeda, *J. Polym. Sci., Polym. Chem. Ed.*, 1974, **12**, 11.
19. R. R. Schrock, paper presented at the 7th International Symposium on Olefin Metathesis, Hull, August 1987.
20. A. Draxler, *Kautsch. Gummi, Kunstst.*, 1983, **12**, 1037.
21. P. A. Patton and T. J. McCarthy, *Macromolecules*, 1987, **20**, 778.
22. T. Ueshima and S. Kobayashi, *Jpn. Plast.*, 1974, **8**, 11.
23. R. Streck, *J. Mol. Catal.*, 1982, **15**, 3.
24. J. G. Hamilton, K. J. Ivin, J. J. Rooney and L. C. Waring, *J. Chem. Soc., Chem. Commun.*, 159 1983; and J. G. Hamilton, K. J. Ivin and J. J. Rooney, *Br. Polym. J.*, 1984, **16**, 21.
25. K. F. Castner and N. Calderon, *J. Mol. Catal.*, 1982, **15**, 47.
26. A. B. Alimuniar, J. H. Edwards and W. J. Feast, *J. Mol. Catal.*, 1985, **28**, 313.
27. P. W. Blackmore and W. J. Feast, *Polymer*, 1986, **27**, 1296; *J. Mol. Catal.*, 1986, **36**, 145; and P. M. Blackmore, W. J. Feast and P. C. Taylor, *Br. Polym. J.*, 1987, **19**, 205.
28. I. F. A. F. El-Saafin and W. J. Feast, *J. Mol. Catal.*, 1982, **15**, 61.
29. W. J. Feast and K. Harper, *J. Mol. Catal.*, 1985, **28**, 293.
30. W. J. Feast and I. S. Millichamp, *J. Mol. Catal.*, 1985, **28**, 331.
31. Plastics Focus, July 1982, **14**, 3 no. 22.
32. R. P. Geer and R. D. Stoutland, Plastics 85, proceedings of the SPE 43rd Annual Technical Conference, ANTEC '85, 1985, 1232.
33. J. H. Edwards, W. J. Feast and D. C. Bott, *Polymer*, 1984 **25**, 359; D. C. Bott *et al.*, *Synth. Met.*, 1986, **14**, 245; P. D. Townsend, *Springer Ser. Solid-State Sci.*, 1985, **63**, 50 and references cited therein.
34. L. R. Gillom and R. H. Grubbs, *J. Am. Chem. Soc.*, 1986, **108**, 733.
35. R. R. Schrock, J. Feldman, R. H. Grubbs and L. Cannizzo, *Macromolecules*, 1987, **20**, 1169.
36. J. S. Murdzek and R. R. Schrock, *Macromolecules*, 1987, **20**, 2640.

# 8

# Polymerization of Acetylene

ALBERTO BOLOGNESI, MARINELLA CATELLANI and SILVIA DESTRI

*Istituto di Chimica delle Macromolecole del CNR, Milano, Italy*



## 8.1 INTRODUCTION

The discovery of the metallic conductivity ($\sigma$) of polyethyne[1] (hereafter polyacetylene) in 1977 disclosed a new field of research on organic polymers, with the aim of combining the properties of two divergent classes of materials: plastics and metals. The characteristics of this material and its technological applications have stimulated the interest of solid-state physicists, electrochemists and chemists, but up to now no industrial development on a medium scale has been reported. Polyacetylene, the simplest of the conducting polymers, is a highly crystalline polymer formed by a conjugated $\pi$-system with *trans* and *cis* configurations,[2] but, depending on the polymerization conditions and subsequent handling, sequences of *cis*-transoid, *trans*-cisoid and *trans*-transoid structures may be present (Figure 1).

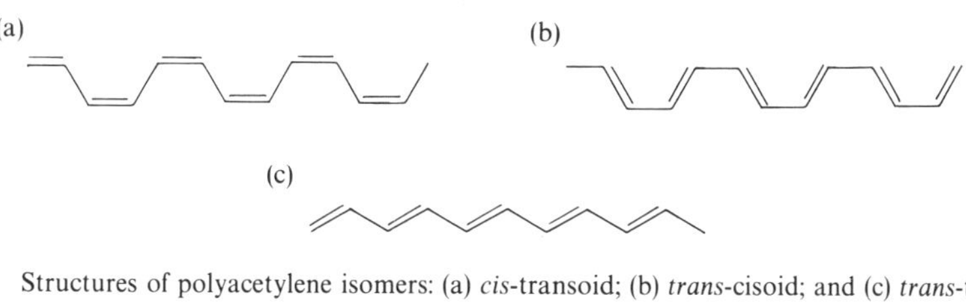

**Figure 1** Structures of polyacetylene isomers: (a) *cis*-transoid; (b) *trans*-cisoid; and (c) *trans*-transoid

*cis*-Polyacetylene can be considered as an insulating material ($\sigma = 4 \times 10^{-9}\ \Omega^{-1}\ cm^{-1}$), while *trans*-polyacetylene ($\sigma = 9 \times 10^{-5}\ \Omega^{-1}\ cm^{-1}$) is a semiconductor.[3] There is an increase in conductivity of 13 orders of magnitude (to $\approx 1.5 \times 10^{3}\ \Omega^{-1}\ cm^{-1}$)[4] when *cis*-polyacetylene is treated with strong oxidizing or reducing agents (doping). The whole process leading to the conductive state may be represented by equation (1),[5] where $X_n$ is a segment of the conjugated chain formed by $n$ carbon atoms which exchanges a full charge with a redox compound or with an electrode of an electrochemical cell. The ion radical formed on the polyacetylene backbone is often called a soliton if the charge carrier is free to move along the chain without passing through an energy barrier, and a polaron if it moves, picking up energy. The process involving a second redox state leads to the formation of another charge carrier called a bipolaron.[6] The nature of the redox reagents (dopants) can be varied by choosing from a variety of substances[7,8] classified as n or p dopants depending on their donor or acceptor character.

$$X_n \underset{\text{reduction}}{\overset{\text{oxidation/A}}{\rightleftarrows}} [X_n^{+\cdot}A^-] \underset{\text{reduction}}{\overset{\text{oxidation/A}}{\rightleftarrows}} [X_n^{2+}2A^-] \tag{1}$$

As indicated by the number of papers that have appeared in the literature in the last decade, the greatest research effort has been devoted to understanding the electrical, electronic and optical behaviour of this polymer. At least three phenomena contribute to the overall electric behaviour: the intrachain transport (single chain), the interchain transport (hopping) and the interparticle transport (percolation).[9] These three conducting pathways form a complicated resistive network, determining the overall mobility of the carriers.

The main factor affecting the overall electrical transport is still a controversial issue. The conductivity of undoped *cis*- and *trans*-polyacetylene and of the $I_2$-doped polymers has been shown[10] to be independent of molecular weight in the range from 400 to 870 000 ($\bar{M}_n$); this suggests that interchain hopping of carriers is the step controlling the phenomenon. The influence of the conjugation length on conductivity has been shown by other workers,[9] who monitored (by Raman spectroscopy) the evolution of long *trans*-segments during *cis–trans* thermal isomerization in undoped polyacetylene. A maximum in conductivity is observed when the highest concentration of long *trans*-segments, together with a rearrangement of the three-dimensional packing (shown by X-ray diffraction data), is reached. Moreover, the prolonged thermal treatment introduces cross-links, decreasing both the conjugation length and the conductivity. More recently,[11] in the BASF laboratories, defect-free polyacetylenes with conductivities higher than $10^5\,\Omega^{-1}\,\text{cm}^{-1}$ have been synthesized (for copper and gold, $\sigma = 7.2 \times 10^4\,\Omega^{-1}\,\text{cm}^{-1}$ and $2.4 \times 10^4\,\Omega^{-1}\,\text{cm}^{-1}$, respectively). In these doped samples the charge transport is supposed to be principally along the conjugated chains.[12]

A complete understanding of the transport mechanism requires a deep knowledge of polymer architecture, because of the close relationship between electrical and structural properties. Free-standing polyacetylene films with different thicknesses consist of polyenic rigid chains held together in fibrils[13] by strong interchain forces, responsible for the insolubility and infusibility of the material. Most of the investigation methods used for studying soluble polymers cannot be adopted for polyacetylene. For example, molecular weight determination cannot be performed by light scattering or viscometry, and unusual methods such as radioquenching,[14] or solubilization by partial hydrogenation,[15] or addition of chlorine[16] have been proposed. Other parameters, such as the distribution of molecular weights, conjugation lengths, number of $sp^3$ defects, *etc*., are not well established and are strongly influenced by reaction conditions (temperature, solvent, catalyst, *etc*.). Nevertheless, the solid polymer has been exhaustively characterized by means of many techniques, such as cross-polarization magic-angle spinning (CP-MAS), transmission electron microscopy (TEM), scanning electron microscopy (SEM) and electron spin resonance (ESR). Relevant literature references can be found in the following sections.

Due to the limited number of X-ray diffraction data specific for unoriented specimens that are available, a variable picture of the crystal packing of polyacetylene isomers has been reported up to now. However, from the large number of structural studies, it can be concluded that the *cis* isomer consistently presents an orthorhombic arrangement,[17] while, for the *trans* isomer, polymorphism has been inferred, depending on polymerization and handling conditions.[18]

Both isomers of polyacetylene absorb in the visible region. The *cis*-isomer absorption has three maxima at 2.1, 2.3 and 2.4 eV. The HOMO–LUMO transition (intrinsic optical gap) is at 1.8 eV. *trans*-Polyacetylene has an absorption maximum at 1.9 eV, its intrinsic optical gap is about 1.4 eV.[19]

Stability is a basic requirement for applications in the fields of electronics, optical switches, solar cells, sensors, etc.[20] The easy oxidation of polyacetylene in air, the rapid decay of its conductivity in an inert atmosphere (4 orders of magnitude in a few days[21]) and its instability under conditions of use (charging and discharging cycles of a battery) are, at present, the biggest limitations to the large-scale application of this fascinating material.

## 8.2 POLYMERIZATION CATALYZED BY ORGANOMETALLIC SYSTEMS

### 8.2.1 Polymerization Catalyzed by $Ti(OBu)_4/AlEt_3$

Natta *et al*., in 1958, were the first to prepare linear high polymers of acetylene using a soluble organometallic catalytic system obtained by reaction of titanium tetrabutoxide with triethyl aluminum [$Ti(OBu)_4/AlEt_3$].[22] Using alkene polymerization techniques, they observed the formation of a black powder with metallic lustre scales and a black shining mirror of polyacetylene on the reactor glass wall. This polymer had a high degree of crystallinity and a *trans* configuration of the polyenic chains.

Later, some researchers noted the formation of uniform polyacetylene film onto concentrated catalyst solutions. Electrical conductivity measurements carried out on thin polyacetylene films showed the semiconducting properties of this material.[23]

In 1974, Shirakawa and co-workers studied the optimum conditions for the simultaneous polymerization and formation of free-standing polyacetylene films on the quiescent catalyst solution.[24] They found that $Ti(OBu)_4/AlEt_3$ yields high-quality films at an aluminum/titanium ratio $\approx 4$ and that macroscopic polyacetylene morphology depends on the catalyst concentration. The minimum concentration for forming films is $3 \times 10^{-3}$ mol l$^{-1}$ of titanium; below this value, the product is a gelatinous mass, while at a lower catalyst concentration, formation of a black powder occurs. Polyacetylene films have densities in the range 0.3–0.6 g cm$^{-3}$.[25]

SEM on polyacetylene and TEM on the polymer grown directly on to an electron microscopy grid show a fibrillar morphology. The thick films consist of an accumulation of randomly disposed fibrils having diameters ranging from 200 to 300 Å. The ultrathin films are formed by very small diameter microfibrils ($\approx$ 50 Å).[13] Electron diffraction shows the polymer-chain axis to be along the fibrillar axis.[26]

When the catalyst concentration is high, the active centres are close to each other and polymerization and crystallization of the polyenic chains are probably nearly simultaneous, producing the characteristic fibrillar structure. The isomeric structure of polyacetylene is strongly dependent on the reaction temperature and, to a lesser extent, on the aluminum/titanium ratio.

Low temperature reaction yields the *cis* isomer, whilst the *trans*-isomer content is higher at elevated temperatures (see Table 1).[24]

**Table 1** Effect of Polymerization Temperature on the *Cis–Trans* Composition of Polyacetylene[a]

| *Temperature* (°C) | *Cis* (%) | *Trans* (%) |
|---|---|---|
| −78 | 98.1 | 1.9 |
| −18 | 95.4 | 4.6 |
| 0 | 78.6 | 21.4 |
| 18 | 59.3 | 40.7 |
| 50 | 32.4 | 67.6 |
| 100 | 7.5 | 92.5 |

[a] T. Ito, H. Shirakawa and S. Ikeda, *J. Polym. Sci., Polym. Chem. Ed.*, 1974, **12**, 11.

The polymerization catalyzed by Ziegler–Natta systems has been studied for 30 years, but the mechanism of this addition polymerization is not fully known. The formation of polyacetylene might occur *via* metathesis or *via* monomer insertion.[27] $^{13}$C NMR studies may indicate that acetylene polymerization occurs by subsequent monomer insertions.[28] It is probable that, with the $Ti(OBu)_4/AlEt_3$ system, the last unit of the growing chain of polyacetylene, as in α-alkene polymerization, is bonded to the transition metal and polymerization proceeds through *cis* insertion of a new coordinated monomer into the titanium–carbon bond (Figure 2).[27] The pristine polymer has a *cis*-transoid configuration.

Different structures have been proposed by several workers for the active polymerization centre.[27,29,30] According to ESR studies, it seems reasonable to suppose that the catalyst species consists of a cluster of titanium ($Ti^{3+}$) and aluminum atoms bridged by butoxy and ethyl groups. Different ESR spectra have been obtained, depending on both aluminum/titanium ratio and temperature, hence the structure of the catalytic complex changes as a function of these parameters.

Radioquenching has been used to determine directly the molecular weight ($\bar{M}_n$) of polyacetylene during the polymerization.[27] With this method, the polymerization is stopped by an excess of tritiated methanol which reacts with all the polymer–metal bonds. By measuring the specific activity of tritium contained in the film, $\bar{M}_n$ ranging from 14000 to 43000 has been calculated at different temperatures.

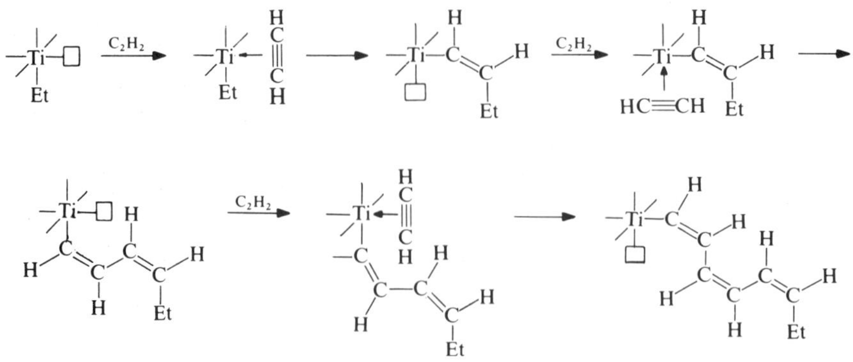

**Figure 2** Proposed mechanism for acetylene polymerization *via cis* insertion; □ = vacant site

*Cis–trans* polymer isomerization is easy because the energies of the two isomeric structures are quite comparable and the activation energy is low.[31] This isomerization occurs *via* thermal, chemical or electrochemical reactions. The process can be monitored by measuring the relative intensity of IR C—H stretching bands of the two isomers, or the electrical resistivity of the sample.[32] *Trans* polymer can usually be obtained by heating *in vacuo* the *cis* isomer synthesized at −78 °C. During this process, short conjugated *trans*-segments are formed at first, followed by progressive lengthening of the *trans*-sequences.[33] The chemical or electrochemical n or p doping of polyacetylene leads to *cis–trans* isomerization. The complete process requires a high doping level.

### 8.2.2 Synthetic Procedure for Polyacetylene Film

In this section, the usual technical procedure for synthesizing a polymer film is reported.

The catalytic system $Ti(OBu)_4/AlEt_3$ produces polyacetylene with different properties depending on concentration, ageing of the catalyst, aluminum/titanium ratio, polymerization temperature and monomer pressure. Under careful preparation conditions, this catalyst yields a high-quality free-standing polyacetylene film,[24] which, upon doping, exhibits high conductivity.

Ziegler–Natta catalysts are easily poisoned by air, water, electron donors, *etc.*, therefore all possible catalyst poisons must be absent in the acetylene polymerization and strict purity criteria for the reagents are recommended.[34] The reactivity of the catalyst, as well as polyacetylene sensitivity to air and water, require the use of high-vacuum apparatus[29] and purification of the inert gas (argon).

The aromatic solvents are dried over magnesium sulfate, refluxed under an argon atmosphere with sodium benzophenone and distilled. Commercial acetylene is purified by flowing through a sodium hydrogensulfite solution ($NaHSO_3$), a dry ice and methanol trap and through three columns containing calcium chloride, phosphorus pentoxide and molecular sieves, respectively. Titanium tetrabutoxide is distilled and stored under argon.

Free-standing *cis*-polyacetylene film is synthesized in a Schlenke-type flask. The reactor with the preformed catalytic system, introduced by a hypodermic syringe, is cooled in a bath of dry ice and methanol, then evacuated. The viscous catalytic solution is homogeneously spread on the reactor wall by shaking the flask, which is rapidly inserted into the cooled bath, and acetylene is finally introduced. Formation of polyacetylene is soon observed on the reactor walls. The polymerization is interrupted by evacuating acetylene from the reactor and by introducing argon.

Purification of the film is carried out at low temperature in order to prevent *cis–trans* isomerization of the polymer. The film is washed with dry cooled toluene or hexane and with methanol several times until the solvents become colourless. The polyacetylene is dried with a stream of argon and stored at low temperature.

The film exhibits a coppery lustre on the side in contact with the glass reactor wall during the polymerization, and a dull brown colour on the surface towards the inside. The morphological aspects of the two faces are different as well (Figure 3).

### 8.2.3 Polymerization with Other Catalyst Systems

Although acetylene has been polymerized by radiation, by anionic, cationic and radical catalysis, and by an enormous variety of transition and rare-earth metal catalysts, many side-reactions,

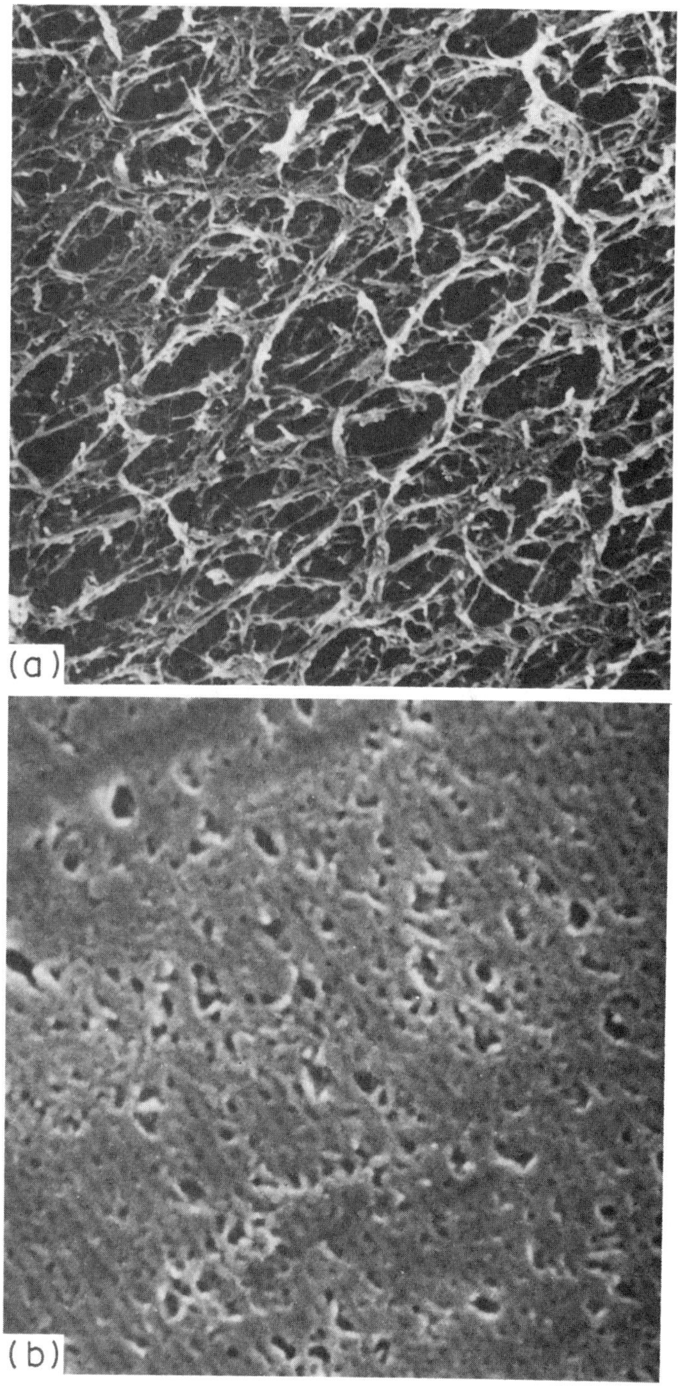


**Figure 3** Scanning electron micrographs of free-standing polyacetylene film: (a) dull face away from the reactor surface; and (b) shiny face toward the reactor surface. Magnification × 50 000 (reproduced by permission of Academic Press from 'Polyacetylene, Chemistry, Physics and Material Science', 1984, p. 121)

depending on the initiation method, disturb the polymerization. The major side-reaction is cyclization to benzene. Other processes lead to low molecular weight oligomers, branched and crosslinked polymers.

The quantitative evaluation of these side reactions, with the exception of cyclization, is very difficult because of the insolubility of the products. On the basis of a comparison between trimerization to benzene and polymerization to linear polymer, a great number of transition metal acetylacetonates have been considered.[35] With $AlEt_3$ as cocatalyst, titanium and vanadium complexes are more reactive than the others in the order Ti, $V \gg Cr > Fe > Co \approx Mn$.

Other titanium-based catalysts are able to polymerize acetylene with different yields of benzene and high polymers. Titanium tetrachloride ($TiCl_4$) and trichloride ($TiCl_3$) (with triethylaluminum as activator) produce, in aromatic or aliphatic hydrocarbon solvents, a black powder with a largely globular morphology, very rich in *trans* isomer.[36]

*Trans* isomer is the main product with $TiCl_4$ and *n*-butyllithium ($LiBu^n$) (lithium/titanium $\approx 4$) at room temperature, while the $TiCl_4/Al(Bu^i)_3$ (triisobutylaluminum) system (with an aluminum/titanium ratio $\approx 2$) yields, at the same temperature, a mixture of both of the isomers.[37] An interesting titanium catalyst is $\mu$-($\eta^1$:$\eta^5$-cyclopentadienyl)-1-tris($\eta$-cyclopentadienyl)dititanium that polymerizes acetylene to a gel product[38] very similar to that obtained with low concentrations of $Ti(OBu)_4/AlEt_3$.

Tetrabenzyltitanium has been used by itself in toluene[39] or mixed with trimethylaluminum in toluene or silicone oil[11] to polymerize acetylene, resulting in a shiny film with fibrillar morphology. The first catalyst affords a distinct fibrillar structure with fibre diameters of 2000 Å. The second yields, in toluene, fibres in the range 120–140 Å, but an unusual transparent film, with a thickness of 800–1000 Å and a density as high as 1.0 g cm$^{-3}$, in silicone oil.[40]

Polymerization of acetylene to high molecular weight insoluble polymer can be performed using nickel halide–tertiary phosphine complexes without an activator. The best yields have been obtained in a poorly coordinative solvent (ethanol/tetrahydrofuran 9/1) at 16–20 °C with both bis(triphenylphosphine)nickel dibromide and bis(tri-*n*-butylphosphine)nickel bromide as well as with bis(triphenylphosphine)nickel diiodide. Polyacetylenes prepared by these catalytic systems are partially crystalline powders, very rich in *trans* isomers.[41] Nickel chloride complexes and cobalt salts (cobalt nitrate) are very good catalysts for acetylene polymerization[42] only with sodium or potassium borohydride in polar solvents. A mechanism for acetylene polymerization with these systems has been suggested by Daniels.[41] It has been demonstrated[43] that it is possible, with the cobalt nitrate/sodium borohydride system and a specific concentration of catalyst, to also obtain a film with fibrillar morphology.[44]

Rare-earth coordination compounds have been used by various Chinese workers[45] to polymerize acetylene. Lanthanide naphthenates, 2-ethylhexyl phosphonates and isopropoxides mixed with triethylaluminum produce, at 30 °C, a 60–80% *cis*-rich crystalline polyacetylene. The lowest percentage of *cis* isomer was obtained using isopropoxide complexes.[45]

Tungsten hexachloride ($WCl_6$) and molybdenum pentachloride ($MoCl_5$), with tetraphenyltin ($SnPh_4$) as cocatalyst, give metathesis polymerization[39] of acetylene. The $WCl_6/SnPh_4$ system provides films, whereas the $MoCl_5/SnPh_4$ catalyst yields powders, both presenting fibrillar morphology. Furthermore, the resulting materials are directly doped during polymerization by catalyst residues.

$MoCl_5$ has also been employed with LiBu to form the most active metathetic system found to date. The resulting polyacetylene is a powder which presents, by electron microscopy, microscopic inclusions with both fibrillar and globular structures.[46] $MoCl_4$, $MoCl_5$ and $WCl_3$ can also be used without cocatalyst, giving highly oxidized polymers.[47] Globular structures can be observed in *trans*-rich polyacetylenes obtained using $WCl_6$ and $MoCl_5$ mixed with water as cocatalyst.[48]

## 8.3 POLYACETYLENE BLENDS AND COPOLYMERS

### 8.3.1 Polyacetylene Blends

Various attempts have been made to prepare modified polyacetylene, in order to obtain materials with improved oxygen resistance and processability.

Composites of low density polyethylene (LDPE) and polyacetylene with different compositions have been prepared[49] and characterized. This matrix has been chosen since it is easily obtained as a film and is inert with respect to the catalyst. LDPE films have been impregnated with $Ti(OBu)_4/AlEt_3$, and acetylene has been polymerized, obtaining a composite possessing intimate mixing at the molecular level. The best results have been obtained at elevated temperatures because the rates of polymerization are high and reflect the high diffusivity of acetylene in the molten matrix. From electron microscopy[50] it has been shown that the characteristic fibrillar morphology is not present, but irregularly shaped entities with dimensions in the range 600–2000 Å have been shown. Moreover, the morphology of polyacetylene–LDPE blends is completely different from that of films obtained by mechanical mixing of the two components. In this latter case, the films are roughly inhomogeneous and generally reach high conductivity only when $I_2$-doped polyacetylene is present in amounts as high as 40% (by weight). The conductivity value of the blends obtained with the *in situ* polymerization are dependent on the polyacetylene concentration as well. There is an increase in conductivity of up to $1\,\Omega^{-1}\,cm^{-1}$ for polyacetylene contents of about 3% (percolation conductivity threshold), which is approximately the same as the loading of the carbon black/polyethylene composites.[51]

A procedure similar to that adopted in ref. 49 has been proposed[52] by other workers for the preparation of polyacetylene/polybutadiene blends: even in this case, the two components of the composite material exist arranged in distinct incompatible phases. However, this material does not show improved resistance to air oxidation.

An increase in polyacetylene stability in the composite has been achieved[53] by using ethylene–propylene–diene terpolymer (EPDM) as a matrix and subsequently crosslinking the blend by radiation. After the *in situ* polymerization, the irradiated blend can be doped with iodine to produce a material having a conductivity of $10^2\,\Omega^{-1}\,cm^{-1}$. The increased stability reflects the high oxygen impermeability of EPDM rubber.

Other experimental works have focused on the possibility of obtaining polyacetylene latices with polystyrene/poly(ethylene oxide) and poly(*t*-butylstyrene)/poly(ethylene oxide) block copolymers.[54]

### 8.3.2 Polyacetylene Copolymers

An understanding of the transport mechanism in polyacetylene is closely connected with the possibility of disentangling the intramolecular or single-chain effects from those induced by intermolecular or chain–chain interactions. Solubilization of polyacetylene has been attempted in order to solve this problem. In fact, it is expected that the interactions among chains in solution should be strongly reduced and the properties of isolated chains would be observed. Various attempts, mainly consisting of monomer modification and copolymer preparation, have been described.

The presence of side-chain groups (alkyls or phenyls), as in poly(methylacetylene), has been reported to solubilize the rigid backbone.[55] However, upon doping, the conductivity values of these materials have been found to be several orders of magnitude lower than those of doped polyacetylene. Steric effects[56] could hinder the planarity of the backbone, therefore decreasing its conjugation length. A free-standing film of acetylene–methylacetylene copolymers (up to a ratio of 1/3) has been prepared. Good conductivity has been reached only for an acetylene-rich sample doped with $AsF_5$ and $I_2$. The resulting copolymer was insoluble in common organic solvents.

Three main methods have been reported for solubilizing polyacetylene, all consisting of preparing copolymers with a different chemical arrangement of polyacetylene and soluble carrier. Copolymer solubility is achieved because of the ability of the carrier to take into solution the insoluble polyenic backbone.

The first method consists of grafting soluble polymeric chains on to the main backbone of polyacetylene.[57] A charge-transfer complex of sodium with polyacetylene is prepared by reaction between sodium dihydronaphthalide in THF and powdered polyacetylene. The unreacted sodium dihydronaphthalide is removed and methyl methacrylate is slowly added dropwise to the stirred suspension of sodium polyacetylene complex. The material, a polyacetylene-*graft*-poly(methyl methacrylate) copolymer (with a comb-like arrangement), is soluble and its UV–visible spectrum indicates a short conjugation length. No conductivity data have been reported, while $^1H$ NMR and $^{13}C$ NMR studies have been unsuccessful in detecting the proton and carbon signals of polyacetylene.

The second synthetic approach consists of the preparation of a copolymer with a flexible main chain (matrix), soluble in organic solvents, and side groups formed by the insoluble polyacetylene. Some workers have used, as matrix, polyisoprene or a random copolymer of polystyrene with a few per cent of butadiene ($\bar{M}_n \approx 200\,000$).[58] A number of different polymers can be used, the main requirement being a relative lack of reactivity toward the catalyst. The carrier polymers are treated to form expoxides or carbonyl groups[59] as grafting sites. Polymerizations are carried out by adding triethylaluminum to a toluene solution of acetylene, titanium tetrabutoxide and the treated carrier polymer. There is a fast change in colour from brown to blue, indicating that polymerization of acetylene takes place. By varying the number of grafting sites on the carrier as well as the molar ratio between the carrier and the catalyst, it is possible to control the number of polyenic chains on the matrix and their length. It has been reported[59] that the graft copolymer is formed due to the quenching of a growing chain of polyacetylene on to a grafting site of the backbone. The quenching reaction is a nucleophilic attack of the carbon atom (bonded to the catalytically active titanium) of the last polymerized acetylene unit on the electrophilic carbon of the grafting site. With this method, only a few chains of polyacetylene per soluble carrier chain can be grafted.[60]

The same kind of graft copolymer has been prepared by growing the polyenic chains directly from the soluble matrix through the synthesis of a matrix-supported catalyst.[61] In one of the proposed synthetic routes, the preformed Ziegler–Natta catalyst reacts with vinyl side-groups of 1,4-*cis*-polybutadiene (containing a few per cent 1,2 units) dissolved in an organic solvent. The reaction between the vinyl group and the catalyst is supposed to lead to a titanium–carbon bond in which the carbon atom belongs to the polymeric backbone. As soon as acetylene bubbles into the solution, polymerization takes place directly by insertion of acetylene into this active titanium–matrix carbon bond. Different ratios of polyacetylene/matrix and *cis–trans* isomers can be obtained by changing the catalyst/vinyl-unit ratio, temperature and solvent. Conductivity values in the range $10^{-2}$–$10\,\Omega^{-1}\,cm^{-1}$ have been measured for 26–50% (by weight) polyacetylene in the cast films.[62] The morphology of the films consists of worm-like aggregates of polyene embedded in a polybutadiene matrix. Light-scattering experiments have shown polyacetylene aggregation in solution. The aggregates change their dimensions depending on the reaction conditions and solvent.[63] They have been studied using both X-ray diffraction from the solution[64] and Raman excitation profile analysis.[63] The $^1H$ NMR spectra of the solution are consistent with the above results because, due to micellar aggregation of polyacetylene, no resonance of the H atoms of the polyenic backbone is observed. A model that seems to be appropriate is a core of polyacetylene chains held together by intermolecular forces, surrounded by polybutadiene, which interacts better with the solvent.

The last synthetic approach for polyacetylene solubilization is based on diblock copolymer preparations. One attempt[65] consists of the synthesis of a living chain of polystyrene, polyethylene or polyisoprene, which can alkylate titanium tetraalkoxide. The alkylated $Ti^{3+}$ is bonded to the final carbon atom of the polymeric chain. Acetylene polymerization occurs in these active titanium–polymer bonds producing the diblock copolymer. By varying the length of the anionic polymer part with respect to the polyacetylene block, materials with different mechanical, physical and electrical characteristics have been prepared. The same kind of copolymer has been obtained with a metathesis catalyst.[66] In both cases the resultant materials are soluble but no characterization of the solutions has been reported up to now.

## 8.4 HIGHER QUALITY POLYACETYLENE

To overcome the drawbacks of polyacetylene mentioned in the introduction, different forms of polymer have been developed by improving both catalyst systems and synthesis methods.

In 1982, some researchers at ENI patented[67] a modified catalyst system able to produce the first highly stretchable polyacetylene. This catalyst is based on: (1) a titanium compound $TiX_mY_{4-m}$ with X = —$NR_2$, —SR, —$OC_6H_4Me$ and Y = —OR, —$O_2CR$, where R are aliphatic or aromatic hydrocarbon residues and *m* ranges from 1 to 4; (2) a reducing agent such as an alkylmetal, namely triethylaluminum.

The polymerization conditions are: (a) aluminum/titanium ratio = 4, (b) hexane as solvent and (c) catalyst ageing at 68–70 °C for 10 h under nitrogen. The procedure to obtain free-standing polymer films from acetylene is very similar to that used by other workers employing almost complete solvent evacuation. Resulting polyacetylene films are characterized by a content of *cis* isomer greater than 40%, depending on the polymerization temperature, and a thickness of up to 5 $\mu$m. The films may be mechanically stretched up to eight times their length with an orientation of the sample fibrils along the drawing direction (*i.e* that of the molecular backbone).[68] The high overall density of this material, 0.9 $g\,cm^{-3}$, makes the sample more resistant to air oxidation than is usual for polyacetylenes. Moreover, the alignment of the fibrils gives the films relevant optical and electrical anisotropies. Doping of stretched and unstretched samples with iodine in the gas phase produces materials which display good conductivity.

The oriented specimens show a conductivity one order of magnitude higher than the unoriented ones (maximum value $2.2 \times 10^3\,\Omega^{-1}\,cm^{-1}$), the conductivity parallel to the drawing direction being one order greater than that of the perpendicular one.[68] The same anisotropy is exhibited by the photoconductivity[69] produced by laser excitation.

Another type of polyacetylene more resistant to air, with a very compact morphology, is obtained by modifying the reducing agent in the catalytic system. The density of films produced using different cocatalysts ranges between 0.62 and 0.86 $g\,cm^{-3}$ depending on the organometallic compound employed, namely $R_2AlOAlR_2$, $R'_2AlSAlR'_2$ or $R'_2AlN(R)AlR'_2$ with R = butyl and R′ = ethyl.[70] The *cis*-isomer content in these materials, synthesized at room temperature, and in the ENI polyacetylene, is higher than that observed in conventional polyacetylenes.

Other workers have developed different synthetic routes or polymerization methods to produce well-oriented polyacetylenes. The first approach, carried out by Feast and co-workers, is based on the thermal elimination of benzene from a soluble stretchable prepolymer prepared by metathesis ring-opening polymerization of a strained monomer, the 7,8-bis(trifluoromethyl)tricyclo[4.2.2.0]-deca-3,7,9-triene.[71] More details about this synthesis are presented in Volume 4, Chapter 7.

The other method consists of polymerizing acetylene with $Ti(OBu)_4/AlEt_3$ (aluminum/titanium 3/1 or 4/1) on an ordered substrate, such as biphenyl[72] and benzene[73] crystals, or in oriented media, such as liquid crystals mostly in nematic phase.[74,75] These polyacetylenes are well ordered, showing fibrillar morphology and very high optical anisotropy. The conductivity parallel to the chain direction has been found to be one order of magnitude greater than that of the perpendicular one.

BASF researchers together with a research group at the University of Montpellier have modified the polymerization method employing the usual systems, namely $Ti(OBu)_4/AlEt_3$ in the ratio 1/4 and $Ti(OBu)_4/LiBu$ with lithium/titanium ratio = 2. These researchers have optimized the morphology and properties of polyacetylene by using a highly viscous complexing solvent together with thermal treatment. Indeed, employing silicone oil as solvent, it is possible to heat the catalyst up to 100–120 °C, at which temperature only one of the four species of $Ti^{3+}$ forming the catalytic system survives.[76] ESR studies[40] have proved that only the species at 1953 G (1 G = $10^{-4}$ T), ascribed to the trimeric complex of catalyst, is living after thermal treatment. This species is responsible for very long conjugated chains almost without $sp^3$ defects (as demonstrated by $^{13}C$ CP-MAS spectra). If the homogeneous black and shining film is produced on a special support of high-density polyethylene or polypropylene films, it is possible to stretch the polyacetylene, thus obtaining a transparent and uniformly oriented film elongated by up to seven times (Figure 4).[11] The conductivity of the oriented material doped with iodine is higher than $2\times10^4\,\Omega^{-1}\,cm^{-1}$ for samples 20 $\mu m$ thick and $8\times10^3\,\Omega^{-1}\,cm^{-1}$ for 0.1 $\mu m$ thick samples.[11] Its very high electrical anisotropy and the temperature independence of the conductivity below 1 K imply genuine metallic behaviour for a high doping level. However, researchers do not disregard a hopping transport mechanism across the fibrils.[12] The conductivity value can be increased if another reducing agent such as LiBu is added to the standard catalyst system in silicone oil. Under these conditions, researchers have synthesized a polyacetylene film with a conductivity higher than $10^5\,\Omega^{-1}\,cm^{-1}$. The samples prepared with this new catalytic system have been stretched on a supporting film by up to five times. This material also exhibits the characteristic fibrillar morphology.[11] We can conclude from all these studies that the achievement of still higher-quality materials from improved syntheses and processing can be expected to give correspondingly better electronic properties useful for technological applications.

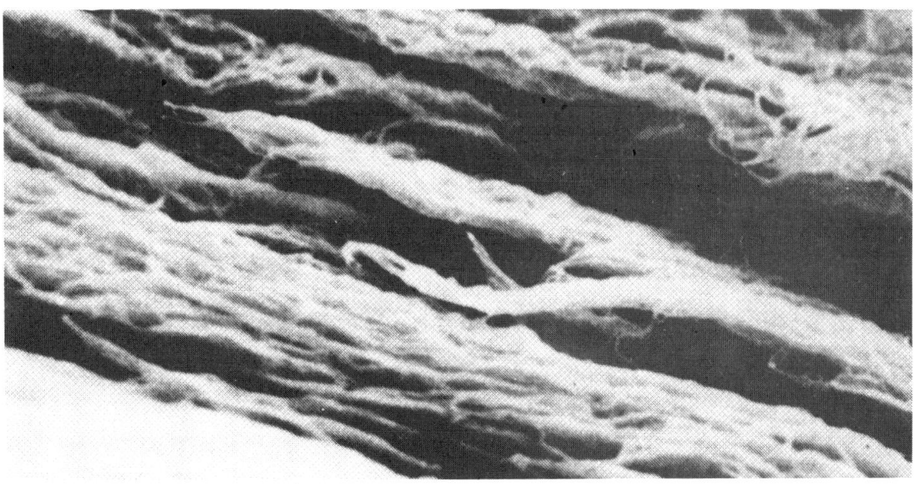

**Figure 4** Scanning electron micrographs of a cross section (low magnification) of undoped polyacetylene film, synthesized with the system $Ti(OBu)_4/Bu^nLi$/silicone oil (reproduced by permission of Marcel Dekker Journals from *J. Macromol. Sci., Chem.*, 1987, **A24**, 797)

## 8.5 REFERENCES


1. H. Shirakawa, E. J. Louis, A. G. MacDiarmid, C. K. Chiang and A. J. Heeger, *J. Chem. Soc., Chem. Commun.*, 1977, 578.
2. J. C. W. Chien, *J. Polym. Sci., Polym. Lett. Ed.*, 1981, **18**, 249.
3. H. Shirakawa, T. Ito and S. Ikeda, *Makromol. Chem.*, 1978, **179**, 1565.
4. J. C. W. Chien, J. M. Warakomski, F. E. Karasz, W. L. Chia and C. P. Lillya, *Phys. Rev. B: Condens. Matter*, 1983, **28**, 6937.
5. G. Wegner, *Makromol. Chem., Macromol. Symp.*, 1986, **1**, 151.
6. R. R. Chance, D. S. Boudreaux, H. Eckardt, R. L. Elsembaumer, J. E. Frommer, J. L. Brédas and R. Silbey, in 'Proceedings of N. A. S. I. School on Quantum Chemistry of Polymers: Solid State Aspects', Braumlange, 1983, ed. J. Ladik and J. M. André, Reidel, Dordrecht, 1984.
7. J. C. W. Chien, in 'Polyacetylene, Chemistry, Physics and Material Science', Academic Press, New York, 1984, p. 336.
8. C. Mathis and B. François, *Makromol. Chem., Rapid Commun.*, 1984, **5**, 359.
9. S. Luzzati, S. Destri, A. Bolognesi, M. Catellani, W. Porzio and R. Tubino, in 'Proceedings of the 4th International School of Condensed Matter Physics, Molecular Electronic, Varna, 1986', ed. M. Borissov, World Scientific, Singapore, 1987, p. 340.
10. M. A. Shen, F. E. Karasz and J. C. W. Chien, *Makromol. Chem., Rapid Commun.*, 1984, **5**, 217.
11. N. Theophilou and H. Naarmann, in 'Proceedings of European Symposium on Polymeric Materials, Polymer for Electronics', Lyon, 1987, European Polymer Federation, p. 1.
12. N. Basescu, Z. X. Liu, D. Moses, A. J. Heeger, A. Naarmann and N. Theophilou, *Nature (London)*, 1987, **327**, 403.
13. F. E. Karasz, J. C. W. Chien, R. Galkiewiez, G. E. Wnek, A. J. Heeger and A. G. MacDiarmid, *Nature (London)*, 1979, **282**, 286.
14. J. C. W. Chien, J. D. Capistran, F. E. Karasz, L. C. Dickinson and M. A. Schen, *J. Polym. Sci., Polym. Lett. Ed.*, 1983, **21**, 93.
15. H. Shirakawa, A. Hamano, S. Kawakami, K. Soga and S. Ikeda, *Macromolecules*, 1980, **13**, 457.
16. V. Enkelmann, G. Lieser, M. Monkenbusch, W. Müller and G. Wegner, *Mol. Cryst. Liq. Cryst.*, 1981, **77**, 111.
17. R. H. Baughman, S. L. Hsu, G. P. Pez and A. J. Signorelli, *J. Chem. Phys.*, 1978, **68**, 5405.
18. L. C. Dickinson, J. A. Hirsch, F. E. Karasz and J. C. W. Chien, *Macromolecules*, 1985, **18**, 2374.
19. C. R. Fincher, Jr., M. Ozaki, M. Tanaka, D. Peebles, L. Lauchlan, A. J. Heeger and A. G. MacDiarmid, *Phys. Rev. B: Condens. Matter*, 1979, **20**, 1589.
20. J. C. W. Chien, in 'Polyacetylene, Chemistry, Physics and Material Science', Academic Press, New York, 1984, p. 566.
21. H. Munstedt, H. Naarmann and G. Köhler, *Mol. Cryst. Liq. Cryst.*, 1985, **118**, 129.
22. G. Natta, G. Mazzanti and P. Corradini, *Atti Accad. Naz. Lincei, Cl. Sci. Fis. Mat. Nat., Rend.*, 1958, **25**, 3.
23. A. Bradley and J. P. Hammes, *J. Electrochem. Soc.*, 1963, **110**, 12.
24. T. Ito, H. Shirakawa and S. Ikeda, *J. Polym. Sci., Polym. Chem. Ed.*, 1974, **12**, 11.
25. H. Shirakawa and S. Ikeda, *Synth. Met.*, 1979/80, **1**, 175.
26. K. Shimamura, F. E. Karasz, J. A. Hirsch and J. W. C. Chien, *Makromol. Chem., Rapid Commun.*, 1981, **2**, 473.
27. M. A. Schen, F. E. Karasz and J. C. W. Chien, *J. Polym. Sci., Polym. Chem. Ed.*, 1983, **21**, 2787.
28. T. C. Clarke, C. S. Yannoni and T. J. Katz, *J. Am. Chem. Soc.*, 1983, **105**, 7787.
29. M. Aldissi, F. Schué, G. L. Giral and M. Rolland, *Polymer*, 1982, **23**, 246.
30. J. A. Shelburne and G. L. Baker, *Macromolecules*, 1987, **20**, 1212.
31. T. Ito, H. Shirakawa and S. Ikeda, *J. Polym. Sci., Polym. Chem. Ed.*, 1975, **13**, 1943.
32. M. Rolland, P. Bernier, S. Lefrant and M. Aldissi, *Polymer*, 1980, **21**, 1111.
33. A. Bolognesi, M. Catellani, S. Destri, G. Morelli, W. Porzio, R. Tubino and M. Zocchi, *Makromol. Chem., Rapid Commun.*, 1983, **4**, 403.
34. J. C. W. Chien, 'Polyacetylene, Chemistry, Physics and Material Science', Academic Press, New York, 1984, p. 25.
35. S. Kambara, M. Hatano and T. Hosoe, *Kogyo Kagaku Zasshi*, 1962, **65**, 720.
36. W. Dietz, P. Cukor, M. Rubner and H. Jopson, *J. Electron. Mater.*, 1981, **10**, 683.
37. W. H. Watson, Jr., C. Warren, W. C. McMordie, Jr. and L. G. Lands, *J. Polym. Sci.*, 1961, **55**, 137.
38. S. L. Hsu, A. J. Signorelli, G. P. Pez and R. H. Baughman, *J. Chem. Phys.*, 1978, **69**, 106.
39. M. Aldissi, C. Linaya, J. Sledz, F. Schué, L. Giral, J. M. Fabre and M. Rolland, *Polymer*, 1982, **23**, 243.
40. N. Theophilou, R. Aznar, A. Munardi, J. Sledz, F. Schué and H. Naarmann, *J. Macromol. Sci., Chem.*, 1987, **A24**, 797.
41. W. E. Daniels, *J. Org. Chem.*, 1964, **29**, 2936.
42. L. B. Luttinger, *Chem. Ind. (London)*, 1960, 1135; *J. Org. Chem.*, 1962, **27**, 159.
43. G. Lieser, G. Wegner, W. Muller and V. Enkelmann, *Makromol. Chem., Rapid Commun.*, 1980, **1**, 627.
44. J. C. W. Chien, in 'Polyacetylene, Chemistry, Physics and Material Science', Academic Press, New York, 1984, p. 127.
45. S. Zhiquan, Y. Mujie, S. Mingxiao and C. Yiping, *J. Polym. Sci., Polym. Lett. Ed.*, 1982, **20**, 411; Y. Cao, R. Qian, F. Wang and X. Zhao, *Makromol. Chem., Rapid Commun.*, 1982, **3**, 687.
46. N. Theophilou, A. Munardi, R. Aznar, G. Sledz, F. Schué and H. Naarmann, *Eur. Polym. J.*, 1987, **23**, 15.
47. M. Voronkov, V. Pukhnarevich, S. Suchchnskaya, V. Z. Annekova, V. M. Annekova and N. Andreeva, *J. Polym. Sci., Polym. Chem. Ed.*, 1980, **18**, 53.
48. J. C. W. Chien, in 'Polyacetylene, Chemistry, Physics and Material Science', Academic Press, New York, 1984, p. 78.
49. M. E. Galvin and G. E. Wnek, *Polymer*, 1982, **23**, 795.
50. M. E. Galvin and G. E. Wnek, *J. Polym. Sci., Polym. Chem. Ed.*, 1983, **21**, 2727.
51. K. Miyasaka, K. Watanabe, E. Jojima, H. Aida, M. Sumita and K. Ishikawa, *J. Mater. Sci.*, 1982, **17**, 1610.
52. M. F. Rubner, S. K. Tripathy, J. Georger, Jr. and P. Cholewa, *Macromolecules*, 1983, **16**, 870.
53. K. I. Lee and H. Jopson, *Makromol. Chem., Rapid Commun.*, 1983, **4**, 375.
54. J. Edwards, R. Fisher and B. Vincent, *Makromol. Chem., Rapid Commun.*, 1983, **4**, 393.
55. P. Cukor, J. I. Krugel and M. F. Rubner, *Polym. Prepr., Am. Chem. Soc., Div. Polym. Chem.*, 1980, **21**, 161.
56. J. C. W. Chien, G. E. Wnek, F. E. Karasz and J. A. Hirsch, *Macromolecules*, 1981, **14**, 479.
57. I. Kmínek and J. Trekoval, *Makromol. Chem., Rapid Commun.*, 1984, **14**, 479.
58. F. S. Bates and G. L. Baker, *Macromolecules*, 1983, **16**, 704.

59. G. L. Baker and F. S. Bates, *Macromolecules*, 1984, **17**, 2619.
60. F. S. Bates and G. L. Baker, *Macromolecules*, 1983, **16**, 1013.
61. S. Destri, M. Catellani and A. Bolognesi, *Makromol. Chem., Rapid Commun.*, 1984, **5**, 353; A. Bolognesi, M. Catellani, W. Porozio and S. Destri, *Polymer*, 1986, **27**, 1128.
62. A. Bolognesi, M. Catellani, S. Destri and W. Porzio, *Makromol. Chem.*, 1986, **187**, 1287.
63. C. Cuniberti, P. Piaggio, G. Dellepiane, M. Catellani, L. Piseri, W. Porzio and R. Tubino, *Makromol. Chem., Rapid. Commun.*, 1986, **7**, 471.
64. R. Dorsinville, R. Tubino, S. Krimchansky, R. R. Alfano, J. L. Birman, A. Bolognesi, S. Destri, M. Catellani and W. Porzio, *Phys. Rev. B: Condens. Matter*, 1985, **32**, 3377.
65. M. E. Galvin and G. E. Wnek, *Mol. Cryst. Liq. Cryst.*, 1985, **117**, 33; M. Aldissi, *Synth. Met.*, 1986, **13**, 87.
66. T. R. Farren, A. J. Amass, M. S. Beevers and J. A. Stowell, *Makromol. Chem.*, 1987, **188**, 2535.
67. U. Pedretti, G. Perego and G. Lugli (ANIC S.p.A), *US Pat.* 3 600 756 (1986) (*Chem. Abstr.*, 1986, **100**, 176 176j); U. Pedretti, G. Perego and G. Lugli (ASSORENI S.p.A.) *Ital. Pat.* 22 722/A82 (1982).
68. G. Lugli, U. Pedretti and G. Perego, *Mol. Cryst. Liq. Cryst.* 1985, **177**, 43.
69. R. Tubino, *Synth. Met.*, 1987, **17**, 509.
70. M. Catellani, S. Destri, A. Bolognesi and E. Albizzati, *Makromol. Chem.*, 1986, **187**, 1345.
71. J. H. Edwards and W. J. Feast, *Polymer*, 1980, **21**, 595.
72. T. Woerner, A. G. MacDiarmid and A. J. Heeger, *J. Polym. Sci., Polym. Lett. Ed.*, 1982, **20**, 305.
73. M. Ozaki, Y. Ikeda and T. Arakawe, *J. Polym. Sci., Polym. Lett. Ed.*, 1983, **21**, 989.
74. K. Araka, A. Mokon, T. Narahara and H. Shirakawa, *Chem. Lett.*, 1984, 1141.
75. M. Aldissi, *J. Polym. Sci., Polym. Lett. Ed.*, 1985, **23**, 167.
76. N. Theophilou, R. Aznar, A. Munardi, J. Sledz, F. Schué and H. Naarmann, *Synth. Met.*, 1986, **16**, 337.

# 9

# Polymerization of Mono- and Di-substituted Acetylenes

GIOVANNA COSTA

*Centro di Studi Chimico-Fisici di Macromolecole Sintetiche e Naturali, Genova, Italy*



## 9.1 GENERAL FEATURES

Substituted polyacetylenes (polyalkynes) are a relatively old class of polymers, which have been widely investigated and reviewed, particularly in the last decade.[1-3] Indeed, the presence of alternating double bonds along the main chain in substituted and unsubstituted polyenes is responsible for some peculiar properties,[2-4] *e.g.* chain stiffness, geometrical isomerism, chemical reactivity, colour, electrical conductivity and paramagnetism, which make them interesting as speciality polymers.

Acetylenic (alkynic) polymers can be obtained either by direct polymerization, starting from the corresponding monomers, or by indirect synthesis from polymeric precursors;[2] in this latter case, the alternating double bond skeleton can be generated by the elimination of some low molecular weight species. Attention will be focused here only on polymers obtained by the first method.

Both ionic and free radical initiators have been used[1,5] to attempt the polymerization of substituted acetylenes, but in most cases low molecular weight products or cyclotrimers were obtained.

When a catalytic system formed by a transition metal derivative and an alkyl or aryl compound of a non-transition metal is used, the reaction of mono- or di-substituted acetylenes results in the addition polymerization of the monomer, as is the case with unsubstituted acetylene. However, unlike acetylene, which can provide high polymers using Ziegler–Natta catalysts under suitable conditions,[6,7] in the case of substituted monomers the resultant products are very often cyclic and/or linear low molecular weight polymers. The relative proportions of these forms depend on both the catalyst–solvent system and the structure of the monomeric unit. Actually, among the monosubstituted monomers, only some primary or secondary alkylacetylenes provide high molecular weight polymers in the presence of Ziegler–Natta catalysts. Oligomers and/or insoluble polymers are obtained starting from acetylenes containing more-branched aliphatic chains or aromatic groups or heteroatoms. Disubstituted monomers do not polymerize at all. Other catalysts, such as a nickel carbonyl/phosphine complex,[8] have been unsuccessfully applied in order to produce high polymers.

Quite recently, a number of catalysts based on transition metals of groups V and VI (Nb, Ta, Mo and W) were found to be effective in the polymerization of specific substituted acetylenes.[9-16]

Indeed, by using these metal derivatives very high molecular weight polymers are obtained, as will be discussed below. A new organometallic complex, based on chromium, has recently been found to induce oligomerization of some acetylenes bearing fluorine-containing substituents.[17] It has to be emphasized that, independently of the substituent, the addition reaction generally proceeds without any evidence of isomerization of the monomeric unit. All the resulting products should therefore maintain the characteristic alternating double bond structure, as shown in equation (1).

$$n\mathrm{RC{\equiv}CX} \longrightarrow \left(\mathrm{RC{=}CX}\right)_n \qquad (1)$$

$$\mathrm{X = H, R^1}$$

Only in the case of poly(phenylacetylene) has a ring structure along the main chain been found; this has been ascribed to thermal isomerization, following the double bond formation.[18]

Unlike polyacetylene, most of the aforementioned polyenes are insulating, amorphous and soluble in common solvents and do not exhibit any appreciable absorption in the visible region. In fact, owing to their twisted conformation, directly related to the steric hindrance of the substituents, the coplanarity of carbon–carbon double bonds in adjacent repeating units is severely reduced with consequent limited conjugation along the main chain and a much weaker overlap between $\pi$ orbitals. Furthermore, whilst polyacetylene is characterized by a high oxygen sensitivity, substituted polyacetylenes are much more stable and their stability seems to increase on increasing the number and/or the steric hindrance of the substituents.

## 9.2 MONOSUBSTITUTED POLYACETYLENES

The polymerization of a great number of monosubstituted acetylenes (alternatively called terminal acetylenes) has been performed by using a wide variety of catalytic systems.[3] In general, high molecular weight polymers can be synthesized either using Ziegler–Natta or Mo- or W-based catalysts. As previously mentioned, monosubstituted polyacetylenes can also be synthesized with free radical, anionic or cationic initiators. However, in this case, only low molecular weight polyenes are obtained.

The choice of catalyst, as well as suitable experimental conditions (*e.g.* temperature, solvent, *etc.*) strongly depend on the chemical nature and steric hindrance of the substituent. In fact, Ziegler–Natta catalysts can be used to achieve polymers only with uncrowded monomers bearing aliphatic groups such as Et, $\mathrm{Bu^n}$ or $\mathrm{Bu^s}$.[7,8,19–21]

Catalytic systems based on Mo and W are much less effective, and give either linear oligomers and cyclotrimers or low molecular weight polymers with *n*- or *s*-alkylacetylenes, respectively.[10,13,16,21,22] So far, no other catalyst is known to provide high polymers when sterically uncrowded acetylenes are polymerized. On the other hand, Mo- or W-based catalytic systems are very effective in the polymerization of highly branched alkylacetylenes,[15,23–25] arylacetylenes[15,26,27] and heteroatom-containing acetylenes.[12,28–35]

In the case of substituents with an electron-withdrawing character, high polymers are generally obtained with Mo catalysts; on the other hand, monomers bearing electron-releasing groups, such as $\mathrm{HC{\equiv}CSiMe_3}$, are reported to polymerize only in the presence of W catalysts.[3] As far as transition metals of group V are concerned, no polymerization takes place when Nb or Ta halides are used with monosubstituted acetylenes. Up to now, only one monomer, *i.e.* 3-(trimethylsilyl)-1-octyne, is known to polymerize when $NbCl_5$ is used as a catalyst.[3] As in the case of acetylene, which usually generates benzene in the presence of Nb or Ta pentachlorides and pentabromides,[36] terminal acetylenes cyclotrimerize to 1,2,4- and 1,3,5-substitued benzenes;[37,38] linear oligomers are produced when Nb or Ta pentafluoride are used.[3] Moreover, $\mathrm{Nb^{III}}$ and $\mathrm{Ta^{III}}$ complexes, such as $M_2Cl_6(SC_4H_8)_3$ (M = Ta, Nb) do not give high molecular weight products and lead to the formation of a mixture of cyclotrimer isomers.[39]

For the sake of simplicity, monosubstituted acetylenes can be roughly divided into three groups, depending on the substituent: (1) sterically unhindered monomers, (2) monomers with bulky substituents, and (3) heteroatom-containing monomers. As is shown schematically in Table 1, this classification is also consistent with the different final reaction products.

The experimental conditions play a significant role in the microstructure of the reaction products, as well as on molecular weights and conversion yields. An interesting example is given by poly-(*t*-butylacetylene), whose geometrical structure can be controlled by choosing a suitable solvent and catalyst.[24] In fact, polymers obtained in non-polar solvents such as toluene in the presence of $WCl_6$

**Table 1** Polymerization of Monosubstituted Acetylenes HC≡CR: Relationship Between Monomer Structure and Catalyst Activity

| *R* | *Catalyst* | *Products* |
|---|---|---|
| Small group | Ziegler–Natta | High polymer |
| | Mo-, W-based | Oligomers and/or cyclotrimers |
| | Nb-, Ta-based | Cyclotrimers — no polymer |
| Bulky or heteroatom-containing group | Ziegler–Natta | No polymer |
| | Mo-, W-based | High polymer |
| | Nb-, Ta-based | Cyclotrimers — no polymer |

as catalyst show both *cis* and *trans* structures in a comparable amount: no selectivity is shown by this system. On the contrary, all-*cis* polymers are obtained when the reaction is carried out in oxygen-containing solvents in the presence of $MoCl_5$ or in nitrogen-containing solvents with $MoCl_5$–$Ph_4Sn$ as catalyst.

In other cases, as, for example, in the polymerization of phenylacetylene, choice of a suitable solvent is an essential prerequisite, not only to achieve high molecular weight polymers with good conversion yields, but also to enhance the polymerization rate.[26,40–42]

## 9.3 DISUBSTITUTED POLYACETYLENES

So far, a smaller number of disubstituted (or internal) acetylenes, as compared with monosubstituted ones, are known to give high polymers in the presence of a suitable catalyst. They include hydrocarbon-substituted monomers, *i.e.* dialkyl, diaryl, alkylaryl acetylenes, and heteroatom-containing acetylenes, among which silyl derivatives seem to exhibit the most unusual and interesting properties. The above monomers do not give high molecular weight products in the presence of Ziegler–Natta catalysts, but can be polymerized by transition-metal-based catalytic systems such as $MoCl_5$, $WCl_6$, $NbCl_5$, $TaCl_5$, $Nb_2Cl_6(SC_4H_8)_3$, $Ta_2Cl_6(SC_4H_8)_3$, *etc.*, sometimes with a cocatalyst, for example $Ph_4Sn$, $Ph_3Bi$ or $Bu^n_4Sn$.

It has been found that both Mo- and W-based catalysts are fairly effective towards most of the hydrocarbon-substituted acetylenes, while Nb- and Ta-based compounds are active with many disubstituted monomers, including those bearing bulky substituents.[3] The general behaviour of these systems is summarized in Table 2.

**Table 2** Polymerization of Disubstituted Acetylenes $R^1C{\equiv}CR^2$: Monomer Structure and Catalyst Activity

| *$R^1$, $R^2$* | *Catalyst* | *Products* |
|---|---|---|
| Small groups | Ziegler–Natta | No polymer |
| | Mo-, W-based | High polymer |
| | Nb-, Ta-based | Cyclotrimer and/or polymer |
| Bulky groups | Ziegler–Natta | No polymer |
| | Mo-, W-based | Polymer |
| | Nb-, Ta-based | High polymer |
| Silicon-containing groups | Ziegler–Natta | No polymer |
| | Mo-, W-based | No polymer |
| | Nb-, Ta-based | High polymer |
| Other heteroatom-containing groups | Ziegler–Natta | No polymer |
| | Mo-, W-based | High polymer |
| | Nb-, Ta-based | No polymer[a] |

[a] Except 1-(pentafluorophenyl)-1-alkynes.

It is worth mentioning here a couple of monomers containing electronegative substituents, namely dicyanoacetylene (NCC≡CCN) and hexafluorobutyne ($CF_3C{\equiv}CCF_3$), which have been polymerized with other initiators.

Poly(dicyanoacetylene) was prepared using anionic catalysts such as butyllithium, butylmagnesium bromide and lithium naphthalene.[43] The resultant products are always low molecular weight polymers. Poly(hexafluorobut-2-yne) was obtained in the presence of $NOAsF_6/BF_3/h\nu$[44] or by fluoride-ion-induced polymerization.[45,46] In this case, because of the insolubility of the polymer both in organic solvents and in mineral acids, its molecular weight has not been determined.

As already mentioned, disubstituted polyacetylenes with high molecular weight (even higher than $10^6$) can be obtained using transition metal catalysts. Molybdenum and tungsten halides, which are effective with monosubstituted monomers, provide very few polymers with internal acetylenes; it is usually necessary to use equimolar amounts of $MoCl_5$ or $WCl_6$ and an organometallic cocatalyst (such as $Ph_4Sn$ or $Bu^n_4Sn$) to get high polymers.[21,47–55] The following metal carbonyl catalysts, $Mo(CO)_6$–$CCl_4$–$h\nu$ and $W(CO)_6$–$CCl_4$–$h\nu$, as well as some isolated metal carbenes (*e.g.* $Ph(MeO)C{=}W(CO)_5$) are also effective.[13,52–54,56] The catalyst activity, as in the case of monosubstituted acetylenes, is generally influenced by the structure of the monomer; in the case of Mo and W derivatives, it has been noticed that acetylenes bearing electron-withdrawing substituents, (*e.g.* 1-chloro-2-phenylacetylene,[51,56] propiolic acid[31] and 1-chloro-1-alkynes[50]) polymerize only in the presence of Mo-based catalysts. On the other hand, polymerization of acetylenes carrying electron-releasing groups has not been reported, and only with 1-(ferrocenyl-1-propyne) has a W-based catalyst been found to be effective.[55]

As far as group V transition metals are concerned, it has been known for a long time that Nb and Ta pentachlorides and pentabromides cyclotrimerize acetylene.[35] Recently, Cotton *et al.*[39] have reported that the catalytic reaction of internal acetylenes with $Nb^{III}$ and $Ta^{III}$ compounds ($M_2Cl_6(SC_4H_8)_3$) can provide polymers and/or cyclotrimers. More recently, Masuda *et al.*[14] have found that $NbX_5$ and $TaX_5$ ($X = Cl$, Br) can give polymers with disubstituted acetylenes. As shown in Table 2 Nb and Ta halides mainly provide cyclotrimers when the steric effect of the substituents is small. On the other hand, when bulky substituents of the monomer are present, high polymers can be obtained.[14,57–61] Also, in this case an equimolar amount of a suitable cocatalyst can be used. The result is an increase in the polymer molecular weight and a faster polymerization reaction. In addition, the use of a cocatalyst can efficiently reduce polymer degradation and affect the relative ratio of cyclotrimers to polymers.[59,61–65]

Heteroatom-containing monomers polymerized by group V transition metals are mostly silicon-containing acetylenes; indeed, only fluorine-containing polyacetylenes have also been synthesized with a Ta-based catalyst.[65] Among silicon-containing polyacetylenes, great interest has arisen regarding poly[1-(trimethysilyl)-1-propyne], for the following reasons. It is the first example of a silicon-containing high molecular weight polyacetylene;[60,66] it exhibits $\bar{M}_w$ as high as *ca.* $4 \times 10^6$, when synthesized in the presence of $TaCl_5$–$Ph_3Bi$;[64] it is air stable and exhibits the highest oxygen permeability of all existing polymers.[66,67]

## 9.4 POLYMERIZATION MECHANISM

Ziegler–Natta-catalyzed polymerization of acetylenes follows the well-established insertion mechanism.[1,3] Many authors have been faced with a problem related to the understanding of the reaction mechanism involved in the polymerization of acetylenes initiated by transition metal halides or transition-metal-based catalysts (W, Mo, Ta, *etc.*).[1,3,15,68]

Some experimental evidence, which will be summarized below, suggests that the polymerization of acetylenes proceeds by a mechanism quite similar to the metathesis polymerization of cycloalkenes. Indeed, it has been established, and is now generally accepted, that alkene metathesis takes place by scission of a carbon–carbon double bond and formation of a metal–carbon double bond, to give a metal–carbene complex (**1**) and a metallacyclobutane intermediate (**2**)[69] as illustrated in equation (2).

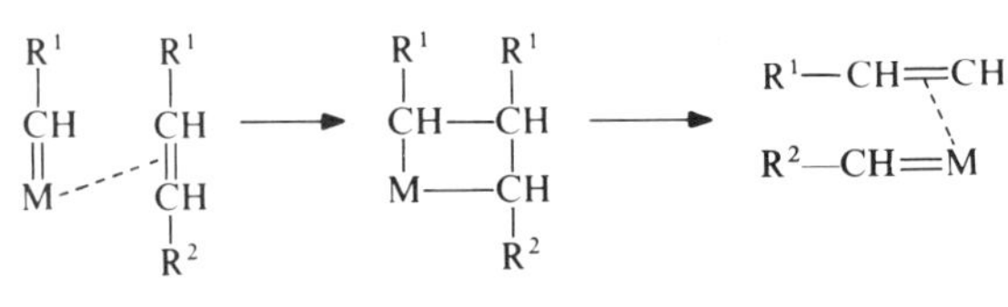

(2)

(**1**) (**2**)

When a cycloalkene is used instead of an alkene, a metathesis polymerization occurs with the same reaction mechanism.[69] This is shown in equation (3). In a similar way when polyacetylenes are obtained by Mo- and W-catalyzed polymerizations, the metal–carbene mechanism shown in equation (4) is proposed.[42]

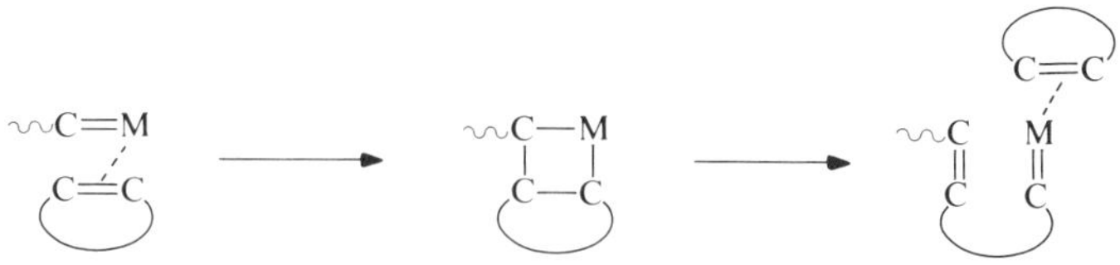
(3)

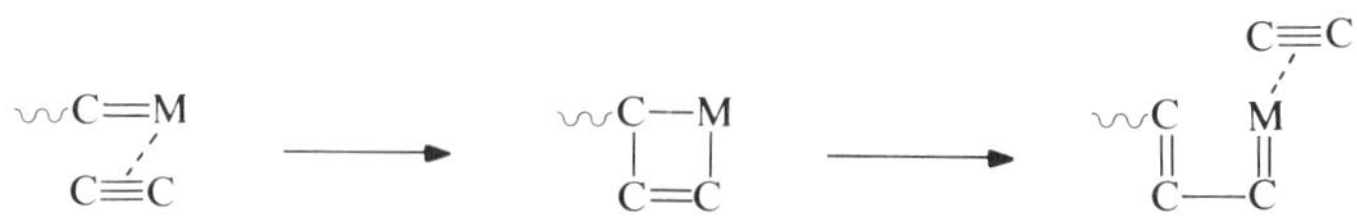
(4)

According to Masuda and Higashimura[3,4,42] this assumption is supported by the following evidence: (i) many transition-metal-based catalysts[70,71] are effective in alkene metathesis and metathesis polymerization of cycloalkenes as well as in acetylene polymerization; (ii) some W-based metal carbenes, isolated by Katz and Lee,[15] can both initiate polymerization of acetylenes (equation 5) and cause alkene metathesis (equation 2); (iii) when $W(CO)_6$–$CCl_4$–$h\nu$ is used in a polymerization system, the presence of a tungsten dichlorocarbene has been pointed out.[72]

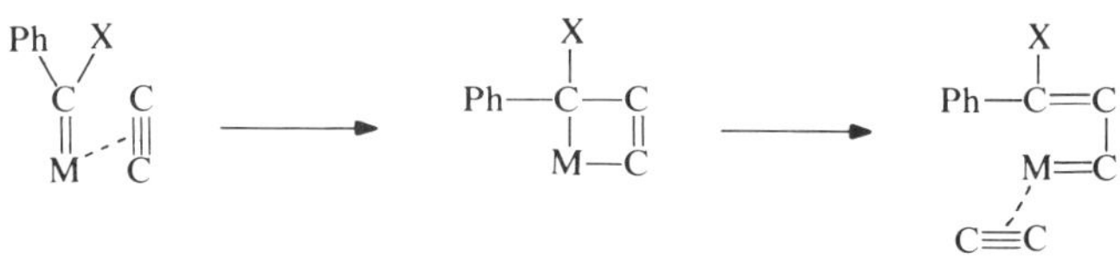
(5)

The mechanism just described has recently been supported by a few additional experimental findings.[73,74] In the case of transition metal halides a similar mechanism has been proposed when $WCl_6$ and $MoCl_5$ catalysts are used.[3] No experimental evidence has been reported for group V transition metals (*i.e.* Nb and Ta), though similar behaviour might also apply to them.

However, the mechanism of carbene formation in the initiation process as well as the nature of the metal carbene is still unknown.[68] Further investigations are needed to clarify this mechanism.

So far, transfer and termination reactions have not been investigated in detail and only one paper dealing with interchain and termination reactions in the polymerization of substituted acetylenes has recently appeared in the literature.[68] Results reported suggest a living character for these polymerizations: indeed the steric hindrance of the substituents seems to prevent both interchain reactions (reduced reactivity of the double bond along the main chain) and termination reactions (backbiting of the propagating metal carbene prevented by steric reasons).

Finally, it must be recalled that, besides polymers, cyclotrimers can be also formed. In this latter case two acetylene molecules must coordinate to the active species.[75,76] Therefore cyclotrimerization reactions should be strongly affected by the presence of bulky substituents.

## 9.5 GENERAL REMARKS

The monomer structure has a strong influence on the type of products obtained from the polymerization reactions of substituted acetylenes. Choice of suitable experimental conditions (*e.g.* catalytic system, solvent and temperature) is a stringent requirement in order to achieve high molecular weight polymers and high conversion yields, to avoid side product formation (cyclotrimers) and to get a polymer of specific microstructure. For several monosubstituted polyacetylenes, many structural investigations have been carried out and in a few cases the conformations of the repeating unit have been determined.[18,22,24] For disubstituted polymers there is still no structural information available.

Steric effects of the substituent groups on the final properties of these polyacetylenes are very important and in fact predominant in most cases. As already mentioned, unlike polyacetylene, the substituted polyenes are amorphous, soluble in many common solvents, colourless, electrically insulating and non-paramagnetic. Indeed, in these compounds very few carbon–carbon double bonds are conjugated and the polyene backbone is non-coplanar, *i.e.* polymer chains take a twisted conformation.

Of course, the differences between substituted and unsubstituted polyacetylenes are enhanced when going from mono- to di-substituted polymers, where steric effects become more severe. On the other hand, whilst polyacetylene is very sensitive to oxygen, substituted polyacetylenes are highly stable and their stability is generally improved by increasing the number and/or the bulkiness of the substituents.[77] This high stability, together with their solubility, can be exploited for practical applications, for example in the membrane separation field. Substituted polyacetylenes generally show a very high gas permeability[78] and special mention must be given to poly[1-(trimethysilyl)-1-propyne] (PTMSP).

It has been found that PTMSP has a very low density and exhibits an extremely high oxygen permeability.[60,66] As compared with poly(dimethylsiloxane) (PDMS), a rubbery polymer which had been, so far, considered the most permeable polymer to oxygen, PTMSP has oxygen permeabilities almost 10 times higher, in spite of its glassy state at room temperature ($T_g > 200\,°C$). On the other hand, although PTMSP exhibits gas diffusivities typical of soft rubbers, the gas solubility is characteristic of high $T_g$ glassy polymers. The combination of the extraordinarily high gas diffusivities and the high penetrant solubility accounts for the unusually high permeability of PTMSP.

This exceptional penetrant–transport behaviour is recognized to be a result of the low molecular packing of the polymer chains, related to both the bulkiness of the trimethylsilyl side groups and to the stiffness of the backbone chains.[79,80]

Another interesting aspect of some film-forming substituted polyacetylenes, such as poly(2-alkynes), which are degraded by $\gamma$-ray irradiation in air, is their potential applications as 'resist materials'.[3] Moreover, although most of these polyenes are electrically insulating materials, they might exhibit appreciable conductivity properties upon doping with a suitable electron acceptor agent. This is the case with poly(phenylacetylene).[81,82]

## 9.6 CONCLUSIONS

In this chapter an attempt has been made to summarize the present status of the chemistry and applications of substituted polyacetylenes. Although this is a relatively young field, there is already a considerable amount of literature. Nevertheless, many fundamental aspects, such as polymerization mechanisms and structure of the polymers, are still open to further investigation. The importance of the applicability of these polymers in various fields has already been fully recognized. The growing interest in these speciality polymers will certainly help to stimulate progress in this area.

## 9.7 REFERENCES


1. C. I. Simionescu and V. Percec, *Prog. Polym. Sci.*, 1982, **8**, 133.
2. H. W. Gibson, in 'Handbook of Conducting Polymers', ed. T. A. Scothein, Dekker, New York, 1986, vol. 1, p. 405.
3. T. Masuda and T. Higashimura, *Adv. Polym. Sci.*, 1987, **81**, 121.
4. T. Masuda and T. Higashimura, *Acc. Chem. Res.*, 1984, **17**, 51.
5. M. G. Chauser, Yu. M. Rodionov and M. I. Cherkashin, *J. Macromol. Sci., Chem.*, 1977, **A11**, 1113.
6. G. Natta, G. Mazzanti and P. Corradini, *Atti Accad. Naz. Lincei, Cl. Sci. Fis., Mat. Nat. Rend.*, 1958, **25**, 3.
7. H. Shirakawa and S. Ikeda, *Polym. J.*, 1971, **2**, 231.
8. L. S. Meriwether, E. C. Colthup, G. W. Kennerly and R. N. Reusch, *J. Org. Chem.*, 1961, **26**, 5155.
9. T. Masuda, K. Hasegawa and T. Higashimura, *Macromolecules*, 1974, **7**, 728.
10. P. S. Woon and M. F. Farona, *J. Polym. Sci., Polym. Chem. Ed.*, 1974, **12**, 1749.
11. F. R. Navarro and M. F. Farona, *J. Polym. Sci., Polym. Chem. Ed.*, 1976, **14**, 2335.
12. M. G. Voronkov, V. B. Pukhnarevich, S. P. Sushchinskaya, V. Z. Annenkova, V. M. Annenkova and N. J. Andreeva, *J. Polym. Sci., Polym. Chem. Ed.*, 1980, **18**, 53.
13. T. J. Katz and S. J. Lee, *J. Am. Chem. Soc.*, 1980, **102**, 422.
14. T. Masuda, T. Takahashi and T. Higashimura, *J. Chem. Soc., Chem. Commun.*, 1982, 1297.
15. T. J. Katz, T.-H. Ho, N.-Y. Shih, Y.-C. Ying and V. I. W. Stuart, *J. Am. Chem. Soc.*, 1984, **106**, 2659.
16. D. Meziane, A. Soum and M. Fontanille, *Makromol. Chem.*, 1985, **186**, 367.
17. Y. Huang, J. Li, Q. Zhou and J. Zhou, *J. Polym. Sci., Polym. Chem. Ed.*, 1985, **23**, 1853.
18. V. Percec, *Polym. Bull. (Berlin)*, 1983, **10**, 1.
19. W. J. Trepka and R. J. Sonnenfeld, *J. Polym. Sci., Part A-1*, 1970, **8**, 2721.
20. F. Ciardelli, S. Lanzillo and O. Pieroni, *Macromolecules*, 1974, **7**, 174.
21. T. Masuda, M. Kawasaki, Y. Okano and T. Higashimura, *Polym. J.*, 1982, **14**, 371.
22. T. Masuda, Y. Okano, K. Tamura and T. Higashimura, *Polymer*, 1985, **26**, 793.
23. T. Masuda, Y. Okano, Y. Kuwane and T. Higashimura, *Polym. J.*, 1980, **12**, 907.
24. Y. Okano, T. Masuda and T. Higashimura, *Polym. J.*, 1982, **14**, 477.
25. Y. Okano, T. Masuda and T. Higashimura, *J. Polym. Sci., Polym. Chem. Ed.*, 1985, **23**, 2527.
26. K. Hasegawa, *Eur. Polym. J.*, 1977, **13**, 315.
27. P. Cukor, J. I. Krugler and M. F. Rubner, *Polym. Prepr., Am. Chem. Soc., Div. Polym. Chem.*, 1980, **21**, 161.

28. Y. Okano, T. Masuda and T. Higashimura, *J. Polym. Sci., Polym. Chem. Ed.*, 1984, **22**, 1603.
29. D. J. Liaw, A. Soum and M. Fontanille, *Makromol. Chem., Rapid. Commun.*, 1985, **6**, 309.
30. H. Tajima, T. Masuda and T. Higashimura, *J. Polym. Sci., Polym. Chem. Ed.*, 1987, **25**, 2033.
31. T. Masuda, M. Kawai and T. Higashimura, *Polymer*, 1982, **23**, 744.
32. Zh. S. Kiyashkina, A. D. Pomogailo, A. I. Kuzayev, G. V. Lagodzinskaya and F. S. Dyachkovskii, *J. Polym. Sci., Polym. Symp.*, 1980, **68**, 13.
33. H. Muramatsu, T. Ueda and K. Ito, *Macromolecules*, 1985, **18**, 1634.
34. T. H. Ho and T. J. Katz, *J. Mol. Catal.*, 1985, **28**, 359.
35. C. Carlini and J. C. W. Chien, *J. Polym. Sci., Polym. Chem. Ed.*, 1984, **22**, 2749.
36. G. Dändliker, *Helv. Chim. Acta*, 1969, **52**, 1482.
37. T. Masuda, Y.-X. Deng and T. Higashimura, *Bull. Chem. Soc. Jpn.*, 1983, **56**, 2798.
38. T. Masuda, T. Mouri and T. Higashimura, *Bull. Chem. Soc. Jpn.*, 1980, **53**, 1152.
39. F. A. Cotton W. T. Hall, K. J. Cann and F. J. Karol, *Macromolecules*, 1981, **14**, 233.
40. T. Masuda, T. Takahashi, K. Yamamoto and T. Higashimura, *J. Polym. Sci., Polym. Chem. Ed.*, 1982, **20**, 2603.
41. T. Masuda, K. Yamamoto and T. Higashimura, *Polymer*, 1982, **23**, 1663.
42. T. Masuda, N. Sasaki and T. Higashimura, *Macromolecules*, 1975, **8**, 717.
43. M. Beneš, J. Peška and O. Wichterle, *J. Polym. Sci., Part C*, 1963, **4**, 1377.
44. S. K. M. Ehrenstorfer, *J. Polym. Sci., Part A-1*, 1969, **7**, 969.
45. J. A. Jackson, *J. Polym. Sci., Polym. Chem. Ed.*, 1972, **10**, 2935.
46. R. D. Chambers, D. T. Clark, D. Kilcast and S. Partington, *J. Polym. Sci., Polym. Chem. Ed.*, 1974, **12**, 1647.
47. T. Masuda, H. Kawai, T. Ohtori and T. Higashimura, *Polym. J.*, 1979, **11**, 813.
48. T. Masuda, Y. Kuwane and T. Higashimura, *Polym. J.*, 1981, **13**, 301.
49. T. Higashimura, Y.-X. Deng and T. Masuda, *Macromolecules*, 1982, **15**, 234.
50. T. Masuda, T. Yoshimura, K. Tamura and T. Higashimura, *Macromolecules*, 1987, **20**, 1734.
51. T. Masuda, M. Yamagata and T. Higashimura, *Macromolecules*, 1984, **17**, 126.
52. T. Masuda, K. Tamura and T. Higashimura, *J. Chem. Soc., Chem. Commun.*, 1985, 1615.
53. M. Kawasaki, T. Masuda and T. Higashimura, *Polym. J.*, 1983, **15**, 767.
54. M. Yamagata, T. Masuda and T. Higashimura, *J. Polym. Sci., Polym. Chem. Ed.*, 1984, **22**, 2275.
55. K. Tamura, T. Masuda and T. Higashimura, *Polym. J.*, 1985, **17**, 815.
56. T. Masuda, Y. Kuwane and T. Higashimura, *J. Polym. Sci., Polym. Chem. Ed.*, 1982, **20**, 1043.
57. T. Masuda, T. Takahashi, A. Niki and T. Higashimura, *J. Polym. Sci., Polym. Chem. Ed.*, 1986, **24**, 809.
58. T. Masuda, T. Takahashi and T. Higashimura, *Macromolecules*, 1985, **18**, 311.
59. A. Niki, T. Masuda and T. Higashimura, *J. Polym. Sci., Polym. Chem. Ed.*, 1987, **25**, 1553.
60. T. Masuda, E. Isobe and T. Higashimura, *Macromolecules*, 1985, **18**, 841.
61. E. Isobe, T. Masuda, T. Higashimura and A. Yamamoto, *J. Polym. Sci., Polym. Chem. Ed.*, 1986, **24**, 1839.
62. T. Masuda, A. Niki, E. Isobe and T. Higashimura, *Macromolecules*, 1985, **18**, 2109.
63. T. Masuda, E. Isobe, T. Hamano and T. Higashimura, *J. Polym. Sci., Polym. Chem. Ed.*, 1987, **25**, 1353.
64. T. Masuda, E. Isobe, T. Hamano and T. Higashimura, *Macromolecules*, 1986, **19**, 2448.
65. T. Yoshimura, T. Masuda, T. Higashimura and T. Ishihara, *J. Polym. Sci., Polym. Chem. Ed.*, 1986, **24**, 3569.
66. T. Masuda, E. Isobe and T. Higashimura, *J. Am. Chem. Soc.*, 1983, **105**, 7473.
67. K. Takada, H. Matsuya, T. Masuda and T. Higashimura, *J. Appl. Polym. Sci.*, 1985, **30**, 1605.
68. J. F. Kunzler and V. Percec, *Polym. Bull. (Berlin)*, 1987, **18**, 303.
69. V. Dragutan, A. T. Balaban and M. Dimonie, 'Olefin Metathesis and Ring-Opening Polymerization of Cyclo-Olefins', Editura Accademici and Wiley, Bucharest, 1985.
70. P. R. Hein, *J. Polym. Sci., Polym. Chem. Ed.*, 1973, **11**, 163.
71. A. B. Alimuniar, J. H. Edwards and W. J. Feast, *J. Mol. Catal.*, 1985, **28**, 313.
72. F. Garnier, P. Krausz and H. Rudler, *J. Organomet. Chem.*, 1980, **186**, 77.
73. M. Clark and M. F. Farona, *Polym. Bull. (Berlin)*, 1982, **7**, 445.
74. T. J. Katz, S. M. Hacker, R. D. Kendrick and C. S. Yannoni, *J. Am. Chem. Soc.*, 1985, **107**, 2182.
75. D. R. McAlister, J. E. Bercaw and R. G. Bergman, *J. Am. Chem. Soc.*, 1977, **99**, 1666.
76. H. Yamazaki and Y. Wakatsuki, *J. Organomet. Chem.*, 1977, **139**, 157.
77. T. Masuda, B.-Z. Tang, T. Higashimura and H. Yamaoka, *Macromolecules*, 1985, **18**, 2369.
78. T. Higashimura, T. Masuda and M. Okada, *Polym. Bull. (Berlin)*, 1983, **10**, 114.
79. Y. Ichiraku, S. A. Stern and T. Nakagawa, *J. Membr. Sci.*, 1987, **34**, 5.
80. L. C. Witchey, H. B. Hopfenberg and R. T. Chern, presented at the Annual Meeting of the AIChE, New York, November 16, 1987.
81. E. T. Kang, A. P. Bhatt, E. Villaroel, W. A. Anderson and P. Ehrlich, *J. Polym. Sci., Polym. Lett. Ed.*, 1982, **20**, 143.
82. P. Cukor, J. I. Krugler and M. F. Rubner, *Makromol. Chem.*, 1981, **182**, 165.

# 10

# Group Transfer and Aldol Group Transfer Polymerization

OWEN W. WEBSTER and DOTSEVI Y. SOGAH

*Du Pont de Nemours & Co., Wilmington, DE, USA*



## 10.1 INTRODUCTION

Two new methods for polymer chain formation involving transfer of trialkylsilyl groups have recently been reported. One gives 'living' polyacrylics at ambient temperatures, the other gives 'living' silyl-capped poly(vinyl alcohol) (PVAL). In the first method, termed group transfer polymerization (GTP),[1-7] a silyl ketene acetal initiator (**1**) reacts with a monomer by a Michael addition. During the addition, the silyl group transfers to the monomer generating a new ketene acetal function (**2**). The new ketene acetal reacts with additional monomer in a repeated fashion to give a living polymer (**3**). The process is illustrated in Scheme 1 for methyl methacrylate (MMA).

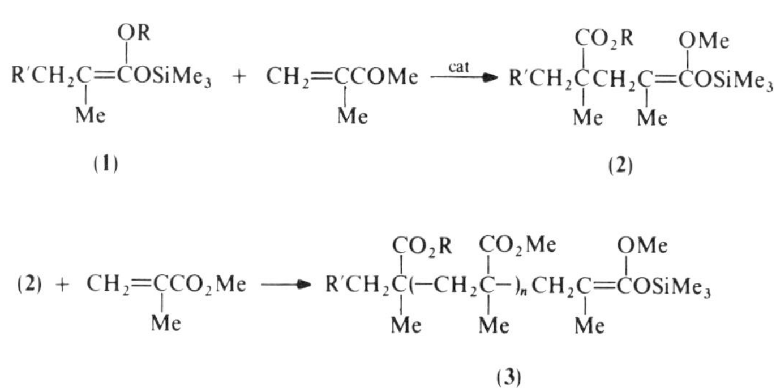

**Scheme 1**

In the second method, termed aldol group transfer polymerization (aldol GTP),[8] an aldehyde initiator reacts with a silyl vinyl ether (**4**) to give silated aldol product (**5**) in which the silyl group has transferred from the silyl vinyl ether to the aldehyde oxygen. Repeated addition of monomer gives living silated PVAL (**6**). The process is illustrated for *t*-butyldimethylsilyl vinyl ether initiated with benzaldehyde in Scheme 2.

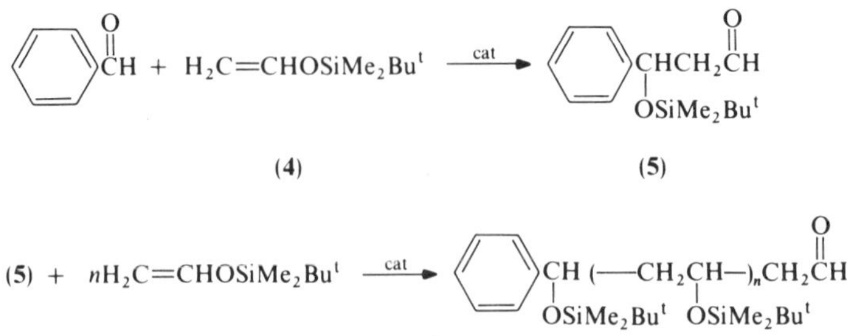

(**4**) (**5**) (**6**)

**Scheme 2**

## 10.2 THE GROUP TRANSFER POLYMERIZATION PROCESS

### 10.2.1 Monomers

Group transfer polymerization works best with methacrylates, to which most of the published data refer.[1,2,6,7] Functional groups sensitive to free radicals, such as 4-vinylbenzyl,[9] allyl and sorbyl, can be present on the ester groups of the monomer and remain unreacted in the polymer. This type of polymer is useful in air-drying finishes.

Active hydrogen compounds interfere with GTP. Thus the polymerization of methacrylic acid or hydroxyethyl methacrylate requires the use of protective groups that can be subsequently removed. Carboxy groups can be introduced by using trimethylsilyl methacrylate (**7**) as the monomer for GTP[7,10] followed by mild hydrolysis. Pendant hydroxy groups can be introduced by using 2-(trimethylsilyloxy)ethyl methacrylate (**8**) as the monomer.[4,7] The trimethylsilyl group is removed from this polymer with methanolic tetrabutylammonium fluoride. Glycidyl methacrylate can be polymerized without involving the epoxy groups provided the temperature is held below 0 °C.

$CH_2{=}C(Me)CO_2SiMe_3$ (**7**)    $CH_2{=}C(Me)CO_2CH_2CH_2OSiMe_3$ (**8**)

Acrylates,[3,11] acrylonitrile,[6] methacrylonitrile,[6] *N,N*-dimethylacrylamide,[6] pentadienoates,[7] and α-methylene-γ-butyrolactones[12] have also been polymerized by GTP. Obtaining high molecular weights with acrylates is difficult, possibly because the silyl function on the growing polymer end migrates to other positions on the chain. Acrylonitrile is extremely active in GTP and gives high molecular weight polymer, but control of molecular weight is difficult.

### 10.2.2 Initiators

To control molecular weight in a living polymerization, it is essential that the rate of initiation be higher or the same as the rate of propagation. Therefore, the best initiators will be similar in structure to the living end of the polymer. Thus for methacrylate GTP, a silyl ketene acetal with an alkyl group in the 2 position is ideal (**1**). Large alkyl groups on the silicon reduce the polymerization rate.[6]

Silyl derivatives that add to methacrylates to produce ketene acetals, for example $Me_3SiCN$ (equation 1),[1,13] $Me_3SiSMe$,[6,7,11] $Me_3SiSPh$,[6,7] $Me_3SiCR_2CN$,[6,7] or $R_2P(O)SiMe_3$[6,7] also initiate GTP.

$$Me_3SiCN + MMA \longrightarrow (\mathbf{1})\ R' = CN,\ R = Me \qquad (1)$$

### 10.2.3 Catalysts

Catalysts for GTP fall into two classes, anionic and Lewis acid. Among the anionic catalysts are fluoride, azide and cyanide,[1,7] and also carboxylates, phenolates, sulfinates, phosphinates, nitrite and cyanate.[14] Even though GTP does not proceed when protonic solvents or monomers are used, some of the better anionic catalysts are derived from the combination of an anion with its conjugate acid, *e.g.* bifluoride[1,7] or biacetate.[14] The anionic catalysts are usually used as their tetraalkylammonium or tris(dimethylamino)sulfonium (TAS) salts. They catalyze GTP by coordination with the silicon atom.[5] GTP works best with low levels of catalyst[13]—as little as 0.01% based on initiator.

Examples of Lewis acid catalysts for GTP are zinc halides, dialkylaluminum halides and tetraalkylaluminoxanes. These catalysts operate by coordination to the monomer.[3] Relatively large amounts of catalyst are required, 10% or more based on initiator. Anionic catalysts work best for methacrylates, Lewis acid catalysts for acrylates.

### 10.2.4 Polymerization Conditions

All reagents and solvents must be dry and pure and the polymerization must be conducted in a dry atmosphere. Typical solvents include toluene, tetrahydrofuran, 1,2-dimethoxyethane, chlorobenzene, and *N,N*-dimethylformamide for nucleophilic catalysis and toluene, dichloromethane and 1,2-dichloroethane for Lewis acid catalysis. Acetonitrile is silated by the initiator under some conditions and is not recommended for GTP.[13]

Temperatures as low as $-100\,^{\circ}C$ and as high as about $120\,^{\circ}C$ can be used but $0\,^{\circ}C$ to $50\,^{\circ}C$ are preferred. For polymerization of acrylates, temperatures at or below $0\,^{\circ}C$ work best. Polymerization is generally very fast and complete and can be controlled by the rate at which the monomer is added during the course of the run. With certain initiators and catalysts, induction periods are noted, for example with trimethylsilyl cyanide.

### 10.2.5 Control of Polymer Structure

The molecular weights of polymethacrylates obtained by GTP are determined by the molar ratio of initiator to monomer. Although this property of GTP allows one to *control* molecular weight accurately in the 1000 to 60 000 range, obtaining a high molecular weight polymer is more difficult since the amount of initiator needed is so low that it begins to match interfering impurity levels. By using very pure monomers, solvents, catalysts and initiators however, one can make polymer in the 100 000 to 200 000 molecular weight range. The molecular weight distribution $M_w/M_n$ for GTP polymers under the best conditions is close to one.

The principle parameters influencing tacticity of poly(methyl methacrylate) (PMMA) GTP are catalyst and temperature. The precise variation of tactic composition of PMMA GTP with polymerization temperature was determined by performing isothermal polymerizations of MMA with $TASHF_2$ as catalyst and THF as solvent.[6,10,16] The isotatic component remains small under all conditions, while the syndiotatic component increases from 50% to 80% as the temperature is lowered from $60\,^{\circ}C$ to $-90\,^{\circ}C$. The diad statistics are essentially Bernouillian. These data are almost identical with those obtained for free radical polymerization of MMA.[15]

Polymers with reactive functionality on one end are useful tools for synthesis of graft polymers, dispersing agents and ABA block polymers. They can be readily made through GTP by merely starting with a functionalized initiator: (**1**) $R'=H$, $R=SiMe_3$ gives a masked carboxylic acid end group; (**1**) $R'=H$, $R=CH_2CH_2OSiMe_3$ gives a masked hydroxyl;[2] and (**1**) $R'=$ CN, SMe, $PO(OSiMe_3)_2$ or $PhCH{=}CH_2$[20] and $R=Me$ gives a polymer with the corresponding end group.[6]

Another way to functionalize polymer chain ends is to react the living silyl ketene end group with electrophilic reagents. Examples are benzaldehyde to give a masked hydroxyl group,[7] bromine to give an α-bromo ester,[2,17] and 4-(bromomethyl)styrene to give a styryl-ended macromonomer.[18]

α,ω-Difunctional (telechelic) polymers are useful for construction of block as well as chain-extended polymers. They can be readily synthesized by coupling living GTP polymers possessing reactive end groups. Suitable coupling agents are *p*-xylylene dibromide or $Br_2$ (Scheme 3). With low molecular weight PMMA (in the 2000 range) one can obtain >99% pure telechelic polymer.[2] However, it is difficult to obtain polymer in the 10 000 to 20 000 molecular weight range of greater than 95% telechelic purity.

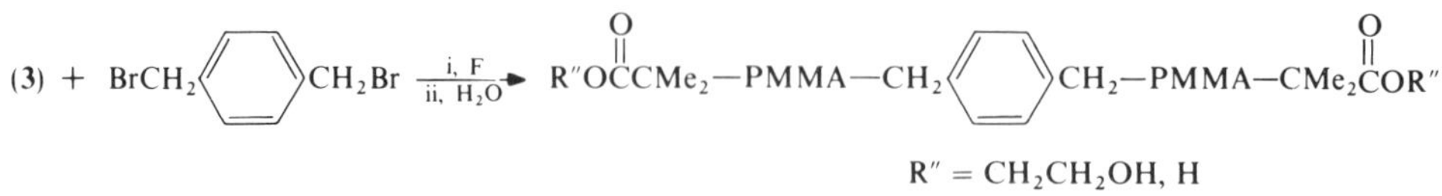

**Scheme 3**

As long as one uses monomers belonging to the same family (all methacrylate or all acrylate, for example), one can make random copolymers simply by adding a mixture of the monomers to the initiator and catalyst. The large difference in reactivity between the various acrylic monomer types (*e.g.* acrylonitrile, methacrylonitrile, acrylates and methacrylates) prevents random copolymer formation by GTP of their mixtures.

A useful attribute of 'living' polymer systems is the ability to prepare block copolymers of predetermined block length and sequence. GTP provides a particularly facile methodology for the sequencing of both homopolymer blocks and random copolymer blocks. One simply adds a new monomer or mixture of monomers when the first batch is depleted. An ABA block polymer was made by sequential addition of methacrylate monomers to the difunctional initiator (**9**).[6]

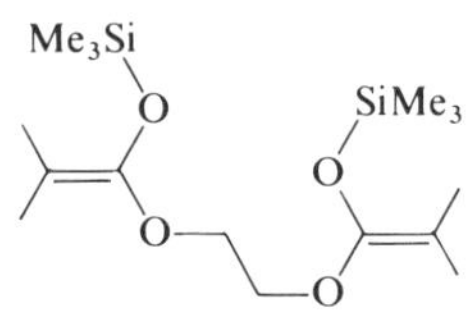

(**9**)

Superior pigment dispersants for acrylic finishes have been made by forming AB block polymers with PMMA or poly(butyl methacrylate) PBMA blocks coupled with short glycidyl methacrylate blocks. The epoxy functions are then converted to polar absorbing groups by reaction with carboxylic acids or amines (equation 2).[19]

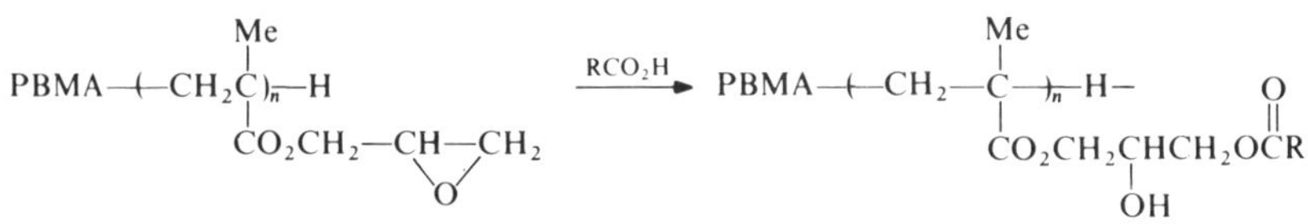 (2)

A method for obtaining acrylic/nonacrylic block polymers is to first treat living GTP polymer with bromine.[2] The resulting α-bromo ester-ended PMMA is then mixed with dimanganese decacarbonyl and irradiated in the presence of styrene to produce a mixture of AB and ABA block polymers (A = PMMA).[17]

Comb polymers have been made by two routes involving GTP. Firstly, methacryl-ended polystyrene,[20] polyphenylene oxide[9] or liquid-crystalline macromonomer[21] were polymerized by GTP. Secondly, GTP was used to make the macromonomer which was then polymerized by free radical initiation. Styryl PMMA was prepared by GTP and was then homopolymerized in 95% conversion by AIBN in benzene at 60 °C.[18]

Initiation of GTP with polyfunctional initiators to produce star polymers presents the problem of synthesis of large, reactive, polyfunctional silyl ketene acetals. An alternative approach is to generate polyfunctional initiators *in situ* by the Michael addition reaction of polyfunctional monomers with silicon reagents. Thus treatment of pentaerythritol tetraacrylate (**10**), with four equivalents of (**1**; R′ = H, R = Me) followed by excess ethyl acrylate (EA), as shown in Scheme 4, gives a four-armed star.[6] Multiarmed star polymers (**12**) can also be made by addition of a difunctional methacrylate, for example ethylene glycol dimethacrylate (**11**), to a living GTP polymer (equation 3).[19]

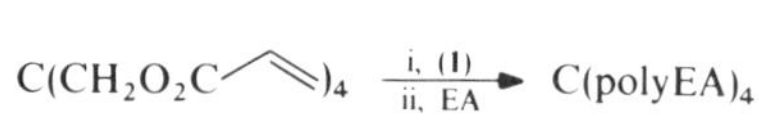

(**10**)

**Scheme 4**

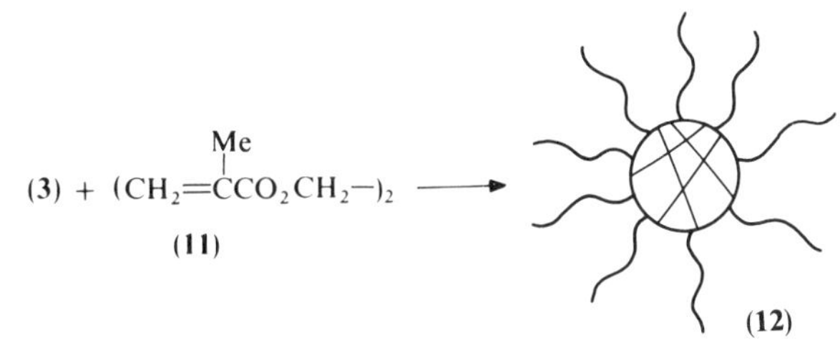

(3)

One end use for these stars is in toughening plastics and coatings. The addition of hydroxy functional stars to hydroxyl resin and then cocrosslinking produces two-phase films which are clear, flexible and hard.[19] Another end use for stars is in rheology control for application of coatings.

### 10.2.6 Mechanism

In GTP, the silyl group does not dissociate from the living ends before transfer to monomer.[5] Labeling studies have shown that trialkylsilyl fluoride does not exchange with living ends under the reaction conditions. Furthermore, the living ends do not exchange with other living ends. The exact nature of the intermediates for associative transfer has not been established, but the sequence of reactions shown in Scheme 5 is emerging as a likely pathway.

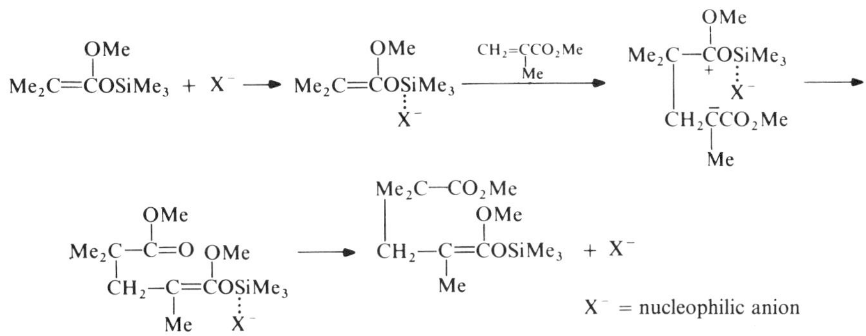

Scheme 5

### 10.2.7 Conclusions

The large number of functional methacrylates available, coupled with the numerous structural variations possible, makes GTP an extremely useful tool for polymer research. In addition, even though GTP is a complex procedure, requiring very pure reagents, dry conditions and much fine tuning, it appears that many industrial applications will be forthcoming.

## 10.3 THE ALDOL GROUP TRANSFER PROCESS

### 10.3.1 Monomers

We find that in aldol GTP the best control of molecular weight is obtained with *t*-butyldimethylsilyl vinyl ether (BDSVE). Scheme 2 outlines the polymerization process.[8] Other monomers that are polymerized include phenyldimethylsilyl, diphenylmethylsilyl, triethylsilyl and isopropyldimethylsilyl vinyl ethers. All of these give polymers in good yield with good control of molecular weight. The polymers are readily converted into PVAL by either refluxing in THF/methanol with $Bu_4NF$, or with hydrochloric acid.

### 10.3.2 Initiators

Most aldehydes will initiate aldol GTP. In general, aromatic aldehydes are preferred over aliphatic aldehydes. Electrophilic reagents which react with silyl enolates to generate an aldehyde also initiate aldol GTP, *e.g.* alkyl halides, acetals, anhydrides and active hydrogen compounds.

### 10.3.3 Catalysts

The catalysts that are operative include Lewis acids such as di(isobutyl)aluminum chloride, titanium tetrachloride, tin tetrachloride, boron trifluoride, zinc bromide, zinc iodide and zinc chloride. The preferred catalysts are the zinc halides. Anion catalysts, such as tris(dimethylamino)sulfonium bifluoride, tetrabutylammonium bifluoride and tetrabutylammonium acetate give less satisfactory results.

### 10.3.4 Polymerization Conditions

Nonreactive chlorinated and nonpolar aromatic solvents, such as dichloromethane, 1,2-dichloroethane, toluene, xylenes and benzene, are preferred for Lewis acid catalysts. Only about $10^{-4}$ to $10^{-2}$ mol % catalyst relative to monomer is needed. Under typical conditions, monomer is added slowly to a mixture of catalyst, initiator and solvent at ambient temperatures. An exotherm is observed and complete conversions are obtained in a few hours.

### 10.3.5 Control of Polymer Structure

As with Michael addition GTP, the molecular weights obtained in aldol GTP are controlled by the molar ratio of monomer to initiator. Good control is obtained in the intermediate molecular weight range of about 20 000 but lower than theoretically predicted molecular weights are obtained in the 100 000 range. In contrast to GTP where impurities tend to increase molecular weights, in aldol GTP impurities generate more living ends and tend to depress molecular weight and broaden the dispersity. Typically, molecular weight dispersities for aldol GTP range from 1.02 to 1.7.

Aldol GTP gives linear silated PVAL chains with no head to head links. The chain stereochemistry appears to be random, like that produced by free radical-initiated vinyl acetate.

Since the living end of the polymer produced by aldol GTP is an aldehyde group, one can convert it into other groups by using one of the classic transformations available in aldehyde chemistry. Aldehyde groups can be obtained on both ends of the polymer by initiation with terephthalaldehyde.[8] A functional group present on the initiator will be on the end of the chain opposite the aldehyde groups. Thus cinnamaldehyde introduces a terminal alkene group while *trans*-retinal introduces a terminal conjugated diene. Before removing the silyl groups to obtain PVAL one must cap the aldehyde function to prevent hemiacetal formation. This can be easily done by reaction with (**1**; R′ = H, R = Me) to form the Reformatski product.[8]

Synthesis of block copolymers by sequential addition of monomers is facile. Thus the block copolymer of BDSVE and trimethylsilyl vinyl ether (TMSVE) is prepared by first polymerizing BDSVE, followed by addition and polymerization of TMSVE. The trimethylsilyl groups can be hydrolyzed selectively to give a poly(vinyl alcohol)/poly(vinyl silyl ether) block polymer. A block polymer of PVAL and MMA was prepared by coupling the aldehyde group of a silylated PVAL (**6**) with living GTP polymer (**3**), followed by hydrolysis of silyl groups (Scheme 6).

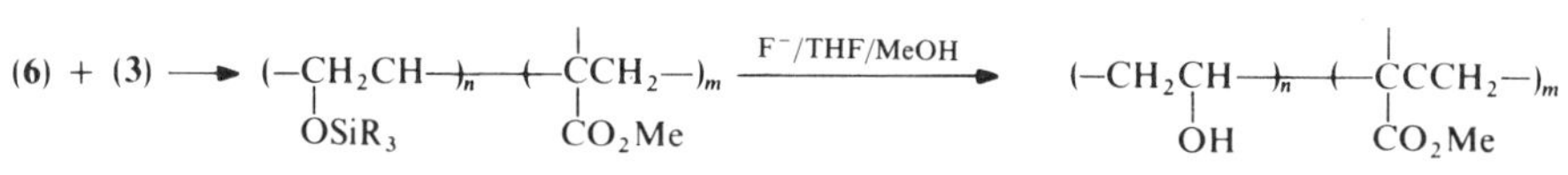

**Scheme 6**

One can also use living polymers prepared by anionic polymerization, in conjunction with aldol GTP, to prepare block copolymers of well-defined segments. Poly(isoprene-*block*-vinyl alcohol) and poly(styrene-*block*-vinyl alcohol) are thus prepared readily by initiation with aldehyde-ended polyisoprene or polystyrene.

Using multifunctional polymeric initiators, one can graft polymer chains to an appropriate backbone. For example, when a polymer with a pendant aldehyde function is used to initiate aldol GTP, one obtains a comb polymer containing a methacrylate backbone and pendant-silylated poly(vinyl alcohol).[8]

When 1,3,5-tris(bromomethyl)benzene and 1,2,4,5-tetrakis(bromomethyl)benzene are used to initiate aldol GTP, star polymers containing three and four arms are obtained.[8]

### 10.3.6 Mechanism

Lewis acids are presumed to catalyze the polymerization by activating the initiator through coordination to the carbonyl functional group. Reaction of the coordinated species with the silyl vinyl ether, followed by the silyl transfer, gives the desired product (Scheme 7). The silyl transfer could occur either in a concerted fashion, *via* the six-membered cyclic $\beta$-silyl-stabilized carbocation intermediate, or as a two-step process.

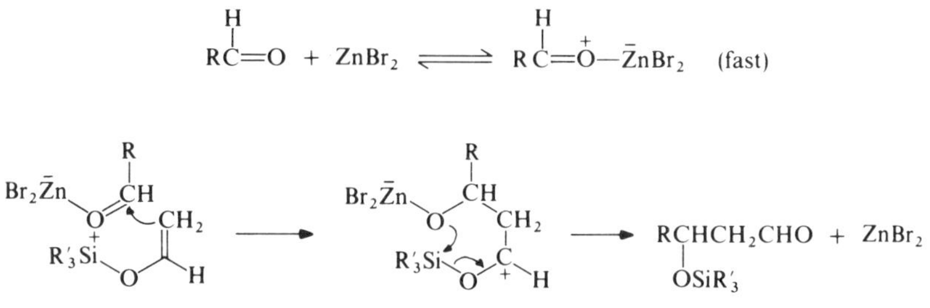

**Scheme 7**

### 10.3.7 Conclusions

The results of aldol GTP demonstrate another instance where organic synthetic methods can be applied to the controlled synthesis of living polymers. The important distinction is that most carbon–carbon bond-forming reactions are not quantitative enough to be used in the formation of high polymers. The aldol GTP clearly has advantages over the previously known methods for synthesis of poly(vinyl alcohol) segments. These advantages include complete monomer conversion over a broad temperature range, living polymer formation, facile block copolymer preparation, good molecular weight control and synthesis of polymers of unusual structures.

## 10.4 REFERENCES


1. O. W. Webster, W. R. Hertler, D. Y. Sogah, W. B. Farnham and T. V. RajanBabu, *J. Am. Chem. Soc.*, 1983, **105**, 5706.
2. D. Y. Sogah and O. W. Webster, *J. Polym. Sci., Polym. Lett. Ed.*, 1983, **21**, 927.
3. W. R. Hertler, D. Y. Sogah, O. W. Webster and B. M. Trost, *Macromolecules*, 1984, **17**, 1415.
4. O. W. Webster, W. R. Hertler, D. Y. Sogah, W. B. Farnham and T. V. RajanBabu, *J. Macromol. Sci., Chem.*, 1984, **21**, 943.
5. D. Y. Sogah and W. B. Farnham, in 'Organosilicon and Bioorganosilicon Chemistry: Structure, Bonding, Reactivity and Synthetic Application', ed. H. Sakurai, Wiley, New York, 1985, chap. 20.
6. D. Y. Sogah, W. R. Hertler, O. W. Webster and G. M. Cohen, *Macromolecules*, 1987, **20**, 1473.
7. O. W. Webster (Du Pont), *US Pat.* 4 417 034 (1983). (*Chem. Abstr.*, 1984, **100**, 86 327); 4 508 880 (1985) (*Chem. Abstr.*, 1985, **102**, 221 336); W. B. Farnham and D. Y. Sogah (Du Pont), *US Pat.* 4 414 372 (1983); (*Chem Abstr.*, 1984, **100**, 68 964); 4 581 428 (1986) (*Chem. Abstr.*, 1986, **105**, 98 133).
8. D. Y. Sogah and O. W. Webster, *Macromolecules*, 1986, **19**, 1775; D. Y. Sogah and O. W. Webster (Du Pont), *US Pat.* 4 544 724 (1985) (*Chem. Abstr.*, 1986, **104**, 89 196).
9. C. Pugh and V. Percec, *Polym. Prepr., Am. Chem. Soc., Div. Polym. Chem.*, 1985, **26**, 303.
10. Y. Wei and G. Wnek, *Polym. Prepr., Am. Chem. Soc., Div. Polym. Chem.*, 1987, **28**, 252.
11. M. T. Reetz, R. Ostarek, K.-E. Piejko, P. Arlt and B. Bömer, *Angew. Chem., Int., Ed. Engl.*, 1986, **25**, 1108.
12. J. Suenaga, D. M. Sutherlin and J. K. Stille, *Macromolecules*, 1984, **17**, 2913.
13. F. Bandermann and H. D. Speikamp, *Makromol. Chem., Rapid Commun.*, 1985, **6**, 335.
14. I. B. Dicker, W. B. Farnham, W. R. Hertler, E. D. Laganis, D. Y. Sogah, T. W. Del Pesco, P. H. Fitzgerald and I. B. Dicker (Du Pont), *US Pat.* 4 588 795 (1986) (*Chem. Abstr.*, 1986, **105**, 98 135).
15. T. G. Fox and H. W. Schnecko, *Polymer*, 1962, **3**, 575.
16. M. A. Mueller and M. Stickler, *Makromol. Chem., Rapid Commun.*, 1986, **7**, 575.
17. G. C. Eastmond and J. Grigor, *Makromol. Chem., Rapid Commun.*, 1986, **7**, 375.
18. R. Asami, M. Takaki and Y. Moriyama, *Polym. Bull. (Berlin)*, 1986, **16**, 125.
19. J. A. Simms and H. J. Spinelli, *J. Coat. Technol.*, 1987, **57**, 125; J. A. Simms and H. J. Spinelli (Du Pont), *PCT Int. Pat. Appl.*, WO86 626 (1986) (*Chem. Abstr.*, 1986, **105**, 24 811).
20. R. Asami, Y. Kondo and M. Takaki, *Polym. Prepr., Am. Chem. Soc., Div. Polym. Chem.*, 1986, **27**, 186.
21. W. Kreuder, O. W. Webster and H. Ringsdorf, *Makromol. Chem:, Rapid Commun.*, 1986, **7**, 5.

# 11

# Polymerization in Emulsions

DONALD H. NAPPER and ROBERT G. GILBERT

*University of Sydney, NSW, Australia*

## 11.1 INTRODUCTION

The early realization that a replacement for rubber might be achieved by dispersing the monomer and/or polymer in an aqueous medium (by analogy with a natural rubber latex) led to the development of the related processes of emulsion polymerization and suspension polymerization. Emulsion polymerization underwent initial development over the period 1930–1950, the driving force being the necessity to find a replacement for natural rubber. The first attempts to develop the process were made in Germany during the First World War. Thereafter Luther and Heuck in 1932[1] presented the first viable emulsion polymerization method. Major developments took place during the Second World War. A historical summary has been given by Blackley.[2] Emulsion polymers are made in huge annual tonnages, and are even replacing metals in some applications. They are used *inter alia* to manufacture synthetic rubber, the latex for several different types of water-based paints, adhesives, and so on.

In bulk or solution polymerization, the reaction is usually carried out in a homogeneous medium. Emulsion polymerizations are carried out in heterogeneous systems; commonly, an aqueous phase and a non-aqueous phase. Usually, the monomer and polymer constitute the non-aqueous phases, although in inverse systems[3] water-soluble monomers are polymerized in a non-aqueous medium. In its simplest form, an emulsion polymerization system is made up of water, monomer(s), emulsifier (surfactant) and initiator; typically 30% monomer, 65% water and the remainder surfactant, modifier (*e.g.* chain transfer agent) and other additives. The monomer must be sparingly soluble in water and generate a water-insoluble polymer which (usually) is swollen by the monomer. One or more monomers may be used, such as styrene, vinyl acetate, methyl methacrylate (*e.g.* automotive lacquer), tetrafluoroethylene (for Teflon), and a huge variety of copolymers such as a 1:4 styrene/butadiene blend (SBR). Normally a water-soluble initiator is used so that polymerization takes place in the latex particles rather than in the emulsion droplets. The latexes exploited in the manufacture of paint usually contain various rubbery and glassy polymers, often with a carefully designed core-shell morphology.

Potassium persulfate can be used as a free-radical initiator for high temperatures, whereas at low temperatures a redox system, such as $Fe^{II}$ and cumene hydroperoxide, is usually employed. Industrially, it is usual to perform emulsion polymerizations at low temperatures; in the case of synthetic rubbers, this reduces the amount of branching of the polymer chain, which in turn reduces the time required to incorporate carbon black into the rubber.

Surfactant (emulsifier) is used to enhance the colloid stability of the polymer latex particles which are the loci of polymerization. There are three types of surfactant: (i) electrostatic stabilizers such as anionic or cationic soaps; (ii) polymeric (steric) stabilizers such as partially hydrolyzed poly(vinyl alcohol); and (iii) electrosteric stabilizers, such as polyelectrolytes which combine both types (i) and (ii). A typical anionic surfactant is sodium dodecyl sulfate (SDS), whilst industrial surfactants are often complex, being typically mixtures of hydrogenated tallow fatty acids and rosen acids. Polymeric stabilizers ('protective agents') are used to provide, for example, good freeze–thaw stability. 'Modifiers' are chain-transfer agents added to control the molecular weight.

A typical laboratory recipe is 270 g styrene, 600 g water and 10.5 g SDS; 1 g $NaHCO_3$ may also be added to control pH. Polymerization is carried out at 90 °C. Initiator ($K_2S_2O_8$) is added at a concentration of *ca.* $10^{-3}$ mol dm$^{-3}$. Emulsification (either ultrasonically or by mechanical agitation) initially results in micelles, swollen with solubilized monomer and surfactant-stabilized monomer droplets ($10^{12}$–$10^{14}$ dm$^{-3}$). The micelles are usually 5–10 nm in size whereas the droplets are of order 1–10 $\mu$m. After the initiator has commenced its (thermal) decomposition, a new phase appears, *i.e.* latex particles. These consist of a number of polymer molecules of fairly high degree of polymerization, stabilized by surfactant and swollen with monomer. These latex particles (usually) are the principle loci of all subsequent polymerization; their number density is typically $10^{16}$–$10^{18}$ dm$^{-3}$. This contrasts with the number density of micelles which is $10^{19}$–$10^{21}$ dm$^{-3}$. In fact, the term 'emulsion polymerization' is a misnomer (arising from the early misconception that polymerization took place in the emulsion droplets where much of the monomer resides, which, as such, is in fact not usually the case).

The advantages of emulsion polymerization include the following. (i) The heat generated by the exothermic free-radical polymerization process can be readily dissipated in the aqueous phase, so that no particularly stringent precautions are required to prevent 'run-away' overheating. (ii) The rate of polymerization is usually considerably faster than in an equivalent bulk process. (iii) The formed polymer often has a considerably higher molecular weight than that from an equivalent bulk process, as well as a different molecular weight distribution (Section 11.8.5). (iv) The polymer is formed as a latex, and can thus be more readily handled; the formed polymer can be obtained by evaporating the water (as occurs when latex paint dries on a wall). (v) The molecular weight is easily controlled by the addition of chain transfer agents which gives ready control of the properties of formed polymer such as mechanical strength. (vi) The process itself, and the resulting polymer latex, is water based rather than solvent based, which minimizes safety and environmental hazards.

The emulsion polymerization process also has certain inherent disadvantages. For example, it is usually difficult (or expensive) to remove the surfactant and the initiator decomposition products, which are therefore usually present in the final polymer product. These may contribute to, for example, undesirable colour on moulding.

## 11.2 CHARACTERISTICS AND CHARACTERIZATION OF EMULSION POLYMERIZATION SYSTEMS

The first reasonably accurate qualitative model of an emulsion polymerization system was given by Harkins,[4] which was shortly thereafter quantified by Smith and Ewart.[5] A problem in the understanding of emulsion polymerization is that it is necessary to take into account a large number of physically reasonable mechanisms involving initiator decomposition, latex particle formation, polymerization within the particles and aqueous phase events. These involve complex phenomena in both colloid and polymer science. Often there are only a limited number of data sets available (such as the steady state polymerization rate as a function of initiator concentration), and theoretical models often contain more unknown parameters than can be properly obtained from the data. A number of assumptions are then required in order to interpret these results, and these assumptions are, in such an approach, quite untestable. Over the years, many notions concerning the mechanism of, for example, latex particle formation, have assumed the status of unquestioned dogma, because any set of data can be fitted by a theory containing a sufficient number of adjustable parameters. A frequently used means of prediction or interpretation of emulsion polymerization characteristics has been through extensive modelling (the 'global' approach); a set of mechanisms is assumed, and the kinetic parameters therein adjusted to ensure agreement with the limited data usually so employed. The presentation given here involves a different approach; where possible, it is designed to ensure that sufficient independent data are available to test mechanistic postulates.

As many as three phases can be present in an emulsion polymerization system: an aqueous phase (containing initiator, surfactant micelles and a small amount of the sparingly soluble monomer); emulsion droplets (containing monomer); and latex particles (containing polymer and some or all unpolymerized monomer). It is therefore convenient to adopt to the following three classifications describing the time evolution of an emulsion polymerization process. These are summarized in Table 1.

*Interval I.* This is the initial stage: the formation of latex particles.

*Interval II.* This is the second stage, which commences when the nucleation of particles is complete. Three phases are now present: aqueous phase, monomer-swollen latex particles and monomer droplets. Polymerization is essentially confined to the latex particles. The particles tend to maintain a constant monomer concentration, because there is a balance between the free energy of mixing and surface energy effects. To maintain this approximately constant monomer/polymer ratio in the latex particles, monomer migrates from the emulsion droplets into the latex particles, through the aqueous phase.

*Interval III.* This is the final stage. As polymerization progresses and monomer is consumed, the monomer droplets eventually disappear, and virtually all remaining monomer is now confined to the latex particles; only latex particles and the aqueous phase (which also contains some dissolved monomer) are present. The concentration of monomer in the latex particles now decreases in time, rather than remaining approximately constant as in Interval II.

These intervals are apparent in a conversion/time plot, as in Figure 1. There is initially an induction period (not shown in the figure) when the primary free radicals arising from the decomposition of initiator consume residual oxygen. There is then a period of increasing rate: Interval I. The reason for the rate increasing is simply that the particle number increases during nucleation. Thereafter follows a period when the rate is approximately constant: Interval II. The rate in this interval is approximately constant because both the number of particles and the monomer concentration within the latex particles are approximately constant. Then follows a period when the rate starts to decrease: Interval III. The decrease in rate occurs because the concentration of monomer within the latex particles is now decreasing monotonically, since droplets are no longer present. Often thereafter there is an increase in the rate. This is still designated Interval III; the

**Table 1** Classifications Describing the Time Evolution of an Emulsion Polymerization

| *Interval* | *% conversion range* | *Micelles* | *Monomer droplets* | *Particle number* | *Comments* | *Particle growth* |
|---|---|---|---|---|---|---|
| I | 0–10 | present | present | increase | nucleation period | increase in size |
| II | 10–40 | absent | present | constant | monomer concentration constant in particles | increase in size |
| III | 40–100 | absent | absent | constant | monomer concentration decreases in particles | size usually decreases slightly |

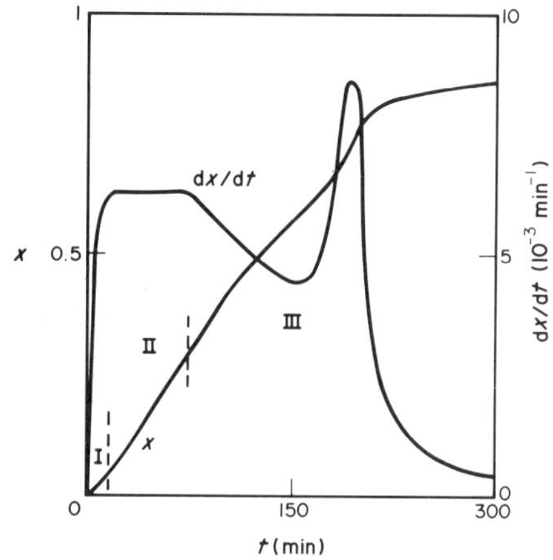


**Figure 1** Typical plot of fraction conversion ($x$) and rate ($dx/dt$) against time in an emulsion polymerization system

increase is due to the gel effect[6,7] (also evident in bulk polymerization systems, see Volume 3, Chapter 12). Eventually, the rate may decrease at very high conversion, if the propagation rate coefficient, $k_p$, becomes diffusion controlled; this is also illustrated in Figure 1.

The following argument shows that the locus of polymerization in an emulsion system is predominantly inside the latex particles, with little polymerization occurring elsewhere. Possible polymerization loci are: (i) the aqueous phase; (ii) the micelles; (iii) the monomer droplets; and (iv) the latex particles.

(i) *Aqueous phase.* The monomers used in emulsion polymerizations are usually only sparingly soluble in water. The polymerization rate per free radical in the aqueous phase is $k_p[M_{aq}]$, where $[M_{aq}]$ is the concentration of monomer in the aqueous phase. Typical values of this product are orders of magnitude less than that observed in an emulsion polymer system; hence the major locus of polymerization cannot be the aqueous phase.

(ii) *Micelles.* Analysis of an emulsion polymerization shows that micelles disappear at less than 10% conversion, so most of the polymer is generated in the absence of micelles.

(iii) *Monomer droplets.* The polymer formed in an emulsion polymer system is finally found to be located in the latex particles. Since polymer is generally soluble in monomer, the polymer could not possibly be first formed in the monomer droplets, which are considerably larger (typically microns) than the latex particles (tenths of microns), since there would be no driving force making the formed polymer then migrate to the latex particles. Moreover, when carrying out an Interval II seeded emulsion polymerization (when both monomer droplets and particles are present and no new particles are formed) one finds the formed polymer is essentially indistinguishable from that in an *ab initio* system. Finally, the number of particles formed in an *ab initio* system is orders of magnitude larger than the number of emulsion droplets, and these latex particles are much smaller in diameter than are the emulsion droplets; thus the surface area of the latter is smaller, and aqueous phase free radicals are much more likely to enter the particles than the micelles.

(iv) *The latex particles.* These remain as the only possible major polymerization locus. There is abundant evidence for this; for example, the polymerization rate is usually proportional to $N_c$, the particle number density (although as will be seen later this dependence is often complicated by variations in the average number of free radicals per particle).

Note that the locus of polymerization may not necessarily be uniformly distributed throughout the latex particles. For example, one may have a 'core-shell' morphology where most of the growth occurs in the peripheral regions of the particles; this is to be expected if one polymer is used as a seed and another monomer, whose polymer is incompatible with the first, is then polymerized onto this seed.[8] In addition certain systems, such as acrylonitrile, are characterized by the monomer and polymer being mutually incompatible, in which case it appears[9–11] that the locus of polymerization is the particle/continuous phase interfacial region.

It has been claimed[12] that chain growth in a styrene system actually occurs only in a shell near the particle surface, resulting in the so-called 'core-shell' morphology. However, it has been demonstrated that this conclusion was based on an erroneous interpretation.[13,14] However, copolymerization systems are commonly designed to have a core-shell type of morphology.

The various quantitative observables that may be used to study an emulsion polymerization system are as follows: rate, particle size distributions, molecular weight distributions, and (for copolymerization systems) the copolymer sequence distributions.

## 11.3 KINETICS OF PARTICLE GROWTH

For both technical and investigative reasons, it is often desirable to separate particle formation from particle growth. In an *ab initio* study, all three intervals may occur. However, in a *seeded* system, one is able to choose conditions so as to start in either Interval II or Interval III, by taking sufficient quantity of a seed latex (formed in a separate *ab initio* run) which is then further polymerized, without renucleation occurring, by the addition of more monomer, surfactant and initiator. In addition to its use in mechanistic investigations, seeded polymerizations are frequently used commercially to control latex or polymer characteristics. Seeded growth obviates the necessity of accounting for the complexities of the nucleation process, and for this reason is discussed first.

A typical recipe for a seeded study of styrene is as follows: 6 g monomer, 2 g of seed particles of unswollen radius 47 nm (the weight refers to the solids content of the latex), 83 g water and 0.35 g of sodium dodecyl sulfate as surfactant. There is a difference between the swollen radius of the monomer-saturated latex particles (which is their radius in the reacting system) and their unswollen radius (which is that after unreacted monomer is removed).

### 11.3.1 Theoretical Description

We now set up the means of computing the rate of polymerization in Intervals II and III (no new particle formation), given the various rate parameters governing this process. This firstly permits the simulation of particle growth (*i.e.* modelling), and secondly creates means of determining these rate parameters (and hence overall mechanisms) from appropriate experimental data.

During Intervals II and III, the particle number density ($N_c$) remains constant. The rate of polymerization in the latex particles is given by equation (1), where $x$ is the fractional conversion of monomer to polymer, $C_M$ is the monomer concentration within the particles (or at the locus of polymerization), $n_M^\circ$ is the number of moles of monomer initially present and $N_A$ is Avogadro's constant. In Interval II, $C_M$ and $k_p$ are both constant, at least to within an excellent approximation. In Interval III, $C_M$ decreases, and one has (for a monomer that is sparingly soluble in water) $x = 1 - (C_M/C_M^0)$, where $C_M^0$ is the concentration of monomer in the particles at the start of Interval III, and where for convenience we set $x = 0$ at the start of Interval III. Thus in Interval III, $\mathrm{d}\ln(1-x)/\mathrm{d}t = -(k_p/N_A V_s)\bar{n}(x, t)$, where $\bar{n}$ is the average number of free radicals per particle and $V_s$ is the swollen volume of the latex particles. This is the same as the previous expression except that $x$ is replaced by $-\ln(1-x)$ and $A$ by $k_p/N_A V_s$, as in equation (2). Thus the analyses of Intervals II and III are the same, except that for the latter instead of $x(t)$ one uses $-\ln(1-x)$, and $k_p/(N_A V_s)$ instead of $A$. Note that $\bar{n}$ and (in Interval III) $k_p$ may vary with conversion.

$$\mathrm{d}x/\mathrm{d}t = k_p N_c C_M/(n_M^\circ N_A)\bar{n} \equiv A\bar{n} \quad \text{(Intervals I, II, III)} \tag{1}$$

$$-\mathrm{d}\ln(1-x)/\mathrm{d}t = (k_p/N_A V_s)\bar{n} \quad \text{(Interval III)} \tag{2}$$

The $\bar{n}$ is given by equation (3), where $N_n$ is the number of latex particles containing $i$ free radicals; for convenience one adopts the normalization $\Sigma N_n = 1$. The time evolution of the $\bar{n}$ is dictated by the following kinetic events within the particles:[5] entry of free radicals into the latex particles (pseudo-first-order rate coefficient $\rho$, *i.e.* the average number of free radicals entering a single latex particle per second); exit (desorption) of free radicals from the particles (first-order rate coefficient $k$, *i.e.* the average fraction of free radicals exiting from a latex particle per second); and bimolecular termination between two free radicals in a particle (pseudo-first-order rate coefficient $c = k_t/N_A V_s$, where $k_t$ is the second-order termination rate coefficient). One thus has the Smith–Ewart[5] equations for the time evolution of the $N_n$, (equation 4) derived as follows. Entry converts a particle containing $n$ free radicals into one containing $n+1$, hence the first term. Exit converts a particle containing $n+1$ free radicals into one containing $n$; moreover, since $k$ is the rate coefficient for exit of a single free radical, then the rate of exit in a particle containing $n+1$ free radicals is $(n+1)k$, hence the second term. Finally, termination converts a particle containing $n+2$ free radicals into one containing $n$; if $c$ is the termination rate coefficient per free radical, the total termination rate (involving the pairwise loss of free radicals) in a particle containing $n+2$ free radicals is proportional to $2({}^{n+2}C_2)=(n+2)(n+1)$, hence the last term.

$$\bar{n} = \sum nN_n \tag{3}$$

$$dN_n/dt = \rho[N_{n-1} - N_n] + k[(n+1)N_{n+1} - nN_n] + c[(n+2)(n+1)N_{n+2} - n(n-1)N_n] \tag{4}$$

These equations are frequently formulated with specific mechanistic assumptions inherent in their form; for example, it is often assumed that $k$ is proportional to the surface area of the particles, or that $\rho$ is proportional to the initiator concentration. Such mechanistic assumptions are eschewed in the present treatment, since they may often be unjustified and their use can therefore lead to misinterpretation. Instead, minimal assumptions are made as to the dependence of any rate parameter on quantities such as particle size, *etc.*, and it will be seen that it is possible to deduce any dependences experimentally.

The Smith–Ewart equations must take into account events in the aqueous phase, which have an important effect in emulsion polymerization kinetics. Even after nucleation has ceased, the kinetics in an emulsion polymerization (as in any heterogeneous system) are extremely complicated. It has been shown[15–18] that these complexities can be condensed into a single quantity termed the 'fate parameter' $\alpha$, resulting in a great simplification. A summary of the reduction is as follows. The steps involved in free radical entry into a latex particle are likely to be: (1) production of primary free radicals from initiator; (2) aqueous phase propagation of the free radicals resulting from process (1) to form surface-active oligomeric free radicals or to precipitate homogeneously from the aqueous phase to form small colloidal particles; (3) homotermination among the free radical species formed in processes (1) and (2) (this leads to the capture efficiency being less than 100%); (4) entry of the oligomeric free radicals (of sufficiently high degree of polymerization) or of colloidal, precipitated entities into the latex particles; (5) aqueous phase free radicals that have excited from the latex particles undergoing subsequent re-entry into a particle; and (6) aqueous phase free radicals which arise from exit undergoing heterotermination with free radicals arising from processes (1) or (2). The overall entry rate coefficient $\rho$ can then be shown[16–18] to be given by equation (5). Here, $\rho_A$ is the component of the entry rate coefficient arising entirely from free radicals generated from initiator, *i.e.* in the absence of exit. The dimensionless quantity $\alpha$ is a 'fate parameter' which takes account of the effect of the aqueous phase kinetics of desorbed free radicals on the entry process; it is confined to the range $-1 \le \alpha \le +1$. It is a complicated function of the various rate coefficients for the aqueous phase processes,[16–18] but its physical implication is that it gives the relative importance of heterotermination ($\alpha = -1$) and re-entry ($\alpha = +1$) in the aqueous phase kinetics. The physical justification of these limits is obvious, as follows. If all exited free radicals undergo re-entry into a latex particle, this must result in an increase in the rate of entry arising solely from initiator by an amount $k\bar{n}$: $\rho = \rho_A + k\bar{n}$: (*i.e.* $\alpha = +1$). If all excited free radicals undergo heterotermination with a free radical arising from initiator, this must result in a decrease in entry by an amount $k\bar{n}$: $\rho = \rho_A - k\bar{n}$: (*i.e.* $\alpha = -1$). Equation (5) encapsulates all the complexities of the aqueous phase kinetics into a simple form. Thus, instead of having to find or fit all the rate parameters for the aqueous phase processes, it is merely necessary to obtain $\alpha$. The extension of the Smith–Ewart treatment (equation 4) by equation (5), seems sufficiently general to describe most of the particle growth kinetics in emulsion polymerization systems (or indeed any compartmentalized heterogeneous free-radical polymerization), and is referred to as the 'generalized Smith–Ewart' mechanism (certain extensions and restrictions are given in Section 11.7).

$$\rho = \rho_A + \alpha k\bar{n} \tag{5}$$

Steady state solutions to equations (3) and (4) have been obtained[19,20] for the case where $\alpha = 0$, in terms of modified Bessel functions of irrational order, $I_m(a)$, as given by equation (6). Here $a=(8\rho/c)^{1/2}$, $m=k/c$. The Bessel functions are best evaluated using the continued fraction method of Ugelstad *et al.*[21] No analytic steady state solution has been found for the general case where $\alpha \neq 0$; however, a readily implemented numerical method has been given.[22] Figure 2 shows the dependence of $\bar{n}_{SS}$ as a function of the dimensionless parameters $\rho_A/c$ and $k/c$, for $\alpha = +1$.

$$\bar{n}_{SS} = aI_m(a)/4I_{m-1}(a) \tag{6}$$

Equations (1)–(5) admit of no general time-dependent solution, *inter alia* because the various rate coefficients may vary with conversion (*e.g.* the value of $k_t$, and hence $c$, in the gel regime). It is however a straightforward matter to obtain numerical solutions although, since the equations are

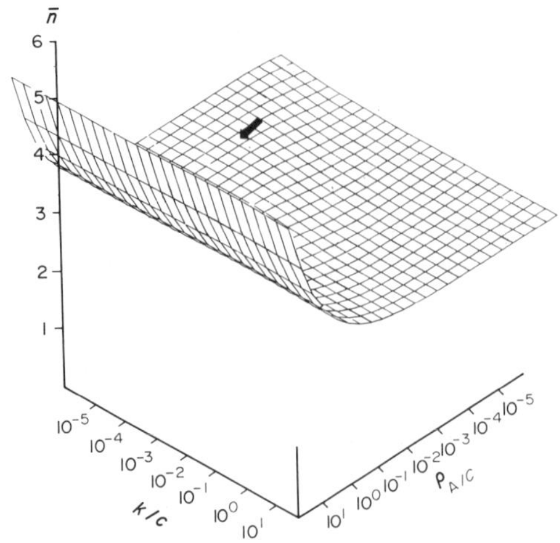


**Figure 2** Steady state $\bar{n}$, as function of dimensionless quantities $\rho_A/c$ and $k/c$, from steady state solution of equations (2) and (3), for $\alpha = 1$; arrow indicates plateau where $\bar{n} = 0.5$ (obtained by method of M. J. Ballard, R. G. Gilbert and D. H. Napper, *J. Polym. Sci., Polym. Lett. Ed.*, 1981, **19**, 533)

stiffly coupled, an appropriate algorithm must be used (*e.g.* the Gear method).[23] The numerical solution requires the truncation of equation (5) at some large value of $n$, denoted $r$. This truncation is made accurate by using the 'instantaneous termination' approximation for $N_{r+1}$; one assumes that entry into a particle containing $r+1$ free radicals results in instantaneous reduction of the free-radical number by two. This results in a slight modification of the generalized Smith–Ewart equations for $N_r$ and $N_{r-1}$, as given in equations (7) and (8). The Gear algorithm solutions of these equations is efficient and numerically stable. Less sophisticated methods of integration (*e.g.* the crude Euler method[24]) are inherently unstable, and will often lead to large errors.

$$dN_{r-1}/dt = \rho N_{r-2} - [\rho + (r-1)k + (r-1)(r-2)c]N_{r-1} + rkN_r + \rho N_r \tag{7}$$

$$dN_r/dt = \rho N_{r-1} - [\rho + rk + r(r-1)c]N_r \tag{8}$$

Various approximate time-dependent solutions of equations (1)–(4) have been discussed by a number of authors.[15–17,25] The most important cases are as follows.

#### *11.3.1.1 High* $\bar{n}$ *limit: the pseudo-bulk equation*

When $\bar{n}$ is comparatively large, the system obeys 'pseudo-bulk' kinetics. Normally, the fact that the free radicals are in isolated latex particles (termed 'compartmentalization') has important kinetic implications. However, in high $\bar{n}$ systems, the effects of compartmentalization can (under certain circumstances) be neglected, *i.e.* the system effectively obeys bulk kinetics. In such a system, equation (4) is replaced by equation (9). This treats the average number of free radicals (per particle) as a variable which is governed by a zeroth-other production (at rate $\rho$), first-order loss (with rate $k\bar{n}$) and second-order loss by termination (with rate $2c\bar{n}^2$). This equation, when valid, obviates the need for determining $\bar{n}$ from the more complicated Smith–Ewart equations. The derivation and conditions of validity of equation (9) have been given by Ballard *et al.*[22,26] There are two regions of validity: when $\bar{n}$ is high (when each individual particle behaves like a pseudo-bulk system); and when both $\alpha \approx +1$ and $k \gg \rho, c$.

$$d\bar{n}/dt = \rho - k\bar{n} - 2c\bar{n}^2 = \rho_A + (\alpha - 1)k\bar{n} - 2c\bar{n}^2 \tag{9}$$

#### *11.3.1.2 Low* $\bar{n}$ *limit: the zero–one system*

In this limit, $c$ is so large that mutual termination occurs instantaneously when a free radical enters a particle which already contains another free radical. No particle contains more than one free radical in such a system, which is termed 'zero–one'; the conditions are $c \gg \rho, k$. Under these circumstances, equation (3) becomes equations (10) and (11). These equations state that entry into a particle containing one free radical results in the formation of a particle containing none; they can be formally derived from equation (4) by making the steady state approximation for $dN_2/dt$ and ignoring higher-order terms.

$$dN_0/dt = -(\rho_A + \alpha k N_1)N_0 + (\rho_A + \alpha k N_1 + k)N_1 \tag{10}$$

$$dN_1/dt = (\rho_A + \alpha k N_1)N_0 - (\rho_A + \alpha k N_1 + k)N_1 \tag{11}$$

Equations (10) and (11) have analytic solutions.[16] Of particular use is the result of these solutions, together with that of equation (1), at long times. One then finds[16] that equation (12) holds, where the long time slope and intercept, $s$ and $i$, are functions of $\alpha$, $\rho_A$ and $k$.[16] These results can be inverted to give the values of $\rho_A$ and $k$ from the slope and intercept of the long time limit of an experimental $x(t)$ curve, as given in equations (13)–(15). Here $\bar{n}_0$ is the initial value of $\bar{n}$. Note that equations (13)–(15) are replaced by simpler expressions for the special case where $\alpha = 0$.[16] Equations (13)–(15) will be seen to furnish useful means of determining rate parameters from experimental conversion/time data. It has been pointed out[17] that, in the limit for which these results are applicable, the values of $\rho_A$ and $k$ deduced from the data can be very sensitive to small errors in the steady state $\bar{n}$.

$$\lim_{t \to \infty} x(t) = i + st \tag{12}$$

$$k = A\ln(F)/(2\alpha i), \qquad \rho_A = Gk \tag{13}$$

$$G = [2\alpha s^2 + A(1 - \alpha)s]/A(A - 2s) \tag{14}$$

$$F = \frac{1}{2} + \{2G + (1 - \alpha) + 4\alpha\bar{n}_0\}/\{2[4G^2 + (\alpha - 1)^2 + 4G(\alpha + 1)]^{1/2}\} \tag{15}$$

The steady state solution of equations (11) and (12) yield equation (16); the subscript SS refers to steady state. This result is particularly useful since it shows that, in the limit as the entry rate becomes large, the steady state $\bar{n}$ achieves the limiting value of 0.5; this limiting case was called 'Case 2' by Smith and Ewart.[5] This plateau can be clearly seen in Figure 2. Note that $\bar{n}$ only assumes this limiting value of 0.5 under special circumstances ($c \gg \rho \gg k$), and is exhibited in relatively few experimental situations; *one should never assume, without firm evidence, that a particular system will exhibit Case 2 kinetics*. Smith and Ewart designated as 'Case 1' the situation when $\bar{n} < 0.5$, and by 'Case 3' that when $\bar{n} > 0.5$.

$$\bar{n}_{SS} = \rho/(2\rho + k) \tag{16}$$

#### *11.3.1.3 Early time limit*

Summation of equation (4) over $n$, after multiplication by $n$, yields the exact form of equation (17),[22] where $\langle n^2 \rangle$ is the second moment of the $N_n$ ($\bar{n}$ being the first moment). Equation (17) cannot be used to find the time evolution of $\bar{n}$, since it is coupled to equations in higher moments. However, if the system starts out initially with $\bar{n} = 0$, then one has the exact result given by equation (18). This also implies that the early time limit of $\bar{n}$ in such a system must obey $\bar{n}(t) = \rho_A(t) + O(t^2)$. This result can be used to estimate $\rho_A$.[26]

$$d\bar{n}/dt = \rho_A - (1 - \alpha)k\bar{n} - 2c(\langle n^2 \rangle - \bar{n}) \tag{17}$$

$$\rho_A = \lim_{t \to 0} \frac{d\bar{n}}{dt} \quad (\text{if } \bar{n}(t = 0) = 0) \tag{18}$$

### 11.3.2 Experimental Means of Determining Rate Parameters

The foregoing theoretical development has outlined how rate data can be calculated from the rate parameters for entry, exit, *etc.* We next consider how rate data can be determined experimentally, and how values of the various rate parameters can be deduced from appropriate data.

A problem which has plagued emulsion polymer science is that postulated mechanisms are often so complex that it is difficult to obtain data which can properly confirm or refute a particular mechanistic postulate and therefore one cannot readily distinguish between quite different mechanisms. A methodology will be outlined which yields unambiguous rate parameters, and hence clear mechanistic information, from experiment.

A simplistic use of insufficient data will be inadequate, and very often misleading. Suppose, for example, that one has only the steady state rate for a range of different initiator concentrations [I] for a particular seeded system in Interval II; this is clearly inadequate to determine all the rate parameters required: $k_p$, $\alpha$, $k$, $c$ and the variation of $\rho$ with [I]. Even use of literature values (for example, that of $k_p$) is problematical, since standard compilations[27] show a very wide range of values. It is obviously essential therefore to have as much data as possible, so as (ideally) to overdetermine the values of the rate parameters for a given microscopic process; if two totally different measurements give the same value of a rate parameter, this provides strong evidence for correctness of the postulated mechanism. This can be done as follows. One studies the polymerization rate of a seeded system, using chemical initiators and also $\gamma$ radiolysis to provide the source of free radicals. The latter has the advantage that it can be switched off instantaneously; this permits one to study the relaxation kinetics, and thus provides new, independent, information on radical loss events. Moreover, one should obtain not only the steady state rate but also the approach to the steady state (both with chemical and $\gamma$ initiation); again, additional information is provided. It has been rigorously proven[28] that such measurements provide, *in toto*, sufficient information to overdetermine all the component rate parameters, and thus to reveal mechanisms unambiguously. This is fundamental to the approach used here, as distinct from the model-fitting 'global' approach discussed above, which is almost always ambiguous. The present technique enables one to find all rate-determining parameters ($k_p$, $k$, $k_t$, $\alpha$ and the dependence of $\rho_A$ on [I]), entirely from emulsion polymerization kinetic data.

An ideal set of data for unambiguous determination of growth rate parameters would consist of: (a) for chemically initiated systems, the steady-state rate and the kinetics of approach to the steady state (slope and intercept), with both of these as functions of initiator concentration [I]; (b) for $\gamma$-radiolysis initiation, the steady state rate in the source, and the kinetics of relaxation following removal from the source, both preferably for different radiative fluxes; and (c) the steady state rate and kinetics of approach to the steady state in a system without any added initiator (to determine the background thermal entry rate coefficient,[29] the value of which is needed for proper evaluation of the values of other kinetic parameters).[18,28] In addition, it is necessary for proper data analysis that no secondary nucleation occur under the chosen conditions. Not all of the desirable data can always be attained from every system, although even in the systems where less information is accessible, considerable qualitative and quantitative mechanistic information can still be deduced.

In order to be able to obtain as much information as possible in kinetic studies of a seeded system, one requires the following. (a) Appropriate equipment for preparation and cleaning of seed latexes. Details can be found, *e.g.*, in Hawkett *et al.*[29] (b) Electron microscope equipment, for particle sizes and testing for secondary nucleation. Details can be found, *e.g.*, in Hawkett *et al.*[29] (c) Equipment for accurate measurement of the time dependence of fractional conversion. The method of choice here is dilatometry;[30] note that this technique can be automated. Dilatometer design must be such that the approach to the steady state can be observed accurately. Other techniques that can be used to monitor conversion include microcalorimetry, densitometry and gravimetry. The last is insufficiently precise to be able to provide essential information such as the approach to the steady state. (d) Access to $\gamma$-radiolysis equipment, designed so that dilatometric studies can be carried out therein. A most important use of this facility (see below) is to re-insert the reacting system back into the source after removal to test for inhibition effects. If $\gamma$-radiolysis equipment is not available, then the information that can be gleaned from the system is limited; however, in the special case of a zero–one system, it is still possible to obtain all the rate-determining parameters from chemical initiation data alone.[28,29,31]

It is desirable to have a monodisperse seed, since rate parameters can be functions of particle size. This often requires preparation at an elevated temperature (for reasons, see Section 11.6.5). The seed must be stable to coagulation over as wide a range of initiator concentration as possible, and initiator and seed concentrations must be chosen so as to avoid secondary nucleation. This requires some trial and error in the latex preparation; a 'bottle polymerizer'[32] is particularly suited for the simultaneous preparation of a number of different latex recipes for such trials.

In interpreting dilatometric data for a monomer that is fairly insoluble in water, the observed changes in volume on polymerization can be related to fractional conversion ($x$) by assuming ideality: $x = \Delta h A_b / C_F g_M^\circ$, where $\Delta h$ is the change in height of the meniscus on the dilatometer bore, $A_b$ is the bore area, $g_M^\circ$ is the initial mass of monomer and the 'contraction factor' $C_F$ is the difference in specific volumes of monomer and polymer, *i.e.* $C_F = d_M^{-1} - d_P^{-1}$ (where $d_M$ and $d_P$ are the densities of monomer and polymer respectively). Additional effects must be taken into account if the monomer is moderately water soluble, *e.g.* methyl methacrylate (MMA). In this case, the partitioning of monomer between the two phases has to be established (by direct measurement). The relation between the observed change in height, $\Delta h$, and conversion, $x$, must then be through a more complex data reduction method, *e.g.* that of Ballard *et al.*[26]

Data interpretation is best carried out for Interval II kinetics, since any changes in $c$ and $k$ due to the gel effect will be absent. Sometimes it may prove impossible to avoid new nucleation in Interval II, and then Interval III data must be exploited. Interval III data must however be treated carefully in the first determination of rate parameters in order to avoid complications arising in the gel regime; these are discussed at a later stage.

### *11.3.2.1 Testing for inhibition effects*

The kinetics of approach to the steady state in a chemically initiated system constitute an important part of the data for deducing unambiguous rate parameters. However, it must be established that the observed kinetics are not some artifact of any inhibitor present. One simple test is to check that the kinetics of approach to the steady state do not alter with differing degrees of deoxygenation.[29] A better test is the following. Using $\gamma$ radiolysis, one operates at a radiation flux such that the time required to reach a steady state upon initial insertion into the source is comparable to that in a chemically initiated system. One re-inserts the reactor back into the $\gamma$ source following its removal (conditions should be chosen to be similar for the initial and second insertions). If the time required to reach the steady state following this second insertion is close to that for the first, then it is likely that no artifacts from inhibitor are present in the chemically initiated data (since the initial exposure to the $\gamma$ flux removes any remaining aqueous phase inhibitor).[26]

We now show how the data may be analyzed to yield unambiguous kinetic parameters. It should be noted that the data, for a given conversion/time curve, can be encapsulated in the observed long time slope and intercept. In principle one could use the entire conversion/time curve rather than just the slope and intercept, however, it is found that experimental accuracy and model sensitivity are such that slope and intercept are sufficient to give the entire conversion curve.[29] This has the important implication that one can obtain no more than two independent pieces of data from a single conversion curve for a chemical initiator. On the other hand, the full $x(t)$ for $\gamma$-relaxation data is required, since the relaxation is often multiple exponential and thus cannot be encapsulated by the two parameters afforded by the long time slope and intercept.

### *11.3.2.2 Zero–one systems*

The data analysis is particularly straightforward if the system can be shown to be zero–one. The general data analysis given in a later section will indeed show clearly if this is in fact the case; however, a simple indication that a system is zero–one is if the steady state rate, when plotted as a function of [I], yields a rate that is independent of [I] at higher [I] values.[29] This can immediately be ascribed to the limit when $\bar{n}=\frac{1}{2}$, from equation (16) and Figure 2. However, the lack of such an observation does not imply that the system is not zero–one, and, as stated, this will be resolved under the complete analysis. An [I]-independent steady state rate immediately gives an accurate value of $A=k_pN_cC_M/(n_M^\circ N_A)$ in equation (1), since $\bar{n}_{ss}=\frac{1}{2}$ at the high-[I] limit. Note that this value is independent of any measurements of $k_p$ in bulk or other systems. Moreover, it gives a value of $A$ (or equivalently of $k_pC_M$) which is applicable to the particular seed latex system being studied. Now, there are literature $k_p$ values for many monomers over a wide temperature range.[27] However, the tabulated values under particular circumstances usually vary by a factor of two or more, depending on the method of measurement and the assumptions made in interpreting the data. It is therefore frequently essential that an independent value for $k_p$ be determined directly from an emulsion polymerization study. Given the value of $A$, the values of $\rho_A$ and $k$ can be obtained from the observed slope and intercept, for an assumed value of $\alpha$, using equations (12)–(15). The choice of $\alpha$ will be discussed below.

For zero–one systems, the value of $\alpha$ can be determined by requiring that the value of $k$ be independent of the initiator concentration, type of initiator and particle number. This is carried out as follows. One takes a series of $x(t)$ curves for different values of [I] and/or $N_c$, and finds the value of $\alpha$ which gives a constant $k$ under all conditions (within experimental error). For a given $x(t)$, the values of $\rho_A$ and $k$ can be obtained, for an assumed value of $\alpha$, from equations (12)–(15). A set of conversion/time curves for different [I] and/or $N_c$ is used to find the values of $\rho_A$ and $k$, for values of $\alpha$ spanning the range that this parameter can assume ($-1\leq\alpha\leq+1$; taking only three values, $\alpha=-1$, 0 and $+1$, is found to be sufficient). The correct value (or range) of $\alpha$ is that which yields a $k$

which is independent of [I] and $N_c$. An example has been given by Whang *et al.*[16] An alternative approach is to determine $\rho_A$ through equation (18) employing the early time $x(t)$ in a chemically initiated system,[26] or alternatively the subsequent insertions of a $\gamma$ initiation,[33] if sufficiently accurate data can be obtained. It has been shown[33,34] that, with appropriate data, this yields the same values for $\rho_A$ as the other methods given here.

#### *11.3.2.3 Thermal contribution to entry*

It must be noted that the interpretation of $\gamma$ relaxation must take account of the 'background thermal entry' rate coefficient, $\rho_T$.[29] This is any component of $\rho_A$ that is present in the absence of any external (chemical or radiative) initiator, and which is presumably due to the thermal production of free radicals from, for example, substances present on the surface of the latex particle. It has been shown[29] that $\rho_T$ can be much larger than any contribution from the thermal polymerization rate of pure monomer. Means of determining $\rho_T$ have been given;[18,29,35] in brief, one can carry out a run in the absence of added initiator and/or find the limiting value of $\rho_A$ as [I] approaches zero. It should be noted that some pioneering work in $\gamma$ radiolysis[36] was originally interpreted without knowledge of the existence of $\rho_T$, and so reached erroneous conclusions as to the order of free-radical loss events; proper accounting for $\rho_T$ brings these data into accord with accepted mechanistic precepts.[35]

#### *11.3.2.4 Systems with any* $\bar{n}$*, 'constant'* k *and* c

Details of the data reduction method have been given elsewhere,[28,31,37] and are summarized below. It is assumed that the data consist of: (1) the slopes and intercepts for a series of chemically initiated runs over a range of [I]; and (2) the time dependence of conversion for a run initiated by $\gamma$ radiolysis, after removal from the source (relaxation mode), for preference with a range of radiation fluxes. The method assumes that $k$ and $c$ are independent of initiator concentration and type, and also independent of conversion except for the circumstances detailed below. All data fitting is done using exact numerical solutions of the generalized Smith–Ewart treatment (equations 1–5). One notes that a single chemically initiated run, since it yields, at most, only two parameters (the slope and intercept) can only be used to deduce two kinetic parameters. (a) In the first step, the chemically initiated data are fitted assuming sets of values of $k_p$ and $\alpha$ which cover the expected range; *e.g.* $k_p$ in the range $10^2$–$10^3$ $dm^3\,mol^{-1}\,s^{-1}$ and $\alpha = -1, 0$ and $+1$ (ten values of $k_p$ and three values of $\alpha$ would give thirty sets of values to be tested). Initially, $k$ is set equal to some reasonable value (*e.g.* that estimated from transfer-diffusion theory,[38] see Section 11.3.7), and $\rho_A$ and $c$ are least squares fitted to the experimental slope and intercept for each [I] (the value of $\rho_A$ depending on [I]). The mean value of $c$ found from all the chemically initiated data is then used to calculate $\rho_A$ and $k$ by re-fitting these data. The mean value of $k$ found from this procedure is then used as in the first step to re-calculate $c$, and the whole procedure iterated until convergence. In each case, the fit consists of minimizing (with respect to $\rho_A$ and $k$ or $\rho_A$ and $c$) the residual $R$ defined by equation (19), where calc and obs refer to the calculated (using equations 1–4) and observed values of the slope ($s$) and intercept ($i$). (b) In the second step, the values of $c$ and $k$ found for each set of values of $k_p$ and $\alpha$ are used to fit the $\gamma$-relaxation data (again with the background thermal entry taken into account, as above). These $\gamma$-relaxation data will be fitted only by a unique pair of $k_p$ and $\alpha$ values (or rather, only by those values over a small range). One thus obtains unique values of all rate parameters.

$$R = [1 - (s_{calc}/s_{obs})]^2 + [1 - (i_{calc}/i_{obs})]^2 \tag{19}$$

There are some important constraints in the fitting procedure which must be taken into account[31] by plotting, for each run, the residual $R$ as a function of whichever pair of parameters ($\rho_A$ and $k$ or $\rho_A$ and $c$) are being fitted. If a very shallow minimum in either $k$ or $c$ is found, this is because this rate parameter is not rate determining for this run (or these conditions); this can occur, for example, if the system is zero–one or if it is pseudo-bulk with $\alpha$ close to $+1$. (i) If this analysis of residuals reveals one of these kinetic parameters as being non-rate determining for all of the runs being used, then the data are simply re-analyzed either as given above (zero–one case) or as described by Ballard *et al.*[26] (pseudo-bulk with $\alpha \triangleq +1$); this plotting of the residuals identifies a system as belonging to either category. (ii) If the residuals show that either $k$ or $c$ is non-rate determining only for certain runs (*e.g.* $k$ may not be rate determining for high [I]) then the value of this rate coefficient given by the least-squares fit for that particular run (or runs) is of course meaningless and must be excluded when finding the mean value.

Finally, care must be taken to avoid using the method when the basic assumption of constant $k$ and $c$ is invalid. For example, the method is inapplicable for Interval III data where $k_t$ (and hence $c$) changes rapidly with weight fraction of polymer; however, this problem will not arise if one operates in Interval II, where the weight fraction of polymer remains fixed. Moreover, this restriction can be easily obviated for certain 'trivial' changes in $k$ and $c$: *e.g.* when $k$ varies inversely as the surface area of the swollen particle (as expected from the transfer-diffusion theory in appropriate circumstances)[38] or when $k_t$ is constant but $c$ varies as the inverse of the swollen particle volume ($c = k_t/N_A V_s$). These 'trivial' variations are readily taken into account by incorporating them into the numerical solution of equations (1)–(4). Means of determining rate parameters under circumstances where the rate parameters change significantly during conversion are given in the next section. The only case when this breakdown is likely to be encountered is for Interval III, when $k_t$ will vary rapidly with $x$.

#### *11.3.2.5 High $\bar{n}$ and changing $k_t$*

As stated, the two preceding methods are inapplicable to Interval III systems where $k$ and/or $c$ changes significantly with conversion (*i.e.* with $w_p$, the weight fraction of polymer) for the conditions of the run (apart from the trivial dependence on swollen size given above). The most likely circumstance for this breakdown is for Interval III data at $w_p$ values where $k_t$ changes rapidly, *e.g.* in the gel regime. In this case, $\bar{n}$ is likely to be high[26] and moreover the diffusion coefficient of oligomeric free radicals in the particle is likely to be low;[33] under either or both of these circumstances, $k$ is unlikely to be rate determining (see Section 11.3.7).[26,33] One may then use the pseudo-bulk relation (equation 9) with $k=0$, as in equation (20). One sees that neither $k$ nor $\alpha$ need then be considered, and the analysis is straightforward. (i) One can assume a functional form for $k_t(w_p)$ containing (say) two unknown parameters which are then least squares fitted to the data by numerical integration of equations (1)–(5). (ii) A better method[26] is to make use of $\gamma$-relaxation data, when it is likely that the system will go from a high $\bar{n}$ value to a very low one (governed by the background thermal entry rate)[26,29,35] over a small range of $w_p$. Over a small range, $k_t$ (and hence $c$) may be treated as constant, in which case equation (20) can be integrated to yield equation (21). Here $\bar{n}_0$ is the initial value of $\bar{n}$ and $\bar{n}_{SS} = (\rho_A/2c)^{\frac{1}{2}}$. This expression can then be least squares fitted to the observed rate to yield an average $k_t$ over that (small) range of $w_p$ involved in that relaxation; details have been given by Ballard *et al.*[26] A major restriction on this method is that a value of $k_p$ is required, in order to convert the observed $x(t)$ to $\bar{n}(t)$ using equation (1). Thus either one must have $k_p$ determined from a run at a lower $w_p$ as described above, or a $k_p$ value available from other sources, as discussed in Section 11.3.4. The same restriction appplies to use of equation (18) to determine $\rho_A$ at any conversion.

$$d\bar{n}/dt = \rho_A - 2c\bar{n}^2 \tag{20}$$

$$\bar{n}(t) = \bar{n}_{SS}\frac{P+1}{P-1}; \quad P = \exp[(8\rho Ac)^{1/2}t]\frac{\bar{n}_{SS}+\bar{n}_0}{\bar{n}_{SS}-\bar{n}_0} \tag{21}$$

### 11.3.3 Monomer Concentration

It has been implicit in the above analysis that the value of $C_M$ is known. In Interval III system for a monomer (such as styrene) that is very insoluble in water, $C_M$ in the monomer-starved particles is trivially calculated from the amount initially added (however, see note below) since essentially all monomer is located inside the latex particles (for more soluble monomers such as MMA, it is necessary to allow for the partitioning of monomer between the particles and the aqueous phase).[26] In Intervals I and II, the monomer concentration within the latex particles achieves an equilibrium saturation value, as a result of the particles minimizing the free energy, which has components arising from interfacial tension from mixing. Consideration of these competing effects leads to equation (22), where $\Gamma$ is the interfacial tension between the latex particle and the water, $\phi$ is the volume fraction of polymer, $V_m$ is the partial molar volume of monomer (to an excellent approximation, $V_m = M_0/d_M$, where $M_0$ is the molecular weight of the monomer), $R$ is the gas constant, $r$ is the unswollen radius of the particle and $\chi$ is the Flory–Huggins interaction parameter. The values of $\phi$

and $w_p$ (the weight fraction of polymer) are related by mass conservation, as given in equation (23). The value of $C_M$ is then found from equation (24). However, the values of $\Gamma$ and $\chi$ at present cannot be predicted for a given system, and indeed the physical basis of equation (23) is open to question. It is therefore necessary to determine the equilibrium monomer concentration in Interval II directly from experiment. There are several means to do this, as follows.

$$\ln(1 - \phi) + \phi + \chi\phi^2 + (2\Gamma V_m/rRT)\phi^{1/3} = 0 \tag{22}$$

$$\phi = [(d_p/d_M)w_p^{-1} + (1 - \{d_p/d_M\})]^{-1} \tag{23}$$

$$C_M = (1 - \phi)/V_m \tag{24}$$

(i) The most straightforward method is by direct analysis, as in the method of Morton *et al.*[39] (ii) The value of $x$ at which the transition from Interval II to Interval III takes place can often be identified as where $dx/dt$ shows a decrease in slope. The value of $C_M$ at saturation of the aqueous phase with monomer is then found by assuming ideal mixing, using equation (25). (iii) The 'static swelling' method[26] involves adding different amounts of monomer to the latex, and one observes when any excess monomer forms a separate phase.

$$C_M = (1 - x)g_M^\circ/(N_c V_s) \tag{25}$$

Given the value of $C_M$ (in any of Intervals I, II and III), the swollen particle volume $V_s$ can be computed at any fractional conversion from equation (26). Here $V$ is the unswollen volume of the system. Equilibrium saturation $C_M$ values have been given in the literature for many systems. They lie typically in the range 4–6 mol dm$^{-3}$.

$$V_s/V = \phi^{-1} = d_M/(d_M - C_M M_0) \tag{26}$$

### 11.3.4 Propagation Rate Coefficients (Low Conversion)

The preceding section gave a number of ways of determining $k_p$ directly from emulsion polymerization kinetics. It is also necessary to consider other techniques, since under certain circumstances (as detailed above) emulsion polymerization conversion/time data alone may be insufficient to yield $k_p$.

The value of $k_p$ should be the same in an emulsion, bulk or any other polymerization system, other conditions being the same. However, $k_p$ can depend on the value of $w_p$, as illustrated in Figure 3. That is, until very high $w_p$, $k_p$ should be essentially constant, because the rate-determining step in propagation is usually 'chemical', *e.g.* the crossing of the activation barrier for reaction between a macroradical and a monomer molecule. This 'low conversion' $k_p$ value will depend on the degree of polymerization of the macroradical up to a few monomer units, whereafter any dependence on end

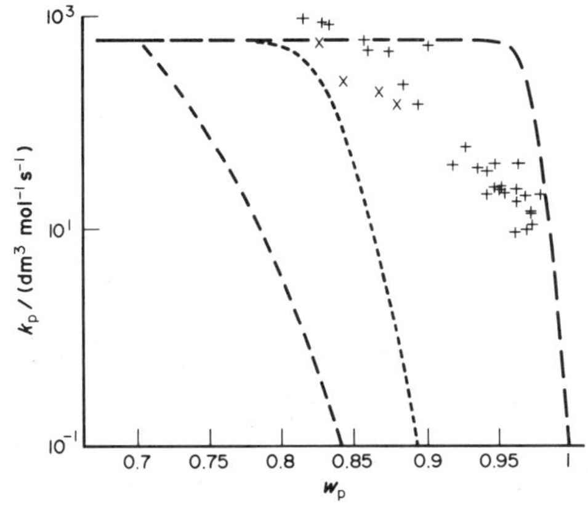


**Figure 3** Propagation rate coefficient for methyl methacrylate at 50 °C, as function of weight fraction polymer: +, values measured directly, using ESR (M. J. Ballard, R. G. Gilbert, D. H. Napper, P. J. Pomery and J. H. O'Donnell, *Macromolecules*, 1986, **19**, 1303); ×, from kinetic method (G. L. Leslie, I. A. Maxwell, M. J. Ballard, R. G. Gilbert and D. H. Napper, *Aust. J. Chem.*, 1988, **41**, 279); lines, predictions of free-volume theories (G. T. Russell, R. G. Gilbert and D. H. Napper, *Macromolecules*, 1988, **21**, 2141)

effects will disappear. At very high conversions (often near, but by no means necessarily at, the glass transition) $k_p$ decreases with $w_p$, as the rate-determining step becomes diffusion of the monomer to the chain end of the macroradical. This decrease will be seen if the system has a high viscosity; *e.g.* it is unlikely to be observed if the temperature is sufficiently great. It is useful to distinguish between the low conversion ('chemical') value, denoted $k_p^\circ$, and that at high conversion.

There are many data available on $k_p$ values in the literature, but inspection of compilations[27] shows that there is a wide range of reported values even under ostensibly the same conditions. The origin of the problem is that data interpretation is usually strongly model dependent. Recent work[40] has shown that good agreement can be obtained between the $k_p$ values reported by different groups using quite different techniques[35,41] as long as the data interpretation involves minimal model-based assumptions. This suggests the correctness of the approaches used by these different groups, which we now discuss briefly. The first is the technique of spatially intermittent polymerization (SIP), a bulk technique,[41] whereas the other uses the emulsion polymer methodology discussed above, in the particular application to a zero–one system.[35]

Figure 4 illustrates the application of the emulsion polymerization technique. This shows the observed steady state rate in the seeded emulsion polymerization of styrene, as a function of initiator concentration. It can be seen that this rate approaches a high [I] limiting value, which from the discussion in Section 11.3.1 must correspond to $\bar{n} = 0.5$ (since $\rho_A$ must increase with [I]; see Figure 2). Now that the value of $\bar{n}$ has been identified, and since one measures the experimental rate (*i.e.* $dx/dt$), and since the value of $C_M$ is known (given $x$) by conservation of mass, all the quantities in equation (1) are known except $k_p$. Thus $k_p$ can be determined quite accurately. One notes that $\gamma$ radiolysis (where the initiation rate is independent of temperature) furnishes a particularly flexible means of applying this technique.[35] This method can be applied only if it is proved conclusively that the system indeed follows Case 2 kinetics ($\bar{n} = 0.5$); the literature contains many incorrect values of $k_p$ (*e.g.* those of Soh)[42] deduced from systems where this assumption appears inapplicable.

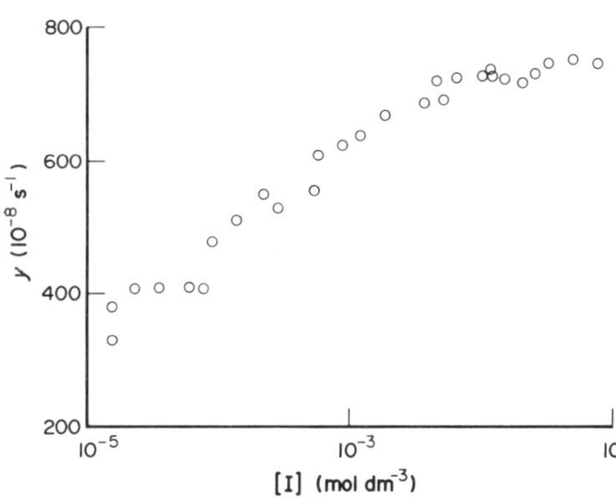


**Figure 4** Steady state polymerization for styrene-seeded emulsion polymerization at 50 °C, as $y = (dx/dt)N_A g_M^\circ / N_c M_0$, as function of initiator concentration (redrawn from *J. Macromol. Sci., Rev. Macromol. Chem.*, 1983, **C23**, 127)

A third technique, which has also been tested against independent measures,[33] is calibrated ESR applied to emulsion polymerization kinetics.[43] This involves essentially no model-based assumptions; it invokes only equation (1), which assumes that the free radicals are uniformly distributed in the polymerizing medium, or, more specifically, that the monomer concentration at the locus of polymerization is uniform. The method requires the concentration of free radicals ($\bar{n}$, measured directly by ESR) and the rate (measured dilatometrically at the same time as the ESR measurement); experimental details have been given by Ballard *et al.*[43] The $k_p$ values from this method for MMA at 50 °C are shown in Figure 3. The technique is difficult to use, but has thus far been applied by two groups.[43,44] It has the particular advantage that it can be used in the diffusion-controlled regime, where $k_p$ changes with $w_p$, and where its precision is the greatest. Another kinetic technique that can be used in this regime will be discussed in Section 11.3.9.

Results for $k_p^\circ$ from the above techniques are collected in Table 2. There is currently no reliable theory for values of $k_p$ at low conversion, although fortunately here it is likely that experimental techniques such as SIP[41] can give accurate values for this quantity.

**Table 2** Some $k_p^\circ$ Values of Relevance to Emulsion Polymerization Systems

| *Monomer* | *Temperature (°C)* | *Method* | *Ref.* | $k_p^\circ$ (dm$^3$ mol$^{-1}$ s$^{-1}$) |
|---|---|---|---|---|
| Styrene | 15–30 | SIP | 41 | $10^{7.0} \exp(-29.5 \text{ kJ mol}^{-1}/RT)$ |
| | 45–65 | EM | 35 | $10^{7.1} \exp(-29 \text{ kJ mol}^{-1}/RT)$ |
| MMA | 15–30 | SIP | 41 | $10^{5.7} \exp(-18.2 \text{ kJ mol}^{-1}/RT)$ |
| BMA | 50 | EM | 37 | $6 \times 10^2$ |
| BA | 50 | EM | 28 | $4.5 \times 10^2$ |

R: gas constant. SIP: spatially intermittent polymerization (bulk method). EM: emulsion polymerization techniques of Section 11.3.1. MMA = methyl methacrylate; BMA = butyl methacrylate; BA = butyl acrylate.

### 11.3.5 Kinetics of Entry: Results and Models

The mechanism whereby free radical activity generated in the aqueous phase is transferred to the latex particles (wherein propagation is primarily located) remains obscure. This section discusses experimental data on $\rho_A$, and uses these to elaborate details of the entry mechanism.

The most extensive data on $\rho_A$ are for styrene, for a wide range of [I], of initiator type and of particle number.[17,18,29,45] Data on $\rho_A$ for styrene, butyl acrylate (BA) and butyl methacrylate (BMA) are given in Table 3. This also contains values of the efficiency $f$ and of $n$, which for purposes of comparison have been calculated at $[\text{I}] = 10^{-3}$ mol dm$^{-3}$and $N_c = 10^{16}$ dm$^{-3}$, using the rate parameters deduced from data at different [I] and $N_c$. In calculating $f$, one uses equation (27), where $k_d$ is the initiator decomposition rate coefficient and $\rho_I$ is the entry rate without the 'background' thermal entry component $\rho_T$, *i.e.* $\rho_I = \rho_A - \rho_T$.

$$f = \rho_I/(2k_d[\text{I}]/N_c) \tag{27}$$

Two important results are immediately apparent: (i) the value of $\bar{n}$ is rarely 0.5; and (ii) the capture efficiency is commonly less than 100%. For the styrene systems in the above work, $f$ can be as low as a few percent. This observation of a low initiator efficiency contradicts an unfortunately common assumption made in interpreting and predicting emulsion polymerization data, which is that many workers assume 100% initiator efficiency. Moreover, the data[17,18,28,29,37,45] show that $f$ is not independent of [I]; the initiator efficiency is not constant. This observation is of course explicable in terms of the occurrence of significant aqueous phase termination.

A simple semi-quantitative treatment has been given[46] to estimate whether capture efficiency will be high or low in a given system. This treatment considers the competing events of capture and aqueous phase termination; it is shown[46] that $f$ will be low if the quantity $\Gamma$ defined by equation (28) is appreciably less than unity. Here $k_{t,\text{aq}}$ is the termination rate coefficient of free radicals in the aqueous phase that are capable of undergoing entry, $z$ is the degree of homogeneous aqueous phase polymerization prior to being capable of undergoing entry, and $[M_{\text{aq}}]$ is the solubility of the monomer in water.

$$\Gamma = 4k_{t,\text{aq}}z^2k_d[\text{I}]/k_p^2[\text{M}_{\text{aq}}]^2 \tag{28}$$

Equation (28) shows that highly soluble monomers are likely to have large values of $\Gamma$, and thus high capture efficiencies. Obviously, too, capture efficiency will also be highest for low initiator concentrations. Table 4 shows calculated values of $\Gamma$ for a number of monomers, the concomitant prediction of whether or not capture efficiency is likely to be high, and (where known) the agreement of this qualitative prediction with experiment. In each case where data are available, these predictions agree with experiment. Note that in this calculation of $\Gamma$, the values of $k_{t,\text{aq}}$ given are estimates based on the likely termination rate of two small oligomers. While the model described here appears to give reliable predictions as to whether the capture efficiency will be high or low, it does depend strongly on the value assumed for $z$.

#### *11.3.5.1 Quantitative theory for entry rate coefficient*

We now turn to detailed models for $\rho_I$, the component of the entry rate coefficient for free radicals arising directly from initiator. The only model applied to this problem and which gives accord with the observed dependence of $\rho_I$ on [I] and $N_c$ has been given by Penboss *et al.*[17,18,45] who extended

**Table 3** Entry Rate Parameters for Various Monomers at 50 °C (see equation 29)

| *Monomer* | *Initiator* | $\beta$ (mol dm$^{-3}$) | $\gamma$ (dm$^3$ mol$^{-1}$ s$^{-1}$) | $k_e$ (dm$^3$ mol$^{-1}$ s$^{-1}$) | $\bar{n}$ | $r_s$ (nm) | $f$ (%) | *Ref.* |
|---|---|---|---|---|---|---|---|---|
| S | $K_2S_2O_8$ | $2.8 \times 10^{-10}$ | $3 \times 10^3$ | $3 \times 10^5$–$4 \times 10^6$ | 0.4 | 69 | 2 | 18 |
| | | $8.0 \times 10^{-11}$ | $5 \times 10^3$ | $5 \times 10^5$–$3 \times 10^4$ | 0.5 | 113 | 2 | 18 |
| | V50 | $6.4 \times 10^{-12}$ | $1 \times 10^4$ | $1.2 \times 10^6$ | 0.4 | 69 | 1 | 18, 45 |
| BMA | $K_2S_2O_8$ | $3 \times 10^{-10}$ | $1.4 \times 10^4$ | $2 \times 10^6$ | 0.7 | 51 | 30 | 31, 37 |
| BA | $K_2S_2O_8$ | $1.2 \times 10^{-10}$ | $6 \times 10^4$ | $4 \times 10^6$ | 7 | 83 | 10 | 28, 31 |

$r_s$ = swollen radius of latex. S = styrene; BMA = butyl methacrylate; BA = butyl acrylate. V50 initiator = 2,2′-azobis(2-amidinopropane) (a cationic initiator). $\bar{n}$ calculated for [I] = $10^{-3}$ mol dm$^{-3}$ and $N_c = 10^{16}$ dm$^{-3}$. $f$ = efficiency for $10^{-3}$ mol dm$^{-3}$ initiator.

**Table 4** Values of Γ from equation (28) at 50 °C

| *Monomer* | [$I$] (mol dm$^{-3}$) | $k_p$ (dm$^3$ mol$^{-1}$ s$^{-1}$) | $k_{t,aq}$ (dm$^3$ mol$^{-1}$ s$^{-1}$) | $z$ | [$M_{aq}$] (mol dm$^{-3}$) | Γ | $f_{est}$ | *Ref.* |
|---|---|---|---|---|---|---|---|---|
| S | $10^{-2}$ | 260 | $7 \times 10^7$ | 3 | $4 \times 10^{-3}$ | 30 | low | 29, 46 |
| | $10^{-5}$ | | | | | $3 \times 10^{-2}$ | high | |
| MMA | $10^{-2}$ | 580 | $3 \times 10^7$ | 40 | 0.15 | 0.3 | 50–100% | 26, 46 |
| | $10^{-5}$ | | | | | $3 \times 10^{-4}$ | high | |
| BA | $10^{-2}$ | 450 | $7 \times 10^7$ | 3 | $6 \times 10^{-3}$ | 4 | < 100% | 28, 46 |
| | $10^{-5}$ | | | | | $4 \times 10^{-3}$ | high | |
| BMA | $10^{-2}$ | 600 | $7 \times 10^7$ | 3 | $3 \times 10^{-3}$ | 14 | low | 46, 37 |
| | $10^{-5}$ | | | | | $1 \times 10^{-3}$ | high | |
| Acrylonitrile | $10^{-2}$ | 500 | $7 \times 10^7$ | 200 | 1.6 | 0.2 | ≙ 100% | 46 |
| | $10^{-5}$ | | | | | $2 \times 10^{-4}$ | high | |
| Vinyl acetate | $10^{-2}$ | $3 \times 10^3$ | $10^8$ | 50 | 0.35 | $10^{-2}$ | high | 46 |
| | $10^{-5}$ | | | | | $10^{-5}$ | high | |

Decomposition rate coefficient ($k_d$) is that for $K_2S_2O_8$ at 50 °C, $1.3 \times 10^{-6}$ s$^{-1}$. [a] $f_{est}$ = initiator efficiency estimated from Γ.
[a] I. M. Kolthoff and I. K. Miller, *J. Am. Chem. Soc.*, 1951, **73**, 3055.

earlier work of Hansen and Ugelstad.[15] This model takes into account the various events leading to entry as described in Section 11.3.1, and gives equation (29), where $\gamma = k_e^2/2k_{t,aq}$, $\beta = k_{t,aq}k_I/k_e^2$, $N = N_c/N_A$, $k_e$ is the second-order rate coefficient for entry of a free radical arising directly from initiator, and $k_I$ is the first-order rate coefficient for production of such a free radical from initiator decomposition ($k_I$ is an effective rate coefficient which takes account of the aqueous phase propagation of a free radical produced directly from the initiator decomposition to a sufficient degree of polymerization to be able to enter the particle). The rate parameters $\beta$ and $\gamma$ can be found by non-linear least squares fitting equation (29) to data on the dependence of $\rho_A$ on [I] and $N_c$; it is found that equation (29) always provides a good fit to experiment. Table 3 exhibits the values of $\beta$ and $\gamma$ so obtained for styrene, BA and BMA. Given an assumed value of $k_{t,aq}$ (as in Table 4), one may then obtain estimates of the values of $k_e$ and $k_I$, also given in Table 3. A sensitivity analysis of the fitting procedure, using extensive data on styrene,[18] yields the uncertainties for $k_e$ for styrene listed in Table 3.

$$\rho_I = \gamma[(N^2 + 4\beta[\mathrm{I}])^{1/2} - N] \tag{29}$$

An extension of equation (29) takes account of the likelihood that $k_I$ has a slight dependence on [I], and uses equation (30), obtained from HUFT theory.[15,47] Here $j_{cr}$ is the critical chain length of an oligomer at which entry can occur. Use of equation (30) instead of an [I]-independent $k_I$ gives essentially the same values for $k_e$. In summary, it is found that equations (29) and (30) are able to fit data excellently.

$$k_I = k_d\left[\frac{k_p[\mathrm{M_{aq}}]}{k_p[\mathrm{M_{aq}}] + (k_{t,aq}k_d[\mathrm{I}])^{1/2}}\right]^{j_{cr}} \tag{30}$$

### *11.3.5.2 Entry mechanism*

The main conclusion to be drawn from the data in Table 3 concerns the value of $k_e$ (the second-order rate coefficient for entry of an oligomeric free radical into a particle), including dependence of $k_e$ on the charge of the species involved. The value of $k_e$ is in the range $10^5 \leq k_e$ $(\mathrm{dm^3\,mol^{-1}\,s^{-1}}) \leq 10^6$. Table 3 shows that $k_e$ is not greatly influenced by the charge on the initiator, and other results[48] show that $k_e$ is significantly affected neither by double layer thickness nor by amount of surfactant on the particle.

The magnitude of $k_e$ immediately eliminates the simple collisional entry mechanism.[49] This hypothesizes that $k_e$ is given by the collisional rate of an oligomer with the latex particle; this predicts[18] a $k_e$ value some ten orders of magnitude larger than experiment, and so is totally inapplicable to these systems. A more realistic model for entry is that of diffusional control from the aqueous phase,[50] coupled with the addition of an activation barrier such as that for displacement of a surfactant molecule from the surface by an entering free radical.[51] Although a physically reasonable activation barrier can be chosen to give agreement with the observed magnitude of $k_e$,[18] this mechanism is in contradiction to the observation that $k_e$ is unaffected by the amount of surface coverage by surfactant.[48] A third mechanism that has been postulated is colloidal entry;[45] this supposes that the entering radical species is of colloidal dimensions (perhaps a few nanometers radius) and that coagulation of this entity with the latex particle is the rate-determining step for radical entry. This is however refuted by the observations[48] that changing the ionic strength and changing the surface charge (by changing the surfactant concentration) does not affect $k_e$ significantly. Moreover, calculations[52] of the rate of coagulation between a colloidal free radical and a particle, using the Barouch–Matijevic[53] solution of the Poisson–Boltzmann equation, gives $k_e$ values of *ca.* $10^9\ \mathrm{dm^3\,mol^{-1}\,s^{-1}}$, orders of magnitude greater than experiment.

The foregoing comparison between various models and experiment shows that no current model is able to give accord with the full range of data. A working hypothesis that is consistent with current knowledge is that entry is controlled by the rate of diffusion of an oligomeric (or colloidal) free radical through some type of barrier in the vicinity of the surface. This would suggest that $k_e$ should be proportional to the swollen surface area of the latex particle (although it must be emphasized that measurements show that the simplistic collision model, which assumes that every particle meeting the surface undergoes entry,[49] is incorrect by many orders of magnitude). Clearly further work is needed to elucidate the nature of the entry mechanism.

### 11.3.6 The 'Fate Parameter'

It will be recalled that the fate parameter $\alpha$ gives the relative importance, for exited free radicals, of complete re-entry ($\alpha = +1$) *vs.* heterotermination ($\alpha = -1$). Table 5 shows the values of $\alpha$ currently available from experiment. For chemical initiators, $\alpha$ is in the range $0 \geq \alpha \geq -1$ for styrene, whereas for MMA, BA and BMA, $\alpha$ appears closer to $+1$. For relaxation studies (*i.e.* in the absence of chemical or $\gamma$-radiolytic initiators), $\alpha$ appears to be $+1$ for all monomers and initiator types. Reliable values for $\alpha$ with $\gamma$ initiation (*i.e.* not in the relaxation mode) are unavailable, but all data can be fitted by assigning the radiolytic $\alpha$ the same value as for that monomer using chemical initiation.

**Table 5** Values for $\alpha$ for Emulsion Polymerization Systems at 50 °C

| *Monomer* | *$\alpha$ ($K_2S_2O_8$ initiator)* | *$\alpha$ ($\gamma$-radiolysis, relaxation mode)* | *Ref.* |
|---|---|---|---|
| S | $-1 \leq \alpha \leq 0$ | +1 | 16 |
| MMA | $0.5 \leq \alpha \leq +1$ | +1 | 26 |
| BMA | $0.5 \leq \alpha \leq +1$ | +1 | 37 |
| BA | $0 \leq \alpha \leq +1$ | +1 | 28 |

The value of $\alpha = +1$ in the relaxation mode (absence of chemical or $\gamma$ initiator) is readily explained. After removal from a radiolysis system there will only be a low concentration of (thermally generated) free radicals in the aqueous phase, and thus the most likely fate for an exited free radical is re-entry, rather than heterotermination.

The dependence of the fate parameter on monomer, with chemical initiators, can be qualitatively understood as follows. The $k_p$ for styrene is significantly lower than that of the other monomers examined. This suggests that the value of $\alpha$ is determined by the length of time spent growing to a sufficient degree of polymerization in the aqueous phase. That is, if the exiting free radical propagates quickly, it is much more likely to re-enter than to heteroterminate. The solubility of the monomer in the aqueous phase is also important in determining both the propagation rate and the number of monomers that must be added to allow entry to occur, both of which might influence $\alpha$.

A general theory for determining $\alpha$ given appropriate rate coefficients has been developed.[18] It is seen that $\alpha$ can be obtained from the relation $\alpha = \partial\rho/\partial(k\bar{n})$, where $\rho$ is given by equation (31), [E] (the concentration of exited free radicals) is given by equation (32), and [R] (the concentration of free radicals derived directly from initiator) is given by equation (33). Here $k_{ht}$ is the rate coefficient for aqueous phase heterotermination between an E and an R species, $k_r$ is that for re-entry, and the other rate coefficients have been defined in Section 11.3.5.

$$\rho = k_e[R] + k_r[E] \tag{31}$$

$$[E] = Nk\bar{n}/(k_rN + k_{ht}[R]) \tag{32}$$

$$k_{t,aq}k_{ht}[R]^3 + (k_{t,aq}k_r + k_{ht}k_e)N[R]^2 + (k_{ht}Nk\bar{n} + k_eN^2k_r - k_I[I]k_{ht})[R] - k_I[I]k_rN = 0 \tag{33}$$

However, this requires values for some rate parameters such as $k_{ht}$ and $k_r$ whose values are uncertain, and a quantitative comparison between theory and experiment has yet to be carried out. Moreover, one sees in Table 5 that the values of $\alpha$ deduced from experiment are subject to what appears to be quite a large uncertainty. However, it is found that (except under rather special circumstances)[26] it is unncessary to know the precise value of $\alpha$. In most cases,[16,26] computed $x(t)$ curves and deduced $\rho_A$ values calculated with $\alpha$ values covering the range of uncertainties given in Table 5 show no significant differences.

One can therefore conclude that the value of $\alpha$ can be obtained, to within the precision needed to interpret and predict experimental data, from an estimate of $k_p$ for the monomer and/or consideration of initiator type.

### 11.3.7 Kinetics of Exit: Results and Models

The importance of exit (desorption) of free radicals from latex particles in emulsion polymerization has only become evident in recent times. Even for a relatively water-insoluble monomer such as styrene, exit of free radicals can be rate determining if the particle size is sufficiently small. In this section, data on exit are examined, suggesting that exit occurs through transfer to monomer followed by diffusion of the resulting small free radical out of the particle before it has time to propagate. Although exit of such an insoluble entity is rare (because of the difference in chemical potential between the particle and aqueous phase), calculations show that it happens with sufficient frequency to make exit kinetically significant in many systems.

A large amount of data on exit, for a range of monomer systems, is now available. Some values have been obtained by the assumption-free methods discussed in Section 11.3.1,[16,29,35,37,54–56] while others have been obtained by indirect (global modelling) techniques.[15,38,57–59] We shall concentrate on the values obtained by the direct methods, which are the most reliable.

Extant exit data are found to be in excellent accord with the transfer/diffusion theory,[15,38] and the results will be discussed in this context. The mechanism involves the following steps: (i) transfer of free radical activity to monomer or to chain transfer agent; (ii) propagation of this new free radical; (iii) diffusion of the free radical before propagation to the boundary of the particle; and (iv) diffusion of the small free radical away from the particle through the aqueous phase. It has been shown that if all of these events are rate determining, then the exit rate coefficient $k$ is given by equation (34). Here $z$ is the degree of polymerization of the exiting free radical, and $D_p$ and $D_w$ are the diffusion coefficients for the exiting species in the particle and in the aqueous phase respectively. The parameter $q$ is the partition coefficient of the exiting species between the (organic) latex particle and the aqueous phase; to a good approximation, $q = C_M/[M_{aq}]$. The transfer/diffusion theory can be simplified in certain limits. For a sparingly soluble monomer such as styrene, $q \gg 6D_w/D_p$, so that equation (34) becomes equation (35). In the limit where diffusion within the particles is rate determining (*i.e.* $6D_w/D_p \gg q$), equation (34) reduces to equation (36).

$$k = (3zD_w/r_s^2)(k_{tr}/k_p)/(q + 6D_w/D_p) \quad (34)$$

$$k \triangleq (3zD_w/r_s^2)(k_{tr}/k_p)([M_{aq}]/C_M) \quad (35)$$

$$k \triangleq z(k_{tr}/k_p)D_p/2r_s^2 \quad (36)$$

Another limit of interest occurs when transfer is significantly slower than either diffusion step, in which case equation (34) is replaced completely by equation (37). In actuality, $k$ will be given by equation (38), where $k_{td}$ is the value calculated from transfer and diffusion (equation 34) and $k_{to}$ is that from transfer only (equation 37).

$$k \triangleq k_{tr}C_M \quad (37)$$

$$k^{-1} = (k_{td})^{-1} + (k_{to})^{-1} \quad (38)$$

Extensive tests of equations (34)–(38) have been carried out on styrene systems. The dependence of $k$ upon $r_s$ (by varying the size of the seed latex)[29] is shown in Figure 5; the variation with $k_{tr}$ (by adding chain transfer agent)[55] is given in Figure 6. The results are in quantitative conformity with the predictions of equation (34), which can be shown to be the appropriate limit for this styrene system; *viz.* $k$ is linear in $r_s^{-2}$ and in the concentration of chain transfer agent (the increase in the value of $k$ in the presence of chain-transfer agents is somewhat diminished by the high propagational reactivity of the transferred free radical). It has also been shown[17] that equation (34) accurately predicts the measured temperature dependence of $k$ for styrene. Adams *et al.*[54] examined the effect

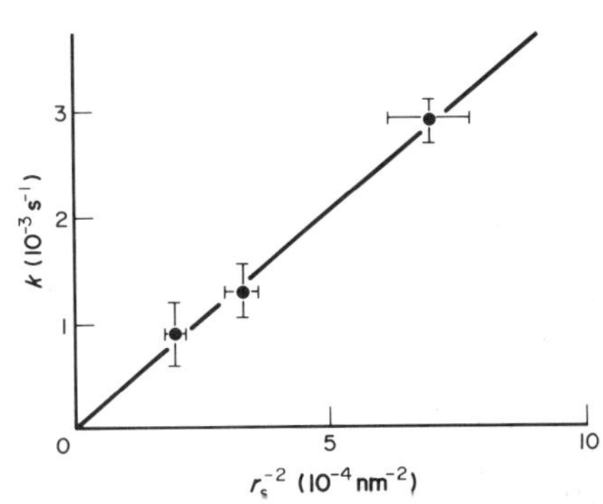


**Figure 5** Dependence of exit rate coefficient $k$ on swollen radius $r_S$, for styrene seed latexes at 50 °C (data from R. G. Gilbert and D. H. Napper, *J. Macromol. Sci., Rev. Macromol. Chem.*, 1983, **C23**, 127)

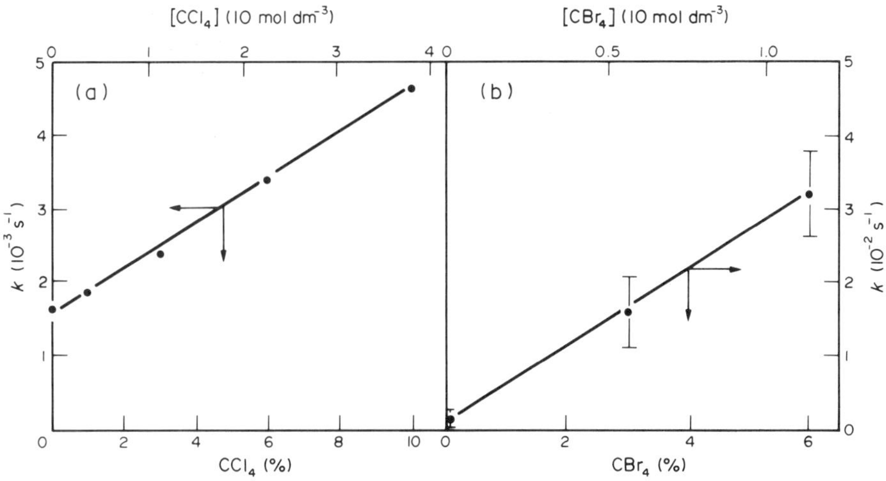


**Figure 6** Dependence of exit rate coefficient $k$ on concentrations of chain transfer agent (reproduced by permission of the Royal Society of Chemistry from *J. Chem. Soc., Faraday Trans. 1*, 1982, **78**, 2129)

on $k$ by adding methanol. This would be expected to increase the solubility of the exiting species in the aqueous phase by decreasing $q$ in equation (34). Excellent accord with theory was again found, including the transition from the transfer/diffusion limit (equation 34) to transfer alone (equation 37), as given in equation (38). Finally, the effect of inert diluents on $k^{56}$ can also be understood in terms of the transfer/diffusion theory and the slow propagation of the transferred free radical. Thus for styrene, the transfer/diffusion theory gives agreement with the data over wide-ranging tests of the prediction of quite different aspects of the theory.

Data for other monomers are much more sparse. The only other 'direct' measurement is for BMA,[37] where the result is again in accord with equation (34) (note that $k$ values reported for BA using the direct technique[28] have subsequently shown to be incorrect).[31] It is felt that results reported by Nomura[38] for many monomers, although probably quite reasonable, may be open to question because they were not obtained by the 'direct' method but instead were deduced from modelling studies. However, it has been shown[58] that $K$ values for styrene deduced from several modelling studies[15,58,59] are in good accord with those obtained by the direct method. This accord between the results obtained by two quite different methods corroborates both.

Finally, we note that there are some results reported for styrene and MMA exit rate coefficients[60,61] which are grossly different from those deduced above by several independent groups. This apparent anomaly can be explained as follows. The styrene results were interpreted using a value of $k_p$ which was a factor of two different from that found by two independent groups[35,40,41] (using separate emulsion and bulk measurements as discussed in Section 11.3.4). The MMA results were inferred from global modelling (*i.e.* agreement with computed conversion/time data) instead of using the 'overdetermined' analysis implicit in using a combination of chemical and $\gamma$-relaxation data. It has been shown[26] that, for MMA, $k$ is often not rate-determining, so that essentially any value can be used to give agreement with chemically initiated conversion/time curves. Hence the anomalous $k$ values[60,61] are probably spurious.

In summary, all extant reliable data on exit can be qualitatively and quantitatively explained in terms of the transfer/diffusion theory, which would therefore seem to be quite reliable for prediction and for modelling studies.

### *11.3.7.1 Vinyl acetate*

The transfer/diffusion theory can be used to explain an apparent difficulty in understanding the mechanism of emulsion polymerization of vinyl acetate, as follows. Conversion/time plots for this monomer are found to be linear over a wide range of $x$ (typically up to 80% conversion), well beyond the transition from Interval II to Interval III.[62] The simple expectation, from equation (1), would be that if $\bar{n}$ were to remain constant, then $x(t)$ would be linear in $t$ in Interval II (as in equation 1), followed by $-\ln(1-x)$ being linear in $t$ in Interval III (as in equation 2). However, the observation is that $x$ remains linear well into Interval III, *i.e.* after the disappearance of monomer droplets. Various hypotheses have been put forward to explain this result,[62] based on cancellations which would not be expected to be applicable over such a wide range as is observed experimentally. However, a simple explanation[63] can be put forward as follows. Estimates of the rate parameters for this system using published values for $k_p$, *etc*., indicate that $\bar{n}$ should be very much less than 0.5, and thus that it should obey zero–one kinetics. If transfer alone is rate determining, equation (37) is applicable. As long as $\alpha$ is not exactly $+1$, and if $\bar{n}$ is small, one has from equations (16) and (5) that $\bar{n} \hat{=} \rho_A/k$, which, when combined with equations (37) and (1), yields equation (39) for the steady state rate. Equation (39) shows that, provided $\rho_A$ is independent of $C_M$, then, in this limit, $dx/dt$ is independent of $C_M$; this result is valid for both Intervals II and III. This then explains the linearity of $x$ with $t$ over such a wide range; the system has a constant $\bar{n}$, and the dependence on the monomer concentration disappears exactly if exit is entirely transfer dominated. The observed zero-order dependence of the rate on monomer concentration can be understood without having to invoke any special peculiarities of the vinyl acetate system apart from a transfer-dominated exit.

$$dx/dt = A\bar{n}_{SS} = k_p N_c \rho_A/(n_M^\circ N_A k_{tr}) \tag{39}$$

### 11.3.8 Kinetics of Termination: Results and Models

Termination processes in emulsion polymerization might be expected to be different from those operative in low-conversion bulk systems. The presence of formed polymer in the latex particles ensures that entanglement effects are almost always present for chains of significant molecular weight. Here we present experimental values for $k_t$ for several systems, and show how they are used to infer the termination mechanism.

The value of $c$ and hence $k_t$ may be obtained using the 'high $\bar{n}$/constant $k$ and $c$' method of Section 11.3.1. The requirement that $k_t$ be constant over the range of conversion used for data reduction (apart from trivial changes due to variation in swollen volume) will usually confine the method to Interval II. This gives the restriction that $k_t$ can only be obtained at the weight fraction corresponding to the equilibrium swelling of the particle by the monomer (typically $w_p \hat{=}$ 30–40%). To date this method has been applied to BMA and BA.[28,31,37] Results are given in Table 6.

**Table 6** Experimental Results and Theoretical Upper and Lower Bounds for $k_t$ ($dm^3\,mol^{-1}\,s^{-1}$) for BMA and BA

| *Quantity* | *BA value* | *BMA value* |
|---|---|---|
| Ref. | 28, 31 | 31, 37 |
| $w_p$ | 0.57 | 0.43 |
| $k_t$ (experiment) | $8 \times 10^2$ | $7 \times 10^3$ |
| $k_t$ (lower bound) | $9 \times 10^2$ | $1.5 \times 10^3$ |
| $k_t$ (upper bound) | $3 \times 10^4$ | $5 \times 10^4$ |

Before going on to discuss another technique which can be used to obtain $k_t$ in Interval III, we note that the 'trivial' dependence of $c$ on particle volume discussed in Section 11.3.1 ($c = k_t/N_A V_s$) can be used to explain an apparent anomaly in the emulsion polymerization of MMA in Interval II; except for very small latex particles, this monomer apparently does not show a true steady state, as there is no linear $x(t)$ region, but rather a gradually increasing rate while monomer droplets are still present.[7,64] The explanation involves the decrease in $c$ caused by the change in swollen volume, as first suggested by Gardon[49] and quantified (as follows) by Ballard *et al.*[26] The gradually increasing rate can be quantified as a dimensionless 'acceleration' $\bar{a}$, defined by equation (40), where $g_M^\circ$ and $g_P^\circ$ are the initial masses of monomer and polymer present, and $\bar{n}^\circ$ is the value of $\bar{n}$ at the start of the quasi-steady state period. The value of $\bar{a}$ can be obtained theoretically by considering the effect of the decrease in $c$ caused by the increase in swollen volume $V_s$ in Interval II. If it is assumed that the system obeys pseudo-bulk kinetics, as in equation (9), and that $\alpha$ in equation (5) is close to $+1$, then the quasi-steady state value of $\bar{n}$ will be given from equation (20) as $\bar{n}_{SS} = (\rho_A N_A V_s/2k_t)$. Consideration of the variation of $V_s$ with $x$ then leads[26] to the theoretical expression given by equation (41), where $\bar{n}^\circ = (\rho_A/2c^\circ)$ and $c^\circ$ is the value when $x = 0$. Comparison of equations (40) and (41) shows that this implies that $\bar{a}$ should take the value of $\frac{1}{2}$ if this explanation is correct. The data of Ballard *et al.*[26] show that, for a latex of radius 37 nm, $\bar{a}$ was found to be $0.6 \pm 0.2$, while for a 46 nm latex, $\bar{a} = 0.5 \pm 0.1$.

The agreement of the data with the predictions of equation (41) supports the suggested explanation, including the supposition that, for this system, $\alpha$ is close to $+1$ and $\bar{n}$ is large.

$$\bar{n}_{SS} = \bar{n}^{\circ}[1 + (\bar{a}g^{\circ}_{M}/g^{\circ}_{P})x] \tag{40}$$

$$\bar{n}_{SS} \triangleq \bar{\bar{n}}^{\circ}(1 + \frac{1}{2}[xg^{\circ}_{M}/g^{\circ}_{P}]) \tag{41}$$

We return to the determination of $k_t$ in emulsion systems. A more widely applicable technique for obtaining $k_t$ from kinetic data is through equation (21). This is valid if $\bar{n}$ is sufficiently high for $k$ not to be rate determining (or, as discussed above for MMA, $k$ is not rate determining because $\alpha \triangleq +1$ and $\bar{n}$ is high). Equation (21) can then be used to least-squares fit $\gamma$-radiolysis relaxation data (given $k_p$, including any dependence on $w_p$). This relation is derived assuming that $k_t$ is constant, which will usually be an excellent approximation over the small range of $w_p$ for which relaxation takes place, and moreover the dependence of $k_t$ on $w_p$ so deduced can then be iteratively improved through exact numerical simulation. The technique is therefore applicable to both Intervals II and III.

There has been an extensive application of this method to MMA by Ballard *et al.*[26,65] The dependence of $k_t$ on $w_p$ was determined in a PMMA seed of high molecular weight, both in the absence of chain-transfer agent (CTA) and with the addition of CTA to observe the effect of change of molecular weight of the growing chain on $k_t$. Results are shown in Figure 7. These $k_t$ data are the best available for any free radical system at high $w_p$, since they are derived essentially free of any mechanistic assumptions; the requisite $k_p(w_p)$ values were obtained directly from ESR measurements[43] (and have been verified by independent kinetic data)[33] and no modelling assumptions are required, since the only free-radical loss event is termination, whose rate is measured directly using $\gamma$-radiolysis in the relaxation mode.

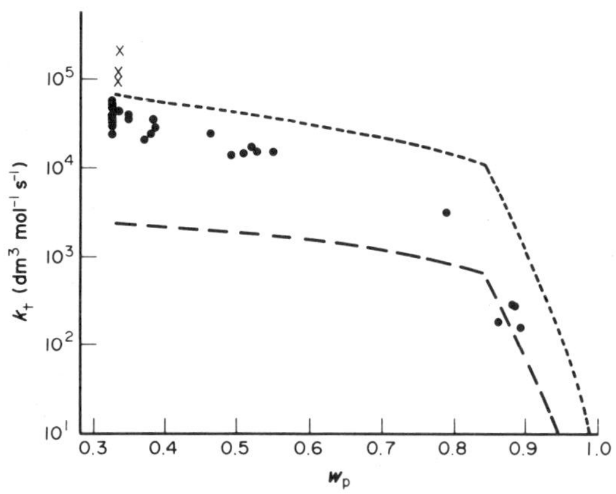


**Figure 7** Dependence of $k_t$ on $w_p$ for MMA at 50 °C: ●, in the absence of chain transfer agent ($\langle M_n \rangle$ of polymer matrix and growing chain both *ca.* $10^6$); ×, in presence of chain transfer agent so that molecular weight of growing chain *ca.* $10^5$; lines, theoretical upper and lower bounds from equations (44) and (45) (redrawn from M. J. Ballard, D. H. Napper and R. G. Gilbert, *J. Polym. Sci., Polym. Chem. Ed.*, 1984, **22**, 3225 and M. J. Ballard, D. H. Napper, R. G. Gilbert and D. F. Sangster, *J. Polym. Sci., Polym. Chem. Ed.*, 1986, **24**, 1027)

### *11.3.8.1 Theoretical models for* $k_t$

We first discuss some general notions of termination in free radical polymerizations. It is generally accepted that termination at low $w_p$ involves three steps:[66] (i) two radical coils gain proximity by translational diffusion; (ii) the free-radical chains ends make contact through segmental diffusion; and (iii) the barriers to chemical reaction are overcome. Since step (iii) is rapid compared to the other two in polymer systems, termination is expected to be entirely diffusion controlled. The value of $k_t$ may be a function of some or all of the following: $w_p$, the molecular weight of the polymer matrix ($M_m$), and the degrees of polymerization of the two terminating species. Since $M_m$ and $w_p$ evolve in time, $k_t$ may also depend implicity on time. The overall loss of free-radical activity by termination is given by equation (42). Here n($p$, $t$) denotes the concentration of chains of degree of polymerization $p$ at time $t$. Equation (42) can be rewritten in terms of an average $\bar{k}_t$ defined by equation (43). Here [R•] is the overall free-radical concentration. However, $\bar{k}_t$ will in general be a function of time and

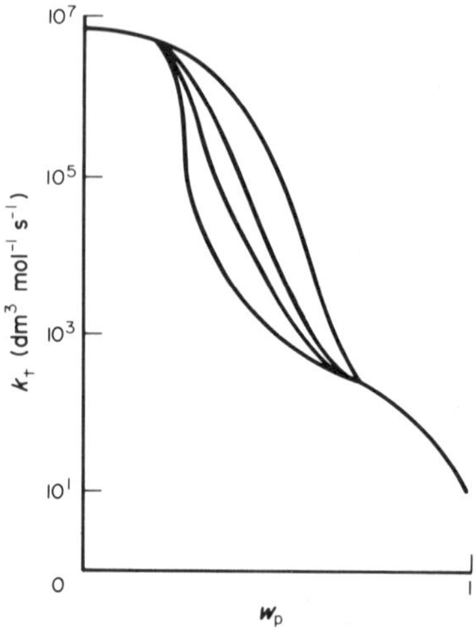


**Figure 8** Notional representation of variation of $k_t$ with $w_p$; actual values and extent of each regime will be system dependent; multiple lines indicate region where $k_t$ governed by centre-of-mass diffusion

indeed of the full history of the system, *e.g.* varying with initiator concentration because [I] may affect $M_m$. A qualitative representation of $k_t$ is given in Figure 8. At low $w_p$, termination is dominated by interactions of unentangled polymer chains; since unentangled chains diffuse rapidly, $k_t$ is governed by segmental diffusion, which will depend only weakly (if at all) on chain length, on $M_m$ and (as long as $w_p$ is sufficiently low) on $w_p$. The value of $k_t$ may even increase slightly as $w_p$ increases, due to a reduction in the solvency of the system on addition of polymer. As the chains become entangled, so the rate-determining step for termination shifts from segmental to centre-of-mass diffusion. At even higher conversions, the chains of the polymer matrix become so entangled that centre-of-mass diffusion ceases, and the chain ends gain proximity solely by 'reaction diffusion'[67] (also called 'residual termination',[49] which although not the term coined by the originator of the mechanism,[67] is convenient and is therefore adopted here). In residual termination, propagational growth causes the chains to diffuse, despite their overall translational immobility. When residual termination is operative, and provided propagation is not diffusion controlled, $k_t$ is expected to depend only very weakly (if at all) on $w_p$, on $M_m$ and on the length of either chain. At intermediate $w_p$, where the termination mechanism is governed by centre-of-mass diffusion of either or both chains, $k_t$ will be a strong function of chain length (of either or both chains), of $w_p$ and of $M_m$. Figure 8 represents this region with multiple lines, indicating that $k_t$ depends on variables additional to $w_p$, and (because of the dependence on $M_m$) on the 'history' of the system (*e.g.* initiator concentration, *etc.*).

$$R|_{\text{termination}} = -\sum_r \sum_s k_t(r, s, w_p, M_m, t)\, n(r, t)\, n(s, t) \tag{42}$$

$$R|_{\text{termination}} = -\bar{k}_t[\text{R}\bullet]^2 \tag{43}$$

The experimental variation of $k_t$ with $w_p$ for MMA shown in Figure 7 shows some of these effects. Note that the values of $k_t$ are several orders of magnitude less than those reported in the literature for $w_p = 0$ (this is of course expected, in view of the change of rate-controlling step with $w_p$). This illustrates the obvious point that use of $k_t$ values for $w_p = 0$ in global modelling (as is not uncommon) will lead to serious error. It can be seen that for $w_p \hat{=} 0.3$, addition of a chain transfer agent (reducing the length of both terminating chains) results in an increase in $k_t$, indicating that here the system is in the intermediate $w_p$ regime of Figure 8. The behaviour at high $w_p$ will be discussed later.

One sees from equations (42) and (43) that, strictly speaking, the kinetics of termination can be represented in terms of a single rate coefficient ($k_t$ or $c$) only at low and high $w_p$; at low $w_p$, this $k_t$ will be almost constant, whereas at high $w_p$, it will depend on $w_p$. In the intermediate regime, a value of the termination rate coefficient deduced from experiment using a single $k_t$, rather than some form of equation (42), will be an apparent rate coefficient which may vary with system characteristics such as [I], *etc.* This may contribute to some of the wide variation of $k_t$ values reported in the literature.

We now discuss theoretical models for $k_t$. First, we consider the residual-termination (high $w_p$) regime. We note that this may in fact be applicable to quite low $w_p$ values in an emulsion polymerization system. For example, with many monomers (such as MMA) which are moderately water soluble, the entering oligomeric free-radical species is expected to be of moderately high degree of polymerization (say, 50–100), and will be likely to undergo rapid propagation on entering the particle, to reach the critical degree of polymerization for entanglement. This would signal the onset of residual termination; if chain transfer agent is absent, this can occur at $w_p$ as low as 0.35. Theories for residual termination have been given by Gardon[49] and by Soh and Sundberg.[68] However, the simplest theory (and that which not only gives the best agreement with experiment but also contains no adjustable parameters) is furnished by the 'rigid chain limit' and 'flexible chain limit' of Russell *et al.*[69] The 'rigid chain limit' provides a *lower bound* for residual $k_t$, which must be applicable at the highest $w_p$ values. It supposes that the growing chain is completely rigid in the polymer matrix on the time scale of propagation. The termination rate coefficient can then be found from the Smoluchowski equation $k_t = 16\pi Dd$, when $D$ is the 'diffusion coefficient' of the propagating chain end and $d$ is the radius of interaction at which termination may be said to occur instantaneously. One can identify $d$ with the Lennard–Jones radius $\sigma$ of the monomer, and $D$ can be found from the diffusion relation $D = \langle r^2 \rangle / 6\Delta t$, where $\langle r^2 \rangle$ is the mean square displacement of the active chain end in time $\Delta t$. The time step $\Delta t$ can be identified as that for propagation, $\Delta t = (k_p C_M)^{-1}$, and $\langle r^2 \rangle^{1/2}$ as the root-mean-square end-to-end distance per monomer unit, denoted by $a$. One thus obtains equation (44). The *upper bound* for $k_t$ in the residual regime can be found by noting that the maximum value for residual termination will occur when the roving chain end is completely flexible. Such a chain has nodes of entanglement every $j_c$ monomer units. This leads,[69] *via* an argument similar to that used to deduce equation (44), to equation (45). To determine $j_c$, the relation[70] $j_c = j_c^0/\phi$ may be used, where $j_c^0$ is the entanglement spacing (in monomer units) of pure polymer and $\phi$ is the volume fraction of polymer.[71] It can be seen that equations (44) and (45) contain no adjustable parameters, and the properties of the polymers required for their evaluation ($j_c^0$ and $a$) can be obtained from physical measurements of the polymer;[71] the value of $\sigma$ can be estimated either from the physical properties of the monomer[72] or from measured $\sigma$ values for analogous substances.[73]

$$k_t(\text{residual, lower bound}) = 4\pi k_p C_M a^2 \sigma / 3 \tag{44}$$

$$k_t(\text{residual, upper bound}) = 8\pi k_p C_M a^3 j_c^{1/2} / 3 \tag{45}$$

Figure 7 compares the predictions of equations (44) and (45) with experimental data for MMA. It can be seen that accord between the extremely simple models and experiment is excellent; the upper and lower bounds neatly encompass the observations. Moreover, the lower bound gives agreement with the observed $k_t$ over a wide range of conversion. It has also been found[34] that equation (44) accurately predicts the high-conversion emulsion polymerization of styrene, providing further evidence for its general applicability.

Table 6 also compares the predictions of these equations with $k_t$ values for BA and BMA. Although in these cases the data are less extensive, they are in accord with the model. It is clear that these simple models are qualitatively and quantitatively applicable to emulsion (and other) polymerizations at higher $w_p$. A current lack is a means of interpolating between these two limits, although it is apparent from Figure 7 that an empirical interpolation should suffice for modelling purposes.

Finally, one notes that the concept of residual termination, as quantified by equations (44) and (45), can be used to understand the sudden drop in $k_t$ observed at higher conversions, as illustrated in Figures 7 and 8. This is due to the drop in $k_p$ as propagation becomes diffusion controlled (see Figure 3). It has also been suggested[69] that a significant contribution to the gel effect in emulsion polymerization is due to the changeover from the upper (flexible chain) to the lower (rigid chain) bounds of $k_t$. Note in this last context that the gel effect in emulsion systems, if observed at all, occurs at much greater $w_p$ than in the equivalent bulk polymerization.

The current status of theories for termination at lower weight fractions, including the dependence of $k_t$ on chain lengths, is considerably less satisfactory. There are many models in the literature, but all contain adjustable parameters which have thus far only been evaluated by comparison with experimental conversion/time curves. As emphasized above, this procedure is unfortunately highly prone to error, and cannot of itself provide unambiguous for the correctness or otherwise of a model. One of the less empirical theories is that of Tulig and Tirrell,[74] based on reptation theory.[75] Friis and Hamielec[76] have given extensive empirical correlations of $k_t$ in terms of $w_p$; however, it must be borne in mind that these were deduced from bulk conversion/time data under the assumption that bulk initiator efficiency $f$ was independent of conversion, an assumption which has subsequently required revision.[40,77] There are also the models of Soh and Sundberg,[68] Gardon,[49] Hamielec and co-workers[78] and of Cardenas and O'Driscoll,[79] but again each suffers from similar deficiencies. A pragmatic approach to modelling the dependence of $k_t$ on $w_p$ would be to take the more reliable of these models at moderate weight fractions and go over to the proven expressions given by equations (44) and (45) when residual termination becomes rate determining. The development of reliable models and sensitive experiments for $k_t$ at moderate $w_p$ is clearly an important area for future development.

Lastly, it is essential to realize that the chain length dependence of $k_t$ expressed in equation (42) can have important effects in emulsion polymerization systems. For example, it has been shown[80] that, for the emulsion polymerization of styrene using large particles (so that termination is rate determining), $\gamma$-radiolysis relaxation data (which are highly sensitive to radical-annihilation events) can only be interpreted at moderate $w_p$ by taking account of two populations of terminating chains: long and short. This approximates the populations $n(p, t)$ in equation (42) by a sum with two terms. The rate of termination between short and long chains is much faster than between two long chains, since the former have a more rapid translational diffusion. Modelling and interpretation of emulsion polymerization kinetics where termination is rate determining, and at $w_p$ values where termination by other than the residual mechanism is possible, must always be carried out bearing this possibility in mind. There is as yet no adequate theory for the dependence of $k_t$ on the length of a short chain.

### 11.3.9 Propagation Rate Coefficients (High Conversion)

We now discuss a method whereby the dependence of $k_p$ on $w_p$ in the diffusion-controlled regime may be obtained entirely from kinetic data, without recourse to ESR measurements.[33] In this regime, monomer diffusion is very slow, and thus equations (34) and (36) show that exit can be ignored. Moreover, $\bar{n}$ is likely to be high, and thus the pseudo-bulk equation (9) may be reliably used. The discussion in Section 11.3.8 shows that, under conditions of high $w_p$, $k_t$ can be accurately found from the rigid chain limit for residual termination (equation 44). Equation (9) with $k = 0$ and equation (44) yield equation (46). Here $q \equiv k_p\bar{n} = -N_A V_s \mathrm{d}\ln(1-x)/\mathrm{d}t$.

$$k_p = 8\pi C_M a^2 \sigma q^2/[3N_A V_s(\rho - \mathrm{d}\bar{n}/\mathrm{d}t)] \tag{46}$$

One sees that the quantity $q = -N_A V_s \mathrm{d}\ln(1-x)/\mathrm{d}t$ may be obtained from experimental conversion/time data. Equation (46) then provides a means of deducing $k_p$ from conversion data, if $\rho$ is known. A value of $\rho$ can be obtained from an observed $\mathrm{d}\bar{n}/\mathrm{d}t$ as $t \to 0$ in a seeded run which starts with $\bar{n} = 0$, from equation (18). Since experiments indicate[34] that $\rho$ depends weakly (if at all) on $w_p$, one can find $\rho$ by starting a kinetic run at a $w_p$ below that at which $k_p$ becomes diffusion controlled. The data for these low $w_p$ values yield $\mathrm{d}\bar{n}/\mathrm{d}t$ from equation (1), given a $k_p^0$ obtained by one of the techniques discussed in Section 11.3.4; $\rho$ can then be found from equation (18). One then takes values of $q$ at $w_p$ where $k_p$ becomes diffusion controlled, and obtains for each $w_p$ a first estimate of $k_p$ from equation (46) with $\mathrm{d}\bar{n}/\mathrm{d}t$ set to zero. Improved values of $k_p$ are obtained using an improved $\mathrm{d}\bar{n}/\mathrm{d}t$ found by solving equations (1)–(5) numerically, with the $k_p(w_p)$ deduced from the first iteration. Iterations are continued until $k_p(w_p)$ converges.

Figure 3 shows the results of the application of this method to data on the seeded emulsion polymerization of MMA.[33] Initiation was by $\gamma$ radiolysis, with multiple insertions into, and removals from, the $\gamma$ source, to eliminate inhibition effects. It can be seen that the agreement between the results from the kinetic method and the ESR results (Section 11.3.4) is excellent, both for the absolute values of $k_p$ and for the decrease in $k_p$ with increased $w_p$. This agreement between two totally different techniques strongly suggests the rectitude of both. Essentially the only assumptions made in the kinetic method are (i) that the free radicals and monomer are uniformly distributed throughout the latex particles and (ii) the validity of the rigid chain limit for $k_t$; a value of $k_p^\circ$ is also required. The kinetic technique has also been applied to styrene emulsion polymerization (where ESR measurements are difficult)[81] to obtain $k_p(w_p)$.[34]

There is at present a dearth of reliable models for $k_p$ at high conversions. There are a number of expressions for $k_p$ in the literature that have been deduced from global modelling of bulk kinetics.[68,78,82] However, all have been based on the assumption that the bulk initiator efficiency is independent of (or only weakly dependent on) $w_p$, a hypothesis which is now realized[40,77] to be in gross error. Indeed, because of this error, expressions for $k_p(w_p)$ currently in the literature[68,78,82] give values that are many *orders of magnitude* less than experiment at high conversion.[43,77] In the region where $k_p$ is diffusion controlled, a simple relationship can be derived[77] to give the diffusive contribution to $k_p$, denoted $k_{p,d}$, in terms of $D_p(w_p)$, the diffusion coefficient of monomer through the polymer matrix as a function of weight fraction polymer. This expression (equation 47), takes account of the contribution to $k_p$ arising from reaction/diffusion, the same dynamical effect which gives rise to equations (44) and (45) for residual termination. Given $k_{p,d}$, the value of $k_p$ at any conversion is found from combining the 'chemical' contribution, $k_p^\circ$, with the diffusive one, as in equation (48).

$$k_{p,d} = 4\pi\sigma N_A(D_P + \frac{1}{6}k_p C_M a^2) \tag{47}$$

$$k_p^{-1} = (k_{p,d})^{-1} + (k_p^\circ)^{-1} \tag{48}$$

$$k_{p,d}^2 + k_{p,d}(k_p^\circ - 4\pi\sigma N_A D_P - \frac{2}{3}\pi(\sigma N_A k_p^\circ C_M a^2) - 4\pi\sigma N_A D_P k_p^\circ = 0$$

$$D_P = k_p[k_p^\circ(4\pi N_A\sigma)^{-1}(k_p^\circ - k_p)^{-1} - (C_M a^2/6)] \tag{49}$$

Equations (47) and (48) can be combined to yield equation (49), which relates $D_P$, $k_p^\circ$ and $k_{p,d}$, the value of $k_{p,d}$, being the positive root of equation (49). Equations (47)–(49) can be used to deduce $D_P(w_p)$ from $k_p$ data such as that in Figure 3. Given $D_P(w_p)$, they also enable one to find $k_p$; however, what is lacking at present is a theory which enables $D_P(w_p)$ to be predicted.

## 11.4 PARTICLE SIZE DISTRIBUTIONS

### 11.4.1 Overview

The latex particles produced in an emulsion polymerization are not strictly monodisperse as they do not all have identical sizes. From measurements of the average size, the full particle size distribution (PSD), and the time evolution of these quantities, considerable mechanistic information can be gained, both about the mechanism of nucleation and about the growth process. In addition, the particle size is important technologically: *e.g.* (i) the rheological properties of the latex are strong functions of the particle size distribution; (ii) the PSD affects properties such as the gloss of the final dried product from an emulsion polymer paint dispersion; and (iii) a highly polydisperse latex flocculates more easily than does a monodisperse one (all other things being equal).

### 11.4.2 Measurement of PSDs

There are a number of techniques for measuring averages of the PSD. These include soap titration, turbidity, ultracentrifugation, simple (angular) light scattering and dynamic light scattering. These only measure averages of the PSD; usually only the first moment, and sometimes the standard deviation, and will not be considered in detail here.

One method which yields information about the higher moments of the PSD is hydrodynamic chromatography (HDC).[83,84] The inversion to deconvolute the observed trace back to a PSD, however, involves technical difficulties which have not yet been fully overcome, and one can only use HDC at present to obtain, at best, the first and second moments of the PSD. The importance of the third moment (whence one can determine the skewness) will become apparent in Section 11.6.

The most direct and reliable method for the measurement of the PSD is calibrated transmission electron microscopy. The latex is first mixed with a suitable standard, usually a monodisperse latex of carefully characterized mean radius, well separated from that of the sample. The latex and standard are then placed upon a microscope grid, dried and photographs of particles are taken. Particular precautions must be taken if the latex is unstable under the electron beam, as is the case with PMMA or PVAC. Such latexes must be hardened, *e.g.* by cross linking with UV or other radiation, or coated with carbon, gold or osmium tetroxide, before measurements are made. If an accurate PSD is required, up to several thousand particles should be counted. Technical requirements to obtain reproducibility have been given, for example, by Lichti *et al.*[85]

Calibrated electron microscopy yields a PSD in terms of the unswollen particle radius, $r$. To gain optimal information, this should then be converted to a PSD in terms of the unswollen particle volume $V$ (for reasons that are given below) as follows. If n($V$) and n($r$) are the PSDs in terms of $V$ and $r$ respectively, then one has equation (50).[85] This has the interesting but often overlooked consequence that a sequence of time-evolving PSDs, if expressed in terms of unswollen volume, may broaden (become more polydisperse) in time, while the same sequence expressed in terms of unswollen radius become more monodisperse. Thus, electron micrographs appear to show that a latex becomes more monodisperse as it grows. The PSD in terms of volume may however show otherwise, because of the $4\pi r^2$ factor. One should therefore specify the size variable ($r$ or $V$) in which a PSD is expressed before commenting on the polydispersity.

$$n(V) = n(r)/4\pi r^2 \tag{50}$$

### 11.4.3 Evolution Equations for the PSD

We now give the equations for the time evolution of the PSD. The PSD is defined as $n(V, t)$; the (relative) number of particles at time $t$ with volume between $V$ and $V + dV$ is $n(V, t)dV$. Define $n_i(V, t)$ as the number of particles containing $i$ free radicals at time $t$ with volume $V$; this is the volume-dependent version of the population $N_i$ of Section 11.3.1. The PSD is then given by equation (51). The $n_i(V, t)$ are defined to be normalized by equation (52).

$$n(V, t) = \sum_i n_i(V, t) \tag{51}$$

$$N_i(t) = \int_0^\infty n_i(V, t)\,dV \tag{52}$$

The time evolution of the $n_i(V, t)$ is given by equation (53). This equation is derived by mass balance in conjunction with the Smith–Ewart kinetic considerations, equations (4) and (5). Here $K_{ii}(V)$ is the rate of volume growth of a particle containing $i$ free radicals, given by equation (54); it is a function of volume $V$ because $C_M$ may depend on $V$. The term $c_i$ is the nucleation rate of a particle containing $i$ free radicals, given by equation (55).

$$\partial n_i/\partial t = \rho[n_{i-1} - n_i] + k[(i + 1)n_{i+1} - in_i] + c[(i + 2)(i + 1)n_{i+2} - i(i - 1)n_i] - \partial(K_{ii}n_i)/\partial V + c_i \tag{53}$$

$$K_{ii}(V) = ik_p M_0 C_M/(N_A d_p) \tag{54}$$

$$c_i = \delta(V - V_m)\delta_{i1}(dN_c/dt)/N_c(t = \infty) \tag{55}$$

Here $\delta_{i1} = 0, i \neq 1$; $\delta_{11} = 1$. It has been assumed that nucleation produces particles whose volume is $V_m$, and that all new particles contain a single free radical (clearly, this can be generalized). The nucleation rate of all particles in equation (55) is denoted $dN_c/dt$. If one considers Intervals II and III, particle sizes are much greater than the size at which particles are 'born', and thus $V_m = 0$. Equations (53)–(55) are an infinite set of coupled partial differential equations, which are in general non-linear, *e.g.* when $\alpha \neq 0$, as given in equation (5). These equations have been derived in various forms by a number of different workers.[86–90]

In Interval II, $C_M$, and hence $K_{ii}$, will be independent of $V$ to a good approximation, whereas in Interval III ($C_M$ decreasing with conversion) and Interval I (very small particles, which may have a smaller $C_M$), $K_{ii}$ would be a function of $V$.

The unswollen volume $V$ (rather than some other size variable such as the swollen or unswollen radius) is in general the optimal variable in which to specify the PSD, because a particle containing a single free radical increases linearly in its unswollen volume in time. While the evolution equations (equations 53–55), are the same in terms of any size variable, the form of the $K_{ii}$ will alter with each size variable. The simplest form of the $K_{ii}$ is that found using $V$ as this variable, when $K_{ii}$ is given by equation (54); that with any other variable will be more complex.

We now consider various means of solving the PSD equations. In general, all of the rate parameters may be complicated functions of volume, and thus numerical solutions must often be used (see below). Nevertheless, analytical solutions for simpler cases have both informative and practical applications.

#### *11.4.3.1 Numerical solutions*

A numerical solution of the full PSD equation can be generated by finite difference, in which one expresses $n_i(V, t)$ as a sum of functions $n_i(V_j, t)$ representing the value of $n_i(V, t)$ at volume $V_j$. This converts the partial differential equations to coupled ordinary differential equations for $dn_i(V_j, t)/dt$, which can be solved numerically by conventional methods. However, in many cases the solution of these equations leads to numerical instabilities. A special variant of the finite difference technique has been derived for the PSD equations[91] which obviates many of these problems.

Another numerical technique, which has been applied to PSDs by a number of authors, is the method of moments. The $m^{th}$ moment of $n_i(V, t)$ is defined by equation (56). Taking the $m^{th}$ moment of equation (53), making use of integration by parts and of the condition that the moments are zero at $V=0$ and $\infty$, and assuming for simplicity that $K_{ii}$ is independent of $V$, one obtains equation (57). Here $\hat{c}_{i,m}$ is the $m^{th}$ moment of the nucleation term and $\Omega_{ij}$ represents the coefficient of $n_j$ in the kinetic terms of the equation for $\partial n_i/\partial t$. Equation (57) is a closed set of coupled ordinary differential equations: *e.g.* the second moments are coupled to the first and zeroth moments, but not to any higher ones. They may thus be readily solved by standard methods (it will also be seen that these equations also have analytic solutions applicable to many cases of interest). Equation (56) can be readily incorporated into a modelling program, for example with the objective of optimizing or controlling a PSD in a CSTR. Some typical applications have been given by Min and Ray[89] and Wood *et al.*[92]

$$M_{i,\mathrm{m}} = \int V^{\mathrm{m}} n_i(V, t)\,\mathrm{d}V \qquad (56)$$

$$\mathrm{d}M_{i,\mathrm{m}}/\mathrm{d}t = \sum \Omega_{ij} M_{j,m} + mK_{ii} M_{i,\mathrm{m}-1} + \hat{c}_{\mathrm{i,m}} \qquad (57)$$

When using the method of moments, it must be borne in mind that a complete distribution function (*e.g.* the PSD) cannot be generated from a finite number of moments, and that one must compare experimental and calculated moments rather than PSDs. Now, PSDs with quite different shapes might have, for example, similar first and second moments, but very different third moments (*i.e.* skewness). Comparison of experimental and theoretical PSDs should therefore involve examination of higher moments. If using the method of moments to obtain rate parameters from observed PSDs (*e.g.* as discussed below) the rate parameters so deduced should be then used to generate a full PSD, for comparison with the original PSD data.

### *11.4.3.2 Analytic solutions*

If a system follows zero–one kinetics (Section 11.3.1), and in addition $\rho$, $k$ and $K_{11}$ are independent of $V$ and $t$ (*e.g.* as will be the case in the Interval II), equation (53) takes the simple form given by equations (58). Here, and in the remainder of this section, we put $K = K_{11}$. Analytic solutions to these have been deduced for two cases of interest: Interval II, when $c_1 = 0$,[90,93] and Interval I where $c_1(t)$ can be represented by two triangular line segments.[93]

$$\partial n_0/\partial t = -\rho n_0 + (\rho + k)n_1; \quad \partial n_1/\partial t = \rho n_0 - (\rho + k)n_1 - K_{11}\partial n_1/\partial t + c_1(t) \qquad (58)$$

Figure 9 shows a plot of the computed time evolution of the PSD using parameters for an Interval II system. One sees the stochastic broadening typical of a compartmentalized system; particles into which a free radical enters (*i.e.* the $n_1$) grow, while those not entered (*i.e.* the $n_0$) remain at fixed volume, thus the PSD broadens as time increases.

Figure 10 shows plots of the computed time evolution of the PSD for two possible types of Interval I systems: one where $c_1$ is a monotonically decreasing function of time (as predicted by the micellar entry[5] and homogeneous nucleation[15,47] theories of particle nucleation) and one where $c_1$ is an initially increasing function of time. It can be seen that the former produces a negatively skewed PSD at early time, while the latter produces a positively skewed one. The explanation of this behaviour is that a decreasing nucleation rate implies that most particles are formed early, whence more particles will have large volumes, while an initially increasing nucleation rate implies that most particles are formed late and hence will have small volumes. Sufficiently long after nucleation has ceased, stochastic effects will mask these differences and the PSDs arising from both forms for $c_1(t)$ will appear more symmetric and become similar, for reasons explained in the next section.

### *11.4.3.3 Analytic solutions for moments*

As shown by Feeney *et al.*,[94] the moment equations for a zero–one system in Interval II with constant $\rho$, $K$ and $k$ have analytic solutions as given by equations (59) and (60). Here $M^{(1)}$ and $M^{(2)}$ are respectively the first and second moments of the PSD ($M^{(1)}$ is the mean volume), $s = (\bar{n} - \bar{n}_0)/\rho$, $r = \bar{n}/\rho$, $\hat{m}$ is the first moment of $\mathrm{n}(V, t=0)$ (*i.e.* the mean volume of the particles at zero time), $\bar{n}_0$ is the initial value of $\bar{n}$, and $m_1$ and $\sigma$ are respectively the first and second moments of $n_1(V, t=0)$.

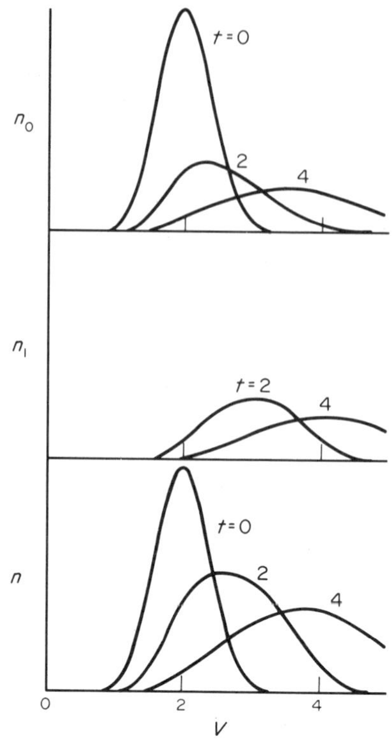


**Figure 9** Theoretical particle size distributions (arbitrary units) as function of volume (dimensionless units) for different times in a zero–one system in Interval II: parameters, $\rho = 1$, $K_{11} = 1$, $k = 0$ (obtained using solutions of G. Lichti, R. G. Gilbert and D. H. Napper, *J. Polym. Sci., Polym. Chem. Ed.*, 1977, **15**, 1957)

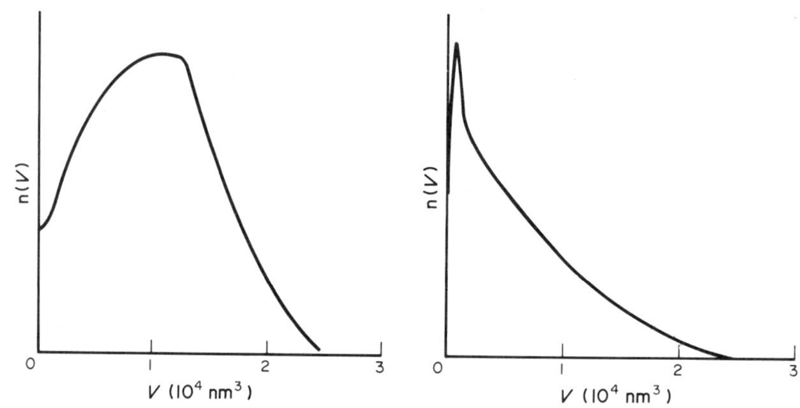


**Figure 10** Theoretical particle size distributions (arbitrary units) as function of volume ($10^4$ nm$^3$), $t = 100$ s, for different forms of nucleation rate. Left, $c_1 = dN_c/dt$ monotonically decreasing ($c_1 = 1 - 10^{-2}t, t < 100$ s; $c_1 = 0, t \geq 100$ s): right, $dN_c/dt$ initially increasing then decreasing ($c_1 = 0.02t$, $t < 50$ s; $c_1 = 2 - 0.02t$, $50 \leq t < 100$, $c_1 = 0$, $t \geq 100$ s). Calculated with $K_{11} = 250$ nm$^3$ s$^{-1}$, $\rho = 10^{-2}$ s$^{-1}$, $k = 10^{-3}$ s$^{-1}$ (obtained using solutions of P. J. Feeney, D. H. Napper and R. G. Gilbert, *Macromolecules*, 1984, **17**, 2520)

Equations (59) and (60) firstly provide a means of determining two rate parameters (*e.g.* $\rho$ and $K$, given $k$) from experimental values for the mean and standard deviation of the volume distribution at a given time. That is, they provide a convenient means of determining rate parameters from the PSD, given one additional piece of information such as the steady state slope (*i.e.* $\bar{n}$). Secondly, equations (59) and (60) enable one to examine how kinetic factors affect the polydispersity of a latex.[94] In the long time limit, these equations can be shown to reduce to equations (61) and (62). From this, the coefficient of variation of the distribution, $CV = [M^{(2)} - (M^{(1)})^2]^{1/2}/M^{(1)}$, at long times is given by equation (63).

$$M^{(1)} = K\bar{n}[t + s(e^{-t/r} - 1)] + \hat{m} \quad (59)$$

$$M^{(2)} = 2K[K\bar{n}^2t(\tfrac{1}{2}t - r) - K\bar{n}^2st + K\bar{n}rt + \hat{m}\bar{n}t - (e^{-t/r} - 1)(K\bar{n}^2r[r + s] - K\bar{n}r^2 + rm_1 - \hat{m}r\bar{n}) - K\bar{n}se^{-t/r}(\bar{n} - 1)(t + r) + K\bar{n}sr(\bar{n} - 1)] + \sigma \quad (60)$$

$$M^{(1)} = K\bar{n}t \quad (61)$$

$$M^{(2)} = 2K(\tfrac{1}{2}K\bar{n}^2t^2 + K\bar{n}rt + \hat{m}\bar{n}t) \quad (62)$$

$$CV = [2t^{-1}(\rho^{-1} + \hat{m}K^{-1}\bar{n}^{-1})]^{1/2} \quad (63)$$

Equation (63) shows that the polydispersity will be decreased by increasing time, $\rho$, $K$ or $\bar{n}$; effects of $k$ are taken into account through $\bar{n}$. If the second term in equation (63) is small, then one has approximately $CV \cong (2/t\rho)^{1/2}$. Hence one sees that the relative polydispersity decreases with increasing time, and also that the polydispersity is smaller, the larger is $\rho$ (*i.e.* the larger is $\bar{n}$: recall that $\bar{n} = \frac{1}{2}$ is the upper limit for a zero–one system). In addition, at sufficiently long times the system becomes independent of its initial PSD (as explained above in connection with Figure 10). The physical explanation of these conclusions is as follows. Increasing polydispersity in a zero–one PSD is caused by stochastic broadening (*vide supra*). The overall decrease in the coefficient of variation with time arises because, although the absolute value of the width of the distribution increases with time, the average volume growth increases even more. The decrease in polydispersity with high $\bar{n}$ (or with $\rho$) arises because non-growing ($n_0$) particles will be converted to growing ($n_1$) ones (and *vice versa*, because of instantaneous termination), and the non-growing particles will lag further and further behind the growing ones, increasing the polydispersity. If on the other hand $\bar{n}$ is high (in a zero–one system), then an individual particle will be rapidly switching between growing and non-growing states, and (since it will then spend half its time growing) the polydispersity will not change in time. Most of these conclusions are also applicable to high $\bar{n}$ systems.

The foregoing discussion has been confined to batch systems. The PSD in CSTR systems is of course described by the same basic equations with the addition of feed and loss terms, and shows its own peculiarities. These have been discussed by Lichti *et al.*[95] and by Poehlein.[96] The principal qualitative effect is a polydispersity that is much greater than in a batch system, because the latex now has a spectrum of residence times.

### 11.4.4 PSDs in Seeded Systems

We now consider applications of the above techniques to obtain rate parameters using the observed time evolution of the PSD.

Figure 11 shows the experimental and fitted PSDs at $t = 0$ and $t = 45$ min for a seeded styrene emulsion polymerization system.[85] The fitted PSD was computed by solution of equations (58) using

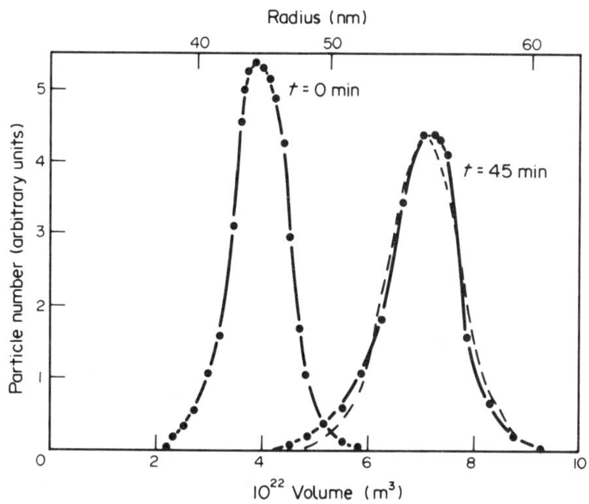


**Figure 11** PSD at time = 0 and 45 min for a styrene-seeded emulsion polymerization, and fit using methods of Section 11.4.3 (reproduced with permission of John Wiley and Sons Inc. from G. Lichti, B. H. Hawkett, R. G. Gilbert, D. H. Napper and S. F. Sangster, *J. Polym. Sci., Polym. Chem. Ed.*, 1981, **19**, 925)

the rate parameters deduced from the PSD data, applying the methods of Section 11.4.3 as follows. The initial distribution was of course that at $t = 0$. The value of $k$ (which in this system is sufficiently small as not to significantly affect the PSD) was set to that found from kinetic studies,[29] *i.e.* $10^{-3}$ $s^{-1}$, and because of the small value of $k$, one can put $\alpha = 0$. The values of $\rho$ and $K$ can be estimated *e.g.* using equations (61) and (62), and then varied to bring about agreement between the observed and calculated PSDs at $t = 45$ min. The values of $\rho$ and $K$ so obtained are in excellent agreement with those obtained from the kinetics. The kinetic parameters determined from PSDs provide a useful check on the validity of those derived by other methods, as given in Section 11.3.2, and *vice versa*.

The foregoing example shows that the PSD in Interval II can be used to deduce kinetic parameters, and alternatively that kinetic parameters, if known with sufficient accuracy, can be used to deduce reliable PSDs.

### 11.4.5 PSDs in *ab initio* Systems

It will be seen that the time evolution of the PSD in an *ab initio* system offers important mechanistic information as to the nucleation process. Figure 12 shows PSDs for a styrene *ab initio* run, obtained at the start of Interval II (just after the cessation of nucleation) and various later times in Interval II. The most striking feature is that the early time PSD is positively skewed, this skewness decreasing markedly at the later times. The observed positive skewness can be qualitatively interpreted without recourse to the ensuing quantitative treatment, although it is emphasized that the latter is necessary to avoid error. The skewness can be immediately ascribed to that responsible for the positive skewness in Figure 10 (Section 11.4.3); the nucleation rate, $dN_c/dt$, must be initially an increasing function of time (although it must later decrease, as nucleation ceases).

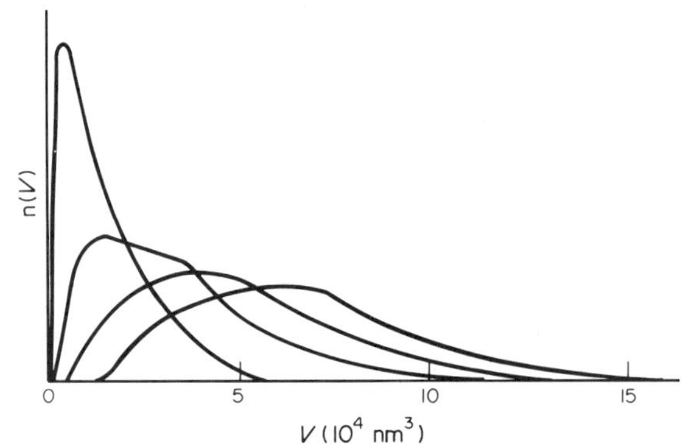


**Figure 12** PSDs (arbitrary units) at $t = 8$, 13, 18 and 23 min after commencement of polymerization in a styrene system (redrawn from G. Lichti, R. G. Gilbert and D. H. Napper, *J. Polym. Sci., Polym. Chem. Ed.*, 1983, **21**, 269)

The quantitative interpretation of the PSD data uses the theory of Section 11.4.3. The data of Figure 12 can be used to determine $c_i(V, t)$, the nucleation rate of particles containing $i$ free radicals, equations (53), (55) and (58). It can be seen that these equations offer (in principle) a means of deducing the time dependence of $dN_c/dt$ from PSD data. However, to do that, it is necessary first to determine the values of $\rho$, $k$ and $K$. For this particular system (styrene), there are a number of simplifications that can be made. Firstly, the system follows 'zero–one' kinetics. Secondly, values for $k_p$ and $C_M$ for similar systems have already been determined (Sections 11.3.3 and 11.3.4). It is thus necessary only to find the values of the rate coefficients for entry ($\rho$) and exit ($k$), and (if required) the dependence of these on $V$. Thirdly, it is found by sample calculation (using the techniques of Section 11.4.3) that the $V$ dependence of these quantities does not in fact need to be considered, since the results for the PSD with both $V$-independent and $V$-dependent $\rho$ and $k$ (as long as the $V$ dependence assumed is in accord with experiment) do not show important differences, at least for the conditions under study. Thus, for the present system, it is only necessary to do the following. (i) Obtain the value for the ratio $\rho/k$ from the steady state slope of the $x(t)$ curve, using equation (16). Note that because the time variation of the PSD is comparatively slow compared to the time required for the system to reach a steady state [*i.e.* $1/(2\rho + k)$] there is no need to separately determine the value of the 'fate parameter' $\alpha$ in this system, this value being incorporated in the value of $\rho$ determined with this method. (ii) From equations (59) and (60), find the value of $\rho$ (and hence $k$) which will produce the observed PSD at the later time from the observed PSD at the earlier time (the PSD data are all in Interval II). Given $\rho$ and $k$ found as described, one then determines the form of $c_i(V, t)$ which will reproduce the observed $n(V, t)$ at early times (at the start of Interval II). This involves choosing an appropriate form for $dN_c(t)/dt$. It is sufficient here to choose a simple but flexible functional form that should be able to give the essentials of the various likely mechanisms of nucleation; the final result does not depend on fine details of $dN_c(t)/dt$, but only on certain overall features. Thus a simple form such as a triangle will suffice, as used for Figure 10.

It is found that no form for $dN_c(t)/dt$ which is initially a decreasing function of time can fit the data, for reasons explained above. The only form of the $dN_c(t)/dt$ that can reproduce the early time PSD is one that is an initially increasing function of time,[91,93] as shown in Figure 10 (exact functional forms have been given elsewhere[91,93]). Thus the observed positive skewness suggests that the nucleation rate is initially an increasing function of time. However, it must also be pointed out that accurate calculations of $dN_c(t)/dt$, as given in Section 11.6.4, show that some (but not the whole) contribution to the positive skewness also arises from the volume dependence of $C_M$, *i.e.* of $K_{11}$ of equation (54).

In conclusion, it can be seen that a knowledge of the growth and nucleation kinetics can either be obtained from, or used to predict, the time evolution of the PSD. In Section 11.6.5 it will be shown that this knowledge can be used to predict and understand effects such as the variation of particle number and polydispersity with quantities such as surfactant concentration and type.

## 11.5 MOLECULAR WEIGHT DISTRIBUTIONS

### 11.5.1 Overview

The molecular weight distribution (MWD) may be an important factor determining the physical properties of a formed polymer such as rate of crack propagation, tensile strength, *etc.* Moreover, certain rate coefficients such as $k_t$ can depend strongly on the molecular weight of the polymer matrix and/or the growing polymer chain.[65,74]

One of the most notable characteristics of many emulsion polymerizations is that the molecular weight of the formed polymer is much greater (*e.g.* by a factor of 10) than that of the same monomer polymerized in bulk. This is due to compartmentalization; the rate coefficient for termination is high, and in bulk systems there is nothing preventing two free radicals from mutually annihilating. However, in emulsion systems that show a low $\bar{n}$, there will not be more than one free radical per particle, and so the rate of termination is greatly reduced; hence the molecular weight is large. Indeed, in many emulsion polymerization systems the principal chain-stopping event is transfer to monomer, which governs the MWD.

This section shows how the MWD can be computed from the rate parameters governing the kinetics of an emulsion polymer system.

The *number molecular weight distribution*, $S(M)$, is defined so that the (relative) number of chains with molecular weight between $M$ and $M + dM$ is $S(M)dM$. The weight MWD, $W(M)$, is given by $MS(M)$. The number average molecular weight $\langle M_n \rangle = \int MS(M)dM / \int S(M)dM$. Some typical values for $\langle M_n \rangle$ are $10^5$ for polystyrene prepared by bulk polymerization and $3 \times 10^6$ for the polymer when prepared by emulsion polymerization; the origin of this important qualitative difference (compartmentalization) was given above. The *weight average molecular weight* $\langle M_w \rangle = \int M^2 S(M)dM / \int MS(M)dM = \int MW(M)dM / \int W(M)dM$. The *polydispersity ratio* $\gamma = \langle M_w \rangle / \langle M_n \rangle$.

There are a number of common values for $\gamma$. (a) An $S(M)$ which is completely monodisperse (*i.e.* a delta function) has $\gamma = 1$. (b) An $S(M)$ curve which consists of a single exponential (*e.g.* the Flory–Schulz most probable distribution)[97] has $\gamma = 2$; this will be seen to apply where the MWD is governed entirely by transfer or (in a bulk or pseudo-bulk system) by disproportionation. (c) An $S(M)$ curve which approximates the distribution $S(M) = Me^{-aM}$ (which is a limiting case of the Poisson distribution) has $\gamma = 1.5$; this is applicable to the case of bulk (or pseudo-bulk) polymerization where the only chain-stopping event is combination. (d) An $S(M)$ containing significant components of two separate exponentials has $\gamma > 2$. It can be seen that $\gamma$ provides a convenient quantitative description of the shape of an $S(M)$ curve, independent of the actual values of $\langle M_n \rangle$ or $\langle M_w \rangle$.

### 11.5.2 Evolution Equations for MWD

The equations giving the time evolution of the MWD must take into account all events which can start and can stop the growth of polymer chains: entry, exit, termination (by both combination and disproportionation) and propagation (as considered in previous sections) and in addition (and, often, most importantly) transfer of free radical activity from a growing chain to another entity.

A complete solution to the problem of determining the MWD in an emulsion polymerization system, given values of the basic rate parameters, has been provided by Lichti *et al.*[95,98] This formulation encompasses less complete expressions developed by Gerrens,[99] Gardon,[49] Lin and Chiu,[100] Katz *et al.*,[101] and Min and Ray.[89,102]

The rate coefficient $a$ giving the increase in molecular weight per unit time is given by equation (64). For the purposes of computing the MWD, the pseudo-first-order bimolecular termination rate coefficient $c$ must be broken up into components arising from bimolecular combination ($c_c$) and disproportionation ($c_d$), as in equation (65). The pseudo-first-order transfer rate coefficient, for transfer both to monomer ($k_{tr,M}$) and to added chain transfer agent (with concentration [A]), $k_{tr,A}$, is denoted by $F$, as in equation (66).

$$a = k_p C_M M_0 \tag{64}$$

$$c = c_c + c_d \tag{65}$$

$$F = C_M k_{f,M} + [A] k_{tr,A} \tag{66}$$

A general MWD theory incorporating the effect of long chain branching and/or cross-linking has yet to be developed; the following treatment assumes that these two effects occur with negligible frequency. The values of $C_M$, $\rho$, $k$, $c_c$, $c_d$, $a$ and $F$ are then sufficient to specify the MWD of a given system; in most cases only one or two of these rate coefficients will actually be important in determining the MWD.

The formulation of Lichti *et al.*[95,98] is based on extensions of the Smith–Ewart concept of the $N_i$, which is the number density of latex particles containing $i$ growing free radicals at time $t$. To determine the MWD, one needs to take explicit account of how long each chain has been growing, *i.e.* of all chain-stopping and chain-starting events. These can be of either first or second order with respect to a growing chain; the first-order events are transfer, entry, exit and bimolecular disproportionation (this last is first order because the molecular weight of the chain in question will be independent of the molecular weight of the second chain with which it undergoes disproportionation); the only second-order event is bimolecular combination (since the molecular weight of the formed chain depends on the molecular weights of both chains which undergo combination).

To take into account the first-order events, Lichti *et al.* defined 'singly distinguished' particles as those containing $i$ free radicals, one of which (the 'distinguishing chain') commenced growth at time $t$ and one of which has been growing for a time $t'$ without any chain-stopping event occurring (see Figure 13).

The number density of the singly distinguished particles can be defined to be $N_i'(t, t')$; since this requires at least one growing chain to be present, one has $i > 0$. At time $t + t'$, the number of

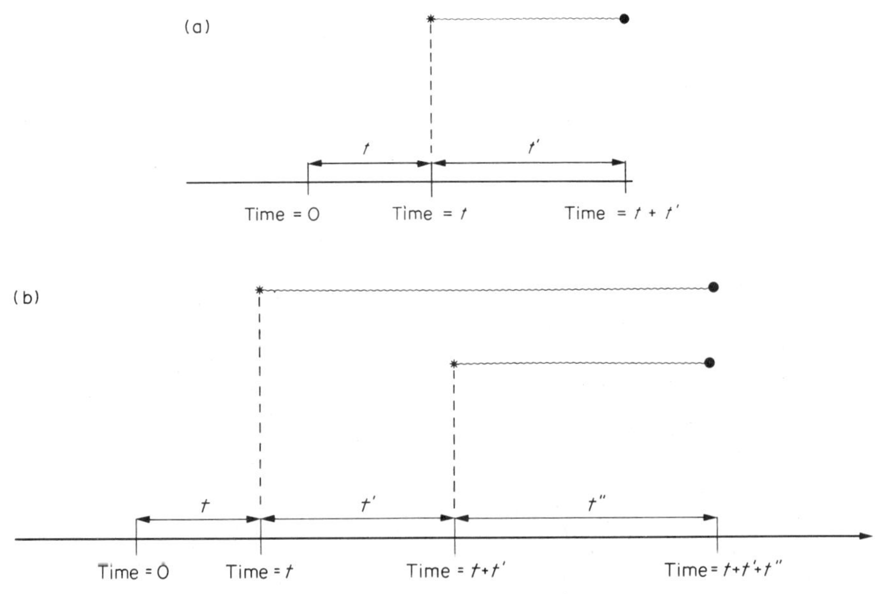


**Figure 13** Various times involved in (a) singly and (b) doubly distinguished chain; * denotes when a chain starts

'distinguished' chains which have been growing for a time between $t'$ and $t'+\mathrm{d}t'$ is $N_i'(t, t')\mathrm{d}t'$. If one knows this distribution function and also the rate of chain-stopping events, then one will know the number of chains that cease growth at any given time and which have up until then been growing for a known time; this will yield the MWD arising from all first-order chain stopping events.

The $N_i'(t, t')$ can be obtained by specifying the evolution equations and the boundary conditions. The appropriate evolution equation can be written down as in equation (67), analogous to the evolution equations for the $N_i(t)$ (equation 4).

$$\partial N_i'/\partial t' = \rho N_{i-1}' - [\rho + ik + F + i(i-1)c]N_i' + ikN_{i+1}' + i(i+1)cN_{i+2}' \quad (67)$$

Equation (67) is deduced[98] by a straightforward extension of that used to deduce the Smith–Ewart equations for the $N_i$. Equation (67) is a set of coupled ordinary linear first order equations for the evolution of $N_i'(t, t')$ in $t'$, Generally, the lifetime of a chain is much less than the time scale over which any of the rate parameters involved can change, and thus the various coefficients in equations (67) may be accurately treated as constant. The solution of equations (67) can then be written down in terms of the eigenvalues of the appropriate matrix,[103,104] as in equations (68). Here $\mathbf{N}'$ is the vector with elements $N_1'$, $N_2'$, . . ., $\lambda_j'$, ${}^{\mathrm{L}}e_j'$ and ${}^{\mathrm{R}}e_j'$ are respectively the eigenvalues, left and right eigenvectors of the matrix defined by the right hand side of equations (67). The initial conditions, $\mathbf{N}'(t, t'=0)$ can be written down immediately from the appropriate chain-starting events, and are given by equation (69). The explicit solutions are given in equation (70). Here ${}^{\mathrm{R}}e_{ij}'$ and ${}^{\mathrm{L}}e_{im}'$ are the $i$th components of the ${}^{\mathrm{L}}e_j'$ and ${}^{\mathrm{R}}e_j'$.

$$\mathbf{N}'(t, t') = \sum \exp(\lambda_j' t')\,{}^{\mathrm{L}}e_j'\,[{}^{R}e_j'\cdot \mathbf{N}'\ (t'=0)] \quad (68)$$

$$N_{i'}(t, t'=0) = \rho N_{i-1} + iFN_i \quad (69)$$

$$N_i'(t') = \sum_j a_{ij}'\exp(\lambda_j' t'),\ a_{ij}' = {}^{R}e_{ij}'\sum_m {}^{\mathrm{L}}e_{\mathrm{im}}'(\rho N_{m-1} + mFN_m) \quad (70)$$

As stated, the $N_{i'}$ give all the necessary information to determine the contribution to the MWD from first-order chain-stopping events. To take account of the second-order contribution (*i.e.* that arising from combination), one defines 'doubly distinguished' particles, with distribution function $N_i''(t, t', t'')$. These are defined as the number of latex particles containing $i$ free radicals, one of which commenced growth at time $t$, grew for a time $t'$, at which moment a new distinguished chain commenced growth (while the original distinguished chain continued to grow), and both chains continued growth for a time $t''$ (see Figure 13). The $N_i''$ are only defined for $i \geq 2$, since only particles containing two or more free radicals can contain two distinguished chains. The evolution equations for the $N_i''(t, t', t'')$, given by equation (71), are deduced[95,98] in a manner similar to those for the $N_i'(t, t')$, except that the evolution will now be in terms of $t''$. The boundary conditions for the $N_i''$ (equation 72), are deduced analogously to those for the $N_i'$ in equation (69). Explicit solutions are given in equations (73).

$$\partial N_i''(t, t', t'')/\partial t'' = \rho N_{i-1}'' - [\rho + 2F + ik + i(i-1)c]N_i'' + (i-1)kN_{i+1}'' + i(i-1)cN_{i+2}'' \quad (71)$$

$$N_i''(t, t', t''=0) = \rho N_{i-1}'(t, t') + (i-1)FN_i'(t, t') \quad (72)$$

$$N_i''(t, t'') = \sum_j\sum_k b_{ijk}\exp(\lambda_k' t' + \lambda_j'' t''),\ b_{ijk} = {}^{R}e_{ij}''\sum_m {}^{\mathrm{L}}e_{km}''[\rho a_{k,m-1}' + (m-1)Fa_{km}'] \quad (73)$$

The $N_i'(t, t')$ and $N_i''(t, t', t'')$ yield the MWD at time $T$ as follows. Since each growing chain can be stopped by only one event, the MWD can be separated into components arising from each type of chain stopping event, *viz.*, exit ($S_{ex}$), transfer ($S_{tr}$), disproportionation ($S_d$) and combination ($S_c$), as in equation (74). Consider first $S_{ex}(M)$. The rate of chain stoppage by exit at time $t_e = t + t'$ can be seen to be given by $\Sigma kN_i'$. To determine the relation between $M$ and $t_e$, one notes firstly that the number of chain stoppages will be very great over the experimental time scale, and thus that $T \gg t'$, and therefore to an excellent approximation, $M = at'$, since equation (64) gives the relation between growth time and molecular weight of the polymer chain. Finally, if the rate coefficients are constant over time $T$, the integral is found by multiplying the integrand by $T$ (if the rate coefficients change over $T$, the overall MWD is simply found by accumulating contributions over appropriate times). The total number of chains of molecular weight $M$ at time $T$ whose growth is terminated by exit is thus given by equation (75). Similarly, for the transfer contribution to the MWD, one has equation (76), and for the contribution from termination by disproportionation, equation (77); again, $M = at'$ in both equations.

$$\mathrm{S}(M) = \mathrm{S}_{\mathrm{ex}} + \mathrm{S}_{\mathrm{tr}} + \mathrm{S}_{\mathrm{d}} + \mathrm{S}_{\mathrm{c}} \tag{74}$$

$$\mathrm{S}_{\mathrm{ex}}(M, T) = \int_0^T \mathrm{d}t_{\mathrm{e}} \sum_{i \geq 1} kN_i'(t, t' = t_{\mathrm{e}} - t) \mathrel{\hat{=}} Tk \sum_{i \geq 1} \sum_j a_{ij}' e^{\lambda_j' t'} \tag{75}$$

$$\mathrm{S}_{\mathrm{tr}}(M, T) = \int_0^T \mathrm{d}t_{\mathrm{e}} \sum_{i \geq 1} FN_i'(t, t') \mathrel{\hat{=}} TF \sum_{i \geq 1} \sum_j a_{ij}' e^{\lambda_j' t'} \tag{76}$$

$$\mathrm{S}_{\mathrm{d}}(M, T) = \int_0^T \mathrm{d}t_{\mathrm{e}} \sum_{i \geq 2} 2(i - 1)c_{\mathrm{d}} N_i'(t, t') \mathrel{\hat{=}} 2c_{\mathrm{d}} T \sum_{i \geq 2} (i - 1) \sum_j a_{ij}' e^{\lambda_j' t'} \tag{77}$$

The contribution to the MWD from combination can be obtained from a knowledge of the $N_i''(t, t', t'')$. Figure 13 shows that the combination of the two distinguished growing chains (one of which has been growing for $t''$ and the other for $t' + t''$) leads to a single chain whose molecular weight $M$ is given by $M = a(t' + 2t'') = at_{\mathrm{e}}$. We can find the total contribution (by combination) to a given molecular weight $M = at_{\mathrm{e}}$ by integrating over all $t''$ which can give this $M$ (this range is from $t'' = 0$ to $t'' = t_{\mathrm{e}}/2$, corresponding to $t' = t_{\mathrm{e}}$ and $t' = 0$, respectively). For the MWD at a time $T$, by reference to Figure 13 one can see that $T = t + t' + t''$. Thus to obtain the contribution to the MWD from combination for $M = at_{\mathrm{e}}$, one integrates over all $N_i''(T - t_{\mathrm{e}} + t'', t_{\mathrm{e}} - 2t'', t'')$, as well as sum over all states $i$ ($\geq 2$) and integrating over all times $t_{\mathrm{e}}$. Thus, the component of the MWD arising from combination is given by equation (78).

$$\begin{aligned} \mathrm{S}_{\mathrm{c}}(M, T) &= 2c_{\mathrm{c}} \int_0^T \mathrm{d}T \sum_{i \geq 2} \int_0^{1/2t_{\mathrm{e}}} \mathrm{d}t'' N_i''(T - t_{\mathrm{e}} + t'', t_{\mathrm{e}} - 2t'', t'') \\ &= 2c_{\mathrm{c}} T \sum_{i \geq 2} \sum_j \sum_k \frac{b_{ijk}[\exp(\frac{1}{2}\lambda_j'' t') - \exp(\lambda_k' t')]}{\lambda_j'' - 2\lambda_k'} \end{aligned} \tag{78}$$

Although the final form of the equations for the MWD so obtained are somewhat complicated algebraically, they may be readily evaluated numerically. In addition, comparatively simple analytic expressions may be derived in a frequently applicable case, that of a 'zero–one–two' system.[98] This description encompasses emulsion polymerization systems wherein the average number of free radicals per particle ($\bar{n}$) does not exceed $\sim 0.7$. These various expressions can be readily incorporated in a modelling program to generate the complete MWD for any system. Equations (74)–(78) can also be used to generate $\langle M_{\mathrm{n}} \rangle$ and $\langle M_{\mathrm{w}} \rangle$. Min and Ray[89, 102] have presented alternative means of incorporating the calculation of $\langle M_{\mathrm{n}} \rangle$ and $\langle M_{\mathrm{w}} \rangle$ into such programs.

### 11.5.3 MWDs of Typical Systems

While the formalism outlined above readily yields numerical solutions, it is useful to examine analytic solutions in limiting cases. For simplicity, we shall assume that all rate parameters are constant over the time period governing the S($M$) in question, *i.e.* we consider the instantaneous MWD.

An especially important case is that for a system which is transfer dominated. The MWD expressions then given above reduce to the bulk/solution formula. That is, in the transfer-dominated limit (*i.e.* where the only chain-stopping event is transfer; $\rho = c = k = 0$) equation (76) reduces to equation (79), *i.e.* single exponential behaviour, with $\gamma = 2$. Equation (79) has the important consequence that experimental MWDs should be plotted as $\ln(S)$ against $M$, to test if the system is indeed transfer dominated. Other tests will be given later.

$$\mathrm{S}_{\mathrm{tr}}(M) \propto \exp[-(M/M_0)(k_{\mathrm{tr}}/k_{\mathrm{p}})] \tag{79}$$

It should be noted that, for most systems, transfer or termination are the dominant chain-stopping events, rather than entry. If the system were zero–one, and if the rate of transfer were comparable to that of entry, then the MWD could be used to determine $\rho$; however, data given later show that such conditions are not satisfied in common systems such as styrene.

Next, we consider the limits of the distinguished particle equations as $\bar{n}$ becomes high, *i.e.* the pseudo-bulk case. These results are applicable to ordinary bulk or solution polymerization, as well as to emulsion systems with high $\bar{n}$. It can be shown (after some algebra) that, in the limit where combination is the only chain-stopping event, one has equation (80), whereas when disproportionation dominates, one has equation (81). Here, for emulsion systems, $b = 2c\bar{n}/(k_{\mathrm{p}} C_{\mathrm{M}} M_0)$, while for bulk systems, $b = 2(k_d f[\mathrm{I}]k_t)^{1/2}/(k_p C_M M_0)$, $f$ being the initiator efficiency. The combination-dominated (pseudo)-bulk limit is the distribution $\mathrm{S}(M) \propto Me^{-bM}$, *i.e.* with $\gamma = 1.5$. The (pseudo)-bulk disproportionation-dominated limit is a single exponential, *i.e.* with $\gamma = 2$.

$$\mathrm{S_c}(M) \quad \propto \quad Me^{-bM} \tag{80}$$

$$\mathrm{S_d}(M) \quad \propto \quad e^{-bM} \tag{81}$$

Equations (80) and (81) show that the test for the system being disproportionation dominated is that a $\ln[\mathrm{S}(M)]$ *vs.* $M$ plot is linear, with a slope proportional to $\bar{n}$, or to [I] for a bulk system; note that this proportionality will not be obtained in the transfer-dominated case, although both will show a linear $\ln[\mathrm{S}(M)]$ plot. The test for combination-dominated case is that a plot of $\ln[\mathrm{S}(M)/M]$ will be linear.

We next consider compartmentalized (*e.g.* emulsion polymer) systems; some polydispersities as functions of $\bar{n}$, generated by solution of equations (74)–(78), are given in Figure 14. Consider a system where termination determines MWD. One sees[95,98] that for compartmentalized transfer and exit-dominated systems, $\gamma = 2$, as in bulk. For the combination-dominated system, $\gamma$(compartmentalized) is found to be significantly greater than $\gamma$(bulk) (*e.g.* 1.9 as compared with 1.5). Similarly, in a disproportionation-dominated system, one finds significant broadening compared with the bulk value of $\gamma = 2$. The bulk value of $\gamma$ is attained only if $\bar{n}$ is sufficiently high.

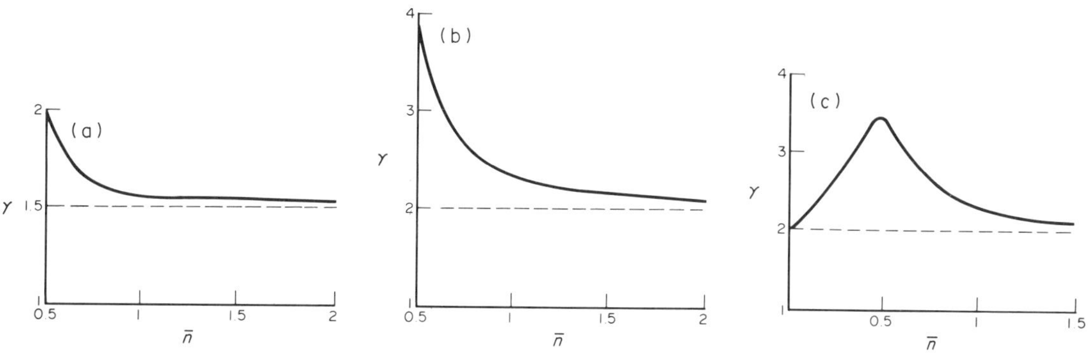


**Figure 14** Plots of polydispersity $\gamma$ as function of $\bar{n}$, for compartmentalized systems: chain stoppage by (a) combination alone, (b) disproportionation alone and (c) exit and disproportionation (redrawn from G. Lichti, R. G. Gilbert and D. H. Napper, in 'Emulsion Polymerization', ed. I. Piirma, Academic Press, New York, 1982)

These compartmentalization effects can be understood as follows. When $\bar{n} = 0.5$, the latex particles contain no more than one free radical at any time, since the entry of a second free radical causes instantaneous termination. In the combination-dominated system, this termination behaves kinetically as does transfer (*i.e.* is first order) and so leaves a contribution to the MWD with $\gamma = 2$. However, with a disproportionation mechanism operative, the entrant free radical leaves in addition a low molecular weight species for each longer dead chain formed. These two contributions to the MWD are additive (*i.e.* one has double exponential behaviour), resulting in a halving of the number-average molecular weight while leaving the weight-average molecular weight essentially unaffected. Thus the overall $\gamma$ changes from 2 to 4 on going from a combination-dominated to disproportionation-dominated system.

### 11.5.4 Experimental Results

The methods for experimental determination of various measures of the MWD are well known. The only reliable method for obtaining the full molecular weight distribution (as distinct from various means thereof) is gel permeation chromatography (see Volume 1). However, this is a relative method, not an absolute one, and needs calibrating with (for example) light scattering data. The GPC trace is *approximately* the weight distribution, $\mathrm{W}(M) = \mathrm{S}(M)M \mathrel{\hat{=}} \mathrm{G}(V_{el})$: the trace at elution volume $V_{el}$, and where $V_{el}$ is related to $M$ by calibration. However, the only precise means of obtaining the $\mathrm{S}(M)$ is by noting that the GPC signal is proportional to the contribution from each molecular weight, leading to the convolution equation (82); a Fredholm integral equation, which must be solved to yield W($M$), and thus S($M$). In equation (82), H($V_{el}$, $M$) is the response function of the instrument and column, which is determined by calibration with a monodisperse sample. A number of means of carrying this out have been presented.[84,105,106] However, recently GPC columns (*e.g.* those from Polymer Laboratories) have become available which reduce the need for this deconvolution.

$$G(V_{el}) = \int_0^\infty H(V_{el}, M)W(M)\,dM \tag{82}$$

In relating theory and experiment, it is clearly insufficient to consider a single measure of the MWD (*e.g.* $\langle M_w \rangle$ or $\gamma$) alone, since interpretation of a single datum is prone to ambiguities. We thus consider only GPC results, which give the full MWD curve.

Friis and Hamielec[107] have presented GPC data on vinyl chloride and vinyl acetate emulsion polymerizations. For these monomers, the main chain-stopping event is thought to be transfer. It was found that the MWDs at early times, when branching is negligible, have a $\gamma$ value close to two, as expected for a transfer-dominated system. A careful experimental study of the MWD produced in a highly compartmentalized system has been carried out by Piirma *et al.*[108] An *ab initio* emulsion polymerization of styrene was used which produced particles small enough (*ca.* 75 nm unswollen radius) to ensure low $\bar{n}$ values in the Interval II region. During this interval, $\gamma$ was found to be essentially constant and have a value slightly larger than two. This value of $\gamma$ implies that the system is dominated either by transfer or some other first-order chain-stopping event. The observed value of $\langle M_w \rangle$, *ca.* $2 \times 10^6$ with oleate as surfactant, in fact corresponds closely to that expected for a system dominated by transfer to monomer. This result is interesting, in view of the fact that the main chain-stopping mechanism for styrene in bulk or solution polymerization is combination. This result is thus a clear example of how compartmentalization can change the dominant chain-stopping mechanism; as stated, this is because free radicals inside the latex particles polymerize in isolation for periods of time long enough to render the MWD transfer dominated. It was found that surfactants such as sodium dodecyl sulfate led to considerably lower $\langle M_w \rangle$ values, indicating that this surfactant acts as a chain transfer agent. This conclusion is supported by the data of Lin and Chiu[100] on styrene systems, who reported $\langle M_w \rangle$ values considerably lower than those found by Piirma *et al.*; this suggests that the surfactant used by the former indeed causes chain transfer. De Graaf and Poehlein[109] also reported the MWD of styrene in a stirred tank reactor (continuous flow). Again, $\gamma$ values of close to two were found. On the other hand, if the reaction temperature is sufficiently high (90 °C or greater), styrene MWDs become combination dominated.

As a further example of experimental and theoretical comparisons involving the MWD, we give in Figure 15 results from our laboratory for the MWD in a styrene emulsion polymerization system at 50 °C, with the sample taken at 25% conversion; the results were obtained by deconvoluting GPC data through solution of equation (82). It is found that $\langle M_n \rangle = 8 \times 10^5$, $\langle M_w \rangle = 2 \times 10^6$, $\gamma = 2.4$. It can be seen that for moderately high molecular weights, the observed MWD follows a single exponential. The slope of this log[S($M$)] *vs.* $M$ plot for all but the lowest molecular weight data yields a value of $k_{tr}/k_p$, from equation (79), of $5 \times 10^{-5}$. This compares with the literature value[27] of $6 \times 10^{-5}$ for this quantity for monomer. Clearly, the system is transfer dominated. At lower molecular weights, however, Figure 14 shows that S($M$) is greater than would be predicted by back extrapolation of the exponential, and indeed seems to show a peak at the lowest ($< 10^4$) molecular

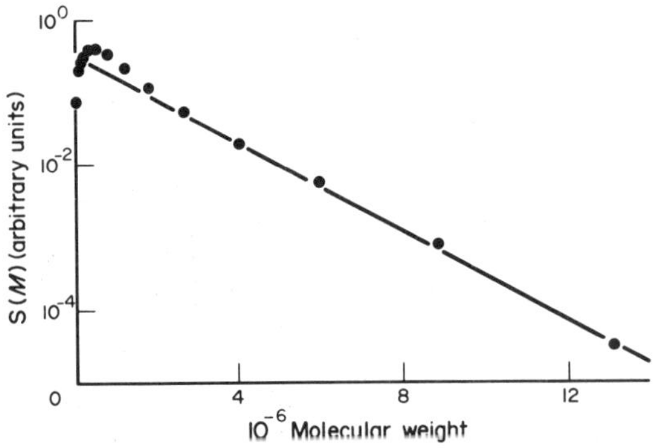


**Figure 15** MWD (arbitrary units) for styrene *ab initio* emulsion polymerization, using oleic acid as surfactant, at 50 °C: sample taken at 25% conversion

weights. This additional polymer may be ascribed to the occurrence of bimolecular termination events, principally combination, during the nucleation period, arising from coagulation of precursor particles (see Section 11.6.3). The apparent decrease in MWD at the very lowest molecular weights, for which there appears no obvious explanation, may be a real effect or may possibly be an artifact of instrumental response and the deconvolution procedure.

There is considerable scope for further work in this area, in particular with regard to the use of theory outlined in the previous section to make mechanistic deductions from observed MWD data, and to use the theory to predict conditions (particularly in CSTR systems) for producing MWDs with desired characteristics.

## 11.6 KINETICS OF PARTICLE FORMATION

### 11.6.1 Preamble

The mechanism of particle formation is particularly difficult to investigate, since there is a plethora of physically reasonable possible contributions to this event, but only a limited number of experimental observables which can be used to test theoretical models. Since all current models contain some parameters whose values are subject to uncertainty, it is possible with limited or inappropriate experimental data to find evidence in support of practically any mechanism. Consider, for example, the pioneering work of Smith and Ewart,[5] who showed that the micellar entry mechanism (*q.v.*) implies that the particle number must be proportional to $[I]^{2/5}[S]^{3/5}$, where [S] is the concentration of surfactant above the critical micelle concentration (CMC). The observation that some systems approximately followed this dependence was then widely quoted as evidence for micellar entry. It was subsequently pointed out by Roe[110] that the same two exponents are also predicted by a homogeneous nucleation mechanism, or indeed any model which *inter alia* assumes that cessation of complete surface coverage by surfactant is one of the governing events. Moreover, Fitch has pointed out[111] that in actuality, a wide range of exponents is observed, and that the data do not always obey a simple power law when plotted accurately. To investigate nucleation, many workers have adopted the 'global' approach of modelling conversion data (*e.g.* see Hansen and Ugelstad's review);[12] it was pointed out that this approach can rarely be used to disprove a supposed mechanism. Others (*e.g.* see Eliseeva's review)[113] have examined different types of data (*e.g.* rates, particle number, particle size, polydispersity and surface properties) to seek clues to the important events contributing to nucleation; because it is not unreasonable to assume that more than one mechanism may govern these properties, physically acceptable conclusions are drawn which are frequently different from, for example, those of Hansen and Ugelstad, for similar systems.

In view of these inconsistencies in the literature, we take a different approach. First, particle growth is separated from particle nucleation; the former was discussed in Section 11.3. Second, sensitive data (such as the time evolution of the particle size distribution function) are examined which can be used to refute postulated mechanisms.

We briefly summarize the principal mechanisms which may be operative in particle nucleation in a given system. (i) Micellar-entry[4,5]—free radicals in the aqueous phase 'sting' micelles containing solubilized monomer to form small particles directly. (ii) Homogeneous nucleation[114]—oligomeric free radicals in the aqueous phase propagate until they attain a sufficiently high degree of polymerization to precipitate, and thereupon yield particles which swell with monomer. (iii) Coagulative nucleation[47,91,112,115]—particles formed by either of the previous two mechanisms are colloidally unstable, and are termed 'precursor particles'; these form colloidally stable latex particles by coagulation and by propagational growth.

### 11.6.2 Experimental Observables

We now consider the experimental observations that can yield information on nucleation.

#### *11.6.2.1 Polymerization rate*

This is a relatively insensitive quantity to mechanistic assumptions, since (Section 11.3.1) it depends not only on $N_c$, but also on $k_p$, $\rho$, $k$, $k_t$ and $\alpha$. Some of these quantities may be functions of the particle volume, which changes dramatically during Interval I. It is thus impossible to use rate data alone to make any mechanistic deductions concerning nucleation, although such data may be used in conjunction with other observables.

#### *11.6.2.2 Particle rate*

This can provide useful data, particularly in the variation of $N_c$ with [S] and [I] and (if technically possible) the time variation of $N_c$;[116] again, however, the interpretation must be carried out in conjuction with other measurements. A related experimental technique is that of competitive nucleation in a seeded system.[112]

#### *11.6.2.3 Particle size distribution*

Particle size distribution data will be seen to provide means of refuting mechanistic suppositions. The full PSD is most informative, rather than just the mean and standard deviation; it will be seen in particular that the sign of the skewness (equivalent to the third moment of the distribution function) is important here.

### 11.6.3 A Case Study: Styrene

We now consider the results of specific studies on styrene nucleation, carried out at 50 °C, using sodium dodecyl sulfate as surfactant.[91,93] Figure 12 shows PSDs obtained at the start of Interval II (just after the cessation of nucleation) and at various later times. As discussed in Section 11.4.5, the early time PSD, expressed as the volume distribution, is positively skewed, this skewness decreasing markedly at the later times; this suggests that the nucleation rate, $dN_c/dt$, must be initially an increasing function of time, of the form shown in Figure 16. The use of the PSD data to deduce the actual form of $dN_c/dt$, through solution of equations (53), was given in Section 11.4.5. It is found that no form for $dN_c(t)/dt$, which is initially a decreasing function of time, can fit the data.[91]

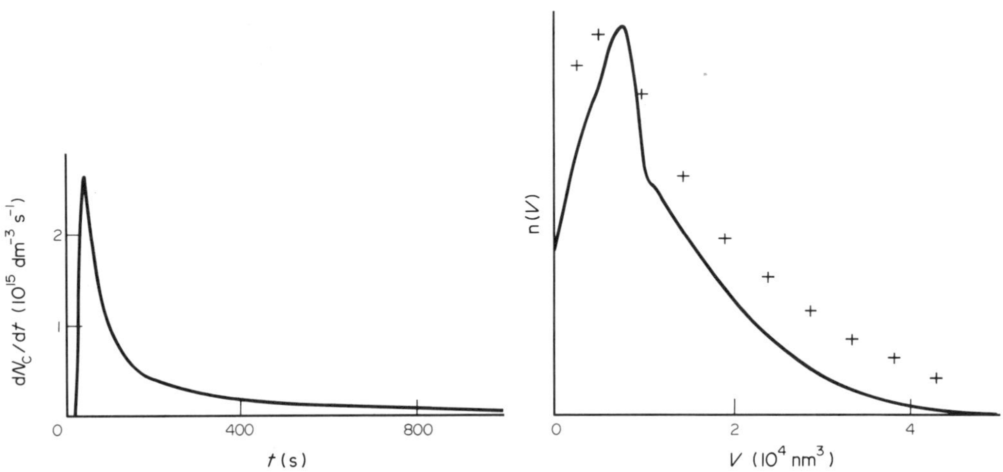


**Figure 16** Calculated and observed PSD (arbitrary units) for the system of Figure 12, and calculated time dependence of $dN_c/dt$ (calculations done using the nucleation model of Section 11.6.4)

Detailed mathematical treatments[112] for micellar entry and homogeneous nucleation mechanisms, without any account of coagulation, show that neither can give this observation; as is physically obvious, both must predict $dN_c(t)/dt$ curves which are initially non-increasing functions of time. The data of Figure 12 therefore refute single-step micellar entry and homogeneous nucleation mechanisms for particle formation, at least in the styrene system under consideration.

We note at this point that, although the micellar entry theory is popularly espoused (presumably for historical reasons), there appears to be absolutely no evidence which unambiguously shows the direct involvement of micelles in the nucleation process; however, there is a growing body of evidence which refutes micellar entry. This mechanism was proposed by Harkins, Smith and Ewart (who are the fathers of our understanding of heterogeneous polymerization) on the basis of the limited data available in the 1940s. The most commonly cited reason for the rectitude of micellar entry is that the observed dependence of $N_c$ on [S] shows a sharp change at the CMC. However, careful studies[117] show that this change can occur above or below the CMC, depending on the system; as will be seen later, this observed dependence on [S] is also consistent with other models for nucleation which are not refuted by the PSD data cited above. The PSD data seem completely incompatible with micellar entry. A further refutation of the micellar entry mechanism is found with the observation of Liesegang rings,[118,119] which (as discussed below) appear to be inexplicable in terms of micellar entry. In addition, it is well known that latex particles can form well below the CMC (indeed, even in the complete absence of added surfactant), demonstrating that there must be a mechanism alternative to micellar entry in these systems. Thus modern data suggest that micellar entry plays little (if any) role in latex particle formation. This mechanism is therefore no longer considered here. Unfortunately, some emulsion polymerization models in current usage[89,120,121] assume micellar entry for the nucleation mechanism, and hence it would appear that use of such models for conditions significantly different from those for which they were parameterized could lead to quantitative and qualitative errors.

The data exemplified in Figure 12 suggest that the nucleation rate must be an increasing function of time. Such behaviour is characteristic of a sequential coagulative mechanism. This observation is not confined to the styrene system. For example, an extensive study by Fitch *et al.*[116] using laser light scattering to deduce (essentially) the particle number as a function of time leads to the same conclusion for the MMA system; however, the data interpretation employed in these studies was model based (using HUFT[15,47] theory) and is thus not as definitive as that given above for styrene. Another experiment, which is strongly suggestive of coagulation, is the formation of Liesegang rings[122] in unstirred emulsion systems.[118,119] These last experiments entail the formation of latex in a spatially inhomogeneous fashion, because of the interplay between reaction and diffusion. Now, it has been established[123] that this spatial inhomogeneity can only arise if the system possesses an autocatalytic step in the mechanism. The coagulative nucleation rate is an increasing function of time, which fulfills this criterion. Quantitative modelling of this Liesegang ring formation[119] shows that accord with the observed spacing and formation time of the bands can be obtained from the equations describing reaction and diffusion, with the reaction kinetics specifically incorporating an initially increasing $dN_c/dt$. The importance of coagulative nucleation in various systems has been suggested by many workers.[47,91,93,115,124,125]

### 11.6.4 Mathematical Model for Particle Nucleation

The foregoing discussion suggests that micellar entry is unlikely to be an important mechanism in particle nucleation in conventional emulsion polymerization systems. We now give details of a very general nucleation model,[93,94,126,127] which incorporates homogeneous nucleation[15,47] and coagulation. The model is in accord with a very wide range of types of data, including those results, discussed above, that refuted micellar entry and suggested the importance of coagulation.

For convenience, one subdivides the species present into two types: 'oligomers' and 'precursors'. The smaller oligomeric free radicals will be extended chains rather than spherical entities. Oligomeric species are defined as being of such low molecular weight that their volume is unimportant. Precursors on the other hand are sufficiently large (say, a few nanometers or more) that they contribute significantly to the mean volume of particle at early times.

#### *11.6.4.1 Qualitative description*

This first step in this multi-step mechanism is homogeneous nucleation.[15,47] Aqueous phase oligomers start life as free radicals generated from initiator decomposition at rate $k_I[I]$. It is assumed that oligomers propagate in the aqueous phase until they reach a degree of polymerization, $J$, at which stage they become primary precursor particles (whose size might be of the order of a nanometer). These particles, once formed, may capture growing oligomers before they homogeneously nucleate, and undergo self termination in the aqueous phase. Primary precursors grow *slowly* by propagation; this slowness is because of the low monomer concentration within these entities, arising from their extreme curvature (see equation 22). 'Primary precursor' denotes newly formed precursor particles which have not undergone significant propagational or coagulational growth; the term 'precursor' particle refers both to these and to particles of any size which are unstable to coagulation. Precursor particles have recently been directly observed using small-angle neutron scattering in special emulsion polymerization systems.[128] Both primary precursors and subsequently formed precursor particles will usually be charged, as a result of end groups arising from initiator (if the latter is ionic) and/or as a result of surfactant adsorption. If the primary precursors have sufficiently low surface charge density, they will be unstable to homo-coagulation.

Once the primary precursor particles have grown sufficiently (mainly by propagation in a system containing large amounts of surfactant, or by both propagation and homo-coagulation in a system with low surfactant concentration), they will start to hetero-coagulate with smaller (more recently formed) precursors. The rapidity of hetero-coagulation between particles with different radii but similar surface charges[53] arises from the rearrangement of the double layer between particles of greatly differing curvatures. Once sufficiently large particles have been formed in significant numbers, they will hetero-coagulate so rapidly with newly formed primary precursors that the latter will not have sufficient time to undergo propagational or coagulative growth to become colloidally stable in their own right; nucleation then ceases. All newly formed oligomeric free radical entities (whether of colloidal dimensions or smaller) then enter directly into colloidally stable ('mature') latex particles.

#### *11.6.4.2 Quantitative description*

The evolution equation for $y(P, t)$, the concentration of oligomers, is given in equation (83). For convenience, their degree of polymerization $P$ is treated as a continuous variable. $P = 1$ corresponds to the free-radical entity formed directly by initiator decomposition (*e.g.* $SO_4^{\dot{}-}$). Here, $k_p$ is the propagation rate coefficient, $C_M(P)$ is the monomer concentration inside an entity of degree of polymerization $P$, $k_t(P, P')$ is the termination rate coefficient between entities of degree of propagation $P$ and $P'$, and $B(V_P, V)$ is the rate coefficient for capture of an oligomeric entity of degree of propagation $P$ (and volume $V_P$) by a precursor or latex particle of volume $V$. $Y(V)$ is the population of precursor and latex particles, as discussed in more detail below. Equation (83) has the boundary condition $y(P = 1, t) = 2k_d[\mathrm{I}]$. The inclusion of termination obviates introducing a radical capture efficiency.

$$\partial y/\partial t = -k_p\partial[C_M(P)y]/\partial P - y\int k_t(P, P')y(P')\mathrm{d}P' - y\int B(V_P, V)Y(V)\mathrm{d}V \tag{83}$$

The time evolution of the concentration of precursors, $Y(V, t)$, is given by equation (84). Here $V_P$ is the volume of a primary precursor, $B(V, V')$ is the rate coefficient for coagulation between precursors of volumes $V$ and $V'$, and the rate of volume growth by propagation, $K(V)$, is given by $K(V) = \bar{n}k_pC_M(V)M_0/(N_A d_p)$, $C_M(V)$ being monomer concentration inside a precursor with volume $V$. Equation (84) is to be solved subject to the boundary condition $Y(V = V_p, t) = k_pC_M(V_p)y(J)$. There is no need to further distinguish Y by the number of free radicals in a particle (*i.e.* the equation is in terms of $Y(V, t)$ rather than $Y_i(V, t)$, $i$ being the number of free radicals), because radical loss will occur to a significant extent in the system. This leads to rapid switching on and off of free-radical activity within each particle; compartmentalization effects can be ignored, and one can instead consider particles to contain only an average number ($\bar{n}$) of free radicals. The approximation will become invalid once coagulation has ceased to be kinetically significant (*i.e.* after the cessation of nucleation); however, one can accurately use the rate of particle formation ($\mathrm{d}N_c/\mathrm{d}t$) generated using the previous approximation in the full set of compartmentalized equations describing the time evolution of the PSD of mature latex particles (Section 11.4.3). The number density of mature (colloidally stable) latex particles is given by equation (85), where $V_m$ is the smallest volume at which particles are deemed to be colloidally stable (in practice, $V_m$ is taken as the volume at which the coagulative lifetime is greater than, say, 10 min).

$$\frac{\partial Y}{\partial t} = \int_{V_P}^{V/2} B(V - V', V')Y(V - V')Y(V')\mathrm{d}V' - Y(V)\int_{V_P}^{\infty} B(V, V')Y(V')\mathrm{d}V' - \frac{\partial}{\partial V}K(V)Y(V) \tag{84}$$

$$N_c(t) = \int_{V_m}^{\infty} Y(V, t)\mathrm{d}V \tag{85}$$

Equations (83)–(85) are the requisite evolution equations, *i.e.* a set of partial integrodifferential equations, which when re-expressed as coupled first-order differential equations are solved numerically by standard means. Explicit forms for the finite-difference equations have been given in the literature.[94,126] Certain parameters (such as the values of $J$ and $V_p$) appearing in these equations are really quantities involved in the simplification of separating the populations into oligomers (of negligible volume) and precursors. In principle, they can be eliminated by taking the limits as $V_p \to 0$, *etc.*, although in practice, lack of knowledge of colloidal properties of extremely small entities means that (for example) a choice of $V_p$ corresponding to a particle radius of less than one or two nanometers becomes questionable, particularly since the smallest species involved will be extended chains rather than spherical entities.

Next we examine the forms of the various rate parameters required for the solution of equations (83)–(85). Because this involves physical properties of very small entities, values of some of these parameters (and indeed, the applicability of assumptions such as their spherical nature) are uncertain. However, adjusting values of unknown parameters to fit limited data (*e.g.* particle number at a given surfactant concentration) will be seen to give quantitative agreement with a very wide range of data, suggesting the reliability of this treatment.

The $C_M(P)$ and $C_M(V)$ are assumed to be given by equation (86), which is in qualitative accord with equation (22) and also approaches the aqueous phase concentration ($C_w$) for very small particle radii. Here $r$ is the particle radius, $C_M^\infty$ is the large radius limiting saturated monomer concentration in the particles, $r_{FH}$ is a radius chosen to express the size dependence of the swelling, and $r^*$ a radius sufficiently small that one expects the value of $C_M$ to be governed by oligomer-like rather than particle-like considerations. The term $\int B(V_P, V)\,Y(V)\,dV$ in equation (84) is replaced by $k_e(r)N_T$, where $k_e(r)$ is the rate coefficient for entry of an oligomeric free radical into a latex or precursor particle, $N_T$ is the total number of precursor plus latex particles, and $r$ is the radius of the largest particle. It is assumed that $k_e(r) = k_{em}r^2$ (*i.e.* free radical entry depends on the particle surface area, as discussed in Section 11.3.5). The coagulation rate coefficients $B(V, V')$ are specified using the Fuchs[129] formulation coupled with DLVO theory,[130] modified as required to encompass hetero-coagulation,[53,131,132] as given in equation (87). Here $k_B$ = Boltzmann's constant, $T$ = absolute temperature, $\eta$ = viscosity of the medium, and $W(V, V')$ is the Fuchs stability ratio, approximated by $W(V, V') = (\kappa\bar{r})^{-1}\exp(E_m/k_BT)$. Here $\kappa^{-1}$ = thickness of the double layer, $E_m$ = height of the potential barrier to coagulation, and $\bar{r} = 2rr'/(r + r')$. The value of $\kappa$ is given by $\kappa = [8\pi^2 N_A I e^2/(4\pi\varepsilon k_B T)]^{1/2}$ where $e$ = electronic charge, $I$ = ionic strength, and $\varepsilon$ = permittivity of the dispersion medium. Note that SI units have been used in this and all subsequent equations. The potential energy barrier to coagulation, $E_m$, can be calculated from DLVO theory with extensions for hetero-coagulation. $E_m$ is the maximum in the total potential energy $V = V_A + V_R$, where the latter quantities are respectively the van der Waals attractive and the electrostatic repulsive potential. $V_R$ can be estimated using equation (88).[132] Here $\zeta$ is the zeta potential of a particle with radius $r$, $\bar{\zeta}^2 = \zeta^2 + \zeta'^2$, and $q = \exp(-\kappa H)$, with $H$ the closest distance between the surfaces of the particles (maximization is with respect to this distance). The constant surface charge density form for $V_R$ is used here because the ions determining the potential are actually incorporated into the latex particles. The zeta potential is calculated from the surface potential by applying the Stern layer corrections: $\zeta = (2k_BT/z_+e)\,\ln[(e^E + 1)/(e^E - 1)]$, where $E = (2z^2e^2n_+/\varepsilon k_BT)^{1/2}\delta + \ln[(e^D + 1)/(e^D - 1)]$, $D = ze\psi_0/2k_BT$, $z$ is the charge on the adsorbed ions, $\delta$ is the thickness of the Stern layer, and $n_+$ is the total number concentration of the counter-ion whose charge is $z_+$.

$$C_M(r) = C_M^\infty \tanh(r/r_{FH}),\quad r > r^* = \exp[\ln(C_w) + (r/r^*)^3\ln\{C_M(r^*)/C_w\}],\quad r \le r^* \tag{86}$$

$$B(V, V') = (4/3)(k_BT/\eta W)(1 + r/r')^2/(2r/r') \tag{87}$$

$$V_R(r, r', H) = \frac{1}{2}\pi\varepsilon\bar{\zeta}^2\bar{r}\left[\frac{2\zeta\zeta'}{\bar{\zeta}^2}\ln[(1 + q)/(1 - q)] + \ln[1 - q^2]\right] \tag{88}$$

Equation (88) can become inaccurate when the two particles have very different radii and very different surface potentials.[53] This can be corrected through use of adjustable parameters fitted to data, and can also in part be corrected by halving the value of $E_m$ found from equation (88) when the radius ration ($r/r'$) differs by more than two, as indicated by exact solutions of the Poisson–Boltzmann equation.[53]

The surface potential ($\psi_0$) for particles of the size and surface charge density considered here, can be estimated from the spherical particle expression $\psi_0 = 4\pi r\sigma/4\pi\varepsilon(1 + \kappa r)$, where $\sigma$ is the surface charge density on the particle. If the surface potential exceeds 50 mV, it is better to use the flat-plate expression $\psi_0 = (2k_BT/z_+e)\sinh^{-1}[2\pi z_+e\sigma/(4\pi\varepsilon k_BT\kappa)]$.

The surface charge density is found by assuming that the surfactant adsorbs onto the particles with a Langmuir isotherm, *i.e.* $N_s = A_t/\{a_S(1+b^{-1}[S_w]^{-1})\}$. Here $N_s$ = total number of surface sites occupied by surface-active species, $A_t = \int 4\pi r^2 Y(V)\,dV$ is the total area of precursors plus latex particles (both of these quantities refer to unit volume of reacting mixture), $[S_w]$ = concentration of the surface-active species in the aqueous phase, $a_s$ = area occupied by a surface-active species when close packed, and $b$ is a constant depending on the surfactant and surface. From mass conservation, $[S_w] = (U^2 + [S_t]b^{-1})^{1/2} - U$, where $U = \frac{1}{2}\{b^{-1} + A_t(N_A a_s)^{-1} - [S_t]\}$ and $[S_t]$ is the total amount of surface-active species present (adsorbed and in solution) per unit volume of the aqueous phase. Surface-active are produced from initiator decomposition, as well as from any added emulsifier, and thus $[S_t] = 2k_d[I]t + [S]$, where [S] is the number of moles of added surfactant (per unit volume of aqueous phase). Theory suggests that the surface charge density should have a depedence on size, which is assumed to have the form $\sigma(r) = \sigma_\infty \min(1, r/r_\sigma)$, where $r_\sigma$ is an adjustable parameter, and the large-radius surface charge density is given by $\sigma_\infty = e/\{a_s(1+b^{-1}[S_w]^{-1})\}$.

The van der Waals attractive potential is given by equation (89), where $s = 2 + (H/\bar{r})$ and $A^*$ is the effective Hamaker constant of the particles in the dispersion medium.

$$V_A(r, r', H) = -(A^*/6)\{2/[s^2 - 4] + 2/s^2 + \ln[(s^2 - 4)/s^2]\} \tag{89}$$

Finally, the values of $B(V, V')$ calculated as above can under certain circumstances (*e.g.* high ionic strength or low [S]) approach the diffusion limit (*ca.* $10^9\ dm^3\,mol^{-1}\,s^{-1}$), which is some orders of magnitude greater than the value of $k_e$ observed under the same conditions (*ca.* $10^6\ dm^3\,mol^{-1}\,s^{-1}$);[18] since this implies an additional barrier such as diffusion through a surface layer (Section 11.3.5) one puts $B(V, V')$ as the smaller of the value calculated above and $k_e$.

The foregoing describes the nucleation model used by Feeney *et al.*[127] Some slight variations have been used by workers at Du Pont;[126] the minor differences between these models are compensated through adjusting one or two parameters to fit data such as the value of $N_c$ at a particular [S] and the time at which nucleation ceases.

### 11.6.5 Comparison of Theory and Experiment

Some results are shown in Figures 16 and 17. Values of the various adjustable parameters used in the calculations of Figure 16 are as follows; $k_{em} = 6 \times 10^{18}\ dm\,mol^{-1}\,s^{-1}$, $r_{FH} = 15$ nm, $r^* = 1$ nm, $r_\sigma = 43$ nm, termination among oligomers in the aqueous phase is assumed negligible, $r_p = 2$ nm, $a_S = 41\ Å^2$, $b = 1200\ dm^3\,mol^{-1}$; detailed justification has been given elsewhere.[126,127] Figure 16 shows the calculated and observed PSD, and calculated $dN_c/dt$, for the styrene system of Figure 12. The accord between theory and experiment is quite acceptable. It can be seen that the calculated $dN_c(t)/dt$ curve has the form implied by the PSD data, *viz.* an initially increasing function of time, followed by a rapid decrease after the surfactant is exhausted.

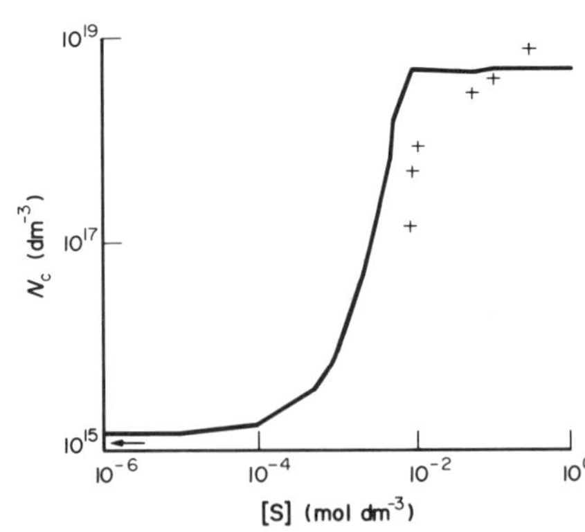


**Figure 17** Calculated and observed dependence of final particle number on concentration of SDS surfactant, for styrene at 80 °C with $[I] = 1.59 \times 10^{-3}\ mol\,dm^{-3}\ (NH_4)S_2O_8$; redrawn from J. R. Richards, J. P. Congalidis and R. G. Gilbert, *J. Appl. Polym. Sci.*, in press. Arrow shows $N_c$ obtained without added surfactant at 70 °C, as given by J. W. Goodwin, J. Hearn, C. C. Ho and R. H. Ottewill, *Colloid Polym. Sci.*, 1974, **252**, 464. Surfactant concentration dependent data from N. Sütterlin, in 'Polymer Colloids II', ed. R. M. Fitch, Plenum Press, New York, 1980

The model is also able to reproduce 'traditional' features of nucleation kinetics. Figure 17 shows the calculated dependence of $N_c(t = \infty)$ on [S] calculated for this model, compared with experiment. It can be seen that, while the dependence on [S] is more complicated than suggested by the Smith–Ewart relation $N_c \propto [S]^{0.6}$, the data are well reproduced, especially the sigmoidal dependence observed with many monomers.[117] It has also been shown that the model agrees excellently[93] with data with [S] = 0, when $N_c$ is three orders of magnitude less than that for high [S]; the model therefore reproduces the fact that $N_c$ changes by many orders of magnitude with [S]. If the calculated $N_c$ are fitted to the functional form suggested by Smith–Ewart theory, $[I]^x[S]^y$, then exponents for [S] of *ca.* 0.6, and for [I] of *ca.* 0.4, are obtained. Thus this more complete theory is also capable of reproducing less sensitive data (the [S] and [I] dependence of $N_c$) as well as data such as the PSDs.

The model can also successfully reproduce the effect of different surfactants on the polydispersity.[127] To give a qualitative example, the observation that Aerosol MA as surfactant yields more monodisperse latexes than does sodium dodecyl sulfate (SDS) can be ascribed to the different adsorption isotherms of these two surfactants; SDS is more strongly adsorbed, hence makes precursors which coagulate more slowly, and thus the nucleation period takes longer and the PSD is more polydisperse. An important conclusion here is that the shorter the nucleation period the more monodisperse is the final latex. The foregoing discussion also shows why emulsion polymerization at higher temperatures tends to produce more monodisperse latexes. This is because the rate of production of precursors is greately increased (*e.g.* through the temperature dependence of $k_d$) and thus more particles are formed earlier; this in turn leads to nucleation ceasing more rapidly, as seen in the last term in equation (83), and thus the final latex is more monodisperse.

Two more effects that the model is able to take into account include the effect on $N_c$ of inert diluent (*e.g.* ethylbenzene) and of indifferent electrolyte (*e.g.* NaCl).[127] It is found the addition of ethylbenzene greatly increases the polydispersity of the PSD and increases $N_c$. This can be seen to arise because the rate of particle formation is reduced through dilution (by inert diluent) and by increasing the exit rate,[56] leading to a much longer nucleation period and hence a more polydisperse PSD and a higher $N_c$. NaCl is seen to decrease $N_c$ and decrease polydispersity. This is because of the increase in rate of coagulation caused by the effect of ionic strength on the double layer. Again, model and experiment are in accord (including the $N_c$ with zero surfactant); increased ionic strength causes greater coagulation, and hence a shorter nucleation period (since the particles become larger more rapidly) and thus a more monodisperse PSD and a lower $N_c$.[94,127]

Further tests to which the model has been subjected[126] include comparison with (i) styrene CSTR conversion, MWD and PSD data[133] and (ii) conversion, copolymer composition and particle number in a styrene/MMA batch copolymerization.[134] Again, the model produces good agreement with these 'traditional' tests of worthiness.

None of the conversion/time curves computed with the model show a rate that decreases with $t$ at any point in Interval I, which is in accord with experiment. However, the computed total particle number is usually briefly increasing at early times, which at first sight would suggest, from equation (1), that $dx/dt$ should also show a maximum. This apparent paradox is resolved when it is realized that the computed maximum in particle number is due to an excess of very small particles, which (because of their extreme curvature) have a low $C_M$ and hence have a slow rate of propagation. The full computation of conversion, using the model, uses equation (1) extended to take this into account, *i.e.* as $dx/dt \propto \int Y(V)C_M(V)\bar{n}(V)\,dV$.

## 11.7 CONCLUSIONS AND FUTURE DIRECTIONS

Many of the basic mechanisms in emulsion polymerization systems are now understood, although modern findings are not always in accord with 'classical' suggestions. Particle formation involves homogeneous nucleation, followed by particle growth by coagulation and by propagation. Measurement of the average number of free radicals per particle ($\bar{n}$) shows that it occasionally takes the 'Case 2' limiting value of 0.5, but values very different from 0.5 are the norm. Free-radical entry into particles is probably controlled by diffusion through the particle–water interface. Free-radical exit is governed by transfer and by diffusion. Termination in the diffusion-controlled regime is understood, although the same cannot be said for propagation. Means of deducing virtually all rate-controlling parameters from appropriate experimental data have been established and are being increasingly applied. Accurate, physically correct modelling of the polymerization process is now readily achieved.

For reasons of space, the development given here has concentrated on the emulsion polymerization of a single monomer. However, extensions of modern knowledge to emulsion copolymerization systems are well in hand. The kinetics of particle growth, Section 11.3, have been extended to copolymerizations by a number of workers, of which the best treatment is that of Nomura and coworkers[135] (who simplified the complete treatment of Ballard *et al.*[136]). The nucleation model of Section 11.6.4 has been extended to copolymerization systems by Richards *et al.*[126]

The material covered in this chapter shows that the present level of understanding of homopolymerizations is moderately high, although there are some important gaps. The mechanism for entry is still poorly understood. Many parameters in nucleation kinetics, especially those involving very small precursor particles, are poorly understood; fortunately, the means of investigating the thermodynamic properties of such small particles (*e.g.* their adsorption isotherms with surfactant) have recently become available.[128] The understanding of the diffusion coefficient of a monomeric species in a polymer at high $w_p$ (which would yield diffusion-controlled $k_p$ values) is still lacking, as is a theory for the chemically controlled component, $k_p^\circ$; another major problem is a viable theory for $k_t$ at intermediate conversions, especially the dependence of this on the chain lengths of the mutually annihilating macroradicals. These two problems are of course not confined to emulsion polymer systems, but to free-radical polymerizations in general.

Despite these deficiencies, modern mechanistic understanding of emulsion polymerizations has made enormous advances since the pioneering work of Harkins, Smith and Ewart. It can be confidently stated that, for much of the emulsion polymerization process, there is good mechanistic knowledge, and the tools are at hand for increasing this; hence reliable mathematical modelling is now possible. With mechanistic understanding, the way is now open for considerable quantitative and qualitative improvement of product and process.

## 11.8 REFERENCES

1. M. Luther and C. Heuck (I.G. Farbenindustrie A.G.), *US Pat.* 1 864 078 (1932) (*Chem. Abstr.*, 1932, **26**, 4505).
2. D. C. Blackley, 'Emulsion Polymerization', Applied Science, London, 1975.
3. J. W. Vanderhoff, E. B. Bradford, H. L. Tarkowski, J. B. Shaffer and R. M. Wiley, *Adv. Chem. Ser.*, 1962, **34**, 32.
4. W. D. Harkins, *J. Am. Chem. Soc.*, 1947, **69**, 1428; *J. Polym. Sci.*, 1950, **5**, 217.
5. W. V. Smith and R. H. Ewart, *J. Chem. Phys.*, 1948, **16**, 592.
6. R. G. W. Norrish and R. R. Smith, *Nature (London)*, 1942 **150**, 336.
7. E. Trommsdorff, H. Köhle and P. Lagally, *Makromol. Chem.*, 1948, **1**, 169.
8. D. R. Bassett and K. L. Hoy, *ACS Symp. Ser.*, 1981, **165**, 371.
9. S. J. McCarthy, E. E. Elbing, I. R. Wilson, R. G. Gilbert, D. H. Napper and D. F. Sangster, *Macromolecules*, 1986, **19**, 2440.
10. L. H. Peebles, in 'Copolymerization', ed. G. E. Ham, Wiley, New York, 1964, vol. 18 of 'High Polymers'.
11. O. G. Lewis and R. M. King, *Adv. Chem. Ser.*, 1969, **91**, 25.
12. M. R. Grancio and D. J. Williams, *J. Polym. Sci., Polym. Chem. Ed.*, 1970, **8**, 2617.
13. J. L. Gardon, *J. Polym. Sci., Polym. Chem. Ed.*, 1973, **11**, 241.
14. J. W. Vanderhoff, *Proc. Water Borne Higher Solids Coating Symp.*, 1976, **3**, Paper No. 6.
15. J. Ugelstad and F. K. Hansen, *Rubber Chem. Technol.*, 1976, **49**, 536.
16. B. C. Y. Whang, D. H. Napper, M. J. Ballard, R. G. Gilbert and G. Lichti, *J. Chem. Soc., Faraday Trans. 1*, 1982, **78**, 1117.
17. R. G. Gilbert and D. H. Napper, *J. Macromol. Sci., Rev. Macromol. Chem. Phys.*, 1983, C**23**, 127.
18. I. A. Penboss, R. G. Gilbert and D. H. Napper, *J. Chem. Soc., Faraday Trans. 1*, 1986, **82**, 2247.
19. W. H. Stockmayer, *J. Polym. Sci.*, 1957, **24**, 314
20. J. T. O'Toole, *J. Appl. Polym. Sci.*, 1965, **9**, 1291.
21. J. Ugelstad, P. C. Møk and J. O. Aasen, *J. Polym. Sci., Part A-1*, 1967, **5**, 2281.
22. M. J. Ballard, R. G. Gilbert and D. H. Napper, *J. Polym. Sci., Polym. Lett. Ed.*, 1981, **19**, 533.
23. C. W. Gear, 'Numerical Initial Boundary Value Problems in Ordinary Differential Equations', Prentice-Hall, New York, 1971.
24. B. W. Brooks, *J. Chem. Soc., Faraday Trans. 1*, 1982, **78**, 3137.
25. D. C. Blackley, 'Emulsion Polymerization', Applied Science, London, 1975.
26. M. J. Ballard, D. H. Napper and R. G. Gilbert, *J. Polym. Sci., Polym. Chem. Ed.*, 1984, **22**, 3225.
27. J. Brandrup and E. H. Immergut, 'Polymer Handbook', 2nd edn., Wiley Interscience, New York, 1975.
28. I. A. Maxwell, D. H. Napper and R. G. Gilbert, *J. Chem. Soc., Faraday Trans. 1*, 1987, **83**, 1449.
29. B. S. Hawkett, D. H. Napper and R. G. Gilbert, *J. Chem. Soc., Faraday Trans. 1*, 1980, **76**, 1323.
30. L. C. Rubens and R. E. Skochdopole, in 'Encyclopedia of Polymer Science and Technology', ed. H. F. Mark, N. G. Gaylord and N. M. Bikales, Interscience, New York, 1966, vol. 5.
31. I. A. Maxwell, E. D. Sudol, D. H. Napper and R. G. Gilbert, *J. Chem. Soc., Faraday Trans. 1*, 1988, **84**, 3107.
32. E. W. Duck, in 'Encyclopedia of Polymer Science and Technology', eds. H. F. Mark, N. G. Gaylord and N. M. Bikales, Interscience, New York, 1966, vol. 5.
33. G. L. Leslie, I. A. Maxwell, M. J. Ballard, R. G. Gilbert and D. H. Napper, *Aust. J. Chem.*, 1988, **41**, 279.
34. G. L. Leslie, R. G. Gilbert and D. H. Napper, to be published.
35. S. W. Lansdowne, R. G. Gilbert, D. H. Napper and D. F. Sangster, *J. Chem. Soc., Faraday Trans. I*, 1980, **76**, 1344.
36. G. J. M. Ley, C. Schneider and D. O. Hummel, *J. Polym. Sci., Part C*, 1969, **27**, 119.
37. L. F. Halnan, D. H. Napper and R. G. Gilbert, *J. Chem. Soc., Faraday Trans. 1*, 1984, **80**, 2851.

38. M. Nomura, in 'Emulsion Polymerization', ed. I. Piirma, Academic Press, New York, 1982; M. Nomura and M. Harada, *J. Appl. Polym. Sci.*, 1981, **26**, 17.
39. M. Morton, S. Kaizermann and M. W. Altier, *J. Colloid Sci.*, 1954, **9**, 300.
40. M. Buback, L. H. Garcia-Rubio, R. G. Gilbert, D. H. Napper, J. Guillot, A. E. Hamielec, D. Hill, K. F. O'Driscoll, O. F. Olaj, Jiacong Shen, D. Solomon, G. Moad, M. Stickler, M. Tirrell and M. A. Winnik, *J. Polym. Sci., Polym. Lett. Ed.*, 1988, **26**, 293.
41. H. K. Mahabadi and K. F. O'Driscoll, *J. Macromol. Sci., Chem.*, 1977, **A11**, 967.
42. S. K. Soh, *J. Appl. Polym. Sci.*, 1980, **25**, 2993.
43. M. J. Ballard, R. G. Gilbert, D. H. Napper, P. J. Pomery, P. W. O'Sullivan and J. H. O'Donnell, *Macromolecules*, 1986, **19**, 1303.
44. W. Lau, D. G. Westmoreland and R. W. Novak, *Macromolecules*, 1987, **20**, 457.
45. I. A. Penboss, D. H. Napper and R. G. Gilbert, *J. Chem. Soc., Faraday Trans. 1*, 1983, **79**, 1257.
46. B. S. Hawkett, D. H. Napper and R. G. Gilbert, *J. Polym. Sci., Polym. Chem. Ed.*, 1981, **19**, 3173.
47. R. M. Fitch and C. H. Tsai, in 'Polymer Colloids', ed. R. M. Fitch, Plenum Press, New York, 1971.
48. M. Trau, M. E. Adams. R. G. Gilbert and D. H. Napper, *Aust. J. Chem.*, in press.
49. J. L. Gardon, *J. Polym. Sci., Part A*, 1968, **6**, 623, 643, 665, 687, 2853, 2859.
50. R. M. Fitch and L. B. Shih, *Prog. Colloid Polym. Sci.*, 1975, **56**, 1.
51. V. I. Yeliseyeva, in 'Emulsion Polymerization', ed. I. Piirma, Academic Press, New York, 1982.
52. P. J. Feeney, D. H. Napper and R. G. Gilbert, *Macromolecules*, 1987, **20**, 2922.
53. E. Barouch and E. Matijevic, *J. Chem. Soc., Faraday Trans. 1*, 1985, **81**, 1797; E. Barouch, E. Matijevic and T. H. Wright, *J. Chem. Soc., Faraday Trans. 1*, 1985, **81**, 1819.
54. M. Adams, D. H. Napper, R. G. Gilbert and D. F. Sangster, *J. Chem. Soc., Faraday Trans. 1*, 1986, **82**, 1979.
55. G. Lichti, D. F. Sangster, B. C. Y. Whang, D. H. Napper and R. G. Gilbert, *J. Chem. Soc., Faraday Trans. 1*, 1982, **78**, 2129.
56. G. Lichti, D. F. Sangster, B. C. Y. Whang, D. H. Napper and R. G. Gilbert, *J. Chem. Soc., Faraday Trans. 1*, 1984, **80**, 2911.
57. M. Nomura and M. Harada, *J. Appl. Polym. Sci.*, 1981, **26**, 17.
58. H. -C. Lee and G. W. Poehlein, *Chem. Eng. Sci.*, 1986, **41**, 1023.
59. M. Nomura, M. Harada, W. Eguichi and S. Nagata, *J. Chem. Eng. Jpn.*, 1971, **4**, 54; M. Nomura, M. Harada, K. Nakagawara, W. Eguichi and S. Nagata, *J. Chem. Eng. Jpn.*, 1971, **4**, 160.
60. B. W. Brooks and B. O. Makanjuola, *J. Chem. Soc., Faraday Trans. 1*, 1981, **77**, 2659.
61. B. W. Brooks, *Colloid Polym. Sci.*, 1987, **265**, 58.
62. V. Stannett, A. Klein and M. Litt, *Br. Polym. J.*, 1975, **7**, 139.
63. M. J. Ballard, P. Hidi, B. J. Chambers, R. G. Gilbert and D. H. Napper, to be published.
64. W. S. Zimmt, *J. Appl. Polym. Sci.*, 1959, **1**, 323.
65. M. J. Ballard, D. H. Napper, R. G. Gilbert and D. F. Sangster, *J. Polym. Sci., Polym. Chem. Ed.*, 1986, **24**, 1027.
66. S. W. Benson and A. M. North, *J. Am. Chem. Soc.*, 1962, **84**, 935.
67. G. V. Schulz, *Z. Phys. Chem. (Frankfurt am Main)*, 1956, **8**, 290.
68. S. K. Soh and D. C. Sundberg, *J. Polym. Sci., Polym. Chem. Ed.*, 1982, **20**, 1299, 1315, 1331, 1345.
69. G. T. Russell, D. H. Napper and R. G. Gilbert, *Macromolecules*, 1988, **21**, 2133.
70. F. Bueche, 'Physical Properties of Polymers', Interscience, New York, 1962.
71. J. D. Ferry, 'Viscoelastic Properties of Polymers', Wiley-Interscience, New York, 1970.
72. R. C. Reid and T. K. Sherwood, 'Properties of Gases and Liquids', McGraw-Hill, New York, 1958.
73. J. O. Hirschfelder, C. F. Curtiss and R. B. Bird, 'Molecular Theory of Gases and Liquids', Wiley, New York, 1964.
74. T. J. Tulig and M. Tirrell, *Macromolecules*, 1981, **14**, 1501.
75. P. G. de Gennes, *J. Chem. Phys.*, 1971, **55**, 572.
76. N. Friis and A. E. Hamielec, *ACS Symp. Ser.*, 1976, **24**, 82.
77. G. T. Russell, D. H. Napper and R. G. Gilbert, *Macromolecules*, 1988, **21**, 2141.
78. S. T. Balke and A. E. Hamielec, *J. Appl. Polym. Sci.*, 1973, **17**, 905; F. L. Marten and A. E. Hamielec, *ACS Symp. Ser.*, 1979, **104**, 43.
79. J. N. Cardenas and K. F. O'Driscoll, *J. Polym. Sci., Polym. Chem. Ed.*, 1976, **14**, 883; 1977, **15**, 1883, 2097.
80. M. E. Adams, D. F. Sangster, D. H. Napper and R. G. Gilbert, to be published.
81. M. J. Ballard, R. G. Gilbert, D. H. Napper, P. J. Pomery and J. H. O'Donnell, *Macromolecules*, 1984, **17**, 504.
82. M. Stickler, *Makromol. Chem.*, 1983, **184**, 2563; F. N. Kelley and F. Bueche, *J. Polym. Sci.*, 1961, **50**, 549.
83. H. Small, *J. Colloid Interface Sci.*, 1974, **48**, 147.
84. C. A. Silebi, Lecture Notes in 15th Annual Short Course 'Advances in Emulsion Polymerization and Latex Technology', Lehigh University, 1984; G. R. McGowan and M. A. Langhorst, *J. Colloid Interface Sci.*, 1982, **89**, 94.
85. G. Lichti, B. S. Hawkett, R. G. Gilbert, D. H. Napper and D. F. Sangster, *J. Polym. Sci., Polym. Chem. Ed.*, 1981, **19**, 925.
86. J. T. O'Toole, *J. Polym. Sci., Part C*, 1969, **27**, 171.
87. D. C. Sundberg and J. D. Eliassen, in 'Polymer Colloids', ed. R. M. Fitch, Plenum Press, New York, 1971, p. 153.
88. L. M. Pis'Men and S. I. Kuchanov, *Vysokomol. Soedin., Ser. A*, 1971, **13**, 1055.
89. K. W. Min and W. H. Ray, *J. Macromol. Sci., Rev. Macromol. Chem.*, 1974, **C11**, 177.
90. G. Lichti, R. G. Gilbert and D. H. Napper, *J. Polym. Sci., Polym. Chem. Ed.*, 1977, **15**, 1957.
91. G. Lichti, R. G. Gilbert and D. H. Napper, *J. Polym. Sci., Polym. Chem. Ed.*, 1983, **21**, 269.
92. D. F. Wood, B. C. Y. Whang, D. H. Napper, R. G. Gilbert and G. Lichti, *J. Polym. Sci., Polym. Chem. Ed.*, 1983, **21**, 985.
93. P. J. Feeney, D. H. Napper and R. G. Gilbert, *Macromolecules*, 1984, **17**, 2520.
94. P. J. Feeney, D. H. Napper and R. G. Gilbert, *J. Colloid Interface Sci.*, 1987, **118**, 493.
95. G. Lichti, R. G. Gilbert and D. H. Napper, in 'Emulsion Polymerization', ed. I. Piirma, Academic Press, New York, 1982.
96. G. W. Poehlein, in 'Encyclopedia of Polymer Science and Engineering', ed. J. I. Kroschwitz, Wiley, New York, 1986, vol. 6, p. 1.
97. P. J. Flory, 'Principles of Polymer Chemistry', Cornell University Press, Ithaca, NY, 1953.
98. G. Lichti, R. G. Gilbert and D. H. Napper, *J. Polym. Sci., Polym. Chem. Ed.*, 1980, **18**, 1297.

99. H. Gerrens, *Fortschr. Hochpolym.-Forsch.*, 1959, **1**, 234.
100. C. C. Lin and W. Y. Chiu, *J. Appl. Polym. Sci.*, 1979, **23**, 2049.
101. S. Katz, R. Shinnar and G. M. Saidel, *Adv. Chem. Ser.*, 1969, **91**, 145.
102. K. W. Min and W. H. Ray, *J. Appl. Polym. Sci.*, 1978, **22**, 89.
103. L. Mirsky, 'Introduction to Linear Albegra', Clarendon Press, Oxford, 1955.
104. R. G. Gilbert and D. H. Napper, *J. Chem. Soc., Faraday Trans. I*, 1974, **70**, 391.
105. C. C. Lin and W. Y. Chiu, *J. Appl. Polym. Sci.*, 1979, **23**, 2049; J. G. Braks and R. Y. M. Huang, *J. Polym. Sci., Polym. Phys. Ed.*, 1975, **13**, 1063; W. W. Yau, H. J. Stoklosa and D. D. Bly, *J. Appl. Polym. Sci.*, 1977, **21**, 1911.
106. K. S. Chang and R. Y. M. Huang, *J. Appl. Polym. Sci.*, 1969, **13**, 1459; A. E. Hamielec, H. J. Ederer and K. H. Ebert, *J. Liq. Chromatogr.*, 1981, **4**, 1697; S. Vozka and M. Kubin, *J. Chromatogr.*, 1977, **139**, 225.
107. N. Friis and A. E. Hamielec, *J. Appl. Polym. Sci.*, 1975, **19**, 97.
108. I. Piirma, V. R. Kamath and M. Morton, *J. Polym. Sci., Polym. Chem. Ed.*, 1975, **13**, 2087; H. L. James and I. Piirma, *ACS Symp. Ser.*, 1976, **24**, 197.
109. A. W. deGraaff and G. W. Poehlein, *J. Polym. Sci., Part A-2*, 1971, **9**, 1955.
110. C. P. Roe, *Ind. Eng. Chem.*, 1968, **60**, 20.
111. R. M. Fitch, *Br. Polym. J.*, 1973, **5**, 467.
112. F. K. Hansen and J. Ugelstad, in 'Emulsion Polymerization', ed. I. Piirma, Academic Press, New York, 1982.
113. V. I. Eliseeva, *Acta Polym.*, 1981, **32**, 355.
114. W. J. Priest, *J. Phys. Chem.*, 1952, **56**, 1077.
115. V. I. Eliseeva, N. G. Zharkova, E. I. Evko and V. M. Luk'yanovitch, *Vysokomol. Soedin., Ser. A*, 1967, **9**, 2478.
116. R. M. Fitch, T. H. Palmgren, T. Aoyagi and A. Zuikov, *Angew. Makromol. Chem.*, 1984, **123/124**, 261.
117. N. Sütterlin, in 'Polymer Colloids II', ed. R. M. Fitch, Plenum Press, New York, 1980.
118. W. Heller, H. B. Klevens and H. Oppenheimer, *J. Dispersion Sci. Technol.*, 1981, **2**, 111.
119. P. J. Feeney, R. G. Gilbert and D. H. Napper, *J. Colloid Interface Sci.*, 1985, **107**, 159.
120. A. Penlidis, J. F. MacGregor and A. E. Hamielec, *AIChE J.*, 1985, **31**, 881; T. O. Broadhead, A. E. Hamielec and J. F. MacGregor, *Makromol. Chem., Suppl.*, 1985, **10/11**, 105.
121. E. P. Dougherty, *J. Appl. Polym. Sci.*, 1986, **32**, 3051, 3079.
122. K. H. Stern, 'Bibliography of Liesegang Rings', 2nd ed., US GPO, Washington, DC, 1967.
123. S. Kai, S. C. Muller and J. Ross, *J. Chem. Phys.*, 1982, **76**, 1392.
124. R. M. Fitch and R. C. Watson, *J. Colloid Interface Sci.*, 1979, **68**, 14.
125. F. K. Hansen and J. Ugelstad, *J. Polym. Sci., Polym. Chem. Ed.*, 1978, **16**, 1953; F. K. Hansen and J. Ugelstad, *J. Polym. Sci., Polym. Chem. Ed.*, 1979, **17**, 3033.
126. J. R. Richards, J. P. Congalidis and R. G. Gilbert, *J. Appl. Polym. Sci.*, in press.
127. P. J. Feeney, M. E. Adams, P. Hidi, D. H. Napper and R. G. Gilbert, *J. Polym. Sci., Polym. Chem. Ed.*, submitted; P. Stolz, R. G. Gilbert and D. H. Napper, *J. Polym. Sci., Polym. Chem. Ed.*, submitted.
128. P. J. Feeney, E. Geissler, R. G. Gilbert and D. H. Napper, *J. Colloid Interface Sci.*, 1988, **121**, 508.
129. N. Z. Fuchs, *Physik*, 1934, **89**, 736.
130. J. T. G. Overbeek, in 'Colloid Science', ed. H. R. Kruyt, Elsevier, Amsterdam, 1960.
131. R. Buscall and R. H. Ottewill, in 'Polymer Colloids', ed. R. Buscall, T. Corner and J. F. Stageman, Elsevier, Amsterdam, 1985.
132. R. Hogg, T. W. Healy and D. W. Fuerstenau, *Trans. Faraday Soc.*, 1966, **62**, 1638; G. R. Wiese and T. W. Healy, *Trans. Faraday Soc.*, 1970, **66**, 490; D. Y. C. Chan and L. R. White, *J. Colloid Interface Sci.*, 1980, **74**, 303.
133. E. E. Badder and B. W. Brooks, *Chem. Eng. Sci.*, 1984, **39**, 1499.
134. M. Nomura and K. Fujita, *Makromol. Chem. Suppl.*, 1985, **10/11**, 25.
135. M. Nomura, K. Yamamoto, I. Horie, K. Fujita and M. Harada, *J. Appl. Polym. Sci.*, 1982, **27**, 2483; M. Nomura, M. Kubo and K. Fujita, *J. Appl. Polym. Sci.*, 1983, **28**, 2767.
136. M. J. Ballard, D. H. Napper and R. G. Gilbert, *J. Polym. Sci., Polym. Chem. Ed.*, 1981, **19**, 939.

# 12

# Polymerization in Micelles and Microemulsions

ALEXANDER S. DUNN

*University of Manchester Institute of Science and Technology, UK*



## 12.1 MICELLAR POLYMERIZATION

While the concentration of emulsifier in the aqueous phase remains above its critical micelle concentration (cmc), monomer solubilized in micelles is necessarily present in emulsion polymerization systems (see Volume 4, Chapter 11), although the amount of this monomer is only significant in comparison with the amount in molecular solution in the aqueous phase for the least water-soluble monomers, of which styrene is typical. However it is possible to conduct micellar polymerization, *i.e.* polymerization of monomer solubilized in micelles, in the absence of a separate phase of monomer present in emulsion droplets. Such systems have been investigated in the hope of obtaining evidence relevant to the mechanism of the emulsion polymerization system, although they have no apparent commercial potential.

Hughes, Sawyer, and Vinograd[1] used X-ray diffraction to study the polymerization of styrene solubilized in very concentrated (14–25%) aqueous solutions of potassium dodecanoate: micelle dimensions were observed to be increased when styrene was solubilized and to revert to their original size when the styrene was polymerized. There was a large excess of emulsifier in these experiments, *i.e.* much more than would have been required to stabilize all the polymer latex particles produced, in contrast to the conditions commonly used in emulsion polymerization in which less emulsifier is present than is required to saturate the surface of the final latex particles with adsorbed surfactant.

Frilette and Hohenstein[2] studied the thermal polymerization of styrene solubilized in 2% potassium 9-octadecenoate (*i.e.* oleate). At 25 °C, this solution solubilized 0.65% styrene whereas the solubility of styrene in water in the absence of soap was 0.022%. The conversion was found to be limited to about 50%. Later work[3] on the bulk thermal polymerization of styrene has shown that this was probably because of competitive reactions which produce cyclic dimers and trimers. The overall activation energy was estimated as $69.5 \pm 6$ kJ mol$^{-1}$, significantly below the value (about 100 kJ mol$^{-1}$) found for the thermal polymerization of styrene in bulk or solution. However doubts have been expressed[4] as to whether the initiation was genuinely thermal in these experiments because potassium oleate is known to form peroxides easily.

Harkins[5] found that a 10% potassium decanoate solution would solubilize 1.9% styrene at 25 °C. When solutions containing 25% or 75% of this amount of styrene were polymerized, it was found that the degree of polymerization of the polystyrene exceeded the number of monomer molecules solubilized in each micelle by a factor of four at the lower concentration and by a factor of 70 at the higher concentration. Later Hyde and Robb[6] obtained a similar result, finding the molecular weight of styrene polymerized in hexadecyltrimethylammonium micelles in the absence of a separate monomer phase to be some 500 times that which would be expected if each polymer molecule was formed by the polymerization of the monomer solubilized in a single micelle. Clearly monomer must diffuse from micelles which have not been initiated to those which have.

Bovey and Kolthoff[4] studied the polymerization of styrene initiated by persulfate in aqueous solution and in both anionic and cationic surfactant solutions. They determined the solubility of styrene in water to be 0.027% at 25 °C and 0.038% at 50 °C, claiming these values to be more accurate than those determined by Lane[7] which are higher. In water or dodecylammonium chloride solution the rate of polymerization was second order in monomer and half order in initiator. The cationic surfactant reduced the rate slightly. In the presence of anionic soaps at the same concentration the rate was much faster, with different kinetics. The rate of polymerization of an aqueous solution of styrene was the same in the presence of 0.019 mol dm$^{-3}$ potassium dodecanoate or hexadecanoate, although this concentration was below the cmc (0.02 mol dm$^{-3}$) of the former and greatly in excess of the cmc (0.001 mol dm$^{-3}$) of the latter. These observations strongly suggest that the function of the emulsifier is to stabilize latex particles nucleated in the true aqueous phase rather than from monomer solubilized in micelles. By contrast, in the emulsion polymerization of styrene when a separate monomer phase is present, the polymerization rates in Interval II (see Volume 4, Chapter 11) have been found to differ at equal total concentration of homologous emulsifiers but to be identical when the concentrations of micellar emulsifiers were equal.[8,9] Robb[10] found the initial rate of polymerization of a styrene emulsion to be eight times faster with sodium dodecyl sulfate than with sodium decyl sulfate at the same micellar concentration, although there would be a larger number of micelles with the latter emulsifier. Sodium dodecyl sulfate micelles are 50% larger than those of sodium decyl sulfate and solubilize more styrene. However the maximum conversion reached with sodium decyl sulfate in this experiment was only 4%: the experiment was ended before Interval II was reached. The reason for the large difference between Interval I rates and the proportionality of the Interval I rate to the amount of monomer solubilized warrants further investigation: it may be that the exit rate of radicals from the smaller micelles is much higher and that the radicals which escape terminate in the water, greatly reducing the rate of nucleation of latex particles.

## 12.2 IMPORTANCE OF SOLUBILIZATION IN EMULSION POLYMERIZATION

Kolthoff and Stricks[11] found that the amount of dimethylaminoazobenzene which could be solubilized was proportional to the micellar concentration of fatty acid soaps and that there was a large increase in their solubilizing power with increasing alkyl chain length. The solubilizing power of some other types of surfactant increased more than in direct proportion to concentration, which might possibly indicate a more rapid increase of micelle size with concentration in those cases (the major effect of increasing the concentration of a surfactant above its cmc is to increase the number of micelles). Kolthoff, Meehan, and Carr[12] found that the amount of styrene solubilized was proportional to the solubilization of dimethylaminoazobenzene and that the initial (*i.e.* Interval I) rate of the butadiene–styrene emulsion copolymerization was also proportional to the solubilization of dimethylaminoazobenzene and hence to the amount of monomer solubilized. Yurzhenko and Kolechkova[13] also found that the initial rate of emulsion polymerization of isoprene was proportional to the amount of monomer solubilized by different surfactants at the same concentration: potassium (sodium) oleate > ammonium oleate > sodium rosinate > sodium butyl α-naphthalene sulfonate.

## 12.3 EFFECT OF VARIATION OF PHASE RATIO

The phase ratio is generally believed not to be an important factor in emulsion polymerization. This was clearly demonstrated for styrene by the meticulous work of Gerrens and co-workers (Table 1). The Interval II rate of polymerization ($R_p$), the final latex particle number concentration ($N$) and the average degree of polymerization ($\bar{P}$) do not vary significantly when the volume of the monomer phase is varied from 5 to 16% at constant initiator and emulsifier concentrations. The average volume of the latex particles ($\bar{V}$) increases proportionally to the volume of monomer. The breadth of the particle size distribution (as measured by the ratio of the standard deviation $\sigma$ to the mean) does decrease as the volume of monomer is increased.

Yurzhenko and Kolechkova[13] observed a much larger effect in the emulsion polymerization of isoprene using an oil-soluble initiator (diazoaminobenzene): a 75% isoprene emulsion polymerized at 3.3% h$^{-1}$, whereas in a more dilute emulsion with 25% isoprene by volume the rate was

**Table 1** Effect of Variation of the Phase Ratio in the Emulsion Polymerization of Styrene at 45 °C[14]

| | | | | | |
|---|---|---|---|---|---|
| Volume % of monomer | 4.9 | 10.7 | 16.0 | 21.4 | 30.8 |
| $R_p \times 10^5$ (mol dm$^{-3}$ s$^{-1}$) | 9.5 | 9.8 | 9.7 | — | — |
| $N \times 10^{-14}$ (per cm$^3$ water) | 1.2 | 1.2 | 1.4 | — | — |
| $\bar{P} \times 10^{-4}$ | 3.1 | 3.5 | 3.1 | — | — |
| $\bar{V} \times 10^{16}$ (cm$^3$) | 2.1 | 4.9 | 8.7 | 13.8 | 19.6 |
| $\sigma/\bar{V}$ (%) | 39.8 | 32.3 | 30.6 | 25.6 | 20.6 |

22.7% h$^{-1}$. Because of the increase of the rate with the volume of the aqueous phase they inferred that monomer solubilized in emulsifier micelles in the aqueous phase was the locus of the reaction. A 3% sodium oleate solution was used in these experiments which was found to solubilize 10% isoprene at room temperature in the presence of 0.04 mol dm$^{-3}$ $Na_2CO_3$. However no experiments were undertaken in the absence of a separate monomer phase.

Konar and co-workers[15] show that the rate of polymerization of a very dilute (0.625%) styrene emulsion is much faster than for a 5% emulsion although they offer no comment on this observation. This effect has been confirmed[16] (Figure 1). The polymerization rate for the 5% emulsion could be reproduced exactly but the polymerization rate of the 0.625% emulsion (which is sufficiently dilute for all the monomer to be solubilized in the emulsifier micelles) is sensitive to the stirring speed. The rates are not in direct proportion to the final particle numbers. Mishal *et al.*[17] have also observed a very high rate (8.6% min$^{-1}$) in the micellar polymerization of styrene: they used 0.5% or 1% solutions of styrene solubilized by a sodium alkyl sulfate emulsifier. The same maximum rate of polymerization was observed whether a water-soluble initiator (persulfate) or an oil-soluble initiator (azodiisobutyronitrile) was used but the induction periods differed in the latter case but not in the former. No comparison was made with emulsion polymerization under similar conditions. They derive an equation for the rate of polymerization in the absence of a separate monomer phase

$$P(t) \simeq 1 - \exp(-\alpha t^2) \tag{1}$$

where P($t$) is the conversion at time $t$ and $\alpha = R_i k_p \Phi_M^0 / 2c_0$, where $R_i$ is the rate of formation of radicals in the aqueous phase, $c_0$ is the initial concentration of monomer (dissolved in water and solubilized by the emulsifier), and $\Phi_M^0$ is the equilibrium volume fraction of monomer in the latex particles which would correspond to $c_0$. This equation provides a reasonable fit for the results obtained with the water-soluble initiator but the value of the propagation rate constant $k_p$ for styrene (80 dm$^3$ mol$^{-1}$ s$^{-1}$ at 50 °C) is very low (the range of literature values is two to four times this value).

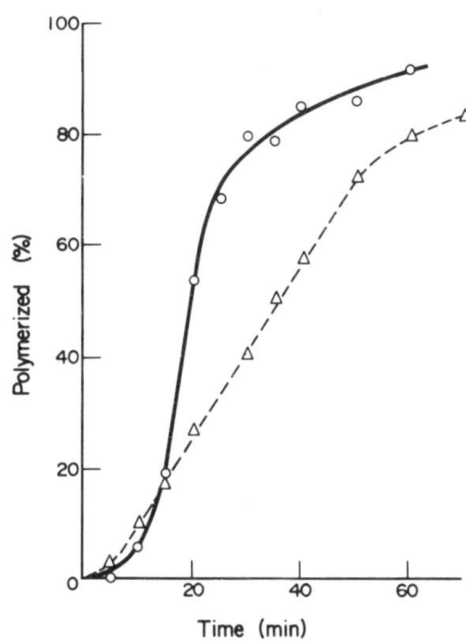


**Figure 1** Conversion–time curves for the polymerization of styrene[16] at 50 °C with 0.1% potassium persulfate and 0.6% sodium dodecyl sulfate in the aqueous solution; (○) micellar solution containing 0.625% v/v styrene, (△) 5% v/v styrene emulsion

The period of constant rate in micellar polymerization is the result of the effect of monomer consumption being compensated for by the effect of continued particle nucleation when emulsifier is in excess, although ultimately ($>70\%$ conversion) the observed rate falls below the calculated rate, probably because of the coalescence of latex particles. Coalescence of latex particles is a factor which has generally been considered to be negligible in emulsion polymerization but the skewness of particle size distribution curves at low conversions[18] provides evidence for its occurrence. High emulsifier:monomer ratios under micellar polymerization conditions may delay the onset of latex particle coalescence, explaining the high rates of polymerization observed.

## 12.4 MICROEMULSION POLYMERIZATION

Emulsions are thermodynamically unstable: an input of energy through vigorous stirring or shaking is needed to form an emulsion because there is a large increase in interfacial area with a consequent increase of the Gibbs surface energy of the system. However, once formed, the emulsion may persist for a long time if the rate of coalescence of the droplets of the dispersed phase is sufficiently slow, *i.e.* the emulsion may have appreciable *kinetic* stability. This is also true of *miniemulsions* in which a smaller droplet size is achieved by the use of a co-emulsifier (a long-chain alkane or alkanol) insoluble in the continuous phase.[19] By contrast, solubilization in micelles and the formation of microemulsions (which have droplet sizes intermediate between those of swollen micelles and those of miniemulsions but which are less than the wavelength of visible light and which are consequently optically transparent) are spontaneous processes which produce thermodynamically stable systems. The reason for this difference is that the van der Waals attractive force between the droplets of the dispersed material increases with droplet size[20] and only becomes significant beyond some minimum size. The droplets of a microemulsion are below this size and are therefore stable. Oil-in-water microemulsions can only be formed over a small range of emulsifier concentration at high emulsifier:monomer ratios and generally require the use of a co-emulsifier. Their optical transparency is advantageous for photochemical experiments. Photoinitiation of polymerization in ordinary, opaque emulsions is possible but all the light is absorbed in a small depth from the illuminated face of the reaction vessel, whereas light can be absorbed more uniformly throughout the vessel if a microemulsion is used.

Nikitina[21] found that if a nonionic emulsifier [a poly(oxyethylene octyl phenol) or a 'Pluronic'—a block copolymer of poly(ethylene oxide) and poly(propylene oxide)] which was much more soluble in water than in oil (70–200 ethylene oxide units: distribution coefficient $c_{oil}/c_{water} < 0.04$) was dissolved in xylene, styrene or vinyl acetate and the solution brought into contact with water, mass transfer of the emulsifier to the water would cause the spontaneous formation of an oil-in-water microemulsion having a narrow particle size distribution with an average particle size of 100 nm. This microemulsion polymerized more rapidly using either a water-soluble or an oil-soluble initiator than when the same amount of emulsifier was dissolved in the water initially. In the case of vinyl acetate the final latex particles were 50% larger than the micro-emulsion droplets: particle size did not vary with conversion. The microemulsion droplets were evidently the principal locus of polymerization, although 2–4% of the original monomer volume was later[22] found to lie in the 20–30 nm size range. It was thought that these smaller latex particles were the product of the polymerization of monomer solubilized in micelles. Similar experiments were also carried out with styrene.[23]

Atik and Thomas[24] produced a microemulsion of a mixture of styrene and divinylbenzene using cetyltrimethylammonium bromide as emulsifier and hexanol as co-emulsifier. Polymerization produced 33 nm diameter latex particles. By contrast, with a toluene microemulsion this latex remained stable even when diluted below the cmc of the emulsifier. Jayakrishnan and Shah[25] question whether this system was a true microemulsion: they studied a more concentrated system using a mixed anionic–nonionic emulsifier (sodium methylpentyl sulfosuccinate and a poly(propylene oxide)–poly(ethylene oxide) block copolymer). Transparent microlatices of polystyrene and poly(methyl methacrylate) were obtained but these were not stable on dilution. The rate of polymerization was somewhat lower than for conventional emulsion polymerization under similar conditions.

Tang, Johnson and Gulari[26] prepared microemulsions containing 3–6% styrene using sodium dodecyl sulfate as emulsifier and pentanol as co-emulsifier in the molar ratio 3.3:1. The average diameter of the original microemulsion droplets was determined as 41 nm by dynamic light scattering (compared with 2.5 nm for a sodium dodecyl sulfate micelle). Using the method of cumulants, dynamic light scattering indicated that the latex particles obtained after polymerization with an oil-soluble initiator had the same size as the initial microemulsion droplets, but the histogram method showed that the particle size distribution was actually bimodal, comprising approximately equal numbers of 22 nm and 72 nm particles. It is possible that the larger particles are produced by the polymerization of microemulsion droplets and the smaller particles by polymerization of monomer solubilized in micelles, but to verify this hypothesis it would be necessary to conduct a micellar polymerization experiment at a lower emulsifier concentration in the absence of microemulsion droplets to see whether only the smaller latex particles were then formed. Johnson and Gulari[27] used a similar system with both oil-soluble and water-soluble initiators but with smaller initial microemulsion droplets (3.6 or 4.9 nm): the final lattices were again bimodal but the larger particles (41–92 nm) were much larger than the original microemulsion droplets.

Photoinitiation of polymerization in a microemulsion has been studied[28] using dibenzyl ketone as photosensitizer. It was found essential to dilute styrene with an equal volume of toluene to prevent turbidity developing during polymerization. Sodium dodecyl sulfate was the emulsifier with pentanol as co-emulsifier. The microemulsion droplets were in the 10–30 nm size range but the final latex particles were larger (35–56 nm). The polymerization rate was found to be of order 0.2 in both the photosensitizer and the light intensity. The rate decreased slightly with increase of conversion which was attributed to a decrease in the number of latex particles as a result of limited coalescence, which would account for the increase of the size of the final particles compared with that of the initial droplets.

Grätzel, Tirousek and Grätzel[29] prepared a surface-active initiator (hexadecyltrimethylammonium persulfate from hexadecylammonium chloride and potassium persulfate)[30] and used it mixed with hexadecyltrimethylammonium chloride and pentanol to prepare transparent microemulsions of styrene, acrylamide, acrolein and methyl methacrylate with and without divinylbenzene as a cross-linking co-monomer. These could be photopolymerized with ultraviolet light or by visible light with either eosin or the ruthenium bipyridyl complex as photosensitizer. Very large latex particles of up to 11.5 $\mu$m were produced, but electron microscopy indicated that these were formed by agglomeration of 10–30 nm particles.

Microemulsion droplets are small enough for termination to be instantaneous when they contain more than one radical. The instantaneous monomer concentration in a polymerizing particle must be below the equilibrium monomer concentration in a monomer–polymer particle of the same size. Consequently, at any time, there must be a tendency for monomer to diffuse from inactive particles to active particles leading ultimately to a reduction in the number of particles and an increase in their size, even in the absence of any coalescence. It is therefore unlikely that polymer latex particles produced from a monomer microemulsion would be identical in size with the initial microemulsion droplets.

## 12.5 POLYMERIZATION OF SURFACE-ACTIVE MONOMERS

Monomers which are also surfactants can also be prepared. Above their cmc values, these form micelles which may affect the course of their polymerization.

Greene, Sheetz and co-workers[31] prepared sodium 9- (or 10-) acryloamidostearate (**1**) which was used as the emulsifier in the preparation of styrene–butadiene lattices primarily with the idea of improving the stability of the latex to electrolyte addition and mechanical agitation. Only a low concentration of the polymerizable emulsifier was used in preparing the initial latex. Additional amounts could be adsorbed and polymerized *in situ* to increase the surface charge density of the particles. It was found that both the mechanical stability and the stability to electrolyte addition were conspicuously better when the polymerized surfactant was compared with the same amount of the same adsorbed surfactant. Adsorbed surfactant molecules desorb when the latex is diluted or when the local concentration is increased during the collison of two particles.

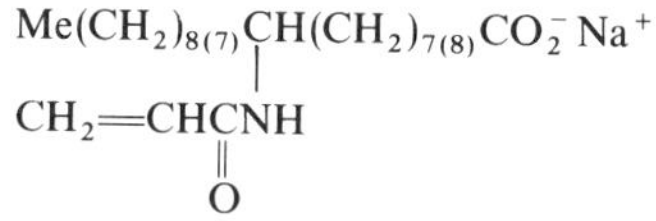


(**1**)

Hamid and Sherrington[32] synthesized a large variety of amphiphilic monomers which were shown to form micelles in aqueous solution. Below their cmc values or in isotropic solution in alcohol, quaternary alkylammonium acrylates and methacrylates formed only dimers and trimers when polymerized because of the effects of charge repulsion and the low concentration. Above the cmc oligomers with a degree of polymerization up to 25 were formed. Although the magnitude of this degree of polymerization is suggestive of the polymerization of individual micelles, it appears to have been actually the result of a facile transfer reaction. The authors argue[33] that polymerization of micelles is unlikely because the rate of exchange of individual molecules between micelle and solution is about twice the rate of addition of monomers to a polymerizing radical.

Nagai *et al.*[34] studied the polymerization of octyl, dodecyl, tetradecyl and octadecyldimethylammonium ethyl methacrylate bromide in water and in organic solvents of varying polarity. They found that micelles (**2**) were formed both in water and nonpolar organic solvents (*e.g.* benzene) but not in polar organic solvents (*e.g.* acetonitrile). The rate of polymerization increased with the alkyl chain length of the monomer and was much faster in water (with persulfate initiation) and in benzene (with azodiisobutyronitrile as initiator) than in acetonitrile at the same initiator concentration.

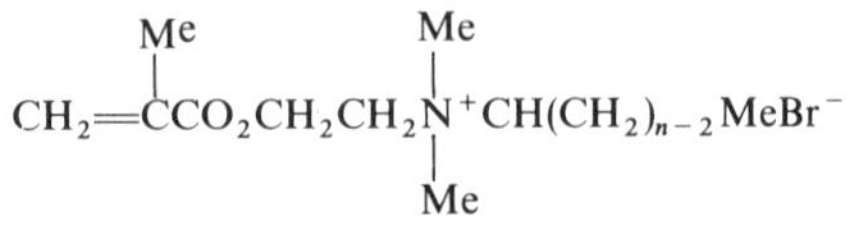

(**2**) $n = 8$–$18$

## 12.6 REFERENCES


1. E. W. Hughes, W. M. Sawyer and J. R. Vinograd, *J. Chem. Phys.*, 1945, **13**, 131.
2. V. J. Frilette and W. P. Hohenstein, *J. Polym. Sci.*, 1948, **3**, 22.
3. F. R. Mayo, *J. Am. Chem. Soc.*, 1968, **90**, 1289.
4. F. A. Bovey and I. M. Kolthoff, *J. Polym. Sci.*, 1950, **5**, 487.
5. W. D. Harkins, *J. Polym. Sci.*, 1950, **5**, 217.
6. A. J. Hyde and D. J. M. Robb, *J. Phys. Chem.*, 1963, **67**, 2089.
7. W. H. Lane, *Ind. Eng. Chem., Anal. Ed.*, 1946, **18**, 295.
8. W. A. G. R. Al-Shahib and A. S. Dunn, *J. Polym. Sci., Polym. Chem. Ed.*, 1978, **16**, 677.
9. A. S. Dunn and W. A. G. R. Al-Shahib, in 'Polymer Colloids II', ed. R. M. Fitch, Plenum Press, New York, 1980, p. 619.
10. I. D. Robb, *J. Polym. Sci., Part A-1*, 1969, **7**, 417.
11. I. M. Kolthoff and W. Stricks, *J. Phys. Colloid Chem.*, 1948, **52**, 915.
12. I. M. Kolthoff, E. J. Meehan and C. W. Carr, *J. Polym. Sci.*, 1951, **6**, 73.
13. A. I. Yurzhenko and M. Kolechkova, *Dokl. Akad. Nauk SSSR*, 1945, **47**, 348.
14. H. Gerrens, *Fortschr. Hochpolym. Forsch.*, 1959, **1**, 234.
15. S. P. Chatterjee, M. Banerjee and R. S. Konar, *J. Polym. Sci., Polym. Chem. Ed.*, 1979, **17**, 2193.
16. T. Farooqui, unpublished work.
17. H. Mishal, G. I. Litvinenko, I. A. Gritskova, V. A. Kaminskii and A. N. Pravednikov, *Vysokomol. Soedin., Ser. B*, 1983, **25**, 139.
18. G. Lichti, R. G. Gilbert and D. H. Napper, *J. Polym. Sci., Polym. Chem. Ed.*, 1983, **21**, 269.
19. J. Ugelstad, M. S. El-Aasser and J. W. Vanderhoff, *J. Polym. Sci., Polym. Lett. Ed.*, 1973, **11**, 503.
20. R. H. Ottewill, in 'Emulsion Polymerization', ed. I. Parma, Academic Press, New York, 1982, chap. 1, *cf.* Figure 9.
21. S. A. Nikitina, in 'Chemistry, Physics, and Practical Applications of Surface Active Agents: Proceedings of the 5th International Congress on Detergents', Ediciones Unidas, Barcelona, 1969, vol. II, part I, p. 331.
22. S. A. Nikitina, V. A. Spiridonova and A. B. Taubmann, *J. Polym. Sci., Part A-1*, 1970, **8**, 3045.
23. I. A. Gritskova, S. S. Medvedev and M. F. Margaritova, *Kolloid-Z.*, 1964, **26**, 168.
24. S. S. Atik and J. K. Thomas, *J. Am. Chem. Soc.*, 1982, **104**, 5868.
25. A. Jayakrishnan and D. O. Shah, *J. Polym. Sci., Polym. Lett. Ed.*, 1984, **22**, 31.
26. H. I. Tang, P. L. Johnson and E. Gulari, *Polymer*, 1984, **25**, 1357.
27. P. L. Johnson and E. Gulari, *J. Polym. Sci., Polym. Chem. Ed.*, 1984, **22**, 3967.
28. P. L. Kuo, N. J. Turro, C. M. Tseng, M. S. El-Aasser and J. W. Vanderhoff, *Macromolecules*, 1987, **20**, 1216.
29. C. K. Grätzel, M. Jirousek and M. Grätzel, *Langmuir*, 1986, **2**, 292.
30. C. K. Grätzel, M. Jirousek and M. Grätzel, *Colloids Surf.*, 1985, **13**, 221.
31. B. W. Greene, D. P. Sheetz and T. D. Filer, *J. Colloid Interface Sci.*, 1970, **32**, 90; B. W. Greene and D. P. Sheetz, *J. Colloid Interface Sci.*, 1970, **32**, 96; B. W. Greene and F. L. Saunders, *J. Colloid Interface Sci.*, 1970, **33**, 393.
32. S. M. Hamid and D. C. Sherrington, *Polymer*, 1987, **28**, 325, 333; *Br. Polym. J.*, 1984, **16**, 39.
33. S. M. Hamid and D. C. Sherrington, *J. Chem. Soc., Chem. Commun.* 1986, 936.
34. K. Nagai, Y. Ohishi, H. Inaba and S. Kudo, *J. Polym. Sci., Polym. Chem. Ed.*, 1985, **23**, 1221.

# 13

# Polymerization in Inverse Microemulsions

FRANÇOISE CANDAU
*Institut Charles Sadron, Strasbourg, France*



## 13.1 STRUCTURE OF MICROEMULSIONS

Although it has been investigated far less than direct emulsion polymerization (see Volume 4, Chapter 9), a new interest in inverse emulsion polymerization has developed, because of the numerous applications of high molecular weight water-soluble polymers in water treatment, paper, mining[1] and, more recently, in enhanced oil recovery.[2] In this process, an aqueous solution of a hydrophilic monomer, such as acrylic acid or acrylamide, is dispersed with a water-in-oil emulsifier in a hydrophobic oil continuous phase. In contrast to many conventional aqueous latexes, which can be highly monodisperse in size and remain stable indefinitely, inverse latexes usually exhibit a large particle size distribution and flocculate quite rapidly.[3,4] The different nature of the electrostatic forces between the two kinds of dispersions is responsible for this behavior.

Recent studies have shown that the polymerization of a water-soluble monomer in a microemulsion medium,[5,6] rather than in an emulsion, could overcome some of these problems. Microemulsions are transparent and thermodynamically stable systems, containing large proportions of oil and water stabilized by an interfacial layer of surface active agents.[7,8] Unlike emulsions, which are essentially of water-in-oil or oil-in-water types, microemulsions can adopt a large variety of labile structural organizations. The latter are characterized by low size domains, the stabilization of which requires a fairly high amount of surfactant ($\sim$10–15%). In the water- or oil-rich regions, microemulsions consist of small globules of uniform size ($\sim 10^{-2}$ μm) swollen with oil and dispersed in water (direct microemulsions) or swollen with water and dispersed in oil (inverse microemulsions). In the phase inversion region, where the mixtures contain comparable amounts of oil and water, the description generally given is that of a bicontinuous structure with randomly connected oil and water domains.[9,10] One remarkable feature is that microemulsions can shift from a water-in-oil to an oil-in-water dispersion without any apparent discontinuity. Lamellar structures have also been observed in systems containing small amounts of surfactants.[11] This variety of structures offers great versatility for choosing the locus of polymerization.

Microemulsion polymerization was first used only around 1980, as a consequence of the extensive studies on microemulsions developed after the oil crisis. In the case of polymerization in direct microemulsions, phase separation often occurs for microemulsions containing 10% or more of monomer, which restricts the applicability of the process.[12] Inverse microemulsion polymerization is more promising for future development. The water-soluble monomers thus far considered are those most widely used in a variety of applications, *i.e.* acrylamide (AM) and sodium acrylate–acrylamide mixtures.

## 13.2 FORMULATION OF POLYMERIZABLE MICROEMULSIONS

The choice of the system to be polymerized is of major importance, since it determines the ultimate properties of the latexes. Acrylamide or acrylamide–sodium acrylate mixtures were polymerized by a free radical process, either inside water-swollen micelles ($d \sim 6 \times 10^{-3}$ $\mu$m) stabilized with an ionic surfactant aerosol AOT (sodium 1,4-bis(2-ethylhexyl)sulfosuccinate),[5,6] or in nonionic microemulsions.[13,14] In both cases, the presence of monomers increases the extent of the microemulsion domain in the phase diagram (Figure 1). The monomer acts as a consurfactant and is partly located at the water–oil interface rather than in the micelles. Confirmatory evidence of such an effect was given for AOT–toluene systems by light scattering and viscometry data, which indicated a strong increase of the interparticular attractive forces with the acrylamide content.[6]

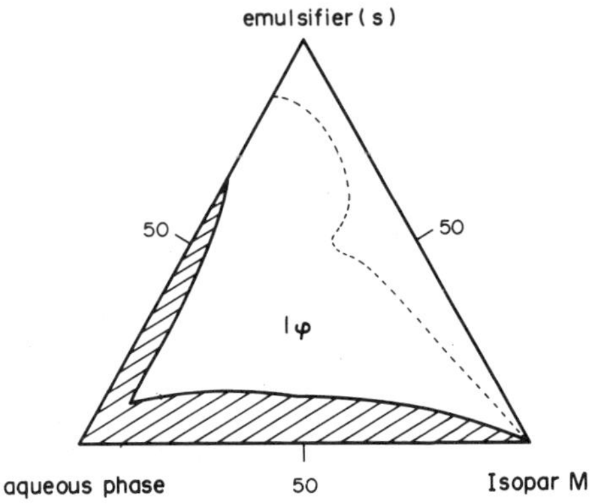


**Figure 1** Pseudo-ternary phase diagram (wt%) obtained with nonionic emulsifiers: hatched area = emulsion domain; ----, in the absence of monomer; —, in the presence of acrylamide + sodium acrylate (water/monomers = 1.25; acrylamide/sodium acrylate = 2.3)

In the case of nonionic emulsifiers, the criteria previously developed for the stability of emulsions[15] have been successfully extended to microemulsions, allowing the optimization of the process.[13,16] In this approach it is assumed that there is a perfect chemical matching between the lipophilic moiety of the emulsifier and oil on one side and the hydrophilic moiety and water on the other side. The corresponding cohesive energy ratio, which determines the nature and the stability of the dispersion, is then directly related to the emulsifier hydrophile–lipophile balance (HLB); this correlates the structure and effectiveness of ethylene-oxide-based surfactants. Thus, it is possible to calculate the required HLB for any given oil whose solubility parameters are known. For example, a blend of sesquioleate sorbitan and poly(oxyethylene) sorbitol hexaoleate was found to be quite efficient for the microemulsification of a 50% aqueous solution of acrylamide[13] or acrylamide–sodium acrylate[14] in an isoparaffinic oil, Isopar M. The minimum emulsifier concentration required is closely associated to the optimum HLB value of the blend. The microemulsions of interest (*i.e.* 25% monomer content and ~11% emulsifiers) exhibit a bicontinuous character. An investigation into the effect of various electrolytes on the solubilization properties of the emulsifier blend has shown that the formation of bicontinuous microemulsions is induced by salts producing a strong salting-out effect, such as sodium acetate[13] or sodium acrylate[14] (the latter being a comonomer for the synthesis of poly(acrylamide-*co*-acrylate)s). Note that these bicontinuous microemulsions have very low interfacial tensions ($\sim 10^{-3}$ dyn cm$^{-1}$; $10^5$ dyn = 1 N).

## 13.3 POLYMERIZATION

The process is characterized by very high reaction rates (minutes rather than hours in inverse emulsion). A common feature found in all experiments is that the final system always consists of a uniform dispersion of spherical latex particles, regardless of the structure of the starting system (globular,[5] bicontinuous[13,14] or lamellar[13]). This could be explained by the high flexibility of the interface and by the conformation of polymeric random coils, which should favor the formation of small nucleated spheres from local bending of the water–oil interface.

In the case of AOT micellar systems, one observes a notable increase in the particle size during the polymerization, so that each final latex particle is the result of the fusion of 60–150 initial micelles. After polymerization there is, as a consequence of the particle growth, an excess of the emulsifier, which forms small hydrated AOT micelles coexisting with the polymer latex particles.[6] The unprecedented discovery was made that each polymer particle contains, on average, one single high molecular weight molecule of polyacrylamide in a collapsed state, compared with thousands of chains in inverse latexes.[17] This suggests kinetics which do not follow the classical Smith–Ewart theory[18] but involve a continuous particle nucleation. In a recent study, a rotating sector technique has been used to follow the polymerization of AM in AOT–toluene systems photoinitiated by azobis-(isobutyronitrile) (AIBN) or by a water-soluble dye, methylene blue/eosin.[19] The rates of polymerization were found to be first order with respect to the incident light intensity and first-order with respect to the initiator concentration. These results confirm that monoradical termination occurs within the droplets through transfer to toluene at the interface, followed by exodiffusion of the benzylic radical into the toluene phase. These radicals are deactivated in the organic phase and do not undergo reentry into other micelles. The molecular weights of the polymers are independent of the rates of polymerization and initiation, which supports the degradative chain-transfer mechanism.

A kinetic study of AM polymerization photoinitiated by benzophenone, or one of its derivatives, in AOT (water + AM)/decane microemulsions was also performed.[20] High quantum yields and biradical termination reactions were found for these systems, as well as high rates of polymerization. It was shown that initiation occurs by competing pathways, either by decane radicals formed by quenching of benzophenone, or through radicals formed *via* electron transfer from benzophenone.

## 13.4 POLYMER CHARACTERISTICS

Polymers and copolymers formed by using this process exhibit high molecular weights[6] and high intrinsic viscosities in aqueous solutions (up to 3700 $cm^3\,g^{-1}$)[21] which make them important for industrial applications. The molecular weight distribution of these samples has not been accurately determined because of the difficulties encountered when dealing with such highly hydrophilic polymers. A sedimentation study[22] has given some evidence for an index of polydispersity much smaller than that found in the classical emulsion process.[23]

Polymerization of AM has also been carried out in water–toluene systems stabilized by amphiphilic block copolymers in the presence of 2-propanol. In that case, the range of attainable molecular weights is limited ($\sim 10^5$) because of chain-transfer reactions to the alcohol.[24]

The microstructure of poly(acrylamide-*co*-sodium acrylate)s was determined by $^{13}$C NMR and compared with statistical models.[25] The average copolymer composition was shown to be independent of the degree of conversion, and the sequence monomer distribution analyzed from triad proportions conforms to Bernouillian statistics within experimental error (Figure 2). The reactivity

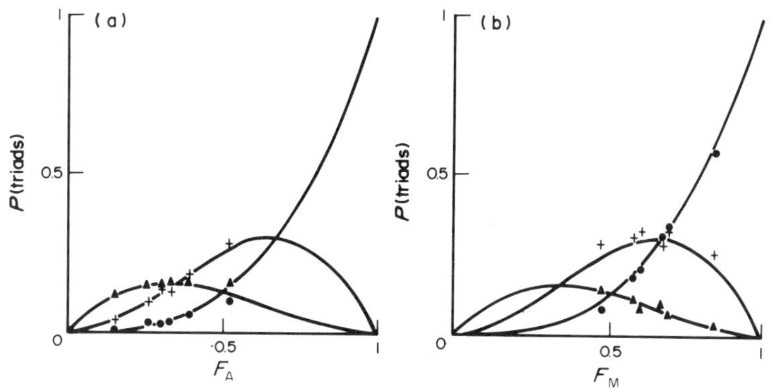


**Figure 2** Comparison of experimental triad proportions (symbols) with those calculated for Bernouillian statistics (curves) for poly(acrylamide-*co*-acrylate) samples at low degrees of conversion; $F_A$ and $F_M$ are the mole fractions in the copolymer of sodium acrylate (A) and acrylamide (M), respectively. (a) ●, AAA; +, AAM; ▲, MAM; (b) ●, MMM; +, MMA; ▲, AMA (reproduced by permission of the American Chemical Society from *Macromolecules*, 1986, **19**, 1895)

ratios of both monomers are therefore close to unity, significantly different from the reported literature values obtained for copolymers prepared in solution ($r_A \sim 0.3$, $r_M \sim 0.95$).[26] This difference was attributed to the microenvironment, which exerts a great influence on the free radical copolymerization of acrylamide with ionogenic monomers, as shown by numerous studies.[27,28] These results confirm a polymerization process by nucleation and interparticular collisions, rather than by monomer diffusion through the continuous medium.

## 13.5 MICROLATEXES

The inverse microlatexes formed after polymerization are clear, fluid and highly stable with no apparent settling even after a period of some months. The dimensions of the particles are rather low ($\sim$40 nm for AOT systems,[6] $\sim$70 nm for nonionic ones[13,14]) with a narrow size distribution, as shown from quasi-elastic light-scattering data. They behave like hard spheres, without strong interactions between the particles. The interparticular forces are indeed reduced after polymerization, as most of the polymer is strongly collapsed within the particles.

The dimensions of the latex particles increase with the acrylamide content, and decrease with increasing emulsifier concentration or with the acrylate content in the comonomer feed (for acrylamide-*co*-acrylate polymers). The latter result was attributed to the combined effects of dehydration of the ethoxylated emulsifier, together with a lowering of the water activity in the particles caused by the sodium acrylate salt.[14]

Additives play an important role in the stabilization of nonionic microlatexes. Stable and transparent microlatexes only form when salting-out-type electrolytes are added to the microemulsions, whereas unstable latexes are produced in the presence of salting-in-type electrolytes or in the absence of any salt. The role of the salts with high salting-out efficiency is to enhance the stability of the starting systems by promoting the formation of a bicontinuous structure characterized by a very low interfacial tension.[13] It is important to point out that polymerization in nonionic microemulsions can be achieved with a monomer to surfactant weight ratio much higher ($\sim$2.5) than in the case of AOT systems ($\sim$0.4). A seeding procedure was applied to the AOT microlatexes in order to increase the polymer content but the resulting swelling of the particles tends to reduce their stability.[29,30]

An important parameter, characteristic of the polymerization mechanism, is the average number of macromolecules $N$ contained in a latex particle. An estimate of this number can be made from the volume of the particle and the molecular weight of the polymer. It has been shown that $N$ is directly related to the acrylamide/emulsifier weight ratio, as illustrated in Figure 3, which gathers data relative to polyacrylamide latex series prepared under various conditions.[31] Note that the limiting case corresponding to one single polymer molecular per particle can be reached by using AOT systems.

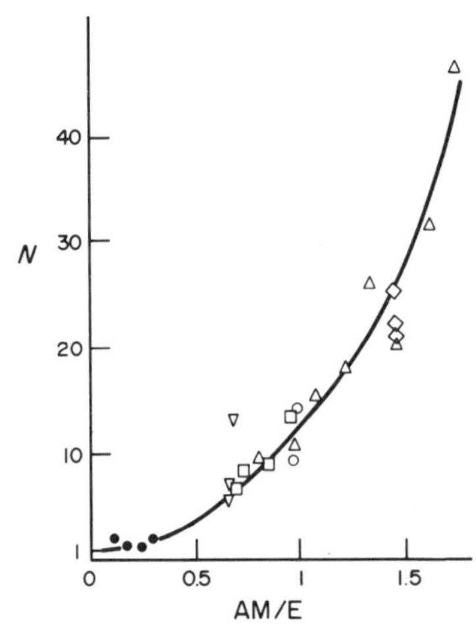


**Figure 3** Variation of the average number of macromolecules per latex particle *vs.* the acrylamide (AM)/emulsifier (E) weight ratio for series of various compositions;[31] ○, △, ▽, □, ◇, nonionic emulsifiers; ●, anionic emulsifier (AOT)

## 13.6 REFERENCES


1. F. Halverson and H. P. Panzer, in 'Kirk-Othmer Encylopedia of Chemical Technology', ed. M. Grayson, Wiley, New York, 1980, vol. 10.
2. W. M. Thomas and D. W. Wang, in 'Encyclopedia of Polymer Science and Engineering', 2nd edn., ed. H. Mark, N. Bikales, C. G. Overberger and G. Menges, Wiley, New York, 1985, p. 169.
3. J. W. Vanderhoff, E. B. Bradford, H. L. Tarkowski, J. B. Shaffer and R. M. Wiley, *Adv. Chem. Ser.*, 1962 **34**, 32.
4. C. Graillat, C. Pichot, A. Guyot and M. S. El Aasser, *J. Polym. Sci., Polym. Chem. Ed.*, 1986, **24**, 427.
5. Y. S. Leong and F. Candau, *J. Phys. Chem.*, 1982, **86**, 2269.
6. F. Candau, Y. S. Leong, G. Pouyet and S. Candau, *J. Colloid Interface Sci.*, 1984, **101**, 167; F. Candau, Y. S. Leong, N. Kohler and F. Dawans (IFP-CNRS), *Fr. Pat.* 2 524 895 (1984) (*Chem. Abstr.*, 1984, DE 3 312 711 $A_1$).
7. K. Shinoda and S. Friberg, *Adv. Colloid Interface Sci.*, 1975, **4**, 281.
8. A. M. Bellocq, J. Biais, P. Bothorel, B. Clin, G. Fourche, P. Lalanne, B. Lemaire, B. Lemanceau and D. Roux, *Adv. Colloid Interface Sci.*, 1984, **20**, 167.
9. L. E. Scriven, *Nature (London)*, 1976, **263**, 123.
10. S. Friberg, I. Lapczynska and G. Gillberg, *J. Colloid Interface Sci.*, 1976, **56**, 19.
11. J. M. Dì Meglio, M. Dvolaitzky, R. Ober and C. Taupin, *J. Phys., Lett, (Orsay, Fr.)*, 1983, **44**, L-229.
12. A. Jayakrishnan and D. O. Shah, *J. Polym. Sci., Polym. Lett. Ed.*, 1984, **22**, 31.
13. C. Holtzscherer, Ph.D. Thesis, University of Strasbourg, 1986 and papers in preparation.
14. F. Candau, Z. Zekhnini and J. P. Durand, *J. Colloid Interface Sci.*, 1986, **114**, 398; J. P. Durand, D. Nicolas, N. Kohler, F. Dawans and F. Candau (IFP), *Fr. Pat.* 2 565 623 and 2 565 692 (1984).
15. A. Beerbower and M. W. Hill, 'McCutcheon's Detergent and Emulsifier Annual', Allured, Ridgewood, NJ, 1971, p. 223.
16. C. Holtzsherer and F. Candau, *Colloids Surf.* (in press).
17. F. Candau, Y. S. Leong and R. M. Fitch, *J. Polym. Sci., Polym. Chem. Ed.* 1985, **23**, 193.
18. W. V. Smith and R. H. Ewart, *J. Chem. Phys.*, 1948, **16**, 592.
19. M. Carver, U. Dreyer, R. Knoesel, F. Candau and R. M. Fitch, *J. Polym. Sci., Polym. Chem. Ed.* (in press).
20. J. P. Fouassier, D. J. Lougnot and I. Zuchowicz, *Eur. Polym. J.*, 1986, **22**, 933.
21. F. Candau, Z. Zekhnini, F. Heatley and E. Franta, *Colloid Polym. Sci.*, 1986. **264**, 676.
22. F. Candau, Y. S. Leong, G. Pouyet and S. Candau, in 'Physics of Amphiphiles: Micelles, Vesicles and Microemulsions', ed. V. Degiorgio and M. Corti, North-Holland, Amsterdam, 1984, p. 830.
23. J. W. Vanderhoff, F. V. di Stefano, M. S. El Aasser, R. O'Leary, O. M. Shaffer and D. L. Visioli, *J. Dispersion Sci. Technol.*, 1984, **5**, 323.
24. Y. S. Leong, G. Riess and F. Candau, *J. Chim. Phys. Phys.-Chim. Biol.*, 1981, **78**, 279.
25. F. Candau, Z. Zekhnini and F. Heatley, *Macromolecules*, 1986, **19**, 1895.
26. S. Ponratnam and S. L. Kapur, *Makromol. Chem.*, 1977, **178**, 1029.
27. K. Plochocka, *Rev. Macromol. Chem.*, 1981, **C20**, 67.
28. V. F. Kurenkov and V. A. Myagchenkov, *Eur. Polym. J.*, 1980, **16**, 1229.
29. C. Holtzcherer, S. Candau and F. Candau, in 'Surfactants in Solution', ed. K. Mittal and P. Bothorel, Plenum Press, New York, 1987, vol. 6, p. 1473.
30. F. Candau, C. Holtzscherer and S. Candau, in 'Physical Optics of Dynamic Phenomena and Processes in Macromolecular Systems', ed. B. Sedlacek, de Gruyter, Berlin, 1985, p. 397.
31. C. Holtzsherer, J. P. Duraud and F. Candau, *Colloid Polym. Sci.*, 1987, **265**, 1067.

# 14

# Aqueous Suspension Polymerizations

JOHN V. DAWKINS

*Loughborough University of Technology, UK*



## 14.1 INTRODUCTION

Polymerization processes in heterogeneous systems are of considerable interest because they offer many practical advantages in industrial applications.[1,2] Among these heterogeneous processes are emulsion and suspension polymerizations of monomers dispersed in an aqueous phase. These two processes have the advantages of the ready control of the heat of polymerization and the possible direct use of the water-based emulsion and suspension polymers.

The term suspension polymerization is defined as a process in which monomer(s) relatively insoluble in water is (are) dispersed as liquid droplets by vigorous stirring which is maintained during polymerization to produce polymer particles as a dispersed solid phase. Initiators soluble in the liquid monomer phase are employed in this polymerization process. The terms pearl and bead polymerization are also used for the suspension polymerization process. The major aim in suspension polymerization is the formation of as uniform a dispersion as possible of monomer droplets in the aqueous phase and the prevention of coalescence of these droplets during the polymerization process. The degree of agitation and the design of the stirrer/reactor system determine the dispersion of monomer droplets, typically with diameters located in the range 10 $\mu$m to 5 mm, in a suspension polymerization which generally also requires the presence of substances known as protective colloids, suspension stabilizers or suspending agents. These substances hinder the coalescence of monomer droplets and the adhesion of partially polymerized particles during the course of polymerization, so that the solid beads may be produced in the same spherical form in which the monomer was dispersed in the aqueous phase. Many (but not all) important commercial suspension polymers have bead sizes above 10 $\mu$m, and so these relatively large particles (compared with emulsion polymers) are simply isolated by filtration and/or sedimentation. Examples of beads of copoly(styrene/divinylbenzene)[3,4] are illustrated in Figure 1.

Suspension polymerization has both common features with and distinguishing features from emulsion polymerization.[1,2] A primary difference is the location of the initiator which generally (but not always) is dissolved in the aqueous phase in emulsion polymerization.[5] A distinction between these polymerizations based on particle size and surfactant can present anomalies, because emulsion polymerization processes are available for the production of polymer particles with diameters[6] as

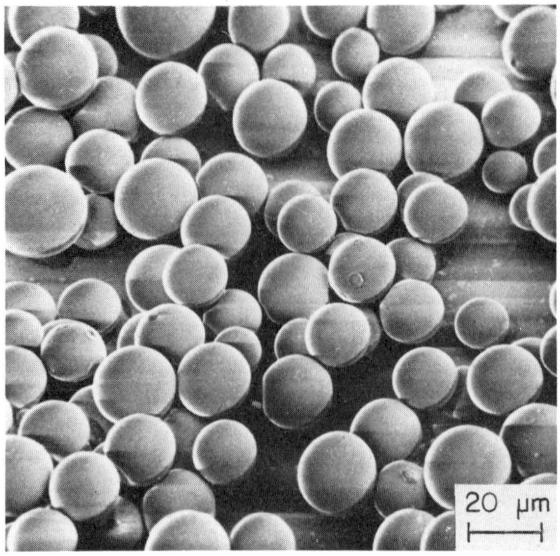


**Figure 1** Scanning electron micrograph of particles of copoly(styrene/divinylbenzene)

high as 20 $\mu$m and suspension polymerization processes may be performed with surfactant at submicellar concentrations.[7] Munzer and Trommsdorff[8] cite the example of protective colloid emulsion polymerization of vinyl acetate using relatively large quantities of a water soluble protective colloid and aqueous phase initiation to produce particle sizes in the range 0.5–10 $\mu$m. This type of polymerization, though probably closely related to emulsion polymerization, is also termed aqueous dispersion polymerization[8] which has been defined as a polymerization process for the preparation of polymer particles in the size range between suspension and emulsion polymers, *i.e.* about 10 $\mu$m size.[7] However, the term dispersion polymerization is also used with reference to the polymerization of a monomer dissolved in organic media containing a macromolecular dispersant which protects the precipitating polymer against flocculation, giving polymer particles ranging from 0.1 to 1.0 $\mu$m in size.[9] This type of process, which initially is homogeneous but transforms to heterogeneous on polymerization, should be contrasted with inverse suspension and inverse emulsion polymerizations where water soluble polymers are produced by dispersing an aqueous phase containing monomer in a hydrophobic organic diluent.[5,8,10,11]

## 14.2 DROPLET SIZE IN DISPERSIONS

A dispersion may be defined as a 'heterogeneous system consisting of at least one immiscible liquid intimately dispersed in another in the form of droplets'.[12] Whereas emulsions which remain unchanged for long periods under favourable storage conditions may be regarded as stable dispersions, suspensions are considered to be unstable dispersions which on cessation of agitation coalesce and separate into two distinct phases. Church and Shinnar[13] introduced the term turbulence-stabilized dispersions for systems in which coalescence is reduced considerably by the combined action of a protective colloid and by turbulent agitation, the droplets remaining stable as long as agitation continues at the same rate.

Application of shear to two immiscible liquid phases reduces the size of the dispersed droplets and produces droplets of varying size. To obtain the limiting droplet size distribution, and therefore mean droplet size, may require a long period of stirring in order to establish the equilibrium between the rates of droplet breakage and coalescence.[14] Theoretical treatments of dispersions calculate the most probable largest droplet diameter which can remain without being broken up in the turbulent flow.[15–17] A drop breaks up under the dynamic forces which are produced by changes in velocity over a distance at most equal to the diameter of the drop, but surface tension forces will tend to hold the drop together. The ratio of shear forces to surface forces is the Weber Number $W_e$ given by

$$W_e = \rho \overline{v^2(D)} D/\sigma \quad (1)$$

where $\rho$ is the mean liquid density, $\overline{v^2(D)}$ is the mean square value of the turbulent velocity fluctuations over a distance $D$, and $\sigma$ is the interfacial tension.

Direct experimental measurements of droplet size have been reported by encapsulation with a thin film of polymer around the droplet surface or by stabilization by mixing the dispersion with a strong soap solution.[18–20] These observations demonstrated that the agitator speed $N$, the interfacial tension $\sigma$ and the volume fraction $\phi$ of the dispersed phase had important effects on droplet size whereas the viscosity of the dispersed phase $\eta_M$ was of secondary importance. Typical changes are represented in Figures 2 and 3, showing how the shape of the distribution curve for droplet size is symmetrical at low $N$ and high $\sigma$ but becomes skewed with a high population of minimum sized droplets at high $N$ and low $\sigma$.[21]

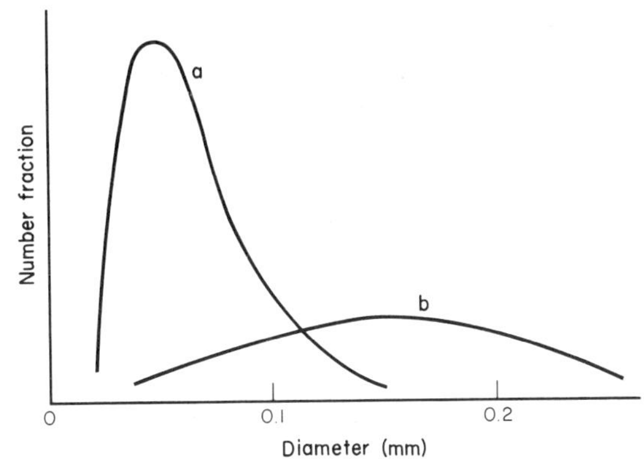


**Figure 2** Distributions of the diameters of dispersed droplets for (a) high speed and (b) low speed agitation

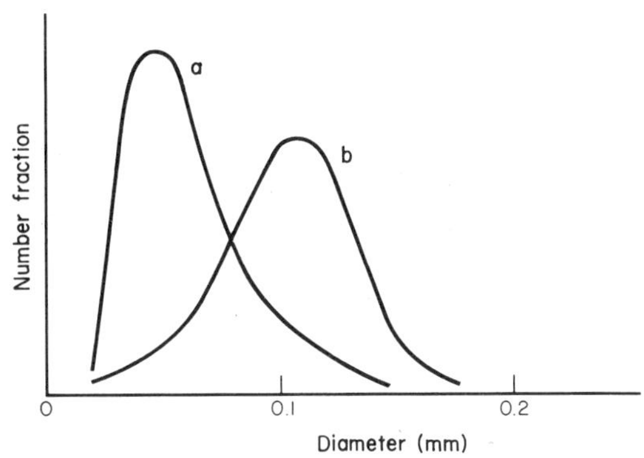


**Figure 3** Distributions of the diameters of dispersed droplets for (a) low interfacial tension and (b) high interfacial tension

The equilibrium droplet size distribution depends on many factors including physical property parameters (density, interfacial tension and viscosity), geometrical parameters (vessel, agitator and baffles) and operational parameters (agitation speed and temperature). A quantitative basis for the dispersion of immiscible liquids was provided by the experimental work of Vermeulen, Williams and Langlois[22] who investigated the variation of interfacial area generated by many liquid–liquid systems in a cylindrical baffled vessel. They found for comparable conditions of vessel and fractions of the two liquids that the volume-to-surface mean diameter (Sauter diameter) of droplets $\bar{D}$ was given by

$$\bar{D}/L = K(N^2 L^3 \rho/\sigma)^{-0.6} \tag{2}$$

where $L$ is the diameter of the agitator and $K$ is a shape factor dependent on the geometry of the vessel and the baffles. A more general equation which takes into account the phase volume of the dispersed phase is

$$\bar{D}/[L\mathrm{f}(\phi)] = K(N^2 L^3 \rho/\sigma)^{-0.6} \tag{3}$$

where $f(\phi)$ is a function of $\phi$ and may be represented by

$$f(\phi) = 1 + 3.3\phi \tag{4}$$

which is valid for values of $\phi$ below about 0.3.[23] This shows that droplet size will increase with phase volume. The term $(N^2L^3\rho/\phi)$ in equations (2) and (3) is a modified Weber number. This is apparent if we replace $D$ and $v(D)$ in equation (1) by $L$ and $NL$ respectively and include the constants of proportionality between $D$ and $L$ and $v(D)$ and $NL$ (raised to the power $-0.6$) into the shape factor constant $K$.

Equations (2) and (3) permit the study of the dependence of the mean droplet size on physical property, geometrical and operational parameters. With $W_e$ it is possible to correlate turbulence stabilization to the experimental dispersion conditions in order to control droplet size. These equations suggest that the viscosities of the dispersed $\eta_M$ and continuous $\eta_w$ phases do not contribute to the correlation between droplet size and the modified Weber number. Rodger, Rushton and Trice,[24] using a similar apparatus to Vermeulen *et al.*,[22] measured the interfacial area of many organic–water systems with a high volume fraction of the organic phase ($\phi = 0.5$) when coalescence is important. They formulated the following generalized equation for the interfacial area $A$ which correlated their experimental results extremely well

$$A = \frac{K'}{L}\left[\frac{N^2L^3\rho_c}{\sigma}\right]^{0.36}\left[\frac{L}{T}\right]^{K'}\left[\frac{\eta_M}{\eta_w}\right]^{0.2}\left[\frac{t}{t_0}\right]^{1/6}\exp\left[\frac{3.6\Delta\rho}{\rho_c}\right]S \tag{5}$$

where $K'$ is an exponent of $L/T$ (a function of Weber Number), $T$ is the vessel diameter, $\rho_c$ is the density of the continuous phase, $\Delta\rho$ is the density difference of the phases, $t$ is the settling time, $t_0$ is the reference settling time (1 min), and $S$ is a scale up function related to $\phi$. The approach of Rodger *et al.*[24] was to correlate $A$ with $W_e$, but unlike the study of Vermeulen *et al.*[22] equation (5) includes the prediction that droplet size should increase with increasing $\eta_w$. The correlations represented by equations (2), (3) and (5) can be simplified for comparison in terms of the reactor variables $N$, $T$ and $L$. Several examples are given in Table 1. The agreement among these correlations is fair considering that there will be differences in baffling and reactor geometry.

**Table 1** Equations for Interfacial Area $A$ in Terms of Agitator and Reactor Parameters

| *Equation*[a] | *Reference* |
|---|---|
| $A = C_1N^{1.2}$ | 22 |
| $A = C_2NT^{-1.2}L^{1.65}$ | 25 |
| $A = C_3N^{1.15}T^{-0.6}L^{1.8}$ | 24 |
| $A = C_4N^{1.2}T^{-1.2}L^2$ | 26 |

[a] $C_1$, $C_2$, $C_3$, $C_4$ combine other variables and constants

The discussion of liquid–liquid dispersion behaviour suggests that the Weber number is a useful representation of several experimental variables. Analysis of the particles from a suspension polymerization will be rather more complex because both suspension- and emulsion-sized particles may be formed in the reactor, the interfacial tension and dispersed phase viscosity will change as monomer is converted to polymer, suspending agents dissolved in the aqueous continuous phase improve dispersion stability and increase $\eta_w$, and the measured particle sizes are lower than the initial droplet sizes due to the contraction occurring on polymerization. This last statement also presumes that the size of the beads is determined at the start of polymerization, which was confirmed by Hopff and co-workers[27,28] who reported the first general experimental study of the dependence of particle size on physical property, geometrical and operational parameters in suspension polymerizations. Polymerizations of several monomers with various suspending agents were performed and the following correlation was proposed

$$\bar{D}/L = AR_e^{0.5}W_e^{-1}(\eta_M/\eta_w)^{0.1} \tag{6}$$

where $A$ is a constant and $R_e$ is the Reynolds number given by

$$R_e = L^2N\rho\eta_w^{-1} \tag{7}$$

This analysis suggests that phase volume is unimportant, though other experimental studies show a dependence of $\bar{D}$ on $\phi$.[23,29] Hopff *et al.*[27,28] performed a large number of experiments demonstrating that particle size increased with increase in dispersed phase viscosity and with decrease in continuous phase viscosity. They also showed that, when the suspending agent, 88 mol% hydrolyzed poly(vinyl alcohol), had a concentration in excess of 1 g 100 $cm^{-3}$ in the aqueous continuous phase, solution viscosity could be varied appreciably with negligible change in $\sigma$. At much lower levels of suspending agent an increase in particle size was observed as the interfacial tension increased. Other workers have investigated correlations of droplet sizes and particle sizes on various experimental parameters.[30-35]

## 14.3 MONOMER PHASE

It is evident that the monomer or mixture of monomers must be sufficiently insoluble in water to form a dispersed phase. The initiator, and chain transfer agent if required, will be added to the monomer phase just before dispersion in water. Since initiator and regulator are generally the same in suspension, bulk monomer and solution polymerizations of a given monomer, it is expected that the polymerization kinetics in these three processes are very similar. Polymerization in each suspension droplet may be considered as a microbulk monomer polymerization. The well-known rate equation for homogeneous polymerization of bulk monomer by the thermal decomposition of a radical initiator is[1,2]

$$R_p = k_p[M](2fk_d[I]/k_t)^{0.5} \tag{8}$$

where $R_p$ is the rate of polymerization, [M] and [I] are the monomer and initiator concentrations, $k_d$, $k_p$ and $k_t$ are the rate constants for initiation, propagation and termination and $f$ is the initiator efficiency. Equation (8) has been shown to be valid for suspension polymerizations of methyl methacrylate.[36] Consequently, the droplet size and suspending agent did not appear to influence $R_p$ and the molecular weight of the polymer. This applies not only for a homogeneous monomer–polymer mixture but also for a polymer precipitating from its own monomer such as poly(vinyl chloride).[37] Differences between bulk monomer and suspension polymerizations might occur if the suspending agent, or impurities in the suspension stabilizer, may act as retarders, inhibitors or regulators.

Copolymerizations in suspension may involve other similar monomers or crosslinking agents.[8] Monomers with substantial solubility in water may be copolymerized. The concentration of the partially water soluble comonomer in droplets will depend on its distribution between the two liquid phases. Furthermore, as polymerization proceeds, the dispersed phase may or may not tend to lose comonomer dependent on comonomer diffusion from the continuous phase. The location of the suspending agent at the interface may influence the comonomer distribution between phases. Comonomer interacting with water may have different reactivity in suspension copolymerization compared with copolymerization of bulk monomers. This is confirmed by monomer reactivity ratios for copolymerizations of methyl methacrylate/methacrylic acid and methyl acrylate/acrylonitrile.[8,38] Because of the comonomer distribution equilibrium and its possible change in polymerization reactivity, controlled monomer addition may be preferred for preparing copolymers homogeneous in composition.

Generally, monomers are polymerized as liquids. Vinyl chloride forms liquid droplets under the pressures employed in industrial suspension polymerizations. Various agents may be added to the monomer phase for specific applications and again distribution between phases depending on solubility must be considered. Typical examples are foaming agents, blowing agents, diluents or swelling agents which are removed after polymerization to form porous beads.[8]

## 14.4 AQUEOUS SUSPENSION PHASE

Generally, turbulence-stabilized dispersions are used for suspension polymerization. When a monomer (or monomer mixture) is agitated in water with a typical weight ratio between monomer and water ranging from 50:50 to 25:75, an unstable dispersion is formed and continuous break-up and coalescence occurs. On cessation of agitation the monomer(s) and water separate into their respective phases. During the suspension polymerization process, the monomer droplets tend to stay in stable dispersion until the polymerization has proceeded to a stage where the droplets become 'tacky' (approximately 30% monomer conversion), but then aggregate. The dispersion of partially polymerized droplets may be maintained throughout polymerization by adding a suspension stabilizer which prevents coalescence and agglomeration. Suspending agents that hinder coalescence may be either water soluble polymers, often called protective colloids, or finely divided insoluble inorganic compounds.

It can be claimed that the suspending agent increases the viscosity of the continuous aqueous phase, thus improving dispersion.[27,28] However, there is evidence that the suspension stabilizer forms a film or skin around the droplet/particle surface,[39] and so it is this layer which prevents coalescence and agglomeration by a mechanism analogous to steric stabilization,[40] which is known to protect particles produced by dispersion and emulsion polymerization[5,9] against flocculation. Some suspension polymerizations use low levels of protective colloid ($\sim 0.1$ wt. %) with little increase in $\eta_w$ for the aqueous phase, and the addition of water insoluble inorganic compounds is not likely to raise $\eta_w$ significantly. Consequently, suspension stabilizers are largely effective when present in the interfacial layers between the water and the monomer droplets, and the tendency for interfacial adsorption is more important than the increase in viscosity of the aqueous phase when choosing a suitable stabilizer.

The thin protecting film is generally regarded as adsorbed on to the surface of the generated droplet. If we consider only stabilization by copolymers, the protective action of this absorbed layer by steric stabilization operates by preventing the approach of the droplets to a separation where the 'tacky' particles can touch and aggregation occur. A good steric stabilizer is one which adsorbs strongly and quickly, is well anchored and provides a thick steric barrier. For a multiblock copolymer $(AB)_n$ in which A is an anchoring component on the droplet surface and B is a steric stabilizer component, A should be organic compatible and B water compatible, giving the train-loop chain arrangement depicted in Figure 4. Stabilization will be improved with longer A and B segments, but the length of the A segment should not prevent solubility of the copolymer in the aqueous phase. Studies by interferometry indicate that the thickness of the surface layer formed by the copolymer and the viscosity of this layer contribute to stabilization by the protective colloid.[39] Little work has been reported on the stabilization mechanism by insoluble particulate suspending agents though it is presumed that surface adsorption is again a dominant factor.[8]

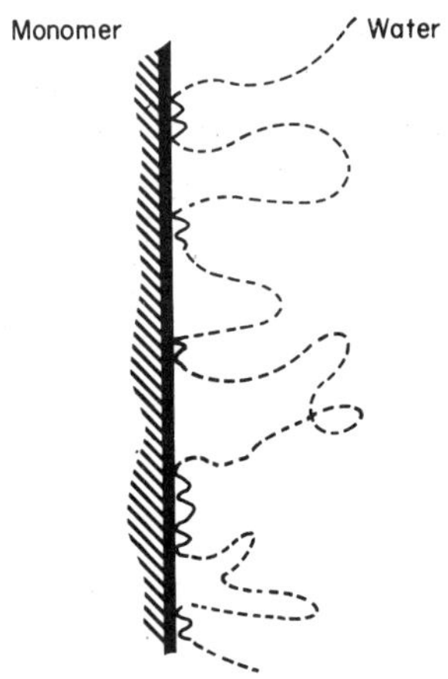


**Figure 4** Train-loop arrangement for multiblock copolymer chain $(AB)_n$ having hydrophobic A (———) and hydrophilic B (– – –) blocks at a droplet–water interface

The suspending agent may not only influence the stability and size of droplet/particle but also the solubility of water in the monomer and/or solubility of monomer in the water. Copolymer suspension stabilizers tend to reduce the interfacial tension between phases which may also result in a foaming problem.

### 14.4.1 Water Soluble Protective Colloids

Early suspending agents involved natural polymers such as gelatin and polysaccharides, and modified carbohydrate polymers such as methyl cellulose, hydroxyalkyl celluloses and carboxymethyl cellulose. Because of the ease of preparation of poly(vinyl acetate) and the control of the degree of hydrolysis, partially hydrolyzed poly(vinyl alcohol) copolymers have become widely used as suspension stabilizers. Other copolymers have been listed,[8] but many of these have only been disclosed briefly in patents and so it is difficult to compare their merits and limitations.

When organic soluble poly(vinyl acetate) is hydrolyzed so that less than about 20 mol% of the acetate groups are left, a water soluble copolymer containing acetate and alcohol groups is produced. Because of water solubility these suspending agents are often grouped under the generic name poly(vinyl alcohol) (PVAL). Their behaviour as protective colloids will be influenced not only by molecular weight and degree of hydrolysis, but also by method of manufacture, positioning of monomer units (whether head-to-head or head-to-tail), stereochemistry, degree of branching and distribution of acetate and hydroxyl groups. Industrially synthesized poly(vinyl acetate) has a wide molecular weight distribution and since the acid hydrolysis process is essentially random, control of block length is impossible. Alkaline hydrolysis using methanol is often preferred industrially, because of easy control of hydrolysis and few waste products.[41] This alkaline product has a more blocky structure where sequences are longer, but it would be expected that anchoring would be poorer because of the shorter sequences of acetate groups. It is presumed therefore that the acetate groups have affinity to the dispersed monomer phase and that the hydroxyl groups are directed toward the aqueous phase. The best suspending agents will require optimization of the degree of hydrolysis together with the number and length of the hydrophilic and hydrophobic sequences. Tadros *et al.*[42] showed that surface layer thickness increases with molecular weight, while Glass *et al.*,[43] investigating the effect of polymeric hydrocolloids on the partial coalescence of liquid drops at liquid–liquid interfaces, showed that droplet stability depended on polymer molecular weight and solution surface tension. The effect of minor structural differences has been shown by Shiraishi and Toyoshima[44] who measured the stability of dispersions using different batches of the same grade of PVAL (degree of polymerization and degree of hydrolysis being constant). These stabilizing properties were shown to vary between batches.

The effect of the degree of hydrolysis was investigated by Lankveld and Lyklema[45] on the liquid paraffin–water system. They measured the adsorption isotherms and also the change in specific surface area as functions of degree of hydrolysis and concentration of stabilizer. For a copolymer with 98 mol% hydrolysis normal behaviour of slowly decreasing droplet size was observed with increasing copolymer concentration, but with 88 mol% hydrolysis they found a minimum in the droplet size–PVAL concentration curve. At low concentrations the 88 mol% PVAL was shown to be a more efficient stabilizer than the 98 mol% grade, although all dispersions were extremely stable. These results are in agreement with the earlier work of Winslow and Matreyek[46] who investigated the effect of a range of stabilizers, including PVAL, on the particles formed during the suspension polymerization of divinylbenzene. They showed that under comparable conditions, (i) high molecular weight grades, (ii) partially hydrolyzed grades, and (iii) high concentrations of stabilizer were associated with beads of lower mean diameter. The decrease of interfacial tension with increasing molecular weight and concentration and decreasing degree of hydrolysis can be used to explain the previous observations. Reference to equations (2), (3) and (5) shows that a decrease in interfacial tension leads to decrease in $\bar{D}$, an overall decrease in droplet diameter and an increase in interfacial area. Dawkins and Stone[3] have shown that the surface tension and interfacial tension behaviour of PVAL (Gelvatol 20–60, degree of polymerization (DP) = 1750, degree of hydrolysis = 88 mol%) are similar, both decreasing with concentration until a critical micelle concentration is reached at 0.158 gm dl$^{-1}$, when no further decrease occurs with increasing concentration. Use of stabilizers above their critical micelle concentration explains the simultaneous formation of both emulsion- and suspension-sized material during the suspension polymerization process, as suggested by Hopff *et al.*[27,28]

Modification of acetate block lengths has been achieved by Hayashi *et al.*[47] and by Shiraishi and Toyoshima.[44] They hydrolyzed poly(vinyl acetate) using methanol/benzene mixtures and produced a degree of blockiness greater than that using normal alkaline hydrolysis with methanol. Associated with the increased block character was better stabilization, lower surface (and thus interfacial) tensions and higher viscosity of aqueous solutions.

The viscosity properties of PVAL have been investigated, chiefly by Naito.[48–50] He discussed the molecular theory of aqueous solutions of completely hydrolyzed grades of PVAL, pointing out that in the viscosity–concentration curve two points of inflexion occur, at concentrations of 2–4 g dl$^{-1}$ and 10–12 g dl$^{-1}$. From this data it was suggested that the solution behaviour of PVAL alters with concentration. Below the first critical concentration each individual macromolecule is dispersed in water in the form of thread-filled spheres containing water. Above the first critical concentration entanglement occurs between the macromolecular spheres, the extent of entanglement increasing with concentration, until above the second critical concentration the entanglement network is almost complete and the solution gels.

It is known that PVAL exhibits non-Newtonian viscosity. Investigating the effect of shear on the viscosity of aqueous solutions of PVAL samples of various concentrations and degrees of polymerization, Naito *et al.*[48–50] showed that dilute solutions ($<0.5\,g\,dl^{-1}$) can be considered to be Newtonian in behaviour. However, at higher concentration the solution becomes non-Newtonian until in the range $2$–$4\,g\,dl^{-1}$ a transition occurs where the apparent viscosity decreases with increasing shear rate, due to the breaking of the entanglements between the molecular spheres and the heavy hydrogen bonding associated with PVAL. This shear effect increases with molecular weight and decreases with an increase in the number of residual acetate groups. Therefore use of PVAL solutions above and below the critical micelle concentration as the continuous phase in suspension polymerizations permits studies of changing interfacial tension at approximately constant viscosity (low PVAL concentration) and changing viscosity at constant interfacial tension (high PVAL concentration).

### 14.4.2 Insoluble Inorganic Particulate Suspending Agents

Dispersion stabilization may be achieved by adding small quantities (0.1 to 1.0%) of finely divided inorganic compounds, such as talc, calcium and magnesium carbonates, and silicates, phosphates, *etc.* A listing of these compounds reported in patents is given elsewhere.[8] Sufficient suspending agent is required to cover the surface of the monomer droplets in order to achieve the required droplet size. Some of these particulate suspending agents require the presence of a surfactant which aids dispersing action of the particles by controlling the wetting angle at the particle surfaces.[51] However, the level of surfactant must be carefully controlled in order to avoid emulsion polymer formation.

## 14.5 AGITATION AND REACTOR SYSTEMS

Laboratory scale suspension polymerization at atmospheric pressure may be performed with three- or four-necked glass reactors equipped with suitable stirrer, reflux condenser and inlet tube and immersed in a thermostatically controlled water bath. Control of reaction temperature is important because thermal variations will influence the mobility of droplets/particles, the solubility of the protective colloid in water and the adsorption of the suspension stabilizer at the droplet/water interface. Temperature variations will also depend on the dissipation of the heat of polymerization and will determine the rate of polymerization $R_p$ in equation (8), the possible occurrence of autoacceleration in $R_p$, and polymer molecular weight and distribution. In larger industrial reactors, temperature is controlled by fluid flow through a jacket, and sometimes by fluid flow through internal piping. However, the use of internal cooling to supplement jacket cooling must allow for the possible deposition of particles on piping with loss of heat transfer capability. Suspension polymerizations are conveniently performed in batch reactors. A gaseous monomer such as vinyl chloride is polymerized in pressure reactors. Continuous reactors for suspension polymerization may be performed with a series of reaction vessels,[52] as long as particle deposition and build-up of polymer can be avoided. Agents to prevent particle agglomeration must be added.

Various procedures are available to charge the reactants, water and suspending agent to the polymerization reactor.[52] If heat dissipation, *e.g.* because of autoacceleration in $R_p$, is likely to be difficult to control by external cooling, then the initiator may be added gradually to balance the rate at which it is depleted. Alternatively, the rate of polymerization during the reaction time may be controlled by employing combinations of initiators of different half-life and temperatures.

Stabilization of the initial dispersion of monomer droplets requires the following conditions: (i) a protective film must be present between the organic droplet and water to prevent immediate coalescence; (ii) agitation must be sufficiently intense so as to separate all adhering droplet pairs and clusters; (iii) in the stable state no notable break-up of droplets should occur owing to agitation; (iv) since the forces affecting the droplets vary with time, the time scale of these fluctuations should be considerably less than the critical coalescence time; and (v) agitation must be sufficiently strong to prevent the separation of the two phases.

Agitation systems, *i.e.* stirrer and reactor, must be carefully designed in order to optimize (i) to (v) above. It is important that the droplets participate in both horizontal and vertical motion to ensure uniform dispersion of droplets throughout the aqueous phase. Reactors must not contain unagitated corners where droplet/particle coalescence may occur.

Agitation systems may be a turbine or a retreat-curved impeller blade. Various baffles are available.[52] Morton and Knott[53] measured the performances of four designs of flask with four-, six- and eight-blade impellers. In all cases a round bottomed flask, modified with four vertical creases and an indented bottom, performed most effectively, particularly with a four-blade impeller. Morton and Redman[54] investigated experimentally the influence of the impeller in high speed stirring, finding that increased effectiveness was attained by: (a) a high speed of rotation; (b) a large diameter impeller; (c) an appropriate blade angle that may preferably be 45°; and (d) a downward bend of the lower edge of the impeller blade.

The use of baffles has been found to be necessary to obtain homogeneous mixing. For the above system, in the absence of any baffling, the liquid will tend to rotate in a mass, *i.e.* in laminar fashion, with the formation of a vortex about the impeller axis. This vortex will increase in depth with an increase in speed of rotation, with, in the case of two immiscible liquids, separation occurring as a function of density. If the impeller is a propeller or pitched-blade turbine, there will be some vertical currents caused by axial flow through the impeller, the extent of these being dependent on the system geometry and the operating conditions. By installing vertical baffles at or near the walls of the reactor, the rotational motion of the fluid is 'broken up' and converted into mean flow currents in the axial and radial directions which results in a 'turnover' of the vessel contents, *i.e.* laminar flow is converted to turbulent flow. The correctly stirred baffled reactor will produce a narrow distribution of monomer droplet sizes in agreement with theoretical predictions.

At the end of the polymerization reaction, residual monomer(s) must be stripped from the particles, e.g. by vacuum or steam distillation. Separation and washing of the polymer particles is performed with continuous centrifuges, sieves or filters. Final drying of the polymer may be accomplished with a rotary air drier or a continuous plate drier.

## 14.6 EXAMPLES

Many monomers may be polymerized by suspension processes.[8,52,55] The more important examples are vinyl chloride, styrene, vinylidene chloride, vinyl acetate, methacrylic esters and acrylic esters. For some applications comonomers such as crosslinking agents are incorporated. Two suspension polymers will be considered to demonstrate typical procedures.

### 14.6.1 Poly(vinyl chloride)

About 80% of poly(vinyl chloride) (PVC) is produced by suspension processes. PVC is a large tonnage polymer required for applications which may or may not require addition of plasticizer. Porosity in suspension particles (about 150 $\mu$m in diameter) develops in polymerization because PVC is insoluble in its monomer and the morphology of porosity will change as polymerization proceeds.[56,57] Porosity aids the stripping of unreacted monomer at the end of polymerization. The processing of PVC depends on the molecular weight of PVC and the morphology of porosity. Easier processing will be possible with lower molecular weights because of lower melt viscosities. Porosity will improve plasticizer uptake for flexible PVC applications. For rigid unplasticized PVC lower porosity is preferred to maximize extruder outputs.

Typical data for PVC porosity have been reported for vinyl chloride monomer polymerized over the range 50–75 °C.[57] Increasing the temperature decreases the molecular weight of the polymer and raises the fusion of primary nuclei, which arise initially from polymer precipitating in the monomer, decreasing porosity. Initiation of polymerization may be performed with lauroyl peroxide, peroxydicarbonate, or a mixed initiator system which generates lower porosity than single initiators. The protective colloid may be poly(vinyl alcohol) (about 75% hydrolyzed), cellulose ether or gelatin. Secondary stabilizers, *e.g.* poly(vinyl alcohol) with a low degree of hydrolysis or a surfactant which lowers the interfacial tension, raise particle porosity. Low agitator speeds give a lower porosity.

Pressure reactors are required with a rating of at least 200 p.s.i. (1 p.s.i. = 703.070 kg m$^{-2}$) which is about 50 p.s.i. greater than the vapour pressure of monomer at the common polymerization temperatures. An anchor-type agitator with a baffled reactor ensures good dispersion. No change in the droplet/particle size distribution between 5% and 100% vinyl chloride monomer conversion has been reported.[58] The factors involving the design and operation of reactors for suspension polymerization in the manufacture of PVC have been discussed.[59] Correlations between particle size for suspension PVC and $W_e$ have been investigated for different reactors.[60,61] PVC formation in radical polymerization is by chain transfer, so higher molecular weight is achieved at lower polymerization temperatures. Because of polymer precipitation in vinyl chloride monomer, polymerization may proceed with autoacceleration in $R_p$.

### 14.6.2 Crosslinked Polystyrene

Porous crosslinked polystyrene beads are widely used for separation processes and chromatography and may be modified *e.g.* for use as ion-exchange resins. The copolymerization of styrene and divinylbenzene with organic diluents dispersed as droplets in water is performed by a suspension process. The diluents are good solvents for the monomers but poor swelling agents for the copolymer, so crosslinked particles with a macroporous morphology are produced. A typical recipe for the dispersed phase[62] is styrene (30%), divinylbenzene (10%), toluene (30%) and *n*-dodecane (30%), with benzoyl peroxide or azobisisobutyronitrile as initiator. The formation of porosity during polymerization has been discussed by Seidl *et al.*[63] Variation of *n*-dodecane relative to toluene determines the pore size distribution.[62]

Dispersion of droplets of styrene with divinylbenzene has been investigated in non-coalescing systems.[64] The models predict the largest surviving droplet size as a function of reactor geometry, agitator speed and physical property parameters. Protective colloids such as PVAL provide adequate suspension stabilization.[3,4] Hydrophobically modified hydroxyethyl cellulose has been examined as a suspension stabilizer,[65] and it was demonstrated that raising the concentration of this suspending agent together with faster agitation decreased the particle size and narrowed the particle size distribution for copoly(styrene/divinylbenzene). Heat dissipation problems may occur in these polymerizations because of the gel effect (*i.e.* rate autoacceleration).[66]

## 14.7 REFERENCES

1. F. W. Billmeyer, 'Textbook of Polymer Science', 3rd edn., Wiley-Interscience, New York, 1984.
2. G. Odian, 'Principles of Polymerization', 2nd edn., Wiley-Interscience, New York, 1981.
3. J. V. Dawkins and T. Stone, unpublished experiments, Loughborough University of Technology, 1974–1977.
4. J. V. Dawkins, T. Stone and G. Yeadon, *Polymer*, 1977, **18**, 1179.
5. D. C. Blackley, 'Emulsion Polymerization', Applied Science, London, 1975.
6. J. Ugelstad, K. Herder-Kaggerud and R. M. Fitch, in 'Polymer Colloids II', ed. R. M. Fitch, Plenum Press, New York, 1980, p. 83.
7. Y. Almog and M. Levy, *J. Polym. Sci., Polym. Chem. Ed.*, 1981, **19**, 115.
8. M. Munzer and E. Trommsdorff, in 'Polymerization Processes', ed. C. E. Schildknecht with I. Skeist, Wiley-Interscience, New York, 1977, vol. 29, chap. 5.
9. K. E. J. Barrett, 'Dispersion Polymerization in Organic Media', Wiley, New York, 1975.
10. J. V. Dawkins and N. P. Gabbott, *Polymer*, 1981, **22**, 291.
11. M. V. Dimonie, C. M. Boghina, N. N. Marinescu, M. M. Marinescu, C. I. Cincu and C. G. Oprescu, *Eur. Polym. J.*, 1982, **18**, 639.
12. P. Becher, 'Emulsions: Theory and Practice', *ACS Monogr.*, 1965, **162**.
13. J. M. Church and R. Shinnar, *Ind. Eng. Chem.*, 1961, **53**, 479.
14. V. M. Martynov, *Kolloid-Z.*, 1948, **10**, 218.
15. A. N. Komogoroff, *Dokl. Akad. Nauk SSSR*, 1941, **30**, 301; 1941, **31**, 538; 1941, **32**, 16.
16. A. N. Komogoroff, *Dokl. Akad. Nauk SSSR*, 1949, **66**, 825.
17. J. O. Hinze, *AIChE J.*, 1955, **1**, 289.
18. A. J. Madden and B. J. McCoy, *Chem. Eng. Sci.*, 1969, **24**, 419.
19. J. W. van Heuven and J. C. Hoevenaar, 'Proceedings of the 4th European Symposium on Chemical Reaction Engineering', Pergamon Press, New York, 1971, p. 217.
20. Y. Mlynek and R. Resnick, *AIChE J.*, 1972, **18**, 122.
21. I. Yamaguchi *et al.*, *Kagaku Kogaku*, 1963, **27**, 576; S. Nagata, 'Mixing Principles and Applications', Halsted Press, New York, 1975, chap. 7.
22. T. Vermeulen, G. M. Williams and G. E. Langlois, *Chem. Eng. Prog.*, 1955, **51**, 85.
23. D. B. Scully, *J. Appl. Polym. Sci.*, 1976, **20**, 2299.
24. W. H. Rodger, V. G. Trice, Jr. and J. H. Rushton, *Chem. Eng. Prog.*, 1956, **52**, 515.
25. I. S. Pavlushenko, N. M. Kostin and F. Matveers, *Zh. Prikl. Khim. (Leningrad)*, 1957, **30**, 1160.
26. F. Rodriguez, L. C. Grotz and D. L. Engle, *AIChE J.*, 1961, **7**, 663.
27. H. V. Hopff, H. Lussi and P. Gerspacher, *Makromol. Chem.*, 1964, **78**, 24, 37.
28. H. Hopff, H. Lussi and E. Hammer, *Makromol. Chem.*, 1965, **82**, 184.
29. M. Narkis, *J. Appl. Polym. Sci.*, 1979, **23**, 2043.
30. J. C. Lee and P. Tasakorn, *Proc. Eur. Conf. Mixing, 3rd*, 1979, **1**, 157.
31. F. Langner, H. U. Moritz and K. H. Reichert, *Chem. Ing. Tech.*, 1979, **51**, 746.
32. M. Konno, K. Arai and S. Saito, *J. Chem. Eng. Jpn.*, 1982, **15**, 131.
33. M. Tanaka and T. Izumi, *J. Chem. Eng. Jpn.*, 1985, **18**, 354.
34. B. Marisi, *J. Vinyl Technol.*, 1986, **8**, 20.
35. A. G. Mikos, C. G. Takoudis and N. A. Peppas, *J. Appl. Polym. Sci.*, 1986, **31**, 2647.

36. M. Jacovic and M. Govedarica, *C. R. Hebd. Seances Acad. Sci., Ser. C*, 1971, **272**, 1295.
37. A. Crosato-Arnaldi, P. Gasparini and G. Talamani, *Makromol. Chem.*, 1968, **117**, 140.
38. Z. Izumi and H. Kitagawa, *J. Polym. Sci., Part A-1*, 1967, **5**, 1967.
39. F. Wolf, B. Hofbauer and S. Eckert, *Plaste Kaut.*, 1972, **19**, 26.
40. B. Vincent, *Adv. Colloid Interface Sci.*, 1974, **4**, 193.
41. C. A. Finch, 'Poly(vinyl alcohol)', Wiley, New York, 1973.
42. M. J. Garvey, Th. F. Tadros and B. Vincent, *J. Colloid Interface Sci.*, 1974, **49**, 57.
43. J. E. Glass, R. D. Lundberg and F. E. Bailey Jr., *J. Colloid Interface Sci.*, 1970, **33**, 491.
44. A. Shiraishi and A. Toyoshima, *Br. Polym. J.*, 1970, **2**, 135.
45. J. M. G. Lankveld and J. Lyklema, *J. Colloid Interface Sci.*, 1972, **41**, 454, 460, 475.
46. F. H. Winslow and W. Matreyek, *Ind. Chem. Eng.*, 1951, **43**, 1108.
47. S. Hayashi, C. Nakano and T. Motoyama, *Kobunshi Kagaku*, 1963, **20**, 303; 1965, **22**, 354.
48. R. Naito, *Kobunshi Kagaku*, 1958, **15**, 191.
49. R. Naito and T. Kominami, *Kobunshi Kagaku*, 1954, **11**, 444.
50. T. Kominami, R. Naito and H. Odanaka, *Kobunshi Kagaku*, 1955, **12**, 218.
51. Von H. Wenning, *Makromol. Chem.*, 1956, **20**, 196; 1956, **22**, 41.
52. E. Farber, in 'Encyclopedia of Polymer Science and Technology', ed. H. F. Mark, N. G. Gaylord and N. M. Bikales, Wiley-Interscience, New York, 1970, vol. 13, p. 552.
53. A. A. Morton and D. M. Knott, *Ind. Eng. Chem.*, 1941, **13**, 649.
54. A. A. Morton and L. M. Redman, *Ind. Eng. Chem.*, 1948, **40**, 1190.
55. H. Warson, *Polym. Paint Colour J.*, 1983, **173**, 541.
56. A. K. Sanderson, *Br. Polym. J.*, 1980, **12**, 186.
57. P. V. Smallwood, *Polymer*, 1986, **27**, 1609.
58. H. Nilsson, C. Silvegren and J. Uustalu, *Polym. Commun.*, 1983, **24**, 268.
59. L. F. Albright and Y. Soni, *J. Macromol. Sci., Chem.*, 1982, **17**, 1065.
60. G. R. Johnson, *J. Vinyl Technol.*, 1980, **2**, 138.
61. M. H. Lewis and G. R. Johnson, *J. Vinyl Technol.*, 1981, **3**, 102.
62. J. C. Moore, *J. Polym. Sci., Part A*, 1964, **2**, 835.
63. J. Seidl, J. Malinsky, K. Dusek and W. Heitz, *Adv. Polym. Sci.*, 1967, **5**, 113.
64. D. E. Leng and G. J. Quarderer, *Chem. Eng. Commun.*, 1982, **14**, 177.
65. S. M. Ahmed, *J. Dispersion Sci. Technol.*, 1984, **5**, 421.
66. K. L. Gardner and G. McNally, *J. Polym. Sci., Polym. Lett. Ed.*, 1978, **16**, 267.

# 15

# Polymerization in Non-aqueous Dispersions

DEREK J. WALBRIDGE

*ICI plc, Slough, UK*



## 15.1 INTRODUCTION

Non-aqueous dispersion polymerization may be defined as the polymerization of a monomer dissolved in an organic liquid to produce an insoluble polymer in the form of a stable colloidal dispersion. The colloidal stability is provided by the adsorption on the surface of the polymer particles of an amphipathic polymeric stabilizer or dispersant which is present in the organic medium. The process may be viewed as a type of precipitation polymerization in which flocculation is prevented and particle size controlled.[1] The dispersion remains fluid and the process can be operated at a very high disperse-phase volume. The technique has been applied to virtually all monomers which undergo free-radical polymerization to produce polymers insoluble in an organic medium and to a more limited range of ionically initiated polymerizations. The early history of the subject and its development up until 1975 has been reviewed in a comprehensive monograph which includes detailed recipes for the preparation of polymer dispersions.[2]

Historically, the development of successful processes for dispersion polymerization in organic media arose from the desire in the surface coatings industry to apply thin films of vinyl and acrylic polymers without resorting to multiple applications of dilute polymer solutions.[3,4] Subsequently, the availability of monodisperse well-characterized polymer colloids stimulated academic studies of colloid stability in non-polar media, and the properties of concentrated suspensions.[5]

Polymer dispersions in organic diluents are stabilized against flocculation by a sheath of lyophilic polymer surrounding the core of lyophobic polymer, by a process which has become known as steric stabilization.[6] The understanding of the mechanism of steric stabilization and the synthesis of effective polymeric steric stabilizers have been key features in the development of reliable dispersion polymerization processes. The alternative mechanism, electrostatic stabilization, which plays such an important role in aqueous emulsion polymerization, would not be expected to be effective in a medium of low dielectric constant[7,8] and certainly no practical system has yet been devised for its use in non-aqueous dispersion polymerization.

The industrial exploitation of this technology has been mainly in the painting of motor cars, but many other applications have been proposed, including reprographics, printing inks, adhesives, foams, polymer reinforcement and polyalkenes.[2]

The technique has been extended to condensation polymers and applied successfully to commercial polyesters and polyamides.[9] An account of this work can be found in Volume 5 of this series.

## 15.2 STERIC STABILIZATION OF POLYMER COLLOIDS IN NON-AQUEOUS MEDIA

### 15.2.1 Conditions for Stabilization by Adsorbed Polymer

In its simplest terms, the colloidal stabilization conferred by adsorbed polymer molecules may be considered as arising from the increase in polymer segment density resulting from the interpenetration or compression of the polymer sheaths when two particles approach one another (Figure 1). The osmotic diffusion of solvent into the region of increased polymer concentration generates a repulsive force, causing the particles to separate. Indeed, the classical method for the measurement of osmotic pressure by recording pressure against concentration in a compression cell has been adapted for the study of sterically stabilized dispersions.[10] The term 'steric stabilization' is now in general use, and is correct in that the effect derives from the spatial or configurational interactions of polymer molecules.[6] It should not be confused with the better known steric effects of organic chemistry, which are concerned with the very short-range repulsions between substituent groups in a molecule.

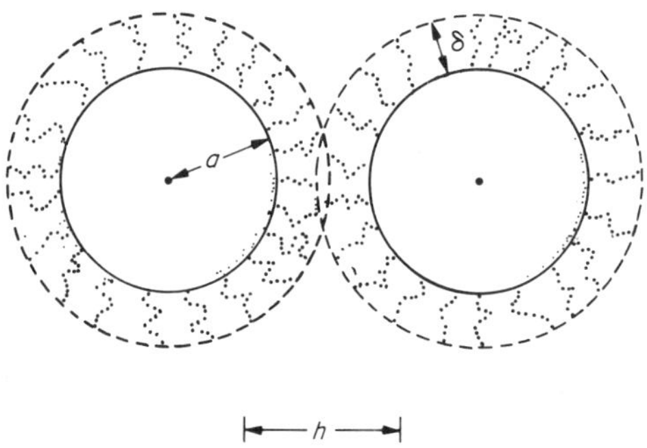


**Figure 1** Schematic representation of the close approach of sterically stabilized particles

For polymer colloids stabilized by firmly attached solvated polymer layers, the thickness of the layer is normally such that, at the point of closest approach, the van der Waals attraction $V_A$ between the particles is reduced to a level at which the particles separate spontaneously by thermal motion. Figure 2 illustrates schematically the potential energy *vs.* distance of separation diagram for two sterically stabilized particles and for different barrier thicknesses.

Unlike electrostatic stabilization, there is no long-range repulsion and the particles are subject only to attractive forces until the outer fringes of the adsorbed molecules are in physical contact. The existence of a secondary minimum, *i.e.* a small net attraction, is therefore quite probable for strongly attracting particles or for shallow steric barriers. The colloid stability is thus dependent on the dimensions and the solution properties of the adsorbed polymer, under the local conditions of temperature, concentration and pressure, and on the size and nature of the particle core.

In non-aqueous media, the heat of solution of a polymer is generally positive. Consequently, the increase in polymer segment concentration in the stabilizing sheaths when particles approach each other would be energetically favoured were it not for the loss in entropy of the stabilizing polymer, which more than compensates for the enthalpic change. Steric stabilization in organic media is thus

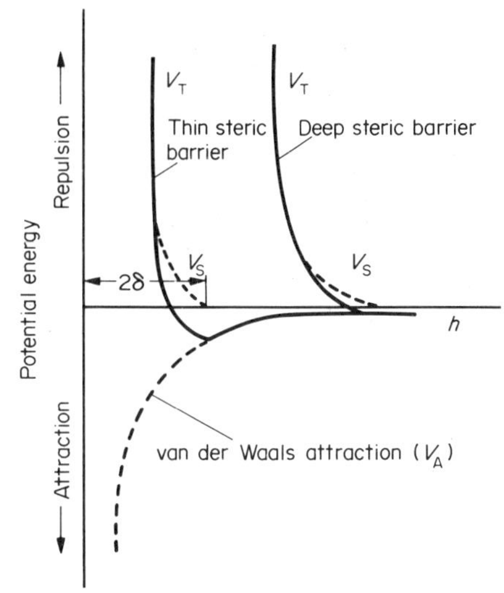


**Figure 2** Potential energy against distance of separation $h$ for sterically stabilized particles with varying depth of steric barrier

primarily dependent on entropy changes and the term 'entropic stabilization' has been employed. A characteristic property of entropic stabilization is the onset of flocculation on cooling a dispersion to a lower critical flocculation temperature (LCFT)[11] at which the entropy term ($T\Delta S$) equals the enthalpic term ($\Delta H$) in the free energy equation. In contrast, dispersions in which enthalpic interaction between the solvent and the stabilizing polymer predominates, as in aqueous dispersions or highly polar media, flocculate on heating to an upper critical flocculation temperature (UCFT).[6,12,13] Critical flocculation points may also be approached by changing the solvent properties of the medium in other ways, *e.g.* by adding a non-solvent for the stabilizing polymer,[6,11] by variation in pressure[14] or by operating near the critical temperature.[20]

Treatments of steric stabilization[6,7,15] have applied various theories of polymer solutions to estimate the steric interaction $V_S$ between two neutral spheres separated by a distance $h$ with an adsorbed polymer layer of depth $\delta$ surrounding each particle (Figure 1). The total interaction potential ($V_T$) may be written

$$V_T = V_A + V_S \qquad (1)$$

where $V_A$ is the van der Waals attraction (Figure 2). When $h < 2\delta$, perturbation of the conformations of the adsorbed chains by compression and by displacement of solvent from the interfacial region will occur. Early theories treated the elastic or volume restriction (compression) effect and the mixing or osmotic effect separately, in expressions of the form[15]

$$V_S = \underbrace{2\pi akTV_2\Gamma_2^2(\tfrac{1}{2}-\chi)S_{mix}}_{\text{mixing term}} + \underbrace{2\pi akT\Gamma_2 S_{el}}_{\text{elastic term}} \qquad (2)$$

where $a$ is the particle radius, $V_1$ the solvent molecular volume, $V_2$ the polymer molecular volume, $\Gamma_2$ the amount of adsorbed polymer (number of chains per unit area) and $\chi$ the Flory polymer/solvent interaction parameter. $S_{mix}$ and $S_{el}$ are geometric functions dependent on the segment concentration profile $\rho(z)$ in the adsorbed layer normal to the interface. The segment concentration profile for individual particles has been determined experimentally by small-angle neutron scattering (SANS). For high surface coverage, the hydrodynamic thickness of the stabilizing sheath is determined largely by the longer extended tails of adsorbed polymer, shorter chains and loops playing a relatively minor role.[16]

More recently, the separate treatment of the mixing and configurational effects has been avoided through the development of self-consistent field theories[6,13,17,18] in which the polymer is assumed to be terminally attached to a surface and the intra- and inter-molecular interactions between segments of all polymers are treated as an excluded volume effect.

The earlier theoretical approach has proved of great practical value in understanding and predicting the behaviour of sterically stabilized polymer colloids. In practice, the mixing term in equation (2) is much the more important, and when it approaches zero, flocculation is usually observed. This condition is met when the term $(\frac{1}{2}-\chi)$ in equation (2) becomes zero at $\chi=\frac{1}{2}$, corresponding to the point at which the second virial coefficient becomes zero and the polymer behaves as an ideal non-interacting solute. For $\chi>\frac{1}{2}$ the polymer coil collapses (poor solvent), while for $\chi<\frac{1}{2}$ the polymer coil swells (good solvent). Since $(\frac{1}{2}-\chi)$ may be rewritten[19] as

$$(\tfrac{1}{2}-\chi)=\psi\left(1-\frac{\theta}{T}\right) \tag{3}$$

where $\psi$ is the entropy parameter and $\theta$ the theta temperature in the Flory–Huggins theory, it follows that $\chi=\frac{1}{2}$ when $\theta=T$, and that the onset of flocculation of a sterically stabilized dispersion should be observed in the vicinity of the theta temperature of the stabilizing polymer in the dispersion medium.[6]

In non-aqueous dispersions, many examples of LCFTs have been recorded on the temperature being lowered through the theta point.[2,6,11] A selection of results for the more common stabilizing polymers is given in Table 1, from which it can be seen that there is a close but not complete correspondence between the LCFT and the theta point. The existence of a UCFT has also been demonstrated for a number of systems. In a non-polar dispersing medium, the UCFT is observed as the critical temperature of the medium is approached.[20–22] In the region of the critical temperature the density of the liquid falls rapidly. As a result, the van der Waals attraction between the particles increases and the solvation of the stabilizing chains becomes less effective. Flocculation at the UCFT is probably a combination of this increase in $V_A$ and decrease in $V_S$.[22] The application of pressure to the liquid medium, which raises the critical temperature, was shown to raise the UCFT for dispersions of polyacrylonitrile in propane stabilized by poly(dimethylsiloxane). The theory of corresponding states has been applied to relate the UCFT to the variation of $\chi$ (Flory interaction parameter) with temperature and pressure.[21,22]

**Table 1** Examples of Critical Flocculation Temperatures (CFT) of Non-aqueous Polymer Dispersions

| *Stabilizing polymer* | *Dispersion medium* | *CFT* (K) | $\theta$ (K) | *Upper (U)/Lower (L)* | *Ref.* |
|---|---|---|---|---|---|
| Poly(12-hydroxystearic acid) | *n*-Heptane | 260 | 240 | L | a |
| Poly(12-hydroxystearic acid) | *n*-Hexanol | 265 | 245 | L | a |
| Poly(lauryl methacrylate) | *n*-Heptane | 215 | 175 | L | a |
| Poly(dimethylsiloxane) | *n*-Heptane/ethanol | 340 | 340 | L | b |
| Poly(α-methylstyrene) | *n*-Butyl chloride | 254 | 263 | L | c |
| Poly(α-methylstyrene) | *n*-Butyl chloride | 403 | 412 | U | c |
| Polystyrene | Cyclopentane | 280 | 293 | L | d |
| Polystyrene | Cyclopentane | 410 | 427 | U | d |
| Polyisobutylene | 2-Methylhexane | 423 | 426 | U | e |

[a] D. H. Napper, *Trans. Faraday Soc.*, 1968, **64**, 1701.
[b] J. V. Dawkins and G. Taylor, *Colloid Polymer Sci.*, 1980, **258**, 79.
[c] M. D. Croucher and M. L. Hair, *Macromolecules*, 1978, **11**, 874.
[d] M. D. Croucher and M. L. Hair, *Colloids Surfaces*, 1980, **1**, 349.
[e] M. D. Croucher and M. L. Hair, *J. Colloid Interface Sci.*, 1981, **81**, 257.

There is thus a wealth of evidence that the solvency of the dispersion medium is a major limiting factor in determining the stability of non-aqueous dispersions. Under the practical conditions of a non-aqueous polymerization, it is prudent to select a stabilizing molecule for which the medium is considerably better than a theta solvent.

Another important factor is high surface coverage (equation 2), which should approximate to close packing of solvated stabilizing polymer chains on the particle surface.[23] Growth of particles without the provision of further stabilizer,[11] or removal of stabilizer by continued replacement of the medium,[24] was found to induce flocculation. The stabilizer must also be firmly attached to the particle surface in order to resist lateral displacement, which would allow the particle cores to touch or closely approach each other. In non-aqueous dispersion polymerizations, the only satisfactory method of satisfying these requirements has been through the use of block or graft copolymer stabilizers. One component provides the solvated moiety, while the second component, which is often referred to as the anchor, should be insoluble in the dispersion medium and should associate strongly or be compatible with the dispersed polymer. Additional stability in the presence of free monomer or strong solvent may be provided by covalently linking the anchor group to the dispersed polymer.[25,26]

### 15.2.2 Effect of Free Polymer in Solution — Depletion Flocculation

In many applications of non-aqueous dispersions in coatings and inks, free polymer is present dissolved in the dispersion medium. The effect of this free polymer on stability is therefore of considerable technological importance, in addition to its theoretical interest.

When free polymer is added to a sterically stabilized dispersion, weak flocculation may be observed. The magnitude of this effect is related to the molecular weight and concentration of the added polymer and to the particle concentration.[27] The higher the molecular weight of the free polymer, the lower the critical concentration at which flocculation is observed. The particle associations are generally reversible and may be broken down under shear or by moving outside the critical concentration limits. An explanation of the effect has been proposed in terms of the behaviour of polymers in the semi-dilute concentration regime. If the dispersed particles are stabilized by close-packed strongly adsorbed polymer, then penetration of the stabilizing layer by free polymer is unlikely. Consequently, when two particles approach to a distance less than the radius of gyration of the free polymer in solution, the latter is excluded from the zone between the particles, which thus becomes depleted of free polymer. Solvent diffuses by osmosis out of this region to equalize the concentration. Application of Flory–Huggins[27] or mean field theory[28] leads to the prediction that, at a critical polymer concentration, the net free energy of mixing can change sign, resulting in a net attraction between the particles. This phenomenon is now generally known as depletion flocculation.[6,13] At higher polymer concentrations, interpenetration of polymer coils takes place and the segment density becomes more uniform. A second critical concentration is then reached, above which the dispersion remains stable.[102] It would be expected that the mixing term (equation 2) would favour interparticle attraction under these conditions, but that the larger contribution of the elastic term should give a net repulsion, resulting in latex stability.

### 15.2.3 Dispersion Stability in Polymer Melts

In the limit, the concentration of polymer in the continuous phase may be raised until the particles are dispersed in a solvent-free melt of the stabilizing polymer.[13,29,30] Under these conditions the mixing term should be negligible, leaving only the elastic contribution. The experimental results suggest that, for small particles with a stabilizer molecular weight of a few thousand, the elastic contribution is sufficient to impart stability in polymer melts. However, evidence was found that, for very low molecular weight stabilizers and large particles, coagulation occurred as a result of the London attraction between the particles.[29]

Under conditions where the molecules of the dispersion medium were comparable in dimensions to the voids between the particles, gelled dispersions were formed, which showed a high yield stress and time-dependent thixotropy.[31] The gelling phenomenon was attributed to the compression of the free polymer molecules within the voids, which resulted in an elastic expansive force pushing outward on the particles, and was compared with the electroviscous effect arising from coulombic repulsions between particles.

### 15.2.4 Stabilizer Barrier Dimensions

The dimensions of the stabilizer barrier have been investigated by a variety of techniques, including the compression of a two-dimensional array of particles at an oil/water interface,[32] three-dimensional packing in a compression cell,[10] viscometry,[31,33,34] photon correlation spectroscopy[35] and SANS.[36] For terminally adsorbed polymers at high surface coverage, all the results supported a picture of polymer chains which were close packed laterally, and highly extended normal to the particle surface. For a low molecular weight stabilizing polymer ($M<10\,000$) the experimental surface layer thickness was close to the fully extended chain length, whereas for $M>10\,000$ the thickness was intermediate between the extended length and twice the root mean-square radius of gyration for the free stabilizer chain in solution.[34] At the higher molecular weights, each stabilizing chain could be visualized as a prolate ellipsoid, the minor axis of which was determined by the spacing of the anchoring points on the particle surface, the major axis by the layer thickness and the total volume by the volume of the same chain as a random coil.

Studies of concentrated non-aqueous dispersions by rheological and optical techniques suggest that particles in which the stabilizing layer thickness is a small fraction of the total (particle + stabilizing shell) radius may be treated as hard spheres.[37–39] As the core diameter is decreased relative to the stabilizer shell, the particles tend to behave as soft spheres. Interpenetration or compression of the stabilizing layers increases progressively as the concentration is raised towards the random-packing limit,[37] with a consequent departure from Newtonian viscosity, a reduction in the particle centre-to-centre separation, and again an approach to hard sphere behaviour.[40]

## 15.3 DESIGN AND SYNTHESIS OF POLYMERIC STERIC STABILIZERS

The most successful types of stabilizer for non-aqueous dispersion polymerization have been based on block or graft copolymers, one component of which is the soluble stabilizing moiety. The 'anchor' component is insoluble in the continuous phase and is adsorbed on, or absorbed into, the disperse phase. The graft stabilizer may be generated in the course of the latex process by grafting reactions between the dissolved stabilizing component and the polymerizing monomer, or it may be preformed and then used in the required amount in the latex process.

It follows from the theoretical discussion in Section 15.2 that the dispersion medium should be better than a theta solvent for the soluble component of the stabilizer under the reaction conditions of the latex process. Table 2 lists a variety of combinations of soluble polymer and dispersion medium which have been successfully employed.

**Table 2** Examples of Polymer Dispersions Prepared by Addition Polymerization

| *Stabilizing polymer* | *Disperse polymer* | *Medium* |
|---|---|---|
| Natural rubber | PMMA, PVAC | Alkane |
| Butyl rubber | PMMA | |
| Polyisobutylene | PMMA | |
| Poly(lauryl methacrylate) | PVC, PMMA, polyacetal | |
| Poly(ethyl hexyl acrylate) | PVAC | |
| Poly(12-hydroxystearic acid) | PMMA, PAN, PVDC, PVAC, PVP | |
| Poly(dimethylsiloxane) | PMMA, PS, PAN, poly($\alpha$-methylstyrene) | |
| Poly(*t*-butylstyrene) | PS | |
| Polystyrene | PAN | Toluene |
| Poly(ethylene phthalate) | Poly(acrylic acid) | Chloroform |
| Poly(vinylpyrrolidone) | PS | Ethanol |
| Hydroxypropylcellulose | PS | Methoxyethanol/ethanol |

The molecular weight of the soluble component is not critical, but for practical reasons[2] the minimum value is in the range 1000–1500. Stability can be provided by molecular weights as low as 300–400,[41] but the problem of constructing an effective stabilizer for polymer dispersions incorporating such low molecular weight groups has not yet been solved.

A minimum molecular weight of 500–1000 would appear to be necessary for the anchor group to be effective, depending on the degree of its insolubility in the medium and the strength of the association with the particle. It is of interest to compare anchoring phenomena in non-aqueous media with the behaviour of surfactants used in the preparation of aqueous dispersions. In aqueous dispersions, the association between the anchoring group of a surfactant and the surface on which it is adsorbed need be only quite weak, because the adsorption arises from the strong rejection of the anchor group by the hydrogen-bonded water structure. In most organic solvents this rejection energy is very much smaller and, in consequence, an anchor group is required which is large enough to be sufficiently insoluble in the medium.[2] A further consequence of this lower limit to the anchor size is that it may be incompatible with the dispersed polymer because of the unfavourable free energy of mixing of dissimilar polymers. For example, when polystyrene was used as the anchor component of a block copolymer for stabilizing a polystyrene dispersion, SANS measurements revealed that it was uniformly distributed in the particle surface. The same stabilizer on poly(methyl methacrylate) (PMMA) clustered into discrete domains in the surface of the particles.[42]

A consequence of the larger anchor group of a non-aqueous stabilizer molecule is the reduction in the rate of dissociation of micellar aggregates into single molecules compared with aqueous surfactants.[22,43,44] A careful balance must therefore be struck between the anchor and soluble components of the stabilizer, (anchor/soluble balance, ASB). The micellar equilibrium may also be adjusted by varying the solvency of the polymerization medium. A convenient method of achieving this is by changing the concentration of polar monomer in solution prior to the commencement of dispersion polymerization.

Examples of stabilizer structures are illustrated in Figure 3. Type (a) represents a stabilizer formed by a random grafting reaction, *e.g.* the grafting of PMMA on to a soluble natural rubber backbone by means of hydrogen abstraction with a peroxide catalyst. The method was employed in the first successful dispersion polymerization process[45] and, more recently, in a series of preparations in which the soluble polymer was polyisobutylene.[46] The reactivity of the soluble group dictates the course of the reaction. Natural rubber was so highly reactive that, although fine particle latexes were produced, there was a tendency to excessive graft formation which led to eventual gelation. At the other extreme, polyisobutylene in comparison with natural rubber is unreactive, giving a low yield of graft and a relatively large particle-size latex. An excess of a relatively unreactive soluble polymer may be used if it has a further role to play in the final application of the product. Thus, etherified melamine formaldehyde resins have been used as stabilizer precursors in the manufacture of hydroxyacrylate copolymer dispersions for thermosetting paints.[47]

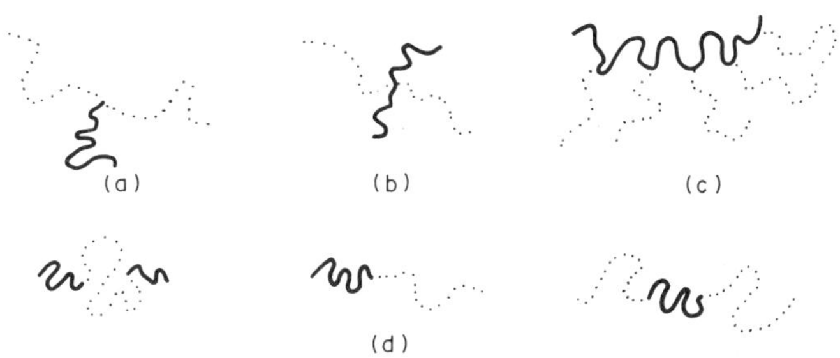


**Figure 3** Possible configurations of copolymer stabilizers: (a) random graft, hydrogen abstraction; (b) random graft, copolymerization; (c) 'comb' graft copolymer; and (d) ABA, AB and BAB block copolymers. ·······, soluble group; ∿∿∿, anchor group

An improvement on the random grafting approach is the incorporation of highly active transfer sites on the soluble chain. An example is the introduction of mercaptoalkylsilyl groups into a poly(dimethylsiloxane) chain. Best results were obtained when two or more mercapto groups per chain were present.[48]

A refinement of this method is shown in Figure 3(b), in which a copolymerizable group is attached to the soluble component, *e.g. via* reaction of acrylic acid with a lauryl methacrylate/glycidyl methacrylate copolymer, and this stabilizer precursor is copolymerized with the anchor component monomer.[49,50] The precursor may be used directly in the dispersion polymerization process or converted to graft copolymer in a prior step.

The 'comb' stabilizer illustrated in Figure 3(c) is simple to synthesize from inexpensive starting materials, and lends itself to considerable variation in the composition of both the soluble and anchor components. In the most widely used version,[51] poly(12-hydroxystearic acid) (PHSA) of molecular weight 1500–2000 is reacted through the terminal carboxyl group with glycidyl methacrylate to form the solvated component. This is then copolymerized, to an overall molecular weight of 10–20 000, with an equal weight of (meth)acrylic monomers, *e.g.* MMA, and a small proportion of glycidyl methacrylate or methacrylic acid to which may be attached polymerizable or other functional groups.

A high yield of graft copolymer is obtained with negligible amounts of ungrafted anchor component.[2] The preparative method provides a molecular structure with alternating anchor and solvated groups of approximately equal size, which is sufficiently flexible to allow the pendant tails of PHSA to pack closely together on the particle surface.[2,23,52]

The composition of the soluble group in the comb stabilizer has been varied by cocondensation of 12-hydroxystearic acid with other hydroxy acids and amino acids, *e.g.* 2,2-bis(hydroxymethyl)propionic acid and aminoundecanoic acid.[53] In a further variation, sufficient monocarboxylic acid was added to esterify residual terminal hydroxyl groups in PHSA to prevent side reactions with other components such as isocyanates.[54,55] Comb stabilizers based on PHSA have found wide industrial application and have provided a simple synthetic route to monodisperse latexes for fundamental studies.[26,33]

Block copolymers of the type shown in Figure 3(d) have also been widely employed in fundamental studies of non-aqueous dispersions in hydrocarbon media. The favoured composition has been a soluble group of poly(dimethylsiloxane) (PDMS; block A) combined with an anchor group of polystyrene (PS; block B) to form AB[44] or ABA[22,30,35,56] block copolymers. Anionic polymerization of styrene was initiated with either *n*-butyllithium or dilithium naphthalene to produce polystyryl anions which were then reacted with a siloxane cyclic trimer, hexamethylcyclotrisiloxane. Block copolymers prepared in this way have well-defined block lengths and a narrow chain-length distribution ($M_w/M_n < 1.25$). The PDMS chains provide terminally anchored tails and, in contrast to PHSA, the molecular weight can be made sufficiently large for an analysis of the thermodynamics of the stabilizing chains in terms of polymer solution-theory.[57] AB block copolymers of polystyrene-*block*-poly(ethylene-*co*-propylene) have been used in similar studies as stabilizers for PMMA and poly(vinyl acetate) (PVAC) dispersions in alkanes.[58]

## 15.4 MECHANISM OF NON-AQUEOUS DISPERSION POLYMERIZATION

### 15.4.1 Nucleation and Growth in Radical Dispersion Polymerization

The kinetics of particle nucleation and growth for a range of radical dispersion polymerizations in organic media were investigated in depth by Barrett and Thomas.[59] Their conclusions may be summarized as follows. In a typical dispersion polymerization, monomer, dispersion stabilizer and initiator form an initially homogeneous solution in the reaction medium. On heating to the reaction temperature, polymerization commences in solution to form oligomeric radicals. These precipitate to form nuclei, in the process associating with the stabilizer and thus becoming sterically stabilized. In the early stages of the polymerization this process of formation of primary particles by homogeneous nucleation may be succeeded by a sudden reduction in particle number due to aggregation. This, in turn, causes a drop in the rate of oligomer capture by existing particles and creates conditions for renucleation to form a fresh crop of particles. The threshold degree of polymerization before precipitation is sufficiently low that nearly all oligomeric radicals form nuclei or are captured by particles before the radicals undergo termination reactions. Monomer is absorbed from solution by the disperse phase once it forms and, subsequently, nearly all the polymerization takes place within the particles.

As polymerization proceeds within the swollen particles, the viscosity within each particle increases and the reaction rate accelerates rapidly due to the reduction in the diffusion-controlled termination rate. Kinetic studies indicated that the extent of the acceleration is closely parallel to the acceleration observed in the corresponding bulk polymerization of monomer at high conversion (see Volume 3, Chapters 6 and 12).

Dispersion polymerization may be regarded as a form of precipitation polymerization (see Volume 4, Chapter 16), in which the presence of stabilizer prevents uncontrolled aggregation of the precipitating particles, thus permitting easy access of monomer to the polymer phase and ensuring good heat transfer to the medium. A comparison of the rates of dispersion, precipitation and solution polymerization of MMA at 80 °C is shown in Figure 4.

A consequence of the microbulk nature of the process is an absence of the dependence of rate on the particle size or number which is a feature of classical aqueous emulsion polymerization[59] (see Volume 4, Chapter 11). A further difference is the absence of a monomer phase requiring separate emulsification. On the other hand, the monomer, which in many cases is a good solvent for its polymer, is present in solution in the dispersion medium at the onset of polymerization and may greatly influence the nucleation stage[60] (see Section 15.5).

The majority of kinetic studies, on which the above picture is based, were conducted with MMA in aliphatic hydrocarbon diluent at 80 °C. At this temperature the polymer was below its glass transition temperature ($T_g$), and the characteristics of dispersion polymerization already mentioned were strongly in evidence.

The appearance of the particles was smooth and spherical, the electron microscope revealing only a fine surface texture, and the stabilizer consumption corresponded to a monolayer covering the surface. In contrast, the dispersion polymerization of vinyl chloride and vinylidene dichloride in alkanes presented the opposite extreme. The process appeared to be one of solution polymerization followed by precipitation or aggregation with existing particles. There was relatively little adsorp-

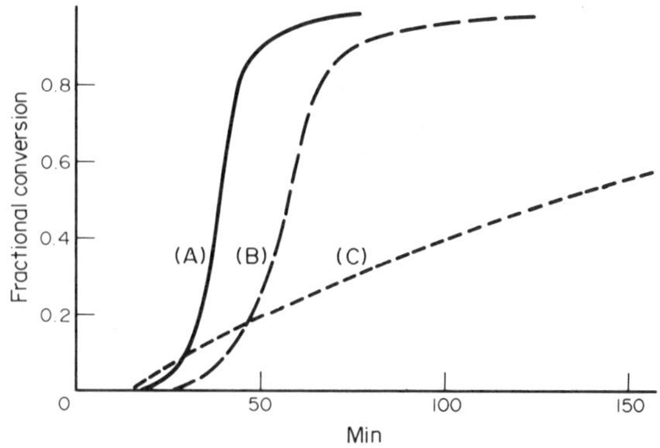


**Figure 4** Comparison of rates of dispersion polymerization (A), precipitation polymerization (B) and solution polymerization (C) of methyl methacrylate at 80 °C (reproduced by permission of John Wiley & Sons Inc. from K. E. J. Barrett and H. R. Thomas, *J. Polym. Sci., Part A-1*, 1969, **7**, 2627)

tion of monomer into the particles and the molecular weight was typical of a solution polymer. The particles were granular in surface texture and were irregular in shape. The stabilizer consumption was excessive and consistent with a mechanism involving continuous formation and aggregation of partly stabilized particles.[2]

Vinyl acetate and ethyl acrylate resembled MMA in behaviour, but as both polymers have much lower $T_g$ values and would have lower particle viscosities, the rate accelerations were much less pronounced at similar reaction temperatures.[61] A strong rate acceleration was observed with acrylonitrile polymerization in alkanes, comparable to that of MMA.

However, the kinetic evidence suggested that particles were heavily swollen with monomer after precipitation, but that, as the monomer concentration dropped with increasing conversion, the particles become more rigid because of strong intermolecular associations, thus limiting the absorption of more monomer. The addition of further monomer at this stage could result in a build-up of monomer concentration in the continuous phase leading to renucleation and an uncontrollable exothermic reaction.

A general expression for the overall rate of polymerization $R_p$, assuming that polymerization takes place predominantly within the particles, is given by[59]

$$R_p = \alpha C_d v^{1/2} k_p (R_i/k_t)^{1/2} \tag{4}$$

where $\alpha$ = monomer partition coefficient, polymer/diluent; $C_d$ = monomer concentration in diluent; $v$ = particle volume fraction; $k_p$ = propagation constant; $k_t$ = termination constant; and $R_i$ = initiation rate.

The polymerization rate is proportional to the total volume of particles. The termination rate does not remain constant throughout the polymerization but falls as the internal viscosity of the particles increases. At high conversions the propagation rate also falls as it becomes diffusion controlled (see Volume 3, Chapter 12). The curve of polymerization rate *vs.* conversion consequently rises to a maximum analogous to that observed in bulk polymerizations. Two limiting cases may be distinguished. In the first, for a low value of $\alpha$, typical of MMA ($\alpha = 1$) and vinyl acetate ($\alpha = 2$) polymer dispersions in aliphatic hydrocarbon, the increase in rate of conversion over the initial part of the polymerization is directly in proportion to time (Figure 5).

In the second case, typical of very polar molecules, such as acrylic acid, acrylonitrile and acrylic acid, the effective value of $\alpha$ is large because the greater part of the monomer partitions at an early stage into the particles, even though the true equilibrium value of $\alpha$ between diluent and unswollen polymer may be no higher than that of methyl methacrylate. The kinetics then approximate to those for a suspension polymerization (see Volume 4, Chapter 14), the rate of conversion rising rapidly to a high level, and thereafter being subject to the competing effects of acceleration, due to the gel effect, counterbalanced by falling monomer concentration (Figure 6).

### 15.4.2 Copolymerization in Radical Dispersion Polymerization

The chemical composition of a copolymer formed by the radical copolymerization of a homogeneous mixture of monomers depends on the relative reactivity of the individual monomers

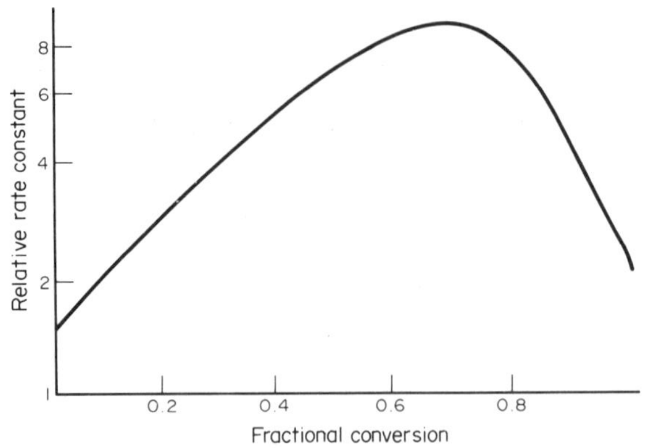


**Figure 5** Variation of relative rate constant with conversion in dispersion polymerization methyl methacrylate at 80 °C (reproduced by permission of John Wiley & Sons Inc. from K. E. J. Barrett and H. R. Thomas, *J. Polym. Sci., Part A-1*, 1969, 7, 2629)

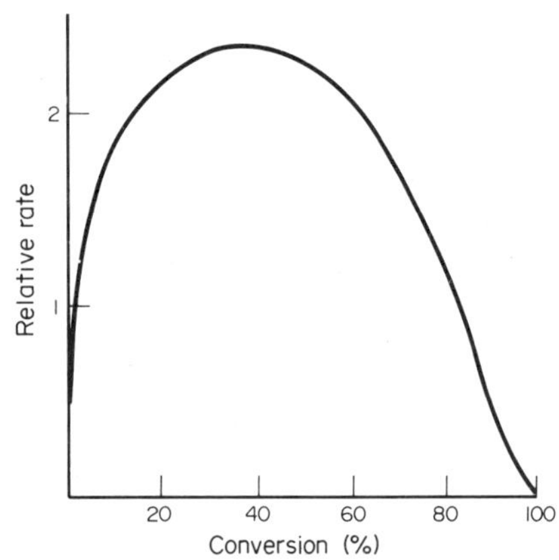


**Figure 6** Variation of relative rate with conversion in dispersion polymerization of acrylonitrile in cyclohexane at 80 °C (reproduced by permission of John Wiley & Sons Inc. from K. E. J. Barrett and H. R. Thomas, in 'Dispersion Polymerization in Organic Media', ed. K. E. J. Barrett, Wiley, London, 1975, p. 186)

towards the propagating polymer radicals.[62] In heterogeneous systems the overall reactivity is also influenced by the relative concentrations of monomer at the site of polymerization. In dispersion polymerization, in which the monomers are freely soluble in the continuous phase, it would be expected that the partition coefficients of the monomers between particle and medium would greatly influence their relative concentrations in the two phases. However, no systematic investigation of the topic has been published. Qualitatively, the effect of differences in partition ratios is to increase the apparent reactivity of more polar molecules, such as acrylic acid or hydroxy(meth)acrylates, and to reduce that of non-polar molecules, such as styrene or the higher alkyl acrylates.

In practical terms, this places a limitation on the fraction of less polar molecules which can be copolymerized, *e.g.* stable dispersions of copolymers of MMA with butyl acrylate containing more than 30% of the latter are difficult to prepare. When excessive amounts of less polar monomer are present, relatively large amounts of solution polymer of marginal solubility are formed, which may bring about coagulation of the latex. Even when such extremes are avoided the composition of the polymer particles is unlikely to be homogeneous. The more polar monomers will polymerize preferentially giving an intitial composition rich in the polar monomer. As the conversion rises, the composition will move towards a higher content of less polar monomer. In the extreme, phase separation may occur within the particles.[63]

### 15.4.3 Control of Molecular Weight

A consequence of the microbulk nature of dispersion polymerization is the formation of very high molecular weight polymer in those examples, such as MMA polymerization, in which diffusion-controlled termination and high exotherms are encountered.[2] In other systems, such as vinyl acetate, transfer to monomer is much more important and molecular weights are relatively low. In most applications of non-aqueous dispersions, the molecular weight of the polymer is not critical and, consequently, relatively little attention has been paid to its control. Chain-transfer agents, most commonly *n*-octanethiol, which have frequently been included in MMA dispersion polymerizations in hydrocarbon to moderate the reaction and prevent a dangerous exotherm, are effective in reducing the molecular weight.[2]

In an investigation of such a system, in which the transfer agent was dodecanethiol, it was concluded that the kinetics were dictated by the escape from the particles of the small radicals formed by transfer to thiol, which then underwent termination in the diluent phase. Under these conditions, the rate was independent of the number of particles and the average number of radicals per particle was very small.[64] The reaction conditions were significantly different from those of Barrett[59] and the conversions were much lower, so it is difficult to make a true comparison.

### 15.4.4 Ionic Initiation of Dispersion Polymerization

Ionically initiated dispersion polymerization of vinyl monomers has not been studied in detail. The published examples are concerned with the anionic polymerization of styrene[65] or α-methylstyrene.[66] Preparations with Ziegler–Natta catalysts have also been described (see Section 15.5), but no mechanistic studies have been reported.

A major difference from the free-radical process is the absence of a termination step involving mutual termination of active species. Consequently, the rate of polymerization would not be expected to accelerate with conversion, neither should it differ greatly from solution polymerization.

The initial polymerization rate of α-methylstyrene in heptane, catalyzed with butyllithium, was found to be faster than the corresponding solution polymerization, but this was ascribed to the higher monomer concentration in the freshly precipitated particles.[66] After the initial stages of the reaction, the rate of conversion was similar to that in solution.

In the polymerization of styrene in *n*-heptane or *n*-dodecane, catalyzed with butyllithium and stabilized with PDMS-*block*-PS copolymers, surprisingly low conversions *ca.* 40% were obtained.[65] This was shown to be the result of a reaction between polystyryllithium and the PDMS blocks (equation 5).

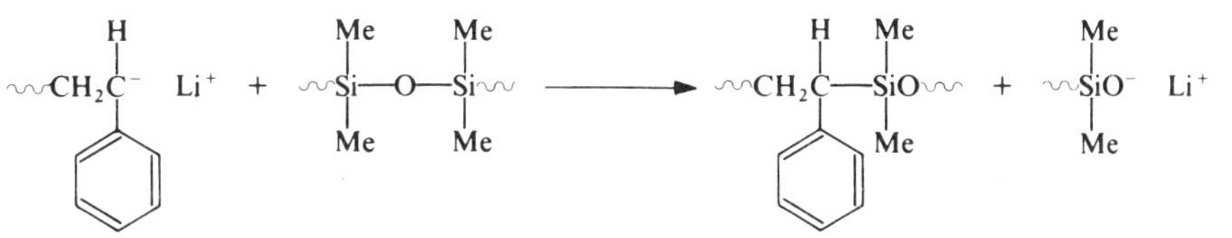


(5)

Nevertheless, the conversion was higher than in a solution polymerization conducted in the presence of PDMS, indicating some protection of the polystyryl anions from the PDMS. Changing the dispersion medium to *n*-dodecane, which was a poorer solvent for PS, in an attempt to reduce the exposure of the active anions to PDMS, gave a two-fold increase in monomer conversion (70%). Finally, when a seed dispersion was prepared in *n*-dodecane consisting of micelles of PDMS-*block*-PS copolymer with a high ASB (in the range 6:1 to 10:1), the conversion was raised to 90%. This was attributed to rapid absorption of monomer and polystyryl anions into the seed micelle. Even so, the molecular weight polydispersity was higher in dispersion systems than in solution, and the molecular weight was about twice the predicted value. Both observations could be explained by the loss of some of the propagating centres as a result of the side reaction with PDMS.

### 15.4.5 Particle Morphology

Particles produced by the microbulk mechanism might be expected to behave as homogeneous spheres of dispersed polymer surrounded by a thin sheath of solvated polymer. Characterization by rheological and optical methods provides support for this picture (see Section 15.2.4). An area of uncertainty is the state of the surface of the core particle, and the extent to which blocks of anchor polymer are uniformly fused into the dispersed polymer. Reference has already been made[42] (see Section 15.3) to the clustering of the incompatible anchor into discrete domains in the particle surface.

Some doubt is cast on this simple picture by a series of studies of fluorescent energy transfer in PMMA particles stabilized with polyisobutylene-*graft*-PMMA copolymers. 1-Naphthylmethyl methacrylate was copolymerized with MMA at levels of 2% and 10% to introduce chromophore groups into the particles.

The phosphorescence spectra and decay times were measured and the phosphorescence quenching was related to the concentration of the naphthalene groups in the particles.[67] In a second series of experiments, anthracene was dissolved in the isooctane medium and the dispersion was irradiated to excite the naphthalene fluorescence. The fluorescence decay curves were analyzed to establish the extent of the energy transfer from the excited naphthalene in the particle core to the anthracene in the medium.[68]

In subsequent experiments, 9-anthrylmethyl methacrylate was copolymerized in dispersion with MMA and the fluorescence polarization was studied.[69] The spectroscopic results led to the conclusion that the particles had a microphase structure in which a continuous network of polyisobutylene interlaced microdomains of PMMA and also formed a stabilizing sheath covering the particle surface. The picture thus differs very significantly from the core-shell structure deduced from other evidence. The particles in this study contained amounts of polyisobutylene (*ca.* 9%) far in excess of that required for a surface layer and there was evidence of extensive grafting of PMMA to the stabilizer. The particle sizes of between 1 μm and 3 μm diameter were rather large, suggesting that coagulation of small particles may have occurred to form aggregates in which trapped stabilizer was buried.

The fluorescence-labelling technique has also been applied to the solvated stabilizing polymer in order to explore the core–stabilizer interface. Ethyl hexyl acrylate was copolymerized with a small amount of phenanthryl methacrylate. This polymer was then used as a random-graft stabilizer precursor in the dispersion polymerization of vinyl acetate in isooctane. Similar experiments were performed with pyrene derivatives. Oxygen quenching of phenanthrene fluorescence and self-quenching and excimer formation of pyrene were used to diagnose the state of the stabilizer–particle interface and the conformation of the stabilizer chains. The results were interpreted as indicating the presence of an extensive interphase region rather than a distinct boundary.[70] It would be of interest to apply these fluorescence methods to the well characterized stabilizers and particles which have been studied by SANS.

## 15.5 PREPARATION OF NON-AQUEOUS POLYMER DISPERSIONS

### 15.5.1 Polymerization Processes

In a typical non-aqueous dispersion polymerization, the reactants are dissolved in the reaction medium to form a homogeneous solution. In the course of the polymerization, insoluble polymer is formed and precipitates. The nucleation process is influenced by a number of factors including: (a) the degree of insolubility of the precipitating polymer; (b) the effect of the monomer on the solvency of the medium, especially at the onset of polymerization when the monomer concentration is at its highest level; (c) the concentration of the dispersion stabilizer; and (d) the ASB of the stabilizer, its micellar equilibrium and the affinity of the anchor for the precipitating polymer.[2] Once a dispersion of stable nuclei has been established, the particle number remains constant unless there is a significant change in the reaction conditions. Such a change would be a rise in the stabilizer concentration to a level at which renucleation took place to form a fresh crop of particles. The greatest demand for stabilizer is at the moment of the first nucleation, after which only a relatively small amount is necessary to stabilize the additional surface area generated as the particles grow.

Two types of experimental procedure have been developed for the free-radical dispersion polymerization of vinyl and acrylic monomers in aliphatic hydrocarbons.[71] In the first, the polymerization of the initially homogeneous solution of monomer, initiator and dispersion stabilizer in hydrocarbon is carried out in a single stage. In a typical process, the temperature is raised to reflux at 80–90 °C, whereupon the solution quickly turns a translucent blue colour and then white, as first nucleation and then growth takes place. The reflux becomes vigorous for some minutes and then dies down as the conversion reaches 80–90%. In the second type of process, the nucleation and growth stages are separated. A seed stage is formed by polymerizing a portion of the monomer and stabilizer, after which the remaining monomer and stabilizer is fed over 1–3 hours to give a controlled reaction rate. A feed process is essential for the preparation of concentrated latex, because above a monomer concentration of *ca.* 35% in the single-stage process the exotherm can be difficult to control, and often only coarse particle latexes can be prepared.

Particle size and number are controlled by the solvency of the medium for the precipitating polymer, by the concentration of the stabilizer and its ASB.[2,44] The number of particles in the seed stage remains virtually constant unless the reaction conditions are altered, *e.g.* by a sudden addition of stabilizer to bring about renucleation. This technique was employed to produce very concentrated dispersions in which the particle diameters and volume fractions were controlled to achieve maximum packing.[9] With a bimodal particle size distribution, a concentration of 85% by weight of PMMA in hydrocarbon was attainable. The process has been operated on an industrial scale.[72]

Although the use of a preformed dispersion stabilizer permits more accurate control of particle size and greater reproducibility, the alternative method of forming the stabilizer *in situ* can be made reliable, particularly when the stabilizer precursor carries a limited number of reactive grafting sites. One advantage of the *in situ* process is that the anchor group of the stabilizer is identical in composition to the precipitated particle, thus avoiding problems of incompatibility. Obviously, in the production of a fine-particle concentrated latex, in which a large surface area is generated, particle nucleation and stabilizer formation must be kept in balance. A characteristic of the preparation of a concentrated non-aqueous dispersion polymerization is the dramatic nature of the coagulation which follows exhaustion of stabilizer. If the polymerization is well advanced, complete solidification of the reaction mixture is likely, while the exothermic reaction continues unabated, without the benefit of the efficient heat transfer of the fluid latex process!

Although the majority of published studies have been concerned with dispersions of the more common homopolymers, PMMA, polyacrylonitrile (PAN), PS and PVAC, many other polymers and copolymers have been prepared.[71] In those cases in which the monomer freely partitions between the medium and the swollen polymer particle, the polymerization is similar to PMMA. Acrylates and vinyl esters, in general, have lower $T_g$ values than PMMA and a greater tendency to transfer to monomer or solvent. Molecular weights consequently tend to be lower and greater difficulty is experienced in reaching high conversions.

Highly polar or hydrogen-bonded monomers, *e.g.* acrylic acid and acrylamide, resemble acrylonitrile. Polymerizations in hydrocarbon are difficult to control, often forming fine particles prone to flocculation on further polymerization.[71] Single-stage and feed processes were reported for PAN in hexane.[35] The single-stage process, in principle, offers a better prospect of control of nucleation, because a high proportion of the monomer is entrapped in the precipitating nuclei. Subsequent feeding of monomer is likely to lead to renucleation, which may result in a polymodal size distribution. Alternatively, new nuclei may be adsorbed by existing particles to give fuzzy or irregularly shaped final particles.

The dispersion polymerization of vinyl chloride in aliphatic hydrocarbon has been carried out on an industrial scale, but the process is difficult to control. The graft dispersing agent was formed *in situ* from a copolymer of lauryl methacrylate and methacrylic acid, reacted with allyl glycidyl ether to provide grafting sites.[71] In the dispersion process, poly(vinyl chloride) precipitated from solution in fine particles as it formed and subsequently aggregated into larger particles (typically 0.5–1.0 $\mu$m diameter) often of irregular shape. Consumption of stabilizer was excessive, the resulting particles being composites of stabilizer and poly(vinyl chloride). Attempts to use preformed graft copolymer stabilizers were not successful.

The majority of examples of non-aqueous dispersion polymerization have been conducted in aliphatic hydrocarbons. The method is nevertheless quite generally applicable to any organic liquid in which a soluble monomer polymerizes to form an insoluble polymer. Thus, for example, dispersions of polystyrene stabilized with hydroxypropylcellulose, have been prepared in mixtures of ethanol and 2-methoxyethanol.[73]

The process is also not limited to radical polymerization, although there are relatively few examples of other types of polymerization. The polymerization of styrene in alkane, initiated with *n*-butyllithium and stabilized with PDMS-*block*-PS copolymer has already been mentioned,[65] as has the anionic polymerization of $\alpha$-methylstyrene.[66]

Block copolymers of *t*-butylstyrene and styrene were shown to act as stabilizers for the anionic dispersion polymerization of styrene.[74] The same workers also described preparations in which a second disperse-phase monomer, such as vinylpyridine or ethylene oxide, was polymerized from the polystyryl anions within the particles to produce particles which were themselves block copolymers.

In a related procedure, the soluble component of a block copolymer was first formed by anionic polymerization followed by growth of the insoluble component to the point of nucleation. Growth was then continued within the particle. Finally, a third monomer was added to produce an ABC block copolymer, in which block A provided the solvated stabilizing moiety and blocks B and C comprised the particle.[75]

A number of ring-opening polymerizations have been reported in the patent literature but no detailed studies have been published.[71] Thus, cyclic ethers including 1,3,5-trioxane have been polymerized in the presence of reactive stabilizer precursors[76] or graft copolymer stabilizers.[77] Similarly, β-propiolactam,[76] caprolactone[78] and caprolactam[79] have been polymerized in dispersion in hydrocarbon media.

### 15.5.2 Preparation of Monodisperse Non-aqueous Polymer Colloids

The relatively short period of nucleation and the subsequent reduction in the solution polymerization rate as a result of diffusion of monomer into the particles are two factors which facilitate the formation of a monodisperse latex in a non-aqueous dispersion polymerization. The availability of monodisperse latexes has greatly assisted the study of steric stabilization and concentrated suspensions.

The polymerization of MMA in *n*-alkanes provides a particularly simple model system. Single-stage formulae have been developed for the preparation of a range of monodisperse latexes with particle diameters between 80 nm and 2.6 μm and coefficients of variation averaging 7%.[76] The recipe given in Table 3, by way of example gave a diameter of 403 nm with a coefficient of variation of 6.9%. The reaction mixture was flushed with nitrogen and stirred under reflux on a steam bath for two hours. When conversion was complete, the hexane was removed and replaced by an equal amount of high boiling hydrocarbon. A tertiary amine catalyst was added and the temperature raised to 120 °C to bring about the reaction between the glycidyl groups in the stabilizer and the carboxylic acid in the latex particle, in order to lock the stabilizer to the particle surface.

**Table 3** Recipe for the Polymerization of MMA

| *Reagents* | *% by weight* |
|---|---|
| Monomer (98.2 MMA/methacrylic acid) | 42.0 |
| Stabilizer [poly(12-hydroxystearic acid-*co*-MMA/glycidyl methacrylate)] comb (40% solution) | 5.25 |
| Azobis(isobutyronitrile) | 0.39 |
| *n*-Octanethiol | 0.2 |
| *n*-Hexane | 35.2 |
| Low aromatic hydrocarbon (b.p. 230–250 °C) | 17.0 |

Monodisperse latexes have also been produced by a seed and feed process with the same reactants.[10] Particle size control with the single-stage process was achieved by adjusting the monomer concentration within the range 35–50% of the total reaction mixture. The relationship between monomer concentration and particle size is illustrated in Figure 7. Stirring the reaction mixture had no significant effect on the particle size. This is typical of preparations of sterically stabilized particles, which, in contrast with aqueous charged stabilized latexes, are very insensitive to shear.[24] With this recipe, stable latexes could not be prepared in the monomer concentration range 8.5–34%. Increase in stabilizer concentration produced a polydisperse latex. Adjustment of the methacrylic acid content, which has a major effect on polymer solubility, would most probably change these concentration limits.

Similar processes have been described for the preparation of narrow distribution PMMA latexes stabilized with PDMS-*block*-PS copolymers in *n*-hexane.[44] In a study of the seed-stage/monomer-feed process, it was found that there was an increase in particle diameter as the monomer content of the seed stage was raised and a reduction in diameter with increasing stabilizer concentration. A reduction in size was also observed as the PDMS soluble group molecular weight was raised from 10 000 to 30 000, whilst maintaining the PDMS concentration constant. This was attributed to the greater area per molecule which could be stabilized by the higher molecular weight chains.

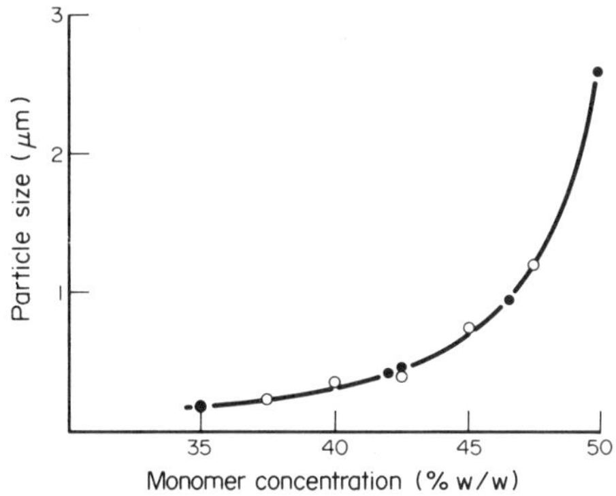


**Figure 7** Particle size against monomer concentration for one-stage dispersion polymerization of methyl methacrylate in aliphatic hydrocarbon: ○, non-stirred preparations; ●, stirred preparations (reproduced by permission of Elsevier Science Publishers from L. Antl *et al.*, *Colloids Surfaces*, 1986, **17**, 73)

The anionic polymerization of styrene in *n*-alkanes[65] was described in Section 15.4.4. The smallest particle diameter was prepared using a PDMS-*block*-PS stabilizer with an ASB sufficiently large to ensure that, when the polymerization was conducted in dodecane, the stabilizer formed irreversible micellar aggregates which acted as polymerization sites. Since all the polymer chains should be initiated simultaneously and all the particles should nucleate at the same time, the system should be ideal for the preparation of monodisperse suspensions.

The preparation of relatively large monodisperse polystyrene particles by free-radical polymerization has also received attention. Reference has been made in Section 15.5.1 to the polymerization of styrene in 2-methoxyethanol/ethanol mixtures stabilized with hydroxypropylcellulose.[73] When the styrene concentration in the reaction mixture was restricted to 15%, monodisperse particles were produced. Increasing the concentration of 2-methoxyethanol in the diluent increased the particle diameter. Suspensions of 3, 7 and 9 μm diameter were produced. Monodisperse PS particles with diameters ranging from 2.5 μm to 6.2 μm have been prepared in ethanol alone with a combination of a poly(vinylpyrrolidone) stabilizer, Aerosol OT (sodium dioctylsulfosuccinate), as a costabilizer, and an azo initiator.[80] The role of the costabilizer is not clear, but in its absence a bimodal distribution was obtained.

## 15.6 APPLICATION OF NON-AQUEOUS POLYMER DISPERSIONS

The special properties of non-aqueous polymer dispersions, such as high polymer content, low viscosity and good shear stability, together with the low heat of evaporation of the diluent and a wide choice of diluent volatility, were the mainspring of their development by the coatings industry. However, the problems encountered in developing satisfactory compositions, in which the polymer in its dispersion form was the major film-forming component, were such that there has been no large scale industrial application of this kind.[2,82]

It was, however, discovered that thermosetting coating compositions, prepared by the dispersion process in aliphatic hydrocarbons and subsequently dissolved in polar solvents, had significant rheological advantages when formulated as automotive paints. These advantages were attributed to the formation of a microgel fraction during polymerization which, when swollen by a stronger solvent, conferred pseudoplastic or thixotropic properties on the coating composition, giving better control of atomization and flow of the coating on spray application.[83] In the original invention, the microgel fraction was generated by attaching two or more polymerizable groups to the anchor group of a PHSA-*graft*-PMMA copolymer comb stabilizer to provide a polyfunctional graft copolymer which entered into crosslinking reactions with the dispersion polymer. Subsequently, processes have been developed for the controlled preparation of crosslinked microgel particles of acrylic polymers, which may then be added to conventionally prepared solution polymers to produce the desired rheology.[54,55,84–86]

In one such process,[84] a seed stage of PMMA was prepared in aliphatic hydrocarbon stabilized with a PHSA-*graft*-PMMA copolymer comb stabilizer. The seed particles provided the locus for the polymerization of the microgel components, comprising MMA, methacrylic acid and glycidyl methacrylate, which were then crosslinked by reaction between the carboxylic acid and glycidyl groups. Finally, a mantle of polymer was grown around the gelled particles. This polymer was similar in composition to the solution polymer in which the microgel particles were to be suspended, and served to control the degree of flocculation of the swollen microgel particles when they were incorporated into the paint.

Many alternative methods for preparing the microgels have been patented, and crosslinking reactions have included the reaction of copolymerized hydroxyl monomers with isocyanates[87] or aminoplast resins[88,89] and reactions between epoxides and carboxyl groups.[84,90]. Automotive coatings incorporating microgels have found extensive use world-wide, particularly for the control of metallic flake alignment within the paint film.[84] As environmental pressures continue to develop to reduce the solvent content of coatings, the use of microgels is likely to increase, in order to control the rheology of high-solids low molecular weight coatings during application.

Coatings based on non-aqueous polymer dispersions may be formulated to coalesce at ambient temperatures in an analogous fashion to aqueous emulsion paints. Thus a coating has been proposed for a wood substrate based on a copolymer of ethyl acrylate, hydroxyethyl acrylate and diallyl phthalate of controlled gel content and extensibility.[91]

After surface coatings, the most frequently cited area of application is in the field of electroreprographics as developers and toners. In many patents, the processes involve heating a higher alkyl (meth)acrylate in a hydrocarbon diluent with a minor quantity of disperse phase monomer and benzoyl peroxide to bring about the simultaneous formation of stabilizer and dispersion. The resulting dispersion may be further compounded with a polyethylene wax to produce a stable wax/acrylic dispersion.[92,93] The rapid drying characteristic of non-aqueous dispersions has led to proposals for their use in printing inks which can be heat-set at lower than conventional temperatures.[94,95] Photoconductive lithographic inks have also been prepared by incorporating a dispersion of poly(*N*-vinylcarbazole).[96]

Polymers prepared as non-aqueous dispersions have also been employed to modify the mechanical properties of bulk materials. Polymer dispersions for the manufacture of unscorched polyurethane foams were prepared by the copolymerization of acrylonitrile and styrene in the glycerol ether of poly(propylene glycol) in the presence of a stabilizer formed from the copolymerization of styrene, acrylonitrile and butoxypoly(propylene glycol) methacrylate.[97] Dispersions of elastomeric polymers are frequently incorporated into coatings or bulk plastics to increase toughness and improve impact resistance. Acrylic microgels, for example, have been patented for use in acrylic lacquers.[103] Epoxy resins have also been modified by first reacting a fraction of the epoxide groups with acrylic acid to provide grafting points and then polymerizing a mixture of 2-ethylhexyl acrylate and glycidyl methacrylate in the molten epoxy resin. Stable dispersions of rubbery polymer were formed in the epoxy matrix.[98] The particle size decreased as the number of acrylic grafting points was increased. Particle diameters from 0.2 $\mu$m to 2.0 $\mu$m were obtained. The surface fracture energy was increased two-fold at the optimum particle size of 1.5 $\mu$m.

The principles of non-aqueous dispersion polymerization have also been applied to Ziegler–Natta catalyzed polymerization of alkenes (see Volume 4, Part 1). An early patent described the polymerization of propylene in a hydrocarbon diluent in the presence of an alkene with six or more carbon atoms, typically 1-octene or 1-hexadecene.[99] More recently, a series of patents described the polymerization of butadiene in hydrocarbon in the presence of block copolymer precursors.[100,101]

## 15.7 REFERENCES


1. K. E. J. Barrett, *Br. Polym. J.*, 1973, **5**, 259.
2. K. E. J. Barrett (ed.), 'Dispersion Polymerization in Organic Media', Wiley, London, 1975.
3. D. W. J. Osmond, *Chem. Ind. (London)*, 1974, **22**, 891.
4. H. R. Thomas, *Farbe + Lack*, 1971, **77**, 525.
5. G. W. Poehlein, R. H. Ottewill and J. W. Goodwin (eds.), 'Science and Technology of Polymer Colloids', Nijhoff, Hague, 1983.
6. D. H. Napper, 'Polymeric Stabilization of Colloidal Dispersions', Academic, London, 1983.
7. G. D. Parfitt and J. Peacock, *Surf. Colloid Sci.*, 1978, **10**, 163.
8. J. Th. G. Overbeek, in 'Colloidal Dispersions', ed. J. W. Goodwin, Royal Society of Chemistry, London, 1982.
9. M. W. Thompson, in 'Polymer Colloids', ed. R. Buscall, T. Corner and J. F. Stageman, Elsevier, London, 1985.
10. R. J. R. Cairns, R. H. Ottewill, D. W. J. Osmond and I. Wagstaff, *J. Colloid Interface Sci.*, 1976, **54**, 45.
11. D. H. Napper, *Trans. Faraday Soc.*, 1968, **64**, 1701.
12. D. H. Napper, in 'Colloidal Dispersions', ed. J. W. Goodwin, Royal Society of Chemistry, London, 1982.
13. R. Buscall and R. H. Ottewill, in 'Polymer Colloids', ed. R. Buscall, T. Corner and J. F. Stageman, Elsevier, London, 1985.

14. R. Evans and D. H. Napper, *J. Colloid Interface Sci.*, 1975, **52**, 260.
15. B. Vincent and S. G. Whittington, *Surf. Colloid Sci.*, 1982, **12**, 69.
16. M. Cohen-Stuart, F. H. W. H. Waajen, T. Cosgrove, B. Vincent and T. L. Crowley, *Macromolecules*, 1984, **17**, 1825.
17. A. K. Dolan and S. F. Edwards, *Proc. R. Soc. London, Ser. A*, 1975, **343**, 427.
18. S. Levine, B. B. Thomlinson and K. Robinson, *Faraday Discuss. Chem. Soc.*, 1978, **65**, 202.
19. P. J. Flory, 'Principles of Polymer Chemistry', Cornell University Press, Ithaca, NY, 1971, p. 523.
20. M. D. Croucher and M. L. Hair, *J. Phys. Chem.*, 1979, **83**, 1712.
21. M. D. Croucher and K. P. Lok, *ACS Symp. Ser.*, 1984, **240**, 317.
22. D. H. Everett, in 'Science and Technology of Polymer Colloids', ed. G. W. Poehlein, R. H. Ottewill and J. W. Goodwin, Nijhoff, Hague, 1983, p. 353.
23. D. W. J. Osmond and D. J. Walbridge, *J. Polym. Sci., Part C*, 1970, **30**, 381.
24. D. J. Walbridge and J. A. Waters, *Discuss. Faraday Soc.*, 1966, **42**, 294.
25. BALM Paints, *Br. Pat.* 1 231 614 (1971) (*Chem. Abstr.*, 1970, **73**, 26 265).
26. L. Antl, J. W. Goodwin, R. D. Hill, R. H. Ottewill, S. M. Owens, S. Papworth and J. A. Waters, *Colloids Surf.*, 1986, **17**, 67.
27. B. Vincent, P. F. Luckham and F. A. Waite, *J. Colloid Interface Sci.*, 1980, **73**, 508.
28. A. P. Gast, C. K. Hall and W. B. Russel, *Faraday Discuss. Chem. Soc.*, 1983, **76**, 189.
29. D. H. Napper and J. B. Smitham, *J. Colloid Interface Sci.*, 1976, **54**, 467.
30. G. N. Choi and I. M. Krieger, *J. Colloid Interface Sci.*, 1986, **113**, 94.
31. G. N. Choi and I. M. Krieger, *J. Colloid Interface Sci.*, 1986, **113**, 101.
32. A. Doroszkowski and R. Lambourne, *J. Polym. Sci., Part C*, 1971, **34**, 253.
33. S. J. Barsted, L. J. Nowakowska, I. Wagstaff and D. J. Walbridge, *Trans. Faraday Soc.*, 1971, **67**, 3598.
34. J. V. Dawkins and G. Taylor, *J. Chem. Soc., Faraday Trans. 1*, 1980, **76**, 1263.
35. D. H. Everett and J. F. Stageman, *Faraday Discuss. Chem. Soc.*, 1978, **65**, 230.
36. D. J. Cebula, J. W. Goodwin, R. H. Ottewill, G. Jenkin and J. Tabony, *Colloid Polym. Sci.*, 1983, **261**, 555.
37. D. W. J. Osmond and I. Wagstaff, in 'Dispersion Polymerization in Organic Media', ed. K. E. J. Barrett, Wiley, London, 1975, p. 243.
38. T. A. Strivens, *Colloids Surf.*, 1986, **18**, 395.
39. P. N. Pusey and W. van Megen, *Nature (London)*, 1986, **320**, 340.
40. I. Marković, R. H. Ottewill, S. M. Underwood and Th. F. Tadros, *Langmuir*, 1986, **2**, 625.
41. A. Doroszkowski and R. Lambourne, *Discuss. Faraday Soc.*, 1978, **65**, 252.
42. J. S. Higgins, J. V. Dawkins and G. Taylor, *Polymer*, 1980, **21**, 627.
43. D. J. Walbridge, in 'Dispersion Polymerization in Organic Media', ed. K. E. J. Barrett, Wiley, London, 1975, p. 74.
44. J. V. Dawkins and G. Taylor, *Polymer*, 1979, **20**, 599.
45. Imperial Chemical Industries Ltd., *Br. Pat.* 893 429 (1962) (*Chem. Abstr.*, 1963, **58**, 11 491).
46. M. D. Croucher and M. L. Hair, *J. Colloid Interface Sci.*, 1981, **81**, 257.
47. Cook Paint and Varnish, *Br. Pat.* 1 134 997 (1968) (*Chem. Abstr.*, 1969, **70**, 30 157).
48. J. C. Saam and C. H. Tsai, *J. Appl. Polym. Sci.*, 1974, **18**, 2279.
49. Imperial Chemical Industries Ltd., *Br. Pat.* 1 052 241 (1966) (*Chem. Abstr.*, 1968, **69**, 10 929).
50. F. A. Waite, *J. Oil Colour Chem. Assoc.*, 1971, **54**, 342.
51. Imperial Chemical Industries Ltd., *Br. Pat.* 1 174 391 (1969) (*Chem. Abstr.*, 1972, **76**, 73 248).
52. R. M. Fitch and Y. K. Kamath, *J. Indian Chem. Soc.*, 1972, **49**, 1209.
53. Imperial Chemical Industries Ltd., *Br. Pat.* 1 325 927 (1973) (*Chem. Abstr.*, 1972, **76**, 128 968).
54. M. S. Chattha and A. N. Theodore, *J. Coat. Technol.*, 1985, **57** (721), 67.
55. M. S. Chattha and J. C. Cassatta, *J. Coat. Technol.*, 1985, **57** (730), 41.
56. D. H. Everett and J. F. Stageman, *Colloid Polym. Sci.*, 1977, **255**, 293.
57. J. V. Dawkins and G. Taylor, *Colloid Polym. Sci.*, 1980, **258**, 79.
58. J. V. Dawkins, G. G. Maghami and S. A. Shakir, *Polym. Prepr., Am. Chem. Soc. Div. Polym. Chem.*, 1985, **26** (1), 234.
59. K. E. J. Barrett and H. R. Thomas, in 'Dispersion Polymerization in Organic Media', ed. K. E. J. Barrett, Wiley, London, 1975, chap. 4.
60. R. M. Fitch and Y. K. Kamath, *J. Colloid Interface Sci.*, 1976, **54**, 6.
61. K. E. J. Barrett, H. R. Thomas and R. J. Tolman, *Kinet. Mech. Polyreactions, Int. Symp. Macromol. Chem., Prepr.*, 1969, **3**, 369.
62. C. H. Bamford, in 'Comprehensive Polymer Science', ed. G. Allen and J. C. Bevington, Pergamon Press, Oxford, 1988, vol. 3, chap. 17.
63. Imperial Chemical Industries Ltd., *Ger. Pat.* 2 140 135 (1970) (*Chem. Abstr.*, 1972, **76**, 154 726).
64. R. M. Fitch and Y. K. Kamath, *Polym. Prepr., Am. Chem. Soc., Div. Polym. Chem.*, 1975, **16**, 228.
65. J. V. Dawkins and G. Taylor, in 'Polymer Colloids II', ed. R. M. Fitch, Plenum Press, New York, 1980, p. 447.
66. G. B. Tampa, *J. Appl. Polym. Sci.*, 1970, **14**, 1227.
67. O. Pekcan, M. A. Winnik and M. D. Croucher, *J. Colloid Interface Sci.*, 1983, **95**, 420.
68. O. Pekcan, M. A. Winnik, L. Egan and M. D. Croucher, *Macromolecules*, 1983, **16**, 699.
69. G. Chiu, M. A. Winnik and M. D. Croucher, *Colloid Polym. Sci.*, 1986, **264**, 25.
70. M. D. Croucher, L. S. Egan and M. A. Winnik, *Polym. Eng. Sci.*, 1986, **26**, 15.
71. K. E. J. Barrett and M W. Thompson, in 'Dispersion Polymerization in Organic Media', ed. K. E. J. Barrett, Wiley, London, 1975, chap. 5.
72. Imperial Chemical Industries Ltd., *US Pat.* 3 686 114 (1972) (*Chem. Abstr.*, 1972, **77**, 140 933).
73. C. K. Ober, K. P. Lok and M. L. Hair, *J. Polym. Sci., Polym. Lett. Ed.*, 1985, **23**, 103.
74. Mobil Oil Corp., *Br. Pat.* 1 359 767 (1974) (*Chem. Abstr.*, 1973, **78**, 44 252).
75. Imperial Chemical Industries Ltd., *Br. Pat.* 1 412 584 (1975) (*Chem. Abstr.*, 1973, **79**, 54 292).
76. Imperial Chemical Industries Ltd., *Br. Pat.* 1 095 931 (1967) (*Chem. Abstr.*, 1965, **62**, 12 022).
77. Imperial Chemical Industries Ltd., *Br. Pat.* 1 095 932 (1967) (*Chem. Abstr.*, 1965, **62**, 12 021).
78. Union Carbide Corp., *US Pat.* 3 632 669 (1972) (*Chem. Abstr.*, 1971, **74**, 42 865).

79. Firestone Tire and Rubber, *Br. Pat.* 1 008 001 (1965) (*Chem. Abstr.*, 1963, **59**, 779).
80. M. S. El-Aasser, Y. Y. Lu, C. M. Tseng and J. W. Vanderhoff, *J. Polym. Sci., Polym. Chem. Ed.*, 1986, **24**, 2995.
81. J. W. Goodwin, J. Hearn, C. C. Ho and R. H. Ottewill, *Br. Polym. J.*, 1973, **5**, 347.
82. R. Dowbenko and D. P. Hart, *Ind. Eng. Chem. Prod. Res. Dev.*, 1973, **12**, 14.
83. PPG Industries Inc., *US Pat.* 4 055 607 (1977) (*Chem. Abstr.*, 1975, **82**, 74 563).
84. A. J. Backhouse, *J. Coat. Technol.*, 1982, **54** (693), 83.
85. Imperial Chemical Industries Ltd., *US Pat.* 4 180 489 (1979) (*Chem. Abstr.*, 1979, **90**, 56 491).
86. Imperial Chemical Industries Ltd., *US Pat.* 4 242 384 (1980) (*Chem. Abstr.*, 1979, **90**, 56 475).
87. Ford Motor Co., *US Pat.* 4 528 317 (1985) (*Chem. Abstr.*, 1985, **103**, 161 933).
88. M. S. Chattha and J. S. Cassatta, *Ind. Eng. Chem. Prod. Res. Dev.*, 1986, **25**, 423.
89. Imperial Chemical Industries Ltd., *US Pat.* 4 340 511 (1980) (*Chem. Abstr.*, 1980, **93**, 48 615).
90. Dai Nippon Ink Chemical K. K., *Jpn. Pat.* 61 051 003 (1986) (*Chem. Abstr.*, 1986, **105**, 210 502).
91. Imperial Chemical Industries PLC, *Eur. Pat. Appl.* 171 235 (1986) (*Chem. Abstr.*, 1986, **105**, 8083).
92. Ricoh Co. Ltd., *Jpn. Pat.* 60 248 711 (1985) (*Chem. Abstr.*, 1986, **104**, 225 414).
93. Ricoh Co. Ltd., *Jpn. Pat.* 60 248 714 (1985) (*Chem. Abstr.*, 1986, **104**, 225 412).
94. Hercules Inc., *Belg. Pat.* 877 607 (1980) (*Chem. Abstr.*, 1980, **92**, 182 686).
95. Nippon Oils and Fats, *Br. Pat.* 2 071 116 (*Chem. Abstr.*, 1982, **96**, 8207).
96. Fuji Photo Film, *Jpn. Pat.* 56 001 970 (1981) (*Chem. Abstr.*, 1981, **95**, 15 992).
97. Union Carbide Corp., *US Pat.* 4 242 249 (1980) (*Chem. Abstr.*, 1981, **94**, 104 365).
98. C. B. Arends, D. K. Hoffman and G. C. Kolb, *Polym. Prepr., Am. Chem. Soc., Div. Polym. Chem.*, 1985, **26** (1), 232.
99. Hercules Inc., *Br. Pat.* 1 165 840 (1969) (*Chem. Abstr.*, 1969, **70**, 88 431).
100. Goodyear Tire and Rubber Co., *US Pat.* 4 418 185 (1983) (*Chem. Abstr.*, 1984, **100**, 52 921).
101. Goodyear Tire and Rubber Co., *US Pat.* 4 435 553 (1984) (*Chem. Abstr.*, 1984, **100**, 175 549).
102. B. Vincent, J. Edwards, S. Emmett and A. Jones, *Colloids Surf.*, 1986, **18**, 261.
103. Imperial Chemical Industries Ltd., *Br. Pat.* 1 242 054 (1971) (*Chem. Abstr.*, 1971, **75**, 130 886).

# 16

# Precipitation Polymerization

ALAIN GUYOT

*CNRS, Lyon Vernaison, France*



## 16.1 INTRODUCTION

This chapter deals with radical polymerization of unsaturated monomers giving rise to the formation of polymers which are insoluble in their monomers. In these systems as soon as the polymerization reaction has started, precipitation, or at least coalescence, takes place. Such an immediate precipitation allows a distinction to be made between these systems and the so-called dispersion polymerization (Volume 4, Chapter 10) in which phase separation occurs during the process when the solvent mixture (monomer plus diluent) becomes progressively a non-solvent for the polymer, owing to the consumption of the monomer. In addition, dispersion polymerization involves the use of block copolymers as non-ionic surfactants which stabilize the dispersion of fine particles. However, both precipitation and dispersion polymerization start with an homogeneous system, in which the initiator (radical generator) is soluble. We will be mainly concerned here with bulk polymerization, although a solvent for the monomer may be added in some cases (for instance water in the case of acrylonitrile polymerization). Some of the polymers formed in these processes are very important commercial polymers: poly(vinyl chloride) (PVC), polyacrylonitrile (PAN), poly(tetrafluorethylene) (PTFE), poly(vinylidene dichloride) (PVDC). They can be directly polymerized in bulk (mainly PVC and PAN) or more often in suspension or in emulsion, since these processes allow a good control of the heat generation associated with polymerization, and also lead to a morphology that is easier to handle. Most of the special features of the precipitation polymerization are valid in the case of suspension polymerization, each polymer bead being considered as a small bulk reactor. However, emulsion polymerization is a special case, which will be treated elsewhere (Volume 4, Chapter 8).

Comprehensive studies of precipitation polymerization involve understanding the consequences of the immediate and full precipitation of the polymer on: (1) the polymerization mechanism; and (2) the morphology of precipitating polymers.

The major differences, concerning the mechanism, between homogeneous and precipitation polymerization deal with the termination process, which, in extreme cases, is practically suppressed, so that living polymers are present, it is then possible to prepare block copolymers through a radical mechanism. However, other peculiarities may be observed for other reactions, mainly transfer reactions.

The morphology of the polymer is governed by the aggregation mechanisms, mostly depending on all the parameters which may affect the stability of colloidal particles formed upon precipitation of the polymers.

Although there are some general features in these polymerizations, such as the increasing polymerization rate in the first periods of the process, each polymer has special characteristics, depending on the reactivity of the growing radical, the solubility of the monomer in its polymer, the glass transition temperature of the polymer, and so on.

Basic studies are chiefly restricted to acrylonitrile and vinyl chloride polymerization, but some other cases will be discussed in which heterogeneity of the medium may lead to living polymers.

## 16.2 SPECIAL FEATURES OF THE POLYMERIZATION MECHANISM

In the bulk polymerization of either VC or AN, the kinetic curve (conversion *vs.* time) show an autocatalytic behavior; such a behavior is generally attributed to the 'so called' gel effect (Chapters 2–5) which decreases the termination rate, because the high viscosity of the medium restricts the growing radical mobility. Such an explanation is chiefly valid for homogeneous systems, but the perturbation of the termination process also seems, here, to be the best explanation for the observed kinetic effect.

In AN polymerization, accumulation of radicals occurs making them easily detectable by ESR techniques (electron spin resonance), which means that their concentration is higher than $10^{-7}$ molar. The radical accumulation was explained by Bamford *et al.*[1] in terms of the 'occlusion theory' already reviewed by Jenkins in 1967.[2]

### 16.2.1 The Occlusion Theory

Since the polymer is totally insoluble in the medium, the growing radicals—after being initiated in the liquid medium, where the conventional radical initiators are soluble—soon precipitate from the liquid phase and becomes tightly coiled. If polymer particles are already present in the medium, the growing radical may be captured by the existing particles and then coalesce before self-precipitation. Whatever the process causing heterogeneity, the radical end is more or less buried inside the remainder of the molecule, and then all the reactants have to diffuse there before reaction may take place; the apparent radical reactivity is decreased compared with solution polymerization. Of course, the decrease in reactivity will first affect the bimolecular termination reaction involving another macromolecule, but it may also affect propagation and transfer reactions involving small molecules. The diffusion of these small molecules towards the trapped radical will then be very dependent on their solubility in the polymer and in this respect, the behavior of AN and VC is very different: AN is considered as insoluble in its polymer, while PVC is swollen by rather large amounts of VC monomer (about 30% of its weight at the usual polymerization temperature of 50 °C).[3]

The ability of a polymer radical to be transfered to a smaller molecule, and especially its own monomer, is also important, as far as the occlusion theory is concerned, because small molecules may escape from the buried zone and allow polymerization to continue. In this respect too, AN and VC polymerizations will be very different, owing to the greater reactivity of PVC radicals towards any kind of molecule but especially towards its own monomer; at 50 °C, the transfer constant to the monomer is $1.1 \times 10^{-3}$ for PVC[4] and $1.7 \times 10^{-5}$ for PAN.[5] A third important parameter governing radical trapping is the difference between the reaction temperature and the glass transition temperature, $T_g$, of the polymer–monomer system (which may be very different from those for pure polymer as pointed out by Hamielec);[6] diffusion rates increase greatly above that temperature, so that occluded radicals may become active again if the temperature is raised. The fact that a trapped radical remains sleeping, but living, at low temperature and active upon swelling its polymer or raising the temperature, has the important consequence of giving the possibility of preparing block copolymers. Then, in the following, we will successively consider the polymerization of acrylonitrile (AN) (and related monomers insoluble in their polymer) and vinyl chloride (VC) (which is able to swell its polymer) and the various cases of block copolymer synthesis from living radicals.

### 16.2.2 Acrylonitrile Polymerization

PAN is soluble in dimethylformamide and ethylene carbonate; in these solvents, the radical polymerization is quite normal (first order in monomer concentration and 1/2 order in initiator concentration) provided the monomer concentration is limited.[7] At a concentration higher than 6 mol $l^{-1}$, precipitation is observed at the onset of polymerization and the polymerization rate is strongly accelerated.[8] In bulk polymerization, the reaction rate progressively increases from the beginning of the process and the initiator exponent is around 0.9.

In photopolymerization, greater after-effects are observed, and, upon successive irradiations, progressively longer after-effects are observed for constant irradiation periods and the precipitated polymer remains active for several hours at room temperature. When the polymerization has been carried out at 25 °C, if the temperature is raised to 60 °C in the dark, a very fast polymerization is observed, followed by a sudden decay in polymerization rate. All these effects can be explained by the existence of free radicals trapped at low temperature, which become active again upon heating. The existence of these radicals was established by ESR or by chemical reaction with 2,2-diphenyl-1-picrylhydrazyl (DPPH).[9] Quantitative estimates from ESR studies[10] have shown that the concentration of trapped radicals may be as high as $10^{-4}$ mol $l^{-1}$, so that the rate of the bimolecular termination process can be reduced by seven orders of magnitude. The concentration of trapped radicals decreases markedly on heating (they are no longer visible by ESR above 60 °C) or upon swelling with a good solvent. However, on swelling with small amounts of DMF, or if methyl acrylate is copolymerized with acrylonitrile, the number of trapped radicals goes through a maximum[11] with increasing amounts of DMF or methyl acrylate; limited swelling allows more radicals to be captured inside the precipitated polymer particles, but if the swelling degree increases, the bimolecular termination reaction becomes more and more important and the stationary concentration of radicals is correspondingly decreased.

Other anomalies concern the molecular weight of PAN. It goes through a maximum at about 65 °C, when the polymerization temperature increases.[7] Before the maximum, the termination of polymer growth mainly occurs by monomolecular trapping, which is more efficient at lower temperatures. When the temperature increases, trapping occurs later, and further, the bimolecular termination process becomes efficient and finally dominating.

When *n*-butanethiol is added to the system as a transfer agent,[12] the acceleration in the polymerization rate is less; the small radicals, resulting from the transfer reaction, can escape from the more occluded regions inside the particles, and can even escape from the particles, then the heterogeneous character of the reaction is less important.

Many kinetic studies have been carried out in bulk polymerization or in aqueous solution (acrylonitrile has a high water solubility of 1.5 mol $dm^{-3}$ at 50 °C). Their interpretation is quite difficult because both nucleation of particles and growth of the existing particles take place. Many contradictory results have been published about the main locus of polymerization (in the water phase[13] or inside the particles[14,15]) and the values of the rate constant $k_p$ and $k_t$. These difficulties have recently been overcome using seeded polymerization in water solution.[16] A very complete kinetic study, using both chemical and $\gamma$-ray initiation has been carried out; in the latter case, successive periods of irradiation and relaxation were studied. Several important conclusions were set: the termination reaction is first order because the radicals become trapped and inaccessible to the monomer; the main locus of polymerization is actually the interface between the particles and the aqueous phase; initiation takes place in the aqueous phase (for chemical initiation) and the growing radicals are captured with 100% efficiency by the particles. For seed particles of about 100 nm in radius, the average number of active radicals in each particle is around 250. Of course, each particle may include many more inactive polymer radicals packed in successive layers in the particle shell. The value of the propagation rate constant $k_p$ is estimated to lie between $3 \times 10^2$ and $12 \times 10^2$ $dm^3\,mol^{-1}\,s^{-1}$, a quite standard value for many monomers at 50 °C. It may be concluded that the main mechanistic perturbation in the acrylonitrile polymerization is restricted to the termination mechanism; it mainly occurs by burial of the growing radicals which become inaccessible to the monomer.

### 16.2.3 Tetrafluorethylene Polymerization

Tetrafluorethylene (see also Volume 3, Chapter 25) is also a monomer which is quite insoluble in its polymer. Trapping of radicals was shown to occur in polymerization carried out in aqueous suspension, using a water soluble initiator.[17] Tournut and Grimaud[18] have been able to reinitiate the polymerization from a polymer prepared in suspension and then isolated and washed. The reinitiation takes place even if the polymer has been in contact with air for a rather long period of time (several weeks). The second polymerization must be carried out at temperatures higher than the initial polymerization temperature, but lower than 110 °C. These facts are in agreement with the occlusion theory and tend to show that the solid polymer occluding the radicals has various degrees of packing; at a given temperature the monomer diffusion is limited to certain regions, with more and more regions being accessible on increasing the temperature. ESR spectroscopy shows weak signals that do not correspond to $CF_2\bullet$ or $CF_2{-}O{-}O\bullet$.[18]

Modeling of TFE polymerization has been carried out recently.[19] It was concluded that the high crystallinity of the polymer causes trapping of the radicals and stops the propagation reaction; this monomolecular termination is considered to operate simultaneously with the usual bimolecular termination mechanism and leads to the presence of a few hundred radicals in the same latex particle.

### 16.2.4 Vinyl chloride Polymerization

The polymerization of vinyl chloride, which has been extensively studied and reviewed,[20–22] presents the characteristic features of precipitation polymerization: immediate precipitation, acceleration of the reaction rate up to rather high conversion, after-effect in irradiation polymerization.[23] However, there are many differences from the case of acrylonitrile, which chiefly come from the two features already mentioned: a high value of the transfer constant to the monomer[4] and swelling of the polymer by a rather large amount of monomer.[3] Due to these features, the mobility of the radicals inside the swollen polymer phases is high enough for the normal termination mechanism to take place; the order of the reaction with respect to the initiator concentration is only slightly higher than 0.5, just showing a trend to increase up to 0.56.[24] Thus accumulation of occluded radicals is probably very limited, and the stationary concentration of radicals in the reaction medium is low enough for them to be undetectable by ESR.

The high value of the transfer constant to the monomer was explained recently. The mechanism is not a direct transfer of the growing radicals to the monomer because neither of the possible chain-end structures corresponding to such a mechanism have been detected. The actual mechanism, shown in Scheme 1, first involves a head-to-head addition propagation reaction, immediately followed by an intramolecular rearrangement, creating a chloromethyl end group. The polymer radical then decomposes to give an unsaturated chain end and a chlorine radical, and finally, that chlorine radical is able to reinitiate the polymerization. The basis of that mechanism was initially suggested by Rigo *et al.*[25] to explain the formation of chloromethyl branches in PVC (through reinitiation after the intramolecular 1,2 migration). It was discussed further and established more firmly by the contribution of many authors;[20,26–29] such an interest is justified because both this mechanism and the mechanism of transfer to polymer are able to explain most of the irregular structures found in PVC,[27,30,31] which are suggested to be at the origin of the thermal degradation of that polymer.

*Head-to-head addition* $\sim\sim CH_2\dot{C}HCl + VC \rightarrow \sim\sim CH_2CHClCHCl\dot{C}H_2$

*1,2 migration* $\sim\sim CH_2CHClCHCl\dot{C}H_2 \rightarrow \sim\sim CH_2CHCl\dot{C}HCH_2Cl$ (ii)

*Decomposition (β-scission)* (ii) $\rightarrow CH_2CH{=}CHCH_2Cl + Cl\bullet$

*Reinitiation* $Cl\bullet + VC \rightarrow CH_2Cl\dot{C}HCl \rightarrow$ dichlorinated chain end

**Scheme 1** Actual mechanism of transfer to monomer in vinyl chloride polymerization

It is interesting to point out here the transient existence of chlorine radicals, which are even more mobile than the monomer radicals, and therefore may easily escape from the polymer particles. In addition, they are very reactive and, upon transfer to the polymer, they are able to explain the formation of small amounts of HCl.

Detailed kinetic studies of the VC polymerization did show a sudden decrease in the rate at very low conversion (between 0.2 and 0.5% in a dilatometer)[24] and then the autoaccelerated process takes place. It was observed that this critical conversion corresponds to a partial coalescence of the already existing particles. Their number is then roughly divided by five to stabilize at around $5 \times 10^{11}$ per gram of monomer. That number remains constant up to about 30% conversion; the size of the particles then grows (Figure 1) and becomes unimodal and very narrow.

The critical conversion was shown to be dependent on the agitation, being lower if the stirring speed is higher (less than 0.05% for a peripheral agitation speed of $2\ m\,s^{-1}$).[32] The rate change is also dependent on the agitation; the rate decreases more if the stirring rate is higher.

More insight into the features associated with precipitation and aggregation of the particles have been obtained from a study of concentrated solution of monomer in poor solvents for PVC.[33] When

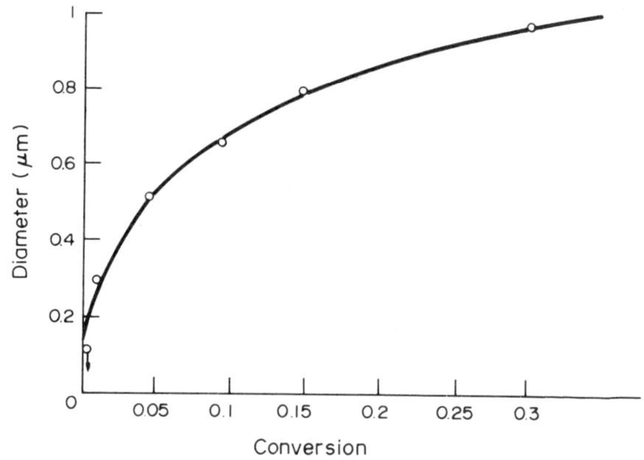


**Figure 1** Particle size of the vinyl chloride polymer as a function of conversion (reproduced from Ref. 24, Figure 9 by permission of John Wiley and Sons Ltd.)

the monomer concentration in diethyl oxalate is lower than 7.5 mol l$^{-1}$, homogeneous polymerization is observed throughout the polymerization. In between 7.5 and 10 mol l$^{-1}$, there is no marked precipitation, but only opalescence appears with increasing conversion. Above 10 mol l$^{-1}$, precipitation is observed after an initial period where the rate increases smoothly, and is associated with a sudden decrease in the rate (Figure 2). This critical conversion is observed at decreasing values of conversion when the monomer concentration increases and must be associated with the aggregation phenomenon observed in bulk polymerization. It is interesting to note that the rate in the initial period is lower than in conditions of homogeneous polymerization, and also the rate decrease associated with the aggregation above the critical conversion increases when the monomer concentration increases.

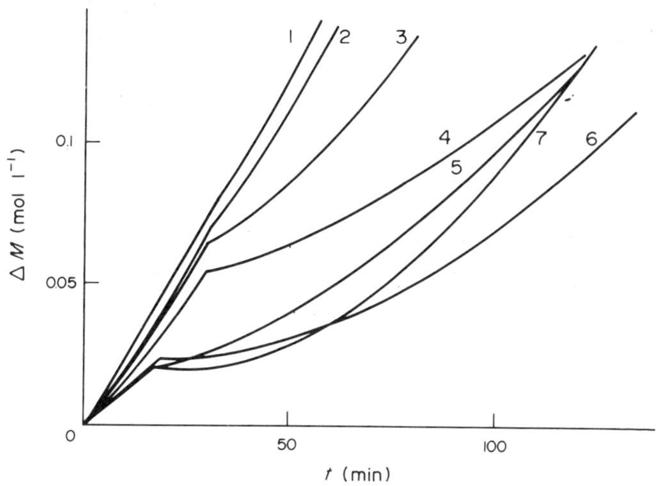


**Figure 2** Polymerization of VC in diethyl oxalate ($t = 25$ °C; initiator [I] $= 3.3 \times 10^{-4}$ mol l$^{-1}$. Monomer concentration (mol l$^{-1}$):(1) 9.81; (2) 10.65; (3) 11.22; (4) 11.63; (5) 12.12; (6) 12.59; (7) 13.05 (reproduced from Ref. 40, Figure 11 by permission of John Wiley and Sons Ltd.)

In the presence of an efficient transfer agent, $CBr_4$, even at low monomer concentration, the initial rate is lowered, but the opalescence is strongly increased. In between 5 and 12 mol l$^{-1}$, for a constant amount of $CBr_4$, the molecular weight increases with the [$CBr_4$]/[VC] ratio, showing an apparent negative value for the transfer constant activation energy. When the monomer concentration increases, the decrease in the initial rate disappears and, in bulk, the initial rate is increased in the presence of $CBr_4$ up to a plateau when the molar ratio $CBr_4$/VC is $8 \times 10^{-4}$.[34] For these conditions, the molecular weight has decreased to a DP of about 25, which seems to correspond to the limit of solubility of the polymer in the monomer.[24] In addition, the critical conversion associated with aggregation, and also the rate acceleration are suppressed in the presence of large amounts of $CBr_4$.

Clearly, the homogeneous reaction is increasingly important when the concentration of $CBr_4$ increases. The experimental data relating to the period before the critical concentration can be rationalized by assuming that the rate of the heterogeneous reaction (responsible for opalescence) is lower than the rate of the homogeneous reaction. This heterogeneous reaction arises from the association of a small number of molecules, and while such an association does not involve the smaller molecules (DP < 25) which remain soluble, it is easier for molecules of moderate molecular weight (effect of $CBr_4$ on the rate and on the opalescence at low monomer concentration). The association involves both dead polymer molecules and growing radicals, the reactivity of which being lowered relative to the propagation reaction and even more so relative to transfer to $CBr_4$, owing to the higher steric hindrance of that molecule. Transfer to $CBr_4$ leads to a small radical coming back to the solution where the more rapid homogeneous reaction takes place, so that on increasing the concentration of $CBr_4$, the rate, initially lowered, may increase again. In the initial period a slight acceleration also takes place; this may be due to a decrease in the rate of termination reaction, caused by the restriction of mobility of the polymer radicals engaged in the association.

At the critical conversion, aggregation of the associated moieties takes place; the severe drop in the reaction rate was interpreted as being mainly due to a fast termination reaction occurring because the concentration of radicals in the small volume of the resulting particle is too high.[24] A new decrease in the propagation rate might also occur but there is no argument to decide if the mobility of the growing radicals is more restricted upon the aggregation of the associated initial particles than in these associated particles. This is because it seems that the initial association should be the basis of the more organized microcrystallite of the polymer (the easier association observed in the presence of $CBr_4$ is in line with the observation that the crystallinity of low molecular weight samples may be higher).[35,36] Bimolecular termination may occur in the particles resulting from aggregation, because they are swollen by monomer; the probability of these reactions is highest when the volume of these particles is lowest.

The number of particles is fixed by the conditions of the aggregation. It depends slightly on the initiation rate, and through that initiation rate, on the polymerization temperature,[32] decreasing with increasing temperature. Using photochemical initiation, Carenza *et al.*[37] also observed a decrease in the particle number on increasing the temperature, and they explained that by the easier coalescence of the swollen particles at higher temperature.

After the aggregation step, the particles just grow and the polymerization rate increases correspondingly with the size of the particles. This acceleration is chiefly explained by a rather slow bimolecular termination mechanism, owing to the high viscosity inside the swollen polymer medium; when the size of the particles increase, the number of radicals simultaneously living inside each particle increases, even if the concentration of radicals inside the polymer phase remains stationary. Such a situation may continue up to about 77% conversion (at 50 °C) when enough monomer remains to saturate the swollen polymer phase. At the end of that period, a sudden increase in the polymerization rate is observed, and is explained as a consequence of the decrease in the termination rate due to an increase in the internal viscosity inside the particles, which results from the decrease in the swelling ratio on consumption of the remaining monomer. This fact is strong support for the statement that the dominating mechanism for the termination reaction continues to be the bimolecular mechanism, obviously being made easier because of the transfer mechanism shown in Scheme 1.

In a study of the morphology of a polymerizing VC droplet suspended in water, Zichy[38] observed the spherical particles at various stages of aggregation. The particle diameter at 2% conversion was observed to be 100 times less than that expected on the basis of the Smoluchowski theory of diffusion controlled flocculation; the conclusion was the existence of a repulsive force to prevent, partly, flocculation. Later on Zichy *et al.*[39] did show that the particles carry negative charges, with a $\zeta$ potential of around $-80$ mV corresponding to about 40 elementary charges per particle. The existence of charged particles was further demonstrated by Davidson and Witenhafer[40] who observed the motion of charged particles in an electric field. Both authors attributed these charges to chloride ions, resulting from the presence of some HCl formed during polymerization. Zichy and Rance,[41] and more recently Törnell *et al.*[42] have used the DLVO theory of electrostatic stabilization of particles to calculate the potential energy against the distance of particle separation in the organic monomer medium, the energy barrier at the maximum is lower than for the commonly studied electrostatically stabilized particles in an aqueous medium (emulsion polymerization) but its decay is very slow with increasing distance. Törnell presented a theory of competitive growth of particles of various sizes by aggregating small particles and was able to explain in that way the very narrow distribution of the particle size observed previously by several authors.[24–43] According to Törnell,[42] at least a proportion of the primary particles precipitating from the liquid phase are charged. A simple charge is too weak to prevent its coagulation with another primary particle, but will cause interactions with larger and more highly charged particles. Thus, the growth of the smaller particles is favored, so that finally all the particles will have about the same size.

The kinetics of polymerization were the subject of many studies.[20,24,34,43–48] Bengough and Norrish[44] showed that the polymer already present in the system can be considered as a catalyst and is responsible for the acceleration of polymerization. They showed that the increase in rate was proportional to the 2/3 power of the amount of polymer already formed and suggested that the polymer is able to absorb the radicals present in the liquid phase proportional to its surface area. The most recent studies[46–48] recognized that the rate is the sum of two terms, one coming from the monomer phase, the second one being associated with the polymer phase. After the nucleation has been completed, the composition of these phases in terms of monomer and polymer remain constant up to the saturation conversion (77% at 50 °C); a very small amount of polymer is dissolved into the monomer and the polymer phase is swollen at saturation by its monomer, hence only the phase volumes change. Talamini[46] was the first to give a complete treatment which is, however, an oversimplification because it does not consider any exchange of radicals between the two phases and admit that the partition coefficient of the initiator between the two phases is unity. He concluded that the ratio $Q$ of the rates between the polymer phase and the monomer phase was constant and estimated it to be 15. More elaborate treatments were given by Ugelstad[48] and Olaj.[47]

For the monomer phase, Olaj considers that radicals were produced in that phase, where the initiator is generally more soluble than in the polymer phase, giving rise to chain propagation in solution. A proportion of the radicals will then be destroyed by termination in solution, while another portion continues to grow as coiled isolated radical chains, which can also be terminated in suspension. However, a large part of these growing radicals are captured by the polymer particles where most of the chain propagation takes place with, finally, a slow termination rate. Olaj also admits that radicals can be produced in the polymer phase, where they grow and terminate. Olaj considers the fact that some radicals can be desorbed from the polymer phase to go into the liquid phase as an infrequent event. Ugelstad takes into account such a possibility, arguing that it mostly involves small radicals generated after a transfer reaction. Ugelstad estimated that the termination rate in the liquid phase is about 50 times larger than in the polymer phase and, on this basis, he proposed that the ratio of rates $Q$ should be about 200. However, the conclusion of Ugelstad was that, even at low conversion, most of the termination reaction occurs inside the particle, despite the fact that its rate was lower than in the liquid phase, and that, due to the fast absorption of growing radicals from the liquid phase and the high mobility of the small chlorine radicals, there is a fast equilibrium distribution of the radicals between the two phases.

While Talamini derived an expression for the rate as

$$-\frac{d[M]}{dt} = a + b[M] \tag{1}$$

both Olaj and Ugelstad derived expressions of the form

$$-\frac{d[M]}{dt} = a + b[M]^{1/2} \tag{2}$$

According to Olaj, the value of $b$ is

$$k_p = \frac{k_i[I]\,d_m \phi_m^{1/2}}{k_{tP} d_p \phi_p} \tag{3}$$

where $k_p$ is the propagation rate constant, $k_i$ the initiation rate constant, [I] the initiator concentration, $k_{tP}$ is the termination rate constant in the polymer phase, $d_m$ and $d_p$ the densities of the monomer and polymer respectively, and $\phi_m$ and $\phi_p$, the volume fraction of the monomer and the polymer phase.

In every case, the contribution of the first term $a$ (which represents the contribution of the liquid phase polymerization) is very small.

The liquid phase is no longer visible beyond about 30% conversion, but still exists in the resin pores. It only disappears at the 'saturation conversion', when all the monomer is dissolved in the swollen polymer. Above that conversion, after a short period where the rate increases, due to the gel effect caused by an increase in viscosity inside the polymer phase, the polymerization rate progressively decreases and vanishes when the conversion is high enough, so that the glass transition temperature of the swollen polymer is equal to the polymerization temperature (around 93% at 50 °C).[6]

That final period of polymerization was mainly studied by Sorvik and Hjertberg[31] in order to understand the formation of irregular structures, suspected to be responsible for the bad thermal stability of the polymer. The Swedish authors simulated the conditions of that final period by carrying out polymerizations at subsaturation pressure in a reactor already containing small PVC particles as a seed. They have estimated the number of branches (chloromethyl, ethyl, butyl and long branches) and of internal double bonds, in different conditions of monomer concentration (VC/PVC ratio) and of temperatures (between 45 and 80 °C). The number of chloromethyl branches increases with both the temperature and the [VC]/[PVC] ratio, while all other kinds of branches also show an increase with the polymerization temperature, but a decrease with increasing values of [VC]/[PVC]. The chloromethyl branches result from a reinitiation of the polymerization after the 1,2 migration step of Scheme 1; its dependence on the [VC]/[PVC] ratio is then expected, while its dependence on the temperature probably reflects the competition between head-to-head and head-to-tail addition in the propagation reaction. All other branches came from intra- or intermolecular transfer to the polymer, which is known to be favored by increasing the temperature and is of course competitive with the propagation, so that it decreases with increasing [VC]/[PVC]. The formation of the internal double bond is more complex; their number decreases with increasing [VC]/[PVC] ratio, but tends to decrease with increasing temperature. According to the Swedish authors, the formation of these internal double bonds can be explained by the reaction shown in Scheme 2.

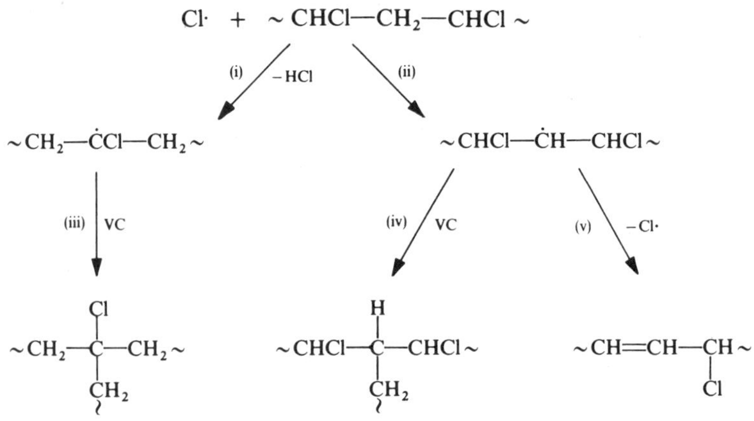

**Scheme 2** Mechanistic scheme for the internal double bonds in vinyl chloride polymerization[31]

Starting from a chlorine radical, there is first a transfer to polymer (reactions i and ii) and then a competition between the decomposition (v) and the formation of long branches (iii and iv). Of course, the transfer to polymer of Cl• radicals is competitive with reinitiation (Scheme 1) and will be favored by a decrease in VC concentration. A possible explanation for the temperature effect might be the selectivity of the transfer reaction (competition between i and ii of Scheme 2); the study of chlorination of PVC pointed out that the methylene group is the preferred site of attack,[49] but on increasing the temperature the other site (CHCl group) would react more frequently, favoring the formation of long branches.

A Monte Carlo simulation of the formation of irregular structures in PVC was recently carried out[29] on the basis of the mechanisms given in Scheme 1 and 2, and also taking into account the accepted mechanisms for transfer to polymer of the normal growing radical, it points out that the experimental number of short and long branches from the data simulating the conditions after the saturation conversion,[50] increases with increasing conversion much more slowly than expected, assuming the propagation reactions to follow a first order law *vs.* the monomer concentration. Actually, the propagation reaction remains much more favorable than expected, this fact tends to show that the mobility of the growing chain end in the polymer phase is very restricted compared with a homogeneous reaction. The restriction in mobility affects the transfer reaction, which needs a polymer chain motion, more than the propagation reaction, which involves more monomer diffusion to the growing macromolecular radical.

### 16.2.5 Block Copolymers from Living Radicals

After having established their occlusion theory, Bamford and Jenkins were the first to predict the possibility of preparation of block copolymers[1] by removing the residual monomer after photoreaction at low temperatures, replacing it with another monomer and heating again at 60 °C. The results greatly depended on the nature of the second monomer; a slow reaction was observed with *n*-propyl or *n*-butyl acrylate without decay of the concentration of trapped radicals; a faster reaction without radical consumption was observed with ethyl acrylate, which probably allows for limited swelling, high enough for diffusion of the monomer towards the radicals, but not sufficient to permit the bimolecular termination process; a very high degree of conversion is observed with methyl acrylate, but with limited decay of the radical concentration. Some additional crosslinking may take place because the final copolymer is not soluble in DMF, a good solvent of both homopolymers. Strong decay of the radical concentration with rapid polymerization was only observed when acrylonitrile again was used as the second monomer.

Later on, Minoura and Ogata[51] produced block copolymers from trapped radicals in polyacrylonitrile, the first step being carried out at 0 °C using $Bu_3B$ as a radical initiator; after drying in vacuum, the second step was carried out at 0 °C again with styrene as a monomer, but using DMF as a swelling agent. In the same paper, the authors describe block copolymers from PVC also prepared at 0 °C, but the second step with styrene and DMF was carried out at 80 °C. Fractional precipitation and elemental analysis (limited to C and H) were given as proof for the formation of block copolymers; it seems, however, that such a conclusion is doubtful in the case of PVC. More credible evidence is given for the PVC–methyl methacrylate system, the two steps being carried out respectively at −78 and +15 °C, with the application of vacuum in between.

Block copolymers were produced from living radicals of PTFE, with either ethylene or VC but not styrene or vinylidene fluoride;[18] most probably nascent PTFE is very tightly coiled, so that the radicals are only accessible to small gaseous molecules.

More recently, a few other systems allowing the preparation of block copolymers from precipitated living polymers were described. Seymour *et al.*[52] first polymerized vinyl acetate (VAC) in silicon oil, a viscous poor solvent in which the poly(vinyl acetate) is precipitated as spherical particles of 4–10 micron diameter; upon polymerization of a second monomer, they obtained in a second step high yields (66–82%) of block copolymers with methyl methacrylate (MMA), acrylonitrile (AN), vinylpyrrolidone, acrylic acid and styrene (S). They also obtained a block terpolymer ABC of VAC–MMA–S from a living block copolymer AB of VAC–MMA. An extensive study of living polymer radicals of *N*-methyl methacrylamide (NMMAm) and their use in the synthesis of block copolymers was published by Otsu.[52] Small particles (100–300 nm) of both NMMAm and NMAm living polymer radicals were obtained upon polymerization in benzene in the presence of di-*tert*-butyl percarbonate, or with AIBN after irradiation. The living radicals were carefully characterized by ESR; quantitative measurements permit the conclusion that up to 92% of the efficient radicals remain alive at moderate temperature. On addition of a second monomer, the ESR signal progressively changes to that characteristic of the second polymer radical. When a copolymerization is carried out in the second step (S–MMA), the composition of the living radicals indicates preferential solvatation inside the particles. The block copolymer of NphenylNAm–MMA shows a very strong phase separation effect in the micron range.

## 16.3 MORPHOLOGY OF THE PRECIPITATING POLYMERS

The morphology of nascent PAN did not receive much attention, because this polymer is mainly used in the fiber industry, where it is wet spun from solutions. The reverse situation is valid for PVC, which is directly used in powder form; depending on the application, the powder must be either very dense (rigid PVC) or highly porous (for easy absorption of plasticizer).

Comparative studies of the initial steps of AN and VC polymerization in quiescent conditions were carried out by Davidson and Witenhafer;[40] optical micrographs of acrylonitrile droplets show that the precipitated polymer quickly agglomerates and forms a tridimensional network without any apparent order (Figure 3).

Under similar conditions, the PVC particles are arranged in very regular arrays (Figure 4). This is a consequence of the electrostatic stabilization of the negatively charged ($Cl^-$) particles in a medium

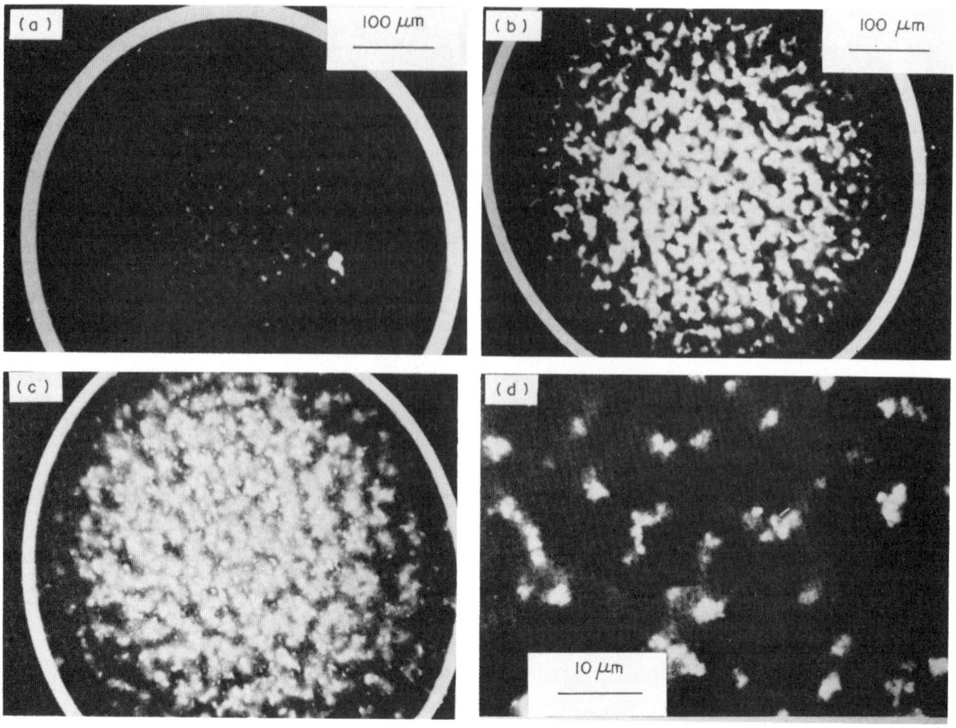


**Figure 3** Single frames of a cine film showing successive stages (a, b, c) in the polymerization of acrylonitrile; dark film optical micrograph (d) showing the nature of the flocs (reproduced from Ref. 40, Figure 11 by permission of John Wiley and Sons Ltd.)

of low dielectric constant. It is possible to modify the electrostatic stabilization by adding an electrolyte which is soluble in VC monomer, such as $Bu_4NBF_4$ or $Bu_4NCl$;[54] the electrical conductivity of the medium is then increased and the particle size may be decreased more than three times, while the particle number is increased by a factor of about forty.

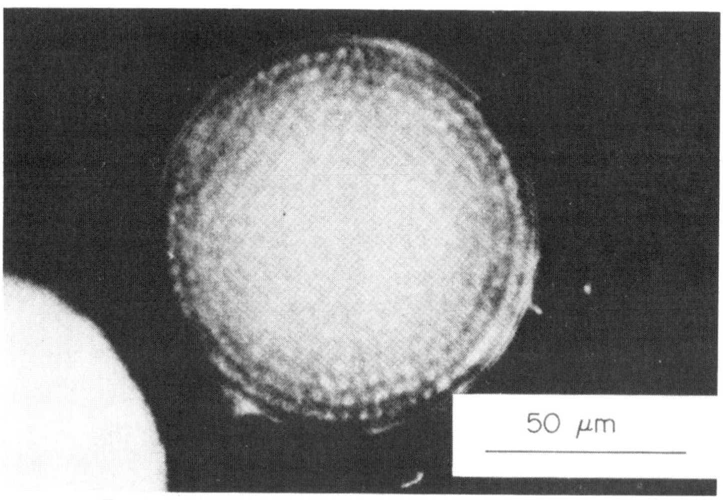


**Figure 4** Single frame of a cine film showing the polymerization of unagitated vinyl chloride droplets (reproduced from Ref. 40, Figure 12 by permission of John Wiley and Sons Ltd.)

Inside a droplet, the agitation plays a major role on the aggregation mode of the primary particles. The aggregation takes place earlier under high speed agitation, and the aggregate tends to have a string structure, as shown in Figure 5. Small charged particles, however, possess a high repulsive electric potential; when a second particle sticks to one, their electric fields superimpose and have a minimum strength at the two end points, so that a third particle will tend to agglomerate at one of these points, and so on. Under mild agitation, the agglomeration takes place later on, when the particles are larger, but then the agglomerates are more spherical (Figure 6). This is possibly because the repulsive potential is lower (at least relative to the particle size) and the electric field less asymmetric. It is also possible that the equilibrium swelling of the polymer by its monomer is not reached before the particle has a size large enough to make the contribution of the curvature to the interfacial energy less dominant.

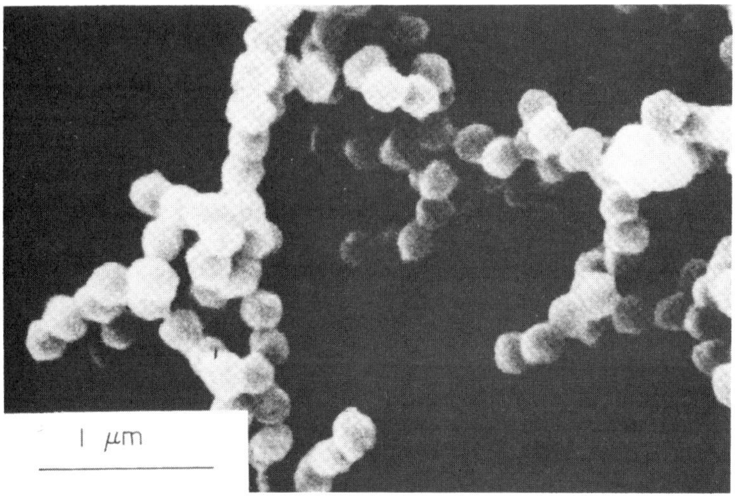


**Figure 5** Agglomerated PVC primary particles as observed at 2% conversion in polymerization experiments with high agitator speed (tip speed 1.5 m s$^{-1}$) (reproduced from Ref. 54, Figure 3 by permission of Butterworth and Co.)

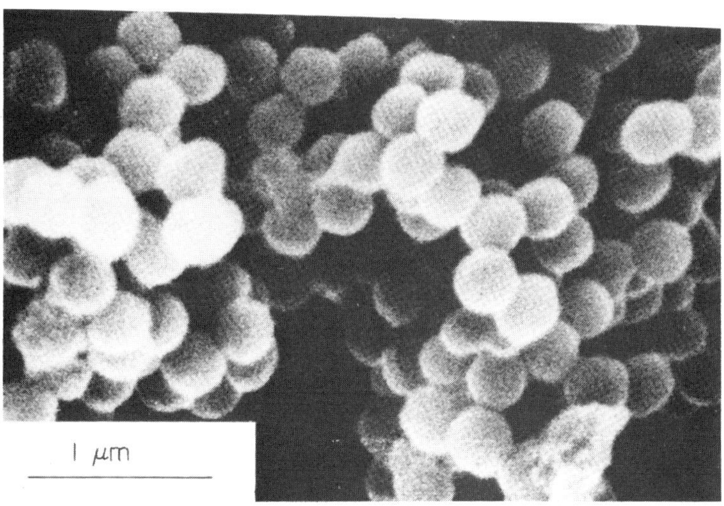


**Figure 6** Agglomerated PVC primary particles as observed at 0.5% conversion in polymerization experiments with no agitation (reproduced from Ref. 54, Figure 4 by permission of Butterworth and Co.)

The formation of the PVC morphology with increasing conversion was described in detail by Allsopp[55] and is summarized in Scheme 3. The polymerization initiation in the liquid VC phase gives growing macroradicals which initially precipitate to give opalescene and undergo a first aggregation step at very low conversion (<0.5%) to form what is called[56] microdomains of 10–20 nm diameter. At conversions around 1–2% (depending on agitation) about 1000 of these microdomains flocculate in particle nuclei of 100–200 nm. Then the number of particles is fixed and large enough so that any new growing macroradical produced in the liquid phase is accreted. These primary particles are large enough to be swollen by the monomer and the microdomains inside coalesce. Furthermore, polymerization takes place mostly inside these particles which grow in size. The swollen particles experience later a third aggregation step, beginning at 4–10%, to give clusters in which the particles remain well separated and just stick together; their individual size being 200–400 nm at that moment. The clustering process continues up to about 20–23% conversion[57] by increasing the size of the clusters to produce a small number of large size clusters. After that agglomeration step, the particles continue to grow to finally reach about 800–1000 nm, and the clusters, more or less fused, reached 2000–10000 nm. These clusters are located inside the grains, which, depending on the process (bulk or suspension) have sizes ranging from 50 to 300 microns. The porosity of the polymer is mainly due to the space between the clusters inside the grains, and also to the space between the particles inside the clusters; the latter tend to be suppressed as the conversion increases, between 75 and 85% conversion, the maximum plasticizer capacity is reduced by 60% while the rate of absorption is reduced by a factor of 2.[58] This seems to be mostly due to the fusing process of the clusters.

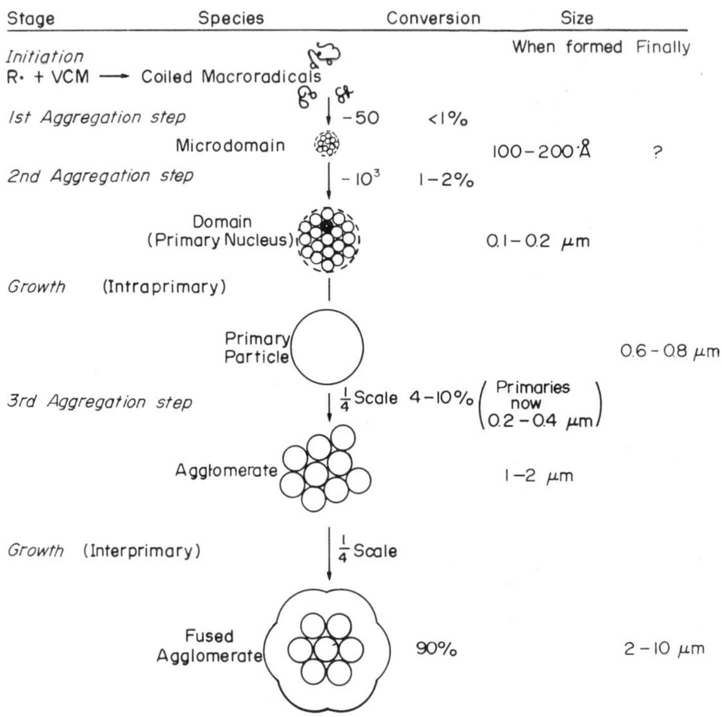


**Scheme 3** Schematic representation of the mechanism of vinyl chloride polymerization: morphology[55]

At present it is uncertain that the memory of the microdomains survives in the final product. One argument for this is the fact that after processing of the polymer, non-deformable microparticles of about 200 nm can be identified more or less linked in a sausage-like object[59] in the whole mass of the polymer; these microparticles are supposed to be composed of internal structure of 20 nm held together by the small crystallites.[60] However, after the second aggregation step, the number of new microdomains coming from nucleation in the liquid phase and then accreted on the existing particles should be limited, because it is believed[20] that polymerization in the liquid phase accounts for a minor part of polymerization, which mainly takes place in the polymer phase inside the swollen particles. So, the understanding of the crystalline texture, and of the formation of the microparticles inside the particles which survive after processing and macrodeformation remains to be determined.

## 16.4 REFERENCES

1. C. H. Bamford and A. D. Jenkins, *J. Chim. Phys. Phys-Chim. Biol.*, 1959, **56**, 798.
2. A. D. Jenkins in 'Vinyl Polymerization', ed. G. E. Ham, Marcel Dekker, New York, 1967, vol. 1, part 1, chap. 6, p. 369.
3. A. R. Berens, *Angew. Makromol. Chem.*, 1975, **47**, 97.
4. G. Vidotto, A. Crosato-Arnaldi and G. Talamini, *Makromol. Chem.*, 1968, **114**, 217.
5. J. Brandrup and E. H. Immergut, 'Polymer Handbook', 2nd edn., Wiley, New York, 1975, p. II 59–61.

6. A. E. Hamielec, R. Gomez–Vaillard and F. L. Marten, *J. Macromol. Sci., Chem.*, 1982, **A17**, 1005.
7. C. H. Bamford and A. D. Jenkins, *Proc. R. Soc. London, Ser. A*, 1953, **216**, 515.
8. G. Vidotto, A. Crosato–Arnaldi and G. Talamini, *Makromol. Chem.*, 1969, **122**, 91.
9. C. H. Bamford and A. D. Jenkins, *Proc. R. Soc. London, Ser. A*, 1955, **228**, 220.
10. C. H. Bamford, J. E. Ingram, A. D. Jenkins and M. R. C. Symons, *Nature (London)*, 1955, **175**, 894.
11. C. H. Bamford, A. D. Jenkins and M. R. C. Symons and M. G. Townsend, *J. Polym. Sci.*, 1959, **34**, 181.
12. C. H. Bamford and E. F. T. White, unpublished results, cited in Ref. 2.
13. F. S. Dainton, P. H. Seaman, D. G. L. James and R. S. Eaton, *J. Polym. Sci.*, 1959, **34**, 209; F. S. Dainton and P. H. Seaman, *J. Polym. Sci.*, 1959, **39**, 279, F. S. Dainton and D. G. L. James, *J. Polym. Sci.*, 1959, **39**, 299; F. S. Dainton and R. S. Eaton, *J. Polym. Sci.*, 1959, **39**, 313.
14. L. H. Pebbles, in 'High Polymer', Wiley, 1964, vol. 18, chap. 9.
15. O. G. Lewis and R. M. King, *Adv. Chem. Ser.*, 1969, **91**, 25.
16. S. J. McCarthy, E. E. Elbing, I. R. Wilson, R. G. Gilbert, H. Napper and D. F. Sangster, *Macromolecules*, 1986, **19**, 2440.
17. S. Sherratt in 'Kirk–Othmer Encyclopedia of Chemical Technology', 2nd edn., ed. Wiley–Interscience, New York, 1966, vol. 9, p. 813.
18. C. Tournut and E. Grimaud, *Ger. Pat.* 2 244 279 (1973) (*Chem. Abstr.*, 1972, **78**, 160 330w) and GFP Symposium, 'Nouveaux Aspects de la Polymérisation radicalaire', Bellevue, 1973, unpublished communication.
19. G. Storti, S. Carra, M. Morbidelli, V. Arcella, G. Broco and E. Gianetti, *J. Fluorine Chem.*, 1987, **35**, 237.
20. J. Ugelstad, P. C. Mork and F. K. Hansenb, *Pure Appl. Chem.*, 1981, **53**, 323.
21. R. H. Burgess, 'Manufacture and Processing of PVC', Applied Science Publishers, London, 1982.
22. G. Butters, 'Particulate Nature of PVC', Applied Science Publishers, London, 1982.
23. M. Magat, *J. Polym. Sci.*, 1955, **16**, 491.
24. J. D. Cotman, Jr., M. F. Gonzales and G. C. Claver, *J. Polym Sci., Part A-1*, 1967, **5**, 1137.
25. A. Rigo, G. Palma and G. Talamini, *Makromol. Chem.*, 1972, **153**, 219.
26. G. S. Park and M. Saleem, *Polym. Bull.*, 1979, **1**, 409.
27. W. H. Starnes, F. C. Schilling, K. B. Abbas, R. E. Cais and F. A. Bovey, *Macromolecules*, 1979, **12**, 556.
28. W. H. Starnes, *Pure Appl. Chem.*, 1985, **57**, 1001.
29. A. Guyot, *Macromolecules*, 1986, **19**, 1090.
30. M. F. Llauro–Darricades, A. Michel, A. Guyot, H. Waton, R. Petiaud and Q. T. Pham, *J. Macromol. Sci., Chem.*, 1986, **A23**, 221.
31. T. Hjertberg and E. M. Sorvik, *J. Polym Sci., Chem.*, 1986, **24**, 1313; *ACS Symp. Ser.*, to be published.
32. J. Boissel and N. Fisher, *J. Macromol. Sci., Chem.*, 1977, **A11**, 1249.
33. M. Ryska, M. Kolinsky and D. Lim, *J. Polym Sci., Part C*, 1967, **16**, 621.
34. J. Breintenbach and A. Schindler, *Monatsh. Chem.*, 1955, **86**, 437.
35. A. Nakajima, H. Namada and S. Hayashi, *Makromol. Chem.*, 1966, **95**, 40.
36. A. Michel and A. Guyot, *J. Polym Sci., Part C*, 1971, **33**, 75.
37. M. Carenza, G. Palma, G. Talamini and M. Tavan, *J. Macromol. Sci., Chem.*, 1977, **11**, 1235.
38. E. L. Zichy, *J. Macromol. Sci., Chem.*, 1977, **A11**, 1205.
39. W. D. Cooper, R. M. Speirs, J. C. Wilson and E. L. Zichy, *Polymer*, 1979, **20**, 265.
40. J. A. Davidson and D. E. Witenhafer, *J. Polym Sci., Polym. Phys. Ed.*, 1980, **18**, 51.
41. D. G. Rance and E. L. Zichy, *Polymer*, 1979, **20**, 266.
42. B. F. Törnell, J. M. Uustalu and B. Jönsson, *Colloid Polymer Sci.*, 1986, **264**, 439.
43. D. N. Bort, Y. Rylov, N. A. Okladnov, B. P. Shtarkman and V. A. Kargin, *Polym. Sci. USSR (Engl. Transl.)*, 1966, **7**, 50.
44. W. I. Bengough and R. G. W. Norrish, *Proc. R. Soc. London, Ser A*, 1950, **200**, 301.
45. H. S. Mickley, A. S. Michaels and A. L. Moore, *J. Polym Sci.*, 1962, **60**, 121.
46. G. Talamini, *J. Polym Sci., Part A-2*, 1966, **4**, 535.
47. O. F. Olaj, *Angew. Makromol. Chem.*, 1975, **1**, 47; *J. Macromol. Sci., Chem.*, 1977, **A11**, 1307.
48. J. Ugelstad, *J. Macromol. Sci. Chem.*, 1977, **A11**, 1281.
49. R. Lukas, J. Svelty and M. Kolinsky, *J. Polym Sci., Polym. Chem. Ed.*, 1981, **19**, 295.
50. T. Hjertberg and E. M. Sorvik, *Polymer*, 1983, **24**, 673, 685.
51. Y. Minoura and Y. Ogata, *J. Polym Sci., Part A-1*, 1969, **7**, 2547.
52. R. B. Seymour and G. A. Stahl, *J. Macromol. Sci. Chem.*, 1977, **A11**, 53.
53. T. Sato and T. Otsu, *Adv. Polym. Sci.*, 1985, **41**, 43.
54. B. F. Törnell and J. M. Uustalu, *Polymer*, 1986, **27**, 250.
55. M. W. Allsopp, *Pure Appl. Chem.*, 1981, **53**, 449, Ref. 2, chap. 7, pp. 151–182.
56. P. H. Geil, *J. Macromol. Sci., Phys.*, 1977, **B14**, 171.
57. A. V. Neimark and L. I. Kheifets, *Vysokomol. Soedin., Ser. A*, 1981, **23**, 1945.
58. J. B. Cameron, A. J. Lundeen, J. H. McCulley, Jr. and P. A. Schab, *J. Appl. Polym. Sci., Appl. Polym. Symp.*, 1981, **36**, 133.
59. O. F. Noel, III and J. F. Bauer, *J. Vinyl Technol.*, 1984, **6**, 143.
60. H. Munstedt, *J. Macromol. Sci., Phys.*, 1977, **B14**, 195.

# 17
# Solid-state Polymerization

BAHATTIN M. BAYSAL
*Technical University of Istanbul, Turkey*



## 17.1 INTRODUCTION

Solid-state polymerization is one of the well-established topics in the field of polymer science explored in the 1960s.The development of solid-state polymerization was largely due to the use of high energy radiation, such as $\gamma$-rays, X-rays, electron beams and sometimes $\alpha$ particles, by polymer chemists. In addition, some polymerizations can be induced by visible light, UV radiation, thermally or with the aid of sensitizers. In particular, since $\gamma$-rays can penetrate into solids, the use of $\gamma$-rays expanded the usable range of temperature to lower temperatures at which monomers exist in the solid state, and led to the production of new types of polymers. The term 'solid-state polymerization' includes all types of polymerization in the solid state, such as crystalline, liquid crystalline and glassy states of pure monomers, mixtures or complexes of monomers with other compounds.

During the last three decades, extensive studies on the solid-state polymerization of many monomers have been reported. Several reviews covering general aspects,[1–5] kinetics and reaction mechanism[6–13] and structural features[14–19] of this subject have already appeared.

Various aspects of solid-state polymerization have been investigated both during irradiation (in-source polymerizations) and after irradiation (post-irradiation reactions). Polymerization may take place during irradiation or on subsequent warming. Free radical, cationic and anionic polymerizations have all been shown to occur. There are various experimental difficulties in following solid-state reactions. For this reason, increasing use is being made of non-destructive techniques, such as optical and electron microscopy, electron spin resonance (ESR), differential thermal analysis, X-ray diffraction, IR and Raman spectroscopy and high-resolution NMR spectroscopy. By using one of the above-mentioned techniques, it is relatively easy to obtain experimental data on molecular structure, molecular packing and molecular vibrations in monomer crystals, as well as on the configuration and conformation of the resulting polymers.

It is well known that the physical state of the crystal, including both the molecular arrangement in the crystal and all kinds of crystal defects, may affect the reaction in the solid state. All reactions and reactions with catalysts in the solids are accelerated at imperfection sites. The crystal imperfections are, commonly, edge dislocations, screw dislocations or some kind of point defect. The inherent imperfections in a monomer crystal are produced by impurities, crystallite sizes and grain boundaries. Additional defects may be produced by high energy radiation.

The data on crystal structure for certain organic monomers which can be polymerized in the solid state are given in Tables 1 and 2 for vinyl and cyclic monomers, respectively. Table 2 includes some other compounds which can polymerize in the solid state. Certain features of the resulting polymeric materials are also given in these two tables.

During the past 20 years, new polymerization reactions that occur in the ordered lattice of appropriate monomeric compounds have been discovered. In favorable cases, macroscopic polymeric single crystals can be prepared and characterized. Physical studies on some of these recently developed materials have resulted in the observation of interesting and potentially useful electrical and optical properties. Consequently, a new aspect of polymer science has been introduced. These new topics in solid-state polymerization are discussed in Volume 4, Chapter 18 and Volume 5, Chapters 14 and 15.

## 17.2 SOLID-STATE POLYMERIZATION OF VINYL COMPOUNDS

The solid-state polymerization of various vinyl compounds has been studied intensively. Acrylamide, methacrylamide, acrylic and methacrylic acids, calcium, barium and alkali metal salts of these acids are among the typical vinyl monomers polymerized in the solid state.

### 17.2.1 Kinetics of Solid-state Polymerizations

Acrylamide is a typical crystalline vinyl monomer and it has been the most thoroughly explored example of solid-state polymerization reactions. Crystalline acrylamide polymerizes in the solid state by $\gamma$-irradiation.[20,21] Detailed kinetic studies have been published on both the in-source and post-irradiation polymerization of this monomer.[22–26] Extensive efforts have been made to clarify the characteristics of the solid-state polymerization of acrylamide, since it is a typical example of a class of monomers. It could be easily crystallized to give good single crystals (an important consideration for related optical studies), and it did not polymerize below $-20\,^{\circ}$C, which made X-ray studies relatively simple.

During the in-source polymerization, the conversion to polymer follows a sigmoid curve with time, *i.e.* there is an induction period, followed by a very rapid increase in polymerization rate and finally a tailing off. This behavior is illustrated in Figure 1, which shows conversion curves for the polymerization of acrylamide at room temperatures during irradiation with $^{60}$Co $\gamma$-rays at different dose rates.[25] The sigmoid curves are typical of many solid-state reactions studied previously, in which reaction is considered to start at specific sites in the crystal (*i.e.* imperfections) and requires the nucleation of a separate product phase in the host crystal.[27]

Various studies on the solid-state polymerization of acrylamide, using various chemical and physical methods, revealed the following experimental findings: (1) the mechanism of polymerization is of a radical type; (2) chain propagation starts with the combination of a monomer molecule and the chain-initiating radical $Me\dot{C}HCONH_2$ at a place, *e.g.* a dislocation, which may still be under the influence of lattice control; (3) chain propagation proceeds at the interfaces between the crystal lattice and polymer phases, where the monomer molecules are released from strict lattice-control; and (4) the resulting polymer is atactic and amorphous.

It was established that the following kinetic relations are consistent with the above experimental findings[25]

$$-\mathrm{d}[\mathrm{M}]/\mathrm{d}t = k_p[\mathrm{M}][\mathrm{R}\cdot] \tag{1}$$

where $k_p$ is the propagation rate constant of the growing radical, [M] is the monomer concentration, and [R·] is the concentration of propagating chains (radicals). If one assumes that $k_p$ is constant, at least at low conversions, the increasing rate can only be explained by an increase in [R·], thus eliminating any steady-state kinetics.

Here, [R·] may be related to the radiation intensity $I$ by the expression

$$[\mathrm{R}\cdot] = KIt \tag{2}$$

where $K$ is a constant composed of an initiation rate constant $k_i$ and an efficiency factor $f$. Instead of equation (1), we can now write

$$-\mathrm{d}[\mathrm{M}]/\mathrm{d}t = k_p k_i f I t[\mathrm{M}] \tag{3}$$

**Table 1** Crystal Data for Some Vinyl Monomers and the Characteristics of their Polymers Obtained by Solid-state Polymerization[15,87,88]

| *Monomer* | *Melting point* (°C) | *Space group* | *Cell dimensions*[a] $a$(Å) | $b$(Å) | $c$(Å) | $\beta$(°) | $Z$[b] | *Polymer* |
|---|---|---|---|---|---|---|---|---|
| MeCHO | −123 | *Pna*2$_1$ | 5.25 | 7.05 | 7.05 | | 4 | Amorphous |
| $CH_2{=}CHCO_2H$ | 11 | *Ibam* | 11.68 | 10.06 | 6.38 | | 8 | Amorphous |
| $CH_2{=}CHCONH_2$ (22 °C) | 64 | *P*2$_1$/*c* | 8.45 | 5.75 | 9.90 | $\beta = 118.7$ | 4 | Amorphous |
| (−100 °C) | | *P*2$_1$/*c* | 8.408 | 5.792 | 9.777 | $\beta = 118.6$ | 4 | |
| (−150 °C) | | *P*2$_1$/*c* | 8.280 | 5.780 | 9.749 | $\beta = 119.62$ | 4 | |
| $CH_2{=}CMeCONH_2$ | 116 | *P*2$_1$/*c* | 8.78 | 6.10 | 10.04 | $\beta = 124.0$ | 4 | Amorphous |
| $(CH_2{=}CHCO_2)_2Ca \cdot 2H_2O$ | | *P*2$_1$/*a* | 14.36 | 6.59 | 11.62 | $\beta = 118.9$ | 4 | Amorphous |
| $(CH_2{=}CMeCO_2)_2Ba \cdot H_2O$ | | *P*2$_1$/*a* | 7.75 | 12.1 | 11.5 | $\beta = 90.5$ | 4 | Amorphous |
|  | 147 | *Pbca* | 7.05 | 9.7 | 25.5 | | 8 | Amorphous |
|  | 178 | *Ima*2 | 8.04 | 5.3 | 28.0 | | 4 | Amorphous |
| $CH_2{=}CHOCO(CH_2)_{16}Me$ | 34.6 ~35.6 | *Pa* | 5.57 | 7.25 | 28.8 | $\beta = 118$ | 2 | Side chain crystallized |
| 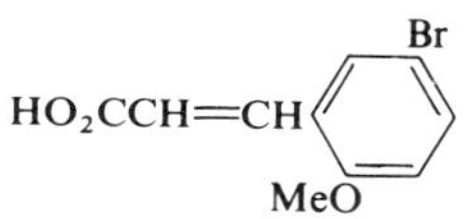 | | *P*2$_1$/*a* | 18.01 | 4.0 | 15.71 | $\beta = 117.9$ | | |
| $NH_2(CH_2)_5CO_2H$ | 207 | *P*2$_1$/*c* | 8.555 | 5.872 | 15.274 | $\beta = 103.1$ | 4 | Oriented crystalline (nylon 6) |
| | | *P*2$_1$/*c* | 8.565 | 5.888 | 15.286 | $\beta = 103.12$ | | |
| $CH_2{=}CHOCOMe$ | −81 | *Pnma* | 12.77 | 6.30 | 6.03 | | 4 | |
| $({-}N{=}CCl{-})_3$ | 146 | *C*2/*c* | 12.92 | 7.46 | 7.57 | $\beta = 60.9$ | 4 | |
| $({-}N{=}PCl_2{-})_3$ | 113.5 | *Pnam* | 14.15 | 12.00 | 6.19 | | 4 | Amorphous (rubber-like) |

[a] 1 Å = $10^{-10}$ m.
[b] Number of molecules per unit cell.

**Table 2** Crystal Data for Some Cyclic and Polymerizable Monomers and the Characteristics of their Polymers Obtained by Solid-state Polymerization[15]

| *Monomer* | *Melting point (°C)* | *Space group* | *a* (Å) | *b* (Å) | *c* (Å) | $\alpha,\beta,\gamma$ (°) | $Z^b$ | *Polymer* |
|---|---|---|---|---|---|---|---|---|
| | | | *Cell dimensions*[a] | | | | | |
| $(\text{—}CH_2S\text{—})_3$ | 215 | $Pm_n$ | 7.668 | 7.003 | 5.285 | | 2 | Oriented crystalline (17/9 helix) |
| $(\text{—}CH_2S\text{—})_3$ | | $Pmn2_1$ | 7.985 | 7.265 | 5.442 | | 2 | Oriented crystalline (2/1 helix) |
| $(\text{—}CH_2O\text{—})_3$ | 64 | $Rc$ | 9.320 | 9.320 | 8.196 | $\gamma = 120$ | 6 | Oriented crystalline (9/5 helix) |
| $(\text{—}CH_2O\text{—})_4$ | 114 | $C2/c$ | 11.455 | 4.160 | 12.232 | | 4 | Oriented crystalline (9/5 helix) |
| $(\text{—}CH_2O\text{—})_5$ | 61 | $Pbcn$ | 8.135 | 10.642 | 7.645 | | 4 | Oriented crystalline (9/5 helix) |
| $(\text{—}CH_2O\text{—})_6$<br>$(\text{—}CH_2O\text{—})_n$ | 72.5<br>178–187 | $R3$<br>$P3_1$ | 7.917<br>4.47 | 7.917<br>4.47 | 11.345<br>17.39 | $\gamma = 120$<br>(fiber axis) | 3 | Oriented crystalline (9/5 helix) |
| 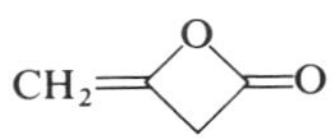 | −6.5 | $P2_1/c$ | 4.00 | 20.67 | 5.11 | $\beta = 101.8$ | 4 | Uniaxially oriented crystalline |
| $(\text{-}SiMe_2O\text{-})_3$ | 64 | $R3m$ | | | | | | |
| 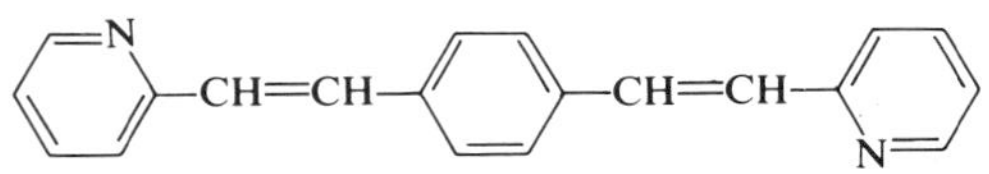 | 232 | $Pbca$ | 21.060 | 9.567 | 7.311 | | 4 | Oriented crystalline |
| 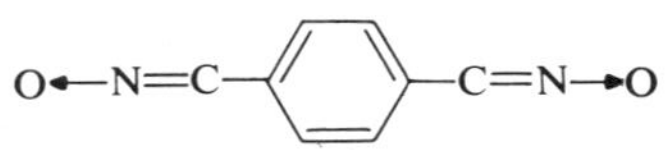 | | | 11.33 | 6.95 | 4.63 | $\alpha = \gamma = 90$<br>$\beta = 104$ | 2 | Amorphous |
| 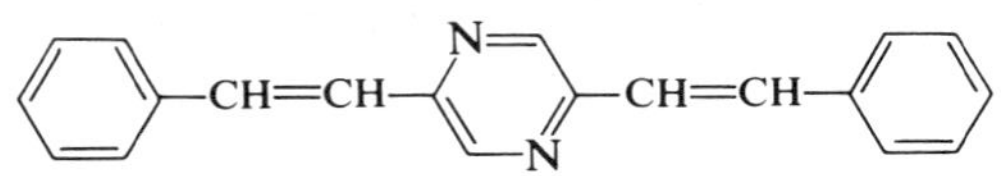 | 236 | $Pbca$ | 20.638 | 9.599 | 7.655 | | 4 | Oriented crystalline |

[a] $1\ \text{Å} = 10^{-10}$ m.
[b] Number of molecules per unit cell.

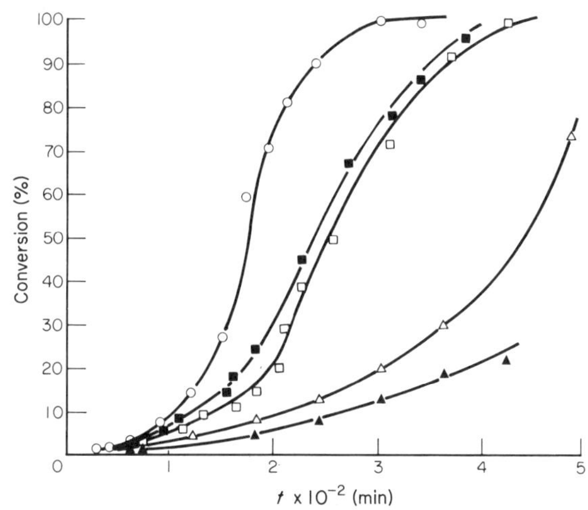


**Figure 1** Polymerization of acrylamide: ○, 0.650 Mrad h$^{-1}$, 27 °C; □, 0.410 Mrad h$^{-1}$, 27 °C; ■, 0.410 Mrad h$^{-1}$, 27 °C, solid solution (10% propionamide); △, 0.221 Mrad h$^{-1}$, 24 °C; ▲, 0.160 Mrad h$^{-1}$, 24 °C (1 rad = $10^{-2}$ J kg$^{-1}$) (reproduced by permission of J. Wiley & Sons from ref. 25)

which in integrated form is

$$\ln([M]_0/[M]) = \frac{1}{2} k_p k_i f I t^2 \tag{4}$$

Figure 2 shows that equation (1) holds up at different temperatures and radiation intensities for the in-source polymerization of acrylamide up to about 20% conversion.[25] An activation energy of 12 kJ mol$^{-1}$ was calculated from the runs at different temperatures using the data at low conversions.

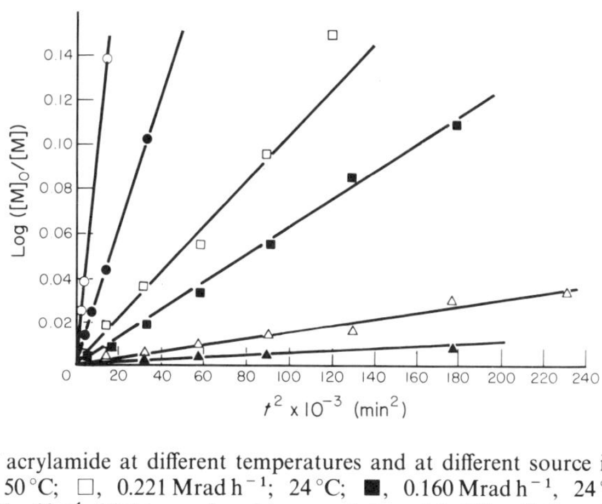


**Figure 2** Polymerization of acrylamide at different temperatures and at different source intensities: ○, 0.221 Mrad h$^{-1}$, 50 °C; ●, 0.160 Mrad h$^{-1}$, 50 °C; □, 0.221 Mrad h$^{-1}$; 24 °C; ■, 0.160 Mrad h$^{-1}$, 24 °C; △, 0.221 Mrad h$^{-1}$, 0 °C; ▲, 0.160 Mrad h$^{-1}$, 0 °C (reproduced by permission of J. Wiley & Sons from ref. 25)

The kinetic picture for the post-irradiation solid-state reaction of acrylamide is somewhat simpler, because [R·] in the equation for propagation is believed to be essentially constant and will be called $[R\cdot]_0$. Here, $[R\cdot]_0$ is the concentration of effective free radicals in the irradiated sample at the time of its removal from the radiation source. The constancy of $[R\cdot]_0$ is not the result of the usual homogeneous polymerization steady-state balance, between initiation and termination, but results from the absence of any effective radical termination step in the post-irradiation reaction. The integrated form of the post-polymerization reaction is

$$\ln([M]_0/[M]) = k_p[R\cdot]_0 t \tag{5}$$

or

$$\ln\left(\frac{1}{1 - x}\right) = k_p[R\cdot]_0 t \tag{6}$$

where $x$ indicates the value of conversion to polymer. Equation (6) is also expressed in the following form[28]

$$\ln\left(\frac{a}{a - x}\right) = k_p[\mathrm{R}\cdot]_0 t \tag{7}$$

where $a$ represents the value of the limiting conversion.

The conversion curves for the post-polymerization of acrylamide at room temperature at different total doses are given in Figure 3. Figure 4 shows that equation (5) holds for total doses from 0.25 to 2.0 Mrad (1 rad $= 10^{-2}$ J kg$^{-1}$). The activation energy for the post-irradiation reaction was calculated from data at low conversions and found to be 42 kJ mol$^{-1}$.[25]

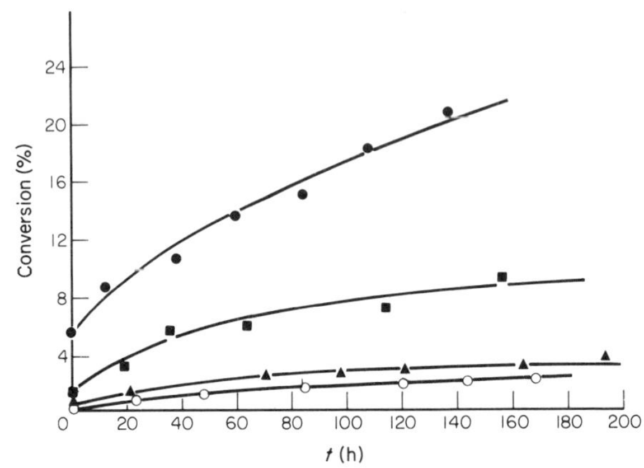


**Figure 3** Post-irradiation polymerization of acrylamide irradiated in a 0.160 Mrad h$^{-1}$ source (irradiation temperature 0 °C, polymerization temperature 24 °C), total dose: ●, 2.0 Mrad; ■, 1.0 Mrad; ▲, 0.50 Mrad; ○, 0.25 Mrad (reproduced by permission of J. Wiley & Sons from ref. 25)

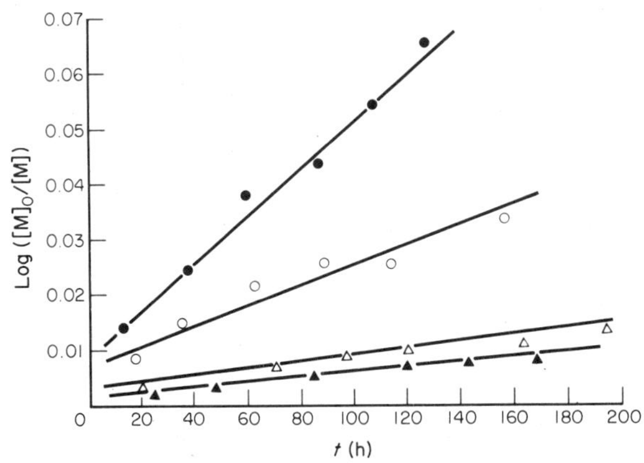


**Figure 4** Post-irradiation polymerization of acrylamide at 0 °C with different total doses: ●, 2.0 Mrad; ○, 1.0 Mrad; △, 0.50 Mrad; ▲, 0.25 Mrad (reproduced by permission of J. Wiley & Sons from ref. 25)

It was demonstrated by ESR that the radical concentration remains high while the rate of polymerization decreases to a very low value.[24,26] It might be argued that the discrepancy between radical concentrations and rates of polymerization is evidence against radical propagation. However, the correlation between changes in the ESR spectrum and rates of polymerization is indicative of free radical propagation, although this does not constitute proof.[11] On the other hand, ionic propagation in solid-state polymerization of acrylamide is unlikely on chemical grounds, since anionic propagation may be expected to lead to the incorporation of $\beta$-alanine residues in the polymer through a proton-migration mechanism as a result of reaction through the amide group;[29] no evidence of this type has been quoted. In other solid-state polymerizations, where reaction almost certainly proceeds through radical intermediates, continuously decreasing rates in the post-irradiation reaction have been observed even though the radical concentration remains effectively constant;[30] the apparent reactivity of the radicals decreases as the reaction proceeds.

It has been demonstrated by detailed and careful studies that a distribution of reaction sites with different properties may be involved in solid-state polymerizations.[31,32] Therefore, it is doubtful if there is any justification for relating parameters in empirical kinetic relations, such as equations (1)–(7), to rate constants for individual processes such as initiation and propagation. The average concentration of reactants may also be of little significance. Changes of temperature will affect the number of defects capable of supporting reaction, and the overall activation energy will contain both physical and chemical contributions.[11] In any particular monomer and crystalline sample, it is difficult to know how the magnitude of the activation energy should be interpreted. The individual rate coefficient and activation energy related to this individual process must be a mean value.[31]

### 17.2.2 Reaction Sites

Formation of polymer from a vinyl monomer involves a change in hybridization of the alkenic carbon atoms of the monomer. This, in turn, requires a change in the geometrical disposition of the subsequent groups. Although the polymers are often denser than the monomers it is fairly obvious, from a general consideration of organic crystal structures, that in most monomers there will be insufficient free volume available in the perfect crystals to allow these configurational changes to occur. Therefore, it is not reasonable to expect polymerization to occur in the perfect regions of the crystal lattice. Consequently, it appears likely that reactions are initiated at imperfections and that polymerization takes place initially in the more imperfect regions of the monomer lattice. The polymeric material obtained is amorphous.

X-Ray studies on $\gamma$-ray induced solid-state polymerization of acrylamide demonstrated that, as polymerization proceeds, there is a weakening of the X-ray diffraction pattern of the monomer and the gradual appearance of diffuse scattering by amorphous polymer.[33,34] However, the original monomer diffraction pattern is still visible at 90% conversion. These results suggest that, throughout the polymerization, amorphous polymer is produced as a separate phase in the crystal[34] (Figure 5).

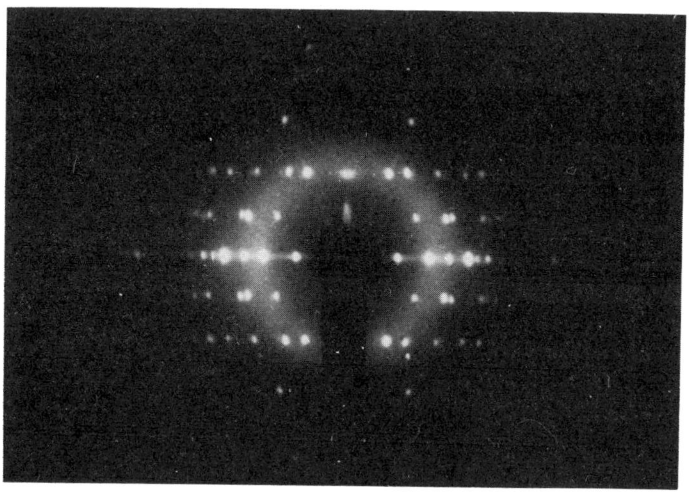

**Figure 5** Rotation diagram of acrylamide crystal irradiated in a $^{60}$Co source, copper radiation, nickel filter[8,34] (reproduced by permission of the American Institute of Physics from ref. 34)

Examination of irradiated and partially polymerized acrylamide crystals with a phase contrast and polarizing microscope reveals that it is possible to follow visually the progress of the polymerizing front. This progress did not take place smoothly at a constant velocity but occurred in jumps.[26,35] It was also observed that polymerization proceeded more rapidly in the vicinity of a scratch on the crystal surface.[26] Studies on acrylamide crystals at various conversions by electron microscopy have shown that polymerization does not occur uniformly throughout the crystal. In the early stages of the reaction, small globular particles of amorphous polymer, of the order of 20 Å diameter (1 Å = $10^{-10}$ m), appear on the surface of the crystals and line up along a preferred crystallographic direction.[36,37] As polymerization proceeds, the particles retain their form but grow in diameter to about 300–400 Å. Eventually, the particles touch each other to give a fibrillar structure.

Studies on the effect of pressure on monomeric crystals indicate that polymerization occurs in dislocations. It was observed that, during the solid-state polymerization of acrylic and methacrylic acid (under UV irradiation), application of a small stress ($\sim 10\,\mathrm{kg\,cm^{-2}}$) on the crystals removes the dislocations from reaction sites, leaving the radicals in an environment where reaction cannot take place.[38,39] For small pressures, the magnitudes of the effects are very sensitive to experimental conditions such as temperature range and the concentrations of small impurities.[38–40]

The influence of high pressures [$5 \times 10^3$–$36 \times 10^3$ atm (1 atm = 101 325 Pa)] on the in-source[41,42] and post-irradiation[43,44] polymerization of crystalline acrylamide and methacrylamide, crystalline and amorphous anhydrous calcium acrylate and crystalline anhydrous barium methacrylate[42,45] have been reported. It was suggested that the main influence of pressure is on propagation. In the polymerization of acrylamide, the lower rates in the early stages are due to the pressure opposing disruptive stresses produced in the lattice by polymerization. However, at higher conversions, the pressure reduces the free volume produced on polymerization of acrylamide and hence increases the rate of propagation. On the other hand, the acceleration produced in the crystalline salts is attributed to the removal of defects, which would be consistent with polymerization occurring mainly in the perfect lattice.[11]

The rates of solid-state in-source polymerization of methacrylonitrile are reduced by a pressure of $5 \times 10^3$ atm.[44] The polymerization of certain fluorocarbons,[46] styrene[47] and methyl methacrylate[47] have been studied under pressures sufficiently high to solidify the monomers at relatively high temperatures; under suitable conditions, using benzoyl peroxide as initiator, styrene and methyl methacrylate polymerized explosively.[47]

### 17.2.3 Nature and Concentration of Free Radicals

The ESR technique has been applied to a number of solid-state polymerizations in order to detect the nature and concentration of radical species. Detailed studies of changes in the ESR spectra and radical concentrations under various experimental conditions provide information related to the mobilities of monomer molecules in the immediate vicinity of initiating and growing radicals.

Free radicals are produced in a number of crystalline vinyl monomers,[48] including acrylamide,[26,31] methacrylamide,[49] acrylic acid,[50] methacrylic acid,[51] itaconic acid (3-carboxy-3-butenoic acid),[52] barium methacrylate,[53] *N*-*t*-butylacrylamide[54] and *N*-vinylcarbazole,[55] when subjected to high energy irradiation at liquid nitrogen temperatures. A detailed account of the application of ESR spectroscopy in polymer research covering the years 1960–1975 has been collected in a comprehensive volume.[56]

Radiolytic hydrogen atoms are produced by interaction of ionizing radiation with the monomer. ESR spectra of the primary radicals are consistent with hydrogen-atom scavenging by the double bond of the monomer.[53,57]

It was found that when a single crystal of acrylamide is irradiated with $\gamma$-rays at $-196\,°\mathrm{C}$ the resulting ESR spectrum consists of two overlapping quartets, which is the pattern for a free radical of structure (**1**). This was confirmed by comparing ESR signals of acrylamide and propionamide crystals irradiated and studied under similar conditions at $-196\,°\mathrm{C}$ (Figure 6). The signal of radical (**1**) is quite stable at low temperatures below about $-125\,°\mathrm{C}$, and it is oriented within the crystal lattice in a manner consistent with the crystal structure.[31] On warming a crystal of acrylamide, irradiated at $-196\,°\mathrm{C}$, the first major change in the ESR spectrum occurs at temperatures above $-125\,°\mathrm{C}$; the original 5-line spectrum changes to a triplet (Figure 7). This new spectrum can be attributed to the addition of a monomer molecule to the original radical producing a species (**2**), formally identical with the propagating radical. No polymerization occurs at these low temperatures, and from crystallographic considerations it has been suggested that the primary radical adds to a nearest-neighbor molecule in the direction of the two-fold screw axis of the crystal.[58]

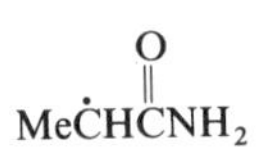

(**1**)

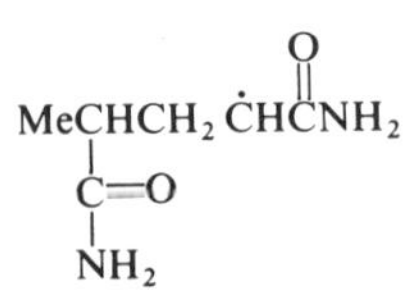

(**2**)

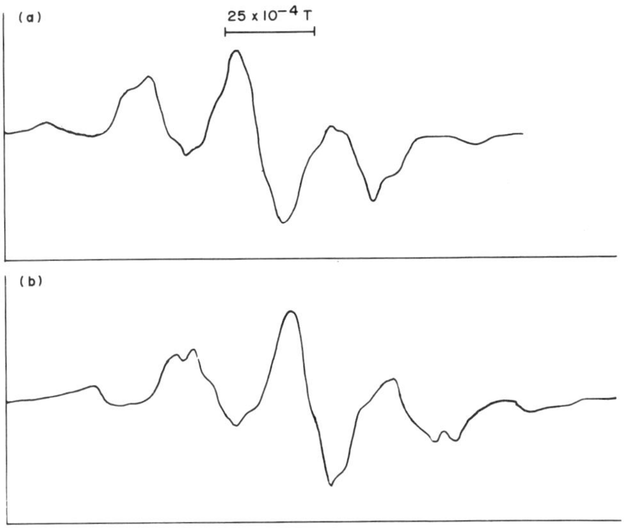


**Figure 6** ESR spectra of (a) acrylamide and (b) proprionamide irradiated and stored at $-196\,^{\circ}C$; radiation dose 20 Mrad (reproduced by permission of the American Chemical Society from ref. 31)

The change of primary radical (**1**) to radical (**2**) occurs at a temperature $T_1$ at which polymerization scarcely takes place in most monomers. Table 3 lists the values of $T_1$ reported for various monomers together with the temperature $T_2$ at which polymerization is found to begin.[13]

**Table 3** Temperatures $T_1$ at which a Primary Radical Adds to the Double Bond of the Monomer and Temperatures $T_2$ at which Polymerization Begins for Various Monomers[13]

| *Monomer* | $T_1$ (°C) | $T_2$ (°C) |
|---|---|---|
| Acrylamide | −125 | −30 to −20 |
| Acrylonitrile (annealed) | −120 | −120 |
| Acrylonitrile (quenched) | −140 | −140 |
| Acrylic acid | −100 | −25 |
| Methacrylic acid | −5 | > −5 |
| Barium methacrylate | −80 | +25 (?) |
| Methyl methacrylate | −130 | |
| Formaldehyde | −196 | Polymerizes by an ionic mechanism |
| Acetaldehyde | −196 | Polymerizes by an ionic mechanism |

Further warming of acrylamide crystals to about $-40\,^{\circ}C$ causes a decrease in radical concentration and a slight change in the spectrum. In the temperature range $-30$ to $-20\,^{\circ}C$ all anisotropy in the ESR spectrum is lost. It is at this stage that the mobility of the molecules in the vicinity of the radical becomes sufficiently great to allow further propagation (**3**), and the nucleation of polymer particles as a second phase.[8,26,35] Post-polymerization is only observed at temperatures above $-20\,^{\circ}C$. Irradiation of acrylamide at $20\,^{\circ}C$ also leads to the formation of free radicals which exhibit an ESR triplet spectrum.[59] The observed change of ESR spectrum which takes place over a fairly wide temperature range[31] is consistent with the existence of a range of imperfections in the solid-state polymerization of methacrylic acid.[32]

$$\sim\!\sim CH_2\dot{C}H\overset{\overset{\displaystyle O}{\|}}{C}NH_2$$

(**3**)

Similar results are obtained with irradiated acrylic acid. The primary $Me\dot{C}HCO_2H$ radicals are oriented in the crystal in two distinct directions, corresponding to the orientations of the unreacted vinyl groups in the monomer crystal lattice.[50] Molecular motions in the crystalline region become detectable at about $-30\,^{\circ}C$, and increase steadily with increasing temperature.

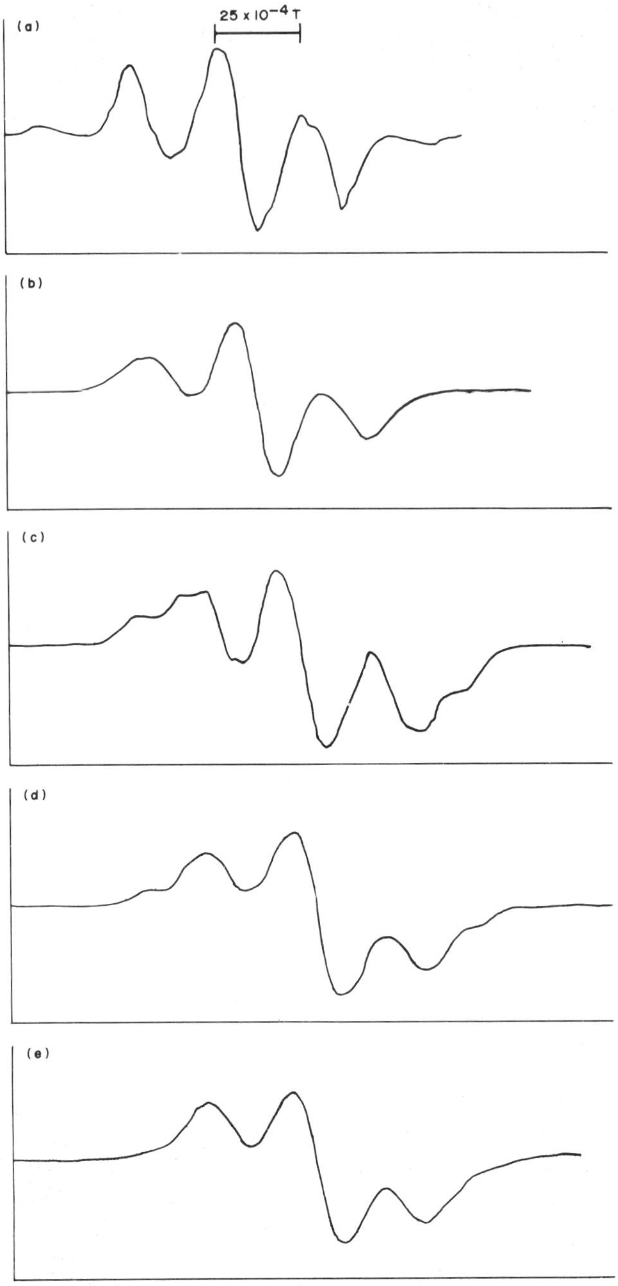


**Figure 7** ESR spectra of acrylamide irradiated at $-196\,°C$ and stored for 16 h at (a) $-145\,°C$, (b) $-80\,°C$, (c) $-40\,°C$, (d) $-30\,°C$ and (e) $-20\,°C$: radiation dose 20 Mrad. The ESR spectra of propionamide at $-145\,°C$ and above are essentially similar to that of acrylamide at $-145\,°C$ (reproduced by permission of the American Chemical Society from ref. 31)

Free radicals are readily observed during the solid-state polymerization of methacrylic acid at all temperatures up to the melting point (16 °C).[30,60] UV radiation of this monomer above 0 °C gives rise to a 9-line ESR spectrum of the propagating radical, which arises from the single radical species ~~C(β)$H_2\dot{C}$(α)$MeCO_2H$ in which there is hindered rotation about the C(α)—C(β) bond. Consequently, certain preferred conformations are adopted in which the methylene protons interact with the unpaired spin to different extents. Only the α-methyl and β-methylene protons contribute to the hyperfine structure. Figure 8 represents a projection along the C(α)—C(β) bond, where H(1) and H(2) are the methylene hydrogens and R the remainder of the growing chain, $\theta_1 + \theta_2 = 120°$.

In the symmetrical conformation, $\theta_1 = \theta_2 = 60°$, $\phi = 0°$ and H(1) and H(2) are equivalent. It was proposed that the 9-line spectrum arises from two preferred conformations, $\phi = 0°$ and $\phi = 15°$,

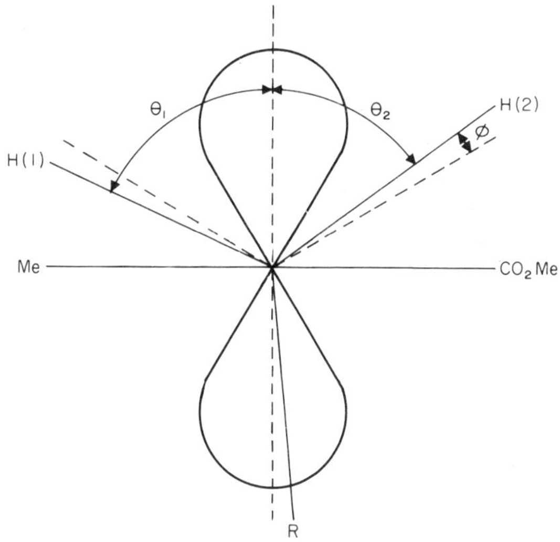


**Figure 8** Radical conformations in methacrylic radicals, showing definition of $\phi$[8,11]

present in different proportions.[61] This view has been confirmed by using monomers containing radicals in solution.[11,63]

Irradiation of methacrylic acid (crystallized below about 0 °C) at lower temperatures (~ − 50 °C) gives rise to a 13-line ESR spectrum which was attributed to the existence of a preferred conformation[60] with $\phi = 5°$ (Figure 8). This restricted conformation results from the lattice forces and the low molecular mobility at these lower temperatures. As the temperature of irradiation is raised to about − 5 °C the spectrum gradually changes from 13 to 9 lines, as the molecular mobility increases.[51,60] Polymerization is more rapid under conditions where the 9-line spectrum is observed.[30,40]

Irradiation of itaconic acid produces the primary radical $Me\dot{C}(CO_2H)CH_2CO_2H$, which gives an ESR spectrum similar to that in Figure 8 (R = —$CO_2H$), with $\phi = 10°$.[52]

The radiation-induced solid-state polymerizations of barium methacrylate dihydrate,[52,66,68] barium methacrylate monohydrate[69,70] and barium methacrylate anhydrate[71] have been studied in detail. The ESR spectrum indicated that the primary radical in the $\gamma$-irradiated barium methacrylate at − 196 °C is $Me_2\dot{C}CO_2^-$ in which both methyl groups are rotating.[53] From single crystal studies, it has been shown that the restricted conformation exhibited by low temperature radicals is stable to a temperature of about 50 °C, below which polymerization is very slow. Above this temperature, the spectrum changes gradually to an isotropic 5-line spectrum corresponding to a propagating radical in a preferred conformation with $\phi = 15°$; under these conditions polymerization is much more rapid. It has also been shown, from ESR spectra of a $D_2O$ salt of barium methacrylate, that the major initiating radicals in the monohydrate were formed by the addition of hydrogen atoms from the water of crystallization to the double bond.[69,70] On the other hand, the yields of high molecular weight polymer in the radiation-induced solid-state polymerization of barium methacrylate anhydrate are negligible if polymerization is prevented during the isolation of residual monomers. However, the crystal lattice was shown by X-ray diffraction to be largely destroyed during reaction. ESR spectra showed that the initial trapped radicals are converted to propagating radicals having a conformation different from that reported for barium methacrylate monohydrate.[69,71]

Low temperature irradiation of other monomers also leads to the formation of primary radicals corresponding to the addition of radiolytic hydrogen atoms to monomer. With acrylonitrile[48,72] and methacrylamide,[48,49] extra lines seems to exist in the ESR spectra which correspond to extra stable radicals at low temperatures.

The solid-state polymerization of various organic monomers is generally assumed to proceed by a radical mechanism. However, it is possible that ionic intermediates play a part in some of these polymerizations. The presence of radical cations, and propagation by a cationic mechanism, has been proposed for the solid-state polymerization of isobutene[1,73] (initiated by $\gamma$-rays at − 196 °C).

It has been suggested that the solid-state polymerization of conjugated dienes[74] (initiated by electron bombardement at − 196 °C) and vinyl carbazole[75] (at 20 °C, initiated by added electron acceptors which form charge-transfer complexes with the monomer) proceeds by a cationic mechanism, while the polymerization of nitroethylene[76] in 2-methyltetrahydrofuran ($\gamma$-irradiated at $-196\,^{\circ}$C) proceeds by an anionic mechanism.

### 17.2.4 Topotactic Reactions in the Solid State

A considerable contribution to the field of solid-state polymerization was made by the systematic studies on chemical reactivities and crystal structures. Crystal structures of various alkenes such as cinnamic, muconic and sorbic acid (3-phenylpropenoic, 2,4-hexadienedioic and (*E,E*)-2,4-hexadienoic acid, respectively), were determined and UV-light-induced cyclodimerization reactions of these compounds were studied.[15,16,77,79]

*trans*-Cinnamic acid shows three crystalline modifications. In the $\alpha$ form (**4**), molecules are related by a center of symmetry and by dimerization (equation 8) give $\alpha$-truxillic acid (**5**); in the $\beta$ form (**6**); molecules related by translation give $\beta$-truxinic acid (**7**; equation 9); on the other hand, the $\gamma$-form cannot be dimerized.[15] In the $\alpha$ and $\beta$ forms, the intermolecular alkenic bonds are 3.7 and 4.0 Å apart, while in the $\gamma$ form the distance is 4.7 Å.[77–79]

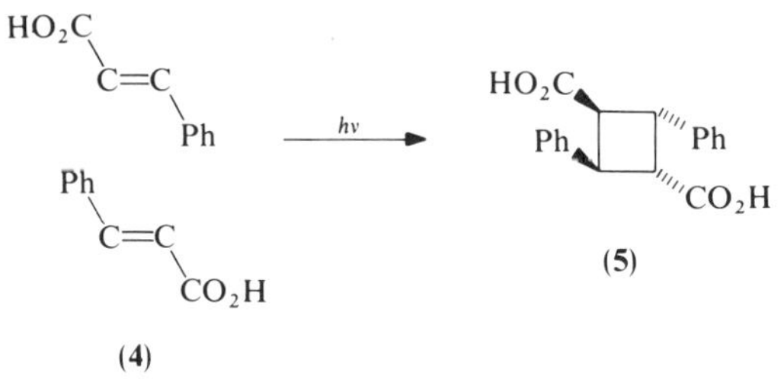


(8)

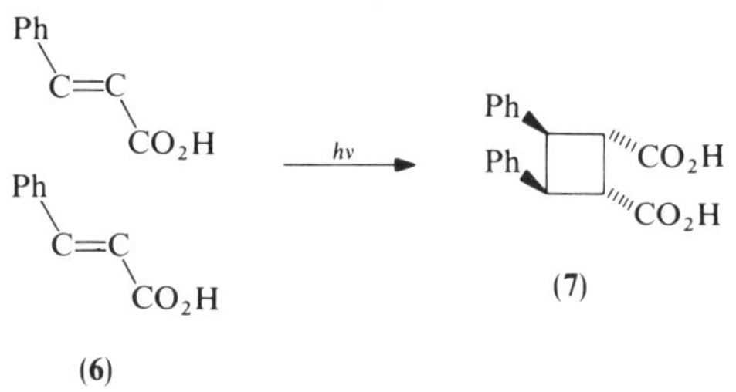


(9)

Generally, only those monomers for which the double bonds lie less than 4 Å apart react in the crystalline state. However, since the C–C bond which is formed in dimerization has a length of only about 1.5 Å, it is necessary to consider the mobility of the molecules in the crystal.

There are several definitions of the term 'topotactic reaction' which are essentially applied to solid-state chemical reactions of small ions or molecules. In the field of solid-state polymerization, small monomer molecules are converted into long-chain molecules. Further, the distinction between lattice control of the polymerization reaction itself or orientation of the polymer is uncertain. Crystallization of the substituent groups in the monomer controls the orientation of the product.

Photopolymerization of 2,5-distyrylpyrazine demonstrates a typical example of lattice-controlled topotactic polymerization.[80,81] This monomer exhibits two crystalline modifications: orthorhombic plate-like and needle-like monoclinic crystals. Plate-like crystals polymerize on irradiation with UV or visible light and produce a four-center-type high molecular weight crystalline polymer, while the needle-like crystals do not exhibit any change on irradiation. Poly(2,5-distylrylpyrazine) is highly crystalline and its crystallites show a high degree of orientation with respect to the monomer single crystal. Figure 9 illustrates the orthorhombic arrangement of several 2,5-distyrylpyrazine (DSP) molecules which will undergo polymerization into one polymer chain along the *c* axis. It can be seen from this figure that only slight translations and rotations of molecules can produce an oriented polymeric crystal, preserving the crystal symmetry of the monomer. 1,4-Bis($\beta$-pyridyl-1-vinyl)-benzene exhibits another example of four-center-type polymerization.[82]

Thermal polymerization of *p*-cyanobenzonitrile (Table 2) gives poly(*p*-phenylene-1,2,4-oxadiazole) in the solid state by cycloaddition polymerization.[83,84] An interesting application of topotactic reactions in the solid state is known as polymerization in clathrates (or canal polymerization) which will be discussed in Volume 4, Chapter 18.

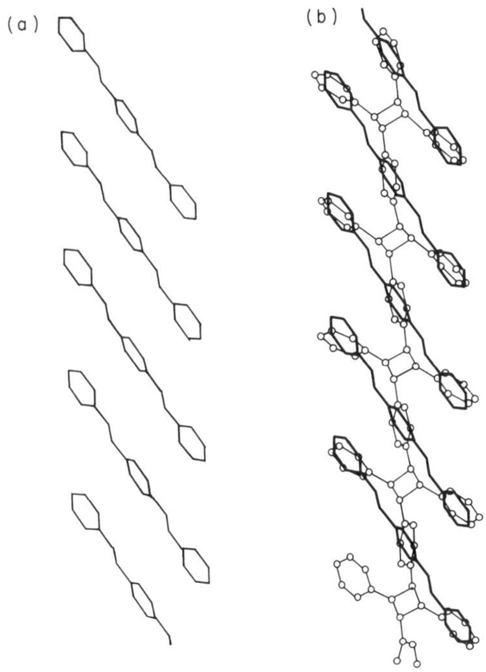


**Figure 9** Polymerization scheme; projection along the *b* axis of DSP: (a) projection of monomer; (b) projection of monomer superimposed on the molecular model of the polymer[15,81] (reproduced by permission of Kodansha–Wiley from ref. 15)

### 17.2.5 Correlation of Solid-state Polymerization with Crystal Structure

The correlation of the experimental results obtained on solid-state polymerization by various techniques during the last three decades with the crystal structures of monomeric crystals provides valuable information.

The crystal structure of acrylamide has been analyzed at room temperature,[15,85] at $-100\,^{\circ}C$[86] and at $-150\,^{\circ}C$.[87,88] It was shown that there was no transition between $-170\,^{\circ}C$ and $84\,^{\circ}C$ (melting point) of the crystals. Crystal lattice data for this monomer are given in Table 1, and its crystal structure is shown in Figure 10. The crystal lattice is constructed from double molecular sheets parallel to the 100 plane. In the sheets, two monomer molecules related by a center of symmetry are connected by N—H· · · · ·O hydrogen bonds and these pairs are connected by other hydrogen bonds, forming a two-dimensional network.[15] Vinyl groups are arranged between such sheets. Consequently, the amide-group layers and the vinyl-group layers stack alternately, parallel to the 100 plane.

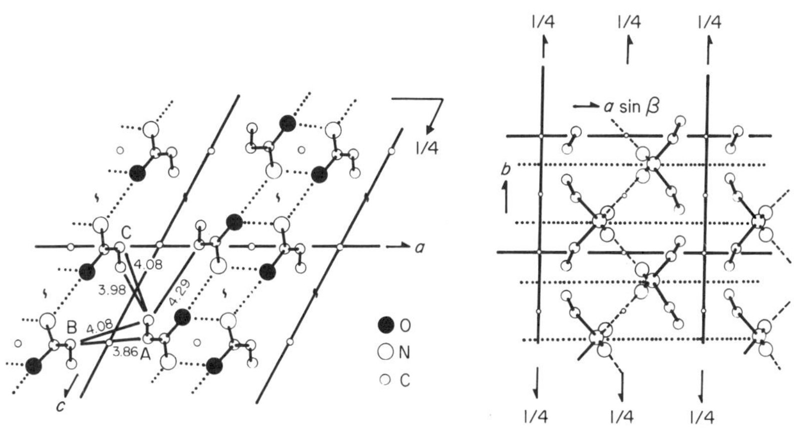


**Figure 10** Crystal structure of acrylamide; dotted lines indicate N—H·····O hydrogen bonds (reproduced by permission of Kodansha–Wiley from ref. 15)

The electron density map of acrylamide projected on the *ac* plane is shown in Figure 11. This map was obtained by using diffraction data measured at room temperature (22 °C), at which solid-state polymerization of acrylamide by X-ray irradiation proceeded rapidly.[34] Therefore, the X-ray intensity data included 'average' information on the crystal structure of the monomer during the polymerization.[15] The anisotropic electron density distribution indicates that molecules vibrate with a large amplitude perpendicular to the 100 plane. It was reported that the root-mean-square amplitude of the $CH_2$ group is 0.32 Å at room temperature,[85] and 0.15 Å at − 100 °C.[86]

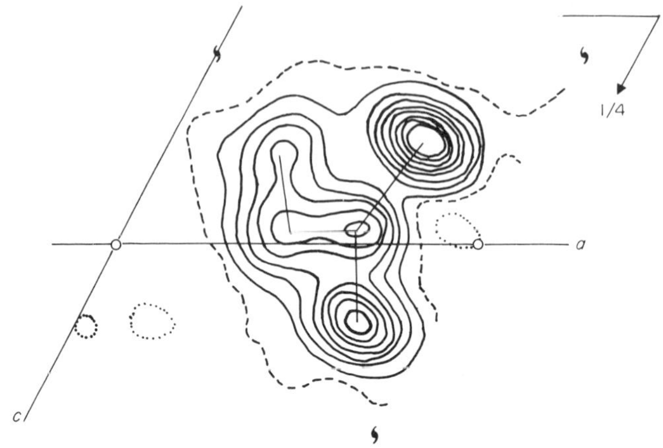


**Figure 11** Electron density map of acrylamide projected on the *ac* plane. Contours are drawn at intervals of 1 e A$^{-2}$, the one-electron line being broken (reproduced by permission of Kodansha–Wiley from ref. 15)

From geometrical considerations, in the initial stages of reaction, the possibility of head-to-head and tail-to-tail additions has been suggested. It was reported that the C(H)---C(H) distance of A and B molecules is 3.7 Å, while the C($H_2$)---C($H_2$) distance of A and C molecules is 3.8 Å at − 100 °C (Figure 10).[85,86] The possibility of a small amount of head-to-head addition has been reported on the basis of NMR and ESR studies.[89] On the other hand, in an earlier ESR study[31] it was supposed that initial combination between two acrylamide molecules related by the two-fold screw axis along the *b* axis (A and C in Figure 10) is preferable, and at a low temperature (∼ − 125 °C), when the molecular mobility is insufficient, the reaction halts at the dimer stage (**2** in Section 17.2.3). However, requirements based on crystal structure indicate that, in the early stage of polymerization, chain propagation may take place in the layer of vinyl groups.[15]

The information provided by the X-ray diffraction data, the anisotropic features of the electron density map, together with the well-known effects of impurities, pressure and scratching of the crystal on the polymer yield, may correlate with the presence of vinyl-group layers. The effect of water may correlate with the hydrogen bonds of amide groups.[15] During polymerization, the growing polymer chain destroys the crystal lattice locally, and the effect is transmitted to the neighboring layers. Polymer nucleation and recrystallization of unreacted monomer must take place simultaneously.[8,34]

The crystal structure of methacrylamide is quite similar to that of acrylamide.[85] However, chemical reactivity and the polymer yield of this monomer are low. Acrylamide and methacrylamide give solid solutions. The copolymers were obtained in high yields from the radiation-induced polymerization of solid solutions of these monomers.[90]

It was found that when polymerization of acrylic acid is initiated with polarized UV-radiation, the molecular disposition has an important influence on the initial stage of polymerization.[91] However, since the intermolecular distance between vinyl carbons is 3.52 Å,[85] chain propagation will be isotropic due to the large molecular motions, and the polymeric product is atactic and amorphous.

A broad-line NMR study was carried out on acrylic acid crystals. The results suggested that the reduction of second moment with temperature is due to the promotion of oscillation of the hydrogen-bonded dimer units about the *a* axis of the crystals.[92] Post-polymerization of acrylic acid irradiated with γ-rays at − 196 °C has also been studied by broad-line NMR and ESR.[93] It was found that the average radical concentration decays by a factor of ten after annealing between − 146 and − 30 °C. However, the local concentrations of radicals which persist after annealing at about − 30 °C actually contribute to the polymerization of this monomer.

If a monomer can exist in more than one crystalline modification, each modification will exhibit a different reactivity in the solid state. There are several examples which have been carried out on monomers in more than one crystalline form.

Crystalline acrylonitrile exhibits a first-order transition at $-113\,^{\circ}C$. High and low temperature forms are stable above and below the transition point, respectively.[72] Samples of the high temperature form can be obtained at $-196\,^{\circ}C$ by shock-cooling liquid monomer, and on subsequent warming this high temperature form reverts to the stable low temperature form at about $-130\,^{\circ}C$. After X-ray irradiation at $-196\,^{\circ}C$, the two forms exhibit diffferent behavior on warming. Post-irradiation polymerization does not occur in acrylonitrile below $-140\,^{\circ}C$.[94] It was found that this polymerization occurs only in the high temperature form, or in association with transition.[72]

The Arrhenius plots of rates of in-source solid-state polymerization of acrylonitrile often show a break between $-140$ and $-110\,^{\circ}C$, corresponding to different activation energies above and below the transition point.[95,99] A number of explanations have been forwarded to account for the abrupt changes in rate of polymerization with temperature. The possibility of different reaction mechanisms, including radical[25,99,100] and ionic[98,100] propagation, collective excitation[7,72,97,101] and thermal energy chains,[4] occurring in different temperature regions, have been considered.

There exist three polymorphic modifications of tributyl(vinyl)phosphonium bromide, and the solid-state polymerization of two forms has been studied.[102] On X-ray irradiation, both forms gave amorphous polymer. Conversion–time curves are sigmoid to 100% conversion throughout the temperature range $-80$ to $+60\,^{\circ}C$. The more unstable form polymerizes faster than the stable form by a factor of two. The difference in rates was attributed to different collisional and steric factors governed by the crystal structures.[102]

Calcium acrylate can be obtained in a number of forms. The yields obtained by the post-irradiation polymerization of this monomer with various modifications are given in Table 4.[11,45] The variations in the yields must reflect the different arrangements of ions and water molecules in the various forms. The rates of the post-irradiation polymerization of the alkali-metal acrylates, at any given temperature, vary considerably with the alkali metal, usually in the order potassium > lithium > sodium.[103]

**Table 4** Post-irradiation Polymerization of Calcium Acrylate at 101 °C[11]

| *Hydrate* | *Form* | *Yield (max.)* (%) |
|---|---|---|
| Dihydrate | Crystalline | 52 |
| Monohydrate | Crystalline | 0 |
| Anhydrous (ex-dihydrate) | Crystalline | 53 |
| Anhydrous (ex-monohydrate) | Crystalline | 14 |
| Anhydrous (ex-dihydrate) | 'Amorphous' | 90 |
| Anhydrous (ex-monohydrate) | 'Amorphous' | 23 |

A similar trend is observed for degrees of polymerization. X-Ray studies on partially polymerized potassium acrylate show that the polymer forms a separate amorphous phase within the monomer crystal.

### 17.2.6 Reaction Mechanism

To propose a general mechanism for solid-state polymerization is not possible, due to the abundance of factors affecting the course of these reactions. However, a considerable amount of data has been accumulated in order to discuss the polymerization mechanism of monomers carrying small substituent groups, such as acrylamide, acrylic and methacrylic acid.

In acrylic acid crystals at $4\,^{\circ}C$, there is considerable molecular mobility,[92] and the mobility is probably more extensive in the imperfections. UV-initiated polymerization of this monomer is rapid, and propagating radicals are randomly oriented in the lattice. However, when polymerization is initiated with polarized UV-radiation, the rate of polymerization in the initial stages shows a strong dependence on the angle between the plane of the electric vector of the polarized UV-light and the crystallographic axes.[91] This result suggests that the vinyl groups of the monomer molecules in the perfect regions of the lattice absorb UV radiation, producing excitons which travel through the lattice and become trapped in imperfections. Some of these excitons give rise to radical formation and initiate the polymerization process.[104] In polymerizations initiated by high energy radiation, the radiolytic products (*e.g.* hydrogen atoms) diffuse through the lattice and initiate polymerization at imperfections.[50]

Detailed ESR studies on the nature of radicals in solid monomers indicate that at low temperatures there is no polymerization, and the radicals are subject to some degree of restriction. In the vicinity of an imperfection, such as a dislocation or grain boundary, some strain will be present in these regions of the lattice and there will be a certain free volume available for molecular motions. At low temperatures these motions may be frozen out. However, even at $-196\,^{\circ}C$ there is presumably sufficient freedom for movement in some defects to allow the addition of a hydrogen atom to the double bond of a monomer molecule.[11,31,50,92] The resulting radical is firmly held in position by intermolecular forces, and these radicals are highly oriented in the crystal lattice.[31,50] At higher temperatures, when there is sufficient molecular movement in the defect, this radical will add to a nearest-neighbor monomer. Accompanying this addition there will be a net shrinkage.[58] However, addition stops at dimer or trimer formation.[31] At this stage, it is probable that substituent groups on the reacted monomers are still under the influence of the original intermolecular forces which maintain the dimer in a preferred orientation and also cause a specific conformation to be adopted, as observed with some monomers. Under these conditions, considerable strain must be developed in the vicinity of the dimer or trimer radical.

Further propagation is apparently not possible until the thermal motions of the molecules increase sufficiently for this strain to be relieved and to allow addition of the radical to another neighboring monomer molecule. This further addition may well not occur in a specific crystallographic direction, and the orientation of the radical with respect to the monomer crystal axes may be lost. It is at this stage that the polymer chain is likely to become detached from the monomer lattice, by breaking strained intermolecular bonds, and nucleate an amorphous polymer phase within the monomer crystal.[8,11,31]

The mechanism suggested above is mainly based on evidence from the solid-state polymerization of monomers carrying small substituent groups. In these cases, there is no evidence that the substituent groups on monomer molecules which have been incorporated into polymer play any further part in the reaction. In the case of monomers carrying large and strongly interacting substituent groups, a considerable amount of modification would be required to take account of the location and propagation of free radicals. In addition, the possibility of side-chain crystallization in the polymer has to be careully studied. Details of the reaction mechanisms of these systems are considered in the following section of this chapter.

### 17.2.7 Solid-state Polymerization of Vinyl Monomers with Long-chain Substituents

Vinyl monomers with long-chain alkanic substituents give interesting crystalline vinyl polymers by post-polymerization with $\gamma$-ray irradiation. During the last two decades, systematic studies have been carried out on ethyl methacrylate,[105] vinyl laurate,[106] vinyl stearate,[107,108,109] allyl palmitate,[110] *N*-hexadecylacrylamide and methacrylamide.[111] The liquid crystalline monomers, such as vinyl oleate[112] and hexadecyl vinyl ether,[113,115] as well as $\beta$-alkyl itaconates,[116] cholesteryl acrylate, cholesteryl methacrylate,[117–119] cholesterylvinyl succinate[120] and *N*-vinylcarbazole,[121,125] were all polymerized. Conversion into polymer is complete for small molecules such as acrylamide, whereas limiting conversions were observed for vinyl monomers having bigger substituents.[109,126,128] These monomers exhibit various crystalline and smectic structures in the condensed state.[129] Radiation-induced solid-state polymerization of various modifications shows large differences depending on the structure of the monomers. Favorable structures are the rectangular monolayers of vinyl groups with either orthorhombic or hexagonal packing, where polymer–monomer solid solutions can be formed. It was suggested that the close-packed arrangement of functional groups in layers makes the ionic initiation more probable.

Monomeric vinyl stearate crystals are monoclinic, in which the vinyl groups are situated in separate layers and the alkanic side-chains in an orthorhombic sub-cell. During solid-state polymerization, the stearate residues attached to the polymer retain their orientation parallel to the side chain of the monomer. However, the packing changes from orthorhombic to hexagonal; at 40% conversion, all stearate residues of the polymer and monomer were packed in the hexagonal arrangement.[107] The polymer formed in this way, although more isotactic than the one prepared in solution, is not stereoregular. The side chains do crystallize and the backbone of the polymeric chain is probably amorphous.[107] Later, a similar situation was observed for the polymerization of vinyl stearate in monomolecular layers.[109]

During the solid-state polymerization of vinyl monomers with long-chain substituents, transformation of the crystal structure of the monomer to form an isomorphous monomer–polymer system is a unique behavior for this class of compounds. This phenomenon is well-documented for the solid-state polymerization of hexadecyl vinyl ether,[113–115] *N*-hexadecylacrylamide and *N*-hexadecylmethacrylamide crystals.[111]

Solid-state polymerization of other side-chain substituted vinyl monomers, such as *p*-acetamidostyrene and *p*-benzamidostyrene, have been studied in detail.[130] Both of these compounds polymerize spontaneously, producing polymers less dense than the crystalline monomers. The retention of IR dichroism has suggested that poly(*p*-benzamidostyrene) is a stereoregular polymer which was obtained by a topotactic reaction.[130]

Valuable information was obtained from studies of the radiation-induced polymerization of binary systems consisting of vinyl monomers and various long-chain compounds in solid–liquid equilibrium states. The radiation-induced polymerization in binary systems consisting of a vinyl monomer (*e.g.* acrylonitrile) and a mixture of inert additives (*e.g.* oily and fatty multicomponent mixtures of glyceride) indicated that the rate of polymerization is significantly reduced.[95,131] In these systems, since the long-chain additives with branched structure can be glassified easily, the in-source polymerization occurs very slowly in the supercooled glassy phase of monomer. On the other hand, if a binary mixture of a vinyl monomer (*e.g.* methyl methacrylate) and a long straight-chain additive (such as an alkane, a long-chain ester or hexadecyl alcohol) was irradiated by $\gamma$-rays, an appreciable amount of post-polymerization was observed.[132] It was assumed that initiation of polymerization takes place as the monomer diffuses to the trapped radical in the crystals of long-chain compounds, and the propagation reaction also proceeds inside the crystal as if acting as a special reaction matrix with a much reduced termination probability.[132,133]

The polymerization of long-chain vinyl compounds has been the subject of much work in the field of solid-state reactions.[107,134–147] Recently, it has been shown that $\gamma$-ray initiated post-polymerization of octadecyl methacrylate is accelerated by particular orientation, suitable packing mode and some freedom of rotational motion of the monomer molecules in the layered structure.[148] Radiation-induced post-polymerizations of polymorphic crystals of octadecyl methacrylate,[148] octadecyl acrylate[149] and the melts of these compounds have been investigated in order to find out the effect of a molecular arrangement of long-chain monomers on polymerization characteristics in the solid state.

Thermal, X-ray and IR analysis show that octadecyl acrylate exhibits three crystalline modifications.[149] The $\alpha$ form is stable in the range 19–32 °C (m.p. of monomer); at lower temperatures the $\beta$ form is stable, whereas the sub-$\alpha$-form appears in $\alpha \rightarrow \beta$ transitions. Experimental data on post-polymerization in the $\alpha$- and $\beta$-form crystals, together with those in the liquid state, are shown in Figure 12. The monomer molecules in triclinic packing ($\beta$ form) show very little tendency to polymerize, while those in hexagonal packing ($\alpha$ form) have high polymerizability that increases with temperature. However, the polymerizability in the molten state at fairly high temperatures is rather low. For octadecyl acrylate, the polymerization rate, limiting (saturated) conversion and polymer molecular weight exhibit maxima just above the melting point of the $\alpha$ form.

Estimations based on X-ray diffractions indicate that polymerization in the $\beta$-form crystal with triclinic packing must be accompanied by a volume expansion of 10% or more.[149] The heat of polymerization seems to be insufficient to cause such unusual expansion in the stable monomer crystal surrounding a propagating radical. This radical is confined to the matrix of closely packed monomer molecules in the $\beta$ form and polymerization is almost impossible.

On the other hand, in the $\alpha$ form, the monomer molecules are arranged hexagonally in the layer structure, with the long molecular axis oriented perpendicularly to the laminar plane and the rotation around the long-chain axis allowed in a statistical sense. This packing mode is similar to that of the side chains of the polymer. In the $\alpha$ form, the reactive functional groups are concentrated in the same plane of the layer structure and situated most effectively for polymerization; that is to say, orientation is favorable. Furthermore, the freedom of rotational motion around the molecular axis favors the propagation of polymerization in all directions within a single layer. In addition, for two-dimensional polymerization of long-chain monomers in the layered structure, the termination reaction may be strongly restricted by the interference of polymer chains already formed.

It was assumed that propagation proceeds in such a way that it will produce isotactic (twisted) or syndiotactic (straight) conformations which are closely related to the microtacticity of the polymer.[149] It was further assumed that the propagating radical cannot pass through the barrier of polymer chain already formed. That means that the radical obstructed by the wall of the polymer chain may propagate in another direction or be trapped in the polymer matrix. This situation seems to be similar to two-dimensional self-avoiding walks. In this case, some monomer molecules should

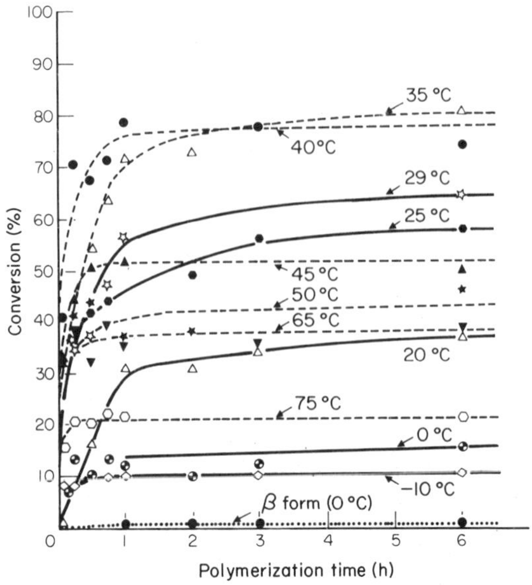


**Figure 12** γ-Ray-initiated post-polymerization of octadecyl acrylate in: (—) the α-form crystal; (---) the molten state with 0.1 Mrad irradiation; and (···) the β-form crystal with 2.0 Mrad irradiation (reproduced by permission of J. Wiley & Sons from ref. 149)

be enclosed in the comb-like polymer chain (Figure 13). The decrease in polymerization probability in the lamellar structure from the initiation point can be expressed by the Gaussian distribution function of distance. Introducing proper approximations that polymerization probability decreases in the cone shape around each initiation point, and assuming that the distribution pattern of the initiation points is of a 'checkerboard' type (that is, every cone that contains one propagating

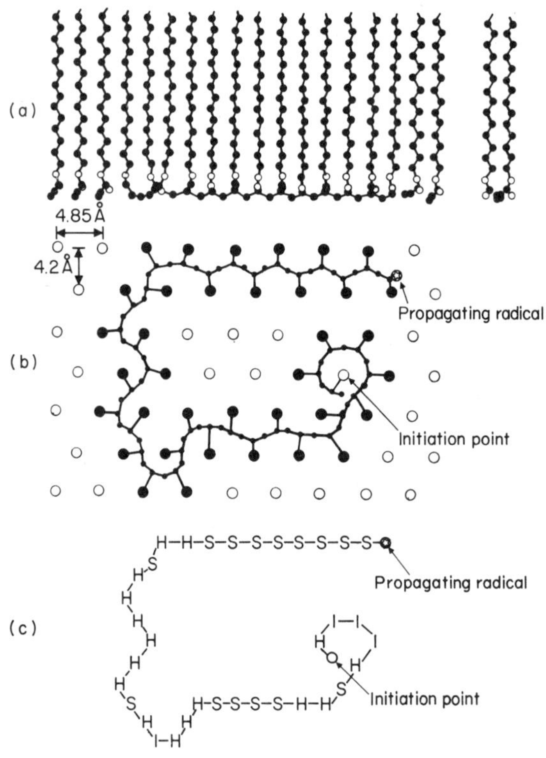


**Figure 13** Schematic illustration of the molecular arrangement in the layered structure and two-dimensional polymerization of octadecyl acrylate in the α form (hexagonal packing): (a) two rectangular side views; (b) plane figure; and (c) corresponding microtacticity of the polymer, I, S and H represent the units of iso-, syndio- and hetero-tactic configurations, respectively (reproduced by permission of J. Wiley & Sons from ref. 149)

polymer chain is surrounded by four neighboring cones), the post-polymerization of long-chain monomers in the layer structure can be interpreted quantitatively.[150]

### 17.2.8 The Effects of Added Substances on Solid-state Polymerizations

Besides polymerization initiators and inhibitors, the influence of other substances in concentrations greater than impurity levels has been studied.

Acrylamide and propionamide form solid solutions in all proportions. The polymerization of acrylamide at room temperature in these solid solutions is an extensively studied binary system. It was observed that the molecular weights of the polyacrylamides which were obtained by in-source[25] and post[24]-polymerizations are reduced by propionamide concentrations as low as ~ 10%. The results on post-polymerization were interpreted in terms of extensive transfer to propionamide in the solid state.[24] However, chain transfer to propionamide is ruled out in this system, since high molecular weight polyacrylamide was extracted from solid solutions containing very low concentrations of acrylamide.[31,58]

Various explanations were offered for the decrease in molecular weight of the polymers obtained from solid solutions containing relatively low concentrations of propionamide.[25,57,58,151,152]

The initial rate of in-source polymerization of acrylamide is increased in solid solutions containing 10% propionamide.[25] On the other hand, polyacrylamides obtained from post-polymerization of solid solutions containing 10% propionamide show a broader distribution in molecular weights compared to pure polyacrylamide obtained under similar conditions.[153,154] If one considers that propionamide is a defect in the crystal lattice, and that the defect density is extremely high with the presence of this additive, a broad distribution with a lower average molecular weight would be a logical consequence.[154]

Studies on the influence of various additives on the polymerization of acrylic acid and acrylamide show that additives which interact strongly with the monomer (*e.g.* forming hydrogen bonds) have the greatest effect.[155]

Crystalline salts of acrylic acid undergo polymerization on mechanical disintegration in the presence of low concentrations of solvents.[156] It was suggested that disintegration produces lattice vacancies which act as traps for electrons, which in turn initiate polymerization under suitable conditions of molecular mobility. A mixture of sodium acrylate and lithium or magnesium chloride crystals, in the presence of traces of solvent, undergo ion exchange accompanied by polymerization.[157]

### 17.2.9 Copolymerization in the Solid State

Frozen mixtures of monomers have been used to obtain copolymers in the solid state. Only homopolymers were produced from monomers which crystallize separately. If the monomers form solid solutions, a copolymer may be formed. The composition of the copolymer will be determined by the nature of the solid solution rather than the reactivity ratios which control copolymer compositions in the liquid state. Solid-state copolymerization reactions have been studied with monomer pairs which either form solid solutions (isomorphous pairs) or eutectic mixtures.

Isomorphous solid solutions of tributylphosphonium bromide and the corresponding iodide were studied in the temperature range − 80 to − 60 °C, and sigmoid conversion–time curves leading to 100% conversion were obtained.[102] It was found that both monomers and all their solid solutions reacted at the same rate at − 60 °C, while the rate changes with composition at − 80 °C.

Solid-state copolymerizations of various binary mixtures of oxetanes were carried out by $\gamma$-irradiation.[90] The composition of solid solutions corresponds to reactivity ratios of unity, which was quite different from those observed in the liquid state.

A series of vinyl monomer pairs leading to eutectic mixtures was found to yield copolymers by irradiation in the solid state. The following monomer pairs have been investigated: *N*-vinylsuccinimide/*N*-vinylpyrrolidone,[158] acrylic acid/vinyl chloroacetate,[159] acrylic acid/acrylamide,[160] styrene/acrylonitrile,[161] vinyl chloroacetate/vinyl laurate,[162] styrene oxide/propylene oxide,[163] monoallyl norbornenedicarboxylates,[164] acrylonitrile/vinyl acetate,[164] binary systems of maleimide and substituted maleimides[165] and *N*-vinylcarbazole in binary systems.[166] The effects of various additives on several binary systems producing copolymers were also studied.

Phase diagrams are available[90] for a number of monomer mixtures. In most cases, eutectics were formed and monomers were polymerized independently during radiation-induced reactions. Maleic anhydride and acenaphthylene, which give a 1:1 copolymer in the liquid state, produce a mixture of polyacenaphthylene and 1:1 copolymer in the solid state. As in many other binary systems, the maximum rate of polymerization occurs at the eutectic composition.[90,167]

Studies on copolymerization in several eutectic systems, composed of α,β-substituted unsaturated compounds which can polymerize by γ-radiation in the solid state, provide valuable information. Examples include maleimide/acenaphthylene,[168] *N*-laurylmaleimide/acenaphthylene,[169] maleimide/succinimide[170] and maleimide/maleic anhydride.[170] For these binary systems, the formation of amorphous by γ-irradiation is explained by a two-phase nucleation mechanism.[8,26]

The conversion to copolymer is generally high in the mixture which contains less maleimide than that in the eutectic composition, where all maleimide should be included as fine crystals in eutectic mixture. Therefore the results can be interpreted by the commonly recognized fact that a crystal defect favors the solid-state polymerization which proceeds by a free radical mechanism, and the participation of a second component (such as succinimide) seems to help the formation of the crystal defect in maleimide crystals by producing eutectics. In contrast to γ-irradiation, crystalline polymeric products were obtained from the above-mentioned mixtures by UV radiation.[168,170] However, there are several binary monomer systems which can form eutectic mixtures but do not exhibit an appreciable copolymerization, such as the *N*-phenylmaleimide/acenaphthylene pair.[169,170]

The acrylic acid/acrylamide mixture is characterized by two eutectics containing 74.5% and 58.7% acrylic acid respectively, and a molecular compound with a composition 2:1 acrylic acid: acrylamide.[167] The resulting copolymers have compositions corresponding to those of the eutectics and the molecular compounds; poly(acrylic acid) is formed under conditions where acrylic acid crystallizes from the monomer mixtures. Similar results were obtained in acrylonitrile/styrene mixtures at $-94\,^{\circ}C$.[161]

Copolymers of vinyl acetate or allyl alcohol with sulfur dioxide can be formed in the solid state.[171]

Radiation-induced copolymerization of multicomponent systems have been studied in solid and supercooled phases.[172–178] In these multicomponent systems, all components are dispersed homogeneously in polymerizing mixtures, including a foaming agent. Eutectic mixtures of acrylamide/itaconic acid were polymerized in the solid state and then foamed by heating at 130–200 °C.[172] The solution systems such as acrylamide/itaconic acid/water and acrylamide/maleic anhydride/urea/water were polymerized in liquid and supercooled phases at higher rates and foamed to a higher degree of expansion. Foams of poly(acrylic acid) and poly(methacrylic acid) were obtained by polymerization and heating of acrylic acid (methacrylic acid)/urea/water systems. The studies related to the multicomponent systems cover the glassification and polymerization of glass-forming monomeric systems. Binary systems consisting of organic acids and amides form stable glassy phases owing to large molecular interactions between the components. These mixtures polymerize at a very high rate in the supercooled phase. It was found that foaming occurred during irradiation in the polymerization of the acrylamide/urea/sulfuric acid system. This behavior was attributed to the promoting effect of sulfuric acid on the decomposition of urea and the heat accumulation by the acceleration of the polymerization.[178]

Block-copolymer preparation is another aspect of solid-state polymerization. It is well established that post-polymerization of acrylamide (and acrylic acid) reaches a certain limiting conversion at a certain temperature. The post-polymerization can be reactivated by increasing the temperature until a limiting conversion is reached[179] (Figure 14). This indicates the reactivity of the buried radicals toward their own monomers. γ-Irradiated acrylamide is used to obtain block copolymers of polyacrylamide–polystyrene and polyacrylamide–poly(methyl methacrylate).[14] The free radical nature of initiation and the change of growing macroradicals from radical type (**3**) to styryl radical throughout the copolymerization were studied by ESR (Figure 15). The diblock structure of the copolymers was determined by torsional shear modulus–temperature measurements.[180]

The living polymeric radicals in micelles which were produced during emulsion polymerization,[181] or radicals in a matrix polymerization by using a complexation agent (*e.g.* chitosan)[182] or free radicals formed in inclusion polymerizations[183] have all been used for block copolymer synthesis. Long-lived propagating radicals of ultra-high molecular weight which were formed in plasma-initiated polymeric systems have also been used for the preparation of block copolymers.[184]

Recently, it was found that acrylamide, methacrylamide and acrylamide derivatives, such as *N*-methylacrylamide and *N*-methylmethacrylamide, were polymerized in adequate systems to form polymer microspheres in high concentrations.[185] Living polymer radicals which were produced in a

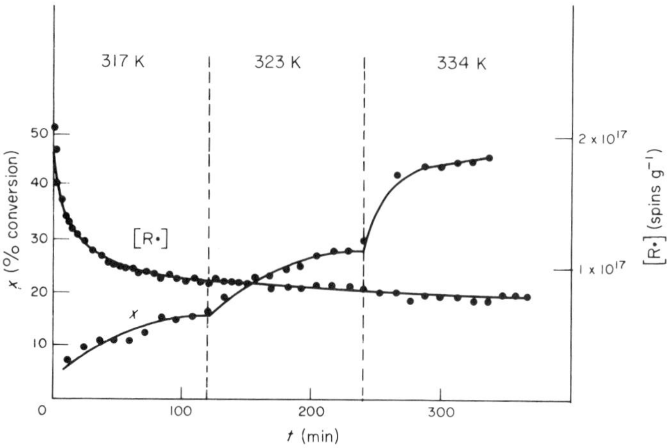


**Figure 14** Reactivation of solid-state polymerization of acrylamide (reproduced by permission of J. Wiley & Sons from ref. 89)

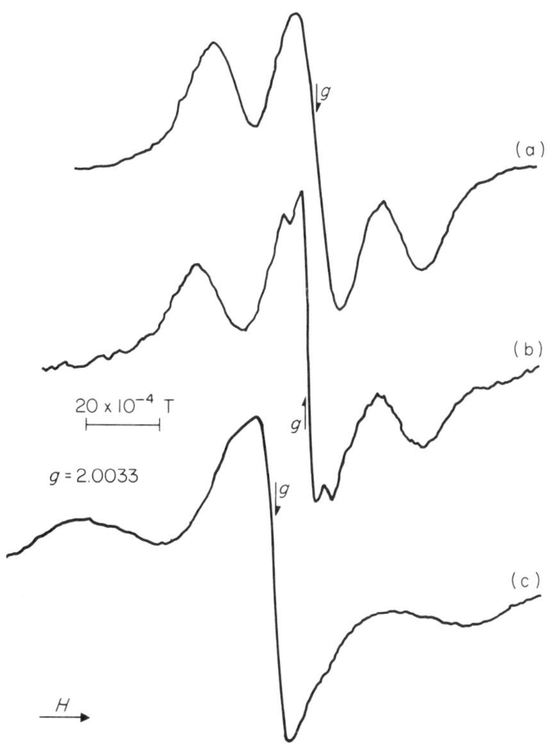


**Figure 15** ESR spectrum of: (a) irradiated acrylamide; (b) irradiated acrylamide/styrene mixture kept at 60 °C or 0.5 h after the addition of styrene; and (c) irradiated acrylamide/styrene mixture kept at 60 °C for 5 h after the addition of styrene[179, 180] (reproduced by permission of J. Wiley & Sons from ref. 180)

suitable polymeric matrix react readily with other vinyl monomers at room temperature, producing block copolymers.[186] The reactions of the living polymer radicals of the amide monomers with vinyl monomers give, in good yields, block copolymers which are amphiphilic, consisting of hydrophilic and lipophilic sequences.[187]

### 17.2.10 Solid-state Polymerization of *N*-*t*-Butylacrylamide

Solid-state in-source and post-polymerizations of *N*-*t*-butylacrylamide, induced by $\gamma$-irradiation, were studied extensively. The experimental results were reported in a recent review.[188] The conversion–time curves for this crystalline monomer are somewhat different to those of acrylamide. However, the rate of post-polymerization follows equation (7).[189] The overall activation energies for post-polymerization of this monomer with total doses of 0.5 Mrad and 1.4 Mrad were reported to be 183 kJ mol$^{-1}$ [189] and 116 kJ mol$^{-1}$,[190] respectively.

In the radiation-induced in-source solid-state polymerization of *N*-*t*-butylacrylamide, about 100% conversion of monomer to polymer was obtained at various irradiation temperatures and dose rates. The variation of intrinsic viscosities as a function of time and conversion is complicated by the extensive branching.

A series of *N*-*t*-acrylamide homologs was selected as a model to study the effects of a gradual structural change of the monomer on its ability to polymerize in the solid state. From the conversion of one monomer as a function of time of irradiation, the kinetic curve of any other monomer of the series can be described if conversion is known for a single time of irradiation with the same dose rate. To describe the dependence of conversion on the time of irradiation, the following simplified relationship is given for a series of *N*-substituted acrylamides.[191]

$$\ln \frac{S}{1 - S} = 3.2803\,[(\theta - 1) + 0.28\,(\theta^3 - 1)] \qquad (10)$$

In this relation, $S$ is the conversion at time $t$, and $\theta$ is the reduced time (*i.e.* the time of irradiation $t$ divided by the half-life $\tau_{1/2}$). For most of the *N*-*t*-alkylacrylamides, the polymerization rate is much higher for shock-cooled crystals than for crystals grown slowly from solution, the difference being especially remarkable for *N*-*t*-alkylacrylamides with large asymmetrical substituents. On the other hand, for *N*-*t*-butylacrylamide the reverse effect was observed, where higher polymer yields required longer times of irradiation.

Recently, solid-state polymerization of complex compounds of *N*-*t*-butylacrylamide with Lewis acids has attracted attention as an interesting topic in this field. Several methods have been developed to prepare complexes of this monomer with Lewis acids.[192] The thermal polymerization of (*N*-*t*-butylacrylamide)$_2$–$ZnCl_2$ in the solid state was studied over the temperature range of 125–135 °C for samples open to the air as well as under vacuum.[193]

The radiation-induced solid-state post-polymerization of (*N*-*t*-butylacrylamide)–$ZnCl_2$ complex was studied in the 80 to 100 °C temperature range and almost 100% conversion was obtained. The polymerization rate of this complex was considerably higher than pure *N*-*t*-butylacrylamide. This system also exhibits an autoacceleration. The radiation-induced in-source polymerization of (*N*-*t*-butylacrylamide)$_2$–$ZnCl_2$ and (*N*-*t*-butylacrylamide)$_2$–$ZrBr_2$ complexes show that $ZnCl_2$ has an accelerating effect on the polymerization rate,[193,194] whereas $ZnBr_2$ has a retarding effect compared with the polymerization rate of uncomplexed *N*-*t*-butylacrylamide at equal temperatures. The lower reactivity of the (*N*-*t*-butylacrylamide)$_2$–$ZnBr_2$ complex was explained by the weaker metal–*N*-*t*-butylacrylamide coordination bond.

## 17.3 SOLID-STATE POLYMERIZATION OF CYCLIC COMPOUNDS

Solid-state polymerization of cyclic compounds was reported in the early 1960s. Highly crystalline polymers, specifically oriented within the monomer crystal, were obtained by radiation-induced solid-state polymerization of crystalline 1,3,5-trioxane (**8**), diketene (**9**; 4-methylene-2-oxetanone), *β*-propiolactone (**10**; 3-propanolide) and 3,3-bis(chloromethyl)oxetane (**11**). The polymers produced were more perfectly oriented than polymers oriented by elongating a fiber. Single crystals of 1,3,5-trioxane were transformed into three-dimensionally oriented poly(oxymethylene) crystals by irradiation with $\gamma$-rays (equation 11, Figure 16).

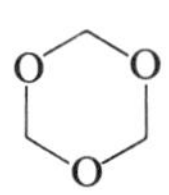

**(8)**

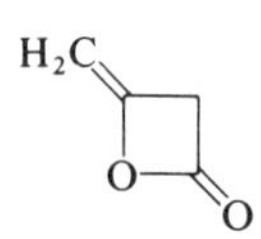

**(9)**

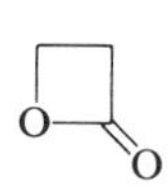

**(10)**

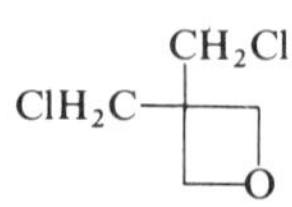

**(11)**

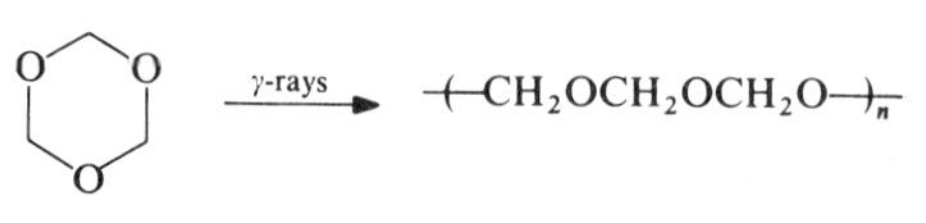

(11)

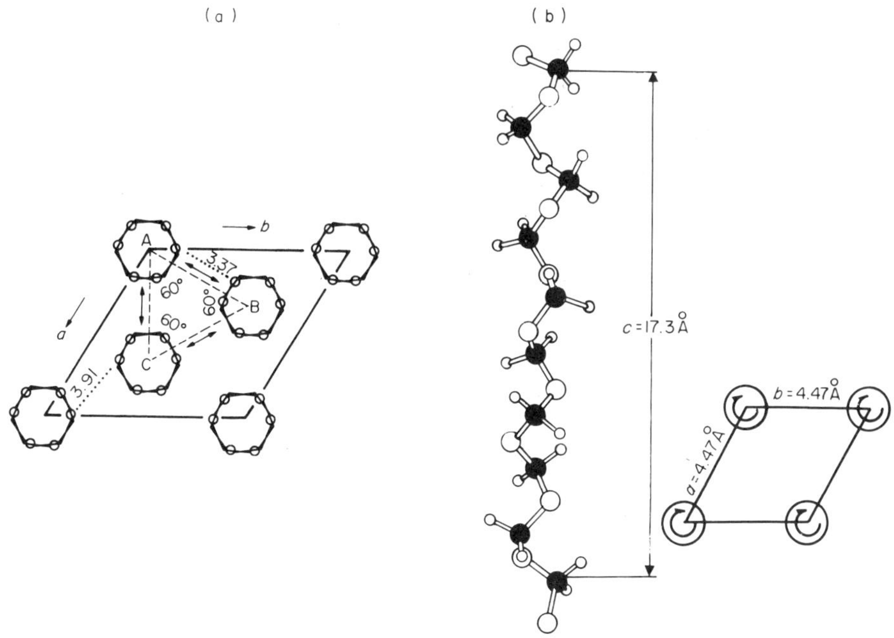


**Figure 16** (a) Crystal structure of 1,3,5-trioxane, white circles represent oxygen atoms, double-headed arrows indicate chain directions of polymer crystallites. (b) Molecular structure of poly(oxymethylene) (reproduced by permission of Kodansha–Wiley from ref. 15)

### 17.3.1 Reactive Species

It is generally accepted that the radiation-induced solid-state polymerization of these cyclic monomers proceeds by a cationic mechanism, since only cationic polymerization is known in solution. Radiolysis of monomer presumably produces cations or radical cations, depending on the nature of the monomer and the presence of impurities. Propagation takes place by addition of monomer to the cation. Solid-state polymerization of these cyclic compounds can be induced by cationic initiators. The polymerization of 1,3,5-trioxane in the solid state initiated by surface catalysis, using conventional cationic initiators such as boron trifluoride and stannic chloride, gave rise to crystalline polymers rather similar to those obtained by radiation-induced initiation. For these initiators, the polymerization must be controlled to some extent by the solid-state polymerization of 1,3,5-trioxane, iodine was included in 1,3,5-trioxane crystals, by freezing a solution of iodine in monomer.[196]

### 17.3.2 Orientation of Polymer Chains in Monomer Crystals

The solid-state polymerization of the cyclic oligomers of formaldehyde, tetroxocane, pentoxecane and hexoxacyclododecane, has been initiated by $\gamma$- or X-ray irradiation. For these oligomers and poly(oxymethylene), X-ray analyses were carried out systematically and the crystallographic data are given in Table 2.[197,198]

The main effort on the solid-state polymerization of cyclic monomers has been essentially directed towards understanding the relationship between the crystal structure of these oligomers and the direction of growth and crystal structure of the poly(oxymethylene). When single crystals of these cyclic oligomers are subjected to $\gamma$- or X-ray irradiation (both in-source and post-irradiation), poly(oxymethylene) is obtained in high yield. It was observed that the polymerization is suppressed at temperatures lower than about 50 °C below the melting points of these oligomers. However, the conversion to polymer increases with polymerization temperature and decreases drastically near the melting points of the oligomers.[197] In each oligomer the single crystal habit of the monomer is preserved at the completion of polymerization. In addition, the preferred orientation of polymer crystallites is characteristic for each oligomer.[197]

The orientations of poly(oxymethylene) (POM) crystals with respect to the crystallographic axes of each oligomer are illustrated in Figure 17. In this figure, hexagonal prisms represent trigonal polymer crystals. The longest axis of the prism is the direction of the polymer chain (the *c* axis), and the six side-faces of the prism are the 100 planes. The size of each prism is a qualitative indication of the relative yields of polymer crystallites in that direction. In the case of hexoxacyclododecane (Figure 17d), in addition to the trigonal polymer crystal, an oriented unstable orthorhombic polymer crystal is also obtained by X-ray induced in-source polymerization at 20 °C (indicated as POM-Mod, square prisms in Figure 17).[15,197]

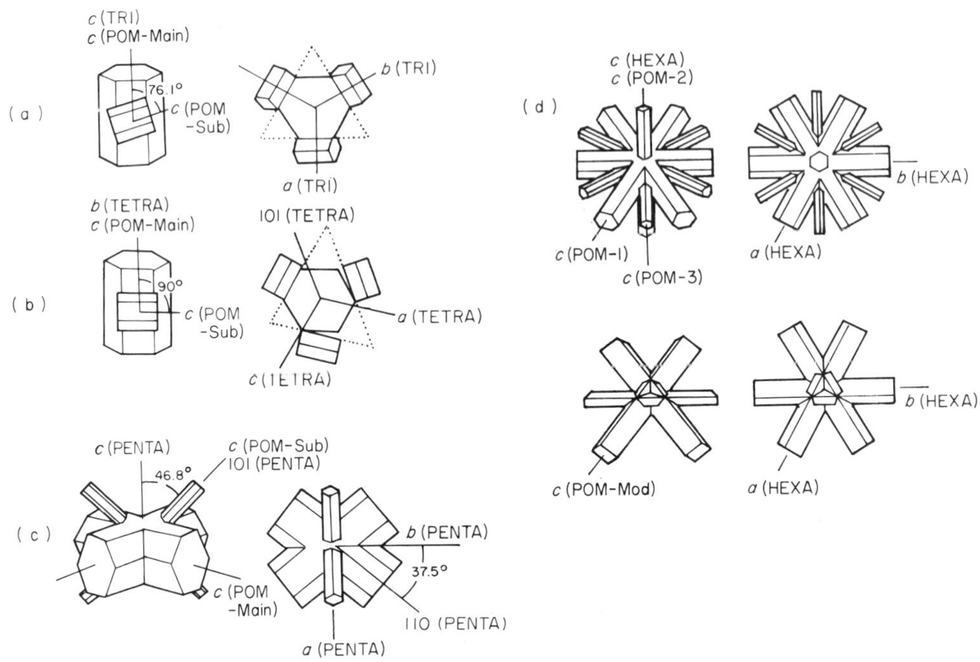


**Figure 17** Relative orientations of poly(oxymethylene) (POM) with respect to the crystallographic axes of the cyclic oligomers of formaldehyde: (a) 1,3,5-trioxane, (b) 1,3,5,7-tetroxocane, (c) 1,3,5,7,9-pentoxecane and (d) 1,3,5,7,9,11-hexoxacyclododecane (reproduced by permission of Kodansha–Wiley from ref. 15)

The relative crystal orientation of the oligomer crystal and the resultant poly(oxymethylene) were determined from X-ray diffraction patterns of oligomer crystals that were partially converted into poly(oxymethylene) by X- or $\gamma$-irradiation. The poly(oxymethylene) formed as a mosaic of main crystals, having the primary orientation, and sub-crystals, having 'twin' orientation. However, the orientations of the sub-crystals of poly(oxymethylene) formed from these oligomers were quite different from one another, as shown from schematic representation of relative crystal orientations in Figure 17.

The chain directions of the main crystal and the sub-crystal were always oriented parallel to specific crystallographic axes of cyclic oligomer crystals. Therefore, these characteristic relative orientations of the oligomer crystal and the resultant poly(oxymethylene) crystal could be interpreted as a consequence of a propagation reaction controlled by the molecular arrangements in the original crystals. These systematic studies show that the solid-state polymerization of cyclic oligomers of formaldehyde might be described as examples of topological polymerization.[197,198]

### 17.3.3 Mechanism and Molecular Processes

Knowledge of the mechanism and the molecular processes involved in solid-state reactions is rather limited. There are many studies on solid-state polymerization kinetics which are difficult to interpret. Since the solid-state transformation seems the main characteristic feature of these reactions, a new approach based on investigating the morphology of the polymeric product is presented as an adequate procedure.

This method is based on the study of the relationship between molecular arrangements in the monomer and polymer phases and on finding out the morphology of the polymeric product as it is polymerized. Two types of solid-state polymerization mechanisms have been proposed:

(1) Firstly, topochemical polymerization, which implies a direct transition from monomer molecules to polymer chains without destruction of the crystal lattice and formation of non-crystalline intermediates. According to this mechanism, the three-dimensional order of the monomer crystal lattice is transferred without loss of periodicity. Therefore, the resulting polymer must possess a three-dimensional well-oriented extended-chain crystal morphology.[3,14,16] A lamellar morphology, as is observed in crystalline polymers, cannot be formed during topochemical polymerization, because the process of chain folding is not consistent with the three-dimensional periodicity of the solid monomer.

(2) The second type of mechanism for solid-state polymerization is concerned with non-crystalline regions. In these regions, the reaction starts preferentially at dislocations, defects and the boundaries of the monomer crystal, where a polymer phase is formed.[11,26,36] At the phase boundary between monomer and polymer, a non-crystalline zone arises which has the character of a liquid phase where polymerization proceeds. Therefore, the chain growth corresponds mostly to a polymerization in the liquid or gas phase. Amorphous polymer is formed which may crystallize in a second separate step.

For these two main types of solid-state polymerization, a clear correlation between the reaction mechanism and the resulting polymer morphology is expected. However, in the course of solid-state polymerizations, relaxation processes have been shown to follow the polymerization step, so that a change of the original morphology is effected.[199] As a result, the correlation between mechanism and morphology in solid-state polymerization becomes more complicated.

The topochemical polymerization mechanism in solid-state reactions attracts great interest for the following reasons: first, study of the strict correlation between monomer and polymer crystals can lead to a deeper insight into the physical properties of organic solids; second, well-oriented extended-chain polymer crystals are of interest for the investigation of physical properties of polymers.

The solid-state polymerization of 3,3-bis(chloromethyl)oxetane (**11**) and its analogs by high energy radiation is a typical example for lattice-controlled solid-state polymerizations of heterocyclic monomers.[9,11,16] These reactions were described in 1962 and were investigated mainly from the kinetic point of view. The polymers were described as highly crystalline and well oriented.[14,16] In addition, the solid-state copolymerization of several parts of oxetanes forming solid solutions was described to be in accordance with the topochemical model of the reaction mechanism.

Detailed studies on the morphology of the polymers resulting from the solid-state polymerizations of various cyclic oxetanes were carried out by using a combination of electron and optical microscopy on polymerizing single crystals and direct observation of the growing polymer phase.[199] These direct morphological studies on the polymerizing single crystals of 3,3-bis(chloromethyl)-oxetane are supported by investigations of the copolymerization behavior of 3,3-bis(chloromethyl)-oxetane and 3-ethyl-3-(chloromethyl)oxetane in the liquid and solid states. The results of these well-designed experiments show that the polymerization gives rise to amorphous polymer in the first step. The poly[3,3-bis(chloromethyl)oxetane] forms irregular platelets which aggregate into larger units without reflecting the crystalline order of the monomer. Subsequent to polymerization, the amorphous polymer crystallizes to the $\beta$ modification of poly[3,3-bis(chloromethyl)oxetane]. If the partially polymerized crystals are extracted by solvents of the monomer, crystallization of the monomer is enhanced, and morphological artifacts arise which were previously mistaken for the true morphology of the as-polymerized polymer.

The proposed mechanism of solid-state polymerization of oxetanes, initiated by high energy radiation, consists of a sequence of three major processes: (a) the lattice-controlled radiochemical initiation step; (b) the polymer chain growth in non-crystalline regions, similar to a polymerization in the liquid phase, where amorphous polymer is formed; and (c) the crystallization giving rise to a certain orientation of the amorphous polymer by swelling in the course of the extraction of unconverted monomer by solvent.[199]

The results of morphological studies clearly indicate that the crystal structure and the orientation of the polymer obtained by solid-state polymerization of oxetanes is not caused by the special reaction mechanism, but by secondary processes taking place after polymerization. However, it should be noted that the polymerization of 3,3-bis(chloromethyl)oxetane by high energy radiation proceeds in the solid state several orders of magnitude more rapidly than in the liquid state.[200] In order to explain this behavior, the special character of the initiation process effected by radiation is suggested.[199]

## 17.4 REFERENCES


1. M. Magat, *Polymer*, 1962, **3**, 449.
2. A. Chapiro, 'Radiation Chemistry of Polymeric Systems', Interscience, New York, 1962.
3. H. Morawetz, in 'Physics and Chemistry of the Organic Solid State', ed. by D. Fox, M. M. Labes and A. Weissberger, Interscience, New York, 1963, vol. I, 1965, vol. II.
4. A. Chalesby, *Rep. Prog. Phys.*, 1965, **28**, 463.
5. N. S. Marans, *Polym. Eng. Sci.*, 1966, **6**, 14.
6. J. H. O'Donnell, *At. Energy Aust.*, 1967, **10**, 9.
7. Y. Tabata, *Adv. Macromol. Chem.*, 1968, **1**, 283.
8. G. Adler and B. Baysal, *Mol. Cryst.*, 1969, **9**, 361.
9. A. Chapiro, in 'Encyclopedia of Polymer Science and Technology', ed. H. F. Mark, N. G. Gaylord and N. M. Bikales, Interscience, New York, 1969, vol. II, p. 702.
10. A. Blumstein, *Adv. Macromol. Chem.*, 1970, **2**, 123.
11. G. C. Eastmond, *Prog. Polym. Sci.*, 1970, **2**, 1.
12. G. M. J. Schmidt, *Pure Appl. Chem.*, 1971, **27**, 647.
13. A. Chapiro, *Isr. J. Chem.*, 1972, **10**, 129.
14. S. Okamura, *Pure Appl. Chem.*, 1972, **30**, 181.
15. Y. Chatani, *Prog. Polym. Sci., Jpn.*, 1974, **7**, 149.
16. M. Nishii and K. Hayashi, *Annu. Rev. Mater. Sci.*, 1975, **5**, 135.
17. Gy. Hardy, F. Cser, K. Nyitrai and N. Fedorova, *Proc. Tihany Symp. Radiat. Chem.*, 1977, **4**, 365.
18. V. Enkelmann, *Colloid Polym. Sci.*, 1978, **256**, 893.
19. Y. Ito and Y. Tabata, *Adv. Chem. Ser.*, 1979, **175**, 109.
20. R. Schulz, A. V. Henglein, H. E. Steinwehr and H. Bambauer, *Angew. Chem.*, 1955, **67**, 232.
21. A. J. Restaino, R. B. Mesrobian, H. Morawetz, D. S. Ballantine, G. J. Dienes and D. J. Metz, *J. Am. Chem. Soc.*, 1956, **78**, 2939.
22. T. A. Fadner, I. D. Rubin and H. Morawetz, *J. Polym. Sci.*, 1959, **37**, 549.
23. H. Morawetz and T. A. Fadner, *Makromol. Chem.*, 1959, **34**, 162.
24. T. A. Fadner and H. Morawetz, *J. Polym. Sci.*, 1960, **45**, 475.
25. B. Baysal, G. Adler, D. Ballantine and P. Colombo, *J. Polym. Sci.*, 1960, **44**, 117.
26. G. Adler, D. Ballantine and B. Baysal, *J. Polym. Sci.*, 1960, **48**, 195.
27. P. G. Garrat, *Polymer*, 1962, **3**, 323.
28. G. F. Parrish, W. A. Trinler and C. K. Burton, *J. Polym. Sci., Part A-1*, 1967, **5**, 2557.
29. D. S. Breslow, G. E. Hulse and A. S. J. Matlack, *J. Am. Chem. Soc.*, 1957, **79**, 3760.
30. C. H. Bamford, G. C. Eastmond and J. C. Ward, *Proc. R. Soc. London, Ser. A*, 1963, **271**, 357.
31. G. Adler and J. H. Petropoulos, *J. Phys. Chem.*, 1965, **69**, 3712.
32. C. H. Bamford, A. Bibby and G. C. Eastmond, *Polymer*, 1968, **9**, 629.
33. G. Adler, *J. Chem. Phys.*, 1959, **31**, 848.
34. G. Adler and W. Reams, *J. Chem. Phys.*, 1960, **32**, 1698.
35. G. Adler, D. Ballantine and B. Baysal, Movie projected at the International Symposium on Macromolecular Chemistry, Moscow, 1960. Typical pictures of this movie are reproduced in: A. Chapiro, in 'Action Chimiques et Biologiques', ed. M. Haissinsky, Masson, Paris, 1966, 10e série, p. 187.
36. C. Sella and J. J. Trillat, *C. R. Hebd. Seances Acad. Sci.*, 1961, **253**, 1511.
37. C. Sella and R. Bensasson, *J. Polym. Sci.*, 1962, **56**, S1.
38. B. Wunderlich, *Angew. Chem., Int. Ed. Engl.*, 1968, **7** (12), 912.
39. G. C. Eastmond, *J. Appl. Chem.*, 1963, **13**, 221.
40. C. H. Bamford, A. Bibby and G. C. Eastmond, *Polymer*, 1968, **9**, 653.
41. Y. Tabata and T. Suzuki, *Polym. Prepr., Am. Chem. Soc., Div. Polym. Chem.*, 1964, **5**, 997; *Makromol. Chem.*, 1965, **81**, 223.
42. P. J. Fydelor and A. Charlesby, *J. Polym. Sci., Part C*, 1969, **16**, 4493.
43. M. Prince and J. Hornyak, *J. Polym. Sci., Part B*, 1966, **4**, 493.
44. M. Prince and J. Hornyak, *J. Polym. Sci., Part A-1*, 1967, **5**, 531.
45. J. B. Lando and H. Morawetz, *J. Polym. Sci., Part C*, 1964, **4**, 789.
46. D. W. Brown and L. A. Wall, *Polym. Prepr., Am. Chem. Soc., Div. Polym. Chem.*, 1964, **5**, 907.
47. G. B. Guarise and G. Talamini, *Corsi Semin. Chim.*, 1967, **5**, 209.
48. R. Bensasson, A. Bernas, M. Bodard and R. Marx, *J. Chem. Phys.*, 1963, **60**, 950.
49. H. Ueda, *J. Polym. Sci., Part A*, 1964, **2**, 2207.
50. Y. Shioji, S. I. Ohnishi and I. Nitta, *J. Polym. Sci., Part A*, 1963, **1**, 3373.
51. C. H. Bamford, A. Bibby and G. C. Eastmond, *J. Polym. Sci., Part C*, 1967, **16**, 2417.
52. M. Fujimoto, *J. Chem. Phys.*, 1963, **39**, 846.
53. J. H. O'Donnell, B. McGarvey and H. Morawetz, *J. Am. Chem. Soc.*, 1964, **86**, 2322.
54. S. Kılıç and B. M. Baysal, *J. Macromol. Sci., Rev. Macromol. Chem. Phys.*, 1986, **C26**, 483.
55. J. Barton, F. Szöcs and J. Nemcek, *Makromol. Chem.*, 1969, **124**, 28.
56. B. Ranby and J. F. Rabek, 'ESR Spectroscopy in Polymer Research', Springer, Berlin, 1977.
57. G. Adler, D. S. Ballantine, T. Davis and R. Rangathan, *J. Phys. Chem.*, 1964, **68**, 2184.
58. G. Adler and W. Reams, *J. Polym. Sci., Part A*, 1964, **2**, 2617.
59. H. Ueda and Z. Kuri, *J. Polym. Sci.*, 1962, **61**, 333.
60. C. H. Bamford, G. C. Eastmond and Y. Sakai, *Nature (London)*, 1963, **200**, 1284.
61. M. C. R. Symons, *J. Chem. Soc.*, 1963, 1186.
62. P. Kourim and K. Vacek, *Tetrahedron Lett.*, 1962, 1051; *J. Polym. Sci., Part C*, 1969, **16**, 3044.
63. H. Fischer, *Z. Naturforsch. A*, 1964, **19A**, 866.
64. J. H. O'Donnell, P. J. Pomery and R. D. Sothman, *Makromol. Chem.*, 1980, **181**, 409.

65. P. Romiszowski and A. Kolinski, *Polymer*, 1982, **23**, 1226.
66. M. J. Bowden and J. H. O'Donnell, *Macromolecules*, 1968, **1**, 499.
67. M. J. Bowden and J. H. O'Donnell and R. D. Sothman, *Makromol. Chem.*, 1969, **122**, 186.
68. M. J. Bowden, J. H. O'Donnell and R. D. Sothman, *Makromol. Chem.*, 1969, **122**, 186.
69. M. J. Bowden and J. H. O'Donnell, *J. Macromol. Sci., Chem.*, 1972, **A6**, 745.
70. M. J. Bowden, J. H. O'Donnell and D. D. Sothman, *Macromolecules*, 1972, **5**, 269.
71. J. H. O'Donnell and R. D. Sothman, *J. Macromol. Sci., Chem.*, 1980, **A14**, 879.
72. R. Bensasson, A. Dworkin and R. Marx, *J. Polym. Sci., Part C*, 1964, **4**, 881.
73. K. Tsuji, H. Yoshida, K. Hayashi and S. Okamura, *J. Polym. Sci., Part B*, 1967, **5**, 313.
74. K. Takeda, H. Yoshida, K. Hayashi and S. Okamura, *J. Polym. Sci., Part A-1*, 1966, **4**, 2710.
75. K. Tsuji, K. Takakura, K. Hayashi and S. Okamura, *J. Polym. Sci., Part A-1*, 1966, **4**, 2028.
76. K. Tsuji, H. Yamaoka, K. Hayashi and S. Okamura, *J. Polym. Sci., Part B*, 1966, **4**, 629.
77. M. D. Cohen, G. M. J. Schmidt and F. I. Soutag, *J. Chem. Soc.*, 1964, 2000.
78. F. L. Hirshfeld and G. M. J. Schmidt, *J. Polym. Sci., Part A*, 1964, **2**, 2181.
79. G. M. J. Schmidt, *Pure Appl. Chem.*, 1971, **27**, 647.
80. M. Hasegawa and Y. Suzuki, *J. Polym. Sci., Part B*, 1967, **5**, 813.
81. M. Hasegawa, Y. Suzuki, H. Nakanishi and F. Nakinshi, *Prog. Polym. Sci., Jpn.*, 1973, **5**, 143.
82. H. Nakanishi, K. Ueno, M. Hasegawa and Y. Sasada, *Chem. Lett.*, 1972, 30.
83. S. Fujimoto, *J. Polym. Sci., Part B*, 1967, **5**, 301.
84. B. M. Culbertson and R. Murphy, *J. Polym. Sci., Part B*, 1966, **4**, 305.
85. I. Nitta, I. Taguchi and Y. Chatani, *Fiber Res. Inst. Osaka Univ. (Japanese)*, No. 12, 89 B1959; Y. Chatani, Doctoral Thesis, Osaka University, 1959.
86. I. V. Isakov, *Zh. Strukt. Khim.*, 1966, **7**, 898.
87. A. Usanmaz, Doctoral Dissertation, Polytechnic Institute of New York, 1974; *Diss. Abstr. Int. B*, 1974, **35** (4), 1610.
88. A. Usanmaz and G. Adler, *Mol. Cryst. Liq. Cryst.*, 1976, **32**, 123.
89. C. Chachaty and A. Forchioni, *J. Polym. Sci., Part A-1*, 1972, **10**, 1905.
90. M. Nishii, H. Tsukamoto, K. Hayashi and S. Okamura, *J. Appl. Polym. Sci.*, 1967, **11**, 1117.
91. G. C. Eastmond, E. Haigh and B. Taylor, *Trans. Faraday Soc.*, 1969, **65**, 2497.
92. B. Arnold and G. C. Eastmond, *Trans. Faraday Soc.*, 1971, **67**, 772.
93. A. Forchioni and C. Chachaty, *J. Polym. Sci., Part A-1*, 1972, **10**, 1923.
94. I. M. Barkalov, V. I. Gol'danskii, N. S. Enikolopyan, S. F. Terekiiova and G. M. Trofimova, *Vysokomol. Soedin.*, 1964, **6**, 92; *Polym. Sci. USSR (Engl. Transl.)*, 1964, **6**, 108.
95. Y. Amagi and A. Chapiro, *J. Chim. Phys. Phys.-Chim. Biol.*, 1962, **59**, 537.
96. I. Barkalov, V. I. Gol'danskii, N. S. Enikolopyan, S. F. Terekhova and G. M. Trofimova, *Dokl. Akad. Nauk. SSSR*, 1962, **147**, 395.
97. Y. Tabama, S. Shu, S. Hayakawa and K. Oshima, *Polym. Prepr., Am. Chem. Soc., Div. Polym. Chem.*, 1964, **5**, 986.
98. A. Chapiro and M. Inoue, *J. Chim. Phys. Phys.-Chim. Biol.*, 1967, **64**, 338.
99. Y. Tabata, K. Ishigure, Y. Fujita and K. Oshima, *J. Polym. Sci., Part C*, 1966, **23**, 663.
100. H. Sobue and Y. Tabata, *J. Polym. Sci.*, 1960, **43**, 459.
101. D. A. Kritkaya, I. K. Larin, A. N. Ponomarev and V. L. Tal'roze, *Vysokomol. Soedin.*, 1964, **6**, 1944; *Polym. Sci. USSR (Engl. Transl.)*, 1964, **6**, 2151.
102. C. S. H. Chen and D. G. Grabar, *J. Polym. Sci., Part C*, 1964, **4**, 849.
103. H. Morawetz and I. D. Rubin, *J. Polym. Sci.*, 1962, **57**, 669.
104. Y. Chatani, Y. Sakata and I. Nitta, *J. Polym. Sci., Part B*, 1963, **1**, 419.
105. Gy. Hardy, K. Nyitrai, G. Kovaks and N. Fedorova, *Acta Chim. Acad. Sci. Hung.*, 1965, **43**, 121.
106. Gy. Hardy, K. Nyitrai, J. Varga and M. Patco, *Period. Polytech., Chem. Eng.*, 1965, **9**, 137.
107. N. Morosoff, H. Morawetz and B. Post, *J. Am. Chem. Soc.*, 1965, **87**, 3035.
108. M. Puterman, T. Fort, Jr. and J. B. Lando, *J. Colloid Interface Sci.*, 1974, **47**, 705.
109. A. Usanmaz and B. M. Baysal, *Middle East Tech. Univ. J. Pure Appl. Sci.*, 1975, **8**, 201.
110. Gy. Hardy, L. Nagy and F. Cser, *Acta Chim. Acad. Sci. Hung.*, 1966, **47**, 211.
111. F. Cser, K. Nyitrai, V. Dévényi and Gy. Hardy, *Magy. Kem. Foly.*, 1976, **82**, 117.
112. Gy. Hardy, N. Fedorova, K. Kovaks and E. Boros-Gyevi, *J. Polym. Sci., Part C*, 1976, **16**, 2675.
113. Gy. Hardy, K. Nyitra, F. Cser, Gy. Cselik and L. Nagy, *Eur. Polym. J.*, 1969, **5**, 133.
114. Gy. Hardy, K. Nyitrai and F. Cser, *Acta Chim. Acad. Sci. Hung.*, 1969, **62**, 253.
115. Gy. Hardy, K. Nyitrai and F. Cser, *Macromol. Synth.*, 1974, **5**, 91.
116. K. Nyitrai, N. L. Nguyen, F. Cser, E. Takaes and Gy. Hardy, *Eur. Polym. J.*, 1976, **12**, 184.
117. Gy. Hardy, F. Cser, A. Kallo, K. Nyitrai, G. Bodor and B. Lengyel, *Acta Chim. Acad. Sci. Hung.*, 1970, **65**, 287.
118. Gy. Hardy, K. Nyitrai and F. Cser, 'Kinetics and Mechanism of Polyreactions, IUPAC International Symposium on Macromolecules, Budapest', Akadémiai Kiadó, Budapest, 1969, Reprint vol. IV, p. 121.
119. Gy. Hardy and K. Nyitrai, *Acta Chim. Acad. Sci. Hung.*, 1970, **65**, 301.
120. K. Nyitrai, F. Cser, B. D. Ngoc and Gy. Hardy, *Magy. Kem. Foly.*, 1976, **82**, 216.
121. J. Kroh and W. Pekala, *Bull. Acad. Pol. Sci., Ser. Sci. Chim.*, 1964, **12**, 419.
122. J. Kroh and W. Pekala, *Bull. Acad. Pol. Sci., Ser. Sci. Chim.*, 1966, **14**, 55.
123. Z. Galdecki, J. Karolak, W. Pekala and J. Kroh, *Bull. Acad. Pol. Sci., Ser. Sci. Chim.*, 1967, **15**, 209.
124. W. Pekala and A. Lesinski, *Nukleonika* 1974, **19**, 769.
125. A. Usanmaz, *Makromol. Chem.*, 1981, **182**, 2351.
126. A. Chapiro and Gy. Hardy, *J. Chim. Phys. Phys.-Chim. Biol.*, 1962, **59**, 993.
127. A. Chapiro and M. Zumer, *J. Chim. Phys. Phys.-Chim. Biol.*, 1965, **62**, 947.
128. F. Tüdos and E. Boras-Gyevi, *Proc. Tihany Symp. Radiat. Chem.*, 1967, **1**, 557.
129. M. D. Cohen and G. M. Schmidt, *J. Chem. Soc.*, 1964, 1996.
130. S. Z. Jakabhazy, H. Morawetz and N. Morosoff, *J. Polym. Sci., Part C*, 1964, **4**, 805.
131. A. Chapiro and Z. Mankowski, *Eur. Polym. J.*, 1971, **7**, 1073.

132. I. Kaetsu, K. Hayashi and S. Okamura, *J. Appl. Polym. Sci.*, 1967, **11**, 1101.
133. I. Kaetsu, A. Ito and K. Hayashi, *J. Appl. Polym. Sci.*, 1973, **17**, 3223.
134. Gy. Hardy, K. Nyitrai, J. Varga, G. Kovacs and N. Federova, *J. Polym. Sci., Part C*, 1963, **4**, 923.
135. M. J. Bowden and J. H. O'Donnell, *J. Polym. Sci., Part A-1*, 1969, **7**, 1657.
136. N. Bredjick and W. J. Burlant, *J. Polym. Sci., Part A-1*, 1970, **8**, 2807.
137. R. Ackermann, O. Inacker and H. Ringsdorf, *Kolloid-Z. Z. Polym.*, 1971, **249**, 1118.
138. A. Cemel, T. Fort, Jr. and J. B. Lando, *J. Polym. Sci., Part A-1*, 1972, **10**, 2061.
139. R. Ackermann, D. Naegele and H. Ringsdorf, *Makromol. Chem.*, 1974, **175**, 699.
140. W. Burlant and A. Adicoff, *J. Polym. Sci.*, 1958, **27**, 264.
141. M. Hatada, M. Nishii and K. Hirota, *Macromolecules*, 1975, **8**, 19.
142. S. A. Letts, T. Fort, Jr. and J. B. Lando, *J. Colloid Interface Sci.*, 1976, **56**, 64.
143. M. Hatada, M. Nishii and K. Hirota, *J. Polym. Sci., Polym. Chem. Ed.*, 1977, **15**, 927.
144. V. Enkelmann and J. B. Lando, *J. Polym. Sci., Polym. Chem. Ed.*, 1977, **15**, 1843.
145. D. Naegele and H. Ringsdorf, *J. Polym. Sci., Polym. Chem. Ed.*, 1977, **15**, 2821.
146. A. Barraud, C. Rosilio and A. Ruaudel-Teixier, *J. Colloid Interface Sci.*, 1977, **62**, 509.
147. Y. Tanaka, F. Yamaguchi, M. Shiraki and A. Okada, *J. Polym. Sci., Polym. Chem. Ed.*, 1978, **16**, 1027.
148. Y. Shibasaki, H. Nakahara and K. Fukuda, *J. Polym. Sci., Polym. Chem. Ed.*, 1979, **17**, 2387.
149. Y. Shibasaki and K. Fukuda, *J. Polym. Sci., Polym. Chem. Ed.*, 1979, **17**, 2947.
150. Y. Shibasaki, *J. Polym. Sci., Polym. Chem. Ed.*, 1980, **18**, 1693.
151. G. Adler, *J. Polym. Sci., Part C*, 1967, **16**, 1211.
152. C. H. Bamford, A. Bibby and G. C. Eastmond, *Polymer*, 1968, **9**, 645.
153. B. Baysal, *J. Polym. Sci., Part C*, 1964, **4**, 935.
154. M. L. Baysal, E. Ergöz and G. Adler, *Middle East Tech. Univ. J. Pure Appl. Sci.*, 1979, **12**, 1.
155. I. Kaetsu, N. Sagane, K. Hayashi and S. Okamura, *J. Polym. Sci., Part A-1*, 1966, **4**, 2241.
156. V. A. Kargin, V. A. Kabanov and G. P. Andrianova, *Vysokomol. Soedin.*, 1959, **1**, 301; *Polym. Sci. USSR (Engl. Transl.)*, 1960, **1**, 106.
157. V. A. Kargin, V. A. Kabanov and N. Ya. Molodtsova, *Vysokomol. Soedin.*, 1961, **3**, 787; *Polym. Sci. USSR (Engl. Transl.)*, 1962, **3**, 657.
158. Gy. Hardy, J. Varga and G. Nagy, *Makromol. Chem.*, 1965, **85**, 58.
159. Gy. Hardy, J. Boros-Gyevi and L. Koronczay, *Period. Polytech., Chem. Eng.*, 1966, **10**, 163.
160. Gy. Hardy and L. Nagy, *J. Polym. Sci., Part C*, 1967, **16**, 2667.
161. A. Chapiro and P. Cordier, *J. Chim. Phys. Phys.-Chim. Biol.*, 1967, **64**, 334.
162. Gy. Hardy and J. Boros-Gyevi, *Acta Chim. Acad. Sci. Hung.*, 1966, **50**, 319.
163. Y. Tabata and H. Sobue, *J. Polym. Sci., Part C*, 1967, **16**, 1693.
164. Gy. Hardy and K. Nyitrai, *Acta Chim. Acad. Sci. Hung.*, 1967, **52**, 105.
165. J. H. O'Donnell and R. D. Sothman, *J. Polym. Sci., Part B*, 1969, **7**, 129.
166. I. Zuchowicz and W. Pekala, *ZfI-Mitt.*, 1981, **43a**, 61.
167. A. Shimizu and K. Hayashi, *J. Polym. Sci., Part A*, 1966, **4**, 1291.
168. H. Yamakita and K. Hayakawa, *J. Appl. Polym. Sci.*, 1969, **13**, 1833.
169. J. E. Herz and V. Stannett, *Macromol. Rev.*, 1968, **3**, 1.
170. K. Hayakawa, H. Yamakita and K. Kawase, *J. Polym. Sci., Part A-1*, 1972, **10**, 1363.
171. Z. Kuri and T. Yoshimura, *J. Polym. Sci., Part B*, 1963, **1**, 107.
172. I. Kaetsu, H. Kamiyama, K. Hayashi and S. Okamura, *J. Macromol. Sci., Chem.*, 1969, **A3**, 1509.
173. I. Kaetsu, Y. Nakase and K. Hayashi, *J. Macromol. Sci., Chem.*, 1969, **A3**, 1525.
174. I. Kaetsu, H. Okubo, A. Ito and K. Hayashi, *J. Polym. Sci., Part A-1*, 1972, **10**, 2203.
175. I. Ishigaki, Y. Watanabe, A. Ito and K. Hayashi, *J. Macromol. Sci., Chem.*, 1977, **A11**, 149.
176. I. Kaetsu, H. Okubo, A. Ito and K. Hayashi, *J. Polym. Sci., Part A-1*, 1972, **10**, 2215.
177. I. Kaetsu, A. Ito and N. Sagane, *J. Appl. Polym. Sci.*, 1973, **17**, 2743.
178. I. Kaetsu, A. Ito and N. Sagane, *J. Appl. Polym. Sci.*, 1973, **17**, 2753.
179. S. Küçükyavuz and B. Baysal, 'Abstracts IUPAC Symposium on Macromolecules, Aberdeen', 1973, p. 68.
180. S. Küçükyavuz and B. Baysal, *J. Polym. Sci., Polym. Chem. Ed.*, 1978, **16**, 2901.
181. E. Bigdeli, R. W. Lenz, B. Oster and R. D. Lundberg, *J. Polym. Sci., Polym. Chem. Ed.*, 1978, **16**, 469.
182. S. Kataoka and T. Ando, *Polymer*, 1984, **25**, 507.
183. M. Miyata, Y. Osaki, K. Takemoto, M. Kamachi and S. Nozakura, *Polym. Prepr., Jpn.*, 1984, **33**, 141.
184. M. Lee, T. Morigami and Y. Minoura, *J. Chem. Soc., Faraday Trans. 1*, 1978, **74**, 1738.
185. Y. Mun, T. Sato and T. Otsu, *Makromol. Chem.*, 1984, **185**, 1507.
186. T. Sato, H. Tanaka and T. Otsu, *J. Polym. Sci., Polym. Lett. Ed.*, 1980, **18**, 189.
187. T. Sato and T. Otsu, *Adv. Polym. Sci.*, 1985, **71**, 41.
188. S. Kılıç and B. M. Baysal, *J. Macromol. Sci., Rev. Macromol. Chem. Phys.*, 1986, **C26**, 483.
189. S. Kılıç and B. M. Baysal, *Makromol. Chem.*, 1983, **184**, 1919.
190. J. Zurakowska-Orszagh and A. Gumulka, *Radiat. Phys. Chem.*, 1980, **15**, 571.
191. J. Zurakowska-Orszagh, *Polymer*, 1978, **19**, 720.
192. J. Zurakowska-Orszagh, K. Mirowski and A. Kaim, *J. Polym. Sci., Polym. Chem. Ed.*, 1981, **19**, 363.
193. S. Kılıç and B. M. Baysal, *J. Polym. Sci., Polym. Chem. Ed.*, 1984, **22**, 453.
194. J. Zurakowska-Orszagh, K. Mirowski and A. Chajewski, *J. Polym. Sci., Polym. Chem. Ed.*, 1982, **20**, 1775.
195. S. Okamura, E. Kobayashi and T. Higashimura, *Makromol. Chem.*, 1966, **94**, 20.
196. A. A. Berlin, L. I. Kuzub and N. S. Yenikolopyan, *Vysokomol. Soedin.*, 1966, **8**, 1204; *Polym. Sci. USSR (Engl. Transl.)*, 1966, **8**, 1325.
197. Y. Chatani, T. Ohno, T. Yamauchi and Y. Miyake, *J. Polym. Sci., Part A-2*, 1973, **11**, 369.
198. J. P. Colson and D. H. Reneker, *J. Appl. Phys.*, 1970, **41**, 4296.
199. B. Keggenhoff and G. Wegher, *J. Polym. Sci., Polym. Phys. Ed.*, 1973, **11**, 769.
200. K. Ueno, H. Tsukamato, K. Hayashi and S. Okamura, *J. Polym. Sci., Part B*, 1967, **5**, 395.

# 18

# Polymerization in Clathrates

GIUSEPPE DI SILVESTRO and PIERO SOZZANI
*Università di Milano, Italy*



## 18.1 AN OVERVIEW

Two or more substances can form adducts showing a single crystalline structure, brought together by different kinds of forces. These forces are, in some cases, relatively strong — such as hydrogen bonds or ion–dipole interactions — or, in other cases, only weak interactions exist, which are essentially due to the shape and dimensions of the molecules involved. This latter characteristic is distinctive of 'clathrates' — or inclusion compounds (IC) — in which the most abundant component (host) packs the molecules of a second component (guest) into its lattice, generating a definite crystalline structure.[1]

Adducts can be formed with high and low molecular weight molecules, in particular monomers, and polymerization carried out inside inclusion compounds is a particular case of polymerization in organized systems. Guest monomers are generally arranged in a line inside a channel-like structure, from which the definition of channel or tunnel polymerization is derived.[2]

Strong analogies can be drawn between ICs and other crystalline adducts, such as hydrates and solvates; their phase diagrams have been extensively studied within the same theoretical framework.[3] However, no attention was given to ICs formed with monomers till 1956, when Clasen achieved the first inclusion polymerization (IP), starting from 2,3-dimethylbutadiene included in thiourea;[4] the process was initiated thermally and proceeded very slowly. A more efficient method for initiating polymerization was introduced by Brown and White in 1960, by using $\beta$-, $\gamma$- or X-rays, adopting urea and thiourea as matrices.[5, 6] In a reasonably short time, high molecular weight products were obtained.

The reaction environment is unusual: neither solvents nor metal catalysts are present, initiation does not need chemical agents and growing chains are isolated one from another, avoiding or markedly depressing transfer and combination reactions. Therefore the method has created much interest because, in principle, it permits a topochemical reaction which transforms a regular row of monomers into a regular sequence of monomer units, possibly retaining the steric relationships of the first structure[7] (Figure 1). Regular polymers showing a degree of constitutional and stereo-

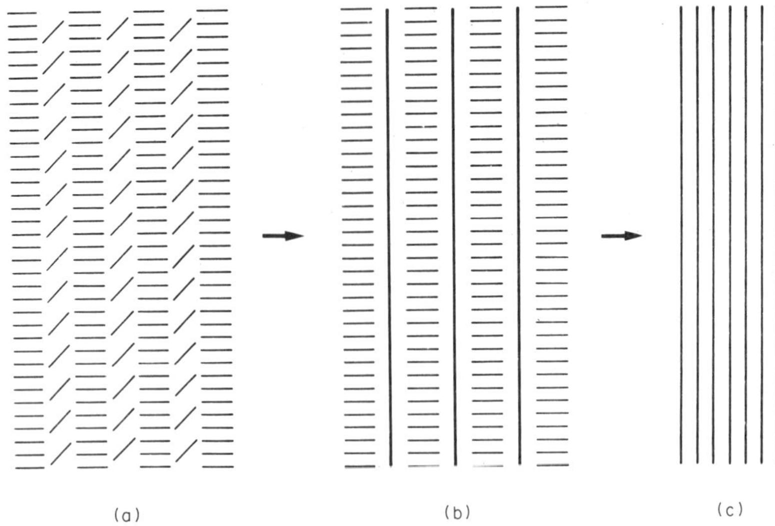


**Figure 1** Schematic representation of inclusion polymerization: (a) inclusion compound with monomers; (b) inclusion compound with polymer; (c) native polymer after extraction

chemical order even higher than those obtained by coordination polymerization can therefore be achieved.

New developments in IP were made by the application, as a host, of the *transoid-trans-transoid-trans-transoid-trans* isomer of perhydrotriphenylene (PHTP), introduced by Farina in 1963.[8] The hydrocarbon matrix of PHTP is suitable for forming inclusion compounds with molecules very different by nature, form and dimensions,[9,10] without the selectivity shown by urea or thiourea molecules.[5,6] Starting from diene monomers, constitutionally pure and highly isotactic polymers are generally obtained.

Furthermore, PHTP being a chiral molecule, asymmetric induction was achieved by using PHTP resolved into its optical isomers; using as a monomer *trans*-1,3-pentadiene, an isotactic optically active polymer was obtained.[11] In 1975, Miyata and Takemoto adopted other chiral hosts: the naturally occurring deoxycholic acid (DCA)[12] and, later, apocholic acid (ACA), the latter being obtained by an easy modification of the former.[13] Asymmetric induction was later demonstrated to be exerted only 'through space' and not through chemical bonds.[14,15]

Other hosts, such as tris(*o*-phenylenedioxy)cyclotriphosphazene (TPP) and similar substituted phosphazenes give good results when used for IP; they were introduced by Finter[16] and Allcock.[17,18] A few examples regarding the application of β-cyclodextrins to IP are also present in the literature.[19] The structures of various hosts are given in Figure 2.

Copolymerization is a different task which can be achieved by modifying the basic techniques of IP, applicable to virtually every matrix mentioned. The only restriction lies in the stability limits of the ternary IC and the possible selectivity towards the monomers. Examples were first presented in the fundamental work by Brown and White,[5,6] but the lack of available means for accurate structural investigations reduced the interest in these products. Only recently have different classes of monomers been tested in this kind of reaction.[20–23] The low probability of termination also permitted the synthesis of block copolymers.[24]

There have been great developments in the study of the fundamental features of IP along with synthetic achievements. At present, a wide research area is being covered. It includes investigation of the stability of the host–monomer IC;[25] proof of the 'living' radical nature of the growing chain-end;[24,26] ESR detection of its structure, conformation and mobility;[26,27] the stability and analysis of the IC with the polymer;[2,28] and, finally, the macroconformation and solid-state properties of the native polymer.[29–31] Moreover, much attention has been devoted to the NMR characterization of the polymer itself,[22,32–34] this bringing indirect information about the statistics and energy of the process.[35,36] The detailed analytical knowledge of polymer and copolymer microstructure, often retaining a controlled degree of order, made them very suitable reference compounds for various spectroscopic studies.[37–40]

In order to predict other possible conditions for IP, the thermodynamic parameters governing host–monomer systems were established, leading to the testing of other monomers and environments.[25,41,42]

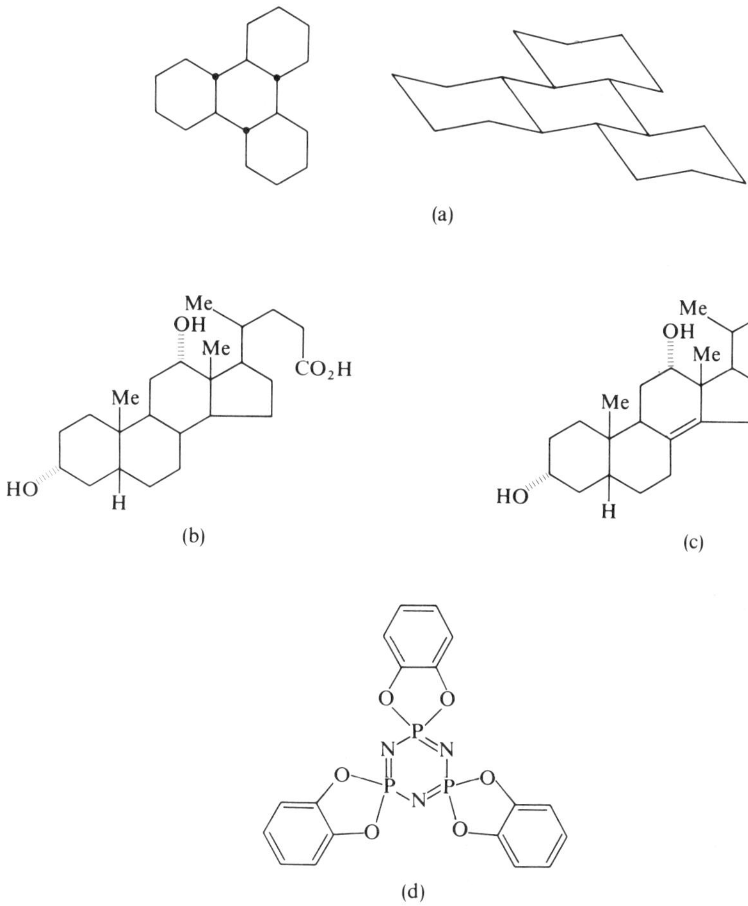


**Figure 2** Examples of the most important hosts: (a) perhydrotriphenylene (PHTP); (b) deoxycholic acid (DCA); (c) apocholic acid (ACA) (d) tris(*o*-phenylenedioxy)cyclotriphosphazene (TPP)

In the following sections, we shall deal with many of these points, paying particular attention to recent results, and trying to construct a rationalization of the whole field. The interested reader can also refer to preceding reviews.[2,7,43,44]

## 18.2 SELECTIVITY PHENOMENA AND TOPOCHEMICAL CONTROL IN INCLUSION POLYMERIZATION

The first aspect which struck researchers was the possibility of obtaining high selectivity and constitutional control by IP. Butadiene, for example, as polymerized in urea, yields a pure and crystalline 1,4-*trans* product;[6] 2,3-dimethylbutadiene and 2,3-dichlorobutadiene, as polymerized in thiourea, yield pure 1,4-*trans* crystalline polymers.[5]

However, extension of the method to other hosts and monomers has shown that it can be applied to a wide range of monomers, and that there is the possibility of obtaining different polymers from the same monomers, when the host or polymerization conditions are changed.

Polymerization results are, however, not easily predictable, but some indications can be drawn from the structure of the IC with monomers. The crystal lattice of the ICs of urea and thiourea is kept together by intermolecular hydrogen bonds and the channel retains a rigidly definite size and structure; crystal packing in the PHTP hydrocarbon matrix involves only relatively weak forces, such as van der Waals interactions, the guest easily fitting into the relatively soft channel. In this last matrix, the selectivity phenomenon is less active through a wide range of monomer dimensions, and even two quite different monomers, such as butadiene and 2,3-dimethylbutadiene, can be copolymerized.[20–22,36] Easy adaptability has no effect on constitutional order, and even copolymers show an almost pure 1,4-*trans* structure. Matrices of low flexibility with respect to PHTP, but less rigid than urea (such as DCA and ACA), are not so selective towards the common diene monomers. Also, dienes substituted by bulky groups have recently been polymerized by using a DCA matrix.[45] However, unlike PHTP, DCA and ACA do not result in high stereochemical control, in most cases.

The wide range of products obtained by IP is summarized in Table 1. The degree of constitutional and configurational regularity is given on the basis of the analytical evidence reported in the literature. Microstructural investigations have been considerably enriched during recent years, due to the progress in analytical techniques, chiefly NMR, but many earlier results still require re-examination. Topochemical control is generally high, but several exceptions are described. In every case, one of the two arrangements prevails (in most cases *trans* against *cis*; head-to-tail against head-to-head and tail-to-tail; and isotactic against syndiotactic) and can be considered as the 'ordered' one, the other being 'defective' due to a selective arrangement.

### 18.2.1 Type of Monomer Addition and Regioselectivity

Butadiene, as polymerized in PHTP and in TPP, produces a pure 1,4-*trans* crystalline polymer,[5,16,17,32] as in the urea matrix; however, polymerization in DCA or ACA produces a variable content (0–30%) of units derived from 1,2 addition, depending on the presence of an inert guest in the clathrate (ethyl acetate, acetone, *etc.*).[63,71] Many different structures can be obtained starting from *cis*- and *trans*-1,3-pentadiene and *trans*-2-methyl-1,3-pentadiene. The monomers polymerized in PHTP produce pure 1,4-*trans* products, also when introduced into copolymer structures.[22] Polymerizations carried out in DCA and ACA on *trans* monomers give rise to a higher content of 1,2- and 1,4-*cis* units, yielding amorphous or partially crystalline polymers with melting points lower than those for polymers synthesized using PHTP as the host.[15,43,58,64,71] In PHTP, the pure *cis* structure was never obtained, probably because it comprises in every case the greatest lateral space-occupying arrangement. In DCA and ACA, only *cis* monomeric units were obtained in the polymerization of 2-*t*-butylbutadiene. However, this result is due to the steric requirements of the monomer and not to the selectivity of the channels. A regular trend is observed for 2-alkyl-substituted butadienes; the *cis* content decreases with decreasing bulkiness of the substituent.[45]

Most diene monomers bearing a group substituted on the 1- or 4- carbon atom produce pure head-to-tail polymers; this is the case for all 1,3-pentadienes in PHTP and most of those in DCA. In other cases, polymers with some degree of regioisomeric defects are obtained. As an example, poly(vinyl fluoride) obtained in urea has head-to-head and tail-to-tail junctions, as demonstrated by $^{19}$F NMR spectroscopy.[48]

Polyisoprene, obtained at different temperatures in PHTP, shows a variable degree of inversion of the monomer unit, detectable to the triad and even tetrad level.[32,35] In addition to the five main peaks, the $^{13}$C NMR spectrum is composed of well-resolved minor peaks in the methylene (diad) and in the unsaturated carbon (triad) ranges, due to sequences containing defective junctions. Their content was 7–16% in syntheses carried out at temperatures in the range $-40$ to $+60\,^{\circ}$C. The process was interpreted as a succession of direct and inverted units with an energy difference of 4 kJ mol$^{-1}$ (Figure 3);[35] 1,4 units prevail in the propagation step over 4,1 units as demonstrated by ESR spectroscopy.[26] The presence of defective junctions was also observed in polyisoprene obtained from polymerization in an ACA matrix.[71] The presence of some sequences containing defective junctions

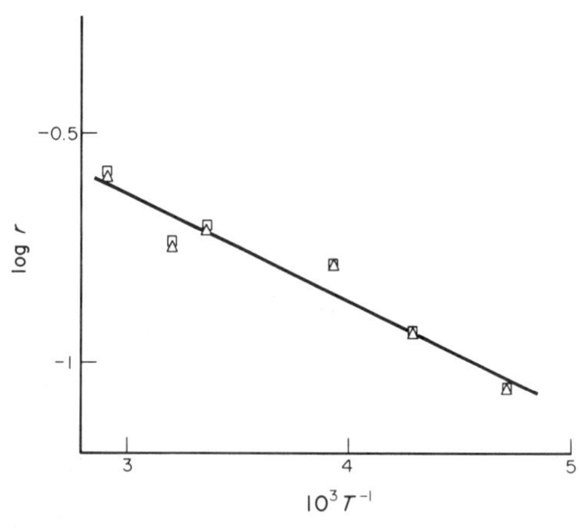


**Figure 3** Arrhenius plot for defects in polyisoprene obtained in PHTP: □, $r = p_{DI}/p_{DD}$; △, $r = p_I/p_D$

**Table 1** Selected Examples of Polymerization in Clathrates

| *Host* | *Monomer* | *Polymer structure* | *Ref.* |
|---|---|---|---|
| Urea | Butadiene | 1,4-*Trans*, crystalline | 6, 46 |
| | Vinyl chloride | Partly syndiotactic | 6, 47 |
| | Vinyl bromide | Partly syndiotactic | 47 |
| | Vinyl fluoride | Partly syndiotactic | 48 |
| | Acrylonitrile | Partly isotactic | 6, 49, 50 |
| | Acrolein | — | 6 |
| Thiourea | 2,3-Dimethylbutadiene | 1,4-*Trans*, crystalline | 5, 51 |
| | 2,3-Dichlorobutadiene | 1,4-*Trans*, crystalline | 5, 52 |
| | Other substituted butadienes | 1,4-*Trans* | 5 |
| | Vinylidene chloride | — | 5 |
| | Several monomer pairs[a] | Statistical copolymers | 5 |
| (±)PHTP | Ethylene | Linear polymer | 42 |
| | Propylene | Partly syndiotactic | 41, 42 |
| | Butadiene | 1,4-*Trans*, crystalline | 32, 53, 54 |
| | Isoprene | 1,4-*Trans*, amorphous | 32, 35, 53, 55 |
| | *trans*-Pentadiene | 1,4-*Trans*, highly isotactic, crystalline | 30, 32, 56, 57, 58, 59 |
| | *cis*-Pentadiene | 1,4-*Trans*, isotactic, crystalline | 32, 56, 59 |
| | 2,3-Dimethylbutadiene | 1,4-*Trans*, crystalline | 32, 53, 55 |
| | *trans*-2-Methylpentadiene | 1,4-*Trans*, highly isotactic, crystalline | 32, 33, 55, 60 |
| | 4-Methylpentadiene | 1,4-*Trans* | 32, 57 |
| | Other substituted butadienes | 1,4-*Trans* | 57 |
| | *trans*-Hexatriene | 1,6-*Trans,trans* | 61 |
| | Other substituted hexatrienes | 1,6-*Trans,trans* | 61 |
| | Several monomer pairs[b] | Block copolymers | 24 |
| | Several monomer pairs[a] | Statistical copolymers | 20, 22, 36 |
| | Vinyl chloride | Partly syndiotactic | 42 |
| | Diacetylene | — | 62 |
| (−)PHTP | *trans*-Pentadiene | 1,4-*Trans*, highly isotactic, crystalline, optically active | 11 |
| | Isoprene + *trans*-pentadiene[b] | Optically active block copolymer | 14 |
| (+)DCA | Butadiene | Mainly 1,4-*trans* | 63 |
| | *trans*-Pentadiene | Mainly 1,4-*trans*, atactic | 15, 58, 64, 65, 66 |
| | *cis*-Pentadiene | Mainly 1,4-*trans*, atactic, optically active | 15, 58, 64, 65, 66 |
| | 2,3-Dimethylbutadiene | 1,4-*Trans* | 12, 64, 67 |
| | 2,3-Dichlorobutadiene | 1,4-*Trans* | 12 |
| | *trans*-2-Methylpentadiene | Mainly 1,4-*trans*, optically active | 64, 65, 66, 68, 69 |
| | *cis*-2-Methylpentadiene | 1,4-*Trans*, optically active | 66, 68, 69 |
| | Other substituted butadienes | Mainly 1,4-*trans* | 43 |
| | 2-*t*-butylbutadiene | Almost pure 1,4-*cis* | 45 |
| | (±)2-*s*-butylbutadiene | Mainly 1,4-*cis*, optically active | 45 |
| | 2-Chloroacrilonitrile | — | 70 |
| (+)ACA | Butadiene | Disordered polymer | 71 |
| | Isoprene | Mainly 1,4 | 71 |
| | *trans*-Pentadiene | Mainly 1,4-*trans*, optically active | 13, 64, 66, 71 |
| | *cis*-Pentadiene | Mainly 1,4-*trans*, optically active | 13, 66, 71 |
| | *trans*-2-Methylpentadiene | Mainly 1,4-*trans*, optically active | 13, 64, 66, 69, 71, 72 |
| | *cis*-2-Methylpentadiene | 1,4-*Trans*, optically active | 13, 66, 71, 72 |
| | Other substituted butadienes | 1,4-*Trans* | 71 |
| | 2-*t*-butylbutadiene | Almost pure 1,4-*cis* | 45 |
| | (±)2-*s*-butylbutadiene | Mainly 1,4-*cis*, optically active | 45 |
| | 2-Chloroacrylonitrile | — | 70 |
| TPP | Vinyl chloride | Mainly syndiotactic | 16 |
| | Butadiene | 1,4-*Trans*, crystalline | 16, 17 |
| | 4-Bromostyrene | — | 17 |
| | Divinylbenzene | Linear non-cross-linked polymers | 17 |
| | Acrylonitrile | Mainly isotactic | 23 |
| | Methyl acrylate | Mainly isotactic | 23 |
| | Methyl methacrylate | Mainly syndiotactic | 23 |
| | Acrylic acid | Mainly syndiotactic | 23 |
| | Methyl vinyl ketone | Mainly isotactic | 23 |
| | Several monomer pairs[a] | Statistical copolymers | 23 |

[a] The two monomers are coincluded at the same time. [b] The second monomer replaces the former after partial polymerization.

is revealed in polypropylene synthesized in PHTP by $^{13}$C NMR analysis. The content of head-to-head junctions is less than 4%.[41]

In the polymerization of unsymmetrical methyl-substituted hexatrienes, defects were detected by ozonization and gas chromatographic analysis of the fragments.[61]

Another interesting effect of the particular reactivity of monomers in the inclusion state was shown by divinylbenzene in TPP. Topological constraints force the monomer to give rise to a linear, not cross-linked, polymer, only one of the double bonds being involved in the polymerization.[17]

### 18.2.2 Tacticity

The configuration of polymers can be modified, depending on the matrix chosen and on the specific conditions. Isotacticity can, in many cases, be even higher than for the same polymers obtained by coordination polymerization, but atactic polymers are also produced. Among other examples of high isotactic content, poly(1,3-pentadiene), synthesized in PHTP at room or a lower temperature starting from the *trans* monomer, shows the highest regularity, as demonstrated by X-ray, IR and DSC analysis.[30,32,57] It contains 97–98% isotactic diads as determined by $^{13}$C NMR.[59]

Although endowed with the same pure 1,4-*trans* structure, a lower *mm* triad content is measured in polypentadienes obtained in DCA and ACA.[13,58] *cis*-Pentadiene in PHTP also leads to a less ordered polymer and, by increasing the polymerization temperature, the syndiotactic content can be considerably enhanced.[59] Therefore it is possible to obtain the expected structure by choosing the host, the temperature or the monomer feed (Figure 4). An analogous distribution of defects was obtained in DCA.[15]

A different tacticity in poly(2-methylpentadiene), as produced in DCA and PHTP, is apparent from the solubility of the polymer. The polymer from PHTP, insoluble in common solvents, was converted by hydrogenation into head-to-tail polypropylene on which the tacticity was measured.[60]

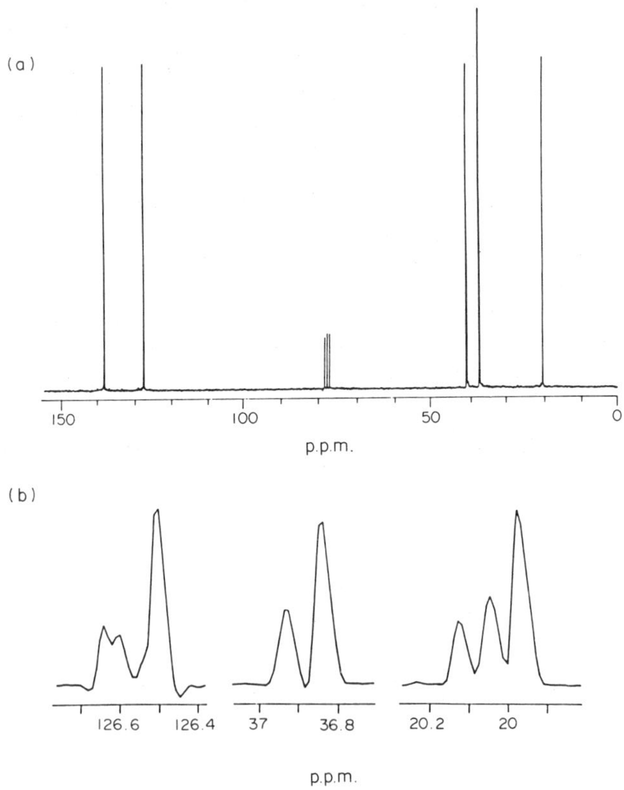


**Figure 4** $^{13}$C NMR spectra of polypentadienes obtained in PHTP, run at 50.3 MHz: (a) polymer synthesized from the *trans* monomer at room temperature (no evidence of splittings are present in the signals); (b) polymer synthesized from the *cis* monomer at +60 °C; it shows the expansion of the signals due to carbons 2,4 and to the methyl, starting from the left; the less intense signals are due to the steric defects

This experiment led to the synthesis of the first example of a new class of polymer stereoisomers for which the name 'hemiisotactic' was proposed.[33,73] They contain two interwoven series of sterically non-equivalent carbon atoms: in one series all the carbons have the same configuration, in the other the carbon configuration is random. Quantitative interpretation of the $^{13}$C NMR spectrum was carried out at the level of decads and undecads.[37] In this way, a steric purity higher than 96%, expressed as isotactic diads, was determined for the initial unsaturated polymer. The general trend was confirmed by $^{1}$H and $^{2}$D NMR spectroscopy.[34]

Good stereochemical control is not obtained for vinyl monomers: polyacrylonitrile is reported as being only partly isotactic when formed in urea[6,49,50] and TPP;[23] polymerization of methyl acrylate and methyl vinyl ketone produces partly isotactic polymers.[23] Prevailingly syndiotactic polymers are formed from vinyl chloride in urea,[6,47] TPP[16] or PHTP,[42] from methyl methacrylate and acrylic acid in TPP,[23] from vinyl fluoride in urea[48] and from propylene in PHTP.[41] Accurate data about the microstructures of many of these polymers are lacking. Only poly(vinyl chloride) and polypropylene have been analyzed at the pentad level. Both polymers synthesized in PHTP contain 70% *r* diads;[41,42] in particular, the stereosequence distribution of polypropylene can be described by a first-order Markov chain with $p_{mr}=0.44$ and $p_{rm}=0.24$.

## 18.3 ASYMMETRIC INDUCTION

The synthesis of optically active 1,4-*trans*-polypentadiene obtained in (−)PHTP was the first example of asymmetric radical polymerization.[11] Enantiomeric excess was determined on methylsuccinic acid, the ozonization product of the polymer. In spite of the low asymmetric induction obtained, this experiment suggested that chirality can be transmitted in an organized medium by weak van der Waals interactions.

The resolution of PHTP into optical isomers is tiresome, so naturally occurring optically active compounds, such as DCA and its derivative ACA, were applied with positive results. *cis*-1,3-Pentadiene polymerized in DCA produced a 1,4-*trans* polymer with a high optical rotatory power, corresponding to an optical purity of 20%, but showing a poor stereoregularity.[15]

In general, it is the chiral shape of the host channel and not the presence of a chiral initiator which is responsible for the preferential formation of the polymeric chains (or monomeric units) with a given configuration. The mechanism of asymmetric induction operates 'through space' and not through chemical bonds. This conclusion was reached from the synthesis of the optically active poly(isoprene-*block*-pentadiene) copolymer.[14] Isoprene was added to irradiated (−)PHTP, and, after partial polymerization, it was stripped away. *trans*-1,3-Pentadiene was then transferred to the living polyisoprenyl radical (Scheme 1).

$$\text{PHTP} \xrightarrow{\gamma\text{-rays}} \text{PHTP}\cdot \xrightarrow{\text{isoprene}} \text{PHTP}\!-\!(\text{isoprene})_n^{\cdot}$$

$$\text{PHTP}\!-\!(\text{isoprene})_n^{\cdot} \xrightarrow{\text{pentadiene}} \text{PHTP}\!-\!(\text{isoprene})_n\!-\!(\text{pentadiene})_m$$

**Scheme 1**

In this way, an achiral polyisoprenic moiety divides the chiral initiator (PHTP radical) from the polypentadiene block; the optical purity of this latter is the same as that of the pentadiene homopolymer obtained under similar conditions. The same conclusion was reached by a statistical analysis of the microstructure of polypentadiene obtained in DCA:[15] the configurational defect distribution was interpreted as an asymmetric Bernoulli process.

Thus, the configuration of each monomeric unit is determined by external factors (the shape and chirality of the cavity) and is independent of the configuration of the preceding unit.

In addition to pentadiene, *cis*- and *trans*-2-methylpentadiene were used; the polymers as synthesized in DCA showed an exceptionally high optical activity.[43,65,66,68,72] The values do not correspond to a comparable optical purity as determined by the optical rotatory power of the ozonization products; however, the optical activity of polymers decreases on increasing the temperature and it reaches a stable value of near zero at +60 °C. The high rotatory power is restored by standing at room temperature for a long time. This particular behaviour was originally interpreted as a conformational change.[74] However, light-scattering and viscosity measurements proved that supermolecular aggregation of polymer chains determine the optical rotatory power at room temperature.[75]

This phenomenon is present only in samples containing a moderate degree of steric disorder; highly regular poly(2-methylpentadiene), as obtained in (−)PHTP, could not be used for solution measurements because it is insoluble at room temperature in common solvents.[32,33]

Optically active poly(2-*s*-butylbutadiene) was obtained by polymerization of racemic monomer in DCA and in ACA matrices.[45] This is not actually a case of asymmetric induction; as the monomer is a racemic mixture, the selection between enantiomers operates in the early stage of formation of the inclusion compound within the chiral host. This is the first example of enantioselective inclusion polymerization.

A different problem arises from the possible presence of the chiral matrix which may, in principle, affect the true optical activity of the polymer. The presence of one host molecule per chain was actually demonstrated in the case of PHTP by using a $^{14}C$-labelled host,[56] but the molecular weights are generally high enough to minimize the influence of the chiral initiator on the optical activity. For polymers obtained in β-cyclodextrin, decomposition of the adduct is much more difficult and a not negligible amount of optical rotation may derive from the residue of several host molecules interacting with the macromolecule like a rotaxane.[19]

## 18.4 INTRODUCTION OF DISORDER AND MODIFICATION OF POLYMERS OBTAINED IN THE INCLUSION STATE

We have previously examined some phenomena involving disorder in homopolymers, both of steric and constitutional origin, which can be modified in relation to the temperature of synthesis. A controlled degree of disorder can also be introduced in IP polymers by adding a second monomer in traces or massive amounts.

Some copolymers were obtained by Brown and White in thiourea at the beginning of the study of IP.[5] However, at that time the concept of high selectivity in IP prevailed, due to a main interest in ordered polymers and the lack of microstructural methods to investigate them. Recently, even monomers markedly different in shape and dimensions, such as butadiene and 2,3-dimethylbutadiene, were copolymerized in PHTP.[20,22] Thus it is not necessary to assume the hypothesis of isomorphism between the two coincluded monomers, which was given as an explanation of the first examples of inclusion copolymerization.[51,55]

Recently, many successful copolymerization results were described in the literature (see Table 1). In particular, previously unstudied pairs of monomers, such as acrylonitrile/acrylic acid and acrylonitrile/methyl methacrylate, were copolymerized in TPP.[23] In some cases, variations in copolymer composition with reaction feed were compared to the results for the same two monomers polymerized in solution by a radical mechanism.[21,36,76] However, the structures of these copolymers are often very different and direct comparison becomes impossible.

Additional information is obtained by $^{13}C$ NMR analysis, by which it is possible to determine the concentration of diads, triads and longer sequences of monomer units. The concentration of each sequence can be compared to values calculated for a particular type of distribution (Bernoulli or first-order Markov).[20,77] In this way, it was shown that two similar monomers lead to ideal azeotropic polymerization ($r_1 = r_2 = 1$), whilst mixtures of very different monomers give longer homopolymeric blocks ($r_1 \cdot r_2 > 1$).[77]

Almost all mono- or di-methyl-substituted 1,3-butadienes can copolymerize in PHTP producing a 1,4-*trans* structure.[22] Only minor amounts of head-to-head and tail-to-tail junctions are sometimes detected in copolymers, although these defects are not present in the respective homopolymers. The constitutional purity of polymers prepared by IP and the variety of possible sequences make them suitable models of more disordered polymers synthesized in another way.

In particular, depending on the monomers chosen and the composition of the feed, it is possible to change sequence types and distributions, thus helping $^{13}C$ NMR spectrum assignment. An extensive study of diene copolymers obtained in PHTP provided a general interpretation of their $^{13}C$ NMR spectra and the elucidation of their microstructure at the level of triads of monomer units. This scheme was tested on a terpolymer (pentadiene, 2-methylpentadiene and 4-methylpentadiene), which showed a nearly statistical distribution of the three monomer units.[22]

A related subject is the introduction, in a subsequent reaction, of controlled disorder into highly ordered IP polymers. This is the case of hemiisotactic polypropylene, in which disorder is introduced by a non-selective hydrogenation of the double bonds.[33]

In addition, polymers endowed with a high level of unsaturation and free from metal catalysts were recently obtained from poly(2-chloroacrylonitrile) (**1**; Scheme 2) prepared by IP.[70] The polymers were directly dehydrogenated by heating at 140 °C, with or without previous extraction

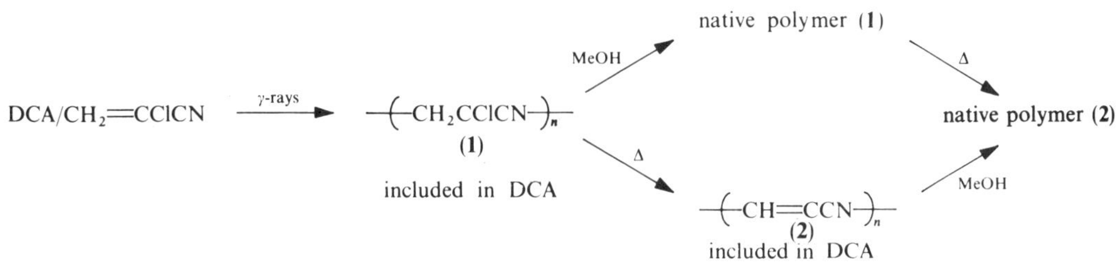


**Scheme 2**

from the matrix, as shown in Scheme 2. The yield of the dehydrochlorination reaction giving poly(1-cyanovinylene) (**2**) reaches 70%.

## 18.5 PROPERTIES OF NATIVE POLYMERS

The topochemical reaction which transforms the host–monomer into the host–polymer crystal produces single polymer chains segregated one from the other. The structure of the channel prevents the chain-folding normally present in crystalline polymers. The polymer itself, when freed from the matrix and avoiding polymer dissolution or melting, can retain these arrangements and, consequently, possesses an extended-chain morphology.

Thus, under mild conditions, it is possible to obtain a similar morphology to that obtained by crystallization under high pressure, as already described for polyethylene.[78] Using a single crystal of host–monomer, the whole process can be performed, retaining the same macroscopic shape, prior to polymer isolation. This is a general result, observed in polybutadiene as polymerized in urea,[6] in polypentadiene synthesized in PHTP,[31] in polydiacetylene as obtained in DCA[64,79] and in PHTP,[62] in poly(2-chloroacrylonitrile) and in poly(1-cyanovinylene) obtained in DCA or ACA.[70]

The first indications of the presence of an extended-chain morphology were given by White.[6] Polybutadiene in that case showed a melting point in the 147–157 °C range. An X-ray diffraction pattern, obtained at low angle, later confirmed a non-folded structure for this polymer.[46]

A thermal study of an extended-chain morphology was carried out on poly(*trans*-1,3-pentadiene) prepared in PHTP.[30] The results are summarized as follows: the native polymer melts at around 100–105 °C, more than 15 degrees higher than that observed during the second melting of the same sample or in samples crystallized from solution; crystallites of the native polymer suspended in the molten polymer are not able to act as crystallization nuclei for extended-chain crystals; a strong dependence of the melting point on the heating rate (superheating) was also observed. It is known that 1,4-*trans*-polybutadiene exhibits a solid–solid enantiotropic transition around 60 °C. The extent of the endothermal effect observed in the DSC curve of samples obtained in clathrates depends on the temperature at which decomposition of the adduct is performed.[46] In polymers freed from the host at low temperature (cold water for urea or thiourea, boiling pentane for PHTP) the transition occurs over a wider temperature range and shows a lower $\Delta H$ than that for polymers extracted at higher temperatures.

## 18.6 MECHANISM AND THERMODYNAMIC RELATIONSHIPS IN INCLUSION POLYMERIZATION

In the preceding sections, it was assumed that IP takes place by a radical mechanism inside the ICs, but no proof was provided. In this section, the thermodynamic basis and mechanism of IP will be considered along with the basis of the process, *e.g.* the existence of the IC with the monomers.

According to a recently introduced terminology, most of the ICs useful for polymerization are 'binary tubulate clathrates'; this definition indicates that two components are present (host and guest), that the space available for the guest is roughly cylindrical and that the interaction between the two components is merely steric in origin, even if stronger forces maintain the host crystal packing. Systems containing two monomers and suitable for copolymerization are 'ternary tubulate clathrates'.[80]

Channel dimensions depend on the type and nature of the host molecule. Urea and thiourea form hydrogen bonds and a rather stiff structure. Only linear monomers can therefore be included in urea channels (5.25 Å, 1 Å$=10^{-10}$ m);[81] cyclic or branched monomers can be included in thiourea channels (6.1 Å).[82] The PHTP lattice is kept together only by van der Waals interactions; the formation of IC with molecules far different in shape, dimension and polarity is promoted by the channel softness.[83] An intermediate behaviour is shown by TPP,[18] DCA and ACA.[84] $\beta$-Cyclodextrins belong to the topological class of rotaxanes, and their adducts are stable even in solution.[85]

In most cases, the pure host does not show empty channels in its own crystalline structure, and the crystalline phases of the host and of the IC are immiscible. This last condition creates an analogy between ICs and hydrates, solvates and racemates.[1,86] A definite decomposition pressure, which obeys the Clapeyron law, can be measured in adducts with volatile guest molecules. Therefore, the pressure–temperature host–monomer phase diagrams are sensitive tools for establishing the field of existence of the IC.[10,87]

Direct examination of a specific host–monomer system could, however, be affected by the occurrence of even a small amount of oligomerization. Assuming saturated molecules as models for monomers, this problem can be overcome. In particular, the phase diagram for *n*-heptane/PHTP was established by vapour pressure measurements and by calorimetric analysis.[10,87] The stability of the IC ranges from the decomposition curve of the IC and the vapour pressure curve of the saturated solution.

Figure 5 shows the pressure–temperature diagram under the hypothesis of non-miscibility of the crystalline phases, ideality of the liquid phase and constancy of the enthalpy of vapourization in the temperature range studied. In order to obtain a complete description of the range of stability of the IC, the interactions in the liquid phase have also to be taken into account, in addition to the solid-state properties. A miscibility gap in the liquid phase lowers the decomposition temperature; thus polarity and steric factors play important roles.[88]

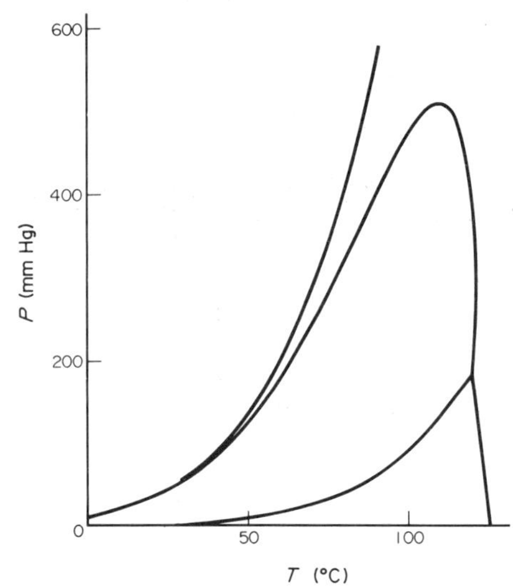


**Figure 5** *n*-Heptane/PHTP phase diagram, 1 mm Hg = 133.322 Pa (see text)

The coincidence of the stability range of the IC with IP boundaries was proved by a sudden drop of the polymerization yield as temperature and pressure exceeded those limits.[55]

The IC between PHTP and polyethylene is isomorphous with the PHTP/*n*-heptane IC and was taken as a model for the host–polymer phase. For this non-volatile guest, the phase diagram was determined by DSC and interpreted on the basis of the Flory–Huggins theory.[28,83]

The results described above have enhanced the knowledge of the two phases existing before and after polymerization. The conversion of a single crystal host–monomer into a single crystal host–polymer was directly observed by X-ray analysis for urea/butadiene[52] and PHTP/butadiene adducts.[54] However, the propagation step was actually followed by ESR techniques, demonstrating the radical nature of the active centre. The polymerization of butadiene was observed in a urea matrix[27] and, later, several diene monomers were observed in PHTP.[26] From these data some otherwise inaccessible conclusions can be drawn about the initiation step, the direction of propagation and, sometimes, the conformational arrangement and mobility of the last few bonds near the propagating chain-end (Figure 6).[26,89]

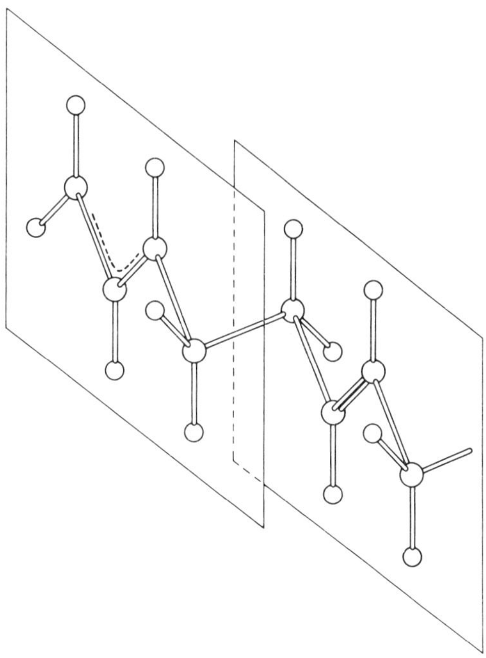

**Figure 6** Conformational arrangement of the growing chain-end of poly(*trans*-1,3-pentadiene) included in PHTP, as determined by ESR spectroscopy

Variable-temperature observations performed in the $-150$ to $+60\,^{\circ}C$ range demonstrated that the conformation of the included polybutadienyl radical is temperature dependent and mobile at room temperature (in relation to the ESR time-scale) in both urea and PHTP. One or two methyl substituents in any position on a butadiene skeleton hinders free rotation of the growing chain-end in PHTP. This was recently also proved true for 2,3-dimethylbutadiene in DCA.[64]

The detection of a radical by ESR spectroscopy does not prove in itself that it is the actual propagating species; the polymer chain might have grown independently by a faster active centre. The problem was solved by performing a block copolymerization experiment monitored by ESR spectroscopy. PHTP-irradiated material shows a complex spectrum due to the superposition of different radicals; a few minutes after the first monomer was added, the specific radical was recorded. After the excess monomer was stripped away, another monomer, which leads to a quite different ESR spectrum for its conjugate radical, was transferred to the tube and the spectrum due to the new radical rapidly appeared.[26] These facts demonstrate both the actual radical nature of this type of polymerization and its living nature.

Even in the presence of the IC, host–monomer polymerization does not always take place in a reasonable time. This happens in the case of *trans,trans*-2,4-hexadiene in PHTP, although a radical formed after irradiation on the diene system was observed by ESR.[26] Two methyls facing one another on a *trans* double bond keep themselves apart (6 Å) thus preventing polymerization.[56] This fact does not hold true in the less selective channels of DCA.[2]

## 18.7 RECENT DEVELOPMENTS

The polymers obtained by IP are a structural reference point for polymers from other sources. Recently, this fact was extensively exploited for the NMR analysis[22,32,33,35,48] and X-ray characterization of the crystalline state.[39,40]

One of the challenging tasks, nowadays, is the synthesis of polymers which contain a high degree of unsaturation and could behave as organic conductors. Native ICs contain isolated polymer chains and the polymers themselves are free of catalysts. These features are present in the previously mentioned dehydrohalogenation of preformed polymers,[70] in the butadiyne reacted directly in DCA[79] or in PHTP.[62]

On the possibility of extending the field of IP to new monomers and systems, alternative methods for initiating polymerization have been tested. A few compounds that can dissociate, and are used for promoting radical polymerization in solution or in mass, have been tried in the solid state. Azobis(isobutyronitrile) (AIBN) and other free radical initiators were included and dissociated by heat, giving rise to polymers with molecular weights of a few thousands.[64]

The quantitative description of the host–guest phase diagram facilitates the determination of the polymerization conditions for host–monomer pairs not previously used; in this way, solid-state polymerization could also be achieved for propylene and ethylene under moderate pressure and at room temperature.[41] In the particular case of propylene, it constitutes the only practicable form of radical polymerization.

## ACKNOWLEDGEMENT

We are indebted to Professor Mario Farina for his useful suggestions and a general discussion of the text.

## 18.8 REFERENCES

1. H. M. Powell, *J. Chem. Soc.*, 1948, 61.
2. M. Farina, in 'Inclusion Compounds', ed. J. L. Atwood, J. E. D. Davies and D. D. MacNicol, Academic Press, London, 1984, vol. 3, chap. 10.
3. J. L. Atwood, J. E. D. Davies and D. D. MacNicol (eds.), 'Inclusion Compounds', Academic Press, London, 1984, vols. 1–3.
4. H. Clasen, *Z. Elektrochem.*, 1956, **60**, 982.
5. J. F. Brown and D. M. White, *J. Am. Chem. Soc.*, 1960, **82**, 5671.
6. D. M. White, *J. Am. Chem. Soc.*, 1960, **82**, 5678.
7. Y. Chatani, *Prog. Polym. Sci., Jpn.*, 1974, **7**, 149.
8. M. Farnia, *Tetrahedron Lett.*, 1963, 2097.
9. M. Farina, G. Allegra and G. Natta, *J. Am. Chem. Soc.*, 1964, **86**, 516.
10. M. Farina, in 'Inclusion Compounds', ed. J. L. Atwood, J. E. D. Davies and D. D. MacNicol, Academic Press, London, 1984, vol. 2, chap. 3.
11. M. Farina, G. Audisio and G. Natta, *J. Am. Chem. Soc.*, 1967, **89**, 5071.
12. M. Miyata and K. Takemoto, *J. Polym. Sci., Polym. Lett. Ed.*, 1975, **13**, 221.
13. M. Miyata, Y. Kitahara and K. Takemoto, *Polym. Bull. (Berlin)*, 1980, **2**, 671.
14. M. Farina, G. Di Silvestro and P. Sozzani, *Makromol. Chem., Rapid Commun.*, 1981, **2**, 51.
15. G. Audisio, A. Silvani and L. Zetta, *Macromolecules*, 1984, **17**, 29.
16. J. Finter and G. Wegner, *Makromol. Chem.*, 1979, **180**, 1093.
17. H. R. Allcock, W. T. Ferrar and M. L. Levin, *Macromolecules*, 1982, **15**, 697.
18. H. R. Allcock, in 'Inclusion Compounds', ed. J. L. Atwood, J. E. D. Davies and D. D. MacNicol, Academic Press, London, 1984, vol. 1, chap. 8.
19. M. Maciejewski, *J. Macromol. Sci., Chem.*, 1979, **A13**, 77.
20. G. Di Silvestro and P. Sozzani, 'Preprints of Convegno Italiano di Chimica delle Macromolecule', Milano, 1981, p. 55.
21. C. Schneider and H. P. Bohlmann, 'Preprints of the IUPAC Symposium on Macromolecules', Florence, 1980, vol. 2, p. 134.
22. P. Sozzani, G. Di Silvestro, M. Grassi and M. Farina, *Macromolecules*, 1984, **17**, 2538.
23. H. R. Allcock and M. L. Levin, *Macromolecules*, 1985, **18**, 1324.
24. M. Farina, *J. Chem. Soc., Chem. Commun.*, 1976, 816.
25. M. Farina and G. Di Silvestro, *J. Chem. Soc., Perkin Trans. 2*, 1980, 1406.
26. P. Sozzani, G. Di Silvestro and A. Gervasini, *J. Polym. Sci., Polym. Chem. Ed.*, 1986, **24**, 815.
27. T. Ohmori, T. Ichikawa and M. Iwasaki, *Bull. Chem. Soc. Jpn.*, 1973, **46**, 1383.
28. M. Farina, G. Di Silvestro and M. Grassi, *Makromol. Chem.*, 1979, **180**, 1041.
29. Y. Chatani and S. Nakatani, *Macromolecules*, 1972, **5**, 597.
30. M. Farina and G. Di Silvestro, *Makromol. Chem.*, 1982, **183**, 241.
31. M. Farina and G. Di Silvestro, *Gazz. Chim. Ital.*, 1982, **112**, 91.
32. P. Sozzani, G. Di Silvestro, M. Grassi and M. Farina, *Macromolecules*, 1984, **17**, 2532.
33. G. Di Silvestro, P. Sozzani, B. Savaré and M. Farina, *Macromolecules*, 1985, **18**, 928.
34. A. De Marco, P. Sozzani, G. Di Silvestro and M. Farina, to be published.
35. G. Di Silvestro, P. Sozzani and M. Farina, *Macromolecules*, 1987, **20**, 999.
36. C. Schneider, personal communication.
37. P. Sozzani and C. Oliva, *J. Magn. Reson.*, 1985, **63**, 115.
38. P. Sozzani and G. Di Silvestro, *Gazz. Chim. Ital.*, in press.
39. S. Brückner, G. Di Silvestro and W. Porzio, *Macromolecules*, 1986, **19**, 235.
40. S. Brückner, S. Luzzati, W. Porzio and P. Sozzani, *Macromolecules*, 1987, **20**, 585.
41. G. Di Silvestro, P. Sozzani and M. Farina, *Polym. Prepr., Am. Chem. Soc., Div. Polym. Chem.*, 1986, **27** (1), 92.
42. G. Di Silvestro, P. Sozzani and M. Farina, *Mol. Cryst. Liq. Cryst.*, to be published.
43. K. Takemoto and M. Miyata, *J. Macromol. Sci., Rev. Macromol. Chem.*, 1980, **C18**, 83.
44. M. Farina and G. Di Silvestro, in 'Encyclopedia of Polymer Science and Engineering', ed. H. F. Mark, N. M. Bikales, C. G. Overberger and G. Menges, 2nd edn., Wiley, New York, 1988, vol. 2, p. 486.
45. M. Miyata, T. Tsuzuki and K. Takemoto, *Makromol. Chem., Rapid Commun.*, 1987, **8**, 501.

46. Y. Chatani and S. Kuwata, *Macromolecules*, 1975, **8**, 12.
47. Y. Chatani, K. Yoshimori and Y. Tatsuta, *Polym. Prepr., Am. Chem. Soc., Div. Polym. Chem.*, 1978, **19** (2), 132.
48. R. F. Cais and J. M. Kometani, *ACS Symp. Ser.*, 1984, **247**, 153.
49. T. Yoshino, H. Kenjo and K. Kuno, *J. Polym. Sci., Polym. Lett. Ed.*, 1967, **5**, 703.
50. K. Matsuzaki, T. Uryu, M. Okada and H. Shiroki, *J. Polym. Sci., Part A*, 1968, **6**, 1475.
51. Y. Chatani and S. Nakatani, *Macromolecules*, 1972, **5**, 597.
52. Y. Chatani, S. Nakatani and H. Tadokoro, *Macromolecules*, 1970, **3**, 481.
53. M. Farina, G. Natta, G. Allegra and M. Löffelholz, *J. Polym. Sci., Part C*, 1967, **16**, 2517.
54. A. Colombo and G. Allegra, *Macromolecules*, 1971, **4**, 579.
55. M. Farina, G. Audisio and M. T. Gramegna, *Macromolecules*, 1972, **5**, 617.
56. M. Farina, U. Pedretti, M. T. Gramegna and G. Audisio, *Macromolecules*, 1970, **3**, 475.
57. M. Farina, G. Audisio and M. T. Gramegna, *Macromolecules*, 1971, **4**, 265.
58. L. Zetta, G. Gatti and G. Audisio, *Macromolecules*, 1978, **11**, 763.
59. M. Farina. in, 'Recent Advances in Mechanistic and Synthetic Aspects of Polymerization', ed. M. Fontanille and A. Guyot. Reidel, Dordrecht, 1987, p. 261.
60. M. Farina, G. Di Silvestro and P. Sozzani, *Macromolecules*, 1982, **15**, 1451.
61. M. Farina, G. Di Silvestro, U. Pedretti and G. Frumusa, *Chim. Ind. (Milan)*, 1973, **55**, 159.
62. B. Jorgensen, R. Liepins and S. Agnew, *Polym. Bull. (Berlin)*, 1986, **16**, 263.
63. M. Miyata and K. Takemoto, *J. Macromol. Sci., Chem.*, 1978, **A12**, 637.
64. M. Miyata, F. Noma, Y. Osaki, K. Takemoto and M. Kamachi, *J. Polym. Sci., Polym. Lett. Ed.*, 1986, **24**, 457.
65. G. Audisio and A. Silvani, *J. Chem. Soc., Chem. Commun.*, 1976, 481.
66. M. Miyata, A. Wada and K. Takemoto, 'Preprints of the IUPAC Symposium on Macromolecules', Strasbourg, 1981, p. 261.
67. M. Miyata and K. Takemoto, *Makromol. Chem.*, 1978, **179**, 1167.
68. M. Miyata and K. Takemoto, *Polym. J.*, 1977, **9**, 111.
69. M. Miyata, K. Shinmen and K. Takemoto, *Polym. Prepr., Am. Chem. Soc., Div. Polym. Chem.*, 1979, **20** (1), 716.
70. M. Miyata, K. Okanishi and K. Takemoto, *Polym. J.*, 1986, **18**, 185.
71. M. Miyata, Y. Kitahara, Y. Osaki and K. Takemoto, *J. Inclusion Phenom.*, 1984, **2**, 391.
72. M. Miyata, Y. Kitahara and K. Takemoto, *Makromol. Chem.*, 1983, **184**, 1771.
73. M. Farina, G. Di Silvestro, P. Sozzani and B. Savaré, *Macromolecules*, 1985, **18**, 923.
74. M. Miyata, Y. Kitahara and K. Takemoto, *Polym. J.*, 1981, **13**, 111.
75. P. Sozzani, G. Di Silvestro, M. Mauritano and M. Farina, 'Preprints of the IUPAC Symposium on Macromolecules', The Hague, 1985, p. 263.
76. C. Schneider and H. P. Bohlmann, 'Preprints of the IUPAC Symposium on Macromolecules', The Hague, 1985, p. 260.
77. P. Sozzani and G. Di Silvestro, unpublished results.
78. B. Wunderlich, 'Macromolecular Physics', Academic Press, Orlando, FL, 1973, vol. 1; 1976, vol. 2; 1980, vol. 3.
79. M. Miyata, H. Tsutsumi and K. Takemoto, 'Preprints of the International Conference on Conductive Polymers', Kyoto, 1986, p. 382.
80. E. Weber and H. P. Josel, *J. Inclusion Phenom.*, 1983, **1**, 79.
81. A. E. Smith, *Acta Crystallogr.*, 1952, **5**, 224.
82. K. Takemoto and N. Sonoda, in 'Inclusion Compounds', ed. J. L. Atwood, J. E. D. Davies and D. D. MacNicol, Academic Press, London, 1984, vol. 2, chap. 2.
83. G. Allegra, M. Farina, A. Immirzi, A. Colombo, U. Rossi, R. Broggi and G. Natta, *J. Chem. Soc., Part B*, 1967, 1020.
84. E. Giglio, in 'Inclusion Compounds', ed. J. L. Atwood, J. E. D. Davies and D. D. MacNicol, Academic Press, London, 1984, vol. 2, chap. 7.
85. W. Saenger, in 'Inclusion Compounds', ed. J. L. Atwood, J. E. D. Davies and D. D. MacNicol, Academic Press, London, 1984, vol. 2, chap. 8.
86. J. E. Ricci, 'The phase rule and heterogeneous equilibrium', Van Nostrand, New York, 1951.
87. M. Farina, G. Di Silvestro and P. Sozzani, *Mol. Cryst. Liq. Cryst.*, 1983, **93**, 169.
88. M. Farina, G. Di Silvestro and A. Colombo, *Mol. Cryst. Liq. Cryst.*, 1986, **137**, 265.
89. M. Farina, G. Di Silvestro and P. Sozzani, *ACS Symp. Ser.*, 1987, **337**, 79.

# 19

# Polymerization by High-energy Radiation

VIVIAN T. STANNETT
*North Carolina State University, Raleigh, NC, USA*
JOSEPH SILVERMAN
*University of Maryland, College Park, MD, USA*
and
JACK L. GARNETT
*University of New South Wales, Kensington, Australia*



## 19.1 INTRODUCTION

### 19.1.1 General

Almost immediately following the discovery of X-rays by Roentgen and of radioactivity by Becquerel, it became clear that absorption of the energetic subatomic emissions associated with these phenomena gave rise to ionization and other endoergic processes, even at low temperatures. The immediate products were usually transient species which followed exoergic paths to form stable products. However, it was not until the mid-1920s that ionizing radiation was used for the room-temperature production of the transient species that initiate chain polymerizations. The reason for the delay is that Staudinger's concept of chain polymerization of vinyl monomers was not advanced until 1920 and not widely accepted until the end of that decade. During that decade, Lind and Mund achieved gas-phase polymerization with $\alpha$ particles and Coolidge achieved similar results with accelerator-produced electrons. In the 1930s, the studies were extended, became more detailed and were applied to liquid monomers. This was made possible by the availability of a few sources of more penetrating ionizing radiation in the form of X-rays, $\gamma$-rays and the newly discovered neutrons. The similarity between radiation polymerization and photopolymerization was apparent from the first, but the differences were not well understood until the early 1950s. It was not until then that relatively inexpensive sources of ionizing radiation became widely available. Also, the radiation chemistry of the polymers that were associated with radiation damage problems in lubricants and insulation in nuclear reactors led, in part, to consideration of radiation methods of synthesizing polymers, curing acrylic coatings and grafting reactions. Both the fundamental and applied aspects have been the subject of many investigations into radiation-induced free-radical and ionic chain polymerizations.

### 19.1.2 Properties of Ionizing Radiation

Radiation is defined as electromagnetic quanta, so that ionizing radiation in the strictest sense would be limited to photons with energies of above 10 eV. In practice, ionizing radiation usually refers to any energetic neutral or ionic fragment with a kinetic energy in the MeV range. Thus fast neutrons, fission recoils, radioactivity-produced $\alpha$ and $\beta$ particles and accelerator-produced electrons and ions are included along with X-rays, $\gamma$-rays and bremsstrahlung as ionizing radiation. In the case of neutral species, fast neutrons and energetic photons, the ionization is produced by charged secondary particles, recoil protons and Compton electrons. For polymerization research and industrial production, the principal radiation sources have been $^{60}$Co and low-energy electron accelerators. The former emits two 1.25 MeV of $\gamma$-rays per decay which have a mean free path of about 250 mm in condensed organic matter. $^{137}$Cs, which emits one 0.66 MeV of $\gamma$-rays per decay, has also been used in polymerization research. Accelerators producing intense beams of electrons in the MeV range have been especially useful for measuring dose-rate effects, but it is the lower-energy electrons, those in the 0.18 0.5 MeV range, which have been employed in the most successful industrial chain-polymerization processes, namely the curing of coatings, finishes and adhesives.

While radiation-source technology is beyond the scope of this chapter, some remarks on the absorption of electrons and $\gamma$-rays are essential to the understanding of radiation polymerization.

The principal means by which the kinetic energy of electrons are absorbed in a monomer are ionization and excitation; electron–nucleus interactions such as Rutherford scattering and bremsstrahlung generation are trivial effects. The electron is not slowed down continuously but makes some 10 000 discrete interactions with the absorber before it is moderated to thermal energy. In a liquid monomer the moderation length is 2–20 nm. Ion pairs and excitations are produced in a sphere of this radius, called a spur. The principal means by which a $^{60}$Co $\gamma$-ray interacts with a monomer is by Compton scattering. The immediate consequence is the production of a Compton electron with a mean energy of 0.6 MeV. The secondary electrons behave much like the accelerator-produced electrons. Thus, in its chemical effects, the primary $\gamma$-ray behaves one or two times as a photon and about 10 000 times as an electron. Another simplifying factor is that the stopping power of a monomer for an electron is almost independent of initial energy in the range of 0.2–15 MeV, as is the mean energy deposited in a spur (about 50 eV) and the energy required to form an ion pair. The consequence is that $\gamma$-rays and electrons produce almost identical chemical effects provided they are delivered at the same absorbed dose and at the same dose rates. These two terms bearing the symbols $D$ and $\dot{D}$ have special units. Dose is given in terms of the rad or the gray (Gy). The former, which is an older unit but still in use, is equivalent to the absorption of 100 eV g$^{-1}$ of absorber. The latter, which is the SI unit, represents the absorption of 1 J kg$^{-1}$.

Most of the work in radiation polymerization involves doses of perhaps 1 Mrad, *i.e.* about 10 kGy. Dose rates usually range from $10^{-2}$ to $10^5$ Gy s$^{-1}$. $\gamma$-Ray sources can readily provide dose rates up to 50 Gy s$^{-1}$. Electron accelerators are used to obtain the higher rates.

The yield of a species produced by radiation is given by the $G$ value, which is defined in the older units as the number of ions, radicals, molecules or other species produced or consumed per 100 eV absorbed. The SI unit, in $\mu$mol J$^{-1}$, is 0.104 times the old value. The product of the dose in Gy and the new $G$ values is the concentration in units of $\mu$mol kg$^{-1}$. The product of the old $G$ value and dose in krad yields the same result within 4%.

An example of the use of the radiation yield is the conversion of methanol to ethylene glycol. $G$(methanol) is 10 molecules per 100 eV but $G$(glycol) is 4 in the same units. Thus 20% of the methanol is converted to other products by radiation. This illustrates the close relationship between the radiation yield and the photochemical quantum yield. Another noteworthy point is that $G$ values for primary effects such as ionization or homolytic scission leading to radical chain initiation are of the order of 1 per 100 eV. On the other hand, $G$(monomer) values can be as high as $10^6$ for the radiation-induced chain-propagation reactions.

Most of the units quoted in this chapter are those used by the authors of the articles from which the material is drawn and therefore follow the older tradition.

Initially, the spur is a small isolated region containing charged and excited species at an elevated temperature, but within the time for a few vibrations, $10^{-13}$ s, thermal equilibrium is reached.

Bimolecular intraspur reactions occur with a rate which decreases as the spur expands by diffusion. After $10^{-6}$ s the chemistry is governed by homogeneous reaction kinetics. Within a spur, neutralization reactions take place in a time which is, to a first approximation, directly proportional to the dielectric constant and the cube of the spur radius and inversely proportional to the mobility of the charge carriers.

Prior to neutralization, the cation radical can initiate a polymerization, especially if the addition of the first monomer unit produces a species with a neutral free-radical site spatially separated from the intramolecular site of the positive charge. The presence of a scavenger for the positive charge can lead to conversion of the cation radical to a neutral free-radical which can undergo the usual chain propagation. Alternatively, the presence of an 'inert' polar species can stabilize charge separation and give rise to cationic polymerization. On the other hand, the capture of an electron by a monomer prior to neutralization yields an anion radical which offers the prospect of anionic polymerization.

The exact details of the intraspur reaction sequence leading to the initiation of polymerization is one of the major research problems in this field. Pulse radiolysis, which permits the study of transient species at submicrosecond levels, and the effects of diluents on the rate and molecular weight of the products of irradiation are the principal techniques used for such studies. Within a few microseconds following the creation of the spurs, which dot the path of the ionizing electron, the usual exothermic reactions of propagation, transfer and termination occur.

The spur model is complicated by several additional pertinent facts. First, the energy deposition by the ionizing electron follows a complex distribution both in space and the amount of energy deposited. Spurs are formed by the deposit of less than 100 eV. If the amount of energy is in the 100–500 eV range, the volume is often treated and referred to as a 'blob', a volume with a greater spatial density of spurs. Deposit of *ca.* 1 kV or more produces an electron sometimes called a $\delta$-ray which gives rise to a short dense track of overlapping spurs. Many of the spurs contain more than one ion pair. Furthermore, at very high dose-rates, there is the problem of spur overlap, particularly in high viscosity systems. Yet, despite the complexity of the radiation physics, the most important areas in the radiation chemistry of polymerization reactions can be described in considerable detail by the simple two-group model, *i.e.* the early intraspur reactions and the homogeneous reactions following spur expansion.

The differences between radiation polymerization and photopolymerization should now be clear. Ionizing radiation is not an MeV 'hammer' which shatters monomer molecules into random fragments. The relatively low-energy depositions in discrete spurs produces geminate pairs of electrons and cation radicals; they also produce excited species of a readily characterized nature, either directly or as a consequence of neutralization. Nevertheless, the effects are less selective than those produced by a narrow band of UV. Consider as an example the effect of oxygen and methanol on the polymerization of styrene initiated by $\gamma$-rays compared with initiation by a mercury arc-lamp in a pyrex envelope. Oxygen can enhance the initiation rate of the free-radical photopolymerization but has no effect on the gamma yield of the initiating free radicals; of course, it can add to the propagating chain as it does in polymerizing chains formed by any free-radical initiator. A few percent of methanol, on the other hand, produces a fivefold increase in the $\gamma$-ray initiation-rate but has no effect on the initiation rate of photopolymerization.

The various effects described in this introductory section have been described well in classical monographs on the field.[1–3] A recent and highly detailed discussion of the fundamentals of the interaction of radiation and matter has also been published.[4]

## 19.2 RADIATION-INDUCED FREE-RADICAL POLYMERIZATION

The events leading to creation of the species giving rise to free-radical, cationic and anionic polymerization have been described in some detail in the previous, introductory, section. Briefly, free radicals are formed by homolytic fission of excited molecules. These may be formed by direct excitation or by positive-molecule-ion-radical–electron recombination, as shown in Scheme 1.

$$\mathrm{M} \xrightarrow{h\nu} \mathrm{M}^* \longrightarrow \mathrm{R}\cdot + \mathrm{R}'\cdot$$

$$\mathrm{M} \xrightarrow{h\nu} \mathrm{M}^{\ddagger} + e^- \longrightarrow \mathrm{M}^* \quad \text{(geminate recombination)}$$

$$\mathrm{M}^* \longrightarrow \mathrm{R}\cdot + \mathrm{R}'\cdot$$

**Scheme 1**

The radiation yields ($G$ values) of free radicals vary with the structure of the monomer from about 0.7 for styrene to about 6 for vinyl acetate. The equations shown in Scheme 1 make it clear that radical yields are always higher than those of free ions. This will be discussed later.

In the case of bulk monomers, it is easy to calculate the rate of initiation from the $G$ values, the dose rate and the physical constants of the monomer. If the $k_p/k_t^{1/2}$ values are known for the monomer in question at the temperature used, the rates of initiation can be calculated from the observed steady-state rates of polymerization. Thus, for the simplest case

$$R_i = (R_p^2 k_t)/(k_p^2[M]^2) \tag{1}$$

$R_i$ combined with the measured dose-rate can then be used to calculate the $G$ value for radicals; this method was developed by Chapiro.[2] The free-radical polymerization of styrene is particularly well behaved and the $k_p/k_t^{1/2}$ values are known over a wide temperature range. It can be used, therefore, to measure the rates of initiation in solution. Other methods, for example, the use of radical scavengers such as 2,2-diphenyl-1-picrylhydrazyl (DPPH), can also be used to obtain a direct value for the rate of radical production. With bulk monomers the $R_i$ values are a function only of the dose rate and the kinetics are quite comparable to those of the benzoyl peroxide-initiated system, for example.

Radiation initiation for solution vinyl polymerization is more complex. A special feature of radiation is that it attacks all components of the system including the solvent. With a monomer–solvent mixture, therefore, the rate of initiation can be represented by a linear equation

$$R_i = \dot{D}[AG_m(i)[M] + BG_s(i)[S]] \tag{2}$$

where $\dot{D}$ is the dose rate and $G_m(i)$ and $G_s(i)$ are the $G$ values for initiation for the monomer and solvent, respectively. [M] and [S] are the monomer and solvent concentrations and $A$ and $B$ are constants to convert the $G$ values into appropriate units, *i.e.* moles, liters, seconds.

The method of Chapiro using known $k_p/k_t^{1/2}$ values for the pure monomer can then be used to calculate the overall $R_i$ if the radical yields are purely additive and ideal kinetics apply. The latter implies that all the solvent radicals react immediately with monomer and the termination step is by the recombination of growing chains. If the purely additive effect, sometimes called the simple dilution effect, applies, then a plot of the overall rates of initiation *vs.* $m$, the mole fraction of monomer, gives a straight line. If the plot curves it indicates that energy transfer is taking place, as will be discussed later. Curvature above the line shows sensitization and below indicates deactivation by the solvent.

A more useful analysis has also been made by Chapiro[2] who showed that

$$\frac{R_p}{R_{p0}} = \left[\frac{m}{m + (1 - m)(V_s/V_m)}\right]^{3/2}\left[1 + \phi_{rel}\frac{(1 - m)}{m}\right]^{1/2} \tag{3}$$

where $R_p$ and $R_{p0}$ are the rates of polymerization in solution and pure monomer, respectively. $\phi_{rel} = \phi_s/\phi_m$, where $\phi_s$ and $\phi_m$ are the molar yield constants for free-radical production per unit exposure dose and unit volume from the solvent and monomer, respectively; (these are almost proportional to the $G$ values). $V_s$ and $V_m$ are the respective molar volumes. Plots of $R_p/R_{p0}$ *vs.* $m$ and a family of curves according to the value of $\phi_{rel}$ are shown in Figure 1. When $\phi_{rel} = 1.0$, as obtained with styrene in benzene or methyl methacrylate in ethyl acetate, for example, the kinetics are similar to those obtained with chemically initiated polymerization systems for which there is no effect of the solvent on the initiation rate.

The curves can usually be fitted with a constant value of $\phi_{rel}$; a number of examples may be found in ref. 2. This is not always the case, however. Values of $\phi_{rel}$ may be calculated from equation (3) at various monomer concentrations. If these are not constant it is usually assumed that energy-transfer processes are operative. The term energy-transfer process in this work is used in its most general sense. The usual meaning implies rapid transfer of excitation from an absorbing site (such as a solvent molecule) over several molecular lengths to an energy sink which becomes reactive (such as a solute which becomes an effective producer of radicals). We use the term to also include fast, but ordinary, chemical reactions which effectively transfer the reactive site. Because these two mechanisms are equivalent, we shall describe the solvent and the transfer of excitation by exchange processes. Regardless of the details of the mechanism, the processes are equivalent and are described below in the conventional sense of radiation-produced excited species that interact with monomer

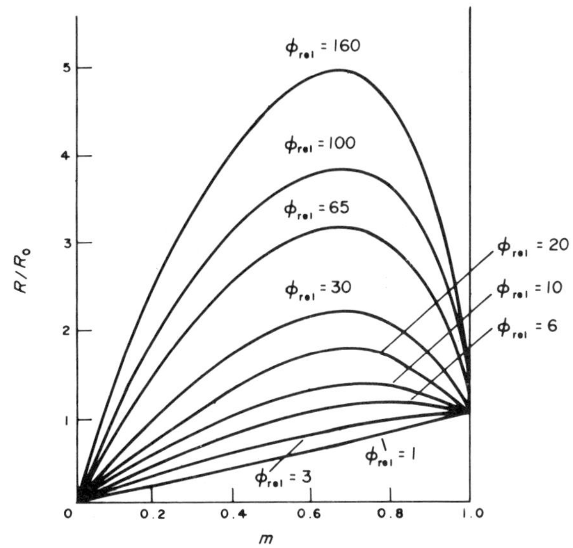


**Figure 1** Relative rates of radiation-induced free-radical polymerization in bulk and in solution at various $\phi_{rel}$ values (reproduced with permission of Wiley-Interscience from ref. 2)

and solvent, thereby producing free radicals or an exchange of energy, thus

$$M^* + S \longrightarrow S^* + M \longrightarrow S\cdot + M \tag{4}$$

$$S^* + M \longrightarrow M^* + S \longrightarrow M\cdot + S \tag{5}$$

An extreme example of energy-transfer behavior was reported by Miller and Stannett[5] with styrene in *n*-dibutyl disulfide (DBS). The results are presented in Table 1 and show a progressive decrease in $\phi_{rel}$ with increasing amounts of disulfide. This implies that energy-transfer processes, predominately from styrene to the disulfide, are taking place. The overall kinetics showed that classic kinetics were otherwise normal. A number of other examples are presented in ref. 2 and by Stannett *et al.*[6]

**Table 1** The Variation of $\phi_{rel}$ with % Solvent for the Radiation-initiated Free-radical Polymerization of Styrene in *n*-Dibutyl Disulfide[a]

| Vol % DBS | 5 | 10 | 20 | 40 | 60 | 80 |
|---|---|---|---|---|---|---|
| $\phi_{rel}$ | 44.0 | 23.8 | 16.6 | 17.3 | 13.6 | 6.9 |

[a] Dose rate 0.005 Mrad h$^{-1}$, at 25 °C.[5]

The kinetics of radiation-induced polymerization in solution in which energy transfer is operative have been developed by Nikitina and Bagdassarian.[7] A term was introduced, $P_{rel}$, representing the relative probability of energy transfer from M to S *vs.* transfer from S to M. This leads to a modified expression for $R_p/R_{p0}$

$$R_p/R_{p0} = \left[\frac{1}{m + (1 - m)V_s/V_m}\right]^{3/2} \left[\frac{1 + \phi_{rel}P_{rel}\dfrac{(1 - m)}{m}}{1 + P_{rel}\dfrac{(1 - m)}{m}}\right]^{1/2} m \tag{6}$$

The development and limitations of this equation have been discussed by Chapiro.[2] The data with styrene and *n*-dibutyl disulfide were analyzed leading to single values of $\phi_{rel} = 8.5$ and $P_{rel} = 3.1$. The degree of fit of the data is shown in Figure 2. Using the value obtained for $\phi_{rel}$ after correcting for energy transfer, the *G*(radical) value (strictly speaking, the *G* value for chain initiation) for dibutyl

$n$-disulfide was estimated to be 3.1. The behavior and values were found to be in good agreement with other data published for various disulfides.

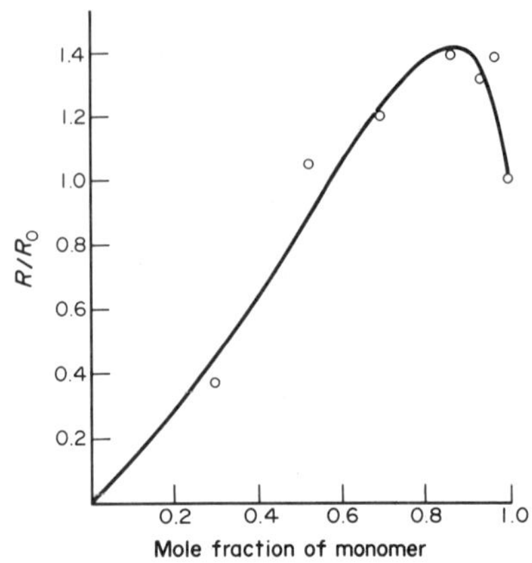


**Figure 2** Relative rates of radiation-induced free-radical polymerization of styrene in dibutyl disulfide at 25 °C. The line corresponds to equation (6) with $\phi_{rel} = 8.5$ and $P_{rel} = 3.1$ (reproduced with permission of J. Wiley & Sons from ref. 5)

When the $\phi_{rel}$ values are less than unity and plots of $R_i$ *vs.* $m$ curve below the $\phi = 1.0$ line, there is deactivation by the solvent, whereas curvature above the line signifies $\phi_{rel} > 1.0$, indicating sensitization of the polymerization by the solvent.

The effects of solvents on the rates of initiation can have considerable implications for radiation grafting. This was clearly shown for the grafting of styrene to cellulose acetate in solution.[6] The use of pyridine with a $\phi_{rel} \sim 0$ compared with dimethylformamide (DMF) with $\phi_{rel} \sim 5.0$ was found to give similar yields of graft copolymer. However, the accompanying homopolymer was only 8% in the case of pyridine compared with 32% with dimethylformamide as the solvent under comparable conditions.

While the energy-transfer model is remarkably successful in summarizing the polymerization kinetics of vinyl monomer solutions, there may be other explanations with certain monomer–solvent systems.[8] The principal observations leading to an alternative explanation are these: (1) the $G$ value for ion-pair formation in gaseous hydrocarbon monomers is $\sim 4$; (2) $G$(free ions) in neat hydrocarbon liquid monomers is $\sim 0.1$; (3) the $G$ value for propagating radical chains in neat hydrocarbon liquid monomers is $< 1.0$; (4) we have yet to find a value greater than 7 for $G$(propagating radicals) in liquid solutions of hydrocarbon monomers. This information is the basis of a model in which the sole function of the additive is a fast reaction which converts into propagating free-radicals the primordial charged species that would otherwise recombine without propagation.

Consider the radiation chemistry of methanol–styrene solutions. From the data of Huang and Chandramouli[9] we calculate $G$(chains) = 0.7 in neat styrene; it rises to 5.7 in 5 vol % methanol in styrene and levels off at 6.5 at higher methanol concentrations. According to the energy-transfer model, these are the steps that take place in styrene containing a small amount of methanol. Energy absorbed in the styrene is efficiently transferred to methanol in quanta well matched to the C—H bond energy. By the alternative model,[8] the principal effect of ionizing radiation is to produce cation and anion radicals. In neat solutions most of them recombine in intraspur reactions before they can diffuse away from each other or before they can add more than a few units by propagation. Neutralization provides the energy for the formation of the small fraction of reactive units that survive and propagate as neutral free-radicals. When a small concentration of methanol is present, however, the charge carriers could undergo rapid diffusion-controlled reactions with the additive that convert them into initiators of the radical polymerization. Further information concerning this approach may be found in ref. 8.

The zero activation energy for radiation initiation, $E_i$, leads to some interesting practical consequences. The overall activation for the rate is given by $E = E_p + 1/2\ E_i - 1/2\ E_t$, where $E_p$ and $E_t$ are the activation energies for propagation and termination, respectively. Since $E_i$ for most catalytic initiation is about 30 and $E_t$ is close to zero, this leads to a value of about 22 kcal mol$^{-1}$ compared with about 7 kcal mol$^{-1}$ for radiation for which $E_i = 0$ (1 cal = 4.184 J). The practical advantage of this is to reduce the exotherm in the case, for example, of runaway polymerizations.

On the other hand, the overall activation energy for the degree of polymerization is equal to $E_p - 1/2E_i - 1/2E_t$. This leads to an approximate value of $-8$ kcal mol$^{-1}$ for chemical and 7 kcal mol$^{-1}$ for radiation initiation. If one wishes to increase the rate by raising the reaction temperature without changing the initiator concentration, this leads to a decrease in molecular weight for chemical initiation but to a corresponding increase in molecular weight for radiation initiation. The practical advantages of this reversal are quite clear.

## 19.3 RADIATION-INDUCED IONIC POLYMERIZATION

### 19.3.1 General

The first direct indication that radiation could induce ionic polymerization was the successful polymerization of liquid isobutene at $-78\,^{\circ}$C with both $^{60}$Co $\gamma$-rays and high-energy electrons.[10] High molecular weights were obtained and, since this monomer is known only to polymerize *via* cations, it was a clear indication that it was ionic in nature. This work was published in 1957 but it is now clear that the radiation polymerization of hexamethylcyclotrisiloxane in the solid state, published in 1956, was also ionic in nature.[11] All the earlier studies, mainly with vinyl monomers, clearly revealed that only free-radical processes were involved. The isobutene results led to a large amount of research into radiation-induced ionic polymerization of a number of monomers. Interesting accounts of the early work were included in the monographs of Charlesby[1] and Chapiro.[2]

The extensive research beginning in 1957 finally led to a rather clear picture of this type of radiation-induced polymerization. Reviews of these, now widely accepted, views have been published by Williams[12] and Hayashi.[13] The elementary steps leading to cationic and anionic initiation have been discussed in the introductory section of this chapter. The exact nature of the positive and negative species which are responsible for the initiation and termination reactions are not clearly known in most cases. In principle, they could be positive or negative ion-radicals. It seems far more likely, however, that normal carbocations or carbanions are the chain carriers. Pulse radiolysis is now beginning to give direct evidence of the exact nature of the active intermediates. Almost all of the ionic polymerizations studied have been cationic in nature, but at least one monomer, nitroethylene, polymerizes anionically and has been studied in detail by Hayashi *et al.*[14]

Typical primary reactions which could occur are shown in Scheme 2. The isobutene mechanism appears plausible and is strongly supported by mass spectrometer studies.[15] The nitroethylene mechanism remains highly speculative, however. Considerable relevant information on electron-transfer polymerizations is available from the work of Szwarc and others.[16–18]

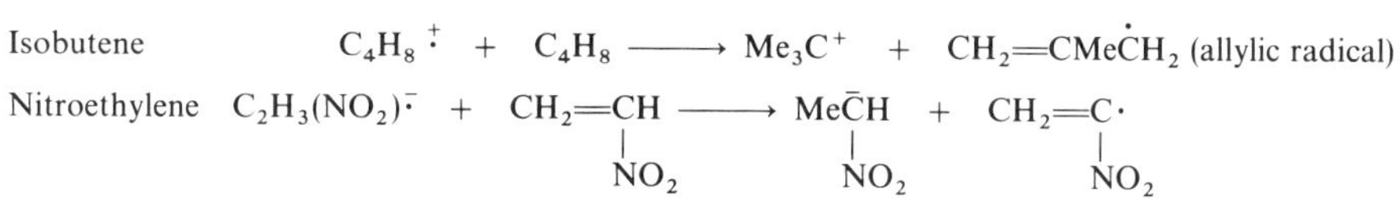

**Scheme 2**

In the case of pure styrene, the mechanism for the formation of the propagating carbocation is not clear. Recent results with picosecond pulse radiolysis have, however, clarified the situation.[8,19] The principal early effect appears to be ionization leading to the cation radical and the anion radical (by electron capture). Within 10 ps, the cation radical adds a monomer to become the dimerized cationic species, with an estimated first-order rate constant of $8 \times 10^9$ l mol$^{-1}$ s$^{-1}$. The dimer cation disappears by a first-order process with a lifetime of 20 ns. This is presumably the trimerization and thus the first step of the cationic polymerization; the calculated rate constant is $4 \times 10^6$ l mol$^{-1}$ s$^{-1}$. This is in excellent agreement with the values estimated from a combination of conductivity and polymerization rate studies with radiation initiation.[20,21] Chemically initiated values are somewhat lower but these are not strictly relevant since they were conducted in chlorinated solvents. It has been well established that solvation and other effects reduce the rate constants with free cationic polymerizations. These results are consistent with the mechanisms shown in Scheme 3.

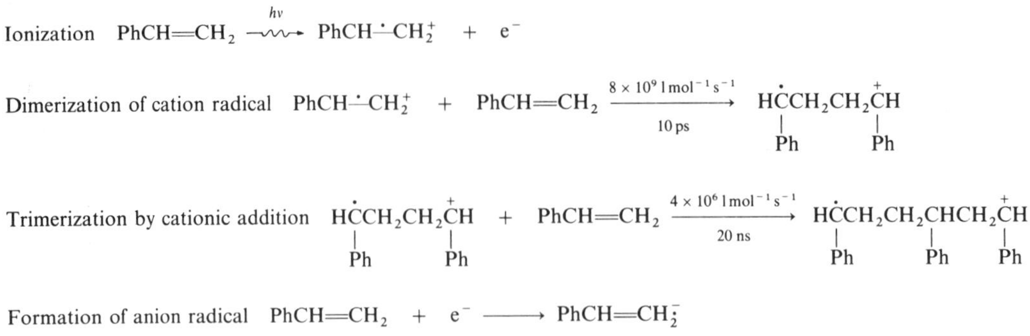


**Scheme 3**

The close proximity of the dimer cation and the anion radical in the spur could lead to some of the high yield of dimers and trimers first reported by Machi, Silverman and Metz.[22]

Pulse radiolysis results were also obtained[8] suggesting that the role of methanol is to serve as a rapid proton-donor to the anion radical converting the latter to a neutral propagating free-radical.[19]

The *G*(i) values for the yield of positive or negative initiating species, which lead to estimates of the rates of initiation, are difficult to detemine. The methods available are the scavenging of the ions with various compounds and the electrical conductivity method using steady-state and pulsed charge collection techniques. They do not always show good agreement; however, the range of values is not too large in most cases. Since the $R_i$ values enter into the rate expression as a square root, the errors become even less significant in evaluating the other kinetic parameters. The free ion yields of only a few monomers have been estimated directly. However, those of a large number of other organic liquids have been determined and have been collected together in an important and useful publication.[23] Apart from the values themselves, interesting trends are presented in ref. 23. The yields increase with temperature to a limited extent; for example, by about sixfold from $-78\,^{\circ}$C to $110\,^{\circ}$C for a number of dienes. The free ion yields also increase with increasing dielectric constant, but with important exceptions, notably the alkanes; for the latter, free ion yields are greater for branched alkanes than for *n*-alkanes although there are no significant differences in the dielectric constant. With polar organic compounds the correlation is often good; see Figure 3.[24] This enables one to make an estimate for methylene chloride, for example, where no values have been determined. It could be asked whether these free ion yields are really equal to the yield of chain initiation. Only one direct comparison has been made,[25] with ethyl vinyl ether in *n*-pentane and neopentane solutions where *G*(free ion) values range 0.16 to 1.0. The rates of polymerization of styrene were found to be proportional to the square roots of the computed overall *G*(i) values. In this case, at least, it appears that all the free ions do indeed participate in the initiation process. It is believed that the free positive and negative species essentially annihilate each other immediately on contact. In fact the $k_t$ values are estimated to be greater than $10^{10}\,\mathrm{l\,mol^{-1}\,s^{-1}}$. These considerations lead to rather simple kinetics when no termination by impurities is involved (Scheme 4).

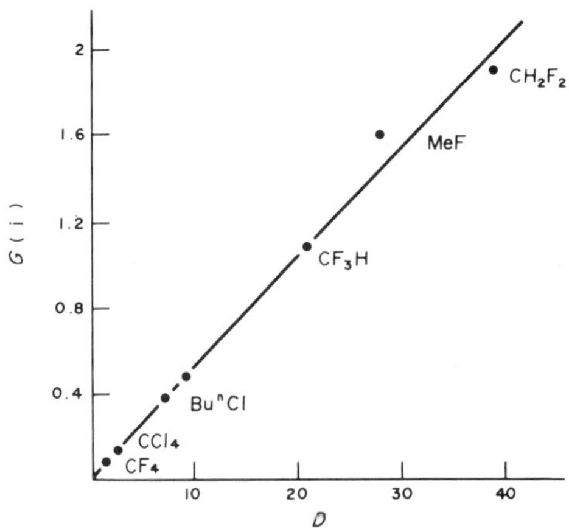


**Figure 3** Correlation between *G*(free ions) and dielectric constant *D* for a series of halogenated compounds (reproduced with permission of IPC Business Press Ltd. from ref. 24)

$$\begin{array}{llll} \text{Initiation} & M \xrightarrow{h\nu} M^{\cdot+} + e^-(Y^-) & & R_i \\ \text{Propagation} & M^{\cdot+} + M \longrightarrow M_2^{\cdot+} \longrightarrow M_n^{\cdot+} & & kR_p \\ \text{Termination} & M_n^+ + Y^- \longrightarrow \text{dead polymer} & & k_t \end{array}$$

**Scheme 4**

Using the normal steady-state method,[12,26] the rate of polymerization $R_p$ is given by

$$R_p = R_i^{1/2} k_t^{-1/2} k_p[M] \tag{7}$$

In conventional units $R_i = \dot{D}G(i)/100N_A$, where $G(i)$ is the radiation yield of free ions as the number per 100 eV of absorbed energy, $\dot{D}$ is the dose rate in eV $s^{-1}$ $ml^{-1}$ and $N_A$ is Avogadro's number.

The termination step, exclusively by ion combination, implies a square-root dependence of the polymerization rate on the intensity. This is, indeed, observed for all monomers studied under strictly pure conditions. Impurities raise the intensity (dose rate) dependence above 0.5 up to 1.0. This would also be true if ion pairs were involved in the propagation step. It is clear therefore that only free, *i.e.* unpaired, ions are the propagating species. Since $[M^+]_{ss} = R_i^{1/2} k_t^{-1/2}$, the low values of $R_i$ and high values of $k_t$ lead to very low steady-state values of the ionic active centers. For example, at 0.2 Mrad $h^{-1}$, the value is only about $10^{-10}$ mol $l^{-1}$ compared with $10^{-8}$ for the corresponding free-radical concentration. This coupled with the high reactivity of free ions leads to the dramatic effect of protonic impurities on the polymerizations, as has been indeed observed. Water is the chief inhibitor and is believed to be regenerated with anions by the reaction

$$H_3O^+ + Y^- \longrightarrow Y + H_2O \tag{8}$$

In the presence of inhibiting impurities the rate of reaction is given by[12,26]

$$R_p = R_i k_p[M]/\{(R_i k_t)^{1/2} + k_{tX}[X]\} \tag{9}$$

where $k_{tX}$ is the rate constant for the reaction of the growing chains with an impurity, X. Since electron accelerators lead to dose rates much much greater than γ-ray sources, it follows that the effect of impurities will be less. This has indeed been shown to be true.[27] However, square-root intensity dependence on the dose rate has not been observed, but first power intensity dependence, at least with styrene, has been observed.

These considerations make radiation-initiated ionic polymerization an excellent method for studying truly free ion polymerizations. Examples of the power of this method have been presented for *p*-methoxystyrene[28] and the vinyl ethers[16,20,21] and are summarized in Table 2.

**Table 2** Estimated $k_p$ and $E_p$ for Selected Monomers Forming Unpaired Cations by Radiation

| *Monomers* | $k_p$ (l $mol^{-1}$ $s^{-1}$) | $E_p$ (kcal $mol^{-1}$)[a] |
|---|---|---|
| Isobutene | $1.5 \times 10^8$ | ~0 |
| Cyclopentadiene | $6.0 \times 10^8$ | <2 |
| α-Methylstyrene | $3.4 \times 10^6$ | ~0 |
| *p*-Methoxystyrene | $3.0 \times 10^6$ | ~0 |
| Styrene | $3.5 \times 10^6$ | ~0 |
| Styrene, pulsed radiolysis | $4 \times 10^6$ | — |
| Free anions | $1 \times 10^5$ | — |
| Free radicals 25 °C | 35 | — |
| Ethyl vinyl ether (25 °C) | $6 \times 10^4$ | >10 |
| Isopropyl vinyl ether (25 °C) | $1.3 \times 10^6$ | 5.3 |

[a] 1 cal = 4.184 J.

The work of Hayashi *et al.*, however, has indicated that an ion-pair component can be introduced by adding strong electron-acceptors such as 1*H*,3*H*-benzo[1,2-*c*:4,5-*c*′]difuran-1,3,5,7-tetrone (pyromellitic dianhydride).[29,30]

Finally, it is clear that certain liquid monomers, notably styrene, can polymerize simultaneously by free-radical, free-anion and free-cation mechanisms. After the extremely rapid reactions leading to the production of the chain initiators, interactions between these different mechanisms appear to be minimal but are unknown. However, it is logical to believe that the anion and cation chains will terminate each other.

There are sufficient quantitative data for net styrene at room temperature to estimate the relative rates at which each process will proceed.

As indicated earlier, $G = 0.7$ for free-radical initiation and about 0.1 for free ions. The termination rate constants are $3 \times 10^7\ \mathrm{l\,mol^{-1}\,s^{-1}}$ for free radicals and $2 \times 10^{11}\ \mathrm{l\,mol^{-1}\,s^{-1}}$ for the free ions. The propagation rate constants have been determined to be 30, $4 \times 10^6$ and about $10^5\ \mathrm{l\,mol^{-1}\,s^{-1}}$ for free-radical, cationic and anionic polymerization, respectively.

The steady-state rates of polymerization, $R_p$, are given by

$$R_p = k_p[\mathrm{M}](R_i/k_t)^{1/2} \tag{10}$$

using the correct rate constants for each case, and

$$R_i = 1.04 \times 10^{-9} G(\mathrm{i}) g \dot{D} \tag{11}$$

where $R_i$ is the initiation rate in $\mathrm{l\,mol^{-1}\,s^{-1}}$ for radicals and ions, respectively, $g$ is the density and $\dot{D}$ the dose rate in rad $\mathrm{s^{-1}}$.

The rates of polymerization calculated by means of equation (10) at 25 °C are: radical, $1.23 \times 10^{-5}\ \mathrm{mol\,l^{-1}\,s^{-1}}$; cation, $7.6 \times 10^{-3}\ \mathrm{mol\,l^{-1}\,s^{-1}}$; and anion, $1.9 \times 10^{-4}\ \mathrm{mol\,l^{-1}\,s^{-1}}$. The ratio is 1:618:15.4 for the free radical, cationic and anionic mechanisms respectively. The overwhelming mechanism at room temperature is clearly cationic under super dry conditions with about 2.5% anionic and a negligible radical contribution.

### 19.3.2 Ring-opening Polymerization

The earliest report of a radiation-initiated ionic polymerization was of hexamethylcyclotrisiloxane in the solid state.[11] Several years later, trioxane was successfully polymerized as a crystalline solid, followed by trithiane. These two monomers gave crystalline polymers with crystal orientations, following those of the monomers, *i.e.* topotactic reactions. All these monomers gave increased yields with increasing temperatures, but these fell to nearly zero above the melting point, *i.e.* in the liquid state. A typical result is shown in Figure 4 with the siloxane cyclic trimer.[31] When the great importance of rigorous drying to promote ionic polymerization was appreciated, the latter system was reexamined by Chawla and St. Pierre.[32] Under these conditions, substantial polymerization was achieved in the liquid state, both with the cyclic siloxane trimer and the tetramer. These monomers and also decamethylcyclopentasiloxane have been studied in considerable detail by Lebrun *et al.*[33] and by Naylor *et al.*[34] The polymerizations were found to be cationic in nature and to follow the expected square-root dependence of the rates on the intensity.[26] Rings were also found to be generated as shown, for example, in Figure 5. A simple ring-opening plus backbiting process was suggested of the type shown in equation (12).

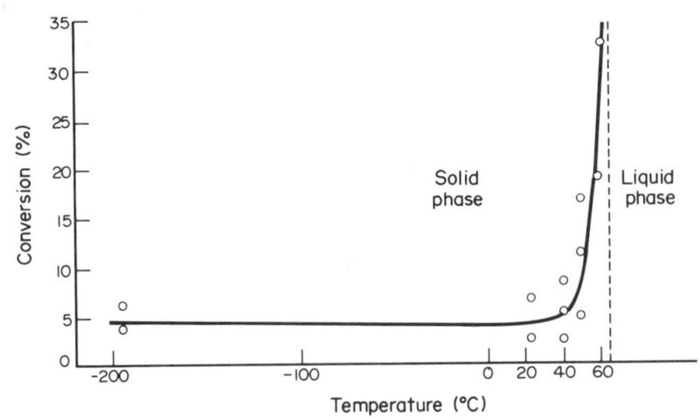


**Figure 4** Rate of polymerization of hexamethylcyclotrisiloxane as a function of temperature. $^{60}$Co $\gamma$-radiation at 23 $\mathrm{kGy\,h^{-1}}$ for 1 h (reproduced with permission of J. Wiley & Sons from ref. 31)

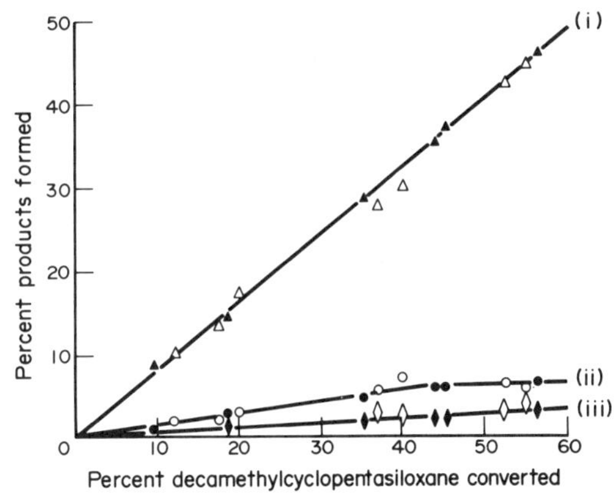


**Figure 5** Yields of (i) polymer, (ii) octamethylcyclotetrasiloxane and (iii) dodecamethylcyclohexasiloxane rings formed by irradiating decamethylcyclopentasiloxane at (▲) 60 °C and (△) 90 °C as a function of conversion, dose rate = 1 kGy h$^{-1}$ (from D. M. Naylor, Ph. D. Thesis, North Carolina State University, 1988)

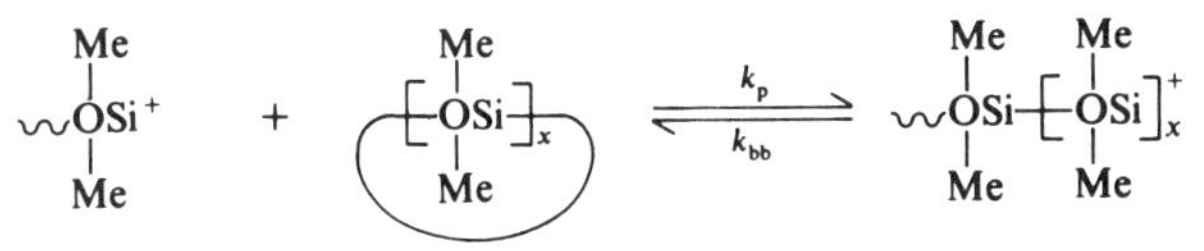


(12)

polymer chain    Ring unit    polymer chain

Various other details of these processes may be found in refs. 33 and 34. A number of expoxy monomers, such as epoxycyclohexane (7-oxabicyclo[4.1.0]heptane),[35] were also shown to polymerize by irradiating in the liquid state. Low yields were obtained; about 0.4 % Mrad$^{-1}$ with comparatively modest drying conditions used. A special type of ring opening was found with β-pinene, which has been studied rather extensively.[36] The effect of drying was demonstrated conclusively and it represents one of the first studies of this kind. All the above were shown to proceed by cationic mechanisms.

For further discussion see also Volume 4, Chapter 17.

## 19.4 GRAFTING AND CURING

### 19.4.1 Introduction

Grafting[1,2,37,38] and curing[39–41] reactions initiated by radiation are related processes and, in many applications, may occur concurrently to the benefit of the finished product. In the present treatment, each process is initially treated separately. Subsequently, the mechanistic significance of the two systems is related.

### 19.4.2 Principles of Grafting

Radiation grafting is a convenient procedure for the modification of both surface and bulk properties of natural and synthetic polymers. Radiation is used to facilitate the covalent bonding of monomeric, oligomeric and polymeric entities to a polymeric substance, referred to as the backbone trunk or base polymer. Bonding may occur along the substrate chains or, much more rarely, at chain ends forming side-chain graft or block copolymers, respectively.

High-energy radiation possesses advantages as an initiator of grafting processes, since the interaction of such radiation with matter produces radicals and ions on the polymer backbone by processes which are virtually independent of temperature. Although the yields may vary greatly, their formation occurs regardless of the molecular structure of the polymeric substrate. Thus any polymeric system when exposed to radiation can be activated for grafting to a degree that is not obtainable by chemical initiation. Strict control over the degree of grafting and also the competing homopolymerization can be exerted by the use of high-energy radiation.

### 19.4.3 Radiation Grafting Methods

Two types of high-energy radiation are currently used for grafting work. These are $\gamma$-rays from a $^{60}$Co (or related) facility and electrons from an electron beam (EB) machine. The differences between the two techniques are related to the penetration and to the dose rates which determine the time required in order to deliver a particular total dose. EB machines usually require only seconds, whereas $^{60}$Co may need hours to deposit the same total dose. However, in terms of flexibility, the greater penetration of the radiation gives advantages to $^{60}$Co. This is of great help when samples include solvents, specific additives or need irradiating under very dry or vacuum conditions. $^{60}$Co is also easier to use when very high or low temperatures are needed.

When the backbone polymer is irradiated to initiate grafting, the process can also lead to simultaneous crosslinking or degradation. Irradiation in the presence of oxygen (air), an effective free-radical scavenger, produces hydroperoxides and other peroxides in the polymer. In the case of glassy or semicrystalline polymers, the radicals formed are sufficiently long-lived that, when the irradiated polymer is placed in contact with an appropriate monomer, reaction occurs yielding graft copolymer. The concept of radical formation in a polymer–monomer system leads to three general methods of radiation grafting. The first procedure utilizes the pre-irradiated backbone polymer before interaction with the monomer. This is known as the pre-irradiation, indirect or consecutive method. The second procedure, known as the mutual, direct or simultaneous method, involves irradiation of the polymer–monomer combination. A third method uses the peroxides formed during irradiation of the substrate polymer in oxygen (air) to later initiate grafting of an added monomer.

### 19.4.4 Pre-irradiation Technique

In this method, the backbone polymer $P_x$ is irradiated *in vacuo* or in the presence of an inert gas prior to exposure to the monomer which may be present as either vapor or liquid. On heating, the radicals formed during irradiation are mobilized and react with monomer giving graft copolymer in high yield. The reaction can also proceed at room temperature if sufficient time is allowed for the monomer to diffuse to the active centers. This is often preferable to heating as it tends to minimize the premature destruction of radicals by recombination. Thus both steps may avoid the application of heat. Solvents or swelling agents may be added to the monomer to facilitate the diffusion process, as discussed later.

### 19.4.5 Mutual Radiation Grafting Technique

In this procedure, the trunk polymer $P_x$ is irradiated whilst in contact with monomer M which may be present as vapor, liquid or in solution. Irradiation may take place in air or, better, in an inert atmosphere and leads directly to the formation of active free-radicals in both polymer $P_x$ and monomer M resulting in graft copolymerization (equations 13 and 14). This is the most efficient method of grafting since the radicals can react as fast as they are produced; however, homopolymer yields can be appreciable (equation 15). Ionic species can also initiate grafting *via* the mutual technique and mechanisms analogous to equation (14) involving positive and negative ionic intermediates can be proposed.[38]

$$P_x \xrightarrow{h\nu} P_x\cdot \qquad M \xrightarrow{h\nu} M\cdot \tag{13}$$

$$P_x\cdot + M \longrightarrow P_xM_n\cdot \longrightarrow \text{graft copolymer} \tag{14}$$

$$M\cdot + M \longrightarrow M_n\cdot \longrightarrow \text{homopolymer} \tag{15}$$

### 19.4.6 Peroxide Technique

Here the polymer is irradiated in the presence of air to produce mainly hydroperoxides. These polymeric peroxides, which are often quite stable, can be decomposed in contact with monomer to produce graft copolymers (equations 16 and 17). If hydroperoxides are the main species, the hydroxy radicals formed lead to homopolymerization (equation 18). Thus

$$P_x \xrightarrow{O_2} P_xO_2\cdot \longrightarrow P_xOOH \xrightarrow[UV]{heat} P_xO\cdot + \cdot OH \tag{16}$$

$$P_xO\cdot + M \longrightarrow P_xOM\cdot \quad \text{graft copolymer} \tag{17}$$

$$\cdot OH + M \longrightarrow HOM\cdot \quad \text{homopolymer} \tag{18}$$

If the initial peroxy radicals react together, peroxides P—OO—P are produced. These are often more stable but do not lead to homopolymer. If a reducing agent is added to the monomer, the production of homopolymer is minimal due to reactions of the type shown in equation (19).

$$POOH + Fe^{2+} \longrightarrow PO\cdot + OH^- + Fe^{3+} \tag{19}$$

It can be seen that the peroxide method is also a pre-irradiation process with similar attendant advantages and disadvantages.

### 19.4.7 Comparison of Pre-irradiation, Mutual and Peroxide Grafting Techniques

One of the advantages of the pre-irradiation grafting method, including the peroxide technique, is its potential suitability for large-scale fast line-speed high-volume production of copolymer using sophisticated electron beam (EB) procedures. Low-energy self-shielded EB machines which have been recently developed and are suitable for many of these grafting applications are capable of delivering 1500 Mrad m $min^{-1}$ to the substrate which can then subsequently be reacted with monomer on-line. When $^{60}Co$ sources are used, the radiation dose-rate is much lower and therefore longer times of irradiation are required. Typical pre-irradiation grafting data with the latter source are shown in Table 3, where styrene and 4-vinylpyridine have been copolymerized to cellulose, a particularly valuable backbone polymer for the pre-irradiation technique, since the trapped radicals formed by radiation in cellulose can be easily stabilized prior to subsequent reaction with monomer.

**Table 3** Grafting of Styrene and 4-Vinylpyridine to Cellulose Using the Pre-irradiation Method[a]

| *Monomer* | *Time* (h)[b] | Graft (%) *20*[c] | *40*[c] | *60*[c] | *80*[c] |
|---|---|---|---|---|---|
| Styrene | 8 | — | — | 2 | 4 |
| Styrene | 20 | — | — | 13 | 16 |
| Styrene | 30 | — | — | 9 | 20 |
| Styrene | 52 | — | — | 25 | 35 |
| Styrene | 70 | 0 | — | 40 | 45 |
| 4-Vinylpyridine | 386 | 0 | 11 | 15 | 15 |

[a] Cellulose pre-irradiated in air at $1.5 \times 10^5$ rad $h^{-1}$ to total dose of $2.5 \times 10^6$ rad, then reaction in monomer methanol solution.
[b] Time of reaction at 60 °C for monomer in methanol.
[c] Monomer concentration in methanol.

The potential disadvantages of the pre-irradiation method originate from the basic concept of pre-irradiation since a minimum threshold radiation-dose is required to rupture sufficient C—H bonds in the backbone polymer to give reasonable yields in the subsequent grafting step during reaction of these trapped radicals with monomer. By contrast, in the simultaneous method, the backbone polymers are irradiated in contact with monomer which structurally may be of a simple alkenic type (*e.g.* vinyl acetate) or a combination of aromatic and alkene bonding (*e.g.* styrene). Alkenic materials similar to the monomers normally act as radiation protection agents in any mixture of compounds, particularly for saturated hydrocarbons. Many of the trunk polymers used in grafting have essentially a saturated hydrocarbon backbone structure. Thus, during radiation grafting by the mutual technique, the monomer may exert radiation protection on the polymer, the mechanism of the process being predominantly by scavenging reactions,[42] particularly involving hydrogen atoms and electrons.[43] In addition, in the mutual method, there are further mechanisms such as energy transfer which lead to grafting but do not involve prior rupture of C—C and C—H bonds in the backbone polymer. Hence grafting by the simultaneous method usually requires a much lower radiation-dose to achieve a particular percentage degree of copolymerization than the corresponding copolymer from the pre-irradiation procedure.

Both pre-irradiation and mutual grafting techniques have been used extensively in the past to synthesize copolymers; however, the mutual technique does possess some practical advantages for certain specific applications and has been utilized in a wider variety of systems. Technically, very much lower radiation-doses are required to graft *via* the mutual technique and this is important for radiation degradation in the backbone polymer, especially for those materials such as PVC which may be radiation sensitive. In addition, for a comparison with curing reactions, the mutual technique is valuable since common mechanisms appear to operate in both processes and, in some applications, curing and grafting occur concurrently to the benefit of the finished copolymer. The emphasis in the remainder of the text will involve the mutual technique. However, reference to pre-irradiation procedures will be made as relevant.

A wide variety of backbone polymers have been utilized in radiation grafting.[2] Ease of radical formation in the backbone polymer relative to radical generation in the monomer is an important factor in determining the efficiency of the process. By contrast, PVC, which readily loses Cl by rupture of the R—Cl bond, easily forms radical sites in the polymer, although molecular dehydrochlorination can be a competing process during irradiation leading to a graft polymer of different backbone structure to that of the original material. Naturally occurring macromolecules such as wool and cellulose are susceptible to facile C—H bond rupture during irradiation and thus radical sites suitable for grafting are easily formed with these materials. A similar situation pertains for the aliphatic-type polymers such as the polyalkenes, particularly polyethylene and polypropylene. By contrast, aromatic-type backbone polymers are generally more radiation stable and may require higher radiation grafting doses than the corresponding aliphatic equivalents.

The presence of solvents in the grafting system can markedly influence the ease of copolymerization and also the properties of the end product. In this respect, the effect of solvent can be related to the structure of the backbone polymer. Thus grafting is enhanced with the appropriate backbone polymer if a swelling solvent is used. With cellulose as an example, representative data for the grafting of styrene in the low molecular weight alcohols (Figure 6) typify this property.[44] In the figure, the wetting and swelling properties of the alcohols are demonstrated, the grafting yield decreasing dramatically with increasing molecular weight of solvent such that, with *n*-butanol, grafting is extremely low under the radiation conditions used. However, electron microscope studies show that the styrene graft in *n*-butanol is essentially confined to the surface of the cellulose, whereas the corresponding copolymer in methanol consists of both surface and bulk graft. The choice of solvent can thus determine the specific nature of the copolymer and its properties.

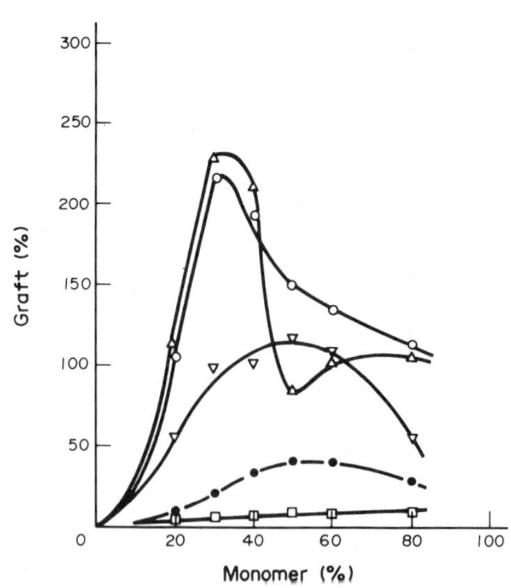


**Figure 6** Grafting of styrene in alcohols to cellulose: dose rate = $8.32 \times 10^3$ rad $h^{-1}$; total dose = 0.20 Mrad; (△) methanol; (○) ethanol; (▽) *n*-propanol; (●) 2-propanol; (□) *n*-butanol

The peak in the grafting yields shown in Figure 6 is generally ascribed to the well-known gel or Trommsdorff effect (see Volume 3, Chapter 12). The experimental conditions needed to observe this effect are important, since not only are the grafting yields highest at this point but the grafted chains are longer. This latter point can be valuable in a preparative context, particularly if specific properties involving the length of the grafted chains are required.

Controlling the swelling of the polymer substrate by the monomer plus solvent is clearly of great importance. If the swelling is too great, termination becomes greater accounting for the peak phenomenon. The radiation chemistry is also an important factor, particularly with respect to the radical yield of the solvent. If this is low, homopolymer is reduced, for example. A more detailed discussion of these and other considerations may be found in the literature.[2,6,44–47]

The grafting yield is inversely proportional to some power of the radiation dose-rate and directly proportional to total dose. This trend is illustrated by the data in Figure 7 for the grafting of styrene in methanol to polyethylene film. The results also show the shift in the Trommsdorff peak from 50% monomer in solvent at 10 000 rad $h^{-1}$ to 70% styrene at 75 000 rad $h^{-1}$. As the radiation dose-rate increases, the magnitude of the gel peak progressively diminishes until it disappears at 112 000 rad $h^{-1}$.

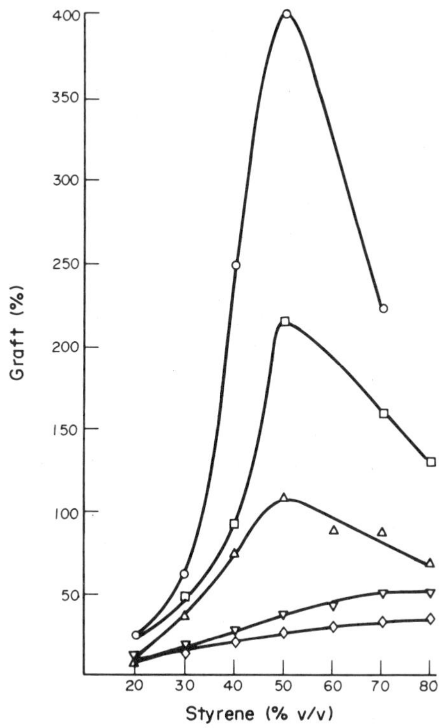


**Figure 7** Effect of dose rate (rad $h^{-1}$) on grafting of styrene in methanol to polyethylene in film at total dose of $2.3 \times 10^5$ rad (○) 10 000; (□) 21 000; (△) 41 000; (▽) 75 000; (◇) 112 000 rad $h^{-1}$

The effects of solvents on the grafting of styrene to polyethylene have also been examined in some detail by Silverman *et al.*[46,47] With styrene–methanol solutions, the rate of grafting increases with the concentration of methanol, passing through a maximum at 70 vol % in the grafting solution. It was determined that the actual methanol concentration in the polyethylene was trivial over the entire range and, therefore, could not precipitate the growing chains and inhibit their termination. On the other hand, dilution of styrene with *n*-octane decreased the grafting rate over the entire range of concentrations. The results with methanol–styrene systems were explained in terms of absorbed monomer only and its effect on the viscosity of the amorphous regions of the polyethylene. The styrene–octane system could be explained by the simple decrease in the available monomer concentration.

### 19.4.8 Effects of Additives

Any technique for increasing yields during radiation grafting is important, especially for those backbone polymers which are sensitive to ionizing radiation. In such instances, it is preferable to use the lowest total radiation-dose to achieve a particular percentage graft. A number of additives have been developed for giving increased copolymer yields. Most of these additives act by enhancing both graft and homopolymerization reactions, the former to a greater extent. Combinations of these additives can also be used to preferentially increase the grafting yields.

Inclusion of polyfunctional monomers, such as divinylbenzene (DVB) and 3-acryloyloxymethyl-1,5-pentylene diacrylate (trimethylolpropane triacrylate; TMPTA), in additive amounts (<1%), significantly enhances radiation grafting.[48] The multifunctional monomers clearly lead to branching and eventually some gelation of the grafted side-chains. This tends to enhance the gel effect and diminish homopolymer formation.

**Table 4** Effect of Sulfuric Acid on Radiation Grafting of Styrene in Methanol and DMF to PVC[a]

| *Styrene (% v/v)* | *Graft (%)* | | | |
|---|---|---|---|---|
| | *1.0 × 10⁵ rad*[b] | *1.0 × 10⁵ rad + H⁺*[b] | *5.0 × 10⁵ rad* | *5.0 × 10⁵ rad + H⁺* |
| 20 | 0 | 10 | 19 | 121 |
| 25 | 0 | 15 | 18 | 141 |
| 27 | 2 | 18 | — | 323 |
| 30 | 13 | 34 | — | 140 |
| 33 | 56 | 65 | 14 | 178 |
| 35 | 58 | 68 | 28 | — |
| 40 | 66 | 70 | 25 | — |
| 50 | 6 | 25 | 15 | — |

[a] Sulfuric acid, $7.5 \times 10^{-2}$ mol l$^{-1}$; dose rate, $1.0 \times 10^5$ rad h$^{-1}$ in air.
[b] Methanol solvent, remainder DMF.

The use of mineral acid as an additive to increase grafting yields has been reported for a number of different backbone polymer systems.[38,49] Typical acid effects observed are shown in Table 4, where addition of mineral acid ($H_2SO_4$) increases the grafting yield of styrene in methanol to PVC at all total radiation-doses studied. An analogous acid enhancement effect is observed when dimethylformamide is used as solvent and also in other solvents,[50,51] so that the concept appears to be of general applicability. When the backbone polymer is polyethylene, analogous acid effects are also found for the styrene system over a wide range of concentrations and dose rates studied, including that corresponding to the Trommsdorff peak as shown in Table 5. The mechanism of the acid effect is complicated. From previous studies of this phenomenon, the enhancement due to this additive has been attributed essentially to two predominant factors, namely the radiolytic yield of hydrogen atoms and also the extent to which grafting polymer (polystyrene) was solubilized in the bulk solution. The hydrogen atom explanation has been somewhat discredited since similar effects were shown to occur with pre-irradiation grafting by Garnett and co-workers.[52] Presumably any hydrogen atoms would have disappeared before grafting took place in the case of polyethylene. Solvent structure has also been considered to be important and those solvents with high radiolytic yields of hydrogen atoms, such as the lower molecular weight alcohols, have, in general, been efficient reagents for grafting in the presence of acids. A number of studies and explanations of the effects of acids have been published. There is clear evidence that there is some synergism between the addition of acids and multifunctional monomers on the grafting yields.[52,53] An example of this may be seen in Figure 8. A number of other additives including inorganic salts and urea have also been shown to lead to increased grafting yields. The exact reasons for all these effects are not clear although some significant explanations have been presented.

In addition to being of fundamental importance to basic radiation grafting theory, the results with the range of additives previously discussed are of value in a preparative context since radiation graft copolymers are now being used for a wide range of applications and any technique for reducing the radiation dose to achieve a particular percentage graft is of both practical and economic significance. The discovery of a range of metal salts and organic compounds,[52,54] such as urea,[55] to complement the acids already known as additives for enhancing grafting is valuable, since the scope of the

**Table 5** Effect of Sulfuric Acid (0.2 mol l$^{-1}$) on Grafting of Styrene in Methanol to Polyethylene Film at Different Dose Rates[a]

| | Graft (%) | | | | | |
|---|---|---|---|---|---|---|
| | 10 000 rad h$^{-1}$ | | 41 000 rad h$^{-1}$ | | 112 000 rad h$^{-1}$ | |
| Styrene (% v/v) | Neutral | Sulfuric acid (0.2 mol l$^{-1}$) | Neutral | Sulfuric acid (0.2 mol l$^{-1}$) | Neutral | Sulfuric acid (0.2 mol l$^{-1}$) |
| 20 | 24 | 32 | 14 | 19 | 7 | 8 |
| 30 | 61 | 82 | 37 | 51 | 14 | 17 |
| 40 | 51 | 344 | 76 | 81 | 23 | 27 |
| 50 | 409 | 543 | 109 | 134 | 25 | 35 |
| 60 | — | — | 89 | 119 | 28 | 36 |
| 70 | 223 | 211 | 89 | 73 | 35 | 37 |

[a] Film 0.12 mm thickness (ex-Union Carbide) to total dose of $2.3 \times 10^5$ rad.

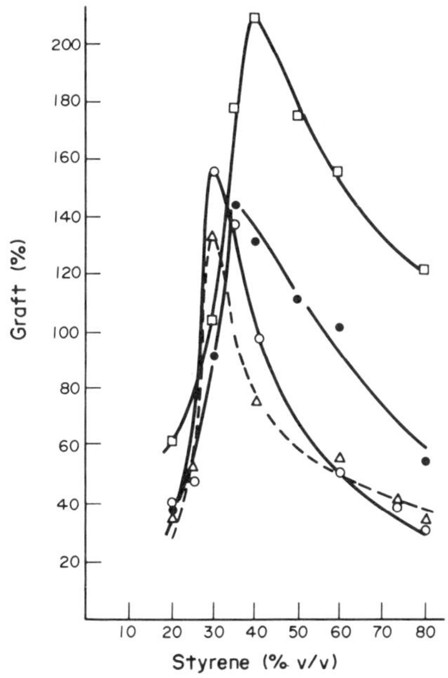


**Figure 8** Effect of sulfuric acid and divinylbenzene on grafting of styrene in methanol to polypropylene film at dose rate of $4.1 \times 10^4$ rad h$^{-1}$ to total dose of $2.4 \times 10^5$ rad: (△) styrene–methanol; (○) styrene–methanol–sulfuric acid (0.2 mol l$^{-1}$); (●) styrene–methanol–divinylbenzene (1% v/v); and (□) styrene–methanol–divinylbenzene (1% v/v)–sulfuric acid (0.2 mol l$^{-1}$)

enhancement technique is very much evident. Thus grafting systems which may have been sensitive to acid now can be treated with metal salts or urea-type materials to achieve the same type of enhanced reactivity. Although the efficiency of the metal salts in increasing grafting yields is higher than for acid, the solubility of certain salts may be limited in the monomer solution, *e.g.* lithium perchlorate and nitrate are amongst the most soluble inorganic salts for the styrene–methanol system over all monomer concentration ranges and thus these two salts have been used predominantly for this early additive work. Because of solubility limitations with the salts, acids and urea, as alternate additives, can be used to replace salts to achieve grafting enhancement in such difficult systems. The nature of the additive used may influence the structure of the end copolymer. Thus chain length of graft, degree of crosslinking and related properties could all be significantly affected by the structure of the additive utilized. Finally, not only are these additive effects important in radiation grafting, the same type of additives and concepts appear to be applicable to the related field of radiation curing.

## 19.4.9 Curing

In practice, radiation curing involves the rapid polymerization of an oligomer/monomer mixture to give a material which can also be crosslinked. Electrons from a self-shielded EB machine with electron energies up to 400 kV are frequently used as high-energy radiation sources to initiate the polymerization.[34–36] The process utilizes 100% solids resin systems, is essentially solvent-free with minimal environmental pollution problems, possesses low energy-consumption with compact equipment, cures at room temperature even in the presence of pigments and fillers and, most importantly, leads to the preparation of new products not capable of being made by other methods.[38,41]

When oligomers and/or monomers are exposed to energetic electrons from a self-shielded EB machine, both radical and ionic species are formed. From basic radiation chemistry work, it has previously been assumed that if curing reactions occurred above 25 °C participation of ions in the process was minimal; however, recent extensive mass-spectrometric studies[56] show that ion lifetimes, particularly negative ions, can be significant even at temperatures in excess of 25 °C. Thus, with appropriate functionality of monomers and oligomers in the prepolymer mixture, the mechanistic roles of cationic and anionic processes in addition to free-radical reactions should be considered in the overall curing of process. The presence of moisture also appears to influence the participation of ions in a radiation polymerization process.[57]

The various features of the formation of cations, anions and free radicals are similar to those discussed earlier in the chapter.

The basic component in any EB curing mixture is the prepolymer which is essentially an oligomer of relatively low molecular weight (~3000) prepared by monomer polymerization or the acrylation of a backbone polymer such as a polyurethane, polyester, polyether or epoxy.[39–41] Acrylate-based oligomers/monomers are the most frequently used in EB curing; however, prepolymer mixtures, oligomers and monomers containing any functionality are theoretically capable of being utilized. One of the major limiting factors is the dose required to cure the resin, usually 2 Mrad for cellulosic-type coatings, any higher dose leading to degradation of the cellulose. The ability of an EB machine to deliver high doses in short periods of time is vital to the success of curing processes. Currently, EB machines delivering 1500 Mrad m $min^{-1}$, *i.e.* 1 Mrad at 1500 m $min^{-1}$ or, more conveniently, 3 Mrad at 500 m $min^{-1}$, are available for curing under an inert gas (generally nitrogen). Extensive research is presently being performed to overcome this oxygen inhibition to give an air-cure system with its obvious advantages.

The dose required to cure is also dependent on the chemical structure of the oligomer/monomer used. This is because, for most resin systems, EB curing is predominantly a radical polymerization reaction (*i.e.* ionic processes are minimized) and high molecular weight polymers are obtained by favoring the propagation process at the expense of termination. Usually propagation is favored by restricting the mobility of the growing polymer chains (gel or Trommsdorff effect) either by raising the viscosity of the oligomer/monomer mixture or by raising the molecular weight of the oligomers used in the curing composition. In general, the former process is preferred since it results in greater flexibility in the application of the oligomer/monomer mixture. Curing is also favored under conditions where polymerization starts at low radiation-doses and results in a Trommsdorff effect. Thus the important chemical properties that affect speed of cure are composition and concentration of monomers and oligomers, the frequency and type of unsaturation in the polyfunctional monomers and oligomers and the molecular weight of the oligomer. The conclusion that cure speed can be increased by increasing both the frequency of unsaturation and the molecular weight of the oligomer has to be treated with caution since, in some instances, higher unsaturation frequency may decrease cure rate due to steric effects, whilst higher molecular weight oligomers may cause solubility problems and difficulty in subsequent application of the mixture.[58]

Multifunctional acrylates can be unsuitable because of high skin-irritancy as determined by the Draize rating,[59] ratings of 1–2 being considered acceptable in routine running. Tripropylene glycol diacrylate is one of the widely used monomers in EB curing work, possessing good solvency for acrylated prepolymers, low Draize rating, relatively low viscosity, and high speed of cure with acceptable viscosity reduction powers (although in the last respect not as efficient as 1,6-hexanediol diacrylate which unfortunately has a relatively high Draize rating).

In order to complete the radiation cure system for practical applications, additives are required for the oligomer/monomer mixture to control factors such as gloss, flow, slip, opacity and UV stability. In addition to incorporating beneficial properties to the finished film, these additives may also simultaneously lead to detrimental effects such as wrinkling, shrinkage and, most importantly, poor adhesion. The role of additives is thus important in curing since, in the presence of additives essential to give the film its basic properties, optimization in film performance becomes a compromise where the maximum properties of each additive may not be realized.

### 19.4.10 Significance of Concurrent Grafting during Curing

The ability to achieve concurrent grafting during radiation rapid-cure processing of films of oligomer on to substrates can be valuable since, under these conditions, a true carbon–carbon bond is formed between coating and substrate. Many important properties of the resulting cured film are thus improved if such concurrent grafting occurs, particularly adhesion and flexibility. For many commercial uses, a carbon–carbon copolymer bond at the interface is not required since interaction energies originating from charge-transfer forces or hydrogen bonding are sufficient to give the required adhesion for a particular purpose. With relatively non-polar substrates, particularly the polyalkenes, satisfactory bonding at the surface during these rapid-cure treatments can be difficult to attain. With EB, concurrent grafting will virtually always occur since the energies of the impinging electrons are sufficiently high to readily break bonds at the surface of the substrate, thus creating sites for copolymerization to occur. The important feature of this concept is to optimize the degree of concurrent grafting during cure, and additives which enhance grafting have relevance to the curing process. Thus certain solvents are shown to enhance grafting to particular backbone polymers and, although EB resin systems are usually solvent free, the functionality in the solvent indicates the type of group which, if present in the oligomer/monomer mixture, may enhance grafting during cure. The multifunctional acrylate reactivity, particularly synergistic effects with additives such as urea (Table 6), is extremely significant in graft enhancement, since these acrylates are also used as reactive diluents in curing processes, so a common beneficial parameter between grafting and curing exists with these materials.

**Table 6** Synergistic Effect of Urea and TMPTA in Grafting of Styrene in Methanol to Polypropylene[a]

| *Styrene (% v/v)* | *Graft (%)* No additive | $U^{b}$ | $U+T^{b}$ |
|---|---|---|---|
| 20 | 14 | 15 | 30 |
| 30 | 64 | 69 | 135 |
| 40 | 72 | 60 | 115 |
| 50 | 44 | 41 | 75 |
| 60 | 32 | 30 | 71 |
| 70 | 26 | 24 | 48 |

[a] Polypropylene film (0.06 mm) irradiated at $7.5 \times 10^4$ rad h$^{-1}$ to total dose of $2.5 \times 10^5$ rad.
[b] U = urea (1% w/v), T = TMPTA (1% v/v).

With respect to curing reactions, silanes and fluorinated alkyl esters are two representative additives which are used to improve slip and flow in the finished film. Recent work on the effect of these two additives on grafting, *per se*, shows that: (i) the silane is a retarder due, presumably, to the repulsion effect of silicon atoms when adsorbed at the grafting interface and (ii) the fluorinated alkyl ester improves grafting yields. Most important is the result when the multifunctional acrylate is added to both silane and fluorinated alkyl ester systems. In each case the presence of the multifunctional acrylate leads to large enhancements in grafting and thus the effect of this acrylate overrides the detrimental effects of even the silanes. Since such multifunctional acrylates are used in curing formulations to control viscosity and achieve crosslinking with speed of cure, judicious choice of such an acrylate in the oligomer/monomer mixture can lead to graft enhancement during cure with beneficial properties to the final product.

Finally, it should be noted that although grafting with simple monomers should at first glance be incapable of sensitizing crosslinking, it has been observed that acrylic acid and acrylonitrile enhance the crosslink yield.[60,61] In the first case, the hydrogen-bonded dimer of the acid behaves as a polyfunctional monomer in a polyethylene matrix, and on heating of the graft to determine gel yield, the interchain dimers form an almost irreversible anhydride. In the case of acrylonitrile–polyethylene grafts, a post-irradiation heat treatment produces interchain links.

## 19.5 REFERENCES


1. A. Charlesby, 'Atomic Radiation and Polymers', Pergamon Press, New York, 1959.
2. A. Chapiro, 'The Radiation Chemistry of Polymeric Systems', Wiley-Interscience, New York, 1962.
3. J. W. T. Spinks and R. J. Woods, 'An Introduction to Radiation Chemistry', 2nd edn., Wiley-Interscience, New York, 1964.
4. Farhataziz and M. A. J. Rodgers, 'Radiation Chemistry. Principles and Applications', VCH Publishers, New York, 1987.
5. L. A. Miller and V. T. Stannett, *J. Polym. Sci., Part A-1*, 1969, **7**, 3159.
6. V. T. Stannett, J. D. Wellons and H. Yasuda, *J. Polym. Sci., Part C*, 1963, **4**, 551.
7. T. S. Nikitina and Kh. S. Bagdasar'yan, *Sb. Rab. Radiats. Khim.*, 1955, 183; see also ref. 2, p. 266.
8. J. Silverman, S. Tagawa, H. Kobayashi, Y. Katsumura, M. Washio and Y. Tabata, *Radiat. Phys. Chem.*, 1983, **22**, 1039.
9. R. Y. M. Huang and P. Chandramouli, *J. Polym. Sci., Part B*, 1969, **7**, 245.
10. W. H. T. Davison, S. H. Pinner and R. Worrall, *Chem. Ind. (London)*, 1957, 1274.
11. E. J. Lawton, W. T. Grubb and J. S. Balwit, *J. Polym. Sci.*, 1956, **19**, 455.
12. F. Williams, in 'Fundamental Processes in Radiation Chemistry', ed. P. Ausloos, Interscience, New York, 1978, p. 515.
13. K. Hayashi, *Polym. J.*, 1980, **12**, 583.
14. H. Yamaoka, F. Williams and K. Hayashi, *Trans. Faraday Soc.*, 1967, **63**, 376.
15. F. W. Lampe, *J. Phys. Chem.*, 1959, **63**, 1986.
16. M. Szwarc, 'Carbanions, Living Polymers and Electron Transfer Processes', Wiley, New York, 1968.
17. M. Szwarc (ed.), 'Ions and Ion Pairs in Organic Reactions', Wiley-Interscience, New York, 1972, vol. I.
18. M. Szwarc (ed.), 'Ions and Ion Pairs in Organic Reactions', Wiley-Interscience, New York, 1974, vol. II.
19. V. T. Stannett and J. Silverman, *ACS Symp. Ser.*, 1983, **212**, 435.
20. F. Williams, Ka. Hayashi, K. Ueno, K. Hayashi and S. Okamura, *Trans. Faraday Soc.*, 1967, **63**, 1501.
21. Ka. Hayashi, K. Hayashi and S. Okamura, *Polym. J.*, 1973, **4**, 426.
22. S. Machi, J. Silverman and D. J. Metz, *J. Phys. Chem.*, 1972, **76**, 930.
23. A. O. Allen, 'Yields of Free Ions Formed in Liquids by Radiation', Reference Data System, NSRD-NBS, vol. 57, 1976.
24. A. Deffieux, W. C. Hsieh, D. R. Squire and V. T. Stannett, *Polymer*, 1982, **23**, 65.
25. H. Kubota, V. Ya. Kabanov, D. R. Squire and V. T. Stannett, *J. Macromol. Sci., Chem.*, 1978, **A12**, 1299.
26. F. Williams, Ka. Hayashi, K. Ueno, K. Hayashi and S. Okamura, *Trans. Faraday Soc.*, 1967, **63**, 1501.
27. D. R. Squire, J. A. Cleaveland, T. M. A. Hossain, W. Oraby, E. P. Stahel and V. T. Stannett, *J. Appl. Polym. Sci.*, 1972, **16**, 645.
28. A. Deffieux, D. R. Squire and V. T. Stannett, *Polym. Bull. (Berlin)*, 1980, **2**, 469.
29. K. Hayashi, M. Irie and Y. Yamamoto, *J. Polym. Sci., Polym. Symp.*, 1977, **56**, 173.
30. Y. Yamamoto, A. Morinaka, M. Irie and K. Hayashi, *Macromolecules*, 1977, **10**, 266.
31. W. J. Burlant and C. Taylor, *J. Polym. Sci.*, 1959, **41**, 547.
32. A. S. Chawla and L. E. St. Pierre, *J. Polym. Sci., Part A-1*, 1972, **10**, 2691.
33. J. J. Lebrun, A. Deffieux, P. Sigwalt, A. Wang and V. T. Stannett, *Radiat. Phys. Chem.*, 1984, **24**, 239.
34. D. M. Naylor, V. T. Stannett, A. Deffieux and P. Sigwalt, *J. Polym. Sci., Polym. Lett. Ed.*, 1986, **24**, 319.
35. D. Cordischi, M. Lenzi and A. Mele, *J. Polym. Sci., Part A-1*, 1965, **3**, 3421.
36. T. H. Bates, J. V. F. Best and T. F. Williams, *J. Chem. Soc.*, 1962, 1531.
37. A. Hebeish and J. T. Guthrie, 'The Chemistry and Technology of Cellulosic Copolymers', Springer, Berlin, 1981.
38. J. L. Garnett, *Radiat. Phys. Chem.*, 1979, **14**, 79.
39. J. L. Gordon and J. W. Prane (eds.), 'Nonpolluting Coatings and Coating Processes', Plenum Press, New York, 1973.
40. G. A. Senich and R. E. Florin, *J. Macromol. Sci., Rev. Macromol. Chem.*, 1984, **C24** (2), 239.
41. J. L. Garnett, *J. Oil Colour Chem. Assoc.*, 1982, **65**, 383.
42. A. Ekstrom and J. L. Garnett, *J. Chem. Soc. A.*, 1968, 2416.
43. A. Chapiro, A. M. Jendrychowska-Bonamour and G. LelieVre, *Faraday Discuss. Chem. Soc.*, 1977, **63**, 134.
44. J. L. Garnett, *ACS Symp. Ser.*, 1977, **48**, 334.
45. S. Dilli and J. L. Garnett, *J. Polym. Sci., Part A-1*, 1966, **4**, 2323.
46. S. Machi, I. Kamel and J. Silverman, *J. Polym. Sci., Part A-1*, 1970, **8**, 3329.
47. I. Kamel, S. Machi and J. Silverman, *J. Polym. Sci., Part A-1*, 1972, **10**, 1019.
48. C. H. Ang, J. L. Garnett, R. Levot and M. A. Long, *J. Appl. Polym. Sci.*, 1982, **27**, 4893.
49. J. L. Garnett and J. D. Leeder, *ACS Symp. Ser.*, 1977, **49**, 197.
50. H. Barker, J. L. Garnett, R. Levot and M. A. Long, *J. Macromol. Sci., Chem.*, 1978, **A12** (2), 261.
51. J. L. Garnett, S. V. Jankiewicz and D. F. Sangster, *J. Polym. Sci., Polym. Lett. Ed.*, 1982, **20**, 171.
52. J. L. Garnett, S. V. Jankiewicz, R. Levot and D. F. Sangster, *Radiat. Phys. Chem.*, 1985, **25**, 509.
53. C. H. Ang, J. L. Garnett, R. Levot and M. A. Long, *J. Polym. Sci., Polym. Lett. Ed.*, 1983, **21**, 257.
54. J. L. Garnett, S. V. Jankiewicz, M. A. Long and D. F. Sangster, *J. Polym. Sci., Polym. Lett. Ed.*, 1985, **23**, 563.
55. P. A. Dworjanyn and J. L. Garnett, *J. Polym. Sci., Polym. Lett. Ed.*, 1988, **26**, 135.
56. S. J. Bett, G. Fletcher and J. L. Garnett, *Radiat. Phys. Chem.*, 1986, **28**, 207.
57. S. V. Nablo and A. S. Denholm, *J. Radiat. Curing*, 1980, **7**, (3), 11.
58. A. S. Hoffman, *At. Energy Rev.*, 1971, **9** (2), 347.
59. R. Holman (ed.), 'UV and EB Curing Formulations for Printing Inks, Coatings and Paints', SITA-Technology, London, 1984.
60. K. Singer, M. S. Joshi and J. Silverman, *J. Polym. Sci., Polym. Lett. Ed.*, 1974, **12**, 387.
61. M. S. Joshi, K. Singer and J. Silverman, *Radiat. Phys. Chem.*, 1977, **9**, 475.

# 20

# Photoinitiated Polymerization

S. PETER PAPPAS
*North Dakota State University, Fargo, ND, USA*



## 20.1 INTRODUCTION

Photoinitiators absorb light in the ultraviolet–visible spectral range, generally 250–450 nm, and convert this light energy into chemical energy in the form of reactive intermediates, such as free radicals and reactive cations, which subsequently initiate polymerization. Photoinitiated polymerization and crosslinking of monomers, oligomers and polymers constitute the basis of important commercial processes with broad applicability, including photoimaging[1–3] and UV curing of coatings[4–7] and inks.[5–8] In addition to photoinitiators, photocrosslinking may also be accomplished with photocrosslinking agents and photocrosslinkable polymers, which are discussed in 'Photocrosslinking', Chapter 5 in Volume 6 of this series.

## 20.2 PHOTOINITIATORS

### 20.2.1 General Considerations

Light absorption by the photoinitiator requires that an emission line from the light source overlaps with an absorption band of the photoinitiator. In this event, the photoinitiator (PI) is raised to an electronically excited state (by promotion of an electron to a higher energy orbital), which is symbolized by PI* in equation (1). Electronically excited state species may be characterized by their orbital nature, such as $n,\pi^*$ or $\pi,\pi^*$, their spin multiplicity, such as singlet ($S$) or triplet ($T$), and their energy level, such as $S_1$ or $S_2$. These considerations are important for an understanding of photoinitiator activity and have been discussed in this context.[9]

$$\mathrm{PI} + \mathrm{Light} \rightarrow \mathrm{PI}^* \tag{1}$$

The lifetime of PI* is short, generally less than $10^{-6}$ s. During this time, PI* may be partitioned among several processes including (1) decay back to PI (with emission of light and/or heat), (2) excited state quenching by $O_2$, monomer or other quenching agent (Q), and (3) a chemical reaction yielding the initiator species, such as I• or $I^+$, as shown in Scheme 1. The initiator species are further partitioned between reactions which either lead or do not lead to initiation.

$$\mathrm{PI} \leftarrow \mathrm{PI}^* \xrightarrow{O_2(Q)} \mathrm{PI}$$
$$\downarrow$$
$$\mathrm{I}\bullet(\mathrm{I}^+)$$

**Scheme 1**

The rate of initiation ($R_i$) may be expressed as the rate of formation of PI*, which corresponds to the number of photons absorbed by PI per unit time and volume ($I_a$), times the fraction ($F$) of PI* that yields initiator species, times the fraction ($f$) of initiator which initiates polymerization, as shown in equation (2).

$$R_i = I_a F f \tag{2}$$

The term $I_a$ is related to the incident light intensity ($I_0$), the number of photons incident to the system per unit time and area, and the absorbance ($A$) (or optical density) of the photoinitiator, as shown in equation (3), where $d$ is the pathlength of light (or thickness of a film).

$$I_a = I_0(1 - 10^{-A})/d \tag{3}$$

The absorbance ($A$) is proportional to the photoinitiator concentration ($c$) and pathlength ($d$), as shown in equation (4), where $e$ is the molar absorptivity of the photoinitiator.

$$A = edc \tag{4}$$

The corresponding energy absorbed by the photoinitiator (per unit time and volume) is $I_a$ times the energy per photon. The energy per mol of photons (a mol of photons is an Einstein) is related to the wavelength ($\lambda$) of the absorbed light by equation (5). The incident light intensity ($I_0$) times the energy per photon equals the energy per unit time (*i.e.* power) incident to the system per unit area at each absorbed wavelength.

$$E(\mathrm{kJ\ mol^{-1}}) = 1.20 \times 10^5/\lambda\ (\mathrm{nm}) \tag{5}$$

Generally, it is desirable that the rate of initiation ($R_i$) is high for efficient utilization of light energy and also uniform throughout the system. Equations (2)–(4) illustrate that $R_i$ increases proportionally with $I_0$, but not with photoinitiator concentration ($c$). Rather, as $c$ (and $A$) increase, the proportion of incident light absorbed decreases exponentially per unit thickness from the initially exposed surface. The optimal absorbance for maximum absorption at the backside of an absorbing system has been calculated to be *ca.* 0.43.[10,11] These considerations are particularly important in curing and imaging applications, and will be discussed in greater detail in 'Photocrosslinking', Chapter 5 in Volume 6 of this work.

## 20.3 PHOTOINITIATORS FOR RADICAL POLYMERIZATION

Current reviews on photoinitiators for radical polymerization are available.[12–14] In general, initiator radicals are photogenerated by (1) *intramolecular* bond cleavage, notably of aryl alkyl ketones, and (2) *intermolecular* H abstraction from a H donor, notably by diaryl ketones.

It is suggested that photoinitiators which undergo intramolecular bond cleavage be classified as $PI_1$, since initiator radicals are produced by a *unimolecular* process. Photoinitiators which undergo intermolecular H abstraction may correspondingly be classified as $PI_2$, since a *bimolecular* reaction is involved.

### 20.3.1 Intramolecular Bond Cleavage of $PI_1$ Photoinitiators

An important criterion for $PI_1$ photoinitiators is the presence of a bond with a dissociation energy lower than the excitation energy of the reactive excited state, on the one hand, and sufficiently high, on the other, to provide adequate thermal stability. Other important criteria are (1) high rates of cleavage, generally evidenced by short excited state lifetimes, to compete with bimolecular quenching, and (2) efficient initiation by the photogenerated free radicals.

#### *20.3.1.1 Aryl alkyl ketones*

Benzoin ethers (BE) are important members of this class and were probably the first photoinitiators used commercially for free radical polymerization. Benzoin ethers undergo photocleavage to produce benzoyl (B) and ether (E) radicals as shown in equation (6).

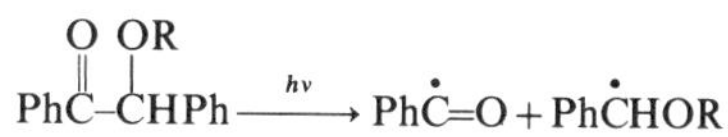

(6)

BE B E

Under conditions of relatively low radical and high monomer concentrations (acrylate and methacrylate), the B and E radicals are reported to be comparably efficient as initiators, based on results with $^{14}$C-labelled benzoin ether derivatives,[15] and independent generation of the E radical.[16] However, under conditions of relatively high radical and low monomer concentrations (styrene), the B radical is primarily responsible for initiation; and the E radical participates, predominately as a chain terminating agent.[17]

Benzoin ethers are reported to undergo cleavage from short-lived ($<10^{-9}$ s) triplet states[18] with quantum yields of 0.4–0.5 in benzene.[19] The less than 100% efficiency of cleavage may be attributed to radical recombination reactions. The short lifetime of the reactive excited state is a direct consequence of facile cleavage ,which competes effectively with bimolecular quenching processes, an important advantage when photoinitiated polymerization is conducted in the presence of oxygen (air) or other potential quenching agents.

Related aryl alkyl ketone photoinitiators include **(1)**–**(4)**, which undergo photocleavage analogously to benzoin ether, as evidenced by various studies,[12–14] including free radical trapping and laser flash photolysis.

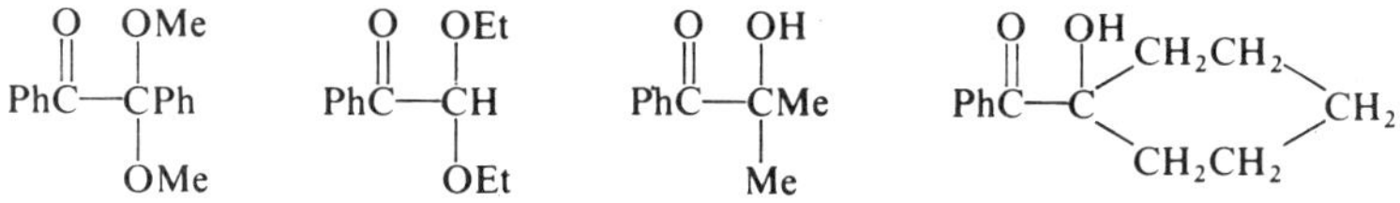

**(1)** **(2)** **(3)** **(4)**

Higher absorptivity in the 350–400 nm spectral region is exhibited by $PI_1$ photoinitiator **(5)**, resulting primarily from conjugation of the *p*-thiomethyl group with the ketone group.[20]

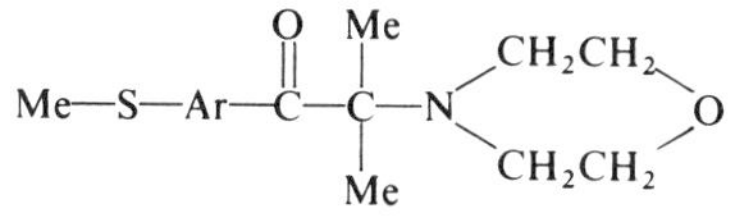

**(5)** Ar = 1,4-Disubstituted benzene

### *20.3.1.2 Cyclic photoinitiators*

Cyclic analogues of benzoin ethers, *e.g.* (**6a**), and benzil ketals, *e.g.* (**6b**), were found to be ineffective as photoinitiators for polymerization of acrylates.[21]

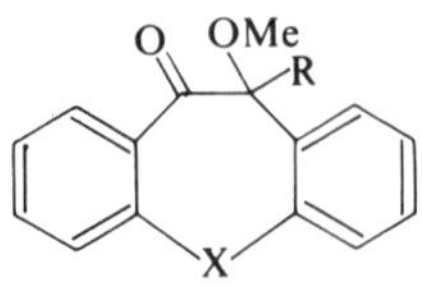

(**6a**) R = H, X = O, S
(**6b**) R = OMe, X = $CH_2$, S

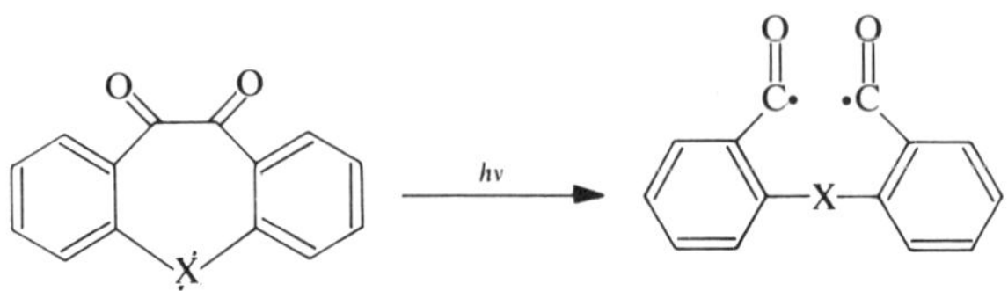 (7)

(7) X = O, S

On the other hand, cyclic benzils (**7**) do function as photoinitiators for polymerization of acrylates,[21] and evidence for photocleavage, see equation (7), was obtained from laser flash photolysis studies.[22]

Photocleavage is also expected to occur with cyclic analogues (**6**); however, rapid recyclization of the resulting biradicals may essentially preclude initiation of polymerization. On the other hand, with cyclic benzils (**7**), recyclization may be retarded by the required *cisoid* configuration of the $\alpha$-dicarbonyl groups.

It is interesting to note that benzil, itself, does not undergo $\alpha$ cleavage in solution (or does so inefficiently).[23,24] Benzil does function as a $PI_2$ photoinitiator by H abstraction from an appropriate donor, as discussed in Section 20.3.22.

### *20.3.1.3 Oximinoketones*

The photochemistry of oximinoketones of general structure (**8**) has recently been reviewed.[12] These $PI_1$ photoinitiators have been proposed to undergo primary C–C or N–O $\alpha$-cleavage depending on the nature of substituent R‴, as shown by (a) and (b), respectively.

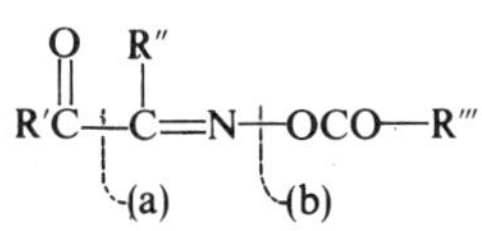

(**8**)

Ultimately, the same products are obtained by secondary cleavage reactions and interpretation of the results is further complicated by *E*, *Z* isomerization.

The oximinoketone chromophore enhances absorptivity in the near UV spectral region, relative to the acetophenone chromophore; however, the relatively weak N–O bond reduces the thermal stability of these photoinitiators and, consequently, their dark storage stability in the presence of reactive monomers.

### *20.3.1.4 Peroxyketones*

Peroxyketones, such as (**9**), are reported to undergo efficient peroxy bond cleavage by photoexcitation of the benzophenone chromophore, as shown in equation (8), (where Ar = 1,4-disubstituted benzene).[25] However, the relatively weak peroxy linkage limits the thermal stability of such photoinitiators.

$$\underset{(\mathbf{9})}{\text{PhCOArCOO–OBu}^t} \xrightarrow{h\nu} \text{PhCOArCOO}\cdot + \text{Bu}^t\text{O}\cdot \qquad (8)$$

#### 20.3.1.5 *Acyl phosphine oxides*

Acyl phosphine oxides, such as (**10**), are reported to exhibit high photoinitiator activity for radical polymerization, thermal and hydrolytic stability, as well as activation by near UV–visible light (350–450 nm).[26] Laser flash photolysis studies have provided evidence for photocleavage, as shown in equation (9).[27] The resulting phosphorus-centred radicals were found to exhibit high reactivity in initiating the polymerization of unsaturated monomers.[28]

(9)

(**10**) Ar = 2,4,6-trimethylphenyl

### 20.3.2 Intermolecular Hydrogen Abstraction of $PI_2$ Photoinitiators

$PI_2$ photoinitiators include diaryl ketones, such as benzophenone, Michler's ketone and thioxanthones, benzils, quinones and 3-ketocoumarins. In contrast to $PI_1$ photoinitiators, which are capable of generating initiator radicals independently, $PI_2$ photoinitiators must undergo a bimolecular reaction, generally with H donors.

For efficient production of initiating free radicals, the bimolecular H abstraction reaction must compete with nonproductive processes, including unimolecular decay of the photoinitiator excited state as well as bimolecular quenching of the excited state, *e.g.* by oxygen as well as by monomers. Furthermore, the reactivity of the radical derived from the H donor is also an important factor. These considerations have been discussed in considerable detail[9,12] and will be highlighted herein.

The facility of H abstraction depends on the reactive excited state photoinitiator, including its excitation energy, orbital nature and spin multiplicity, as well as on the H donor. Particularly reactive H donors are those in which the abstractable H is (1) positioned $\alpha$ to oxygen, as in ethers ($R_2\underline{CH}$–OR), or to nitrogen, as in tertiary amines ($R_2\underline{CH}$–$NR_2$), or (2) directly attached to sulfur, as in thiols ($R\underline{SH}$).

Excited state $PI_2$ photoinitiators are generally more reactive with tertiary amines than with alcohols, ethers or thiols by two to three orders of magnitude, which constitutes a particularly important advantage for amine H donors when photoinitiated polymerization is conducted in air, owing to competitive quenching of the $PI_2$ excited state by oxygen, and/or in the presence of monomers which quench the $PI_2$ excited state. For example, styrene, *N*-vinylpyrrolidone and methyl methacrylate quench excited state (triplet) benzophenone with rate constants of $3.3 \times 10^9$, $3.6 \times 10^8$ and $6.9 \times 10^7$ $M^{-1} s^{-1}$, respectively, in benzene.[29] Furthermore, methyl acrylate quenches benzophenone excited state with a rate constant of $7.5 \times 10^6$ $M^{-1} s^{-1}$, *i.e.* an order of magnitude slower than methyl methacrylate.[30]

The high reactivity of tertiary amines with excited state $PI_2$ photoinitiators (with rate constants on the order of $10^9$ $M^{-1} s^{-1}$, closely approaching rates of diffusion in fluid media) has been attributed to rapid formation of an excited state complex (exciplex) stabilized by charge transfer contribution from the electron-rich amine and electron-accepting excited state $PI_2$. (Note that excited state species are both better at accepting and donating electrons than their ground states, corresponding to the magnitude of their excitation energy.)

#### 20.3.2.1 *Diaryl ketones*

H abstraction by excited state benzophenone (BP*) from a tertiary amine ($R_2NCH_2R$), *via* a charge transfer stabilized exciplex, is shown in Scheme 2. Initiation of polymerization of an acrylate monomer by the resulting $\alpha$-amino radical is also depicted. The capability of the $\alpha$-amino radical to initiate polymerization is important, since the accompanying benzophenone ketyl radical has been shown to be a relatively inefficient initiator species[31,32] and may function primarily as a chain terminating agent.[32] The low initiating reactivity of benzophenone ketyl radical probably applies to diaryl ketyl radicals in general, owing to their steric bulk and resonance stabilization. Whereas $\alpha$-amino radicals are efficient initiators for acrylates,[33] they appear to be inefficient initiators for styrene.[34]

$$\mathrm{PhC(=O)Ph} \xrightarrow{h\nu} \mathrm{BP^*}$$

$$\mathrm{BP^* + R_2NCH_2R \rightarrow [\overset{\delta^-}{BP} \ldots \overset{\delta^+}{NR_2}{-}CH_2R]^*}$$
$$\text{Exciplex}$$

$$\text{Exciplex} \xrightarrow{\text{H Transfer}} \mathrm{Ph_2\dot{C}{-}OH + R_2N\dot{C}HR}$$

$$\mathrm{R_2N\dot{C}HR + CH_2{=}CHCOOR \rightarrow R_2NCH(R)CH_2{-}\dot{C}HCOOR}$$

**Scheme 2**

Bis(4,4′-dimethylamino)benzophenone (Michler's ketone) (MK), possessing both diaryl ketone and tertiary amine groups, undergoes photoinduced H abstraction from ground state MK, although the efficiency is low.[35] MK* also undergoes H transfer with benzophenone (BP), probably *via* an exciplex in which MK* functions as the H donor,[36] as shown in Scheme 3, where Ar = 1,4-disubstituted benzene. A comparison of Schemes 2 and 3 illustrates that exciplex formation between a H donor and an acceptor may be preceded by excitation of either partner.

$$\mathrm{MK} \xrightarrow{h\nu} \mathrm{MK^*}$$

$$\mathrm{MK^* + BP \rightarrow [MK \ldots BP]^*}$$
$$\text{Exciplex}$$

$$\text{Exciplex} \xrightarrow{\text{H Transfer}} \mathrm{Ph_2\dot{C}{-}OH + Me_2N{-}Ar{-}C(=O){-}Ar{-}N(Me)CH_2^{\bullet}}$$

**Scheme 3**

Thioxanthones (TX), such as the 2-isopropyl analogue (**11**), function as $PI_2$ photoinitiators by H abstraction from tertiary amines, systems which have been investigated extensively by spectroscopic and flash photolysis techniques.[37,38] Strong interest in thioxanthones results from high absorptivity in the near UV–visible together with low coloration, in contrast to MK, making TX derivatives attractive as photoinitiators for thin and white pigmented films.

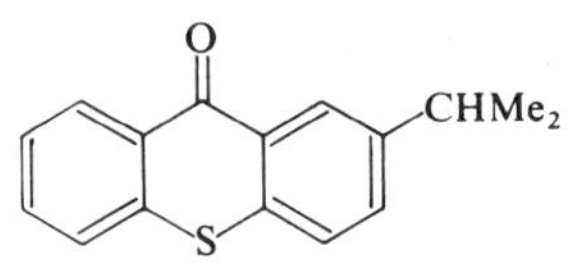

(**11**)

### 20.3.2.2 *Benzil and quinone derivatives*

As noted in Section 20.3.1.2 (cyclic photoinitiators), strong evidence against benzil undergoing photocleavage in solution has been obtained from radical scavenger studies[23] and reisolation of benzil after irradiation in oxygen-free benzene.[24] These results are intuitively reasonable in view of the low excitation energy of benzil (*ca.* 224 kJ mol$^{-1}$), which is insufficient for overcoming the CO–CO bond energy, estimated at >270 kJ mol$^{-1}$, as previously discussed.[9] Furthermore, benzil does not function effectively as a photoinitiator in the absence of H donors.[39]

On the other hand, benzil is known to undergo photoinduced H abstraction[40] and functions as a $PI_2$ photoinitiator by H abstraction from tertiary amines.[41]

However, in apparent contradiction to these considerations, the water-soluble benzil derivative (**12**) has been shown to undergo facile photolysis in aqueous solution.[42]

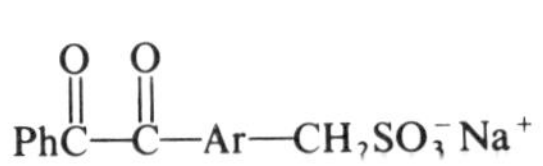

(**12**) Ar = 1,4-Disubstituted benzene

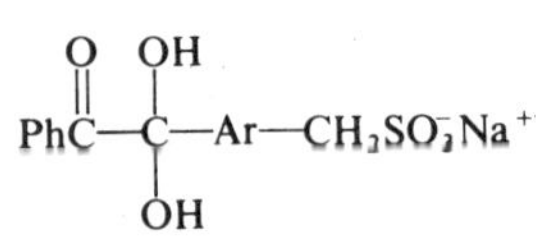

(**13**) Ar = 1,4-Disubstituted benzene

A possible explanation which reconciles the above findings is that benzil (**12**) exists in aqueous solution in equilibrium with the corresponding hydrate (**13**) (or isomer). Hydrate (**13**) is structurally related to benzil ketal (**1**) and is expected to undergo facile photocleavage analogously, as shown in equation (10).

$$(\mathbf{13}) \xrightarrow{h\nu} \mathrm{Ph\dot{C}{=}O + Na^{+}\ {}^{-}O_3SCH_2\text{–}Ar\text{–}\dot{C}(OH)_2} \qquad (10)$$

This explanation is supported by photochemical studies on biacetyl (MeCO–COMe).[43] Like benzil, biacetyl is a poor photoinitiator for radical polymerization, except in the presence of H donors. However, biacetyl photoinitiates the polymerization of monomers, such as vinyl acetate and *N*-vinylpyrrolidone, in the absence of H donors when irradiation is conducted in methanol or water. Evidence is presented that solvated biacetyl (**14**) is the active photoinitiator, which presumably undergoes cleavage, as shown in equation (11).

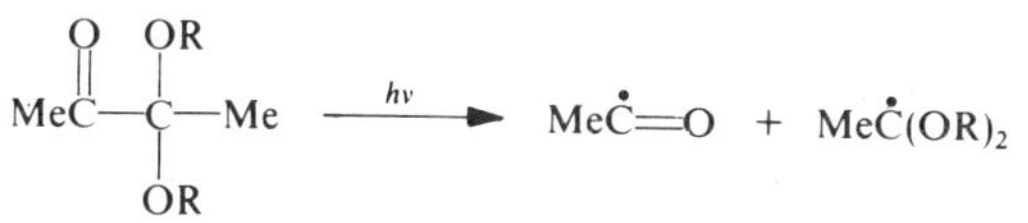


(11)

(**14**) R = H or Me

Nevertheless, the benzil issue remains open. Recent flash photolysis studies on aromatic ketone-containing photoinitiators, including benzil, have been interpreted in support of CO–CO photocleavage of benzil as the primary step for initiator radical generation.[44]

Anthraquinone and 2-*tert*-butylanthraquinone are reported to function as efficient photoinitiators for MMA (methyl methacrylate) polymerization in tetrahydrofuran solution (H donor) in the presence or absence of oxygen.[45] Camphor quinone appears to be a photoinitiator of choice together with tertiary amines for dental applications, utilizing visible light.[46]

#### *20.3.2.3 3-Ketocoumarins*

3-Ketocoumarins of general structure (**15**), which were developed primarily as photosensitizers for imaging applications,[47] also function as $PI_2$ photoinitiators by H abstraction from tertiary amines, as well as from α-phenoxy- and α-thiophenoxy-acetic acid.[48,49]

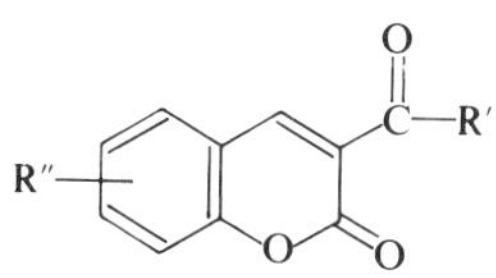


(**15**)

3-Ketocoumarins exhibit high absorptivity ($50\text{–}90 \times 10^3\ M^{-1}\,cm^{-1}$) which can be varied throughout the UV–visible spectral region by proper selection of substituents R′ and R″. Photoinitiator efficiencies of 3-ketocoumarin/H donor compared to MK/BP have been reported.[48]

### 20.3.3 Photoinduced Intramolecular Bond Cleavage Followed by Intermolecular Hydrogen Abstraction of Bis(imidazolyl)s

An important photoinitiator system for free radical polymerization in imaging applications utilizes bis(triarylimidazolyl)s of general structure (**16**) together with H donors, such as thiols (RSH).[50] Photoinduced bond cleavage of the bis(imidazolyl)s yields relatively stable (noninitiating) imidazolyl radicals (Im•), which subsequently yield initiating radicals by abstraction from the H donor, *e.g.* RSH, as shown in Scheme 4. The spectral response of bis(imidazolyl)s may be extended deep into the visible region by photosensitization,[51] as discussed in Section 20.4.2.

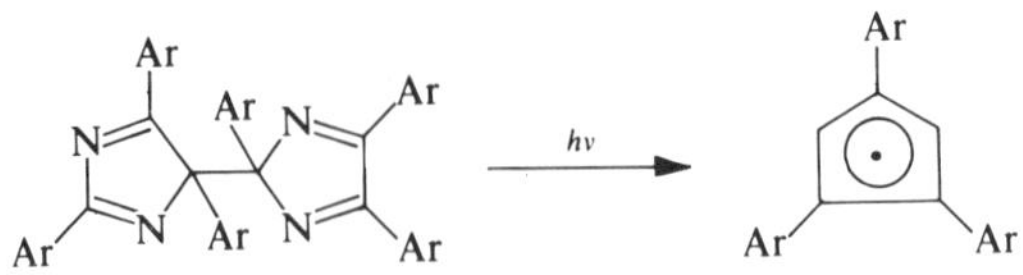


**(16)**

$$\text{Im}\bullet + \text{RS—H} \longrightarrow \text{ImH} + \text{RS}\bullet$$

$$\text{RS}\bullet + \text{Monomer} \longrightarrow \text{Polymer}$$

**Scheme 4**

## 20.4 PHOTOSENSITIZATION OF RADICAL POLYMERIZATION

In the broadest sense, a photosensitizer sensitizes or activates a system towards light regardless of the mechanism. More narrowly, a photosensitizer is regarded as an energy transfer agent which absorbs light, transfers the energy to an acceptor and reverts to its original state. In the first reported mechanistic study on photosensitization of onium salts for cationic polymerization (see Section 20.6.1), a broader definition of photosensitization was adopted, including electron transfer as well as exciplex-induced bond cleavage.[52] In common with energy transfer, these processes may also result, ultimately, in regeneration of the photosensitizer or at least its basic chromophoric group, which allows catalytic activity. This broader definition of photosensitization, including all processes which may proceed catalytically, is also adopted herein. Within this scope, the exciplex-induced H transfer between MK and BP (see Scheme 6) may be considered photosensitization, since the MK chromophore remains intact.

### 20.4.1 Energy Transfer

Photosensitization by energy transfer is illustrated in Scheme 5. Following light absorption by the photosensitizer (PS), the participating excited state (PS*) transfers its energy to the acceptor (A). The resulting excited state acceptor (A*) should be capable of generating initiating radicals, generally by intramolecular bond cleavage or intermolecular H abstraction.

$$\text{PS} \xrightarrow{h\nu} \text{PS}^*$$

$$\text{PS}^* + \text{A} \rightarrow \text{PS} + \text{A}^*$$

$$\text{A}^* \rightarrow \text{I}\bullet$$

**Scheme 5**

The energy transfer step may occur by resonance energy transfer, generally from the first excited singlet state ($S_1$), or by exchange energy transfer, generally from the first excited triplet state ($T_1$).[53] In either case, efficient energy transfer requires that the excitation energy of the photosensitizer, $E^*$(PS), is equivalent to or greater than that of the acceptor, $E^*$(A). In other words, the energy transfer process should not violate the first law of thermodynamics.

General requirements for energy transfer photosensitizers are (1) desirable absorption characteristics (to efficiently utilize the available light), (2) relatively high energy and long lifetime of the participating excited state (to promote energy transfer), and (3) photostability (to provide catalytic activity).

The high absorptivity of thioxanthones, such as (**11**), in the near UV–visible spectral region, their relatively high energy (*ca.* 273 kJ mol$^{-1}$) and long lifetime of excited triplet states, together with photostability in the absence of H donors make them particularly attractive as photosensitizers for radical polymerization in the presence of competing absorbing substances, such as white and colored pigments. Acceptors have included various aromatic and heteroaromatic sulfonyl and methylene chlorides,[9,54] as well as the *p*-thiomethylacetophenone $PI_1$ photoinitiator (**5**).[20]

### 20.4.2 Electron Transfer

Photosensitization by electron transfer is illustrated in Scheme 6. Following light absorption by the photosensitizer (PS), the participating excited state (PS*) undergoes electron transfer with an electron acceptor (A) or donor (D). One or more of the resulting redox products should be capable of generating initiating radicals.

$$\mathrm{PS} \xrightarrow{h\nu} \mathrm{PS}^*$$

$$\mathrm{PS}^* + \mathrm{A} \rightarrow \underset{\text{Exciplex}}{(\mathrm{PS} \ldots \mathrm{A})^*} \rightarrow \underset{\text{Redox products}}{\mathrm{PS}^{\dot{+}}\,\mathrm{A}^{\dot{-}}}$$

or

$$\mathrm{PS}^* + \mathrm{D} \rightarrow \underset{\text{Exciplex}}{(\mathrm{PS} \ldots \mathrm{D})^*} \rightarrow \underset{\text{Redox products}}{\mathrm{PS}^{\dot{-}}\,\mathrm{D}^{\dot{+}}}$$

$$\text{Redox products} \rightarrow \mathrm{I}\bullet$$

**Scheme 6**

Efficient electron transfer requires that the excitation energy of the photosensitizer, $E^*(\mathrm{PS})$, is sufficient to effect the electron transfer process, including separation of the redox products. This energy requirement may be expressed as the thermodynamic free energy change ($\Delta G$) for the reaction in terms of half-wave redox potentials ($E_{1/2}$), as shown in equation (12), where $E_{1/2}^{\mathrm{ox}}(\mathrm{D/D^+})$ and $E_{1/2}^{\mathrm{red}}(\mathrm{A^-/A})$ represent redox potentials (in volts) for oxidation of the donor (D) and reduction of the acceptor (A), respectively; $E^*(\mathrm{PS})$ represents the excitation energy of the photosensitizer in kJ mol$^{-1}$, and the value 97 converts the redox potentials from volts into kJ mol$^{-1}$. The last term represents the coulombic attraction energy between products of unlike charge at an equilibrium distance $R$ and may be considered as the energy required to separate such products to infinite distances. In this term, $e$ is the electronic charge and $\varepsilon$ is the dielectric constant of the medium. Equation (12) applies specifically to redox reactions in acetonitrile, in which case the coulombic energy term is essentially negligible. However, in less polar media where the energy for charge separation is enhanced, a modified equation may be used.[55]

$$\Delta G = 97[E_{1/2}^{\mathrm{ox}}(\mathrm{D/D^+}) - E_{1/2}^{\mathrm{red}}(\mathrm{A^-/A})] - E^*(\mathrm{PS}) - e^2/R\varepsilon \tag{12}$$

Photosensitization by electron transfer is thermodynamically favourable when the free energy change ($\Delta G$) in equation (12) is a negative value. With reference to Scheme 6, the photosensitizer may function as the donor (D) with acceptor (A) or as the acceptor (A) with donor (D). Both electron accepting and donating capability of the excited state photosensitizer are enhanced, relative to the ground state, by the magnitude of the excitation energy, as indicated in equation (12).

Photosensitization of bis(imidazolyl)s of general structure (**16**) (see Section 20.3.3) with visible light absorbing dyes, which possess relatively low excitation energies, probably occurs by electron transfer, as shown in Scheme 7.

$$\mathrm{PS} \xrightarrow{h\nu} \mathrm{PS}^*$$

$$\mathrm{PS}^* + (\mathbf{16}) \rightarrow \underset{\text{Exciplex}}{(\mathrm{PS} \ldots \mathbf{16})^*} \rightarrow \mathrm{PS}^{\dot{+}} + (\mathbf{16})^{\dot{-}}$$

$$(\mathbf{16})^{\dot{-}} \rightarrow \mathrm{Im}\bullet + \mathrm{Im}^-$$

$$\mathrm{Im}\bullet + \mathrm{RSH} \rightarrow \mathrm{ImH} + \mathrm{RS}\bullet$$

$$\mathrm{RS}\bullet + \text{Monomer} \rightarrow \text{Polymer}$$

$$(\mathbf{16})^{\dot{-}} + \mathrm{PS}^{\dot{+}} \rightarrow (\mathbf{16}) + \mathrm{PS} \tag{13}$$

**Scheme 7**

Following light absorption by the photosensitizer (dye) (PS), electron transfer to bis(imidazolyl) (**16**) yields the corresponding radical anion $(\mathbf{16})^{\dot{-}}$, which is expected to cleave spontaneously into the imidazolyl radical (Im•) and anion (Im$^-$). The relatively stable (noninitiating) imidazolyl radical undergoes H abstraction from a thiol (RSH) to produce the initiating thiyl radical (RS•).

The facility of spontaneous cleavage of the bis(imidazolyl) radical anion $(\mathbf{16})^{\dot{-}}$ is important since this process competes with thermodynamically favored back electron transfer, which regenerates the starting materials, as shown in equation (13) in Scheme 7. The driving force for back electron transfer (disregarding the coulombic energy term) may be expressed by equation (14), where the redox potentials correspond to those in equation (12).

$$\Delta G = 97[E_{1/2}^{\mathrm{ox}}(\mathrm{A^-/A}) - E_{1/2}^{\mathrm{red}}(\mathrm{D/D^+})] \tag{14}$$

The free energy change ($\Delta G$) for back transfer will necessarily be favorable (*i.e.* negative), unless the forward redox reaction occurs spontaneously in the absence of light. In general, efficient generation of initiating radicals by photoinduced electron transfer requires that a subsequent irreversible process occurs which competes with back transfer. Recent reviews on control of back electron transfer in photoinduced electron transfer processes in organic systems,[56] and on dye sensitized photopolymerization,[57] are available.

$$\mathrm{Im}^- + (\mathbf{16}) \rightarrow \mathrm{Im}\bullet + (\mathbf{16})^{\overline{\bullet}} \tag{15}$$

The possibility that the imidazolyl anion ($\mathrm{Im}^-$) may also reduce the bis(imidazolyl) (**16**), as shown in equation (15), is particularly intriguing since the resulting radical anion (**16**$^{\overline{\bullet}}$) would be expected to regenerate the anion ($\mathrm{Im}^-$) by spontaneous cleavage (see Scheme 7). Thereby, these reactions would constitute a chain reaction for the generation of initiating radicals. That is, one absorbed photon would result in the formation of many initiating radicals. Since each radical initiates the polymerization of many monomers, the process would constitute an amplication of photons.

## 20.5 PHOTOINITIATORS FOR CATIONIC POLYMERIZATION

Current reviews on photoinitiators for cationic polymerization are available.[58–60] (See also Chapter 39 in Volume 3 of this work.) The two major classes are onium salts[58,59] and mixed ligand arene cyclopentadienyl metal salts[60] with complex metal halide anions.

### 20.5.1 Onium Salts

#### *20.5.1.1 Aryldiazonium salts*

The first commercial system for photoinitiated cationic polymerization in UV curing and imaging applications utilized arenediazonium salts of complex metal halide anions together with epoxy functional monomers and oligomers.[61] Photolysis of these salts yields Lewis acids, as shown in equation (16), which may initiate cationic polymerization of epoxides directly and/or produce protic (Brönsted) acids by reaction with adventitious (or purposefully added) hydroxyl substances, such as water or alcohols, as shown in equation (17).

$$\mathrm{Ar{-}N_2^+\,PF_6^-} \xrightarrow{h\nu} \mathrm{Ar{-}F + N_2 + PF_5} \tag{16}$$

$$\mathrm{PF_5 + ROH \rightarrow H^+\,PF_5(OR)^-} \tag{17}$$

The spectral response of diazonium salts may be varied throughout the near UV–visible range by appropriate substituents on the aryl ring.[62] Furthermore, both the absorptivity and thermal stability of the diazonium salts is enhanced by electron-donating groups (one of the few exceptions to Murphy's laws). Nevertheless, the inherent thermal instability of arenediazonium salts limits their practical utility as photoinitiators in crosslinking systems.

#### *20.5.1.2 Diaryliodonium and triarylsulfonium salts*

These deficiencies prompted investigations on alternative photoinitiators for cationic polymerization, which led to the development of diaryliodonium (**17**) and triarylsulfonium (**18**) salts with complex metal halide anions.[58,63] In contrast to arenediazonium salts, the iodonium and sulfonium salts exhibit high thermal stability, as well as hydrolytic stability, and may be considered second generation photoinitiators for crosslinking by cationic polymerization.

$$\underset{(\mathbf{17})}{\mathrm{Ar_2I^+X^-}} \qquad \underset{(\mathbf{18})}{\mathrm{Ar_3S^+X^-}}$$

$$\mathrm{X^- = PF_6^-,\ AsF_6^-\ and\ SbF_6^-}$$

Irradiation of these salts results in homolytic cleavage of Ar–I (or Ar–S) bonds to yield reactive radical cations, as shown in equations (18) and (19), respectively.

$$Ar_2I^+X^- \xrightarrow{h\nu} ArI^{\cdot+}X^- + Ar\cdot \tag{18}$$

$$Ar_3S^+X^- \xrightarrow{h\nu} Ar_2S^{\cdot+}X^- + Ar\cdot \tag{19}$$

Direct evidence for homolytic bond cleavage has been obtained from laser flash photolysis studies.[64] Furthermore, the radical cations have been found to be highly reactive with nucleophiles, including cyclohexene oxide and vinyl ethers.[64–66]

The phenyliodinium radical cation ($PhI^{\cdot+}$) was found to be more than ten times more reactive with cyclohexene oxide relative to other oxygen nucleophiles, including methanol, tetrahydrofuran (THF) and water. These results provide experimental support for direct initiation of cyclohexene oxide polymerization by $PhI^{\cdot+}$ even in the presence of other oxygen nucleophiles as well as H donors (*e.g.* THF).

A possible reaction mechanism for photoinitiated polymerization of epoxides utilizing a diphenyliodonium salt (**17**), as photoinitiator, which is consistent with the laser flash photolysis studies, as well as with polymer end group analysis studies (see below), is provided in Scheme 8. (Note that the anion is not included.)

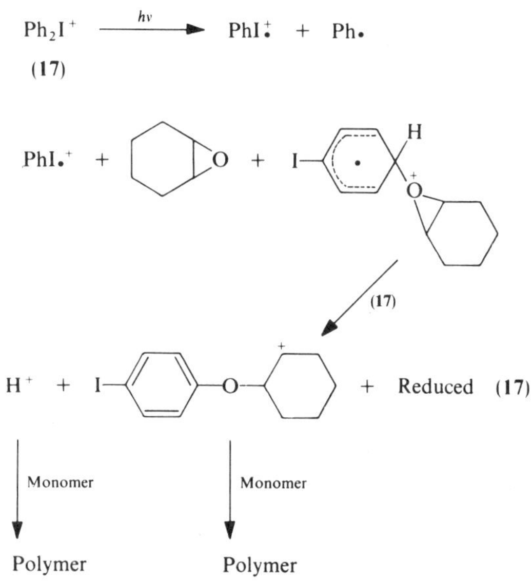


**Scheme 8**

Key features of this scheme are (1) direct reaction of $PhI^{\cdot+}$ with the epoxide, (2) oxidation of the resulting adduct by a second diphenyliodonium salt, and (3) initiation of polymerization by both the resulting proton ($H^+$) and activated epoxide. This mechanism predicts both hydroxyl and 4-alkoxyphenyl iodide polymer end groups.

Support for photoinitiator fragment incorporation comes from studies with the triarylsulfonium salt (**19**).[67] Utilizing (**19**) as photoinitiator for cationic polymerization of cyclohexene oxide, spectroscopic evidence (absorption and emission) has been obtained for the incorporation of 4-alkoxy diphenyl sulfide end groups in poly(cyclohexene oxide). These results are readily explicable in terms of initiation by diphenylsulfinium radical cation ($Ph_2S^{\cdot+}$) in accordance with Scheme 8.

Furthermore, direct evidence for formation of $Ph_2S^{\cdot+}$ from irradiation of (**19**), as shown in equation (20), has been obtained from laser flash photolysis studies.[64] Thus, the laser flash photolysis and end group analysis studies are complementary and support direct initiation of cationic polymerization (of cyclohexene oxide) by the reactive phenyliodinium and diphenylsulfinium ions, derived from the corresponding iodonium and sulfonium salt photoinitiators, in accordance with Scheme 8.

$$Ph_2\overset{+}{S}\text{–Ar–SPh} \xrightarrow{h\nu} Ph_2S^{\cdot+} + \cdot\text{Ar–SPh} \tag{20}$$

(**19**) Ar = 1,4-disubstituted benzene

In the absence of reactive monomer, photogenerated $ArI^{+\cdot}$ and $Ar_2S^{+\cdot}$ are expected to react with nucleophiles (NuH) and possibly H donors (RH) to produce protons, as shown in Scheme 9. The reactions in Scheme 9 account for polymerization of monomer introduced after irradiation,[58,59] as well as for photoinduced acid catalyzed reactions utilized in photoimaging applications.[68,69] The electrophilicity of diarylsulfinium ions ($Ar_2S^{+\cdot}$) is well documented.[70]

$$ArI^{+\cdot}\,(Ar_2S^{+\cdot}) + NuH \rightarrow H^+ + \text{Unidentified Products}$$

$$ArI^{+\cdot}\,(Ar_2S^{+\cdot}) + RH \rightarrow H^+ + ArI(Ar_2S) + R\cdot$$

**Scheme 9**

In addition to the proposals for photogeneration of cationic initiators in Schemes 8 and 9, both of which originate with homolytic bond cleavage, evidence for photoinduced heterolytic bond cleavage of diaryliodonium and triarylsulfonium salts in methanol has been reported,[71] as shown in Scheme 10. However, the reported substantial yields of fluorobenzene appear to be inconsistent with earlier studies on the photochemistry of diaryliodonium[72] and triarylsulfonium salts,[73] carried out under similar conditions. Possibly, fluorobenzene arises from secondary photochemical reactions.

$$Ph_2I^+(Ph_3S^+)PF_6^- \xrightarrow{h\nu} PhI(Ph_2S) + Ph^+\,PF_6^-$$

$$Ph^+\,PF_6^- \rightarrow PhF + PF_5$$

**Scheme 10**

### 20.5.1.3 *Phenacylsulfonium and 4-hydroxyphenylsulfonium salts*

Dialkyl phenacylsulfonium[74] and 4-hydroxyphenylsulfonium[75] salts, such as (**20**) and (**21**), respectively, also function as photoinitiators for cationic polymerization. However, in contrast to triarylsulfonium salts, polymerization is reported to cease following light exposure. Furthermore, polymerization does not occur when (**20**) and (**21**) are irradiated in the absence of monomer followed by monomer addition. These results, together with additional evidence (including deuterium exchange studies) have been interpreted in terms of reversible photogeneration of acid, as shown in equations (21) and (22).

$$PhC(=O)CH_2\overset{+}{S}Me_2\;PF_6^- \xrightarrow{h\nu} PhC(=O)CH{=}SMe_2 + H^+PF_6^- \qquad (21)$$

(**20**)

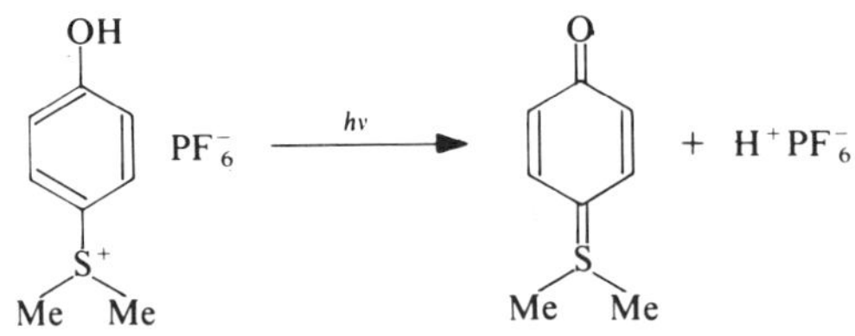


(**21**) (22)

In the case of phenacylsulfonium salts, such as (**20**), ylide ($R_2C{=}SR_2$) formation proceeds by intramolecular H abstraction by the ketone from the $\gamma$ carbon.[74] In both cases, monomer can apparently compete for protons with the highly basic ylide products. However, the lifetime of the protons is reduced in the presence of the basic ylides.

### 20.5.1.4 *Sulfoxonium salts*

Aryloxysulfoxonium salts of general structure (**22**) are reported[76] to be efficient photoinitiators for cationic polymerization with activities comparable with or greater than diaryliodonium and triarylsulfonium salts. They apparently undergo irreversible photolysis and acid generation.

Sulfoxonium salts of general structure (**23**) are also reported to be efficient photoinitiators for cationic polymerization.[76] With these photoinitiators, cationic polymerization is reported to

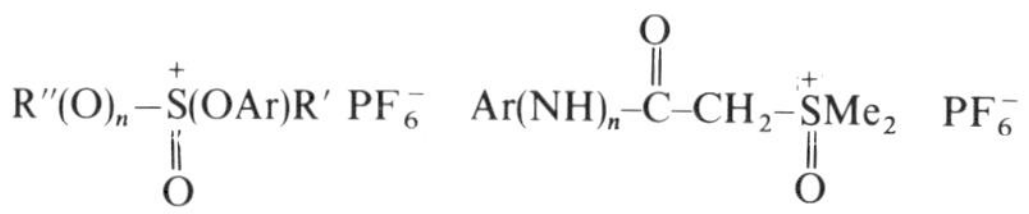

(22) $n=0$ or 1 (23) $n=0$ or 1

cease following light exposure, as observed with phenacyl- and 4-hydroxyphenyl-sulfonium salts (see above).

### 20.5.2 Mixed Ligand Arene Cyclopentadienyl Metal Salts

A newer class of photoinitiators for cationic polymerization are mixed ligand arene cyclopentadienyl metal salts of complex metal halide anions,[77,78] such as ($\eta^6$-benzene)($\eta^5$-cyclopentadienyl)$Fe^{II}$ hexafluorophosphate (**24**). A mechanism for photoinitiated polymerization of epoxides is provided in Scheme 11.[60,79]

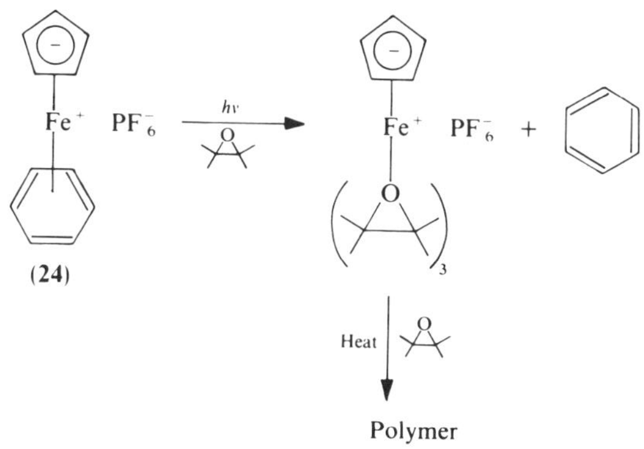


**Scheme 11**

The proposed photoinduced ligand exchange of the arene by three epoxide groups is based on earlier ligand exchange studies.[80] With bicyclic epoxides, such as cyclohexene oxide, polymerization is spontaneous under ambient conditions. However, moderate heating (50 to 150 °C) is required with less reactive epoxides, such as glycidyl ethers. The thermal requirement may be reduced substantially in the presence of oxidizing agents, such as peroxides, which apparently oxidize $Fe^{II}$ to the more reactive $Fe^{III}$ state. It is noteworthy, in this regard, that photoinitiated polymerization of epoxides, utilizing triarylsulfonium salts, is also enhanced by moderate heating, which also reduces the adverse effects of high humidity.[61]

The spectral response of the mixed arene cyclopentadienyl metal salts may be varied throughout the near UV–visible region by proper selection of the arene component and also by photosensitization.[81]

## 20.6 PHOTOSENSITIZATION OF CATIONIC POLYMERIZATION

The spectral response of onium salts[58,59] as well as mixed ligand arene cyclopentadienyl metal salts[81] can be extended to longer wavelengths by photosensitization.

### 20.6.1 Onium Salts

Representative photosensitizers for cationic polymerization with diaryliodonium ($Ar_2I^+$), triarylsulfonium ($Ar_3S^+$), dialkyl phenacylsulfonium ($DPS^+$) and dialkyl 4-hydroxyphenylsulfonium ($DHS^+$) salts are provided in Table 1. The results were obtained from photoinduced polymerization of 3,4-epoxycyclohexylmethyl 3′,4′-epoxycyclohexanecarboxylate (**25**), as evaluated by photogelation[59] and photocalorimetry.[58]

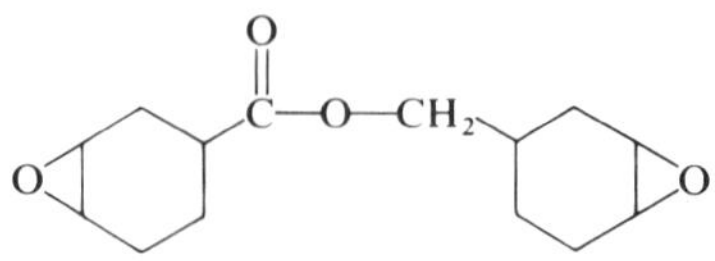

**(25)**

**Table 1** Photosensitizer/Onium Salt Combinations for Cationic Polymerization of Di-epoxide **(25)**[a]

| *Photosensitizer* | $Ar_2I^+$ [82] | $Ar_3S^+$ [82] | $DPS^+$ [91] | $DHS^+$ [91] |
|---|---|---|---|---|
| Anthracene | yes | yes | yes | no |
| Perylene | yes | yes | yes | no |
| Phenothiazine | yes | yes | yes | no |
| Michler's ketone | yes | yes | b | b |
| Xanthone | yes | no | b | b |
| Thioxanthone | yes | no | yes | yes |
| Benzophenone | yes[c] | no | no | yes |
| Acetophenone | yes[c] | no | no | yes |
| Fluorenone | d | e | f | yes |
| 9,10-Anthraquinone | d | e | f | yes |

[a] Yes signifies that polymerization was observed. [b] Not reported; predicted to occur. [c] Requires a H donor. [d] Not reported; predicted to occur with a H donor. [e] Not reported; predicted not to occur. [f] Not reported; predicted to be borderline with a H donor.

### *20.6.1.1 Triarylsulfonium salts*

Parallel studies on (1) photosensitized polymerization of di-epoxide (**25**), (2) quantum yields of photosensitized photolysis, and (3) luminescence quenching experiments (to identify the photosensitizer excited states) with triphenylsulfonium hexafluoroarsenate, coupled with calculations on the thermodynamic favorability of energy and electron transfer, have provided strong evidence for the electron transfer mechanism, as shown in Scheme 12.[82] Also see Section 20.4.2.

$$PS \xrightarrow{h\nu} PS^*$$

$$PS^* + Ar_3S^+(Ar_2I^+) \rightarrow \underset{\text{Exciplex}}{(PS\ldots Ar_3S^+)^*} \text{ or } \underset{\text{Exciplex}}{(PS\ldots Ar_2I^+)^*}$$

$$\text{Exciplex} \rightarrow PS^{\dot{+}} + Ar_2S(ArI) + Ar\cdot \qquad (23)$$

$$PS^{\dot{+}} + \text{Monomer} \rightarrow \text{Polymer}$$

**Scheme 12**

The electron transfer process (*via* the exciplex), equation (23), is shown to yield $Ar_2S$ and Ar•, directly, in accordance with recent studies which provide evidence for concomitant bond cleavage with one-electron reduction of aryl alkyl sulfonium salts.[83] The reported irreversible potential for reduction of triphenylsulfonium ion is $-1.2$ V (*vs.* SCE).[84] The accompanying photosensitizer radical cation ($PS^{\dot{+}}$) is expected to initiate cationic polymerization directly, as shown, or indirectly by interaction with nucleophiles (or H donors) to yield Brönsted acids, which initiate polymerization.

Photosensitization by electron transfer is further supported by spectroscopic evidence for the formation of *N*-ethylphenothiazine radical cation on irradiation of *N*-ethylphenothiazine in the presence of triarylsulfonium (as well as diaryliodonium) salts. Furthermore, phenothiazine and *N*-ethylphenothiazine radical cations are reported to initiate cationic polymerization.[85]

The participating excited states of the photosensitizers were determined by quenching of their emission with triphenylsulfonium hexafluoroarsenate. In the case of anthracene, perylene, phenothiazine and Michler's ketone, the observed quenching is thermodynamically favorable by electron transfer, but not by energy transfer. Thus, photosensitized polymerization of di-epoxide (**25**) (see Table 1) and sensitized photolysis of the sulfonium salt may be attributed to electron transfer, as shown in Scheme 12.

The triplet excited states $^3(PS)^*$ of xanthone and acetophenone were also quenched by the sulfonium salt, for which electron transfer is thermodynamically unfavorable. In these cases, quenching may be attributed to favorable triplet energy transfer, as shown in equation (24). However, neither photosensitized polymerization of di-epoxide (**25**) (see Table 1) nor sensitized photolysis of the sulfonium salt is observed with xanthone or acetophenone, signifying that the resulting excited triplet state of $Ph_3S^+$ does not participate in the photogeneration of reactive ions, or does so inefficiently.

$$^3(PS)^* + Ph_3S^+ \rightarrow PS + {}^3(Ph_3S^+)^* \tag{24}$$

These results also signify that excited singlet states of $Ph_3S^+$ are primarily responsible for photoinitiator activity in direct irradiation, as well, and that intersystem crossing to triplet states represents an energy wastage process.

These conclusions reasonably follow from the results with xanthone and acetophenone. However, quenching of xanthone and acetophenone phosphorescence by the sulfonium salt is relatively inefficient, probably resulting from similar triplet excitation energies of the ketones and sulfonium salt. Consequently, these results do not preclude the possibility that more efficient photosensitized photolysis of triphenylsulfonium salts may be observed with higher energy (or longer lived) triplet photosensitizers.

An alternative mechanism for photosensitization, which may be considered, is exciplex-induced bond cleavage, shown in Scheme 13.

$$PS \xrightarrow{h\nu} PS^*$$

$$PS^* + Ph_3S^+ \rightarrow (PS \ldots Ph_3S^+)^*$$
$$\text{Exciplex}$$

$$\text{Exciplex} \rightarrow PS + Ph\bullet + Ph_2S^{\dot{+}} \tag{25}$$

$$Ph_2S^{\dot{+}} + \text{Monomer} \rightarrow \text{Polymer}$$

**Scheme 13**

In this case, the shared excitation energy within the exciplex may be utilized to cleave the Ph–S bond of the sulfonium salt, as shown in equation (25), to yield the reactive $Ph_2S^{\dot{+}}$ initiating ion (which is also formed by direct irradiation of the sulfonium salt). This mechanism requires sufficient energy in the exciplex for homolytic Ph–S bond cleavage. However, the apparent nonreactivity of triplet state sulfonium salt with an excitation energy $>290\ kJ\ mol^{-1}$ (from emission studies) casts doubt on the likelihood of this mechanism for photosensitizers with lower excitation energies, including perylene and phenothiazine.[59]

A feature which clearly distinguishes the electron transfer mechanism (Scheme 12) from exciplex induced bond cleavage (Scheme 13) is the contrasting role of the photosensitizer in generation of the active initiator. In the electron transfer mechanism, the reactive cationic species ($PS^{\dot{+}}$) is derived from the photosensitizer; by contrast, in the bond cleavage mechanism, the reactive cationic species ($Ar_2S^{\dot{+}}$) is derived from the onium salt.

#### 20.6.1.2 *Diaryliodonium salts*

Photosensitization studies analogous to those discussed above with triphenylsulfonium hexafluoroarsenate were also carried out with diphenyliodonium hexafluoroarsenate.[82] The results support the electron transfer mechanism (Scheme 12) for all effective photosensitizers in Table 1, except benzophenone and acetophenone.

With benzophenone and acetophenone, a H donor is required and the results may be explained by an indirect electron transfer process, as outlined in Scheme 14 for benzophenone and 2-propanol.

Key features of this proposal are (1) photogeneration of ketyl radicals by H abstraction and (2) oxidation of the ketyl radicals by the iodonium salt to produce carbocations (protonated ketones), which dissociate to yield protons for initiation of cationic polymerization.

Reduction of the iodonium salt (by the ketyl radicals) is shown to occur with concomitant dissociation into PhI and Ph•, in accordance with electrochemical reduction of diphenyliodonium salts for which an irreversible potential of −0.2 V (*vs.* SCE) is reported.[86]

H abstraction by phenyl radical from 2-propanol (equation 28) regenerates the corresponding ketyl radical. Thus, equations (27) and (28) constitute a radical chain reaction leading ultimately to protons.

$$Ph_2C{=}O \xrightarrow{h\nu} (Ph_2C{=}O)^*$$

$$(Ph_2C{=}O)^* + Me_2CH{-}OH \rightarrow Ph_2\dot{C}{-}OH + Me_2\dot{C}{-}OH$$

$$Ph_2\dot{C}{-}OH + Ph_2I^+ \rightarrow Ph_2\overset{+}{C}{-}OH + PhI + Ph\bullet \tag{26}$$

$$Me_2\dot{C}{-}OH + Ph_2I^+ \rightarrow Me_2\overset{+}{C}{-}OH + PhI + Ph\bullet \tag{27}$$

$$Ph\bullet + Me_2\overset{+}{C}H{-}OH \rightarrow Me_2\dot{C}{-}OH + Ph{-}H \tag{28}$$

$$Ph_2\overset{+}{C}{-}OH \rightarrow Ph_2C{=}O + H^+$$

$$Me_2\overset{+}{C}{-}OH \rightarrow Me_2C{=}O + H^+$$

$$H^+ + \text{Monomer} \rightarrow \text{Polymer}$$

**Scheme 14**

The proposal in Scheme 14 predicts that quantum yields of formation of protons (as well as PhI) may exceed one. That is, more protons may be produced than photons absorbed by benzophenone. This predicted amplification of photons in the formation of protons has been observed and evidence for a radical chain reaction has been obtained.[82]

Radicals produced photochemically with free-radical photoinitiators (as well as thermally) have also been shown to participate in the generation of reactive cations by the reduction of diaryliodonium[87] (as well as aryldiazonium)[87,88] salts, as illustrated in Scheme 15.

$$PhC({=}O){-}C(OMe)_2{-}Ph \xrightarrow{h\nu} Ph\dot{C}{=}O + Ph\dot{C}(OMe)_2$$

$$Ph\dot{C}(OMe)_2 + Ph_2I^+ \rightarrow Ph\overset{+}{C}(OMe)_2 + PhI + Ph\bullet$$

$$Ph\overset{+}{C}(OMe)_2 + \text{Monomer} \rightarrow \text{Polymer}$$

**Scheme 15**

Triarylsulfonium salts do not participate efficiently in these processes (Schemes 14 and 15),[67,88,89] which undoubtedly reflects their less favorable reduction potentials. The free energy change ($\Delta G$) for the redox reactions in Scheme 14, equations (26) and (27), may be estimated from equation (29), where the value 97 converts the redox potentials from volts into kJ mol$^{-1}$. Utilizing oxidation potentials ($E^{ox}_{1/2}$) of $-0.8$ to $-1.0$ V for acetone and benzophenone ketyl radicals, respectively,[90] together with reduction potentials ($E^{red}_{1/2}$) of $-0.2$ and $-1.2$ V for diphenyliodonium and triphenylsulfonium salts, respectively, $\Delta G$ values are seen to be highly favorable (exergonic) for reduction of the iodonium salt ($-58$ to $-78$ kJ mol$^{-1}$) and unfavorable for reduction of the sulfonium salt ($+19$ to $+39$ kJ mol$^{-1}$).

$$\Delta G = 97[E^{ox}_{1/2} - E^{red}_{1/2}] \tag{29}$$

### *20.6.1.3 Phenacylsulfonium salts*

Photosensitization of dialkyl phenacylsulfonium salts by anthracene, perylene, phenothiazine and thioxanthone (see Table 1) is reported to occur by the electron transfer mechanism, in accordance with Scheme 12, and spectroscopic evidence for the perylene radical cation has been obtained.[91] Irreversible reduction potentials of phenacylsulfonium salts are *ca.* $-0.8$ V (*vs.* SCE),[83] intermediate between diaryliodonium and triarylsulfonium salts. The more favorable reduction potential for phenacylsulfonium salts accounts for photosensitization by thioxanthone, which is not observed with triarylsulfonium salts.[58]

Whereas direct irradiation of phenacylsulfonium salts results in reversible acid formation, as discussed in Section 20.5.1.3, photosensitization produces acid irreversibly. The photosensitization results are in accord with electron transfer, which is expected to occur irreversibly, as shown in equation (30).

$$PS^* + PhC(O)CH_2{-}\overset{+}{S}Me_2 \rightarrow PS^{+\bullet} + PhC(O)CH_2^{\bullet} + Me_2S \tag{30}$$

### *20.6.1.4 4-Hydroxyphenylsulfonium salts*

Photosensitization of dialkyl 4-hydroxyphenylsulfonium salts by thioxanthone, benzophenone, fluorenone and anthraquinone (see Table 1) has been attributed to H abstraction from the phenol group followed by electron transfer between the resulting ketyl radical and phenol radical cation.[91] This mechanism is outlined in Scheme 16, in which the excited state photosensitizer is represented by $(Ar_2C{=}O)^*$.

Photosensitization by aromatic ketones results in reversible acid formation, as observed in direct irradiation of hydroxyphenylsulfonium salts (see Section 20.5.1.3). Reversible acid generation is consistent with Scheme 16, since formation of the highly basic ylide is proposed, which is expected to limit the lifetime of protons (see equation (31) in Scheme 16).

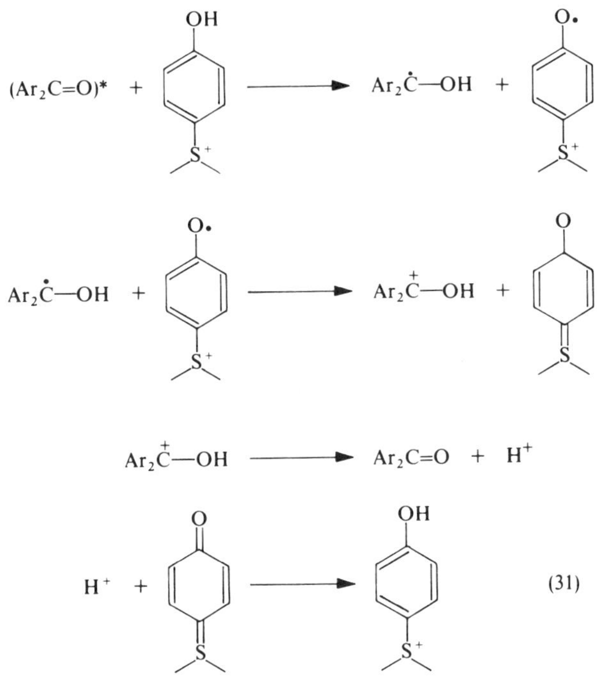

Scheme 16

Polynuclear aromatic hydrocarbons are reported to be inactive as photosensitizers for 4-hydroxyphenylsulfonium salts,[91] in contrast to their effectiveness with phenacylsulfonium as well as with triarylsulfonium salts (see Table 1). The lack of photosensitization by polynuclear aromatics, such as anthracene and perylene, which should be capable of undergoing electron transfer to 4-hydroxyphenylsulfonium salts, in accordance with Scheme 12, is an interesting observation, which does not appear to have an obvious explanation.

**Table 2** Reversible and Irreversible Photogeneration of Cationic Initiators by Direct and Sensitized Photolysis of Onium Salts

| *Onium salt* | *Direct irradiation* | *Sensitized irradiation* |
|---|---|---|
| Diaryliodonium | irreversible | irreversible |
| Triarylsulfonium | irreversible | irreversible |
| Phenacylsulfonium | reversible | irreversible |
| 4-Hydroxyphenylsulfonium | reversible | reversible |

### 20.6.2 Mixed Ligand Arene Cyclopentadienyl Metal Salts

Photosensitizers, including anthracene and anthrone,[92] have been claimed for photoinitiated cationic polymerization with mixed ligand arene cyclopentadienyl metal salts. Studies on the mechanism of photosensitization have, apparently, not been reported.

## 20.7 REVERSIBLE AND IRREVERSIBLE PHOTOGENERATION OF ACID

Direct and sensitized photolysis of onium salts constitute versatile systems for reversible and irreversible photogeneration of cationic initiators. The systems are summarized in Table 2.

## 20.8 REFERENCES

1. L. F. Thompson, C. G. Willson and M. J. Bowden (eds.) *ACS Symp. Ser.*, 1983, **219**.
2. G. E. Green, B. P. Stark and S. A. Zahir, *J. Macromol. Sci., Rev. Macromol. Chem.*, 1981–2, **C21**, 187.
3. G. A. Delzenne, *Adv. Photochem.*, 1979, **11**, 1.
4. S. P. Pappas (ed.), 'UV Curing: Science and Technology', Technology Marketing Corp., Norwalk, CT, 1985, vol. II.
5. G. A. Senich and R. E. Florin, *J. Macromol. Sci., Rev. Macromol. Chem. Phys.*, 1984, **C24**, 239.
6. C. G. Roffey, 'Photopolymerization of Surface Coatings,' Wiley–Interscience, New York, 1982.
7. S. P. Pappas (ed.), 'UV Curing: Science and Technology', Technology Marketing Corp., Norwalk, CT, 1978.
8. S. G. Wentink and S. D. Koch (eds.), 'UV Curing in Screen Printing for Printed Circuits and the Graphic Arts', Technology Marketing Corp., Norwalk, CT, 1981.
9. S. P. Pappas and V. D. McGinniss, in ref. 7, pp. 1–22.
10. G. A. Thommes and V. J. Webers, *J. Imaging Sci.*, 1985, **29**, 112.
11. A. R. Gutierrez and R. J. Cox, *Polym. Photochem.*, 1986, **7**, 517.
12. H. J. Hageman, *Prog. Org. Coat.*, 1985, **13**, 123.
13. S. P. Pappas in 'New Trends in the Photochemistry of Polymers', eds. N. S. Allen and J. F. Rabek, Elsevier, London, 1985, pp. 99–111.
14. S. P. Pappas, *Radiat. Phys. Chem.*, 1985, **25**, 633.
15. L. H. Carlblom and S. P. Pappas, *J. Polym. Sci., Polym. Chem. Ed.*, 1977, **15**, 1381.
16. S. P. Pappas and R. A. Asmus, *J. Polym. Sci., Polym. Chem. Ed.*, 1982, **20**, 2643.
17. H. J. Hageman, F. P. B. van der Maeden and P. C. G. M. Janssen, *Makromol. Chem.*, 1979, **180**, 2531.
18. S. P. Pappas and A. K. Chattopadhyay, *J. Am. Chem. Soc.*, 1973, **95**, 6484.
19. F. D. Lewis, R. T. Lauterbach, H.-G. Heine, W. Hartmann and H. Rudolph, *J. Am. Chem. Soc.*, 1975, **97**, 1519.
20. W. Rutsch, G. Berner, R. Kirchmayr, R. Huesler, G. Rist and N. Buehler, *Org. Coat.*, 1986, **8**, 175.
21. S. P. Pappas and D. C. Sutton, unpublished results.
22. S. P. Pappas and J. C. Scaiano, unpublished results.
23. A. Ledwith, P. J. Russell and L. H. Sutcliffe, *J. Chem. Soc., Perkin Trans. 2*, 1972, 1925.
24. C. W. Porter, H. C. Ramsperger and C. Steel, *J. Am. Chem. Soc.*, 1923, **45**, 1827.
25. S. N. Gupta, I. Gupta and D. C. Neckers, *J. Polym. Sci., Polym. Chem. Ed.*, 1981, **19**, 103.
26. M. Jacobi and A. Henne, *J. Radiat. Curing*, 1983, **10(4)**, 16.
27. T. Sumiyoshi, W. Schnabel, A. Henne and P. Lechtken, *Polymer*, 1985, **26**, 141.
28. T. Sumiyoshi and W. Schnabel, *Makromol. Chem.*, 1985, **186**, 1811.
29. R. Kuhlmann and W. Schnabel, *Polymer*, 1976, **17**, 419.
30. S. P. Pappas, D. Berger and T. Kilp, unpublished results.
31. J. F. Kinstle and S. L. Watson, Jr., *J. Radiat. Curing*, 1975, **2(2)**, 7.
32. J. Hutchison, M. C. Lambert and A. Ledwith, *Polymer*, 1973, **14**, 250.
33. M. R. Sandner, C. L. Osborn and D. J. Trecker, *J. Polym. Sci., Polym. Chem. Ed.*, 1972, **10**, 3173.
34. S. P. Pappas and R. A. Asmus, unpublished results.
35. T. H. Koch and A. H. Jones, *J. Am. Chem. Soc.*, 1970, **92**, 7503.
36. C. C. Wamser, G. S. Hammond, C. T. Chang and C. Baylor, Jr., *J. Am. Chem. Soc.*, 1970, **92**, 6362.
37. G. Amirzadeh and W. Schnabel, *Makromol. Chem.*, 1981, **182**, 2821.
38. N. S. Allen, F. Catalina, P. N. Green and W. A. Green, *Eur. Polym. J.*, 1986, **22**, 793.
39. J. Hutchison and A. Ledwith, *Polymer*, 1973, **14**, 405.
40. B. M. Monroe, *Adv. Photochem.*, 1971, **8**, 77.
41. P. K. Sengupta and S. K. Modak, *Makromol. Chem.*, 1985, **186**, 1593.
42. R. A. Bottom, J. Guthrie and P. N. Green, *Polym. Photochem.*, 1985, **6**, 59.
43. M. V. Encinas, J. Garrido and E. A. Lissi, *J. Polym. Sci., Polym. Chem. Ed.*, 1985, **23**, 2481.
44. N. S. Allen, F. Catalina, P. N. Green and W. A. Green, *Eur. Polym. J.* 1986, **22**, 49.
45. A. Ledwith, G. Ndaalio and A. R. Taylor, *Macromolecules*, 1975, **8**, 1.
46. E. C. Dart and J. Nemcek (ICI Ltd.), *US Pat.* 4 071 424 (1978) (*Chem. Abstr.*, 1973, **79**, 79 526).
47. D. P. Specht, P. A. Martic and S. Farid, *Tetrahedron*, 1982, **38**, 1203.
48. J. L. R. Williams, D. P. Specht and S. Farid, *Polym. Eng. Sci.*, 1983, **23**, 1022.
49. D. P. Specht, C. G. Houle and S. Y. Farid (Eastman Kodak Co.), *US Pat.* 4 289 844 (1981) (*Chem. Abstr.*, 1981, **95**, 106 408).
50. R. Dessauer and C. Looney, *Photogr. Sci. Eng.*, 1979, **23**, 287.
51. T. E. Dueber and W. J. Link (E. I. DuPont de Nemours and Co.), *US Pat.* 4 454 218 (1984) (*Chem. Abstr.*, 1985, **101**, 46 322).

52. S. P. Pappas and J. H. Jilek, *Photogr. Sci. Eng.*, 1979, **23**, 140.
53. N. J. Turro, 'Modern Molecular Photochemistry', Benjamin/Cummings, CA, 1978, pp. 296–361.
54. V. D. McGinniss, *ACS Symp. Ser.*, 1976, **25**, 135–49.
55. A. Weller, *Z. Phys. Chem.*, 1982, **133**, 93.
56. M. A. Fox, *Adv. Photochem.*, 1986, **13**, 237.
57. D. F. Eaton, *Adv. Photochem.*, 1986, **13**, 427.
58. J. V. Crivello, *Adv. Polym. Sci.*, 1984, **62**, 1.
59. S. P. Pappas, *Prog. Org. Coat.*, 1985, **13**, 35.
60. F. Lohse and H. Zweifel, *Adv. Polym. Sci.*, 1986, **78**, 61.
61. W. R. Watt, in ref. 4, pp. 247–82.
62. S. I. Schlesinger, *Photogr. Sci. Eng.*, 1974, **18**, 387.
63. J. V. Crivello, in ref. 7, pp. 23–77.
64. S. P. Pappas, B. C. Pappas, L. R. Gatechair, J. H. Jilek and W. Schnabel, *Polym. Photochem.*, 1984, **5**, 1.
65. S. P. Pappas, B. C. Pappas, L. R. Gatechair and W. Schnabel, *J. Polym. Sci., Polym. Chem. Ed.*, 1984, **22**, 69.
66. M. Tilley, Y. Yagci, W. Schnabel and S. P. Pappas, unpublished results.
67. A. Ledwith, *Photogr. Sci. Eng., Prepr. 22nd Fall Symp.*, Arlington, VA, Nov. 1982, pp. 44–5.
68. J. V. Crivello, *ACS Symp. Ser.*, 1984, **242**, 3–10.
69. H. Ito and C. G. Willson, *ACS Symp. Ser.*, 1984, **242**, 11–23.
70. A. J. Bard, A. Ledwith and H. J. Shine, *Adv. Phys. Org. Chem.*, 1976, **13**, 155.
71. R. S. Davidson and J. W. Goodin, *Eur. Polym. J.*, 1982, **18**, 589.
72. J. W. Knapczyk, J. J. Lubinkowski and W. E. McEwen, *Tetrahedron Lett.*, 1972, 3739.
73. J. W. Knapczyk and W. E. McEwen, *J. Org. Chem.*, 1970, **35**, 2539.
74. J. V. Crivello and J. L. Lee, *Macromolecules*, 1983, **16**, 864.
75. J. V. Crivello and J. H. W. Lam, *J. Polym. Sci., Polym. Chem. Ed.*, 1980, **18**, 1021.
76. B. P. Stark and E. Irving, *Ind. Chem. Bull.*, 1982, **1**, 164.
77. K. Meier, N. Buehler, H. Zweifel, G. Berner and F. Lohse (Ciba-Geigy AG), *Eur. Pat. Appl.* 94 915 (1983) (*Chem. Abstr.*, 1984, **100**, 105 179).
78. M. C. Palazzotto and W. A. Hendrickson (3M Co.), *Eur. Pat. Appl.* 109 851 (1984) (*Chem. Abstr.*, 1985, **101**, 153 584).
79. K. Meier and H. Zweifel, *J. Radiat. Curing*, 1986, **13**(4), 26.
80. T. P. Gill and K. R. Mann, *J. Organometal. Chem.*, 1981, **216**, 65.
81. K. Meier and H. Zweifel, *J. Imaging Sci.*, 1986, **30**, 174.
82. S. P. Pappas, L. R. Gatechair and J. H. Jilek, *J. Polym. Sci., Polym. Chem. Ed.*, 1984, **22**, 77.
83. F. D. Saeva and B. P. Morgan, *J. Am. Chem. Soc.*, 1984, **106**, 4121.
84. P. S. McKinney and S. Rosenthal, *J. Electroanal. Chem. Interfacial Electrochem.*, 1968, **16**, 261.
85. A. Ledwith, *Polym. Prepr., Am. Chem. Soc., Div. Polym. Chem.*, 1982, **23**(1), 323.
86. H. E. Bachofner, F. M. Beringer and L. Meites, *J. Am. Chem. Soc.*, 1958, **80**, 4269.
87. A. Ledwith, *Makromol. Chem., Suppl.*, 1979, **3**, 348.
88. H. Baumann and H.-J. Timpe, *J. Prakt. Chem.*, 1984, **326**, 529.
89. S. P. Pappas and R. L. Daskin, unpublished results.
90. P. S. Rao and E. Hayon, *J. Am. Chem. Soc.*, 1974, **96**, 1287.
91. J. V. Crivello and J. L. Lee, *Macromolecules*, 1981, **14**, 1141.
92. K. Meier and H. Zweifel (Ciba-Geigy AG), *Eur. Pat. Appl.* 152 377 (1985) (*Chem. Abstr.*, 1986, **104**, 59 420).

# 21
# Plasma Polymerization

H. YASUDA and Y. IRIYAMA
*University of Missouri-Rolla, Rolla, MO, USA*



## 21.1 INTRODUCTION

Plasma polymerization refers to the formation of polymeric materials by using either the plasma state of monomers, or reactive species created in the plasma state. Plasma is the fourth state of matter, consisting largely of ionized gas which maintains overall electrical neutrality. The plasma used in the formation of polymeric materials is the so-called 'low temperature plasma' or 'low pressure plasma', in which thermal equilibrium is not maintained among electrons, ions and neutral species. The kinetic energy ($kT$) of electrons, expressed in the electron temperature, is in the vicinity of 60 000 K, while corresponding values for ions and neutral species are often in the range of 300–500 K.

Low temperature plasmas can be created in a variety of ways; the most practical, for the formation of polymeric materials in laboratories and in industrial applications, is the electric glow discharge created by using power sources as varied as DC and microwave. Therefore, the term 'glow discharge polymerization' is used synonymously with 'plasma polymerization', although glow discharge plasmas represent only one aspect of plasmas. The use of high temperature plasma for organic materials is not practical, because of their heat labile characteristics. On the other hand, plasma polymerization is not restricted to organic materials, and the plasma synthesis of inorganic materials should, in principle, be included in plasma polymerization. Some details on plasma polymerization which are not covered by this article may be found in general references.[1-6]

Although the plasma state is generally identified by the ionization of gases, and ionizaton is the essential step in creating plasmas, the population of ions in a low pressure plasma is small compared with that of neutral species. Typical glow discharges used for plasma polymerization of organic monomers contain nearly a million times more neutral species than ions.[7] Therefore, it is naive to consider that a direct correlation exists between the ionization process and the formation of polymeric materials (polymerization). In this respect, it is interesting to compare the role of ionization in radiation polymerization and plasma polymerization.

In the polymerization of vinyl monomers by ionizing radiation, such as $\gamma$ radiation, a direct correlation seems to exist between the ionization of molecules and the initiation of polymerization, although the instantaneous concentration of ionized species is extremely small. Ionization is the first step of polymerization in radiation polymerization.[8] In contrast, the instantaneous concentration of ions formed in plasma polymerization is much greater, but the formation of polymer is not directly linked to the ionization. The majority of reactive species which participate in polymer formation are formed in processes parallel to the ionization.

Formation of the reactive species which are used in forming polymeric materials takes place in the plasma. Depending on the location of subsequent polymer-forming processes, plasma polymerization can be divided into two major categories: (i) plasma-state polymerization and (ii) plasma-induced polymerization. In simple terms, plasma-state polymerization is the vacuum deposition of plasma polymers; plasma-induced polymerization is the condensed phase (liquid and solid) polymerization of monomers initiated by the action of a plasma. The latter is often referred to as plasma-initiated polymerization; however, 'plasma-induced polymerization' seems to be more appropriate, because the polymerization is initiated by reactive species created under the influence of a plasma, not by plasma itself.

Deposition of polymers from monomers in vacuum does not occur in conventional polymerization reactions.[9] Due to a low ceiling temperature for polymerization in vacuum, most polymerizations, such as addition polymerization of vinyl monomers, do not yield materials with high enough molecular weights to be recognized as polymers. Plasma-state polymerization seems to be an exception in which polymeric materials deposit from vacuum. Another notable example of vacuum deposition polymerization is the thermally initiated vacuum deposition of derivatives of paracyclophene known as 'parylene polymerization'. Recent studies on the fundamentals of parylene polymerization[10–14] have shown a striking similarity between plasma polymerization and parylene polymerization, and both processes seem to constitute the major portion of the vacuum deposition polymerization. It is not surprising that the same basic principles of polymer formation in vacuum apply to both processes.

## 21.2 PLASMA-STATE POLYMERIZATION

### 21.2.1 Growth Mechanism of Plasma Polymerization

In plasmas, vinyl monomers and their saturated counterparts polymerize in nearly the same manner. Table 1 compares deposition rates from various vinyl monomers and the corresponding saturated compounds. The results indicate that the contribution of the vinyl double bond is

**Table 1** Polymerization Parameter: Vinyl *vs.* Saturated Vinyl Compounds[a]

| *Vinyl compounds* | $k^b$ ($10^4$ cm$^{-2}$) | *Saturated vinyl compounds* | $k^b$ ($10^4$ cm$^{-2}$) |
|---|---|---|---|
| 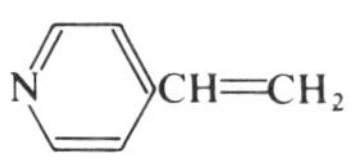 | 7.59 | 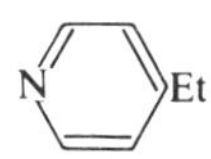 | 4.72 |
| 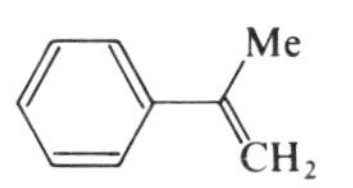 | 5.33 | 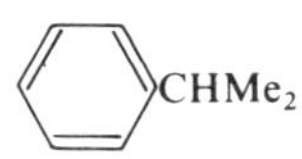 | 4.05 |
| 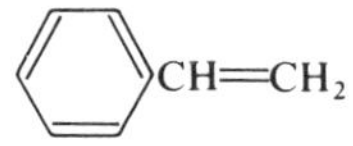 | 5.65 |  | 4.52 |
| 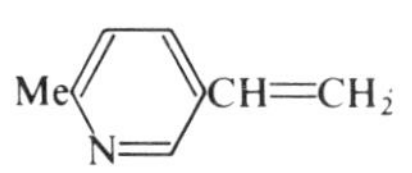 | 7.65 | 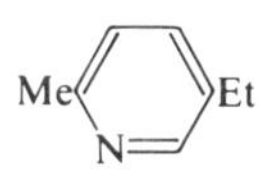 | 7.38 |
| 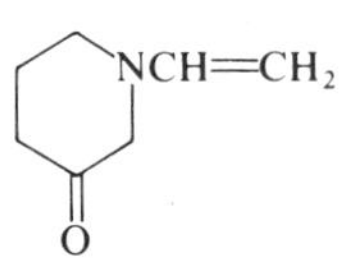 | 7.55 | 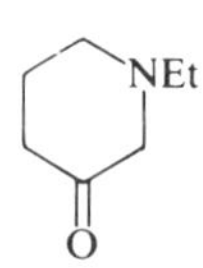 | 3.76 |
| $H_2C{=}CHC{\equiv}N$ | 5.71 | $EtC{\equiv}N$ | 4.49 |
| $H_2C{=}CCl_2$ | 5.47 | $MeCHCl_2$ | 2.98 |
| $H_2C{=}CHCH_2NH_2$ | 2.86 | $EtCH_2NH_2$ | 2.52 |

[a] Adapted from ref. 9.
[b] $r = kF_w$, where $r$ is the rate of polymer deposition (g cm$^{-2}$ min$^{-1}$) and $F_w$ is the weight-based monomer flow rate (g min$^{-1}$).

negligible, and suggest that plasma polymerization proceeds *via* a completely different growth mechanism compared to the addition chain-growth mechanism by which vinyl monomers polymerize. To explain polymer formation in vacuum, a rapid step-growth polymerization (RSGP) mechanism has been proposed.[15] The overall polymerization mechanism based on the RSGP mechanism is shown in Figure 1.

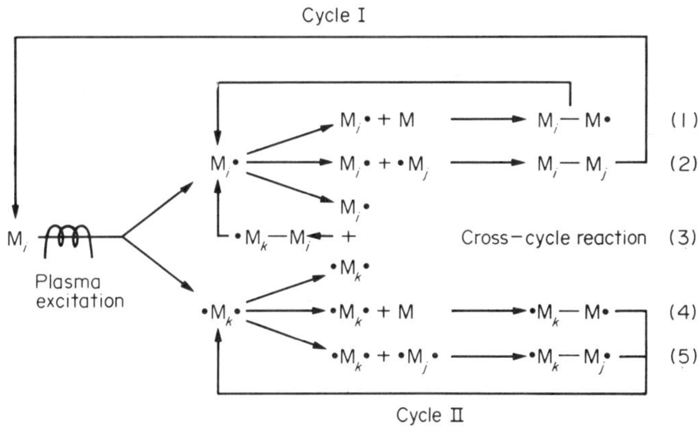


**Figure 1** Schematic representation of bicyclic step-growth mechanism of plasma polymerization (adapted from ref. 15)

In the schematic representation, $M_x$ refers to an unexcited neutral species that can be the original monomer molecule or any dissociation product, including molecules such as hydrogen, halogens and hydrogen chloride. Activated species capable of participating in chemical reactions to create a covalent bond are designated M•. Difunctional activated species are shown as •M•. Subscripts *i*, *j* and *k* merely indicate differences in the sizes of the species involved. Although activated species are represented by the symbol for a free radical, for simplicity and reasons to be explained later, any activated species can be considered in the reactions shown in Figure 1. For instance, the cationic radical formed by the ionization of styrene is represented by •M• in the figure to indicate that the species has two active sites.

An important aspect of the polymer-forming mechanism is that the major reactions which lead to an increase in molecular size are the reactions between activated species, rather than the reactions between a reactive species and a molecule which constitute the propagation reaction in chain-growth polymerization. In this respect, plasma polymerization involves many reactions which can be viewed as polyrecombination reactions. It seems to be a natural consequence that recombination (termination) reactions prevail over propagation reactions when the population of activated species increases beyond a critical point. In a sense, plasma polymerization may provide a prediction of what would happen if the dose rate of a radiation process was increased to a level comparable to that of plasma polymerization. According to Westwood[16] the dose rate used in plasma polymerization, estimated from the *G* value of polymerization, is roughly $10^6$ times greater than that used in polymerization induced by $\gamma$ rays.

As shown in Figure 1, the overall reaction contains two major routes to rapid step-growth. Cycle I is *via* repeated activation of reaction products from monofunctional activated species and the repeated reaction of these species. Cycle II is *via* difunctional or multifunctional activated species, similar to that found in parylene polymerization. Chain-growth polymerization carried out by activated species, particularly free radicals, of certain monomers is possible under certain conditions; however, such reactions as shown in (1) and (4) do not propagate in vacuum, in most cases, due to the low ceiling temperature. Reaction (3) is a cross-cycle reaction which converts a difunctional activated species into a monofunctional activated species, and plays an extremely important role if the monomer structure is such that the contributions of cycle I and cycle II are similar. The importance of reaction (3) is seen in pulsed glow discharge polymerization, as will be discussed later.

Activated species which participate in cycle I are formed by a variety of chemical reactions that occur in plasmas; however, by far the most important mechanism seems to be the detachment of hydrogen from hydrocarbons and the elimination of simple molecules such as hydrogen chloride or hydrogen fluoride, if halogens are present in the monomer structure. This suggestion is supported by the fact that dissociations such as dehydrogenation, dehydrochlorination and dehydrofluorination require much less energy than that required to ionize a monomer.[7]

Difunctional activated species are formed mainly by the opening of a double bond, a triple bond or a cyclic structure. An aromatic structure easily breaks up in plasma to form difunctional activated species; one molecule of benzene is nearly equivalent to three molecules of acetylene in many aspects of plasma polymerization.

### 21.2.2 Monomer Structures and Polymerization Characteristics

The rates at which polymers deposit from plasmas of organic vapors can be correlated to monomer structure. In the case of hydrocarbons, the presence or absence of double bonds, triple bonds and cyclic structures determines the deposition rate. As mentioned above, the detachment of hydrogen from a hydrocarbon is important in plasma polymerization and, consequently, by comparing the hydrogen yield (number of hydrogen molecules liberated per monomer molecule participating in polymer formation) of monomers in plasma polymerization, it is possible to classify hydrocarbon monomers into three major groups.[17] Monomers in a group behave more or less similarly, as shown in Figures 2 and 3. In addition to the three groups identified for hydrocarbons by the hydrogen yield, oxygen-containing compounds can be recognized as group D.[18]

Thus the plasma polymerization behavior of organic compounds can be categorized by the four groups of monomers: group A—compounds containing triple bond(s), aromatic or heteroaromatic structures; group B—compounds containing double bond(s) and/or cyclic structures; group C—linear and branched compounds without aforementioned structures; and group D—oxygen-containing (carbonyl, ether, carboxylic acid) compounds. (Compounds which contain oxygen directly linked to an aromatic structure and a nitrogen atom, such as an amide, do not show typical group D behavior.)

Because of the nonspecific nature of plasma polymerization, illustrated by the data in Table 1, the structure of the monomer appears to have relatively little influence on the polymerization characteristics of organic compounds.[9] This is largely true in the context of selective reactivity due to the structure of monomers found in conventional polymerizations. However, the influence of monomer structure, as categorized by groups A–D, tends to be accentuated when plasma polymerization conditions are varied.

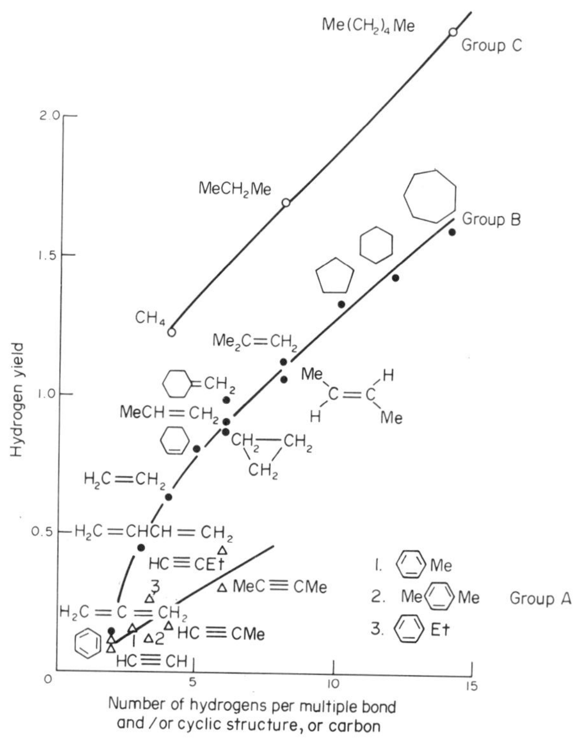


**Figure 2** Numbers of hydrogen molecules evolved per molecule of monomer when it polymerizes (hydrogen yield) as a function of chemical structure (adapted from ref. 17)

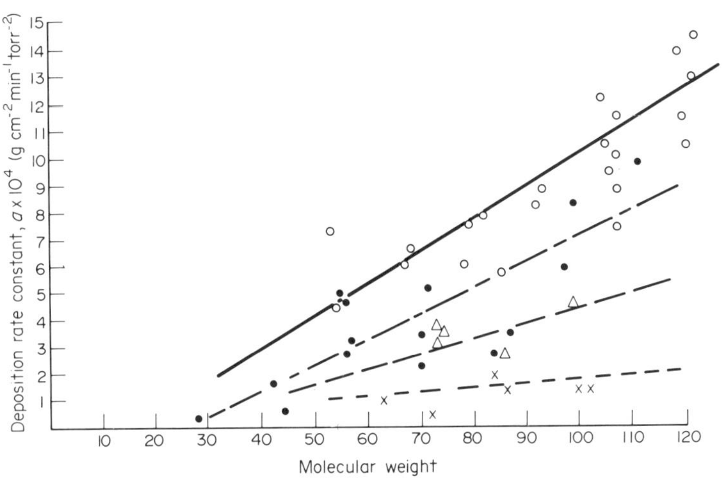


**Figure 3** Dependence of polymer deposition rate constant $a$ on molecular weight of monomer: ○, group A, triple-bond-containing, aromatic and heteroaromatic compounds; ●, group B, double-bond-containing and cyclic compounds; △, group C, compounds without aforementioned structures; and ×, group D, oxygen-containing compounds (adapted from ref. 18)

Due to the complex nature of the mechanism described in Figure 1, which also indicates the occurrence of a considerable degree of destruction of monomer structure during the process of polymer formation, plasma polymerization is highly system dependent, *i.e.* the polymerization reaction as well as the properties of the polymer formed depend on how the polymerization is carried out. In other words, a monomer does not yield a single polymer but a variety of polymers, depending on the type of reactor and the operational parameters, particularly the level and mode of energy input.

The system dependent nature of plasma polymerization can be demonstrated by the different polymer characteristics depicted by the electron spectroscopy for chemical analysis (ESCA) C 1*s* spectra of plasma polymers of tetrafluoroethylene collected in different locations within a reactor. The reactor and the locations of sample collection are shown in Figure 4, and ESCA C 1*s* spectra of the corresponding samples are shown in Figure 5.[19] It is important to recognize that there is no polymeric material which can be fully described and identified as the plasma polymer of tetrafluoroethylene, whereas the conventional polymer of the same monomer can be well described and identified as poly(tetrafluoroethylene) (PTFE).

The change of polymerization characteristics that can be recognized to be most sensitive to the structure of monomer may be seen in the influence of pulsed discharge.[20] A pulsed discharge may be viewed as the analogue of the rotating sector in photoinitiated polymerization. The ratio $r$ of 'off' time $t_2$ to 'on' time $t_1$, $r=t_2/t_1$, is expected to influence the polymerization rate depending on the

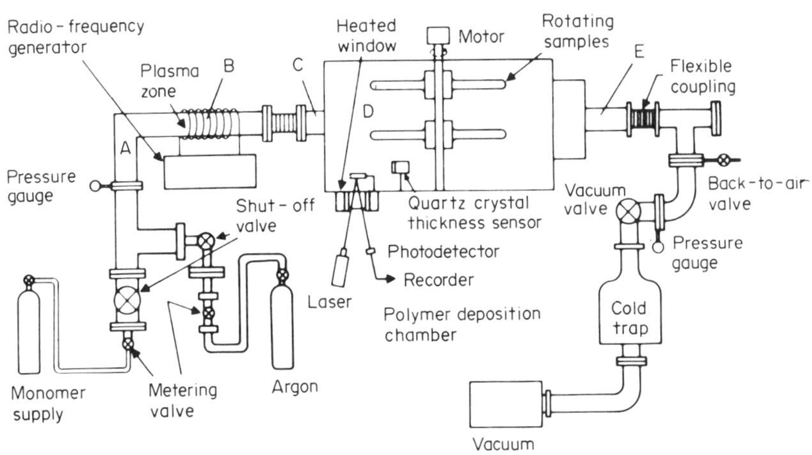


**Figure 4** Glow discharge thin polymer film deposition system (adapted from ref. 19)

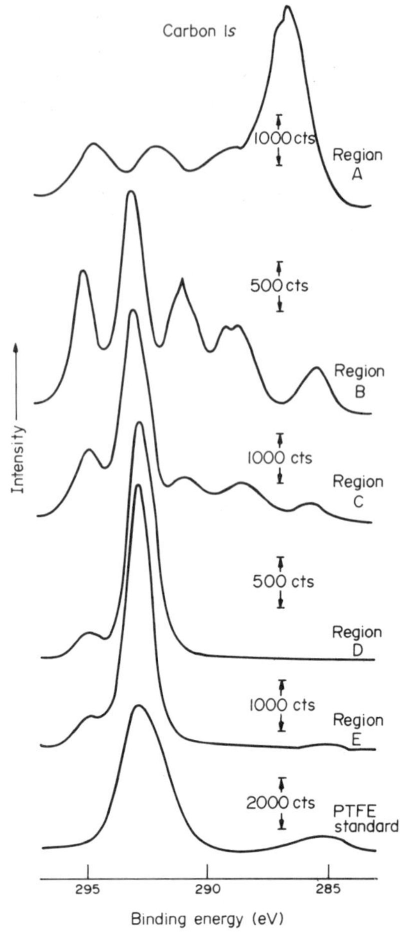


**Figure 5** ESCA C 1*s* spectra of plasma polymer of tetrafluoroethylene from different regions of the reactor depicted in Figure 2 (adapted from ref. 19)

relative time scale of $t_2$ to the lifetime of free radicals in free-radical addition polymerization of a monomer. The rate of vinyl polymerization decreases to a rate in the range $1/(1+r)$ (if $t_2 \gg$ average lifetime) to $1/(1+r)^{1/2}$ (when $t_2$ is close to the average lifetime).[21] In contrast, pulsed glow discharge exhibits rather abnormal effects of the pulse. Some monomers show a more or less anticipated decrease in polymer deposition rates, but some show increased deposition rates. The most significant effect of pulsed discharge can be seen in the concentration of free radicals trapped in plasma polymers, which reflects the unique mechanism of polymer formation in plasmas.

As expected from the mechanism described in Figure 1, particularly cycle II that leads to free radical 'living polymer', plasma polymers generally contain many free radicals in as-polymerized polymers. The free radical concentration can be correlated to the monomer types based on structural characteristics.[9] The greater the contribution of cycle II in the overall polymerization mechanism, the more free radicals are found in the plasma polymers.

Free radicals are also found in substrate materials on which plasma polymers deposit.[22] Free radicals in substrates are caused primarily by the radiation effects of a plasma, which usually emits highly energetic vacuum UV as well as the characteristic visible glow. An inverse relationship exists between the concentrations of free radicals in the plasma polymer and in the substrate.

The effects of pulsed discharges may be understood by examining the roles of cycle I and cycle II in Figure 1, for the four groups of monomers. For group A monomers, cycle II dominates in plasma polymerization. Group C monomers polymerize largely *via* cycle I, and group B monomers polymerize *via* both routes. Oxygen in group D monomers tends to be liberated in plasmas and blocks cycle II, and to a lesser extent, cycle I; this effect is termed the 'poisoning effect' of oxygen in group D monomers. The influence of pulsed discharges for some typical monomers, investigated using a cycle of 100 $\mu$s 'on' and 900 $\mu$s 'off', is shown in Table 2. The effects are expressed by the

**Table 2** Effect of Pulsing on Plasma Polymers of Some Monomers[a]

| *Monomer* | *Types of reaction involved*[b] | *Cycles in plasma polymerization mechanism* | $\delta$ | *Changes due to pulsed discharge*[c] *Deposition rate* | $C_s^0$ | $C_g$ | $\cos\theta(H_2O)$ |
|---|---|---|---|---|---|---|---|
| Tetramethyldisiloxane | Detachment of H | I | (−9) | (−47) | (−90) | 0 | −0.28 |
| | Detachment of Me | I | | | | | |
| Propionic acid | Detachment of H | I | (−18) | (−110) | 0 | (−81) | 0.53 |
| | Detachment of $CO_2H$ | I and poisoning | | | | | |
| Acrylic acid | Detachment of H | I | (−29) | (120) | (140) | (−100) | 0.49 |
| | Detachment of $CO_2H$ | I and poisoning | | | | | |
| | Opening of double bond | II | | | | | |
| Cyclohexane | Detachment of H | I | (21) | (−90) | (−100) | (−100) | −0.09 |
| | Ring opening | II | | | | | |
| Ethylene | Detachment of H | I | (19) | (2) | (970) | (−79) | −0.09 |
| | Opening of double bond | II | | | | | |
| Acetylene | (Detachment of H) | (I) | (200) | (−23) | (81) | 0 | −0.01 |
| | Opening of triple bond | II | | | | | |
| Benzene | Detachment of H | I | (150) | (−8) | (−50) | 0 | −0.76 |
| | Aromatic ring opening | II | | | | | |
| Styrene | Detachment of H | I | (87) | (−16) | (−86) | 0 | −0.09 |
| | Aromatic ring opening | II | | | | | |
| | Opening of double bond | II | | | | | |
| Vinyl acetate | Detachment of H | I | (−5) | (−48) | (−21) | (−71) | −0.03 |
| | Detachment of OCOMe | I and poisoning | | | | | |
| | Opening of double bond | II | | | | | |
| Ethylene oxide | Detachment of H | I | (−5) | (−7) | (−33) | (−76) | −0.09 |
| | Ring opening | II and poisoning | | | | | |
| Tetrafluoroethylene | Detachment of F | I and ablation | (−11) | (106) | (−35) | (−84) | −0.01 |
| | Opening of double bond | II | | | | | |
| Hexafluorobenzene | Detachment of F | I and ablation | (50) | (−22) | (−27) | 0 | −0.045 |
| | Aromatic ring opening | II | | | | | |
| Vinyl fluoride | Detachment of H | I | (47) | (−45) | (480) | (−100) | 0.07 |
| | Detachment of F | I and ablation | | | | | |
| | Opening of double bond | II | | | | | |
| Vinylidene fluoride | Detachment of H | I | (−6) | (−47) | (100) | (−64) | −0.36 |
| | Detachment of F | I and ablation | | | | | |
| | Opening of double bond | II | | | | | |

[a] Adapted from ref. 9.
[b] Possible reactions in continuous plasma.
[c] Numbers in parentheses are percentages based on values of continuous plasma; $\delta = p_g/p_m$; $C_s^0$, spin concentration in plasma polymer; $C_g$, spin concentration in glass substrate; $\theta$, contact angle of water. A negative value for $\cos\theta$ means that the polymer becomes more hydrophobic: a positive value indicates that it is more hydrophilic.

percent change in parameters due to the pulse, relative to the corresponding value for the continuous discharge. The parameters used to describe changes due to the pulse are described below:

The pressure parameter, $\delta$, is the change of system pressure due to inception of glow discharge, given by $\delta = p_g/p_m$, where $p_m$ is the system pressure of monomer flow before glow discharge is initiated, and $p_g$ is the steady state pressure during plasma polymerization. Values of $\delta$ for group A and B monomers are generally smaller than unity; the pressure drops during plasma polymerization. Values for group C and D monomers are generally larger than unity; the total number of gas molecules increases due to the evolution of gaseous by-products. Numbers listed in Table 2 are changes due to the pulsing, and consequently the signs of changes do not indicate the original values, *i.e.* $>1$ or $<1$.

$C_s^0$ is the concentration of free radicals in the plasma polymer; it is measured by electron spin resonance (ESR) as a function of time after the sample is exposed to air, and the initial concentration is estimated by extrapolation of data to time zero. The polymer signal is 'residual free radicals' from the polymerization process.

$C_g$ is the surface concentration of free radicals in a glass substrate; the ESR signal is measured after the polymer deposit is removed physically. The glass signal is a measure of the 'radiation effect' (mainly of vacuum UV) of plasma from which plasma polymers deposit on the substrate (glass rod).

The most striking effect of pulsed discharges is the conspicuous increase in free radicals in some plasma polymers, which seems to indicate the characteristic polymerization mechanisms illustrated in Figure 1. In so far as the concentration of polymer free-radicals is concerned, there seem to be two major factors that one should keep in mind in order to interpret the data shown in Table 2. These are: (i) the major sources of polymer free-radicals are the chemical reactions involved in cycle II; and (ii) the 'cross-cycle' reaction, reaction (3) in Figure 1, reduces polymer free-radicals during the continuous plasma polymerization; the influence of reaction (3) would be most pronounced if the contributions of cycle I and cycle II were nearly equal. If the contribution of either cycle I or cycle II is small, reaction (3) has little influence on the overall polymerization.

Data shown in Table 2 clearly reflect the polymerization mechanisms represented by the rapid step-growth polymerization. The most pronounced increase in polymer free-radicals due to the pulsed discharge is found for plasma polymerization of ethylene, for which cycle I, based on the detachment of hydrogen, and cycle II, based on the opening of the alkenic double bond, play nearly equal roles in the continuous discharge polymerization. Consequently, a significant portion of difunctional activated species are converted to monofunctional activated species by reaction (3) during the 'on' period of pulsed glow discharge. During the 'off' period, however, the cycle I reactions as well as reaction (3) cease and, during this period, difunctional activated species, once created, are largely conserved by the continued reactions within cycle II. As a net effect of the pulsed glow discharge, the contribution of cycle II to the overall polymerization increases, in spite of the rather long resting period involved in the pulsed glow discharge; the increase in the contribution of cycle II results in an increase in polymer free-radicals.

### 21.2.3 Characteristics of Plasma Polymers and Conditions of Polymerization

Plasma polymerization is a system dependent process and a well defined polymer cannot be obtained from a monomer. Consequently, properties of a plasma polymer cannot be discussed in terms of the monomer used and no plasma polymer of styrene can be defined in the same manner as conventional polystyrene is defined. Plasma polymers of styrene could form a wide spectrum, starting from a polymer which is close to polystyrene in chemical structure, but not quite close enough to consider as polystyrene, to a graphitic or diamond-like carbon film, depending on the conditions of plasma polymerization.

Perhaps the most important of the operational parameters which influence the nature of plasma polymers is the level of energy input into a plasma of monomer vapor. Since most plasma polymerizations are carried out in flow systems in vacuum, the density of input energy to a plasma system can be expressed by the composite parameter $W/FM$, where $W$ is the discharge wattage, $F$ is the molar or volume flow rate, and $M$ is the monomer molecular weight. $FM$ is the mass flow rate. The value of $W/FM$ represents the energy input per unit mass, and the SI units of this composite parameter are J kg$^{-1}$.

How $W/FM$ influences the properties of plasma polymers can be examined by simple estimates of required energy for some hypothetical plasma polymerizations of styrene. Assuming that free-radical addition polymerization of styrene occurs exclusively, it requires an energy input sufficient to break a $\pi$ bond of the alkenic double bond, *i.e.* the bond energy of the $\pi$ bond divided by the molecular weight of styrene, which is $2.6 \times 10^6$ J $kg^{-1}$. For higher energy inputs, there is a high probability that more than one $\pi$ bond of the styrene molecule would be broken. If we consider the extreme case of breaking all the bonds in styrene to form a carbon film, the energy level required in the same hypothetical calculation is $7.6 \times 10^7$ J $kg^{-1}$. Similarly, to form carbon films from methane, it is necessary to increase the (hypothetical) energy input to $1.04 \times 10^8$ J $kg^{-1}$.

The ionization energy of an organic molecule is nearly an order of magnitude higher than the C–H bond energy. Therefore, the ionization process in plasma polymerization is a rather nondiscriminatory destructive process. It is easy to recognize that the energy input level will play a crucially important role in determining the structure of plasma polymers. In practical plasma polymerization, energy inputs of $10^7$ to $10^{11}$ J $kg^{-1}$ are employed. Considering the energy necessary to ionize the molecule, and the inevitable energy loss associated with the process, the energy input level of plasma polymerization is in a reasonable range based on the hypothetical calculation based on bond energies. From this consideration, it can easily be conceived that polystyrene-like plasma polymer could be prepared by employing the lower energy input, while a highly crosslinked three-dimensional network or a carbon film would require considerably higher energy input manifested by the $W/FM$ parameter.

While the parameter $W/FM$ is important in controlling plasma polymerizations, its absolute value is not enough to judge or control the nature of polymer obtained. It is necessary to know the domain of plasma polymerization and the relative position of a set of conditions employed in the domain. The domain of plasma polymerization may be visualized by analyzing the dependence of polymer deposition rate on operational parameters.

Deposition rates, at fixed $F$, are generally linear in $W$, but start to deviate from linearity, and finally become independent of $W$. The wattage at which the line deviates from the initial linear dependence, and becomes wattage independent, varies with flow rate as shown in Figure 6. Deposition rates, at fixed $W$, also show an initial linear increase with $F$ and deviate from linearity as $F$ increases. A further increase in $F$ leads to a decrease in deposition rate depending on the wattage used, as shown in Figure 7. These two dependencies can be summarized by using the parameter $W/FM$. A plot of the data shown in Figures 6 and 7 against the corresponding $W/FM$ is shown in Figure 8.

This general dependence can be used to demonstrate the domains of plasma polymerizations, as shown schematically in Figure 9. At a low $W/FM$, the deposition rate is linearly dependent on this parameter. On the right-hand side of the dashed line connecting values of $(W/FM)_c$ for each flow rate the deposition rate is independent of $W/FM$. This region (iii) is the monomer-deficient region, where sufficient discharge power is available but the monomer feed is the rate-determining factor. To the left of the dashed line (region ii) is the power-deficient transient region, where ample monomer is available but the power input rate is not sufficient. In an extreme case, the deposition rate becomes linearly dependent on $W/FM$, as shown by the initial straight line (region i, where the power input rate is rate determining).

As shown in Figure 9, values of $(W/FM)_c$ depend on monomer flow rate and its dependency is related to the change of gas phase before and during plasma polymerization. In the case of

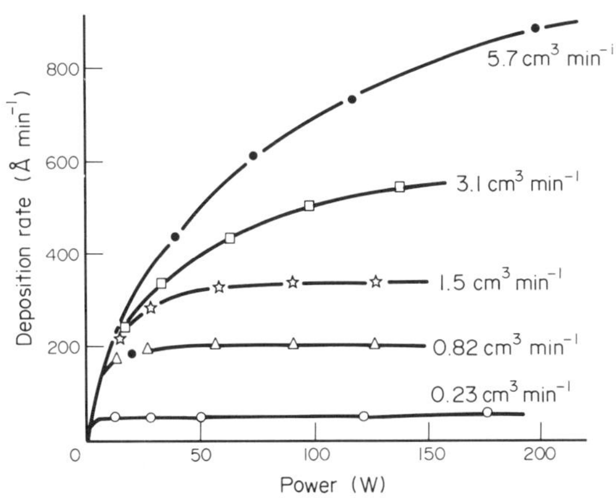


**Figure 6** Dependence of deposition rate of the plasma polymer of tetramethyldisiloxane on discharge wattage at various monomer flow rates $cm^3$ (STP) $min^{-1}$: from top; 5.7, 3.1, 1.5, 0.82 and 0.23 (adapted from ref. 9)

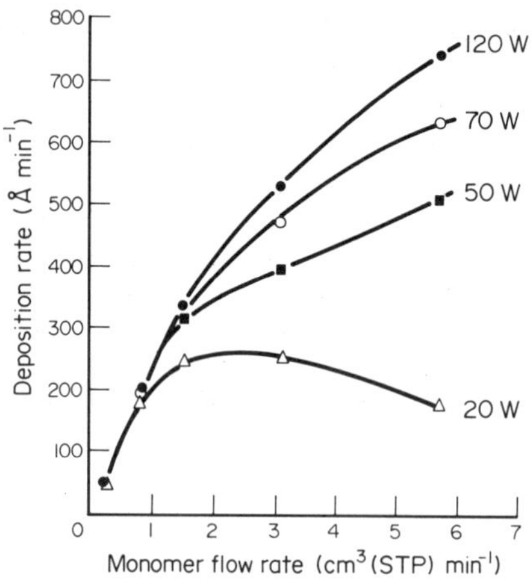


**Figure 7** Dependence of deposition rate of the plasma polymer of tetramethyldisiloxane on flow rate of the monomer at various discharge wattages (watts): from top; 120, 70, 50 and 20 (adapted from ref. 9)

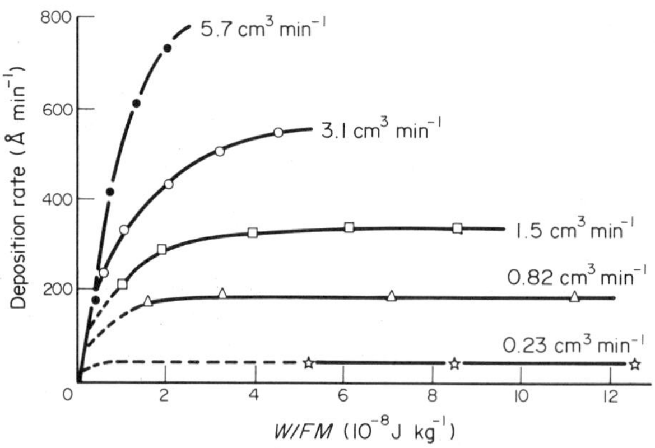


**Figure 8** Dependence of deposition rate of the plasma polymer of tetramethyldisiloxane on the composite parameter $W/FM$; the same data shown in Figure 5 are plotted against $W/FM$ in this figure (adapted from ref. 9)

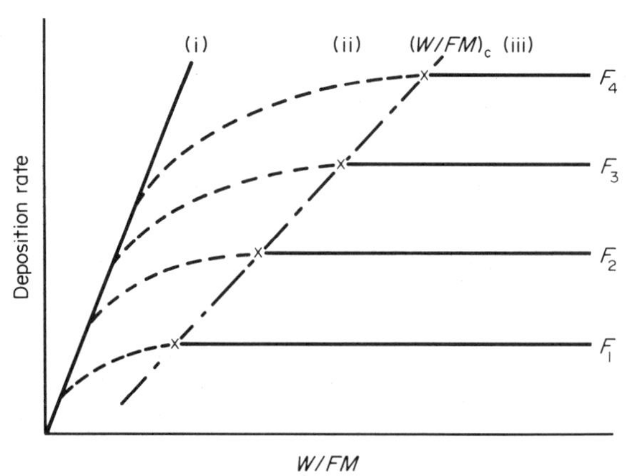


**Figure 9** Dependence of plasma polymerization domains on deposition rate and $W/FM$ at various flow rates; a schematic representation (adapted from ref. 9)

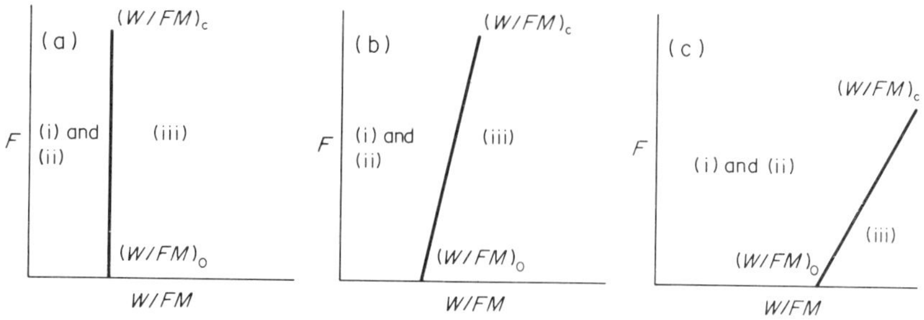


**Figure 10** Dependence of plasma polymerization domains on monomer flow rate and *W/FM*: (a) group A monomers; (b) group B monomers; (c) group C monomers (adapted from ref. 9)

hydrocarbons, the dependency is related to the hydrogen yield and, consequently, the nature of flow rate dependency can be generally expressed by the groups of monomers as shown schematically in Figure 10.[9]

The rate at which a plasma polymer deposits on to a strategically placed substrate can be represented by deposition rates in different units, depending on the practical needs for such parameters. For the kinetic aspect of plasma polymer deposition, the deposition rate $k_1$, based on [$\mathrm{kg\,m^{-2}\,s^{-1}}$], can be used. The deposition rate $k_1$ obtainable for a monomer is dependent on monomer flow rate and discharge wattage. Consequently, it is difficult to compare plasma polymerization characteristics of different monomers by comparing values obtained under arbitrarily chosen sets of conditions. For a more systematic comparison of plasma polymerization characteristics of various monomers, the specific deposition rate, $k_0 = k_1/FM$, can be used. $k_0$ is the polymer conversion ratio per unit surface and has dimensions of $\mathrm{m^{-2}}$. The overall conversion ratio is given by $k_0$ multiplied by the total surface area on which plasma polymers deposit. Since $k_1$ and $k_0$ are generally not average deposition rates, an accurate estimate of the overall polymer yield requires information on the distribution of deposition rates within a reactor.

The temperature dependence of polymer deposition for various monomers under different discharge conditions is best expressed in terms of the specific deposition rate $k_0$ rather than the deposition rate $k_1$. Specific deposition rates as functions of temperature are shown in Figure 11 with the best fits obtained by regression analysis.[15] As shown in the figure, the temperature dependence of $k_0$ is given by $k_0 = Ae^{-bT}$, where $A$ represents the extrapolated specific deposition rate at $T = 0$ K. Thus, the polymer deposition rate can be expressed by two parameters; the pre-exponential factor $A$ and the temperature dependence factor $b$. Values of $b$ for perfluorocarbon and hydrocarbon monomers, which represent gaseous monomers which do not condense easily and liquid monomers which condense relatively easily, respectively, are tabulated in Table 3. The results indicate that the

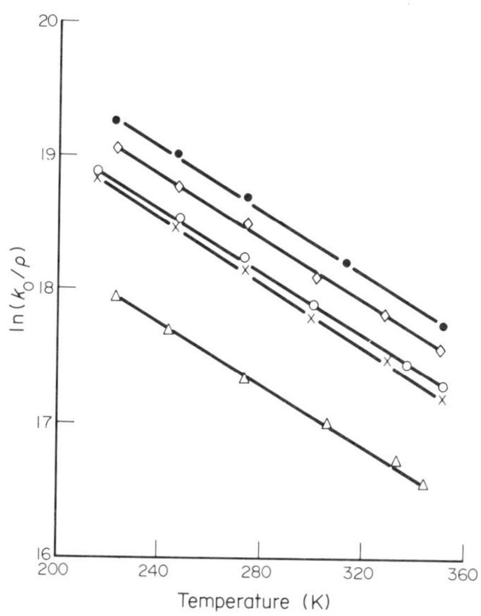


**Figure 11** Effect of substrate temperature on the specific deposition rates for tetrafluoroethylene; $\rho$ is the specific weight of polymer (adapted from ref. 15)

**Table 3** Values of Temperature Dependence Factor $b$ for Some Monomers[a]

| *Monomer* | *Molecular weight* | *Boiling point* (°C) | $b$ |
|---|---|---|---|
| Ethylene | 28 | −104 | 0.0045±0.0003 |
| Styrene | 104 | 145 | 0.0060±0.0003 |
| Tetrafluoroethylene | 100 | −76 | 0.0112±0.002 |
| Perfluoro-2-butyltetrahydrofuran | 416 | 107 | 0.0143±0.004 |

[a] Adapted from ref. 15.

adsorption or condensation of monomer plays a relatively minor role in plasma polymerization, *i.e.* $b$ is nearly independent of the condensability of monomers but more dependent on their chemical nature.

Plots of log $A$ *vs.* $W/FM$, Figure 12, show similar trends, indicating the domain where $A$ becomes independent of $W/FM$. The value of $(W/FM)_c$, the critical value above which plasma polymeriz-

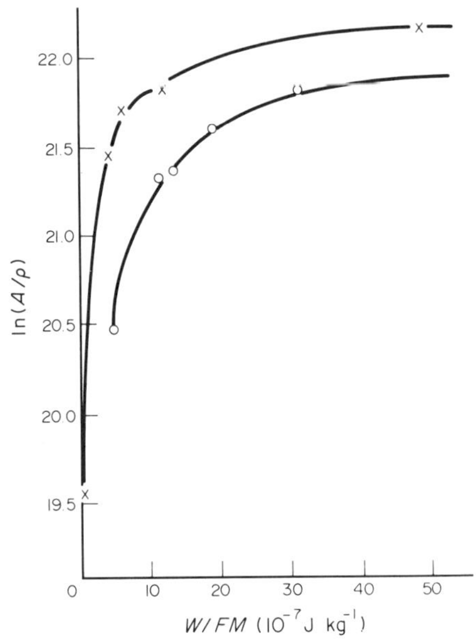


**Figure 12** Dependence of the pre-exponential factor $A$ on the composite parameter $W/FM$; $\rho$ is the specific weight of polymer (from ref. 15)

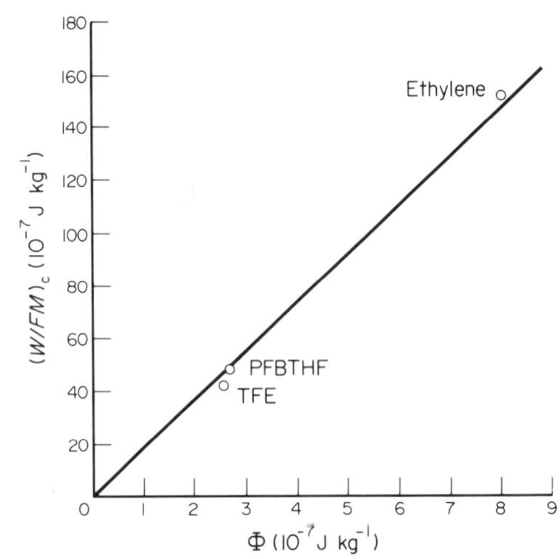


**Figure 13** Correlation between $(W/FM)_c$ and the specific bond energy of monomers (from ref. 15)

ation can be considered in the monomer-deficient region, can be correlated to the nature of the monomers. In Figure 13, values of $(W/FM)_c$, estimated from plots of log $A$ *vs.* $W/FM$, are plotted against specific bond energies, (total bond energies)/(molecular weight), of monomers. Thus, in the monomer-deficient region, nearly all the chemical bonds in an original monomer molecule are likely to be broken in a plasma; the polymer formation deviates significantly from 'molecular' reactions and the building blocks of plasma polymers become atoms of the original monomer. In this domain of plasma polymerization, polymer formation reactions are more atomic or nonmolecular, in contrast to strictly molecular polymerization in most conventional polymer formation reactions.

### 21.2.4 General Characteristics of Plasma Polymers

Since plasma polymerization is highly system dependent, it is often meaningless to discuss properties of plasma polymers based strictly on types or structures of monomers employed, as is done for conventional polymers. On the other hand, several aspects can be considered as common-denominator characteristics of plasma polymers, and these are briefly described in this section.

(a) Plasma polymers have no discernible repeating units and, consequently, have few linear chain segments exceeding five carbons or other atoms. The characteristics of a plasma polymer are difficult to predict from the nature and structure of the monomer used.

(b) Three-dimensional networks formed by plasma polymerization can be viewed as highly crosslinked and/or branched networks. However, the use of terms such as crosslinks and branching (developed for conventional long-chain polymers) for a network of plasma polymer may result in misleading interpretations of polymer structures and characteristics. Some plasma polymers formed under certain conditions are no more than oligomers, in a strict sense, and are found to be readily soluble in many solvents, despite their highly crosslinked nature. Terms used to describe plasma polymers, including such terms as polymerization, polymers, branching and crosslinking, require reexamination of their definition and meaning.

(c) Most plasma polymers contain many free radicals in as-polymerized forms. Consequently, most polymers after exposure to air contain considerable amounts of oxygen-containing groups, even if the original monomer had no oxygen-containing functions. Free radicals trapped in a plasma polymer are generally rather stable and difficult to destroy by chemical reactions, due to the relatively tight network structure. The reduction of trapped free radicals can best be achieved by the selection of monomer and polymerization conditions which minimize formation of free radicals during plasma polymerization.

(d) The polymer-formation steps discussed above are simultaneous processes of polymer formation and polymer deposition; reactive species impinge continuously on as-developing polymer surfaces. Consequently, a plasma polymer film tends to develop internal stress (expansive force at the interface). While such an internal stress is highly desirable in certain applications, an excessive build-up of internal stress, which increases with thickness of deposition, often causes cracking of the film and partial separation from the substrate. The extent of internal stress depends on the nature of monomer used and the deposition conditions. This aspect is not well understood; however, successful utilization of plasma polymerization seems to depend on the total control and manipulation of the internal stress in plasma polymers.

(e) The surfaces of some well-executed plasma polymers are found to be imperturbable in many respects; *i.e.* chemically, physically and surface dynamically. Consequently, nearly totally passive surfaces can be formed by plasma polymerization.

## 21.3 PLASMA-INDUCED POLYMERIZATION

### 21.3.1 Growth Mechanism of Plasma-induced Polymerization

In contrast to plasma-state polymerization, in which gas-phase reactions (gas–gas and gas–surface reactions) play major roles, plasma-induced polymerization occurs in condensed phases, usually in liquid phases of monomers the surfaces of which are exposed to a plasma. Polymerization-initiating species are formed under the influence of plasma either in the monomer liquid or vapor phase and on the surfaces that contact with plasma. In the latter cases, activated species will be transferred to the liquid phase spontaneously or by mixing which follows plasma irradiation. Polymerization occurs in the condensed phase after plasma irradiation is terminated. While the duration of plasma irradiation is short, from 30 s to a few minutes, the subsequent polymerization time is relatively long, from a few days to a few months. Monomers used in plasma-induced polymerization are monomers for free-radical addition polymerization.

In view of the activated species in plasmas as discussed in previous sections, the most logical chain-carrying species would seem to be free radicals. Many experimental data tend to prove that the growth mechanism of plasma-induced polymerization is free-radical chain-growth polymerization. The tacticity of poly(methyl methacrylate) (PMMA) formed by plasma-induced polymerization was identical to that for PMMA formed by free-radical polymerization.[23,24] Copolymerization characteristics of plasma-induced polymerizations of various monomer pairs also confirm that the mechanism of plasma-induced polymerization is free-radical chain-growth.[25–28] Plasma-induced polymerization of acrylamide (AAM) was depressed by the presence of a free radical inhibitor, hydroquinone.[29]

The formation of free radicals in a liquid by plasma irradiation can be studied by the use of 2,2-diphenyl-1-picrylhydrazyl (DPPH). DPPH is a stable free radical in a solution and its concentration can be quantitatively determined by UV absorption spectroscopy. DPPH reacts quantitatively with other reactive free radicals, and the decay of its concentration is used to determine rates of free-radical formation in various free-radical initiating systems. Figure 14 depicts spectral changes of DPPH in dimethyl sulfoxide (DMSO) solution (a) by plasma irradiation and (b) by thermal decomposition of a free-radical initiator, azobis(isobutyronitrile) (AIBN).[30] Estimated rates of free-radical formation from these results were $3.3 \times 10^{-5}$ mol $l^{-1}$ $s^{-1}$ for plasma irradiation and $2.5 \times 10^{-9}$ mol $l^{-1}$ $s^{-1}$ for thermal decomposition of AIBN. Although the absolute values are not of specific importance, due to the system dependent aspect of plasmas, the comparison certainly indicates that free radicals are formed in the solution exposed to plasma, and that rates of free-radical formation by plasma irradiation are nearly four orders of magnitude greater than that in a typical thermal decomposition of free-radical initiator.

Polymerization of water-soluble monomers is often influenced by the presence of water.[31–33] The influence of water on their polymerization rates is often observed in radiation-induced polymerization of such monomers; this phenomenon can be taken as supporting evidence for the normal chain-growth mechanism rather than any unusual aspect of plasma-induced polymerization.

### 21.3.2 Formation of Polymerization-initiating Species by Plasma

Although the mechanism of plasma-induced polymerization is free-radical chain-growth, the overall polymerization is significantly different from conventional free-radical addition polymerizat-

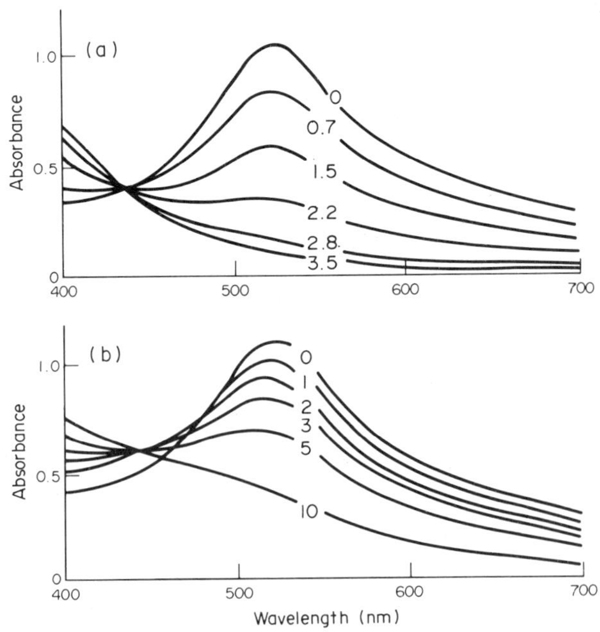


**Figure 14** Change in spectra of DPPH in DMF by (a) plasma irradiation and (b) thermal decomposition of AIBN. Numbers shown in (a) denote plasma irradiation time in seconds, and numbers in (b) thermal decomposition time in hours at 50 °C: $[DPPH] = 10^{-4}$ mol $l^{-1}$; $[AIBN] = 10^{-3}$ mol $l^{-1}$; plasma discharge at 100 W (adapted from ref. 30)

ion. An important difference is that the polymer molecular weight increases with reaction time (*i.e.* with conversion) in plasma-induced polymerization, while in conventional free-radical polymerization the molecular weight of the polymer is independent of conversion, except at extremely high conversion where diffusion-controlled termination increases the molecular weight of the polymer. Figure 15 depicts the increase in molecular weight as a function of reaction time (and with percent conversion) observed for plasma-induced polymerization of methyl methacrylate (MMA); 1 minute (monomer vapor) plasma irradiation, and post-plasma polymerization at 25 °C.[34] At high conversion, the molecular weight of PMMA formed by plasma-induced polymerization reached $3 \times 10^7$ g mol$^{-1}$. Such an extraordinarily high molecular weight (by free-radical polymerization), *e.g.* $1 \times 10^7$ g mol$^{-1}$, was also obtained in plasma-induced polymerization of AAM.[35]

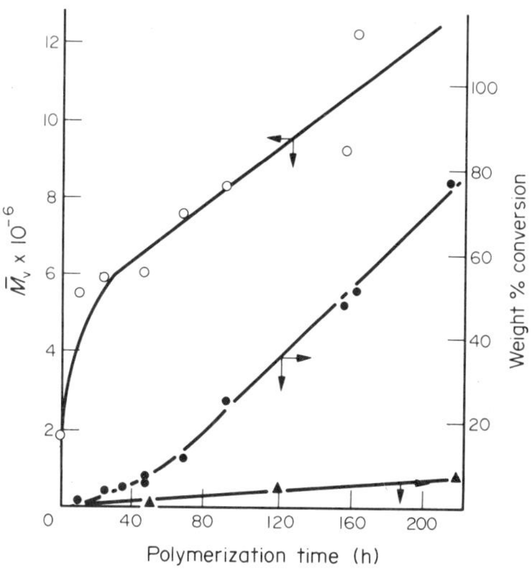


**Figure 15** Polymer conversion and molecular weight of polymers formed by plasma-induced polymerization of methyl methacrylate (adapted from ref. 34)

These unusual characteristics may be attributed to the nature of the initiating species created by the plasma. In other words, plasma-induced polymerization could be visualized as the free-radical addition polymerization initiated by unusual initiators formed under the influence of plasma. From the view point of plasma-state polymerization (see Figure 1 for the mechanism), it is most likely that the diradicals would contribute to such abnormal phenomena. Once diradicals initiate free-radical polymerization (an ordinary addition polymerization in this case, in contrast to the rapid step-growth polymerization in plasma-state polymerization), no termination can occur by the recombination of free radicals and the reaction system leads to a 'free-radical living polymer' system. The presence of free-radical living polymer systems was confirmed by the formation of block copolymers by adding the second monomer after sufficient amount of the first monomer was polymerized.[31,36]

In Table 4, plasma-induced polymerizations reported in the literature are summarized. It is evident that not all monomers that can be polymerized by free radicals (compounds with alkenic double bonds) polymerize by plasma-induced polymerization. It is interesting to note that all monomers which have been polymerized by plasma-induced polymerization *i.e.* methacrylates and water-soluble monomers, are known to behave unpredictably in water; usually they do not follow the kinetics predicted by theory. Therefore, some explanations for mechanistic features of plasma-induced polymerization are related to the unusual behavior of these monomers.

Unusual characteristics which might be related to the formation of the reactive species responsible for the initiation of plasma-induced polymerization are: (i) the capability of monomers to polymerize thermally (without a free-radical initiator) is related to whether a monomer can be polymerized by plasma-induced polymerization; (ii) plasma irradiation does not necessarily require liquid monomer in contact with the vapor phase in which the plasma is generated. In other words, a solvent which is irradiated with plasma (of its own vapor) can initiate polymerization of certain monomers; and (iii) polymerization kinetics of some monomers (particularly water-soluble monomers) are influenced by the nature of solvent and by the presence of water.

The similarity of polymers formed by thermal and plasma-induced polymerizations of MMA is depicted in Figure 16, in which their gel permeation chromatography (GPC; a) and thermo-

**Table 4** List of Conditions of Plasma-induced Polymerization

| *Monomer* | *Solvent*[a] | *Vapors in plasma* | *Polymerization rate*[b] | *Ref.* |
|---|---|---|---|---|
| Methyl methacrylate | — | Monomer | 0 | 23, 25, 32, 34, 41 |
| | MIB | Monomer and solvent | 0 | 41 |
| Ethyl methacrylate | — | Monomer | X | 25, 41 |
| | MIB | Monomer and solvent | 0 | 41 |
| *n*-butyl methacrylate | — | Monomer | X | 25, 41 |
| | — | Monomer | 0 | 32 |
| | MIB | Monomer and solvent | 0 | 41 |
| Methyl acrylate | — | Monomer | X | 25 |
| Ethyl acrylate | — | Monomer | X | 32, 50 |
| *n*-butyl acrylate | — | Monomer | X | 25, 32 |
| Styrene | — | Monomer | X | 25, 51 |
| | $CCl_4$ | Monomer and solvent | X | 25 |
| | Benzene | Monomer and solvent | X | 25 |
| α-methylstyrene | — | Monomer | X | 25 |
| Acrylic acid | — | Monomer | X | 25 |
| | — | Monomer | 00 | 50, 52 |
| | $H_2O$ | Monomer and solvent | 00 | 25 |
| | MIB | Monomer and solvent | 0 | 52 |
| Methacrylic acid | — | Monomer | X | 25, 31, 50 |
| | $H_2O$ | Monomer and solvent | 00 | 25, 31 |
| | DMF | Monomer and solvent | X | 31 |
| | MIB | Monomer and solvent | 0 | 51 |
| Acrylamide | — | Monomer | X | 25, 33 |
| | tr. water | Monomer and solvent | 0 | 33 |
| | $H_2O$ | Monomer and solvent | 00 | 25, 33 |
| | $H_2O$ | Monomer and solvent | 0 | 50 |
| | DMF | Monomer and solvent | X | 31 |
| Methacrylamide | — | Monomer | X | 25 |
| | $H_2O$ | Monomer and solvent | 00 | 25 |
| | — | Monomer and solvent | X | 31 |
| Hydroxyethyl methacrylate | $H_2O$ | Monomer and solvent | 00 | 31 |
| | DMF | Monomer and solvent | X | 31 |
| *N*-vinylcarbazole | — | Monomer | X | 53 |
| | Benzene | Monomer and solvent | X | 50 |
| Acrylonitrile | — | Monomer | X | 29 |
| Methacrylonitrile | — | Monomer | X | 53 |
| Vinylpyrrolidone | — | Monomer | X | 29 |
| Vinyl acetate | — | Monomer | X | 53 |
| 4-vinylpyridine | — | Monomer | X | 53 |

[a] MIB = methyl isobutylate, DMF = *N*,*N*-dimethylformamide, tr. = trace.
[b] 00, polymerize fast; 0, polymerize slowly; X, polymerize very slowly or no polymerization.

gravimetric–differential scanning calorimetry (TG–DSC; b) data are compared with PMMA formed by thermal decomposition of AIBN.[37] This similarity might be linked to the polymerization mechanisms in which diradicals play important roles in the initiation.

Several mechanisms have been proposed for the thermal polymerization of MMA, *i.e.* polymerization in the absence of added initiator (see Volume 3, Chapter 10). The exact species which are responsible for the thermal polymerization are still unknown; however, all the current proposals invoke the formation of diradicals.[34,38–40] In contrast to the formation of diradicals in the thermal polymerization of MMA, styrene (that cannot be polymerized by plasma-induced polymerization) does not seem to form diradicals but does form monoradicals.[40]

Mass-spectroscopic analysis of the gas-phase products of plasma treatment of MMA vapor indicated the presence of allylic radicals.[41] Although the contribution of such a species to the initiation of polymerization is questionable, the structural feature which yields allylic radicals under the influence of plasma seems to be related to the formation of reactive species which can initiate polymerization of MMA. It was reported that the plasma effluent gases of methyl isobutylate (MIB) and 4-methylpent-1-ene, both of which yield allylic radicals under the influence of plasma, initiated polymerization of MMA.[41] It was also reported that an oily nonvolatile product of plasma polymerization of MMA can initiate polymerization of MMA.[39] Thus, it seems clear that formation of polymerization-initiating species in plasmas requires some favored structures, and that those

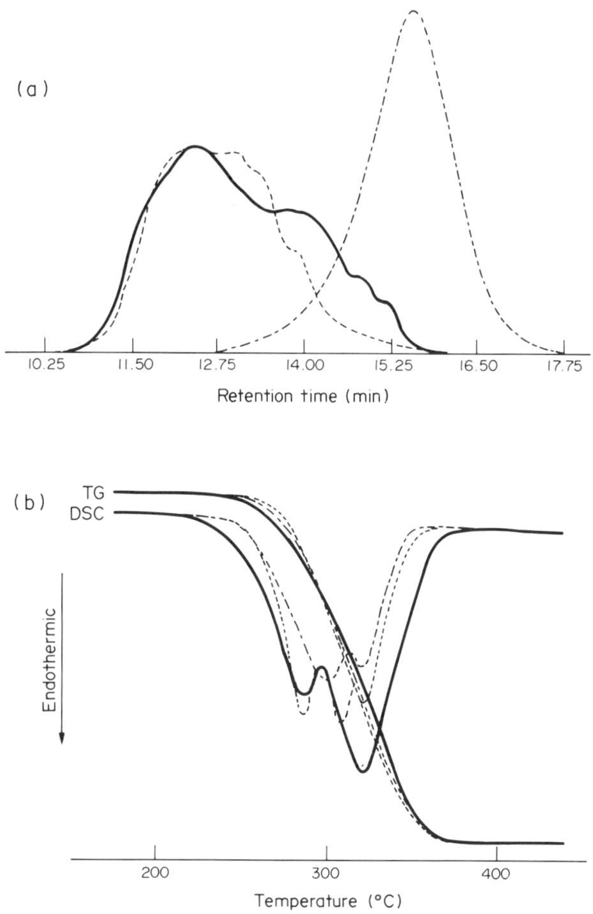


**Figure 16** Comparison of polymer characteristics: (a) gel permeation chromatograms and (b) TG–DSC thermograms; for poly(methyl methacrylate) formed by plasma-induced polymerization (——), by thermal decomposition of AIBN (– – –), and by thermal polymerization (without an initiator) (adapted from ref. 37)

compounds are not necessarily monomers. Plasma-exposed *N*,*N*-dimethylformamide (DMF) is able to initiate polymerization of some water-soluble monomers.[29]

Questions which arise are: (a) where are polymerization-initiating species formed; and (b) how are those species transferred to the bulk liquid phase where polymerization takes place? Theoretically there seem to be three possibilities: (1) the reactive species are created in the gas phase and can be deposited directly onto the liquid surface or onto the wall of the reactor and subsequently dissolved into the liquid on mixing; (2) the reactive species are created at the interface of liquid and plasma; and (3) the reactive species are formed in the bulk liquid, probably by high energy vacuum UV emitted by the plasma.

Considering that gas-phase reactions involve highly reactive and energetic species such as plasma reactions, the reactions at the surface and pure gas-phase reactions cannot be completely separated and, particularly in vacuum, the interaction with surface increases in importance as the system pressure decreases; the first two cases may be considered as the major combined mechanisms for initiation plasma-induced polymerization. Experimental data seem to support these schemes. Although the possibility of the third mechanism cannot be eliminated, experimental evidence has not yet been reported. Perhaps because of the relatively short duration of plasma exposure, this possibility has been neglected in the experimental design of most studies.

Experiments performed without liquid monomer during plasma irradiation seem to confirm the combined initiation schemes. The experimental procedures of these experiments are as follows: first, a sealed ampoule of vapor of either MMA or MIB (without liquid) is used to create plasma, and the ampoule is opened and its vapor is replaced by dry nitrogen. Then MMA is placed in the ampoule and polymerized at a constant temperature after the ampoule is resealed in vacuum. Although the contribution of glass free-radicals is suggested,[42] the contribution of plasma polymer which deposits on the surface of glass during the first plasma irradiation process seems to be a more likely explanation.[43,44]

Because of the very short duration of plasma irradiation, there seems to be a trend to consider that plasma-induced polymerization is completely different from plasma-state polymerization. Experimental data shown above, however, indicate that plasma-induced polymerization is an extension of plasma-state polymerization. The unique chemical reactions within plasmas continue in the liquid phase during polymerization; however, the growth mechanism is completely different, as polymerization takes place in the liquid phase and is a post-plasma process, *i.e.* there is no influence of the plasma during polymerization.

The direct influence of plasma, as well as the direct deposition of reactive species on the monomer, are more evident in the case of solid state polymerization, in which solid instead of liquid monomer is employed. It has been reported that 1,3,5-trioxane[45,46] and other cyclic compounds such as cyclophosphazene[47] and cyclosiloxane[48] can be polymerized in the solid state by plasma irradiation. In certain cases, plasma-irradiated solid monomer polymerized only after the solid monomer was dissolved by adding a solvent.[49] These examples clearly show that, in plasma-induced polymerization, the phases, and often locations, for formation of polymerization-initiating species, initiation of polymerization and propagation are all different and can be controlled by selecting (i) monomer and monomer phase; (ii) plasma gas or vapor; and (iii) the conditions for each step, *i.e.* plasma irradiation, initiation and propagation.

## 21.4 REFERENCES

1. R. F. Baddour and R. S. Timmins (eds.), 'The Application of Plasma to Chemical Processing', MIT Press, Cambridge, MA, 1967.
2. J. H. Hollahan and A. T. Bell (eds.), 'Techniques and Applications of Plasma Chemistry', Wiley, New York, 1974.
3. D. T. Clark, A. Dilks and D. Shuttleworth, in 'Polymer Surfaces', ed. D. T. Clark and W. J. Feast, Wiley, London, 1978, chap. 9.
4. H. V. Boening, 'Plasma Science and Technology', Cornell University Press, Ithaca, NY, 1982.
5. D. T. Clark, *Pure Appl. Chem.*, 1982, **54**, 415.
6. D. T. Clark, *Org. Coat.*, 1984, **7**, 113.
7. H. Kobayashi, M. Shen and A. T. Bell, *J. Macromol. Sci., Chem.*, 1974, **A8**, 1345.
8. K. Hayashi, *Polym. J.*, 1980, **12**, 583.
9. H. Yasuda, 'Plasma Polymerization', Academic Press, Orlando, FL, 1985.
10. P. Kramer, A. K. Sharma, E. E. Hennecke and H. Yasuda, *J. Polym. Sci., Polym. Chem. Ed.*, 1984, **22**, 475.
11. M. Gazicki, G. Surendran, W. James and H. Yasuda, *J. Polym. Sci., Polym. Chem. Ed.*, 1985, **23**, 2255.
12. M. Gazicki, G. Surendran, W. James and H. Yasuda, *J. Polym. Sci., Polym. Chem. Ed.*, 1986, **24**, 215.
13. M. Gazicki, W. J. James and H. K. Yasuda, *J. Polym. Sci., Polym. Lett. Ed.*, 1985, **23**, 639.
14. Y.-S. Yeh, G. Surendran, P. W. Kramer, W. J. James and H. Yasuda, *J. Polym. Sci., Polym. Chem. Ed.*, in press.
15. H. Yasuda and C. R. Wang, *J. Polym. Sci., Polym. Chem. Ed.*, 1985, **23**, 87.
16. A. R. Westwood, *Eur. Polym. J.*, 1971, **7**, 363.
17. H. Yasuda, M. O. Bumgarner and J. J. Hillman, *J. Appl. Polym. Sci.*, 1975, **19**, 531.
18. H. Yasuda, *Contemp. Top. Polym. Sci.*, 1979, **3**, 103.
19. D. F. O'Kane and D. W. Rice, *J. Macromol, Sci., Chem.*, 1976, **A10**, 567.
20. H. Yasuda and T. Hsu, *J. Polym. Sci., Polym. Chem. Ed.*, 1977, **15**, 81.
21. B. Vollmert, 'Polymer Chemistry', Springer, New York, 1973.
22. H. Yasuda, *J. Macromol. Sci., Chem.*, 1976, **A10**, 383.
23. Y. Osada, A. T. Bell and M. Shen, *J. Polym. Sci., Polym. Lett. Ed.*, 1978, **16**, 309.
24. G. Akovali, G. Demirel and S. Ozkar, *J. Macromol. Sci., Chem.*, 1983, **A20**, 887.
25. Y. Osada, M. Shen and A. T. Bell, *ACS Symp. Ser.*, 1979, **108**, 253.
26. B. C. Simionescu, A. Natansohn and C. I. Simionescu, *Polym. Bull.*, 1980, **2**, 809.
27. B. C. Simionescu, A. Natansohn, M. Leanca, C. Ananiescu and C. I. Simionescu, *Polym. Bull.*, 1980, **3**, 247.
28. B. C. Simionescu, A. Natansohn, M. Leanca, C. Ananiescu and C. I. Simionescu, *Polym. Bull.*,1981, **4**, 569.
29. Y. Osada, M. Takase and Y. Iriyama, *Polym. J.*, 1983, **15**, 81.
30. Y. Osada and Y. Iriyama, *J. Phys. Chem.*, 1984, **88**, 5951.
31. Y. Osada, Y. Iriyama and M. Takase, *Kobunshi Ronbunshu*, 1981, **38**, 629.
32. Y. Osada and M. Takase, *J. Polym. Sci., Polym. Lett. Ed.*, 1983, **21**, 643.
33. G. Akovali and B. Orhan, *J. Polym. Sci., Polym. Chem. Ed.*, 1984, **22**, 3351.
34. D. R. Johnson, Y. Osada, A. T. Bell and M. Shen, *Macromolecules*, 1981, **14**, 118.
35. Y. Osada and M. Takase, *Nippon Kagaku Kaishi*, 1983, 439.
36. C. I. Simionescu, B. Simionescu, M. Leanca and C. Ananiescu, *Polym. Bull.*, 1981, **5**, 61.
37. M. Kuzuya, K. Kamiya, T. Kawaguchi and T. Okuda, *J. Polym. Sci., Polym. Lett. Ed.*, 1983, **21**, 509.
38. M. Stickler and G. Meyerhoff, *Polymer*, 1981, **22**, 928.
39. C. W. Paul, A. T. Bell and D. S. Soong, *Macromolecules*, 1985, **18**, 2312.
40. W. A. Pryor and L. D. Lasswell, *Adv. Free-Radical Chem.*, 1975, **5**, 27.
41. M. Kuzuya, T. Kawaguchi, T. Daikyo and T. Okuda, *J. Polym. Sci., Polym. Lett. Ed.*, 1983, **21**, 515.
42. M. Kuzuya, T. Kawaguchi, S. Mizutani and T. Okuda, *J. Polym. Sci., Polym. Lett. Ed.*, 1985, **23**, 69.
43. M. Kuzuya, T. Kawaguchi, M. Nakanishi and T. Okuda, *J. Chem. Soc., Faraday Trans. 1*, 1986, **82**, 1441.
44. M. Kuzuya, T. Kawaguchi, Y. Yanagihara, S. Nakai and T. Okuda, *J. Polym. Sci., Polym. Chem. Ed.*, 1986, **24**, 707.

45. Y. Osada, M. Shen and A. T. Bell, *J. Polym. Sci., Polym. Lett. Ed.*, 1978, **16**, 669.
46. A. Odajima, Y. Nakase, Y. Osada, A. T. Bell and M. Shen, *ACS Symp. Ser.*, 1979, **108**, 263.
47. Y. Osada, M. Hashidzume and E. Tsuchida, *Nature (London)*, 1980, **286**, 693.
48. Y. Osada and M. Hashidzume, *J. Polym. Sci., Polym. Lett. Ed.*, 1981, **19**, 369.
49. Y. Osada and A. Mizumoto, *Macromolecules*, 1985, **18**, 302.
50. B. C. Simionescu, M. Leanca, C. Ananiescu and C. I. Simionescu, *Polym. Bull.*, 1980, **3**, 437.
51. B. C. Simionescu, M. Leanca, S. Ioan and C. I. Simionescu, *Polym. Bull.*, 1981, **4**, 415.
52. M. Kuzuya, T. Kawaguchi, T. Daikyo and T. Okuda, *J. Polym. Sci., Polym. Chem. Ed.*, 1985, **23**, 77.
53. Y. Osada, in 'Plasma Zyugoh', ed. Y. Osada, Tokyo Kagaku Dojin, Tokyo, 1986, chap. 4.

# 22
# Alternating Copolymerization

JOHN M. G. COWIE
*University of Stirling, UK*

## 22.1 INTRODUCTION

While homopolymerization of vinyl monomers has produced many useful and commercially successful polymers, the addition of a second monomer to produce a copolymer immediately increases the range of possible structures that can be formed, and also, therefore, the range of properties obtained, thus providing the investigator with a new dimension for the production of novel materials. The types of structure which can be formed from two monomers A and B range from the block copolymers (**1**) and (**2**), through the tapered block (**3**), to the statistical copolymer (**4**) and, finally, the alternating structure (**5**)

AAAAAAAAAAAABBBBBBBB (**1**) AAAAAABBBBBBAAAAAAAAAAABBBBAAABBB (**2**)

AAAAAAAABAAABAABBABBBABBBBBB (**3**) AABABBBABABBAABABBABBBAABAAABA (**4**)

ABABABABABABABABABABABABABAB (**5**)

Structures (**1**)–(**3**) result from carefully controlled polymerization conditions and are normally synthesized by deliberate design. When considering structures (**4**) and (**5**), however, the final monomer sequence tends to depend much more on the nature of the two monomers. Thus, two possible extreme cases can be envisaged, one in which copolymerization of the two monomers is so strongly favoured that a copolymer is formed with the monomer units A and B alternating in a strictly regular fashion in the chain, and the other in which A and B, when polymerized together, simply form a mixture of the homopolymers of A and B. The latter is not often met with in practice, but the former occurs quite frequently and will be discussed in detail in this chapter.

### 22.1.1 Definition

If a copolymer is, by convention, considered to be the product of a reaction between two monomers, each of which can form (separately) a homopolymer, then step-growth polymers formed by the reaction of two difunctional monomers can be excluded. This discussion will be restricted, therefore, to systems prepared by radical or ionic addition polymerizations which result in the incorporation of two monomers in a strictly alternating manner along the whole chain.

### 22.1.2 Alternating Tendency

When two monomers, $M_A$ and $M_B$, are copolymerized, the composition of the product is rarely the same as that of the initial monomer feed ratio. This suggests that there are factors governing the incorporation of each monomer into the copolymer chain which are not obvious in homopolymerization reactions.

The first successful kinetic analysis of a copolymerization reaction was put forward by Mayo and Lewis,[1] based on Dostal's hypothesis[2] that the rate of addition of a monomer to a radical chain-end depends only on the nature of the radical end group and not on the length or chemical composition of the chain to which the radical is attached.

The Mayo and Lewis treatment, which is usually referred to as the 'terminal model', involves the four reaction steps shown in equations (1)–(4).

$$\sim\sim M_A\cdot + M_A \xrightarrow{k_{AA}} \sim\sim M_A\cdot \qquad (1)$$

$$\sim\sim M_A\cdot + M_B \xrightarrow{k_{AB}} \sim\sim M_B\cdot \qquad (2)$$

$$\sim\sim M_B\cdot + M_B \xrightarrow{k_{BB}} \sim\sim M_B\cdot \qquad (3)$$

$$\sim\sim M_B\cdot + M_A \xrightarrow{k_{BA}} \sim\sim M_A\cdot \qquad (4)$$

This treatment assumes that the composition of the resulting copolymer will depend mainly on the relative rates of each of these reactions, and that, for effective copolymerization to take place, the cross propagation reactions (2) and (3) should be more favoured than the self propagation steps (1) and (4). The rate constants can then be expressed in terms of the monomer reactivity ratios $r_A = k_{AA}/k_{AB}$ and $r_B = k_{BB}/k_{BA}$, which provide a measure of the preference for a propagating chain-end radical to add its own monomer rather than the other monomer.

If $k_{AA} = k_{AB}$ and $k_{BB} = k_{BA}$, $r_A = r_B = 1$, then the rate of addition of $M_A$ or $M_B$ to the chain end will be determined by their relative concentrations, and the composition of the copolymer will be equal to the ratio of the two monomers in the reaction mixture. This is generally termed an 'ideal' copolymerization and is represented in Figure 1 by the broken diagonal line (curve i), where the mole fraction of $M_A$ in the copolymer ($F_A$) is plotted against its mole fraction in the feed ($f_A$).

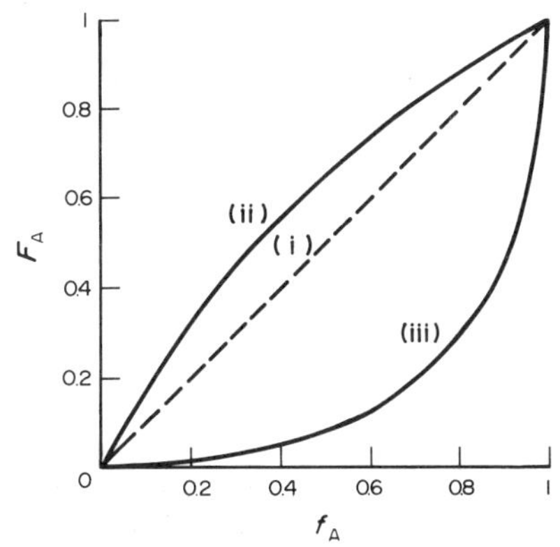


**Figure 1** Mole fraction of monomer ($M_A$) in the copolymer ($F_A$) plotted against its mole fraction in the feed ($f_A$) for an 'ideal' copolymerization (curve i), and values of $r_A$ ($=1/r_B$) of 2 and 0.1 (curves ii and iii, respectively)

The reactivity ratios will, therefore, provide a guide to the copolymer composition and, as 'ideal' conditions are rarely found, the composition dependence of the copolymer on the feed ratio usually deviates significantly from a linear dependence. This is illustrated for different values of $r_A$ and $r_B$ in Figure 1.

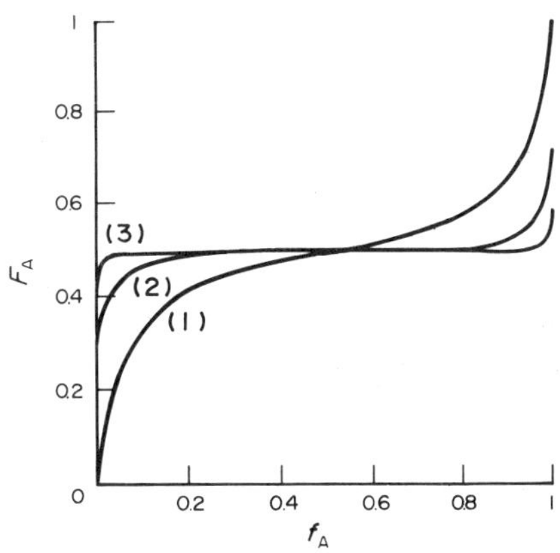


**Figure 2** Mole fraction of monomer $M_A$ in the copolymer ($F_A$) against its mole fraction in the feed ($f_A$); curve (1), $r_A = r_B = 0.1$; curve (2), $r_A = r_B = 0.01$; and curve (3), $r_A = r_B = 0.001$

Values of $k_{AB}$ and $k_{BA}$ must be much larger than $k_{AA}$ and $k_{BB}$ if alternating structures are to be obtained, consequently, $r_A$ and $r_B$ should both tend towards zero as the alternating sequences become more perfect. This is shown schematically in Figure 2, where the reactivity ratios are equal and vary from 0.1 to 0.01 then to 0.001. Thus, for a highly alternating copolymer, the composition remains essentially constant at a 1:1 mole ratio, irrespective of the initial feed.

### 22.1.3 Sequence Length Distribution

The values of the reactivity ratios will, therefore, give some indication of the regularity of the alternating structure in a given copolymer, but they do not immediately provide information about the distribution of monomer units in the chain. This can be estimated using statistical arguments, which can be developed using the reaction scheme outlined for the terminal model.

If the probability that a monomer $M_A$ will add onto its own radical is $P_{AA}$, while $P_{AB}$ is the probability that $M_B$ will be favoured instead, then, as the radical chain-end must add either $M_A$ or $M_B$, the probabilities of these events can be expressed by equations (5) and (6).

$$P_{AA}=\frac{k_{AA}[M_A\bullet][M_A]}{k_{AA}[M_A\bullet][M_A]+k_{AB}[M_A\bullet][M_B]}=\frac{r_A[M_A]}{r_A[M_A]+[M_B]} \quad (5)$$

$$P_{AB}=(1-P_{AA})=\frac{[M_B]}{r_A[M_A]+[M_B]} \quad (6)$$

The probability of forming a sequence of $m$ units of $M_A$ in the chain is then

$$P_A(m)=P_{AA}^{(m-1)}P_{AB} \quad (7)$$

The average sequence length can be estimated from these equations if the reactivity ratios are known, *i.e.* if $r_A=r_B=0.1$, there will be only nine alternating units for every non-alternating pair, whereas if $r_A=r_B=0.01$ then an average of 99 $[M_A{-}M_B]$ links in sequence will be obtained in every 100. The structure becomes more perfectly alternating as $r_A$ and $r_B$ both approach zero.

### 22.1.4 Monomer Reactivity

Values of $r_A$ and $r_B \approx 0$ imply rapid cross propagation reactions, therefore formation of an alternating copolymer will be favoured if the monomers show little inclination to undergo homopolymerization. This is not the only criterion, however, and the tendency towards alternation was commented on by Mayo and co-workers,[3-5] who observed that either there were specific interactions between the comonomers, or else steric effects were responsible. A more acceptable rationale for the tendency towards alternation considers, instead, the polarity of the double bond in the monomers. Mayo and Walling[6] reached the significant conclusion that two monomers will tend to form an alternating copolymer when there is a large difference between the polarities of each double bond. A vinyl monomer with an electron-withdrawing substituent, such as –CN or $-CO_2H$, which decreases the electron density on the double bond, will thus favour alternation with monomers possessing electron-donating groups, such as –OR or $-O_2CMe$, which have electron-rich double bonds.

Pertinent examples are systems such as maleic anhydride (MAH) and divinyl ether,[7] or α-methylstyrene (α-MS) and 2,4-dicyanobut-1-ene (DCB),[8] both comprising strong donor–acceptor pairs. In addition, these individual monomers undergo radical homopolymerization very slowly, or with considerable difficulty. This latter characteristic is, to some extent, a bonus, as the basis of the Mayo–Lewis treatment is that the cross propagation reactions are favoured because of the stabilization of the transition state by electrostatic interactions between the propagating radical and the double bond of the incoming monomer.

The general conclusion is, then, that a free radical copolymerization between a strong donor monomer and a strong acceptor monomer will result in a copolymer with a predominantly alternating structure. Typical examples of the two types are listed in Table 1, and copolymerization of a monomer from one column with a monomer from the other should give a copolymer with a high tendency towards alternation.

**Table 1** Typical Donor and Acceptor Monomers[34]

| *Donors* | *Acceptors* |
|---|---|
| Alkyl vinyl ethers | Acrylonitrile |
| Alkyl vinyl sulfides | Citraconic anhydride |
| Alkyl phenylvinyl ethers | $\beta$-Cyanoacrolein |
| Benzoselenophene | Diethyl fumarate |
| Butadiene | Diethyl maleate |
| Cyclopentene | Divinyl sulfone |
| 1,1-Diphenylethylene | Fumaronitrile |
| Dimethylbutadiene | *N,N'*-Fumaroyldipyrrolidone |
| Furan | Maleic anhydride |
| Indole | Maleimide |
| Indene | Methyl acrylate |
| Isoprene | Methylglutaronitrile |
| $\alpha$-Methylstyrene | Methyl maleate |
| Methyl methacrylate | Picryl methacrylate |
| 2-Naphthyl methacrylate | *N*-Phenylmaleimide |
| Phenylacetylene | Sulfur dioxide |
| Phenyl vinyl éthers | Tetracyanoquinodimethane |
| 2-Isopropenylnaphthalene | Tricyanoethylene |
| Stilbene | Vinylidene cyanide |
| Styrene | |
| Thiophene | |
| *N*-Vinylcarbazole | |
| Vinyl acetate | |
| Vinyl chloride | |
| 1-Vinylindole | |
| Vinylnaphthalene | |
| Vinylpyridine | |
| *N*-Vinylpyrrolidone | |

### 22.1.5 Deviations from the Terminal Model

The Mayo–Lewis terminal model has proved remarkably successful in predicting the copolymerization behaviour of many of these donor–acceptor combinations, but not all of them. Several systems appear to deviate from this simple model, for example styrene (S)–fumaronitrile[9] and S–MAH.[10] One explanation, proposed originally by Bartlett and Nozaki,[11] is that a donor–acceptor charge-transfer complex forms between the two participating monomers, and subsequently enters into the propagation step of the copolymerization as an identifiable unit. This idea was based on the observation that solutions of the donor and acceptor were often coloured, *e.g.* a S–chloromaleic anhydride mixture is yellow, indicating the formation of such complexes.

The evidence for the active participation of such donor–acceptor complexes in alternating systems is mixed, and it is still a controversial subject, as will be seen in subsequent sections, although the existence of such complexes is not denied.

## 22.2 DONOR–ACCEPTOR COMPLEXES

### 22.2.1 Binary Complexes

The term 'donor–acceptor complex' covers a wide spectrum of intermolecular complexes in which the strength of the intermolecular interaction ranges from weak dispersion forces, through dipole–dipole interactions, to complete transfer of electronic charge between donor and acceptor molecules. In certain alternating systems a strong charge-transfer complex is generated by a one electron 'T-class' reaction,[12] which involves the interaction between a donor radical cation and an acceptor radical anion, as shown in structure (**6**) for S and MAH. In many systems much weaker interactions may be observed, and so the more general donor–acceptor (D–A) terminology will be adhered to. As the D–A is assumed to exist in equilibrium (equation 8) with the unassociated species, where $x$ and $y$ normally have values of 1 or 2, the magnitude of the equilibrium or formation constant $K$ can be regarded as a measure of the strength of the complex. This can be measured using spectroscopic

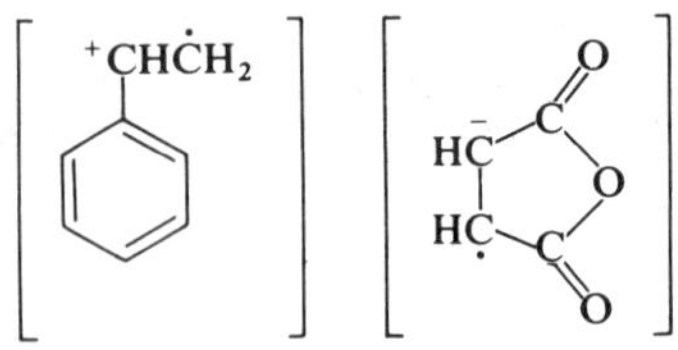

**(6)**

techniques; evidence for complex formation comes from the appearance of a new absorption band in the UV/visible spectrum of a comonomer mixture or solution, as long as this new band does not overlap the characteristic absorptions of either of the monomers. If this happens, evidence of complex formation may be obtained from alternative techniques, such as $^1$H NMR. Here, a change in the chemical shift of a resonance for one monomer in the mixture can indicate complex formation.

$$xD + yA \rightleftharpoons D_xA_y \tag{8}$$

Due consideration must also be given to the role played by the solvent (if present) in the complexation, as some solvents may be capable of forming a donor–acceptor complex with one of the monomers. Thus, it is preferable to use an inert solvent, which will not complex significantly with either monomer, when determining the formation constant $K$ for a donor–acceptor pair. Solvents such as hexane or carbon tetrachloride fulfil this requirement, but are not always miscible with both monomers.

### 22.2.2 Complexes Involving Lewis Acids

The situation is complicated further by the work of Hirooka *et al.*,[13] who opened up a new area when they reported that the tendency towards alternation of many comonomer systems could be enhanced by the addition of a Lewis acid. They deduced that the Lewis acid complexed with monomers with either pendant carbonyl or cyano groups, and so increased the difference in polarity between these monomers and donor molecules. On this basis they divided potential monomers into two groups, as shown in Table 2. Those in class A are monomers with electron-rich double bonds, which act as donors, whereas those in class B normally have their double bond conjugated with an

**Table 2** Classification of Monomers According to Hirooka[108]

| *Class A (donors)* | *Class B (acceptors)* |
|---|---|
| Ethylene | Methyl acrylate |
| Propylene | *n*-Butyl acrylate |
| Isobutene | Methyl methacrylate |
| 1-Hexene | Methyl α-chloroacrylate |
| 1-Octadecene | *n*-Butyl methacrylate |
| Styrene | Methyl crotonate |
| α-Methylstyrene | *n*-Butyl crotonate |
| 2-Butene | Acrylonitrile |
| 2-Methyl-2-butene | α-Chloroacrylonitrile |
| Butadiene | Methacrylonitrile |
| Isoprene | Acrylic acid |
| Cyclopentene | *n*-Octyl acrylate |
| 2-Norbornene | Acryloyl chloride |
| β-Methylstyrene | *N*-*n*-Octylacrylamide |
| *trans*-Stilbene | *N*,*N*-Di-*n*-butylacrylamide |
| Dihydronaphthalene | Methyl vinyl ketone |
| Vinyl chloride | |
| Vinyl acetate | |
| Vinyl benzoate | |
| Vinylidene chloride | |
| Allyl chloride | |
| Allyl acetate | |

electron-withdrawing group, such as a carbonyl or cyano group. Class B monomers can have their accepting properties enhanced by complexation with a Lewis acid, which reduces further the electron density of the vinyl group and thus strengthens the acceptor character of the monomer significantly. This does depend somewhat on the nature of the Lewis acid. It has been found that acrylic monomers complexed more effectively with $Et_3Al_2Cl_3$ and $Bu^iAlCl_2$, and were less active with other Lewis acids. It is also worth noting that conjugation is an important feature of class B monomers; non-conjugated monomers, such as vinyl acetate, are not included in this classification.

Complexation in these systems can thus be regarded as a multistage process. If the Lewis acid (MX) is considered to be a means of strengthening the donor–acceptor complex, then the formation of a ternary complex can be postulated, which will proceed *via* the three stages outlined in equations (9)–(11), where step (9) is the formation of the Lewis acid–acceptor complex, which then forms the ternary complex in step (10). The reaction (11) recognizes that normal D–A complex formation will also be a feature of the system. These stoichiometries can be established spectrophotometrically.

$$\mathrm{MX} + n\mathrm{A} \underset{}{\overset{K_A}{\rightleftharpoons}} \mathrm{MX(A)}_n \tag{9}$$

$$\mathrm{MX(A)}_n + p\mathrm{D} \overset{K_n}{\rightleftharpoons} \mathrm{MX(A)}_n\text{----}\mathrm{D}_p \tag{10}$$

$$\mathrm{A} + m\mathrm{D} \overset{K_0}{\rightleftharpoons} \mathrm{A}\text{----}\mathrm{D}_m \tag{11}$$

## 22.3 COMPLEX STOICHIOMETRY

### 22.3.1 Binary Complexes

The values of $x$ and $y$ in equation (8) can be established spectroscopically using the 'continuous variation method' proposed by Job.[14] Both UV/visible spectroscopy and $^1$H NMR data are amenable to analysis by this technique, and can be illustrated with the following examples.

When the donor–acceptor pair 2,3-dihydro-1,4-dioxin (1,4-dioxene) and MAH are dissolved in chloroform, an absorption band appears with a maximum at $\lambda = 366$ nm, which is characteristic of the D–A complex, as shown in Figure 3. It can be seen from the figure that this band displays a maximum absorption at a 1:1 mole ratio of the donor and acceptor, when determined as a function of the mixture composition. This is interpreted as the stoichiometry of the D–A complex.

This method has been extended by Sahai *et al.*[15] to encompass $^1$H NMR data. They have shown that, for a 1:$n$ complex, a plot of $\Delta\delta^A_{obs}[A]_0$ against $X_D$ will give a maximum at $X_D = n/(n+1)$. Here $X_D$ is the mole fraction of the donor, $[A]_0$ is the initial concentration of the acceptor and $\Delta\delta^A_{obs}$ is the observed chemical shift in the NMR spectrum of the acceptor protons in a complexing medium.

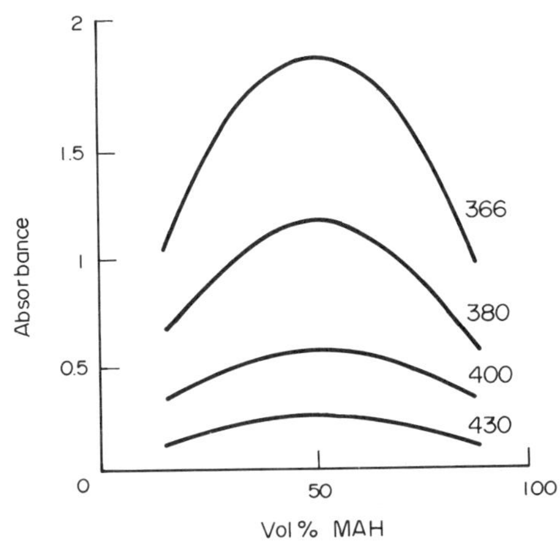


**Figure 3** Analysis of the charge transfer complex formed between MAH and 2,3-dihydro-1,4-dioxin in chloroform solution, using the continuous variation method. Curves have been determined at wavelengths of 366, 380, 400 and 430 nm (reproduced by permission of Huthig and Wepf from S. Iwatsuki and Y. Yamashita, *Makromol. Chem.*, 1965, **89**, 205)

It has been observed that both α-MS and isoprene (IP) form donor–acceptor complexes with DCB, but their characteristic absorption bands cannot be resolved by UV spectroscopy. The Sahai method has proved successful, however, and the chemical shift data from $^{1}H$ NMR are shown in Figure 4. These curves provide evidence for the formation of 1:1 complexes in each case.

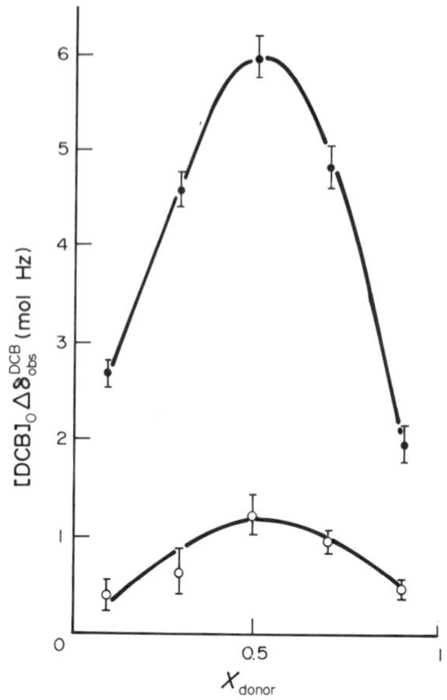


**Figure 4** Stoichiometric determination of the donor–acceptor complex by the NMR continuous variation method. $X_{donor}$ = mole fraction of donor molecules in donor–DCB mixtures; ●, α-MS–DCB; ○, IP–DCB

It was shown in Section 22.2.2 that binary complexes can form between a Lewis acid and a class B monomer by coordination of the carbonyl or cyano group to the central metal atom in the Lewis acid; see, for example, structures (**7**) and (**8**).

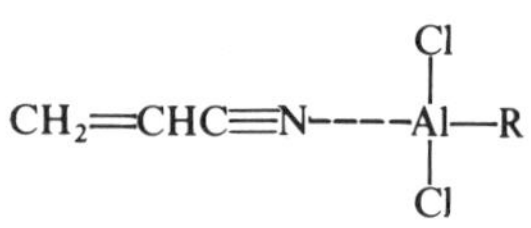


(**7**)

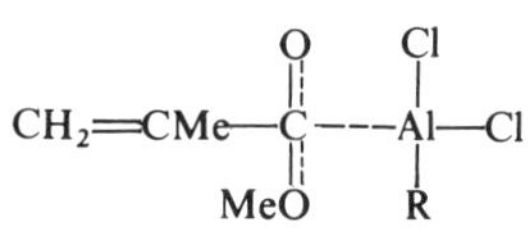


(**8**)

These complexes are often sufficiently stable to allow them to be isolated, usually as crystalline complexes which are very hygroscopic[16,17] and also polymerize readily. The large shifts in the IR bands associated with the stretching of the cyano or carbonyl groups substantiate the existence of these complexes.[18,19]

$^{1}H$ and $^{13}C$ NMR have also provided useful evidence for 1:1 complexation.[20–22] These complexes are not always 1:1 and 1:2 have also been identified.

### 22.3.2 Ternary Complexes

When $SnCl_4$ is added to methyl methacrylate (MMA), 1:1 and 1:2 complexes can form, as shown in equations (12) and (13).

$$SnCl_4 + MMA \underset{}{\overset{K_1}{\rightleftharpoons}} SnCl_4MMA \tag{12}$$

$$SnCl_4MMA + MMA \underset{}{\overset{K_2}{\rightleftharpoons}} SnCl_4(MMA)_2 \tag{13}$$

These tin complexes are either trigonal, bipyramidal or octahedral. The 1:2 $SnCl_4(MMA)_2$ appears to be octahedral, whereas, in the 1:1 complex, two Raman lines at 355 and 396 $cm^{-1}$ indicate the bipyramidal form; $Zn^{II}$, $B^{III}$ and $Al^{III}$ complexes tend to be tetrahedral. The stoichiometry was established using gravimetric and elemental analysis.[23,24]

While the formation of binary complexes between a Lewis acid and a class B monomer is firmly established, ternary complexation between the binary system and a donor monomer is more tenuous. The strength of interaction is probably much weaker but, by careful selection of solvent, temperature and measuring conditions, some evidence for the existence of ternary complexes can be obtained.

In the $SnCl_4$–MMA system, a 1:2 'binary' complex can be formed (equation 13). When a donor such as S is added to this system, a 1:1 interaction between MMA and S can be identified[21,25] using the continuous variation method based on $^1H$ NMR data, as shown in Figure 5.

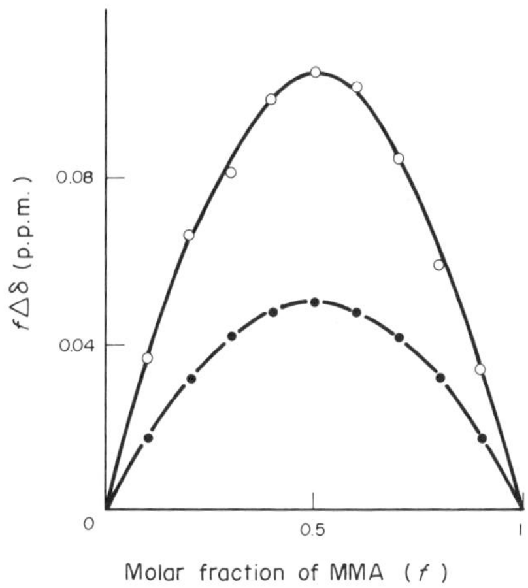


**Figure 5** Continuous variation plots of chemical shifts for the $SnCl_4(MMA)_2$–S system in *n*-hexane at 253 K; ○, methoxy protons; ●, α-methyl protons (reproduced from ref. 21 with permission from Wiley)

Similarly, when the 1:2 $ZnCl_2(AN)_2$ complex (AN = acrylonitrile) is mixed with S, a charge transfer band appears in the UV spectrum at 320–350 nm. Application of the Job method indicated a 1:1:1 complex between $ZnCl_2$, AN and S,[26] and this type of complex formation was confirmed in the $ZnCl_2$–$(MMA)_2$–S system, and $ZnCl_2$–$(AN)_2$–divinyl ether.[27]

Furukawa *et al.*[28] examined some $EtAlCl_2$–acrylic monomer–butadiene systems. Addition of butadiene to the $EtAlCl_2$/MMA complex led to the formation of a charge transfer complex with a characteristic absorption at 340 nm, which was shown to be a 1:1:1 complex. It was also proposed that, when a Lewis acid complexes with an acrylic monomer, the polarizability of the monomer is increased. NMR examination of the complexed monomer indicated an apparent delocalization of the π electrons, which was believed to be the driving force necessary for the formation of the ternary complex with a donor molecule.[20]

## 22.4 MEASUREMENT OF FORMATION CONSTANTS

The formation constants are measured predominantly by UV and NMR spectroscopy, but can also be studied using vapour pressure, calorimetry and dielectric methods.[29–34]

### 22.4.1 UV/Visible Absorption Spectra

In a binary system composed of a donor D and an acceptor A, with conditions such that $[D] \gg [A]$, the Benesi–Hildebrand[35] relation, or one of its several modifications, can be used to establish the formation constant $K$ of the D–A complex. These equations have linear forms

Benesi–Hildebrand
$$\frac{\mathscr{L}[A]_0}{d} = \frac{1}{\varepsilon} + \frac{1}{\varepsilon K[D]_0} \tag{14}$$

where $\mathscr{L}$ is the optical path length, ε is the extinction coefficient, $d$ is the absorbance at a given wavelength and $[A]_0$ and $[D]_0$ are the initial concentrations of acceptor and donor, respectively.

Rearrangement of equation (14) gives equation (15).

Scott
$$\frac{\mathscr{L}[A]_0[D]_0}{d} = \frac{[D]_0}{\varepsilon} + \frac{1}{\varepsilon K} \qquad (15)$$

The Benesi–Hildebrand equation can only be applied to 1:1 complexes and it is not always the best way to measure $K$, particularly when the D–A complex is weak. It can give zero or negative intercepts and may exhibit curvature at high donor concentrations, depending on the solvent used in the measurements. These drawbacks have been commented on by Person,[36] Deranleau,[37] and Guidry and Drago,[38] who stress that the initial concentration of the acceptor must be kept low and constant at approximately 0.1 mol dm$^{-3}$, while the initial concentration of the donor is varied in the range 1.0 to 9.0 mol dm$^{-3}$.

The Scott[39] equation is an improvement, but both equations (14) and (15) tend to give unequal weight to the experimental points and are insensitive to more than one stoichiometry in solution. For these reasons the Scatchard–Foster modification is frequently used (equation 16), where $n$ is included to allow for the possibility of a departure from 1:1 complexation. The advantage of using equation (16) is that a wider range of experimental values can be considered without undue emphasis being placed on data measured over a limited concentration range. If the charge transfer band characteristic of the D–A complex tends to overlap one of the acceptor absorptions, then the Ketelaar equation (equation 17) can be used, where $d^* = d - \varepsilon_A \mathscr{L}[A]_0$ and $\varepsilon_{DA}$ is the extinction coefficient of the donor–acceptor complex. An example of a Ketelaar plot is shown in Figure 6 for the MAH–vinyl acetate (VA) complex formation in acetonitrile,[40] where $\lambda_{max}$ for the complex could not be resolved because of strong absorptions of the vinyl acetate at $\lambda < 258$ nm. These UV methods have been critically reviewed by Polle.[41]

Scatchard–Foster
$$\frac{d}{[D]_0^n} = K\mathscr{L}\varepsilon[A]_0 - Kd \qquad (16)$$

Ketelaar
$$\frac{\mathscr{L}[A]_0}{d^*} = \frac{1}{(\varepsilon_{DA} - \varepsilon_A)} + \frac{1}{(\varepsilon_{DA} - \varepsilon_A)K[D]_0} \qquad (17)$$

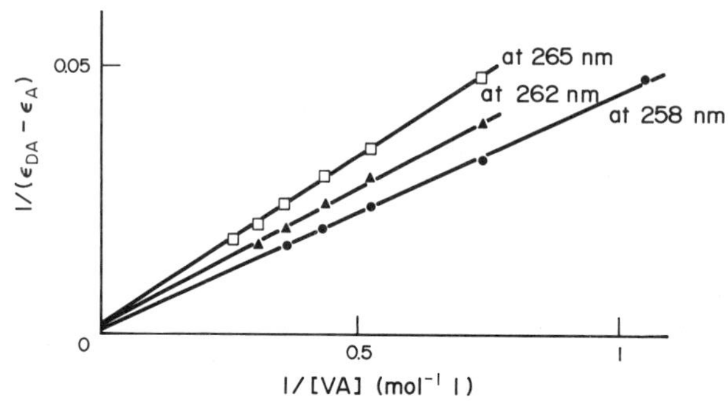


**Figure 6** Ketelaar plot for the determination of the formation constant of complexation between MAH and vinyl acetate (VA) in acetonitrile at 300 K, [MAH] = $3.437 \times 10^{-3}$ mol dm$^{-3}$ (reproduced from ref. 40 with permission from Springer)

### 22.4.2 NMR Analysis

In certain D–A systems no suitable UV charge transfer band can be detected, and some of these may be amenable to $^1$H NMR analysis. Ideally, these systems should possess the following properties: (a) the components of the mixture should be completely miscible with one another; (b) the $^1$H NMR resonances of donor or solvent should not overlap the signals of the acceptor (and *vice versa*); and (c) both donor and acceptor molecules should contain proton(s) which give rise to a single sharp signal.

In weak D–A complexes there is a rapid exchange between the complexed and uncomplexed states, and so the chemical shifts of the protons of one of the monomers ($\delta_{OBS}$) are each observed as a single peak, corresponding to the weighted average of the shift due to the free molecules ($\delta_f$) and that due to the complexed monomers ($\delta_{DA}$), as shown schematically in Figure 7. If it is assumed that the

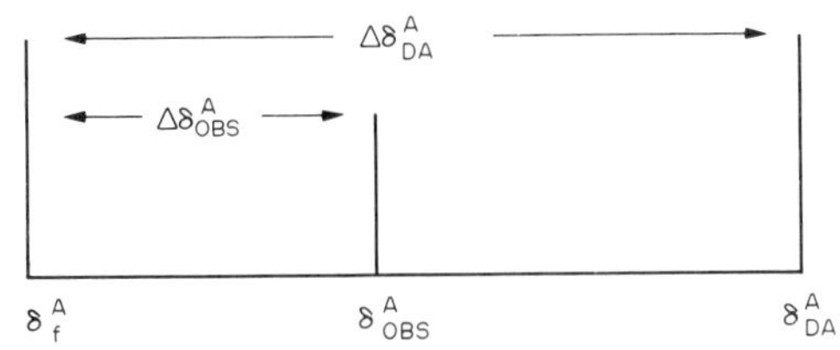


**Figure 7** Schematic representation of the relation between chemical shift observed for acceptor protons in the complexing medium ($\delta^{A}_{OBS}$), the chemical shift of acceptor protons in the pure complex ($\delta^{A}_{DA}$) and the shift due to the free (uncomplexed) molecules ($\delta^{A}_{f}$)

equilibrium represented by equation (18) exists, then the corresponding equation takes the form of equation (19).

$$K = \frac{[DA]}{([A]_0 - [DA])([D]_0 - [DA])} \tag{18}$$

$$\delta_{OBS} = \delta_p \frac{[A]}{[A]+[DA]} + \delta_{DA} \frac{[DA]}{[A]+[DA]} \tag{19}$$

The chemical shifts for either donor or acceptor protons can be used, whichever is most convenient, but if the donor is always kept in excess, as required by the Benesi–Hildebrand approach, then the simplified relation (20) can be used, where $\Delta\delta_{OBS} = \delta_{OBS} - \delta_f$ and $\Delta\delta_{DA} = \delta_{DA} - \delta_f$. Modification of equation (20) gives the NMR equivalent of the Scatchard equation (equation 21).

Hanna–Ashbaugh

$$\frac{1}{\Delta\delta_{OBS}} = \frac{1}{\Delta\delta_{DA}} + \frac{1}{\Delta\delta_{DA}[D]_0 K} \tag{20}$$

Scatchard

$$\frac{\Delta\delta_{OBS}}{[D]_0} = \Delta\delta_{DA} K - \Delta\delta_{OBS} K \tag{21}$$

From these equations, values of $K$ can be obtained by plotting $1/\Delta\delta_{OBS}$ against $1/[D]_0$, and a number of systems have been examined in this way. MAH has a $^1$H NMR spectrum which is particularly suited to this treatment, as the protons appear as a singlet at 7.02 p.p.m. Addition of a donor molecule causes the singlet to shift; the $K$ values can be estimated for MAH complexes with furan,[42] S[44] and many others, which are listed in Table 3.

The method is not confined to MAH complexes, however, and this is illustrated for the DCB–IP system. The $^1$H NMR spectra for various concentrations of this donor–acceptor pair are shown in Figure 8, which demonstrates the shift of the vinyl protons of the DCB with added IP at a temperature of 306 K. Treatment of the data using the Hanna–Ashbaugh[45] equation gave linear plots at three temperatures, as shown in Figure 9(a). It has been demonstrated[46] that the Benesi–Hildebrand relations will give linear plots even if a $D_2A$ complex is present, whereas the Scatchard equivalent plot[47] is more sensitive and is curved if the stoichiometry of the complex is not 1:1. Replotting of the data using equation (21) also gave a linear dependence (see Figure 9b) and both plots gave identical values of $K$ which are listed in Table 4.

The formation constants for ternary complexes can also be measured. For the MMA–S system in the presence of $SnCl_4$, Hirai and Komiyama[21] have measured $K_1$ and $K_2$ values for equations (12) and (13) in hexane solution, using the proton chemical shift of the MMA component. Inspection of the $K$ values in Table 5 shows that $SnCl_4$ enhances complex formation when compared with the binary complex; it also shows toluene to be a better donor than styrene.

Komo *et al.*[22] have also used this method to determine ternary complex formation involving complexes of ($AlEtCl_2AN$----toluene) and ($AlEt_3AN$----toluene). Further examples of the constants for formation of the ternary complexes involving $SnCl_4$ are shown in Table 6, where values obtained in both the presence and absence of Lewis acids can be compared.

Usually, conditions are selected such that $[D] \gg [A]$, in order to produce linear plots; the advent of computerized non-linear curve-fitting techniques should now allow a much wider range of concentrations to be investigated. Values of $K$ obtained must still be examined critically, however, as there are inconsistencies in the literature. It should be remembered that $K$ is often a concentration constant and not necessarily a true thermodynamic constant, unless the activity coefficients $\gamma$ are

**Table 3** Equilibrium Constants $K$ for Charge Transfer Complexes Between MAH and Electron Donor Molecules[34]

| *Electron donor* | $10^2 K$ ($dm^3\,mol^{-1}$) | *Solvent* | $T$(°C) | *Method* | *Ref.* |
|---|---|---|---|---|---|
| Styrene | 24.0 | $CHCl_3$ | 25 | UV | 50 |
| Styrene | 34.0 | *n*-Hexane | 30 | NMR | 46 |
| α-Methylstyrene | 7.0 | $CDCl_3$ | — | NMR | 98 |
| Vinyl acetate | 6.6 | $CHCl_3$ | 25 | UV | 50 |
| Vinyl acetate | 6.0 | — | — | UV | 59 |
| Vinyl acetate | 56.0 | *n*-Hexane | 25 | NMR | 101 |
| Furan | 6.9 | $CHCl_3$ | RT | NMR/UV | 42 |
| Furan | 47.0 | Heptane | 24 | UV | 60 |
| Methylfuran | 5.8 | $CHCl_3$ | RT | UV | a |
| Dimethylfuran | 2.0 | $CHCl_3$ | RT | UV | a |
| Tetrahydrofuran | 8.6 | Heptane | RT | UV | 60 |
| Dihydrofuran | 44.5 | Hexane | 36 | NMR | 44 |
| Benzofuran | 1.0 | Cyclohexane | 20 | NMR | 43 |
| Dihydropyran | 11.0 | Heptane | 24 | UV | 60 |
| Dihydropyran | 2.0 | Dichloromethane | — | UV | b |
| Indole | 28.0 | $CHCl_3$ | 20 | NMR | 43 |
| Benzothiophene | 30.0 | $CHCl_3$ | 20 | NMR | 43 |
| Divinyl ether | 0.6 | $CHCl_3$ | — | UV | a |
| Divinyl ether | 18.0 | Heptane | 24 | UV | 49 |
| Ethyl vinyl ether | 15.0 | Heptane | 24 | UV | 60 |
| Ethyl vinyl ether | 3.6 | $CHCl_3$ | 23 | UV | c |
| 2-Chloroethyl vinyl ether | 29.0 | Hexane | 30 | NMR | 44 |
| 2-Chloroethyl vinyl ether | 10.0 | Benzene | 30 | NMR | 76 |
| 2-Chloroethyl vinyl ether | 10.5 | $CHCl_3$ | 30 | NMR | 76 |
| 2,3-Dihydro-1,4-dioxin | 6.9 | Benzene | 25 | UV | 76 |
| 2,3-Dihydro-1,4-dioxin | 4.0 | Toluene | 25 | UV | 76 |
| 2,3-Dihydro-1,4-dioxin | 5.1 | Chlorobenzene | 25 | UV | 76 |
| 2,3-Dihydro-1,4-dioxin | 5.4 | $CHCl_3$ | 25 | UV | 76 |
| 2,3-Dihydro-1,4-dioxin | 4.7 | Acetone | 25 | UV | 76 |
| PhCH=CHOR | | | | | |
| R = Ethyl | 28.0 | $CDCl_3$ | 30 | NMR | 48 |
| R = *n*-Butyl | 20.0 | $CDCl_3$ | 30 | NMR | 48 |
| R = 1-Methylpropyl | 30.0 | $CDCl_3$ | 30 | NMR | 48 |
| R = 2-Methylbutyl | 38.0 | $CDCl_3$ | 30 | NMR | 48 |
| PhCH=CHSR | | | | | |
| R = Ethyl | 25.0 | $CDCl_3$ | 30 | NMR | 48 |
| R = *n*-Propyl | 38.0 | $CDCl_3$ | 30 | NMR | 48 |
| R = Isopropyl | 28.0 | $CDCl_3$ | 30 | NMR | 48 |
| R = 1-Methylpropyl | 40.0 | $CDCl_3$ | 30 | NMR | 48 |

[a] Y. A. Ragab and G. B. Butler, *J. Polym. Sci., Polym. Chem. Ed.*, 1981, **19**, 1175. [b] G. B. Butler and K. C. Joyce, *J. Polym. Sci., Part C*, 1968, **22**, 45. [c] G. B. Butler and A. F. Campus, *J. Polym. Sci., Part A-1*, 1970, **8**, 523.

**Table 4** Formation Constants for the DCB–IP System at Three Temperatures

| *Temperature* (K) | $K$ ($dm^3\,mol^{-1}$) *Equation (20)* | *Equation (21)* |
|---|---|---|
| 273 | 0.09 | 0.09 |
| 288 | 0.08 | 0.08 |
| 306 | 0.07 | 0.07 |

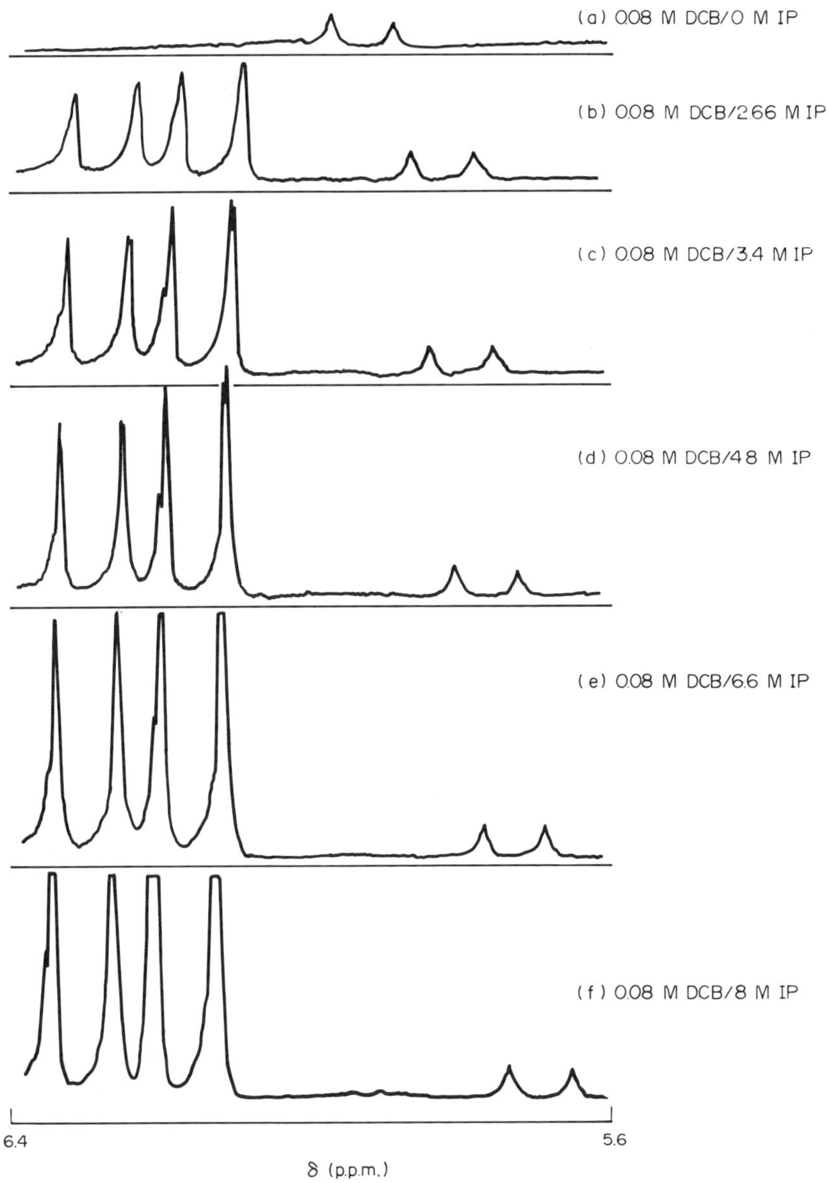


**Figure 8** $^1$H NMR spectra of DCB/IP solutions in $CCl_4$ at 306 K illustrating the high field shifts of the DCB vinylic resonances upon complexation with IP

**Table 5** Formation Constants[a] in the $SnCl_4$–MMA–S System in Toluene at −40 °C

| *Molecular complex* | $K$ ($dm^3 mol^{-1}$) |
|---|---|
| $SnCl_4$MMA . . . . S | 0.6 |
| $SnCl_4(MMA)_2$ . . . . Toluene | 1.22 |
| MMA . . . . S | 0.23 |
| MMA . . . . Toluene | 0.27 |

[a] Data taken from ref. 21.

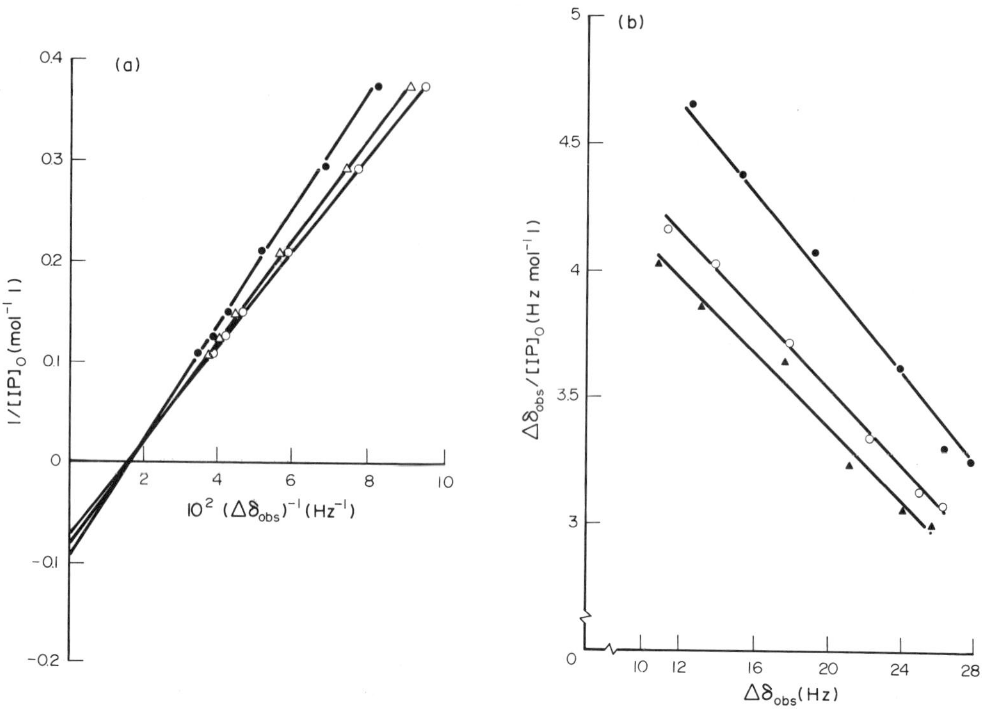


**Figure 9** (a) $^1$H NMR determination of the formation constant for DCB–IP complexation in $CCl_4$ using the Hanna–Ashbaugh relation, equation (20); ●, 273 K; △, 288 K; and ○, 306 K; (b) same data plotted according to the equivalent NMR Scatchard relation, equation (21), using same temperatures as in Figure 9(a)

**Table 6** Equilibrium Constants for Formation of Ternary Complexes. Lewis Acid, when Present, Stannic Chloride[108]

| *Complex* | *Temperature* (°C) | *Solvent* | ($K$ mol$^{-1}$ dm$^3$) | *Ref.* |
|---|---|---|---|---|
| MMA----S | −50 | *n*-Hexane | 0.27 | 104 |
| MMA----S | −40 | *n*-Hexane | 0.23 | 21 |
| MMA----S | −20 | *n*-Hexane | 0.18 | 21 |
| MMA----S | 0 | *n*-Hexane | 0.14 | 21 |
| MMA----S | 24 | *n*-Hexane | 0.11 | 21 |
| MMA----S | 40 | Toluene | 0.152 ± 0.013 | a |
| MMA----S | 40 | Tetrachloroethylene | 0.073 ± 0.017 | b |
| MMA----Tol | −50 | Toluene | 0.33 | b |
| MMA----Tol | −40 | Toluene | 0.27 | 21 |
| MMA----Tol | −20 | Toluene | 0.20 | 21 |
| MMA----Tol | 0 | Toluene | 0.15 | 21 |
| MMA----Tol | 24 | Toluene | 0.12 | 21 |
| MMA----Tol | 40 | Toluene | 0.024 ± 0.007 | a |
| MMA----Tol | 35 | Cyclohexane | 0.27 | 110 |
| MMA----TCE | 40 | Tetrachloroethylene | 0.131 ± 0.058 | a |
| $SnCl_4(MMA)_2$----S | −50 | *n*-Hexane | 0.75 | 21 |
| $SnCl_4(MMA)_2$----S | −40 | *n*-Hexane | 0.60 | 21 |
| $SnCl_4(MMA)_2$----S | −20 | *n*-Hexane | 0.40 | 21 |
| $SnCl_4(MMA)_2$----S | 0 | *n*-Hexane | 0.28 | 21 |
| $SnCl_4(MMA)_2$----Tol | −50 | *n*-Hexane | 1.65 | 21 |
| $SnCl_4(MMA)_2$----Tol | −40 | *n*-Hexane | 1.22 | 21 |
| $SnCl_4(MMA)_2$----Tol | −20 | *n*-Hexane | 0.70 | 21 |
| $SnCl_4(MMA)_2$----Tol | 0 | *n*-Hexane | 0.42 | 21 |
| $SnCl_4$MMA----S | −50 | *n*-Hexane | 0.99 | 104 |
| $SnCl_4$MMA----Tol | −50 | *n*-Hexane | 1.29 | 104 |

[a] J. R. Ebdon, private communication. [b] W. Gerrard, M. F. Lappert, H. Pyszora and J. W. Wallis, *J. Chem. Soc.*, 1960, 2182.

known, or it is assumed that the relation $\gamma_{DA}/\gamma_{D}\delta_{A} = 1$ is valid or remains constant over the concentration range studied.[48]

### 22.4.3 Solvent Effects

The influence of the solvent cannot be ignored when attempting to measure $K$, as it may compete as a donor or acceptor molecule itself. Butler and Fujimori[49] studied the complexing of MAH with divinyl ether in solvents of increasing dielectric constant and found a tenfold decrease in $K$ (see Table 7). The problem has been studied by Caze and Loucheux[50] who developed the concept of solvent competition, first proposed by Merrefield and Phillips,[51] which considers equilibria between the donor, the acceptor and the solvent.

**Table 7** Effect of Solvent on the Equilibrium Constant $K$ for Complex Formation between MAH and Divinyl Ether

| *Electron donor* | *Solvent* | *Dielectric constant* | *$10^2 K$* ($dm^3\ mol^{-1}$) | *Ref.* |
|---|---|---|---|---|
| Divinyl ether | Heptane | 1.9 | 10.8 | 49 |
| | $CCl_4$ | 2.2 | 9.8 | 49 |
| | $CHCl_3$ | 4.8 | 3.6 | a |
| | $CH_2Cl_2$ | 9.1 | 1.4 | 49 |
| | $CH_2Cl_2$ | 9.1 | 1.4 | b |
| | $CH_2ClCH_2Cl$ | 10.4 | 1.3 | 49 |

[a] G. B. Butler and A. F. Campus, *J. Polym. Sci., Part A-1*, 1970, **8**, 523. [b] G. B. Butler and K. C. Joyce, *J. Polym. Sci., Part C*, 1968, **22**, 45.

Litt and Wellinghoff[52] have also made a careful examination of the effect of solvent and observed that, when several systems were plotted according to equation (14), curvature developed at high donor concentrations. They developed a new model in which the reaction occurs in two stages, solvation followed by complexation, and they concluded that the ability of the donor to compete with the solvent during the first step was one of the most important factors. Thus the variations in $K$ were often due to these competitive interactions.

## 22.5 RADICAL INITIATED ALTERNATING COPOLYMER SYSTEMS

### 22.5.1 Maleic Anhydride

The most extensively studied acceptor molecule is MAH, which forms alternating copolymers readily with a wide range of donor molecules. The monomer homopolymerizes with some difficulty, but often reacts spontaneously with donors[53–59] and a selection of these systems, which form 1:1 alternating copolymer structures, is given in Table 8. Not all of the copolymers incorporating MAH have the expected 1:1 composition and more complex structures have been identified.

#### *22.5.1.1 Maleic anhydride and dienes*

Conjugated dienes copolymerize readily with MAH and often form Diels–Alder adducts. This reaction may compete with the alternating copolymerization. When 1,3-dienes are the comonomers, the composition of the copolymer tends to be 1:1, with the diene unit incorporated predominantly in the *cis* form; Scheme 1 has been suggested as a possible route to the copolymer. This postulates the formation of a charge transfer complex which either forms the Diels–Alder adduct or undergoes copolymerization *via* the exciplex.[60] If the cyclic adduct is very stable, high temperatures may be required to form the alternating copolymer, and it has been suggested that retrograde dissociation of the adduct must take place before copolymer can be formed.

**Table 8** Alternating Copolymer with MAH[a]

| | *Polymerization* | *CT complex detected* | *Property* | *Structure* | *Ref.* |
|---|---|---|---|---|---|
| Styrene | At 80 °C by BPO, 70 °C by AIBN | UV | $M_n = 10\,000–1300$ | 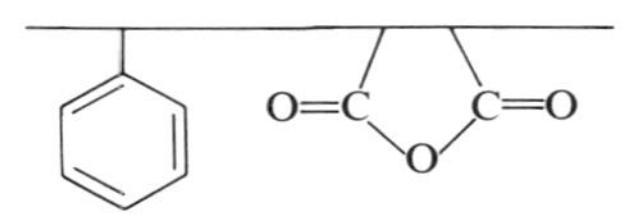 | 10, 53 |
| α-Methylstyrene | At 80 °C in decalin by AIBN | NMR and UV | $T_m = 210\,°C$ | 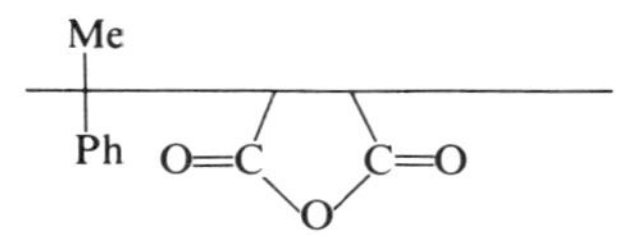 | 54 |
| Phenylethyne | At 98 °C in but-2-one or acetophenone by AIBN or UV | | $M_n \leq 9.1 \times 10^4$ | 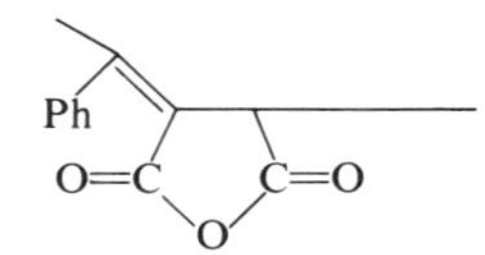 | 55 |
| Furan | At 70 °C in benzene by AIBN | NMR and UV | $[\eta] \leq 0.12\,dl\,g^{-1}$ | 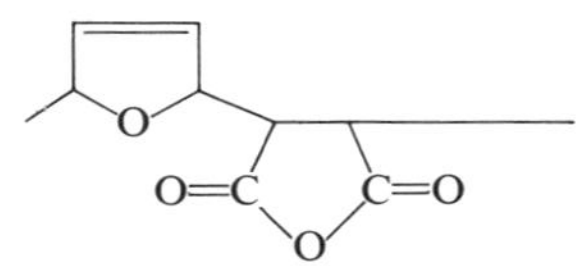 | 56, 57 |
| 2-Methylfuran | Under UV light or AIBN, MPP in benzene, dioxane or THF at 30–70 °C | | $[\eta] \leq 0.33\,dl\,g^{-1}$ | 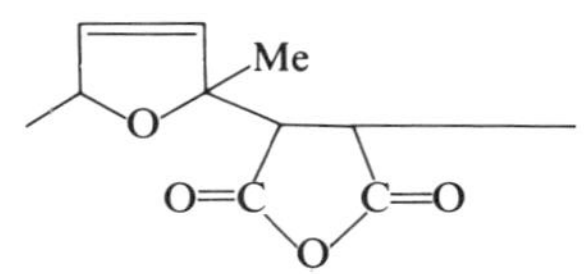 | 56 |
| Benzofuran | At 60 °C in cyclohexane by AIBN | NMR | $\eta_{red} = 0.06 \sim 0.11\,dl\,g^{-1}$ $C = 0.5\,g\,dl^{-1}$ | 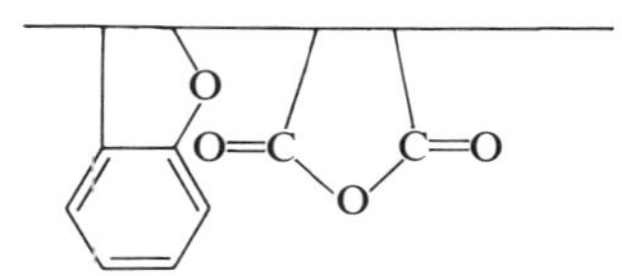 | 43 |

| | | | | | |
|---|---|---|---|---|---|
| Indole | At 60 °C in $CHCl_3$ by AIBN | NMR | $\eta_{red} = 0.07 \sim 0.09\ dl\ g^{-1}$ $C = 0.5\ g\ dl^{-1}$ | 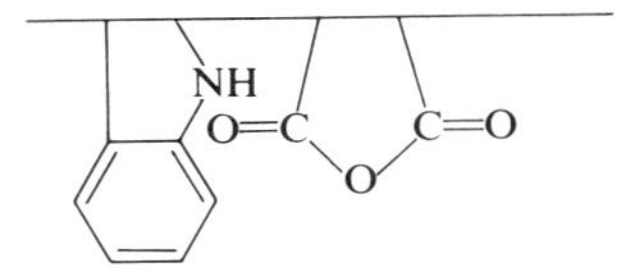 | 43 |
| Benzothiophene | At 60 °C in $CHCl_3$ by AIBN | NMR | $\eta_{red} = 0.09 \sim 0.16\ dl\ g^{-1}$ $C = 0.5\ g\ dl$ | 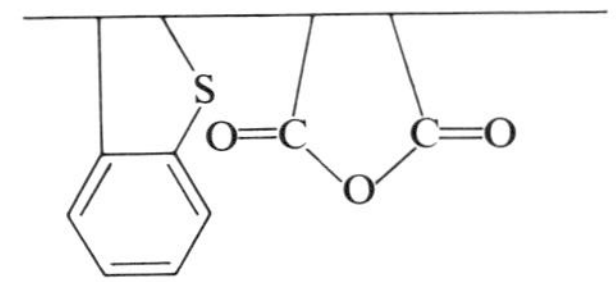 | 43 |
| Ethyl vinyl ether | At 60 °C in $CHCl_3$ by BPO | UV | $\eta_{rel} < 1.418$ $C = 0.5\ g\ dl^{-1}$ | 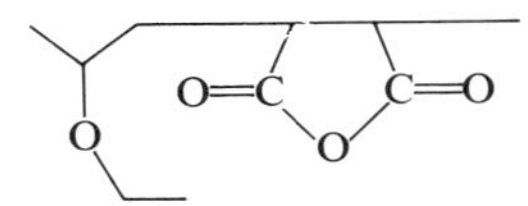 | 49 |
| 2,3-Dihydropyran | At 60 °C in $CHCl_3$ by BPO | UV | $\eta_{rel} \approx 1.025$ $C = 0.5\ g\ dl^{-1}$ | 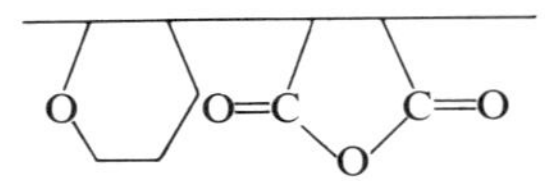 | 58 |
| Vinyl acetate | | NMR and UV | | 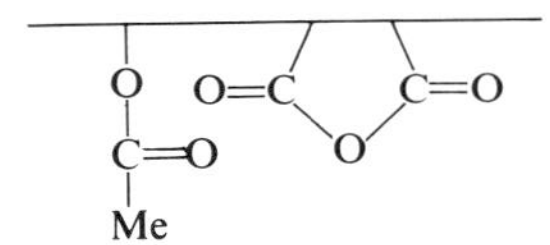 | 59 |

[a] Reproduced with permission from *Rubber Chem. Technol.*, 1978, **51**, 600.

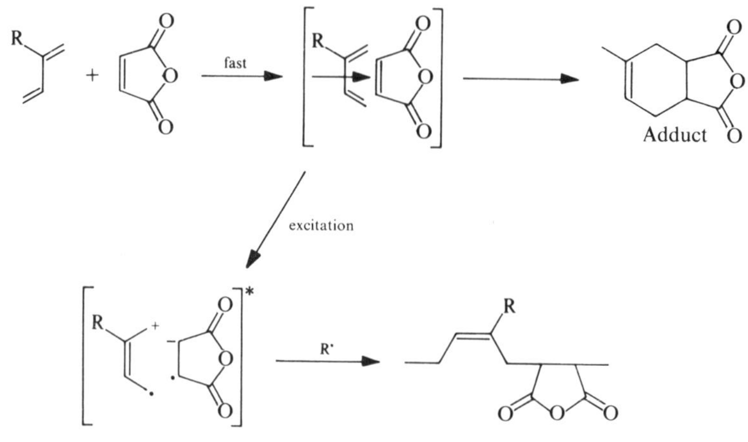


**Scheme 1**

The reactions of MAH with 1,4-dienes have been studied in some detail by Butler and co-workers,[7,27,49,61–64] who have established that copolymers with a 1:2 diene:MAH composition are more commonly produced than 1:1 structures. Thus, MAH and divinyl ether form the complex structures shown in Scheme 2.

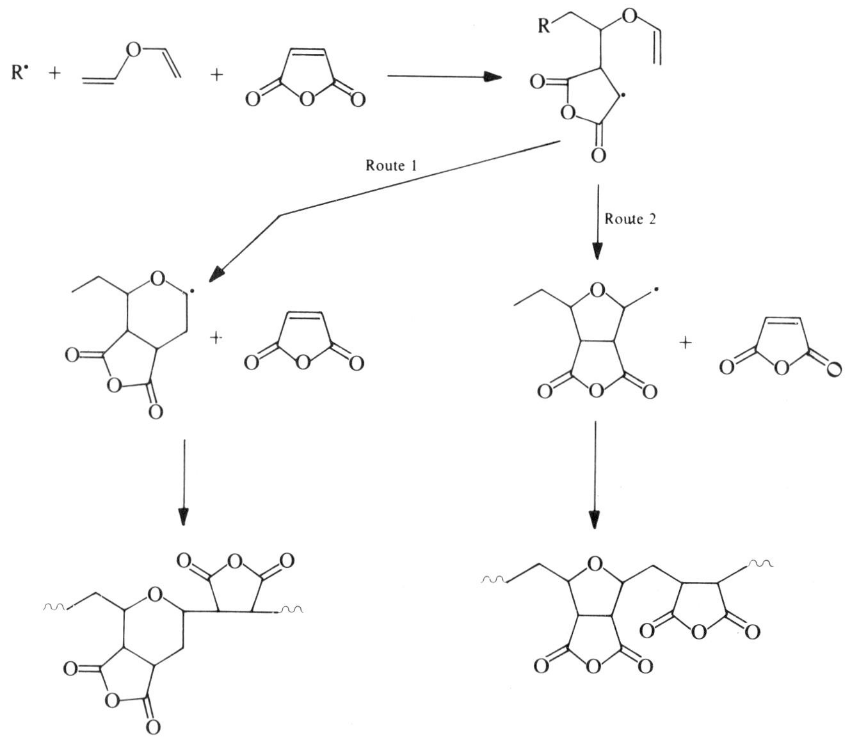


**Scheme 2**

Multiring copolymers are also formed when 1,5- and 1,6-dienes are used, as illustrated in Scheme 3.

## 22.5.2 Other Systems

There are many other examples of alternating systems, some of which will be covered in subsequent sections, and the majority of these tend to form 1:1 structures. Of these, fumaronitrile copolymerized with S[65,66] and *N*-vinylcarbazole (NVC),[67,68] *N*-vinylpyrrolidone with methyl

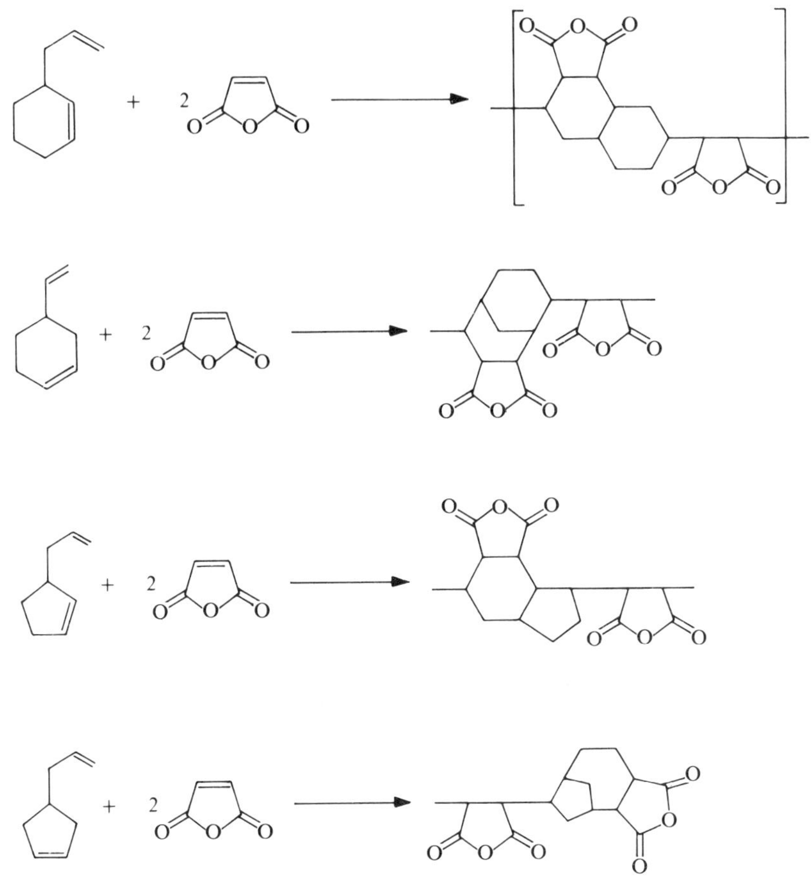

**Scheme 3**

α-cyanocinnamate,[69] vinyl acetate with ethyl α-cyanocinnamate,[70] together with many of the MAH and substituted-MAH-containing systems, have been studied with the aim of elucidating the exact mechanism of alternation. Much of the attention in this area is now focussed on this aspect.

## 22.6 MECHANISMS OF RADICAL INITIATED ALTERNATING COPOLYMERIZATIONS

### 22.6.1 Radical Initiated Binary Systems

The identification of D–A complexes in many alternating copolymer systems has generated the idea that the complex may act as a unique species capable of entering into the propagation steps in the reaction. While this concept has been investigated in some detail, there is, as yet, no firm consensus on the part played by such complexes. The idea has been tested by examining copolymerization reactions using three models, which have been developed to explain the mechanism; these are (a) the terminal model (Mayo and Lewis), (b) the penultimate model and (c) the complex-participation model. The applicability of each can be assessed, either by determining the variation of the copolymer composition with monomer feed composition over the whole composition range, or by studying the kinetics of the reaction as a function of monomer feed composition. It is also advantageous to be able to determine the sequence distribution, but this is not always possible.

#### *22.6.1.1 The terminal model*

The terminal model is the simplest approach to describing alternating copolymerization and the reaction scheme has been outlined in Section 22.1.2. The model predicts that alternation will be enhanced by strong polar or other attractive forces between the monomers which favour the cross propagation reactions to the virtual exclusion of the homopolymerization steps.

The model has been criticized by several workers who have found it gives an inadequate description of their systems,[71] and a number of factors contributing to its failure have been postulated.

(a) The penultimate unit in the propagating chain may influence the reactivity of the chain-end radical.[72]

(b) Many of the monomers are capable of forming D–A complexes and these should be considered as a third polymerizable species.[11]

(c) If one of the monomers is a diene it can add to the propagating radical in several ways.

(d) If polymerization is carried out at or near a ceiling temperature, a depropagation reaction might have to be part of the kinetic scheme.

While some of the reported discrepancies between the results and the model predictions may be analytical, and experimental error cannot be ruled out as a source of deviation from a simple kinetic scheme, two, more elaborate, models have been advanced which appear to give improved descriptions in certain systems.

### 22.6.1.2 *The penultimate model*

A scheme based on criticism (a) assumes that the penultimate unit in the growing chain has an influence on the reactivity of the terminal radical, thereby affecting its ability to react with an incoming monomer. This is called the 'penultimate model'[72] and requires the eight propagation steps shown in equations (22)–(29) to be considered.

$$\sim M_AM_A\cdot + M_A \xrightarrow{k_{AAA}} \sim M_AM_AM_A\cdot \tag{22}$$

$$\sim M_AM_A\cdot + M_B \xrightarrow{k_{AAB}} \sim M_AM_AM_B\cdot \tag{23}$$

$$\sim M_AM_B\cdot + M_A \xrightarrow{k_{ABA}} \sim M_AM_BM_A\cdot \tag{24}$$

$$\sim M_AM_B\cdot + M_B \xrightarrow{k_{ABB}} \sim M_AM_BM_B\cdot \tag{25}$$

$$\sim M_BM_A\cdot + M_A \xrightarrow{k_{BAA}} \sim M_BM_AM_A\cdot \tag{26}$$

$$\sim M_BM_A\cdot + M_B \xrightarrow{k_{BAB}} \sim M_BM_AM_B\cdot \tag{27}$$

$$\sim M_BM_B\cdot + M_A \xrightarrow{k_{BBA}} \sim M_BM_BM_A\cdot \tag{28}$$

$$\sim M_BM_B\cdot + M_B \xrightarrow{k_{BBB}} \sim M_BM_BM_B\cdot \tag{29}$$

Thus, there are four reactivity ratios, defined as $r_A = k_{AAA}/k_{AAB}$, $r_B = k_{BBB}/k_{BBA}$, $p_A = k_{BAA}/k_{BAB}$ and $p_B = k_{ABB}/k_{ABA}$. The instantaneous copolymer composition can then be derived from equation (30), which is much more complex than that for the terminal model and needs an iterative non-linear least-squares curve-fitting procedure to test the data. If $p_A = r_A$ and $p_B = r_B$ then there is no penultimate effect, but when there is no such equivalence the reaction scheme can be simplified by making certain assumptions.[10] The reactivity ratios can be reduced to $r_A$ and $p_A$ by assuming that one of the monomers has little tendency to homopolymerize, so that reactions (25), (28) and (29) can be neglected, limiting the scheme to five steps.

$$\frac{d[M_A]}{d[M_B]} = \frac{1+\{(p_A[M_A]/[M_B])(r_A[M_A]/[M_B]+1)/(p_A[M_A]/[M_B]+1)\}}{1+\{p_B(p_B+[M_A]/[M_B])/[([M]_A/[M_B])(p_B+[M_A]/[M_B])]\}} \tag{30}$$

### 22.6.1.3 *Complex-participation model*

Perhaps the biggest source of disagreement is the extent to which the D–A complexes, observed in many systems, participate in the alternation mechanism. Evidence cited to support this participation is based on several types of observation.

Initial rates in many copolymerizations exhibit maxima around equimolar feed ratios at which it is assumed the concentration of the D–A complex will be greatest. This behaviour is illustrated for the DCB–α-MS system[8] in Figure 10, but such maxima can be accommodated within the scope of the terminal model, making this proof inconclusive. Examples of spontaneous initiation have been quoted as supporting evidence, but, as before, doubt concerning the exact source of the initiation, which may be caused by trace impurities, makes this case uncertain.

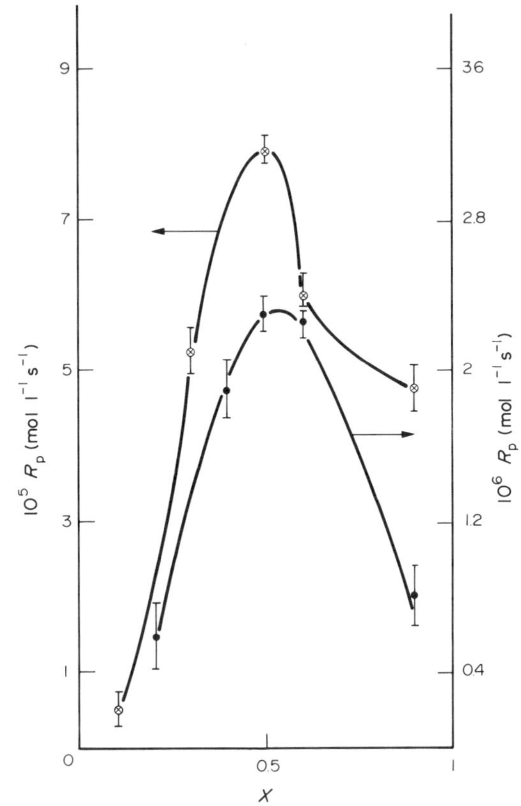


**Figure 10** Plots of initial copolymerization rate against mole fraction of DCB, for the DCB–IP system (⊗) and DCB–α-MS system (●)

Kelen and Tudos[71] have used a graphical analysis of the reactivity ratios as a diagnostic tool. When data plotted according to their relations are found to be curved, then they postulate that the terminal model is no longer applicable. However, O'Driscoll[73] has examined the effect of random errors on their plot, and has shown that deviations from linearity can be induced by random errors of only 3% in the data; consequently, care must be exercised in the interpretation. Nevertheless, kinetic schemes involving complex participation have been derived and applied with some success to certain systems. Early work[74–76] proposed an extreme mechanism, in which propagation was entirely by homopolymerization of the D–A complex which added to the chain in the appropriate head–tail orientation. This was not entirely satisfactory and an alternative approach, which postulates simultaneous propagation by the complex and free monomers, is now favoured. This is typified by the generalized model of Seiner and Litt[77] comprising eight propagation steps and an equilibrium constant, as shown in equations (31)–(39), where $\overline{M_AM_B}$ represents the complex which

$$\sim M_A\cdot + M_A \xrightarrow{k_{AA}} \sim M_AM_A\cdot \tag{31}$$

$$\sim M_A\cdot + M_B \xrightarrow{k_{AB}} \sim M_AM_B\cdot \tag{32}$$

$$\sim M_A\cdot + \overline{M_BM_A} \xrightarrow{k_{A\overline{BA}}} \sim M_AM_BM_A\cdot \tag{33}$$

$$\sim M_A\cdot + \overline{M_AM_B} \xrightarrow{k_{A\overline{AB}}} \sim M_AM_AM_B\cdot \tag{34}$$

$$\sim M_B\cdot + M_A \xrightarrow{k_{BA}} \sim M_BM_A\cdot \tag{35}$$

$$\sim M_B\cdot + M_B \xrightarrow{k_{BB}} \sim M_BM_B\cdot \tag{36}$$

$$\sim M_B\cdot + \overline{M_BM_A} \xrightarrow{k_{B\overline{BA}}} \sim M_BM_BM_A\cdot \tag{37}$$

$$\sim M_B\cdot + \overline{M_AM_B} \xrightarrow{k_{B\overline{AB}}} \sim M_BM_AM_B\cdot \tag{38}$$

$$M_A + M_B \rightleftharpoons \overline{M_AM_B} \text{ (or } \overline{M_BM_A}\text{)} \tag{39}$$

can approach the chain end in two orientations. Seiner and Litt had to define six reactivity ratios, and obtained a complex and rather intractable equation for the instantaneous copolymer composition which, as Pittman and Rounsefell[78] later demonstrated, can be solved using Marquandt's algorithm. They commented, however, that the data in the literature were insufficiently accurate to benefit from this detailed treatment.

Simplifying assumptions can be made but, alternatively, the reactivity ratios can be redefined, and this approach has been adopted by Cais *et al.*[79] and by Natansohn *et al.*,[80] although each group has offered different definitions. Cais and co-workers derived a simpler expression for the copolymer composition, equation (40), by defining a 'transition probability' $P$ for each propagation step, where $P$ was defined by equation (41). In this latter equation $x$ can be A, B, AB or BA, and a similar expression can be derived for $P_{Bx}$ when the radical end is $\sim M_B^{\bullet}$. The probabilities can be expressed in terms of the six reactivity ratios: $r_A = k_{AA}/k_{AB}$; $r_B = k_{BB}/k_{BA}$; $p_A = k_{A\overline{AB}}/k_{A\overline{BA}}$; $p_B = k_{B\overline{BA}}/k_{B\overline{AB}}$; $s_A = k_{A\overline{BA}}/k_{AB}$; and $s_B = k_{B\overline{AB}}/k_{BA}$.

$$\frac{d[M_A]}{d[M_B]} = \frac{(1-P_{AB})(P_{BA}+P_{B\overline{BA}})+(1-P_{BB})(P_{AB}+P_{A\overline{AB}})}{(1-P_{AA})(P_{BA}+P_{B\overline{BA}})+(1-P_{BA})(P_{AB}+P_{A\overline{AB}})} \tag{40}$$

$$P_{Ax} = \frac{\text{Rate of addition of substrate } x \text{ to } \sim M_A^{\bullet}}{\Sigma\ (\text{Rates of addition of all } x \text{ to } \sim M_A^{\bullet})} \tag{41}$$

In the approach used by Natansohn *et al.* the reactivity ratios are modifications of the Cais definitions, with $r_A$ and $r_B$ the same but with $\alpha_{AA} = s_A p_A$, $\alpha_{AB} = s_A$, $\alpha_{BA} = s_B$ and $\alpha_{BB} = s_B p_B$. This gives equation (42) for the copolymer composition and can also be solved by Marquandt's algorithm.

$$\frac{d[M_A]}{d[M_B]} = \frac{1+r_{AB}([M_A]/[M_B])+K[\overline{M_AM_B}](\alpha_{AA}\alpha_{BC}+\alpha_{BB}\alpha_{AC})+\alpha_{BC}([\overline{M_AM_B}]/[M_A])+(\alpha_{AA}+\alpha_{AC}+\alpha_{BB}r_{AB})([\overline{M_AM_B}]/[M_B])}{1+r_{BA}([M_B]/[M_A])+K[\overline{M_AM_B}](\alpha_{AA}\alpha_{BC}+\alpha_{BB}\alpha_{AC})+\alpha_{AC}([\overline{M_AM_B}]/[M_B])+(\alpha_{BB}+\alpha_{BC}+\alpha_{AA}r_{BA})([\overline{M_AM_B}]/[M_A])} \tag{42}$$

#### 22.6.1.4 *Complex-dissociation model*

A variation of the complex-participation model has emanated from the original suggestion by Tsuchida and Tomono[81] that complex disruption might take place during the propagation steps. Karad and Schneider[82] have extended this idea and formulated a complex-dissociation scheme. The constrained form of the model, in which one monomer does not homopropagate through either free monomer or complex, leads to the same expression for copolymer composition as the constrained form of the Seiner–Litt approach. The general model can also be used to account for non-dissociation of the complex, and thus also embraces the complex-participation model.

The model has been described by Hill *et al.*[83] who replaced equations (33), (34), (37) and (38) from the Seiner–Litt sequence by equations (43)–(46).

$$\sim M_A^{\bullet} + \overline{M_BM_A} \xrightarrow{k_{A\overline{BA}}} \sim M_B^{\bullet} + M_A \tag{43}$$

$$\sim M_A^{\bullet} + \overline{M_AM_B} \xrightarrow{k_{A\overline{AB}}} \sim M_A^{\bullet} + M_B \tag{44}$$

$$\sim M_B^{\bullet} + \overline{M_BM_A} \xrightarrow{k_{B\overline{BA}}} \sim M_B^{\bullet} + M_A \tag{45}$$

$$\sim M_B^{\bullet} + \overline{M_AM_B} \xrightarrow{k_{B\overline{AB}}} \sim M_A^{\bullet} + M_B \tag{46}$$

Using transition state probabilities, they formulated the copolymer composition as equation (47), where $r_A$, $r_B$, $s_A$ and $s_B$ are as defined previously, but now $q_A = k_{A\overline{AB}}/k_{AB}$ and $q_B = k_{B\overline{BA}}/k_{BA}$. It is possible to calculate 'best values' for the reactivity ratios and the complex formation constant, from either composition analysis or triad fraction analysis of the sequence distribution of the products.

$$\frac{d[M_A]}{d[M_B]} = \frac{([M_B]+s_A[\overline{M_AM_B}])(r_B[M_B]+[M_A]+s_B[\overline{M_AM_B}]+q_B[\overline{M_AM_B}])}{([M_A]+s_B[\overline{M_AM_B}])(r_A[M_A]+[M_B]+s_A[\overline{M_AM_B}]+q_A[\overline{M_AM_B}])} \tag{47}$$

### 22.6.2 Model Discrimination Using Composition Data

Model testing does depend on the accuracy of the composition analysis, and this must be considered when attempting to decide which model is most appropriate.

### 22.6.2.1 *Maleic anhydride–styrene system*

The most intensively studied alternating copolymer system is undoubtedly that involving MAH and S.[10,44,81,84–86] There is a strong donor–acceptor interaction which, when measured in hexane, has a formation constant $K = 0.34\ dm^3\ mol^{-1}$. Participation of the D–A complex is suggested by curvature in the Kelen–Tudos plot and the inability of the terminal model to fit the composition data. Tsuchida *et al.*[44,81] noted a maximum in the rate of copolymerization at 1:1 monomer compositions in solvents such as benzene and carbon tetrachloride, which disappeared when competitive donor solvents, such as tetrahydrofuran and acetone, were used. They suggested that this indicated complex participation in the propagation. Barb[53] applied the penultimate model to the system, and an exhaustive study by Dodgson and Ebdon[10,87] confirmed that this model was the most satisfactory description. The latter carried out reactions in a wide range of solvents and tested all three models, using the assumptions that MAH does not homopolymerize with either free monomer or complex and that there was no complexation with solvent. Both penultimate and complex-participation models gave better fits with the experimental data than the terminal model, as shown in Figure 11.

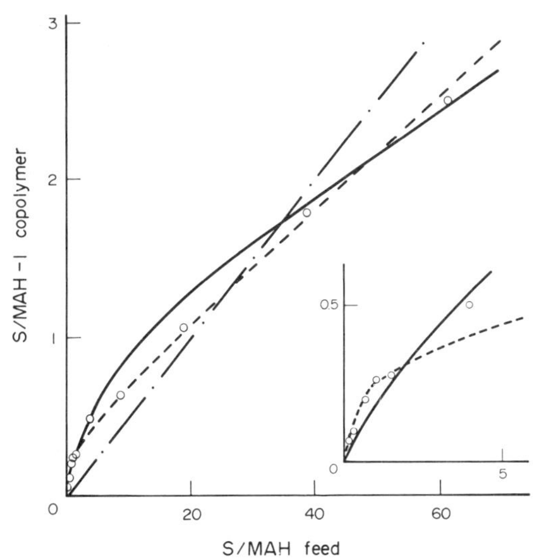


**Figure 11** 'Barb' plot for the copolymerization of S and MAH in methyl ethyl ketone at 333 K; ——, penultimate model: -----, complex participation model; and —·—·—, terminal model (reproduced from ref. 10)

As further diagnostic evidence, they added $ZnCl_2$ to the system and found this actually diminished the alternating tendency; dilution experiments also appeared to have little effect on the copolymer composition and, as this should have altered the D–A complex concentration, they concluded that the penultimate model was best. Recent work by Rätzsch and Steinert[85] supports this conclusion; they demonstrated that at high S:MAH ratios of 80:1, when nearly all the MAH will be complexed, addition of cyclohexyl radicals in reducing media gave the hydrogenated anhydride moiety before S could add onto the radical. This indicated that propagation did not involve addition of a tightly bound complex.

Farmer *et al.*[86] subjected Dodgson and Ebdon's data for bulk copolymerization to a rigorous data-fitting analysis using the complex-participation model without any constraint. It can be seen from Figure 12 that this gave an excellent fit, and the analysis gave a value of $K = 0.19\ dm^3\ mol^{-1}$ which is in close agreement with the experimental value $K = 0.25\ dm^3\ mol^{-1}$. This group also tested the complex-dissociation model without restriction[48] using a search method developed by Chandler.[88] The best fit data were located using a 'goodness of fit' parameter.

The copolymer compositions predicted by this method differ only slightly from the complex-dissociation model, and both provide a better fit for the experimental data than the penultimate model. Hill *et al.* are doubtful about this discrimination procedure, however, and have suggested that sequence distribution analysis may be more sensitive. Several groups have used $^{13}C$ NMR for sequence distribution analysis.[89–92] Roth *et al.*[91] found it difficult to measure the composition

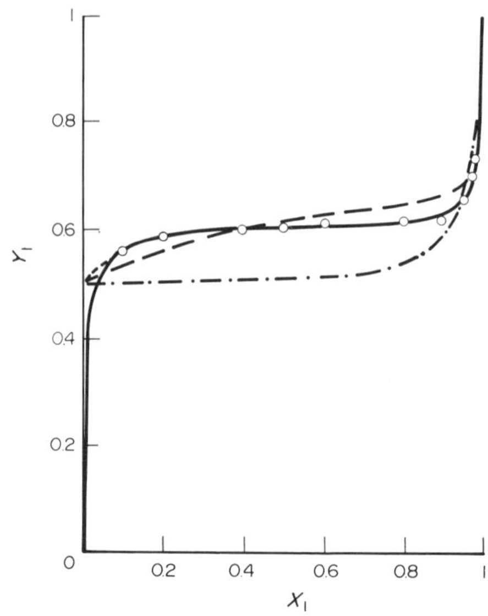


**Figure 12** Composition curves for the copolymerization of MAH and S ($Y_1$ = mole fraction of S in the copolymer, $X_1$ = mole fraction of S in the feed); ——, complex model with no restriction on the MAH sequences; – – – – –, complex model with only single MAH sequences; —·—·—, terminal model; and ———, penultimate model[10] (reproduced from ref. 86 with permission of Marcel Dekker)

accurately, but found the terminal model to be superior. However, Farmer *et al.*[86] have shown that, if the number fraction of single or double diad units in the copolymer is calculated for the various methods, the discrimination is large, as can be seen in Figures 13 and 14, and this could be the most useful diagnostic tool. From the preceeding series of conflicting conclusions, it is clear that this technique must be investigated further.

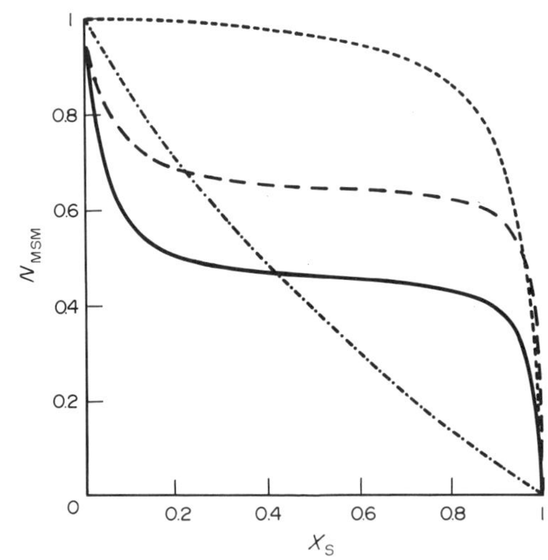


**Figure 13** Number fraction of single S sequences ($N_{MSM}$) in MAH–S copolymer ($X$ = mole fraction of S in the feed); ——, complex model; ———, complex dissociation model; —·—·—, penultimate model; and – – – – –, terminal model (reproduced from ref. 83 with permission from the American Chemical Society)

### 22.6.2.2 *Isoprene–2,4-dicyanobut-1-ene system*

While there is some doubt concerning the mechanism of the MAH–S system, this is not unusual. Cowie and Cree[8] have examined several systems which alternate, involving DCB and a number of donors. The data for DCB–IP were tested for the complex-participation model using the Cais

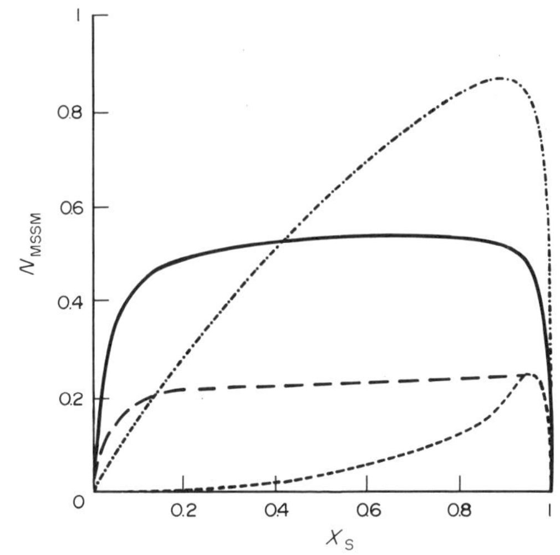


**Figure 14** Number fraction of double S sequences ($N_{MSSM}$) in MAH–S copolymer; legend as in Figure 13 (reproduced from ref. 83 with permission from the American Chemical Society)

method (equation 40), and a comparison of experimental data with both terminal and complex-participation models is shown in Figure 15. The penultimate model was also tested using equation (30) and is also compared with the terminal model in Figure 16. As both models appear to be better than the terminal model, another criterion for distinguishing between the complex and penultimate schemes must be used. This was assessed by means of an objective statistical $F$-test,[93] as suggested by Hill and O'Donnell, where the ratio $F$ is the residual sum of the squares for the models being compared as defined by equation (48).

$$F = \frac{(SS_i - SS_j)/(P_i - P_j)}{SS_i/(n - P_i)} \tag{48}$$

Here $P_i$ and $P_j$ are the number of variable parameters in each model, $n$ is the number of experimental data points and $SS$ is the residual sum of squares. The value of $F$ is then compared with a critical value found in statistical tables to determine whether there is a statistically significant improvement in the fit and that this is not just a result of using a larger number of variables in the model.

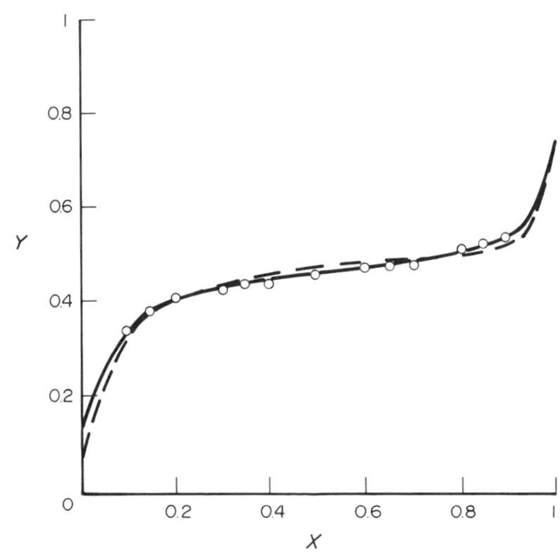


**Figure 15** Copolymer composition curves for the solution copolymerization of DCB and IP at 333 K ($Y$ = mole fraction of DCB in the copolymer and $X$ = mole fraction of DCB in the feed mixture); ○, experimental data; ——, complex participation model; and ---, terminal model

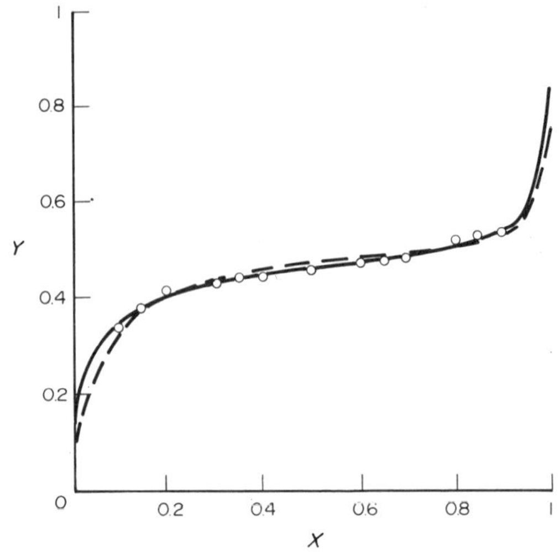


**Figure 16** Copolymer composition curves for the solution copolymerization of DCB and IP at 333 K ($Y$ = mole fraction of DCB in the copolymer and $X$ = mole fraction of DCB in the feed mixture); ○, experimental data; ——, penultimate model; and ----, terminal model

Application of the $F$-test to the data shown in Figures 15 and 16 led to the conclusion that there was no statistically significant improvement using the complex-participation model, but that the penultimate model was certainly a better description of the data than either of the others. Unfortunately, one of the calculated reactivity ratios had a value which was somewhat unrealistic, and the apparent improvement could be affected by the diene propagating in more than one configuration. Thus, while the $F$-test may not be wholly conclusive, it is worth investigating.

Whenever possible, composition analysis should be supplemented by kinetics studies.

### 22.6.3 Polymerization Kinetics

The kinetic approach offers an alternative way of determining the extent, if any, of complex participation in the propagation. Two methods are worthy of consideration.

#### *22.6.3.1 Shirota method*

A simplified complex-participation model was outlined by Shirota and co-workers[94] and is given in equations (49)–(54). However, it has been observed that in certain systems, such as MAH–divinyl ether, the termination reaction seems to be a first order reaction and does not involve significant chain coupling; therefore the monomolecular termination steps in equation (55) were proposed.

$$\left.\begin{array}{lcl} I & \longrightarrow & 2R\cdot \\ R\cdot + M_A & \longrightarrow & \sim M_A\cdot \\ R\cdot + M_B & \longrightarrow & \sim M_B\cdot \\ R\cdot + \overline{M_AM_B} & \longrightarrow & \sim M_B\cdot \\ R\cdot + \overline{M_BM_A} & \longrightarrow & \sim M_A\cdot \end{array}\right\} \text{Initiation} \qquad (49)$$

$$\sim M_A\cdot + M_B \xrightarrow{k_{AB}} \sim M_B\cdot \qquad (50)$$

$$\sim M_B\cdot + M_A \xrightarrow{k_{BA}} \sim M_A\cdot \qquad (51)$$

$$\sim M_A\cdot + \overline{M_BM_A} \xrightarrow{k_{A\overline{BA}}} \sim M_A\cdot \qquad (52)$$

$$\sim M_B\cdot + \overline{M_AM_B} \xrightarrow{k_{B\overline{AB}}} \sim M_B\cdot \qquad (53)$$

$$\left.\begin{array}{lll}\sim\!\sim M_A\cdot + \sim\!\sim M_A\cdot & \xrightarrow{k_{tAA}} & \text{polymer} \\ \sim\!\sim M_A\cdot + \sim\!\sim M_B\cdot & \xrightarrow{k_{tAB}} & \text{polymer} \\ \sim\!\sim M_B\cdot + \sim\!\sim M_B\cdot & \xrightarrow{k_{tBB}} & \text{polymer}\end{array}\right\}\ \text{Bimolecular termination} \tag{54}$$

$$\left.\begin{array}{lll}\sim\!\sim M_A\cdot & \xrightarrow{k_{tA}} & \text{polymer} \\ \sim\!\sim M_B\cdot & \xrightarrow{k_{tB}} & \text{polymer}\end{array}\right\}\ \text{Monomolecular termination} \tag{55}$$

It was then considered that the overall rate of polymerization $R_p$ was the sum of the rates for the free monomer $R_p(\mathrm{f})$ and the complex $R_p(\mathrm{DA})$. If $X = [M_B]/[M_A]$, $R_i$ is the rate of initiation and $K$ is the formation constant of the D–A complex, then, under steady-state conditions, equations (56) to (58) are obtained.

$$\frac{R_p}{[M_A]} = A(X)F(X)[M_A] + A(X) \tag{56}$$

$$A(X) = \frac{2k_{BA}k_{AB}XR_i^{\frac{1}{2}}}{[k_{tAA}k_{BA}^2 + 2k_{tAB}k_{BA}k_{AB}X + k_{tBB}k_{AB}^2X^2]^{\frac{1}{2}}} \tag{57}$$

$$F(X) = K\left\{\frac{k_{A\overline{BA}}}{k_{AB}} + \frac{k_{B\overline{AB}}}{k_{BA}}X\right\} \tag{58}$$

By measuring $R_p$ for various concentrations of $M_A$ at a fixed feed ratio $X$, a plot of $R_p/[M_A]$ against $[M_A]$ should be linear and give $A(X)$ and $F(X)$. Hence, the initial rates of polymerization $R_p(\mathrm{f})$ and $R_p(\mathrm{DA})$ follow from equations (59) and (60).

$$R_p(\mathrm{f}) = A(X)[M_A] \tag{59}$$

$$R_p(\mathrm{DA}) = A(X)F(X)[M_A]^2 \tag{60}$$

For the monomolecular processes, the corresponding relations are equations (61) and (62).

$$\frac{R_p}{[M_A]} = B(X)F(X)[M_A] + B(X) \tag{61}$$

$$B(X) = \frac{2k_{AB}k_{BA}XR_i}{k_{tA}k_{BA} + k_{tB}k_{AB}X} \tag{62}$$

Copolymerization of the NVC–diethyl fumarate (DEF),[95] MAH–alkyl phenylvinyl ethers,[96] vinyl acetate–ethyl α-cyanocinnamate,[70] MAH–phenyl vinyl ether[97] and MAH–α-MS[98] systems have been successfully treated using this approach. These studies show that there is participation of both the free monomers and, to a lesser extent, the D–A complex in the NVC–DEF system, but much greater participation of the D–A complex in the MAH–α-MS system, as can be seen from Figure 17.

The MAH–S system was recently reexamined by Deb and Meyerhoff,[99] who added the homopolymerization steps to the Shirota mechanism. They developed a more complicated rate equation which is not amenable to a graphical solution unless simplifications are made. Their analysis does not separate out the individual rates for the complex and free monomers, but they concluded that the reaction was best described by incorporating complex-participation steps in the terminal model. This is consistent with the analysis based on their composition data.[100]

The theory that both the complex and free monomers contribute to the propagation is gaining support from kinetic evidence. This is also the case when an alternative kinetic analysis is used, as outlined in the following section.

### 22.6.3.2 Georgiev–Zubov model

A bimolecular termination model, similar to that proposed by Shirota, was used by Georgiev and Zubov,[101] who defined three reactivity ratios: $\beta_A = k_{ABA}/k_{AB}$, $\beta_B = k_{BAB}/k_{BA}$ and $\beta_C = k_{AB}/k_{BA}$, and

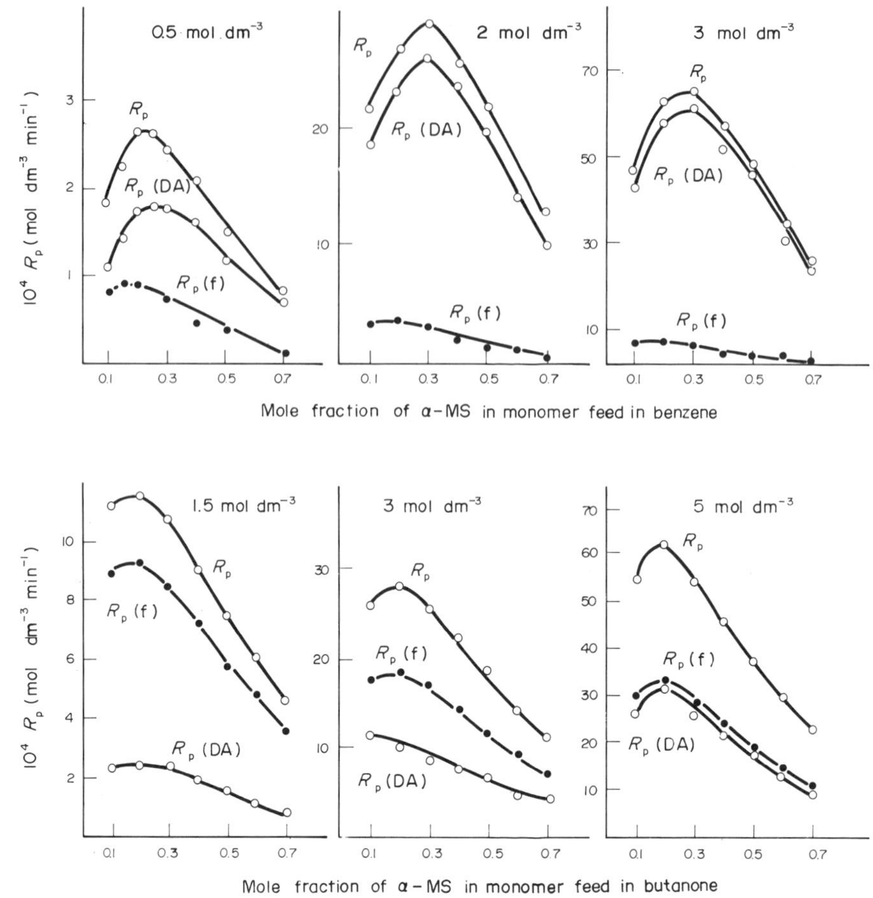


**Figure 17** Relative participation of the free monomer and the donor–acceptor complex in the copolymerization of α-MS and MAH at various feed concentrations and two solvents (reproduced from ref. 98 with permission from Wiley)

derived equation (63) to analyze the rate of copolymerization.

$$R_p = k_{AB}\left\{\frac{R_i}{k_t}\right\}^{\frac{1}{2}}\left\{\frac{([M_A][M] - [M_A] + 2\beta_A[M_AM_B] + ([M] - [M_A])([M] + 2\beta_B[M_AM_B])}{\beta_C([M] - [M_A]) + [M_A]}\right\} \tag{63}$$

Here [M] is the total monomer concentration and $k_t$ is the total rate constant for all the termination reactions. Large values of $\beta_A$ and $\beta_B$ indicate substantial participation of the D–A complex, whereas small or zero values would suggest a free monomer mechanism. This model has been used by Arnaud *et al.*,[102] who examined the MAH–vinyl acetate system in benzene and obtained values of $\beta_A = 7.0$ and $\beta_B = 0$, a result which was interpreted as indicating participation of both the complex and free monomer in the propagation. This interpretation has been supported by Fujimori *et al.* for MAH–vinyl acetate in methyl ethyl ketone[40] and MAH–isobutyl vinyl ketone in chloroform.[103]

In neither system was the maximum rate of polymerization observed to coincide with a 1:1 feed composition; rather, it moves towards this value, as shown in Figure 18, as the total monomer concentration increases. This is consistent with some complex participation, because the concentration of the complex is expected to grow as the total monomer concentration rises.

If the rate is seen to be a maximum at monomer concentrations $[M_A]$ and $[M_B]$, then the conditions for the maximum are as expressed in equation (64), where $a = 2K[M_A][M_B]$; $b = K[M_B]^2 - [M_A][M_B]$; $c = [M_B]$; $d = [M_B]^{-1}K[M_A]^2([M_B] - [M_A])$; $e = -2K[M_A]^2$; and $f = [M_A]^2/[M_B]$. Equation (64) can be solved for $\beta_A$, $\beta_B$ and $\beta_C$ by determining $[M_A]/[M_B]$ for three different total monomer concentrations. The values of $\beta$ obtained for the two systems were $\beta_A = 35$, $\beta_B = 2.2$ and $\beta_A - 65.5$, $\beta_B - 4.42$, respectively, thereby implying substantial complex participation in both systems.

$$a\beta_A\beta_C + b\beta_B\beta_C + c\beta_C + d\beta_A + e\beta_B = f \tag{64}$$

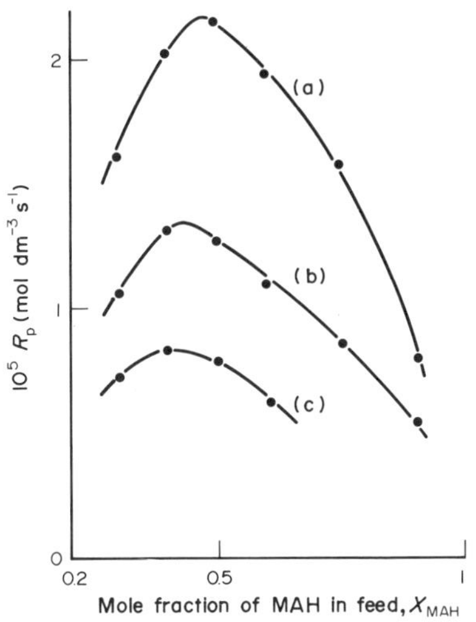


**Figure 18** Initial rate ($R_p$) of the copolymerization of VA with MAH in methyl ethyl ketone at 323 K: (a) [VA + MAH] = 2.25 mol dm$^{-3}$, (b) [VA + MAH] = 1.50 mol dm$^{-3}$ and (c) [VA + MAH] = 0.75 mol dm$^{-3}$ (reproduced from ref. 40 with permission from Springer)

## 22.7 FREE RADICAL COPOLYMERIZATION IN THE PRESENCE OF A LEWIS ACID

When a Lewis acid is added to a comonomer mixture, both binary and ternary complexes can form and may play a part in the propagation reactions. Three main mechanisms for alternation have been proposed to explain the behaviour of these systems; they are (i) ternary complex homopolymerization, (ii) enhanced cross propagation and (iii) a radical complex mechanism, in which propagation involves a complex of the two monomers which is formed at the growing chain ends.

One way of attempting to discriminate among these mechanisms is to establish the rate of polymerization as a function of monomer composition, and quite detailed kinetic studies have been carried out on selected systems. As there are distinct differences in the results reported for different systems, several examples will be summarized under the heading of the mechanism attributed to the copolymerization.

### 22.7.1 Ternary Complex Mechanism

The formation of a ternary complex involving the donor–acceptor and Lewis acid has been discussed in Section 22.3.2. Homopolymerization of this ternary complex is one method of this generating an alternating copolymer.

#### *22.7.1.1 Methyl methacrylate–styrene–$SnCl_4$ system*

Hirai and co-workers[104–106] made a close study of the copolymerization of MMA and S in the presence of $SnCl_4$ using both thermal and photoinitiation. This is not a reaction which undergoes spontaneous initiation. The solvents used were toluene at 223 K and 1,2-dichloroethane at 253 K; conversions were kept below 7.5%. In toluene (Tol) solution, 11 complex equilibria were postulated and, as the formation constants of several of these could be measured (see Table 6), the concentrations of binary and ternary complexes such as $SnCl_4$MMA, $SnCl_4(MMA)_2$, ($SnCl_4(MMA)_2$---S) and ($SnCl_4(MMA)_2$---Tol) could be estimated. As concern with the possible participation of the ternary complex was paramount, their kinetic analysis focussed on this aspect, and they demonstrated that the initial rates of copolymerization were proportional to ($SnCl_4$MMA---S) and ($SnCl_4$MMA---S)$^{3/2}$ for the thermal and photoinitiated reactions, respectively. Scheme 4 was proposed, which gives a classical free-radical copolymerization rate equation if it is assumed that ($SnCl_4$MMA---S) is the monomeric species. This gave equation (65), where $x = 1.0$ for thermal initiation with $I = fk_d/2$, and 3/2 for the photoinitiation with $I = \Phi I_0$ (the product of the quantum yield and the initial light intensity).

$$\text{Initiation}\quad \begin{cases} \mathrm{SnCl_4MMA{-}{-}{-}{-}S} \xrightarrow{h\nu} \mathrm{R\cdot} \\ \mathrm{R\cdot + SnCl_4MMA{-}{-}{-}{-}S} \longrightarrow \mathrm{R_1\cdot} \end{cases}$$

$$\text{Propagation}\quad \begin{cases} \mathrm{R_1\cdot + SnCl_4MMA{-}{-}{-}{-}S} \xrightarrow{k_p} \mathrm{R_2\cdot} \\ \mathrm{R_n\cdot + SnCl_4MMA{-}{-}{-}{-}S} \xrightarrow{k_p} \mathrm{R_{n+1}\cdot} \end{cases}$$

$$\text{Termination}\quad \mathrm{R_m\cdot + R_n\cdot} \xrightarrow{k_t} \text{polymer}$$

**Scheme 4**

$$R_p = k_p\left(\frac{I}{k_t}\right)^{\frac{1}{2}}[\mathrm{SnCl_4MMA{-}{-}{-}{-}S}]^x \tag{65}$$

The copolymers had an alternating structure irrespective of the initial feed ratio. The data were also tested using equations derived from a cross propagation and a complex-radical mechanism, but gave poor fits with both. Consequently, it was proposed that homopolymerization of the ternary complex was responsible for the alternating structure. It was concluded that the reactivity of this ternary complex was overwhelmingly larger than that of either the free monomers or any of the other complexes present.

While this mechanism seems to be the best explanation of the data, it may be an unusual system. Reference to Table 9 shows that the concentration of the ternary complex is quite large relative to the free monomers, and perhaps larger than is normally obtained in other systems. Such conditions could well favour propagation predominantly through the ternary complex species.

**Table 9** Concentrations of Free Monomers and Ternary Complex for the MMA–S–$SnCl_4$ System in 1,2-Dichloroethane at 253 K[108]

| *Initial concentration* ($\mathrm{mol\,dm^{-3}}$) | | | *Ternary complex* | *Free monomers* ($\mathrm{mol\,dm^{-3}}$) | |
|---|---|---|---|---|---|
| *MMA* | *S* | *$SnCl_4$* | ($\mathrm{mol\,dm^{-3}}$) | *MMA* | *S* |
| 2.40 | 1.00 | 1.00 | 0.41 | 0.38 | 0.59 |
| 2.40 | 2.00 | 1.00 | 0.74 | 0.35 | 1.25 |
| 4.80 | 1.00 | 2.00 | 0.60 | 0.77 | 0.40 |

### 22.7.1.2 *Butadiene with methyl methacrylate or acrylonitrile in the presence of $AlEtCl_2$–$VOCl_3$*

The ternary complex mechanism has also been proposed by Furukawa *et al.*[107] for reactions involving butadiene (Bd) with either MMA or AN to which alkyl aluminum halides have been added. The rate of the reaction could be accelerated by addition of $VOCl_3$, because the redox reaction between $AlEtCl_2$ and $VOCl_3$ produces radicals.

NMR analysis showed that strongly alternating structures were formed, and that the diene was almost exclusively in the *trans*-1,4 configuration. It was observed that the molecular weight of the copolymer increased with reaction time, which seemed to suggest a step-growth mechanism was operative. This is difficult to reconcile with the expected free radical mechanism, and Bamford[108] considers it more likely that the efficiency of the initiating species decays with time and that the chain length increases as the effective radical concentration drops.

The kinetic behaviour can be summarized in equations (66) and (67), where the concentration of the active species [C*] is given by equation (68). Furukawa found that the kinetic observations could be explained on the basis of equations (69) to (71). First, a complex equilibrium is established, as in equations (69) and (70) where A is the acrylic monomer. The concentration of the complex is then given by equation (72) where $[AlEtCl_2]_0$ is the initial concentration. It follows that, for $K_2 \gg K_1$, equation (72) reduces to equation (73) and, when $K_1 \gg K_2$ with $n=3$, it reduces to equation (74).

$$R_p \propto [\mathrm{AlEtCl_2}]^{3/2}[\mathrm{VOCl_3}]^{\frac{1}{2}}[\mathrm{MMA}]^0[\mathrm{Bd}]^0 \tag{66}$$

$$R_p \propto [\mathrm{AlEtCl_2}]^{3/2}[\mathrm{VOCl_3}]^{\frac{1}{2}}([\mathrm{Bd}]^0/[\mathrm{AN}]^2) \tag{67}$$

$$[C^*] \propto [AlEtCl_2]^{\frac{1}{2}}[VOCl_3]^{\frac{1}{2}} \quad (68)$$

$$AlEtCl_2 + nA \overset{K_{n1}}{\rightleftharpoons} AlEtCl_2A_n \quad (69)$$

$$AlEtCl_2 + A + Bd \overset{K_2}{\rightleftharpoons} AlEtCl_2A{-}{-}{-}Bd \quad (70)$$

$$\text{Propagation} \sim\!\underset{\substack{| \\ EtAlCl_2}}{Bd}{-}A\cdot + \underset{\substack{| \\ EtAlCl_2}}{Bd}{-}{-}{-}A \rightarrow \sim\!Bd{-}A{-}\underset{\substack{| \\ EtAlCl_2}}{Bd}{-}\underset{\substack{| \\ EtAlCl_2}}{A}\cdot \quad (71)$$

$$[AlEtCl_2A{-}{-}{-}Bd = \frac{K_2[AlEtCl_2]_0[A][Bd]}{1 + K_1[A]^n + K_2[A][Bd]} \quad (72)$$

$$[AlEtCl_2A{-}{-}{-}Bd] = [AlEtCl_2][A]^0[Bd]^0 \quad (73)$$

$$[AlEtCl_2A{-}{-}{-}Bd] = (K_2/K_1)[AlEtCl_2][Bd]^0/[AN]^2 \quad (74)$$

The data did not fit the cross propagation or radical complex mechanisms but complied more closely with the idea of a ternary molecular complex propagation in both systems.

Bamford[108] has expressed some reservations about this interpretation. In particular, the observed maximum in the copolymerization rate is difficult to reconcile with zero order monomer dependences. There may also be formation of a Diels–Alder adduct, which could complicate the kinetics, and, for conditions implying a large $K_2$, nearly all of the $AlEtCl_2$ would have to enter into complex formation, which is not substantiated by experiment. Nevertheless, using current data, the ternary complex mechanism is still the most compelling.

### 22.7.2 Cross Propagation

This is perhaps the simplest mechanism, where it is assumed that addition of the Lewis acid increases the differences in monomer polarities and thereby enhances the cross propagation reactions.

#### *22.7.2.1 Methyl acrylate–diene mixtures with $Et_3Al_2Cl_3$*

Copolymerizations involving dienes (butadiene and IP) and methyl acrylate (MA) complexed with triethylaluminum sesquichloride ($Et_3Al_2Cl_3$) in toluene at 298 K were examined by Bamford and Han.[109] Photoinitiation using $Mn_2(CO)_{10}$ in carbon tetrachloride yielded strongly alternating copolymers, with the diene units predominantly in the *trans*-1,4 configuration. The reactions were complicated at low MA concentrations by the formation of Diels–Alder adducts (Scheme 5), but this side reaction was negligible when $[MA] > 0.1\ mol\,dm^{-3}$. The kinetic data can be summarized in equation (75) and the rate of adduct formation is given by equation (76), where [D] is the diene concentration. This shows that adduct formation can be minimized by having a low concentration of diene and a rapid initiation process.

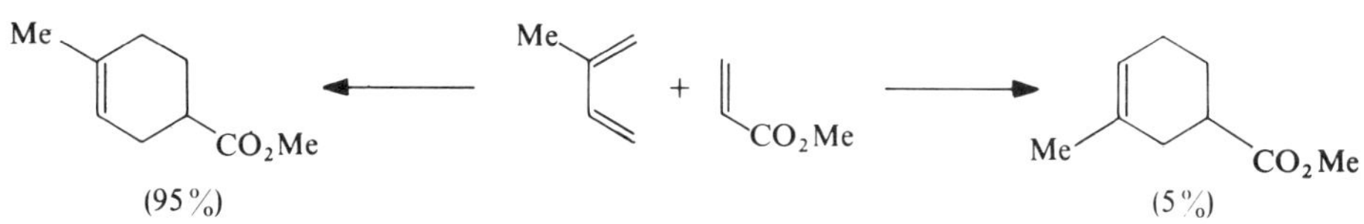


**Scheme 5**

$$R_p = k[MA{-}Et_3Al_2Cl_3][D]^0[MA]^0_{free} \quad (75)$$

$$k_A[MA{-}Et_3Al_2Cl_3][D] \quad (76)$$

Bamford and Han estimated that there was a 40- to 80-fold enhancement of the cross propagation rate constants in the presence of $Et_3Al_2Cl_3$ compared with the reaction in the absence of a Lewis acid. Their kinetic analysis favoured the cross propagation mechanism.

#### *22.7.2.2 Methyl methacrylate–isoprene–$Et_3Al_2Cl_3$ system*

Evidence for a cross propagation mechanism in the MMA–IP–$Et_3Al_2Cl_3$ system has been obtained from a comprehensive study by Ebdon and Gabbott.[110] Four complex-forming equilibria were considered, the binary complex between MMA and the Lewis acid and three ternary complexes, the most important of which, according to the formation constant, was ($Et_3Al_2Cl_3$MMA ---IP). The kinetic data can be summarized by equation (77).

$$R_p = k[Et_3Al_2Cl_3]^{1.35}[MMA]^{0.26}[IP]^{0.03} \tag{77}$$

An increase in [IP] led to an increase in the concentration of the ternary complex but gave no corresponding increase in the rate. This, together with the almost zero order dependence of [IP] in the rate equation, precludes a major participation of the ternary complex in the reaction, which is better described by a cross propagation mechanism.

While there are similarities to the MMA–S–$SnCl_4$ system (Section 22.7.1.1), in so far as thc rate is proportional to the ternary complex concentration, there is a significant difference in that the concentration of ternary complex in this system was low, and probably never exceeded 6%.

### 22.7.3 Radical Complex Mechanism

This mechanism is perhaps the most controversial and there are suggestions that it is really a mixed radical complex and enhanced cross propagation reaction.

#### *22.7.3.1 Methyl acrylate–styrene–$Et_3Al_2Cl_3$ system*

The methyl acrylate (MA)–S system was originally investigated by Hirooka[111] in the presence of triethylaluminum sesquichloride in toluene and 1,2-dichloroethane solutions. The kinetic data from a series of experiments can be summarized by equation (78).

$$R_p = k[MA\text{–}Et_3Al_2Cl_3]^{1.0}[MA]^0[S]^0 \tag{78}$$

While Hirooka considered that the presence of an MA–Lewis acid complex had been established, no evidence for the existence of ternary complex formation could be obtained from either cryoscopic measurements or UV spectroscopy. Neither were any chain transfer reactions detected when carbon tetrachloride was added. Replacement of S by allyl-type monomers such as propylene or isobutene all gave alternating copolymers in good yield, suggesting that degradative chain transfer reactions were also absent. These observations suggested that the mechanism of copolymerization was different from conventional reactions, and led to the rejection of a cross propagation and a ternary complex mechanism. The radical complex mechanism, Scheme 6, was proposed instead.

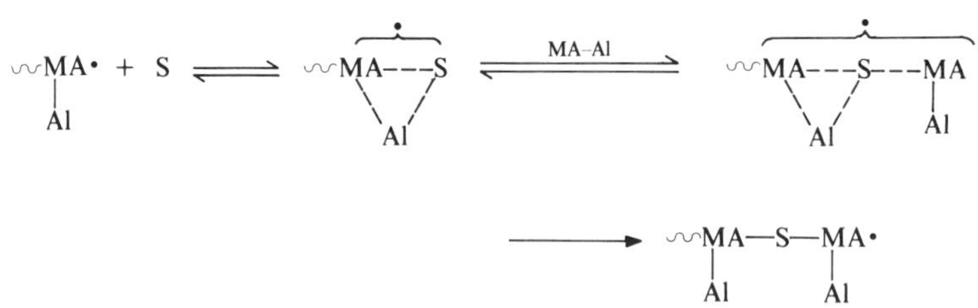

**Scheme 6**

Bamford and Malley[112,113] reexamined the system using well-characterized initiating systems. While in agreement with some parts of Hirooka's work, they differed in several points. They found that, for low S concentrations (< 0.2 mol $dm^{-3}$), the kinetics changed from zero to first order in [S]. Their data also suggested the absence of a cross termination reaction and the presence of chain transfer reactions when carbon tetrabromide was added. The latter effect was reduced by Lewis acid complexation of the monomer. The cross propagation rate coefficients were not only larger than the homopropagations but were also larger than the cross propagation rate constants in the absence of a

Lewis acid. The observed enhancements were 140 times for the ~MA–S• + MA → ~MA• reaction and 11 times for the ~S–MA• + S → ~S• reaction. The rise in the former reaction was most important, as the latter is already quite fast in the absence of a Lewis acid. Bamford and Malley concluded that a cross propagation reaction was the most probable mechanism.

A kinetic study of MA–S in the presence of $Et_2AlCl$ has been carried out by Zubov *et al.*,[114] who based their analysis on the Shirota complex-participation scheme in which two of the propagation reactions involved addition of the ternary complex (AlMA – – – – S). They concluded that propagation was by a mixed mechanism, with the dominant contribution coming from monomer addition to S• and to complexed ~MA• radical end groups, with only a minor contribution from ternary complex addition.

It is clear from the foregoing that no single mechanism will cover all copolymerizations carried out in the presence of a Lewis acid. Indeed, reaction conditions may be important in this respect, as it appears that these can influence the extent to which a species may participate in the reaction. The suggestion that a mixed mechanism may occur is worth exploring further but, if operative, makes comparisons among systems difficult.

## 22.8 ALTERNATION IN THE PRESENCE OF ZIEGLER-TYPE CATALYSTS

### 22.8.1 General Observations

Catalysts formed by mixing vanadium or titanium compounds with alkyl aluminum compounds can be used to prepare alternating structures of propylene and dienes. The feasibility of this reaction was first demonstrated by Furukawa and co-workers,[115–117] who noted that the catalytic activity was dependent on the temperature of preparation of the catalyst. When this was aged at room temperature, homopolymerization of the diene was obtained, but if the catalytic mixtures were made at −70 °C, alternating copolymers were formed. The original studies involved propylene, IP and butadiene, but this has now been extended to included other substituted alkenes and substituted dienes. Acetylene has proved more difficult to incorporate, but limited success has been achieved using propyne and butadiene.[118] However, the inclusion of a substituent introduces the possibility of steric hindrance during propagation, and the alternating tendency for the alkenes decreases in the order $CH_2{=}CH_2 > CH_2{=}CHMe > CH_2{=}CHEt > CH_2{=}CHCH_2CHMe_2 > CH_2{=}CHCMe_3$ whereas for the dienes the order is $CH_2{=}CHCH{=}CH_2$ > $CH_2{=}CHCMe{=}CH_2$ > *trans*-$CH_2{=}CHCH{=}CHMe$ > *cis*-$CH_2{=}CHCH{=}CHMe$ > $CH_2{=}CHCH{=}CMe_2$ > $CH_2{=}CMeCMe{=}CH_2$ > $MeCH{=}CHCH{=}CHMe$. Thus an alternating structure is obtained if 4-methyl-1,3-pentadiene is copolymerized with ethylene, but a block copolymer results if propylene is used.

### 22.8.2 Kinetics

Furukawa's group[119,120] have studied the polymerization of propylene (PP) and butadiene (Bd) using catalysts prepared from $VOCl_3/AlBu^i_3$ and $TiCl_4/AlBu^i_3$ at low temperatures in toluene. The intrinsic viscosity of the product was found to increase steadily throughout the course of the reaction, as shown in Figure 19, and this was interpreted as indicating a step growth mechanism with slow chain transfer. The reaction was first order with respect to monomer, and the magnitude of the derived rate constants, listed in Tables 10 and 11, show that they are comparable to those for reactions using normal Ziegler catalysts, except for a noticeably higher value for the chain transfer

**Table 10** Variation of the Rate Constants and Concentration[a] of the Active Species with Temperature of Polymerization for the System Propylene [3.71]/Butadiene [3.71]/$TiCl_4$ [$6.9\times10^{-3}$]/$AlBu^i_3$ [$1.57\times10^{-2}$]

| *Temperature* (°C) | $10^4K$ (min$^{-1}$) | [$C^*$] (mol dm$^{-3}$) | $10^3(k_f/k_p)$ | $k_p$(dm$^3$ mol$^{-1}$ min$^{-1}$) | $10^2k_f$ (dm$^3$ mol$^{-1}$ min$^{-1}$) |
|---|---|---|---|---|---|
| −30 | 26.4 | 1.56 | 2.57 | 16.9 | 4.34 |
| −40 | 12.5 | 1.32 | 2.04 | 9.45 | 1.93 |
| −50 | 3.64 | 1.03 | 1.58 | 3.53 | 0.54 |

[a] Concentrations in mol dm$^{-3}$. Reproduced with permission from *Rubber Chem. Technol.*, 1978, **51**, 600.

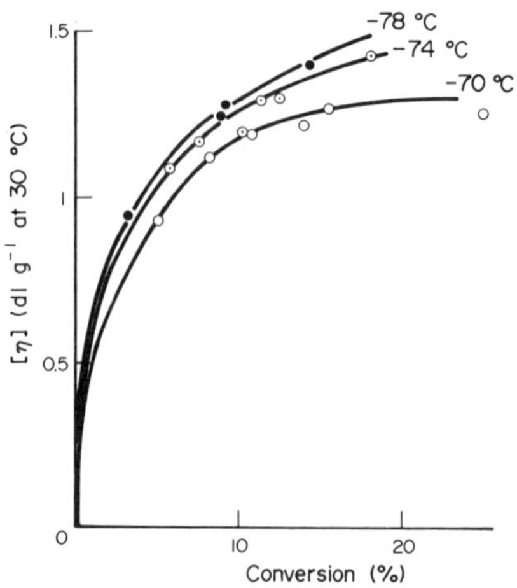


**Figure 19** Relationship between the intrinsic viscosity $[\eta]$ of alternating PP–Bd copolymers and conversion: $[PP]=2.76\ mol\,dm^{-3}$, $[Bd]=2.76\ mol\,dm^{-3}$, $[VOCl_3]=3.89\times10^{-3}\ mol\,dm^{-3}$ and $[Bu^i{}_3Al]=1.95\times10^{-2}\ mol\,dm^{-3}$; solvent = toluene (reproduced with permission from *Rubber Chem. Technol.*, 1978, **51**, 600.)

**Table 11** Effect of Temperature on the Concentration[a] of the Active Species and Rate Constants for the Polymerization of Propylene [2.76]/Butadiene [2.76]/$VOCl_3$ $[3.89\times10^{-3}]$/$AlBu^i{}_3[1.95\times10^{-2}]$

| *Temperature* (°C) | $10^4 K$ ($min^{-1}$) | $[C^*]$ ($mol\,dm^{-3}$) | $10^3\,(k_f/k_p)$ | $k_p$ ($dm^3\,mol^{-1}\,min^{-1}$) | $10^2\,k_f$ ($dm^3\,mol^{-1}\,min^{-1}$) |
|---|---|---|---|---|---|
| −70 | 9.55 | 1.56 | 2.64 | 6.12 | 1.62 |
| −74 | 6.61 | 1.56 | 2.12 | 4.24 | 0.899 |
| −78 | 3.74 | 1.50 | 1.88 | 2.49 | 0.468 |

[a] Concentrations in $mol\,dm^{-3}$. Reproduced with permission from *Rubber Chem. Technol.*, 1978, **51**, 600.

process; $k_f$ is a rate coefficient for chain transfer. It was reasoned that this might explain why a low temperature had to be used if an alternating structure was to be obtained.

They also found that the best way to analyze the data was to invoke a two step propagation, one step involving the addition of butadiene to the propylene end ($k_{PB}$) and the other in which propylene added to the butadiene end group ($k_{BP}$). The rate constants were estimated using equation (79), where $[C^*]$ is the concentration of active chain ends and $K_B$ and $K_P$ are coordination constants for

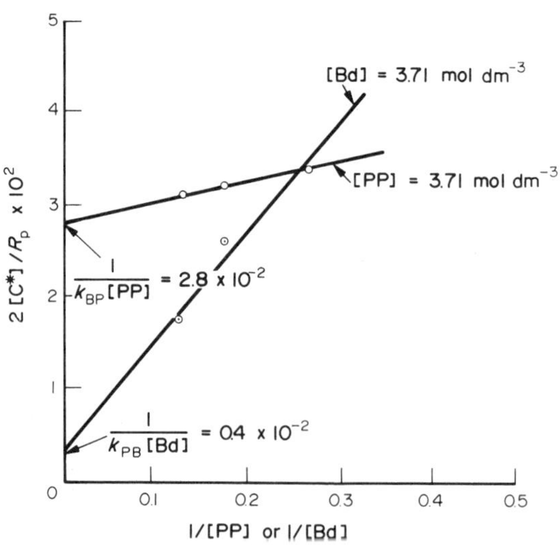


**Figure 20** Relationship between $2[C^*]/R_p$ and $1/[PP]$ or $1/[Bd]$, as given by equation (80) (reproduced with permission from *Rubber Chem. Technol.*, 1978, **51**, 600)

butadiene and propylene, respectively. Consideration of steady-state conditions leads to equation (80), and, from the derived data for conditions of either constant butadiene or constant propylene, the constants ($k_{BP}K_P$) and ($k_{PB}K_B$) can be estimated, as shown in Figure 20. The different values suggest that there are different reactivities at the two types of chain end.

$$R_p = k_{PB}K_B[C^*][Bd] + k_{BP}K_P[C^*][PP] \tag{79}$$

$$2[C^*]/R_p = (k_{PB}K_B[Bd])^{-1} + (k_{BP}K_P[PP])^{-1} \tag{80}$$

The alternating structure of the products was established using NMR and ozonolysis. A typical 220 MHz NMR spectrum is shown in Figure 21; the assignments are summarized in Table 12.

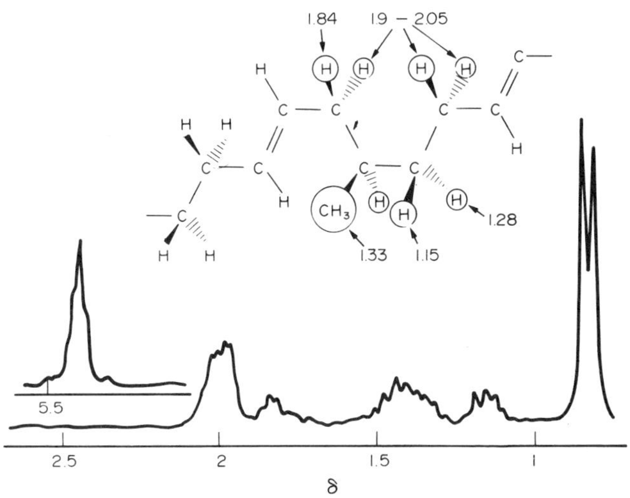


**Figure 21** 220 MHz NMR spectrum of butadiene–propylene alternating copolymer (reproduced from T. Suzuki, Y. Takegami, J. Furukawa and R. Hirai, *J. Polym. Sci., Part B*, 1971, **9**, 931, with permission from Wiley)

**Table 12** 220 MHz NMR Spectrum Data for a Propylene (PP)–Butadiene (Bd) Alternating Copolymer[117]

| δ (p.p.m.) | *Assignment* |
|---|---|
| 0.85 | $CH_3$ (PP) |
| 1.15 | $CH_2$ (PP) |
| 1.28 | $CH_2$ (PP) |
| 1.45 | CH(PP) |
| 1.73 | $CH_3$(end group) |
| 1.84 | $CH_2$(Bd) |
| 1.90–2.05 | $CH_2$ (Bd) |
| 2.65 | $CH_2$(end group) |
| 4.68 | $=CH_2$ |
| 5.45 | =CH (Bd) |

Ozonolysis supports the idea of addition according to equation (81), which illustrates the IP–propylene propagation.

$$\overset{\delta+}{C}H_2{=}CH\underset{|}{\overset{Me}{C}}{=}\overset{\delta-}{C}H_2 + {}^{\delta+}\overset{Me}{CH}{=}\overset{\delta-}{C}H_2 \rightarrow \sim\!\!\!\sim\!(CH_2CH{=}\underset{Me}{C}CH_2\underset{Me}{C}HCH_2)_n\!\sim\!\!\!\sim \tag{81}$$

### 22.8.3 Mechanism of Alternation

There is no evidence to support monomer complex involvement in the alternating copolymerization of alkenes or dienes, and so a cross propagation mechanism seems most likely. It is suggested that the transition metal in the catalyst may control monomer addition in one of two possible ways: (i) there could be unidentate and bidentate coordination of the different incoming monomers, or (ii) selective coordination may be controlled by the steric restrictions at the growing end of the chain. The former mechanism is favoured by Furukawa and is outlined in Scheme 7 for IP and propylene.

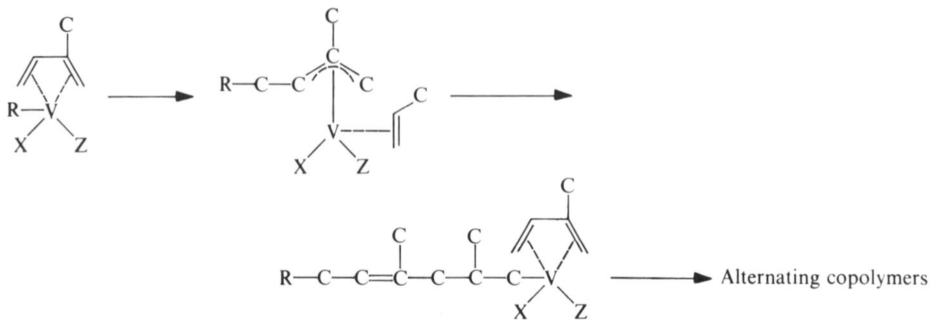


**Scheme 7**

The growing chain end, therefore, has two types of structure; either a propylene terminal unit or an IP terminal unit. The former coordinates only one site on the vanadium, whereas the latter exists as a $\pi$-allylic form and occupies two coordination sites on the metal. Thus, when IP is the terminal unit it is coordinated by two sites, leaving only one site which the propylene can occupy but not another IP. This encourages alternation in the system.

## 22.9 SPONTANEOUS ALTERNATING COPOLYMERIZATIONS

In certain systems, when the comonomers are powerful electron donors and acceptors, a spontaneous reaction occurs, resulting in the formation of a highly alternating copolymer. In some systems the initiation is probably *via* a radical mechanism, whereas in others a zwitterion is involved.

### 22.9.1 Radical Initiation

Sulfur dioxide is a strong electron donor and has been observed to react spontaneously with bicyclo[2.2.1]hept-2-ene to produce a copolymer, as shown in equation (82).

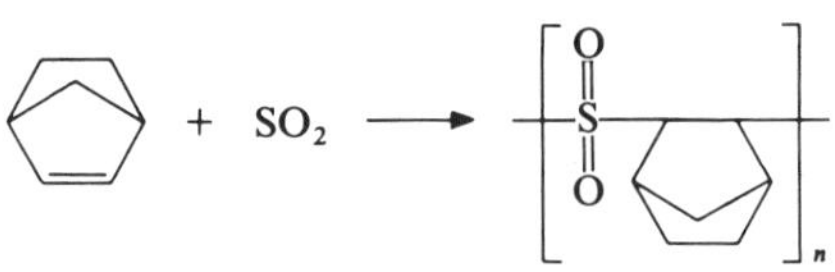

(82)

Mixtures of MAH with cyclic ethers[58] and phenyl vinyl sulfide[121] exhibit spontaneous alternating copolymerization, and many other strong donor–acceptor comonomer pairs also show this characteristic. While several authors have shown that impurities have an influence on these reactions, and may be the source of initiation,[122] others believe it is a consequence of charge-transfer complex formation. A recent study by Nagai *et al.*[123] on the spontaneous copolymerization of S and $\alpha$-chloromaleic anhydride suggests that initiation occurs through the radicals resulting from charge-transfer complex formation. The polymerization rate was accelerated when the mixture was exposed to UV radiation and this was attributed to photoexcitation of the charge transfer complex which increases the charge separation.

Spin-trapping techniques[124,125] have been used to study the mechanism of spontaneous initiation, and an intramolecular proton migration, Scheme 8, has been proposed, forming two radicals which then undergo cross initiation.

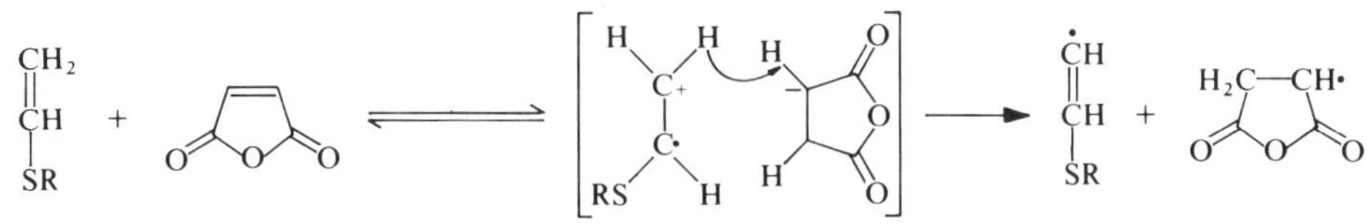

**Scheme 8**

The addition of Lewis acids to certain comonomers also leads to spontaneous polymerization in certain systems. Thus, addition of ethyl(phenyl)aluminum dichloride to butadiene–S mixtures, and ethylboron dichloride or triethylaluminum sesquichloride to S–MA mixtures all lead to spontaneous copolymerizations. Recent work has cast doubt on some of these observations, and Bamford considers that many cases of spontaneous copolymerization can be attributed to the presence of impurities such as peroxides, oxygen and possibly light.[108] When extreme care is taken to ensure an oxygen-free system, the rate of these 'spontaneous' reactions is very slow in the absence of light. Systems containing Lewis acids may be particularly susceptible because they are easily oxidized.

While the question of spontaneity in radical initiated systems remains unresolved, there is no doubt about the spontaneous nature of copolymerizations which occur *via* zwitterion intermediates.

## 22.9.2 Spontaneous Alternating Copolymerization *via* Zwitterion Intermediates

A second class of spontaneously initiated alternating copolymer systems has been described by Saegusa and co-workers,[126–129] who demonstrated that this can occur through the generation of a zwitterion structure. The initial reaction is between two monomers, one with nucleophilic reactivity ($M_N$) and one with electrophilic activity ($M_E$), which, when they interact, give rise to a dimeric dipolar species, equation (83). This is the primary zwitterion which is responsible for initiation and propagation, as in equation (84).

$$M_N + M_E \rightarrow {}^+M_N\text{–}M_E^- \tag{83}$$

$$n({}^+M_N\text{–}M_E^-) \rightarrow {}^+M_N\text{–}(M_E\text{–}M_N)_{n-1}\text{–}M_E^- \tag{84}$$

Initially the propagation is achieved by addition of the zwitterions formed in step (83) to each other to create the macrozwitterion in step (84), but, as the reaction proceeds and the concentration of macrozwitterions builds up, these larger structures may combine. This is manifest in a marked increase in molecular weight, as illustrated in Figure 22 for the alternating copolymerization of 2-oxazoline and $\beta$-propiolactone,[130] which proceeds according to Scheme 9. The initiation involves nucleophilic attack on the lactone by the lone pair on the nitrogen to form the primary zwitterion.

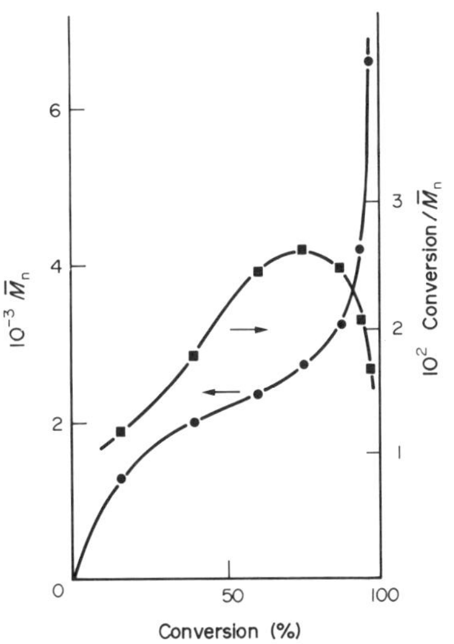


**Figure 22** Alternating copolymerization of 2-oxazoline and $\beta$-propiolactone in acetonitrile (reproduced from ref. 126 with permission from Marcel Dekker)

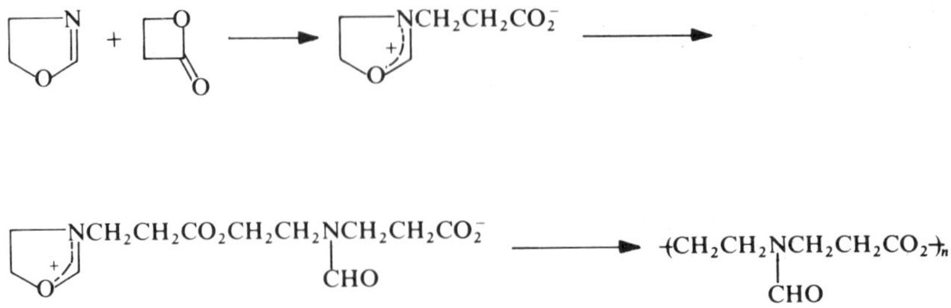

Scheme 9

The chain is propagated by attack of the carboxylate anion of one zwitterion on the 5-position of the oxazoline ring of another zwitterion, leading eventually to macrozwitterionic structures. Ultimately, the latter will also interact, as their concentration increases, resulting in high molecular weight polymer at high conversions. These characteristics are similar to step growth polymerizations.

While the alternating addition appears to be maintained in most cases, there is the possibility that homopropagation might take place *via* Scheme 10, and this should not be totally ignored. Another feature of the reaction path outlined in Scheme 9 that was noted was that alternating copolymers were obtained only when the composition of the feed was in excess of 50 mol % of 2-oxazoline. This behaviour is illustrated in Figure 23. The yield and molecular weight of the copolymer also tended to vary with solvent polarity, but molecular weights were always relatively low, increasing from about 2000 to 4000 as the reaction temperature was raised from 313 K to 393 K. Evidence of the ionic 'living' polymer nature of the copolymerization was obtained when addition of more monomer to a system which had undergone maximum conversion led to a further increase in molecular weight.

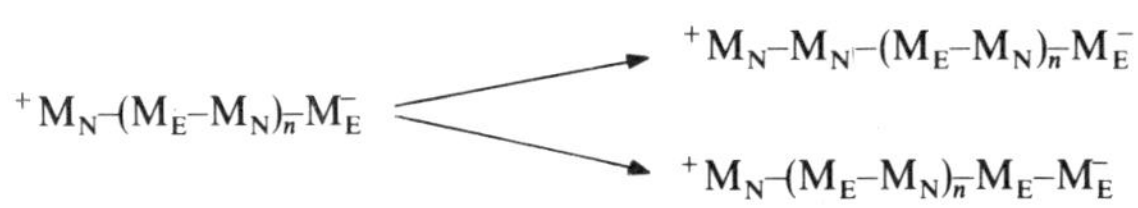

Scheme 10

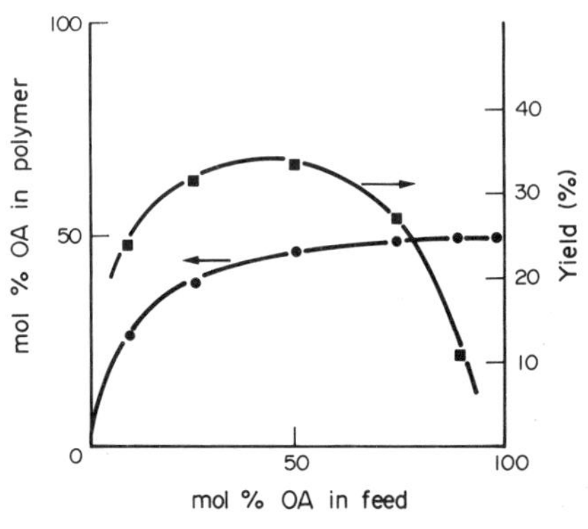


**Figure 23** Composition curve (●) and conversion against composition (■) for the copolymerization of 2-oxazoline (OA) and β-propiolactone at 313 K in dimethylformamide (redrawn from ref. 130 with permission from the American Chemical Society)

This type of reaction has now been reported for a wide range of compounds which have the appropriate electrophilic or nucleophilic characteristics. Reactions proceed according to the same general reaction scheme in which the formation of a primary zwitterion is the key step.

### *22.9.2.1 Ammonium-type zwitterions*

A number of systems involving an ammonium-type zwitterion have been reported, and in Table 13 there are examples of $M_N$ monomers and the corresponding $M_E$ monomers with which they will react.

**Table 13** $M_N$ Monomers Giving an Ammonium Site and $M_E$ Monomers[a]

*$M_N$ monomers*

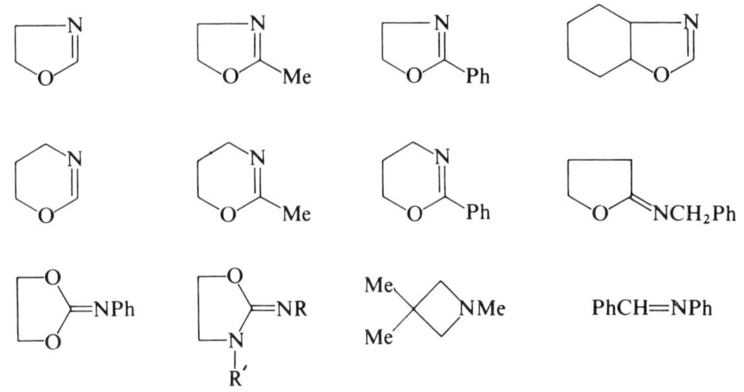

PhCH=NPh

*$M_E$ monomers*

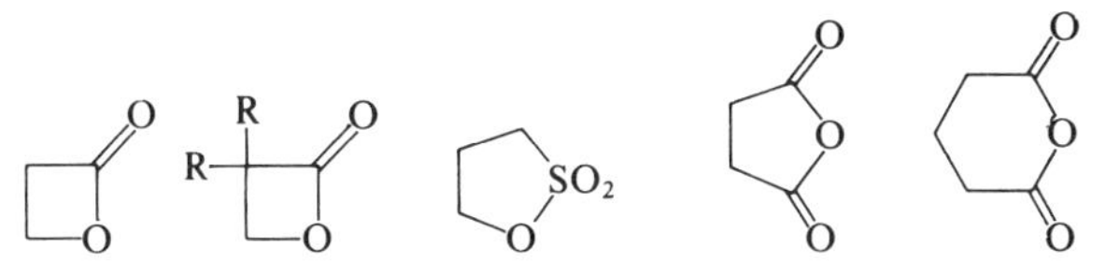

$CH_2{=}CHCO_2H$

$CH_2{=}C(Me)CO_2H$ $\quad$ $CH_2{=}C(X)CO_2H$ (X=Cl, Br) $\quad$ $CH_2{=}CHCO_2R$

$CH_2{=}CHCONH_2$ $\quad$ $CH_2{=}CHSO_2NH_2$ $\quad$ (R= $CH_2CH_2OH$, $CH_2CH_2CH_2OH$)

[a] Reproduced from ref. 34, with permission from Plenum Publishing Co. Ltd.

In addition to $\beta$-propiolactone, the cyclic imino ethers will copolymerize spontaneously with the $\alpha$-methyl- and $\alpha,\alpha$-dimethyl-$\beta$-propiolactones, although the rate of reaction tends to decrease with increasing ring substitution in the lactone. The effect of solvent polarity on molecular weight is illustrated using the system $\beta$-propiolactone combined with *N*-phenyliminocarbonate, which react according to equation (85).

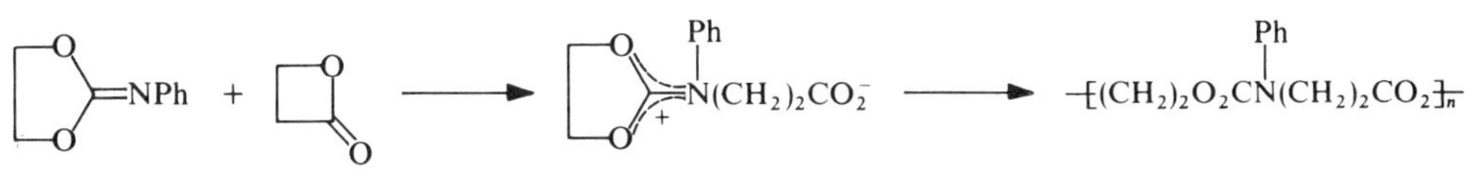

(85)

The data are gathered in Table 14, and show that the molecular weight of the product becomes larger as the dielectric constant of the solvent increases. This suggests that the formation of the primary zwitterion is the rate determining step and will be favoured by more polar solvents.

**Table 14** Effect of Solvent Polarity on the Number Average Molecular Weight ($M_n$) of the Copolymer Formed from $\beta$-Propiolactone and Ethylene *N*-Phenyliminocarbonate

| *Solvent* | *Dielectric constant* | $M_n$ |
|---|---|---|
| PhCl | 5.7 | 4000 |
| $CH_2Cl_2$ | 9.1 | 6970 |
| $PhNO_2$ | 35.7 | 8080 |
| $MeNO_2$ | 35.9 | 13 200 |

Cyclic imino ethers also react readily with acrylic acid, which behaves as an $M_E$ monomer,[131] as shown in equation (86). This is believed to proceed by way of a Michaels-type adduct with subsequent proton transfer to give a zwitterion. The general reaction path illustrated in Scheme 11 is, therefore, the suggested route to the copolymer.

(86)

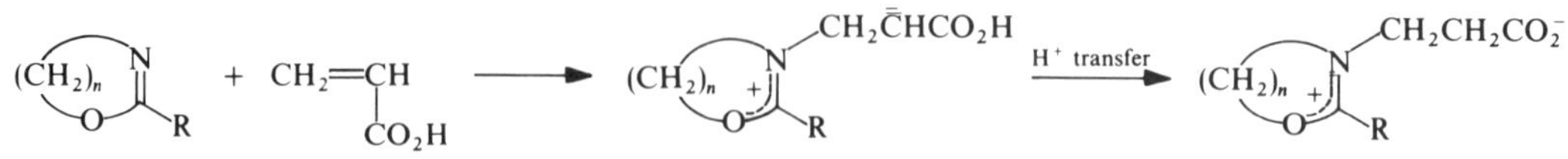

$n = 2, 3 \quad R = H, Me, Ph$

**Scheme 11**

The copolymerization of 2-methyl-2-oxazoline with α-methyl-, α-chloro- and α-bromo-substituted acrylic acids proceeds in a similar fashion[132] and yields alternating structures with molecular weights in the range 2000–4000, giving the repeat unit represented by structure (**9**).

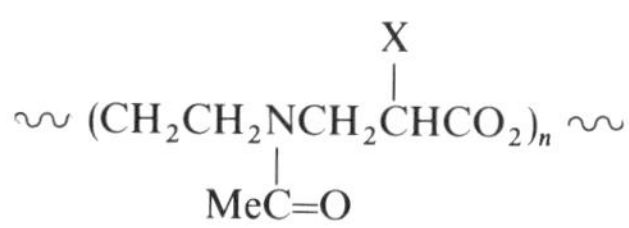

(**9**) X = Me, Cl or Br

Acrylic acid also forms alternating structures with 1,3,3-trimethylazetidine[133] and *N*-benzylideneaniline,[134] as in equations (87) and (88), but the molecular weights are again very low. The cyclic imino ethers also react with other derivatives of acrylic acid, such as acrylamide,[135] ethylenesulfonamide[136] and hydroxyalkyl acrylates,[137] producing the structures shown in equations (89)–(91).

(87)

$$CH_2{=}CHCO_2H + PhCH{=}NPh \longrightarrow \left[CHPhNPh(CH_2)_2CO_2\right]_n \quad n \sim 3 \tag{88}$$

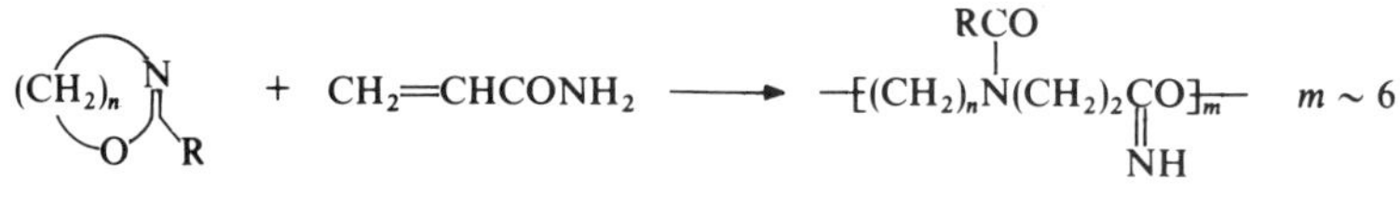
(89)

(90)

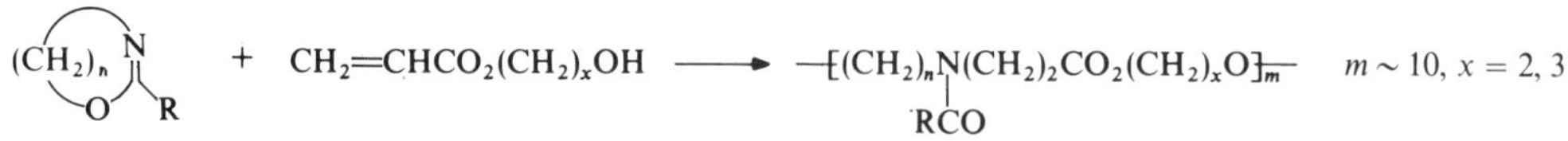
(91)

### 22.9.2.2 *Phosphonium-type zwitterions*

Saegusa has extended the range of alternating structures attainable by this method, by making use of the nucleophilic characteristics of trivalent phosphorus. This was stimulated by the observation that, in the cationic ring-opening polymerization of 2-phenyl 1,3,2-dioxaphospholane, some phosphonium intermediates could be detected.[138] Typical $M_N$ and $M_E$ monomers for this type of reaction are shown in Table 15.

**Table 15** $M_N$ Monomers Giving a Phosphonium Site and $M_E$ Monomers[34]

*$M_N$ monomers*

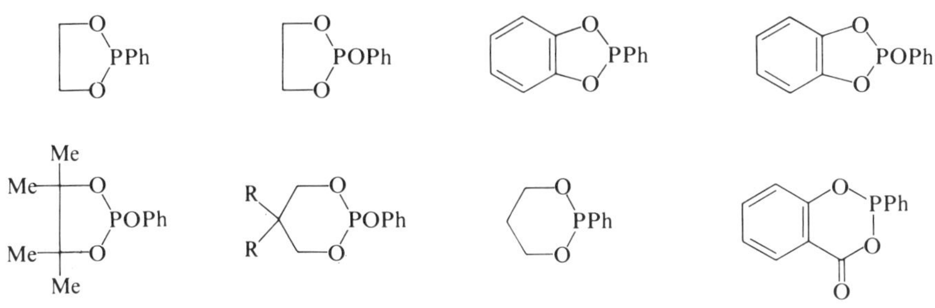

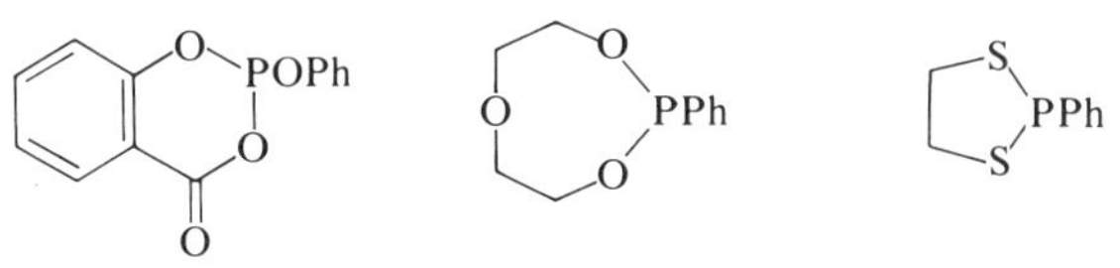

$(PhO)_2PPh$ $(PhO)_3P$

*$M_E$ monomers*

$CH_2{=}CHCO_2H$ $CH_2{=}CMeCO_2H$ $RCH{=}CHCO_2H$ $CH_2{=}CHCO_2CH_2CH_2CO_2H$

$CH_2{=}CHCO_2CH_2CH_2OH$ $CH_2{=}CHC({=}O)X$ $CH_2{=}CHCONH_2$ $CH_2{=}CHSO_2NH_2$

$RC({=}O)CO_2H$

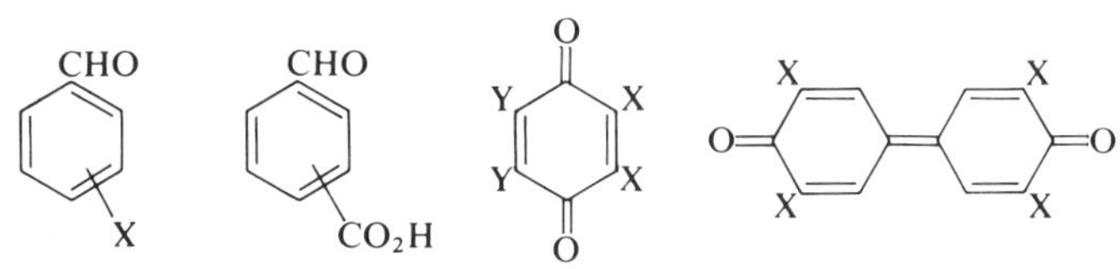

The reactions of the 2-phenyldioxaphospholane with acrylic acid and with $\beta$-propiolactone lead to the same alternating structure; the proposed mechanism is outlined in Scheme 12. Again, the degree of polymerization of the products tends to be small. Ambient temperature reactions lead to the formation of a crystalline solid (**10**) which, from X-ray analysis, was found to be a pentavalent phosphorus compound, but not a zwitterion.[139] This compound is quite stable and copolymer can only be formed when it is heated to a higher temperature (423 K) in a polar solvent.

Other crystalline pentavalent structures can be isolated when 2-phenyldioxaphospholane reacts with MA and phenyl vinyl ketone[140] in equimolar quantities according to Scheme 13. This reaction can be regarded as an equilibrium between the pentavalent phosphorus entity and the primary zwitterion containing an enolate anion through which propagation takes place.

Electron deficient alkenes, such as MA, methyl vinyl ketone and acrolein, copolymerize with benzophosphorinones in the manner shown in Scheme 14 to give alternating structures. Propagation is believed to occur *via* the enolate oxygen and to progress in a regiospecific fashion at the ambidentate anionic sites formed in the primary zwitterion.[141]

### *22.9.2.3 Redox copolymerizations*

Reactions which change the oxidation state of the phosphorus from $P^{III}$ to $P^V$ have been called 'redox' copolymerizations by Saegusa.[142] These involve the reaction of cyclic phosphorus mono-

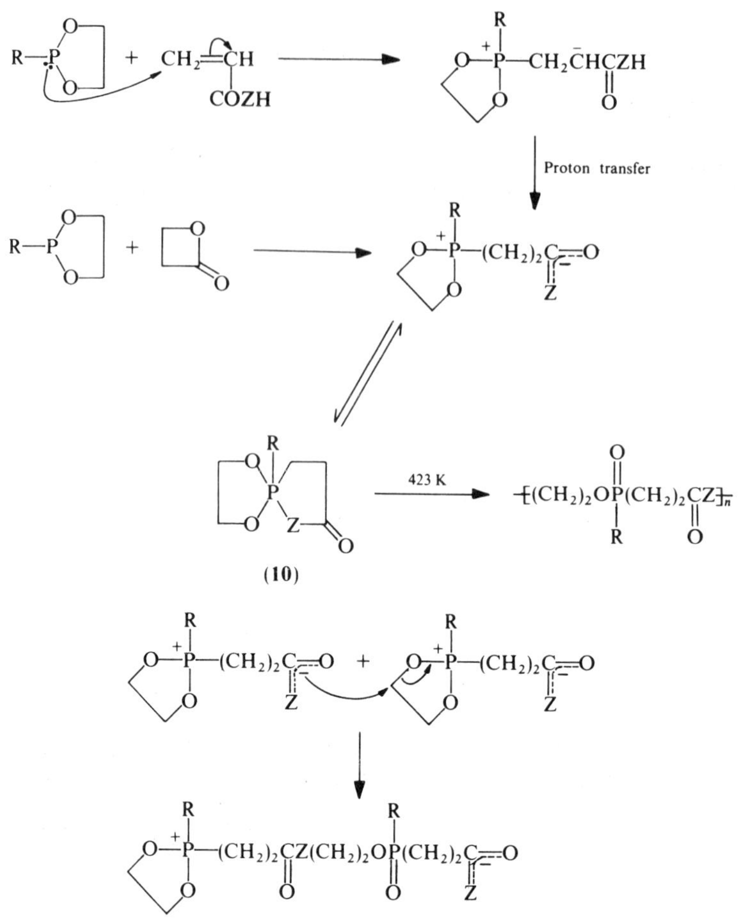


**Scheme 12**

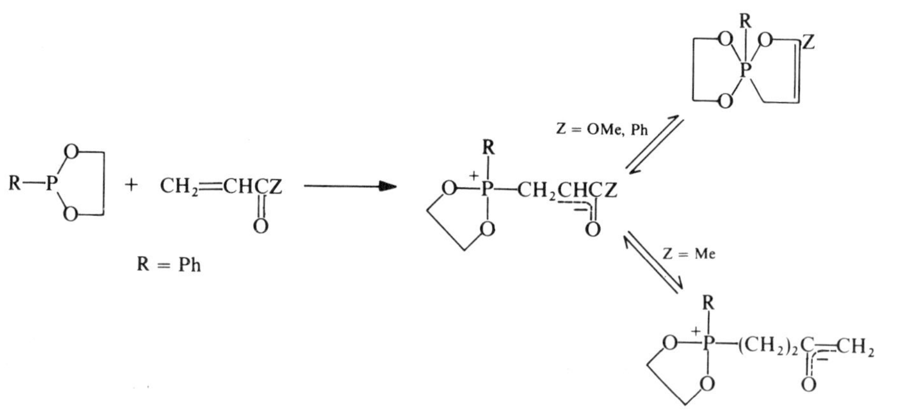


**Scheme 13**

mers with quinones of α-keto acids. Alternating structures are formed at elevated temperatures by the nucleophilic attack of the phosphorus on the carbonyl unit, as outlined in Scheme 15.[143,144]

When cyclic acyl phosphonite was mixed with 4,4′ diphenoquinones, copolymers with a phosphonate structure were produced according to equation (92). The former also produces 1:1 alternating structures with cyclic disulfides.[145]

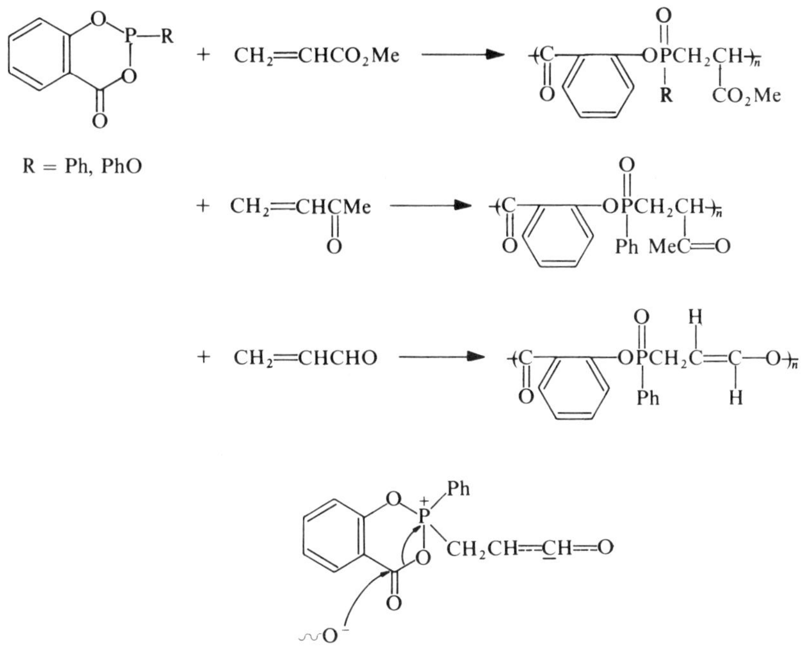

**Scheme 14**

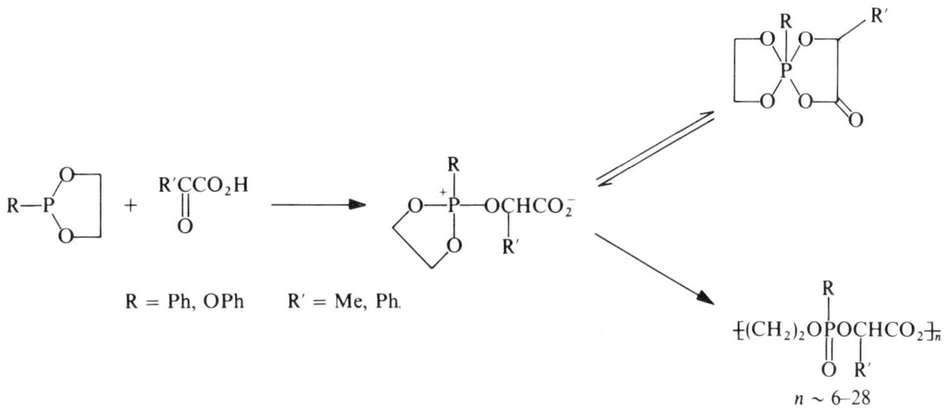

**Scheme 15**

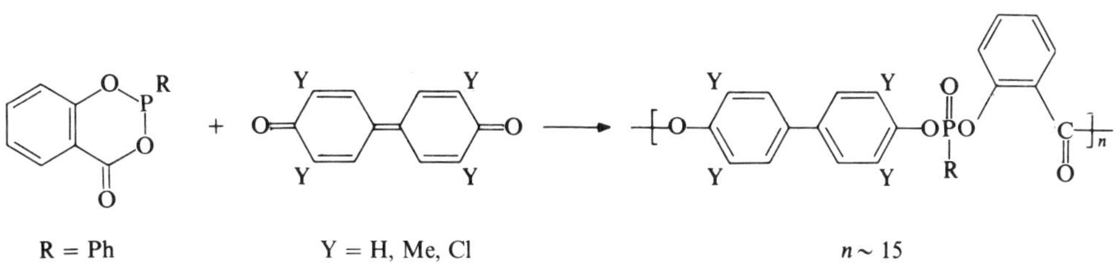

(92)

### *22.9.2.4 Observations*

It is clear that, while the formation of alternating structures *via* zwitterions is an interesting phenomenon, the molecular weights of the products reported so far are too low to be of much practical use themselves. Future work in this area should focus on the nature of the termination process, probably through end group analysis, which might improve understanding of the system and lead to the production of higher molecular weight products or block copolymers.

## 22.10 REFERENCES


1. F. R. Mayo and F. M. Lewis, *J. Am. Chem. Soc.*, 1944, **66**, 1594.
2. H. Dostal, *Monatsh. Chem.*, 1936, **69**, 424.
3. F. M. Lewis, F. R. Mayo and W. F. Hulse, *J. Am. Chem. Soc.*, 1945, **67**, 1701.
4. C. Walling and F. R. Mayo, *Discuss. Faraday Soc.*, 1947, **2**, 295.
5. C. Walling, E. R. Briggs, K. B. Wolfstirn and F. R. Mayo, *J. Am. Chem. Soc.*, 1948, **70**, 1537.
6. F. R. Mayo, F. M. Lewis and C. Walling, *J. Am. Chem. Soc.*, 1948, **70**, 1529.
7. G. B. Butler, *J. Macromol. Sci., Chem.*, 1971, **A5**, 219.
8. J. M. G. Cowie and S. Cree, to be published.
9. G. E. Ham, *J. Polym. Sci.*, 1960, **45**, 169.
10. K. Dodgson and J. R. Ebdon, *Eur. Polym. J.*, 1977, **13**, 791.
11. P. D. Bartlett and K. Nozaki, *J. Am. Chem. Soc.*, 1946, **68**, 1495.
12. L. P. Ellinger, *Adv. Macromol. Chem.*, 1968, **1**, 169.
13. M. Hirooka, H. Yabuuchi, S. Morita, S. Kawasumi and K. Nakaguchi, *J. Polym. Sci., Part B*, 1967, **8**, 47.
14. P. Job, *Ann. Chim. (Paris)*, 1928, **9**, 113.
15. R. Sahai, G. L. Loper, S. H. Lin and H. Eyring, *Proc. Natl. Acad. Sci. USA*, 1974, **71**, 1499.
16. M. Imoto, T. Otsu and Y. Harada, *Makromol. Chem.*, 1963, **65**, 180.
17. H. Hirai, T. Ikegami and S. Makishima, *J. Polym. Sci., Part A-1.*, 1969, **7**, 2059.
18. H. Hirai, S. Okuzawa, T. Ikegami and S. Makishima, *J. Fac. Eng. Univ. Tokyo, Ser. B*, 1967, **29**, 115.
19. T. Ikegami and H. Hirai, *J. Polym. Sci., Part A-1*, 1970, **8**, 463.
20. J. Furukawa, E. Kobayashi, S. Nagata and T. Moritani, *J. Polym. Sci., Polym. Chem. Ed.*, 1974, **12**, 1799.
21. H. Hirai and H. Komiyama, *J. Polym. Sci., Polym. Chem. Ed.*, 1974, **12**, 2701.
22. Y. Komo, S. Kondo and K. Iimura, *J. Polym. Sci., Polym. Lett. Ed.*, 1974, **12**, 153.
23. S. Okuzawa, H. Hirai and S. Makishima, *J. Polym. Sci., Part A-1*, 1969, **7**, 1039.
24. H. Hirai and T. Ikegami, *J. Polym. Sci., Part A-1*, 1970, **8**, 2407.
25. H. Hirai, M. Komiyama and N. Toshima, *J. Polym. Sci., Part B*, 1971, **9**, 883.
26. T. Ikegami and H. Hirai, *J, Polym. Sci., Part A-1*, 1970, **8**, 195.
27. K. Fujimori and G. B. Butler, *J. Macromol. Sci., Chem.*, 1973, **A7**, 387.
28. J. Furukawa, E. Kobayashi and Y. Iseda, *J. Polym. Sci., Part B*, 1970, **8**, 47.
29. G. Briegleb, 'Electronen-Donator–Acceptor-Komplexe', Springer, Berlin, 1961.
30. L. J. Andrews and R. M. Keefer, 'Molecular Complexes in Organic Chemistry', Holden–Day, San Francisco, 1964.
31. R. Foster, 'Organic Charge-transfer Complexes', Academic Press, London, 1969.
32. 'Molecular Complexes', ed. R. Foster, Crane Russak, New York, 1974, vol. 2.
33. F. R. Hartley, C. Burgess and R. M. Alcock, 'Solution Equilibria', Ellis–Horwood, Chichester, 1980.
34. J. M. G. Cowie, 'Alternating Copolymers', Plenum Press, New York, 1985.
35. H. A. Benesi and J. H. Hildebrand, *J. Am. Chem. Soc.*, 1949, **71**, 2703.
36. W. B. Person, *J. Am. Chem. Soc.*, 1965, **87**, 167.
37. D. A. Deranleau, *J. Am. Chem. Soc.*, 1969, **91**, 4050.
38. R. M. Guidry and R. S. Drago, *J. Am. Chem. Soc.*, 1973, **95**, 6645.
39. R. L. Scott, *Recl. Trav. Chim. Pays-Bas*, 1956, **75**, 787.
40. K. Fujimori and A. S. Brown, *Polym. Bull. (Berlin)*, 1986, **15**, 223.
41. E. G. Polle, *Russ. Chem. Rev. (Engl. Transl.)*, 1974, **43**, 631.
42. G. B. Butler, J. T. Badgett and M. Sharabash, *J. Macromol. Sci., Chem.*, 1970, **A4**, 51.
43. E. J. Goethals, A. Cardon and R. Grosjean, *J. Macromol. Sci., Chem.*, 1973, **A7**, 1265.
44. E. Tsuchida, T. Tomono and H. Sano, *Makromol. Chem.*, 1972, **151**, 245.
45. M. W. Hanna and A. L. Ashbaugh, *J. Phys. Chem.*, 1964, **68**, 811.
46. G. D. Johnson and R. E. Bowen, *J. Am. Chem. Soc.*, 1965, **83**, 1655.
47. R. Foster and C. A. Fyfe, *Trans. Faraday Soc.*, 1965, **61**, 1626.
48. D. J. T. Hill, J. H. O'Donnell and P. W. O'Sullivan, *Prog. Polym. Sci.*, 1982, **8**, 215.
49. G. B. Butler and K. Fujimori, *J. Macromol. Sci., Chem.*, 1972, **A6**, 1533.
50. C. Caze and C. Loucheux, *J. Macromol. Sci., Chem.*, 1973, **A7**, 991.
51. R. E. Merrefield and W. D. Phillips, *J. Am. Chem. Soc.*, 1958, **80**, 2778.
52. M. H. Litt and J. Radovic-Wellinghoff, *J. Phys. Chem.*, 1977, **81**, 2644.
53. W. G. Barb, *Trans. Faraday Soc.*, 1953, **49**, 143; W. G. Barb, *J. Polym. Sci.*, 1953, **11**, 117.
54. R. B. Seymour and D. P. Garner, *Polymer*, 1976, **17**, 21.
55. H. Block, M. A. Cowd and S. M. Walker, *Polymer*, 1972, **13**, 549.
56. N. G. Gaylord, S. Maiti, B. K. Patnaik and A. Takahashi, *J. Macromol. Sci., Chem.*, 1972, **A6**, 1459.
57. N. G. Gaylord and S. Maiti, *J. Macromol. Sci., Chem.*, 1972, **A6**, 1481.
58. K. Fujimori, *J. Macromol. Sci., Chem.*, 1975, **A9**, 495.
59. C. Caze and C. Loucheux, *J. Macromol. Sci., Chem.*, 1975, **A9**, 29.
60. N. G. Gaylord, M. Stolka, A. Takahashi and S. Maiti, *J. Macromol. Sci., Chem.*, 1971, **A5**, 867.
61. K. Fujimori and G. B. Butler, *J. Macromol. Sci., Chem.*, 1972, **A6**, 1609.
62. B. Zeegers and G. B. Butler, *J. Macromol. Sci, Chem.*, 1973, **A7**, 349.
63. K. Fujimori and G. B. Butler, *J. Macromol. Sci., Chem.*, 1973, **A7**, 415.
64. J. J. van Heiningen and G. B. Butler, *J. Macromol. Sci., Chem.*, 1974, **A8**, 1175.
65. M. H. Litt and J. Radovic-Wellinghoff, *Polym. Prepr., Am. Chem. Soc., Div. Polym. Chem.*, 1976, **17**, 596.
66. P. Karad and C. Schneider, *J. Polym. Sci., Polym. Chem. Ed.*, 1978, **16**, 1137.
67. Y. Shirota, M. Yoshimura, A. Matsumoto and H. Mikawa, *Macromolecules*, 1974, **7**, 4.
68. Y. Shirota, A. Matsumoto and H. Mikawa, *Polym. J.*, 1972, **3**, 643.
69. G. B. Kharas and M. H. Litt, *Polym. Bull. (Berlin)*, 1984, **12**, 65.
70. G. B. Kharas and D. H. Kohn, *J. Polym. Sci., Polym. Chem. Ed.*, 1983, **21**, 1457.

71. F. Tüdös, T. Kelen and T. Földer-Berezhnikh, *J. Polym. Sci., Polym. Symp.*, 1975, **50**, 109.
72. K. Merz, T. Alfrey and G. Goldfinger, *J. Polym. Sci.*, 1946, **1**, 75.
73. R. C. McFarlane, P. M. Reilly and K. F. O'Driscoll, *J. Polym. Sci., Polym. Lett. Ed.*, 1980, **18**, 81.
74. D. Booth, F. S. Dainton and K. J. Ivin, *Trans. Faraday Soc.*, 1959, **55**, 1293.
75. W. G. Barb, *Proc. R. Sóc. London, Ser. A*, 1952, **212**, 66, 177.
76. T. Kokubo, S. Iwatsuki and Y. Yamashita, *Macromolecules*, 1968, **1**, 482.
77. J. A. Seiner and M. H. Litt, *Macromolecules*, 1971, **4**, 308.
78. C. U. Pittman and T. D. Rounsefell, *Macromolecules*, 1975, **8**, 46.
79. R. E. Cais, R. G. Farmer, D. J. T. Hill and J. H. O'Donnell, *Macromolecules*, 1979, **12**, 835.
80. A. Natansohn, D. Galea, V. Percec and C. I. Simionescu, *J. Macromol. Sci., Chem.*, 1981, **A15**, 393.
81. E. Tsuchida and T. Tomono, *Makromol. Chem.*, 1971, **141**, 265.
82. P. Karad and C. Schneider, *J. Polym. Sci., Polym. Chem. Ed.*, 1978, **16**, 1137.
83. D. J. T. Hill, J. H. O'Donnell and P. W. O'Sullivan, *Macromolecules*, 1983, **16**, 1295.
84. C. H. Bamford and W. G. Barb, *Discuss. Faraday Soc.*, 1953, **14**, 208.
85. M. Ratzsch and Y. Steinert, *Makromol. Chem.*, 1984, **185**, 2411.
86. R. G. Farmer, D. J. T. Hill and J. H. O'Donnell, *J. Macromol. Sci., Chem.*, 1980, **A14**, 51.
87. K. Dodgson and J. R. Ebdon, *Makromol. Chem.*, 1979, **180**, 1251.
88. J. P. Chandler, *Quantum Chem. Prog. Exch.*, 1976, **11**, 307.
89. J. Schaefer, *Macromolecules*, 1971, **4**, 107.
90. K. Arita, T. Ohtomo and Y. Tsurumi, *J. Polym. Sci., Polym. Lett. Ed.*, 1981, **19**, 211.
91. H. Roth, M. Arnold and M. Ratzsch, *Acta Polym.*, 1981, **32**, 277.
92. B. E. Buchak and K. C. Ramey, *J. Polym. Sci., Polym. Lett. Ed.*, 1976, **14**, 401.
93. E. Wilson, 'An Introduction to Scientific Research', McGraw-Hill, New York, 1952.
94. M. Yoshimura, H. Mikawa and Y. Shirota, *Macromolecules*, 1975, **8**, 713.
95. M. Yoshimura, H. Mikawa and Y. Shirota, *Macromolecules*, 1978, **11**, 1085.
96. R. Vukovic, V. Kuresevic and D. Fles, *J. Polym. Sci., Polym. Chem. Ed.*, 1979, **17**, 3835.
97. A. Smirnov, G. I. Deryabina, T. L. Petrova, G. S. Georgiev, V. B. Golubev and V. P. Zubov, *Vysokomol. Soedin., Ser. A*, 1981, **23**, 407.
98. I. Javni, D. Fles and R. Vukovic, *J. Polym. Sci., Polym. Chem. Ed.*, 1982, **20**, 977.
99. P. C. Deb and G. Meyerhoff, *Eur. Polym. J.*, 1984, **20**, 713.
100. P. C. Deb, *J. Polym. Sci., Polym. Lett. Ed.*, 1985, **23**, 233.
101. G. S. Georgiev and V. P. Zubov, *Eur. Polym. J.* 1978, **14**, 93.
102. R. Arnaud, C. Caze and J. Fossey, *J. Macromol. Sci., Chem.*, 1980, **A14**, 1269.
103. K. Fujimori, P. P. Organ, M. J. Costigan and I. E. Craven, *J. Macromol. Sci., Chem.*, 1986, **A23**, 647.
104. H. Hirai, *J. Macromol. Sci., Chem.*, 1975, **A9**, 883.
105. H. Hirai and M. Komiyama, *J. Polym. Sci., Polym. Chem. Ed.*, 1975, **13**, 2419.
106. M. Komiyama and H. Hirai, *J. Polym. Sci., Polym. Chem. Ed.*, 1976, **14**, 307.
107. J. Furukawa, Y. Iseda, K. Haga and N. Kataoka, *J. Polym. Sci., Part A-1*, 1970, **8**, 1147.
108. C. H. Bamford, in 'Alternating Copolymers', ed. J. M. G. Cowie, Plenum, New York, 1985.
109. C. H. Bamford and X-Z. Han. *J. Chem. Soc., Faraday Trans. 1*, 1982, **78**, 855, 869.
110. J. R. Ebdon and N. P. Gabbott, *Polymer*, 1983, **24**, 565.
111. M. Hirooka, *J. Polym. Sci., Part B*, 1972, **10**, 171.
112. C. H. Bamford and P. J. Malley, *J. Polym. Sci., Polym. Lett. Ed.*, 1981, **19**, 239.
113. C. H. Bamford and P. J. Malley, *J. Chem. Soc., Faraday Trans. 1*, 1982, **78**, 2497.
114. V. P. Zubov, M. B. Lachinov, E. V. Ignatova, G. S. Georgiev, V. B. Golubev and V. A. Kabanov, *J. Polym. Sci., Polym. Chem. Ed.*, 1982, **20**, 619.
115. A. Kawasaki, I. Maruyama, M. Taniguchi, R. Hirai and J. Furukawa, *J. Polym. Sci., Part B*, 1969, **7**, 613.
116. J. Furukawa, H. Amano and R. Hirai, *J. Polym. Sci., Polym. Chem. Ed.*, 1972, **10**, 681.
117. J. Furukawa, *Angew. Makromol. Chem.*, 1972, **23**, 189.
118. Maruzen Petrochemical Co. Ltd., *Jpn. Pat.* 72 42 883 (1972) (*Chem. Abstr.*, 1973, **78**, 137 049).
119. J. Furukawa, *Acc. Chem. Res.*, 1980, **13**, 1.
120. M. Taniguchi, Dissertation, Kyoto University, 1976.
121. T. Otsu and H. Inoue, *Makromol. Chem.*, 1969, **128**, 31.
122. S. Aoki, R. F. Tarvin and J. K. Stille, *Macromolecules*, 1970, **3**, 472.
123. K. Nagai, K. Akiyama and N. Kuramoto, *Makromol. Chem.*, 1985, **186**, 1863.
124. T. Sato, K. Hibino and T. Otsu, *J. Macromol. Sci., Chem.*, 1975, **A9**, 1165.
125. T. Sato, M. Abe and T. Otsu, *J. Macromol. Sci., Chem.*, 1981, **A15**, 367.
126. T. Saegusa, S. Kobayashi, Y. Kimura and H. Ikeda, *J. Macromol. Sci., Chem.*, 1975, **A9**, 641.
127. T. Saegusa, S. Kobayashi and Y. Kimura, *Pure Appl. Chem.*, 1976, **48**, 307.
128. S. Kobayashi and T. Saegusa, *Pure Appl. Chem.*, 1981, **53**, 1663.
129. S. Kobayashi and T. Saegusa, in 'Alternating Copolymers', ed. J. M. G. Cowie, Plenum Press, 1985, chap. 5.
130. T. Saegusa, H. Ikeda and H. Fujii, *Macromolecules*, 1972, **5**, 354.
131. T. Saegusa, S. Kobayashi and Y. Kimura, *Macromolecules*, 1974, **7**, 139.
132. T. Balakrishnan and M. Periyasamy, *Polymer*, 1982, **23**, 1372.
133. T. Saegusa, Y. Kimura, S. Sawada and S. Kobayashi, *Macromolecules*, 1974, **7**, 956.
134. T. Saegusa, S. Kobayashi and J. Furukawa, *Macromolecules*, 1975, **8**, 703.
135. T. Saegusa, S. Kobayashi and Y. Kimura, *Macromolecules*, 1975, **8**, 374.
136. T. Saegusa, S. Kobayashi and J. Furukawa, *Macromolecules*, 1976, **9**, 728.
137. T. Saegusa, Y. Kimura and S. Kobayashi, *Macromolecules*, 1977, **10**, 239.
138. H. J. Harwood and N. K. Patel, *Macromolecules*, 1968, **1**, 233.
139. T. Saegusa, Y. Kimura, N. Ishikawa and S. Kobayashi, *Macromolecules*, 1976, **9**, 724.
140. T. Saegusa, S. Kobayashi and Y. Kimura, *Macromolecules*, 1977, **10**, 64.

141. T. Saegusa, T. Kobayashi and S. Kobayashi, *Polym. Bull. (Berlin)*, 1979, **1**, 535.
142. T. Saegusa, *Pure Appl. Chem.*, 1981, **53**, 691.
143. T. Saegusa, T. Yokoyama and S. Kobayashi, *Polym. Bull. (Berlin)*, 1978, **1**, 55.
144. T. Saegusa, S. Kobayashi and T. Yokoyama, *Polym. Bull. (Berlin)*, 1978, **1**, 91.
145. S. Kobayashi, T. Y. Chow and T. Saegusa, *Polym. Bull. (Berlin)*, 1983, **9**, 588.

# 23
# Cyclopolymerization

GEORGE B. BUTLER
*University of Florida, Gainesville, FL, USA*



## 23.1 INTRODUCTION

Cyclopolymerization is defined as any type of chain-growth addition polymerization that leads to formation of cyclic structures in the main chain of the polymer. Its scope is broad, including, at least conceptually, not only symmetrical diene counterparts of all known monomers capable of undergoing addition polymerization but also the unsymmetrical ones. Many of the latter have not been investigated.

It is the purpose of this review to extend coverage of the subject to recent developments beyond those included in the most recently published summary.[1]

Cyclopolymerization had its inception in an observation that diallyl quaternary ammonium bromides failed to yield crosslinked polymers upon radical-initiated polymerization.[2] Since the early investigations, it has been demonstrated that by use of appropriate initiators almost all 1,6-dienes can be polymerized to yield soluble saturated polymers containing rings along the linear chain.

Cyclopolymerization then is an exception to one of the most widely accepted general principles of modern polymer science, *i.e.* the one established by Staudinger[3] (Nobel Laureate, 1953) that polymerization of nonconjugated dienes leads to crosslinked, and therefore insoluble, nonlinear polymers or copolymers. It was observed that a variety of diallyl quaternary ammonium salts polymerized to yield soluble, and hence linear, polymers containing little or no residual unsaturation (Scheme 1). A polymerization mechanism that involves an alternating intramolecular–intermolecular chain propagation was proposed to account for these unexpected results. The term 'cyclopolymerization', although less descriptive and accurate, has grown to be the most widely used term to describe these phenomena.

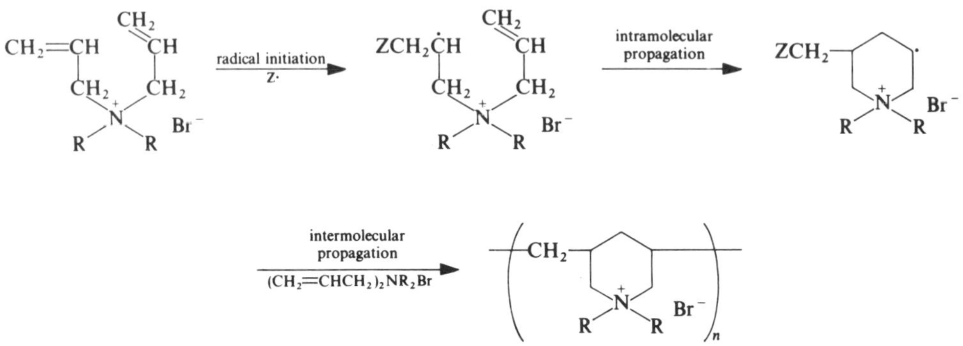


**Scheme 1**

To substantiate the proposed structures, representative polymers were degraded to show conclusively the presence of cyclic structures in the polymer chain.[4] However, these studies did not unequivocably establish the size of the ring formed during the intramolecular step. The piperidinium structure proposed was reasonable based upon the generally accepted hypothesis advanced by Flory[5] (Nobel Laureate, 1974) regarding the predominance of the more stable radical in controlling the course of vinyl polymerization. More recent studies have shown that in numerous cases cyclopolymerizations do not conform to this hypothesis but cyclize to structures derived *via* propagation through the less stable intermediate, *i.e.* the reactions appear to be proceeding under kinetic rather than thermodynamic control. A discussion of these results, including supporting evidence, has been published in an earlier review.[6]

Certain of the diallylamines and quaternary ammonium salts yield cyclopolymers consisting largely of five-membered rings. A $^{13}$C NMR study of poly(*N*-methyl-*N*,*N*-diallylamine)[7] showed formation of five-membered rings; however, when the allyl groups are substituted in the 2-position, mixtures of both five- and six-membered rings form. Also, it has been shown by $^{13}$C NMR and model-compound studies that poly(diallyldimethylammonium chloride)[8] consists mostly of five-membered rings linked mainly in a 3,4-*cis* configuration.

## 23.2 NONCONJUGATED SYMMETRICAL DIENES

### 23.2.1 Carbocyclic Rings

Numerous diene monomers have been studied and those which are functionally capable of generating five- or six-membered rings almost invariably cyclopolymerize quite cleanly. Certain monomers undergo cyclopolymerization *via* all of the well-known methods of initiation of polymerization. 2,6-Diphenyl-1,6-heptadiene represents an ideal monomer which undergoes initiation by radical, cationic and anionic methods, as well as by thermal means and Ziegler–Natta methods, to yield essentially the same cyclopolymer in each case (equation 1).[9]

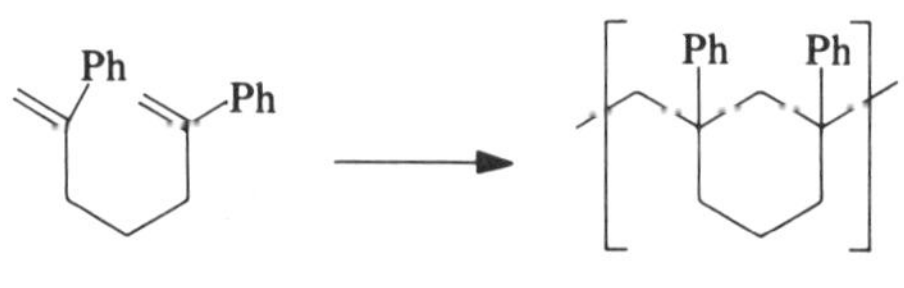

(1)

Soluble polymers (**1a–d**) essentially free of unsaturation[10] have been obtained by free-radical initiation of 1,6-heptadiene-2,6-dicarboxylic acid, its methyl and ethyl esters, and nitrile, respectively (equation 2). Two polymers (**1a** and **1c**) have been partially dehydrogenated with potassium perchlorate. Spectral data indicated the presence of aromatic rings in the dehydrogenated material. The predominant structural unit of these polymers thus was concluded to be (**1**). The polymers of 4-substituted 2,6-dichloro-1,6-heptadienes undergo up to 70% dehydrochlorination to polymers containing 1,4-cyclohexadiene rings in the main chain.

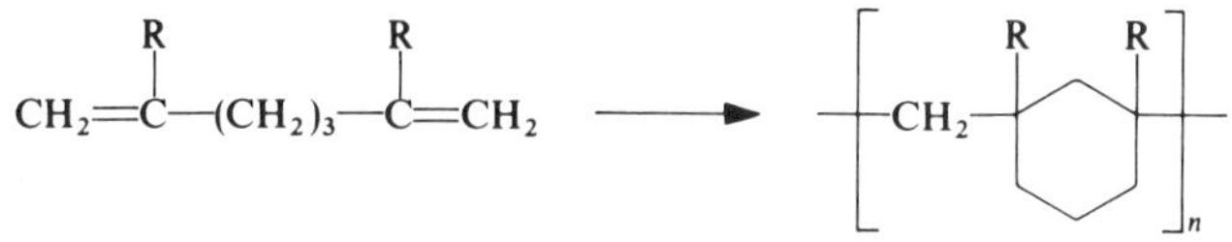

(2)

(**1**) **a**: R = $CO_2H$
**b**: R = $CO_2Me$
**c**: R = $CO_2Et$
**d**: R = CN

Cyclopolymerization of *cis*-1,3-divinylcyclohexane has been studied by the use of metal-coordination initiation.[11] A cyclopolymer containing bicyclo[3.3.1]nonane units with little residual unsaturation was obtained. The cyclopolymerization of this monomer by Friedel–Crafts and other cationic initiators has been studied.[12] $Et_2AlCl/TiCl_4$ and similar systems were also effective. The polymers were composed almost entirely of the cyclized unit.

Perfluoro-1,6-heptadiene and 4-chloroperfluoro-1,6-heptadiene have been polymerized and their glass transition temperatures ($T_g$) determined.[13] The polymers were prepared by exposing monomer at high pressure and temperature to radiation from $^{60}Co$. The polymers were highly saturated as shown by their IR spectra. The cyclic structures (**2**) and (**3**), respectively, are most probable from the evidence available. Polymers of the perfluoro-$\alpha,\omega$-dienes have much lower $T_g$ values than polymers of perfluoro-$\alpha$-alkenes.

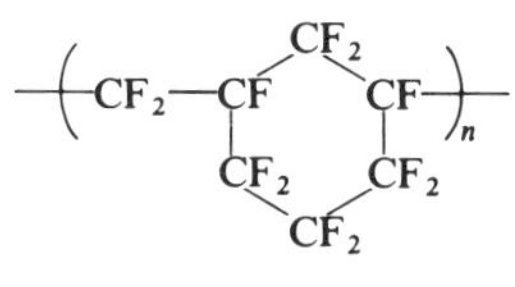

(**2**)

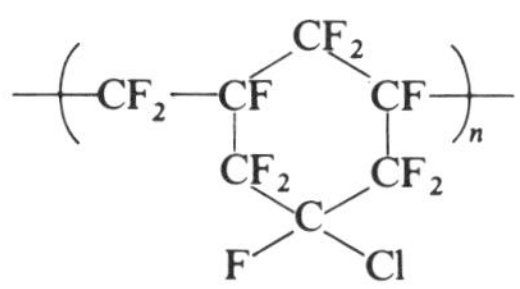

(**3**)

### 23.2.2 Heterocyclic Rings

In the early studies on radical-initiated polymerization of unsaturated quaternary ammonium salts,[14] it was found that monomers containing three allyl double bonds generally produced cross-linked water-insoluble polymers, that monomers containing two allyl double bonds produced linear water-soluble polymers and that monomers containing only one allyl double bond failed to polymerize. The production of linear soluble polymers from monomers containing two allyl double bonds was explained[15] by the cyclopolymerization mechanism shown in Scheme 1. Two of the polymers were degraded by well-known methods.[4] Both poly(diallylammonium bromide) and poly(diallyldimethylammonium bromide) gave products resulting from the cleavage of nitrogen-containing rings present in the structures. These results confirmed the cyclopolymerization mechanism, but not the size of the ring.

Cyclopolymerization of diallylamines has been reviewed by several authors,[16] and the papers presented at two symposia on this topic have been published.[17] The effect of $\beta$-substituents on the cyclopolymerization of diallylamines has been studied.[18] Electron spin resonance (ESR) studies and analysis of products (**4**) and (**5**) show that the piperidine ring (**5**) content increases with the bulk of the

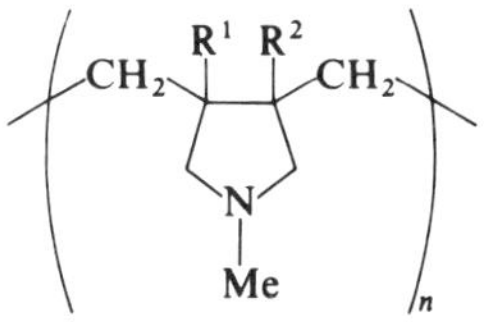

(**4**)

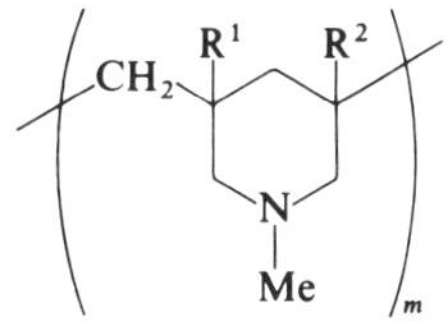

(**5**)

β-substituent, which causes a small decrease in the basicity of the polymers. Diallylamines with a variety of polar and potential metal-chelating groups as substituents on the nitrogen atom have been prepared.[19] Their ability to form cyclopolymers as well as the polymer structure and properties have been investigated.

Diacrylamide has been cyclopolymerized to a polymer which is soluble in hot dimethylformamide and which contains only a small number of double bonds. A six-membered ring structure was presumed since the polymer showed an IR peak which corresponded approximately to the characteristic peak of the carbonyl group in glutarimide. Dimethacrylamide and *N*-substituted dimethacrylamides have been polymerized yielding soluble and fusible polymers which contain practically no unsaturation (**6** and **7**; Scheme 2). The formation of two different cyclic structures is possible; either a five- or six-membered ring, depending on the orientation of the second vinyl group at the intramolecular propagation step, to head-to-tail (**6**) or head-to-head structure (**7**). These studies have been discussed in greater detail in an earlier review.[1]

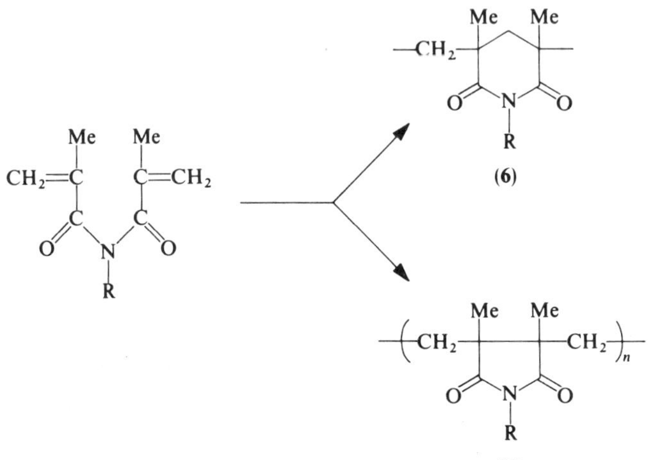

**Scheme 2**

A comparison of IR spectra of polymers obtained by polymerization of monomers of the dimethacrylic series with those of polymers formed by partial thermal deamination of the corresponding polymethacrylamides indicated that important differences exist in the carbonyl absorption region. By this approach, polymerizations of *N*-substituted dimethacrylamides were shown to yield cyclopolymers containing predominantly five-membered ring repeating units, while deamination of the corresponding polymethacrylamides gave polymers with six-membered cyclic units.

Radiation-induced cyclopolymerization of *N*-substituted dimethacrylamides has been studied[20] in the liquid, supercooled-liquid, glassy and crystalline states. Cyclopolymerization occurred completely, predominantly to five-membered rings, in all states except the glassy state, where no polymerization was observed.

An increase in polarity of the medium for cyclopolymerization of dimethacrylamides increases the ratio of five- to six-membered rings in the polymer.[21] The probable conformations of the imide group in dimethacrylamides and their cyclic polymers (**6** and **7**) were determined by IR spectroscopy.

Radical cyclopolymerization of *N*-propyldimethacrylamide has been reinvestigated. Kinetic orders of the polymerization rate with respect to the monomer and the initiator were 1.0 and 0.5, respectively, at concentrations greater than 1.6 $mol\,l^{-1}$. Absolute values of rate constants of propagation and termination reactions were determined by the rotating sector method in bulk and in 50% acetonitrile solution at 30 °C. The polymer consists of five-membered cyclic repeating units almost exclusively. The rate constants for addition of polymer radical to vinyl monomers and those for addition of polymer radicals to *N*-propylmethacrylamide were also evaluated. On the basis of the rate constants obtained, the rate-determining step of propagation was the intermolecular addition of the growing radical to the monomer. The growing radical, formed by intramolecular addition to C=C of the same molecule, was as reactive as polymer radicals from resonance-stabilized monomers. The five-membered ring formation was highly favored by the conformation of the monomer in which coplanarity of the newly formed bond and the semi-occupied orbital of the new radical center is readily achieved. Relatively low reactivity of the polymer radical was accounted for by delocalization of the unpaired electron on the three coplanar carbon atoms. Radical addition of

2-methyl-2-propanethiol to the monomer gave only the substituted succinimide. The five-membered ring formation in the addition reaction was due to the favorable conformation of the monomer.[22]

Other studies on dimethacrylamide and *N*-substituted dimethacrylamides have provided evidence that, in the solid state, the monomers preexist in a conformation favorable for cyclization.[23,24] The thermal or photochemical polymerization of $RN(COCMe{=}CH_2)_2$ (R = Me or Ph) in benzene gave polymers containing five-membered rings and some unsaturation.

A novel monomer, 4-(diallylamino)pyridine, has recently been reported and its cyclopolymerization studied. The polymers were water soluble. The absence of NMR peaks due to vinyl groups in the spectrum of the polymer indicated a high degree of intramolecular reaction during polymerization. The monomer was also copolymerized with diallylamine hydrochloride and dimethyldiallylammonium chloride.[25]

Another novel monomer, *N*,*N*-diacryloylmetribuzin, synthesized by treating the herbicide, metribuzin (4-amino-3-methylthio-6-*t*-butyl-1,2,4-triazin-5(4*H*)-one), with acryloyl chloride in the presence of $Et_3N$, has been cyclopolymerized using free-radical initiators to form soluble polymers containing no residual unsaturation measurable by IR.[26]

Numerous recent references deal with the synthesis of the commercially important monomer, diallyldimethylammonium chloride (DADMAC) or other salts,[27–30] its polymers [poly(DADMAC)] or copolymers,[31–44] kinetic and mechanistic studies on the polymerization process[45–47] and analytical procedures.[48] Also, a study of the polymerization process for mono- and diallylamine derivatives has been reported.[49]

γ-Ray initiated polymerization of formaldehyde divinyl acetal at 0 °C in the liquid state, or at −78 °C by solution polymerization, gives polymers which contain cyclic units. $^{13}$C NMR spectroscopy was used to show that the polymers contain a *cis*-4,5-disubstituted 1,3-dioxolane ring as the predominant unit. Other divinyl acetals have also been studied.[1]

Carbon suboxide (**8**; $C_3O_2$) has been reported to undergo an unusual type of cyclopolymerization. A mechanism involving a zwitterionic intermediate with competing propagation or termination steps has been proposed to explain the details of the mechanistic pathway (Scheme 3).[50]

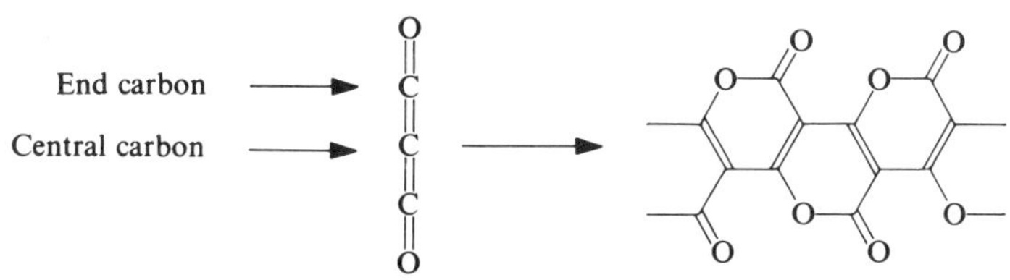


**(8)**

**Scheme 3** Poly(carbon suboxide): a polycyclic chain of six-membered lactone rings with polyconjugated double bonds

Optical fibers are prepared by melt spinning together a polymer containing poly(acrylic anhydride) or poly(methacrylic anhydride) units, unsaturated carboxylic acid units, aromatic vinyl compound units and methyl methacrylate units as the core and a polymer having a lower refractive index as the sheath. Such a copolymer as the core and poly(methyl methacrylate) as the sheath have been melt-spun together and drawn to give flexible optical fibers with good resistance to heat and cracking.[51]

A cyclopolymerization study and regioselective synthesis of vinyl itaconates has been published (itaconic acid = 2-methylenebutanedioic acid). A selective synthesis for both 1- and 4-alkyl vinyl itaconate regioisomers was presented. Polymerization experiments with these two monomers revealed that only 4-vinylitaconic acid and 1-alkyl 4-vinylitaconates lead to cyclopolymers. A synthesis of the γ-lactone repeat unit found in the cyclopolymer confirmed the structure and stereochemistry of the polymer.[52]

Polymerizations of a number of unsaturated silanes that are functionally capable of polymerizing by the cyclopolymerization mechanism have been investigated.[53] Allyl-substituted silanes studied included diallylmethylphenylsilane, diallyl(cyclotetramethylene)silane, and diallyl(cyclopentamethylene)silane.

A limited number of organometallic monomers capable of undergoing cyclopolymerization have been studied. 1,1-Divinylferrocene, the most extensively studied monomer of this type, has been reported to form soluble polymers[54] *via* both radical and cationic initiation with formation of three-carbon bridged ferrocene units in the chain. Recent investigation of these polymers by Mossbauer spectroscopy,[55] however, has shown that they possess significantly different structures. By use of model compounds, it was shown that radical initiation yields the three-carbon bridged cyclopolymer, but that polymers produced by cationic initiation do not have this structure.

Several diunsaturated phosphonium salts have been synthesized and their cyclopolymerization characteristics studied.[56] Divinylphenylphosphine copolymerizes with acrylonitrile in accord with the cyclocopolymerization mechanism.

### 23.2.3 Larger Ring Sizes

Poly(diallyl phthalate) has been commercially important for many years. The monomer is functionally capable of undergoing cyclopolymerization, and early studies[57] led to the conclusion that soluble polymers could be obtained prior to gelation that contained 31–41% of a cyclic structure presumed to consist of 11 or 13 members.

Cyclopolymerization of the hydrocarbons $CH_2{=}CH(CH_2)_nCH{=}CH_2$, in which $n$ is equal to 4–9, 11, 12, 14 and 18, has been studied by use of coordination initiation.[58] In all cases, both cyclic and acyclic recurring units were formed, the largest extent of cyclization (25%) ocurring at calculated ring size = 7. Large rings have also been synthesized from many other monomers. Cyclopolymers having [3.3]paracyclophane repeating units (**9**; equation 3) are obtained by cationic initiation[59] but not by other initiators. Tetraethylene glycol divinyl ether (**10**; 3,6,9,12,15-pentaoxa-1,16-heptadecadiene) was reported to yield a polymer containing [16]-crown-5 units (**11**; equation 4) that is effective as a phase-transfer catalyst.[60] The divinyl monomer, 1,2-bis[(2-vinyloxy)-ethoxy]benzene was shown[61] to cyclopolymerize *via* cationic initiation to yield the 13-membered ring polymer.

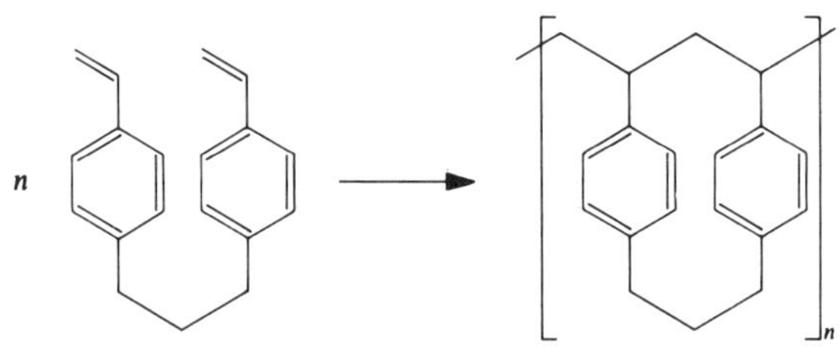

1,3-bis(4-vinylphenyl)propane (**9**) (3)

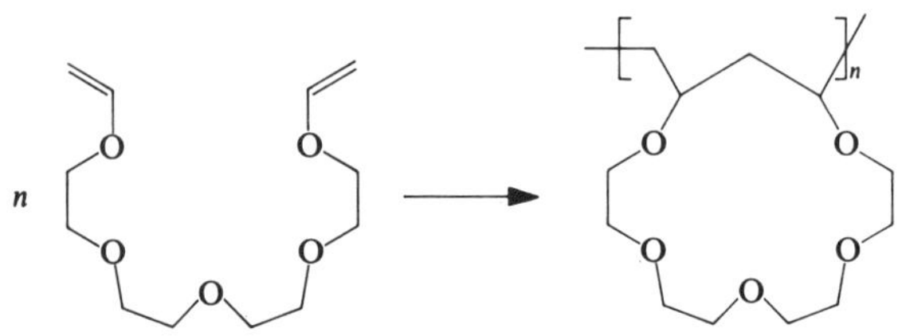

(**10**) (**11**) (4)

Cyclopolymerization of diallyl ether with sulfur dioxide gives soluble copolymers containing 1–2 mol diallyl ether : 1 mol sulfur dioxide. The formation of rings from diallyl ether and sulfur dioxide and of 3,4-disubstituted tetrahydrofuran (THF) rings by cyclized diallyl ether units on the main chain was proposed, based on NMR spectra.[62]

To clarify the possibility of preparing a polymer that contained head-to-head (HH) methyl methacrylate units, radical cyclopolymerization of *o*-di(methacryloyloxy)benzene was investigated. The intramolecular propagation proceeded mainly by a head-to-tail (HT) mechanism.[63] The preparation, thermal properties and reaction of HH cyclopolymers from polymerization of *N*-substituted dimethacrylamide derivatives to soluble HH cyclopolymers have been studied.[64] A recent study included preparation and characterization of poly(methyl acrylate) and poly(methyl methacrylate) consisting of HH and HT units through cyclopolymerization of acrylic and methacrylic anhydrides. The content of HH units of these HH/HT polymers was determined by $^1H$ NMR and $^{13}C$ NMR spectra.[65]

The kinetics of polymerization of *N,N'*-methylenebisacrylamide have been studied using a $Ce^{IV}$–thiourea redox system as initiator. The rate of polymerization was proportional to $[Ce^{IV}]^{1/2}$, $[\text{thiourea}]^{1/2}$ and $[\text{monomer}]^{3/2}$. A cyclopolymerization mechanism fits in with the experimental results.[66]

During studies of cyclopolymerization in the presence of alkylaluminum chlorides, cyclopolymerizations of 2-(*o*-allylphenoxy)ethyl methacrylate and the higher homologs in the 11–19-membered ring region were investigated. The addition of alkylaluminum chlorides to these monomers was effective for increasing the extent of cyclization. The effect of alkylaluminum chlorides was remarkable in formation of the 11-membered ring, but was less for lower-membered rings.[67]

Preparation and characterization of head-to-head cyclopolymers through radical cyclopolymerization of *N*,*N'*-dimethyl-*N*,*N'*-dimethacryloylurea led to a completely soluble polymer comprised of a seven-membered cyclized unit.[68]

Direct observation of the propagating species in the cationic polymerization of 1,3-bis(*p*-vinylphenyl)propane in 1,2-dichloroethane has been accomplished. Stopped-flow spectroscopy was used to study the cationic propagating species in the cyclopolymerization of 1,3-bis(*p*-vinylphenyl)propane and 1,2-bis(*p*-vinylphenyl)ethane by trifluoromethanesulfonic acid ($CF_3SO_3H$) in 1,2-dichloroethane.[69] Cationic cyclopolymerization of 1,3-bis(4-vinylnaphthyl)propane and the polymer structure have been studied. The cyclopolymers contained predominantly cisoid[3.3](1,4)naphthalenophane units in the main chain.[70]

Cyclopolymerization of ethylene glycol divinyl ether and the photocrosslinking reaction of its polymer have been studied. Soluble polymers with methylene-1,4-dioxane-2,3-diylmethylene and 2-[2-(vinyloxy)ethoxy]ethylene structural units were obtained.[71]

Polymerization of *N*,*N'*-methylenebisacrylamide has been carried out with a redox initiator. A polymerization mechanism involving cyclopolymerization in the propagation step was suggested.[72]

Cyclopolymerization of the divinyl ethers, (−)-(*S*)-2,2'-bis[2-(vinyloxy)ethoxy]-1,1'-binaphthyl or (−)-(*S*)-2,2'-bis{2-[2-(vinyloxy)ethoxy]ethoxy}-1,1'-binaphthyl produced optically active polymeric products having a crown ether structure.[73]

Poly(4,7-dioxa-1,10-decylene dimethacrylate) obtained by microhetereogeneous three-dimensional polymerization at 293 K in the presence of 2,2,6,6-tetramethylpiperidine-1-oxyl (TEMPO) inhibitor, introduced at various stages of the polymerization, exhibited higher compression elasticity modulus (*E*) than uninhibited samples at the same conversion.[74] Polymerization of 4,7-dioxa-1,10-decylene dimethacrylate gave a soluble polymer of cyclolinear structure containing $\leq 1$–2 long-chain branches per macromolecule. Gel formation was observed on polymerization of other dimethacrylate compounds.[75]

ESR studies of solid-state polymerization of irradiated *N*-(*p*-bromophenyl)dimethacrylamide and *sym*-dimethacryloyldimethylhydrazine have been published. The intermediate radicals formed in the cyclopolymerization of both monomers were studied. An initiation radical could be observed in each case.[76] The polymerization of *sym*-dimethacryloyldimethylhydrazine was not affected by solvents.[77]

The mechanism of free-radical cyclopolymerization of acryloyl- and methacryloyl-methylallylamines has been studied. The difference in cyclopolymerizabilities between the two monomers could be interpreted on the basis of the hypothesis that the lower the polymerizabilities of the monofunctional counterparts of unconjugated dienes, the higher their cyclopolymerizabilities.[78]

Fully identified cyclic structures formed *via* radically initiated cyclizations and cyclopolymerizations are explained using frontier orbital theory. This theory, briefly introduced for general application to organic polymer chemistry, works very well in explaining the ring size in cyclopolymerization except when steric hindrance is predominant in controlling the course of cyclization.[79]

MINDO/3 and $^{13}C$ NMR results have been used to predict the free-radical polymerizability of allyl monomers. Polymerizable mono- and di-allyl monomers exhibited high negative eigenvalues for the C—H bond $\alpha$ to the allyl group and small $\Delta\delta$ values (distance between the $\beta$- and $\gamma$-carbon NMR peaks) as correlated by MINDO/3 calculations and $^{13}C$ NMR results, respectively. Monomers with low negative eigenvalues (corresponding to weak $\alpha$-C—H bonds) underwent degradative chain-transfer instead of polymerization. Monomers with intermediate eigenvalues or $\Delta\delta$ values polymerized with difficulty.[80]

Styrene–methyl methacrylate copolymers derived from styrene–methacrylic anhydride copolymers have been studied. The structures of copolymers of styrene and methyl methacrylate prepared from styrene–methacrylic anhydride polymers were calculated from NMR. The $CH_3O$ proton resonances of the derived copolymers are more intense in the highest field area than would be expected if the resonance were due only to cosyndiotactic styrene–methyl methacrylate–styrene triads. An explanation for the discrepancies was proposed.[81]

Electrophoretic mobilities have been determined for $(CH_2{=}CHCH_2)_2N^+Me_2Cl^-$, $CH_2{=}CMeCO_2CH_2CH_2NMe_2$, $CH_2{=}CMeCO_2CH_2CH_2N^+Me_2EtCl^-$, $CH_2{=}CMeCO_2$-$CH_2CH_2N^+Me_2CH_2CH_2CO_2^-$ and the corresponding polymers. Low electrophoretic mobility of the polymers of $(CH_2{=}CHCH_2)_2N^+Me_2Cl^-$, $CH_2{=}CMeCO_2CH_2CH_2NMe_2$ and $CH_2{=}CMe\text{-}CO_2CH_2CH_2N^+Me_2EtCl^-$ supported the contention that insensitivity of the polymerization kinetics of $(CH_2{=}CHCH_2)_2N^+Me_2Cl^-$, $CH_2{=}CMeCO_2CH_2CH_2NMe_2$ and $CH_2{=}CMe\text{-}CO_2CH_2CH_2CH_2N^+Me_2EtCl^-$ in aqueous solutions to pH, ionic strength of the medium and the nature of counterions reflects the formation of intramolecular complexes (*via* $N^+$ and ether O) having high affinity to counterions and giving stable ion-pairs. These intramolecular complexes exist in growing macroradicals and polymers, and render them effectively neutral in spite of formal electric charges.[82]

The Monte Carlo method has been used to model radical polymerization of polyfunctional monomers with regard to intramolecular cyclization, *i.e.* conformational changes in the growing chain which allow the reaction between two radicals or between a double bond and one radical of the same chain.[83]

In a continuing study of head-to-head polymers, *N*-phenyldimethacrylamide was cyclopolymerized to poly(*N*-phenyldimethacrylamide) containing 94% five-membered rings, but more importantly containing 6% six-membered rings in the cyclopolymer.[84]

Poly(DADMAC) has been prepared by $Na_2S_2O_8$-initiated polymerization of the respective monomer in the presence of disodium versenate (versene acid = ethylenediaminetetraacetic acid), and the effect of the concentration of monomer, $K_2S_2O_8$ and disodium versenate, temperature and pH of the reaction mixture on the yield and intrinsic viscosity of the polymer was determined.[85]

A paper relevant to this discussion of cyclopolymerizations deals with computer simulation of gelation. The influence of cyclization on the extent of gelation of telechelic linear prepolymers with tetrafunctional crosslinking agents at the critical point was investigated by computer simulation using the upper–lower bound methods and scaling methods.[99]

Binding of methyl orange and its homologs by polyion complexes in aqueous solution has been studied. Polyion complexes of sodium polymethacrylate and poly(DADMAC), which are insoluble in water and have an equal number of positive and negative charges, bind organic anions (methyl orange, ethyl orange, propyl orange, butyl orange and pentyl orange) in aqueous solution. The significance of the hydrophobic and the electrostatic interactions which accompany the binding have been described.[87]

Complexation of poly(DADMAC) and 1-anilino-8-naphthalenesulfonic acid in water, aqueous NaCl and methanol–water mixture has been studied by measurements of fluorescence, UV spectroscopy and potentiometric titration.[88] The electrostatic interaction of the components results in the formation of a 1:32 stoichiometric complex between the reactants.[89]

Interaction of poly(DADMAC) with cyanoferrate(II) and cyanoferrate(III) anions has been studied, and the conclusion was drawn that the binding of the anions proceeded by an anionic mechanism and was accompanied by a spectral shift of the cyanide bands.[90]

Complexing of eosine in analysis of cationic flocculants has received some attention. The spectrophotometric determination of cationic fluocculants such as poly(DADMAC) in the presence of eosine is affected by the pH of the aqueous solutions used, with pH 3.4–3.6 being optimal for the determination.[91]

Association of polyelectrolytes with oppositely charged mixed micelles has been studied. Phase diagrams were obtained for mixtures of a strong polycation and poly(DADMAC) with a nonionic surfactant (Triton X-100) and an anionic surfactant sodium dodecanesulfonate (SDS) in the presence of NaCl. The results are consistent with intrapolymer condensation of micelles.[92]

In a continuing study of polyelectrolyte–micelle complexes,[3] solutions of poly(DADMAC) in NaCl and in Triton X-100 in the presence of SDS were characterized by the quasi-elastic light-scattering technique. The effect of solution variables on the phase boundary was discussed.[93]

The effect of the salt content on formation and structure of polyelectrolyte complexes (symplexes) has been studied. The particle size of complexes of poly(DADMAC) with sodium cellulose sulfate in water was shown by light scattering to increase with increasing NaCl concentration. Further salt addition results in limited aggregation.[94] The adsorption of an aqueous saline solution of the polymer on colloidal silica was investigated as a function of pH, NaCl concentration and molecular weight. These adsorption characteristics were discussed in connection with the charge and conformation of the polymer in water or salt solution and also with the surface charge of colloidal silica.[95]

Characteristics of the behavior of *N*-hexadecyl-*N*,*N*-diallyl(dodecyloxycarbonylmethyl)ammonium bromide in water have been studied. The ultrasonic dispersion of the monomer in water gave vesicular colloids, which can be used for immobilization of biological objects and modeling membrane processes.[96]

Salt linkage formation of poly(DADMAC) with acidic groups in the polyion complex between human carboxyhemoglobin and potassium poly(vinyl alcohol) sulfate has been observed.[97]

Polysalts have been prepared from poly(acrylic acid), poly(itaconic acid), maleic anhydride-α-methylstyrene copolymer or maleic anhydride–propylene copolymer with poly(DADMAC), polyethyleneimine or two quaternized acrylic polymers. The results were discussed in terms of the acid strength and distribution of the anionic groups, both of which affect the composition of the polysalts.[98] Symplex polysalts have also been prepared by treating sodium polyphosphates with poly(DADMAC), polyethylenimine and cationically modified polyacrylamides. Morphological studies indicated that the ratio of the components in the complexes was controlled more strongly by electrostatic charge than in the case of symplexes involving cellulose derivatives as anionic components.[99] Sparingly soluble metal salts of poly(DADMAC) or its copolymer are obtained instantaneously at ambient temperature at pH 1–8 by treating with a dilute aqueous solution of metal-containing anions. This reaction is utilized for detoxification and purification of solutions and waste water, metal concentration and dressing or preservation of wood surfaces.[100]

Further discussion of the polyelectrolytic titration, method, information and limits has been published. In the determination of polyacrylamide, anionic and cationic starch, and polyethyleneimine in pulp suspensions or their filtrates by titration with potassium poly(vinyl alcohol) sulfate or poly(DADMAC) as anionic or cationic standard reagent, respectively, the end point of titration was traced with high sensitivity using a streaming current detector.[101]

The conditions of formation and some properties of compositions from polyelectrolyte complexes and latexes have been studied. Films were formed by mixing polyelectrolyte solutions, *e.g.* poly(DADMAC) in aqueous $NH_3$ or 50% $HCO_2H$ with 15% acrylate latexes, pouring on a polyethylene substrate and drying at ambient temperature. The physical properties of the composites depended on composition.[102]

Thermal degradation of poly(DADMAC) and 1:1 DADMAC–methacrylic acid copolymer gave products containing methyl chloride (main component), allyl chloride, trimethylamine, allyldimethylamine and (from DADMAC–methacrylic acid copolymer) HCl and $CO_2$. The phase transition temperatures were determined from elongation as a function of temperature.[103]

Problems of interfacial phenomena associated with application to determining the amount of cationic polyelectrolyte [poly(DADMAC)] adsorbed on cellulose fibers have been studied. Potassium poly(vinyl alcohol) sulfate was used as an anionic polymer.[104] A theoretical model for the electric double layer of fibers with an adsorbed layer of cationic polyelectrolyte has been proposed to account for the structural changes by the adsorbed monolayer. This model could also be applied to general cases in which the reversal of polarity of the charge occurs by an adsorbed monolayer of cationic polyelectrolyte in equilibrium with inorganic salts having counterions with relatively low polarization energies.[105]

In studies of electric conduction properties of cellulose fibers with adsorbed layers of cationic polyelectrolyte, the effect of multilayer formation of poly(DADMAC) on the electric conduction behavior of carboxymethyl cellulose (CMC) fiber was investigated. The excluded volume effect and protonic transition mechanism accounted for the conduction behavior of CMC fiber with absorbed multilayers.[106] Electric conduction properties of cellulose fibers with adsorbed layers of cationic polyelectrolyte have been determined. The electric conduction of a poly(DADMAC)-containing dissolving-pulp pad was determined as a function of the amount of adsorbed poly(DADMAC), KCl concentration and pH, and the conduction data were analyzed. The results suggested that mobile ions in the electric double layer of the pad are replaced by immobile poly(DADMAC) ions as adsorption of poly(DADMAC) proceeds.[107] In a continuing study of electric conduction properties of cellulose fibers with adsorbed layers of cationic polyelectrolyte and to provide theoretical interpretations for the empirical relation $R = GL + W$($R$ = total pad resistance, $L$ = pad length, $G$ and $W$ = constants), a mathematical model was proposed for the electrical conduction of dissolving pulp, cotton linters and CMC containing poly(DADMAC) as polyelectrolyte. The model correlated the electrical conduction with the hydrodynamic properties of the cellulosic pad and permitted estimation of specific conductance values of fiber surfaces using the hydrodynamic data obtained by the streaming current method.[108]

## 23.3 NONCONJUGATED UNSYMMETRICAL DIENES

Methyl allyl maleate and methyl allyl fumarate were perhaps the earliest examples of unsymmetrical dienes to be studied,[109] but numerous other such monomers have now been studied and discussed in detail.[110] Cyclopolymerization[109] of allyl and methallyl crotonates appear to represent the first examples in which it was shown that the less stable radical predominated in the cyclization step in contrast to the prediction of the Flory hypothesis.[5]

### 23.3.1 Carbocyclic Rings

Cyclopolymerization of *o*-allylstyrene has been studied by free-radical and Friedel–Crafts initiators; all of the vinyl groups participated in polymerization, and the allyl groups partly remained as residual double bonds.[111] Other unsymmetrical dienes reported in the early literature to produce cyclopolymers are *o*-allyl-α-methylstyrene, 1-allylindene and *o*-isopropenylstyrene.[1]

### 23.3.2 Heterocyclic Rings

Cyclopolymerization of 1-vinyluracil by free-radical initiated polymerization has been studied.[112] The polymer contained substituted dihydrouracil rings believed to have been formed by the cyclopolymerization mechanism.

The synthesis of polyampholytes (**12**) and (**13**) by cyclopolymerization of *N*-(2-phenylallyl)-acrylamide and *N*-ethyl-*N*-(2-phenylallyl)acrylamide followed by hydrolysis of the cyclic amides (equation 5) has been reported.[113]

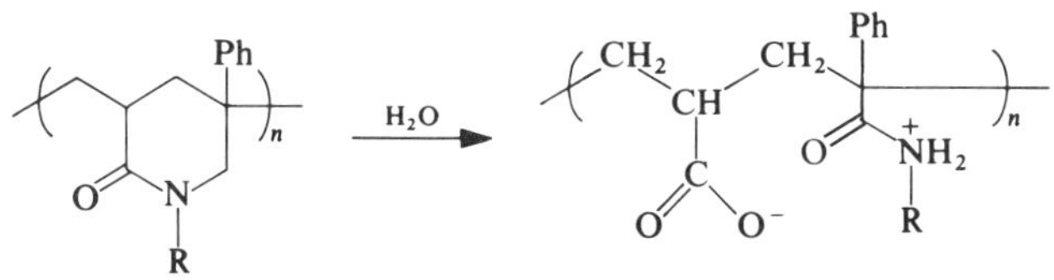


(**12**) (**13**) (5)

Cyclopolymerization of *N*-allyl-*N*-methylmethacrylamide has been investigated.[114] Radical initiation resulted in polymers with mainly five-membered rings with some six-membered rings and pendant methacryl groups. *N*-Allylacrylamide and *N*-allylmethacrylamide were cyclopolymerized in toluene.[115] More recent cyclopolymerization studies of *N*-allylacrylamide and *N*-allylmethacrylamide[116] have shown that soluble polymers can be obtained only under special conditions.

Cyclopolymerization of *N*-methacryloyl-*N*-methylcrotonamide has been studied. While *N*,*N*-disubstituted methacrylamide and crotonamide, which can be considered as structural moieties of *N*-methacryloyl-*N*-methylcrotonamide, do not homopolymerize, *N*-methacryloyl-*N*-methylcrotonamide nevertheless yielded a homopolymer consisting of five-membered cyclic repeating units. The coplanarity of the three carbon atoms and reluctance of the reacting double bonds to homopolymerize seem to be the minimum requirements for forming a polymer consisting of solely five-membered cyclic repeating units.[117]

Linear polymers from the radical cyclopolymerization of the monoallyl esters of maleic and citraconic acids have been obtained.[118] On the basis of residual unsaturation determinations, it was concluded that the polymers contained 63–65% cyclic units. Similarly, the polymerization of diallyl maleate[119] yielded a cyclopolymer containing 33.4% unsaturation (one double bond per monomer unit), of which 31.4% was due to allyl double bonds and 2% to maleate double bonds. Polymerization of allyl acrylate[120] to conversions up to 60% gave a soluble polymer that contained 59% of the predicted γ-valerolactone structure, along with residual amounts of both types of double bonds.

Vinyl *trans*-cinnamate[121] has been polymerized to linear soluble polymers containing γ-lactone rings that have only 10–30% residual unsaturation. Vinyl methacrylate and allyl methacrylate have been polymerized in toluene.[115] Vinyl methacrylate cyclopolymerized, and the cyclization content of the soluble polymers obtained at low conversions was estimated by IR spectroscopy to be 50–60%. In contrast, allyl methacrylate produced no soluble polymer, even at low conversion. However, in another study it was shown that allyl methacrylate yielded soluble polymers containing appreciable amounts of cyclic units when polymerized under suitable conditions.[122] By polymerizing a solution of allyl methacrylate in benzene, it was found that soluble polymers containing about 20% of γ-lactone-type structural units were obtained at conversions lower than 10%.[123] Soluble polymers have also been obtained by polymerization of allyl methacrylate in carbon tetrachloride, dioxane and diallyl ether solutions.[124] It was suggested that the propagation and cyclization reactions proceed only *via* addition to the methacrylyl groups of the monomer. A more recent study of cyclopolymerization of methallyl methacrylate has been published. The formation of five-membered rings occurred at higher polymerization temperatures.[125]

A free-radical and anionic initiated homopolymerization study of vinyl acrylate and vinyl methacrylate gave soluble polymers containing vinyl group residual unsaturation.[126] Poly(vinyl methacrylate) prepared with radical initiators contained 10–20% cyclic units, and poly-(vinyl acrylate) similarly prepared contained 50–60% cyclic units.

The mechanism of cyclopolymerization of vinyl acrylate, methacrylate and $\alpha$-chloroacrylate has been studied more recently. The residual double bond in the polymers was confirmed to be mainly a vinyl group by IR spectroscopy and ICl titrimetry.[127]

Polymerization of allyl acrylate gave partially cyclized soluble poly(allyl acrylate).[128] The cyclization constant (ratio of propagation rate to cyclization rate) was 2.42. The heat of activation was 15.4 kcal mol$^{-1}$ (1 cal = 4.184 J), and the difference between the activation energies of inter- and intra-molecular propagation was 1.4–1.8 kcal mol$^{-1}$. Controlled ring closure has been shown to be possible in the radical cyclopolymerization of allyl methacrylate. The cyclopolymerization kinetics and mechanism have been discussed.[129] Radical-initiated cyclopolymerization of isopropenyl acrylate and methacrylate has been studied at 60 °C in benzene. The residual double bond in the polymers was determined to be mainly isopropenyl groups by spectroscopy and titrimetry.[130] *o*-Isopropenylphenyl vinyl ether has been polymerized[131] by cationic and radical initiation to soluble cyclopolymers. IR spectroscopy showed that cyclization occurred *via* head-to-head addition, leading to five-membered rings.

Cyclopolymerization of methyl 2-methallyl maleate has been studied and has been shown to lead preferentially to five-membered lactone ring formation. This monomer has a stronger tendency to cyclopolymerize than allyl methyl maleate.[132] The kinetics of radical-initiated cyclopolymerization of methyl 2-methallyl fumarate have been investigated and compared with the polymerization results of allyl methyl fumarate. In the former, the extent of cyclization and five-membered ring formation were enhanced.[133] Other unsymmetrical monomers which have been studied are vinyl *o*-isopropenylbenzoate, *o*-allylphenyl acrylate, *o*-vinylphenyl acrylate and 2-(*o*-allylphenoxy)ethyl acrylate.[1]

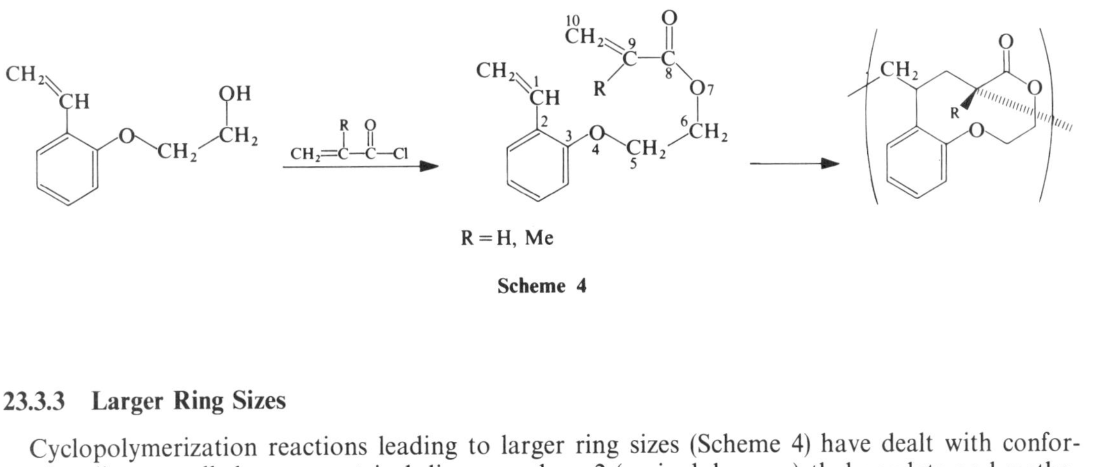

R = H, Me

**Scheme 4**

### 23.3.3 Larger Ring Sizes

Cyclopolymerization reactions leading to larger ring sizes (Scheme 4) have dealt with conformationally controlled unsymmetrical dienes, such as 2-(*o*-vinylphenoxy)ethyl acrylate and methacrylate[134] and their cyclopolymerization, utilizing both radical and photoinitiated polymerization. The polymers obtained were essentially completely cyclized.

A method of synthesis of polymers containing crown lactone units *via* cyclopolymerization in the presence of alkylaluminum chlorides has been developed. Polymerization of the acrylates $CH_2{=}CHCH_2(OCH_2CH_2)_yO$-*o*-$C_6H_4O(CH_2CH_2O)_xCOCH{=}CH_2$ ($x$ = 1–2, $y$ = 1–2) or *o*-$CH_2{=}CHCH_2C_6H_4O(CH_2CH_2O)_xCOCH{=}CH_2$ ($x$ = 2–5) by alkylaluminum chlorides gave cyclopolymers containing crown ether lactone units. The cation-binding ability of the polymers was demonstrated.[135]

Studies on cyclopolymerization in the presence of alkylaluminum chlorides have been extended to acrylates and methacrylates containing oligo(oxyethylene) units in the 11–20-membered ring region. 2-[2-(*o*-Allylphenoxy)ethoxy]ethyl acrylate and higher homologs containing oligo(oxyethylene) units capable of forming 14-, 17- and 20-membered rings, respectively, were studied. Although the effect of alkylaluminum chlorides was gradually reduced with increasing ring size, it was remarkable in the formation of 14-membered rings for acrylates.[136]

## 23.4 DIENES WHICH YIELD POLYCYCLIC SYSTEMS

Polycyclic systems have been introduced into polymers *via* cyclopolymerization in a number of ways. For example, *cis*-1,3-divinylcyclohexane and *o*-divinylbenzene have been cyclopolymerized to polymers containing bicyclic units.[137] However, in these cases as well as in many others discussed in previous reports,[1,11] only one ring system is developed during the propagation step. Thus, examples presented in this section will be limited to those monomers which introduce at least a bicyclic unit from open-chain or branched monomers, or the initial ring structure present is altered during the propagation step. The first case has been defined as a succession of ring closures, and such a phenomenon was proposed as early as 1961.[138]

For example, triallylethyl- (**14**) and tetraallyl-ammonium bromides have been converted to soluble polymers containing small amounts of residual unsaturation. Also, a trialkenic monomer, 3-vinyl-1,5-hexadiene, has been converted to a soluble polymer having little residual unsaturation using a Ziegler catalyst. These results indicate an extensive linear chain structure containing 2,6-methylene-linked bicyclo[2.2.1]heptyl rings (**15**).

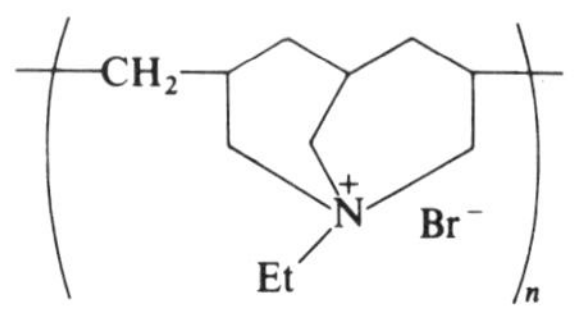

(**14**)

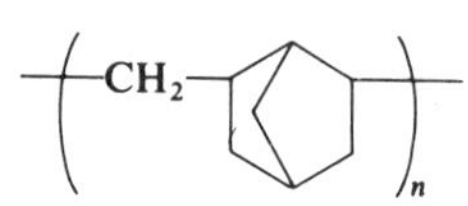

(**15**)

2-Ethoxycarbonylbicyclo[2.2.1]hepta-2,5-diene polymerized readily with free-radical initiators to give high-molecular-weight soluble essentially saturated polymers.[139] The IR data suggested that the polymer consisted primarily of tricyclo[2.2.1.0$^{2,6}$]heptane (nortricyclene) repeating units (**16**) and (**17**). Poly(bicyclo[2.2.1]hepta-2,5-diene) obtained by $\gamma$-radiation-induced polymerization[140] was shown to contain the nortricyclene structure (**18**) as the main unit in the polymer.

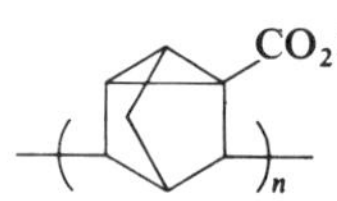

(**16**)

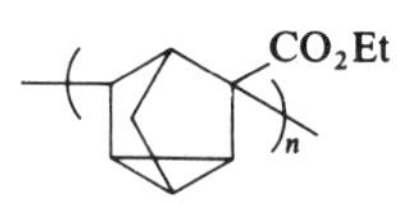

(**17**)

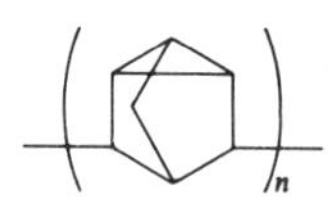

(**18**)

It has also been observed[141] that *cis,cis*-1,5-cyclooctadiene, under the influence of Ziegler-type initiators, yields a substantially saturated soluble polymer through a bicyclic ring-closure polymerization. Spectral and other physical properties suggest it to be poly(2,6-bicyclo[3.3.0]octane). A more recent study has shown that polybicyclic structures are obtained by polymerization of 4-vinylcyclohexene at $-15$ to $-10\,^{\circ}$C in a solution containing acetyl chloride and $AlCl_3$ with acetyl chloride–monomer ratio = 1 and $AlCl_3$–monomer ratio = (1.2–1.3):1.[142]

A wide variety of cyclocopolymerization reactions involving a single 1,4-diene monomer have been observed. For example, radical-initiated polymerization of *N,N*-divinylaniline in solution led to formation of a polymer made up of bicyclic units.[143] A soluble polymer containing about 20% residual unsaturation had been reported previously, produced *via* homopolymerization of divinyl ether.[114] Assuming that radical initiation takes place at the $\beta$-methylene group of the monomer, structures (**19**)–(**22**) were postulated as the predominant units.

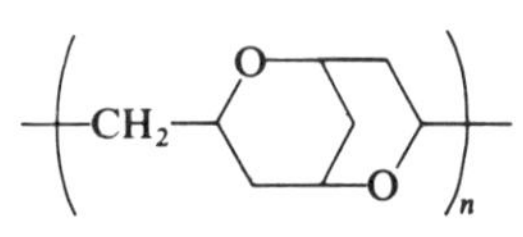

(**19**)

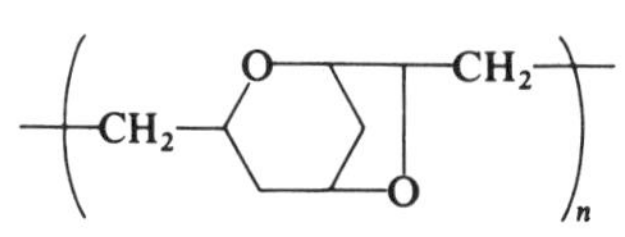

(**20**)

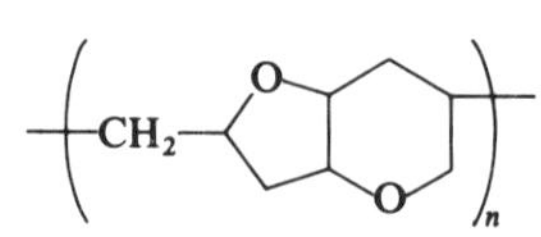

(**21**)

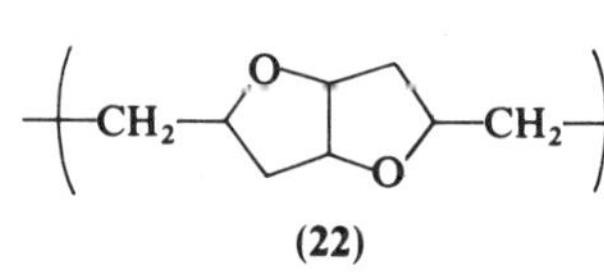

(**22**)

An interesting and novel synthesis of poly(metacyclophane) (**23**), a new polymer, has been described. 8,16-Dihydroxy[2.2]metacyclophane dimerized spontaneously in $CHCl_3$ containing oxygen in the presence of sunlight to give the dimer (**24**) which polymerized at 300–400 °C to give the title polymer (**23**).[145]

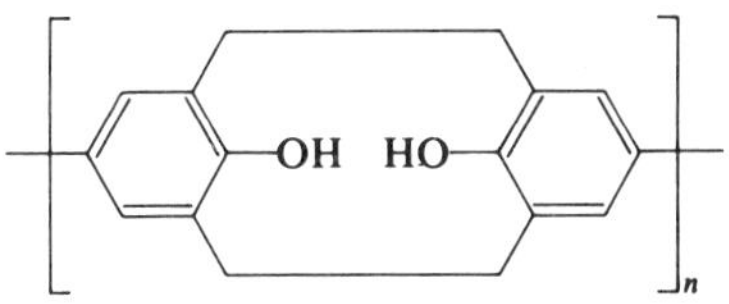

(**23**)

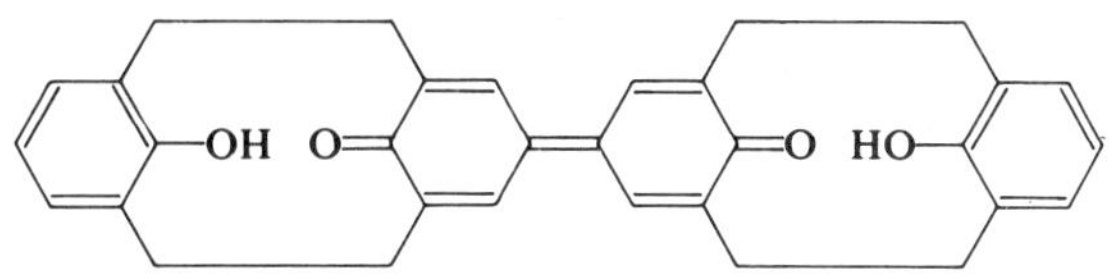

(**24**)

Cyclopolymerization of (*E,E*)-[6.2]paracyclophane-1,5-diene proceeds by an intramolecular–intermolecular mechanism to give a polymer containing [3.2]paracyclophane in the repeat unit. Thermal analysis of the polymer indicates a complex thermooxidative behavior in the presence of oxygen, while depolymerization occurs above about 400 °C in an inert atmosphere.[146]

1,2,3-Triisocyanatopropane has been polymerized to give a linear polymer for which a bicyclic recurring unit (**25**) was proposed. *cis*-1,3-Diisocyanatocyclohexane has been cyclopolymerized under similar conditions. Evidence was presented for structure (**26**) in the polymer. Details have been discussed in a previous publication.[1]

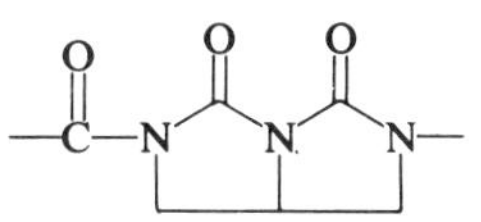

(**25**)

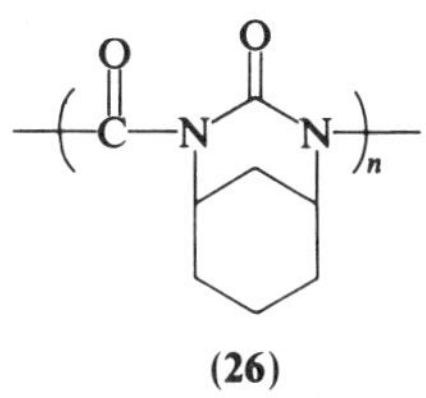

(**26**)

## 23.5 SYMMETRICAL MONOMERS CONTAINING OTHER MULTIPLE BONDS

Other types of monomers that have been shown to undergo cyclopolymerization are diynes, dialdehydes, diisocyanates, diepoxides and dinitriles, as well as unsymmetrical monomers with respect to the functional groups referred to here. In addition, a variety of monomers are reported to lead to bicyclic structures.[110]

### 23.5.1 Diynes

Diacetylenes were studied as early as 1961[1] and have been studied more recently as a result of the present interest in semiconductors. 1,6-Heptadiyne reportedly led to polymer structure (**27**). A more recent investigation of this monomer was initiated as the result of strong interest in electrically conducting films. The results of this investigation show that 1,6-heptadiyne can be converted to a free-standing film which can be doped with iodine to yield an electrically conducting polymer film. The conclusion was that the cyclopolymer possessed predominantly structure (**27**), in accord with that proposed earlier.

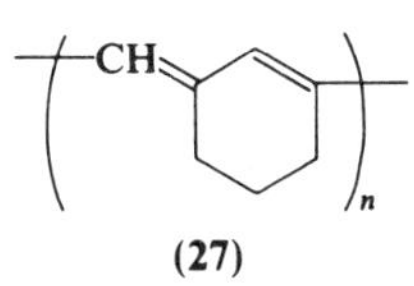

(**27**)

The effect of free radicals on the kinetics of oxidation and doping of polyenes has been studied. Oxidation of the polyenes is related to radical formation, oxygen insertion at radicals being the primary mechanism.[147] The cyclopolymerization of propiolic anhydride in the presence of weak nucleophiles has been reported. Polymerization of the 2-propyn-1-yl esters of several unsaturated acids has also been reported. These polymer structures have been discussed in a previous review.[1]

The polymerization of *N*-substituted di(2-propyn-1-yl)amines in boiling pyridine or DMF containing $PdCl_2$ as the catalyst also gave soluble conjugated cyclic polymers.[148]

### 23.5.2 Dialdehydes

As early as 1908, it was reported that glutaraldehyde gave soluble solid products on standing at room temperature.[149] However, the structures of the products were not elucidated. Both the spontaneous and cationic-initiated polymerization of glutaraldehyde were reinvestigated and the anticipated results obtained. The ring structure was inferred to be six membered, since the IR spectra of the polymers were similar to those of substituted tetrahydropyrans.[1] Other aliphatic dialdehydes which have been studied include succindialdehyde, malealdehyde and *cis*- and *trans*-diformylcyclohexane.[1]

*o*-Phthalaldehyde has perhaps been studied more extensively than any other dialdehyde. This monomer homopolymerizes readily to cyclopolymer using $\gamma$-ray irradiation, cationic catalysts, anionic catalysts and coordinated anionic catalysts. *p*-Phthalaldehyde and *m*-phthalaldehyde resisted polymerization. It was proposed that dipole–dipole interactions between the two adjacent aldehyde groups promote the cyclopolymerization process through an intermediate or in a concerted manner rather than the stepwise manner.[1]

The cationic-initiated copolymerization of *o*-, *m*- and *p*-phthalaldehydes ($M_2$) with styrene ($M_1$) has been studied.[150] The monomer reactivity ratios were $r_1 = 0.77$, $r_2 = 0$ for the *meta* isomer and $r_1 = 0.60$, $r_2 = 0$ for the *para* isomer. Neither the *meta* nor the *para* isomer participated in the polymerization and acted simply as a compound having an electron-withdrawing group, thus reducing the cationic reactivity of these monomers.

### 23.5.3 Diisocyanates

The cyclopolymerization of several aliphatic 1,2-diisocyanates has been reported using sodium cyanide in DMF as initiator below −30 °C. Soluble high molecular weight products were obtained which were reported to have cyclic structures. 1,2,3-Triisocyanatopropane also polymerized to give a linear polymer for which a bicyclic recurring unit (**25**) was proposed. An alternating intramolecular–intermolecular chain-growth mechanism analogous to that postulated for 1,6-dienes was assumed. *cis*-1,3-Diisocyanatocyclohexane has been cyclopolymerized under similar conditions.[1]

A pulsed NMR study of cyclopolymerization of 4,4′-diphenylmethane diisocyanate reported kinetics of cyclopolymerization of the monomer in the presence of $P(NEt_2)_3$ catalyst at 323.15–343.15 K. IR spectra of the polymer indicated that cyclotrimerization of the monomer was accompanied by cyclodimerization of the isocyanate groups.[151] Cyclopolymerization of isocyanates in the presence of other organophosphorous compounds has been observed. The kinetics and mechanism of polymerization of 2,4-tolylene diisocyanate, 1,6-hexamethylene diisocyanate and 1,4-tetramethylene diisocyanate in the presence of triethyl phosphite or triisopropyl phosphate catalysts at 20–80 °C have been discussed.[152] The homo- and co-polymerization of $OCN(CH_2)_4NCO$ and $OCN(CH_2)_6NCO$ in bulk in the presence of $(Et_3N)_3P$ catalyst occurred with cyclization, giving heterochain polymers containing isocyanurate and uretidinedione rings.[153] Cyclopolymerization of 2,4-tolylene diisocyanate (2,4-TDI) in the presence of phosphorus triamides has been studied. $(R_2N)_3P$ (R = Me, Et, Bu) catalyzed both the cyclodimerization and cyclotrimerization of 2,4-TDI at 23–60 °C to give oligomers containing uretidinedione and isocyanurate units as a soluble white powder, stable to 280 °C and decomposing at 400–420 °C.[154]

The effects of intramolecular reactions on network properties have been studied. The correlations between the modulus of polyurethane networks formed at complete reaction and the extent of intramolecular reactions at gelation were interpreted *via* mathematical equations. The occurrence of post-gel intramolecular reaction prevents attainment of the perfect network.[155]

### 23.5.4 Diepoxides

Diepoxides of suitable structure also yield cyclopolymers. 1,2,5,6-Diepoxyhexane has been polymerized under the influence of a variety of initiators to afford soluble polymers for which tetrahydropyran recurring units were proposed.[156] The polymerization of *o*-di(epoxyethyl)benzene has also been investigated. *N,N*-Bis(2,3-epoxypropyl)aniline undergoes cyclopolymerization to give linear polymers containing 2,4,6-substituted morpholindiyl units using potassium *t*-butoxide as initiator. These and many other studies on a variety of diepoxides and their participation in cyclopolymerization have been reported in recent reviews.[1,156]

During three-dimensional polymerization of diglycidyl phthalate (glycidol = 2-hydroxymethyloxirane) with aromatic diamines, side reactions involving cyclization and aminolysis can result in network defects. The presence of such reactions and their relative contribution to the overall reaction have been established by gel chromatographic analysis of model systems of diglycidyl phthalate with aniline or *o*-anisidine.[157]

The effects of cyclization reactions on topological structure and properties of crosslinked polymers prepared from diglycidyl ethers and aromatic amines have been studied. Ring formation in the crosslinking of epoxy resins was studied using a statistical model. Systems based on ortho diglycidyl ethers or esters exhibited a marked tendency to form cyclic defects in the linear network.[158]

The synthesis of polymers with thiacrown ether units *via* cyclopolymerization of 1,2-bis(2,3-epithiopropoxy)benzene and 1,2-bis[2-(2,3-epithiopropoxy)ethoxy]benzene gave polymers with thiacrown ether units. The higher metal-binding capacity of the 1,2-bis[2-(2,3-epithiopropoxy)ethoxy]benzene homopolymer compared to the 1,2-bis(2,3-epithiopropoxy)benzene homopolymer was related to a larger size of the thiacrown ether rings of the former polymer.[159]

Cyclization in the reaction between diglycidylaniline and amines has been studied. The reaction of diglycidylaniline with *N*-methylaniline or aniline was investigated by HPLC, and the products were identified by mass spectrometry. An eight-membered ring formed by the intramolecular reaction of the secondary amino group and the epoxy group in the diglycidylaniline–aniline monoadduct was the main cyclic product.[160]

### 23.5.5 Dinitriles

Succinonitrile, as discussed in a recent review,[1] has been reported by a number of investigators to undergo cyclopolymerization to a black solid polymer, the spectra of which were consistent with structure (**28**). Anionic initiators such as sodium methoxide at atmospheric pressure, free-radical and cationic initiators have been reported. The polymer is reported to possess semiconductor properties.

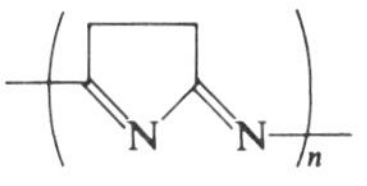

**(28)**

Homopolymers and copolymers of fumaronitrile, maleonitrile, succinonitrile, diphenylmaleonitrile, *meso*- and DL-diphenylsuccinonitrile, *o*-phthalonitrile and 1,2,4,5-tetracyanobenzene have been prepared using a variety of initiators. On the basis of the physical appearance of the polymers, their solubility and analytical data, as well as chemical and spectroscopic evidence, completely conjugated cyclic structures were proposed. The polymers possess good thermal stability.[1]

Cyclopolymerization of aromatic nitriles to polymers with ring–chain structures has been observed. A review, with 13 references, of the preparation, properties and uses of heat-resistant polymers obtained by the triaryl-1,3,5-triazine ring-forming reaction of terminal nitrile groups on polyimides, polybenzimidazoles and polyimidazopyrrolones has summarized this work.[161]

A non- or partially crosslinked triaryltriazine ring-containing polymer or copolymer is prepared by cyclopolymerizing an aromatic nitrile-modified oligomer precursor in the presence of an excess of aromatic nitrile (chain extender) at $>100\,^{\circ}C$ and $>200\ lbf\,in^{-2}$ ($1\ lbf\,in^{-2} \simeq 6.895 \times 10^3$ Pa).[162]

## 23.6 MONOMERS CONTAINING DIFFERENT MULTIPLE BONDS (FUNCTIONALLY UNSYMMETRICAL MONOMERS)

A variety of monomers having two kinds of functional groups have been studied in cyclopolymerization. For example, *o*-vinylbenzaldehyde has been cyclopolymerized *via* cationic initiation to yield polymers containing both of the predicted cyclized structures shown in eqution (6). The cyclopolymer also contained 20% pendant vinyl groups and 5% pendant aldehyde groups.[156]

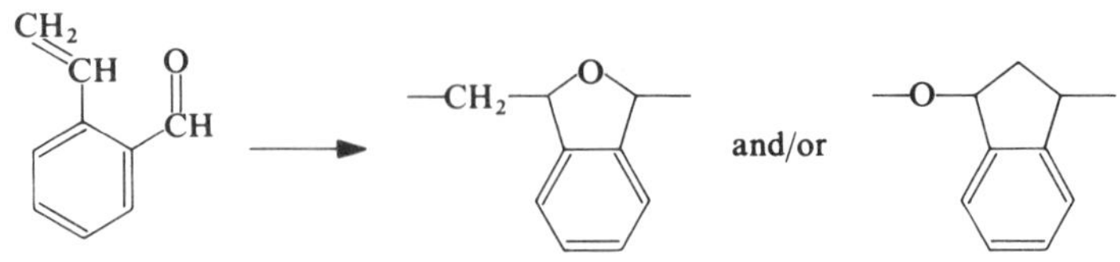 (6)

Glycidyl acrylate and methacrylate have been reported to cyclopolymerize, using $BF_3OEt_2$ as initiator. Although some disagreement still exists as to the exact mechanism, there appears to be general agreement that both functional groups of the monomers participate in a cyclopolymerization mechanism.[1]

Polymerization of acrolein dimer with an initiator system such as $AlEt_3 \cdot H_2O$ or $AlEt_2$ (equation 7) gave a polymer consisting in part of 6,8-dioxabicyclo[3.2.1]octane units (**29**; R = H).[163] Methacrolein dimer was cyclopolymerized, giving a soluble polymer having only 1,4-dimethyl-6,8-dioxabicyclo[3.2.1]octane (**29**; R = Me) recurring units using $BF_3OEt_2$ as initiator.[164]

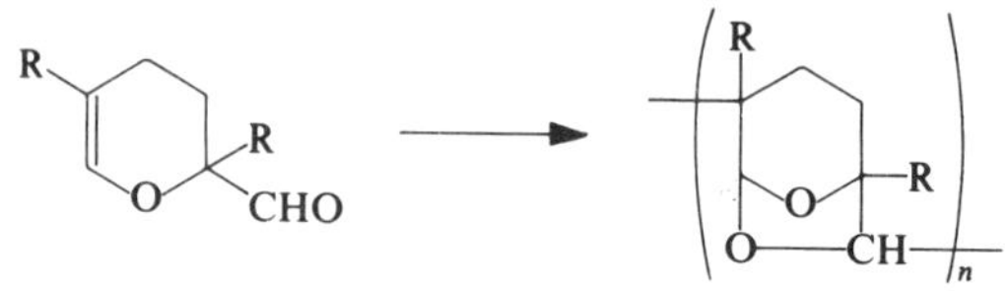 (7)

**(29)**

## 23.7 COPOLYMERIZATION OF NONCONJUGATED DIENES WITH VINYL MONOMERS

Copolymerization studies of 1,6-dienes with conventional vinyl monomers have been carried out. Copolymerization involving cyclopolymerization can proceed by two distinctly different pathways: (a) the nonconjugated diene alone can form the cyclic structure followed by reaction of the resulting cyclic radical (or other reactive intermediate) with the comonomer or (b) the comonomer can contribute to formation of the cyclic structure along with the nonconjugated diene. The latter type is generally defined as cyclocopolymerization, and will be discussed later.

One of the earliest investigations carried out was that of copolymerization of several diallyl monomers with acrylamide and acrylonitrile.[165] Diallyldimethylammonium chloride, diallylmethylamine hydrobromide, diallylamine, diallylmethylamine, diallylmethylamine oxide, and the zwitterion obtained by quaternizing diallylmethylamine with α-haloacetic acid were among those compounds studied. Soluble copolymers of high molecular weight were obtained.

Copolymerizations of methacrylic anhydride and a variety of common types of vinyl monomers have been described in a previous review.[156] Methyl methacrylate, styrene, ethyl acrylate, vinyl acetate, 2-chloroethyl vinyl ether and allyl chloroacetate were among the comonomers studied. Both bulk and solution polymerizations at 60–80 °C have been used. In general, soluble copolymers were formed under the following conditions: (a) the less reactive the comonomer in free-radical copolymerizations, (b) the greater the dilution, (c) the greater the difference in the total moles of the two components in the charge and (d) the lower the conversion.

Copolymers of ethylene and 1,5-hexadiene, prepared through use of a modified Ziegler catalyst, are crystalline over the compositional range of 15–93 mol % of ethylene.[166] An unusual mutual synergistic effect on polymerization rate in the system has also been reported. The addition of ethylene to the 1,5-hexadiene monomer feed resulted in an increase of 1,5-hexadiene conversion up to about 10 wt % ethylene. An increased rate of ethylene polymerization by addition of 1,5-hexadiene was also observed.

Copolymerization of diallylamine hydrochloride with sulfur dioxide gives copolymers the composition of which is always 1:1 regardless of the feed ratio.[165] Conversions, when calculated on the basis of the 1:1 diallylamine hydrochloride–sulfur dioxide complex, range up to 100%.

This limited discussion deals only with representative examples of copolymerization of nonconjugated dienes with conventional vinyl monomers. Numerous additional examples are discussed in recent reviews.[1,156] Recently published copolymerization studies deal predominantly with the commercially important monomer diallyldimethylammonium chloride (DADMAC).

Grafted copolymers of DADMAC have been used in developing organic pigments useful in loading paper. Thus, an acrylamide–DADMAC prepolymer was grafted with styrene to give a graft copolymer which was crosslinked with glyoxal.[167] Copolymers of DADMAC with methacrylic acid which exhibit flocculating and desalting properties have been patented in the USSR.[168] Copolymerization of *N,N*-dimethylaminoethyl methacrylate with DADMAC gave perfectly soluble copolymers and no vinyl groups were observed by physical methods.[169]

A stable microbiocidal aqueous dispersion comprising methylene bisthiocyanate and a self-inverting emulsion of a copolyampholyte containing DADMAC has been found to be suitable for inhibiting the growth of slime-forming bacteria, fungi and algae.[170] Emulsions containing a saponified DADMAC–vinyl acetate copolymer or allyltrimethylammonium chloride–vinyl acetate copolymer have been prepared and mixed with thickeners, freeze–thaw stabilizers, rustproofing agents, preservatives, defoamers, emulsifiers or dispersants and water.[171] Polyampholytes and their salts have been prepared from anionic, cationic and some nonionic monomers and used as scale and corrosion inhibitors, in drilling fluids and in water-flood oil recovery. A typical example of such a polyampholyte is an acrylic acid–DADMAC–*N,N*-dimethylacrylamide terpolymer (70:20:10).[172]

Cationic polyelectrolytes from acrylamide and DADMAC have been prepared in aqueous solutions and their flocculating capacities were tested in aqueous suspensions of $TiO_2$. Acrylamide homopolymer of molecular weight $<10^6$ had virtually no flocculating capacity. The flocculating capacity increased and the optimum dose decreased with increasing molecular weight of the copolymers, but larger doses than those of the homopolymer were required for producing the same sedimentation rates.[173]

A copolymer of DADMAC with acrylic or methacrylic acid has been used as a bactericide for water–oil-base cutting fluids.[174] The effect of polyelectrolyte flocculants on microorganisms in receiving streams has been studied. Tests of polyelectrolytic toxicity to aquatic microorganisms demonstrated that poly(DADMAC) did not inhibit the cumulative oxygen-uptake of a culture of mixed aquatic microorganisms, as measured with an electrolyte respirometer. Although oxygen-uptake curves showed that the polymer could not support microbial activity by the mixed aquatic microbe, a *Pseudomonas* species was isolated from the St. Lawrence River which could biodegrade and assimilate the polyelectrolyte. This bacterial species was also able to utilize the monomer as a source of both carbon and nitrogen.[175]

A graft copolymer of cellulose and DADMAC has been prepared by impregnating cellulose with aqueous monomer containing $(NH_4)_2S_2O_8$, removing the excess and heating the impregnated cellulose.[176] Reactivity ratios ($r$) have been determined for copolymerization of DADMAC ($M_1$) and acrylamide ($M_2$). Both $r_1$ and $r_2$ depended on DADMAC concentration in the initial comonomer solution, and increased with decreasing DADMAC concentration. An increase in total monomer concentration increased $r_1$, but $r_2$ remained nearly constant. The results are explained by association of the cationic monomer and electrostatic interactions.[177] Terpolymers containing *N,N*-diallylacetamide for use as high-temperature oil-well drilling fluids have been synthesized.[178]

Certain copolymers of diallylamine have been used in developing dye-fastness improvements. Copolymers having structural units (**30**) and (**31**) (where $R^{1-4}$ = H, Me) in 95–50:5–50 ratio are prepared and used to improve dye fastness of textiles.[179]

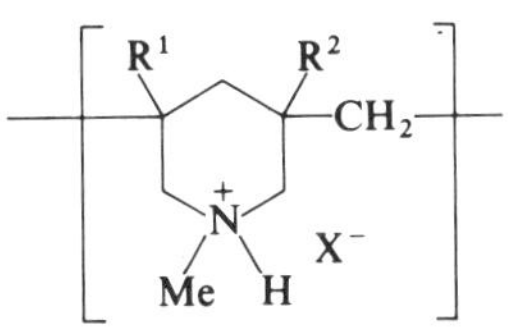

(**30**)

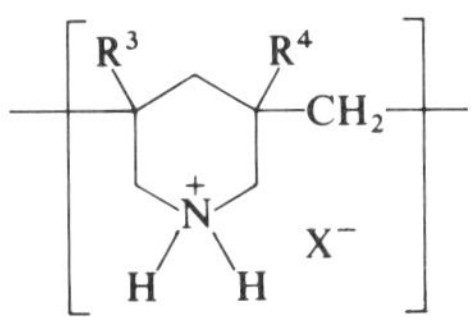

(**31**)

Cyclopolymerization of *N,N*-diacryloylmetribuzin with hydrophilic comonomers to yield polymeric herbicides has been studied. Copolymerization with acrylamide and acrylic acid were conducted to increase hydrophilicity. Studies in aqueous buffer systems indicated that hydrolysis of the labile bond to release metribuzin was occurring.[180]

Copolymers of allyl methacrylate with styrene with high gel rigidity have been developed for negative electron-beam resists which do not swell during development or flow in thermal processing.[181] Cyclization has been observed in a study of methyl methacrylate–methylene dimethacrylate copolymers and the effect of dilution and the dimethacrylate structure on the gel point determined. The relations between crosslinking and cyclization in the systems are discussed.[182]

Cationic block copolymers of DADMAC have been claimed in a recent patent. Water-soluble cationic block copolymers, useful for flocculating agents and dewatering assistants for papermaking, are prepared which contain a polymer having amine groups that are predominantly quaternized and a polymer having amine groups that are predominantly primary or secondary.[183]

## 23.8 CYCLOCOPOLYMERIZATION OF DIENES WITH SELECTED MONOMERS

A type of copolymerization distinctly different from that described above now constitutes a significant portion of the cyclopolymerization literature.[184] This process, referred to as cyclocopolymerization, incorporates both comonomers into the developing cyclic structure. Perhaps the most extensively studied example of this unusual type of copolymerization is the cyclocopolymer of divinyl ether and maleic anhydride (**32**; Scheme 5). This copolymer (DIVEMA) has been extensively studied for its biological properties; it exhibits a broad spectrum of biological activities,[184] including antitumor, antiviral, antibacterial, anticoagulant and antiarthritic properties, as well as being capable of generating interferon, inhibiting inverse transcriptase, activating macrophages and eliminating plutonium.

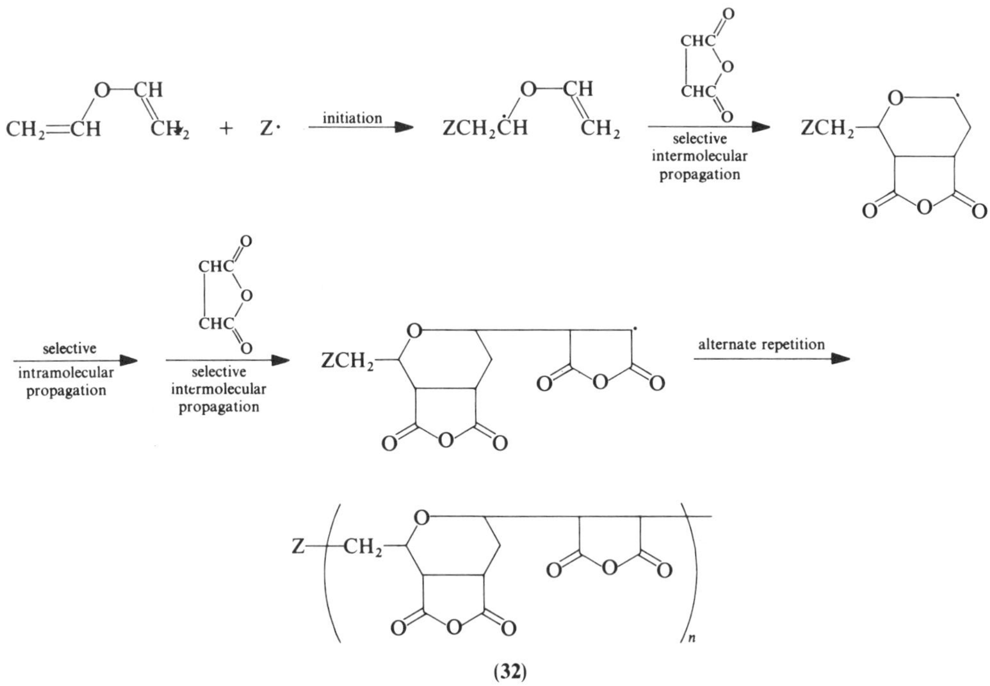


**(32)**

**Scheme 5**

Cyclocopolymerization of a wide variety of comonomers has been studied, and it has been concluded that cyclocopolymerization is most likely to occur when one monomer possesses strong electron-donor properties and the other strong electron-acceptor properties.[156,184] Thus, it appears reasonable to suspect that an intermediate charge-transfer complex (or donor–acceptor complex) plays a significant role in controlling the regularly alternating cyclocopolymerization.

Divinyl ether and maleic anhydride had been observed to yield soluble linear copolymers before the cyclopolymerization mechanism had been postulated.[156,184] However, it was not until this mechanism had become accepted that a plausible explanation for this copolymerization was forthcoming. The postulated mechanism (Scheme 5) accounted for the observed properties of the copolymer and has now become widely accepted.

Cyclocopolymerization of sulfur dioxide with *cis,cis*-1,5-cyclooctadiene using radical initiation in dilute solution has been reported.[185] The copolymer having structure (**33**) was obtained in film- and

(8)

**(33)**

fiber-forming molecular weights, had no residual unsaturation, and contained 2 mol sulfur dioxide per 1 mol cyclooctadiene (equation 8).

Cyclocopolymerization of (+)-limonene with maleic anhydride in THF was reported recently. The composition of the copolymer showed that (+)-limonene readily underwent cyclocopolymerization giving 1:2 alternating copolymers which were optically active.[186]

The participation of donor–acceptor complexes in the radical polymerization of maleic anhydride with divinyl ether has been evaluated by a method based on polymerization rate constants experimentally determined at two different total monomer concentrations and two different monomer concentration ratios. The participation of the complex in the copolymerization was only 1–17% despite a significantly higher complex reactivity compared to monomer reactivity.[187]

$^{13}C$ NMR studies of the structure of DIVEMA indicated that copolymer samples prepared in toluene or benzene contain tetrahydrofuran (THF) and tetrahydropyran (THP) rings in a ratio of 0.8:1.[188] An experimentally reproducible procedure for the alternating cyclopolymerization of maleic anhydride and divinyl ether has been published. The polymer was characterized by IR, NMR and gel permeation chromatography.[189] Calcium salts of DIVEMA have been shown to have much reduced toxicity. Solid cyclic ~1:2 divinyl ether–maleic anhydride polymer calcium salts are as effective as the corresponding alkali metal salts in treating tumors or viruses, and are much less toxic.[190] Many additional examples of cyclocopolymerization of 1,4- and 1,5-dienes have been discussed in detail in published reviews.[1,156,184]

## 23.9 LADDER POLYMERS

Cyclopolymerization has been used (Scheme 6) to synthesize ladder polymers from preformed polymers of general structure (**34**). Perhaps the most important example of this mode of polymer modification is 'Black Orlon', a polymer obtained by heat treatment of polyacrylonitrile and precursor in one process for the manufacture of graphite fibers.[1,184] Partial-ladder polymer synthesis from conjugated dienes has been reported and discussed in several reviews.[1,191,192] Copolymers of ethylene and isoprene having a monomer mole ratio of 25:1, and in which the isoprene units were incorporated into the chain through the 3,4 double bond, when treated with phosphorous oxychloride in tetrachloroethylene, yielded soluble copolymers containing fused cyclohexane units.

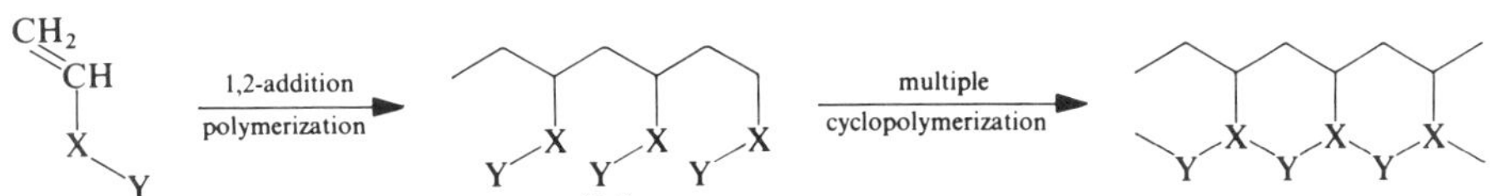


(**34**)

—X—Y = —CH=CH$_2$, —C=O, —C≡N or —N=C=O

**Scheme 6**

A method of cyclized rubber production has been developed in which polyisoprene is treated with *p*-toluenesulfonic acid in an aromatic hydrocarbon.[193] Cyclized rubber has been used in the development of correction-fluid compositions. These fluids are applicable to water- and oil-based printing and writing inks and also contain $TiO_2$, rubber binders and hydrocarbon solvents.[194] A developer for cyclized rubber photoresists has been described. The method provides high-resolution negative type photoresists and permits carrying out simultaneous developing and rinsing processes.[195]

Cyclopolymerization during the synthesis of poly(vinyl chloride-*co*-1,3-butadiene) has been observed. The presence of other important structural defects in butadiene–vinyl chloride copolymers, including a cyclohexyl branch arrangement that results from a cyclopolymerization reaction, were also observed.[196]

High-purity cyclized polyisoprene has been prepared by living polymerization of isoprene in the presence of an organometallic catalyst to give cyclized rubber and bringing the polymer into contact with a catalyst composed of $BF_3$–ether complex and a carboxylic acid under mild conditions.[197] Protective coatings based on cyclized rubber have been developed. Formation, physical properties and coating applications of cyclized rubber have been discussed with an emphasis on its use in antifouling and marine points, and some formation aspects have been explored, including the choice of solvents, pigments and plasticizers.[198]

Acrolein, an $\alpha,\beta$-unsaturated aldehyde, contains a conjugated double bond system and can be considered as a 1,3-diene. In contrast to 1,3-dienes, the double bonds in acrolein possess unequal reactivities. Polymerizations initiated with water-soluble redox systems produce polymers almost exclusively through the vinyl group.[199] Ionic initiators are effective agents for initiating polymerization of acrolein through the carbonyl double bond; however, the polymers also contain free carbonyl groups which show that polymerization also occurs to some extent through the vinyl double bond. When sodium naphthalene is used in a nonaqueous medium, more than 70 mol % of the copolymer consists of bond formation through the carbonyl group.[200] Radical-initiated polymerization of acrolein[201] yields an insoluble white powdery polymer which contains 2,6-disubstituted tetrahydropyran units in support of the cyclopolymerization mechanism.

Vinyl isocyanate has been polymerized through the isocyanate group by using sodium cyanide as initiator at $-55\,^{\circ}C$ to yield *N*-vinyl-1-nylon.[202] When this polymer, in dimethylacetamide solution, was treated with AIBN and UV light at room temperature, the product was soluble in solvents of high dielectric strength and was shown to consist of 90% ladder polymer. The results of other studies on this monomer essentially confirm these findings.[203]

Macrocyclic ladder polymers (**35**; equation 9) are postulated to arise from the crosspolymerized product of 1,11-dodecadiyne. Electron diffraction patterns were obtained from the macromonomer and crosspolymerized crystals of the polymer. The unit cell of both macromonomer and crosspolymerized material was monoclinic, space group $P2_1/n$.[204]

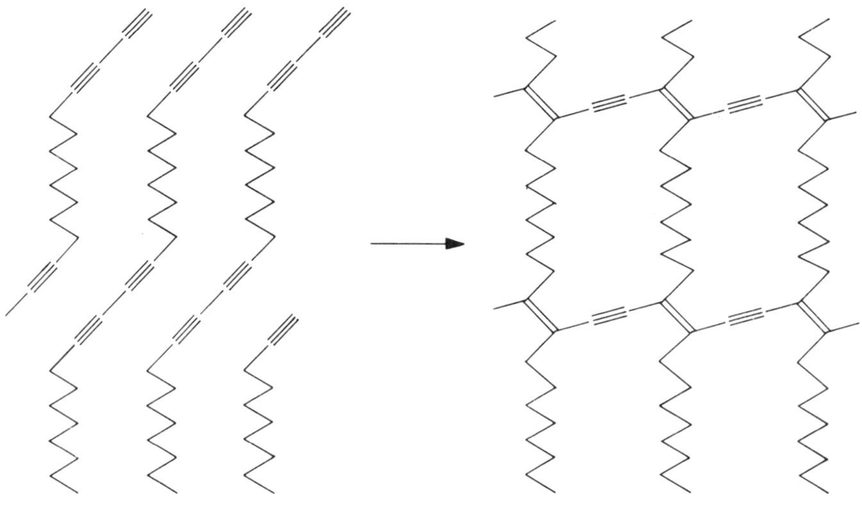

(9)

**(35)**

## 23.10 MECHANISM OF CYCLOPOLYMERIZATION

### 23.10.1 Kinetic Studies

Allyl monomers are classically known for their degradative chain-transfer characteristics. The ability of the diallyl quaternary ammonium monomers to form high polymers apparently without interference from degradative chain-transfer appears to be an exception. An explanation for the lack of degradative chain-transfer on the monomer during radical polymerization of *N*,*N*-diallyl-*N*,*N*-dialkylammonium halides, has been offered; formation of the intermediate (**36**; R = methyl, ethyl) in chain transfer on the monomer in radical polymerization of *N*,*N*-diallyl-*N*,*N*-diethylammonium chloride and DADMAC, and its rearrangement to the radical (**37**), was postulated as an explanation of experimental data indicating formation in the chain transfer of a radical more active than the allyl radical (*e.g.* **36** prior to rearrangement ) and participation of this active radical in chain propagation giving a linear polymer, rather than a branched or crosslinked polymer that would be formed if (**36**) acted as a chain initiator. The presence of one, rather than two, terminal double bonds, and of a methyl terminal group, indicated by $^{1}H$ and $^{13}C$ NMR spectroscopy in the isolated polymers, supported the postulated mechanism.[205]

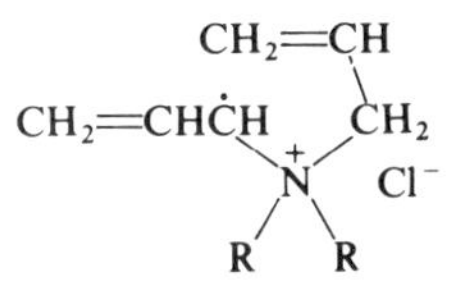

**(36)**

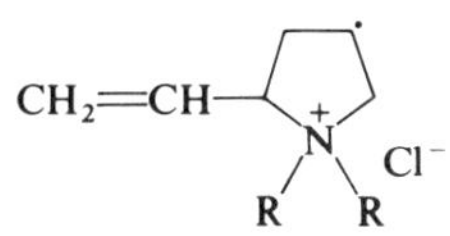

**(37)**

Mechanistic studies dealing with various aspects of the cyclopolymerization process are extensive and cannot be presented in detail here. However, important details of the process are embodied in a kinetic study carried out on methacrylic anhydride.[206] This study showed that the energy of activation for the intramolecular cyclization step is higher than that for the intermolecular step by about 2.6 kcal $mol^{-1}$ (1 cal = 4.184 J). The rate of cyclization, however, was found to be considerably greater than the intermolecular propagation step, in support of a very high steric factor favoring cyclization. The value $k_c/k_{11}$ was found to be 2.4 mol $l^{-1}$, and the Arrhenius frequency factor ratio was found to be 256 mol $l^{-1}$ in favor of the cyclization step. Also, it was shown that increasing steric factors, higher temperature, lower monomer concentration, poorer solvent and higher conversion favor cyclization.

Numerous other kinetic studies have reported $k_c/k_{11}$ values to vary from 2.8 for radical-initiated polymerization of *o*-divinylbenzene in benzene at 50 °C up to 200 for radical-initiated polymerization of formaldehyde divinyl acetal in benzene at 50 °C. Arrhenius frequency factor ratios have been reported to vary from 50 mol $l^{-1}$ for radical-initiated polymerization of *o*-divinylbenzene in benzene at 50 °C up to $2.2 \times 10^4$ for cationic-initiated polymerizaion of *o*-divinylbenzene in toluene. The values for $E_c - E_{11}$ have been reported to vary from approximately zero for radical-initiated polymerization of vinyl *trans*-cinnamate at 70 °C in benzene to 5.3 kcal $mol^{-1}$ for cationic-initiated polymerization of *o*-divinylbenzene in toluene.[156]

Many recently published kinetic studies related to cyclopolymerization have concentrated on the commercially important monomer DADMAC and its copolymerization characteristics. The polymerization of DADMAC initiated by $(NH_4)_2S_2O_8$ followed kinetics of 2.3 order in the monomer and 0.47 order in the initiator, with an overall activation energy of 15.4 kcal $mol^{-1}$. Comparison of the experimental kinetic equation for polymerization of DADMAC with the theoretical equations for polymerization of unconjugated symmetrical dienes suggested that the polymerization of the monomer occurred with partial cyclization and chain termination by recombination of cyclized radicals. The $k$ and $\alpha$ constants of the Mark–Houwink equation for poly(DADMAC) (molecular weight 13 500–22 500, 25 °C, solution in 1 mol $l^{-1}$ NaCl) were determined ($12.6 \times 10^{-4}$ and 0.51, respectively).[207] New results on cyclopolymerization kinetics of the same monomer have been published. During the cyclopolymerization of pure monomers using oxygen-free water as solvent and $K_2S_2O_8$ as initiator, the rate of polymerization was 0.75 order in $K_2S_2O_8$ concentration, second order in DADMAC concentration, 0.5 order in DADMAC cation concentrations and 0.5 order in $Cl^-$ concentration.[208]

Studies of the kinetics of radical polymerization of DADMAC have been continued. The rate of free-radical initiation, determined from integral evaluation of the time dependence of conversion and inhibition, was of 1.5 and second order in initiator and monomer, respectively. The overall rate of polymerization was proportional to the 0.75 and third power of initiator and monomer, respectively, owing to the effect of the rate of initiation.[209] A study of the kinetics of radical polymerization of this monomer in water–methanol, and methanol solutions with or without the presence of low molecular weight electrolytes revealed that the rate of polymerization sharply and nonlinearly increases with increasing initial concentration of monomer.[210] The mechanism of the initiation and termination reactions with $K_2S_2O_8$ as initiator has been studied. In aqueous solution, primary radicals are formed simultaneously by activated decomposition of ion pairs ($S_2O_8^{2-}$/monomer associations), forming sulfate ion radicals, and redox reaction of $S_2O_8^{2-}$ with $Cl^-$ in ion pairs with monomer, forming sulfate anion radicals and Cl·. The partial reactions were shown by quantitative analysis to proceed at equal rates. Cl· formed in the second reaction acts as both initiator and terminator, the additional chain termination being regarded as transfer to monomer in formal kinetic terms.[211]

The kinetics and mechanism of the radical cyclopolymerization of DADMAC has been reviewed. The kinetic analysis, taking into consideration nearly complete cyclization, a linear increase of $k_p k_t^{0.5}$ with [M] and a different mechanism of initiation depending on the nature of the initiator, leads to rate equations which fit the experimental data well. Initiation with $S_2O_8^{2-}$ has the following peculiarities: formation of primary radicals by redox reaction with chloride ions and interaction with the monomer cation, additional termination by chlorine atoms and an experimental chain-transfer constant to monomer which includes transfer to monomer and termination by chlorine radicals.[212]

### 23.10.2 Ring Size Studies

It was recognized quite early that less-favored ring sizes may be formed in cyclopolymerization and that the less stable radical may predominate in the cyclization.[109] It was shown that the less stable radical predominated in the cyclization step during polymerization of allyl and methallyl crotonates. Justification of this preference for five-membered ring formation has been based on both electronic and steric factors. A suggestion that there may be an electronic interaction between the initially formed radical and the neighboring double bond of the diene has received considerable attention.[192,213]

Extensive work dealing with small molecule cyclizations has been done, and much of this work has been summarized in a review article.[214] In a study of cyclization of 1-substituted 4-hexenyl radicals, it was shown that, as substitution in the 1-position by radical-stabilizing groups (one or two CN or C(O)OEt groups) was increased, the mixture of cyclized products changed from nearly pure cyclopentane to nearly pure cyclohexane derivatives. Higher temperatures favored cyclohexane formation. Also, it was shown that the cyclization reactions were reversible when C(1) was disubstituted, although it had been shown[215] earlier that in the case of the primary hexenyl radical, the cyclization step is irreversible. On this basis, it was proposed that the cyclopentane product is the kinetically preferred product while the cyclohexane product is preferred *via* thermodynamic control, with the energy of activation for cyclization being higher for the cyclohexane derivatives.

The majority of the radical cyclizations reported have been accomplished by permitting alkyl radicals to react with an unsubstituted double bond, thus producing the intermediate nonresonance-stabilized primary or secondary radical.[216] As stabilization of the generated radical increased, the ratio of six-membered ring increased. A more recent study[217] has shown that when the radical generated after cyclization of the six-membered ring can be stabilized more by resonance than the competing primary radical, the ratio of six-membered to five-membered rings is markedly increased. The results of this study strongly indicate that radical stabilization results in greater selectivity and thus an increase in the more thermodynamically stable six-membered ring. The cyclization rate constant ratios at 40 °C and the derived Arrhenius parameters showed that the radical-stabilizing effect of the methyl group was not sufficient to prevent the predominance of the tetrahydrofuran (THF) ring. The $k_1/k_2$ ratio at this temperature was 43; however, the corresponding ratio for the phenyl case was only 0.55. The Arrhenius parameter ratios, $A_2/A_1$, were $0.35 \pm 0.08$ for the methyl case and $1.8 \pm 0.2$ for the phenyl case. An analogy can be drawn between five-membered ring formation in cyclopolymerization of 1,6-dienes and head-to-head polymerizaion in a classical vinyl monomer.

By means of IR, $^1$H NMR, and $^{13}$C NMR spectroscopy, it has been shown[218] that the ratio of five- to six-membered ring formation in cyclopolymerization of acrylic anhydride at 50 °C varies with the dipole moment ($\mu$) of the solvent from 5% five-membered ring at $\mu = 0$ to 30% at $\mu = 4$. A pronounced temperature effect was also observed. On the basis of the previously described studies on acrylic anhydride, it was postulated[219] that, if methacrylic anhydride were polymerized above the ceiling temperature of its monoene counterpart, methacrylic isobutyric anhydride, the poly-(methacrylic anhydride) should consist largely of five-membered ring anhydride structures. Previous studies[218] had shown that cyclic poly(methacrylic anhydride) consisted exclusively of six-membered rings under the lower temperature conditions. As postulated, polymerization of methacrylic anhydride at 160 °C in benzonitrile was found to give 53% five-membered rings, and polymerization at 180 °C gave almost exclusively the five-membered ring structure. Furthermore, in accord with the postulate, five-membered ring formation was found to increase with increasing temperature and dilution.

Radiation-induced cyclopolymerization of *N*-substituted dimethacrylamides has been studied extensively in the liquid, supercooled-liquid, glassy and crystalline states.[220] In most of these cases, cyclopolymerization occurred exclusively, predominantly to five-membered rings, except in the glassy state where no polymerization was observed. The ratio of six-membered rings was higher in the crystalline state than in the liquid states. A single-crystal X-ray diffraction and solid-state polymerization study[221] of *N*-(*p*-bromophenyl)dimethacrylamide led to the conclusion that the monomer crystallizes in a conformation that slightly favors the intramolecular over the intermolecular polymerization reaction.

Certain of the diallylamines and quaternary ammonium salts also yield cyclopolymer consisting largely of five-membered rings. On the basis of a $^{13}$C NMR study of poly(*N*,*N*-diallyl-*N*-methylamine),[227] it was concluded that five-membered rings were formed; however, when the allyl groups were substituted in the 2-position, mixtures of both five- and six-membered rings were formed. Also, by $^{13}$C NMR and model-compound studies, it was shown that poly(DADMAC)[228] consisted predominantly of five-membered rings linked mainly in a 3,4-*cis* configuration.

Many attempts have been made to account for the otherwise unpredictable ring size or preference for head-to-head propagation in cyclopolymerization and other radical cyclizations.[109] It has been suggested that I strain[213] in the cyclic structures being formed during cyclopolymerization may shed some light on the competition between five- and six-membered ring systems and their relative ease of formation. In five-membered ring systems, when hybridization at any one carbon is changed from $sp^3$ to $sp^2$, I strain is decreased. The transition from cyclopentane to cyclopentanone would reduce I strain. In the six-membered ring, change of hybridization from $sp^3$ to $sp^2$ increases I strain. Consideration of the fact that five-membered ring formation in the dimethacrylamides requires introduction of $sp^2$ hybridization at two carbons, which should reduce I strain relative to cyclopentane, whereas $sp^2$ hybridization at two carbons in the six-membered ring should increase I strain relative to cyclohexane, may offer a partial explanation for the five-membered ring preference in these cases.[222] Similar comparisons have been referred to in previous reviews.[156] On the other hand, the relative differences in the calculated heats of formation[223] of the cyclic anhydrides and their open-chain dicarboxylic acids indicate that the six-membered ring should be favored over the five-membered ring from these considerations.[219]

It is now well established that the ring size formed in free-radical cyclization reactions of 5-hexenyl radicals is influenced according to whether the reaction is rate or equilibrium controlled.[224] With many alkenyl radicals, the formation of a five-membered ring is kinetically favored over formation of a six-membered ring.[225] However, with 5-substituted 5-hexenyl radicals, the formation of a six-membered ring can be increased due to either the steric effect of the substituent causing a retardation in the rate of 1,5 cyclization[226] or a resonance-stabilizing effect that favors 1,6 cyclization.[217] 1,5 Ring-closures of substituted 5-hexenyl and related radicals are stereoselective: 1- or 3-substituted systems afford mainly *cis*-disubstituted products, whereas 2- or 4-substituted systems give mainly *trans* products.[225] Studies of the cyclopolymerization of variously substituted diallylamines suggest that these monomers also exhibit the characteristics of kinetic *vs.* thermodynamic control.[227]

The stereoelectronic requirements of the transition state for radical addition reactions have been proposed to explain the kinetic preference for the formation of five-membered ring products.[228,229] This concept demands maximum overlap of the half-filled $p$ orbital with the vacant $\pi^*$ orbital of the double bond. Approach of the radical must be along a vertical line from one of the carbon atoms of the double bond with the orbitals holding the three electrons being in the same plane throughout the reaction. Such conditions can be met by 1,5 cyclization of 5-hexenyl radicals but not by 1,6 cyclization. It has also been suggested that the predominance of *cis* addition in some cyclopolymerizations may be the result of hyperconjugative maxima of the $p$ orbital with the alkyl CH $\alpha^*$ orbitals, leading to a delocalized orbital of similar symmetry to the $\pi^*$ orbital, which results in an attractive interaction between the alkyl substituent and the double bond (Figure 1). This model for the cyclization reaction can explain the increasing preference for cyclopolymers with six-membered rings as bulky 2-substituents are introduced to one or both allyl groups of diallylamines.

**Figure 1**

Support for this hypothesis to explain the cyclization mechanism by examination of molecular models, calculations of geometrical probability factors and molecular orbital evaluations have been discussed in a recent review.[156] The ring size in cyclopolymers will be influenced by the nature of the cyclization reaction; it is therefore important to understand the energetics and kinetics of model reactions. However, there are indications that the correlation between models and polymers is unsatisfactory.[227]

The cyclolinear structure of polymers of $N,N$-diallyl-$N,N$-dialkylammonium halides has been further studied. The nature of the $N$-alkyl substituents, the anion of the quaternary ammonium salt and the solvent affect the structure and configuration of cyclolinear poly($N,N$-dialkyl-$N,N$-diallylammonium halides). An increase in monomer concentration from 0.25 mol l$^{-1}$ to 3 mol l$^{-1}$ in preparation of poly(DADMAC) and poly($N,N$-diethyl-$N,N$-diallylammonium chloride) in aqueous solutions and of poly(DADMAC) in methanol solutions led to no changes in the polymer structure.[34]

The present understanding of cyclopolymerization would indicate that the structures of cyclopolymers are highly dependent on the nature of the monomers and the reaction conditions. Predictions of structures based upon thermodynamic control of the cyclization reaction can be misleading since kinetic products are often found. An important driving force for the cyclization reaction may be an intramolecular interaction, particularly in some cases where a charge-transfer interaction can occur or is induced, and in cationic-initiated reactions. However, the predominance of cyclization over intermolecular propagation can be ascribed to a smaller decrease in activation entropy, which compensates for the unfavorable difference in activation energies. Further studies on the kinetics and energetics of the cyclization reaction during polymerization are necessary before a completely acceptable mechanistic explanation is forthcoming.

### 23.10.3 Interpretations of Mechanistic Proposals

A satisfactory theory for cyclopolymerization must adequately account for the following observations: (a) almost all nonconjugated dienes and other similar polymerizable structures studied to date undergo cyclopolymerization in preference to crosslinking or other modes of propagation; (b) such dienes invariably undergo polymerization at a higher rate than their respective monoene counterparts; (c) the cyclization step has invariably been shown to be of higher activation energy than propagation without cyclization; and (d) the cyclization step often leads to propagation *via* the less thermodynamically stable reaction intermediate, leading to the less thermodynamically stable ring structure. Many examples in which the rates of cyclopolymerization are significantly greater than those for homopolymerization of their corresponding monoalkenic counterparts have been published. Relative rates up to 590 have been reported.[222]

An early theory suggested that an electronic interaction may occur between the nonconjugated double bonds of 1,6-dienes or between the intramolecular double bond and the reactive center after initiation. Such an interaction would provide an energetically favorable pathway from monomer to cyclic product and be the major driving force for cyclopolymerization. This theory has been investigated extensively, with conflicting results. Spectroscopic studies have produced evidence for interspatial interactions in some monomers that cyclopolymerize but not all.[156] However, kinetic evidence has been obtained[230] which shows that the reactivity of each double bond in the 3-allyl-5-hexenyl radical, a model for cyclopolymerization, is about five times greater than the double bond in the 5-hexenyl radical under similar conditions. Presently, no satisfactory explanation for this enhanced reactivity is available. It cannot be attributed solely to homoconjugation. Support for this proposal is also provided by a photoelectron spectroscopic study which has shown that the conformation in which the two double bonds of a 1,5-diene are crossed is more stable by 2.3 kcal mol$^{-1}$ than the open-chain conformation.[231]

Stabilization of the reactive center would be expected to occur in the case of cationic initiation. 1,3-Bis(4-vinylphenyl)propane can be polymerized, by use of cationic initiators, to a cyclopolymer having [3.3]paracyclophane units in the chain.[59] An interaction was proposed between the styryl cation and the intramolecular styryl group that stabilizes the transition state leading to cyclopolymer.

Studies have invariably shown that the intramolecular step leading to cyclization has a higher activation energy than the intermolecular step. Many of these studies point toward a thermodynamic explanation for cyclopolymerization. Although the differences in activation energies between the intramolecular and intermolecular propagation reactions suggest that the former is a slightly less favorable reaction,[206] the higher tendency for cyclization has been explained[232] as being due to a smaller decrease in entropy for such a reaction compared to intermolecular addition. Thus, the Gibbs free energy of activation will be smaller for cyclopolymerization than for intermolecular propagation.

This hypothesis has been extended, leading to the proposal that highly cyclized polymers may be obtained from monomers whose monofunctional counterparts do not homopolymerize. This has been successfully applied to a number of systems, including *N*-substituted dimethacrylamides.[220] Thus, *N*-isobutyryl-*N*-propylmethacrylamide did not homopolymerize. However, complete cyclization occurred on homopolymerization or copolymerization of *N*-propyldimethacrylamide. It was proposed that the loss of internal rotation degrees of freedom is less effective in decreasing entropy than a loss of translational and rotational degrees of freedom and the activation entropies favoring cyclization.

## 23.11 MECHANISM OF CYCLOCOPOLYMERIZATION

### 23.11.1 Kinetic Studies

The first example of cyclocopolymerization in which both comonomers contribute to the cyclic structure was that of the alternating copolymerization with cyclization of divinyl ether and maleic anhydride.[156,184] The mechanistic steps postulated for this copolymerization are shown in Scheme 5. This mechanism accounts for the failure of the system to crosslink in accord with the widely accepted theory of Staudinger.[3] The presence of cyclic ether groups was confirmed by chemical evidence which involved cleavage by hydroiodic acid and incorporation of iodine into the copolymer. A wide variety of 1,4-dienes and alkenes have since been shown to undergo cyclocopolymerization in a similar manner.

A fundamental study of this mode of copolymerization[233] has been carried out. A series of copolymer composition equations were derived that are consistent with the addition of monoalkene to diene radicals by a concerted bimolecular step proceeding through a cyclic transition state to produce the cyclic repeating unit. It was recognized quite early[156,184] that the above requirements could be satisfied if the monomer pair formed a charge-transfer complex that participated in the copolymerization. Considerable evidence has now been presented for charge-transfer complex participation in a variety of highly alternating cyclocopolymerizations.[234]

### 23.11.2 Ring Size Studies

A review paper has summarized much of the published literature dealing with ring size in cyclocopolymerization.[184] The early work[235] had led to the suggestion that the more thermodynamically stable six-membered ring was formed. However, it is now apparent that cyclocopolymerization could lead to the thermodynamically favored six-membered ring *via* the more stable radical, to the five-membered ring *via* the less stable radical, or to a mixture of both.

Recent studies,[188] which included model compound and deuterated monomer syntheses, $^{1}$H and $^{13}$C NMR investigations of the resulting copolymers and solvent effects, have led to the tentative conclusion that the copolymer includes both THF and THP structures. The ratio of the two appears to vary in favor of the THF ring with increasing polarity of the medium. A copolymer prepared in benzene was reported to contain THF–THP structures in a ratio of 0.8:1.0. The stereochemistry of the chiral centers in the copolymer is uncertain, although evidence has been presented for *trans* junctures at the bridgeheads in the THP structure and *cis* junctures for the corresponding centers in the THF structure.

Large-ring-containing cyclocopolymers have recently been reported. Cyclocopolymerization of the divinyl monomer 1,2-bis[2-(vinyloxy)ethoxy]benzene with maleic anhydride *via* radical initiation has been shown[236] to produce the 1:2 alternating cyclocopolymer. The monomer pair was shown by UV spectroscopy to form a charge-transfer complex having an equilibrium constant ($K$) of 0.280, suggesting participation of the complex in the copolymerization. Also consistent with this proposal was the observation that the maximum rate of copolymerization occurred at the monomer molar ratio of 1:2.

### 23.11.3 Interpretations of Mechanistic Proposals

The present understanding of the mechanistic aspects of cyclocopolymerization has been summarized in a review article.[156] Although cyclopolymerization and cyclocopolymerization are certainly related mechanistically, the unusual nature of the latter reaction, in which bimolecular ring-closure is highly favored, requires further explanation. The early kinetic treatment,[233] which suggested that the bimolecular ring-closure step is concerted, led to the proposal that the propagating species is a charge-transfer complex.[237] This proposal was supported by evidence for the existence of complexes in several comonomer pairs. Solvent effects on the cyclocopolymerization of maleic anhydride and divinyl ether have provided further evidence to support a charge-transfer complex as the active species. However, it has been concluded from kinetic data that the mechanism can be explained without invoking a charge-transfer complex.[156]

Current concepts of cyclocopolymerization would indicate that the structures of cyclocopolymers are highly dependent on the nature of the monomers and the reaction conditions. Predictions of structures based upon thermodynamic control of the cyclization reaction can be misleading, since kinetic products are often found. An important driving force for the cyclization reaction may be an intramolecular interaction, particularly in some cases where a charge-transfer interaction can occur or is induced, and in cationic-initiated reactions. However, the predominance of cyclization over intermolecular propagation can be ascribed to a smaller decrease in activation entropy which compensates for the unfavorable difference in activation energies. Further studies on the kinetics and energetics of the cyclization reaction during polymerization are necessary before a completely acceptable mechanistic explanation is forthcoming.

## 23.12 REFERENCES


1. G. B. Butler, in 'Encyclopedia of Polymer Science and Engineering', 2nd edn., Wiley, New York, 1986, vol. 4.
2. G. B. Butler and F. L. Ingley, *J. Am. Chem. Soc.*, 1951, **73**, 895.
3. H. Staudinger and W. Heuer, *Chem. Ber.*, 1934, **67**, 1164.
4. G. B. Butler, A. Crawshaw and W. L. Miller, *J. Am. Chem. Soc.*, 1958, **80**, 3615.
5. P. J. Flory, *J. Am. Chem. Soc.*, 1937, **59**, 241.
6. G. B. Butler, in 'Polymeric Amines and Ammonium Salts', ed. E. J. Goethals, Pergamon Press, New York, 1981, p. 125.
7. D. G. Hawthorne, S. R. Johns, D. H. Solomon and R. I. Willing, *J. Chem. Soc., Chem. Commun.*, 1975, 982; *Aust. J. Chem.*, 1976, **29**, 1955.
8. J. E. Lancaster, L. Baccei and H. P. Panzer, *J. Polym. Sci., Polym. Lett. Ed.*, 1976, **14**, 549.
9. N. D. Field, *J. Org. Chem.*, 1960, **25**, 1006.
10. C. S. Marvel and R. D. Vest, *J. Am. Chem. Soc.*, 1959, **81**, 984.
11. G. C. Corfield and A. Crawshaw, *J. Macromol. Sci., Chem.*, 1971, **5** (1), 1873.
12. C. Aso, T. Kunitake, R. K. Khattak and N. Sugi, *Makromol. Chem.*, 1970, **134**, 147.
13. D. W. Brown and L. A. Wall, *Polym. Prepr., Am. Chem. Soc., Div. Polym. Chem.*, 1968, **9**, 1401.
14. G. B. Butler and R. L. Bunch, *J. Am. Chem. Soc.*, 1949, **71**, 3120.
15. G. B. Butler and R. J. Angelo, *J. Am. Chem. Soc.*, 1957, **79**, 3128.
16. R. M. Ottenbrite and W. S. Ryan, *Ind. Eng. Chem. Prod. Res. Dev.*, 1980, **19**, 528.
17. D. H. Solomon and D. G. Hawthorne, *J. Macromol. Sci., Rev. Macromol. Chem.*, 1976, **C15**, 143.
18. D. G. Hawthorne and D. H. Solomon, *J. Macromol. Sci., Chem.*, 1975, **A9**, 149; *J. Macromol. Sci., Chem.*, 1976, **A10**, 923.
19. J. H. Hodgkin and D. H. Solomon, *J. Macromol. Sci., Chem.*, 1976, **A10**, 893.
20. W. E. Gibbs and R. L. Van Deusen, *J. Polym. Sci.*, 1961, **54**, SI.
21. Y. M. Boyarchuk, *Vysokomol. Soedin., Ser. A*, 1969, **11** (10), 2161.
22. B. Yamada, T. Saya and T. Otsu, *Makromol. Chem.*, 1982, **183** (3), 627.
23. C. Jakus and G. Smets, in 'Proceedings of the I. U. P. A. C. Macromolecular Symposium, 28th', 1982, p. 171 (*Chem. Abstr.*, 1983, **99**, 105 814r).
24. C. Jakus, G. Smets and T. Zeegers-Huyskens, *Bull. Soc. Chim. Belg.*, 1984, **93** (3), 175.
25. L. J. Mathias, R. A. Vaidya and R. H. Bloodworth, *J. Polym. Sci., Polym. Lett. Ed.*, 1985, **23** (6), 289.
26. C. L. McCormick, Z. B. Zhang and K. W. Anderson, *Polym. Prepr., Am. Chem. Soc., Div. Polym. Chem.*, 1983, **24** (2), 364.
27. Nippon Kayaku Co., Ltd., *Jpn. Pat.* 81 16448 (1981) (*Chem. Abstr.*, 1981, **95**, 43 911b).
28. A. I. Kartashevskii, D. F. Varfolomeev, L. G. Sementsova, T. S. Kirillov, L. Y. Skundina and I. E. Izmailov, *Khim. Tekhnol. Topl. Masel*, 1981, **6**, 37 (*Chem. Abstr.*, 1981, **95**, 208 941b).
29. S. Saeki, M. Suehiro and Y. Fukui, *Jpn. Pat.* 85 184 052 (1985) (*Chem. Abstr.*, 1986, **104**, 69 296j).
30. D. Ballschuh, J. Rusche and R. Ohme, *Ger. (East) Pat.* 158 783 (1983) (*Chem. Abstr.*, 1983, **99**, 38 975d).
31. D. A. Topchiev, G. T. Bikasheva, A. I. Martynenko, N. N. Kaptsov, L. A. Gudkova and V. A. Kabanov, *Vysokomol. Soedin, Ser. B.*, 1980, **22** (4), 269.
32. A. Wyroba, *Polimery (Warsaw)*, 1981, **26** (2), 50.
33. A. Wyroba, *Przem. Chem.*, 1981, **60** (11–12), 531.
34. D. A. Topchiev, G. T. Nazhmetdinova, A. M. Krapivin, V. A. Shreider and V. A. Kahanov, *Vysokomol. Soedin., Ser. B.*, 1982, **24** (6), 473.
35. M. Hahn, W. Jaeger, F. Seehaus, C. Wandrey and G. Reinisch, *Ger. (East) Pat.* 156 979 (1982) (*Chem. Abstr.*, 1983, **98**, 107 993b).
36. D. Neigel and C. D. Szymanski, *Br. Pat.* 2 125 051 (1984) (*Chem. Abstr.*, 1984, **101**, 7798d).
37. D. A. Topchiev, G. T. Nazhmetdinova, A. I. Kartashevskii, A. V. Nechaeva and V. A. Kabanov, *Izv. Akad. Nauk SSSR, Ser. Khim.*, 1983, **10**, 2232.
38. G. T. Nazhmetdinova, V. A. Shreider, D. A. Topchiev and V. A. Kabanov, *Izv. Akad. Nauk SSSR, Ser. Khim.*, 1984, **5**, 1024.
39. D. Neigel, *Eur. Pat. Appl.* 103 698 (1984) (*Chem. Abstr.*, 1984, **101**, 73 258d).
40. R. J. Schaper, *US Pat.* 4 439 580 (1984) (*Chem. Abstr.*, 1984, **100**, 210 998r).
41. C. D. Szymanski and D. Neigel, *US Pat.* 4 517 351 (1985) (*Chem. Abstr.*, 1985, **103,** 54 890d).
42. N. A. Golubkova, A. I. Martynenko, N. A. Babaev, A. V. Nechaeva, A. A. Efendiev, D. A. Topchiev and V. A. Kabanov, *Izv. Akad. Nauk SSSR, Ser. Khim.*, 1986, **2**, 485.
43. K. Ishimi, *Jpn. Pat.* 86 126 114 (1986) (*Chem. Abstr.*, 1986, **105**, 173 268h).
44. E. Dumitriu, S. Oprea and M. Dima, *Mater. Plast. (Bucharest)*, 1981, **18** (4), 202.
45. C. Wandrey, W. Jaeger and G. Reinisch, *Acta Polym.*, 1982, **33** (2), 156.
46. W. Jaeger, M. Hahn, C. Wandrey, F. Seehaus and G. Reinisch, *J. Macromol. Sci., Chem.*, 1984, **A21** (5), 599.
47. M. Hahn, W. Jaeger, C. Wandrey and G. Reinisch, *Acta Polym.*, 1984, **35** (5), 350.
48. Y. A. Klyachko, M. A. Shmaider, M. L. Korshunova, I. V. Kolganova and D. A. Topchiev, *Zh. Vses. Khim. Ova.*, 1984, **29** (1), 111 (*Chem. Abstr.*, 1984, **100**, 176 887s).
49. K. Kageno, T. Ueda and S. Harada, *Eur. Pat. Appl.* 140 309 (1985) (*Chem. Abstr.*, 1985, **103**, 105 434t).
50. N. Yang, H. Haubenstock and A. Snow, *ACS Symp. Ser.*, 1982, **195**, 417.

51. Asahi Chemical Industry Co., Ltd., *Jpn. Pat.* 85 115 902 (1985) (*Chem. Abstr.*, 1985, **103**, 179 368w).
52. D. W. Bjorkquist, R. D. Bush, F. S. Ezra and T. W. Keough, *J. Org. Chem.*, 1986, **51** (16) 3192.
53. N. C. Billingham, A. D. Jenkins, E. B. Kornfli and D. R. M. Walton, *J. Polym. Sci., Polym. Chem. Ed.*, 1977, **15**, 675.
54. S. L. Sosin, V. A. Jashi, V. A. Antipova and V. V. Korshak, *Vysokomol. Soedin., Ser. B*, 1970, **12**, 669; *Vysokomol. Soedin., Ser. B*, 1974, **16**, 347; V. V. Korshak and S. L. Sosin, in 'Organometallic Polymers', ed. C. E. Carraher, J. E. Sheats and C. U. Pittman, Academic Press, New York, 1977, p. 26.
55. G. C. Corfield, J. S. Brooks and S. Plimley, *Polym. Prepr., Am. Chem. Soc., Div. Polym. Chem.*, 1981, **22**, 3.
56. G. B. Butler, D. L. Skinner, W. C. Bond, Jr. and C. L. Rogers, *J. Macromol. Sci., Chem.*, 1970, **4** (6), 1437.
57. W. Simpson, T. Holt and R. J. Zetie, *J. Polym. Sci.*, 1953, **10**, 489.
58. C. S. Marvel and W. E. Garrison, *J. Am. Chem. Soc.*, 1959, **81**, 4737.
59. J. Nishimura and S. Yamashita, *Polym. Prepr., Am. Chem. Soc., Div. Polym. Chem.*, 1981, **22**, 46.
60. L. J. Mathias and J. B. Canterberry, *Polym. Prepr., Am. Chem. Soc., Div. Polym. Chem.*, 1981, **22**, 38.
61. G. B. Butler and Q. S. Lien, *Polym. Prepr., Am. Chem. Soc., Div. Polym. Chem.*, 1981, **22**, 54.
62. K. Fujimori, *J. Polym. Sci., Polym. Chem. Ed.*, 1983, **21** (1), 25.
63. T. Ohya and T. Otsu, *J. Polym. Sci., Polym. Chem. Ed.*, 1983, **21** (11), 3169.
64. T. Otsu and T. Ohya, *J. Macromol. Sci., Chem.*, 1984, **A21** (1), 1.
65. T. Ohya and T. Otsu, *J. Polym. Sci., Polym. Chem. Ed.*, 1983, **21** (12), 3503.
66. S. Paulrajan, A. Gopalan, N. R. Subbaratnam and K. Venkatarao, *Polymer*, 1983, **24** (7), 906.
67. K. Yokota, T. Kakuchi, A. Nanasawa, J. Iwata and Y. Takada, *Polym. J.*, 1982, **14** (7), 509.
68. T. Otsu and T. Ohya, *Polym. Bull. (Berlin)*, 1983, **9** (8–9), 355.
69. J. Nishimura, Y. Ishida, S. Yamashita, H. Hasegawa, M. Sawamota and T. Higashimura, *Polym. J.*, 1983, **15** (4), 303.
70. J. Nishimura, M. Furukawa, S. Yamashita, T. Inazu and T. Yoshino, *J. Polym. Sci., Polym. Chem. Ed.*, 1981, **19** (12), 3257.
71. T. Nishikubo, T. Iizawa, A. Yoshinaga and M. Nitta, *Makromol. Chem.*, 1982, **183** (4), 789.
72. A. Gopalan, S. Paulrajan, N. R. Subbaratnam and K. V. Rao, *J. Polym. Sci., Polym. Chem. Ed.*, 1985, **23** (7), 1861.
73. T. Kakuchi and K. Yokota, *Makromol. Chem., Rapid Commun.*, 1985, **6** (8), 551.
74. V. M. Lagunov, I. V. Golikov and G. V. Korolev, *Vysokomol. Soedin., Ser. A*, 1982, **24** (1), 131.
75. N. M. Bol'bit, A. L. Izyumnikov, E. D. Rogozhkina, N. K. Faizi and Y. A. Chikin, *Vysokomol. Soedin., Ser. A*, 1985, **27** (8), 1621.
76. T. Kodaira and G. B. Butler, *J. Macromol. Sci., Chem.*, 1985, **A22** (2), 213.
77. T. Kodaira, S. Tsuji, K. Okuyama and Y. Hosoki, *Polym. J.*, 1984, **16** (12), 901.
78. T. Kodaira and Y. Sumiya, *Makromol. Chem.*, 1986, **187** (4), 933.
79. Y. Lai and G. B. Butler, *J. Macromol. Sci., Chem.*, 1984, **A21** (11–12), 1547.
80. R. A. Vaidya and L. J. Mathias, *J. Polym. Sci., Polym. Symp.*, 1986, **74**, 243.
81. D. L. Neumann and H. J. Harwood, *ACS Symp. Ser.*, 1982, **195**, 43.
82. A. I. Martynenko, A. M. Krapivin, A. B. Zezin, D. A. Topchiev and V. A. Kabanov, *Vysokomol. Soedin., Ser. B*, 1982, **24** (8), 580.
83. I. I. Romantsova, O. V. Pavlova, S. M. Kireeva and Yu. M. Sivergin, *Dokl. Akad. Nauk SSSR, Ser. Khim.*, 1986, **289** (2), 422.
84. F. Xi and O. Vogl, *J. Macromol. Sci., Chem.*, 1983, **A20** (3), 321.
85. A. Wyroba, *Przem. Chem.*, 1981, **60** (11–12), 531.
86. L. Y. Shy, Y. K. Leung and B. E. Eichinger, *Polym. Prepr., Am. Chem. Soc., Div. Polym. Chem.*, 1984, **25** (1), 278.
87. T. Takagishi, H. Kozuka and N. Kuroki, *J. Polym. Sci., Polym. Chem. Ed.*, 1981, **19** (12), 3237.
88. E. A. Bekturov, R. E. Khamzamulina, G. Reinisch, W. Jaeger and Z. K. Belgibaeva, *Acta Polym.*, 1984, **35** (7), 521.
89. R. E. Khamzamulina, Z. K. Bel'gibaeva, G. Reinisch, E. A. Bekturov and W. Jaeger, *Izv. Akad. Nauk Kaz. SSR, Ser. Khim.*, 1985, (3), 33.
90. E. A. Bekturov, S. E. Kudaibergenov, V. Z. Ushanov and S. S. Saltybaeva, *Makromol. Chem.*, 1985, **186** (1), 71.
91. Y. A. Klyachko, M. A. Shnaider, M. L. Korshunova, I. V. Kolganova and D. A. Topchiev, *Zh. Vses. Khim. Ova.* 1985, **30** (6), 573 (*Chem. Abstr.*, 1986, **104**, 170 570u).
92. P. L. Dubin and R. Oteri, *J. Colloid Interface Sci.*, 1983, **95** (2), 453.
93. P. L. Dubin, D. W. McQuigg and D. R. Rigsbee, *Polym. Prepr., Am. Chem. Soc., Div. Polym. Chem.*, 1986, **27** (1), 420.
94. H. Dautzenberg and H. Dautzenberg, *Acta Polym.*, 1984, **36** (2), 102.
95. E. Kokufuta and K. Takahashi, *Macromolecules*, 1986, **19** (2), 351.
96. V. V. Egorov, E. V. Batrakova, L. V. Titkova, V. V. Demin, V. P. Zubov and A. N. Barnakov, *Vysokomol. Soedin., Ser. B.*, 1982, **24** (5), 370.
97. E. Kokufuta, H. Shimizu and I. Nakamura, *Macromolecules*, 1981, **14** (5), 1178.
98. B. Philipp, L. T. Hong, K. J. Linow and J. M. G. Cowie, *Eur. Polym. J.*, 1981, **17** (6), 615.
99. B. Philipp, L. T. Hong, W. Dawydoff, K. J. Linow and U. Schuelke, *Z. Anorg. Allg. Chem.*, 1981, 479, 219.
100. J. Rusche, D. Ballachuh and R. Ohme, *Ger. (East) Pat.* 147 949 (1981) (*Chem. Abstr.*, 1981, **95**, 187 949t).
101. W. Schempp and H. T. Tran, *Wochenbl. Papierfabr.*, 1981, **109** (19), 726, 728, 731.
102. G. B. Shalbaeva, (USSR), *Deposited Document*, 1982, VINITI 3676–83, 20–4 (*Chem. Abstr.*, 1984, **101**, 193 433d).
103. O. V. Vasil'eva, A. V. Davydov, I. V. Kolganova, A. I. Mikaya, V. G. Zaikin, Y. G. Yanovakii, D. A. Topchiev and G. V. Vinogradov, *Dokl. Akad. Nauk SSSR*, 1984, **276** (3), 621.
104. F. Onabe, *Mokuzai Gakkaishi*, 1982, **28** (7), 445 (*Chem. Abstr.*, 1982, **97**, 111 466d).
105. F. Onabe, *Mokuzai Gakkaishi*, 1983, **29** (9), 593 (*Chem. Abstr.*, 1983, **99**, 196 809g).
106. F. Onabe, *Mokuzai Gakkaishi*, 1983, **29** (8), 513 (*Chem. Abstr.*, 1983, **99**, 196 798c).
107. F. Onabe, *Mokuzai Gakkaishi*, 1983, **29** (7), 467 (*Chem. Abstr.*, 1983, **99**, 124 286m).
108. F. Onabe, *Mokuzai Gakkaishi*, 1983, **29** (7), 459 (*Chem. Abstr.*, 1983, **99**, 124 285k).
109. M. D. Barnett and G. B. Butler, *J. Org. Chem.*, 1960, **25**, 309.
110. G. B. Butler, G. C. Corfield and C. Aso, *Prog. Polym. Sci.*, 1974, **4**, 71.
111. K. Yokota and Y. Takada, *Kobunshi Kagaku*, 1969, **26** (288), 317.
112. H. Kaye, *Macromolecules*, 1971, **4** (2), 147.
113. H. L. Panzik and J. E. Mulvaney, *J. Polym. Sci., Polym. Chem. Ed.*, 1972, **10**, 3469.
114. T. Kodaira and O. Murata, *J. Polym. Sci., Polym. Chem. Ed.*, 1979, **17**, 319.

115. W. Kawai, *J. Polym. Sci., Part A-1*, 1966, **4**, 1191.
116. F. F. Shcherbina and I. P. Fedorova, *Ukr. Khim. Zh. (Russ. Ed.)*, 1967, **33**, 394.
117. B. Yamada, T. Saya, T. Ohya and T. Otsu, *Makromol. Chem.*, 1982, **183** (4), 963.
118. I. A. Arbuzova, S. A. Plotkina and O. V. Sokolova, *Vysokomol. Soedin.*, 1962, **4**, 843.
119. I. A. Arbuzova and S. A. Plotkina, *Vysokomol. Soedin.*, 1964, **6**, 662.
120. R. C. Schulz, M. Marx and H. Hartmann, *Makromol. Chem.*, 1961, **44–46**, 281.
121. J. Roovers and G. Smets, *Makromol. Chem.*, 1963, **60**, 89.
122. L. Trossarelli, M. Guaita and A. Priola, *Ric. Sci.*, 1965, **35** (11-a), 429.
123. L. Trossarelli, M. Guaita and A. Priola, *Ann. Chim. (Rome)*, 1966, **56**, 1065.
124. J. P. J. Higgins and K. E. Weale, *J. Polym. Sci., Part A-1*, 1968, **6**, 3007.
125. A. Matsumoto, M. Tano, H. Ishido and M. Oiwa, *J. Polym. Sci., Polym. Chem. Ed.*, 1983, **21** (2), 609.
126. W. Fukuda, M. Nakao, K. Okumura and H. Kakiuchi, *J. Polym. Sci., Part A-1*, 1972, **10** (1), 237.
127. W. Fukuda, Y. Yamano, M. Tsuriya and H. Kakiuchi, *Polym. J.*, 1982, **14** (2), 127.
128. M. Raetzsch and L. Stephan, *Plaste Kautsch.*, 1971, **18** (8), 572.
129. A. Matsumoto, H. Ishido, M. Oiwa and K. Urushido, *J. Polym. Sci., Polym. Chem. Ed.*, 1982, **20** (11), 3207.
130. W. Fukuda, H. Sato, T. Imai and H. Kakiuchi, *Polym. J.*, 1986, **18**, (9), 631.
131. K. Yokota and Y. Takada, *Kobunshi Kagaku*, 1973, **30**, 71; *Kobunshi Kagaku*, 1973, **30**, 217.
132. K. Urushido, Y. Takizawa, A. Matsumoto and M. Oiwa, *Makromol. Chem., Rapid Commun.*, 1983, **4** (1), 21.
133. K. Urushido, F. Iwasaki, A. Matsumoto and M. Oiwa, *Makromol. Chem.*, 1986, **187** (4), 711.
134. S. C. Chu and G. B. Butler, *J. Polym. Sci., Polym. Lett. Ed.*, 1977, **15**, 277.
135. K. Yokota, T. Kakuchi, Y. Taniguchi and Y. Takada, *Makromol. Chem., Rapid Commun.*, 1985, **6** (3), 155.
136. K. Yokota, T. Kakuchi, T. Iiyama and Y. Takada, *Polym. J.*, 1984, **16** (2), 145.
137. C. S. Marvel and E. J. Gall, *J. Org. Chem.*, 1960, **25**, 1784.
138. D. S. Trifan and J. J. Hoglen, *J. Am. Chem. Soc.*, 1961, **83**, 2021.
139. P. J. Graham, E. L. Buhle and N. Pappas, *J. Org. Chem.*, 1961, **26**, 4658.
140. R. H. Wiley, W. H. Rivera, T. H. Crawford and N. F. Bray, *J. Polym. Sci.*, 1962, **61**, S38.
141. B. Reichel, C. S. Marvel and R. Z. Greenley, *J. Polym. Sci., Part A*, 1963, **1**, 2935.
142. U. K. Agaev, I. A. Guseinov, I. P. Tverdokhleb and S. Z. Rizaeva, *Soviet Pat.* 975 721 (1982) (*Chem. Abstr.*, 1983, **98**, 144 002y).
143. E. Y. C. Chang and C. C. Price, *J. Am. Chem. Soc.*, 1961, **83**, 4650.
144. C. Aso, S. Ushio and M. Sogabe, *Makromol. Chem.*, 1967, **100**, 100.
145. S. Mizogami and S. Yoshimura, *J. Chem. Soc., Chem. Commun.*, 1985, 1736.
146. D. T. Longone and D. T. Glatzhofer, *J. Polym. Sci., Polym. Chem. Ed.*, 1986, **24** (8), 1725.
147. J. M. Pochan and H. W. Gibson, *Org. Coat. Plast. Chem.*, 1980, **43**, 872.
148. G. V. Ambartsumyan, S. B. Gevorkayn, V. G. Kharatyan, V. B. Gavalyan, A. A. Saakyan, S. G. Grigoryan and L. A. Ak'opyan, *Arm. Khim. Zh.*, 1984, **37** (3), 188.
149. C. Harries and L. Tank, *Ber. Dtsch. Chem. Ges.*, 1908, **41**, 1701.
150. C. Aso, S. Tagami and T. Kunitake, *J. Polym. Sci., Part A-1*, 1970, **8** (6), 1323.
151. V. P. Prokop'ev, M. I. Bakhitov, I. G. Davlethaev, T. I. Kupreishvili and E. V. Kuznetsov, *Vysokomol. Soedin., Ser. B*, 1981, **23** (9), 688.
152. M. I. Bakhitov, L. S. Zainutdinova, E. V. Kuznetsov and F. L. Kigman in 'Sint. Poliuretanov', ed. S. I. Omel'chenko, Izd. Naukova Dumka, Kiev, USSR 1981, p. 114 (*Chem. Abstr.*, 1982, **97**, 72 867p).
153. M. I. Bakhitov, I. G. Davlethaev, V. P. Prokop'ev and E. V. Kuznetsov, *Vysokomol. Soedin., Ser. A*, 1984, **26** (6), 1153.
154. N. K. Emelyanova, M. I. Bakhitov and E. V. Kuznetsov, *Vysokomol. Soedin., Ser. A*, 1985, **27** (7), 1434.
155. R. F. T. Stepto, *Polym. Prepr., Am. Chem. Soc., Div. Polym. Chem.*, 1985, **26** (2), 46.
156. G. B. Butler, *Acc. Chem. Res.*, 1982, **15**, 370.
157. L. M. Chepel, A. N. Zelenetskii, O. B. Salamatina, N. P. Zaitseva, G. M. Trofimova and D. D. Novikov, *Vysokomol. Soedin., Ser. A*, 1983, **25** (2), 410.
158. L. M. Chepel, V. A. Topolkaraev, A. N. Zelenetskii, E. V. Prut, G. M. Trofimova, D. D. Novikow and A. A. Berlin, *Vysokomol. Soedin., Ser. A*, 1982, **24** (8), 1646.
159. K. Yokota, H. Hashimoto, T. Kakuchi and Y. Takada, *Makromol. Chem., Rapid Commun.*, 1984, **5** (11), 767.
160. L. Matejka, M. Tkaczyk, S. Pokoray and K. Dusek, *Polym. Bull. (Berlin)*, 1986, **15** (5), 389.
161. L. C. Hsu and W. H. Philipp, in 'Proceedings of the I. U. P. A. C. Macromolecular Symposium, 28th', 1982, p. 170 (*Chem. Abstr.*, 1983, **99**, 88 589b).
162. L. C. Hsu, *US Pat.* 4 555 565 (1985) (*Chem. Abstr.*, 1986, **104**, 149 671t).
163. Y. Kitahama, H. Ohama and H. Kobayashi, *J. Polym. Sci., Part C*, 1969, **23** (2), 785.
164. Y. Kitahama, *J. Polym. Sci., Part A-1*, 1968, **6**, 2309.
165. W. H. Schuller, J. A. Price, S. T. Moore and W. M. Thomas, *J. Chem. Eng. Data*, 1959, **4**, 273.
166. H. S. Makowski, B. K. C. Shim and Z. W. Wilchinsky, *J. Polym. Sci., Part A*, 1964, **2**, 4973.
167. Hercules Inc., *Belg. Pat.* 881 397 (1980) (*Chem. Abstr.*, 1981, **94**, 193 965r).
168. D. A. Topchiev, N. N. Kaptsov, L. A. Gudkova, V. A. Kabanov, A. I. Martynenko, B. N. Trushin and E. S. Parkhamovich, *Soviet Pat.* 910 664 (1982) (from Otkrytiya, Izobret., *Prom. Obraztsy, Tovarnye Znaki*, 1982, **9**, 91; *Chem. Abstr.*, 1982, **97**, 73 635y).
169. E. Dumitriu, S. Oprea and M. Dima, *Mater. Plast. (Bucharest)*, 1981, **18** (4), 202.
170. G. A. Weber, R. H. Pettebone and M. J. D'Errico, *Eur. Pat. Appl.* 69 573 (1983) (*Chem. Abstr.*, 1983, **98**, 174 866a).
171. Nippon Synthetic Chemical Industry Co., Ltd., K. K. Samso, Ube Industries, Ltd., *Jpn. Pat.* 83 15 546 (1983) (*Chem. Abstr.*, 1983, **99**, 54 544t).
172. C. A. Costello, G. F. Matz, N. S. Sherwood, B. P. Boffardi, M. A. Yorke and Z. Amjad, *Eur. Pat. Appl.* 82 657 (1983) (*Chem. Abstr.*, 1983, **99**, 125 373z).
173. A. Wyroba, *Przem. Chem.*, 1983, **62** (12), 681.
174. A. M. Ginberg, L. D. Dremyatakaya, T. A. Ivanova, R. A. Vladiminski, Y. A. Klyachko, M. A. Shnaider, E. V. Kamenskaya, V. A. Konstantinov, V. V. Gorodilin and B. A. Khokhlov, *Soviet Pat.* 1 071 630 (1984) (from *Otkrytiya, Izobret., Prom. Obraztsy, Tovarnye Znaki*, 1984, **5**, 80; (*Chem. Abstr.*, 1984, **100**, 177 625s).

175. D. Mourato and R. Gehr, *Sci. Tech. Eau*, 1983, **16** (4), 323, 327 (*Chem. Abstr.*, 1984, **100**, 108 802k).
176. I. S. Shkurnikova, M. A. Penenzhik, A. D. Virnk and D. A. Topchiev, *Izv. Akad. Nauk SSSR, Ser. Khim.*, 1984, (4), 928.
177. C. Wandrey and W. Jaeger, *Acta Polym.*, 1984, **36** (2), 100.
178. D. M. Giddings, D. G. Ries and A. R. Syrinek, *US Pat.* 4 502 966 (1985) (*Chem. Abstr.*, 1985, **102**, 187 828r).
179. M. Iwata, T. Saka, T. Ikeda and N. Kondo, *Jpn. Pat.* 86 133 213 (1986) (*Chem. Abstr.*, 1986, **105**, 210 367n).
180. C. L. McCormick, Z. B. Zhang and K. W. Anderson, *J. Controlled Release*, 1986, **4** (2), 97 (*Chem. Abstr.*, 1986, **105**, 185 753a).
181. J. S. Shu, W. Lee, G. I. Varnell and J. I. Bartelt, *US Pat.* 4 318 976 (1982) (*Chem. Abstr.*, 1982, **96**, 182 930r).
182. J. C. Rabadeux, D. Durand and C. Bruneau, *Makromol. Chem., Rapid Commun.*, 1984, **5** (4), 191.
183. J. E. Boevink, A. P. Derrick and J. A. Moodie, *US Pat.* 4 347 339 (1982) (*Chem. Abstr.*, 1982, **97**, 183 424m).
184. G. B. Butler, in 'Anionic Polymeric Drugs', ed. L. G. Donaruma, R. M. Ottenbrite and O. Vogl, Wiley-Interscience, New York, 1980, p. 49.
185. A. H. Frazer and P. O'Neill, *J. Am. Chem. Soc.*, 1963, **85**, 2613.
186. T. Doiuchi, H. Yamaguchi and Y. Minoura, *Eur. Polym. J.*, 1981, **17** (9), 961.
187. G. S. Georgiev, *God. Sofii. Univ., Khim. Fak.*, 1977 (pub. 1981), **72** (1), 131 (*Chem. Abstr.*, 1982, **96**, 123 404z).
188. W. J. Freeman and D. S. Breslow, *ACS Symp. Ser.*, 1982, **186**, 243.
189. G. B. Butler and C. C. Wu, *Macromol. Synth.*, 1982, **8**, 89.
190. R. L. Wolgemuth, *US Pat.* 4 223 109 (1980) (*Chem. Abstr.*, 1983, **98**, 35 183n).
191. G. B. Butler, G. C. Corfield and C. Aso, *Prog. Polym. Sci.*, 1975, **4**, 71.
192. G. C. Corfield and G. B. Butler, in 'Developments in Polymerization—3', ed. R. N. Haward, Allied Sciences, Essex, 1982.
193. A. Y. Vainer and T. A. Dryakina, *Soviet Pat.* 730 696 (1980) (from *Otkrytiya, Izobret., Prom. Obraztsy, Tovarnye Znaki*, 1980, **16**; 86; (*Chem. Abstr.*, **93**, 115 706x).
194. Carbon Paper Co., Ltd., *Jpn. Pat.* 84 24 765 (1984) (*Chem. Abstr.*, 1984, **101**, 40 014t).
195. Fujitsu, Ltd., *Jpn. Pat* 84 02 043 (1984) (*Chem. Abstr.*, 1984, **101**, 31 146k).
196. W. H. Starnes, Jr., G. M. Villacorta, F. C. Schilling, G. S. Park and A. H. Saremi, *Polym. Prepr., Am. Chem. Soc., Div. Polym. Chem.*, 1983, **24** (1), 253.
197. M. Todoko and H. Watanabe, *Eur. Pat. Appl.* 110 356 (*Chem. Abstr.*, 1984, **101**, 74 112p).
198. J. Van der Klooster, *J. Oil Colour Chem. Assoc.*, 1984, **67** (4), 105.
199. R. C. Schulz, *Angew. Chem., Int. Ed. Engl.*, 1964, **3**, 416.
200. R. C. Schulz, in 'Encyclopedia of Polymer Science and Technology', ed. N. M. Bikales, H. F. Mark and N. G. Gaylord, Wiley, New York, 1964, vol. 1, p. 160.
201. R. C. Schulz, *Kunststoffe*, 1957, **47**, 303.
202. C. G. Overberger, S. Ozaki and H. Mukamal, *J. Polym. Sci., Polym. Lett. Ed.*, 1964, **2**, 627.
203. R. C. Schulz and R. Stenner, *Makromol. Chem.*, 1964, **72**, 202.
204. J. B. Lando and M. Thakur, *Macromolecules*, 1983, **16** (1), 143.
205. V. A. Kabanov, D. A. Topchiev and G. T. Nazhmetdinova, *Vysokomol. Soedin., Ser. B.*, 1984, **26** (1), 51.
206. T. F. Gray, Jr. and G. B. Butler, *J. Macromol. Sci., Chem.*, 1975, **A9** (1), 45.
207. A. Wyroba, *Polimery (Warsaw)*, 1981, **26** (4), 139.
208. G. Reinisch, W. Jaeger, M. Hahn and C. Wandrey, in 'Proceedings of the I. U. P. A. C. Macromolecular Symposium, 28th', 1982, p. 83 (*Chem. Abstr.*, 1983, **99**, 158 900x).
209. M. Hahn, W. Jaeger and G. Reinisch, *Acta Polym.*, 1983, **34**, (6), 322.
210. D. A. Topchiev and G. T. Nazhmetdinova, *Vysokomol. Soedin., Ser. A*, 1983, **25** (3), 636.
211. M. Hahn, W. Jaeger, C. Wandrey and G. Reinisch, *Acta Polym.*, 1984, **35** (5), 350.
212. W. Jaeger, M. Hahn, C. Wandrey, F. Seehaus and G. Reinisch, *J. Macromol. Sci., Chem.*, 1984, **A21** (5), 599.
213. G. B. Butler, 'Proceedings of the International Symposium on Macromolecules, Rio de Janiero, 1974', ed. E. B. Mano, Elsevier, Amsterdam, 1975, p. 57.
214. M. Julia, *Acc. Chem. Res.*, 1971, **4**, 386; *Pure Appl. Chem.*, 1967, **15**, 167; *Rec. Chem. Prog.*, 1964, **25**, 1.
215. R. C. Lamb, P. W. Ayers and M. K. Toney, *J. Am. Chem. Soc.*, 1963, **85**, 3483.
216. C. Walling, J. H. Cooley, A. A. Ponaras and E. F. Racah, *J. Am. Chem. Soc.*, 1966, **88**, 5361.
217. T. W. Smith and G. B. Butler, *J. Org. Chem.*, 1978, **43**, 6.
218. G. B. Butler and A. Matsumoto, *J. Polym. Sci., Polym. Lett. Ed.*, 1981, **19**, 167.
219. A. Matsumoto, T. Kitamura, M. Oiwa and G. B. Butler, *Makromol. Chem., Rapid Commun.*, 1981, **2**, 683.
220. T. Kodaira and M. Saki, *Polym. J.*, 1979, **11**, 595.
221. S. A. Stone-Elander, G. B. Butler, J. H. Davis and G. J. Palenik, *Macromolecules*, 1982, **15**, 45.
222. G. B. Butler and S. Kimura, *J. Macromol. Sci., Chem.*, 1971, **A5**, 181.
223. 'Handbook of Chemistry and Physics', 49th edn., Chemical Rubber Co., Cleveland, OH, 1968, p. D-184.
224. A. Matsumoto, K. Iwanami and M. Oiwa, *J. Polym. Sci., Polym. Chem. Ed.*, 1981, **19**, 213.
225. A. L. J. Beckwith, C. J. Easton and A. K. Serelis, *J. Chem. Soc., Chem. Commun.*, 1980, 482.
226. A. L. J. Beckwith, I. A. Blair and G. Phillipou, *Tetrahedron Lett.*, 1974, 2251.
227. D. G. Hawthorne and D. H. Solomon, *J. Polym. Sci., Polym. Symp.*, 1976, **55**, 211.
228. A. L. J. Beckwith, G. E. Gream and D. S. Struble, *Aust. J. Chem.*, 1972, **25**, 1081.
229. A. L. J. Beckwith, *Tetrahedron*, 1981, **37**, 3073.
230. A. L. J. Beckwith, T. Lawrence and A. K. Serelis, *J. Chem. Soc., Chem. Commun.*, 1980, 484.
231. J. C. Bunzli, A. J. Burak and D. C. Frost, *Tetrahedron*, 1973, **29**, 3735.
232. O. Chiantore, G. Camino, A. Chiorino and M. Guaita, *Makromol. Chem.*, 1977, **178**, 125.
233. J. M. Barton, G. B. Butler and E. C. Chapin, *J. Polym. Sci., Part A*, 1965, **3**, 501.
234. G. B. Butler and A. F. Campus, *J. Polym. Sci., Part A-1*, 1970, **8**, 545.
235. G. B. Butler, *J. Polym. Sci.*, 1960, **48**, 279; G. B. Butler, *US Pat.* 3 320 216 (1967) (*Chem. Abstr.*, 1967, **67**, P22 329z) (*US Pat. Reissue* 26 407 (1968)); G. B. Butler, *J. Macromol. Sci., Chem.*, 1971, **A5** (1), 219.
236. G. B. Butler and Q. S. Lien, *ACS Symp. Ser.*, 1982, **195**, 149.
237. G. B. Butler and K. C. Joyce, *J. Polym. Sci., Part C*, 1968, **22**, 45.

# 24
# Electroinitiated Polymerization

GIULIO D. GUERRA
*CNR, Pisa, Italy*



## 24.1 INTRODUCTION

The use of electrochemical reactions to produce species able to initiate polymerization processes began in the years around 1950, with some pioneering work being radical polymerizations initiated by cathodic reduction of inorganic redox couples,[1] cathodic reduction of $H^+$ to atomic hydrogen,[2–4] and anodic oxidation of carboxylate ions.[5] Some years later, the first electroinitiated anionic polymerization was obtained by direct transfer of electrons to styrene in pyridine at a magnesium cathode, in the presence of NaI as supporting electrolyte,[6] and in 1960 the first evidence of cationic polymerizations initiated by electrolysis was reported.[7] The progress in the field of electroinitiated polymerization up to the beginning of 1971 has been extensively reviewed by Breitenbach.[8]

## 24.2 INITIATION PROCESSES

The main question to be answered in investigating electroinitiated polymerizations concerns the actual initiation reactions, that is which electrochemical processes occurring at the electrodes are able to produce directly either the active centres or some initiating species. Indeed, whereas the nature of the propagation (radical or ionic) in electroinitiated polymerizations can be easily ascertained as in conventional polymerizations by means of suitable reagents, the initiation process can only be ascertained by sophisticated electrochemical techniques, which, moreover, need to be adapted for use with polymerizable compounds. For these reasons, the nature of the electrochemical reactions leading to initiation has been uncertain or misunderstood for many years; in most cases, the electrolytic production of some initiating species was simply inferred by the results of the electrolysis at constant current-intensity of the supporting electrolyte solution in the absence of a monomer,[4,9,10] and no investigation was made into the species (electrolyte, solvent or some adventitious impurity) actually reacting at the electrode. Even when the electrodic reaction was investigated by means of polarographic and voltammetric measurements, the electropolymerization was generally carried out under constant current-intensity, with no control of the electrodic potential,[8] so that a multiple process could occur at the electrode. These facts led many authors to postulate initiation mechanisms which were questioned later when the initiator-formation processes were reinvestigated by more appropriate electrochemical techniques. For example, 'indirectly' electroinitiated cationic polymerizations, in the presence of perchlorates and tetrafluoroborates as supporting electrolytes, are initiated, respectively, by $HClO_4$ and $BF_3$ plus HF formed at the anode.[11] The acids responsible for the polymerization were first supposed to originate from $ClO_4^{\bullet}$[12,13] and $BF_4^{\bullet}$[10,14] radicals, which would abstract H• from some hydrogen-donating species present in the solution to give the corresponding acids. Experiments carried out by electrolyzing $Bu_4NClO_4$ solutions at constant anodic potentials showed that, with 1,2-dichloroethane as solvent,[15] $H^+$ formation occurs at potentials lower than that of $ClO_4^-$ oxidation (~2 V *vs.* Ag/$AgClO_4$ saturated) and is due to the anodic oxidation of adventitious moisture. At higher potentials, $ClO_4^-$ undergoes an oxidative decomposition giving only $ClO_2$, detected in the anolyte by its absorption band at 355 nm.[16] With acetonitrile as solvent, potentiostatic $Bu_4NClO_4$ electrolysis showed that acidity, in addition to that arising from moisture oxidation, is produced above the anion decomposition potential by reactions between the anodically formed $ClO_2$ and the solvent, giving rise to a mixture of $HClO_4$ and weaker acids.[17] In both cases, no $ClO_4^{\bullet}$ formation was observed. In the electrolysis of $BF_4^-$ solutions, acid formation *via* the $BF_4^{\bullet}$ radical must also be excluded; indeed, in an electrolysis of a 0.075 mol $l^{-1}$ $Bu_4NBF_4$ solution in 1,2-dichloroethane, carried out at the anion oxidation potential (+2.7 V *vs.* Ag/$AgClO_4$ saturated), both evolution of white fumes of $BF_3$ and etching of the platinum anode were observed.[18] Although no accurate investigation of the anodic reactions was made, this fact indicates that the anodically formed fluorine radicals attack the anode, with formation of soluble fluoroplatinates, rather than the solvent, with formation of HF.

Only when potentiostatic electrolysis techniques, in which the working-electrode potential is kept constant *vs.* a reference electrode during the electrolysis, began to be used to carry out electroinitiated polymerizations, was it possible to obtain electropolymerizations with a well-defined electrodic process giving rise to the electrosynthesis of an initiator[19–21] or to direct electrochemical initiation.[22–24] This fact made possible the kinetic analysis of several polymerization systems.[19–22,24–26] Although the direct correlation between quantities of electricity and active species assumed by some authors has been criticized,[27] the kinetic parameters evaluated have a mechanistic significance, and, at least in one case,[24] have been confirmed by computer simulation.[28] Potentiostatic electrolysis has also permitted the formation of solid perylene cation radical, which can be recovered and reacted separately with various monomers to give polymerization by anodic electrocrystallization.[29] Further progress in ascertaining the electrochemical initiation process was achieved by means of the rotating ring-disk electrode; in this way, it has been possible to identify transient species, such as radical cations,[30] radical anions[31] and initiating inorganic redox couples.[32,33]

The attention paid to electrodic processes also allowed the electrolytic component of the so-called 'electric field effect' to be ascertained, *i.e.* the rate increase observed when a strong DC electric field is applied to ionic polymerizations.[34] Such a rate increase had been attributed by Ise's school to electrostatic causes, such as the second Wien effect and the desolvation of the growing free ions.[35] Results not explicable in terms of electrostatic effects were found by Giusti *et al.* on reversing periodically the field applied to an iodine-initiated polymerization of anethole (1,4-$CHMe{=}CHC_6H_4OMe$),[36] and were correctly interpreted by Funt *et al.* in terms of electrolytic phenomena.[37] That the 'electric field effect' is mainly an electroinitiated polymerization induced by the field in addition to the chemically initiated polymerization, was proved for the iodine-initiated polymerizations of anethole and vinyl ethers[34,38] by comparison with the electroinitiated polymerizations of the same monomers in the presence of $Bu_4NI_3$,[39–41] and for the $BF_3 \cdot Et_2O$-initiated polymerization of anethole and styrene by voltammetric measurements.[42] The 'negative field effect', *i.e.* the retardation of the polymerization rate by the field observed in the $BF_3 \cdot Et_2O$-initiated polymerizations of some aromatic vinyl monomers,[42,43] is also due to electrochemical reactions, as well as the influence of the electric field on the rates and reactivity ratios of radical homo- and copolymerizations,[44–46] which obviously cannot be interpreted in terms of ion-pair dissociation or free-ion desolvation.

## 24.3 COMPARISON WITH OTHER TECHNIQUES

A comparison between electroinitiation and other polymerization techniques shows that the former has both advantages and disadvantages. Among the advantages are: the possibility, in electroinitiated 'living' anionic polymerization, of controlling the molecular weight distribution by varying the intensity, the application time and the polarity of the current;[47,48] the possibility of varying the monomeric composition of copolymers with the potential of the working electrode;[49–51] and the possibility of synthesizing products which cannot be obtained by conventional techniques.

The main disadvantages are: the presence of the supporting electrolyte, which, for the ionic electropolymerizations, results in large concentrations of counterions, which depress polymerization rates and molecular weights;[24,27] and the fact that the propagating species in electroinitiated polymerizations are very similar to those in 'conventional' ones means that electrochemical polymerization techniques have not found a major practical use; but, in several cases, they can help to solve problems which are otherwise difficult.[11]

## 24.4 PRACTICAL APPLICATIONS

A very important part of the electropolymerization field is constituted by those electrochemical techniques which can be adapted to practical industrial applications.[52] One of these is the previously mentioned electrosynthesis of the cationic-initiator perylene cation radical perchlorate, produced on the anode in the solid state; the possibility of its being reacted separately with the monomer avoids the disadvantages of 'pre' and '*in situ*' generation.[29] Another application is the use of electrolysis to induce changes in the structure of electroactive macromolecules. The anodic electrolysis of salts of macromolecular carboxylic acids can produce, by the Kolbe reaction, radical sites on the chain able to initiate graft copolymerization.[53,54] By carrying out the same reaction in the presence of water, hydroperoxidic side-groups have been formed on the chains of poly(acrylic acid)s and poly(methacrylic acid)s, which, after decomposition, initiated graft copolymerization with higher yields than direct grafting.[53,55] In the presence of *N*-acetylaminocaproic and cyanoacetic acids, pentylacetamino and cyanomethylene groups were grafted on to methacrylic acid by a crossed Kolbe synthesis.[56,57] Copolymers of acrylic and methacrylic acid with various comonomers were crosslinked, internally modified and reacted with various compounds by Kolbe electrolysis.[53] Poly(styrene-*co*-vinylbenzophenone) was cathodically electrolyzed to form radical anionic sites on the benzophenone groups, which add acrylonitrile or methyl methacrylate, leading to graft copolymers, probably through an anionic mechanism.[58] As regards electrochemical modification of natural macromolecules, electrochemical grafting of methyl methacrylate on to cellulose occurs[59] and anodic electropolymerization of the blood-protein fibrinogen yields fibres resembling those of noncrosslinked fibrin,[60,61] as shown by electron microscopy.

Another important aspect is that of electropolymerization occurring in aqueous media. It comprises most of the earliest studies of electroinitiated polymerization,[1-4] and deals with polymerizations propagating by a radical mechanism, so that the common ion rate and molecular weight depression present in ionic electropolymerizations is avoided. As regards the electroinitiation techniques, radical production by Kolbe electrolysis has been used to polymerize vinyl acetate,[62] methyl methacrylate,[62] acrylic acid[63,64] and acrylamide.[64-68] In a polymerization of acrylamide carried out using tartaric acid as electrolyte, the peculiar nature of the electroactive acid causes the formation of initiating species different from those formed in the conventional Kolbe electropolymerization.[69] An unusual application of the Kolbe reaction is the electropolymerization of dicarboxylic acids, a polymerization not only 'electroinitiated', but also 'electropropagated'.[70-73] Anodic electropolymerizations in aqueous media of acrylonitrile by electrogenerated $HSO_4^{\bullet}$ radicals,[74] methyl methacrylate by monomer oxidation in the presence of $Fe^{2+}$ ions[75] and acrylamide by electrogenerated iridium oxides[76] have been reported. Cathodic electropolymerizations of methacrylic acid in aqueous media, by reduction of a bypyridilium salt in the presence of small quantities of oxygen,[77] and of methyl methacrylate[2-4,25,78-81] have been obtained. As regards the latter monomer, the initially supposed initiation by cathodically produced $H^{\bullet}$ radicals[2-4] was later excluded, and the role of monomer peroxides[78-80] and of added $Fe^{3+}$ ions[79] was pointed out; more recent studies seemed to exclude the peroxide hypothesis and attributed the initiation to monomer reduction; the presence of $Fe^{2+}$ or $Fe^{3+}$ ions, depending on the nature of the cathode materials, reduces the inhibition period.[25,81]

Other electroinitiated polymerizations are of practical application because of the structure of the polymers obtained. These include the electropolymerization of alkynic derivatives;[82,83] the electrochemical copolymerization of $SO_2$ and conjugated dienes to give polysulfones;[84] the electropolymerizations of acrylonitrile to give polymers containing ring structures;[85-87] the electroinitiated 'three-dimensional' polymerization of oligoester acrylates performed by Ukrainian workers;[88-91] and the electrochemical polycondensations obtained by Phillips *et al.*,[92-94] who also made a thorough investigation of the electropolymerization of donor–acceptor complexes.[95-97]

Electrocoating by means of *in situ* electropolymerization is perhaps the most important application of electroinitiated polymerization, as shown by the great number of patents existing in this field.[98-104] Although electrode coating has also often been obtained in normal electropolymerization experiments,[105-108] the most interesting studies in this field are those carried out in order to obtain surfaces coated with polymeric films for special purposes. Trifonov *et al.* polymerized acrylonitrile on a cathodically polarized germanium optical prism to follow the reaction by internal reflection spectroscopy.[86,87,109–111] Electropolymerization on graphite fibres gives a material used in reinforced polymer composites, with improved adhesion between the fibres and the matrix polymer,[112–114] and a fire-retardant film around the fibres.[115] This work was carried out mainly by Subramanian, who also reviewed the electropolymerizations made on this and other electrode materials.[112]

Another very important application of *in situ* electropolymerization is its use to protect metals from corrosion. Apart from the early pioneering works, such as the anionic polymerization of caprolactam obtained by Gilch and Michael on an iron cathode at 160 °C in the melt,[116] results of possible technological interest can only be reached with *in situ* electropolymerizations taking place in aqueous media, a fact which considerably restricts the choice of monomers and polymerization processes.[117] Mengoli *et al.* obtained corrosion-resistant films by electropolymerizing phenolic compounds[117–120] and other monomers[121,122] on iron, copper and zinc anodes; protective coatings of poly(oxyphenylene)s were deposited on to mild steel by Otero *et al.* by means of the electropolymerization of phenyl salicylate by cyclic potential sweeps in methanol solution[123] and by Lucchesi *et al.* by electropolymerization of various phenolic compounds on copper and iron anodes in aqueous solution.[124–126]

The technique of *in situ* electropolymerization has also been widely used to obtain 'chemically modified' electrodes; that is electrodes covered with a layer of an electroactive polymer, thus giving special properties to the electrode itself.[127] Such electrodes, initially made only by chemically attaching the reactive layer to the electrode surface,[128] are now increasing in number and their field of application is expanding, owing to the use of appropriate electropolymerization techniques.[129] Some examples of the possible uses of the polymer-modified electrodes are given here. Energy-storage batteries have been designed with electrodes made with polythiophene,[130,131] polyazulene,[131] polyaniline,[132,133] poly(4-vinylpyridinium hydrobromide perbromide)[134] and polypyrrole.[135] Organic conjugate polymers have been grafted on to semiconductors by electropolymerization in order to protect them against photodegradation and to obtain new 'organic-on-inorganic' electronic devices of potential interest.[136] Electrodes coated with polyanilines, a class of electroactive polymers showing a great difference in conductivity between their oxidized and reduced forms,[137] have been employed in electrochromic displays and in electrocatalysis.[137–144] Similar applications have been found for homo- and co-polymers of thiophenes[145–149] and pyrroles[150–154] electrosynthesized on electrodes.

One may conclude that electrochemical initiation is very important, both from a theoretical and from a practical point of view. In particular, its use in obtaining materials for special purposes has opened the doors to a wide expansion of this polymerization technique.

## 24.5 REFERENCES


1. I. M. Kolthoff and L. L. Ferstandig, *J. Polym. Sci.*, 1951, **6**, 563.
2. E. Dineen, T. C. Schwan and C. L. Wilson, *J. Electrochem. Soc.*, 1949, **96**, 226.
3. G. Parravano, *J. Am. Chem. Soc.*, 1951, **73**, 628.
4. W. Kern and H. Quast, *Makromol. Chem.*, 1953, **10**, 202.
5. S. Goldschmidt and E. Stöckl, *Chem. Ber.*, 1952, **85**, 630.
6. J. Y. Yang, W. E. McEwen and J. Kleinberg, *J. Am. Chem. Soc.*, 1957, **79**, 5833.
7. J. W. Breitenbach, Ch. Srna and O. F. Olaj, *Makromol. Chem.*, 1960, **42**, 171.
8. J. W. Breitenbach, O. F. Olaj and F. Sommer, *Adv. Polym. Sci.*, 1972, **9**, 47.
9. B. L. Funt, D. Richardson and S. N. Bhadani, *Can. J. Chem.*, 1966, **44**, 711.
10. B. M. Tidswell and A. G. Doughty, *Polymer*, 1971, **12**, 431.
11. G. Silvestri, S. Gambino and G. Filardo, *Adv. Polym. Sci.*, 1981, **38**, 27.
12. S. Nakahama, S. Hino and N. Yamazaki, *Polym. J.*, 1971, **2**, 56.
13. S. N. Bhadani and A. Ghose, *Trans. SAEST*, 1974, **9**, 170.
14. B. L. Funt and T. J. Blain, *J. Polym. Sci., Part A-1*, 1971, **9**, 115.
15. P. Cerrai, G. Guerra and M. Tricoli, *Eur. Polym. J.*, 1979, **15**, 153.
16. G. E. Glass and R. West, *Inorg. Chem.*, 1972, **11**, 2847.
17. F. Andruzzi, P. Cerrai, G. Guerra, L. Nucci, A. Pescia and M. Tricoli, *Eur. Polym. J.*, 1982, **18**, 685.
18. P. Cerrai, G. D. Guerra and M. Tricoli, unpublished results.
19. G. Mengoli, G. Farnia and E. Vianello, *Eur. Polym. J.*, 1969, **5**, 61.
20. G. Mengoli and G. Vidotto, *Makromol. Chem.*, 1971, **150**, 277.
21. B. L. Funt and V. Verigin, *Can. J. Chem.*, 1974, **52**, 1643.
22. U. Akbulut, J. E. Fernandez and R. L. Birke, *J. Polym. Sci., Polym. Chem. Ed.*, 1975, **13**, 133.
23. F. Bruno, M.-C. Pham and J.-E. Dubois, *J. Chim. Phys. Phys.-Chim. Biol.*, 1975, **72**, 490.
24. P. Cerrai, G. Guerra, M. Tricoli and L. Nucci, *Eur. Polym. J.*, 1980, **16**, 867.

25. A. Lucchesi, in 'Meeting of the International Society of Electrochemistry, [Abstracts], 31st, Venice', ed. E. Vecchi, Photo-offset, Padua, 1980, preprint E13, vol. 2, p. 652.
26. U. Akbulut, S. Eren and L. K. Toppare, *Polymer*, 1984, **25**, 1028.
27. A. Gandini, H. Cheradame, *Adv. Polym. Sci.*, 1980, **34–35**, 215, 224.
28. G. D. Guerra, P. Cerrai and M. Tricoli, in 'Convegno Italiano di Scienza delle Macromolecole, [Atti], 8th, Milan', ed. A. I. M. Photo-offset, Pisa, 1987, vol. II, p. 355.
29. E. Oberrauch, T. Salvadori and S. Cesca, *J. Polym. Sci., Polym. Lett. Ed.*, 1978, **16**, 345.
30. B. L. Funt, W. Severs and A. Glasel, *J. Polym. Sci., Polym. Chem. Ed.*, 1976, **14**, 2763.
31. L.-S. R. Yeh, *J. Electroanal. Chem. Interfacial Electrochem.*, 1977, **84**, 159.
32. B. L. Funt and L.-C. Hsu, in 'European Symposium on Electric Phenomena in Polymer Science, [Preprints], Pisa', ed. P. Giusti, Photo-offset, Pisa, 1978, preprint B4, p. 79.
33. L.-C. Hsu, A. Glasel, J. F. Skinner and B. L. Funt, *J. Electrochem. Soc.*, 1979, **126**, 939.
34. P. Giusti, *Makromol. Chem.*, 1974, **175**, 1157.
35. N. Ise, *Adv. Polym. Sci.*, 1969, **6**, 347.
36. P. Giusti, P. Cerrai, M. Tricoli, P. L. Magagnini and F. Andruzzi, *Makromol. Chem.*, 1970, **133**, 299.
37. J. Tanner, V. Hornof and B. L. Funt, *Makromol. Chem.*, 1972, **157**, 227.
38. P. Cerrai, P. Giusti, G. Guerra and M. Tricoli, *Eur. Polym. J.*, 1974, **10**, 1141.
39. P. Cerrai, P. Giusti, G. Guerra and M. Tricoli, *Eur. Polym. J.*, 1974, **10**, 1195.
40. P. Cerrai, P. Giusti, G. Guerra and M. Tricoli, *Eur. Polym. J.*, 1975, **11**, 101.
41. G. Attalla, P. Cerrai, P. Giusti, G. Guerra and M. Martelli, *Chim. Ind., Milano*, 1978, **60**, 197.
42. P. Cerrai, G. Guerra, L. Nucci and M. Tricoli, *Eur. Polym. J.*, 1985, **21**, 505.
43. U. Akbulut, R. L. Birke and J. E. Fernandez, *Makromol. Chem.*, 1978, **179**, 2507.
44. M. Rätzsch, H. Pretzsch and K. Chomiakow, *Wiss. Z. Tech. Hochsch. 'Carl Schorlemmer' Leuna-Merseburg*, 1977, **19**, 365.
45. M. Albeck, A. Levinger and K. Shiff, *J. Chem. Soc., Faraday Trans. 1*, 1978, **74**, 1488.
46. M. Albeck and S. Shaik, *J. Chem. Soc., Faraday Trans. 1*, 1978, **74**, 1496.
47. B. L. Funt and D. R. Richardson, *J. Polym. Sci., Part A-1*, 1970, **8**, 1055.
48. B. L. Funt and V. Hornof, *J. Polym. Sci., Part A-1*, 1971, **9**, 2171.
49. L. Toppare, S. Eren and U. Akbulut, *J. Polym. Sci., Polym. Chem. Ed.*, 1984, **22**, 2941.
50. L. Toppare, S. Eren and U. Akbulut, *J. Polym. Sci., Polym. Chem. Ed.*, 1985, **23**, 303.
51. S. Eren, L. Toppare and U. Akbulut. *Polym. Commun.*, 1987, **28**, 36.
52. B. L. Funt, in 'Encyclopedia of Polymer Science and Engineering', ed. H. F. Mark, N. M. Bikales, C. G. Overberger and G. Menges, Wiley, New York, 1986, vol. 5, p. 587.
53. G. Smets, R. Van Gorp and G. Van Haeren, *Eur. Polym. J.*, 1969, (Suppl.) 215.
54. G. Guerra, in 'IUPAC International Symposium on Macromolecules [Proceedings], Dublin', ed. D. C. Pepper, Photo-offset, Dublin, 1977, preprint III-27, vol. 2, p. 511.
55. G. Smets, A. Poot, M. Mullier and J. P. Bex, *J. Polym. Sci.*, 1959, **34**, 287.
56. G. Smets, X. Van der Borght and G. Van Haeren, *J. Polym. Sci., Part A*, 1964, **2**, 5187.
57. G. J. G. Smets and J. P. Bex, *US Pat.* 3 330 745 (1967) (*Chem Abstr.*, 1967, **67**, 64 370g).
58. B. L. Funt and L.-C. Hsu, *J. Polym. Sci., Polym. Chem. Ed.*, 1980, **18**, 1957.
59. J. A. Epstein and A. Bar-Nun, *J. Polym. Sci., Part B*, 1964, **2**, 27.
60. G. Stoner and L. Walker, *J. Biomed. Mater. Res.*, 1969, **3**, 645.
61. G. D. Guerra, M. Palla and G. Soldani, in 'Giornate della Elettrochimica Italiana 1986, Società Chimica Italiana, Divisione di Elettrochimica, Riunione Scientifica, [Atti], 6th, Bologna', ed. S. Trasatti, Photo-offset, Bologna, 1986, preprint C1, p. 110.
62. W. B. Smith and H.-G. Gilde, *J. Am. Chem. Soc.*, 1960, **82**, 659.
63. W. B. Smith and D. T. Manning, *J. Polym. Sci.*, 1962, **59**, S45.
64. A. Kunugi, T. Tanimoto and S. Nagaura, *Electrochim. Acta*, 1972, **17**, 2397.
65. Z. Ogumi, I. Tari, Z. Takehara and S. Yoshizawa, *Bull. Chem. Soc. Jpn.*, 1974, **47**, 1843.
66. Z. Ogumi, I. Tari, Z. Takehara and S. Yoshizawa, *Bull. Chem. Soc. Jpn.*, 1976, **49**, 841.
67. Z. Ogumi, Z. Takehara and S. Yoshizawa, *Bull. Chem. Soc. Jpn.*, 1976, **49**, 2883.
68. S. Yoshizawa, Z. Takehara, Z. Ogumi and C. Nagai, *J. Appl. Electrochem.*, 1976, **6**, 147.
69. P. K. Sengupta and R. Chakraborty, *J. Macromol. Sci., Chem.*, 1984, **A21**, 701.
70. A. Gozlan, G. Agam, S. Vardi and A. Zilkha, *Eur. Polym. J.*, 1984, **20**, 759.
71. A. Gozlan and A. Zilkha, *Eur. Polym. J.*, 1984, **20**, 1199.
72. A. Gozlan and A. Zilkha, *Eur. Polym. J.*, 1987, **23**, 133.
73. A. Gozlan and A. Zilkha, *Eur. Polym. J.*, 1987, **23**, 143.
74. G. Pistoia, B. Scrosati and M. A. Voso, *Eur. Polym. J.*, 1976, **12**, 53.
75. A. Lucchesi, P. Giusti and P. Cerrai, in 'Convegno Italiano di Scienza delle Macromolecole, [Atti], 4th, Colleferro', ed. A. I. M., Photo-offset, Genoa, 1979, preprint A7, p. 94.
76. T. Fernández Otero, *Makromol. Chem., Rapid Commun.*, 1984, **5**, 125.
77. R. Arnold and D. A. Swift, *Aust. J. Chem.*, 1969, **22**, 859.
78. A. I. Fedorova, Li Kuo-tung and I. V. Shelepin, *Russ., J. Phys. Chem. (Engl. Transl.)*, 1964, **38**, 920.
79. S. M. Jovanović, D. B. Šepa and M. V. Vojnović, *Makromol. Chem.*, 1970, **139**, 289.
80. G. Pistoia, A. Ricci and M. A. Voso, *J. Appl. Polym. Sci.*, 1976, **20**, 2441.
81. A. Lucchesi, P. Cerrai and P. Giusti, in 'IUPAC International Symposium on Macromolecules, [Preprints], 26th, Mainz', ed. I. Süderwald and R. Weis, Photo-offset, Mainz, 1979, preprint A1-10, vol. I, p.35.
82. V. Farafonov, M. Grovu and C. Simionescu, *J. Polym. Sci., Polym. Chem. Ed.*, 1977, **15**, 2041.
83. C. Simionescu, V. V. Farafonov and M. Grovu, *Rev. Roum. Chim.*, 1979, **24**, 379.
84. M. Delamar, P.-C. Lacaze, B. Lemiere, J.-Y. Dumousseau and J.-E. DuBois, *J. Polym. Sci., Polym. Chem. Ed.*, 1982, **20**, 245.
85. F. Beck and H. Leitner, *Angew. Makromol. Chem.*, 1968, **2**, 51.
86. A. Trifonov, B. Iordanov and M. Poneva, *Izv. Otd. Khim. Nauki (Bulg. Akad. Nauk.)*, 1971, **4**, 131.
87. A. Z. Trifonov and I. D. Shopov, *J. Electroanal. Chem.*, 1972, **35**, 415.

88. G. S. Shapoval, E. S. Shevchuk and T. E. Lipatova, *Sint. Fiz.-Khim. Polim.*, 1975, **17**, 3.
89. G. S. Shapoval, T. E. Lipatova and E. S. Shevchuk, *Vsb., Fiz.-Khim. Svoistva Sint. Vysokomol. Soedin.*, 1976, 17.
90. G. S. Shapoval and T. E. Lipatova, in 'European Symposium on Electric Phenomena in Polymer Science, [Preprints] Pisa', ed. P. Giusti, Photo-offset, Pisa, 1978, preprint B8, p. 94.
91. G. S. Shapoval and T. E. Lipatova, in 'European Symposium on Electric Phenomena in Polymer Science, [Preprints] Pisa', ed. P. Giusti, Photo-offset, Pisa, 1978, preprint B9, p. 100.
92. D. C. Phillips, J. F. Meier and J. A. Jackson, *J. Polym. Sci., Polym. Chem. Ed.*, 1974, **12**, 1193.
93. D. C. Phillips, S. Spewock and W. M. Alvino, *J. Polym. Sci., Polym. Chem. Ed.*, 1976, **14**, 1137.
94. D. C. Phillips and W. M. Alvino, *J. Polym. Sci., Polym. Chem. Ed.*, 1976, **14**, 1151.
95. D. C. Phillips, D. H. Davies and J. D. Smith, *Makromol. Chem.*, 1972, **154**, 317.
96. D. H. Davies, J. D. B. Smith and D. C. Phillips, *Macromolecules*, 1973, **6**, 163.
97. J. D. B. Smith, D. C. Phillips and D. H. Davies, *J. Polym. Sci., Polym. Chem. Ed.*, 1979, **17**, 1411.
98. G. Reinhard and E. Mersiowsky, *East Ger. Pat.* 236 117 (1986) (*Chem. Abstr.*, 1987, **106**, 121 455t).
99. M. Haruta, Y. Kuwae, H. Munakata, S. Yuasa and E. Sugawa, *Jpn. Pat.* 61 143 192 (1986) (*Chem. Abstr.*, 1987, **106**, 129 419e).
100. K. Sakata and J. Umetsu, *Jpn. Pat.* 61 193 379 (1986) (*Chem. Abstr.*, 1987, **106**, 53 272d).
101. A. Otani, T. Ichinose, T. Sasaki and M. Abe, *Jpn. Pat.* 61 225 214 (1986) (*Chem. Abstr.*, 1987, **106**, 85 252p).
102. R. Bittihn and G. Ely, *Ger. Pat.* 3 519 090 (1986) (*Chem. Abstr.*, 1987, **106**,123 097b).
103. T. Kojima, S. Tsuchiya, Y. Kudo, S. Yoshimura and J. Ozaki, *Jpn. Pat.* 61 240 625 (1986) (*Chem. Abstr.*, 1987, **106**, 187 577r).
104. S. Suzuki, Y. Asami and F. Honma, *Jpn. Pat.* 62 35 456 (1987) (*Chem. Abstr.*, 1987, **106**, 179 885z).
105. D. Laurin and G. Parravano, *J. Polym. Sci., Part C*, 1968, **22**, 103.
106. R. C. Schultz and W. Strobel, *Monatsh. Chem.*, 1968, **99**, 1742.
107. D. C. Phillips, D. H. Davies and J. A. Jackson, *Makromol. Chem.*, 1976, **177**, 3049.
108. U. Akbulut, *Makromol. Chem.*, 1979, **180**, 1073.
109. A. Trifonov, P. Nikolov and Ts. Popov, *J. Mol. Struct.*, 1973, **17**, 143.
110. A. Trifonov, Ts. Popov and B. Iordanov, *J. Mol. Struct.*, 1973, **15**, 257.
111. A. Trifonov, *Z. Phys. Chem. (Leipzig)*, 1973, **254**, 156.
112. R. V. Subramanian, *Adv. Polym. Sci.*, 1979, **33**, 33.
113. R. V. Subramanian, *Pure Appl. Chem.*, 1980, **52**, 1929.
114. J. R. MacCallum and D. H. MacKerron, *Eur. Polym. J.*, 1982, **18**, 717.
115. J. J. Jakubowski and R. V. Subramanian, *Composites*, 1980, **11**,161.
116. H. Gilch and D. Michael, *Makromol. Chem.*, 1966, **99**, 103.
117. G. Mengoli, *Adv. Polym. Sci.*, 1979, **33**, 1.
118. M. M. Musiani, G. Mengoli, B. Pelli and C. Folonari, *Makromol. Chem.*, 1982, **183**, 1869.
119. G. Mengoli, M. M. Musiani, B. Pelli, M. Fleischmann and I. R. Hill, *Electrochim. Acta*, 1983, **28**, 1733.
120. G. Mengoli and M. M. Musiani, *Makromol. Chem., Macromol. Symp.*, 1987, **8**, 265.
121. G. Mengoli, M. M. Musiani, S. Daolio and C. Folonari, *Makromol. Chem.*, 1980, **181**, 1909.
122. M. M. Musiani, G. Mengoli and F. Furlanetto, *J. Appl. Polym. Sci.*, 1984, **29**, 4433.
123. T. F. Otero, M. T. Ponce and Y. J. Jiménez, *Makromol. Chem., Macromol. Symp.*, 1987, **8**, 255.
124. A. Lucchesi, G. Maschio and P. Cerrai, in 'Meeting of the International Society of Electrochemistry, [Abstracts], 32nd, Dubrovnik–Cavtat', ed. B. Lovreček, Photo-offset, Bor, 1981, preprint VI-43, vol. II, p. 1061.
125. A. Lucchesi, G. Maschio, P. Cerrai and G. Angeli, in 'Convegno Italiano di Scienza delle Macromolecole, [Atti], 5th, Milan', ed. A. I. M., Photo-offset, Genoa, 1981, preprint 3S, p. 390.
126. E. Baicchi, Ph.D. Thesis, University of Pisa, 1983.
127. R. W. Murray, A. G. Ewing and R. A. Durst, *Anal. Chem.*, 1987, **59**, 379A.
128. R. W. Murray, *Acc. Chem. Res.*, 1980, **13**, 135.
129. F. Beck and H. Guder, *Makromol. Chem., Macromol. Symp.*, 1987, **8**.
130. J. H. Kaufman, T.-C. Chung, A. J. Heeger and F. Wudl, *J. Electrochem. Soc.*, 1984, **131**, 2092.
131. R. J. Waltman, A. F. Diaz and J. Bargon, *J. Electrochem. Soc.*, 1984, **131**, 740.
132. A. Kitani, M. Kaya and K. Sasaki, *J. Electrochem. Soc.*, 1986, **133**, 1069.
133. E. M. Geniès, M. Lapkowski, C. Santier and E. Vieil, *Synth. Met.*, 1987, **18**, 631.
134. G. Mengoli, M. M. Musiani, M. Fleischmann and N. Garrard, *Electrochim. Acta*, 1987, **32**, 55.
135. R. Bittihn, G. Ely, F. Woeffler, H. Münstedt, H. Naarmann and D. Naegele, *Makromol. Chem., Macromol. Symp.*, 1987, **8**, 51.
136. F. Garnier and G. Horowitz, *Makromol. Chem., Macromol. Symp.*, 1987, **8**, 159.
137. A. Thyssen, A. Borgending and J. W. Schultze, *Makromol. Chem., Macromol. Symp.*, 1987, **8**, 143.
138. A. Kitani, J. Yano and K. Sasaki, *J. Electroanal. Chem.*, 1986, **209**, 227.
139. A. Kitani, M. Kaya, J. Yano, K. Yoshikawa and K. Sasaki, *Synth. Met.*, 1987, **18**, 341.
140. S. H. Glarum and J. H. Marshall, *J. Phys. Chem.*, 1986, **90**, 6076.
141. S. H. Glarum and J. H. Marshall, *J. Electrochem. Soc.* 1987, **134**, 142.
142. S. Gottesfeld, A. Redondo and S. W. Feldberg, *J. Electrochem. Soc.*, 1987, **134**, 271.
143. U. Hayat, P. N. Bartlett, G. H. Dodd and J. Barker, *J. Electroanal. Chem.*, 1987, **220**, 287.
144. J.-C. LaCroix and A. F. Diaz, *Makromol. Chem., Macromol. Symp.*, 1987, **8**, 17.
145. R.-K. Yuan, D. Peramunage, M. Tomkiewicz and D. S. Ginley, *J. Electrochem. Soc.*, 1987, **134**, 886.
146. H. An, M. Seki and R. Yosomiya, *Makromol. Chem., Rapid Commun.*, 1987, **8**, 325.
147. M. Bragadin, P. Cescon, A. Berlin and F. Sannicolò, *Makromol. Chem.*, 1987, **188**, 1425.
148. S. Osawa, M. Ito, K. Tanaka and J. Kuwano, *Synth. Met.*, 1987, **18**, 145.
149. F. Chao, J. L. Baudoin, M. Costa and P. Lang, *Makromol. Chem., Macromol. Symp.*, 1987, **8**, 173.
150. B. L. Funt, E. M. Peters and J. D. Van Dyke, *J. Polym. Sci., Polym. Chem. Ed.*, 1986, **24**, 1529.
151. M. Okuno, K. Itoh, A. Fujishima and K. Hondo, *J. Electrochem. Soc.*, 1987, **134**, 837.
152. P. Audebert, G. Bidan and M. Lapkowski, *J. Electroanal. Chem. Interfacial Electrochem.*, 1987, **219**, 165.
153. A. Merz, R. Baumann and A. Haimerl, *Makromol. Chem., Macromol. Symp.*, 1987, **8**, 61.
154. F. Beck and M. Oberst, *Makromol. Chem., Makromol. Symp.*, 1987, **8**, 97.

# 25

# Polymerization of Cyclosiloxanes

THOMAS C. KENDRICK
*Dow Corning Ltd., Barry, UK*
and
BHUKAN M. PARBHOO and JAMES W. WHITE
*Dow Corning Ltd., University of Sussex, Brighton, UK*

## 25.1 INTRODUCTION

The cyclodiorganopolysiloxanes (**1**) are one of the two principal classes of siloxane intermediate used in the manufacture of siloxane polymers. The other is short chain linear $\alpha,\omega$-polydisiloxanols (**2**). Both are formed simultaneously during the hydrolysis of the appropriate diorganodichlorosilane. The cyclosiloxane fraction can be separated from the linear polysiloxanols by simple flash distillation. Ring-opening polymerization of the cyclosiloxanes, which can be effected by both basic and acidic catalysts, produces high molecular weight linear polymers. Molecular weight control is obtained by the use of hexaorganodisiloxanes (**3**), which act as chain transfer agents. The polymerization with both catalyst systems is a complex equilibrium polymerization and comprises a series of competing reactions involving both cyclic and linear species as represented below (equations 1–5) for octamethylcyclotetrasiloxane ($D_4$) and hexamethyldisiloxane ($M_2$).

$\overline{\lfloor (SiR_2O)_n \rfloor}$ (**1**)  $HO(SiR_2O)_nSiR_2OH$ (**2**)  $R_3SiOSiR_3$ (**3**)

Initiation: $$I + D_4 \rightleftharpoons P_4^* \quad (1)$$

Propagation: $$P_n^* + D_4 \rightleftharpoons P_{n+4}^* \quad (2)$$

Depropagation: $$P_n^* \rightleftharpoons P_{n-x}^* + D_x \quad (3)$$

Redistribution: $$P_n^* + P_m \rightleftharpoons P_{n+x} + P_{m-x}^* \quad (4)$$

Transfer: $$P_n^* + M_2 \rightleftharpoons P_n + M^* \quad (5a)$$

$$M^* + D_4 \rightleftharpoons P_4^* \quad (5b)$$

The conversion of cyclosiloxanes to polysiloxanes takes place without the formation of any additional chemical bonds and since the siloxane bond energy in the cyclic compound is generally the same as that in the polymer, then the reactions are thermodynamically controlled through entropy changes. Thus factors that control the configurational entropy of the cyclosiloxane ring, such as the size of the organic groups attached to silicon and the size of the ring itself, also influence the rate of polymerization and the position of equilibrium.

The cyclotrisiloxanes (**4**) constitute a special class of siloxane monomers. The siloxane bond in these monomers is different from that in the higher cyclosiloxanes and indeed in the polysiloxane itself. This gives rise to a strain energy of up to 38 kJ mol$^{-1}$ in the cyclotrisiloxanes[1] and the polymerization is no longer controlled by purely entropic factors. The result is that the cyclotrisiloxanes are far more reactive to both basic and acidic catalysts than the cyclotetrasiloxanes. Hence polymerization conditions can be chosen so that the rates of the depropagation and redistribution reactions are negligibly small compared to the rate of polymerization. Under these conditions the complexity of the polymerization is much reduced.

$\overline{\lfloor (SiR_2O)_3 \rfloor}$

(**4**)

## 25.2 RING–CHAIN EQUILIBRIA IN THE SILOXANE SYSTEM

In their pioneering work on the kinetics of the base-catalyzed polymerization of octamethylcyclotetrasiloxane ($D_4$) Grubb and Osthoff[2] concluded that the rate of polymerization of $D_4$ was governed by competing polymerization and depolymerization reactions

$$P_n + D_4 \overset{K_1}{\rightleftharpoons} P_{n+4} \quad (6)$$

and determined that the equilibrium concentration of $D_4$ in the bulk polymerization catalyzed by potassium hydroxide at 156.2 °C was 0.165 mol l$^{-1}$. However, it was Scott[3] who first reported that the equilibrium in equation (6) involved cyclic species other than $D_4$. He measured the equilibrium concentration of $D_4$ and $D_5$ formed in the acid-catalyzed polymerization of $D_4$ and postulated the participation of $D_6$, $D_7$ and $D_8$ in the equilibrium process but was unable to detect their presence at equilibrium. Carmichael and Winger[4] employed gas/liquid chromatography (GLC) techniques to measure the equilibrium concentrations of the cyclic species $D_n$ for $n = 3–10$ formed in the potassium silanolate and sulfuric-acid-catalyzed polymerizations of $D_4$. They found that the equilibrium concentration of cyclics was independent of the nature of the catalyst and for cyclic species other than $D_3$ was also independent of temperature in the range 25–178 °C. Brown and Slusarczuk[5] used gel permeation chromatography (GPC) and GLC techniques to confirm the participation of cyclic species beyond $D_{10}$ in the equilibrium process. They isolated individual cyclic species up to $D_{25}$ and used fractionation techniques and high resolution GPC to demonstrate the presence of a continuous population of cyclic species extending up to at least $D_{400}$ at equilibrium.

The results cited above clearly demonstrate that the cyclic polymer equilibrium is independent of the nature of the catalyst and involves a range of cyclic species extending to very high molecular weights. Furthermore for cyclic species other than $D_3$ the position of equilibrium is determined by purely entropic factors. A theory for cyclic formation in such polymer systems has been developed by Jacobson and Stockmayer.[6]

Consider the following equilibrium between siloxane rings $D_x$ and polymeric chains $-D_n-$ and $-D_{n-x}-$

$$-D_n- \overset{K_x}{\rightleftharpoons} -D_{n-x}- + D_x \tag{7}$$

The equilibrium constant $K_x$ is given by

$$\begin{aligned} K_x &= [-D_{n-x}-]\,[D_x]/[-D_n-] \\ &= [D_x]/p^x \end{aligned} \tag{8}$$

where $p$ is defined by the relationship $(1-p)$ = the fraction of Si–O units that are present in the polymer as end groups.

The Jacobson–Stockmayer theory separates equation (7) into an equilibrium between linear species

$$-D_n- \overset{K_3}{\rightleftharpoons} -D_{n-x}- + -D_x- \tag{9}$$

and an equilibrium between the linear species $-D_x-$ and the corresponding cyclic molecule $D_x$

$$-D_x- \overset{K_4}{\rightleftharpoons} D_x \tag{10}$$

For $x \geq 4$ the enthalpy changes for reactions (9) and (10) cancel out and the free energy change for reaction (7) is the net of the entropy changes for reactions (9) and (10). Jacobson and Stockmayer estimate these entropy changes theoretically and calculate $K_x$ from the relationship

$$K_x = \exp(\Delta S_x^\circ/R) \tag{11}$$

where $\Delta S_x^\circ = \Delta S_3^\circ - \Delta S_4^\circ$ is the standard entropy change for reaction (7).

In equilibrium (9) consider one of the skeletal atoms that forms the siloxane bond that breaks. In the bonded form it is confined to a small volume element $v_s$ at bonding distance $l$ from the other. After the bond breaks it can be found anywhere in the total volume $V$ of the system. The entropy change associated with this process is therefore

$$\Delta S_3 = k \ln(V/\sigma_a v_s) \tag{12}$$

where $\sigma_a = 2$ is a symmetry number for the siloxane chain.

In equilibrium (10) the probability that one chain end is located in the volume element $v_s$ at bonding distance $l$ from the other is $\Sigma^l W(\boldsymbol{r})d\boldsymbol{r}$, where $W(\boldsymbol{r})$ is the radial distribution function for the end to end vector $\boldsymbol{r}$ of the $x$-mer chain. For sufficiently large $x$, $W(\boldsymbol{r})$ can be represented by a normalized Gaussian distribution[7] and

$$\Sigma^l W(\boldsymbol{r})d\boldsymbol{r} = [3/2\pi(\bar{\boldsymbol{r}}^2)]^{3/2} v_s \tag{13}$$

The entropy change $\Delta S_4$ for chain closure to a ring then becomes

$$\Delta S_4 = k \ln[3/2\pi(\bar{\boldsymbol{r}}^2)]^{3/2} v_{s\sigma_a}/\sigma_b \tag{14}$$

where $\sigma_b = 2x$ is a symmetry number for the $x$-mer ring. For sufficiently large $x$ the bonds broken in reaction (9) and formed in reaction (10) are identical and the equilibrium constant $K_x$ for equilibrium (7) is given by

$$\begin{aligned} K_x &= K_3/K_4 \\ &= \exp[(\Delta S_3^\circ - \Delta S_4^\circ)/R] \\ &= [3/2\pi\langle \bar{r}^2\rangle]^{3/2}/2N_a x \text{ mol l}^{-1} \end{aligned} \tag{15}$$

where $N_a$ is Avogadro's number. In the bulk polymer $\bar{r}^2 = \bar{r}_0^2 = 2C_x x l^2$ where $C_x$, the characteristic ratio for the linear $x$-mer, is essentially constant for values of $x > 30$ and $l = 1.64 \times 10^{-8}$ cm is the length of the Si–O bond

$$K_x = (3/\pi)^{3/2}/16 l^3 C_x^{3/2} x^{5/2} N_a \tag{16}$$

The Jacobson and Stockmayer theory predicts independence of the molar cyclization constant $K_x$ with dilution and the dependence of $K_x$ on $x$ to the power of $-5/2$. However, for large $n$

$$\begin{aligned} K_x &= [D_x]/p^x \\ &= [D_x] \end{aligned} \tag{17}$$

Thus a further prediction of the theory is the existence of a critical dilution beyond which the system is composed entirely of cyclics.

A rigorous comparison of theory with experiment has been carried out by Semlyen and Wright.[8] They find $K_x$ to be independent of dilution for cyclic species $D_x$, where $x$ lies in the range 11–40. For cyclic species where $x$ is 4–10, $K_x$ increases with dilution. There is a critical dilution beyond which only cyclics are formed but the weight fraction of cyclics as a function of dilution departs from theory due to the dependence of $K_x$ on dilution for the smaller rings (see Figure 1).

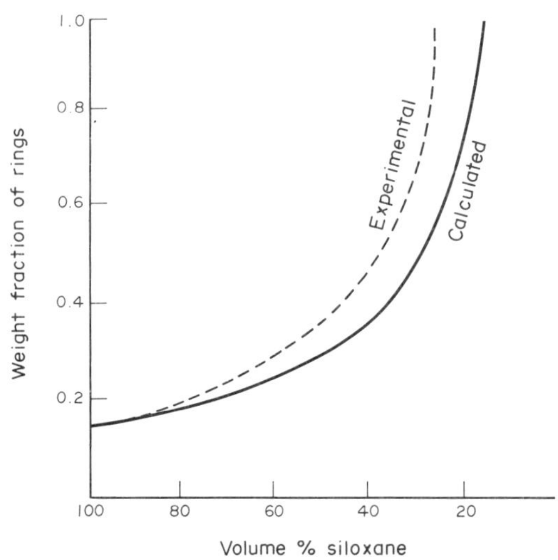


**Figure 1** Weight fraction of cyclics in poly(dimethylsiloxane) equilibrates as a function of solvent dilution (reproduced by permission of Butterworths Scientific Ltd. from ref. 8)

Siloxane chains with more than 30 units obey Gaussian statistics and the plot of log $K_x$ against log $x$ for $x > 20$ has a slope of 2.57, almost identical with that predicted by theory. However, the theory fails to account for the detailed features of the plot for values of $x < 20$ units since the assumption that such chains obey Gaussian statistics is no longer valid (Figure 2). The cyclosiloxanes $(RMeSiO)_x$ behave in a similar fashion to the dimethylcyclosiloxanes, though deviations from theory for the lower molecular weight cyclics is more pronounced the larger the size of the R group (Figure 3).

Bevers and Semlyen[10] have employed the Jacobson and Stockmayer theory without assuming Gaussian statistics to calculate $K_x$ for the cyclics $D_x$ where $x = 4$–9. They have used the Flory, Crescenzi and Mark rotational isomeric state model[11] to calculate the fraction of conformations of

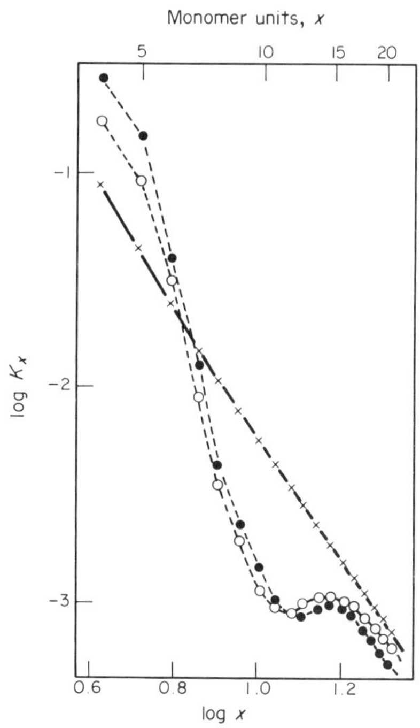


**Figure 2** Molar cyclization equilibrium constants for poly(dimethylsiloxane)s as a function of ring size: ●, solution equilibration; ○, bulk equilibration; ×, calculated (reproduced by permission of Butterworths Scientific Ltd. from ref. 8)

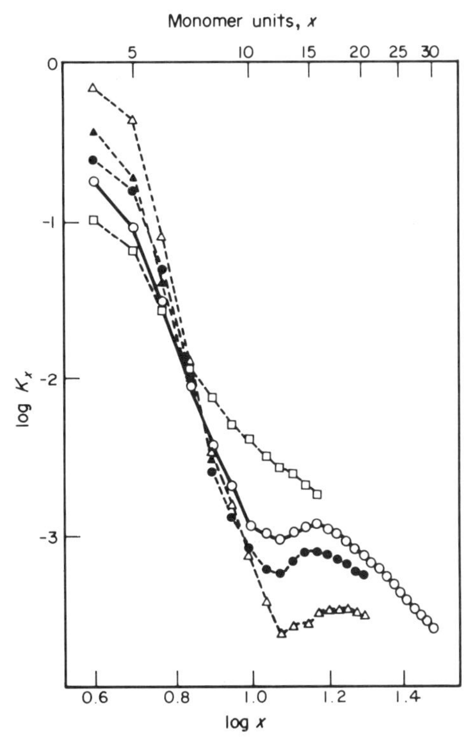


**Figure 3** Molar cyclization equilibrium constants for siloxanes $(RMeSiO)_n$: R = H, □; R = Me, ○; R = Et, ●; R = Pr$^n$, ▲; R = $CF_3CH_2CH_2$, △, (reproduced by permission of Butterworths Scientific Ltd. from ref. 9)

the linear $x$-mer $-D_x-$ that have terminal atoms in a position where ring closure can be effected. They find good agreement with theory for $K_8$ and $K_9$. Scales and Semlyen[12] have extended these calculations to $D_x$ having values of $x$ in the range $7<x<13$. Agreement with theory is excellent for $K_8$, $K_9$ and $K_{10}$ but the theoretical value for $K_{11}$ is three times that measured experimentally. The authors conclude that whilst the rotational isomeric state model gives a good qualitative understanding of the relationship between $K_x$ and $x$, the discrete nature of the model (three rotational states and a fixed bond angle at oxygen) is responsible for the limitations of the theory when applied to these lower molecular weight cyclics. A more detailed description of the ring–chain equilibria in the siloxane system can be found in reviews by Wright[13] and Wright and Beevers.[14]

## 25.3 ANIONIC POLYMERIZATION OF CYCLIC SILOXANES

### 25.3.1 Introduction

The use of alkali metal hydroxides to catalyze the polymerization of cyclodiorganopolysiloxanes was first disclosed by Hyde[15] in 1949. He observed that initially the metal hydroxide reacts with the cyclosiloxane to form a metal silanolate, which he found to be capable of catalyzing the polymerization. Grubb and Osthoff[2] demonstrated that the rates of polymerization of octamethylcyclotetrasiloxane ($D_4$) catalyzed by equimolar amounts of potassium hydroxide and potassium silanolate were identical, and concluded that the catalytic species was the metal silanolate and that the polymerization was an addition reaction where the growing polymer chain adds discrete $D_4$ units to form high molecular weight polymers. In their comprehensive study of the bulk polymerization of $D_4$ catalyzed by potassium hydroxide these authors found that the rate of polymerization was first order in monomer, that the pseudo first-order rate constant had a one-half order dependency on the catalyst concentration, and that the conversion of cyclic monomer to polymer was a reversible process. These general features of the polymerization have been confirmed by several groups of workers,[16,17,18] some of whom have observed a pronounced induction period before first-order kinetics are realized.[18]

There is general agreement in the literature that the fractional order in monomer arises from interchange reactions involving dormant and active catalyst species, though there is some conflict as to their nature. The half order in catalyst has been variously described as arising from ionization of the metal silanolate,[2,16] where the active species is the silanolate anion, or by deaggregation of a dormant ion pair dimer to yield active ion pairs.[17,18] Dissociation into free ions will only be significant at low concentrations or in media of very high dielectric constant.[19] Such conductivity measurements as have been carried out have been done so under conditions where the metal silanolate is almost certainly present as multi-ionic structures like ion pairs or ion pair aggregates. Nevertheless, there is no concrete evidence for the existence of the discrete silanolate anion under conditions where polymerization can occur; for example conductivity measurements on sodium silanolate in the inactive trifluoropropylmethylcyclotetrasiloxane $(D^F)_4$ with and without activating solvents[20] confirm that free ions are absent under conditions where $(D^F)_3$ polymerization readily takes place. Conversely there is much evidence in the literature[21,22,23] for the association of metal silanolates into aggregates in non-polar solvents. In boiling toluene the degree of aggregation for sodium trimethylsilanolate is 8.6 and that for lithium trimethylsilanolate is 4.9, both values being constant over a 0.05–0.5 $mol\,l^{-1}$ concentration range.[24] Similar measurements have not been carried out on the higher alkali metal silanolates since they are less inclined to form aggregates and therefore their solubility in non-polar solvents is very low. Likewise similar measurements on quaternary ammonium and phosphonium silanolates have not been reported in the literature despite their high solubility, high catalytic activity and industrial importance as transient catalysts for the polymerization of cyclosiloxanes.

In solution the primary aggregates will exist in dynamic equilibrium with smaller aggregates and we may consider the following sequence of equilibria (equations 18–21) to be representative of the metal silanolate (AM) in non-polar solvents

$$(AM)_n \rightleftharpoons (AM)_{n-2} + (AM)_2 \tag{18}$$

$$(AM)_{n-2} \rightleftharpoons (AM)_{n-4} + (AM)_2 \tag{19}$$

$$(AM)_2 \rightleftharpoons 2AM \tag{20}$$

$$AM \rightleftharpoons A^- + M^+ \tag{21}$$

Several of the metal silanolate species involved in these equilibria have the potential to catalyze the polymerization of cyclosiloxanes. However, it is probable that the primary catalytic species is the ion pair, but this is not firmly established. If the rate of reaction of the ion pair with the cyclosiloxane

$$AM + D_n \rightleftharpoons PM \tag{22}$$

is slower than the reverse reactions of the various equilibria, then equilibrium will be established and the concentration of ion pairs will be proportional to the $1/n$th power of the $n$-meric catalyst species. It is possible for the polymeric siloxanolates (PM) to become involved in the aggregation–deaggregation equilibria

$$AM + PM \rightleftharpoons (AMPM) \tag{23}$$

and this may effect the position of the various equilibria and lead to a non steady state concentration of ion pairs at least in the early stages of the polymerization. Induction periods where the rate of polymerization increases before first-order kinetics are established have been frequently observed for the polymerization of cyclosiloxanes in non-polar solvents. If more than one catalytic species is active in the polymerization, then the rate of polymerization is given by

$$R_p = \{k_d + f(2k_m - k_d)\}[C]_0[D]/2 \tag{24}$$

where $f$ is the mole fraction of catalyst present as the ion pair, $k_d$ is the propagation rate constant for the ion pair dimer, $k_m$ that for the ion pair and $[C]_0$ and $[D]$ are the initial concentration and concentration at time $t$ of catalyst and monomer respectively. A similar expression describes the situation where the ion pair and the free ion are the active catalytic species.

The reactivity of the metal silanolate catalyst is dependent on the nature of the metal counter ion, the larger metal ions giving rise to more active catalysts. For example in the metal silanolate series the order of catalyst activity is $Li < Na < K < Rb < Cs$,[25] with quaternary ammonium and quaternary phosphonium silanolates having the same order of activity as caesium silanolate. Lithium and sodium silanolates are not very powerful catalysts for cyclosiloxane polymerization unless used in conjunction with an activating solvent such as tetrahydrofuran (THF) or dimethyl sulfoxide (DMSO).

The large dipole moment of ion pairs causes them to interact strongly with polar molecules with the result that small amounts of polar solvents profoundly affect the course of the polymerization. Two types of interaction can be distinguished:[26] (i) where the interacting solvent does not solvate the ions and lies outside the ion pair; and (ii) where one of the ions is strongly solvated to yield a solvent-separated ion pair. Thus in polar solvents the following dynamic equilibrium will be set up involving ion pairs, solvent-separated ion pairs and free ions

$$\underset{\text{ion pair}}{(A^-M^+)S} \rightleftharpoons \underset{\text{solvent-separated ion pairs}}{A^-|S|M^+} \rightleftharpoons \underset{\text{free ions}}{A^- + SM^+} \tag{25}$$

The species which predominates will depend on several factors, including the nature of the counter ion, the solvating power of the solvent and the dielectric constant of the medium.

### 25.3.2 Kinetics of the Base-catalyzed Polymerization of Cyclotetrasiloxanes

There is general agreement in the literature on a number of features of the anionic polymerization of octamethylcyclotetrasiloxane ($D_4$) *viz.*: (i) the polymerization exhibits an induction period which is more pronounced at lower temperatures; (ii) the rate of consumption of monomer (the internal order in monomer) obeys first-order kinetics and the dependence of the apparent first-order rate constant on initial monomer concentration (the external order in monomer) is unity; and (iii) the order in catalyst is fractional and often 0.5.

The propagation/depropagation reaction is the rate-determining step in the polymerization and is represented below for the polymerization of $D_4$

$$D_4 + P_n^* \underset{k_1'}{\overset{k_1}{\rightleftharpoons}} P_{n+4}^* \tag{26}$$

where $[P_n^*]$ is the concentration of active species based on the polymer molecule of $n$ dimethylsiloxane units

$$\begin{aligned} -d[D_4]/dt &= k_1[P^*][D_4] - k_1'[P^*] \\ &= k_1[P^*]([D_4] - [D_4]_e) \end{aligned} \tag{27}$$

where [P*] is now the total concentration of active species and $[D_4]_e$ is the concentration of monomer at equilibrium. If $[D_4]_0$ is the initial monomer concentration then

$$-\ln\{([D_4]-[D_4]_e)/([D_4]_0-[D_4]_e)\}=kt \tag{28}$$

which at low conversions of monomer to polymer reduces to

$$([D_4]_0-[D_4])/([D_4]_0-[D_4]_e)=kt \tag{28a}$$

Data generated according to equations (28) or (28a) yield reliable values for the pseudo-first-order rate constant *k*.

Kucera and co-workers[203–208] have postulated a reaction order of 0.5 in monomer. They propose the following scheme (equations 29–31) for the polymerization where the rate-controlling step is rearrangement of a pentacovalent silicon anion to a silanolate anion.

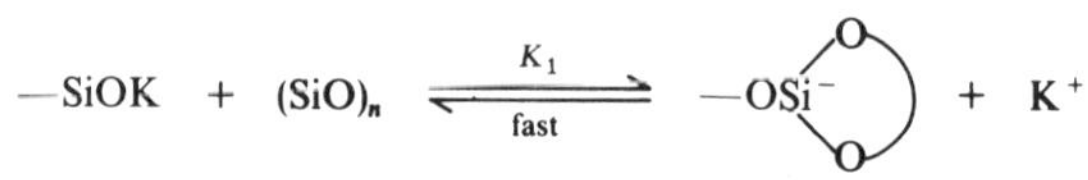


(29)

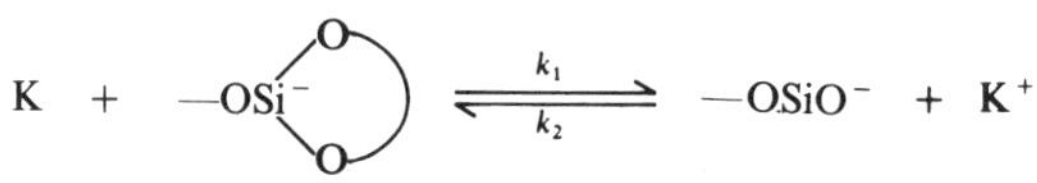


(30)

$$-OSiO^- + K^+ \overset{K_2}{\rightleftharpoons} -OSiOK \tag{31}$$

They derive the following expression for the rate of polymerization, where $[C]_0$ is the initial catalyst concentration, $[M]_0$, $[M]_e$ and [M] are the concentrations of monomer initially, at equilibrium and at any time *t* respectively

$$-d[M]/dt=k_1K_1^{1/2}[C]^{1/2}(\{[M]-[M]_e\}/[M]^{1/2}) \tag{32}$$

which integrates to

$$\frac{([M]_0^{1/2}-[M]^{1/2})}{[M]_e^{1/2}}+\frac{1}{2}\ln\frac{([M]_0^{1/2}-[M]_e^{1/2})([M]^{1/2}+[M]_e^{1/2})}{([M]_0^{1/2}+[M]_e^{1/2})([M]^{1/2}-[M]_e^{1/2})}=\frac{k_1(K_1[C]_0)^{1/2}}{2[M]_e^{1/2}}t \tag{33}$$

Our experience is that the integrated form of this rate equation does not adequately represent the polymerization kinetics at high conversions of monomer to polymer.

The most comprehensive kinetic investigation is that of Grubb and Osthoff.[2] They followed the polymerization by measuring changes in the vapour pressure of $D_4$ during the course of the polymerization and determined *k* from the relationship

$$-\ln(1.06\varphi_0-0.06)=kt \tag{34}$$

where $\varphi_0$ is the volume fraction of monomer and $0.06=[D_4]_e/[D_4]_0$, the volume fraction of monomer at equilibrium. Equation (34) reduces to

$$-\ln(1.06/[D_4]_0)([D_4]-[D_4]_e/1.06)=kt \tag{35}$$

which approximates to equation (28) if $[D_4]_e/1.06=[D_4]_e$ and $1.06=([D_4]_0+[D_4]_e)/[D_4]_0\cong[D_4]_0/([D_4]_0-[D_4]_e)$. Values of *k* calculated according to equation (28) are identical to those calculated from equation (34) in the $D_4$/KOH system studied by Grubb and Osthoff.

The mechanism proposed by Grubb and Osthoff for the bulk polymerization of $D_4$ catalyzed by KOH is based on the assumption that the potassium silanolate dissociates into silanolate ions which are the active catalytic species in the polymerization

$$-(OSiMe_2)_nOK \rightleftharpoons -(OSiMe_2)_nO^- + K^+ \tag{36}$$

$$-(OSiMe_2)_nO^- + D_4 \rightleftharpoons -(OSiMe_2)_{n+4}O^- \tag{37}$$

If both the ion pair and the silanolate anion are active catalytic species in the polymerization, then the rate of polymerization of $D_4$ is given by

$$-d[D_4]/dt = ([D_4]-[D_4]_e)(k_p[SiOK]+k_i[SiO^-]) \quad (38)$$

$$k/[C]_0 = k_p+(k_i-k_p)K/[C]_0^{1/2} \quad (39)$$

where $K$ is the dissociation constant for the ion pair, $[C]_0$ the initial catalyst concentration, $k$ the observed first-order rate constant and $k_i$ and $k_p$ are the propagation rate constants for the free ion and the ion pair respectively. Plots of $k/[C]_0$ *vs.* $1/[C]_0^{1/2}$ are linear but pass through the origin (see Figure 4), signifying the participation of one species only in the propagation reaction, which quite probably is the ion pair. A similar result is obtained for polymerizations in 50% volume THF.[27]

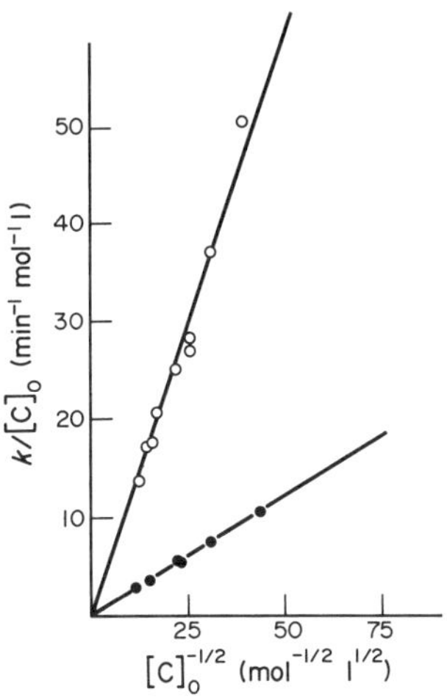


**Figure 4** Variation of $k/[C]_0$ with $(1/[C]_0)^{1/2}$ for the KOH-catalyzed polymerization of $D_4$ in bulk: ○, 152.6 and ●, 125 °C

Thus in the $D_4$/KOH system $k/[C]_0^{1/2}$ is constant over a wide range of catalyst concentrations and at a number of temperatures (Table 1).[2,16,28,29]

**Table 1** Variation of $k$ with $[C]_0$ for the KOH-catalyzed Polymerization of $D_4$

| *Temperature* (°C) | $[C]_0 \times 10^3$ (mol kg$^{-1}$) | $k/[C]_0^{1/2}$ (min$^{-1}$ mol$^{-1/2}$ kg$^{1/2}$) |
|---|---|---|
| 152.6[a] | 0.39–7.7 | $1.02 \pm 0.13$ |
| 140[b,29] | 1.00–6.8 | $0.46 \pm 0.08$ |
| 125[29] | 0.53–7.4 | $0.25 \pm 0.02$ |

[a] Data reproduced by permission of the American Chemical Society from ref. 2.
[b] Data reproduced by permission of John Wiley and Sons Inc. from ref. 16.

Attempts to confirm the presence of free silanolate anions by conductivity measurements have been inconclusive.[24,25] If indeed free ions are the active catalyst species in the polymerization, then increasing the dielectric constant ($D$) of the polymerization medium should enhance the dissociation of the potassium silanolate and at sufficiently high $D$, the order in catalyst should be unity and the propagation rate constant $k_3 = k/[C]_0$ independent of the nature of the metal counter ion. Morton, Deisz and Bostick[16] have measured the rate of polymerization of $D_4$ catalyzed by KOH at 140 °C in the presence of varying amounts of nitrobenzene, a polar solvent that does not have the capacity for solvating the metal counter ion (Table 2). Modest changes in the dielectric constant do significantly increase the rate of polymerization.

Yuzhelevskii and his co-workers[28] have studied the polymerization of mixed propylmethyl-/dimethyl-cyclotetrasiloxanes ($D_n^{Pr}D_{4-n}$) catalyzed by potassium silanolate in acetone solution at

**Table 2** Variation of $k$ with Nitrobenzene Concentration for Polymerization of $D_4$ at 140 °C[a,b]

| *Nitrobenzene* (vol %) | 0 | 2 | 5 | 9 | 17 |
|---|---|---|---|---|---|
| $k \times 10^2$ (min$^{-1}$) | 2.0 | 3.1 | 4.9 | 8.4 | 17 |
| $D_{soln.}$ *at 140 °C* | 2.4 | 2.7 | 3.0 | 3.5 | 4.4 |

[a] [KOH] = $1.79 \times 10^{-3}$ mol kg$^{-1}$. [b] Data reproduced by permission of John Wiley and Sons Inc. from ref. 16.

monomer concentrations where $D$ of the medium (17) is close to values where the concentration of free ions should be significant (Table 3). The polymerization was first order in catalyst and the propagation rate constants for the potassium silanolate and sodium silanolate catalyzed polymerizations were of similar magnitude. In the absence of solvent, potassium silanolate is 40–60 times more effective a catalyst than sodium silanolate in this system. These results suggest that the silanolate anion is the active catalyst in acetone. However, since acetone does have the capability for specific solvation of (*e.g.* lithium) cations,[30] this single result is not in itself sufficient proof that the silanolate anion is the active species in this polymerization.

**Table 3** Polymerization of $D^{Pr}_n D_{4-n}$ in Acetone[a]

| *Monomer* | $D_4$ | $D^{Pr}D_3$ | $D^{Pr}D_3$ | $D^{Pr}_2 D_2$ | $D^{Pr}_3 D$ | $D^{Pr}_4$ |
|---|---|---|---|---|---|---|
| *Catalyst* | SiOK | SiOK | SiONa | SiOK | SiOK | SiOK |
| $k/C$ (min$^{-1}$ mol$^{-1}$ l) | 360 | 8.5 | 5.1 | 5.0 | 2.6 | 1.2 |

[a] $[C]_0 = 0.76 \times 10^{-3}$ mol l$^{-1}$, temperature 40 °C, [monomer] 0.84 mol l$^{-1}$. Data from ref. 28.

### *25.3.2.1 Polymerization of cyclotetrasiloxanes containing substituents other than methyl on silicon*

Changing the substituent on silicon has little effect on the overall kinetics of the polymerization. The order in monomer and catalyst remain unchanged and the activation energy for the polymerization is almost identical with that for the dimethylcyclosiloxanes. The rate of polymerization, though, does depend on the inductive effect of the groups attached to silicon. Andrianov[31] and his co-workers have studied the polymerization kinetics for a series of monosubstituted cyclotetrasiloxanes of general formula (**5**), where R is ethyl, vinyl, phenyl and chlorophenyl. The polymerization in bulk, catalyzed by potassium hydroxide, follows first-order kinetics and the activation energies for phenyl- and chlorophenyl-substituted cyclics are similar to that for $D_4$. Yuzhelevskii[28,32] and his co-workers have studied the polymerization of propyl- and trifluoropropyl-substituted cyclotetrasiloxanes of general formula $D^{Pr}_n D_{4-n}$ and $D^F_n D_{4-n}$, where $D^{Pr}$ represents a propylmethylsiloxane unit and $D^F$ a trifluoropropylmethylsiloxane unit. They find that the polymerization in bulk catalyzed by potassium silanolate is first order in monomer and half order in catalyst. Once again activation energies are similar to that for $D_4$. Yuzhelevskii[33] has also studied the anionic polymerization of $\beta$-cyanoethyl- and $\gamma$-cyanopropyl-substituted cyclotetrasiloxanes catalyzed by potassium silanolate, sodium silanolate and lithium silanolate. The polymerization is first order in monomer and half order in lithium silanolate catalyst. Cations are poorly solvated in alkyl cyanides therefore rate enhancement by specific cation solvation should be absent in this system. However, these cyanoalkylcyclotetrasiloxanes have significantly higher dielectric constants than $D_4$, which will accelerate the rate of polymerization. Thus the enhanced rate of polymerization observed with these cyanoalkylcyclotetrasiloxanes is the result of this factor plus the inductive effect of the cyano group itself. By carrying out the polymerization of $D_4$ in the presence of equivalent concentrations of trimethyl($\gamma$-cyanopropyl)silane. Yuzhelevskii has been able to isolate the rate enhancement due to the purely inductive effect of the cyanoalkyl group itself. The kinetic results are collected together in Table 4.

$$\overline{\lfloor (RMeSiO)(Me_2SiO)_3 \rfloor}$$

(**5**)

**Table 4** Relative Rates of Polymerization for Substituted Cyclotetrasiloxanes Catalyzed by Potassium Silanolate[a]

| *Monomer* | *Catalyst* (mol l$^{-1}$ × 10$^3$) | *k* (min$^{-1}$ × 10$^3$) *at the indicated temperature* (°C) *90* | *110* | *120* | *130* | *Activation energy* (kJ mol$^{-1}$) |
|---|---|---|---|---|---|---|
| $D_4$ | 1.64 | | | | 4.58 | 81.9 |
| $D^ED_3$ | 1.64 | | | | 4.09 | |
| $D^VD_3$ | 1.64 | | | | 5.73 | |
| $D^PD_3$ | 1.64 | | | | 6.97 | 79.0 |
| $D^{ClP}D_3$ | 1.64 | | | | 7.79 | 74.8 |
| $D_4$ | 2.0 | | 2.25 | | 8.95 | 80.3 |
| $D^{Pr}D_3$ | 2.0 | | 1.16 | | 4.16 | 81.5 |
| $D_2^{Pr}D_2$ | 2.0 | | 0.62 | | 1.92 | 72.3 |
| $D_3^{Pr}D$ | 2.0 | | 0.28 | | 0.88 | 73.2 |
| $D_4^{Pr}$ | 2.0 | | | | 0.35 | |
| $D_4$ | 1.6 | | | 4.3 | 8.0 | 81.5 |
| $D^FD_3$ | 1.6 | | 3.0 | 6.7 | 10.8 | 81.9 |
| $D_2^FD_2$ | 1.6 | | 4.0 | 7.7 | 14.4 | 81.9 |
| $D_3^FD$ | 1.6 | | 5.0 | 9.5 | 17.6 | 80.3 |
| $D_4^F$ | 1.6 | 2.0 | 6.7 | 11.9 | | 70.2 |
| $D^{\beta CN}D_3$ | 2.6 | 137.2[b] | | | | |
| $D^{\gamma CN}D_3$ | 2.3 | 89.5[b] | | | | |
| $D_4$+RCN | 2.2 | 9.1[b] | | | | |

[a] Data from refs. 28, 31, 32 and 33. [b] Polymerizations carried out at 70 °C.

Within a single class of mixed cyclosiloxane the introduction of electron-donating substituents reduces the rate of polymerization, whilst electron-withdrawing substituents increase it. If the assumption is made that the inductive effect of multiple substituents is additive, then there is a good correlation between log $k/k_{D4}$ and the Taft $\sigma^*$ within a cyclosiloxane series (Figure 5). The activation energy remains relatively constant as the substituent on silicon is changed, suggesting that the major

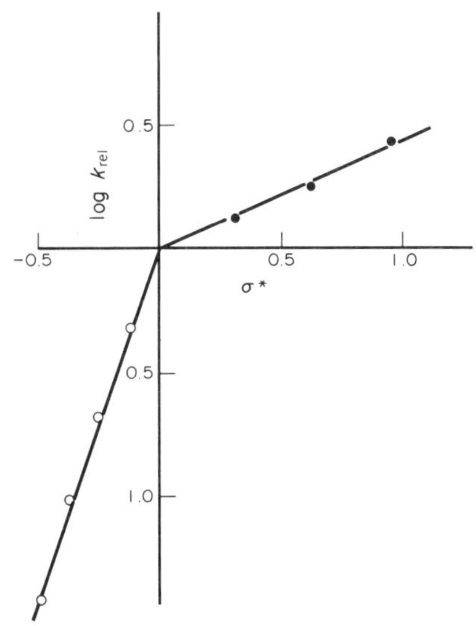


**Figure 5** The inductive effect of the substituent on silicon and its impact on rate of the polymerization of substituted cyclotetrasiloxanes: ○, ($D_n^{Pr}D_{4-n}$) and ●, ($D_n^FD_{4-n}$)

contributor to the apparent activation energy is not influenced by the nature of the substituent on silicon. For example if this is the component from the propagation step, then the major effect of the substituent is to increase or decrease the number of active centres involved in the polymerization.

### 25.3.3 Polymerization of Cyclotrisiloxanes

In the cyclotrisiloxane series the complexity of the polymerization is greatly reduced since conditions can be chosen where the rates of the depropagation and redistribution reactions are negligible compared to the rate of propagation. Kopylov, Prikhodko and Kovyazin[34] have studied the polymerization of hexamethylcyclotrisiloxane ($D_3$) in toluene catalyzed by $\alpha,\omega$-bistetramethyl-ammonium siloxanolate in toluene solution at 22–50 °C. The polymerization is first order in monomer and catalyst and the propagation rate constant $k_3$ is 200 $min^{-1}$ $mol^{-1}$ l. The activation energy for the propagation reaction is 124 kJ $mol^{-1}$. Andrianov, Temnikovskii, Khananashvili and Lyapina[35] report similar results for the polymerization of *cis*-1,3,5-trimethyl-1,3,5-triphenylcyclo-trisiloxane catalyzed by $\alpha$-hydroxy-$\omega$-tetramethylammonium methylphenylsiloxanolate in benzene solution at 12–17 °C. The polymerization is first order in monomer and catalyst and the activation energy is 103 kJ $mol^{-1}$. The propagation rate constant at 17.5 °C is 2500 $min^{-1}$ $mol^{-1}$ l. With these quaternary ammonium silanolate catalysts the catalyst is present as a single catalytic entity, which is probably the ion pair.

Mazurek and Chojnowski[36] find that the polymerization of $D_3$ in heptane/dioxane 95/5 volume catalyzed by potassium silanolate is half order in catalyst. Kopylov, Prikhod'ko, Kovyasin and Dubchak[37] find that the polymerization of $D_3$ in toluene at 80 °C catalyzed by sodium trimethyl-silanolate is first order in monomer and first order in catalyst. This is an unusual result, though the concentration range used by these workers (0.05 to 0.2 mol $l^{-1}$) is higher than normal and we may assume that the polymerization is catalyzed by ion pair dimers only. In the presence of added trimethylsilanol, which will aid dissociation of the aggregates to ion pairs, the order in catalyst reverts to the more commonly observed figure of 0.5 and the rate of polymerization is increased ($k$, the observed first-order rate constant, increases by more than 30 times when the silanol is present in a 15 times excess over the catalyst). In contrast the same authors[34] find that when the catalyst is $\alpha,\omega$-bis(tetramethylammonium) oligodimethylsilanolate the addition of a 15 times excess of the silanol reduces $k$ by a factor of 12 and the order in catalyst with and without the silanol is unity.

### 25.3.4 Mechanism of the Initiation Process

There is some dispute as to the nature of the active catalyst in the metal-silanolate-catalyzed polymerization of cyclosiloxanes.[2,16,17,18] The existence of dormant and active species has been invoked to explain the frequently observed fractional order in catalyst and there is general agreement on this aspect of the mechanism, though the exact nature of the dormant and active species has not been resolved. In fact the nature of the dormant and active species will depend on the reaction conditions and in particular on: (i) the nature of the counter ion; (ii) the dielectric constant of the medium; (iii) the presence or absence of polar solvents; and (iv) the temperature of the polymerization and the concentration of the catalyst.

Chojnowski and Mazurek[18] have attempted to resolve this issue of the nature of the active species by studying the reaction of phenyldimethylsilanolates with cyclosiloxanes under conditions where siloxane bond redistribution reactions involving the polymer and the catalyst are suppressed. For the reaction of sodium phenyldimethylsilanolate (**6**) with 2,2,5,5-tetramethyl-1-oxa-2,5-disilacyclo-pentane (**7**) they find that the rate of disappearance of (**6**) in an excess of (**7**) follows first-order kinetics. The observed first-order rate constant $k$ varies with the initial monomer concentration $[M]_0$ and the initial catalyst concentration $[C]_0$ according to the following expression

$$k = K[M]_0/[C]_0^{1/4} \tag{40}$$

The propensity for sodium silanolates to form aggregates is well known[21,23] and the authors propose the following mechanism based on a series of aggregation/deaggregation equilibria between the phenyldimethylsilanolate ion pair ($A_1$) and the polymeric phenyldimethylsiliconate ion pair ($B_1$) formed by reaction of (**6**) with (**7**)

$$nA_1 \overset{K_0}{\rightleftharpoons} A_n \tag{41}$$

$$(n-m)A_1 + mB_1 \rightleftharpoons A_{(n-m)}B_m \tag{42}$$

$$nB_1 \rightleftharpoons B_n \tag{43}$$

They assume that the reaction of $A_1$ with M is the rate-determining step

$$A_1 + M \overset{k_5}{\rightleftharpoons} B_1 \tag{44}$$

and derive the following expression for the rate of the reaction of (**6**) with (**7**)

$$-dA/dt = k_5[M][A]/\{(nK_0)^{1/n}[A]_0^{(n-1)/n)}\} \tag{45}$$

where [A] is the concentration of (**6**) at time $t$, $[A]_0$ is the initial concentration of (**6**) and $K_0$ is the equilibrium constant for reaction (41). For $n=2$ this is identical to the expression derived by Szwarc[26] for the addition of living polystyryl to 1,1-diphenylethylene. Equation (45) predicts the general features of the kinetics but does not support an order of 1/4 in initial (**6**). Chojnowski and Zietera[38] have studied the reaction of sodium polystyryl (**8**) with cyclosiloxanes. The reaction does not follow simple first-order kinetics, presumably because cross-aggregation of (**8**) with sodium silanolates would produce aggregates with dissociation constants intermediate in value between that for (**8**) and sodium silanolate aggregates and assumptions concerning the interrelationships of the equilibrium constants for reactions (41) to (43), which are necessary to derive equation (45), are no longer valid. However, the general features of the reaction are the same and $k$ shows an inverse dependence on the initial concentration of (**8**). The kinetics of the reaction of potassium phenyldimethylsilanolate (**9**) has similar features.[36,39] Deviation from first-order kinetics is again observed and for reactions of $D_6$ with (**9**) the slope of $\log k$ *vs.* $\log A_0$ is close to $-0.5$. Potassium trimethylsilanolate is not very soluble in non-polar solvents but cross-aggregates with lithium or sodium silanolates are soluble. This suggests that the potassium silanolate, unlike its lithium and sodium counterparts, is not highly aggregated in solution. The kinetic results point to a maximum aggregation number for potassium siloxanolate of 2. In this respect it more resembles sodium polystyryl than sodium silanolate.

### 25.3.5 Polymerization in the Presence of Activator Solvents

The ability of solvents to increase the rate of the base-catalyzed polymerization of cyclosiloxanes was first reported by Hyde[15] who observed increased rates of polymerization when acetone was used as a solvent with alkali metal hydroxide catalysts. The considerable rate enhancement that can be obtained by using these dipolar aprotic solvents is well illustrated in the table below taken from the data of Yuzhelevskii[20] for the $(D^F)_3$/sodium siloxanolate system (Table 5). $k_0$ is the observed first-order rate constant in the absence of activator solvent. The data indicate two broad classes of dipolar aprotic solvent: (i) those like nitrobenzene and alkyl cyanides where the cation is poorly solvated and the anion is essentially naked and where the rate of polymerization increases smoothly with the dielectric constant of the medium; and (ii) those like hexamethylphosphorotriamide

**Table 5** Relative Rates of Polymerization of $D_3^F$ in the Presence of Activators using Sodium Siloxanolate as Catalyst ($2 \times 10^{-3}$ mol $l^{-1}$)[a]

| *Solvent (complexant)* | *[Solvent]* | *Temperature* (°C) | $k/k_0$ |
|---|---|---|---|
| Nitrobenzene | 0.1 | 40 | 2.1 |
| Tetrahydrofuran | 0.1 | 40 | 10.5 |
| Acetonitrile | 0.1 | 40 | 34 |
| Acetone | 0.1 | 40 | 95 |
| Dimethylformamide | 0.01 | 30 | 150 |
| (Dimethyl sulfoxide) | 0.01 | 30 | 155 |
| (Tributyl phosphate) | 0.01 | 40 | 300 |
| (Hexamethylphosphorotriamide) | 0.001 | 30 | 28 |
| (Diethylene glycol dimethyl ether) | 0.001 | 20 | 28 |

[a] Data reproduced by permission of Plenum Publishing Corp. from ref. 20.

(HMPT) and dimethyl sulfoxide (DMSO) where the cation is highly solvated and which are highly activating at low concentrations generally of the order of the catalyst concentration.

The polymerization of cyclosiloxanes catalyzed by metal silanolates in solvents from both classes have been studied by several groups of workers. The metal silanolates like sodium and lithium silanolates are highly aggregated and have no catalytic activity in non-polar solvents and thus we can reasonably conclude that these aggregates are dormant catalytic species in the polymerization of cyclosiloxanes. The polymerization of $D_3$ catalyzed by lithium silanolate in toluene/THF at a [THF] of 3.2 $mol\,l^{-1}$ has been studied by Gladkova, Durgar'yan and Nametkin.[40] The polymerization is first order in monomer and of fractional order in catalyst. Similar results were obtained by Morton and Bostick for the polymerization of $D_4$ catalyzed by potassium siloxanolate[27] in THF. The potassium silanolate catalyzed polymerization of $D_4$ and mixed methylpropyl-/dimethyl-cyclosiloxanes in acetone has been referred to previously.[28] In this medium of high dielectric constant the polymerization is first order in both catalyst and monomer. Yuzhelevskii, Kagan and Fedoseeva[20] have studied the polymerization of $(D^F)_3$ catalyzed by sodium siloxanolate at 30 °C in the presence of dimethylformamide (DMF) at concentrations varying from 1 to 10 times that of the catalyst. The polymerization is first order in monomer and the first-order rate constant $k$ is directly proportional to the catalyst and the solvent concentrations at constant catalyst to solvent molar ratios. However, the variation of $k$ with catalyst concentration at constant solvent concentration is complex. $k$ increases with increasing catalyst concentration up to a maximum and then decreases with further increases in the catalyst concentration. Lee, Frye and Johannson[41] have studied the polymerization of $D_3$ catalyzed by butyllithium ($4\times10^{-3}$ $mol\,l^{-1}$) in *o*-xylene at 25 °C and in the presence of DMSO at concentrations 50–100 times that of the catalyst. The rate of polymerization increases with the $[DMSO]^{1.5}$. Holle and Lehan[42] have polymerized $D_3$ with lithium silanolate catalyst in the presence of HMPT and find that the observed first order rate constant is proportional to $[HMPT]^2$ and $[catalyst]^{1/2}$.

In the presence of those activating solvents that solvate the cation the metal silanolates are present in appreciable concentrations as solvent-separated ion pairs.[26] The solvent-separated ion pair will be in equilibrium with contact ion pairs, solvated ion pair dimers and free ions. If the active catalytic species are the ion pair and the free ion, then the observed first-order rate constant will be a composite term involving the propagation rate constant for the ion pair ($k_p$) and that for the free ion ($k_i$)

$$k = k_p[\equiv SiOM] + k_i[\equiv SiO^-] \tag{46}$$

Consider the equilibrium

$$SiOMA + nA \overset{K}{\rightleftharpoons} SiO^- + AMA_n^+ \tag{47}$$

If $f$ = mole fraction of catalyst as the ion and the remaining catalyst is present as the solvated ion pair then

$$K = f^2[C]_0/(1-f)[A]^n \tag{48}$$

$$[A]_0 = [A] + [C]_0(1+nf) \tag{49}$$

and

$$f = \{K(1-f)[A]^n/[C]_0\}^{1/2} \tag{50}$$

If $f \ll 1$ then

$$[A] = [A]_0 - [C]_0 \tag{51}$$

$$f = \{(K[A]_0 - [C]_0)/[C]_0\}^{1/2} \tag{52}$$

and

$$k/[C]_0 = k_p + (k_i - k_p)\{K([A]_0 - [C]_0)^n/[C]_0\}^{1/2} \tag{53}$$

where $[A]_0$ and $[C]_0$ are the initial solvent and catalyst concentrations respectively. The data for the rates of polymerization in various activator solvents are in accord with this equation.

Polymerizations in THF where the solvent is in a large excess over the catalyst give linear plots for $k/[C]_0$ *vs.* $1/[C]_0^{1/2}$ for both lithium and potassium siloxanolate catalyzed polymerizations, that for the latter catalyst indicating the participation of a single catalytic species only in the polymerization (Figure 6). $k_p$ for $D_3$ and lithium siloxanolate is 24 $min^{-1}\,mol^{-1}\,l$.

The polymerization of $D_3$ catalyzed by lithium silanolate at constant catalyst concentration in the presence of DMSO gives a linear plot for $k/[C]_0$ *vs.* $[DMSO]^{3/2}$, signifying that for this activator

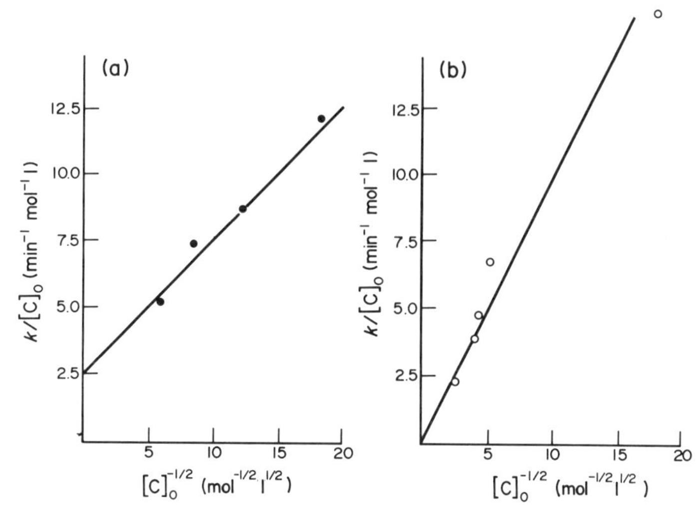


**Figure 6** Variation of $k/[C]_0$ with $(1/[C]_0)^{1/2}$: (a) polymerization of $D_4$ with potassium silanolate; (b) polymerization of $D_3$ with lithium silanolate

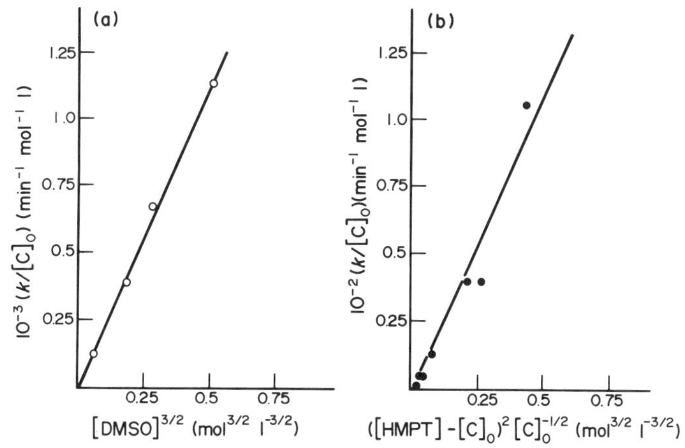


**Figure 7** Variation of $k/[C]_0$ with $\{([A]_0-[C]_0)^n/[C]_0\}^{1/2}$ for the polymerization of $D_3$ with lithium silanolate in (a) DMSO and (b) HMPT

solvent $n=3$. The polymerization of $D_3$ catalyzed by lithium silanolate in the presence of HMPT also fits the equation for $n=4$. Both plots pass through the origin, indicating the participation of a single catalytic species in the polymerization (Figure 7).

The polymerization of $(D^F)_3$ catalyzed by sodium siloxanolate in the presence of DMF at concentrations similar to that of the catalyst is highly complex. If $n=1$, then at constant [DMF] $([A]_0-[C]_0)$ will decrease as $[C_0]$ increases and $k$ will therefore increase with $[C]_0$ to a maximum and then decrease with further increases in $[C]_0$. This is observed. At constant $[A]_0/[C]_0$ $k$ will vary directly with $[C]_0$ and this also is observed.

The work of Boileau[43] with lithium catalysts and highly specific cation complexing solvents is particularly significant. The author describes the use of the macrobicyclic ligand or cryptate (**10**), which forms multicontact complexes with the metal counter ion of the catalyst. These complexes, which are often metal ion specific give rise to large increases in the rate of polymerization by

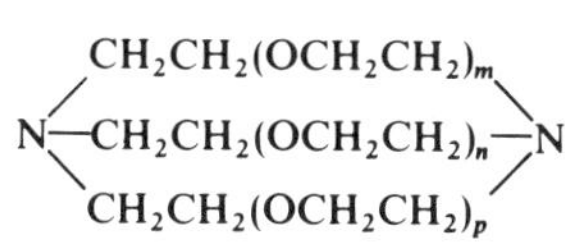


**(10)**

assisting deaggregation of the ion pair aggregates. The [2.1.1] cryptate where $m=2$ and $n$ and $p=1$ is highly specific for the lithium counter ion. Polymerization of $D_3$ with the lithium [2.1.1] cryptate is first order in catalyst. Deactivation of the living poly(dimethylsiloxane) shows no significant change in viscosity of the polymer solution and confirms the absence of aggregates in this system. The author assumes the active catalytic species to be the lithium cryptate ion pair and estimates the ion pair propagation rate constant to be 78 $min^{-1}$ $mol^{-1}$ l. With the [2.2.1] cryptate the rate of polymerization is much lower and deactivation of the living polymer gives a substantial reduction in viscosity. The order in catalyst is 0.25, confirming the presence of ion pair aggregates.

### 25.3.6 The Effect of Additives on the Kinetics of Polymerization of Cyclosiloxanes

The polymerization of $D_4$ with KOH exhibits an induction period. Small amounts of water reduce the induction period and increase the rate of polymerization, presumably by behaving as a protic solvent and aiding the dissociation of the potassium silanolate. However, at higher concentrations the water acts as a transfer agent and decreases the rate of polymerization. The effect of water on the kinetics of the polymerization of $(D^P)_3$ (0.46 $mol\,l^{-1}$) catalyzed by tetramethylammonium siloxanolate ($1.65 \times 10^{-4}$ $mol\,l^{-1}$) in benzene at 20 °C has been studied by Andrianov and his co-workers.[44] As the water concentration increases, so does the induction period, whilst the rate of polymerization and the polymer molecular weight decrease. The proposed mechanism is

$$\text{-(SiO)M} + (D^P)_3 \rightleftharpoons \text{-(SiO)}_3\text{M} \quad (54)$$

$$\text{-(SiO)M} + \text{HOH} \rightleftharpoons \text{-(SiO)OH} + \text{MOH} \quad (55)$$

$$\text{MOH} + (D^P)_3 \rightleftharpoons \text{-(SiO)}_3\text{M} \quad (56)$$

The reduction in the rate of polymerization and the extended induction period are an indication of a reduction in the number of active centres, which the authors propose is due to rapid formation of the less active complex (**11**), which then establishes an equilibrium with the silanol-terminated polymer and the metal hydroxide, the latter being a more active catalyst for the polymerization. The extended induction period is seen as a period leading up to steady state concentration of the complex. Alternatively the induction period could be due to reaction of the catalyst with the water to form silanols and the reduced rate of polymerization to the known effect of silanols on rate of base-catalyzed polymerizations.[34,37] The effect of silanols on the kinetics of the polymerization of $D_3$ catalyzed by quaternary ammonium siloxanolate[34] and sodium trimethylsilanolate[37] has been studied by Kopylov and co-workers. With the quaternary ammonium silanolate catalyst the polymerization in the absence of trimethylsilanol is first order in monomer and catalyst. The addition of trimethylsilanol reduces the rate of polymerization and at constant [silanol] the observed first-order rate constant $k$ exhibits a non-linear relationship with [catalyst]. At high concentrations of silanol the order in catalyst is one. At a constant molar ratio of [silanol] : [catalyst] the order in catalyst is unity. The authors postulate the formation of complexes between the silanol and the ion pair which are less active than the non-complexed ion pair.

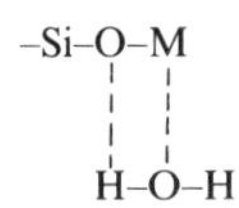


(**11**)

One would expect the trimethylsilanol to perform two functions, *viz.* to help break down the aggregates into ion pairs and to behave as a protic solvent and solvate the anion giving a less active catalytic species. This may be represented according to the following scheme where $[A]_0$ and $[C]_0$ are the initial silanol and catalyst concentrations respectively

$$(\equiv\text{SiOM})_2 + 2\text{A} \rightleftharpoons 2[\equiv\text{SiOMA}] \quad (57)$$

$$\equiv\text{SiOMA} + \text{A} \overset{K}{\rightleftharpoons} \equiv\text{SiOAMA} \quad (58)$$

$$[\text{A}]_0 = [\text{A}] + [\equiv\text{SiOMA}] + 2[\equiv\text{SiOAMA}] \quad (59)$$

The quaternary ammonium silanolate catalyst will be present as ion pairs and if $f$ is the mole fraction of catalyst present as SiOAMA, then $(1-f)$ is the mole fraction present as SiOMA and

$$[A]_0 = [A]+[C]_0+f[C]_0 \tag{60}$$

For $f[C]_0 \ll [A]$

$$[A]_0 = [A]+[C]_0$$
$$k/[C]_0 = k_p+(k_{sp}-k_p)K([A]_0-[C]_0)/\{1+K([A]_0-[C]_0)\} \tag{61}$$

$k_p$ and $k_{sp}$ are the propagation rate constants for the ≡SiOMA and ≡SiOAMA ion pairs respectively, $K$ is the equilibrium constant for their interconversion and $k$ is the observed first-order rate constant. This equation predicts: (i) $k$ will increase in a non-linear fashion with $[C]_0$ at constant [silanol]; (ii) the ratio $R=K([A]_0-[C]_0)/\{1+K([A]_0-[C]_0)\}$ at constant $[C]_0$ will tend to a constant value as $[A]_0$ is increased and $k$ will become independent of $[A]_0$ at high $A_0$; (iii) at constant $[A]_0/[C]_0$ the order in catalyst will be unity; and (iv) for situations where $[C]_0>[A]_0$ the ratio $R$ will again tend to a constant value and the order in catalyst will be unity. These features are observed.

For polymerizations with sodium silanolate as catalyst, $k$ increases significantly with increasing [silanol] : [catalyst] and this is almost certainly due to breakdown of aggregates by the silanol as is proposed by the authors. In this respect the trimethylsilanol is behaving more as an activator solvent than as a complexing agent for the active catalyst species. Kucera and his co-workers[45] find the inhibiting effect of water to be more apparent at low temperatures than at high. They explain this by invoking the intervention of complexes of the type proposed by Andrianov, which are dormant at 100 °C but which decompose to give active species at higher temperatures. Kucera has examined the effect of a number of additives on the polymerization. The most significant observation is that the KOH-catalyzed polymerization is retarded by the addition of lithium or sodium hydroxides. This effect probably results from the known ability of the metal silanolates to form cross-aggregates, which would reduce the concentration of KOH in its active deaggregated form. Andrianov[46] and co-workers have studied the polymerization of cyclosiloxanes with organic diols. The diol is completely consumed during the reaction, presumably by a mechanism similar to that proposed for water.

### 25.3.7 The Polymerization of Cyclosiloxane Species Containing More than Four Siloxane Units

The preceeding discussion has focused on the polymerization of cyclotrisiloxanes and cyclotetrasiloxanes and ignored the behaviour of the higher members of the cyclosiloxane series. The relative rates of polymerization of the higher dimethylcyclosiloxanes have been reported by a number of groups of workers, the most complete study being that of Laita and Jelinek.[47] They determined the relative rates of polymerization for cyclic species up to $D_7$ in the bulk polymerization catalyzed by KOH at 150 °C. Thomas[29] has measured the relative rates of polymerization of $D_4$ and $D_5$ catalyzed by KOH over a range of temperatures and Yuzhelevskii and co-workers[28] that for $D_3$ and $D_4$ at 110 °C. The various results, converted to a common temperature of 150 °C, are quoted below (Table 6). In the presence of dipolar aprotic solvents this reactivity sequence is not retained. Mazurek and Chojnowskii[39] quote relative rates for the polymerization of $D_4$ and $D_7$ catalyzed by KOH in a number of solvent systems at 30 °C and Boileau[43] for the polymerization of $D_3$, $D_4$, $D_5$

**Table 6** Relative Rates of Polymerization of Dimethylcyclosiloxanes Catalyzed by KOH in Bulk[a]

| *Monomer* | *Relative rate* | *Activation energy* (kJ mol$^{-1}$) |
|---|---|---|
| $D_3$ | 53, 34[28] | 73.2, 69.4[28] |
| $D_4$ | 1 | 81.5, 81.9[29] |
| $D_5$ | 1.6, 1.4[29] | 81.5, 90.3[29] |
| $D_6$ | 11.3 | 81.5 |
| $D_7$ | 316 | 81.5 |

[a] Data from refs. 28 and 47.

**Table 7** Relative Rates of Polymerization of Dimethylcyclosiloxanes in Various Solvents[a,b]

| *Solvent (complexant)* | $D_3$ | $D_4$ | $D_5$ | $D_6$ | $D_7$ |
|---|---|---|---|---|---|
| 95/5 heptane/dioxane | | 1 | | | 432 |
| 95/5 heptane/diglyme | | 1 | | | 8 |
| 95/5 heptane/(HMPT) | | 1 | | | 0.7 |
| THF | | 1 | | | 15 |
| THF/(18-DBC-6)[b] | | 1 | | | 0.9 |
| Li(THF) (211 cryptate) | 280 | 1 | 0.4 | 0.06 | |

[a] Data reproduced by permission of Huthig and Wepf Verlag from ref. 39, and by permission of the American Chemical Society from ref. 43. [b] 18-DBC-6 is dibenzo-18-crown-6.

and $D_6$ at room temperature catalysed by lithium silanolate ([2.1.1] cryptate). The results are collected in Table 7.

The reaction of potassium phenyldimethylsilanolate with dimethylcyclosiloxanes in the non-solvating solvent system 95/5 heptane/dioxane at 30 °C has been studied by Mazurek and Chojnowskii.[39] Their results are presented in Table 8.

**Table 8** Relative Rates of Reaction of Potassium Phenyldimethylsilanolate with Dimethylcyclosiloxanes at 30 °C in 95/5 Heptane/Dioxane[a]

| *Monomer* | $D_3$ | $D_4$ | $D_5$ | $D_6$ | $D_7$ | $D_8$ | $D_9$ |
|---|---|---|---|---|---|---|---|
| *Relative rate* | 88 | 1 | 1.1 | 12 | 161 | 182 | 124 |

[a] Data reproduced by permission of Huthig and Wepf Verlag from ref. 39.

The authors propose that there is an interaction between the silanolate and the cyclosiloxane which facilitates the cleavage of the siloxane. They visualize the interaction to involve the cation of the ion pair in a multipoint interaction with the cyclosiloxane similar to the polydentate interactions between cations and crown ethers (Scheme 1). Thus species like crown ethers and HMPT which form stronger complexes with the cation will disrupt this multipoint interaction. Thus the relative rate of the polymerization of $D_4$:$D_7$ is reduced from 1:161 to 1:0.7. It should be realized however that the relative rates are determined from observed first-order rate constants which are themselves composite terms involving the equilibrium constant for the formation of active catalytic species from the catalyst. If the higher cyclic species favour the formation of active species by some mechanism such as is proposed by Mazurek, then the relative rates for the actual propagation reaction will be more in accord with the data of Boileau where the catalyst is present as a single active species, namely the lithium cryptate/silanolate ion pair.

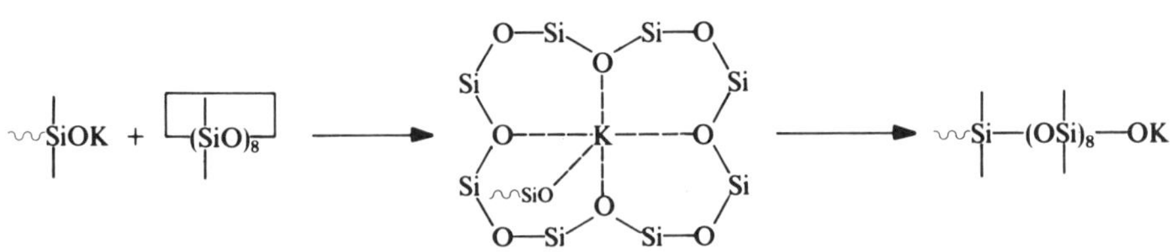

**Scheme 1**

### 25.3.8 Molecular Weight Distributions and the Nature of the Redistribution Reactions Involving the Siloxane Chain

The polymerization of cyclotrisiloxanes can be effected under conditions where monomer addition to the active centre occurs without termination or transfer to polymer. If the exchange between active and dormant catalytic species is fast compared to the rate of propagation, then growth of all the polymer molecules proceeds simultaneously and the polymer will have a Poisson distribution.[7] The number average degree of polymerization $x_n$ and the ratio of weight average degree of polymerization $x_w$ to $x_n$ is given by the following equations

$$x_n = 1 + n \tag{62}$$

$$x_w/x_n = 1 + n/(1+n)^2 \tag{63}$$

where $n$ is the number of monomer units reacted per polymer molecule . There have been a number of studies where polymers of narrow molecular weight distribution have been prepared. For example Holle and Lehan[42] have measured the molecular weights of polymers prepared by polymerizing $D_3$ with lithium siloxanolate using HMPT as the activator solvent. Their results show good correspondance with theory for both $x_n$ and $x_w/x_n$ (Table 9.)

**Table 9** Molecular Weights and Molecular Weight Distributions in the Polymerization of $D_3$ Catalyzed by Lithium Silanolate/HMPT[a]

| $[D_3]$ (mol l$^{-1}$) | $[C]_0$ (mol l$^{-1}$) | *Conversion* (%) | $X_n$ *Observed* | $X_n$ *Theory* | $x_w/x_n$ *Observed* | $x_w/x_n$ *Theory* |
|---|---|---|---|---|---|---|
| 0.39 | 0.027 | 98 | 15 | 13 | 1.11 | 1.080 |
| 0.52 | 0.002 | 25 | 66 | 59 | 1.03 | 1.015 |
| 0.52 | 0.001 | 56 | 292 | 284 | 1.02 | 1.003 |
| 0.23 | 0.000 17 | 50 | 677 | 671 | 1.03 | 1.002 |
| 0.26 | 0.0001 | 65 | 1690 | 1815 | 1.09 | 1.001 |
| 0.43 | 0.000 067 | 70 | 4494 | 4685 | 1.16 | 1.000 |

[a] Data from ref. 42.

If the polymerization is carried out under conditions where chain transfer to polymer can occur and if the siloxane bonds in the polymer chain are of equal reactivity towards the active chain end, then the molecular weight distribution of the polymer will be broader and at equilibrium will have the most probable distribution[7] where

$$x_n = 1/(1-p) \tag{64}$$

$$x_w/x_n = 1 + p \tag{65}$$

The ratio $x_w/x_n$ will tend to a value of 2 as the extent of reaction $p$ approaches unity. Davydova[48] and co-workers found that the polydispersity of PDMS prepared from $D_4$ and potassium siloxanolate becomes broader as the polymerization proceeds. Changes that do occur during the polymerization are observed in the range of high molecular weights only and number average molecular weights remain fairly constant throughout the polymerization (Table 10).

Mazurek and co-workers[49] have studied the concentration of linear oligomers in fully equilibrated living poly(dimethylsiloxane)s. Polymerizing $D_3$ with potassium trimethylsilanolate should give at equilibrium chains with trimethylsilyl groups at each end (I), chains terminating in one trimethylsilyl group and one silanolate group (II) and chains terminated in silanolate groups only (III). By neutralizing the living silanolate chain ends with trimethylchlorosilane the authors found that the concentration of $MD_nM$ oligomers exceeded that predicted from the most probable distribution by a factor of 10. By neutralizing the living polymer with ethyldimethylchlorosilane the authors were

**Table 10** Variation of $x_w/x_n$ as a Function of the Degree of Polymerization of $D_4$ for Polymerizations With and Without $M_2$ Chain Transfer Agent[a]

| $x_n$ | *Without $MD_2M$* *Conversion* (%) | $x_w/x_n$ | $x_n$ | *With $MD_2M$* *Conversion* (%) | $x_w/x_n$ |
|---|---|---|---|---|---|
| 2095 | 27 | 2.6 | 2027 | 30 | 2.98 |
| 2095 | 45 | 3.2 | 2216 | 52 | 3.20 |
| 1920 | 65 | 4.4 | 2256 | 75 | 3.24 |
| 2432 | 85 | 4.7 | 2540 | 84 | 3.23 |

[a] Data from ref. 48.

able to estimate the amounts of oligomers of type I, II and III present at equilibrium. They found that these low molecular weight oligomers are derived from species of type III, *i.e.* chains terminating in silanolate groups only. A maximum in the oligomer concentration occurs at $n=4$–5. The explanation offered is that the silanolate chain ends aggregate to form ion pair dimers and the ease of their doing so is probably related to the entropy change for chain closure to a ring such as developed by Jacobson and Stockmayer.[6] Such phenomena might explain the broadening of the molecular weight distribution and the skewing of the distribution to high molecular weights at high conversions observed by Davydova.[49] The introduction of an end block molecule such as $MD_2M$ reduces the opportunity for finding silanolate end groups on the same chain with the result that the molecular weight distribution is sharpened and the tendency to favour high molecular weights at high conversions disappears. Under conditions where aggregation does not occur Boileau[43] finds that the polymerization of $D_4$ with butyl lithium [2.1.1] cryptate gives a bimodal molecular weight distribution with the $x_w/x_n$ values for the high molecular weight peak being always equal to 2.

The polymerization of $D_3$ with catalysts that yield narrow molecular weight distributions proceeds by successive addition of $D_3$ units to the growing polymer chain. Zavin[50] and his co-workers have polymerized $D_3$ with lithium butyl/HMPT and equimolar amounts of the chain transfer agents $Me_nSi(OMe)_{4-n}$. In the presence of dimethyldimethoxysilane, oligomers of the general formula $MeO(Me_2SiO)_nSiMe_2OMe$ (**12**), where $n=3$, 6, 9 and 12, are formed in preference to oligomers of intermediate chain length. Such redistribution reactions as do occur in this system are specific to the terminal siloxane bond connected to the dead chain end of the oligomer. Thus oligomers having exclusively trimethoxysilyl or exclusively methoxy end groups are not formed, signifying that the redistribution reactions for the oligomers (though not necessarily for the high molecular weight polymer) are not random along the oligomer chain. Fessler and Juliano[51] have studied the reaction of equimolar amounts of $D_3$ and lithium *n*-butyldimethylsilanolate (BDLi) and characterized the oligomers formed. The principal product formed is $BD_2Li$, which is formed as rapidly as BDLi is consumed and faster than $D_3$ is consumed. The $BD_4Li$ which is formed by the primary reaction

$$BDLi + D_3 \rightleftharpoons BD_4Li \tag{66}$$

must react very rapidly with a second molecule of BDLi to give $BD_2Li$ and $BD_3Li$ which will react with a further molecule of BDLi to yield two molecules of $BD_2Li$. The specificity of this redistribution can be understood if one invokes a rapid exchange between ion pair and ion pair dimers, such as has been used to explain the kinetics of $D_3$ polymerization with lithium catalysts. When aggregates are not present, this specificity should disappear.

### 25.3.9 Summary

The anionic polymerization of cyclosiloxanes is a complex process. For the alkali metal silanolate catalysts the weight of experimental evidence supports a mechanism based on growth from the metal silanolate ion pair. The ion pair is in dynamic equilibrium with ion pair dimers, which for the smaller alkali metal ions like lithium and sodium are themselves in dynamic equilibrium with ion pair dimer aggregates. The fractional order in catalyst which is observed is a direct result of the equilibria between ion pairs, ion pair dimers and ion pair dimer aggregates. Polar solvents and species that form complexes with the cation have a marked accelerating effect on the rate of polymerization. The polar solvents break down the aggregates and increase the concentration of ion pair dimers and hence the concentration of ion pairs. Species like crown ethers and the [2.1.1] cryptate which form strong complexes with the metal cation increase the dissociation of ion pair dimers into ion pairs. In the case of the lithium [2.1.1] cryptate dissociation into ion pairs is complete and the order in catalyst is unity.

Many apparent anomalies in the polymerization are explained on the basis of the equilibrium between ion pair dimers and ion pairs. Thus the higher than anticipated concentrations of cyclic oligomers observed in the $\alpha,\omega$-dipotassium-siloxanolate-catalyzed polymerization of permethylcyclosiloxanes is thought to be due to the formation of internal ion pair dimers by silanolate ion pairs on the chain ends of the same polymer molecule. The wide differences in the relative rates of polymerization of permethylsiloxanes in the presence and absence of polar solvents also has its origins in this association/dissociation equilibrium. In non-polar solvents the cyclosiloxane is thought to form multicontact complexes with the metal cation which would assist the deaggregation reaction and favour the production of ion pair dimers and ion pairs. Some cyclics will be more effective than others and this gives rise to their relative reactivities in non-polar systems. With polar solvents or species which form complexes with the metal cation the cyclosiloxane is no longer able to impact the aggregation/deaggregation equilibria and the relative rates of polymerization reflect those for the propagation reaction itself.

The polymerization is accelerated in solvents of high dielectric constant which do not have the capability to solvate the ion pair. Under these conditions, which would favour the dissociation of the ion pair into the free ions, it is perhaps reasonable to conclude that the polymerization is catalyzed by both the ion pair and the free silanolate anion. Similar considerations apply to the industrially important quaternary ammonium and phosphonium silanolates whose large diffuse cations should favour ionization. However, there is no hard kinetic data on either system to support this conclusion.

The ion-pair–ion-pair-dimer equilibrium is set up extremely rapidly and exchange between the ion pair and the dimer is far more rapid than reaction of the ion pair with the cyclosiloxane even when it is a cyclotrisiloxane. Thus the polymerization of cyclotrisiloxanes under conditions where redistribution reactions involving the siloxane bonds in the polymer are absent leads to monodisperse polymers. All of the chains grow at the same rate despite the fact that at any instant both dormant (ion pair dimers) and active (ion pairs) species exist in equilibrium with one another.

## 25.4 CATIONIC POLYMERIZATION

### 25.4.1 Introduction

The polymerization of cyclosiloxanes using both Brönsted and Lewis acid catalysts has been known since the earliest days of the silicone industry[52] and some of the first commercially available high molecular weight siloxane polymers were prepared in this way.[53] Until recently there have been few attempts to understand the fundamental chemistry and mechanisms underlying these reactions;[52,54,55] however, within the last decade the cationic polymerization of cyclosiloxanes has received much attention, which is particularly appropriate in view of the increasing importance of cationic polymerization in the manufacture of some of the newer commercial silicone products. It is well established that the products formed and the course of the polymerization process is governed to a large extent by the nature of the starting monomers.[56] If only permethylsiloxanes are considered, then the activity towards acid catalysts decreases in the order $D_3 > MM > MDM > MD_nM > D_4$, so that under normal conditions polymerization of dimethylcyclosiloxanes proceeds steadily to reach an equilibrium point beyond which no further change can be detected (curve A, Figure 8). By contrast the activity towards basic catalysts decreases in the order $D_3 > D_4 > MD_nM > MDM > MM$ and the characteristic viscosity/time profile observed for a similar polymerization catalyzed by basic catalysts (curve B, Figure 8) reflects this opposite order. The exception is $D_3$, which is the most reactive cyclosiloxane towards both basic and acidic catalysts due to its relative instability introduced by the strain energy of the six-membered ring system.

Catalysts that have been reported to initiate cationic polymerization of siloxanes include most strong protonic acids like $H_2SO_4$[1,52,53,57] or $HClO_4$[58] and a variety of Lewis acids.[1,55,57] Other sources of exchangeable $H^+$ are also important, particularly in commercial processes. These may be

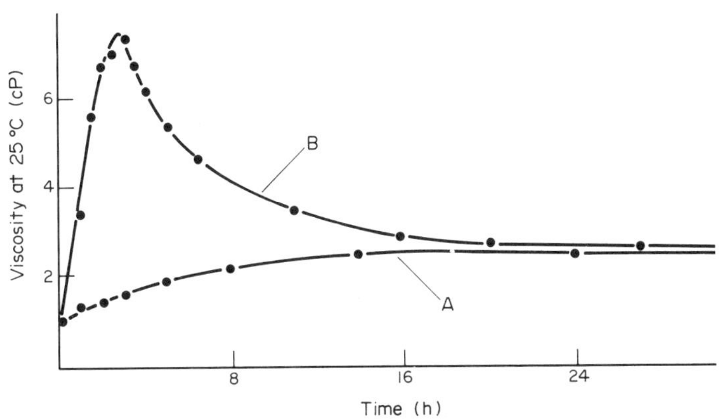


**Figure 8** Viscosity *vs.* time for 1 mol MM and 1 mol $D_4$: curve A with 4% $H_2SO_4$; curve B with 0.1% $Me_4NOH$ (curves drawn using data from ref. 56 by permission of the American Chemical Society)

cation exchange resins[59] such as sulfonated polystyrene, activated clays like $H_2SO_4$-treated montmorillonite[28,60] or, more recently, graphite interchelate compounds of Lewis acids.[61,62] Even the solid state polymerization of $D_3$ using a $^{60}Co$ source is reported to proceed *via* a cationic route.[63,64] Many of the newer commercial processes, particularly those for the synthesis of organofunctional siloxanes,[65,66] utilize strong sulfonic acids such as $CF_3SO_3H$. It is this catalyst that has been singled out for particular attention in the most recent kinetic and mechanistic studies[67,68] and the implications of these studies on the synthesis of cyclosiloxanes have been extensively reviewed by Wright.[13] However despite a great deal of effort the precise nature of the polymerization process has yet to be established.

### 25.4.2 The Mechanism of the Cationic Polymerization of Cyclosiloxanes Catalyzed by $CF_3SO_3H$

Historically two general mechanisms have been proposed for the cationic polymerization of cyclosiloxanes–addition polymerization and acidolysis/condensation.

In addition polymerization the growth occurs by reaction of monomer with some active propagating centre –Si*, which might be an oxonium ion, $SiMe_2\text{–}\overset{+}{O}\text{–}(SiMe_2)_n$,[55,69] or a discrete silicenium ion.[1,70]

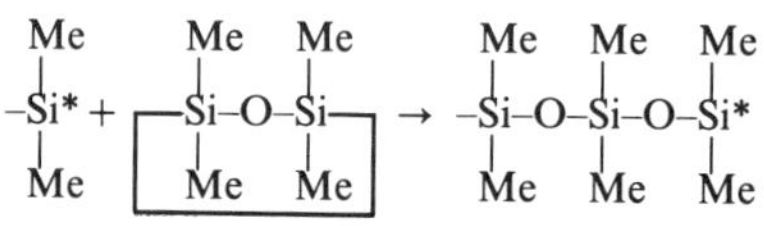

(67)

The following scheme (equations 68–75) proposed for the polymerization of $D_4$ with the catalyst/cocatalyst system $Fe_2Cl_6/HCl$ illustrates the principal features of the addition polymerization mechanism.[55] The actual catalytic species is the complex acid $HFeCl_4$, which is represented below as HA. For the sake of clarity the $AHA^-$ counter ion has been left out of the scheme.

Ionization of the acid
$$2HA \overset{K_1}{\rightleftharpoons} H^+\, AHA^- \tag{68}$$

Initiation
$$H^+ + D_4 \rightleftharpoons D_4^+H \tag{69}$$
$$D_4^+H \overset{k_3}{\rightleftharpoons} HD_4^+ \tag{70}$$

Propagation
$$D_4 + HD_n^+ \overset{k_4}{\rightleftharpoons} HD_n^+D \quad \text{siloxonium ion} \tag{71}$$
$$HD_4^+D \rightleftharpoons HD_{n+4}^+ \quad \text{silicenium ion} \tag{72}$$

Termination
$$HD_n^+ + HCl \overset{k_6}{\rightleftharpoons} HD_nCl + H^+ \tag{73}$$
$$HD_n^+ + H_2O \overset{k_7}{\rightleftharpoons} HD_nOH + H^+ \tag{74}$$
$$HD_n^+ + D_4 \overset{k_8}{\rightleftharpoons} HD_nMe + D_4^+ \tag{75}$$

Based on reaction (71) being the rate-controlling step, the following expression for the apparent first-order rate constant $k$ is derived from the scheme as

$$k = \frac{K_1k_3k_4[HA]^2}{k_8[D_4]_0 + k_7[H_2O] + k_6[HCl]} \tag{76}$$

At low [HCl] and hence low $[H_2O]$ and constant [HA], $1/k$ is proportional to $[D_4]_0$, the initial monomer concentration. However, the variation of $k$ with [HCl] is complex because of its involvement in the equilibria

$$\equiv SiOH + HCl \rightleftharpoons \equiv SiCl + H_2O \tag{77}$$
$$Fe_2Cl_6 + 2HCl \rightleftharpoons 2HFeCl_4 \tag{78}$$

which control both the [HA] and $[H_2O]$.

In acidolysis/condensation initiation is thought to occur *via* decomposition of an intermediate siloxonium complex formed by electrophilic attack on the siloxane oxygen by a proton and which affords silanol (SiOH) and silyl ester (SiA) groups[52,54]

$$HA + D_3 \rightarrow HO\overset{Me}{\underset{Me}{Si}}\text{–O–}\overset{Me}{\underset{Me}{Si}}\text{–O–}\overset{Me}{\underset{Me}{Si}}A \quad (79)$$

The acidolysis product may then undergo hetero- or homo-polycondensation and hydrolysis reactions to form oligomers and high polymer. These equilibrium reactions are represented below by equations (80) through (82)

$$\text{–SiOH} + \text{–SiOH} \rightleftharpoons \text{–SiOSi–} + H_2O \quad (80)$$

$$\text{–SiA} + H_2O \rightleftharpoons \text{–SiOH} + HA \quad (81)$$

$$\text{–SiA} + \text{–SiOH} \rightleftharpoons \text{–SiOSi–} + HA \quad (82)$$

In the case of strong protonic acids these equilibria strongly favour the formation of the silyl ester, for example in a molar solution of $D_4$ catalyzed by $CF_3SO_3H$ ($5 \times 10^{-3}$ mol l$^{-1}$), NMR studies show that 93% conversion of the monomer is acheived in less than 30 s.[71] More recently two groups of researchers in Lodz[67,72–82] and Paris[68,71,83] have tackled the problem of elucidating the complex mechanisms involved in the cationic polymerization by slightly different approaches. Chojnowski *et al.* have chosen to study the polymerization of $D_3$ initiated by trifluoromethanesulfonic acid (triflic acid) because the initial reactions proceed cleanly to almost complete conversion driven by the strain energy of *ca.* 15 kJ mol$^{-1}$ in the six-membered siloxane ring.[84] This makes the primary polymerization extremely fast and distinct from any secondary redistribution reactions[75–77] and enables them to be studied in isolation from the secondary reactions. On the other hand Sigwalt and co-workers have chosen to concentrate on the polymerization of $D_4$ with the same catalyst because of its greater relevance to industrial processes.[68] Both groups have shed much light on the nature of the cationic polymerization and despite some differences in interpretation and conclusions their experimental observations are largely in agreement with one another.[71]

#### *25.4.2.1 Some general features of the polymerization of cyclosiloxanes catalyzed by $CF_3SO_3H$*

Recent studies have shown clearly that the polymerization of siloxanes initiated by strong acids such as $CF_3SO_3H$ follows a complex rate law (see Figure 9) in which it is necessary to invoke both of the mechanisms represented by equations (67) and (80) through (82) in an interrelated manner.[67,68,76] At equilibrium cyclic oligomers coexist with both high and low molecular weight

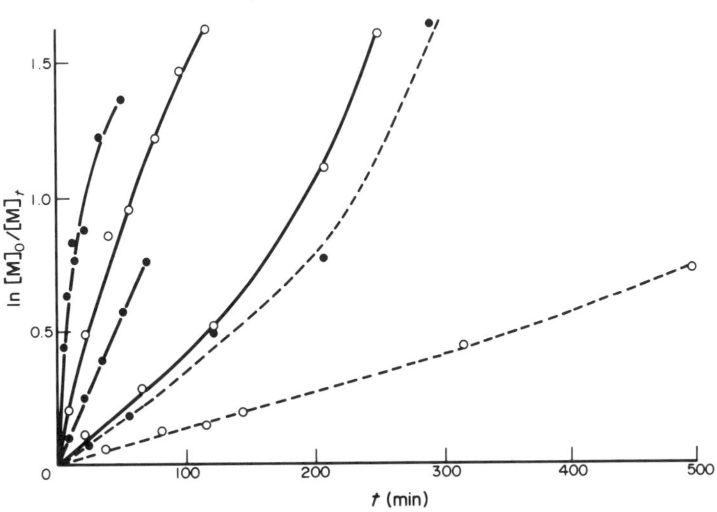


**Figure 9** First-order plots for the polymerization of $D_3$ in $CH_2Cl_2$ (——) or heptane (---) at varying initial catalyst and monomer concentrations (reproduced by permission of Huthig and Wepf Verlag from ref. 76)

linear polymers. The formation of the cyclic oligomers has been studied separately by gas chromatography[73,75] and has yielded valuable mechanistic information. The cyclic population, especially that obtained in the polymerization of $D_3$, is discontinuous under kinetic conditions and comprises a series of cyclosiloxanes $D_{3n}$, where $n$ is an integer and each cyclosiloxane differs from its nearest neighbours by three dimethylsiloxane units (Figure 10).

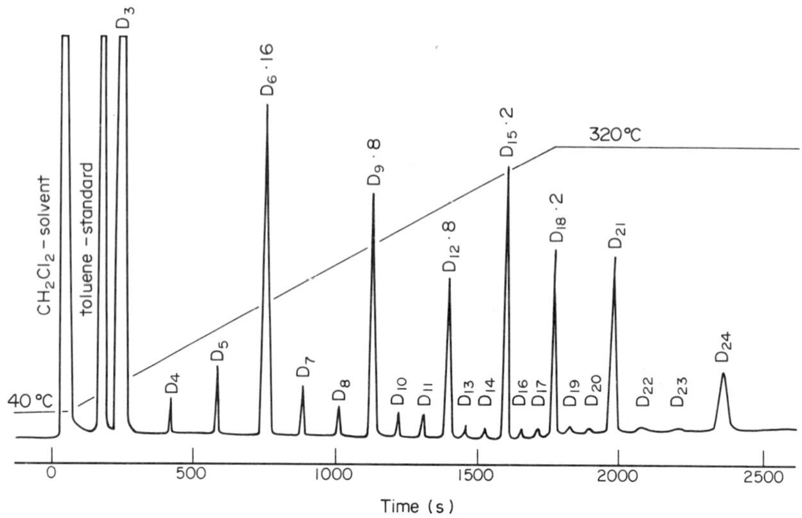


**Figure 10** GLC of the volatile fraction from the reaction of $D_3$ (20% in $CH_2Cl_2$) with $SnCl_4$ catalyst at 74% conversion (reproduced by permission of Huthig and Wepf Verlag from ref. 73)

Small quantities of additives, particularly water, have a profound influence on the kinetics of the polymerization. For example the reaction of $D_3$ with $CF_3SO_3H$ has a negative apparent activation energy under anhydrous conditions. This type of phenomenon is apparently quite common for processes in which a complex but more stable intermediate is formed prior to activation; however, the activation energy changes to the expected positive value on addition of water[76] as shown in Figure 11. The effects are less distinct in the case of $D_4$ but the overall character of the rate dependence on [water] is similar.[67]

The clearer understanding of the cyclosiloxane/$CF_3SO_3H$ system which is emerging from these studies has led to a reexamination of other apparently less complex cationic polymerizations.

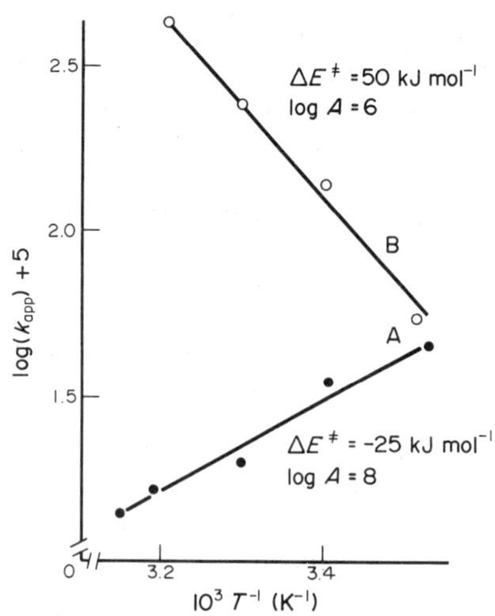


**Figure 11** The effect of temperature on log($k_{app}$) for the polymerization of $D_3$ (2 M in $CH_2Cl_2$) initiated by $CF_3SO_3H$ ($5.9 \times 10^{-4}$ M): curve A, no additional water; curve B, $3 \times 10^{-2}$ M water added (reproduced by permission of Huthig and Wepf Verlag from ref. 76)

Several observations suggest that the same general mechanism may be operative in a large number of cationically initiated siloxane polymerizations. For example, in the polymerization of $D_3$ a variety of catalysts including strong protonic acids, Lewis acids or carboxonium salts either in solution or in bulk give qualitatively similar results.[73] The products have very characteristic compositions of their volatile fractions, which in some cases feature almost exclusive formation of the $D_{3n}$ oligomers (compare Figures 10 and 12). Other common features have been noted in the cationic polymerization of hexamethylcyclotrisiloxane and octamethylcyclotetrasiloxane.[67] These include similarities in the general character of the effect of additional water on the system and a similar dependence of the rate on initial monomer and acid concentrations. For the polymerization of $D_4$ a large increase (from *ca.* 24 to 72 kJ mol$^{-1}$) in the apparent activation energy with added water was also observed and in both systems there is considerable departure from simple first-order kinetics.

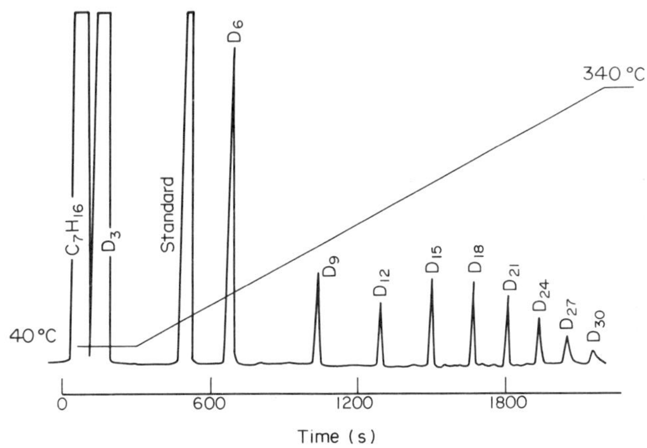


**Figure 12** GLC of the cyclic oligomer fraction isolated from the reacion of $D_3$ (40%) with $CF_3SO_3H$ in heptane quenched at 42% conversion of the monomer (reproduced by permission of Huthig and Wepf Verlag from ref. 75)

#### 25.4.2.2 *Kinetic studies of the cationic polymerization of cyclosiloxanes*

The kinetics of the acid-catalyzed polymerization of cyclosiloxanes have been followed using dilatometry and GLC[67,76] or GPC[68] techniques. The rate of consumption of monomer does not follow simple first-order kinetics as illustrated by the data for the polymerization of $D_3$ initiated by $CF_3SO_3H$ shown in Figure 9. A similar complex dependence of the rate of polymerization on monomer concentration was observed for the corresponding $D_4$ polymerization. Rate constants and activation parameters derived from this data were obtained in most instances by measurement of the initial or maximum rate of polymerization. The addition of water not only increases the rate of polymerization of $D_3$ and to a lesser extent $D_4$ but it also fundamentally alters the kinetics of polymerization. Thus the rate of polymerization of $D_3$ in $CH_2Cl_2$ in the absence of water actually decreases with increasing initial monomer concentration to the power of 0.7 but when water is present the external order in monomer reverts to the expected first-order dependence. A negative value of the activation energy was obtained for this polymerization ($E_A = -25$ kJ mol$^{-1}$) together with an unusually low preexponential factor ($A = 10^{-8}$ s$^{-1}$). As shown in Figure 11 these values also change dramatically on the addition of water to yield the more conventional values of $E_A = 50$ kJ mol$^{-1}$ and $A = 10^6$ s$^{-1}$.

In the absence of an end-blocking species four distinct groups of products are observed by GPC;[68,76] small cyclics, macrocyclic oligomers, low molecular weight polymer and high polymer. The oligomers and low polymer rapidly reach steady state concentrations, whilst the high polymer continues to increase both its molecular weight and relative proportion at the expense of the starting monomers until equilibrium is attained. These assignments are made on the basis that the theoretical number of end groups calculated for the added acid is sufficient to account only for the end groups of the high polymer and linear oligomers. The remaining distributions must therefore be cyclic materials. Qualitatively similar GPC results are obtained at equilibrium whether starting from $D_4$, $D_5$ or PDMS.[68] In the case of $D_3$ it is also notable that the high molecular weight polymer fraction appears at the beginning of the reaction.[67,71] Figure 13 shows a typical GPC trace for the composition at equilibrium obtained from the polymerization of $D_4$ initiated by $CF_3SO_3H$.

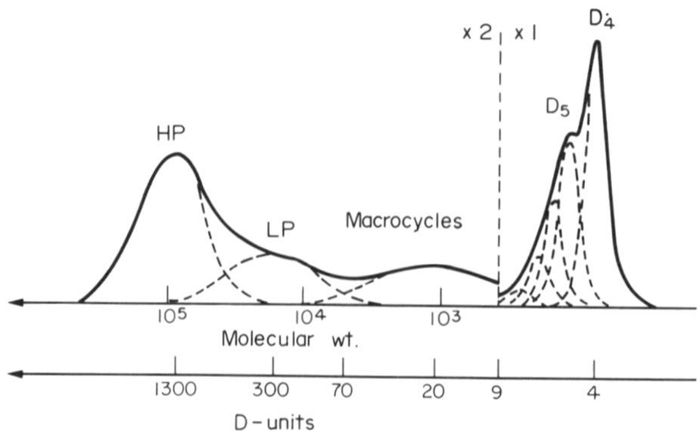


**Figure 13** Gel permeation chromatogram obtained at equilibrium from $D_4$ (0.966 mol $l^{-1}$ in $CH_2Cl_2$) catalyzed by $CF_3SO_3H$ ($5.85 \times 10^{-3}$ mol $l^{-1}$) at 20 °C (reproduced by permission of Academic Press from ref. 68)

By varying the initial concentration of reactants Sigwalt[68] *et al.* found that the initial rate of polymerization $(R_p)_0$ of $D_4$ in $CH_2Cl_2$ was given by

$$(R_p)_0 = -(d[D_4]/dt)_0 = k([D_4]_0)^{0.7}([CF_3SO_3H]_0)^{2.2} \tag{83}$$

They assumed the propagation step to be

$$D_4 + P^* \underset{k_{-1}}{\overset{k_1}{\rightleftharpoons}} P^* \tag{84}$$

which gives the integrated rate equation

$$\ln\frac{([D_4]_0 - [D_4]_e)}{([D_4] - [D_4]_e)} = k_1\int_0^t [P^*]\,dt \tag{85}$$

Analysis of the linear portions of the left side of this expression against time gave values for the apparent rate constant ($k_{app}$) which were also dependent on both the initial acid and the inverse of the initial monomer concentrations in agreement with Chojnowski's observations. For the same reaction the Lodz group found the order in acid was 2.0; however this fell to a value of 1.3 on addition of water and they proposed that if sufficient water could be added, the order in acid would become unity, the limiting factor being its solubility in the reaction medium.

### 25.4.3 Oligomer Formation in the Polymerization of Cyclosiloxanes with $CF_3SO_3H$

GLC analysis of the primary products from the polymerization confirms that the cyclic siloxanes initially enter the polymer as discrete monomer units[67,73,75,85] (either $D_3$ or $D_4$). Thus in the polymerization of $D_3$ catalyzed by $CF_3SO_3H$ the volatile fraction contains a predominance of cyclic oligomers $D_{3n}$, where the number of dimethylsiloxane units in each oligomer is a multiple of three.[73,75,86] The chromatogram in Figure 12 illustrates that in specific cases such oligomers are formed almost exclusively; however, more commonly small amounts of other cycles $D_4$, $D_5$, $D_7$, *etc.* are produced concurrently and are attributed to the onset of the depolymerization process (contrast Figures 10 and 12). This kind of remarkable kinetic enhancement is less obvious for similar experiments using $D_4$,[68,83] although under certain conditions multiples of the $D_4$ unit can be detected.[67,72,85] When the end-blocking units MM are present, no high polymer is formed but the linear oligomers $MD_{3n}M$ are observed and are also far in excess of their equilibrium values.[74] In this case the distribution of the $MD_{3n}M$ species decreases in a highly regular manner as shown in Figure 14, the molar ratio, $u = MD_{3n}M/MD_{3(n+1)}M$ of adjacent homologues being almost constant and independent of $n$. This feature is characteristic of a chain reaction in which redistribution processes play a minor role.[73,74] Such a chain reaction might be ring opening of the monomer by some active centre giving chain growth (equation 67) and termination by reaction of the propagating centre with MM. However, the same regularity is not observed for the corresponding $D_{3n}$ series where the ratio $u = [D_{3n}]/[D_{3(n+1)}]$ is strongly dependent on the reaction conditions such as solvent and type of catalyst and on the relative thermodynamic stabilities of the cyclic products.

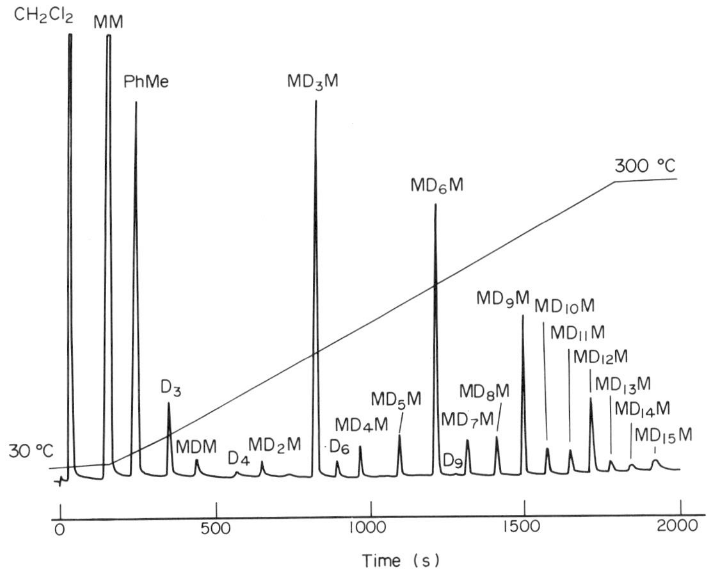


**Figure 14** GLC of the products of polymerization of $D_3$ (20%) and MM (20%) with $CF_3SO_3H$ in $CH_2Cl_2$ quenched at 93% conversion (reproduced by permission of Huthig and Wepf Verlag from ref. 74)

Similar experiments using MDM as end-blocking agent showed very low specificity in the distribution of oligomers.[74] The authors argue that a mechanism involving a five-coordinate silicon transition state provides the best explanation of these results.

### 25.4.3.1 *Mechanism of the cyclization reaction in the polymerization of cyclosiloxanes*

The mechanism of formation of macrocyclics in the cationic polymerization of $D_3$ has been the subject of extensive studies, mainly by GLC analysis of the volatile fraction.[72,75,76,81] This work has shown that in the absence of an end-blocking species intramolecular or intermolecular condensation of chain ends effectively competes with the linear growth mechanism, *i.e.* the addition of monomer units to an active centre

$$\underset{\text{cyclic}}{\boxed{D_x}} \leftarrow \underset{\text{linear}}{-(D_x)-} \xrightarrow{D_3} \text{high polymer} \qquad (86)$$

Two possible mechanisms were considered for the cyclization process: (i) linear growth of the macromolecule followed by end to end ring closure (Scheme 2); and (ii) ring expansion polym-

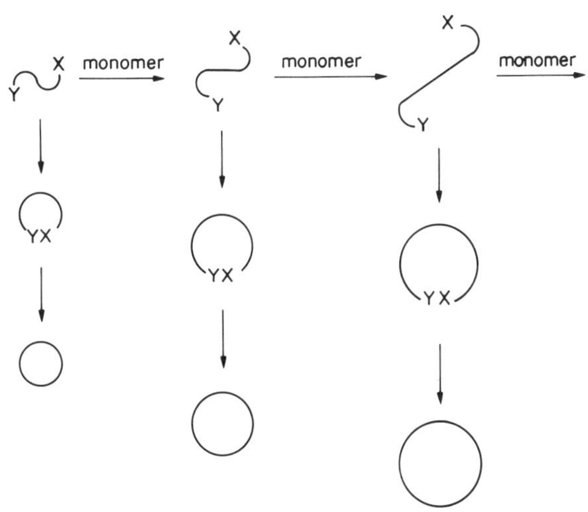


**Scheme 2**

erization by insertion of monomer into the activated cyclic (Scheme 3). A third possibility of ring formation by depolymerization was rejected in this case because the depolymerization process is a random process and therefore could not account for the characterestic distributions and the almost exclusive formation of certain $D_{3n}$ oligomers.

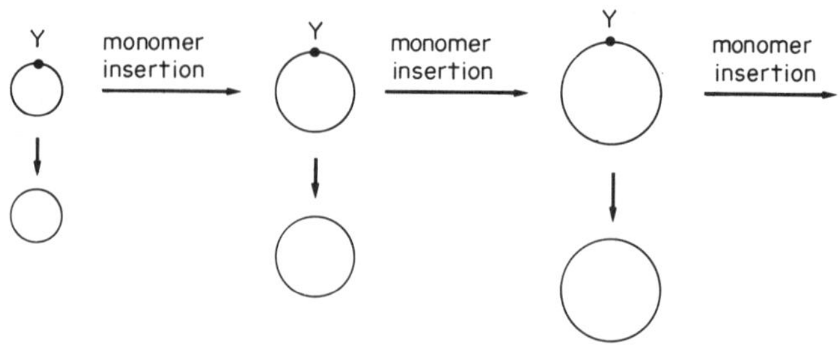


**Scheme 3**

Chojnowski *et al.* prefer the former ring closure mechanism because it can account for most of their experimental observations. At first sight this is perhaps rather surprising for large molecules containing as many as 100 atoms in the chain; however, it should be noted that the mean end to end distance of a macromolecule increases with the square root of its real length. For instance conformational analysis of a 50 unit siloxane polymer predicts that on average the termini will be about 40 Å apart,[11,87] but probability dictates that a significant fraction will have end groups close enough together for reaction to occur. The proposed end to end ring closure mechanism[75] assumes a linear growing species $B(D_3)_nA$ composed of a terminal group B and an unspecified propagating centre A which can undergo competitive reactions leading either to linear propagation or cyclization according to the general scheme shown below (Scheme 4).

$$\begin{array}{ccc} B(D_3)_nA & \xrightarrow[k_p]{D_3} & B(D_3)_{n+1}A \rightarrow \\ \downarrow k_n & & \downarrow k_{n+1} \\ D_{3n} + AB & & D_{3(n+1)} + AB \end{array}$$

**Scheme 4**

The rates of formation of the various oligomers would then be

$$d[D_{3n}]/dt = k_n[A(D_3)_nB] \tag{87}$$

$$d[D_{3(n+1)}]/dt = k_{n+1}[A(D_3)_{n+1}B] \ etc. \tag{88}$$

If propagation is much faster than cyclization ($k_p[D_{3n}] \gg k_n$), such that high polymer with $DP \gg n$ is formed and intermolecular reactions between A and B are neglected, the existence of a pseudo-stationary state for a particular set of reaction conditions can be assumed. This hypothesis was later confirmed by detailed studies of equilibria in acid/siloxane model systems.[82] Under these conditions the instantaneous concentrations of intermediates in the above model are equal, *i.e.* $[A(D_3)_nB] = [A(D_3)_{n+1}B]$ *etc.* and the relative rate of formation of the cyclic oligomers is independent of the concentrations of their respective linear intermediates

$$\frac{d[D_{3n}]}{d[D_{3(n+1)}]} = \frac{k_n}{k_{n+1}} = \frac{[D_{3n}]_t}{[D_{3(n+1)}]_t} \tag{89}$$

where $[D_{3n}]_t$ is the concentration of the 3*n*th cyclic at time *t*.

If the proposed scheme is correct, then for a particular reaction the ratio, *u*, should also be independent of monomer and catalyst concentrations, so the same $[D_{3n}]_t/[D_{3(n+1)}]_t$ values are expected for different initial monomer concentrations and conversions. This condition also requires that a logarithmic plot of the distributions of kinetically controlled $D_{3n}$ cyclics at different conversions should have the same shape and be parallel to each other. This is supported by experiment as shown in Figure 15. The data shows that the distribution pattern near the beginning of the kinetically controlled polymerization is almost identical to that when the reaction is nearing completion. Additional experiments indicate that this characteristic pattern does not vary with initial monomer concentration.

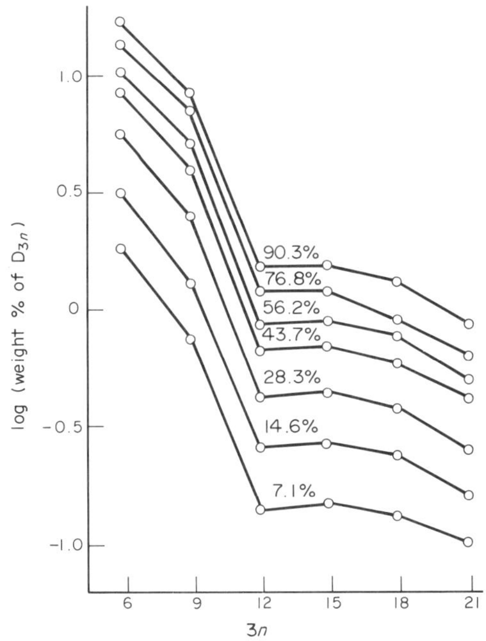


**Figure 15** Log[weight of $D_{3n}$ oligomers (w.r.t. total siloxanes)] in the polymerization of $D_3$ (40% in $CH_2Cl_2$) with $CF_3SO_3H$ at various conversions (reproduced by permission of Huthig and Wepf Velag from ref. 75)

The molar free energy of activation for the cyclization reaction in the scheme given above is

$$\Delta G_n^\circ = \Delta H_n^\circ - T\Delta S_n^\circ \tag{90}$$

However, since the same reaction between end groups A and B is involved at each stage and the rings formed are strainless, the activation enthalpy, $\Delta H_n^\circ$, should be independent of ring size, $n$. This is confirmed by the observation that under the same conditions but at different temperatures the characteristic distribution of oligomers does not change, hence oligomer formation in this scheme is controlled by the activation entropy of cyclization $\Delta S_n^\circ$. Furthermore this may be separated into component parts $(\Delta S_n^\circ)_{AB} + (\Delta S_n^\circ)_c$, where $(\Delta S_n^\circ)_{AB}$ is the activation entropy for reactions between the end groups A and B and $(\Delta S_n^\circ)_c$ is the difference between the entropy of the ground state conformation of the $A(D_{3n})B$ species and that where the molecules which have terminal groups in suitable position for closure. For sufficiently large values of $n$, $(\Delta S_n^\circ)_{AB}$ should also be independent of ring size, therefore it is the conformational entropy $(\Delta S_n^\circ)_c$ that governs the ratio of oligomers and the following expression holds

$$k_n/k_{n+1} = \ln\,([D_{3n}]_t/[D_{3(n+1)}]_t \tag{91}$$

$$= (\Delta S_n^\circ)_c/R - (\Delta S_{n+1}^\circ)_c/R \tag{92}$$

*i.e.* the distribution of the $D_{3n}$ oligomers is dependent solely on the loss of conformational entropy in the cyclization transition state.[75]

The equilibrium distribution of cyclic oligomers is also governed by conformational entropy considerations as first described by Jacobson and Stockmayer for the formation of macrocyclics.[5,6,8,88] However the mechanism of cyclization in an equilibrium process, *i.e.* the redistribution of siloxane bonds by polymerization and depolymerization and elimination of cyclics is quite different from that in the kinetically controlled scheme as proposed by Chojnowski. The Jacobson–Stockmayer theory predicts that $[D_x]$ varies as $x^{-5/2}$ and the double logarithm plot of $[D_x]$ *vs.* ring size should have a slope of $-5/2$ for sufficiently large values of $x$. This theory, which has been extended to some non-equilibrium systems by Kilb,[89] and further developed by others,[90–92] predicts that for kinetically controlled reactions $[D_x]$ is proportional to $x^{-3/2}$.

The observed experimental values for rings with more than 18 siloxane units in the polymerization of $D_3$ by $CF_3SO_3H$ in heptane were $-1.7$ for the kinetic distribution and $-2.7$ for the equilibrium distribution as shown in Figure 16. The striking similarity between the equilibrium and kinetically controlled distributions, especially in the small cyclic region, and the difference of unity in the

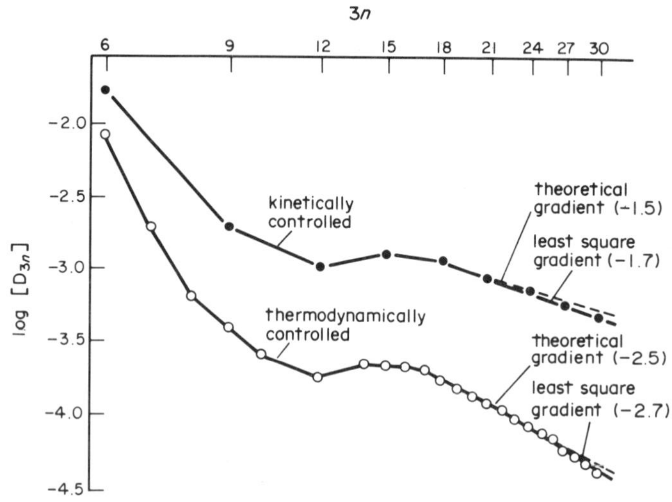


**Figure 16** Comparison of the kinetically controlled and equilibrium distributions of oligomers in the $D_3/CF_3SO_3H$ system in heptane (reproduced by permission of Huthig and Wepf Verlag from ref. 75)

exponent of the limiting slopes is taken as strong evidence in favour of the end to end ring closure mechanism (note that in the example given a minimum is observed at $n=4$ in both cases).

It is also clear that certain parallels exist between the proposed end to end ring closure mechanism and an intermolecular condensation reaction which could also account for the formation of high polymer. However, these macrocylic polymers always constitute a separate population and GPC studies show clearly that high molecular weight linear polymer is formed in the early stages of the reaction. This is consistent with the conclusion that the growth of the high polymer is more likely to occur *via* the direct addition of monomer to an activated end group in competition with intra- and inter-molecular condensations as previously mentioned.

The ring expansion mechanism was dismissed on the grounds that polymerization cannot occur by insertion of an activated monomer into larger cyclics because the composition of the cyclic population would be largely dependent on the conversion of monomer (*cf.* Figure 15). The alternative of ring expansion by insertion of monomer *via* an active centre on a growing cyclic species is theoretically possible but the structure of the active centre would vary with catalyst type and reaction conditions. The only possibility for chain termination would be by transfer to monomer in which case the relative rates of formation of oligomers is given by

$$d[D_{3n}]/d[D_{3(n+1)}] = k_{tr}/k_p + 1 \tag{93}$$

where $k_p$ and $k_{tr}$ are the rate constants of propagation and transfer. In this scheme the growing species preserves its ring structure so that the formation of oligomers is not dependent on the probability of ring closure and any similarity between the kinetic and equilibrium distributions would be purely fortuitous. Since the ratio of propagation to transfer rate $k_p/k_{tr}$ should also be independent of ring size no deviation in the distribution of the smaller rings is expected. This interpretation is in direct contrast to the experimental observations depicted in Figures 15 and 16.

### 25.4.4 The Acidolysis/Condensation Equilibria

The contribution of the condensation process to the kinetics and mechanism of acid polymerization has been determined by studies of the equilibria in cyclic and model linear siloxanes using both $CF_3SO_3H$ and $CF_3CO_2H$ as catalyst.[68,78,82] Assuming that each ring formed corresponds to the consumption of one monomer unit by acidolysis and neglecting intermolecular condensations, it is estimated that this process accounts for *ca.* 10% of the conversion of monomer in the polymerization of a 40% solution of $D_3$ in dichloromethane.[76]

The processes depicted in equations (79) through (82), which give rise to silanol and silyl ester functions, are well established and can be readily monitored by $^{19}F$ and $^{1}H$ NMR spectroscopy.[68,76,93] GLC techniques[82] and IR spectroscopy[78] were employed to investigate the homocondensation reaction (80) and to determine the effects of H bonding. The acidolysis of MM was found to be similar to that for cyclosiloxanes but without added complications arising from polymerization of the monomer. With strong protic acids such as $CF_3SO_3H$ the acidolysis step is rate determining and the equilibria favouring the silyl ester are established very rapidly. However the interaction of siloxanes with the moderately strong trifluoroacetic acid (TFA) depicted in equations (94) through (97) are more amenable to kinetic analysis.

$$CF_3CO_2H + D_3 \rightleftharpoons HOSi(Me_2){-}O{-}Si(Me_2){-}O{-}Si(Me_2)O(O)CCF_3 \tag{94}$$

$$-SiOH + -SiOH \rightleftharpoons -SiOSi- + H_2O \tag{95}$$

$$-SiOOCCF_3 + H_2O \rightleftharpoons -SiOH + CF_3CO_2H \tag{96}$$

$$-SiOOCCF_3 + -SiOH \rightleftharpoons -SiOSi- + CF_3CO_2H \tag{97}$$

The initial rate can be expressed in terms of the initial concentrations as follows

$$R_0 = -d[\text{acid}]_0/dt = d[\text{ester}]_0/dt$$
$$= k[\text{acid}]_0^n\,[\text{monomer}]_0^m \tag{98}$$

The reaction can be followed by measuring changes in the acid or ester concentration as a function of time; however, this approach gives rise to some unusual results. Low apparent orders in monomer (*e.g.* $-1.4$ for $D_3$ and zero for $D_4$) and high apparent orders in acid (3.1–3.5) are observed and in common with the polymerization of $D_3$ by $CF_3SO_3H$ several other inconsistencies were also noted.[78] For example, the apparent value of the activation energy for acidolysis of the SiO bond is higher for $D_3$ (56 kJ mol$^{-1}$) than for the strainless $D_4$ (45 kJ mol$^{-1}$) and the reactions proceed at similar rates although the solvolytic cleavage of $D_3$ in alcohol was as much as $10^4$ times faster.[93] Also the addition of water enhances the rate of conversion to ester whereas proton acceptors such as THF decrease the rate of reaction. Equation (98) fails to account for H-bonding phenomena and the actual concentrations of uncomplexed acid and monomer are considerably less than their initial concentrations.

H bonding and self-association of TFA with siloxanes and water were studied independently by IR spectroscopy and the following equilibria were found to be the most important

$$TFA + TFA \rightleftharpoons (TFA)_2 \quad \text{(acid dimer)} \tag{99}$$

$$TFA + D_4 \rightleftharpoons (D_4{\cdot}TFA) \quad \text{(binary acid base complex)} \tag{100}$$

Values of the initial concentrations of free acid $[\text{acid}]_f$ and monomer $[\text{monomer}]_f$ were determined experimentally and the kinetics reinterpreted in terms of the expression

$$R_0 = k[\text{acid}]_f^p\,[\text{monomer}]_f^q \tag{101}$$

The exponents $p$ and $q$ determined using equation (101) are now more closely related to the molecularity of the reaction. In the case of $D_3$ the values of $p$ and $q$ are 3.1 and 1.0 respectively, whereas for $D_4$ they are 3.5 and 1.5. Chojnowski and co-workers offer an interpretation based on the number of molecules participating in the transition state. They propose that an H-bonded complex of $p$ molecules of acid and one or more monomer units, formed by a series of rapid reversible reactions, is capable of ring opening. This complex is also in equilibrium with other non-active H-bonded species and uncomplexed acid according to Scheme 5. In the $D_3$ case the active species may be thought of as a complex of one monomer and three acid molecules which can transfer charge easily. The addition of amphiprotic (proton donor or acceptor) species such as water results in the formation of an even more reactive H-bonded complex, which can take part in extremely fast charge reorganization due to cooperative H bond migration.[78] In discussing these systems Sigwalt *et al.* agree that association phenomena are important and suggest that hydrated acid species containing

$$pCF_3CO_2H + qM \rightleftharpoons [(CF_3CO_2H)_p{\cdot}M_q]_{\text{active}} \xrightarrow[\text{opening}]{\text{ring}}$$

$$CF_3C(O)OSi(Me_2){-}O{-}Si(Me_2){-}OSi(Me_2)OH + [(CF_3CO_2H)_{p-1}{\cdot}M_{q-1}]_{\text{inactive}}$$

**Scheme 5**

up to seven molecules of water may be responsible for the cleavage of siloxane bonds. Thus water molecules may act as synergists by increasing the acidity of the catalyst and the basicity of the siloxane in transient H-bonded polymeric complexes of the type shown in Figure 17.

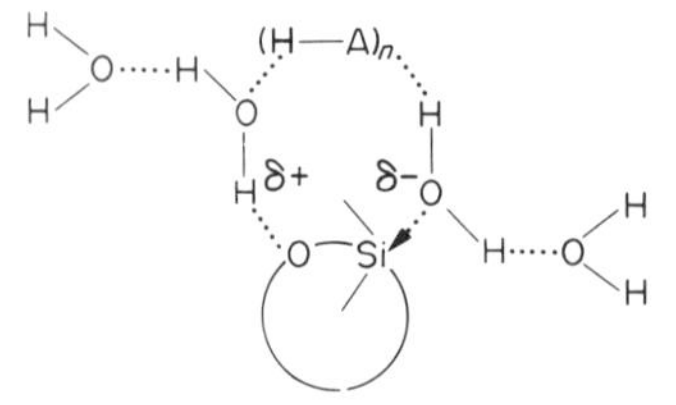

**Figure 17** Transient H-bonded polymer complex

The dramatic changes observed on the addition of small quantities of water can now be explained in terms of its effect on the ease of formation of the H-bonded complex cyclic transition state. With increasing water content a rapid increase in the rate of monomer consumption occurs, which passes through a maximum and ultimately levels off. Rate increases of up to three orders of magnitude were noted for the polymerization of $D_3$ in the presence of $CF_3SO_3H$, although the effect is either not observed[68] or is less obvious in similar reactions[67] involving $D_4$. These features are also catalyst dependent and are illustrated by a plot of apparent first-order rate constant against concentration of added water in Figure 18. Silanol (as $HOMe_2SiOSiMe_2OH$), when added in equivalent amounts, has an identical influence on the kinetics of polymerization of $D_3$[76] as shown in Figure 19. This supports previous arguments that the dynamic equilibrium between silanol and ester groups in equations (80) through (82) is rapidly established. However, neither silanol nor silyl esters are effective catalysts for the ring opening of siloxanes alone[54] and the data in Figures 18 and 19 suggest that the rate extrapolates to zero for the totally anhydrous system (in practice this is never achieved because siloxane cleavage by acidolysis generates water). Thus as well as facilitating charge transfer in the H-bonded transition state, water also enhances the rate by increasing the stationary state concentration of free acid. The H-bonded complex is also invoked to explain the negative apparent order in monomer observed in the polymerization of anhydrous $D_3$ because at higher levels of monomer the formation of other inactive or less active complexes is encouraged.

Various other additives have been evaluated by Sauvet *et al.*[68] and the results may also be interpreted in terms of their effect on the stationary state concentrations of active species, for example triflic (trifluoromethanesulfonic) anhydride inhibits the polymerization, whereas the silyl ester of $CF_3SO_3H$ has an accelerating effect. However, there are discrepancies, for instance weak acids are said to promote $D_3$ but retard $D_4$ reactions[58,69,78] and in particular the large effects of water or silanol are not substantiated by Sauvet's results.

Attempts to relate the conclusions drawn from the experiments on the weaker TFA to strong acids such as $CF_3SO_3H$ were made by studying a model transesterification equilibrium.[82] As expected the

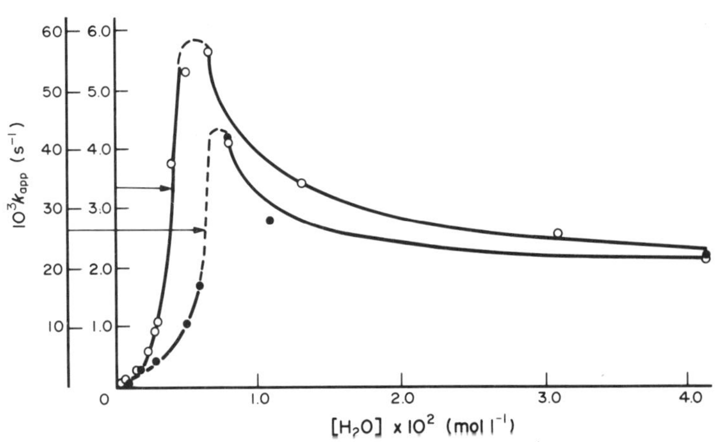


**Figure 18** The effect of water on the apparent rate of polymerization of $D_3$ (2 mol l$^{-1}$ in $CH_2Cl_2$) by $CF_3SO_3H$ at 30 °C (reproduced by permission of Huthig and Wepf Verlag from ref. 76)

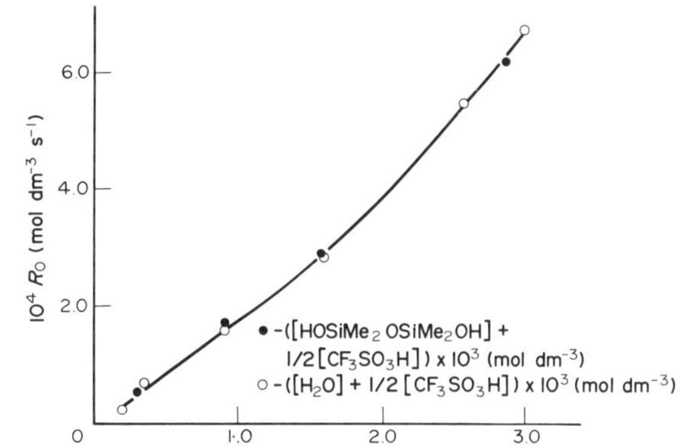


**Figure 19** The effect of equivalent quantities of water and silanol on the initial rate of polymerization of $D_3$ (reproduced by permission of Huthig and Wepf Verlag from ref. 76)

stronger acids show a higher tendency towards silyl ester formation and their rates of reaction are much higher. The $^{19}$F and $^1$H NMR spectra of the $CF_3SO_3SiMe_3/MeCO_2H$ system for example shows only a single ester species corresponding to the $CF_3SO_3H$ ester even at relatively high concentrations of acetic acid. Additionally, formation of the multi-H-bonded complex is favoured with weaker acids and monomers of low basicity but is not expected to be so significant in the polymerization with $CF_3SO_3H$. With strong acids and more nucleophillic monomers (*e.g.* $D_3$) protonation followed by reaction of a second siloxane unit, affording a siloxonium active centre, is more likely to occur.[72,78] However, despite these differences the overall character of the interdependence of the equilibrium concentrations of reactive species in the two cases has many common features.[82]

### 25.4.5 Summary

The preceding discussion indicates the exceedingly complex nature of the cationic process with several points of detail still to be resolved. Nevertheless most workers concur that the two mechanisms, *viz.* addition polymerization and acidolysis/condensation are operative and are likely to occur simultaneously. Chojnowski and Sigwalt differ in the emphasis placed on each mechanism and in their description of the precise nature of the active centre, which remains the subject of much debate. For example the former describes activation by a silanol/acid complex (equation 102) and concludes that condensation accounts for approximately 10% of the polymerization of $D_3$, whereas the latter favours initiation *via* the silyl ester group leading to a tertiary siloxonium ion and a homoconjugate triflate complex[94] (equation 103) and suggests that polycondensation plays a major role in the polymerization of $D_4$.

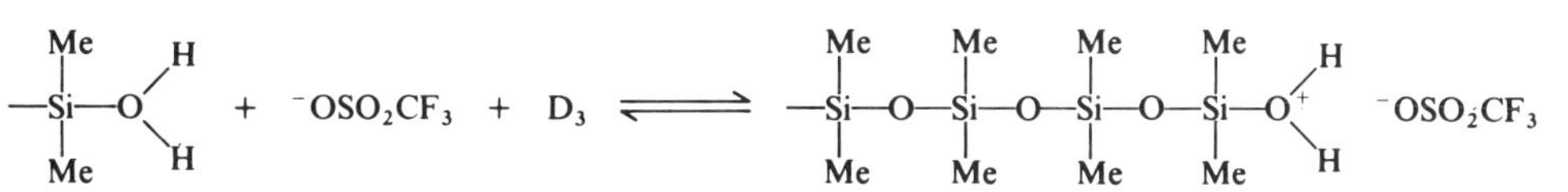 (102)

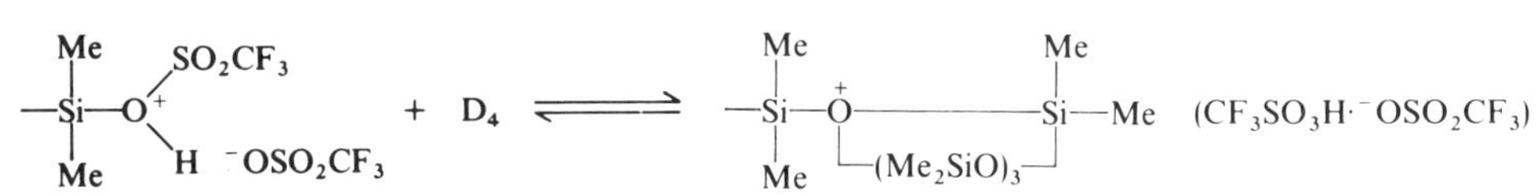 (103)

The kinetics of polymerization of $D_3$ and $D_4$ are quite different but each system produces qualitatively similar results with cyclic oligomers being predominant in both cases. The discontinuous size distribution of both cyclic and linear oligomers in the $D_3$ case is not observed for $D_4$ but it is possible to select polymerization conditions where kinetic enhancement of $D_{4n}$ can be detected. Although the results of the two groups are broadly in agreement, both prefer not to invoke the previously proposed[1,70] discrete silicenium ion as the active species. Attempts to prove the existence of such intermediates have been numerous[95] but it is probably at best a fleetingly transient species in the presence of siloxane bonds. Nevertheless this possibility cannot be excluded. For instance Lambert and his coworkers[96] have recently reported the observation of long lived $Ph_3Si^+$ and $Me_3Si^+$ ions in solution. Their conductivity results and cryoscopic molecular weight measurements support the formation of ions and the solutions are known to initiate styrene polymerization.[96] On the other hand NMR and crystallographic data presented by Olah's group[97] suggest that the silyl perchlorates are covalently bonded in both the solid state and in solution. It is significant, however, that despite the continuing controversy surrounding their stability, the existence of transient silicenium ions is no longer disputed, and their participation in the cationic polymerization process should perhaps be reexamined.

## 25.5 COPOLYMERIZATION OF CYCLOSILOXANES AND THE CHARACTERIZATION OF SILOXANE COPOLYMERS

### 25.5.1 Introduction

The poly(dimethylsiloxane)s (PDMS) are the most important class of commercially available silicone polymers. Their commercial acceptance is a result of their inertness towards chemicals and radiation, their low surface tension, their excellent thermal stability and their non-conductance of heat and electricity. These desirable properties are the result of a unique combination of the electronic and geometric properties of the silicon/oxygen and silicon/carbon bonds, which impart to the macromolecule its specific physical, chemical and mechanical properties. Keeping the siloxane backbone intact and substituting methyl groups along the chain or at the chain end by other inorganic or organic groups generates a wide variety of new siloxane macromolecules. As a consequence the known properties of PDMS are slightly changed and genuinely new properties appear, thereby broadening the materials base of the silicone industry. For example, phenyl groups alter the crystallinity and the mobility of the polymer at low temperatures, whilst the trifluoropropyl group reduces swelling of polysiloxanes in solvents. The result is that siloxane copolymers are used in many applications such as elastomers, coatings, biomaterials, antifoams, substrates for controlling the release of drugs, in microlithography, aerospace, *etc.*

Despite their considerable industrial importance, our understanding of the structure–properties relationships of siloxane copolymers is less than satisfactory. However, the emergence of new characterization techniques is reawakening interest in this aspect of siloxane polymer chemistry.

### 25.5.2 General Classification of Siloxane Copolymers

Siloxane copolymers may be classified in several different ways depending upon the structure and the microstructure of the different siloxane units.[57] The symbols M, D, T and Q will be used to represent the mono-, di-, tri- and tetra-functional siloxane structural units respectively. Siloxanes where one or more methyl groups have been replaced by another organic group or inorganic atom will be identified by the use of superscripts $M^{X,Y,Z}$, $D^{X,Y}$, $T^{X}$, *etc.*

#### *25.5.2.1 Molecular macrostructure* (see also Volume 4, Chapter 3)

Siloxane copolymers may be described according to their macrostructure, *i.e.* how the M, D, T and Q units are combined in the polymer. Several classes of copolymer can be recognized.

*Linear Chain polysiloxanes* ($MD_nM$). These comprise linear polymers where M and D units may be replaced by $M^X$, $M^{X,Y}$, $M^{X,Y,Z}$ and by $D^X$, $D^{X,Y}$ units respectively. Several combinations of these units will generate telechelic and chain-functionalized linear copolymers. Since almost any organic or inorganic group can be attached to the siloxane backbone, such copolymers exhibit a wide range of interesting properties and are used in many applications.

*Branched chain polysiloxanes.* These are linear chains containing T, $T^X$ or Q units as branching centres.

*Cyclic polysiloxanes.* Various combinations of D and T units can generate macrocyclics that may be classified into four subgroups, *viz.* monocyclic, polycyclic, spirocyclic and spherocyclic polysiloxanes. The three latter groups can be interpreted as arising from a cross-linking of monocyclic ring systems.[57]

*Cross-linked polysiloxanes.* These molecular structures are three-dimensional networks where T and/or Q units are combined with D and/or M units. A variety of polymers of complex structures are thus produced such as ladder-structured polymers and silicone resins.

The linear chain polysiloxanes are the subject of this section.

#### 25.5.2.2 *Molecular microstucture*

In the simplest case of a linear chain copolymer of infinite length made from two different units, three different microstructures may be distinguished.

(i) Random copolymers where the two monomers in the chain are statistically distributed:

–AAABABBABABBBABAABBBABABAABAABBABAAAABABABB–

This situation is the most favoured thermodynamically.

(ii) Alternating copolymers where the two monomers alternate along the chain:

–ABABABABABABABABABABABABABABABABABABABABA–

Due to the nature of the siloxane bond this situation is rarely encountered.

(iii) Block copolymers where each monomer forms relatively long sequences which are linked together:

–AAAAAAAAAABBBBBBBAAAAAAAAABBBBBBBBBAAAAAAAAA–

In these schemes A and B may represent any $D^{X,Y}$ siloxane unit. Subtle changes in the sequencing of the different siloxane units may generate significant changes in the properties of the copolymer and the need for characterization of the copolymer at the molecular level is self-evident.

The advent of highly sophisticated analytical instruments assisted by computers has improved the precision, accuracy and resolution of the analysis of siloxane copolymers. Direct observation of silicon nuclei embedded in different electronic and geometric environments is now possible in the liquid, elastomeric, amorphous and crystalline states.

### 25.5.3 Characterization of Siloxane Copolymers

The characterization of polysiloxanes in general and of copolymers in particular first requires the determination of the molecular weight distribution (MWD) and its related average molecular weights and of the number average degree of polymerization (DP). These measurements are then followed by the analysis of functional groups with regard to their nature, their concentration and their relative positions in the macromolecular chain.

#### 25.5.3.1 *Measurement of the molecular weights and molecular weight distributions of siloxane copolymers*

The molecular weight distribution (MWD) and the different molecular weights ($M_n$, $M_w$, $M_z$) are most frequently determined by the size exclusion chromatography (SEC) technique (see Volume 1, Chapter 12) also known as gel permeation chromatography (GPC).[98,99,100] For homopolysiloxanes, the concentration of the eluted polymer fraction is measured with a differential refractometer. This type of detector is suitable for linear or branched homopolymers. However, a calibration curve must be constructed in order to convert the elution volume into molecular weight and the chromatogram into the MWD.[98] This is done with standard polymers, *viz.* monodisperse polystyrenes or poly(dimethylsiloxane)s. The calibration procedure is very important for accuracy in the determination of the MWD and there is still considerable effort in the literature to find better calibration methods.[101–105] Universal calibration curves are based on the assumption that the solvodynamic volume is a function only of molecular size. Although this is true for homopolymers when concentration effects are absent,[106–108] it is not necessarily valid for copolymers, since the solvodynamic volume and SEC retention volume may be a function of the polymer composition as well as of its molecular weight. In the case of copolymers, when the composition is invariant with molecular weight, the universal calibration procedure can be applied.[109] However, when it changes with molecular weight, measurements of the MWD presents additional problems. The dependency of elution volume of the different species on the molecular composition and structure is one such problem. Another problem arises from the fact that specific organic groups have different refractive indices. If the copolymer composition changes with the degree of polymerization,[110] then the measured differential refractive index is not directly related to the amount of copolymer being eluted. Finally the introduction of groups such as hydroxyls, amines, *etc.* will increase the potential for intermolecular association, which will influence the elution time. Large discrepancies have been observed between number average molecular weight $M_n$ determined by SEC and by $^{29}$Si NMR, which have been attributed to differential spreading of the chromatogram by various disiloxanol species.[111]

These particular problems have not been solved and efforts to do so are still being pursued. The breakthrough will be the possibility of specifically detecting the different eluting copolymer molecules. The development and combination of new techniques of detection have recently been proposed that overcome these problems.[110] An evaporative detector (ED), a differential refractive index detector (DRI) and a low-angle laser light-scattering detector (LALLS) are simultaneously used to monitor the refractive index difference between the solvent and the solution, the concentration of the eluting species and the molecular weight of the eluting species. This latter development has been applied to organic copolymers.

The number average degree of polymerization, DP, is the other parameter which is very important in the characterization of polymers. Any analytical technique that can quantitate the end group concentration provides absolute values of DP and $M_n$. Spectroscopic methods such as FT–IR and FT–NMR are preferred to chemical titrations of reactive groups since the spectroscopic techniques are more rapid, specific, precise and accurate.[112] Furthermore chemically non-reactive groups such as the trimethylsilyl group can be determined. The importance of MWD, $M_n$ and DP on the performance of the polymer and its chemical reactivity has been emphasized.[113]

### *25.5.3.2 Analysis of the chemical functions in the copolymer*

Information concerning the global content of various inorganic and organic functional groups obtained by classical methods of chemical analysis is well documented.[57,98,111,114,115] Such chemical analytical methods have several disadvantages. The chemical reaction involved must be driven to completion. Each chemical function requires a specific reagent and since the reactivity of a given function might be dependent on its molecular environment, so the copolymer microstructure might influence reactivity of a given group. Interference of a second functional group may complicate the analysis. In the determination of silanol for example, the water and carbinol concentration must be taken into account.[98] IR and NMR spectroscopies have been shown to be powerful analytical tools for the analysis of siloxane copolymers.[98] In the past IR spectroscopy has been used mainly for qualitative and semi-quantitative analysis[57,98] such as the identification of substituents like methyl, phenyl, hydrogen and alkoxy. Recently the advent of fast computers and the Fourier transformation (FT) technique[116] has raised the limit of detection and improved the resolution of the spectroscopic techniques. Hydroxyl groups have been analyzed using the computer-assisted substraction technique to isolate the Si–OH band at 3900–3450 $cm^{-1}$. Interference of water and of siloxane frequencies are thus removed.[117] IR spectroscopy has been used to measure the DP of linear siloxane polymers.[116–119] For polymers with $M^H$ terminal units the strong and isolated Si–H stretching band at 2135 $cm^{-1}$ is easily measurable.[120] However the overlapping peaks for $CH_3$ in M and in D siloxane units makes their respective quantitative determination extremely difficult.[118] The technique of Fourier self-deconvolution can be applied to the observed resolution of FT–IR spectra and it is very efficient in resolving the overlapping peaks. This technique consists of taking the FT of the IR spectrum, enhancing the contributions of the high frequency Fourier components by convoluting the transformed data and retransforming the result to construct a deconvoluted spectrum which displays IR peaks with the natural resolution of the instrument (Figure 20).[117,118] Using this technique chain lengths of up to 200 siloxy units in PDMS molecules can be determined.[118,121–123]

NMR spectroscopy is a powerful technique for compositional and structural analysis of silicones.[57,98] Proton NMR has been used when differentiation between protons attached on different functional groups is possible[57,98] and the technique has been applied to the analysis of methyl, phenyl, vinyl, hydroxyl and hydrogen groups in polysiloxanes. Various functional groups attached to the *n*-propyl group, such as CN, SR, $NR_2$, Br, Cl, $^+NR_3$, OR, OOCR, may be distinguished from the methylene proton chemical shifts.[124,125] However, the methyl protons of M and D siloxane units cannot be differentiated. When chemical functions such as OH cannot be

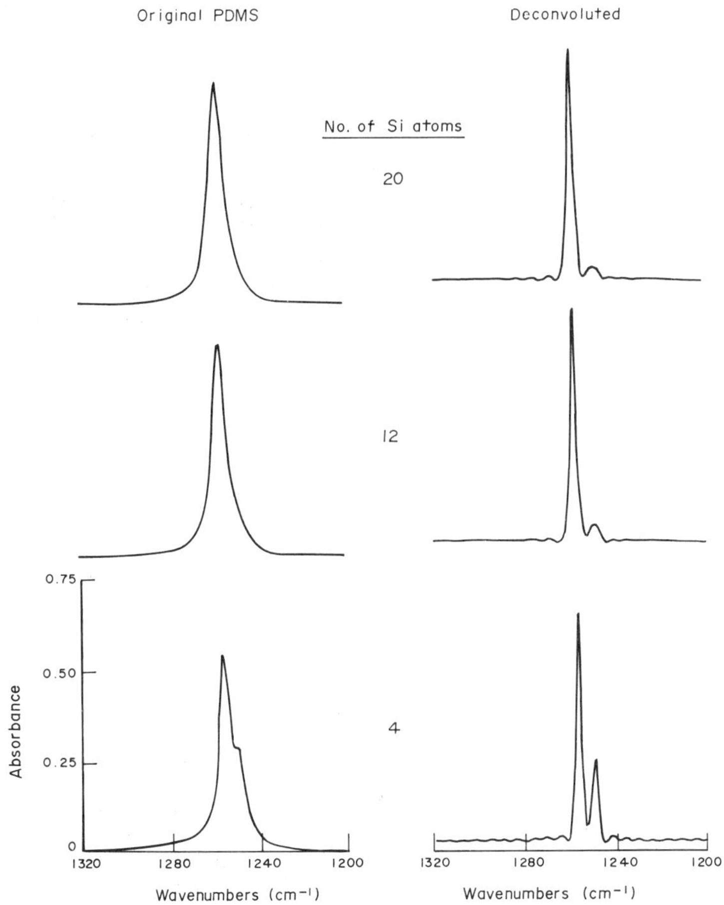


**Figure 20** The FT–IR of the symmetric methyl deformation for three representative PDMS oligomers are shown on the left; the deconvoluted results are shown on the right (reproduced by permission of the Society for Applied Spectroscopy from ref. 118)

observed, derivatization sometimes overcomes the problem.[98] The relatively small range of chemical shifts for proton nuclei (10 p.p.m) and the increased complexity of the spectra due to scalar coupling limits the applicability of proton NMR for elucidating the microstructure of siloxane copolymers. In principle $^{13}$C NMR can be applied and characterization of the degree of polymerization of poly(dimethylsiloxane)s has been reported recently.[111,126]

$^{29}$Si NMR allows direct observation of the silicon nuclei that form the backbone of polysiloxanes and brings a new and unrevealed dimension in their structural description. High resolution $^{29}$Si NMR at high frequency using superconducting magnets has no rival in determining the changes in the electronic, atomic and molecular environment of silicon atoms. Excellent reviews have been published on the subject.[127–132] The sensitivity of the chemical shift is such that the M, D, T and Q units are well separated. Any substitution of the methyl group is clearly differentiated (Figure 21).

The sensitivity of chemical shift is such that the silicon nucleus is influenced by substitution at up to six bonds away from the observed nucleus (Figure 22). This allows characterization of up to heptad sequences.[133] It is then a relatively simple matter to analyze the macrostructure in terms of the number average degree of polymerization of linear chains or even number average molecular weights as chain and chain end groups are simultaneously observed.[112,127–132] The only problem associated with this technique in determining the average DP is the rapidly decreasing sensitivity of chain ends as the molecular weight increases. Whilst polarization transfer pulse sequences might be a viable alternative,[134,135] FT–IR is superior in this respect.[118]

From the integrated intensities of the different resonance signals attributed to different siloxane units $M^{X,Y,Z}$, $D^{X,Y}$, $T^{X}$, calculation of their relative molar percentage and of the concentration of the specific X, Y and Z chemical functions can be performed. A typical $^{29}$Si NMR spectrum of a methyl/vinyl branched fluid is shown in Figure 23.

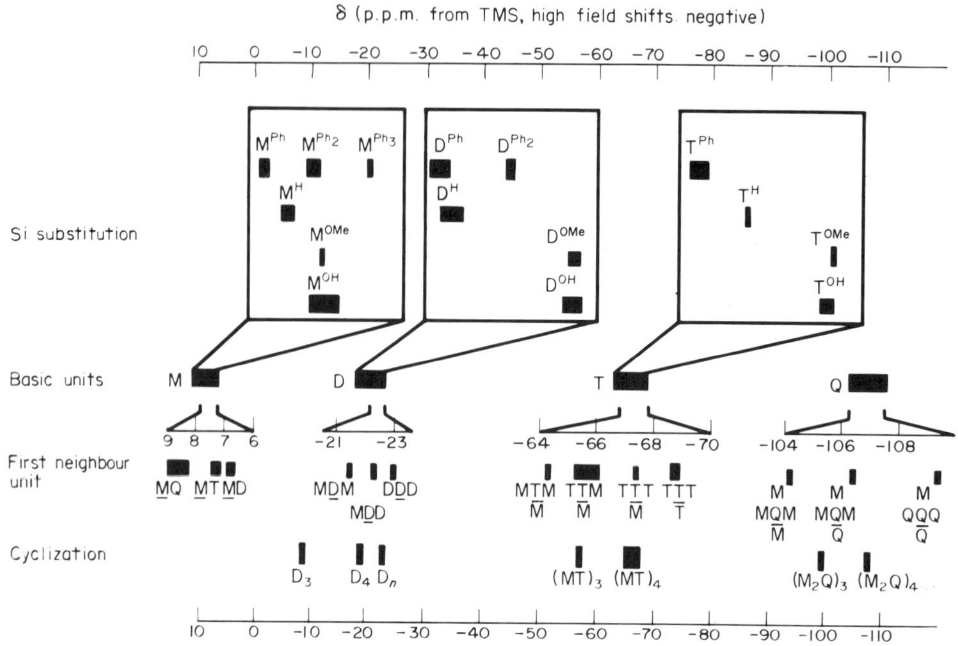


**Figure 21** Structure correlations of $^{29}Si$ chemical shifts: effect of substitution and of the first neighbour unit on the chemical shift of basic units M, D, T, Q (reproduced by permission of Springer Verlag and the authors from ref. 162)

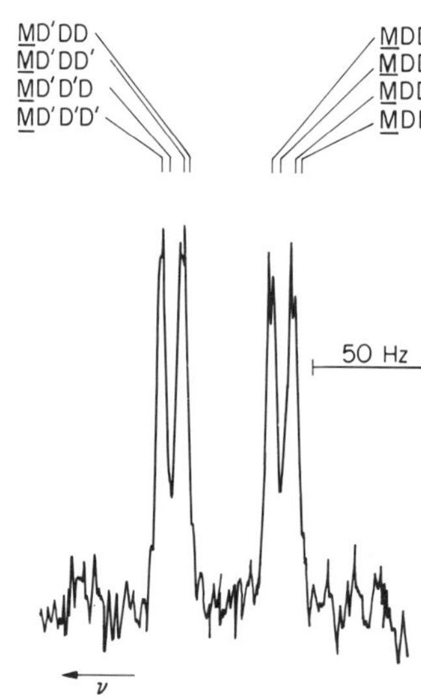


**Figure 22** $^{29}Si$–$^{1}H$ Spectrum of $Si_{18}H_8$ ($MD_8D_8^HM$) showing the microstructure for the M units (reproduced by permission of Elsevier Sequoia SA from ref. 133)

As can be seen, all main units can be detected. The normalized integrated intensities give the relative molar percentage of the different silicone units. T and $T^{Vi}$ (Vi = vinyl) branching as well as potential sites of branching ($D^{OH}$, $D^{Vi,OH}$) are observed and expressed as molar percentages. From these results an average model structure of the polysiloxane is drawn. When the structure is known, the average degree of polymerization and the average number molecular weight can be calculated. The concentration of hydroxyl functions as $M^{OH}$, $D^{OH}$, $D^{Vi,OH}$ siloxy units are easily deduced and expressed as global and specific contents (Table 11). The concentration of vinyl functions as $D^{Vi,OH}$ and $T^{Vi}$ siloxy units or any other organofunctional group can be evaluated by this process. The

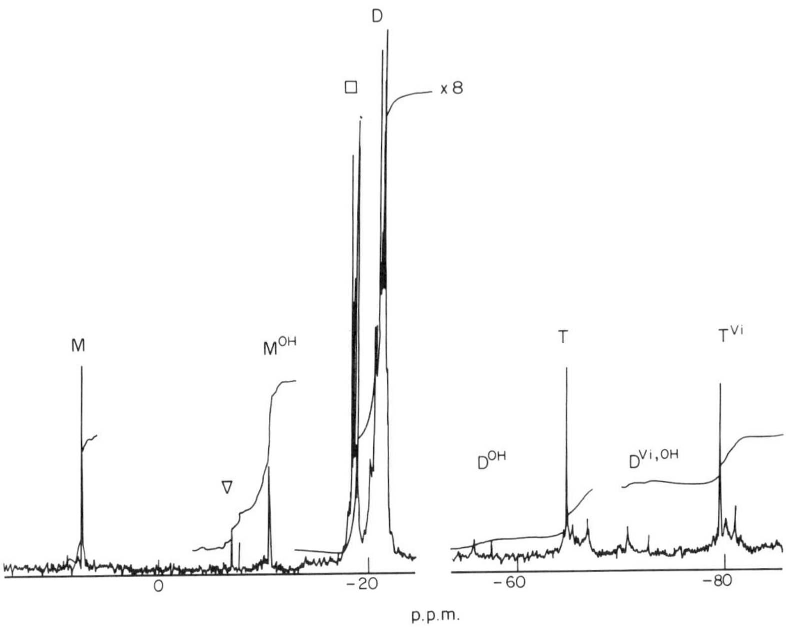


**Figure 23** $^{29}$Si–$^{1}$H NMR spectrum of a methyl/vinyl T-branched silicone liquid: ▽, substituted cyclotrisiloxanes; □, substituted cyclotetrasiloxanes (reproduced from ref. 112)

**Table 11** Molar Percentage of Various Siloxane Units of a Methyl/Vinyl Branched Fluid as Determined by $^{29}$Si NMR

| *Unit* | M | $M^{OH}$ | X | D | $D^{OH}$ | T | $D^{V,OH}$ | $T^{V}$ | Total |
|---|---|---|---|---|---|---|---|---|---|
| *I*(%) | 2.7 | 2.8 | 19.1 | 56.2 | 1.5 | 8.1 | 1.8 | 7.8 | 100 |

[a] T and $T^{V}$ are the specific methyl and vinyl branching levels. X represents mixed cyclotetrasiloxanes.

equations are

$$\% \text{OH} = (I^{OH} \times \text{MW}^{OH} \times 100)/\text{MW}^{tot} \tag{104}$$

$$\% \text{Vi} = (I^{Vi} \times \text{MW}^{Vi} \times 100)/\text{MW}^{tot} \tag{105}$$

$I^{X}$ refers to the normalized integrated intensity of the specified X unit (Table 12), $\text{MW}^{X}$ refers to the molecular weight of the X chemical function. $\text{MW}^{tot}$ is determined from normalized integrated intensities ($I_{u}^{X}$) and the MW of the silicone units

$$\text{MW}^{tot} = \Sigma(I^{X} \times \text{MW}^{X}) \tag{106}$$

The summation is made over all observed silicone units.

**Table 12** Specific and Total [Hydroxyl] and [Vinyl] Determined from Values Reported in Table 11

| | | | |
|---|---|---|---|
| % *OH* (w/w) | $M^{OH}$ | 0.64 ± 0.04 | |
| | $D^{OH}$ | 0.34 ± 0.02 | |
| | $D^{V,OH}$ | 0.34 ± 0.02 | |
| | Total | 1.32 ± 0.08 | $^{29}$Si NMR |
| | Total | 1.7 ± 0.3 | Chemical |
| % *vinyl* (w/w) | $D^{V,OH}$ | 0.54 ± 0.03 | |
| | $T^{V}$ | 2.81 ± 0.15 | |
| | Total | 3.3 ± 0.2 | $^{29}$Si NMR |
| | Total | 3.3 ± 0.3 | Chemical |

The NMR results are certainly more accurate than any other chemical method as chemical reactions are not involved. There are no side reactions, there is no need to ensure complete reaction and there is no interference with other reactive functions as is observed in chemical analysis.[57,98] When a good signal to noise ratio is obtained, the precision (relative error percentage) is 5%. For the chemical analysis of hydroxyl the relative error is 5–15%. A single NMR spectrum can yield information concerning a very complex silicone fluid that would otherwise require several different analytical methods. Information obtained from this NMR study would have required the simultaneous use of GLC to show the presence of mixed cyclics and the use of chemical methods of analysis to determine the OH and vinyl content and the level of branching. Furthermore NMR allows identification and quantification of specific types of OH and vinyl groups as well as different types of branching.

Over the last 10 years there has been a significant number of publications dealing with $^{29}$Si NMR studies on organisilicon polymers.[127–132,136–147] However, there are NMR techniques that present obvious advantages that have not yet been fully exploited for siloxane copolymer characterization. The low sensitivity of $^{29}$Si nuclei can be boosted and resolution increased by the multipulse sequence techniques like insensitive nuclear enhancement *via* polarization transfer (INEPT)[136] and distortionless enhancement by polarization transfer (DEPT).[137] These pulse sequence techniques transfer the proton nuclear spin polarization to the $^{29}$Si spins *via* the $^1$H–$^{29}$Si second-order scalar coupling. Furthermore this transfer depends only on proton relaxation and thus significantly decreases the experimental time.[138] A new technique, recycle-flow fourier transform NMR has been recently proposed[126,138–140] and has been applied to the $M_n$ determination of siloxane polymers with $^{13}$C and $^{29}$Si nuclei. This technique has several advantages. It reduces the total experimental time in getting a good signal to noise ratio; an improvement of three to five times is possible with the flow technique over the conventional $^{29}$Si NMR enhanced by the use of the Cr(acac)$_3$ relaxation reagent. Sample contamination is avoided as is the adverse effect of the reagent on the assignment of chemical shifts and on the resolution. Furthermore samples that cannot solubilize the chromium compound can be run. Its major limitation is the implementation of additional hardware within the cryomagnet.[126] Figure 24 shows the spectra obtained using this technique.

Quite often siloxane copolymers attain such molecular weights or have so many cross-links that they become insoluble in virtually any solvent (Figure 25). In such cases classical techniques of analysis are not applicable any more. In simple NMR experiments, dipolar interactions between nuclear spins and the anisotropy of the $^{29}$Si chemical shift lead to strong line broadening and overlapping of peaks with already poor signal to noise ratios.[148] The various silicone structural units can no longer be differentiated. The recent progress made in the NMR field has brought new spectra acquisition principles. $^{29}$Si–$^1$H cross-polarization, high power proton decoupling and

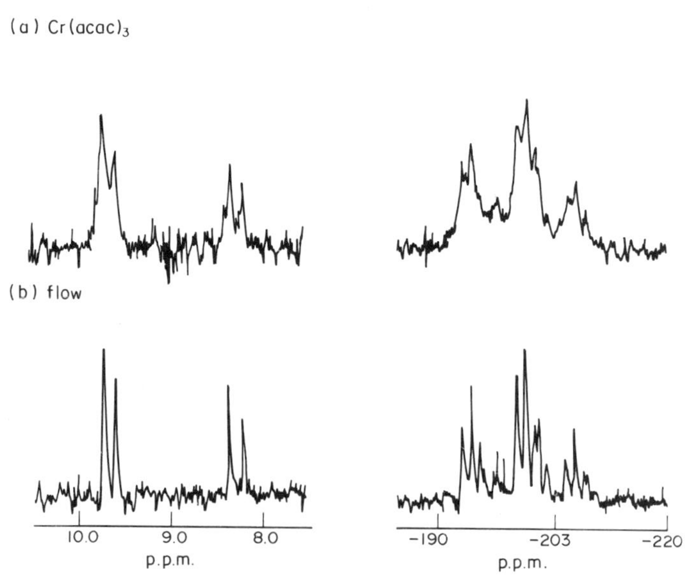


**Figure 24** Spectra of the end and middle group regions of the DC 550 phenylmethylsiloxane copolymer blend demonstrate the reduced spectral quality derived from (a) the 0.05 M Cr(acac)$_3$ sample compared to (b) the flow measurement (reproduced by permission of the American Chemical Society from ref. 126)

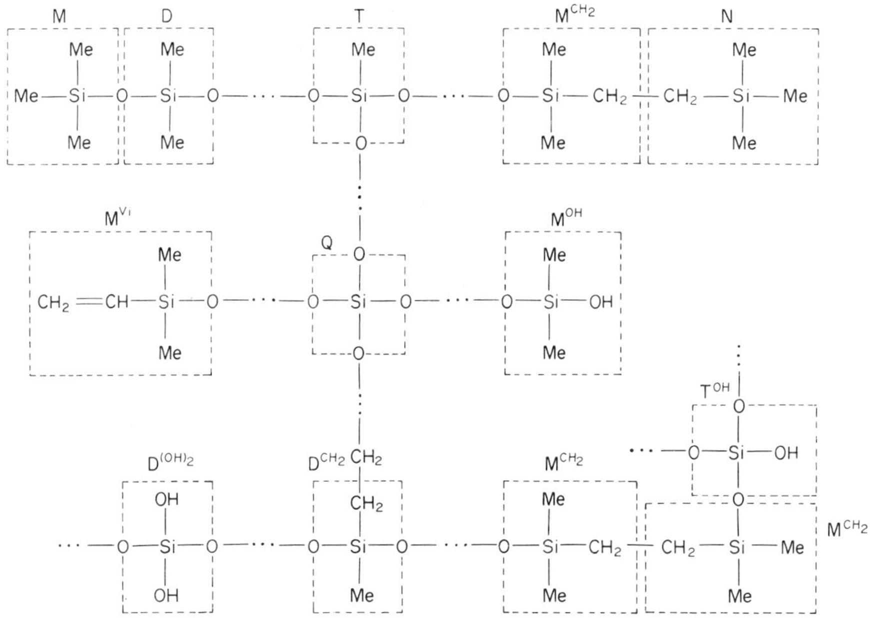


**Figure 25** Hypothetical PDMS network illustrating various sites of silicon atoms and their corresponding chemical functionalities (reproduced by permission of John Wiley and Sons Inc. from ref. 158)

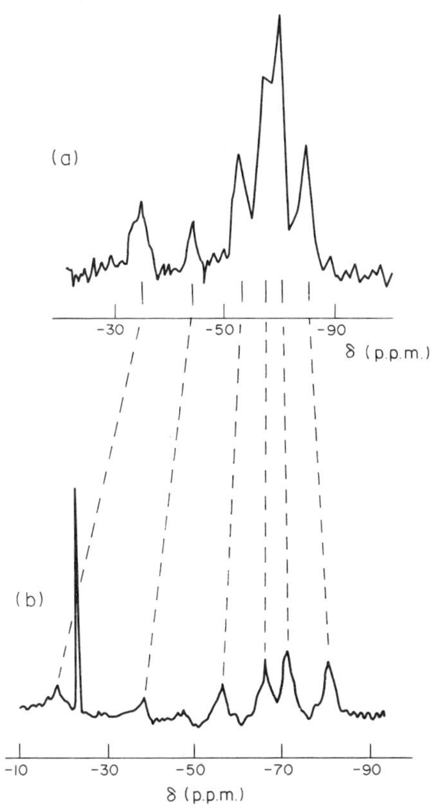


**Figure 26** $^{29}$Si NMR spectrum of methylphenylsilicon resin: (a) solid powder, 1900 scans, pulse repetition rate 1 s, contact time 3 ms (total time 32 min); (b) 15% solution in $CCl_4$, 16 100 scans, pulse repetition time 6 s (total time 26.8 h) (reproduced by permission of Elsevier Sequoia SA from ref. 148)

sample spinning at magic angle have made possible the observation of $^{29}$Si nuclei of solid samples. Resonance peaks are narrow and well resolved.[148–151] Engelhardt *et al.*[148] have shown that the detailed information on siloxane structural units available in liquid state NMR is also revealed in solid state NMR and the relationships between $^{29}$Si chemical shifts and molecular structure are unchanged. An illustration is given in Figure 26, based on the NMR of a methylphenylsilicone resin.[148]

The complex structure of his resin can be discerned immediately. Resonance signals are attributed to D, T, $T^P$ units and to $M^{P_2,OH}$, $D^{OH}$, $D^{P,OH}$ arising from unreacted silanols. The sharp line at −21.9 p.p.m. is assigned to $D_n$ units in the chain. This peak cannot be observed in the solid state when the cross-polarization technique is used because of the high flexibility of siloxane chains. However under classical FT-NMR conditions these mobile D units are detected.[148,152] Several other examples of the application of this technique can be found in the literature.[148–162]

Blends of different compositions of a methyl resin with PDMS and PDMDPS gums have been studied as a function of temperature and compared with studies of the pure components.[155] $^{29}$Si NMR spectra of resin/poly(dimethyl-*co*-diphenylsiloxane) (PDMDPS) gum are shown in Figure 27.

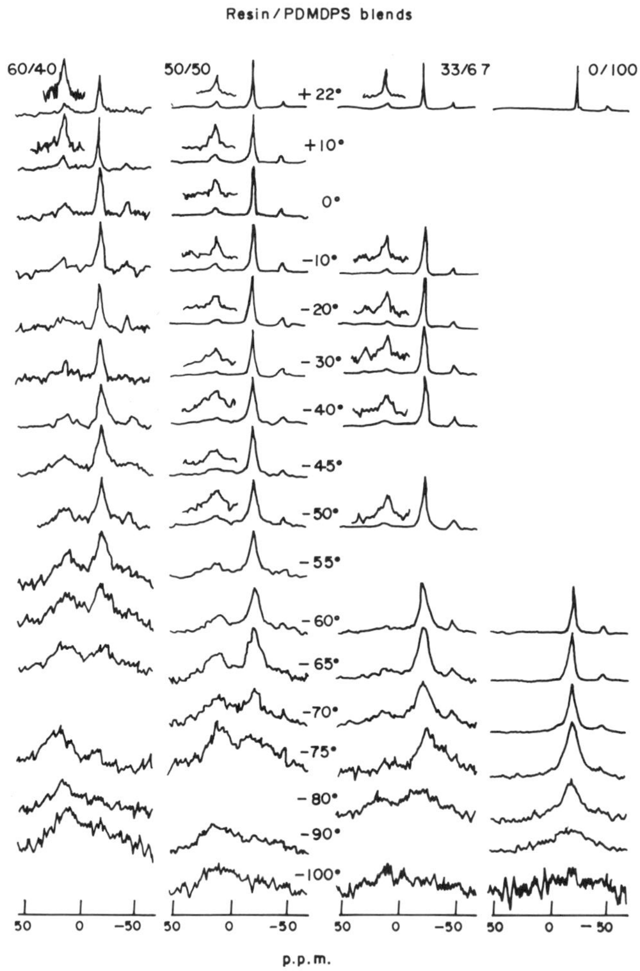


**Figure 27** $^{29}$Si variable temperature spectra of resin/PDMDPS Blends; the spectra are plotted at different vertical gain (reproduced by permission of the American Chemical Society from ref. 155)

This study shows that the glass transition temperature can be detected in solid state NMR. Relaxation times and nuclear Overhauser enhancement factors indicate substantial segmental motion at ambient temperature. A model is proposed for these blends which assumes the blend is composed of two phases, *viz.* a resin phase that contains a small amount of low molecular weight gum and a gum phase that contains some resin. Solid state $^{29}$Si NMR provides exceptional capability in the qualitative chemical analysis of siloxane networks.[157–159] Microscopic network properties such as junction functionality and number density or network imperfections such as dangling chain ends can be examined. A siloxane network was synthesized[157] by reacting vinyl-terminated oligomers with tetrafunctional $Si(OSiMe_2H)_4$. In the Q unit region (*ca.* 104 p.p.m.) one would expect a single Q resonance. Instead, three peaks are observed, the two extra peaks being attributed to 'butterfly' and 'bow tie' structures which are considered as defects in the network (Figure 28).

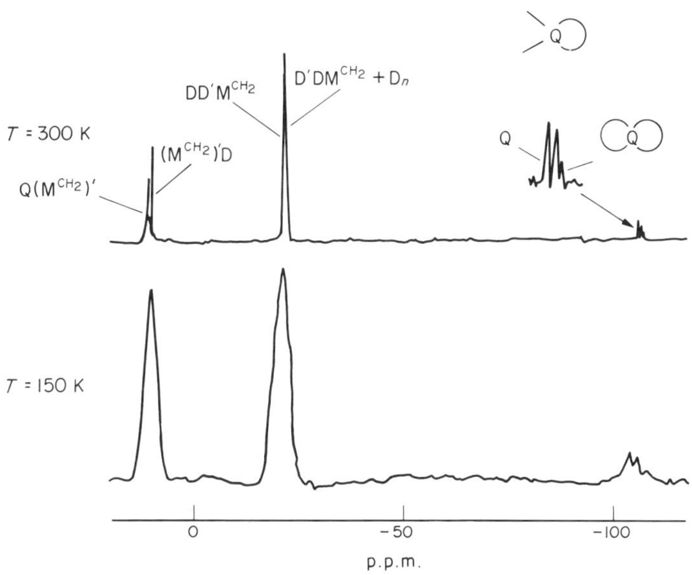


**Figure 28** Model vinyl end-linked network: $^{29}Si$ cross-polarization/magic angle spinning spectra of an unextracted tetrafunctional vinyl end-linked PDMS network synthesized from chains of $M_n = 520$, DP = 7 (reproduced by permission of the American Chemical Society from ref. 157)

### 25.5.4 Linear Siloxane Copolymers

The macrostructure description of homopolymers is relatively simple. They are well characterized by the chemical composition of chain and chain end groups, the number average degree of polymerization, the molecular weight distribution and the related average molecular weights. The theory of homopolymerization is also relatively simple as few parameters are needed to describe the chain growth.

The description of the copolymer macrostructure will also involve characterization of these molecular parameters. These in turn can be related to the chemical, physical and mechanical properties of the copolymer. However, different properties were observed for a given copolymer macrostructure and it was recognized that the ordering of the different comonomer units in the chain, which could be non-uniform, would account for the apparently inconsistent observed properties.[161–171] Early studies on copolymer microstructures are still being pursued today.[172–185] As there were no direct methods that could reveal the sequencing of the monomer units in the polymer chain, these efforts concentrated on theoretical models of copolymerization based upon kinetic and statistical considerations and were designed to predict the copolymer microstructure formed at any conversion and from any comonomer composition.[166] Attempts have been made to chemically analyze organic copolymers,[186, 187] but it is the advent of NMR that has enabled the determination of sequence distributions to be achieved and to be compared with theory.

The sequence distribution in copolymers is extremely important since it may affect the copolymer properties in a subtle yet significant way. It depends on the methods of synthesis and on reaction conditions and is governed by the kinetics and the thermodynamics of the polymerization process. Three general methods of synthesis are known. Sequential polymerization, condensation copolymerization and mixed cyclic copolymerization.

### 25.5.4.1 *Sequential polymerization*

Sequential polymerization consists of first polymerizing a given cyclosiloxane to the desired chain length, whereupon the comonomer cyclosiloxane is then introduced into the reaction medium and proceeds to grow from the living chain end. The reaction is terminated at the desired conversion.[188,189] The catalyst must be selective towards the cyclosiloxane in order to have a living polymer.[189] Thus the active reaction centre must only react with the monomer and not with the linear chains already formed. Under these conditions pure block copolymers will be produced. Otherwise a random distribution of monomer units will appear as the result of redistribution reactions involving the polymer chain.[1,57] The concentration of the polymer chains in the system will be equal to the concentration of the catalyst and, if impurities are absent, will remain constant throughout the course of the reaction. The number average degree of polymerization will be described by expression (107) if total conversion of the comonomers is assumed

$$\mathrm{DP} = (C_{\mathrm{A}}^{0} + C_{\mathrm{B}}^{0})/C_{\mathrm{P}} \tag{107}$$

In equation (107) $C_{\mathrm{A}}^{0}$ and $C_{\mathrm{B}}^{0}$ represent the initial molar concentration of the comonomers A and B, $C_{\mathrm{P}}$ represents the concentration of the copolymer chains. Depending on the functionality of the catalyst, either $A_xB_y$ or $B_yA_xB_y$ block type copolymers can be made. In multisequential polymerization, these A and B runs alternate along the polymer chain.

### 25.5.4.2 *Condensation polymerization*

Prepolymers are first prepared by ring-opening polymerization. These homopolymers, which are generally hydroxyl end blocked, are then condensed to yield block type copolymers. The concentration and the functionalities of the different prepolymers will determine the final microstructure of the copolymers. An $A_xB_yA_x$ type copolymer, for example, will be obtained by condensing monofunctional $A_x$ and difunctional $B_y$ prepolymers.[1,57,190] Prediction of the copolymer macrostructure and of the lengths of the different blocks is straightforward. Problems only arise when redistribution reactions occur which scramble the order previously established. This process can lead to any combination of prepolymers provided that the homopolymers can be prepared.

### 25.5.4.3 *Mixed cyclics copolymerization*

The process of synthesizing high molecular weight polymers from cyclics is well established and widely used in the silicone industry. Polymerizing different cyclosiloxanes will yield copolymers where the macrostructure can be well controlled. However, the microstructure, which depends on several reactions parameters, is not easily predictable. The sequences of the siloxane units are determined by the relative reactivities of the monomers and the active chain ends. These reactivities depend on inductive, mesomeric and steric effects and each copolymer molecule is both the end result and a record of the complex interplay of these factors. In this process the different cyclics are mixed together and copolymerized. The reaction is initiated by basic or acidic catalysts and a stepwise addition polymerization kinetic scheme is followed. Cyclotrisiloxanes are most frequently used in these copolymerizations since the chain growth mechanism dominates the kinetics and redistribution reactions involving the polymer chain are of negligible importance. Several different copolymers may be obtained by this process and, moreover, they will be monodisperse and free from cyclics. The microstructure can be varied from pure block to pure random copolymers. Anionic copolymerization of the cyclotetrasiloxanes octamethylcyclotetrasiloxane and octaphenylcyclotetrasiloxane has been studied and it has been shown that rate of the chain transfer *via* redistribution reactions is already appreciable even in the early stages of copolymerization.

### 25.5.4.4 *Theoretical models for the polymerization of mixed cyclics*[161–183]

Consider the copolymerization of two monomers $D_3$ and $(D')_3$, representing two different cyclotrisiloxanes A and B. In the stepwise addition polymerization scheme the individual steps will be initiation, propagation and termination. Initiation and termination bear much resemblance to those same steps in homopolymerization, but it is the propagation step that gives mixed copolymerization its special character by determining the composition of the copolymer. Therefore in this treatment only this propagation step is considered and the chain lengths are assumed to be infinite. Ring-opening reactions will initiate the copolymerization process by creating two active propagation centres D* and D′*

$$D_3 + XY \rightarrow X\text{–}D\text{–}D\text{–}D^*\,Y^* \tag{108}$$

$$(D')_3 + XY \rightarrow X\text{–}D'\text{–}D'\text{–}D'^*\,Y^* \tag{109}$$

When experimental conditions are chosen so that the two active centres selectively attack the monomers, four propagation reactions can be considered

$$\text{–}D\text{–}D\text{–}D^* + D_3 \xrightarrow{k_{AA}} \text{–}D\text{–}D\text{–}D\text{–}D\text{–}D\text{–}D^* \tag{110}$$

$$\text{–}D\text{–}D\text{–}D^* + (D')_3 \xrightarrow{k_{AB}} \text{–}D\text{–}D\text{–}D\text{–}D'\text{–}D'\text{–}D'^* \tag{111}$$

$$\text{–}D'\text{–}D'\text{–}D'^* + D_3 \xrightarrow{k_{BA}} \text{–}D'\text{–}D'\text{–}D'\text{–}D\text{–}D\text{–}D^* \tag{112}$$

$$\text{–}D'\text{–}D'\text{–}D'^* + (D')_3 \xrightarrow{k_{BB}} \text{–}D'\text{–}D'\text{–}D'\text{–}D'\text{–}D'\text{–}D'^* \tag{113}$$

where $k_{AA}$ and $k_{BB}$ represent the homopropagation rate constants and $k_{AB}$ and $k_{BA}$ represent the cross-propagation rate constants. These four reactions compete with each other. In this model it is assumed that these reactions are independent of the number and nature of attached units.[171] The mechanism of the reaction is not considered as it does not influence the conclusions drawn from the model.[178] The rates of propagation are independent of the conversion and remain constant regardless of the change in monomer composition.[178] They are assumed to be dependent only on the nature and concentration of the terminal active unit regardless of the growing chain's previous history.[160,161,170] This assumption is important in the mathematical derivation since the way the chains grow may then be described by the four reactions (110)–(113). When the growth of the chains is affected by the nature of the monomer unit immediately preceding the active centre, eight reactions will describe the copolymerization process.[185]

Mayo and Lewis[166] and Alfrey and Goldfinger[167] using a kinetic approach and Goldfinger and Kane[168] using a probabilistic approach derived the equation that links the copolymer composition to the comonomer composition. Whilst the kinetic development assumes steady conditions, this assumption is not necessary in developing the copolymer composition from probabilistic considerations of sequence distributions.

If A and B represent the two cyclotrisiloxanes, then the rates of consumption of monomers A and B can be expressed as

$$-d[A]/dt = k_{AA}[A^*][A] + k_{BA}[B^*][A] \tag{114}$$

$$-d[B]/dt = k_{BB}[B^*][B] + k_{AB}[A^*][B] \tag{115}$$

The ratio of these two equations combined with the steady state condition ($k_{AB}[A^*][B] = k_{BA}[B^*][A]$) yields an equation independent of the concentration of the active centres but a function of the monomer reactivity ratios $r_1 = k_{AA}/k_{AB}$ and $r_2 = k_{BB}/k_{BA}$

$$d[A]/d[B] = [A](r_1[A] + [B])/[B](r_2[B] + [A]) \tag{116}$$

For the initial stages of the polymerization this equation can be rewritten in a form that relates the mole fraction of A in the copolymer $X_A^{cp}$ as a function of the mole fraction $X_A$ and $X_B$ of the monomers

$$X_A^{cp} = (r_1 X_A^2 + X_A X_B)/(r_1 X_A^2 + 2X_A X_B + r_2 X_B^2) \tag{117}$$

where the mole fraction of A in the copolymer formed during a differential time interval is

$$X_A^{cp} = [\Delta A]/(\Delta A + \Delta B) \tag{118}$$

ΔA and ΔB are the amounts of the corresponding comonomers which enter the polymer in a differential time interval. Equation (117) describes the copolymer composition. The reactivity ratios can be obtained from a non-linear regression analysis[170,192] or from its linear rearranged form

$$\frac{X_A(1 - 2X_A^{cp})}{(1 - X_A)X_A^{cp}} = \frac{X_A^2(X_A^{cp} - 1)}{(1 - X_A)^2 X_A^{cp}} r_1 + r_2 \tag{119}$$

This equation neither requires nor yields information about specific rates of polymerization. The reactivity ratios determine the direction of the chain growth and hence the microstructure of the copolymer. They can account for pairs of monomers that do or do not readily form block copolymers.[170]

If one considers a situation where the ring opening of monomer A is easier than that of B, then $k_{AA}$ is greater than $k_{AB}$ and $k_{BA}$ is greater than $k_{BB}$. If the active chain end is a silanolate and the substituent on the siloxane unit is electronegative, then $k_{BB}$ is greater than $k_{AB}$ and $k_{BA}$ is greater than $k_{AA}$.

$$\begin{array}{ccc} k_{BA} & > & k_{AA} \\ \vee & & \vee \\ k_{BB} & > & k_{AB} \end{array} \qquad (120)$$

These qualitative comparisons of rate constants predict that in the initiation step, once A* is formed, it will preferentially react with A, but if it reacts with B, the new B* formed will react with A to reform an A*. As $k_{AA} > k_{AB}$ the chance of A* reacting with monomer B can be considered as being small. The preferential consumption of A is thus promoted both by the high rate of its homopolymerization and by the even higher rate of interaction between monomer A and the macroanion B* since $k_{BA} > k_{AA}$. On the other hand, monomer B, which is less active than monomer A, is more inclined to interact with an active centre of its own nature (B*) than the macroanion A* ($k_{AB} < k_{BB}$). Further into the copolymerization a state is reached where reaction (110) strongly competes with reaction (111) since

$$-\mathrm{d}[A]/\mathrm{d}t = k_{AA}[A^*][A] \qquad (121)$$

$$-\mathrm{d}[B]/\mathrm{d}t = k_{AB}[A^*][B] \qquad (122)$$

where $k_{AA} > k_{AB}$ and [A] is now $\ll$[B]. At this stage A and B add in a random fashion to the growing copolymer chains. When A has completely reacted, the final stage is attained where only B is left and a pure B block is formed. The size of the random AB block depends on the relative reactivities of the monomers. The dual role of the electroactive substituents on silicon in the anionic copolymerization of cyclosiloxanes is well illustrated by this analysis. Electronegative substituents facilitate attack of the catalyst on the silicon atom. Once the ring is opened and the macroanion formed, the nucleophilic activity of the anion is lowered relative to a permethylcyclosiloxane. Yet, upon substituting a methyl group by an electronegative group, the rate of anionic polymerization still increases. Thus the nature of the cyclosiloxane, which dictates the rate of the ring-opening reaction, has more influence on the whole process than the nature of the chain end.[192]

### 25.5.4.5 *Copolymer composition as a function of the degree of polymerization*

The copolymer composition equation (117) can be used to show graphically the initial copolymer composition as a function of initial monomer composition (Figure 29).[163,193] Curve I shows the

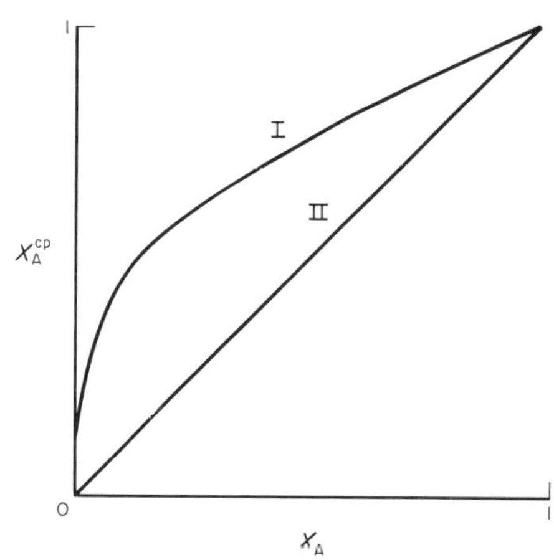


**Figure 29** Initial mole fraction of component A (diphenyl) in the copolymer as a function of its mole fraction in initial comonomer mixture: Copolymerization of diphenylcyclotrisiloxane (A) and dimethylcyclotrisiloxane (B); $r_1 = 2.14$, $r_2 = 0.045$ (reproduced by permission of the Division of Polymer Chemistry of the American Chemical Society from ref. 197)

initial composition for an AB copolymer where $r_1 > 1$ and $r_2 < 1$. The fraction of A monomer units will be higher in the copolymer than in the mixture of the two monomers as a consequence of A being more reactive than B in the ring-opening polymerization. One may also expect that curve I corresponds to a copolymer with 'block' character. Curve II shows the initial composition when $r_1 = r_2 = 1$ and represents the composition of a pure random copolymer. The effect of comonomer composition on the initial copolymer composition is well illustrated.

The description of the copolymer composition as a function of the conversion or of time is of much more interest. The first treatment was undertaken by Skeist who drew an analogy between copolymer composition and distillation of a binary mixture of liquids.[175] In the system above, if $M$ is the total amount of A and B monomers and $\mathrm{d}M$ represents the number of moles of monomers that polymerize over a small time increment, the number of moles of component A in the copolymer is then $X_A^{cp}\,\mathrm{d}M$. At the same time the number of moles of component A in the monomer has been reduced to

$$(M - \mathrm{d}M)(X_A - \mathrm{d}X_A) \tag{123}$$

The material balance of A is

$$X_A M - (M - \mathrm{d}M)(X_A - \mathrm{d}X_A) = X_A^{cp}\,\mathrm{d}M \tag{124}$$

If the product of the two differentials is neglected, one has

$$\mathrm{d}M/M = \mathrm{d}X_A/(X_A^{cp} - X_A) \tag{125}$$

By integrating, one obtains the composition conversion equation

$$\ln(M/M_0) = \int_{X_A^0}^{X_A} \mathrm{d}X_A/(X_A^{cp} - X_A) \tag{126}$$

This equation combined with the copolymer composition equation (117) gives the instantaneous copolymer composition as a function of polymerization conversion. An analytical solution of equation (126) is not possible and numerical computation has to be used to solve the equation. The measured copolymer composition is the average composition from the beginning to a given conversion. The instantaneous copolymer must be integrated once again to provide the average copolymer composition. Figure 30 shows the variation of the instantaneous copolymer composition and the average copolymer composition as a function of the conversion for the system shown in Figure 29 (initial comonomer composition $X_A = 0.5$).

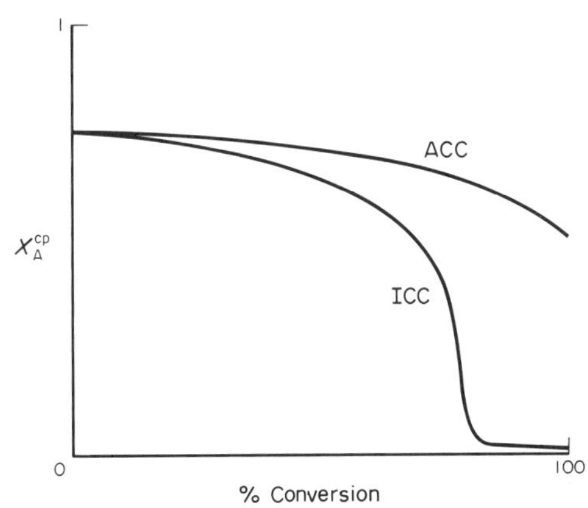


**Figure 30** Instantaneous (ICC) and average (ACC) copolymer composition as a function of conversion for the poly(diphenyl-*co*-dimethylsiloxane); initial monomer mixture is $X_A^0 = 0.50$ (reproduced by permission of the Division of Polymer Chemistry of the American Chemical Society from ref. 197)

## 25.5.5 Copolymer Microstructure

So far the theory of copolymerization has provided a description of the copolymer macrostructure as a function of the conversion. The copolymer microstructure is described by the distribution of the different siloxane units along the macromolecular chain. Sequences of the monomer units can be considered theoretically and observed experimentally at the dyad, triad, $n$-ad levels. Two approaches to the determination of microstructure may be considered: one derived from the kinetic scheme and the other from NMR experimental data of copolymers at equilibrium. Both approaches are based on probabilistic considerations. The most simple statistical approach to predict the *n*-ad sequence distribution is the Bernoulli trial process. When the copolymerization is non-Bernoullian, simple Markovian chain statistics generally apply.

### 25.5.5.1 Determination of copolymer microstructure by analysis of kinetic chains

Price[180] and Bovey[191] were the first to develop the statistical analysis of polymers to study tacticity using Bernoullian and Markovian models. Randall and Koenig have published comprehensive surveys of the subject.[194,195] In the Bernoullian trial distribution the monomer is added to the growing chain independently of the outcome of the previous addition. In the Markovian trial distribution the nature of the preceding unit does influence the chain growth. Price and Bovey have approached the problem through equivalent statistical approaches but applied different states to define the chain ends and the adding units.

Price used a one state model (A or B unit) and the previous event corresponds to the active chain end. The Price–Bernoullian model is then described by two probabilities as the nature of the active chain end will not influence the addition of the monomer units

$$\text{–X}^* + \text{A} \rightarrow \text{–A}^* \qquad P_{\text{A}} \tag{127}$$

$$\text{–X}^* + \text{B} \rightarrow \text{–B}^* \qquad P_{\text{B}} \tag{128}$$

The Price first-order Markovian model is described by four propagation reactions defined by four conditional probabilities

$$\text{–A}^* + \text{A} \rightarrow \text{–AA}^* \qquad P_{\text{A/A}} \tag{129}$$

$$\text{–A}^* + \text{B} \rightarrow \text{–AB}^* \qquad P_{\text{B/A}} \tag{130}$$

$$\text{–B}^* + \text{A} \rightarrow \text{–BA}^* \qquad P_{\text{A/B}} \tag{131}$$

$$\text{–B}^* + \text{B} \rightarrow \text{–BB}^* \qquad P_{\text{B/B}} \tag{132}$$

In Bovey's approach adapted to silicone copolymers a two-state model is used and the preceding event is the nature of the monomer unit immediately preceding the active chain end. The Bovey–Bernoullian model is now described at dyad levels by four reactions defined by the four probabilities of occurrence of the four dyads

$$\text{–A}^* + \text{A} \rightarrow \text{–AA}^* \qquad P_{\text{AA}} \tag{133}$$

$$\text{–A}^* + \text{B} \rightarrow \text{–AB}^* \qquad P_{\text{AB}} \tag{134}$$

$$\text{–B}^* + \text{A} \rightarrow \text{–BA}^* \qquad P_{\text{BA}} \tag{135}$$

$$\text{–B}^* + \text{B} \rightarrow \text{–BB}^* \qquad P_{\text{BB}} \tag{136}$$

The Bovey first-order Markovian model requires eight reactions to describe the copolymerization

$$\text{–AA}^* + \text{A} \rightarrow \text{–AAA}^* \qquad P_{\text{AA/A}} \tag{137}$$

$$\text{–AA}^* + \text{B} \rightarrow \text{–AAB}^* \qquad P_{\text{AB/A}} \tag{138}$$

$$\text{–BA}^* + \text{A} \rightarrow \text{–BAA}^* \qquad P_{\text{AA/B}} \tag{139}$$

$$\text{–BA}^* + \text{B} \rightarrow \text{–BAB}^* \qquad P_{\text{AB/B}} \tag{140}$$

$$\text{–AB}^* + \text{A} \rightarrow \text{–ABA}^* \qquad P_{\text{BA/A}} \tag{141}$$

$$\text{–AB}^* + \text{B} \rightarrow \text{–ABB}^* \qquad P_{\text{BB/A}} \tag{142}$$

$$\text{–BB}^* + \text{A} \rightarrow \text{–BBA}^* \qquad P_{\text{BA/B}} \tag{143}$$

$$\text{–BB}^* + \text{B} \rightarrow \text{–BBB}^* \qquad P_{\text{BB/B}} \tag{144}$$

It is essential that the approach taken is stated in order to know the model that is sustained by the experimental observations. We shall explore in detail the Bovey–Bernoullian and the Price–Markov models as they have been applied to anionic polymerization of cyclotrisiloxanes. As A and B each symbolizes three siloxane units, the Price–Bernoullian and Bovey–Markovian models can be discarded. The correctness of any approach to a given copolymer system can be tested and confirmed by experimental observation. In general when Bernoullian statistics do not describe the sequence distribution, Markovian statistics do.

### 25.5.5.2 Bovey–Bernoulli trial distribution

In the Bovey–Bernoullian statistics the growth of a chain ended with either A* or B* depends on a random selection of the A and B adding comonomers. The probability for an A* active chain end to attack an A or a B monomer is given by

$$P_{AA} = k_{AA}C_A/(k_{AA}C_A + k_{AB}C_B) \tag{145}$$

$$P_{AB} = k_{AB}C_B/(k_{AA}C_A + k_{AB}C_B) \tag{146}$$

and similarly the probability that a B* active chain end will attack an A or B monomer is given by

$$P_{BA} = k_{BA}C_A/(k_{BA}C_A + k_{BB}C_B) \tag{147}$$

$$P_{BB} = k_{BB}C_B/(k_{BA}C_A + k_{BB}C_B) \tag{148}$$

where $P_{XY}$ is the probability that an X active chain end attacks a Y monomer.
These probabilities are related by

$$P_{AA} + P_{AB} = 1 \quad \text{and} \quad P_{BA} + P_{BB} = 1 \tag{149}$$

and a single parameter $P_{AA}$ or $P_{AB}$ needs to be considered for the growth of a chain ended by A*. The influence of the nature of the chain end is implicitely taken into account by the rate constants. Equation (145) can be rewritten as a function of the mole fraction of B in solution

$$P_{BB} = r_2X_B/[1 + (r_2 - 1)X_B] \tag{150}$$

If one selects a B unit in the copolymer chain, then A and B units can be arranged around it in three different ways in a triad model[178]

$$\text{ABA} \qquad \text{ABB} = \text{BBA} \qquad \text{BBB} \tag{151}$$

The mole fractions or probabilities of occurrence of these triads are easily deduced. Similarly the probabilities of occurrence of pentad sequences can be calculated.

Lee and Marko[197] applied the Bovey–Bernoullian statistics to siloxane copolymers and compared the theoretical predictions with proton NMR. The comonomers hexaphenylcyclotrisiloxane and hexamethylcyclotrisiloxane here defined as A and B monomers are copolymerized under non-equilibrium conditions and with anionic catalysis. (Figure 31) shows the change in mole fraction of the triads as a function of the B mole fraction in monomer. The mole fraction of triads can be related

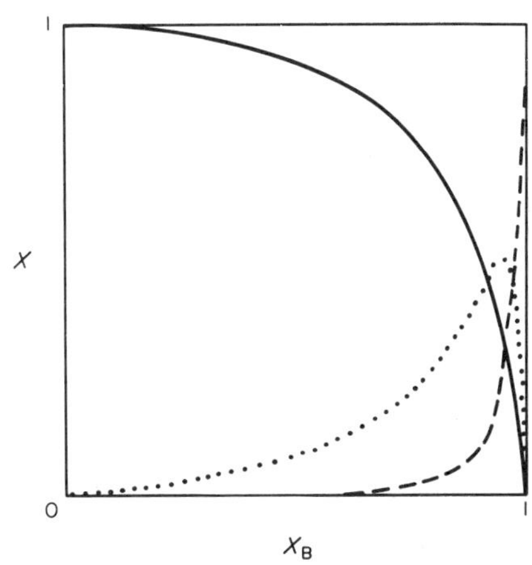


**Figure 31** Mole fractions of the triads ABA (——), ABB (· · ·) and BBB (---) as a function of mole fraction $X_B$ for the hexaphenylcyclotrisiloxane (A) and hexamethylcyclotrisiloxane (B) copolymerization system; Bovey–Bernoulli approach (reproduced by permission of the Division of Polymer Chemistry of the American Chemical Society from ref. 197)

to the mole fraction of B in the comonomer composition. This will provide three differential sequence distribution curves that represent the probability of occurrence of these triads at a given extent of reaction (Figure 32). These differential distribution curves must be integrated in order to obtain the overall averaged mole fraction of the different triads already formed as a function of the conversion (Figure 33). The data can be converted to theoretical NMR spectra and compared with NMR spectra of the copolymer. Methyl proton resonance signals were analyzed and ratios of integrated intensities for different signals could be evaluated at different conversions and compared with those predicted by the theory (Figure 34).

### 25.5.5.3 Price–Markov trial distribution

In Price–Markov statistics the probability of forming any dyad AA, AB, BA or BB is dependent on the chemical nature of the active siloxane end unit. This model involves four transition or

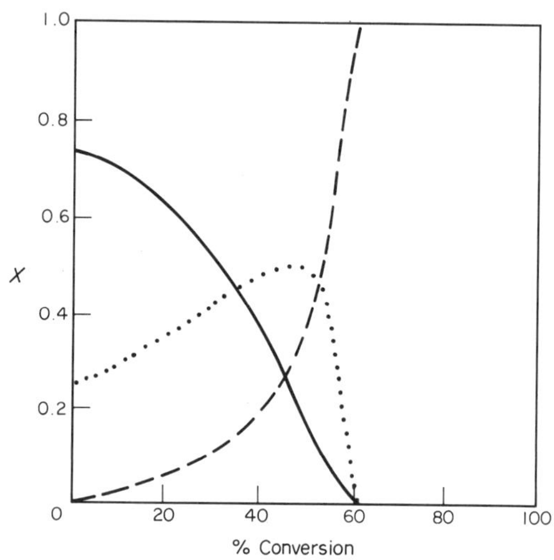


**Figure 32** Differential sequence distribution of the triads ABA (——), ABB (· · ·) and BBB (–––) as a function of conversion; $X_A^0=0.72$, $r_2=0.45$; Bovey–Bernoulli approach (reproduced by permission of the Division of Polymer Chemistry of the American Chemical Society from ref. 197)

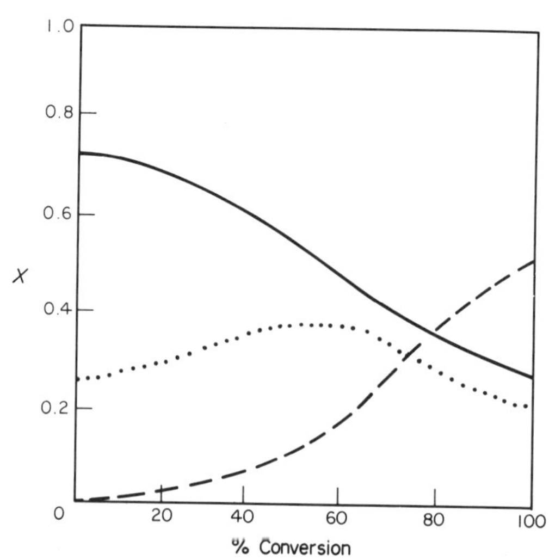


**Figure 33** Integrated sequence distribution of the triads ABA (——), ABB (· · ·) and BBB (–––) as a function of conversion; $X_A^0=0.72$, $r_2=0.045$; Bovey–Bernoulli approach (reproduced by permission of the Division of Polymer Chemistry of the American Chemical Society from ref. 197)

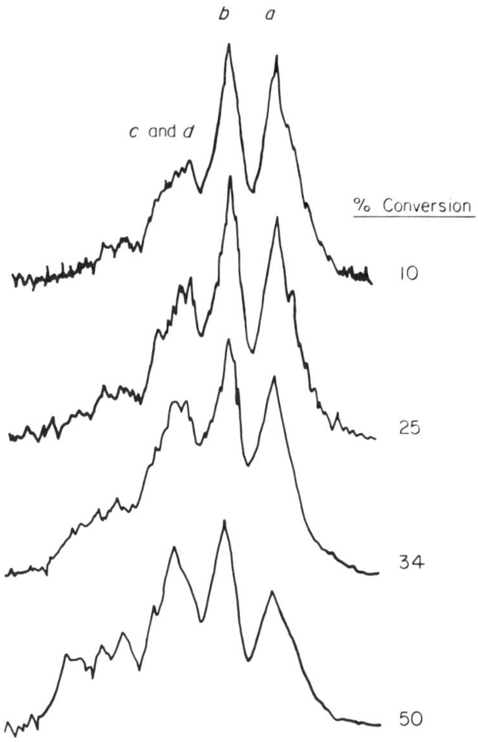


**Figure 34** Methyl proton NMR spectra of dimethylphenyl copolymer at various stages of polymerization; chemical shift values: $a = -0.236$ p.p.m. $b = -0.173$ p.p.m., $c$ and $d = -0.08$ and $-0.02$ p.p.m. (reproduced by permission of the Division of Polymer Chemistry of the American Chemical Society from ref. 197)

ABA = D″D″D″DDDD″D″D″ ABB = D″D″D″DDDDDD

*aba* *bbccdd*

BBB = only *c* and *d* type methyl protons were assumed

conditional probabilities related by

$$P_{A/A} + P_{B/A} = 1 \tag{152}$$

$$P_{A/B} + P_{B/B} = 1 \tag{153}$$

where $P_{X/Y}$ means the probability that a chain ending in Y* will add an X monomer or in other words the probability of occurrence of a YX dyad provided that the event Y* has occurred. Two independent parameters need to be considered to describe the process of copolymerization. These transition probabilities are $P_{B/A}$ and $P_{A/B}$. At this stage one has to define the probability of finding a particular initial state A* or B* as $P_{A^*}$ and $P_{B^*}$. As there are only two possibilities the sum of these is equal to unity

$$P_{A^*} + P_{B^*} = 1 \tag{154}$$

and

$$P_{A^*} = P_{A^*}P_{A/A} + P_{B^*}P_{A/B} \tag{155}$$

$$P_{B^*} = P_{B^*}P_{B/B} + P_{A^*}P_{B/A} \tag{156}$$

The probabilities of finding A and B units can now be expressed in terms of transition probabilities only

$$P_{A^*} = P_{A/B}/(P_{A/B} + P_{B/A}) \tag{157}$$

$$P_{B^*} = P_{B/A}/(P_{A/B} + P_{B/A}) \tag{158}$$

The conditional probabilities give access to the probabilities of occurrence of the 4 dyads

$$P_{AA} = P_{A/A}P_{A^*} = P_{A/B}(1 - P_{B/A})/(P_{A/B} + P_{B/A}) \tag{159}$$

$$P_{AB} + P_{BA} = 2P_{A/B}P_{B^*} = 2P_{A/B}P_{B/A}/(P_{A/B} + P_{B/A}) \tag{160}$$

$$P_{BB} = P_{B/B}P_{B^*} = P_{B/A}(1 - P_{A/B})/(P_{A/B} + P_{B/A}) \tag{161}$$

In the statistical analysis it is essential to get sufficient observations to establish true conformity to a particular model.[194,196] For dyad distribution only two independent observations are available for determining two unknowns, therefore the dyad level is not suited to test this conformity. It is necessary to analyze the copolymer at triad level to get enough degrees of freedom for first-order Markov analysis. The triad distribution is given by six observables with two independent variables $P_{A/B}$ and $P_{B/A}$ and five dependent variables

$$P_{AAA} = P_{A/B}(1-P_{B/A})^2/(P_{A/B}+P_{B/A}) \tag{162}$$

$$P_{AAB}+P_{BAA} = 2P_{A/B}P_{B/A}(1-P_{B/A})/(P_{A/B}+P_{B/A}) \tag{163}$$

$$P_{BAB} = P_{B/A}^2 P_{A/B}/(P_{A/B}+P_{B/A}) \tag{164}$$

$$P_{ABA} = P_{B/A}P_{A/B}^2/(P_{A/B}+P_{B/A}) \tag{165}$$

$$P_{ABB}+P_{BBA} = 2P_{A/B}P_{B/A}(1-P_{A/B})/(P_{A/B}+P_{B/A}) \tag{166}$$

$$P_{BBB} = P_{B/A}(1-P_{A/B})^2/(P_{A/B}+P_{B/A}) \tag{167}$$

Baratova, Mileskevich and Gurari have applied the Price–Markovian statistics to the anionic copolymerization of hexamethylcyclotrisiloxane (HM) with the *cis* and *trans* isomer of 1,3,5-trimethyl-1,3,5-triphenylcyclosiloxane (TMTP)[198] and of the *cis*-TMTP with pentamethylphenylcyclotrisiloxane (PMP) and pentamethylvinylcyclotrisiloxane (PMV).[192] The catalyst used was sodium poly(methylsiloxanediolate). They have determined the kinetic and the probabilistic microstructure parameters of the copolymerization. The monomer reactivity ratios $r_1$ and $r_2$ have been calculated from the differential Mayo–Lewis equation. By independently following the homopropagation reactions, the associated rate constants $k_{AA}$ and $k_{BB}$ are calculated. Thus the rate constants of the cross-propagation reactions $k_{AB}$ and $k_{BA}$ can be deduced. The kinetic data are used to calculate the probabilistic microstructure parameters of the copolymers, *viz.* the probabilities of dyad sequence distribution, the nominal probabilities of transition $P_{A/B}$ and $P_{B/A}$, the coefficient of microheterogeneity (defined as the sum of these transition probabilities) and the number average sequence lengths $l_A$ and $l_B$. The results are reported in Tables 13–16.

As noted above comparison between the theoretical prediction and experimental observation must be undertaken with great care for dyad distribution. Furthermore proton NMR spectra are so complex that fitting theoretical to experimental data could lead to misinterpretation. However, a qualitative description is informative.

Let us consider the *cis*-TMTP–PMP cyclosiloxane pair in detail. It can be seen from the rate constants that *cis*-TMTP cyclosiloxane (A) is preferentially consumed in its homopolymerization ($k_{AA}$) and in the copolymerization ($k_{BA}$) resulting from the attack of a macroanion with a terminal fragment B* or more precisely –D$^P$–D–D*. The less active heterogeneous cyclic monomer (B) is

**Table 13** Parameters of Anionic Homo- and Co-polymerization of Certain Diorganylcyclotrisiloxanes

| *A* | *B* | $r_1$ | $r_2$ |
|---|---|---|---|
| *trans*-TMTP | HM | 70.2 | 0.074 |
| *cis*-TMTP | HM | 59.9 | 0.02 |
| *cis*-TMTP | PMP | $18.6 \pm 0.2$ | $0.22 \pm 0.004$ |
| *cis*-TMTP | PMV | $18.2 \pm 0.2$ | $0.11 \pm 0.003$ |

[a] Catalyst sodium poly(methylphenylsiloxanediolate) (0.01% by weight), 110 + 0.5 °C. Data from refs. 192 and 198.

**Table 14** Rate Constants of Elementary Growth Reactions ($k \times 10^3$, min$^{-1}$)[a]

| *A–B* | $k_{AA}$ | $k_{BA}$ | $k_{BB}$ | $k_{AB}$ |
|---|---|---|---|---|
| | | ($k \times 10^3$, min$^{-1}$) | | |
| *cis*-TMTP–PMP | $16.0 + 0.2$ | $23.0 \pm 0.1$ | $5.1 \pm 0.1$ | $0.9 \pm 0.4$ |
| *cis*-TMTP–PMV | $16.0 \pm 0.2$ | $20.0 \pm 0.1$ | $2.2 \pm 0.1$ | $0.9 \pm 0.4$ |

[a] Data from refs. 192 and 198.

**Table 15** Probabilistic Characteristics of Copolymerization and the Structures of the Macromolecular Chains of the Copolymers[a]

| *A* | *B* | $X_A^0$ | *Conversion* (%) | $P_{AA}$ | $P_{AB}+P_{BA}$ | $P_{BB}$ |
|---|---|---|---|---|---|---|
| *trans*-TMTP | HM | 0.28 | 13.2 | 0.90 | 0.09 | 0.007 |
| *cis*-TMTP | HM | 0.31 | 19.4 | 0.89 | 0.10 | 0.002 |
| *cis*-TMTP | PMP | 0.52 | 11.0 | 0.90 | 0.09 | 0.01 |
| | | 0.51 | 27.0 | 0.87 | 0.11 | 0.02 |
| | | 0.51 | 38.0 | 0.83 | 0.14 | 0.03 |
| *cis*-TMTP | PMV | 0.35 | 30.5 | 0.70 | 0.25 | 0.05 |

[a] $X_A^0$ initial molar fraction of monomer A. $P_{ij}$ probability of occurrence of dyad *ij* in Price–Markov approach. Data from refs. 192 and 198.

**Table 16** $P_{ij}$ Transition Probability (Price–Markov)[a]

| *A–B* | $P_{B/A}$ | $P_{A/B}$ | $K_m$ | $l_A$ | $l_B$ |
|---|---|---|---|---|---|
| *trans*-TMTP–HM | 0.05 | 0.88 | 0.93 | 18.90 | 1.13 |
| *cis*-TMTP–HM | 0.05 | 0.95 | 1.00 | 18.87 | 1.05 |
| *cis*-TMTP–PMP | 0.05 | 0.81 | 0.86 | 21.0 | 1.23 |
| | 0.06 | 0.73 | 0.79 | 17.0 | 1.37 |
| | 0.08 | 0.70 | 0.78 | 13.0 | 1.43 |
| *cis*-TMTP–PMV | 0.15 | 0.71 | 0.86 | 6.7 | 1.40 |

[a] $K_M = P_{A/B} + P_{B/A}$ = microheterogeneity coefficient. $l_i$ number average sequence length (see Table 15). Data from refs. 192 and 198.

more inclined to react with a B* (–D$^P$–D–D*) terminated chain ($k_{BB}$) than with an A* (–D$^P$–D$^P$–D$^{P*}$) terminated chain. The dual role of the electronegative substituent on the silicon atom in the anionic polymerization is well illustrated. The substituent facilitates ring opening by the nucleophilic attack of the catalyst on the substituted silicon atom. However, the newly formed macroanion has a relatively lower nucleophilic activity. Both the nature of the cyclotrisiloxane and of the macroanion have an influence on the whole copolymerization process. However, the type of cyclic is determinant since the overall rate of copolymerization increases on substituting a methyl group by a relatively more electronegative group. Figure 35 and the data collected in Tables 15 and 16 show that for the *cis*-TMTP–PMP cyclotrisiloxanes pair there is a tendency for block copolymer formation. The

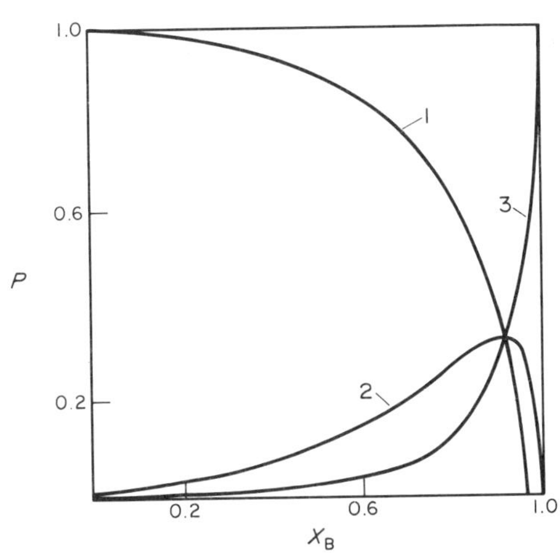


**Figure 35** Instant values of the probabilities of formation of dyads $P_{AA}$ (1), $P_{AB}+P_{BA}$ (2) and $P_{BB}$ (3) as a function of the composition of the comonomer mixture for systems *cis*-TMTP (A)–PMP (B) cyclotrisiloxanes; Price–Markov approach (reproduced from ref. 192)

probability of homosequence (–A–A–) occurrence is relatively more important at the start of the process and then decreases as the monomer is consumed. The probability of transition $P_{A/B}$ decreases with conversion and the related number average sequence length increases. Since A symbolizes a three siloxane unit monomer, the number average sequence length of methylphenylsiloxane block reaches 60 units in the *cis*-TMTP–PMP copolymer. Figure 35 displays the instantaneous probabilities of formation of the dyads AA, AB + BA, BB as a function of the initial composition of the comonomers. The mean values of probabilities of occurrence of these dyads are plotted against conversion percentage (Figure 36).

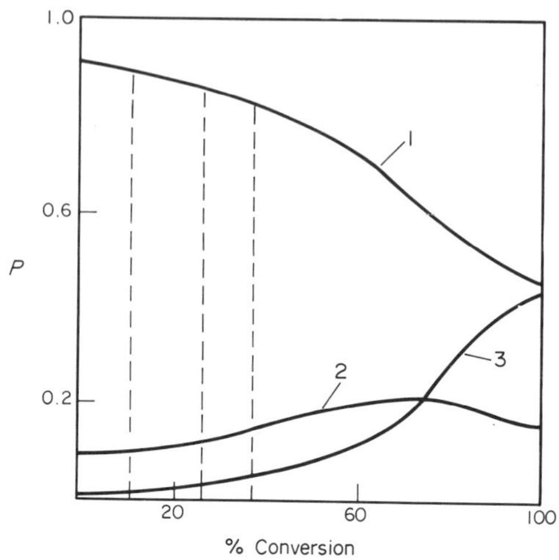


**Figure 36** Mean values of the probability of the formation of dyads $P_{AA}$ (1), $P_{AB}+P_{BA}$ (2) and $P_{BB}$ (3) as a function of conversion for systems *cis*-TMTP (A)–PMP (B) at $X_A^0 = 0.51$; Price–Markov approach (reproduced from ref. 192)

An interesting problem raised in this study concerns the site of ring opening of the cyclotrisiloxane. Three possible variants of ring opening can be considered for the PMP cyclic

$$\text{(I)} \quad -D^P-D-D^* \tag{168}$$

$$\text{(II)} \quad -D-D^P-D^* \tag{169}$$

$$\text{(III)} \quad -D-D-D^{P*} \tag{170}$$

Only the single variant (I) is evidenced by proton NMR, namely the attack by the most active chain end on the most electropositive silicon atom ($-D^{Ph}-$). This confirms the known mechanism of anionic polymerization of cyclic polysiloxanes.[1,192] From these considerations a copolymer with a regular microstructure of mainly $(-D^{Ph}-D-D-)_n$ block may be synthesized. The kinetic data and the Markovian statistics show that pseudo-block copolymers are the result of the copolymerization of the *cis*-TMTP (A) with PMP (B) symbolized in the following scheme

$$\underset{\text{block}}{B\text{–}(B)_z}\text{–}\underset{\text{random}}{(AB)_y}\text{–}\underset{\text{block}}{(A)_x}\text{–}\underset{\text{random}}{(AB)_y}\text{–}\underset{\text{block}}{(B)_z\text{–}B} \tag{171}$$

The central block is not a pure A block as B units are incorporated in the early stages of the process.

### 25.5.5.4 *Average sequence length*

The average sequence length (ASL) is a measure of sequential homogeneity in copolymers. The basic definition of the average sequence length is the total number of siloxane units divided by the total number of sequences. Classically, it is calculated using the assumption that the chains are infinitely long[172,176,199] using probabilistic arguments. The probability of finding exactly $m$ units of A in a row and $n$ units of B in a row is given by

$$P_{Am} = P_{AA}^{m-1} P_{AB} \tag{172}$$

$$P_{Bn} = P_{BB}^{n-1} P_{BA} \tag{173}$$

These probabilities $P_{Am}$ and $P_{Bn}$ are the fractions of all $A_i$ and $B_j$ sequences of lengths $m$ and $n$, respectively, where $i$ and $j$ vary from 1 to $\infty$. These summations are writen as

$$\sum_{m=1}^{\infty} P_{Am} = 1 \quad \text{and} \quad \sum_{n=1}^{\infty} P_{Bn} = 1 \tag{174}$$

The ASL values for infinite chain length ($l_A, l_B$) are given by

$$l_A = \sum_{m=1}^{\infty} m P_{AA}^{m-1} P_{AB} \tag{175}$$

$$l_B = \sum_{n=1}^{\infty} n P_{BB}^{n-1} P_{BA} \tag{176}$$

or

$$l_A = (P_{AB})^{-1} \quad \text{and} \quad l_B = (P_{BA})^{-1} \tag{177}$$

The average sequence length is, as expected, the inverse of the probability of changing the nature of the run $R$. For copolymers made by condensation or by polymerization stopped at low conversion the assumption of infinite chain length no longer holds. Galvan and Tirrell[184] developed an expression for the ASL as a function of the chain length. In their scheme they visualized finite copolymers terminated by four types of chain ends, depending on the initial and final type of unit. Thus chains may be started and ended either by A or B units. They arrive at the following expression for the ASL of A as a function of the chain length

$$\langle \bar{l}_A \rangle = 2/\{[(1 - P_{AB} - P_{BA}L^l + Y_A/T_A)/l] + P_{AB} + P_{BA}L^l\} \tag{178}$$

where $T_A$ is the probability of finding an A unit in a chain of length $l$; $L^l = T_B/T_A$; and $Y_A$ is the probability that the first monomer unit in a chain is an A unit. The bar over $l_A$ means the average and the $\langle \ \rangle$ sign refers to the arithmetical mean. The expression for ASL in chains of infinite length is a particular case of expression (178) where $l_A^{\infty} = l_A$ when $l$ tends to $\infty$ and $L^l$ tends to $P_{AB}/P_{BA}$. This treatment will yield the instantaneous chain length distribution, the relative population of monomer A, and the ASL. For monomer unit A in the copolymer chain these parameters are plotted as a function of chain length $l$ using kinetic data generated from the Ziegler–Natta copolymerization of ethylene and propylene.[174] Three different cases (I, II, III) have been studied, where the values of $X_A$ were respectively 0.5, 0.7, 0.9. It can be seen that in all cases very few long chains are formed. $\langle l_A \rangle$ approaches its asymptotic value at large $l$. Figure 37 shows that when $P_{AA}$ is increased (case III) the chain has to be longer to reach the asymptotic value. Table 17 provides the results for the three cases according to the infinite and finite chain length approaches. The discrepancy between the values determined according to the finite and the infinite chain length conditions is most important when the mole fraction of A is higher than that of B and long runs of a given monomer appear. Up to the present time this approach has not been applied to siloxane copolymers. This microstructure parameter may be obtained experimentally from NMR data. Lee and Marko[197] determined the sequence lengths of diphenylsiloxane units in the poly(dimethyl-*co*-diphenylsiloxane) by analyzing model compounds with proton NMR. The differential sequence distribution and the proton NMR data showed that in the first 20% of conversion, the sequence length of diphenyl and dimethyl-siloxy units are mostly six and three, respectively. Over the last 40% conversion the terminal segments of copolymer chains become pure dimethyl blocks.

**Table 17** Comparison Between ASL Predicted in the Infinite Chain Approximation (ASL–INF) and ASL Predicted in the Finite Chain Case[a]

| *Case* | $X_A$ | *ASL* | *Mean ASL* | *Difference* |
|---|---|---|---|---|
| I | 0.5 | 7.14 | 6.95 | 2.7% |
| II | 0.7 | 15.62 | 14.89 | 4.7% |
| III | 0.9 | 62.50 | 54.87 | 12.2% |

[a] Data reproduced by permission of John Wiley and Sons Inc. from ref. 184.

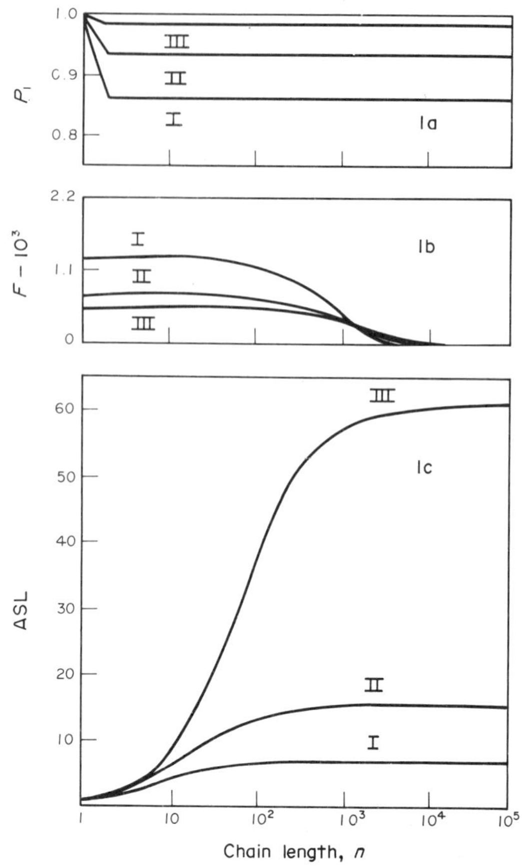


**Figure 37** Average sequence length (ASL) in copolymers as a function of chain length: 1a, monomer A mole fraction $T_A$; 1b, chain length distribution $F$ (*i.e.* relative concentration of chains of lengths $l$); 1c, ASL of A as a function of chain length $l$; (I) $X_A = 0.5$; (II) $X_A = 0.7$; (III) $X_A = 0.9$ (reproduced by permission of John Wiley and Sons Inc. from ref. 184)

### *25.5.5 Determination of copolymer microstructure by analysis of chains at equilibrium*

The microstructure of the siloxane copolymers has so far been deduced from kinetic and probabilistic considerations of the ring-opening polymerization of highly strained cyclosiloxanes under non-rearranging anionic catalysis and under non-equilibrium conditions. The breakthrough leading to the approach where the microstructure could be obtained from a direct analysis of the copolymer regardless of its history came from Jancke, Engelhardt and Kriegsmann who combined $^{29}$Si NMR data and statistical concepts.[200] They used the earlier concept of run number proposed by Harwood and Ritchey.[183] In a copolymer of infinite chain length, the run number $R$ is defined as the average number of monomer sequences ($R$) occurring in a copolymer per 100 monomer units. $R$ can be related to any sequence distribution sensitive physicochemical property.

In Figure 38 a 30 unit section of a cyclic copolymer chain of infinite length is arranged in 15 alternating runs or sequences. Every run of As is followed by an A–B linkage. There are as many A runs as B runs. The $^{29}$Si NMR spectrum of such a copolymer will show two main groups of signals corresponding to A and B siloxane units. Each of these patterns will be a triplet since unit A or B is influenced in three different ways by its first A and B neighbours. These triplets correspond to three different triads. As the magnetic field increases, the influence of the second neighbours can be detected. As a result each of the signals constituting the triplet will be further split into three. The nine signals correspond to a pentad sequence (Figure 39). When the distribution of A and B siloxane units changes from pure random to pure block microstructure or when their molar concentrations are not equal, the relative intensities in the NMR pattern will gradually change from that shown in Figure 39. Furthermore a splitting of signals can arise due to non-additivity of chemical shifts for different arrangements of neighbouring siloxane units in the pentad sequences where the central A is observed, *viz.* AAABB and ABAAB.

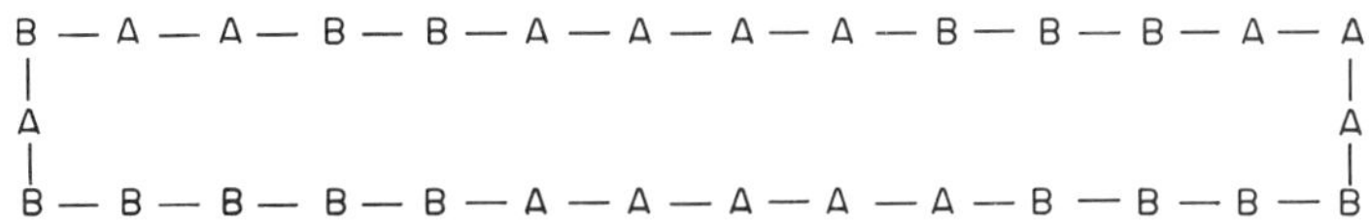

**Figure 38** Schematic representation of runs or sequences for an AB copolymer

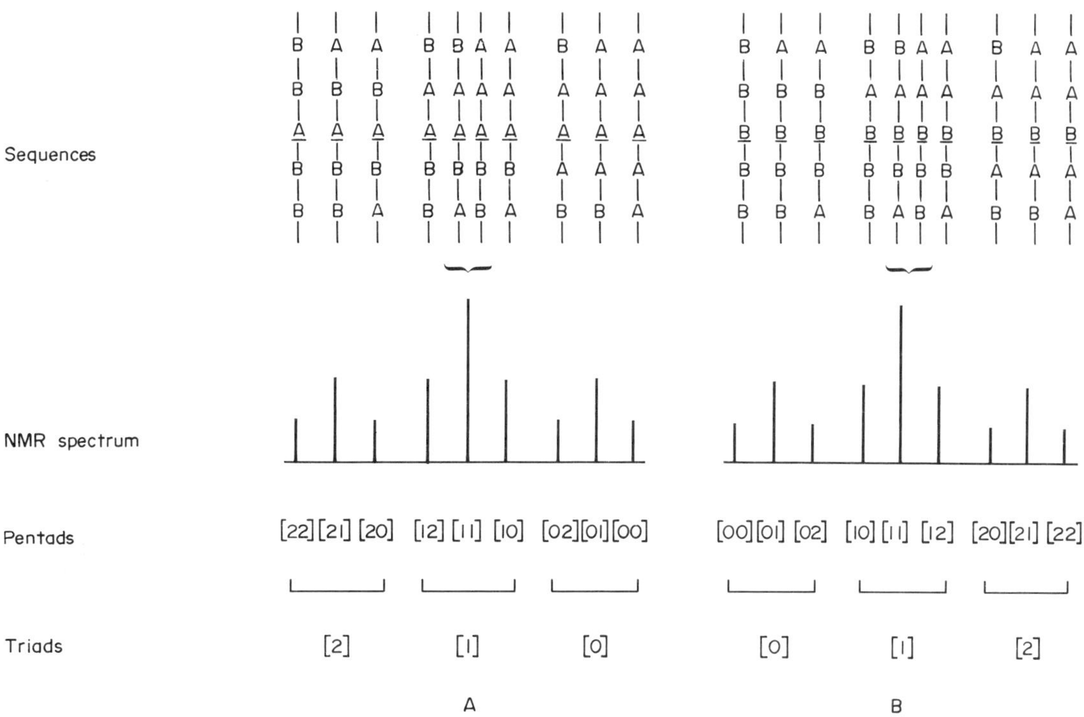


**Figure 39** Pentad sequences and their corresponding A and B $^{29}$Si NMR Signals for a Random $A_xB_x$ copolymer (reproduced by permission of VEB Deutscher Verlag for Grundstoffindustrie from ref. 200)

A $^{29}$Si NMR spectrum of a methylphenylsilicone oil is shown in Figure 40 and illustrates the pattern described above. The first step in the determination of the microstructure that corresponds to this spectrum is to derive the probabilities of occurrence of each signal corresponding either to a triad or to a pentad sequence on either an A or B monomer unit. These probabilities are directly proportional to the NMR signal intensities attributed to these sequences. In the kinetic situation $P_{ij}$ is the instantaneous probability which could be related to the kinetic rate constants. At equilibrium, $P_{ij}$ is the average probability of finding a dyad *ij* in the copolymer chain and in essence is simply the mole fraction of an *i* species in a possible *ij* dyad sequence. According to the theory the probability of finding the triad BAB in the copolymer where the central A is observed is given by

$$P_{\mathrm{BAB}} = P_{\mathrm{AB}}^2 P(\mathrm{A}) \tag{179}$$

Similarly, the pentad sequence ABABB will have a probability of occurrence of

$$P_{\mathrm{ABABB}} = 2P_{\mathrm{AB}}^2 P_{\mathrm{BA}} P_{\mathrm{BB}} P(\mathrm{A}) \tag{180}$$

The coefficient 2 appears as there are two ways of putting A and B at the end of the triad. $P(\mathrm{A})$ is the probability of finding A in the chain and is the mole fraction $X_{\mathrm{A}}$ (reported as a percentage). Table 18 contains the probability expressions for all triad and pentad sequences.

A consequence of the equality of A and B runs is that the percentages of various linkages can be related to the run number and to percentages of A and B units in the copolymer ($100X_{\mathrm{A}}$, $100X_{\mathrm{B}}$). The run number concept shows that

$$\%\,\text{A–A links} = \%\,\text{A} - \%\,\text{AB links} = 100X_{\mathrm{A}} - R/2 \tag{181}$$

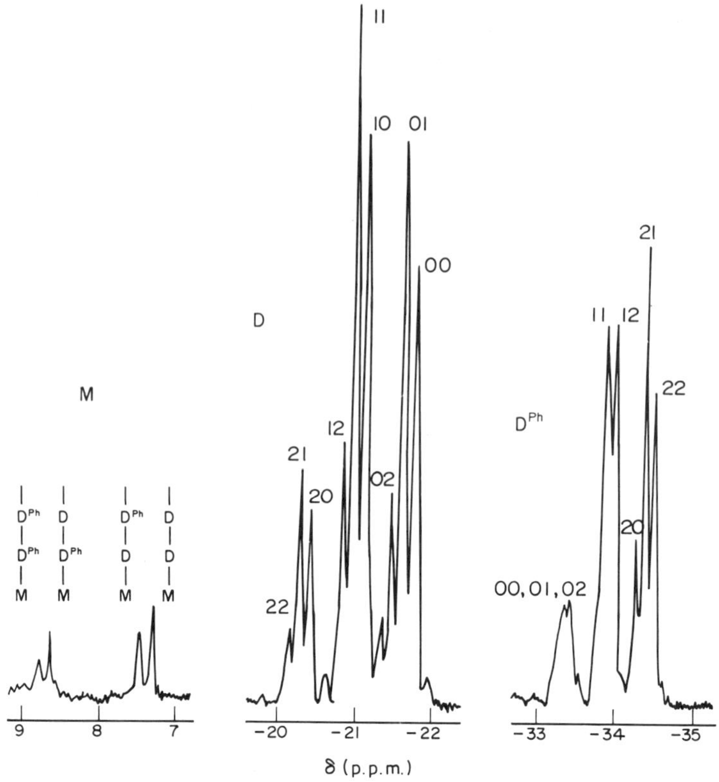


**Figure 40** $^{29}$Si NMR spectrum of a methylphenyl silicone oil. OV-7 (reproduced by permission of Springer Verlag and the authors from ref. 162)

**Table 18** Probabilities of Occurrence of Triad and Pentad Sequences

| *Triad* | *Probability* | *Intensity* | *Pentad* | *Probability* | *Intensity* |
|---|---|---|---|---|---|
| | | | 00 | $P^4_{AA}P(A)$ | $I_{00}$ |
| 0 | $P^2_{AA}P(A)$ | $I_0$ | 01 | $2P^3_{AA}P_{AB}P(A)$ | $I_{01}$ |
| | | | 02 | $P^2_{AA}P^2_{AB}P(A)$ | $I_{02}$ |
| | | | 10 | $2P^2_{AA}P_{AB}P_{BA}P(A)$ | $I_{10}$ |
| | | | 11 | $2P^2_{AA}P_{AB}P_{BB}P(A)$ | $I_{11}$ |
| 1 | $2P_{AB}P_{AA}P(A)$ | $I_1$ | 11 | $2P_{AA}P^2_{AB}P_{BA}P(A)$ | $I_{11}$ |
| | | | 12 | $2P_{AA}P^2_{AB}P_{BB}P(A)$ | $I_{12}$ |
| | | | 20 | $P^2_{AB}P^2_{BA}P(A)$ | $I_{20}$ |
| 2 | $P^2_{AB}P(A)$ | $I_2$ | 21 | $2P^2_{AB}P_{BA}P_{BB}P(A)$ | $I_{21}$ |
| | | | 22 | $P^2_{AB}P^2_{BB}P(A)$ | $I_{22}$ |

$$\text{\% B–B links} = \text{\% B} - \text{\% BA links} = 100X_B - R/2 \tag{182}$$

$$\text{\% (AB + BA) links} = 2(\text{\% AB links}) = 2(\text{\% BA links}) = R \tag{183}$$

The equations that follow give the different probabilities of linkages or dyads in terms of the mole fraction of A and B monomer units and to the run number $R$

$$P_{AA} = \frac{100X_A - R/2}{100X_A} \qquad P_{BB} = \frac{100X_B - R/2}{100X_B} \tag{184}$$

$$P_{AB} = \frac{R}{2X_A 100} \qquad P_{BA} = \frac{R}{2X_B 100} \tag{185}$$

$$\text{where } P_{AA} + P_{AB} = 1 \quad \text{and} \quad P_{BA} + P_{BB} = 1 \tag{186}$$

The relative NMR intensities of triad sequences are functions of the run number and the mole fraction of the A or B siloxy units. Appropriate combinations of different NMR intensities provide expressions that lead to the run number. For example the ratio between intensities of triads [1] and [2] will be

$$\frac{I_1}{I_2} = 2\frac{P_{AA}P_{AB}P(A)}{P_{AB}^2 P(A)} = 2\frac{P_{AA}}{P_{AB}} \tag{187}$$

Substituting the expressions for $P_{AA}$ and $P_{AB}$ from equations (184) and (185) in equation (186), one obtains the expression

$$R = \frac{4I_2}{I_1 + 2I_2} X_A 100 \tag{188}$$

or

$$R = 100 q_A X_A \tag{189}$$

Combinations of the three NMR intensities $I_1$, $I_2$, $I_3$ of the triad will give $q_A$ and similar treatment will give $q_B$. The run number is also given by

$$R = 100 q_B X_B \tag{190}$$

The constants $q_A$ and $q_B$ are calculated with the following expressions

$$\begin{aligned} q_i &= 4I_2/(2I_2 + I_1) \qquad q_i = 2I_1/(2I_0 + I_1) \\ q_i &= 2 - 2(I_0/I_q)^{1/2} \qquad q_i = 2(I_2/I_q)^{1/2} \end{aligned} \tag{191}$$

Considering that the A and B units are statistically distributed along the copolymer chain, the run number will be

$$R_{random} = 200 X_A X_B \tag{192}$$

Thus the experimental run number relative to the random run number will show the average microstructure in terms of its block, random or alternate character. Indeed when $R > R_{random}$, the siloxane comonomer units are found to alternate, whilst when $R < R_{random}$, they are found to agglomerate in blocks. Obviously, when $R = R_{random}$, the two different siloxane units are randomly distributed along the macromolecular chain. Since there are as many A runs as B runs in the copolymer

$$q_A X_A = q_B X_B \tag{193}$$

As $X_A + X_B = 1$, it follows that $X_A$ can be calculated from the appropriate ratio of peak intensities

$$X_A = q_B/(q_A + q_B) \tag{194}$$

The other important microstructure parameter, the number average sequence length (ASL) $l_A$ and $l_B$, can be calculated from this experimental run number and the mole fraction of A and B

$$l_A = 200 X_A/R \quad \text{and} \quad l_B = 200 X_B/R \tag{195}$$

The product of the reactivity ratio under non-equilibrium conditions can also be obtained

$$r_A r_B = (l_A - 1)(l_B - 1) \tag{196}$$

Knowledge of $X_A$, $X_B$ and $R$ thus permits one to calculate a set of microstructure parameters for a given copolymer. The $^{29}$Si NMR analysis of the end groups will provide a description of the three last siloxane units of the copolymer chain and the copolymer microstructure is fully revealed at the molecular level. The microstructure parameters corresponding to the poly(dimethyl-*co*-methylphenylsiloxane) shown in Figure 40 are collected in Table 19.

Finally, stereoisomerism of poly(diorganosiloxane)s $(RR'SiO)_n$ may be expected as a consequence of the pseudo-asymmetry of the silicon atoms. These polysiloxanes could be considered as configur-

**Table 19** Microstructure Parameters of Poly (dimethyl-*co*-methylphenylsiloxane) Determined by $^{29}$Si NMR[a]

| | |
|---|---|
| $X_A = 0.589$ | $X_B = 0.411$ |
| $l_A = 2.4$ | $l_B = 1.6$ |
| $P_{AA} = 0.585$ | $P_{BB} = 0.405$ |
| $P_{AB} = 0.415$ | $P_{BA} = 0.595$ |
| $R_{exp} = 48.9$ | $R_{ran} = 48.4$ |

[a] Average composition $M_2D_{58}D_{42}^{Ph}$.

ational copolymers where the two different siloxane units adopt two different stereochemical configurations.[131] Polymers obtained under non-equilibrium conditions from *cis* and *trans* isomers of 1,3,5-trimethyl-1,3,5-triphenylcyclotrisiloxane[192,198,201] as well as copolymers ($D_x^P D_y$; $x = 1$ or $2, y = 1$, 2 or 3) obtained from heterofunctional condensation of methylphenyldichlorosilane with 1,3-dihydroxy-1,1,3,3-tetramethyldisiloxane[201] have been studied by $^1$H NMR at high frequency. The stereoregularity in these macromolecules can be represented as a set of iso-, syndio- and hetero-tactic triads.[202] This tacticity can be observed by $^1$H high resolution NMR[195] of poly(*cis*-$D_3^P$) and poly(*trans*-$D_3^P$) (Figure 41). The presence of only two signals in the spectrum of poly(*cis*-$D_3^P$) indicates the iso- and hetero-tactic triads, whereas the three signals in the case of poly(*trans*-$D_3^P$) are assigned to the three cases of tactic triads. Thus the degree of stereospecificity of the polymerization of the isomers of cyclotrisiloxanes can be determined.[201]

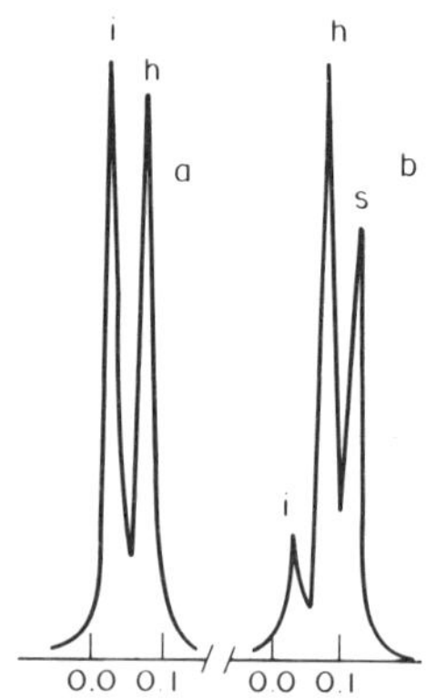


**Figure 41** Proton NMR spectra of the homopolymers of poly(*cis*-$D_3^P$) (a) and poly(*trans*-$D_3^P$) (b) (90 MHz); i = iso, h = hetero, s = syndio triads (reproduced from ref. 198)

The homopolymer poly(methylhydrosiloxane) was first revealed by $^{29}$Si NMR as being a co-polymer where the two different siloxane units adopt two different stereochemical configurations.[131] A fine structure in the $^{29}$Si NMR spectrum is observed due to asymmetric effects. Figure 42 shows that D′ signals of $MD'_5M$ oligomer. The three main signals are assigned to $D_1$, $D_2$, $D_3$. The splitting of $D_1$ into a doublet is due to the possibility of $D_2$ being either (+) or (−) thus giving rise to two different environments for $D_1$. The resonances of $D_2$ and $D_3$ each consist of three lines of 1:2:1 relative intensities ratio, each unit having two possible asymmetric neighbours. These features are still observed in a poly(methylhydrosiloxane) of higher molecular weight $MD_{50}M$ (Figure 42). The relative intensities of the triplet peaks suggest that the polymer is either atactic or that chains of different tacticities are randomly distributed in the mixture.

These configurational copolymers exhibit interesting properties such as no tendency to crystallization[201] and different optical anisotropies as observed for the *cis* and *trans* isomers of the $D_3^P$.

### *25.5.5.6 Correspondence between the kinetic and thermodynamic approaches*

In the first part of this chapter, we have discussed the copolymerization model that leads to a copolymer when starting from different cyclotrisiloxanes. Run numbers of polymer formed in-

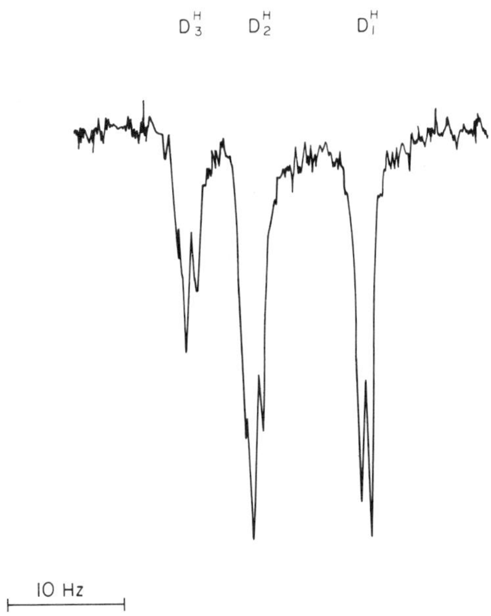


**Figure 42** $^{29}$Si NMR spectrum of $MD_5^H M$ silicone ($D^H$ region); the three groups of lines are due to the three $D^H$ units and the fine splitting due to asymmetry effects (reproduced by permission of the Royal Society of Chemistry from ref. 131)

stantaneously for a given comonomer composition can be calculated.[183] Thus the rate of incorporation of A and B units into the copolymer is

$$d[A+B]/dt = k_{AA}[A^*][A]+k_{AB}[A^*][B]+k_{BA}[B^*][A]+k_{BB}[B^*][B] \tag{197}$$

and the rate of sequences $S$ generated is

$$dS/dt = k_{AB}[A^*][B]+k_{BA}[B^*][A] \tag{198}$$

The ratio of these two expressions in the steady state condition leads to the average number of monomer units per run in the polymerization at a given time

$$d[A+B]/dS = 1+\frac{r_1X_A^0}{2X_B^0}+\frac{r_2X_B^0}{2X_A^0} \tag{199}$$

where $X_I$ refers to the molar fraction of I in the comonomer mixture.
Since $R = 100dS/d[A+B]$, it follows that

$$R = \frac{200}{2+r_1X_A/X_B+r_2X_B/X_A} \tag{200}$$

Finally, the copolymerization can be stopped at any time or allowed to attain thermodynamic equilibrium. The copolymers at equilibrium can then be analyzed. The linear form of equation (200) can be rewritten as

$$2\left(\frac{100}{R}-1\right)\frac{X_B}{X_A} = \frac{X_B^2}{X_A^2}r_2+r_1 \tag{201}$$

This linear relationship yields the 'instantaneous' monomer reactivity ratios.

From the kinetic model, linkage probabilities were related to monomer reactivity ratios and feed composition data.[166] These probabilities were also determined from sequence distribution analysis of the copolymer at equilibrium. The comparison shows the possibility of determining 'instantaneous' monomer reactivity ratios from a knowledge of the compositions and run numbers of instantaneous polymers formed at various feed compositions.[175] Table 20 shows the linkage probabilities according to the kinetic and equilibrium models.

**Table 20** Linkage Probabilities for Dyads Determined from the Kinetic Model and from the Microstructure of the Copolymer at Equilibrium[a]

$$P_{AA} = \frac{r_1 X_A}{r_1 X_A + X_B} = \frac{X_A - R/200}{X_A}$$

$$P_{AB} = \frac{X_B}{r_1 X_A + X_B} = \frac{R/200}{X_A} = /l_A$$

$$P_{BB} = \frac{r_2 X_B}{X_A + r_2 X_B} = \frac{X_B - R/200}{X_B}$$

$$P_{BA} = \frac{X_A}{X_A + r_2 X_B} = \frac{R/200}{X_B} = /l_B$$

[a] Data reproduced by permission of John Wiley and Sons Inc. from ref. 183.

### 25.5.6 Summary

In the case of copolymerization using catalysts that do not rearrange the siloxane chain, and under non-equilibrium conditions, a theoretical model can be developed to describe the microstructure. This same microstructure can be analyzed by $^1H$ or $^{29}Si$ NMR. The concordance between theory and experiment enables confirmation of the correctness of either the theoretical mathematical model or the assignment of NMR signals. High quality quantitative information on the microstructure of copolymers as a function of time or at any conversion or for any comonomer composition can be obtained. For copolymerization under less precise conditions, the microstructure can be derived from the siloxane copolymer itself by direct observation of the magnetic resonance signals of $^{29}Si$ nuclei.

The theoretical formalisms and the characterizations of both macrostructure and microstructure provide the most complete description of the copolysiloxane macromolecules. This will lead to a better fundamental understanding of copolymerization processes and of the mechanism of polymerization at the molecular level. An in-depth study of the effect of reaction conditions on the sequence distribution becomes feasible. Finally the microstructure quantitative parameters allow correlation between microstructure and properties, which is extremely important from an industrial point of view.

## 25.6 REFERENCES

1. M. G. Voronkov, V. P. Mileshkevich and Yu. A. Yushelevskii, 'The Siloxane Bond', The Consultants Bureau, New York, 1974, chap. 1.
2. W. T. Grubb and R. C. Osthoff, *J. Am. Chem. Soc.*, 1955, **77**, 1405.
3. D. W. Scott, *J. Am. Chem. Soc.*, 1946, **68**, 2294.
4. J. B. Carmichael and R. Winger, *J. Polym. Sci., Part A*, 1965, **3**, 971.
5. J. F. Brown, Jr. and G. M. J. Slusarczuk, *J. Am. Chem. Soc.*, 1965, **87**, 931.
6. H. Jacobson and W. H. Stockmayer, *J. Chem. Phys.*, 1950, **18**, 1600.
7. P. J. Flory, 'Principles of Polymer Chemistry', Cornell University Press, New York, 1953, chaps. 8 and 10.
8. J. A. Semlyen and P. V. Wright, *Polymer*, 1969, **10**, 543.
9. P. V. Wright and J. A. Semlyen, *Polymer*, 1970, **11**, 462.
10. M. S. Beevers and J. A. Semlyen, *Polymer*, 1972, **13**, 385.
11. P. J. Flory, V. Crescenzi and J. E. Mark, *J. Am. Chem. Soc.*, 1964, **86**, 146.
12. L. E. Scales and J. A. Semlyen, *Polymer*, 1976, **17**, 601.
13. P. V. Wright, in 'Ring Opening Polymerization', ed. K. J. Ivin and T. Segusa, Elsevier, London, 1984, vol. 2.
14. P. V. Wright and M. S. Beevers, in 'Cyclic Polymers', ed. J. A. Semlyen, Elsevier, London, 1986.
15. J. F. Hyde, *US Pat.* 2 490 357 (1949) (*Chem. Abstr.*, 1950, **44**, 4490d).
16. M. Morton, M. A. Deisz and E. E. Bostick, *J. Polym. Sci., Part A*, 1964, **2**, 513.
17. M. Shinohara, *Polym. Prepr., Am. Chem. Soc., Div. Polym. Chem.*, 1973 **14** (2), 1209.
18. J. Chojnowski and M. Mazurek, *Makromol. Chem.*, 1975, **176**, 2999.
19. C. A. Kraus, *J. Phys. Chem.*, 1956, **60**, 129.
20. Yu. A. Yuzhelevskii, E. G. Kagan and N. N. Fedoseeva, *Dokl. Akad. Nauk SSSR*, 1970, **190**, 647.
21. Von. H. Schmidbaur, J. A. Perez-Garcia and H. S. Arnold, *Z. Anorg. Allg. Chem.*, 1964, **328**, 105.
22. W. S. Tatlock and E. G. Rochow, *J. Org. Chem.*, 1952, **17**, 1555.

23. H. Schmidbaur and S. Waldmann, *Angew. Chem., Int. Ed. Engl.*, 1964, **3**, 697.
24. R. H. Baney, O. K. Johannson, F. M. Koski and G. E. Vogel (Dow Corning Corp.), unpublished work.
25. D. T. Hurd, R. C. Osthoff and M. L. Corrin, *J. Am. Chem. Soc.*, 1954, **76**, 249.
26. M. Szwarc, 'Carbanions, Living Polymers and Electron Transfer Processes', Wiley, New York, 1968, chaps. 5 and 8.
27. M. Morton and E. E. Bostick, *J. Polym. Sci., Part A*, 1964, **2**, 523.
28. Yu. A. Yuzhelevskii, Ye. G. Kagan, N. P. Timofeyeva, T. D. Doletskaya and A. L. Klebanskii, *Vysokomol. Soedin., Ser. A*, 1971, **13**, 183.
29. D. R. Thomas, private communication.
30. A. J. Parker, *Q. Rev., Chem. Soc.*, 1962, **16**, 163.
31. K. A. Andrianov, S. Ye. Yakushkina and N. N. Terent'eva, *Vysokomol. Soedin., Ser. A*, 1968, **10**, 1721.
32. Yu. A. Yuzhelevskii, Ye. G. Kagan, E. V. Kogan, A. L. Klebanskii and N. N. Nikiforova, *Vyskomol. Soedin., Ser. A*, 1969, **11**, 1539.
33. Yu. A. Yuzhelevskii, E. B. Dmokhovskaya, A. L. Klebanskii and N. V. Kozlova, *Vysokomol. Soedin., Ser. A*, 1969, **11**, 432.
34. V. M. Kopylov, P. L. Prikhod'ko and V. A. Kovyazin, *Vysokomol. Soedin., Ser. A*, 1982, **24**, 1751.
35. K. A. Andrianov, V. A. Temnikovskii, L. M. Khananashvili and N. A. Lyapina, *Vysokomol. Soedin., Ser. A*, 1972, **14**, 2235.
36. M. Mazurek and J. Chojnowski, *Macromolecules*, 1978, **11**, 347.
37. V. M. Kopylov, P. L. Prikhod'ko, V. A. Kovyazin and I. L. Dubchak, *Vysokomol. Soedin., Ser. A*, 1984, **26**, 1882.
38. J. Chojnowski and J. Zietera, *Bull. Acad. Pol. Sci., Ser. Sci. Chim.*, 1979, **27**, 555.
39. M. Mazurek and J. Chjonowski, *Makromol. Chem.*, 1977, **178**, 1005.
40. N. K. Gladkova, S. G. Durgar'yan and N. S. Nametkin, *Dokl. Akad. Nauk SSSR*, 1979, **244**, 86.
41. C. L. Lee, C. L. Frye and O. K. Johannson, *Polym. Prepr., Am. Chem. Soc., Div. Polym. Chem.*, 1969, **10**, 1361.
42. H. J. Hölle and B. R. Lehen, *Eur. Polym. J.*, 1975, **11**, 663.
43. S. Boileau, *ACS Symp. Ser.*, 1985, **286**, 23.
44. K. A. Andrianov, V. M. Kopylov, V. A. Temnikovskii and L. M. Khananashvili, *Vysokomol. Soedin., Ser. A*, 1976, **18**, 1714.
45. M. Kucera, M. Jelinek, J. Lanikova and K. Vesely, *J. Polym. Sci.*, 1961, **53**, 311.
46. K. A. Andrianov, V. M. Kopylov, E. I. Khubulava, L. M. Khananashvili, A. I. Nogaideli and T. E. Stakhrovskaya, *Dokl. Akad. Nauk SSSR*, 1977, **233**, 854.
47. Z. Laita and M. Jelinek, *Vysokomol. Soedin.*, 1962, **4**, 1739.
48. V. P. Davydova, Z. S. Lebedeva and I. A. Gryaznykh, *Vysokomol. Soedin., Ser. A*, 1967, **9**, 2226.
49. M. Mazurek, M. Scibiorek, J. Chojnowskii, B. G. Zavin and A. A. Zhdanov, *Eur. Polym. J.*, 1980, **16**, 57.
50. B. G. Zavin, A. A. Zhdanov, M. Scibiorek and J. Chojnowski, *Eur. Polym. J.*, 1985, **21**, 135.
51. W. A. Fessler and P. C. Juliano, *Polym. Prepr., Am. Chem. Soc., Div. Polym. Chem.*, 1971, **12**, 150.
52. W. Patnode and D. F. Wilcock, *J. Am. Chem. Soc.*, 1946, **68**, 358.
53. R. R. McGregor and E. Warrick, *US Pat.* 2 437 204 (1948) (*Chem. Abstr.*, 1948, **42**, 5265h).
54. D. T. Hurd, *J. Am. Chem. Soc.*, 1955, **77**, 2998.
55. T. C. Kendrick, *J. Chem. Soc.*, 1965, 2027.
56. S. W. Kantor, W. T. Grubb and R. C. Osthoff, *J. Am. Chem. Soc.*, 1954, **76**, 5190.
57. W. J. Noll, 'Chemistry and Technology of Silicones', Academic Press, New York, 1968.
58. K. A. Andrianov, M. I. Shkol'nik, V. M. Kopylov and N. Bravina, *Vysokomol. Soedin., Ser. B*, 1974, **16**, 893.
59. S. Schindler and K. Ruehlmann, *Plaste Kautsch.*, 1978, **25**, 384.
60. N. N. Baglei and M. T. Bryk, *Ukr. Khim. Zh.*, 1981, **47** (4), 409.
61. I. B. Rashkov, I. Gitsov and I. Panaiotov, *Polym. Bull. (Berlin)*, 1983, **10** (11–12), 487.
62. I. B. Rashkov and I. Gitsov, *J. Polym. Sci., Polym. Chem. Ed.*, 1986, **24**, 155.
63. A. S. Chawla and L. E. St. Pierre, *J. Polym. Sci.*, 1972, **10**, 2691.
64. J. J. Lebrun, A. Deffieux, P. Sigwalt, A. Wang and V. Stannett, *Radiat. Phys. Chem.*, 1984, **24**, 239.
65. I. Yilgor and J. E. McGrath, *Polym. Prepr., Am Chem. Soc., Div. Polym. Chem.*, 1985, **26**, 57.
66. J. E. McGrath, J. S. Riffle, A. K. Banthia, I. Yilgor and G. L. Wilkes, *ACS Symp. Ser.*, 1983, **212**, 145.
67. L. Wilczek, S. Rubinsztajn and J. Chojnowski, *Makromol. Chem.*, 1986, **187**, 39.
68. G. Sauvet, J. J. Lebrun and P. Sigwalt, 'Cationic Polymerization and Related Processes', ed. E. J. Goethals, Academic Press, London, 1984, p. 237.
69. K. A. Andrianov and S. E. Yakushkina, *Vysokomol. Soedin.*, 1959, **1**, 613; 1960, **2**, 1508.
70. E. V. Kogan, A. G. Ivanova, V. O. Reikhsfel'd, N. I. Smirnov and V. N. Gruber, *Vysokomol. Soedin.*, 1963, **5**, 1183 (*Polym. Sci. USSR, Engl. Transl.*), 1963, **5**, 249.
71. P. Sigwalt, *Actual. Chim.*, March, 1986, 45.
72. J. Chojnowski, S. Rubinsztajn and L. Wilczek, *Actual. Chim.*, March, 1986, 56.
73. J. Chojnowski, M. Mazurek, M. Scibiorek and L. Wilczek, *Makromol. Chem.*, 1974, **175**, 3299.
74. J. Chojnowski and M. Scibiorek, *Makromol. Chem.*, 1976, **177**, 1413.
75. J. Chojnowski, M. Scibiorek and J. Kowalski, *Makromol. Chem.*, 1977, **178**, 1351.
76. J. Chojnowski and L. Wilczek, *Makromol. Chem.*, 1979, **180**, 117.
77. M. Scibiorek and J. Chojnowski, *Eur. Polym. J.*, 1981, **17**, 413.
78. L. Wilczek and J. Chojnowski, *Macromolecules*, 1981, **14**, 9.
79. J. Chojnowski and Z. Lasocki, *Proc. IUPAC Macromol. Symp.*, 28th, 1982, 118.
80. J. Chojnowski, L. Wilczek and S. Rubinsztajn, 'Cationic Polymerization and Related Processes', ed. E. J. Goethals, Academic Press, London, 1984, p. 253.
81. J. Chojnowski, S. Rubinsztajn and L. Wilczek, *J. Chem. Soc., Chem. Commun.*, 1984, 69.
82. L. Wilczek and J. Chojnowski, *Makromol. Chem.*, 1983, **184**, 77.
83. J. J. Lebrun, G. Sauvet and P. Sigwalt, *Makromol. Chem., Rapid Commun.*, 1982, **3**, 757.
84. W. A. Piccoli, G. G. Haberland and R. L. Merker, *J. Am. Chem. Soc.*, 1960, **82**, 1883.
85. J. B. Carmichael and J. Heffel, *J. Phys. Chem.*, 1965, **69**, 2213.
86. J. F. Hampton, *US Pat.* 3 465 016 (1969) (*Chem. Abstr.*, 1969, **71**, 81 517q).

87. V. Crescenzi and P. J. Flory, *J. Am. Chem. Soc.*, 1964, **86**, 141.
88. P. J. Flory and J. A. Semlyen, *J. Am. Chem. Soc.*, 1966, **88**, 3209.
89. R. W. Kilb, *J. Phys. Chem.*, 1958, **62**, 969.
90. N. Goodman and H. Morawetz, *J. Polym. Sci., Part C*, 1970, **31**, 177.
91. R. F. T. Stepto and D. R. Waywell, *Makromol. Chem.*, 1972, **152**, 263.
92. M. Gordon and W. B. Temple, *Makromol. Chem.*, 1972, **152**, 277.
93. Z. Lasocki, J. Kulpinski and W. Gador, *Polimery (Warsaw)*, 1970, **15**, 442.
94. D. Souverain, A. Leborgne, G. Sauvet and P. Sigwalt, *Eur. Polym. J.*, 1980, **16**, 861.
95. R. J. P. Corriu and M. Henner, *J. Organomet. Chem.*, 1974, **74**, 1.
96. J. B. Lambert, J. A. McConnell and W. J. Schulz, Jr., *J. Am. Chem. Soc.*, 1986, **108**, 2482; W. J. Schulz, private communication.
97. G. K. S. Prakash, S. Kenyaniyan, R. Aniszfeld, L. Heiliger, G. A. Olah, R. C. Stevens, H. K. Choi and R. Beau, *J. Am. Chem. Soc.*, 1987, **109**, 5123.
98. A. L. Smith, 'Analysis of Silicones', Wiley, New York, 1974, vol. 41.
99. A. C. Ouano, E. M. Barrall, II, J. F. Johnson, 'Techniques and Methods of Polymer Evaluations', ed. P. E. Slade, Jr, 1975, vol. 4, part II.
100. N. C. Billingham, 'Molar Mass Measurement in Polymer Science', Kogan Page, London, 1977.
101. R. R. Vrijbergen, A. A. Soeteman and J. A. M. Smit, *J. Appl. Polym. Sci.*, 1978, **22**, 1267.
102. C. J. B. Dobbin, A. Rudin and M. F. Tchir, *J. Appl. Polym. Sci.*, 1980, **25**, 2985.
103. K. K. Chee, *Polymer*, 1985, **30**, 4137.
104. D. C. Bugada and A. Rudin, *J. Appl. Polym. Sci.*, 1985, **30**, 4137.
105. H. A. Andreetta, I. H. Sorokin and R. V. Figini, *Makromol. Chem., Rapid. Commun.*, 1985, **6**, 419.
106. Z. Grubisic, P. Rempp and H. Benoit, *J. Polym. Sci., Part B*, 1967, **5**, 753.
107. H. Kh. Mahabadi and A. Rudin, *Polym. J.*, 1979, **11**, 123.
108. W. L. Elsdon, J. M. Goldwasser and A. Rudin, *J. Polym. Sci., Polym. Lett. Ed.*, 1981, **19**, 483.
109. J. M. Goldwasser and A. Rudin, *J. Liq. Chromatogr.*, 1983, **6**, 2433.
110. V. Grinshpun and A. Rudin, *J. Appl. Polym. Sci.*, 1986, **32**, 4303.
111. R. W. La Rochelle, J. D. Cargioli and E. A. Williams, *Macromolecules*, 1976, **9** (1), 85.
112. D. Butler and B. Parbhoo (Dow Corning Ltd.), unpublished results.
113. J. E. Herz, A. Belkebir-Mrani and P. Rempp, *Eur. Polym. J.*, 1973, **9**, 1165.
114. F. A. Bovey, 'High Resolution NMR of Macromolecules', Academic Press, New York, 1972.
115. W. Kemp, 'Qualitative Organic Analysis', McGraw-Hill, London, 1986.
116. R. S. McDonald, *Anal. Chem.*, 1986, **58**, 1906.
117. G. W. Griffith, *Ind. Eng. Chem., Prod. Res. Dev.*, 1984, **23**, 590.
118. E. D. Lipp, *Appl. Spectrosc.*, 1986, **40**, 1009.
119. P. J. Madec and E. Maréchal, *J. Polym. Sci., Polym. Phys. Ed.*, 1980, **18**, 2417.
120. A. L. Smith and N. C. Angelotti, *Spectrochim. Acta*, 1959, **15**, 412.
121. J. K. Kauppinen, D. J. Moffatt, H. H. Mantsch and D. G. Cameron, *Appl. Spectrosc.*, 1981, **35**, 271.
122. J. K. Kauppinen, D. J. Moffatt, H. H. Mantsch and D. G. Cameron, *Appl. Opt.*, 1981, **20**, 1866.
123. J. K. Kauppinen, D. J. Moffatt, H. H. Mantsch and D. G. Cameron, *Anal. Chem.*, 1981, **53**, 1454.
124. E. Liepins, I. Zicmane and E. Lukevics, *J. Organomet. Chem.*, 1986, **306**, 167.
125. W. Simmler, H. Walz and H. Niederprum, *Chem. Ber.*, 1963, **96**, 1495.
126. D. A. Laude and C. L. Wilkins, *Macromolecules*, 1986, **19**, 2295.
127. R. K. Harris and B. J. Kimber, *Appl. Spectrosc. Rev.*, 1975, **10** (1), 117.
128. R. K. Harris, in 'NMR and the Periodic Table', ed. A. K. Harris and E. B. Mann, Applied Science Publishers, London, 1978, p. 309.
129. E. A. Williams and D. J. Cargioli, *Annu. Rep. NMR Spectrosc.*, 1979, **9**, 221.
130. E. A. Williams, *Annu. Rep. NMR Spectrosc.*, 1983, **15**, 235.
131. R. K. Harris and B. J. Kimber, *J. Chem. Soc., Chem. Commun.*, 1974, 559.
132. E. A. Williams, J. D. Cargioli and R. W. Larochelle, *J. Organomet. Chem.*, 1976, **108**, 153.
133. R. K. Harris and B. J. Kimber, *J. Organomet. Chem.*, 1974, **70**, 43.
134. J. M. Dereppe and B. Parbhoo, *Anal. Chem.*, 1986, **58**, 2641.
135. J. Schraml, V. Blechta, M. Kvicalova, L. Nondek and V. Chvalovsky, *Anal. Chem.*, 1986, **58**, 1892.
136. G. A. Morris and R. Freeman, *J. Am. Chem. Soc.*, 1979, **101**, 760.
137. D. M. Doddrell, D. T. Pegg, W. Brooks and M. R. Bendall, *J. Am. Chem. Soc.*, 1981, **103**, 727.
138. D. A. Laude, Jr., R. W. K. Lee and C. L. Wilkins, *J. Magn. Reson.*, 1984, **60**, 453.
139. D. A. Laude, Jr., R. W. K. Lee and C. L. Wilkins, *Anal. Chem.*, 1985, **57**, 1281.
140. D. A. Laude, Jr., R. W. K. Lee and C. L. Wilkins, *Anal. Chem.*, 1985, **57**, 1286.
141. R. K. Harris, B. J. Kimber, M. D. Wood and A. Holt, *J. Organomet. Chem.*, 1976, **116**, 291.
142. R. K. Harris and M. L. Robins, *Polymer*, 1978, **19**, 1123.
143. D. J. Burton, R. K. Harris, K. Dodgson, C. J. Pellow and J. A. Semlyen, *Polym. Commun.*, 1983, **24**, 278.
144. E. A. Williams, J. D. Cargioli and P. E. Donahue, *J. Organomet. Chem.*, 1980, **192**, 319.
145. Y.-M. Pai, W. P. Weber and K. L. Servis, *J. Organomet. Chem.*, 1985, **288**, 269.
146. G. Engelhardt, H. Jancke, M. Magi, T. Pehk and E. Lippmaa, *J. Organomet. Chem.*, 1971, **28**, 293.
147. H. G. Horn and H. C. Marsmann, *Makromol. Chem.*, 1972, **162**, 255.
148. G. Engelhardt, H. Jancke, E. Lippmaa and A. Samoson, *J. Organomet. Chem.*, 1981, **210**, 295.
149. E. T. Lippmaa, M. Alla, T. J. Pehk and G. Engelhardt, *J. Am. Chem. Soc.*, 1978, **100**, 1929.
150. E. T. Lippmaa, M. Magi, A. Samoson, G. Engelhardt and A. R. Grimmer, *J. Am. Chem. Soc.*, 1980, **102**, 4889.
151. T. Saluvere, J. Puskar, J. Past, E. T. Lippmaa and G. Engelhardt, *Proc. Congr. Amperè 19th*, 1976, 336.
152. L. W. Jelinski, J. J. Dumais, P. I. Watnick, S. V. Bass and L. Shepherd, *J. Polym. Sci., Polym. Chem. Ed.*, 1982, **20**, 3285.
153. H. Jancke, G. Engelhardt and H. Grosse-Ruyken, *J. Organomet. Chem.*, 1983, **247**, 139.
154. G. E. Maciel, M. J. Sullivan and D. W. Sindorf, *Macromolecules*, 1981, **14**, 1607.
155. R. A. Newmark and B. C. Copley, *Macromolecules*, 1984, **17**, 1973.
156. B. C. Copley, *Org. Coat. Plast. Chem.*, 1983, **48**, 121.

157. K. Beshah, J. E. Mark, A. Himstedt and J. L. Ackerman, *Polym. Prepr., Am. Chem. Soc., Div. Polym. Chem.*, 1985, **26** (2), 52.
158. K. Beshah, J. E. Mark, J. L. Ackerman and A. Himstedt, *J. Polym. Sci., Polym. Phys. Ed.*, 1986, **24**, 1207.
159. K. Beshah, J. E. Mark and J. L. Ackerman, *Macromolecules*, 1986, **19**, 2194.
160. R. G. W. Norrish and E. F. Brookman, *Proc. R. Soc. London, Ser. A*, 1937, **163**, 205.
161. R. G. W. Norrish and E. F. Brookman, *Proc. R. Soc. London, Ser. A*, 1939, **171**, 147.
162. G. Engelhardt and H. Jancke, *Polym. Bull. (Berlin)*, 1981, **5**, 577.
163. F. T. Wall, *J. Am. Chem. Soc.*, 1940, **62**, 803.
164. F. T. Wall, *J. Am. Chem. Soc.*, 1941, **63**, 1862.
165. F. T. Wall. *J. Am. Chem. Soc.*, 1944, **66**, 2050.
166. F. R. Mayo and F. M. Lewis, *J. Am. Chem. Soc.*, 1944, **66**, 1594.
167. T. Alfrey and G. Goldfinger, *J. Chem. Phys.*, 1944, **12**, 205.
168. G. Goldfinger and T. Kane, *J. Polym. Sci.*, 1948, **3**, 462.
169. M. Fineman and S. D. Ross, *J. Polym. Sci.*, 1950, **5**, 259.
170. C. C. Price, *J. Polym. Sci.*, 1946, **1**, 83.
171. E. Merz, T. Alfrey and G. Goldfinger, *J. Polym. Sci.*, 1946, **1**, 75.
172. M. Szwarc, M. Levy and R. Milkovitch, *J. Am. Chem. Soc.*, 1956, **78**, 2656.
173. M. Szwarc, *Nature (London)*, 1956, **178**, 1168.
174. T. Alfrey, Jr., J. Bohrer and H. Mark, 'Copolymerization', Interscience, New York, 1952.
175. I. Skeist, *J. Am. Chem. Soc.*, 1946, **68**, 1781.
176. R. Simha and H. Branson, *J. Chem. Phys.*, 1944, **12**, 253.
177. W. H. Stockmayer, *J. Chem. Phys.*, 1945, **13**, 199.
178. R. L. Miller and L. E. Nielsen, *J. Polym. Sci.*, 1960, **46**, 303.
179. A. Miyake and R. Chujo, *J. Polym. Sci.*, 1960, **46**, 163.
180. F. P. Price, *J. Chem. Phys.*, 1962, **36**, 209.
181. B. D. Coleman and T. G. Fox, *J. Polym. Sci., Part A*, 1963, **1**, 3183.
182. W. Ring, *J. Polym. Sci., Part B*, 1963, **1**, 323.
183. H. J. Harwood and W. M. Ritchey, *J. Polym. Sci., Part B*, 1964, **2**, 601.
184. R. Galvan and M. Tirrell, *J. Polym. Sci., Polym. Chem. Ed.*, 1986, **24**, 803.
185. G. E. Ham, 'Copolymerization', Interscience, New York, 1964, vol. XVIII.
186. C. S. Marvel, G. D. Jones, T. W. Mastin and G. L. Schertz, *J. Am. Chem. Soc.*, 1942, **64**, 2356.
187. T. Alfrey, Jr., H. C. Haas and C. W. Lewis, *J. Am. Chem. Soc.*, 1951, **73**, 2851.
188. E. E. Bostick, *Polym. Prepr., Am. Chem. Soc., Div. Polym. Chem.*, 1965, **10**, 877.
189. H. R. Allcock, 'Contemporary Polymer Chemistry', F. W. Lampe, Prentice Hall, Englewood Cliffs, NJ, 1981.
190. R. L. Merker, M. J. Scott and G. G. Haberland, *J. Polym. Sci., Part A*, 1964, **2**, 31.
191. F. A. Bovey, *Acc. Chem. Res.*, 1968, **1**, 175.
192. T. N. Baratova, V. P. Mileshkevich and V. E. Gurari, *Polym. Sci. USSR (Engl. Transl.)*, 1983, **25**, 2899.
193. R. L. Merker, *Br. Pat.* 916 135 (1963) (*Chem. Abstr.*, 1963, **59**, 10 120h).
194. J. C. Randall, 'Polymer Sequence Determination. C-13 NMR Method', Academic Press, New York, 1977.
195. J. L. Koenig, 'Chemical Microstructure of Polymer Chains', Wiley, New York, 1980.
196. G. S. Georgiev, *Polym. Bull. (Berlin)*, 1983, **10**, 527.
197. C. L. Lee and O. W. Marko, *Polym. Prepr., Am. Chem. Soc., Div. Polym. Chem.*, 1978, **19**, 250.
198. V. P. Mileshkevich, T. N. Baratova and V. E. Gurari, *Polym. Sci. USSR (Engl. Transl.)*, 1982, **24**, 27.
199. K. A. Andrianov, B. G. Zavin and G. F. Sablina, *Polym. Sci. USSR (Engl. Transl.)*, 1979, **20**, 1240.
200. H. Jancke, G. Engelhardt, H. Kriegsmann and F. Keller, *Plaste Kautsch*, 1979, **26**, 612.
201. L. S. Bresler, V. P. Mileshkevich, Yu. A. Yuzhelevskii and N. P. Timofeeva, *Zh. Strukt. Khim.*, 1978, **19**, 453.
202. K. A. Andrianov, S. E. Yakushkina, V. S. Svistunov and B. A. Lavrukhin, *Dokl. Akad. Nauk SSSR*, 1969, **188**, 345.
203. M. Kucera, *Khim. Prakt. Primen. Kremneorg. Soedin. Tr. Konf., Leningrad*, 1958, **6**, 194.
204. M. Kucera, *Collect. Czech. Chem. Commun.*, 1960, **25**, 547.
205. M. Kucera and M. Jelinek, *Vysokomol. Soedin.*, 1960, **2**, 1860.
206. M. Kucera and M. Jelinek, *Collect. Czech. Chem. Commun.*, 1960, **25**, 536.
207. M. Kucera, *J. Polym. Sci.*, 1962, **58**, 1263.
208. J. Laita, P. Hlozek, B. Bucek and M. Jelinek, *J. Polym. Sci., Part C*, 1967, **16**, 669.

# 26

# Phosphazene High Polymers

HARRY R. ALLCOCK
*The Pennsylvania State University, University Park, PA, USA*



## 26.1 INTRODUCTION

Most known polymers are organic macromolecules derived either from living organisms or from petrochemical monomers. However, in principle, many elements other than carbon could form part of a macromolecular backbone, and the incorporation of 'inorganic' elements into a polymer structure should broaden the range of properties accessible in polymeric substances.[1]

The first indications of what was possible in this area came from the discovery and development of the poly(organosiloxanes) ('silicones') in the 1940s, and from the existence since the late 1800s of a rubbery, laboratory curiosity of formula $(NPCl_2)_n$.[2] This latter species was the first known polyphosphazene. Although it was too unstable to moisture to allow detailed study or application, its rubbery properties belied the conventional wisdom that organic polymers are flexible, tough or elastomeric, while inorganic materials are brittle and rigid. In the past 20 years, the field of phosphazene chemistry has developed to the point that different polyphosphazenes now span the range of properties from low temperature elastomers to fiber or film-forming thermoplastics and glasses, from water-soluble to highly hydrophobic polymers, and from bioactive biomedical materials to electrical semiconductors or solid electrolytes. Virtually all known polyphosphazenes are stable to water, most are stable to acids, bases and many solvents, and a substantial number are stable to heat and resistant to oxidation or burning. At least 300 examples of different polyphosphazenes are known, with different side groups or combinations of side groups, and the number continues to grow steadily. They now rival many areas of organic polymer chemistry in molecular diversity, scientific interest and prospective applications.

## 26.2 THE BASIC SYNTHESIS METHOD

The method of synthesis used for nearly all linear or macrocyclic polyphosphazenes was established by Allcock, Kugel and Valan in the mid-1960s.[3,4] It is based on the chemistry outlined in Scheme 1.

The starting material is hexachlorocyclotriphosphazene (**1**) (formerly known as 'phosphonitrilic chloride'), now prepared on an industrial scale from phosphorus pentachloride and ammonium

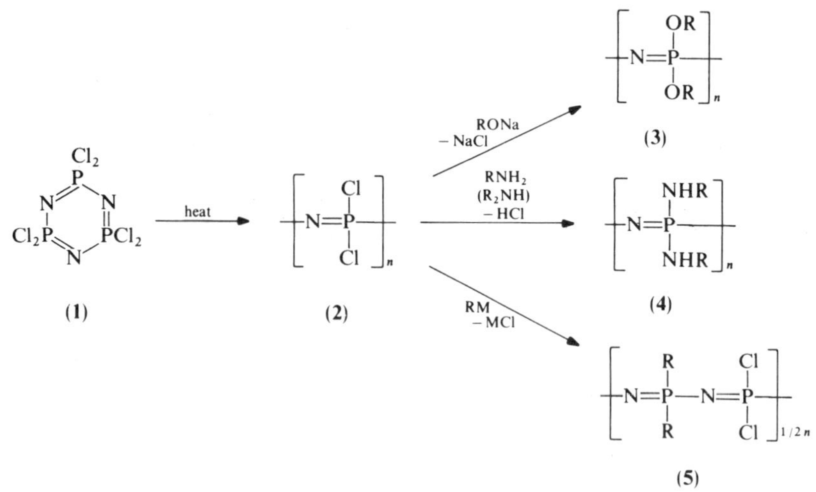

$n \simeq 15\,000$, M = a metal such as Li

**Scheme 1**

chloride. Compound (**1**) is heated in the melt at 210–250 °C to induce polymerization to uncross-linked essentially linear poly(dichlorophosphazene) (**2**). Failure to control this reaction results in the conversion of (**2**) into an insoluble cross-linked material identical to the 'inorganic rubber' reported a century ago.

Uncross-linked poly(dichlorophosphazene) (**2**) dissolves in a variety of anhydrous organic solvents and, in solution, reacts with a wide range of organic and organometallic nucleophiles to give polymers of structure (**3**), (**4**) or (**5**). In most cases, all the chlorine atoms in (**2**) can be replaced to yield hydrolytically stable, unreactive polymers. Thus, the key feature of this chemistry is the use of (**2**) as a highly reactive macromolecular intermediate. Given the broad range of organic or organometallic nucleophiles available, and the fact that two or more different side groups can be introduced concurrently or sequentially, it will be clear that the polymers known at the present time represent only a small fraction of those that could be obtained.[1,3–8]

For reactions with certain nucleophiles (particularly organometallic reagents), it is sometimes preferable to use the closely related compound poly(difluorophosphazene) (**7**) as a reaction substrate.[9] This polymer can be prepared as a fluorocarbon-soluble species by polymerization of the trimer, $(NPF_2)_3$ (**6**; equation 1), itself obtained by reaction of $(NPCl_2)_3$ with sodium fluoride.

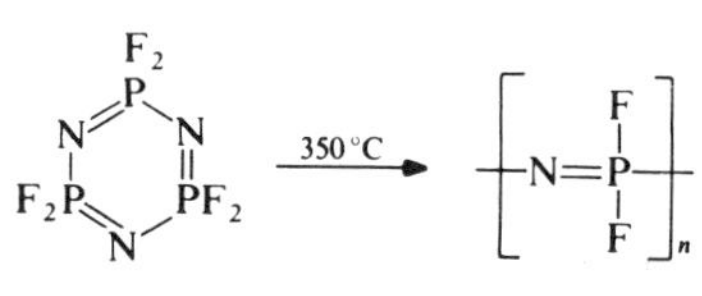

(1)

## 26.3 DIVERSITY OF THE SYNTHESIS METHOD

The substitutive synthesis method outlined above allows an exceedingly broad range of side groups to be attached to a polyphosphazene chain. A few representative examples are illustrated in Table 1.

In general, the limits of this method appear to be defined by the following three factors. First, bulky organic side groups attached to the chain can hinder the replacement of the remaining chlorine atoms. Thus, reaction of $(NPCl_2)_n$ with diethylamine results in a facile replacement of one chlorine atom at each phosphorus, but the remaining chlorine atoms are resistant to reaction with the same reagent,[10] presumably because of steric shielding effects. However, the partly substituted species will react with less-hindered nucleophiles to yield fully substituted derivatives (**8**; equation 2). Similar steric effects are evident when steroid oxides react with (**2**).[11] Forcing reaction conditions (high temperatures) are generally required in order to bring about complete replacement of the

**Table 1** Representative Polyphosphazenes and Their Properties

| *R in* $\left[ -N=PR_2- \right]_n$ | $T_g$(°C) | $T_m$(°C) | *Solubility* |
|---|---|---|---|
| *Elastomers* | | | |
| $OPr^n$ | −100 | | THF |
| OEt | −84 | | THF, ethanol |
| $OCH_2CH_2OCH_2CHOMe$ | −84 | | Water |
| OMe | −76 | | Methanol |
| $OCH_2CF_3/OCH_2(CF_2)_xCF_2H$ | −60[a] | | Fluorocarbons |
| $OC_9H_{19}$ | −56 | | THF |
| $OPh/OC_6H_4Et$ | −10 | b | Hot aromatic hydrocarbons |
| $OCH_2CF_3/CH_2SiMe_3$ | −40 to −60[a] | | |
| $NHCH_2CO_2Et$ | −23 | | THF |
| F | −96 | −68, −40 | Fluorocarbons |
| Cl | −66 | −7.2, (+39°) | Benzene, THF |
| Br | −15 | | |
| *Microcrystalline thermoplastics* | | | |
| $OCH_2CF_3$ | −66 | +242 | Acetone |
| OPh | −8 | +360 | Hot aromatic hydrocarbons |
| *Amorphous thermoplastics (glasses)* | | | |
| NHMe | +14 | | Water |
| NHPh | +91 | | Benzene, THF |
| $C_6H_{10}O_6$ (glucosyl) | +92 | | Water |

[a] Value varies with ratio of two substituents or the length of the $CF_2$ chain. [b] Complex melting behavior. [c] For the stretched polymer.

chlorine in (**2**) by aryl oxides, including phenoxide.[4] These reactivity differences allow both gross and subtle structural tailoring of the molecular structure to be achieved.

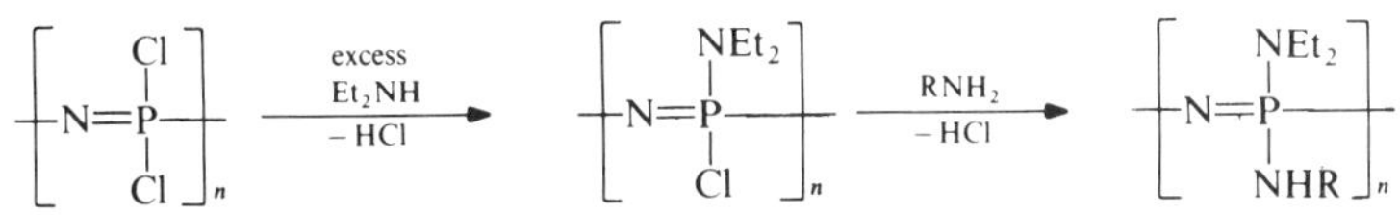


(**2**) (**8**) (2)

Second, the substitutive method of synthesis restricts the reaction to the use of monofunctional nucleophiles. Di- or tri-functional reagents cross-link the chains. Thus, the introduction of multifunctional side groups, such as sugar,[12] amino acid[13] or other units of biological interest,[14–17] usually requires an elaborate sequence of blocking and deblocking reactions.

Finally, the replacement of halogen atoms in $(NPCl_2)_n$ or $(NPF_2)_n$ by organometallic reagents is limited, at present. Such reactions are of considerable interest, because they offer a route for the synthesis of species with alkyl or aryl groups bonded directly to the skeleton through C–P bonds (by the use of Grignard, organolithium or organocopper reagents)[18–21] or with transition metals linked to the skeleton through M–P bonds (by the use of organometallic anions). Such species are described later. However, organometallic reactions of this type suffer from competing side reactions such as skeletal cleavage, metal–halogen exchange and (with alkyl groups present) from metal–hydrogen exchange and cross-linking. In general, $(NPF_2)_n$ is a better substrate for organometallic reactions than is $(NPCl_2)_n$.[9]

## 26.4 ALTERNATIVE SYNTHESIS ROUTES

The problems associated with the reactions of $(NPCl_2)_n$ with Grignard or organolithium reagents have stimulated the development of an alternative route to polyphosphazenes that bear alkyl or aryl side groups linked to the chain through C–P bonds. This route is based on the observation that it is nearly always easier to carry out organometallic substitution reactions on phosphazene cyclic trimers, such as (**1**) or (**6**), than on high polymers, such as (**2**) or (**7**). Thus, a number of different methods have been explored for the synthesis of cyclic trimers or tetramers that bear alkyl or aryl side groups.[21–39] Such trimers might be induced to polymerize to the analogous high polymer, in a similar manner to the polymerization of $(NPCl_2)_3$ to $(NPCl_2)_n$.

This approach is practical only if the cyclic trimer bears *both* organic and halo side groups. Trimers such as $[NP(Me)_2]_3$ or $[NP(Ph)_2]_3$ undergo ring–ring equilibration reactions, but they do not yield high polymers.[40] In general, cyclic trimers that bear one alkyl or aryl group plus five halogen atoms (**9**), will polymerize. Similar behavior is found for species with two or three non-geminally disposed organic groups (**10** or **11**), but geminal organic substituents (**12**) generally retard polymerization. Compounds such as (**9**)–(**12**) can be obtained *via* the reactions of (**1**) or (**6**) with organolithium, Grignard or organocopper reagents, or by the reactions of cyclophosphazene anions with alkyl halides.[21–39]

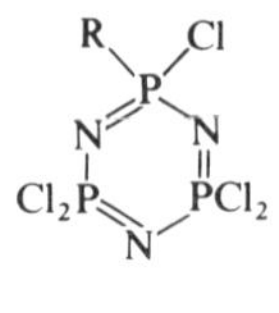
(**9**)

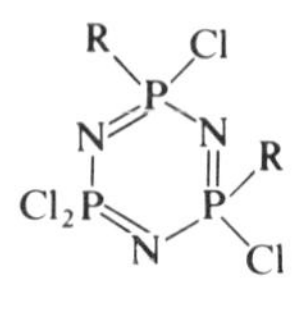
(**10**)

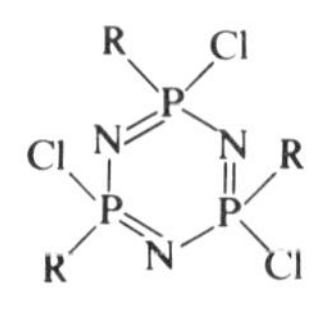
(**11**)

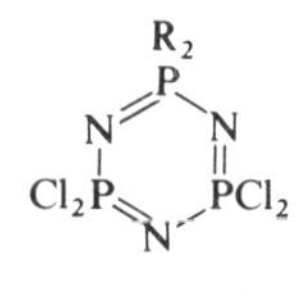
(**12**)

Polymerization of (**9**)–(**11**) yields organohalogenophosphazene polymers, in which the chlorine atoms can be replaced readily by treatment with conventional nucleophiles, such as alkoxides or amines. Such mixed substituent polymers are generally non-crystalline elastomers.

A second, alternative, synthesis of alkyl or arylphosphazene polymers makes use of an entirely different approach. This method, developed by Neilson and Wisian–Neilson, involves the pyrolytic condensation and polymerization of silylphosphinimines, such as (**13**), to give polymers of type (**14**), as in equation (3).[41,42] Polyphosphazenes prepared by this method have lower molecular weights than those obtained by the polymer substitution route, but they contain no halo, alkoxy, aryloxy or amino cosubstituents. Thus, they are valuable molecules for direct comparison with poly(organosiloxanes).

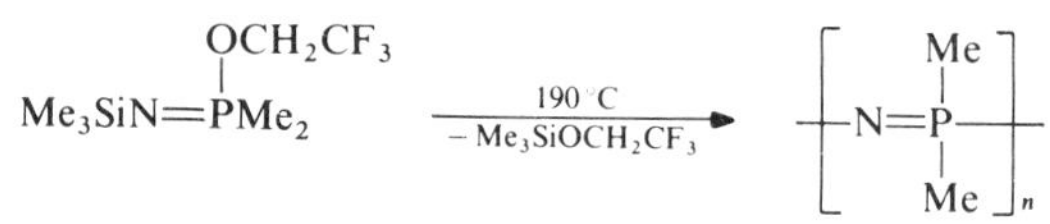
(3)

(**13**) (**14**)

## 26.5 POLYPHOSPHAZENES OF BIOLOGICAL INTEREST

It will be obvious from the foregoing comments that the methods available for the synthesis of polyphosphazenes allow a wide variety of different organic groups to be linked directly to the chain. Such groups may be bioactive agents, such as steroids, local anesthetics, antibacterial agents, *etc.* However, the additional possibility exists that organic groups already attached to the skeleton may serve as spacer units for the attachment of bioactive units. Such a method requires the presence of a second functional site on the spacer unit. The possibility also exists that the biological activity can be mediated by the choice of 'non-active' cosubstituent groups. Thus, water solubility, biodegradability or hydrophobicity can be designed into the macromolecule by the choice of the appropriate cosubstituent. Some of the possibilities that have been examined are shown in Table 2.

Biomedical evaluation work is under way on many of these systems, and individual polymers show promise as polymeric drugs, enzyme reactor substrates, inert biomedical materials and immobilized receptor stimulators. Other applications are being investigated in membrane science and biomedical transducers.

## 26.6 TRANSITION METAL DERIVATIVES

The newest class of polyphosphazenes bear transition metals in the side group structure. Such polymers are of interest as polymer-bound catalysts, electrode mediators, optical fibers and semiconducting polymers. Three main approaches have been developed for the incorporation of

**Table 2** Bioactive Polyphosphazenes 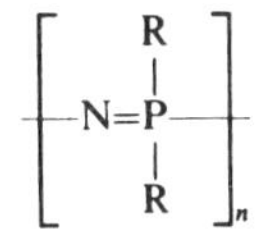

| *Water solubilization* | *Biological activity* |
|---|---|
| R = MeNH<br>glucose residues<br>Me<br>$MeOCH_2CH_2O$ | R =<br>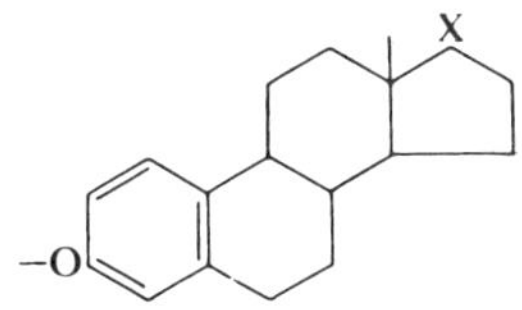 |
| *Water destabilization*<br>R = $EtO_2CCH_2NH$<br>imidazolyl | 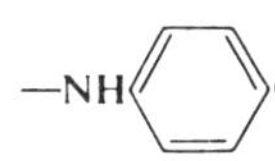 (procaine)<br>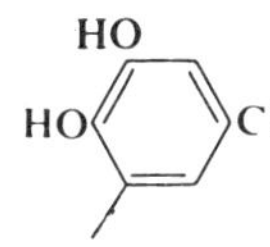 (dopamine) |
| *Water stabilization*<br>R = $OCH_2CF_3$ *etc.*<br>OPh *etc.*<br>Ph<br>Me | 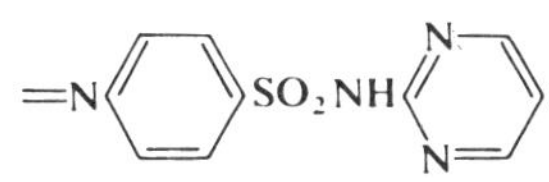 (sulfadiazine)<br>Heparin<br>$PtCl_2$ (as *N*-coordination adduct)<br>Enzymes |

transition elements. In the first approach, organic side groups bearing coordination sites for metals are attached to the chain. An example is shown as structure (**15**). Such species are prepared by the introduction of *p*-bromophenoxy substituent groups, followed by lithiation and treatment with diphenylchlorophosphine.[43] The reaction of this polymer with iron, rhodium and osmium organometallic units has been studied.[44] In addition, the same carrier polymer has been examined as a carrier for cobalt carbonyl hydroformylation catalysts.[45]

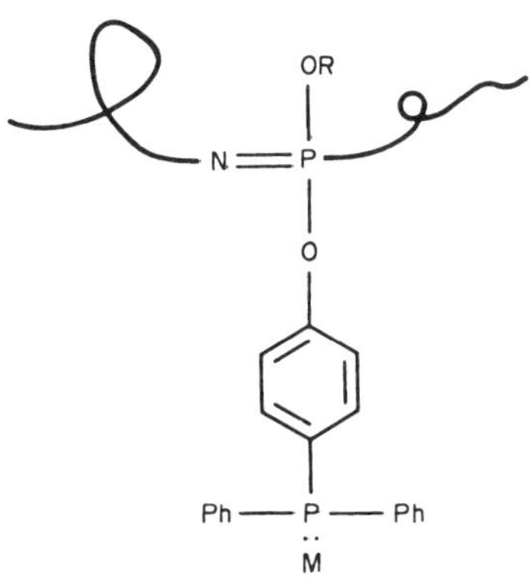

(**15**) M = a metal carbonyl

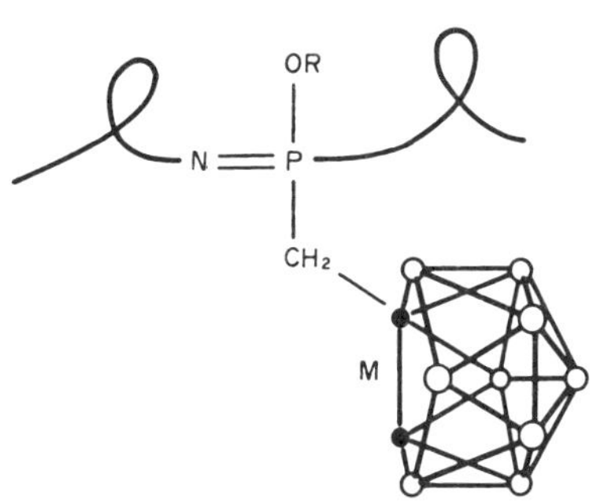

(**16**)

Secondly, polyphosphazenes have been prepared in which transition metals are $\pi$ bonded to carboranyl[46] and cyclopentadienyl residues attached to the polymer backbone. For example, species (**16**) has been prepared in which a rhodium hydrogenation catalyst occupies the open face of the *nido*-carboranyl side group. Species (**17**) have been synthesized by the reactions of $(NPF_2)_3$ with dilithioruthenocene or dilithioferrocene.[47] Ring strain associated with the presence of the trans-annular organometallic bridge increases the ease of polymerization relative to $(NPF_2)_3$.[48] The resultant high polymer (**18**; equation 4), when treated with sodium trifluoroethoxide, yields a stable macromolecule. The ferrocenyl derivative is an amber-colored film-forming material. It can be doped with iodine to form a weak semiconductor. Copolymerization of the ferrocenyl and ruthenocenyl forms of (**17**) yields a polyphosphazene with both organometallic side groups.[48] Such species have interesting polarographic properties.

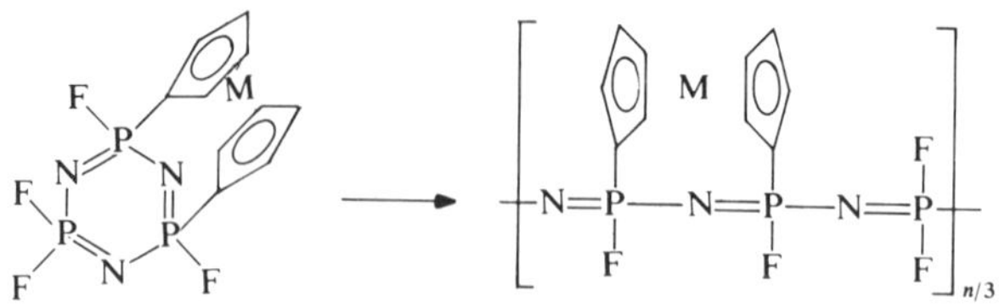

M = Fe or Ru (4)

(**17**) (**18**)

Thirdly, transition metals can be linked to the skeleton through P–M bonds by the chemistry shown in Scheme 2.[49] Polymer (**19**) is first prepared by the reaction of $(NPCl_2)_n$ with a stoichiometric deficiency of sodium trifluoroethoxide, in order to reduce the functionality of the polymer. Treatment with $KFe(C_5H_5)(CO)_2$ (pathway a) brings about replacement of most of the chlorine atoms by the organometallic unit. A final treatment with trifluoroethoxide removes the remaining chlorine. In pathway (b), species (**19**) is converted first to a polyphosphazene anion by treatment with lithium triethylborohydride. These anionic sites then react with the electrophile, $Fe(C_5H_5)(CO)_2I$, to yield compound (**20**). Again, any traces of remaining P–Cl units are removed by treatment with sodium trifluoroethoxide.

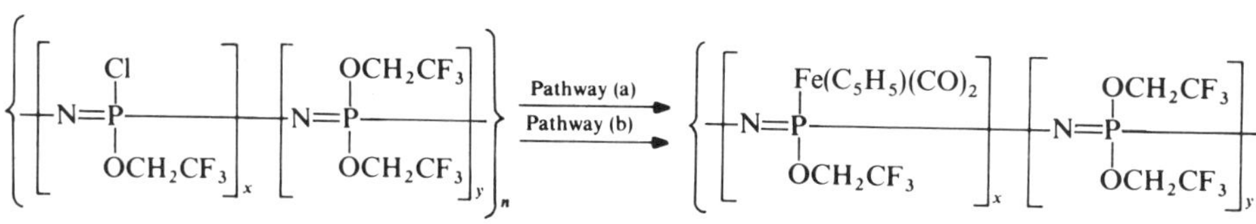

(**19**) (**20**)

Pathway (a): i, $[Fe(C_5H_5)(CO)_2]^-$; ii, $NaOCH_2CF_3$
Pathway (b): i, $LiBEt_3H$; ii, $Fe(C_5H_5)(CO)_2I$; iii, $NaOCH_2CF_3$

**Scheme 2**

## 26.7 STRUCTURAL CHEMISTRY

In what ways are polyphosphazenes different from classical organic polymers and how can these differences be understood in terms of the inorganic backbone and the organic or organometallic side groups' structure?

Many physical properties of polymers can be correlated with chain flexibility or the lack of it. In general, elastomers have flexible chains, while rigid polymers have relatively inflexible chain structures. Chain flexibility can be traced to the influence of two main fundamental molecular factors—the intrinsic torsional freedom of the skeletal bonds and the role played by the side groups in restricting torsional motions of the skeletal bonds. In these terms there should be no fundamental differences between organic and inorganic polymers, other than the inherent variations in skeletal bond length and angles, bond polarity and $\pi$ bonding.

Phosphazenes possess an unusual skeletal structure, the electronic nature of which is still not fully understood.[50] After electron pairs have been assigned to each of the $\sigma$ bonds, four electrons in each repeating unit remain to be accounted for. Two of these are assigned to a lone pair orbital at nitrogen (structure **21**), while the remaining two are believed to occupy a $d_\pi$–$p_\pi$ orbital—a form of stabilized excited state derived from the 3*d* orbitals of phosphorus and a 2*p* orbital of nitrogen. Unlike organic $\pi$ bonds, a $d_\pi$–$p_\pi$ bond imparts no significant torsional barrier to the N–P bond, presumably because up to five of the diffuse *d* orbitals on each phosphorus can 'switch' into a bonding position as the skeletal bonds undergo torsion. Moreover, the symmetry of the $d_\pi$–$p_\pi$ system is such that a node occurs at each phosphazene. Thus, the $\pi$ bonding can be considered as three-center 'islands' of $\pi$ character, essentially insulated from each oher, as shown in structure (**22**).[51]

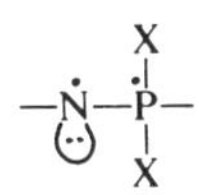

**(21)**

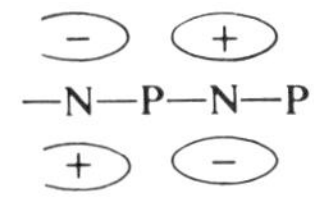

**(22)**

These descriptions explain both the high flexibility of the phosphazene backbone and the lack of 'metallic' character of the type found in poly(sulfur nitride) or doped polyacetylene.

Thus, several polyphosphazenes have extremely low glass transition temperatures ($T_g$) (see Table 1). These are cases where the side groups are insufficiently bulky, rigid or polar to inhibit chain flexibility. On the other hand, rigid bulky side groups raise the torsional barrier and bring about a marked increase in $T_g$.

Some polyphosphazenes are microcrystalline (see Table 1), and the conformations adopted by the chains in the crystalline domains are of some interest. X-Ray diffraction studies of $(NPCl_2)_n$,[52] $(NPF_2)_n$,[53] $[NP(OCH_2CF_3)_2]_n$,[4] $[NP(OPh)_2]_n$,[4] and $[NP(OC_6H_4Cl\text{-}m)_2]_n$[54] suggest a chain repeat distance of $\simeq$4.9 Å and a near *cis–trans* planar conformation of the backbone. This is the conformation that minimizes intramolecular repulsions associated with the side groups for all the above structures, except $(NPF_2)_n$.[53] This latter polymer also assumes a spiral helical conformation at low temperatures.

In general, very little detailed information is currently available for polyphosphazenes, and X-ray structures of cyclic, trimeric and short chain linear phosphazenes are the main source of bond angle and bond length estimates.[50,55]

## 26.8 APPLICATIONS

The main applications of polyphosphazenes, so far, have been as high performance elastomers.[6–8] Mixed substituent polymers of formula (**23**) are non-crystalline, oil and fuel resistant elastomers with very low glass transition temperatures (see Table 1). They also resist burning and are stable at 200 °C. For these reasons, they are used as O-rings, oil seals, pipelines and in other oil and fuel related applications. Mixed substituent aryloxyphosphazene polymers (**24**) are also elastomers, but have higher $T_g$ values. Their main use is as non-burning expanded foam materials in thermal and electrical insulation.

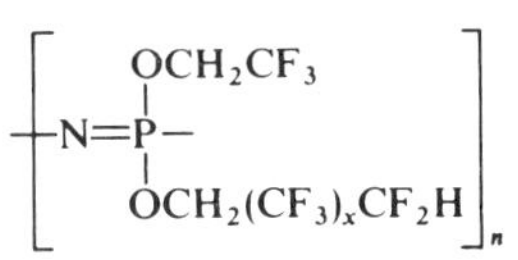

**(23)**

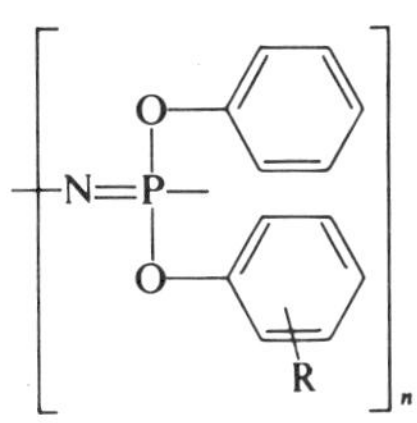

**(24)**

A number of other applications for polyphosphazenes are in the developmental stage. A variety of polymers, some with bioactive groups, are being evaluated as biomedical materials, membranes and immobilized drug formulations. Others, particularly compound (**25**), are candidates for solid battery electrolyte applications. Species (**25**) functions as a solid solvent for lithium triflate, and the resultant solid solution shows a high ionic electrical conductivity at room temperature.[56,57]

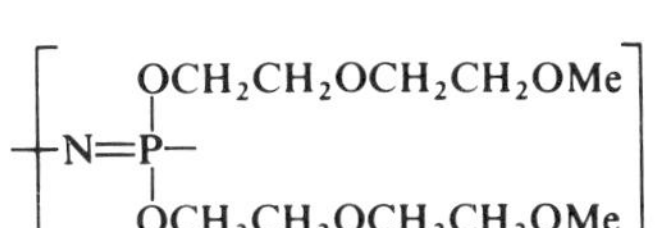

**(25)**

## 26.9 REFERENCES


1. H. R. Allcock, *Chem. Eng. News*, 1985, **63**, 22.
2. H. N. Stokes, *Am. Chem. J.*, 1897, **19**, 782.
3. H. R. Allcock and R. L. Kugel, *J. Am. Chem. Soc.*, 1965, **87**, 4216.

4. H. R. Allcock, R. L. Kugel and K. J. Valan, *Inorg. Chem.*, 1966, **5**, 1709; H. R. Allcock and R. L. Kugel, *Inorg. Chem.*, 1966, **5**, 1716.
5. G. Allen, C. J. Lewis and S. M. Todd, *Polymer*, 1970, **1**, 31.
6. R. E. Singler, G. L. Hagnauer and R. W. Sicka, *ACS Symp. Ser.*, 1982, **193**, 229.
7. D. P. Tate, *J. Polym. Sci., Polym Symp.*, 1974, **48**, 33.
8. H. R. Penton, *Kautsch. Gummi, Kunstst.*, 1986, **39**, 304.
9. T. L. Evans and H. R. Allcock, *J. Macromol. Sci., Chem.*, 1981, **A16**, 409.
10. H. R. Allcock, W. J. Cook and D. P. Mack, *Inorg. Chem.*, 1972, **11**, 2584.
11. H. R. Allcock and T. J. Fuller, *Macromolecules*, 1980, **13**, 1338.
12. H. R. Allcock and A. G. Scopelianos, *Macromolecules*, 1983, **16**, 715.
13. H. R. Allcock, T. J. Fuller, D. P. Mack, K. Matsumura and K. M. Smeltz, *Macromolecules*, 1977, **10**, 824.
14. H. R. Allcock and P. E. Austin, *Macromolecules*, 1981, **14**, 1616.
15. H. R. Allcock, P. E. Austin and T. F. Rakowsky, *Macromolecules*, 1981, **14**, 1622.
16. H. R. Allcock, P. E. Austin and T. X. Neenan, *Macromolecules*, 1982, **15**, 689.
17. H. R. Allcock, W. C. Hymer and P. E. Austin, *Macromolecules*, 1983, **16**, 1401.
18. H. R. Allcock, D. B. Patterson and T. L. Evans, *J. Am. Chem. Soc.*, 1977, **99**, 6095.
19. H. R. Allcock and C. T.-W. Chu, *Macromolecules*, 1979, **12**, 551.
20. H. R. Allcock, T. L. Evans and D. B. Patterson, *Macromolecules*, 1980, **13**, 201.
21. H. R. Allcock, A. G. Scopelianos, J. P. O'Brien and M. Y. Bernheim, *J. Am. Chem. Soc.*, 1981, **103**, 350.
22. H. R. Allcock and G. Y. Moore, *Macromolecules*, 1975, **8**, 377.
23. P. J. Harris and H. R. Allcock, *J. Am. Chem. Soc.*, 1978, **100**, 6512.
24. P. J. Harris and H. R. Allcock, *J. Chem. Soc., Chem. Commun.*, 1979, 714.
25. R. J. Ritchie, P. J. Harris and H. R. Allcock, *Macromolecules*, 1979, **12**, 1014.
26. H. R. Allcock and P. J. Harris, *J. Am. Chem. Soc.*, 1979, **101**, 6221.
27. H. R. Allcock, *Polymer*, 1980, **21**, 673.
28. H. R. Allcock, R. J. Ritchie and P. J. Harris, *Macromolecules*, 1980, **13**, 1332.
29. H. R. Allcock, A. G. Scopelianos, J. P. O'Brien and M. Y. Bernheim, *J. Am. Chem. Soc.*, 1981, **103**, 350.
30. H. R. Allcock, P. J. Harris and M. S. Connolly, *Inorg. Chem.*, 1981, **20**, 11.
31. H. R. Allcock and P. J. Harris, *Inorg. Chem.*, 1981, **20**, 2844.
32. P. J. Harris, J. L. Desorcie and H. R. Allcock, *J. Chem. Soc., Chem. Commun.*, 1981, 852.
33. H. R. Allcock, M. S. Connolly and P. J. Harris, *J. Am. Chem. Soc.*, 1982, **104**, 2482.
34. H. R. Allcock, J. L. Desorcie and P. J. Harris, *J. Am. Chem. Soc.*, 1983, **105**, 2814.
35. H. R. Allcock, M. S. Connolly and R. R. Whittle, *Organometallics*, 1983, **2**, 1514.
36. H. R. Allcock, J. L. Desorcie and L. J. Wagner, *Inorg. Chem.*, 1985, **24**, 333.
37. H. R. Allcock and M. S. Connolly, *Macromolecules*, 1985, **18**, 1330.
38. H. R. Allcock, D. J. Brennan, J. M. Graaskamp and M. Parvez, *Organometallics*, 1986, **5**, 2434.
39. H. R. Allcock, J. L. Desorcie and G. H. Riding, *Polyhedron*, 1987, **6**, 119.
40. H. R. Allcock and D. B. Patterson, *Inorg. Chem.*, 1977, **16**, 197.
41. P. Wisian–Neilson and R. H. Neilson, *J. Am. Chem. Soc.*, 1980, **102**, 2848.
42. P. Wisian–Neilson, R. R. Ford, R. H. Neilson and A. K. Roy, *Macromolecules*, 1986, **19**, 2089.
43. H. R. Allcock, T. J. Fuller and T. L. Evans, *Macromolecules*, 1980, **13**, 1325.
44. H. R. Allcock, K. D. Lavin, N. M. Tollefson and T. L. Evans, *Organometallics*, 1983, **2**, 267.
45. R. A. Dubois, P. E. Garrou, K. D. Lavin and H. R. Allcock, *Organometallics*, 1986, **5**, 460.
46. H. R. Allcock, A. G. Scopelianos, R. R. Whittle and N. M. Tollefson, *J. Am. Chem. Soc.*, 1983, **105**, 1316.
47. H. R. Allcock, K. D. Lavin, G. H. Riding, P. R. Suszko and R. R. Whittle, *J. Am. Chem. Soc.*, 1984, **106**, 2337.
48. H. R. Allcock, K. D. Lavin and G. H. Riding, *Macromolecules*, 1985, **18**, 1340.
49. H. R. Allcock, M. N. Mang, G. H. Riding and R. R. Whittle, *Organometallics*, 1986, **5**, 2244.
50. H. R. Allcock, 'Phosphorus–Nitrogen Compounds', Academic Press, New York, 1972, chap. 2.
51. M. J. S. Dewar, E. A. C. Lucken and M. A. Whitehead, *J. Chem. Soc.*, 1960, 2423.
52. H. R. Allcock, R. A. Arcus and E. G. Stroh, *Macromolecules*, 1980, **13**, 919.
53. H. R. Allcock, R. L. Kugel and E. G. Stroh, *Inorg. Chem.*, 1972, **11**, 1120.
54. S. M. Bishop and I. H. Hall, *Br. Polym. J.*, 1974, **6**, 193.
55. H. R. Allcock, N. M. Tollefson, R. A. Arcus and R. R. Whittle, *J. Am. Chem. Soc.*, 1985, **107**, 5166.
56. P. M. Blonsky, D. F. Shriver, P. E. Austin and H. R. Allcock, *J. Am. Chem. Soc.*, 1984, **106**, 6854.
57. P. M. Blonsky, D. F. Shriver, P. E. Austin and H. R. Allcock, *Polym. Mater. Sci. Eng.*, 1985, **53**, 118.

# 2

# Monoalkene Polymerization: Mechanisms

PETER J. T. TAIT and NEIL D. WATKINS
*University of Manchester Institute of Science and Technology, UK*



## 2.1 INTRODUCTION

In spite of the intense research activity that has centred around Ziegler–Natta catalyst systems since their discovery in the early 1950s, no definitive, unequivocal polymerization mechanism has yet been devised to fully describe their behaviour. One reason for this is the complexity of the systems employed. For instance, the catalyst itself may react as a solid or may be soluble in the reaction medium; activation by an organometallic compound is usually required but not always so; the monomers that are homo- or co-polymerized may be gases or liquids; electron donors may be added to improve stereoregulation; and the whole process may be performed under slurry, solution or gas phase conditions. Since the actual number of propagating polymerization centres present on a solid catalyst is usually only a small proportion of the total number of surface sites, it is evident why direct spectroscopic studies of structure have thus far been of limited value. Although developments in NMR spectroscopy of solids, EXAFS (extended X-ray absorption fine structure) and SIMS (secondary ion mass spectroscopy) may hold promise for the future, most evidence so far reported in the scientific literature has come from indirect studies of polymerization kinetics, molecular weights and molecular weight distributions, polymer structure and active centre concentrations.

It has nevertheless been established that two key steps in the polymerization process are complexation between the alkene and the active centre followed by insertion into the growing polymer chain. Any proposed mechanism should include these steps and also be able to account for chain transfer reactions and the stereoregulating ability of the catalyst.

A discussion of some of the more important mechanisms that have been suggested for α-alkene polymerization will be given in this chapter followed by a brief summary of mechanisms for diene polymerization. A more detailed review has been made by Boor.[1] Finally, a summary of some models for active centres in heterogeneous and homogeneous catalysts will be presented. The question of the origin of stereoregulation will be considered in a subsequent chapter.

## 2.2 MECHANISMS FOR α-ALKENE POLYMERIZATION

Various mechanisms have been proposed which differ according to whether the growth centre is a bound radical, a bound anion or a metal–alkyl bond. Of these the last is by far the most widely accepted group and can be subdivided according to whether the growth centre is an activator–alkyl bond or a transition metal–alkyl bond. All of these mechanisms will be considered together with supporting and contradictory experimental evidence.

### 2.2.1 Early Mechanisms

A number of early mechanisms were based on a radical mode of polymerization propagation. Various sources of initiation were proposed including the transition metal centre, the metal alkyl centre, or an alkyl radical produced by reduction of an alkylated transition metal. For example, in the mechanism of Nenitzescu *et al.*[2] the reduction of a titanium alkyl compound from an oxidation state of four to three produces alkyl radicals which were believed to initiate polymerization, as shown in Scheme 1.

$$R_3Al + TiCl_4 \longrightarrow R_2AlCl + RTiCl_3$$

$$RTiCl_3 \longrightarrow R\cdot + TiCl_3$$

$$R\cdot + C_2H_4 \longrightarrow RCH_2\dot{C}H_2 \longrightarrow \xrightarrow{C_2H_4} \text{polymer}$$

**Scheme 1**

In such mechanisms the growing polymer chains are envisaged as existing in the form of 'bound radicals'. Thus, in the scheme of Friedlander and Oita[3] initiation occurs when an electron is transferred from the transition metal compound surface to an adsorbed alkene molecule (Scheme 2).

Cross section

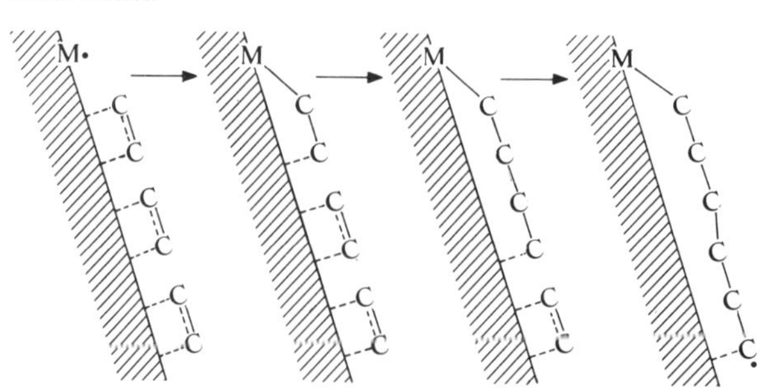

**Scheme 2**

Topchiev *et al.*[4] proposed a mechanism in which a chemisorbed aluminum alkyl molecule alkylates a surface titanium atom. Reduction of the titanium from an oxidation state of three to two produces an alkyl radical which adds to a chemisorbed alkene molecule; both the growing radical and the reacting alkene molecules remain bound to the $TiCl_3$ surface (Scheme 3).

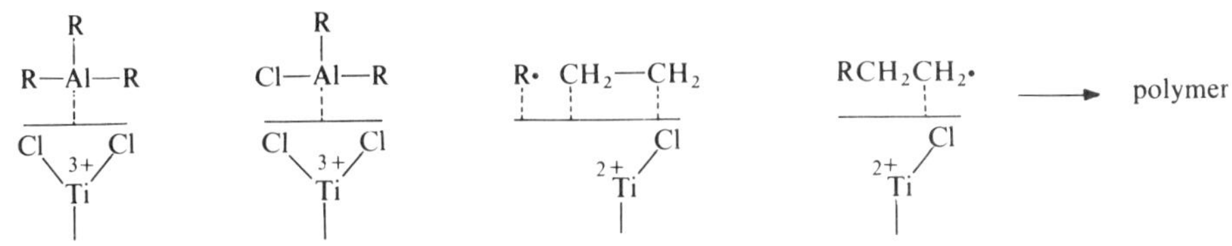

**Scheme 3**

The aluminum alkyl compound has also been proposed as the source of the initiator radical.[5] A different early suggestion was of an anionic mechanism in which the anionic end of the growing polymer chain is not bound to any metal centre.[6] Initiation occurs by transfer of an alkyl anion from adsorbed metal alkyl to adsorbed alkene molecules; propagation proceeds as the anionic end adds to further adsorbed monomer molecules.

Both radical and anionic-type mechanisms have now been discounted since, as will be described in the next section, there is considerable experimental evidence available which indicates that polymer growth occurs at a metal–alkyl bond. Nevertheless, under appropriate conditions some monomers such as vinyl chloride will polymerize by a free radical mechanism in the presence of some or all of the components of a Ziegler–Natta catalyst. Such cases are, however, exceptional; polyethylene produced by Ziegler–Natta systems has a higher density and higher crystallinity than that produced by a free radical process due to the absence of chain branching.

### 2.2.2 Propagation at the Metal–Alkyl Bond

The weight of experimental evidence argues in favour of propagation occurring at a metal–alkyl bond. Three crucial experimental considerations lead to this conclusion: (i) the use of aluminum alkyls labelled with $^{14}C$ leads to the formation of polymer containing labelled carbon atoms; (ii) quenching of polymerizations using a tritiated alcohol (*e.g.* $MeO^3H$) results in the introduction of tritium atoms into the final polymer; and (iii) hydrogen is often added to polymerizations catalyzed by Ziegler–Natta systems where it acts as a transfer agent and lowers the molecular weight of the polymer formed. When deuterium (as $^2H_2$) or tritium (as $^3H_2$) is added, incorporation of the appropriate label into the polymer occurs (Scheme 4).

$$MCH_2CH_2P + X_2 \longrightarrow MX + XCH_2CH_2P$$
$$X = H,\ ^2H \text{ or } ^3H$$

**Scheme 4**

Metal-alkyl growth centre mechanisms can be classified according to whether chain propagation occurs at the activator compound or the transition metal compound.

#### *2.2.2.1 Propagation at the activator–alkyl bond*

Mechanisms which involve propagation at the activator–alkyl bond are often termed 'bimetallic' since complexes such as that shown in structure (**1**) are envisaged as the growth centres; the growing polymer chain is partially bonded between the activator and transition metal atoms. Polymer growth occurs *via* insertion of monomer into the aluminum–carbon bond.

The mechanism of Natta[7] was an important early example of this type. Natta's previous work on the Aufbau reaction suggested that polymer growth occurred *via* insertion of an alkene molecule into an Al—C bond. The incoming alkene molecule forms a $\pi$ bond with the transition metal atom, simultaneously cleaving the bridging bond. The alkene molecule is polarized by bonding to the transition metal and then inserts into the aluminum–alkyl bond. Finally the bridging bond reforms (Scheme 5).

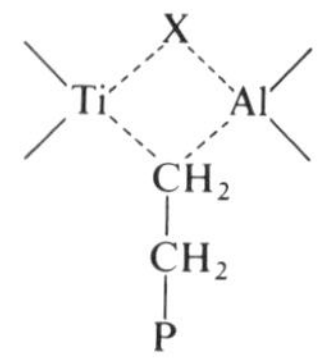

(1) X = halogen ligand,
P = growing polymer chain

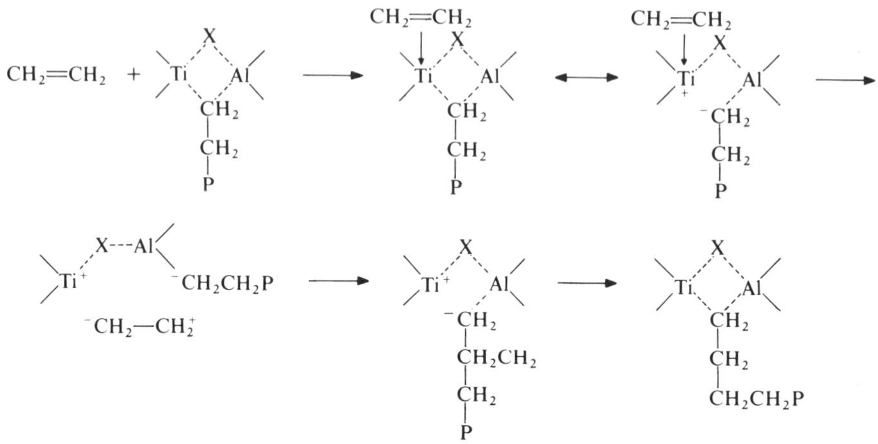

**Scheme 5**

A mechanism essentially similar to Natta's was proposed by Patat and Sinn;[8] an alkene molecule becomes partially bonded between the transition metal atom and the methylene group of the last unit added to the growing polymer chain. A $\sigma$ bond is formed between the alkene and the methylene group, following which the latter becomes detached from the aluminum atom; finally a new bridging bond is formed (Scheme 6).

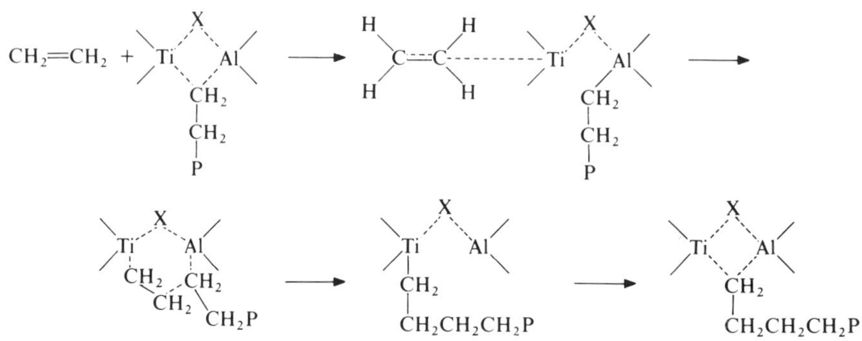

**Scheme 6**

The mechanism of Uelzmann[9] is also bimetallic and features the formation of an ion pair from the reaction between $TiCl_3$ and the activator. The cation produced $(TiCl_2)^+$ attracts and activates an alkene molecule which reorientates and then inserts into an aluminum alkyl bond, $(AlR_3Cl)^-$ acting as the growth centre (Scheme 7).

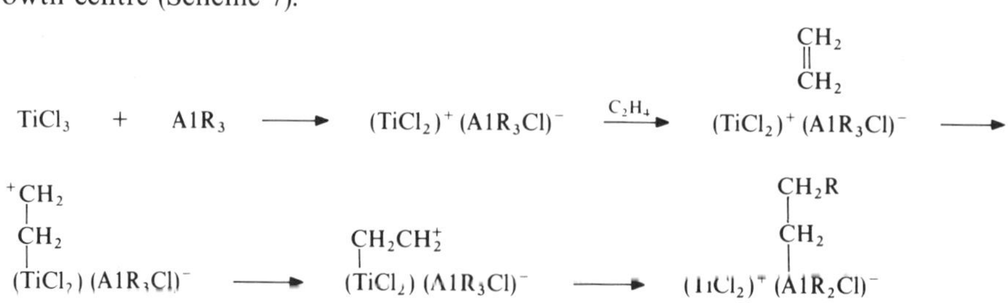

**Scheme 7**

Boor[10] has proposed a bimetallic mechanism in which insertion into an aluminum alkyl bond occurs *via* a concerted process (Scheme 8).

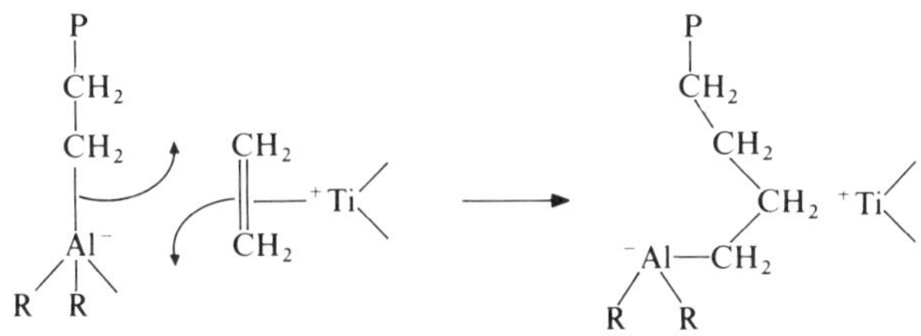

**Scheme 8**

In mechanisms of the type described above, the role of the $TiCl_3$ surface is to complex with alkene molecules and, by polarization, to promote insertion into the activator–alkyl bond. Other mechanisms have been devised in which the $TiCl_3$ surface acts simply to activate adsorbed aluminum alkyl; polymerization proceeds *via* migration of the growing polymer chain from an aluminum alkyl molecule to an alkene molecule which is complexed to an adjacent aluminum alkyl. A simultaneous electron transfer reaction through the $TiCl_3$ surface maintains electroneutrality.[11]

The popularity of mechanisms involving polymer growth at an activator–alkyl bond among early workers in the field was probably largely due to the influence of the Aufbau reaction, in which ethylene (ethene) is polymerized to low molecular weight polymer by group I–III alkyl compounds alone. Furthermore, the observation that many metal alkyl compounds such as $AlEt_3$ exist in dimeric form (structure **2**) supports arguments for the type of complexes proposed in such bimetallic mechanisms as those of Natta, and Patat and Sinn.

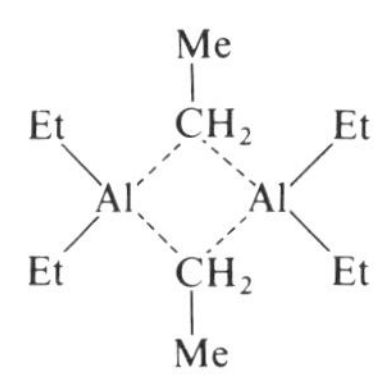

**(2)**

Additional evidence comes from the synthesis, and characterization by X-ray spectroscopy, of soluble bimetallic complexes of the sort shown in structure (**3**).[7,12]

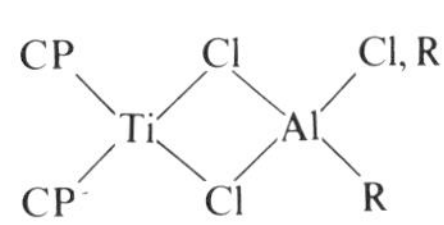

**(3)** R = Me or Et,
Cp = $\pi$-cyclopentadienyl

Such complexes were found to be active for the polymerization of ethylene. An analysis of the end groups of the polymers prepared using different metal alkyls (Table 1) together with the observation that $Cp_2TiPh_2$ alone will not polymerize ethylene led to the deduction that polymer growth occurs at an aluminum–alkyl bond.

However, this conclusion has been criticized on the basis of the following[13] considerations: (i) when a mixture of $Ph_3Al$ and $Et_3Al$ is used with $Cp_2TiCl_2$ only *ethyl* groups are found in the

**Table 1** End Groups of Polyethylenes Produced with $Cp_2TiR_2$–$AlR_3$ Catalyst Systems.

| *Catalyst* | *End group* |
|---|---|
| $Cp_2TiCl_2 + AlPh_3$ | Ph— |
| $Cp_2TiPh_2 + AlEt_3$ | Et— |
| $Cp_2TiPh_2 + AlPh_3$ | Ph— |

polymer; and (ii) a facile exchange of Ph and Et groups occurs when $Cp_2TiPh_2$ and $AlEt_2Cl$ are mixed.

The results of (i) and (ii) seem to indicate that polymerization occurs predominantly at titanium–ethyl bonds owing to the higher alkylating ability of $AlEt_3$.

A number of studies have been made which show that catalytic activity and stereoregulating ability depend on such factors as the size, ionic nature and complexing ability of the metal alkyl used. For example, Natta[14] used combinations of $\alpha$, $\gamma$ and $\delta$ forms of $TiCl_3$ with various metal alkyls to polymerize propylene and found that the isotacticity of the polymer produced depended only on the structure of the activator, and not on the crystalline modification of $TiCl_3$ used. It was concluded that polymer growth occurs at the activator–alkyl bond since it is the activator that has the greatest influence on the polymerization. Steric effects of the alkyl ligands (which become the end groups of the final polymer chain) have been discounted since their effect will be lost after a few subsequent insertions of monomer.

Studies have shown that if ethylene polymerization is conducted in the presence of deuterium ($D_2$) then the relative amounts of CHD and $CH_2D$ units found in the final polymer are dependent on the structure of the activator compound used.[15] Scheme 9 shows how the two different units arise. Such a finding has been interpreted as indicating that either the metal alkyl is the chain growth centre or that a bimetallic complex is formed which is strongly influenced by the metal alkyl used.

$$MCH_2CH_2P + D_2 \xrightarrow{\text{transfer}} MD + DCH_2CH_2P$$

$$MCH_2CH_2P + D_2 \xrightarrow{\text{exchange}} MCH_2CHDP + HD$$

**Scheme 9**

Kinetic studies[16] and quantum mechanical calculations[17] have also been performed which support a bimetallic scheme.

In spite of the examples just given, it is nevertheless true that in general supporting evidence for propagation at activator–alkyl bonds is inconclusive. Experiments that are based on an examination of the changes in polymerization behaviour that occur when the activator compound is varied can usually be explained by differences in ability to alkylate or to reduce the transition metal. At present, opinion has shifted firmly in favour of propagation at the transition metal.

### 2.2.2.2 *Propagation at the transition metal–alkyl bond*

As early as 1956 a transition metal–alkyl bond was suggested as being the growth centre for polymerization. Nenitzescu *et al.*[2] proposed a mechanism in which the alkene was polymerized by an alkylated tetravalent form of titanium (Scheme 10). However this scheme was rejected in favour of the radical mechanism described in Section 2.2.1.

$$RTiCl_3 \longrightarrow R^-TiCl_3^+ \xrightarrow{C_2H_4} RCH_2CH_2^-TiCl_3^+$$

**Scheme 10**

The observation that the rate of polymerization of ethylene using the system $AlR_3$–$TiCl_4$ (R = Et or $Bu^i$) reaches a maximum when the average valence state of the titanium is two has been interpreted as evidence for a divalent alkyltitanium halide being the active site.[18] A mechanism was proposed which envisaged that an ethylene molecule becomes complexed with the transition metal and is then inserted into the titanium–alkyl bond (Scheme 11).

$$ClTi(C_2H_4)_nR + C_2H_4 \xrightarrow{\text{slow}} \underset{ClTi(C_2H_4)_nR}{\overset{CH_2{=}CH_2}{|}} \xrightarrow{\text{fast}} ClTi(C_2H_4)_{n+1}R$$

**Scheme 11**

It should be noted, however, that whilst divalent titanium catalysts are known which are highly active for ethylene polymerization, activities for $\alpha$-alkene polymerizations are generally low.

The most widely accepted of the mechanisms involving growth at the transition metal alkyl is that devised by Cossee and Arlman.[19] Arlman recognized that chlorine vacancies must exist on the surface of $TiCl_3$ in order for electroneutrality of the crystal to exist,[20] and from this Cossee devised an active centre model based on a titanium atom having an octahedral configuration. In this model the metal is surrounded by four chlorine ligands from the crystal lattice, an alkyl group introduced by the activator and a coordination vacancy (structure **4**).

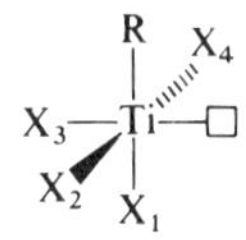

(**4**) □ = ligand vacancy, $X_1$–$X_4$ = halide ligand

Polymerization occurs *via* complexation of an alkene molecule at the coordination vacancy followed by transfer of the alkyl group to the bound alkene. A new $\sigma$ bond is formed between the growing polymer chain and the transition metal atom. The new structure formed is equivalent to the original except that the positions of the alkyl group and vacancy have exchanged. A final re-exchange of the alkyl and vacancy position is required (Scheme 12) in order to explain the stereoregulating ability of heterogeneous catalysts in $\alpha$-alkene polymerization.

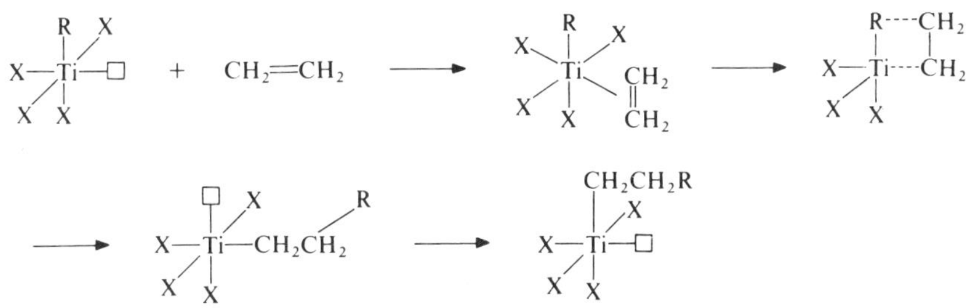

**Scheme 12**

Cossee and Arlman provided strong support for their mechanism from molecular orbital calculations. Figure 1 shows the overlap of the molecular orbitals which are involved in the formation of a $\pi$ bond between the alkene and the transition metal.

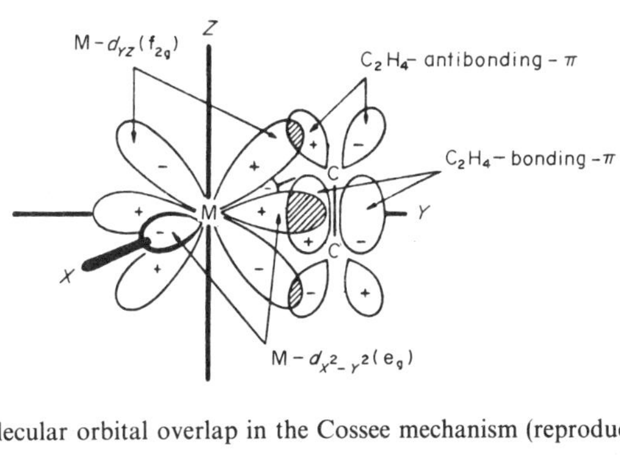


**Figure 1** Molecular orbital overlap in the Cossee mechanism (reproduced from ref. 19)

The metal–alkyl bond of the $RTiCl_4$—□ complex is stable in the uncomplexed state; however, coordination of an alkene molecule results in a mixing of *d* orbitals from the metal with $\pi^*$ antibonding orbitals from the alkene. The resultant orbital is sufficiently lower in energy relative to the original metal 3*d* orbitals that an electron can easily be excited into it from the metal–alkyl bond; the resultant alkyl radical attaches itself in a concerted way to the nearest carbon atom of the bound alkene.

As mentioned previously the final step of the Cossee and Arlman mechanism requires a 'back-jump' of the Ti—R bond in order to fulfil stereochemical requirements; such a step has been considered a weakness of the scheme. Allegra[21] has considered the situation arising from cleavage of $TiCl_3$ crystals and has proposed a different mode of alkene $\pi$-complexation (structure **5**).

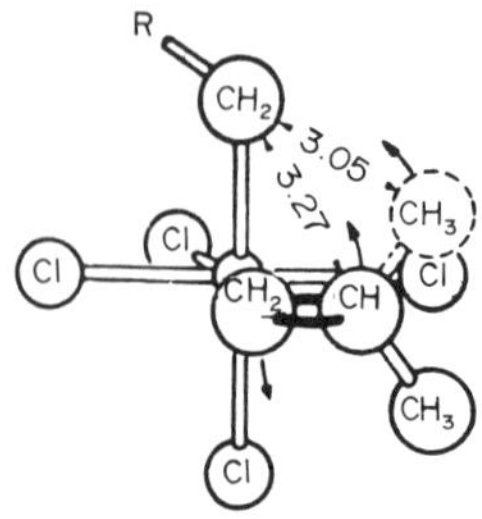


**(5)**

As in the Cossee–Arlman mechanism, active titanium atoms near the cleavage surface are surrounded by a free octahedral site, one singly bonded chlorine atom and four chlorine atoms making bridges to other titanium atoms. Reaction with the activator results in the replacement of the singly bonded chlorine with an alkyl group. Owing to steric interactions, coordination of alkene is believed to occur such that the complexed C=C bond is perpendicular to the Ti—R bond rather than parallel to it as in Cossee–Arlman active centre. A *trans*-like orientation of the C—Me bond is preferred to a *cis*-like mode, since it does not involve any unacceptable interatomic distances. Furthermore the two Cl atoms are equivalent; therefore, following alkene addition and insertion, the Ti—R bond and vacancy will have exchanged position but a further alkene molecule may coordinate to the free titanium vacancy with the same chirality as before and a back-jump step is not necessary to explain stereoregulation.

The mechanism of Rodriguez and Van Looy[22] is based on the Cossee–Arlman scheme but is bimetallic in that an activator molecule is complexed to the transition metal; chain propagation is however monometallic. The nature of the metal alkyl determines the stereoregularity of the polymer formed due to steric interaction with the complexed alkene (Scheme 13).

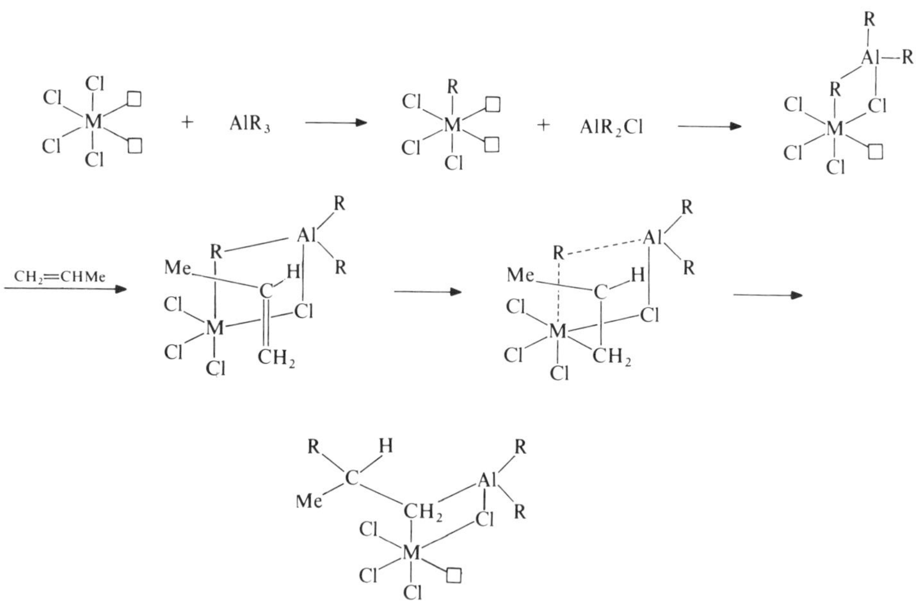


**Scheme 13**

A novel polymerization mechanism has been proposed by Ivin *et al.*[23] in which propagation occurs *via* carbene and metallocyclobutane intermediates rather than by insertion (Scheme 14). α-Elimination of hydrogen from the carbon atom bonded to the transition metal produces a metal hydride and a carbene; the latter reacts with a coordinated alkene molecule to form a metallocyclobutane intermediate which subsequently rearranges.

Of the experimental evidence available in favour of propagation at the transition metal–alkyl bond the strongest is the existence of systems which will polymerize alkenes in the absence of transition metal compounds. Such systems consist of transition metal compounds or salts to which other reagents (but not metal alkyls) may or may not be added; examples include the Phillips catalyst ($CrO_2$ supported on silica) and tetrabenzylzirconium or tetrabenzyltitanium on oxide

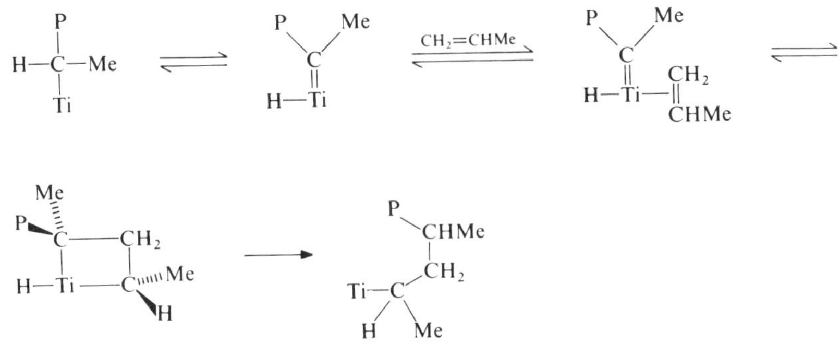

**Scheme 14**

supports. In terms of polymerization behaviour they are similar to conventional two-component catalysts; thus isotactic polypropylene and copolymers of ethylene may be prepared, hydrogen may be added as a transfer agent to control the molecular weight, and the addition of tritiated alcohols leads to the introduction of tritium atoms into the polymer.

Considerable experimental evidence has been obtained concerning the alkylation of the transition metal. It has been demonstrated that $TiCl_4$ is easily alkylated by $AlMe_3$ at 25 °C to give a mixture of $MeTiCl_3$ and $MeTiCl_2$, the relative quantities of the products formed depending on the ratio of reactants used.[24] The vapour phase reactions of $TiCl_4$ with $AlMe_3$ and $ZnMe_2$ have been studied by IR spectroscopy and the catalytic activity of the solid phase correlated with its methyl content.[25] A comprehensive study of the reactions of various metal alkyls ($AlMe_3$, $AlMe_2Cl$ and $AlEt_2Cl$) with α- and β-$TiCl_3$ at high concentrations and in the absence of solvents has been made.[22,26] The results showed that half of the Ti atoms present on the solid surface were alkylated; the majority of these alkylated complexes decompose, but those present on the lateral faces escape decomposition, and it is at these sites that polymerization was assumed to occur.

Kinetic evidence for propagation at a transition metal alkyl complex has been reported although the findings are not conclusive. It has been found that the relative reactivities of propylene and ethylene with $ZrCl_4$, $VCl_4$, $VOCl_3$ and $TiCl_4$ catalysts depend only on the transition metal salt and not on the structure of the metal alkyl.[27] The conclusion that the metal alkyl is not the growth centre is, however, only valid if alkene complexation at the transition metal is *not* rate-determining, since otherwise the copolymer composition would only depend on the adsorption equilibria of each of the monomers at the reaction site. A consideration of the activation energy observed for polymerization indicates that atomic rearrangement following monomer complexation is the rate-determining step and so the above conclusion appears valid.

Natta *et al.*[28] established from kinetic measurements based on the change in viscosity average molecular weight with polymerization time that the propagation rate constant, $k_p$, depends only on the transition metal salt and not on the metal alkyl structure. The alkylated transition metal was considered to be the growth centre, although complexed metal alkyl is present which stabilizes the transition metal–carbon bond and influences the stereochemistry of the system. Complexation of the alkene molecule at the transition metal centre was considered to be the rate-determining step.

Burfield and Tait[29] have established that in the polymerization of 4-methyl-1-pentene the propagation rate constant is independent of the particular aluminum alkyl compound used.

Significant evidence for a monometallic mechanism has been provided by kinetic studies on soluble complexes such as that formed from $Cp_2TiCl_2$ and $AlEt_2Cl$. Spectroscopic methods have shown that alkylation occurs in such systems by exchange of a chloride for an alkyl ligand (Scheme 15).[30] The initial rate of ethylene polymerization using this catalyst was correlated with the concentration of the $Cp_2TiClEt \cdot AlEtCl_2$ complex.

$$Cp_2TiCl_2 + AlEt_2Cl \xrightleftharpoons{\text{complexation}} Cp_2TiCl_2 \cdot AlEt_2Cl \xrightleftharpoons{\text{exchange}} Cp_2TiClEt \cdot AlEtCl_2$$

**Scheme 15**

The cationic species $[Cp_2TiEt]^+$ formed from this complex has been proposed as being responsible for polymerization.[31] During electrodialytic studies, polymerization of ethylene was found to take place at the cathode, where $[Cp_2TiEt]^+$ was present, but not at the anode, where $[AlEtCl_3]^-$ was present.

The addition of $^{14}C$-labelled alkylating agents to similar soluble systems leads to the formation of labelled polymer, again reinforcing the conclusion that insertion of monomer occurs at a transition metal–alkyl bond (Scheme 16).[32]

$$\text{`}Cp_2Ti\cdot THF\text{'} + {}^{14}CH_3I \longrightarrow Cp_2Ti\cdot {}^{14}CH_3I + THF \xrightarrow{CH_3AlCl_2} [Cp_2Ti\cdot {}^{14}CH_3]^+$$

THF = tetrahydrofuran

**Scheme 16**

Finally, Henrici-Olivé and Olivé[33–36] have shown that the active species in the systems $Cp_2TiCl_2$—$AlEtCl_2$ and $Cp_2TiEtCl$—$AlEtCl_2$ is a tetravalent titanium atom since polymerization rates decrease with the extent of reduction of $Ti^{IV}$ to $Ti^{III}$ The model shown in structure **(6)** was proposed for the active centre.

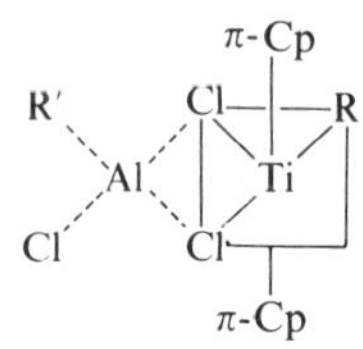

**(6)**

In summary, the concept of polymerization propagation occurring at a transition metal–alkyl bond has been generally accepted in preference to propagation at an activator–alkyl bond. Although the evidence available to support such a conclusion is not totally unequivocal, it is nonetheless more convincing than any supporting evidence to the contrary. In particular, the mechanism of Cossee together with its subsequent modifications has been widely adopted.

## 2.3 MECHANISMS FOR DIENE POLYMERIZATION

Although some workers have assumed that mechanisms for diene polymerization are similar to those of α-alkenes, the situation is undoubtedly more complex since a wider range of compounds may perform as catalysts and the stereochemistry of polymerization is more complex; 1,2-syndiotactic and 1,2-isotactic polymers may be produced together with polymers having *cis*- and *trans*-1,4 structures.

The subject of diene polymerization has been comprehensively reviewed elsewhere.[37, 38]

### 2.3.1 Proposals for the Growth Centre

Just as for α-alkene polymerization, free radical mechanisms can be discounted for various reasons: (i) typical overall activation energies for diene polymerizations ($\sim 15\ \text{kcal mol}^{-1}$) are considerably lower than those in free radical polymerization ($\sim 25\ \text{kcal mol}^{-1}$); (ii) the polymer produced by Ziegler–Natta catalysts has high *cis*-1,4, *trans*-1,4 or 1,2 structural content; and (iii) the polymer produced is generally free of crosslinking.

Conventional ionic mechanisms can also be discounted on the basis of observations (ii)–(iii) above, together with the fact that Lewis bases can often be added without the polymerization being terminated.

Experimental evidence in favour of polymer growth at a metal–alkyl bond is similar in kind to that proposed for α-alkene polymerization, but is not as conclusive. For example, although the use of labelled activators [*e.g.* $Al(^{14}C_2H_5)_3$] with titanium catalysts has been shown to lead to the production of labelled polymer,[39] no incorporation of $^{14}C$ was reported when $Al(^{14}C_2H_5)_2Cl$ was used with a cobalt based system.[40]

Models for diene polymerization have been proposed in which growth occurs at the activator–alkyl bond; in these schemes the transition metal is required for coordination of the monomer, and thus the growth centres are bimetallic. Two models that have been proposed for bimetallic growth centres are shown in structures **(7)** and **(8)**.

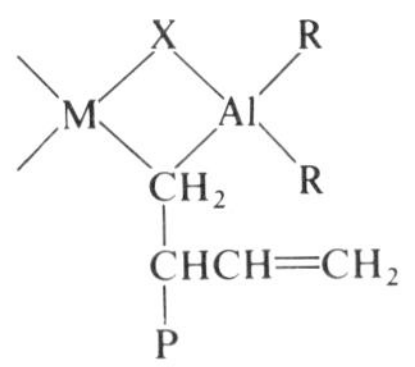

(7)

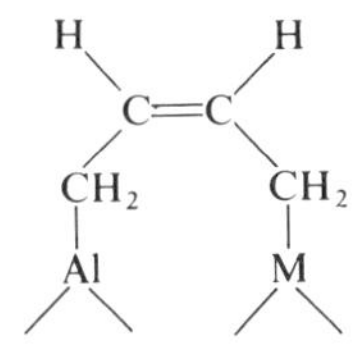

(8)

Experimental evidence for such mechanisms is scarce and is based on the observation that catalytic activity and stereoregulating ability may depend upon the metal alkyl structure. For example, Bressler *et al.*[41] observed different activities for the polymerization of isoprene when $\beta$-$TiCl_3$ was used with $Al(Bu^i)_3$ or $Al(Bu^i)_2Cl$, and furthermore the polymer produced had different molecular weight characteristics. However such findings can be explained by the two metal alkyls producing different numbers of active centres.

Various models have been suggested for growth centres involving polymerization at a transition metal–alkyl bond; several different modes of bonding of the monomer to the transition metal have been proposed, *viz.* a localized $\sigma$ type bond (structure **9**), a stabilized $\sigma$ bond (structure **10**) and a delocalized $\pi$-allyl bond (structure **11**).

$MCH_2CH{=}CHCH_2P$

(9)

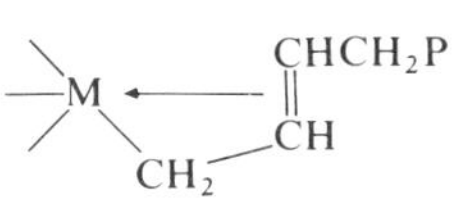

(10)

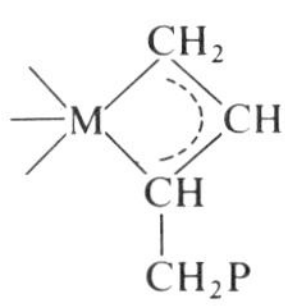

(11)

Structures (**10**) and (**11**) can be considered as essentially the same. Strong evidence for the involvement of a $\pi$-allyl intermediate comes from the existence of activator-free butadiene polymerization catalysts based on $\pi$-allyl compounds such as bis($\pi$-allyl)nickel. The $\pi$-allyl bonding mode is believed to be more likely with metals on the right-hand side of the transition metal block; those having fewer *d* electrons, such as titanium, are more likely to form $\sigma$ bonds. However, it is possible that a catalyst may adopt both bonding modes during polymerization, insertion occurring into the $\sigma$ bonded form.

## 2.4 POLYMERIZATION GROWTH SITES AND ACTIVE CENTRE MODELS

Numerous proposals have been made for the structure of the transition metal centres responsible for polymerization. Some of the more important models will be described here.

### 2.4.1 The Active Centre in $TiCl_3$-based Catalysts

A number of models for the active centres present in $TiCl_3$ catalysts have been proposed including those of Cossee and Arlman, Allegra, and Rodriguez and Van Looy as described in Section 2.2. A determinant factor in the formulation of such structures is the actual locus of polymerization on the $TiCl_3$ surface. For example, in the mechanism of Cossee and Arlman[19] polymerization is suggested to occur at exposed titanium atoms present on the lateral faces of $TiCl_3$ crystals formed by the edges of the close-packed anion planes. Evidence for such a hypothesis comes from electron microscopy studies[21,42] which have clearly demonstrated polymer growth occurring along spiral dislocations and on surface defects of the $\alpha$-$TiCl_3$ crystal. The geometry of the Cossee–Arlman active site is shown in Figure 2; the alkyl ligand and three of the chlorines around the titanium atom are square coplanar, the plane through them forming an angle of 55° to the close-packed anion planes. The environments of the alkyl ligand and coordination vacancy are not equivalent, necessitating the 'back-jump' step of the Cossee–Arlman mechanism.

Theoretical calculations have also supported polymerization occurring at edge sites. Layer edges are present in all of the layer forms of $TiCl_3$, and an appreciable amount of titanium atoms are located at edge sites in high surface area catalysts.

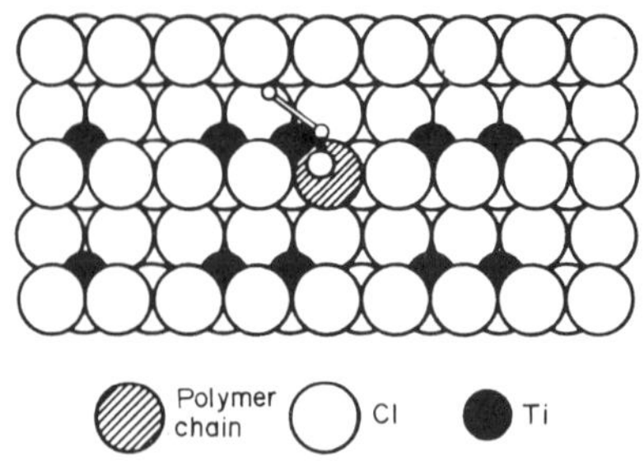


**Figure 2** Geometry of the active centre in the Cossee–Arlman mechanism according to ref. 19

Kissin *et al.*[43] have proposed a monometallic active centre present in heterogeneous catalysts having two vacant coordination positions and which is located on the basal planes of the lattice crystals. The centres are produced by removal of chlorine atoms from the basal surfaces of the transition metal halide by alkylation reactions, and thus contain metal–carbon bonds. The active site model is shown in Figure 3.

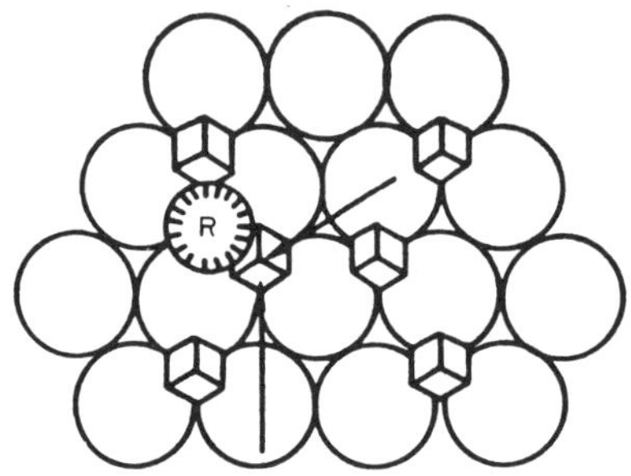


**Figure 3** Model of active centre for Ziegler–Natta polymerization according to Kissin and Chirkov. R is an alkyl group of small volume; the arrows correspond to the direction of monomer approach (reproduced from ref. 43)

Evidence for the Kissin model was provided by active centre concentration determinations using a methanol inhibition technique. The value obtained for the polymerization of propylene by α-$TiCl_3$–$AlEt_3$ was $3 \times 10^{18}$ sites $m^{-2}$ and suggested that the active sites practically cover the whole surface of the catalyst. Since the basal planes account for some 95% of the total surface area then the majority of catalytic centres must be located on these planes. It should be noted that the Kissin theory relies on the validity of the methanol inhibition method and furthermore takes no account of catalyst modification and disintegration during polymerization.

Fuji[44] has studied three different catalyst systems for the polymerization of ethylene: (i) $TiCl_4$ reduced by $AlEt_2Cl$; (ii) $TiCl_4$ reduced by $AlEt_2Cl$ and $TiCl_4$ added as cocatalyst; and (iii) $TiCl_4$ reduced by $AlEt_2Cl$ and $AlEt_2Cl$ added as cocatalyst.

The titanium present in all three catalysts was in the $Ti^{III}$ state. However, differences in polymerization rates and the numbers of branches and double bonds in the polymer formed were attributed to the physical state around the titanium atoms. Investigations of concentrations of alkyl groups unstable to heat treatment at 140 °C, assumed to be Ti—R linkages, and those that were thermally stable, assumed to be either bonded Al—R or Ti - - - - - R - - - - - Al, led Fuji to propose the structures shown in Figure 4 for the active centres present in catalysts (i), (ii) and (iii).

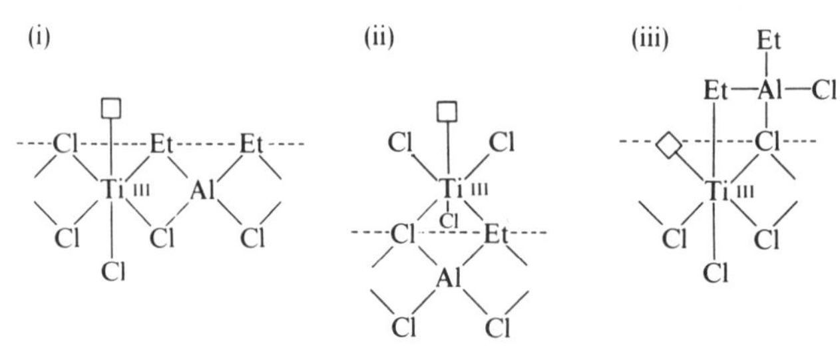


**Figure 4** Structures proposed for the active centres present in the catalyst systems described by Fuji (reproduced from ref. 44)

### 2.4.2 Models for Active Centres in Magnesium Chloride Supported Catalysts

Commercially important catalysts for the polymerization of propylene have been developed which are based on the use of magnesium chloride as a support. In a typical preparation anhydrous magnesium chloride is ball-milled in the presence of an electron donor and then reacted with a titanium compound such as $TiCl_4$. Although many studies have been made concerning such parameters as the effect of ball-milling on the support, interactions with the electron donor, *etc.*, few models have been proposed for the active centres present in these complex systems.

Chien *et al.*[45] have made an extensive study of a 'high mileage' catalyst prepared by ball-milling magnesium chloride, pre-treated with HCl, in the presence of ethyl benzoate and then treating with *p*-cresol, $AlEt_3$ and $TiCl_4$. The oxidation states of the titanium atoms present were reported as $Ti^{II}$ (8%), $Ti^{III}$ (38%) and $Ti^{IV}$ (54%). Electron paramagnetic studies on the final catalyst revealed a single observable $Ti^{III}$ species which was strongly attached to the catalyst surface (structure **12**). This $Ti^{III}$ species is coordinatively unsaturated; the vacant coordination position may be occupied by a weak ligand or solvent molecule.

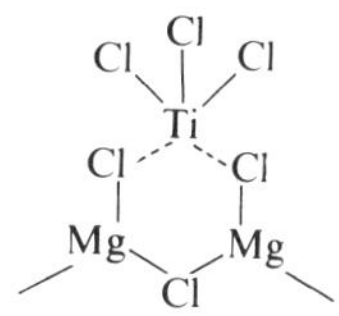

**(12)**

Only approximately 20% of the $Ti^{III}$ present was attributed to the EPR observable species. It was concluded that the remainder of the $Ti^{III}$ had adjacent sites occupied by one or more $Ti^{III}$ ions. Chlorine bridge structures such as that shown in structure (**13**) were suggested as being responsible for the EPR silence.

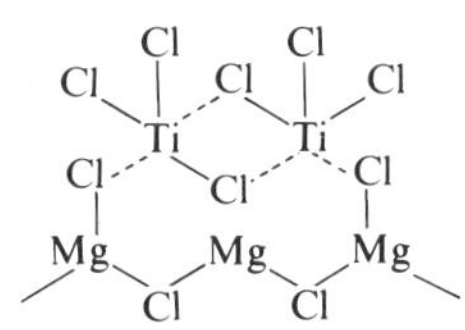

**(13)**

Activation of the catalyst with a mixture of $AlEt_3$ and methyl *p*-toluate (3:1) leads to a reduction of 90% of the $Ti^{IV}$ to lower oxidation states. Furthermore, a single $Ti^{III}$ species with rhombic symmetry is produced with a structure believed to be that shown in structure (**14**). The species was found to be unstable, and on ageing was converted to a $Ti^{III}$ species with axial symmetry.

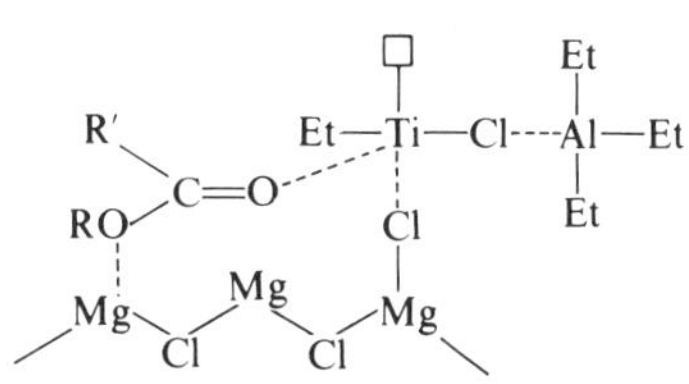

**(14)**

Structure (**14**) is coordinatively unsaturated and contains a $Ti^{III}$ atom with a vacant site and an asymmetric field, and thus fulfills the requirements of a site capable of stereospecific polymerization.

Kakugo *et al.*[46] have studied the microstructure of polypropylenes produced by $TiCl_3$-based and $MgCl_2$-supported catalysts using $^{13}C$ NMR spectroscopy. Two types of isospecific centres were identified in the catalysts studied which produced polymer with high and low stereospecificities. By analogy with $TiCl_3$-based systems, Scheme 17 was proposed for the formation of active centres in $MgCl_2$-supported catalysts.

Kakugo *et al.*[46] proposed that the addition of electron donors converts the non-stereospecific centres to ones of high isospecificity, whereas the low isospecific centres are rendered inactive as shown in Scheme 18.

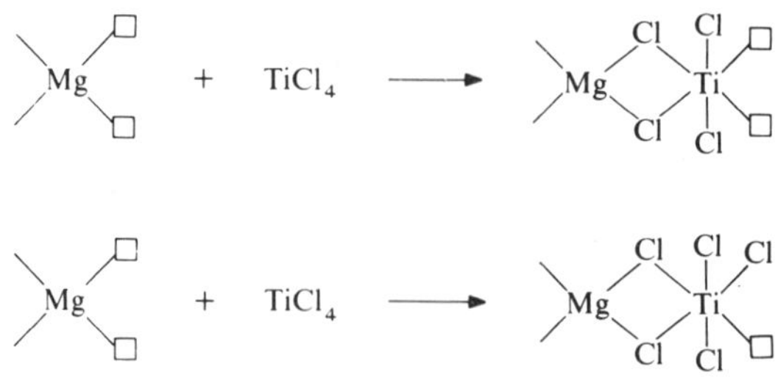


**Scheme 17**

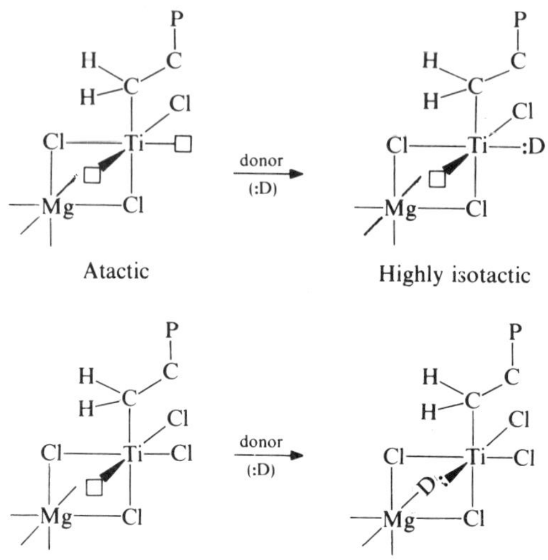


**Scheme 18**

Doi *et al.*[47] have carried out similar $^{13}C$ NMR studies on the structures of polypropylenes produced by various catalyst systems, including $TiCl_4/AlEt_3$, $\delta$-$TiCl_3/AlEt_3$, $MgCl_2/TiCl_4$/ethyl benzoate–$AlEt_3$ and $TiCl_4/MnCl_2$/ethyl benzoate–$AlEt_3$. Two types of active centre were proposed, one producing highly isotactic polypropylene and the other producing atactic polymer composed of isotactic and syndiotactic stereoblocks of short sequence length. The structures of the centres are shown in Figure 5. In site A the surface metal (Mg, Mn or Ti) is coordinatively saturated, whereas in site B it has a chlorine vacancy. Reversible migration of the alkylaluminum compound in site B was suggested as the cause of the frequent changes in polymer structure.

### 2.4.3 Models for Active Centres in the Phillips Catalyst

An important catalyst for the commercial polymerization of HDPE is the Phillips Cr/silica catalyst.

The Phillips system comprises an oxide support, usually silica or alumina–silica, impregnated with chromium(VI) oxide to a loading of 0.5–5.0% chromium, typically 1%. Supports are often modified by compounds such as titania.

The catalysts produce high polymerization rates and in contrast to true Ziegler–Natta systems do not require an activator or cocatalyst compound.

The support plays an important role in the stabilization of the $Cr^{VI}$ species; a high temperature treatment is required to activate the catalyst and to remove any moisture from the surface. Heating of unsupported $CrO_3$ in the range 473–703 K leads to decomposition to $Cr^{III}$ with the evolution of oxygen. However, the use of a support such as silica stabilizes the $Cr^{VI}$ and prevents reduction to $Cr^{III}$ during the heat treatment.

The stabilization of Cr is believed to be due to the interaction of $CrO_3$ with hydroxyl groups on the support surface. Hogan[48] showed that a $CrO_3/SiO_2$–$Al_2O_3$ catalyst containing 2.49 wt% Cr

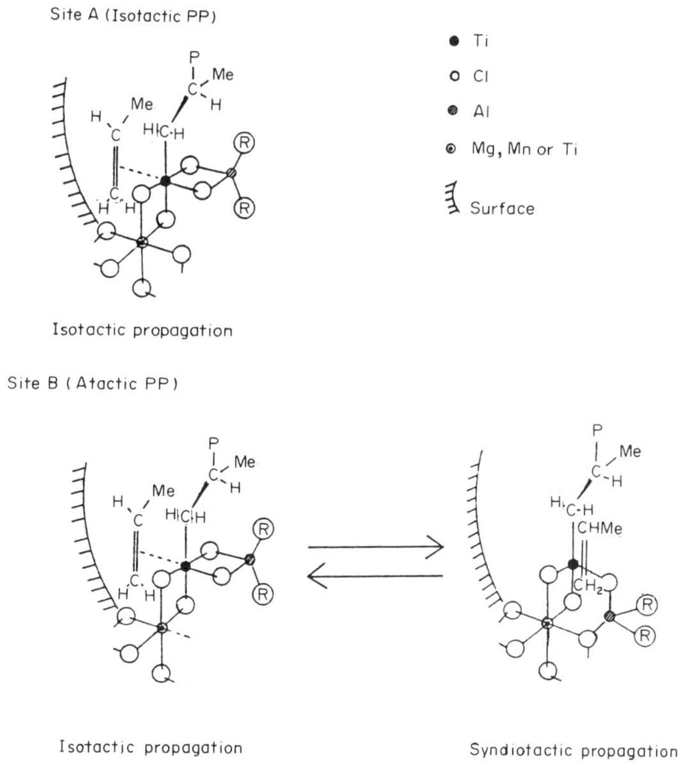


**Figure 5** Structures for active centres proposed by Doi *et al.*, site A producing highly isotactic polypropylene, site B producing atactic polypropylene (reproduced from ref. 47)

contains 96% as $Cr^{VI}$ after calcination at 813 K in dry air, and he concluded that the stabilization was due to the formation of surface chromate and dichromate species (Scheme 19).

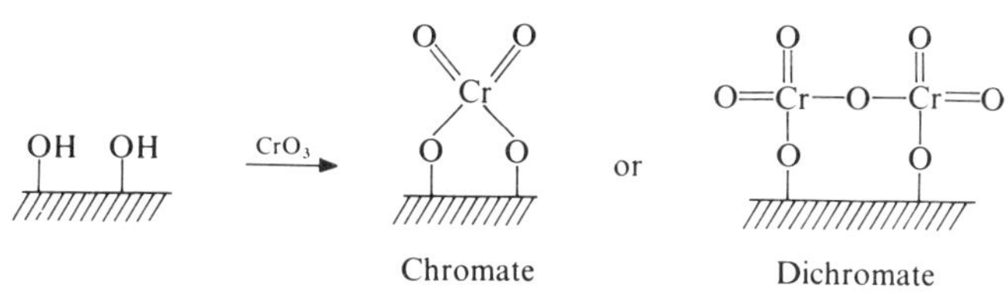


**Scheme 19**

The valence state of Cr during polymerization has been the subject of considerable dispute; every valence state between $Cr^{II}$ and $Cr^{VI}$ has in fact been proposed as the active species. Although some early workers proposed that $Cr^{V}$ was responsible for the polymerization of ethylene, few workers now believe that higher oxidation states are the active species. Most researchers at the present time believe that the active species is $Cr^{II}$ or $Cr^{III}$, or perhaps a combination of both. However, it is also possible that the coordinative unsaturation of the Cr is as important as its valency.

### 2.4.4 Active Centres in Soluble Catalyst Systems

Much work has been performed using homogeneous Ziegler–Natta catalyst systems with the aim of establishing a general mechanistic scheme; such systems are easier to study than heterogeneous systems in that each metal atom is supposed to form an active centre. The studies of Natta using the $Cp_2TiCl_2$ catalyst for the polymerization of propylene has been described earlier; the model for the active centres of the catalyst $Cp_2TiCl_2$–$AlEtCl_2$ proposed by Henrici-Olivé was shown in structure (**6**).

More recently a fascinating discovery has been made whereby a very highly active catalyst for the polymerization of ethylene can be prepared by pre-treating $AlEt_3$ with water and adding to dialkylbis(cyclopentadienyl)titanium(IV) catalysts.[49] Earlier work had shown that the addition of water to homogeneous chloride-containing systems based on bis(cyclopentadienyl)titanium(IV) catalysts increased the activity. Cihlar *et al.*[50] proposed the stabilized complexes shown in Scheme 20 from similar studies.

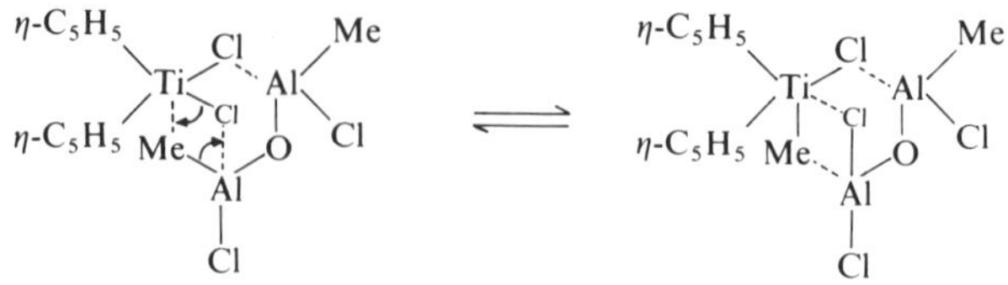

**Scheme 20**

The addition of water to $AlMe_3$ leads to the formation of an oligomeric aluminoxane with the structure [—OAl(Me)—]$_n$. Further condensation with elimination of $AlMe_3$ is believed to give rise to cyclic structures of the type suggested by Kaminsky and Sinn,[51] shown in Scheme 21.

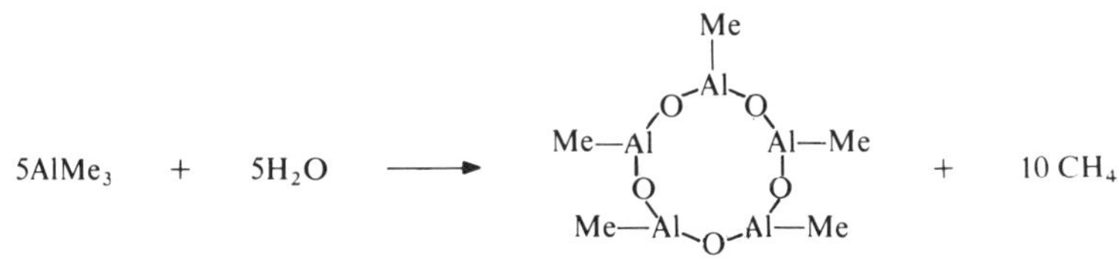

**Scheme 21**

Kaminsky *et al.*[52] have shown that cyclopentadienyl derivatives of zirconium [*e.g.* bis(cyclopentadienyl)dimethylzirconium] in conjunction with alkylaluminoxanes produce exceptionally active catalysts for ethylene polymerization, $> 10^8$ g PE/g Zr. A large excess of aluminoxane was required and the rate was found to be proportional to the Zr concentration and to depend quadratically on that of the aluminoxane. The polymerization rate increased after an induction period to a maximum value and thereafter remained constant, indicating a 'living' polymerization system. The complex shown in structure (**15**) was proposed. The system could also polymerize propylene to atactic polymer and copolymerize ethylene and propylene.

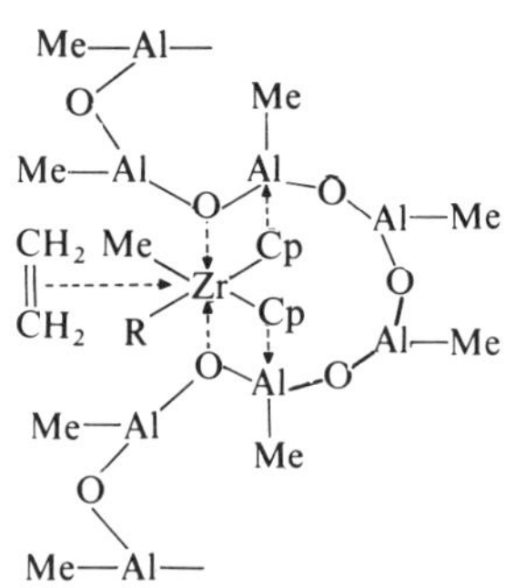

(**15**)

In later papers, however, Kaminsky *et al.*[53,54] have reported that highly isotactic polypropylene can be prepared using stereorigid chiral zirconocenes such as a racemic mixture of ethylenebis(indenyl)zirconium dichloride or ethylene(tetrahydroindenyl)zirconium dichloride together with methylaluminoxane cocatalyst. *rac*-Ethylene(bistetrahydroindenyl)zirconium dichloride is shown in structure (**16**).

A separation of racemic mixtures of the zirconocene compounds has been performed to yield an optically pure enantiomer. The use of such a compound has led to the first ever preparation of optically active polypropylene, since the growing polymer chain is formed predominantly in only one screw sense.

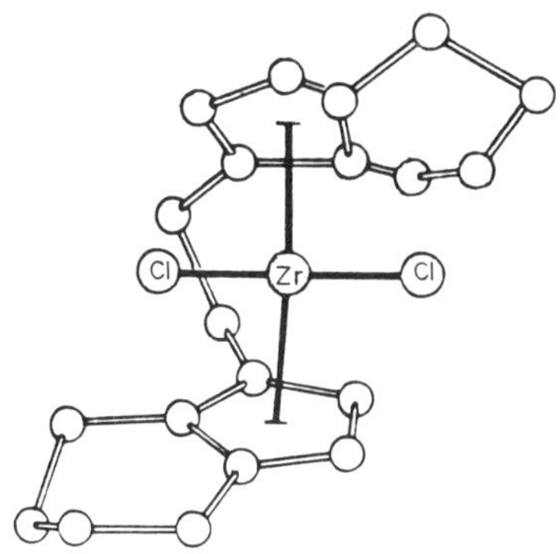


**(16)**

## 2.5 KINETICS OF ZIEGLER–NATTA POLYMERIZATION

### 2.5.1 Introduction

It is well established that the overall kinetic behaviour of any catalytic polymerization process results from the overlap of three potentially time-dependent reactions: the time dependence of the intrinsic catalyst activity; the time dependence of the overall polymerization reaction; and the time dependence of any diffusion processes which may also operate. The kinetic behaviour of Ziegler–Natta catalysts is no exception to such reactions. Furthermore the kinetic behaviour of heterogeneous Ziegler–Natta catalysts is complicated by adsorption reactions of monomer, of cocatalyst, of donors (internal and external) and of by-products formed in reactions leading to the formation of active centres. For these reasons the kinetics of Ziegler–Natta polymerization reactions are often exceedingly complicated and present a fascinating and exciting challenge to investigators.

### 2.5.2 Phenomenological Behaviour

#### *2.5.2.1 Kinetic rate–time profiles*

A comparison of kinetic data on alkene polymerization allows classification of the typical kinetic curves in coordinates of yield and time, or, better, of rate and time. Such kinetic rate–time profiles are significant in Ziegler–Natta polymerizations in that their particular shape may be characteristic of a particular catalyst or catalyst–monomer system. The importance of such profiles has been discussed by Keii[55] and Tait.[56–58] A number of different types of kinetic rate–time profiles are encountered in studies of Ziegler–Natta catalysts and some typical examples are shown in Figure 6.

In general such profiles can be considered to consist of three periods, *viz.* an acceleration period, a stationary period and a decay period, and as can be seen some catalyst systems show all three types.

Type (a) behaviour is shown by many first generation catalyst systems, *e.g.* $\alpha$-$TiCl_3$, $VCl_3$, *etc.*, when used with dialkylaluminum halides as cocatalysts for the polymerization of propylene in hydrocarbon media.[59] It can be seen that these polymerizations show an acceleration or settling period during which the rate increases to reach a more or less steady value, the duration of this period being of the order of 20–60 min for many propylene polymerizations when carried out at 1 atm pressure in the temperature range 50–70 °C.

The nature of this acceleration period was for a long time the subject of debate. Early research by Natta and Pasquon[59] using $\alpha$-$TiCl_3$ catalysts established that breakdown of the $\alpha$-$TiCl_3$ matrix took place due to the growth of polymer chains during the early stages of the polymerization, thus exposing fresh titanium atoms and leading to the formation of new active centres and an increase in the polymerization rate. After breakdown of the $\alpha$-$TiCl_3$ matrix to primary crystallites had been established, a steady state rate was observed for a significant period of time. The effect is illustrated in Figure 7.

The attainment of a steady rate of polymerization allowed Natta and his co-workers[59] to carry out a very systematic investigation and analysis of the polymerization of propylene by $\alpha$-$TiCl_3$–$AlR_2Cl$ catalysts.

Mechanical grinding of the $\alpha$-$TiCl_3$ catalysts shortens the acceleration period considerably but changes the character of the catalyst, the polymerization process becoming unstable (*cf.* Figure 7), and finally tending to the same rate value as before. Whilst breakdown of the catalyst matrix

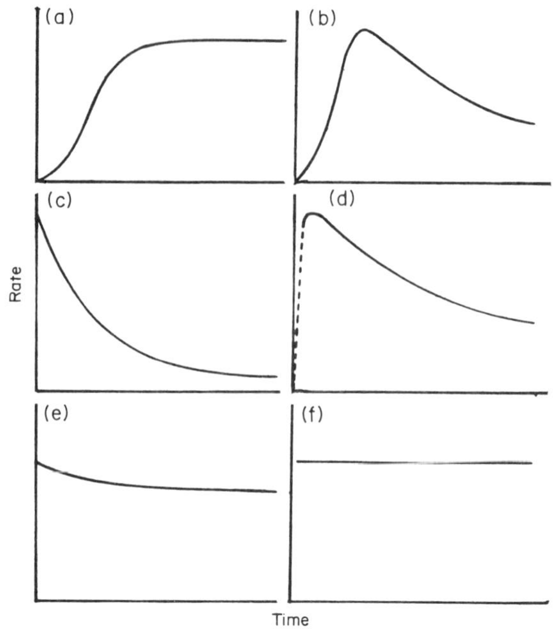


**Figure 6** Some typical kinetic rate–time profiles (reproduced from ref. 58)

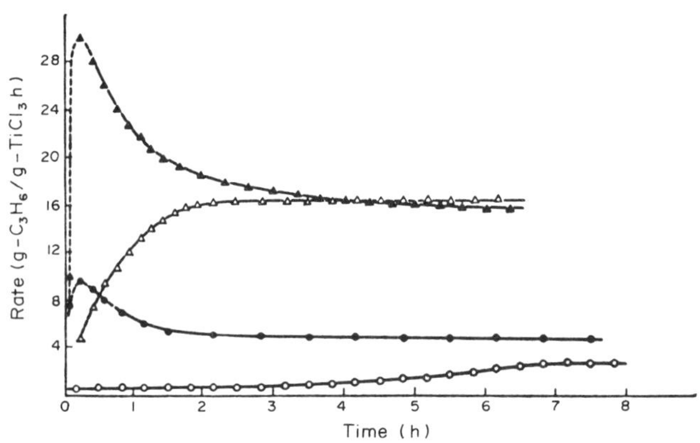


**Figure 7** Effect of grinding on the shapes of the kinetic curves. (▲) and (●) show ground $TiCl_3$ (sample A), particle size $\leqslant 2\ \mu m$; △ and ○ show unground $TiCl_3$ (sample A), particle size 1–10 $\mu m$; P = 1450 mmHg; triangles, $T = 70\ °C$; circles, $T = 32\ °C$ (reproduced from ref. 59)

certainly takes place, it is difficult nevertheless to visualize a breaking process which always gives the same level of crystal dispersion.

Type (b) behaviour is often observed when trialkylaluminum compounds are used as cocatalysts. The use of a trialkylaluminum compound usually produces a more active but less stable catalyst. In type (b) behaviour the rate of polymerization increases to reach a maximum and then decreases with increase in polymerization time.

Many high activity supported catalyst systems[60–64] show either (c) or (d) type behaviour, in which the rate may either start at a maximum value or rise very rapidly to a maximum value and then decrease rapidly with time, *e.g.* $MgCl_2/EB/TiCl_4$–$AlEt_3$ catalyst systems when used for either ethylene or propylene polymerization (EB = ethyl benzoate).

Type (c) behaviour was observed in the early polymerization of styrene by catalysts derived from $TiCl_4$ and $AlEt_3$.[65,66] Such behaviour is also shown by many homogeneous catalyst systems, *e.g.* $Cp_2TiEtCl$–$AlEtCl_2$ in ethylene polymerization,[36,67] and by modified catalyst systems, such as $VOCl_3 \cdot 2THF$–$AlR_3$, in vinyl chloride polymerization.[68]

Type (e) behaviour is shown by Solvay & Cie ether-treated catalysts in propylene polymerization.[57,69,70] Breakdown of the porous catalyst particles is practically instantaneous on treatment with a dialkylaluminum compound and the settling period is more or less eliminated. Thereafter the polymerization rate decreases very gradually with time and the catalyst system shows good stability.

Type (f) behaviour, for which the polymerization system shows no settling or adjustment period but whose rate remains strictly constant with time, is rare and so is not often encountered. Nevertheless such behaviour is shown by $MgCl_2$-supported catalysts containing phthalate esters when used for the polymerization of 4-methyl-1-pentene (4-MP-1). Here no settling period is observed and the polymerization rate remains almost completely constant with time.[58,64,71,72] Catalyst systems $Cp_2TiCl_2$–aluminoxane also shown constant polymerization rates apart from a short settling period.[52]

The principal factors giving rise to the observed types of kinetic rate–time profiles are as follows.

### *(i) Catalyst matrix type and preparation*

Solvay & Cie ether-treated type catalysts have highly porous structures and catalyst matrix breakdown takes place very rapidly on treatment with dialkylaluminum chloride to produce primary catalyst particles. Alkylation of exposed Ti atoms is thus completed over a short time interval. These catalyst systems show reduced settling periods in comparison to either $\alpha$-$TiCl_3$–$AlEt_2Cl$ or $\gamma$-$TiCl_3 \cdot 0.33AlCl_3$–$AlEt_2Cl$ systems.[69,70] A typical plot of rate *vs.* time is shown in Figure 8.

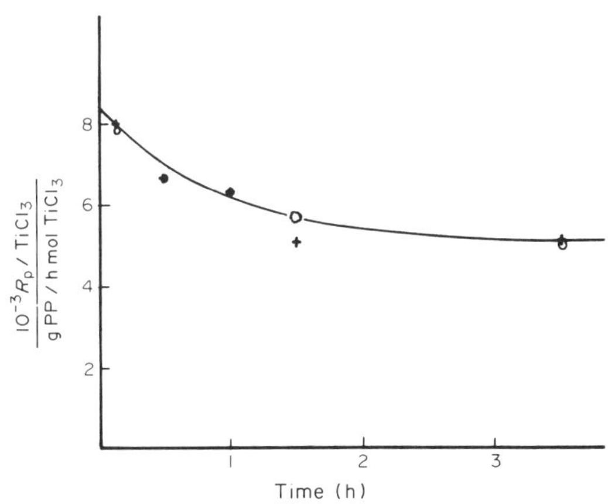


**Figure 8** Plot of rate of polymerization *vs.* time for the polymerization of propylene at 60 °C using a Solway & Cie type catalyst; $[TiCl_3] = 5.0$ mmol dm$^{-3}$; $[AlEt_2Cl] = 10.0$ mmol dm$^{-3}$; 1 atm pressure

When the breakdown of the catalyst matrix takes place over a longer time scale, as in the polymerization of propylene using spheroidal $MgCl_2$/phthalate/$TiCl_4$ catalysts,[73] more complex kinetic rate–time profiles can be generated, as shown in Figure 9.

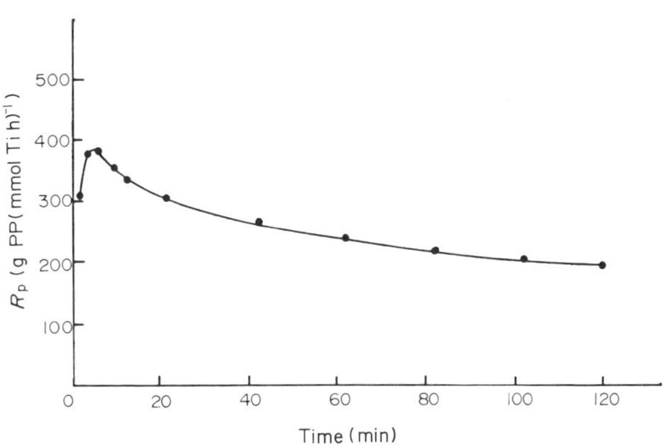


**Figure 9** Plot of rate of polymerization of propylene *vs.* time using the catalyst $MgCl_2$/DIOP/$TiCl_4$, $[Ti] = 0.08$ mmol dm$^{-3}$; $[Ti]:[AlEt_3]:[DPDMS] = 1:169:16.9$. Pressure = 1 atm, solvent = pentamethylheptane, temperature = 60 °C

### (ii) Alkylaluminum

Kinetic rate–time profiles are affected by both the type and concentration of alkylaluminum compound used as cocatalyst. The observed behaviour is complicated and is the result of alkylation, reduction, extraction (*i.e.* $AlCl_3$ from $TiCl_3 \cdot 0.33AlCl_3$ catalysts) and catalyst modification reactions.[64]

The effect of variation of alkylaluminum concentration in the case of $MgCl_2$/electron donor/$TiCl_4$ catalyst systems is illustrated in Figure 10. This behaviour is believed to arise from adsorption reactions and has been analyzed by Keii.[61] The presence of an electron donor (internal and/or external) also affects the rate–time behaviour.

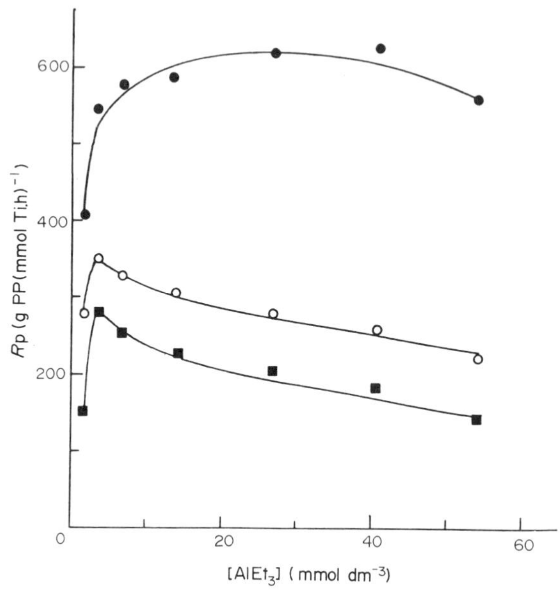


**Figure 10** Variation of rate of polymerization of propylene with concentration of $AlEt_3$: ● = polymerization rate at 4 min; ○ = polymerization rate at 60 min; □ = polymerization rate at 120 min; [Ti] = 0.08 mmol dm$^{-3}$; pressure = 1 atm; solvent = pentamethylheptane; temperature = 60 °C

### (iii) Monomer pressure

The pressure of monomer may also affect the observed kinetic rate–time profiles.[55] Many catalyst systems which have steady kinetic rate–time profiles for polymerizations at atmospheric pressure show decay type profiles when used at significantly higher pressures of monomer, *i.e.* at 5–10 bar.

### (iv) Temperature

The overall effect of increased temperature when this is raised beyond a certain critical value appears to be destabilization of the catalyst system. Catalysts showing types (a) or (e) behaviour can thus show types (b), (c) or (d) behaviour at higher temperatures. Although most published results relate to the temperature range 30–100 °C some industrial processes operate at temperatures above 200 °C.

### (v) Monomer

The monomer used can affect the observed kinetic rate–time profiles. It is found that α-alkenes such as 4-MP-1 show rate–time profiles which are different from those obtained in either ethylene or propylene polymerizations.[64, 71]

### (vi) Polymerization media

Ziegler–Natta polymerizations are most often carried out in hydrocarbon media. Polymerizations carried out in aromatic solvents can show very different rates and kinetic rate–time profiles.[58]

It is believed that the complicated behaviour as outlined above arises from both chemical and physical factors. The effective particle size and particle size distribution as functions of time will determine the numbers of exposed transition metal atoms which can be involved in the formation of active centres at any given time. Chemical reactions which control the formation and deactivation of active centres will control the instantaneous concentration of potential or actual active centres.

### *2.5.2.2 Molecular weight–time profiles*

The change of molecular weight with time in Ziegler–Natta polymerizations can be conveniently discussed in terms of two different types of behaviour which are shown schematically in Figure 11.

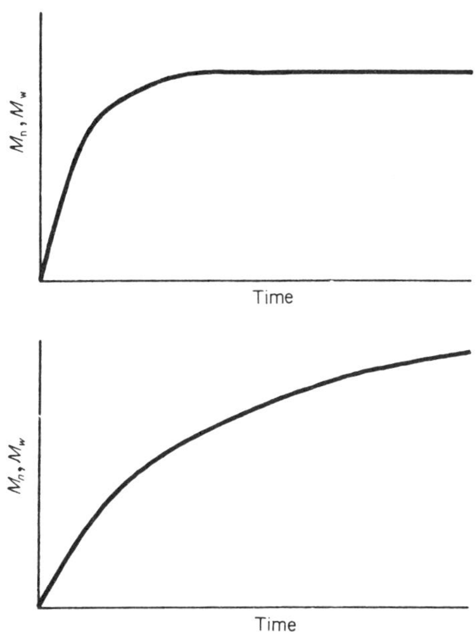


**Figure 11** Dependence of molecular weight of polymer on time for polymerizations using heterogeneous Ziegler–Natta catalysts

In the first case the molecular weight increases for an initial period of time and then reaches a constant value. A number of polymerization systems show this type of dependency and include $\gamma$-$TiCl_3$–$AlEt_3Cl$,[74] $\alpha$-$TiCl_3$–$AlEt_3$,[59] $VCl_3$–$Al(Bu^i)_3$,[75] *etc.* The duration of the first stage in Figure 9 depends strongly on the polymerization temperature and the concentration of monomer. When the polymerization is carried out at low temperatures the molecular weight growth can last for a few hours.[59] For polymerizations carried out at 70–80 °C and high concentrations of monomer[76] the molecular weight becomes independent of time at an early stage, *i.e.* $< 30$ min. In the second case the molecular weight continues to increase throughout the duration of polymerization although the rate of this increase diminishes with time. Such a dependency has been reported for the polymerization of ethylene,[77] propylene[78] and 4-MP-1[79] by the catalyst system $\delta$-$TiCl_3$–$AlEt_2Cl$. Such behaviour can be explained in principle when the duration of polymerization was not long enough to allow equilibrium between chain initiation and chain termination.

## 2.5.3 Kinetic Models

### *2.5.3.1 Early kinetic models*

A number of simple kinetic models[59, 80] have been developed for catalysts which have relatively low activities and are characterized by kinetic rate–time profiles of the type shown in Figure 6 (a and f). These models are based on the assumption that the total concentration of active centres, $C^*$, remains constant throughout the polymerization.

*(i) Chain initiation*

This step is assumed to be the insertion of the first monomer molecule, M, into a transition metal–carbon bond in an active centre, Cat—R

$$\text{Cat—R} + \text{M} \xrightarrow{k_i} \text{Cat—P}_1 \tag{1}$$

*(ii) Chain propagation*

The chain propagation step is regarded as the insertion of a monomer molecule into a transition metal–carbon bond in a propagation centre, Cat—P

$$\text{Cat—P}_1 + \text{M} \xrightarrow{k_p} \text{Cat—P}_2 \tag{2}$$

$$\text{Cat—P}_2 + \text{M} \xrightarrow{k_p} \text{Cat—P}_3 \tag{3}$$

$$\text{Cat—P}_n + \text{M} \xrightarrow{k_p} \text{Cat—P}_{n+1} \tag{4}$$

In this scheme all propagation centres are regarded as equally active and having the same propagation constant, $k_p$, independent of their geometric location and of their degree of polymerization. However since it has been established that all centres do not have equal activities, values of $k_p$ must be regarded as average values ($\bar{k}_p$).

*(iii) Chain transfer*

Where no specific chain transfer agent has been added to the polymerization system three transfer reactions are usually considered.

*(a) Transfer with monomer.*

$$\text{Cat—P}_n + \text{M} \xrightarrow{k_m} \text{Cat—R}' + \text{P}_n \tag{5}$$

*(b) Transfer with alkylaluminum.*

$$\text{Cat—P}_n + \text{AlR}_x \xrightarrow{k_a} \text{Cat—R}'' + \text{P—AlR}_{x-1} \tag{6}$$

Chain transfer reactions with monomer and with alkylaluminum are introduced into the polymerization scheme because of the well established experimental observations that the molecular weight of the polymer produced decreases as the concentrations of monomer and/or alkylaluminum compound increases.

*(c) Spontaneous transfer.*

$$\text{Cat—P}_n \xrightarrow{k_{sp}} \text{Cat—R}''' + \text{P}_n \tag{7}$$

The three new initiation centres formed in reactions (5)–(7) have different chemical structures. Hence for the polymerization of propylene using either $AlEt_3$ or $AlEt_2Cl$ as cocatalyst, R′ = *n*-propyl, R″ = ethyl and R‴ = H. For most catalyst systems and under normal conditions of monomer and alkylaluminum concentrations, spontaneous chain transfer reactions can be neglected.

*(d) Transfer with transfer agents.* Some compounds, particularly hydrogen, are effective as chain transfer agents and lead to a reduction of molecular weight. For this reason hydrogen is usually added in the commercial production of polyethylene and polypropylene.

A kinetic analysis may be carried out using the above reaction scheme.[80]

The total concentration of active centres, $C^*$, consists of the concentration of propagating centres, $C_p^*$, together with the sum of the concentrations of different initiation centres, $\Sigma C_i^*$, at which an initiation reaction takes place at that particular moment

$$C^* = C_p^* + \sum_j C_i^* \qquad (8)$$

Under stationary state conditions only the steady state concentration with respect to the concentration of propagation centres need be considered and

$$\frac{dC_p^*}{dt} = 0 \qquad (9)$$

also $\sum_j R_{\text{initiation}} = \sum_j R_{\text{transfer}}$.

Thus in the above reaction scheme

$$k_i'C_i'[M] = k_m C_p^*[M] \qquad (10)$$

$$k_i''C_i''[M] = k_a C_p^*[A] \qquad (11)$$

$$k_i'''C_i'''[M] = k_{sp} C_p^* \qquad (12)$$

where [M] and [A] are the concentrations of monomer and alklyaluminum respectively, and $k_i'$, $k_i''$ and $k_i'''$ are the rate constants for chain initiation involving the Cat—R′, Cat—R″ and Cat—R‴ species respectively (*cf.* equation 1).

It can now be shown that

$$C_p^* = C^*(1 + k_m/k_i' + k_a[A]/k_i''[M] + k_{sp}/k_i'''[M])^{-1} \qquad (13)$$

Hence the rate of polymerization

$$R_p = k_p C_p^*[M] \qquad (14)$$

$$= \frac{k_p C^*[M]^2}{[M] + k_m[M]/k_i' + k_a[A]/k_i'' + k_{sp}/k_i'''} \qquad (15)$$

and the number-average degree of polymerization

$$\bar{P}_n = \frac{k_p}{k_m[M] + k_a[A] + k_{sp}} \qquad (16)$$

Some important limiting expressions can be obtained from equations (13), (15) and (16) when [M] is high, *i.e.*

$$C_p^* = C^*(1 + k_m/k_i')^{-1} \qquad (17)$$

and

$$R_p = \frac{k_p C^*[M]}{(1 + k_m/k_i')} \qquad (18)$$

when

$$k_i' \approx k_p \gg k_m; \quad C_p^* \approx C^* \qquad (19)$$

and

$$R_p = k_p C^*[M] \qquad (20)$$

Also

$$\bar{P}_n = k_p/k_m \qquad (21)$$

*i.e.* $\bar{P}_n$ at high [M] becomes independent of all polymerization variables.

Under more normal experimental conditions

$$\bar{P}_n^{-1} = k_m/k_p + k_a[A]/k_p[M] + k_{sp}/k_p[M] \qquad (22)$$

Hence a plot of $\bar{P}_n^{-1}$ as $1/[M]$ for polymer samples prepared at constant [A] should be a straight line having $k_m/k_p$ as the intercept and $(k_a[A]/k_p + k_{sp}/k_p)$ as the slope. Similarly when $\bar{P}_n^{-1}$ is plotted *vs.* [A] for polymer samples prepared at constant [M], a straight line should be obtained having $(k_m/k_p + k_{sp}/k_p[M])$ as the intercept and $k_a/k_p[M]$ as the slope. Consequently all $k_t/k_p$ ratios may be evaluated,[80] and typical values for a number of catalyst systems are listed in Tables 2 and 3.

**Table 2** $k_m/k_p$ Ratios for Alkene Polymerization with Various Heterogeneous Ziegler–Natta Catalysts

| *Monomer* | *Catalyst system* | *Temperature* (°C) | $k_m/k_p$ | *Ref.* |
|---|---|---|---|---|
| Ethylene | $\delta$-$TiCl_3$–$AlEt_3$ | 50 | $0.9–1.8 \times 10^{-5}$ | 81 |
| | $\delta$-$TiCl_3$–$AlEt_3$ | 80 | $0.35–0.7 \times 10^{-4}$ | 82 |
| | $\gamma$-$TiCl_3$–$AlEt_2Cl$ | 40 | $1.38 \times 10^{-4}$ [a] | 83 |
| | $VCl_3$–$Al(Bu^i)_3$ | 60 | $2.1–4.3 \times 10^{-4}$ | 84 |
| Propylene | $\alpha$-$TiCl_3$–$AlEt_3$ | 70 | $5.7–11.5 \times 10^{-4}$ | 84 |
| | $\alpha$-$TiCl_3$–$Al(Bu^i)_3$ | 70 | $19–38 \times 10^{-4}$ | 84 |
| | $\alpha$-$TiCl_3$–$BeEt_2$ | 30 | $1.7–3.5 \times 10^{-4}$ | 85 |
| | $\alpha$-$TiCl_3$–$BeEt_2$ | 60 | $8.8–17.5 \times 10^{-4}$ | 85 |
| | $\alpha$-$TiCl_3$–$ZnEt_2$ | 50 | $20–40 \times 10^{-4}$ | 86 |
| | $\alpha$-$TiCl_3$–$ZnEt_2$ | 60 | $10–20 \times 10^{-4}$ | 86 |
| | $\delta$-$TiCl_3$–$AlEt_3$ | 70 | $3.6–7.3 \times 10^{-4}$ | 82 |
| | $VCl_3$–$Al(Bu^i)_3$ | 60 | $10 \times 10^{-4}$ | 84 |
| | $VCl_3$–$Al(Bu^i)_3$ | 70 | $2.8 \times 10^{-4}$ | 84 |
| 4-Methyl-1-pentene | $VCl_3$–$Al(Bu^i)_3$ | 30 | $4.4–8.8 \times 10^{-4}$ | 87 |

[a] These estimations are based on $\bar{M}_n$ data. In all other cases $\bar{M}_w$ was calculated from viscosity data and $\bar{M}_w/\bar{M}_n$ was assumed to be in the range of 5–10.

**Table 3** $k_a/k_p$ and $k_{sp}/k_p$ Ratios for Alkene Polymerization with Various Heterogeneous Ziegler–Natta Catalysts

| *Monomer* | *Catalyst system* | *Temperature* (°C) | $k_a/k_p$ | $k_{sp}/k_p$ | *Ref.* |
|---|---|---|---|---|---|
| Ethylene | $\gamma$-$TiCl_3$–$AlEt_2Cl$ | 40 | — | $3.8 \times 10^{-6}$ [a] | 83 |
| | $\delta$-$TiCl_3$–$AlEt_3$ | 75–80 | $1.7 \times 10^{-4}$ [a] | — | 82 |
| | $\delta$-$TiCl_3$–$Al(Bu^i)_3$ | 75–80 | $6 \times 10^{-4}$ [a] | — | 82 |
| Propylene | $\alpha$-$TiCl_3$–$AlEt_3$ | 70 | $17–34 \times 10^{-4}$ | — | 84 |
| | $\alpha$-$TiCl_3$–$AlEt_3$ | 70 | $15–30 \times 10^{-4}$ | $1.2–2.5 \times 10^{-4}$ | 88 |
| | $\alpha$-$TiCl_3$–$Al(Bu^i)_3$ | 70 | $19–38 \times 10^{-4}$ | — | 84 |
| | $\alpha$-$TiCl_3$–$Al(Bu^i)_3$ | 50–70 | $38–75 \times 10^{-4}$ | — | 88 |
| | $\alpha$-$TiCl_3$–$BeEt_2$ | 70 | $17–34 \times 10^{-4}$ | $1.5–3 \times 10^{-4}$ | 88 |
| | $\alpha$-$TiCl_3$–$ZnEt_2$ | 50 | $11–31 \times 10^{-2}$ | 0 | 86 |
| | $\alpha$-$TiCl_3$–$ZnEt_2$ | 60 | $14–29 \times 10^{-2}$ | — | 86 |
| | $\alpha$-$TiCl_3$–$ZnEt_2$ | 70 | $12–24 \times 10^{-2}$ | 0 | 88 |
| | $\delta$-$TiCl_3$–$AlEt_3$ | 70 | $4.5 \times 10^{-4}$ | — | 82 |
| | $VCl_3$–$Al(Bu^i)_3$ | 60 | $5.6 \times 10^{-4}$ [b] | $2.5 \times 10^{-5}$ [b] | 89 |

[a] The estimations are based on $\bar{M}_n$. In all other cases $\bar{M}_w$ was calculated from viscosity data and $\bar{M}_w/\bar{M}_n$ was assumed to be in the range of 5–10.
[b] $\bar{M}_w/\bar{M}_n = 4$.

It should be noted that the values of $k_a$ for $AlEt_3$, $BeEt_2$ and $AlEt_2Cl$ are the effective values and include the equilibrium constant for dissociation into monomeric species, *i.e.*

$$(k_a)_{real} = k_a/K_{diss}^{1/2} \tag{23}$$

### 2.5.3.2 *Adsorption models*

As early as 1956 Erich and Mark[90] pointed out that since most Ziegler–Natta catalyst systems were heterogeneous in nature that it was most likely that adsorption reactions were involved in such polymerizations. Since this time adsorption processes have featured in many kinetic schemes.

Adsorption phenomena were used by Saltman[91] in an attempt to derive a general kinetic scheme for Ziegler–Natta polymerizations. He proposed that monomer and metal alkyl were reversibly adsorbed on to the surface of a crystalline transition metal halide and that chain propagation occurred between the adsorbed metal alkyl and monomer, either adsorbed or in solution. Under conditions of high metal alkyl concentration the rate of propagation, assuming that this was with unadsorbed monomer, was given by

$$R_p = k_p[M][S] \tag{24}$$

where [S] is the concentration of active surface centres, [M] the monomer concentration and $k_p$ the rate constant for propagation.

From their studies on the polymerization of propylene using a $\delta$-$TiCl_3 \cdot 0.33\,AlCl_3$ catalyst, Keii *et al.*[92] proposed a Langmuir-type dependence of the rate of polymerization during the initial stage of polymerization on the aluminum alkyl concentration and expressed the rate of polymerization by

$$R = kP^2 \frac{K[A]}{1 + K[A]} t \tag{25}$$

where $R$ is the rate of polymerization, $k$ a rate constant, $P$ the pressure of propylene, [A] the concentration of aluminum alkyl, $t$ the time of polymerization and $K$ the equilibrium adsorption constant for adsorption of alkyl on an active centre. The stationary state rate, $R$, was represented by

$$R = k_p P \frac{K[A]}{1 + K[A]} \tag{26}$$

Vesely *et al.*[93] proposed a kinetic scheme where propagation was considered to occur between adsorbed metal alkyl and adsorbed rather than unadsorbed monomer. In this case the rate of polymerization is represented by

$$R_p = k_p \theta_M \theta_A S \tag{27}$$

where $\theta_M$ and $\theta_A$ are the fractions of surface, $S$, covered by adsorbed monomer and metal alkyl respectively. It should be noted that this equation predicts a fall-off in rate at high metal alkyl concentrations.

Reich and Stivala[94] have also considered the kinetic situation amongst others where adsorbed metal alkyl constitutes the active centres and in which propagation takes place between adsorbed metal alkyl and monomer, either adsorbed or in solution.

The above reaction schemes are based on the assumption that the active centres are adsorbed metal alkyl species. The active centres are however more likely to be alkylated transition metal entities as proposed by Cossee,[19] and as considered by Otto and Parravano.[95]

#### 2.5.3.3 *Model of Burfield, McKenzie and Tait*

Burfield, McKenzie and Tait considered the active centres to be alkylated vanadium entities,[96,97] such as $VCl_2R$, on the surface of the vanadium trichloride crystals, rather than metal alkyl molecules merely adsorbed on to the surface. The kinetic model proposed was found to give good agreement for the polymerization of 4-methyl-1-pentene (4-MP-1) by the catalyst system $VCl_3/AlR_3$, although the scheme is obviously applicable to a much wider range of polymerization systems. The polymerization system $VCl_3/AlR_3$/4-MP-1 was chosen since all the components could be obtained in a high state of purity, and because this system shows very stable kinetic rate–time profiles of the type shown in Figure 12.

The main proposals of this model are as follows:

(a) Chain initiation arises from the following reactions:

$$VCl_3 + AlR_3 \xrightarrow{\text{fast}} VCl_2R \cdot AlR_2Cl \tag{28}$$

$$VCl_2R \cdot AlR_2Cl + AlR_3 \underset{}{\overset{\text{slow}}{\rightleftharpoons}} VCl_2R + Al_2R_5Cl \tag{29}$$

$$VCl_2R + M \longrightarrow VCl_2P \tag{30}$$

$VCl_2P$ is equivalent to $C^*$ and equation (29) describes the initiation process.

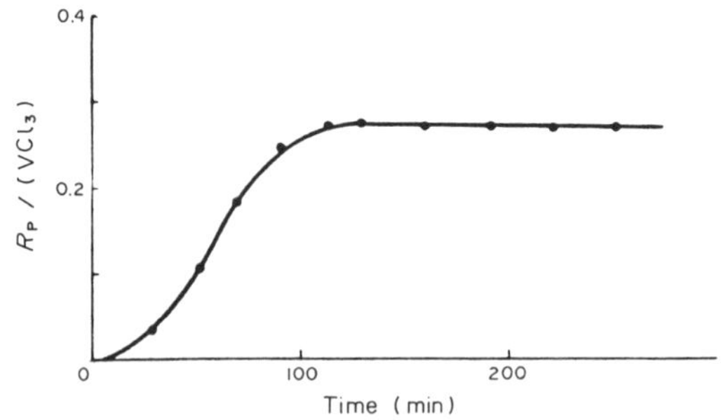


**Figure 12** Plot of $R_p/[VCl_3]$ *vs.* time. $[VCl_3] = 18.5$ mmol l$^{-1}$; $\{Al(Bu^i)_3\} = 37.0$ mmol l$^{-1}$; [4-MP-1] = 2.0 mol l$^{-1}$; solvent = benzene; temperature = 30 °C (reproduced from ref. 56)

(b) Chain propagation is regarded as a two-stage process, involving initially the adsorption of monomer on to an active centre followed by insertion of adsorbed monomer into a transition metal carbon bond, *i.e.* the process is monometallic, although this does not exclude the possibility of adsorbed alkylaluminum at a neighbouring site. The overall rate equation, describing the steady state polymerization period, is thus given by

$$R_p = k_p \theta_M C^* \tag{31}$$

where $k_p$ is the propagation rate constant with respect to adsorbed monomer, $\theta_M$ is the fraction of surface centres covered by adsorbed monomer, and $C^*$ is the concentration of active centres during this polymerization period.

(c) Monomer and also alkylaluminum molecules are involved in competitive reversible adsorption reaction with active centres, *viz.*

$$VCl_2P + M \xrightleftharpoons{K_M} VCl_2P \cdot M \tag{32}$$

$$VCl_2P + A \xrightleftharpoons{K_A} VCl_2P \cdot A \tag{33}$$

The adsorption of monomer and alkylaluminum is thus regarded as adsorption or complexation reactions. The term adsorption relates to weak chemisorption of species on to the active centres on the catalyst surface and only species which are capable of this type of specific interaction with the active centres are considered as adsorbed, any physisorption, which is much weaker, being neglected. Monomer may be adsorbed through complex formation of a π-donor complex (**17**). Alkylaluminum compounds are more probably adsorbed through the formation of bridge structures such as (**18**). Complexes of type (**18**) are considered to be inactive.

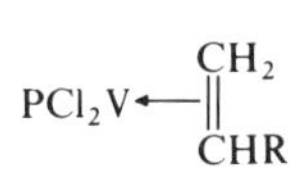


(**17**)

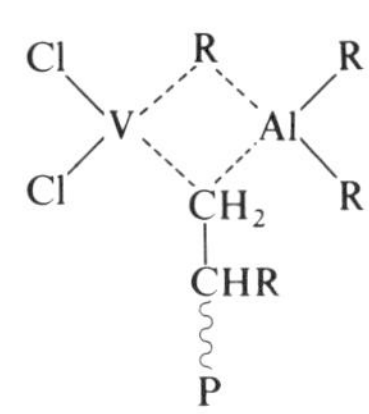


(**18**)

(d) Adsorption of both monomer and alkylaluminum are described by Langmuir–Hinshelwood isotherms, *viz.*

$$\theta_M = \frac{K_M[M]}{1 + K_M[M] + K_A[A]} \tag{34}$$

and

$$\theta_A = \frac{K_A[A]}{1 + K_M[M] + K_A[A]} \tag{35}$$

where [M] and [A] are the equilibrium concentrations of monomer and metal alkyl, and $K_M$ and $K_A$ are the equilibrium constants for the respective adsorption equilibria defined in equations (32) and (33). The alkylaluminum thus affects the rate of polymerization through the transition metal alkylation reaction (equations 28 and 29) and by reversible adsorption on the active centres with the formation of temporarily inactive centres.

(e) The overall rate of polymerization is given by the expression

$$R_p = \frac{k_p K_M [M] C^*}{1 + K_M [M] + K_A [A]} \tag{36}$$

(f) Chain transfer reactions take place between adsorbed alkylaluminum and the growing chain attached to the transition metal and also between adsorbed monomer and the growing polymer chain (equation 37). Transfer with adsorbed monomer may proceed *via* a $\beta$-hydride transfer reaction (equation 38).

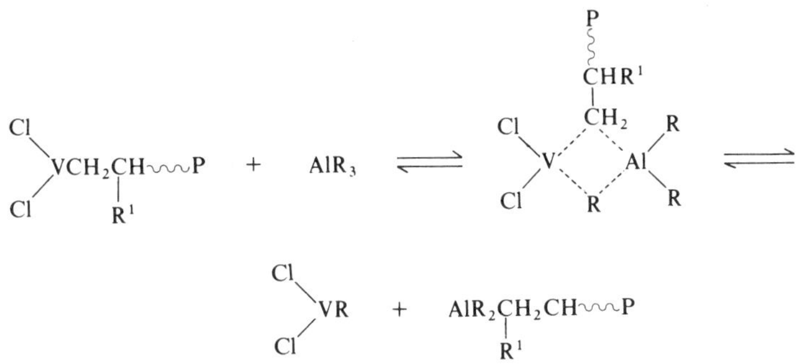

(37)

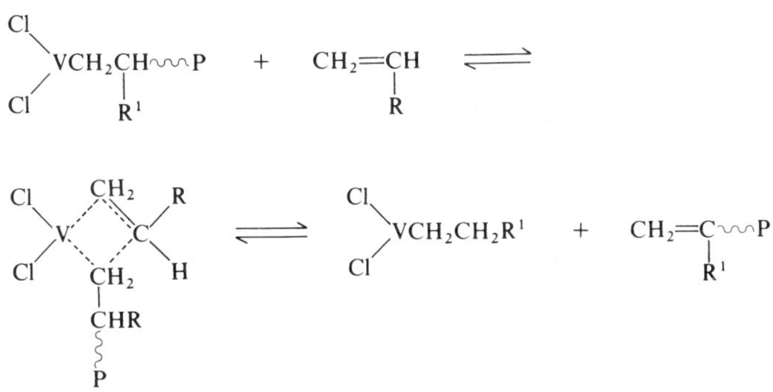

(38)

The rates of chain transfer with adsorbed metal alkyl, $R_{ta}$, and with adsorbed monomer, $R_{tm}$, are given by

$$R_{ta} = k_a \theta_A C^* \tag{39}$$

and

$$R_{tm} = k_m \theta_M C^* \tag{40}$$

where $k_a$ and $k_m$ are the rate constants for transfer with adsorbed alkylaluminum and adsorbed monomer, respectively.

(g) The number-average degree of polymerization $\bar{P}_n$, is given by

$$P = \frac{\int_0^t k_p \theta_M C^* \, dt}{C^* + \int_0^t k_m \theta_M C^* \, dt + \int_0^t k_a \theta_A C^* \, dt} \tag{41}$$

In the use of the above kinetic model to interpret their extensive kinetic results on the polymerization of 4-MP-1 by $VCl_3$–$AlR_3$ catalyst systems, Tait and co-workers used an independent method (tritium quenching) to determine $C^*$. The dependence of the overall rate of polymerization during the steady period on the alkylaluminum concentration, both at constant and varying [A]:[M] ratios, was found to be in quantitative agreement with the proposed model. A value of $3.10 \times 10^3$ min$^{-1}$ at 30 °C was obtained for the propagation rate constant and this was independent of the nature of the alkylaluminum compound. This same kinetic model has been used[99] successfully in an analysis of kinetic data for the polymerization of 4-MP-1 using a high activity catalyst system, *viz.* $MgCl_2/EB/TiCl_4–Al(Bu^i)_3$. Comparative values for some of the principal kinetic parameters are listed in Table 4.

**Table 4** Some Comparative Values for Kinetic Parameters for the Polymerization of 4-MP-1 with Various Catalysts at 40 °C

| *Catalyst system* | $k_a$ (min$^{-1}$) | $K_A$ (dm$^3$ mol$^{-1}$) | $k_M$ (dm$^3$ mol$^{-1}$) | $10^{-3}\,k_p$ (min$^{-1}$) |
|---|---|---|---|---|
| $\delta$-$TiCl_3 \cdot 0.33\,AlCl_3–Al(Bu^i)_3$ | 0.20 | 16 | 0.38 | 1.29 |
| $MgCl_2/EB/TiCl_4–Al(Bu^i)_3$ [a] | 4.0 | 30 | 1.5 | 4.26 |
| $VCl_3–AlBu^i)_3$ | 0.067 | 1.5 | 0.16 | 3.10 |

[a] Catalyst contains 2.6 wt% Ti.

### 2.5.3.4 *Model of Zakharov, Yermakov and co-workers*

Yermakov *et al.*[100–102] have also proposed a kinetic model for chain propagation and chain transfer reactions in Ziegler–Natta polymerization involving adsorption reactions. The model is detailed below, and is somewhat similar to that proposed by Tait *et al.*[96,97]

Chain propagation

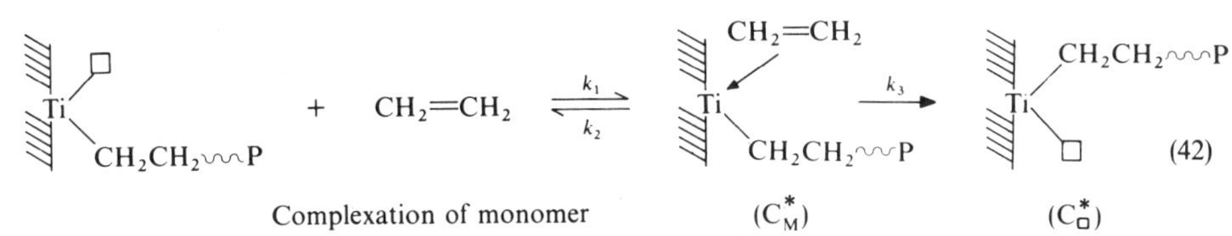

(42)

Complexation of alkylaluminum

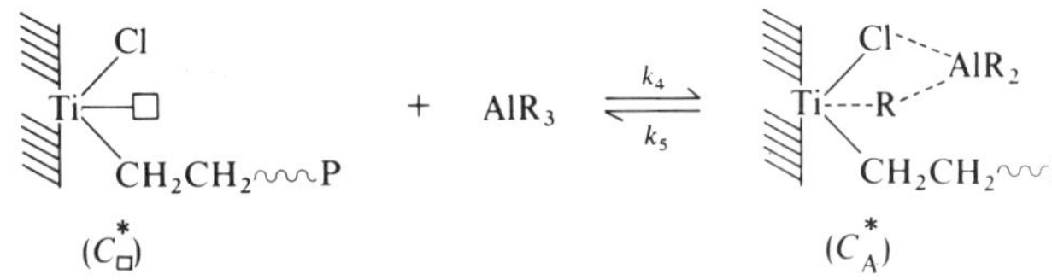

(43)

Temporary deactivation of propagation centres

Chain transfer with adsorbed alkylaluminum

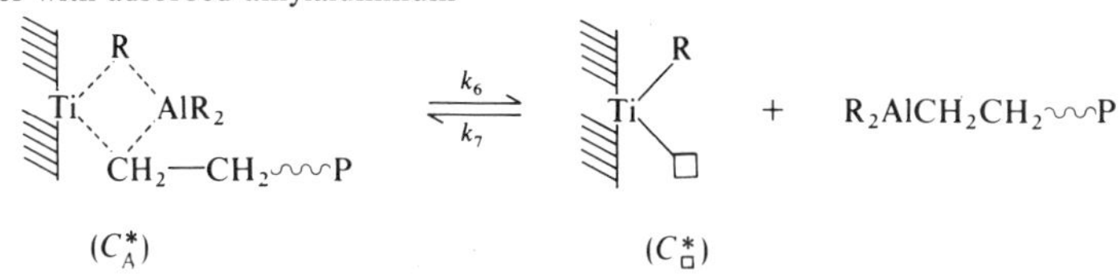

(44)

Chain transfer with adsorbed monomer

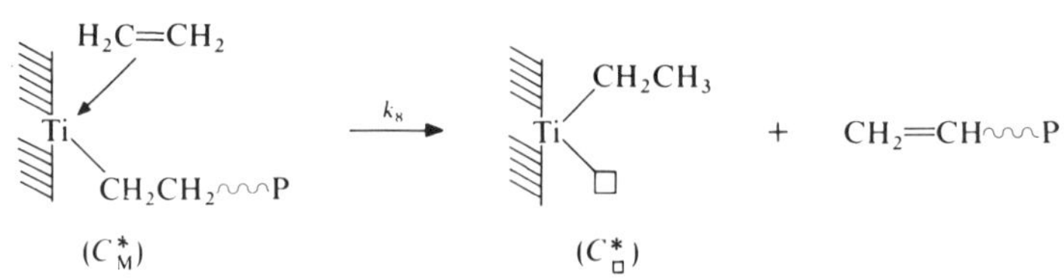

(45)

$\beta$-Hydride transfer

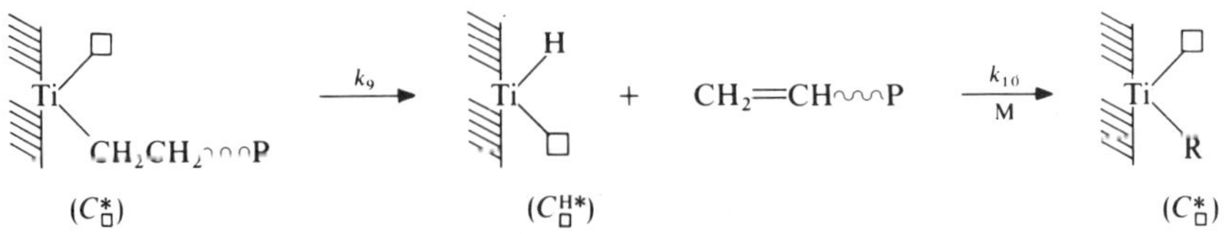

(46)

The important features of this model are as follows.

(a) Chain propagation occurs at titanium–carbon bonds and is monometallic — the alkylaluminum cocatalyst is not incorporated as such into the propagation centre. This is consistent with the appropriate $k_p$ values being independent of the type of alkylaluminum compound used as cocatalyst.

(b) Chain propagation is a two-stage reaction which proceeds through prior coordination of monomer at the titanium atom in the active centre followed by insertion of the coordinated monomer into the titanium–carbon bond.

(c) Both monomer and alkylaluminum are involved in chain transfer reactions, these reactions also taking place through adsorption reactions between monomer and active centres and alkylaluminum and active centres respectively.

(d) The alkylaluminum cocatalyst affects the overall rate of polymerization and the stereoregularity of the resulting polymer due to variation in the actual numbers of centres which are formed and also by variation in the ratio of centres of different stereospecificities, since the complexation reactions alter the number of propagating centres and the ratio of isospecific to aspecific centres.

(e) The overall rate of polymerization is given by

$$R_p = \frac{k_1 k_3}{k_2 + k_3 + k_1[M]} C_p^*[M] \tag{47}$$

*i.e.*

$$k_p = \frac{k_1 k_3}{k_2 + k_3 + k_1[M]} \tag{48}$$

(*cf.* equation 14).

(f) The total number of active centres, $C^*$, is given by

$$C^* = C_{\square}^* + C_M^* + C_A^* + C_{\square}^{H*} \tag{49}$$

According to Yermakov

$$C_p^* = C_{\square}^* + C_M^* \tag{50}$$

And neglecting $C_{\square}^{H*}$

$$C^* = C_p^* + C_A^* \tag{51}$$

When $k_1[M] \ll (k_2 + k_3)$ it can be shown that

$$C_p^* = \frac{C^*}{1 + K[A]} \tag{52}$$

where

$$K_A = k_4/k_5 \tag{53}$$

For dimeric forms of alkylaluminum compounds the corresponding equation is

$$C_p^* = \frac{C^*}{1 + K_A K_D^{1/2}[A_2]^{1/2}} \tag{54}$$

where $K_D$ is the equilibrium constant for the alkylaluminum dissociation.

Yermakov *et al.*[101] have used this kinetic model in analyses of their extensive experimental studies on the polymerization of ethylene and propylene using a number of catalyst systems, *e.g.* $TiCl_2$, $TiCl_2$–$AlR_3$ δ-$TiCl_3 \cdot 0.33\ AlCl_3$–$AlR_3$, and as a result have been able to evaluate values for $k_p$, $k_a$, $K_A$, $K_D$, $C^*$ and $C_p^*$. Values of $C^*$ and $C_p^*$ were determined[103] using tritium-labelled methanol and $^{14}CO$ radio-tagging (see later). Values of $k_p$ were shown to be independent of the nature of the alkylaluminum compound and of the composition of the $TiCl_3$. The value of $k_a$ was found to vary with the type of monomer, being two orders higher for ethylene than propylene. For ethylene polymerization the $k_a$ value was an order less for $Al(Bu^i)_3$ than for $AlEt_3$.[104]

### *2.5.3.5 Model of Böhm*

Böhm[105] has published a significant kinetic model for Ziegler–Natta polymerization which is claimed to include all reactions considered relevant in both heterogeneous and homogeneous

Ziegler–Natta catalysts. For heterogeneous systems the Böhm model is a particular case of the Rideal mechanism.[106] First the different reactants, *i.e.* monomer, hydrogen and cocatalyst, interact by complexation with the active centre. Then in further steps these complexed entities interact with the polymer (or hydrogen) group attached to the active centre by a covalent bond; the catalytic process taking place in such a way that the same covalent bond is re-established or else a new one is formed. The adsorbed molecule is assumed to be held in a deep van der Waals trough which may be identified with the free coordination site on the active centre. The various reactions are shown in Scheme 22.

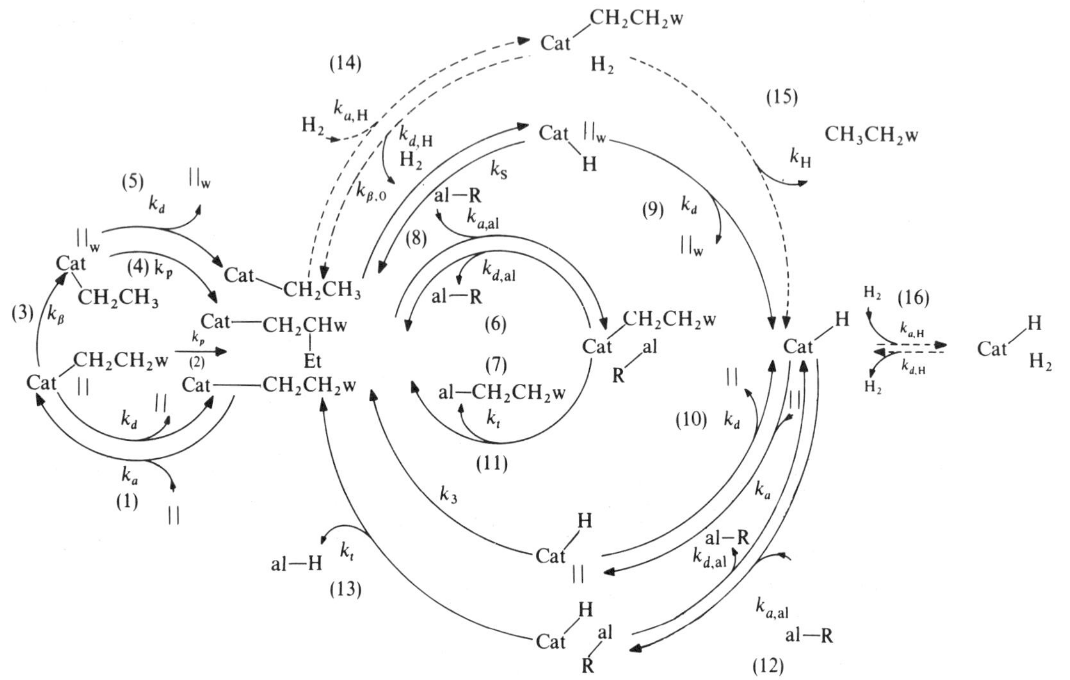


**Scheme 22.** Reaction scheme of Ziegler–Natta polymerization process; Cat, catalytic active centre; || ethylene molecule; W, polymer chain; || w, polymer chain with a vinyl group; al, 1/3 Al; R, alkyl group (for example, Et); H, $H_2$, hydrogen atom or hydrogen molecule (reproduced from ref. 105)

The following features concerning this kinetic model should be noted.

(a) Complex formation, reaction (1) in Scheme 22, between the active centre and the alkene is assumed to be reversible (*cf.* Tait and Yermakov).

(b) The polymer molecules are formed by a subsequent insertion reaction.

(c) $\beta$-H elimination takes place *via* a six-centre cyclic transition state (Scheme 22, reaction 3).

(d) Active centres and alkylaluminum interact reversibly (Scheme 22, reaction 6), and chain transfer may result (*cf.* Tait and Yermakov).

(e) Spontaneous transfer (Scheme 22, reaction 8) to form a Cat—H bond is allowed for in this scheme (*cf.* Ziegler and Natta).

(f) Chain transfer with hydrogen is included (Scheme 22, reactions 14–16), and may explain the rate-lowering effects which have sometimes been observed in the presence of hydrogen.

(g) By assuming steady state condition for all intermediates and that all rate constants are independent of the degree of polymerization, Böhm has deduced that

$$R_p = \frac{k_p k_a [M]_0}{k_p + k_d + k_a [M]_0} \frac{1}{\left(1 + \frac{b}{a} + \frac{c}{a}\right)} \frac{n^*}{n_k} \qquad (55)$$

where

$$a = \frac{k_p + k_d + k_a [M]_0}{k_p + k_d} \qquad (56)$$

$$b = \frac{k_{\beta,0}}{k_s + k_d}\left\{1 + \frac{k_d + \dfrac{k_d k_a[M]_0}{k_s + k_d} + \dfrac{k_d k_{a,al}[al]_0}{k_{d,al} + k_t}}{\dfrac{k_s k_a[M]_0}{k_s + k_d} + \dfrac{k_t k_{a,al}[al]_0}{k_{d,al} + k_t}}\right\} \tag{57}$$

$$c = \frac{k_{a,al}[al]_0}{k_t + k_{d,al}} \tag{58}$$

In this scheme as detailed by Böhm, $n^*$ is the quantity of active centres (*cf.* $C^*$). Equations (55)–(58) are somewhat cumbersome but if $k_{\beta,0} = 0$ and $k_i = k_p$ and if it is assumed that there are equilibria due to monomer and alkylaluminum, equation (55) reduces to that given by Tait *et al.*[96,97]

(h) An expression can also be derived to relate the number-average degree of polymerization to the monomer and alkylaluminum concentrations, *i.e.*

$$\frac{1}{P_{n,\infty}} = A + B\frac{1}{[M]_0} + C\frac{[al]_0}{[M]_0} \tag{59}$$

where

$$A = \frac{k_d k_\beta}{k_p(k_p + k_d + k_\beta)} \tag{60}$$

$$B = \frac{k_d k_{\beta,0}(k_p + k_d)}{(k_d + k_s)k_p k_a} \tag{61}$$

$$C = \frac{k_t k_{a,al}(k_p + k_d)}{(k_{d,al} + k_t)k_p k_a} \tag{62}$$

(i) Finally the Böhm treatment allows the derivation of an important equation which allows the evaluation of the number of active centres from molecular weight data, *i.e.*

$$\frac{KA}{M_n} = \frac{n^*}{k_K} + \frac{1}{M_{n,\infty}}KA \tag{63}$$

where

$$M_{n,\infty} = P_{n,\infty} M_m \tag{64}$$

The Böhm kinetic model has been successfully applied[107,108] to the polymerization of ethylene using a highly active catalyst derived from $Mg(OEt)_2$ and $TiCl_4$.

## 2.6 ACTIVE CENTRE DETERMINATION

### 2.6.1 Objectives

The determination of active centre concentrations in Ziegler–Natta polymerization systems has presented an exciting challenge ever since the early recognition that only a small percentage of the transition metal atoms present in these systems was active. Such determinations are crucial to any fundamental characterization of Ziegler–Natta catalyst systems and initial investigations were carried out as early as 1959.[59] The importance of active centre concentration determination has been highlighted elsewhere but the following observations are considered to be of importance.

(a) A knowledge of the number of propagating centres, $C_p$, allows evaluation of the appropriate values for the rate constant for chain propagation, $k_p$, since the overall rate of polymerization, $R_p$, is given by the equation

$$R_p = k_p C_p^*[M] \tag{14}$$

where [M] is the concentration of monomer at the propagating centres. Since this concentration may not be the same as the bulk concentration in solution the following relationship is more strictly applicable.

$$R_p = k_p C_p^* \theta_M \tag{31}$$

where $\theta_M$ is the fraction of centres complexed by monomer and $k_p$ is the rate constant for chain propagation with adsorbed monomer. A knowledge of [M] (or $\theta_M$) and $R_p$ together with $C_p$ allows calculations of values of $k_p$.

(b) Where more than one type of active centre is present

$$R_p = \sum_i k_{p,i} C^*_{p,i}[M] \tag{65}$$

and

$$R_p = \sum_i k_{p,i} C^*_{p,i}\theta_M \tag{66}$$

The simplest case where equations (65) and (66) are applicable is for the polymerization of propylene by catalyst systems producing atactic and isotactic polymer. Some methods for active centre concentration allow evaluation of $C^*_p$ (atactic) and $C^*_p$ (isotactic) leading to the appropriate values of $k_p$ (atactic) and $k_p$ (isotactic). Otherwise it should be realized that only average values of $C^*_p$ and $k_p$ will be obtained, *i.e.* $\bar{C}^*_p$ and $\bar{k}_p$ respectively.

(c) Consideration of the particular type of kinetic rate–time profile shown by the catalyst system being investigated should be made since the rate of polymerization may change drastically with time, as in some catalyst systems of the type $MgCl_2$/ethyl benzoate/$TiCl_4$. Since the rate of polymerization can be expected to be related, directly or otherwise, to the active centre concentration, it is evident that values of active centre concentration relate to the times of their determination.

(d) It should also be realized that it is not only $C^*_p$ which may change with time; the intrinsic activity of the catalytic centre may also vary with time, and this will be reflected in a change in the value of the propagation rate constant $k_p$.

Over the years a wide range of techniques have been used in the determination of active centre concentrations and the subject has received considerable attention in the literature.[69,109–113] For this reason only the most commonly used methods will be examined in this account. More detailed information may be found elsewhere.

### 2.6.2 Kinetic and Molecular Weight Methods

Values of active centre concentration may be evaluated from kinetic and molecular weight data.

#### *2.6.2.1 Variation of the number-average degree of polymerization with time*

It can be shown[109] that

$$C^*_p = \frac{\tau R_p}{\bar{P}_n} \tag{67}$$

and

$$\tau = \frac{d(1/\bar{P}_n)}{d(1/t)}\bar{P}_n \tag{68}$$

where $\tau$ is the average lifetime of the growing polymer chain and $\bar{P}_n$ the number-average degree of polymerization.

The method requires evaluation of $\bar{P}_n$ as a function of time together with the corresponding $R_p$ values. From a plot of $1/\bar{P}_n$ *vs.* $1/\tau$ together with a knowledge of $\bar{P}_n$, and $R_p$, $C^*_p$ may be evaluated.

This method was first used by Natta but has been used by many subsequent workers. Scknecko and Kern[109,110] have discussed the applicability of this method, drawing attention to the fact that only average values of $\tau$ and $C^*_p$ are obtained.

#### *2.6.2.2 Variation of yield with time*

This procedure has been frequently used in estimations of $C^*_p$ values. In particular, Haward,[114,115] Böhm,[107] Kashiwa[116] and other workers have made use of the method using equations of the form

$$\frac{Y}{\bar{M}_n} = (1 + k_{tr}[X]t)C_p^* \tag{69}$$

$$= C_p^* \quad \text{when } t = 0$$

where $Y$ is the yield of polymer at time $t$ and $k_{tr}$ is the rate constant for chain transfer with some agent X (*e.g.* hydrogen or metal alkyl).

The method requires determination of yield as a function of time together with the measurement of appropriate $\bar{M}_n$ values. A plot of $Y/\bar{M}_n$ *vs.* $t$ allows the limiting value of $Y/\bar{M}_n$ when $t \rightarrow 0$ to be evaluated.

The method is best applicable to polymerization systems which show little variation of $C^*$ with time, such as catalyst systems derived from $TiCl_4$ and $Mg(OR)_2$.

The use of equation (69) is also the basis of methods used for the determination of values of $C_p^*$ after very short polymerization times.

Some typical $C^*$ values determined by kinetic and molecular weight methods are listed in Table 5.

### 2.6.3 Quenching Methods

The basic assumption of this method is that the growing polymer chain has a relatively basic carbon atom bound to the transition metal atom and which can react with the labelling group of the quench reagent, QL, as shown in equation (70).

$$\text{CatCH}_2\underset{\substack{|\\ \text{R}}}{\text{CH}}\!\sim\!\text{P} + \text{QL} \longrightarrow \text{CatQ} + \text{LCH}_2\underset{\substack{|\\ \text{R}}}{\text{CH}}\!\sim\!\text{P} \tag{70}$$

Over the years a number of quench reagents have been used, including radioactive iodine,[126] inactive iodine,[127] tritiated alcohols,[29,78,128–137] tritiated water,[138] deuterated water and deuterated methanol.[22] However most recent investigations have involved tritiated alcohols when equation (71) is followed.

$$\text{CatCH}_2\underset{\substack{|\\ \text{R}^1}}{\text{CH}}\!\sim\!\text{P} + \text{ROT} \longrightarrow \text{CatOR} + \text{TCH}_2\underset{\substack{|\\ \text{R}^1}}{\text{CH}}\!\sim\!\text{P} \tag{71}$$

From the radio-content of the decontaminated polymer sample, the metal–polymer bond concentration in the catalyst system, [MPB], may be evaluated using the equation

$$[\text{MPB}] = \frac{kAG}{a} \tag{72}$$

where $k$ = correction factor for the isotope effect, $A$ = specific activity of the polymer, $G$ = polymer yield at time of quenching and $a$ = specific activity of the methanol.

There is a major complication in most Ziegler–Natta polymerization systems in that chain transfer with alkylaluminum occurs leading to the production of centres of the type $R_2AlCH_2CHR^1\sim P$ which are non-propagative yet which will react with tritiated alcohols to yield labelled polymer chains. Thus the total concentration of metal–polymer bonds, $[MPB]_t$, will increase with time and is given by the equation

$$[\text{MPB}]_t = C_0^* + N_{tr} \tag{73}$$

where $N_{tr}$ is the concentration of transferred metal–polymer bonds at time $t$, and $C_0^*$ is the concentration of active centres at $t = 0$.

To evaluate $C_0^*$ it is necessary to use plots of either [MPB] *vs.* time, or [MPB] *vs.* conversion respectively.[29] It should be noted that the method gives values for the active centre concentration in terms of [MPB] whether the centres are actually propagating or not. Hence it is appropriate to use the symbol $C_0^*$ rather than $C_p^*$. Also since the method requires determination of [MPB] over a period of time it is most usefully applied to polymerization systems showing steady kinetic rate–time profiles.

**Table 5** Some Typical Active Centre Concentrations for the Polymerization of Propylene using Kinetic and Molecular Weight Methods

| *Author* | *Catalyst* | *Temperature* (°C) | *$10^3[C^*]$* (mol (mol Ti)$^{-1}$) | $k_p$ ($dm^3\,mol^{-1}\,s^{-1}$) | *Method*[a] | *Ref.* |
|---|---|---|---|---|---|---|
| Natta | $\alpha$-$TiCl_3$ | 70 | 2–6.3 | 2.5 | MW/K | 143 |
| Tanaka and Morikava | $\delta$-$TiCl_3 \cdot 0.33\,AlCl_3$–$AlEt_3$ | 70 | 10 | 5.6 | K | 117 |
| Chirkov | $VCl_3$–$Al(Bu^i)_3$ | 70 | 4.7 | 46.5 | K | 118 |
| | $\alpha$-$TiCl_3$–$Al(Bu^i)_3$ | 70 | 5.4 | 2.9 | K | 119 |
| Schnecko and Kern | $TiCl_3$–$AlEt_2Cl$[b] | 60 | 7.5 | 6.2 | K | 109, 110 |
| | $TiCl_3$–$AlEt_2Cl$[c] | 60 | 19 | 7.5 | K | 109, 110 |
| Doi *et al.* | $\delta$-$TiCl_3$–$AlEt_3$ | 41 | 15 | 23 | MW | 120 |
| | $\delta$-$TiCl_3$–$AlEt_3$ | 40 | 15 | 31 | MW | 121 |
| Haward *et al.* | $TiCl_4/MgC_8H_{17}Br$–$Al(C_8H_{17})_3$ | 40 | 30 | 28 | MW | 122 |
| Keii *et al.* | $TiCl_4/MgCl_2$/EB–$AlEt_3$ | 41 | < 23 | > 440 | MW | 123 |
| | $TiCl_4/MgCl_2$ | 60 | 200–600 | 240–730 | MW | 124 |
| | $TiCl_4/MgCl_2$/EB–$AlEt_3$–EB | 60 | 28 | 2700 | MW | 124 |
| Bukakov *et al.* | $TiCl_4/MgCl_2$/EB–$Al(Bu^i)_3 \cdot$ EMB[d] | 70 | 70 | 1000 | MW | 125 |

[a] MW = molecular method, K = kinetic method.
[b] $TiCl_4$ reduced by Ti.
[c] $TiCl_4$ reduced by Al.
[d] EMB = ethyl *p*-methoxybenzoate.

An advantage of using quenching techniques of this type is that the rate constant for chain transfer with alkylaluminum may be obtained, since

$$[\mathrm{MPB}]_t = C_0^* + \int_0^t R_t\,dt \tag{74}$$

and

$$\frac{C_0^*}{R_{ta}} = \frac{1 + K_M[M]}{k_a K_A[A]} + \frac{1}{k_a} \tag{75}$$

One serious drawback to the method is that since the alcohol used for quenching is only partially tritiated a kinetic isotope effect will be present. It is thus necessary to establish the value of $K$ in equation (72). In practice $K$ can be determined[136,137] either by variation of the $^3H$ content or by variation in the rate of addition. In both cases the relative availability of the two isotopes will vary and so $K$, equal to the ratio $k_H/k_{^3H}$, can be evaluated. The value of $K$ has been found to vary with the particular catalyst, cocatalyst, monomer and temperature and so must be determined for each particular polymerization system.

Also it is important to establish that main-chain labelling does not take place,[139] and that effective decontamination of the polymer has been achieved.

Some typical values of $C^*$ obtained using quenching techniques are listed in Table 6.

### 2.6.4 Radio-tagging Methods

$^{14}C$-labelled alkylaluminum compounds were first used by Natta[59,143] and Chien[126] in order to determine values of active centre concentrations for the polymerization of propylene. Most recent investigations have involved the use of radio-labelled poisons such as $^{14}CO$ and $^{14}CO_2$ which can insert into active transition metal–carbon bonds and become incorporated into the polymer chain. The use of such compounds has its origin in research carried out at the Hercules Powder Co., Welmington, USA, and ICI, Welwyn Garden City, UK.

By way of practical operation the method is very simple and for this reason has been adopted by many workers. The polymerization reaction is first treated with $^{14}CO$ or $^{14}CO_2$, left to react for a specific time (contact time $t_c$) and then quenched by addition of acidified alcohol. The overall reaction is believed to be represented by the sequence of reactions (76) to (78).

Coordination 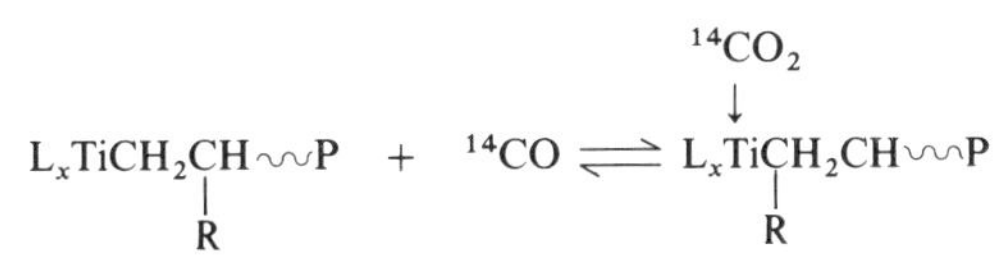 (76)

Insertion  (77)

Quenching  (78)

After quenching the radio-tagged polymer is isolated, carefully decontaminated, and its $^{14}C$ content determined either by combustion analysis and/or gel scintillation counting.

Although simple in operation the method has been the subject of considerable debate, some workers being concerned lest the insertion step should be reversible and therefore the insertion reaction incomplete.[137,144–146]

It is now realized that selection of the appropriate contact time, $t_c$, is of major importance as is the use of acid in the quenching step.

**Table 6** Some Typical Active Centre Concentrations for the Polymerization of Propylene using Radio-tagging methods

| *Author* | *Catalyst* | *Temperature* (°C) | *$10^3[C^*]$* (mol (mol Ti)$^{-1}$) | $k_p$ (dm$^3$ mol$^{-1}$ s$^{-1}$) | *Method* [a] | *Ref.* |
|---|---|---|---|---|---|---|
| Natta | $\alpha$-$TiCl_3$–$AlEt_2Cl$ | 70 | 3 | — | $^{14}C$ | 143 |
| Chien | $\alpha$-$TiCl_3$–$AlEt_2Cl$ | 70 | 36 | 1.1 | $^{14}C$ | 148 |
| Zakharov *et al.* | $TiCl_2$ | 70 | 0.0013 | 76 | $^{14}CO_2$ | 149 |
| | $\delta$-$TiCl_3\cdot 0.3AlCl_3$–$AlEt_3$ | 70 | 0.36 | 70 | $^{14}CO$–$PH_3$ | 149 |
| | $\delta$-$TiCl_3\cdot 0.33AlCl_3$–$AlEt_2Cl$ | 70 | 1.4 | 73 | $^{14}CO$–$PH_3$ | 149 |
| Chien and Hsieh | $Bz_3TiCl/Mg(OH)Cl$–$AlEt_2Cl$ | 50 | 0.9 | 32 | $^{14}CO$–$PH_3$ | 150 |
| Burns | $\delta$-$TiCl_3$–$AlEt_2Cl$ | 60 | 10–20 | 8.2 | $^{14}CO$ | 151 |
| | Solvay $TiCl_3$–$AlEt_2Cl$ | 60 | 7.9–15.2 | 17–47 | $^{14}CO$ | 151 |
| Giannini | Montedison type | 65 | — | 500–140[b] | $^{14}CO$ | 60 |
| | $\delta$-$TiCl_3\cdot 0.33AlCl_3$–$Al(Bu^i)_3$ | 65 | — | 130–100[c] | $^{14}CO$ | 60 |
| Zakharov *et al.* | $TiCl_m(Mg)$[d] | 70 | 0.14[e] | 470[e] | $^{14}CO$ | 125 |
| | $TiCl_n(Al)$[e] | 70 | 0.04[e] | 100[e] | $^{14}CO$ | 125 |
| | $MgCl_2/TiCl_4$–$AlEt_3$ | 70 | 5.2[e] | 740[e] | $^{14}CO$ | 125 |
| | $MgCl_2/TiCl_4\cdot EB$–$AlEt_3$ | 70 | 2.7[e] | 870[e] | $^{14}CO$ | 125 |

[a] $^{14}C$ refers to labelled aluminum alkyl; $^{14}CO$—$PH_3$, $^{14}CO$ refer to the inhibition agents used.
[b] $t_c = 5$–$180$ min.
[c] $t_c = 15$–$180$ min.
[d] $TiCl_4$ reduced by $MgBu_2$.
[e] Refers to isotactic value.
[f] $TiCl_4$ reduced by aluminum alkyl.

Tait *et al.*[147] have investigated the applicability of the methods to soluble $Cp_2ZrCl_2$–methylaluminoxane catalyst systems in ethylene polymerization and have demonstrated that the expected $C$ values are obtained for short $t_c$ values as is shown in Figure 13, a result also confirmed by tritiated alcohol determinations.

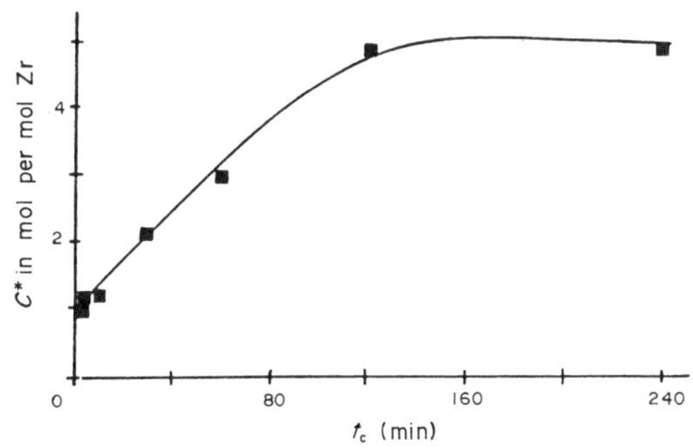


**Figure 13** Polymerization system bis($\eta$-cyclopentadienyl)zirconium(IV) dichloride/methylaluminoxane/ethylene. Variation of active centre concentration $C^*$ with contact time $t_c$ (reproduced from ref. 147)

An advantage of the $^{14}CO$ radio-tagging method is that for propylene polymerization it allows differentiation between centres which produce stereoregular and those producing non-stereoregular polymer.[57] Separation of the polymer into fractions consisting of mainly isotactic polypropylene from mainly atactic polymer can be achieved using boiling $n$-heptane or xylene. Analysis for $^{14}C$ on both fractions allows evaluation of $(C_p)_{insol}$ and $(C_p)_{sol}$. Some data for the polymerization of propylene is shown in Table 7. Some typical values for $C_p$, obtained using this method, are shown in Table 8.

**Table 7** Values of $C_p^*$ and $k_p$ for Propylene Polymerization at Different Temperatures Using a Ball-milled $MgCl_2/EB/TiCl_4$ Catalyst and $AlEt_3$[a]

| *Temperature* (°C) | $10^2C_p^*$ (mol (mol Ti)$^{-1}$) *Total*[b] | *Insol.*[c] | *Sol.*[d] | $k_p$(dm$^3$ mol$^{-1}$ s$^{-1}$) *Total* | *Insol.* | *Sol.* |
|---|---|---|---|---|---|---|
| 40 | 7.14 | 3.44 | 3.70 | 369 | 703 | 59 |
| 50 | 8.03 | 4.93 | 3.10 | 669 | 1034 | 89 |
| 60 | 10.30 | 6.68 | 3.61 | 858 | 1256 | 122 |
| 70 | 11.31 | 7.47 | 3.84 | 1341 | 1952 | 163 |

[a] Polymerization time before injection of $^{14}CO$ was 2 min; contact time = 4 min.
[b] $C_p^*$ (total) was determined using the unextracted polymer.
[c] $C_p^*$ (insol) was determined using the insoluble fraction from polymer extracted with boiling $n$-heptane for 24 h.
[d] $C_p^*$ (sol) = $C_p^*$ (total) − $C_p^*$ (insol)

**Table 8** Some Typical Active Centre Concentrations for the Polymerization of Propylene using Quenching Methods

| *Author* | *Catalyst* | *Temperature* (°C) | $10^3[C^*]$ (mol (mol Ti)$^{-1}$) | $k_p$ (dm$^3$ mol$^{-1}$ s$^{-1}$) | *Method*[a] | *Ref.* |
|---|---|---|---|---|---|---|
| Schnecko *et al.* | $TiCl_3$–$AlEt_2Cl$ | 60 | 3 | — | $I_2$ | 109, 141 |
| Coover *et al.* | $TiCl_3$–$AlEt_2Cl$[b] | 70 | 1 | 40 | TQ | 132 |
| Caunt | $\alpha$-$TiCl_3$–$AlEt_2Cl$ | 60 | 4 | — | $I_2$ | 127 |
| Kohn *et al.* | $\alpha$-$TiCl_3$–$AlEt_2Cl$ | 70 | 28 | 124 | TQ | 130 |
| Bier *et al.* | $TiCl_3\cdot 0.3AlCl_3$–$AlEt_2Cl$ | 50 | 10 | 1.2 | TQ | 78 |
| Schnecko *et al.* | $\delta$-$TiCl_3\cdot 0.33AlCl_3$–$AlEt_2Cl$ | 60 | 5.3 | 18 | TQ | 110 |
| Yermakov *et al.* | $\delta$-$TiCl_3\cdot 0.3AlCl_3$–$AlEt_2Cl$ | 70 | $60\alpha$[c] | $1.2/\alpha$[c] | TQ | 135 |
| Baulin *et al.* | $TiCl_4/MgO$–$AlEt_3$ | 70 | 330 | 4.8 | TQ | 141 |
| | $TiCl_4/Al_2O_3\cdot SiO_2$–$AlEt_3$ | 70 | 220 | 1.0 | TQ | 142 |

[a] $I_2$ = iodine quenching; TQ = tritium quenching.
[b] Stauffer HA type catalyst.
[c] $\alpha$ = kinetic isotope effect.

### 2.6.5 Simultaneous Adsorption and Rate Measurements

A method which promises to be of some importance is that developed by Caunt, Tait and co-workers,[57,69,151–153] which involves the injection of a small quantity of a suitable inhibitor into a polymerization system and the simultaneous measurement of the amount of inhibitor adsorbed and the drop in the overall rate of polymerization. Both allene and carbon monoxide have been used as probes, with some measure of success, in polymerization systems showing stable kinetic rate–time profiles; some results are reported in Table 9.

**Table 9** Some Typical Active Centre Concentrations for the Polymerization of Propylene using Simultaneous Adsorption Measurements

| *Author* | *Catalyst* | *Temperature* (°C) | $10^3[C^*]$ (mol (mol Ti)$^{-1}$) | $k_p$ (dm$^3$ mol$^{-1}$ s$^{-1}$) | *Method*[a] | *Ref.* |
|---|---|---|---|---|---|---|
| Tait *et al.* | $\delta$-$TiCl_3 \cdot 0.33AlCl_3$–$AlEt_2Cl$ | 60 | 13.7[b] | 5.9 | SARM | 151 |
| | Solvay type/$AlEt_2Cl$ | 60 | 22.7[b] | 7.6 | SARM | 151 |
| | Solvay type/$AlEt_3$ | 60 | 121[b] | 10.2 | SARM | 151 |

[a] SARM = simultaneous adsorption and rate measurement using allene inhibition.
[b] Average values of $C^*$ for a series of determinations.

For this method to be successful the following criteria need to be satisfied: (a) the poison adsorbed on active centres should remain there long enough for its equilibrium concentration to be determined, *i.e.* it should not be consumed, or at least not rapidly, in side reactions; (b) all centres must be poisoned at the time of the determinations carried out in (a); (c) the poison should be adsorbed only on centres active in polymerization and not on other non-propagative centres; (d) only one molecule of the poison should be adsorbed per active centre, or else the exact stoichiometry should be established; and (e) the probe molecule, *i.e.* the poison, should be of similar molecular dimensions to the monomer studied.

The validity of the method has been examined and it is apparent that for some polymerization systems, when using allene and CO as poisons, significant consumption of these probes takes place leading to values for $C^*$ which are too high.[152,153] However one attractive advantage of the method is that it can provide useful information concerning the distribution of centres of differing activities.

### 2.6.6 Concluding Remarks

Ziegler–Natta polymerization systems have been shown to be very complex and the model of a fixed geometric centre which has a definable identity and activity invariant with time is far too simplistic a model. Thus, since the various methods as outlined in this chapter for the determination of active centre concentration depend on the measurement of different parameters, and since most Ziegler–Natta catalyst systems have centres of widely different activities and geometric locations, some variation in the actual values obtained for $C^*$ must be expected. Where a fuller analysis of $C^*$ values is necessary determinations should be carried out using more than one method.

## 2.7 REFERENCES


1. J. Boor, Jr., in 'Ziegler–Natta Catalysts and Polymerizations', Academic Press, New York, 1979.
2. C. D. Nenitzescu, C. Huch and A. Huch, *Angew. Chem.*, 1956, **68**, 438.
3. H. N. Friedlander and K. Oita, 'Abstracts, 130th meeting of the American Chemical Society', 1956, p. 138.
4. A. V. Topchiev, B. A. Krentsel and L. G. Sidarova, *Dokl. Akad. Nauk SSSR*, 1959, **128**, 732; A. V. Topchiev, B. A. Krentsel and L. L. Stotskaya, *Russ. Chem. Rev. (Engl. Transl.)*, 1961, **30**, 192.
5. E. W. Duck, *J. Polym. Sci.*, 1959, **34**, 86.
6. A. Gilchrist, *J. Polym. Sci.*, 1959, **34**, 49.
7. G. Natta and G. Mazzanti, *Tetrahedron*, 1960, **8**, 86.
8. P. Patat and H. Sinn, *Angew. Chem.*, 1958, **70**, 496.
9. H. Uelzmann, *J. Polym. Sci.*, 1958, **32**, 457; *J. Org. Chem.*, 1960, **25**, 671.
10. J. Boor, *J. Polym. Sci., Part C*, 1963, **1**, 257.
11. G. Bier, *Kunststoffe*, 1958, **48**, 354; A. Gumbolt and H. Schmidt, *Chem.-Ztg.*, 1959, **83**, 636.
12. G. Natta, P. Pino, G. Mazzanti and R. Lanzo, *Chim. Ind. (Milan)*, 1957, **39**, 1032; G. Natta, P. Pino, G. Mazzanti, V. Giannini, E. Mantica and M. Peraldo, *J. Polym. Sci.*, 1957, **26**, 120.

13. G. L. Karapinka and W. L. Carrick, *J. Polym. Sci.*, 1961, **55**, 145.
14. G. Natta, I. Pasquon, A. Zambelli and G. Gatti, *J. Polym. Sci.*, 1961, **51**, 387; G. Natta, *Chim. Ind.* (*Milan*), 1960, **42** (11), 1207.
15. A. Schindler, *J. Polym. Sci., Part B*, 1965, **3**, 147.
16. T. Keii, *J. Indian Chem. Soc.*, 1979, **56**, 269.
17. O. A. Ponomarev, K. S. Minsker, V. Pshenichnikov and Yu. A. Sangalov, *Vysokomol. Soedin., Ser. A*, 1975, **17**, 309.
18. D. B. Ludlum, A. W. Anderson and C. E. Ashby, *J. Am. Chem. Soc.*, 1959, **80**, 1380.
19. P. Cossee, *Tetrahedron Lett.*, 1960, **17**, 12; P. Cossee, 'Proceedings of the 6th International Congress on Coordination Chemistry, 1961', Macmillan, New York, 1961, p. 241.
20. J. Arlman, *Proc. Int. Congr. Catal., 3rd*, 1964, **2**, 957.
21. G. Allegra, *Makromol Chem.*, 1971, **145**, 235.
22. L. A. M. Rodriguez and H. M. Van Looy, *J. Polym. Sci., Part A-1*, 1966, **4**, 1951, 1971.
23. K. J. Ivin, J. J. Rooney and C. D. Stewart, *J. Chem. Soc., Chem. Commun.*, 1978, 603; K. H. Ivin, J. J. Rooney, C. D. Stewart, M. L. H. Green and R. Mahtab, *J. Chem. Soc., Chem. Commun.*, 1978, 604; M. L. H. Green, *Pure Appl. Chem.*, 1978, **30**, 27.
24. C. Beerman and H. Bestian, *Angew. Chem.*, 1959, **71**, 618.
25. A. P. Gray, A. B. Callear and F. H. Edgecombe, *Can. J. Chem.*, 1963, **41**, 1502.
26. L. A. M. Rodriguez, H. M. Van Looy and J. A. Gabant, *J. Polym. Sci., Part A-1*, 1966, **4**, 1905, 1917, 1927.
27. F. J. Karol and W. L. Carrick, *J. Am. Chem. Soc.*, 1961, **83**, 2654; F. J. Karol, W. L. Carrick, G. L. Karapinka and J. J. Smith, *J. Am. Chem. Soc.*, 1960, **82**, 1502.
28. G. Natta, A. Zambelli, I. Pasquon and G. M. Giongo, *Chim. Ind. (Milan)*, 1966, **48**, 1298, 1307; I. Pasquon, *Pure Appl. Chem.*, 1967, **15**, 465.
29. D. R. Burfield and P. J. T. Tait, *Polymer*, 1972, **13**, 315.
30. W. P. Long and D. S. Breslow, *J. Am. Chem. Soc.*, 1960, **82**, 1953; W. P. Long, *J. Am. Chem. Soc.*, 1959, **81**, 5312.
31. A. E. Shilov, A. K. Shilova and B. N. Vobkov, *Vysokomol. Soedin.*, 1962, **4**, 1688.
32. E. A. Nevel'skii and F. S. Dyachkovskii, *Vysokomol. Soedin., Ser A*, 1962, **11**, 797.
33. G. Henrici-Olivé and S. Olivé, *Adv. Polym. Sci.*, 1969, **6**, 421.
34. G. Henrici-Olivé and S. Olivé, *J. Polym. Sci., Part C*, 1969, **22**, 965.
35. G. Henrici-Olivé and S. Olivé, *Makromol. Chem.*, 1969, **121**, 70.
36. G. Henrici-Olivé and S. Olivé, *Angew. Chem., Int. Ed. Engl.*, 1968, **7**, 821.
37. W. Cooper, *Ind. Eng. Chem. Prod. Res. Dev.*, 1970, **9** (4), 457.
38. Ph. Teyssié, A. Devaux, P. Wadjiandresu, M. Julémont, I. M. Thomassin, E. Walckiers and R. Warin, in 'Preparation and Properties of Stereoregular Polymers', ed. R. W. Lenz and F. Ciardelli, Reidel, Dordrecht, 1980, p. 131.
39. G. Natta and L. Porri, 'Polymer Chemistry of Synthetic Elastomers', ed. J. P. Kennedy and E. G. M. Tornqvist, Interscience, New York, 1969, vol. 23, p. 597.
40. G. Natta, L. Porri and A. Carbonara, *Atti Accad. Naz. Lincei, Cl. Sci. Fis., Mat. Nat., Rend.*, 1960, **29**, 491 (*Chem. Abstr.*, 1962, **56**, 1477).
41. L. S. Bressler, I. Ya. Poddubnyi and V. N. Sokolov, *J. Polym. Sci., Part C*, 1969, **16**, 4337.
42. L. A. M. Rodriguez and J. A. Gabant, *J. Polym. Sci., Part C*, 1963, **4**, 125.
43. Y. V. Kissin, S. M. Mezhikovsky and N. M. Chirkov, *Eur. Polym. J.*, 1970, **6**, 267; *Eur. Polym. J.*, 1970, **6**, 525.
44. S. Fuji, in 'Coordination Polymerization; A Memorial to K. Ziegler', ed. J. C. W. Chien, Academic Press, New York, 1975, p. 748.
45. J. C. W. Chien, J.-C. Wu and C.-I. Kuo, *J. Polym. Sci., Polym. Chem. Ed.*, 1982, **20**, 2019; *J. Polym. Sci., Polym. Chem. Ed.*, 1982, **20**, 2461.
46. M. Kakugo, T. Miyatake, Y. Naito and K. Mizunuma, *Macromolecules*, 1988, **21**, 314.
47. Y. Doi, E. Suzuki and T. Keii, 'Transition Metal Catalyzed Polymerizations: Alkenes and Dienes', ed. R. P. Quirk, Harwood, New York, 1983, part A, p. 97.
48. J. P. Hogan, *J. Polym. Sci., Part A-1*, 1970, **8**, 2637.
49. A. Andersen, H. G. Cordes, J. Herwig, W. Kaminsky, A. Merk, R. Mottweiler, H. Sinn and H.-J. Vollmer, *Angew. Chem., Int. Ed. Engl.*, 1976, **15**, 630.
50. J. Cihlar, J. Mejzlik and O. Hamrik, *Makromol. Chem.*, 1978, **179**, 2553.
51. W. Kaminsky and H. Sinn, *Adv. Organomet. Chem.*, 1980, **18**, 99.
52. W. Kaminsky, in 'Transition Metal Catalyzed Polymerizations: Alkenes and Dienes', ed. R. P. Quirk, Harwood, New York, 1983, vol. 4, p. 225.
53. W. Kaminsky, 'Catalytic Polymerization of Olefins, Studies in Surface Science and Catalysis', ed. T. Keii and K. Soga, Kodansha, Tokyo, 1986, vol. 25, p. 293.
54. H. H. Brintzinger, in 'Transition Metals and Organometallics as Catalysts for Olefin Polymerization', ed. W. Kaminsky and H. Sinn, Springer-Verlag, Berlin, 1988, p. 249.
55. T. Keii, 'Kinetics of Ziegler-Natta Polymerization', Kodansha, Tokyo, 1972.
56. D. R. Burfield, I. D. McKenzie and P. J. T. Tait, *Polymer*, 1976, **17**, 130.
57. P. J. T. Tait, in 'Preparation and Properties of Stereoregular Polymers', ed. R. W. Lenz, Reidel, Dordrecht, 1980, p. 85.
58. P. J. T. Tait in 'Catalytic Polymerization of Olefins, Studies in Surface Science and Catalysis', ed. T. Keii and K. Soga, Kodansha, Tokyo, 1986, vol. 25, p. 304.
59. G. Natta and I. Pasquon, *Adv. Catal.*, 1959, **11**, 1.
60. U. Giannini, *Makromol. Chem., Suppl.*, 1981, **5**, 216.
61. T. Keii, *Makromol. Chem.*, 1982, **183**, 2285.
62. J. C. W. Chien, 'International Symposium on Transition Metal Catalyzed Polymerization, Akron, OH', 1986.
63. P. J. T. Tait, in 'Transition Metals and Organometallics as Catalysts for Olefin Polymerization', ed. W. Kaminsky and H. Sinn, Springer-Verlag, Berlin, 1988, p. 309.
64. P. J. T. Tait, in 'History of Polyolefins', ed. R. B. Seymour and T. Cheng, Reidel, New York, 1986, p. 213.
65. F. Danusso, D. Sianesi and B. Caleagno, *Chim. Ind.* (*Milan*), 1958, **40**, 628.
66. G. M. Burnett and P. J. T. Tait, unpublished data; P. J. T. Tait, Ph.D. Thesis, University of Aberdeen, 1959.
67. L. F. Borisova, E. A. Fushman, E. I. Vizen and N. M. Chirkov, *Eur. Polym. J.*, 1973, **9**, 953.

68. A. G. Chesworth, R. N. Haszeldine and P. J. T. Tait, *Polymer*, 1973, **14**, 24.
69. P. J. T. Tait, in 'Transition Metal Catalyzed Polymerization: Alkenes and Dienes', Harwood, New York, 1983, vol. 4, p. 115.
70. A. L. Burns, Ph. D. Thesis, University of Manchester, 1976.
71. P. J. T. Tait and A. E. Enenmo, to be published.
72. P. J. T. Tait, ACS Meetings, Chicago.
73. P. J. T. Tait and N. D. Watkins, to be published.
74. B. M. Grieveson, *Makromol. Chem.*, 1956, **84**, 93.
75. V. I. Tsvetkova, A. N. Plusnin, B. F. Bolshakova, B. A. Uvarov and N. M. Chirkov, *Vysokomol. Soedin., Ser. A.* 1969, **11**, 1817.
76. N. Kashiwa and J. Yoshitake, *Makromol. Chem., Rapid Commun.*, 1982, **3**, 211.
77. G. Bier and G. Lehmann in 'Copolymerization', ed. G. E. Ham, Interscience, New York, 1964.
78. G. Bier, W. Hoffmann, G. Lehmann and G. Seydel, *Makromol. Chem.*, 1962, **58**, 1.
79. V. S. Shteinbak, V. V. Amerik, F. I. Yakobson, Y. V. Kissin, D. V. Ivanyukov and B. A. Krentsel, *Eur. Polym. J.*, 1975, **11**, 457.
80. Y. V. Kissin, 'Isospecific Polymerization of Olefins', Springer-Verlag, New York, Berlin, 1985.
81. H. Schnecko, M. Reinmöller, W. Lintz, K. Weirauch and W. Kern, *Makromol. Chem.*, 1965, **84**, 156.
82. V. A. Zakharov, N. B. Chumaevsky, G. B. Bukatov and Y. I. Yermakov, *Kinet. Katal.*, 1977, **18**, 848.
83. B. M. Grieveson, *Makromol. Chem.*, 1965, **84**, 93.
84. N. M. Chirkov, 'Kinetics and Mechanisms of Polyreactions, IUPAC International Symposium on Macromolecular Chemistry, Plenary and Main Lectures: Budapest', 1969, p. 297.
85. A. P. Firsov, N. D. Sandomirskaya, V. I. Tsvetkova and N. M. Chirkov, *Vysokomol. Soedin.*, 1961, **3**, 1352.
86. A. P. Firsov, A. D. Ter-Gazaryan and N. M. Chirkov, *Vysokomol. Soedin.*, 1964, **6**, 357.
87. D. R. Burfield and P. J. T. Tait, *Polymer*, 1974, **15**, 87.
88. A. P. Firsov, V. I. Tsvetkova and N. M. Chirkov, *Dokl. Akad. Nauk SSSR, Ser. Khim*, 1964, **11**, 1956.
89. L. A. Novokshoneva, V. I. Tsvetkova and N. M. Chirkov, *Vysokomol. Soedin.*, 1965, **7**, 898.
90. F. Eirich and H. F. Mark, *J. Colloid Sci.*, 1956, **11**, 748.
91. W. H. Saltman, *J. Polym. Sci.*, 1960, **46**, 375.
92. T. Keii, K. Soga and N. Saiki, *J. Polym. Sci., Part C*, 1967, **16**, 1507.
93. K. Vesely, J. Ambroz, R. Vilim and O. Hamrick, *J. Polym. Sci.*, 1961, **55**, 25.
94. L. Reich and S. S. Stivala, *J. Polym. Sci., Part A*, 1963, **1**, 203.
95. F. D. Otto and G. Parravano, *J. Polym. Sci., Part A*, 1964, **2**, 5131.
96. D. R. Burfield, I. D. McKenzie and P. J. T. Tait, *Polymer*, 1972, **13**, 302.
97. P. J. T. Tait, in 'Coordination Polymerization: A Memorial to K. Ziegler', ed. J. C. W. Chien, Academic Press, New York, 1975, p. 155.
98. D. R. Burfield, P. J. T. Tait and I. D. McKenzie, *Polymer*, 1972, **13**, 321.
99. P. J. T. Tait and M. AbuEid, 'Proceedings 9th Iberoamerican Symposium on Catalysis, Lisbon', 1984, vol. 1, p. 163.
100. Y. A. Zakharvo, G. D. Bukatov and Y. I. Yermakov, *Makromol. Chem.*, 1975, **176**, 1959.
101. V. A. Zakharov, G. D. Bukatov, N. B. Chumaevski and Y. I. Yermakov, *Makromol. Chem.*, 1977, **178**, 976.
102. G. D. Bukatov, V. A. Zakharov and Y. I. Yermakov, *Makromol. Chem.*, 1978, **179**, 2093.
103. V. A. Zakharov, N. B. Chumaevskii, G. D. Bukatov and Y. I. Yermakov, *Makromol. Chem.*, 1976, **177**, 763.
104. G. D. Bukatov, N. B. Chumaevski, V. A. Zakharov, G. I. Kuznetsova and Y. I. Yermakov, *Makromol. Chem.*, 1977, **178**, 953.
105. L. L. Böhm, *Polymer*, 1978, **19**, 545.
106. E. K. Rideal, *Proc. Cambridge Philos. Soc.*, 1939, **35**, 130.
107. L. L. Böhm, *Polymer*, 1978, **19**, 553.
108. L. L. Böhm, *Polymer*, 1978, **19**, 562.
109. H. Schnecko and W. Kern, 'Kinetics and Mechanisms of Polyreactions, IUPAC, International Symposium on Macromolecular Chemistry, Plenary and Main Lectures, Budapest', 1969, p. 365; *Chem.-Ztg.*, 1970, **94**, 229.
110. H. Schnecko, K. A. Jung and W. Kern in, 'Coordination Polymerization: A Memorial to K. Ziegler', ed. J. C. W. Chien, Academic Press, New York, 1975, p. 91.
111. Y. I. Ermakov and V. A. Zakharov, *Russ. Chem. Rev. (Engl. Transl.)*, 1972, **41** (3), 1972.
112. J. Mejzlik, *Makromol. Chem., Macromol. Symp.*, 1986, **3**, 359.
113. J. Mejzlik, M. Lesna and J. Kratochvila, *Adv. Polym. Sci.*, 1987, **81**, 83.
114. D. G. Boucher, I. W. Parsons and R. N. Haward, *Makromol. Chem.*, 1974, **175**, 3461.
115. D. G. Boucher, I. W. Parsons and R. N. Haward, *J. Polym. Sci., Polym. Chem. Ed.*, 1978, **16**, 1683.
116. T. Tsutsui and N. Kashiwa, *Polym. Commun.*, 1988, **29**, 180.
117. S. Tanaka and H. Morikava, *J. Polym. Sci., Part A*, 1965, **3**, 3147.
118. N. Chirkov, L. Novokshonova, L. Berseneva and V. Tsvetkova, *Polym. Sci. USSR (Engl. Transl.)*, 1967, **9**, 631.
119. Y. Kissin, S. Mezhikovsky and N. Chirkov, *Eur. Polym. J.*, 1970, **6**, 267.
120. Y. Doi, A. Moringa and T. Keii, *Makromol. Chem., Rapid Commun.*, 1980, **1**, 193.
121. E. Suzuki, M. Manura, Y. Doi and T. Keii, *Makromol. Chem.*, 1979, **180**, 2235.
122. K. Gardner, I. W. Parsons and R. N. Haward, *J. Polym. Sci., Polym. Chem. Ed.*, 1978, **16**, 1683.
123. E. Suzuki, M. Tamura, Y. Doi and T. Keii, *Makromol. Chem.*, 1979, **180**, 2235.
124. N. Kashiwa and J. Yoshitake, *Makromol. Chem., Rapid Commun.*, 1983, **4**, 41.
125. G. Bukatov, S. Shepetiv, V. Zakharov, S. Sergeev and Y. Yermakov, *Makromol. Chem.*, 1982, **183**, 2657.
126. J. C. W. Chien, *J. Am. Chem. Soc.*, 1959, **81**, 86.
127. A. D. Caunt, *J. Polym. Sci., Part C*, 1963, **4**, 49.
128. H. Schnecko, K. A. Jung and L. Grosse, *Makromol. Chem.*, 1971, **148**, 67.
129. C. F. Feldman and E. Perry, *J. Polym. Sci.*, 1960, **46**, 217.
130. E. Kohn, H. J. L. Schuurmans, J. V. Cavender and R. A. Mendelson, *J. Polym. Sci.*, 1962, **58**, 681.
131. W. Cooper, D. E. Eaves, G. D. T. Owen and G. Vaughan, *J. Polym. Sci., Part C*, 1966, **4**, 211.
132. H. W. Coover, J. E. Guillet, R. L. Combs and F. B. Joyer, *J. Polym. Sci., Part A*, 1966, **4**, 2583.
133. E. A. Youngman and J. Boor, *J. Polym. Sci., Part B*, 1966, **4**, 913.

134. K. A. Jung and H. Scknecko, *Makromol. Chem.*, 1972, **154**, 227.
135. Y. I. Yermakov and V. A. Zakharov, in 'Coordination Polymerization: A Memorial to K. Ziegler', ed. J. C. W. Chien, Academic Press, New York, 1975, p. 91.
136. M. A. Schen, F. E. Karasz and J. C. W. Chien, *J. Polym. Sci.*, 1983, **21**, 2787.
137. J. C. W. Chien and Chi Kuo, *J. Polym. Sci.*, 1985, **23**, 731.
138. D. G. H. Ballard, E. Jones, R. J. Wyatt, R. T. Murray and P. A. Robinson, *Polymer*, 1974, **15**, 169.
139. D. R. Burfield and C. M. Savaviar, *Macromolecules*, 1979, **12** (2), 243.
140. H. Schnecko, W. Lintz and W. Kern, *J. Polym. Sci., Part A-1*, 1967, **5**, 205.
141. A. Baulin, A. Radionov, S. Ivanchev and N. Domareva, *Eur. Polym. J.*, 1980, **16**, 937.
142. S. S. Ivanchev, A. A. Baulin and A. G. Rodionov, *J. Polym. Sci., Polym. Chem. Ed.*, 1980, **81**, 2045.
143. G. Natta, *J. Polym. Sci.*, 1959, **34**, 21.
144. J. Mejzlik and M. Lesna, *Makromol. Chem.*, 1977, **178**, 261.
145. M. Lesna and J. Mejzlik, *React. Kinet. Catal. Lett.*, 1978, **9**, 99.
146. J. Mejzlik, M. Lesna and J. Kratochvila, *Adv. Polym. Sci.*, 1987, **81**, 84.
147. P. J. T. Tait, B. L. Booth and M. O. Jejelowo, *Makromol. Chem., Rapid Commun.*, 1988, **9**, 393.
148. J. W. C. Chien, *J. Polym. Sci., Part A*, 1963, **1**, 425.
149. V. A. Zakharov, N. B. Chumayevskii, S. I. Makhtarulin, G. D. Bukatov and Y. I. Yermakov, *React. Kinet. Catal. Lett.*, 1975, **2**, 329.
150. J. W. C. Chien and J. Hsieh, *J. Polym. Sci.*, 1976, **14**, 1915.
151. A. L. Burns and P. J. T. Tait, to be published; A. L. Burns, Ph. D. Thesis, University of Manchester, 1976.
152. A. D. Caunt and P. J. T. Tait, *Br. Polym. J.*, 1981, **13**, 22.
153. A. D. Caunt, S. Davies and P. J. T. Tait, in 'Transition Metal Catalyzed Polymerization: Alkenes and Dienes', ed. R. P. Quirk, Harwood, New York, 1983, vol. 4, p. 149.

# 4

# Monoalkene Polymerization: Copolymerization

PETER J. T. TAIT and IAN G. BERRY
*University of Manchester Institute of Science and Technology, UK*



## 4.1 INTRODUCTION

Ever since their original discovery Ziegler–Natta catalysts have been used for the copolymerization (*i.e.* the simultaneous polymerization of two or more monomers) as well as for the homopolymerization of alkenes and dienes. Such copolymerization reactions cannot only offer an intriguing insight into the nature of the catalyst and the polymerization reaction, but also yield polymers of alternative and useful properties. Indeed, copolymers are arguably of far greater commercial and industrial significance than homopolymers.

## 4.2 COPOLYMERIZATION KINETICS

### 4.2.1 Copolymer Composition Equation

#### *4.2.1.1 Derivation*

Each monomer in a copolymerization reaction competes for the available active catalytic centres, and the composition, structure and molecular weight of the copolymer produced reflect this competition. The mechanistic features of the copolymerization reaction are assumed to be essentially similar to those of the homopolymerization reaction, although certain modifications need to be made to allow for the added complexity of the propagation step which can proceed in at least four ways (Scheme 1).

$$T_M{-}M_1{-}P \;+\; M_1 \xrightarrow{k_{11}} T_M{-}M_1{-}M_1{-}P$$

$$T_M{-}M_1{-}P \;+\; M_2 \xrightarrow{k_{12}} T_M{-}M_2{-}M_1{-}P$$

$$T_M{-}M_2{-}P \;+\; M_1 \xrightarrow{k_{21}} T_M{-}M_1{-}M_2{-}P$$

$$T_M{-}M_2{-}P \;+\; M_2 \xrightarrow{k_{22}} T_M{-}M_2{-}M_2{-}P$$

**Scheme 1**

Scheme 1 is a modification of the propagation steps proposed for free radical copolymerization and from it an expression for the copolymer composition in terms of comonomer feed concentrations (equation 1) can be obtained[1]

$$\frac{d[M_1]}{d[M_2]} = \frac{[M_1](r_1[M_1] + [M_2])}{[M_2](r_2[M_2] + [M_1])} \tag{1}$$

where $d[M_1]$ = concentration of comonomer 1 in copolymer; $d[M_2]$ = concentration of comonomer 2 in copolymer; $[M_1]$ = concentration of comonomer 1 in reaction feed; $[M_2]$ = concentration of comonomer 2 in reaction feed; $r_1 = k_{11}/k_{12}$ = reactivity ratio of comonomer 1; and $r_2 = k_{22}/k_{21}$ = reactivity ratio of comonomer 2.

#### 4.2.1.2 *Reactivity ratios*

The reactivity ratio is a measure of the tendency for a comonomer to show a preference for insertion into a growing chain in which the last inserted unit was the same, rather than the other comonomer. Values are obtained either by kinetic measurement, namely fitting copolymer composition data at several comonomer ratios to a suitable form of the copolymer composition equation (1);[2] or by microstructural analysis of the copolymer, usually $^{13}$C NMR determination of monomer sequence distributions.[3] Copolymers of high comonomer content (>8 mol %) or copolymers prepared using $^{13}$C enriched comonomer are generally used in the latter case.[4]

The product of the reactivity ratios, $r_1 \cdot r_2$, is usually between zero and one. When $r_1 \cdot r_2$ is approximately one, the last added unit in the chain does not influence the next addition and Bernoullian statistics govern the formation of a random copolymer. When $r_1 \cdot r_2$ tends to zero, there is some influence from the last inserted unit when first order Markovian statistics operate, or from the penultimate inserted unit when second order Markovian statistics operate. An alternating copolymer structure is favoured in this case. Finally, when $r_1 \cdot r_2$ is greater than one there is a tendency for comonomers to form long segments, or perhaps even undergo homopolymerization, and block copolymer formation predominates. This may be illustrated for an ethylene (E), propylene (P) copolymer as follows

| | | | |
|---|---|---|---|
| $r_1 \cdot r_2$ | = | 1 random | EEPEPEPPEP |
| $r_1 \cdot r_2$ | < | 1 alternating | EPEPEPEPEP |
| $r_1 \cdot r_2$ | > | 1 block | EEEEPEPPPE |

#### 4.2.1.3 *Ziegler–Natta copolymerization*

Analysis of copolymer structure and the evaluation of reactivity ratios can therefore act as a valuable mechanistic probe as the structure effectively provides a permanent record of the polymerization reaction. Examples of the application of reactivity ratios will be given later but the insights which the determination of reactivity ratio products may afford into copolymer formation and structure should be regarded with caution as their evaluation and interpretation may give rise to uncertainty. Firstly, discrepancies exist in the values of reactivity ratio product obtained by $^{13}$C NMR and those obtained by kinetic measurement.[5] This anomaly has been reported for both soluble[6] and heterogeneous[7,8] systems and for vanadium and titanium based catalysts. Secondly, in deriving the copolymer composition equation (1) certain assumptions are made which may render it inapplicable to Ziegler–Natta catalyst systems. The features of these systems which may give rise to problems of applicability and interpretation are the time dependent changes which occur in polymerization kinetics and also the nature of the active catalytic centre itself.

*(i) Changes in polymerization kinetics*

Time dependent changes in catalytic activity are well known in Ziegler–Natta systems. Such changes, whether due to changes in active centre concentration, changes in the rate constant or constants, or to diffusion phenomena, will make it difficult to apply the copolymer composition equation (1) unambiguously. Changes in any of these as comonomer concentration in the reaction feed is varied will also add further complications.

*(ii) The nature of the active centre*

*(a) Multiplicity of active centres.* The derivation of the copolymer composition equation (1) assumes only one type of active centre to participate in chain propagation. This may be true in certain homogeneous Ziegler–Natta catalyst systems but in most heterogeneous systems multiple centres often exist and hence quantities derived from the copolymer composition equation (1) merely represent average values for the different sites present. An underestimation of the true blockiness present in a copolymer may therefore result if a single centre model is assumed. A more rigorous mathematical treatment accounting for multiple centres can be undertaken and may explain the discrepancy in reactivity ratio products determined by $^{13}$C NMR and kinetic measurement.[5] The greater accuracy obtained by $^{13}$C NMR microstructural analysis may also account for such reactivity ratio differences as quantities derived from the copolymer composition equation (1) are very sensitive to small errors in compositional determination.[2]

*(b) Steric factors.* The influence of the last inserted unit upon subsequent insertions is crucial in copolymerizations as the last unit may influence which comonomer is next inserted and also the mode of that insertion.[9] The derivation of the copolymer composition equation (1) assumes the active centre to be totally regiospecific; Scheme 1 may therefore require modification to account for regiospecific primary and secondary insertion, although in most heterogeneous systems primary insertion is believed to predominate and no modification is necessary. Homogeneous vanadium based catalysts on the other hand, which are capable of producing syndiotactic polymer, may be incompletely described by Scheme 1. Comonomer structure and the stereochemistry of the active centre can therefore profoundly influence copolymer composition.

## 4.2.2 Controlling Copolymer Composition

Both physical and chemical reaction conditions can influence copolymerization kinetics and these influences manifest themselves not only in their effect upon the rate of reaction but also in the structure and composition of the copolymer produced. Copolymer properties are closely related to copolymer composition, which is a complex function of: (i) total comonomer content, the distribution of comonomer units along an individual chain (intramolecular composition) and the variation in comonomer content and distribution between different chains (intermolecular composition); (ii) molecular weight and molecular weight distribution; and (iii) the relative content of head–tail additions to head–head or tail–tail additions (*i.e.* the degree of regiospecificity).

A degree of control can be exercised over these factors and therefore over copolymer properties by paying careful attention to physical and chemical influencing factors, the most important of which are as follows, although they may be interdependent: (i) operating conditions; (ii) chemical composition of the catalyst; (iii) physical nature of the catalyst; and (iv) structure of the comonomers.

### *4.2.2.1 Operating conditions*

Whichever polymerization process is chosen — solution, gas phase or slurry — a drift in comonomer feed composition may result in a copolymer of non-uniform composition. In batch copolymerizations, especially those conducted to low yield, a constant comonomer ratio may be approximated but this may prove more difficult to achieve in continuous processes. Diffusion effects may exacerbate the problem. In slurry processes the monomers dissolved in the diluent must be kept in equilibrium with a gaseous phase of known composition and high stirring rate, low solution viscosity and a reactant flow well above the copolymerization rate are necessary to achieve this.

Temperature can exert an influence *via* its effect upon monomer solubility and also upon the overall reaction rate and the rate of chain transfer reactions. Additionally, if temperature influences monomer coordination it may affect copolymer composition. A slight decrease in reactivity ratios with increasing temperature has been reported during ethylene–butene-1 copolymerization (Table 1),[10] although no dependence upon temperature was found up to 75 °C for ethylene and propylene copolymerization in the $VCl_4$–$Al(C_6H_{13})_3$ system.[11]

**Table 1** Reactivity Ratio Variation with Temperature for Ethylene–Butene-1 Copolymerization[10]

| $T$ (°C) | $r_1$ (*Ethylene*) | $r_2$ (*Butene-1*) |
|---|---|---|
| 120 | $51 \pm 5$ | $0.04 \pm 0.02$ |
| 85 | $67 \pm 8$ | $0.08 \pm 0.04$ |
| 65 | $110 \pm 20$ | $0.1 \pm 0.05$ |
| 50 | $130 \pm 25$ | $0.1 \pm 0.05$ |

Finally, the order of addition of reaction components may be important. For example, in the $VCl_4$–$AlEt_3$–ethylene, propylene copolymerization system it was found that by mixing the catalyst components in the presence of pure propylene a copolymer richer in propylene was produced compared with the copolymer obtained when the components were mixed in pure ethylene or in an ethylene–propylene mixture.[12]

### 4.2.2.2 *Catalyst composition*

The structure of the transition metal component and sometimes the nature of the metal alkyl component can influence copolymer composition. It has been shown that the relative reactivities of comonomers can be influenced by the electronegativity and the oxidation state of the transition metal centre. Propylene reactivity for example increases in ethylene–propylene copolymerization with increasing transition metal electronegativity and ethylene reactivity is generally greater in titanium rather than vanadium based catalysts.[13] Reactivity also decreases with decreasing oxidation state in both titanium and vanadium based systems[14,15] and propylene reactivity was found to be lost completely during ethylene–propylene copolymerization by over-reduction of the catalyst; pure polyethylene formed rather than a copolymer.[16]

In general, the trends observed in overall polymerization rate with variation in catalyst–cocatalyst ratio are as observed in homopolymerization but copolymerization systems may be more sensitive to over-reduction by cocatalyst.

### 4.2.2.3 *Nature of catalyst*

The presence of multiple centres, particularly in heterogeneous catalysts, is believed to be responsible for the production of a molecular weight distribution ($M_w/M_n = 10$) in Ziegler–Natta systems. Although diffusion phenomena may be important[17] such a multiplicity of active centres is also capable of producing a distribution in the extent of comonomer incorporation—copolymer composition—as the variation in ligand environment at different centres influences the ease of comonomer coordination at the centre. It should be noted that copolymerization is a particularly effective method of demonstrating the presence of a distribution of active centres of different steric behaviour in Ziegler–Natta catalysts.

The compositional distribution may additionally be superimposed upon the molecular weight distribution to add to the structural complexity of the copolymer. Control of the molecular weight by transfer agents, variation in operating procedure and transfer reactions, which may also of course involve the comonomer, may all further increase compositional heterogeneity.

In comparison to copolymers prepared using heterogeneous catalysts, those obtained from homogeneous systems generally possess a narrow molecular weight distribution ($M_w/M_n = 2$) showing little compositional variation.[18] Such catalysts generally possess a single type of active site and produce random or alternating copolymers.

#### 4.2.2.4 *Effect of comonomer structure*

In Ziegler–Natta copolymerization reactions reactivity ratio values for each comonomer that differ in magnitude by an order of three are often obtained.[10] In contrast, those values obtained in radical copolymerization vary in magnitude by just several factors, clearly implying a close interaction between the active centre and the monomer in Ziegler–Natta systems. This fact is well appreciated from homopolymerization studies and following the trend in homopolymerization, higher α-alkenes become less and less readily copolymerized as their size increases.[19,20,21] Despite the sometimes poor correlation between copolymer structure and reactivity ratio values (Section 4.2.1.3), such values can be used more confidently to compare relative comonomer reactivity in a copolymerization system. A systematic study[22] of $TiCl_3$–$AlEt_3$ and $MgCl_2$, $TiCl_4$–$AlEt_3$ catalysts for ethylene copolymerization with linear α-alkenes from $C_6$ to $C_{36}$ showed the α-alkene reactivity ratio value to decrease as chain length increased from $C_6$ to $C_{20}$, but thereafter it remained almost constant.

A further illustration of steric influences upon copolymerization behaviour, which also highlights the value of copolymerization reactions as a mechanistic probe, is provided by the stereoselective heterogeneous copolymerization of ethylene with racemic α-alkenes. Copolymers of either the —R(*R*) or —R(*S*) comonomer are formed (Scheme 2).[23]

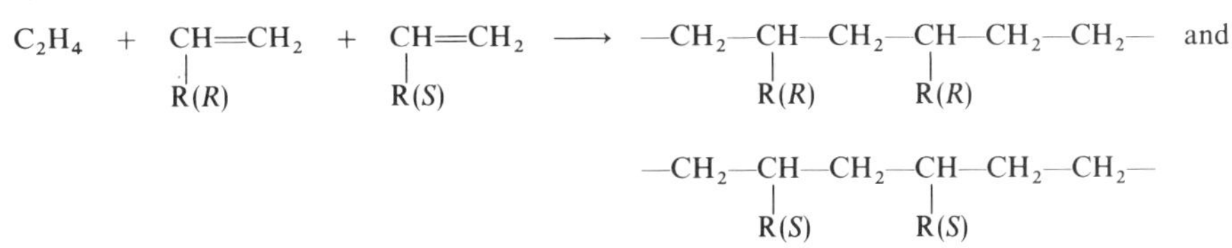

**Scheme 2**

Finally, the comonomer can influence copolymer structure by acting as a chain transfer agent during copolymerization, although as the bulkiness increases, its efficiency in this respect decreases. Comonomer presence can therefore broaden the molecular weight distribution, and decreases in the average molecular weight and the average degree of polymerization have been reported in the presence of comonomer.[24,25]

#### 4.2.2.5 *Rate enhancement*

The crucial significance of steric factors in Ziegler–Natta catalysis might lead to the conclusion that copolymerization proceeds at a lower rate than homopolymerization. This has indeed been reported for an ethylene–propylene[26] copolymerization but an increasing number of workers are finding the reverse to be true in ethylene–α-alkene copolymerization. An increase in ethylene concentration in the $VCl_4$, $Al(C_6H_{13})_3$–ethylene, propylene system produced an increase in rate beyond that expected from the known greater reactivity of ethylene.[27] The existence of active centres which could homopolymerize or copolymerize ethylene with α-alkenes but not homopolymerize α-alkenes was suggested by way of explanation.

An increase in ethylene–propylene copolymerization rate over the ethylene homopolymerization rate has been reported for the $MgCl_2$, $TiCl_4$–$AlEt_3$ system[15] and rate enhancement has now been observed in many other systems for a wide range of catalysts and comonomers. These include $Cp_2ZrCl_2$–methyl aluminoxane–$C_2H_4$, $C_3H_6/C_6H_{12}$;[28] $MgH_2$, $TiCl_4$–$AlEt_3$–$C_2H_4$, $C_3H_6/C_4H_8/C_8H_{16}$;[22] $TiCl_4$, Grignard reagent–$C_2H_4$, $C_6H_{12}$;[29] and $TiCl_3$, $0.33AlCl_3$–$AlEt_2Cl$–$C_2H_4$, $C_8H_{16}$.[30]

Recent kinetic and active centre studies may explain this puzzling phenomenon.[30,31] A significant rate enhancement which depended upon comonomer concentration was found during ethylene and

4-methylpentene-1 (4-MP-1) copolymerization for the following catalyst systems; (a) $TiCl_3$, $0.33AlCl_3$–$Al(Bu^i)_3$; (b) $MgCl_2/EB/TiCl_4$–$Al(Bu^i)_3$ (EB = ethylbenzoate); and (c) $MgCl_2$/DIBP/$TiCl_4$–$Al(Bu^i)_3$ (DIBP = diisobutyl phthalate).

Active centre concentration measurements using the $^{14}CO$ radiolabelling technique showed the enhanced rate of ethylene polymerization in the presence of 4-MP-1 to result from an increased number of active centres. Such an increase could arise from several causes: (a) fragmentation of the catalyst matrix to expose previously unexposed active sites; (b) active centre formation reactions involving the comonomer; (c) displacement of adsorbed or complexed molecules from the active site by the comonomer; and (d) diffusion phenomena.

## 4.3 COPOLYMER STRUCTURE

### 4.3.1 Alkene–Alkene Copolymers

#### *4.3.1.1 Ethylene–propylene*

Ethylene–propylene copolymers are the most widely investigated of the Ziegler–Natta copolymers[32] and illustrate the profound effect upon properties that comonomer incorporation brings. Both high density polyethylene and polypropylene are highly crystalline polymers but the disruption of long sequences of one monomer by the introduction of a comonomer lowers the ability of the copolymer to crystallize. The properties of both homopolymers can be improved in this way. Small amounts of propylene incorporated into high density polyethylene are necessary to lower the crystallinity slightly and give a less brittle product. Larger amounts of propylene incorporated into a random copolymer (EPM, 'ethylene–propylene monomer' copolymer) produces an elastomeric material, small quantities of which (5–20%) if blended with isotactic polypropylene dramatically improve the flexibility properties near 0 °C.

Successful elastomeric materials require a low glass transition temperature to give good low temperature behaviour and also good elasticity. The absence of unsaturation in EPM elastomers precludes the possibility of vulcanization or curing to introduce crosslinking as a means of improving elasticity, but this deficiency can be overcome by copolymerization with a third comonomer. Copolymerization with a diene[33] such as 1,4-hexadiene or ethylidenenorbornene or a triene[34] produces a terpolymer (EPDM or EPTM, 'ethylene–propylene–diene or triene monomer') possessing unsaturated branches along the main chain which can act as sites for crosslinking. Vanadium based catalysts are generally preferred for the production of these random copolymers and EPDM rubbers now find wide application in many fields but particularly in the automobile industry.

#### *4.3.1.2 Ethylene–α-alkene*

A variety of catalysts and a number of different α-alkene comonomers other than propylene have been used for copolymerization with ethylene. This type of copolymerization has been extensively investigated, especially for ethylene–butene-1 copolymers,[35] and has recently assumed major commercial significance.

Small amounts of α-alkene comonomer (typically 1.5–5.0 mol %) incorporated in an ethylene copolymer have the effect of producing side chains at points where the α-alkene is inserted into the linear polyethylene backbone. Thus, copolymerization with butene-1 produces ethyl branches but butyl, isobutyl and hexyl branches are also commonly introduced. The branches disrupt the crystallinity of high density polyethylene (HDPE) and lower the density of the material so that it simulates many of the properties of low density polyethylene (LDPE) manufactured by free radical polymerization of ethylene. Ethylene–α-alkene copolymers are therefore collectively termed linear low density polyethylenes (LLDPE). Comonomer content and distribution together with branch length can be used to control both the density and melting point[4] of the copolymer product.

The origin of LLDPE can be traced back to the early 1950s when Du Pont began to produce 'Sclair' and 'Rotothene' resins. Phillips Petroleum Co. manufactured LLDPE in the 1960s in a solution process using silica–alumina supported $CrO_3$ catalysts, but it was not until 1977 and the claims made by Union Carbide for the Unipol gas phase process that widespread attention was attracted to LLDPE. The Unipol process uses a high activity supported catalyst to copolymerize ethylene with butene-1, pentene-1 or hexene-1, and large energy and plant-cost savings over high pressure LDPE manufacture are possible. Solution processes and low pressure slurry processes are also now successfully producing a variety of different LLDPEs.

The success of LLDPE lies not only in its comparative ease of manufacture but also in the control over the structure that can be gained by careful manipulation of manufacture conditions and in the superior physical properties of the product. LDPE contains a combination of short and long chain branches; the short branches disrupt the crystal structure while the fewer long branches affect properties such as the melt strength. LLDPE in comparison has no long chain branches just chain branches of constant length; this molecular structure leads to a higher crystallinity than LDPE at the same density and therefore a higher melting point and a higher rigidity. Additionally, control of the reaction conditions allows a narrower molecular weight distribution to be obtained in LLDPE and this leads to improved shear behaviour.

The difference in structure between LDPE and LLDPE brings advantages both in ease of processability and in mechanical performance of the polymer. The improved viscosity and extrusion properties in LLDPE combined with its better seal strength, particularly if an LLDPE terpolymer is used, mean LLDPE is suitable for many film and packaging applications. Additionally, better impact, tensile and tear strength, improved puncture resistance, excellent optical properties and the scope for using thinner films of equivalent strength (downgauging) have all contributed towards the rapid success of LLDPE over LDPE.

The relatively recent interest in LLDPE and its huge commercial significance has precluded extensive publication of structural analyses in the scientific literature. The complex structure, as illustrated by the appearance of several peaks in differential scanning calorimetry (DSC), has also hindered structural analysis. The DSC behaviour suggests a non-uniform composition which might be explained by the presence of multiple sites in heterogeneous catalysts[36] or by diffusion phenomena, or both (Section 4.2.1.3.ii.a). Understanding the structure of copolymers with such a highly heterogeneous compositional nature as LLDPE is not easy and some form of fractionation, and preferably cross-fractionation,[37,38] is necessary for a fuller appreciation of the structure. Cross-fractionation involves separation according to compositional structure (by temperature rising elution fractionation, TREF, for example) and then separation of these fractions by molecular weight (using gel permeation chromatography, GPC, for example). The reverse process is also possible.[39]

The distribution of comonomer contents and molecular weights and the inter-relationship between these can all influence copolymer properties and so the interest in LLDPE structure is great. Despite the complex structure some level of general understanding has been achieved and this may be summarized as follows:[40] (i) LLDPE is a blend of molecules with different comonomer contents; (ii) there is variation in comonomer distribution over the entire range of molecular weights; (iii) molecules with the same length are found to differ in comonomer content, perhaps ranging from LDPE to HDPE type molecules; and (iv) the average comonomer content rises as molecular weight decreases.

The low level of comonomer incorporation in LLDPE might lead to the expectation of a random intramolecular distribution of comonomer units. This has indeed been confirmed for homogeneous catalysts but in heterogeneous catalysts there is some evidence to suggest that incorporation is not random, especially at higher comonomer content. $^{13}$C NMR of LLDPEs has shown sequences of two (diads) or more (triads, *etc.*) α-alkene comonomer units to be significant in concentration.[21] This observation suggests that once a comonomer unit is inserted the insertion of another α-alkene molecule is encouraged. Terpolymerization studies seem to confirm this. The presence of short chain α-alkenes (butene-1 to octene-1) is found to support the insertion of bulky α-alkenes such as hexadecene but, conversely, the insertion of the longer chain α-alkene does not increase the incorporation of the shorter chain molecule.[21] The observation that the short chain branches are located near the longer chain branches has led to the suggestion that a 'synergistic effect' operates in which the steric hinderance at the active centre is temporarily reduced by the short chain α-alkene.[41]

#### 4.3.1.3 Other copolymerizations[42]

There are many other combinations of comonomer pairs which have been studied and the scope illustrates the true versatility of Ziegler–Natta catalysts.

##### (i) α-Alkene–α-alkene

The modification in physical properties which comonomer incorporation can bring by disruption to polymer crystallinity can be reduced or prevented by the copolymerization of two comonomers of similar structure, for example 3-methylbutene-1 and 4-methylbutene-1. Another example of copolymerization in this class is propylene–butene-1 copolymerization in which both comonomers, by virtue of their similar homopolymerization rates, have similar reactivity ratio values. Depending upon the catalyst used random or non-random copolymers can be produced.

*(ii) Vinyl aromatic monomers*

The low reactivity of styrene and substituted styrenes necessitates the use of high concentrations to achieve incorporation in a copolymer. Nevertheless, copolymers of styrene with ethylene, heptene-1, octadiene and 4-methylpentene-1 have been reported. Block structures of high crystallinity tend to form in such copolymers.

*(iii) α-Alkene–internal alkene*

The ability of a comonomer in terpolymerization to improve the incorporation of long chain α-alkenes has been mentioned previously (Section 4.3.1.2). Similarly, internal alkenes (cycloalkenes or non-α-alkenes) can be copolymerized with ethylene although they cannot undergo homopolymerization themselves. Ethylene seems to minimize steric interactions at the active centre and thus copolymers of ethylene with butene-2, cyclopentene, cyclohexene, cycloheptene and cyclooctene, have all been reported.

*(iv) α-Alkene–polar monomers*

Examples which have been reported include ethylene and *t*-butyl acrylate, 6-chlorohexene- and 4-methylpentene-1, and styrene and 2-methyl-5-vinylpyridine.

### 4.3.2 Diene Copolymers

#### *4.3.2.1 Alkene–diene copolymers*

The use of diene comonomers in EPDM elastomers has been mentioned already but dienes can also be incorporated as comonomers in copolymerization as well as terpolymerization. The copolymerization of styrene with isoprene or butadiene,[42] isobutylene with butadiene,[43] and cyclopentene with dicyclopentadiene[44] amongst others have all been reported. By far the simplest and most investigated systems in this class are the ethylene–diene (EDM) and propylene–diene (PDM) copolymers however.

*(i) Ethylene–diene copolymers*[45]

Vanadium based catalysts such as $V(acac)_3$–$AlEt_2Cl$ have been well studied in ethylene–butadiene (EBD) copolymerization. Low rates, low molecular weight product and poor utilization of butadiene are common problems however, and the acidity of the active centre appears to be crucial for successful coordination of the diene and its subsequent insertion into the V—C bond. The use of a Lewis acid such as water improves diene incorporation and a good aromatic solvent, high Al:V ratio and low temperature to improve selectivity all aid copolymer formation. In these systems there is a tendency towards blockiness; diene coordination is strong and insertion of a subsequent diene unit is preferred over ethylene insertion. This was found even in a supported titanium based catalyst which produced a more random distribution of comonomer units than a vanadium catalyst.

Butadiene is present almost exclusively as *trans*-1,4 units in EBD copolymers, presumably due to the nature of the coordination through one double bond.

*(ii) Propylene–diene copolymers*[46]

The experimental conditions necessary for successful ethylene–diene copolymerization also appertain to propylene–diene copolymerization and the addition of carbonyl compounds or weak Lewis acids to both vanadium and titanium based catalysts to control active centre acidity is adopted. Alternating or random copolymers are formed, most interest being centred around the alternating copolymers.

#### 4.3.2.2 Diene–diene copolymers[47]

There has been little incentive to thoroughly investigate such copolymers but butadiene-2-phenylbutadiene, piperylene and isoprene copolymers have been prepared.

### 4.3.3 Block Copolymers

The improvement in the performance of isotactic polypropylene at low temperature which can be achieved by blending with small amounts of EPM elastomer has been mentioned already (Section 4.3.1.1). Polyisobutylene, polyisoprene and polybutadiene can also be used as blending material yet what is gained in toughness by blending is often lost in tensile strength. The desire to overcome this incompatibility by incorporating elastomeric lengths and isotactic lengths in the same chain has been the driving force behind the interest in block copolymerization.

In attempting to synthesize block copolymers the aim is to form two or more long segments of different composition within the same polymer chain. Possible ideal structures which can be envisaged for ethylene (E), propylene (P) and copolymer (PE) sequences include: P—E, P—EP, EP—(EP)′, P—EP—P, P—E—P—E, P—EP—P—EP, (EP)′—EP—(EP)′—EP.

The successful synthesis of block copolymers requires that a number of conditions are satisfied. (a) All polymerization centres must be activated at the same time. (b) All polymerization centres must stay alive during the polymerization, *i.e.* all polymer chains must remain attached to the active centres. (c) All polymerization centres must have equal activity towards the various monomers— this also implies that all alkene molecules have equal accessibility to the active centres. (d) The crossover propagation rates ($T_M$—AAAA + B and $T_M$—BBBB + A) must be high for all centres. (e) In the case of ethylene–propylene block copolymers the centres must be isospecific for propylene polymerization and produce linear polyethylene segments.

It is essential to distinguish chains which are 'alive' from active centres which are 'alive'. A centre can remain 'alive' after being involved in the production of possibly thousands of individual macromolecules, but the chain itself is only considered to be 'alive' if it remains attached to the active centre and is capable of further growth.

The above requirements were formulated for typical anionic initiators and it is evident that many of them do not hold for heterogeneous Ziegler–Natta catalysts, where a number of transfer reactions have been established and where as a consequence the average lifetime of a growing chain may be relatively short. Early estimates for the lifetime of growing chains in propylene polymerization were of the order of several minutes.[48,49] Kissin[50] quotes values of between 11.9 and 226 min for polymerization of propylene using a $\delta$-$TiCl_3$–$AlEt_2Cl$ catalyst at temperatures from 90–30 °C respectively. However considerably longer lifetimes have been claimed for some polymerization systems.[51] The existence of living chains has been illustrated by the increase in polymer yield and number average molecular weight with time but in order to achieve this a low temperature (−78 °C) and elimination of chain transfer reactions must be achieved.[52,53] Many typical supported catalysts are unstable (the activity for propylene polymerization may drop by a factor of at least two after 20 min at 60 °C) and have somewhat higher $k_P$ values which could lead to a reduction in the average lifetime.

In spite of such difficulties many patents have been filed claiming the synthesis of block copolymers.[54] Commercial polymerization procedures for producing propylene–ethylene block copolymers are specifically designed to maximize the production of block copolymer molecules, and typically would involve the use of twin reactors. Whilst it has not been possible to prepare, under normal polymerization conditions, clean block copolymers similar to styrene–diene (butadiene or isoprene) copolymers prepared by the use of lithium alkyls,[55] there can be no doubt that the use of heterogeneous Ziegler–Natta catalysts in such a polymerization will produce a complex mixture of *in situ* produced macromolecules which will contain, besides homopolymer molecules, a small fraction of block copolymer material. So far neither fractionation nor thermal studies have been able to provide a reliable estimation of the block copolymer content of such materials.

These propylene–ethylene copolymers because of their attractive blend of physical properties are produced commercially on a very large scale and are sold as value added products. Research and development in this area is likely to continue to be of major commercial and industrial interest for many years to come.

## 4.4 REFERENCES


1. J. Boor, in 'Ziegler–Natta Catalysts and Polymerizations', Academic Press, New York, 1979, chap. 20.
2. J. L. Koenig, in 'Chemical Microstructure of Polymer Chains', Wiley, New York, 1980, chap. 5.
3. J. L. Koenig, in 'Chemical Microstructure of Polymer Chains', Wiley, New York, 1980, chaps. 3 and 8.
4. S. D. Clas, D. C. McFadden, K. E. Russell and M. V. Scammell-Bullock, *J. Polym. Sci., Polym. Chem. Ed.*, 1987, **25**, 3105.
5. S. Floyd, *J. Appl. Polym. Sci*, 1987, **34**, 2559.
6. C. Cozewith, *Macromolecules*, 1987, **20**, 1237.
7. N. Kashiwa, A. Mizuno and J. Minami, *Polym. Bull. (Berlin)*, 1984, **12**, 105.
8. K. Soga, T. Shiono and Y. Doi, *Polym. Bull. (Berlin)*, 1983, **10**, 168.
9. J. Boor, in 'Ziegler–Natta Catalysts and Polymerizations', Academic Press, New York, 1979, p. 576.
10. L. L. Böhm, *J. Appl. Polym. Sci.*, 1984, **29**, 279.
11. G. Natta, G. Mazzanti, A. Valvassori, G. Sartori and A. Barbagallo, *J. Polym. Sci.*, 1961, **51**, 429.
12. T. Suminoe, N. Yamazaki and S. Kambara, *Kobunshi Kagaku*, 1963, **20**, 461.
13. W. L. Carrick, F. J. Karol, G. L. Karapinka and J. J. Smith, *J. Am. Chem. Soc.*, 1960, **82**, 1502.
14. G. Natta, A. Valvassori and G. Sartori, in 'Polymer Chemistry of Synthetic Elastomers', ed. J. P. Kennedy and E. G. M. Törnqvist, Wiley, New York, 1969, part 2, p. 679.
15. N. Kashiwa and J. Yoshitake, *Makromol. Chem.*, 1984, **185**, 1133.
16. K. Soga, T. Sano and R. Ohnishi, *Polym. Bull. (Berlin)*, 1981, **4**, 157.
17. W. H. Ray, T. W. Taylor, K. Y. Choi and H. Yean, in 'Transiton Metal Catalyzed Polymerizations', ed. R. P. Quirk, Harwood Academic Press, New York, 1983, p. 191.
18. C. Cozewith and G. Ver Strate, *Macromolecules*, 1971, **4**, 482.
19. Y. V. Kissin, in 'Transition Metal Catalyzed Polymerizations', ed. R. P. Quirk, Harwood Academic Press, New York, 1983, p. 597.
20. Y. V. Kissin, *Adv. Polym. Sci.*, 1974, **15**, 92.
21. T. A. Ojala and G. Fink, *Makromol Chem., Rapid Commun.*, 1988, **9**, 85.
22. Y. V. Kissin and D. L. Beach, *J. Polym. Sci., Polym. Chem. Ed.*, 1984, **22**, 333.
23. F. Ciardelli, P. Locatelli, M. Marchetti and A. Zambelli, *Makromol. Chem.*, 1974, **175**, 923.
24. L. L. Böhm, *Makromol. Chem.*, 1981, **182**, 3291.
25. N. Kashiwa, T. Tsutsui and A. Toyota, *Polym. Bull. (Berlin)*, 1984, **12**, 111.
26. S. N. Gan, S. I. Chen, R. Ohnishi and K. Soga, *Polymer*, 1987, **28**, 1391.
27. A. Valvassori, G. Sartori, G. Mazzanti and G. Pajaro, *Makromol. Chem.*, 1963, **61**, 46.
28. T. Tstusui and N. Kashiwa, *Polym. Commun.*, 1988, **29**, 180.
29. A. Munoz-Escalona, H. Garcia and A. Albornoz, *J. Appl. Polym. Sci.*, 1987, **34**, 977.
30. P. J. T. Tait, G. W. Downs and A. A. Akinbami, *Int. Symp. Transition Met. Catal. Polym.*, Akron, Ohio, 1986.
31. P. J. T. Tait, in 'Transition Metals and Organometallics as Catalysts for Olefin Polymerization', ed. W. Kaminsky and H. Sinn, Springer-Verlag, Berlin, 1988, p. 309.
32. F. P. Baldwin and G. Ver Strate, *Rubber Chem. Technol.*, 1972, **45**, No. 3, 709.
33. G. E. Ham (ed.), 'Copolymerization', Wiley, New York, 1964, chap. 4.
34. S. Cesca, in 'Transition Metal Catalyzed Polymerizations', ed. R. P. Quirk, Harwood Academic Press, New York, 1983, p. 588.
35. J. Boor, in 'Ziegler–Natta Catalysts and Polymerizations', Academic Press, New York, 1979, p. 577.
36. T. Usami, Y. Gotoh and S. Takayama, *Macromolecules*, 1986, **19**, 2722.
37. L. Wild, T. R. Ryle and D. C. Knobeloch, *Polym. Prepr., Am. Chem. Soc., Div. Polym. Chem.*, 1982, **23**, 133.
38. L. Wild, T. R. Ryle, D. C. Knobeloch and I. R. Peat, *J. Polym. Sci., Polym. Phys. Ed.*, 1982, **20**, 441.
39. P. Schouterden, G. Groeninckx, B. Van der Heijden and F. Jansen, *Polymer*, 1987, **28**, 2099.
40. V. B. F. Mathot, Polycon '84-LLDPE, Plenary Session A.
41. J. V. Seppälä and N. Oy, *J. Appl. Polym. Sci.*, 1985, **30**, 3545; 1986, **31**, 657 and 699.
42. J. Boor, in 'Ziegler–Natta Catalysts and Polymerizations', Academic Press, New York, 1979 p. 580.
43. S. Cesca, in 'Transition Metal Catalyzed Polymerizations', ed. R. P. Quirk, Harwood Academic Press, New York, 1983, p. 583.
44. R. J. Minchak, H. Tucker and J. Hugh Macey, in 'Transition Metal Catalyzed Polymerizations', ed. R. P. Quirk, Harwood Academic Press, New York, 1983, p. 863.
45. S. Cesca, *Macromol. Rev.*, 1975, **10**, 1.
46. J. Furukawa, *J. Polym. Sci., Polym. Symp.*, 1974, **48**, 19; J. Furukawa, *Polymer*, 1974, **15**, 162.
47. J. Boor, in 'Ziegler–Natta Catalysts and Polymerizations', Academic Press, New York, 1979, p. 581.
48. G. Natta and I. Pasquon, *Adv. Catal.*, 1959, **11**, 1.
49. S. Tanaka and H. Morikawa, *J. Polym. Sci., Part A*, 1965, **3**, 3147.
50. Y. V. Kissin, in 'Transition Metal Catalyzed Polymerizations', ed. R. P. Quirk, Harwood Academic Press, New York, 1983, p. 52.
51. G. Bier, G. Lehmann and H. J. Leugering, *Makromol. Chem.*, 1961, **44/46**, 347.
52. Y. Doi, S. Suzuki, F. Nozawa and K. Soga, in 'Catalytic Polymerization of Olefins, Studies in Surface Science and Catalysis', ed. T. Keii and K. Soga, Kodansha, Tokyo, 1986, vol. 25, p. 257.
53. G. G. Evens and E. M. J. Pijpers, in 'Transition Metal Catalyzed Polymerizations', ed. R. P. Quirk, Harwood Academic Press, New York, 1983, p. 245.
54. T. G. Heggs, in 'Block Copolymers', ed. D. C. Allport and W. H. Janes, Wiley, New York, 1973, chap. 4 and 8D.
55. J. Boor, in 'Ziegler–Natta Catalysts and Polymerizations', Academic Press, New York, 1979, p. 588.

# Subject Index